Elementary Statistics

A Step by Step Approach

SIXTH EDITION

Elementary Statistics

A Step by Step Approach
Reinforced Binding

Allan G. Bluman

Professor Emeritus
Community College of Allegheny County

Boston   Burr Ridge, IL   Dubuque, IA   Madison, WI   New York   San Francisco   St. Louis
Bangkok   Bogotá   Caracas   Kuala Lumpur   Lisbon   London   Madrid   Mexico City
Milan   Montreal   New Delhi   Santiago   Seoul   Singapore   Sydney   Taipei   Toronto

Higher Education

ELEMENTARY STATISTICS: A STEP BY STEP APPROACH, SIXTH EDITION

Published by McGraw-Hill, a business unit of The McGraw-Hill Companies, Inc., 1221 Avenue of the Americas, New York, NY 10020. Copyright © 2007 by The McGraw-Hill Companies, Inc. All rights reserved. No part of this publication may be reproduced or distributed in any form or by any means, or stored in a database or retrieval system, without the prior written consent of The McGraw-Hill Companies, Inc., including, but not limited to, in any network or other electronic storage or transmission, or broadcast for distance learning.

Some ancillaries, including electronic and print components, may not be available to customers outside the United States.

This book is printed on acid-free paper.

3 4 5 6 7 8 9 0 DOW/DOW 10 09 08 07

ISBN-13   978–0–07–327160–6
ISBN-10   0–07–327160–8

About This Text
Reinforced Binding

What Does It Mean?

For adopting schools this means these texts can be expected to be more durable and last longer when subjected to daily classroom use in a school environment where textbooks are adopted for multiple years.

This text has been adopted by colleges and universities yet it is often used in high schools for teaching Advanced Placement*, honors, elective, and college prep courses. Because advanced high school programs adoption periods often last several years and a text must stand up to usage by multiple students, McGraw-Hill has elected to manufacture this text in a manner compliant with the "Manufacturing Standards and Specifications for Textbooks" (MSST) published by the "National Association of State Textbook Administrators" (NASTA).

The MSST manufacturing guidelines provide guidance and minimum standards for the binding, paper type, and other physical characteristics of a text with the goal of making it more durable. These manufacturing standards are in common use for manufacturing of basal level texts. For full production specification detail visit: www.NASTA.org

*Pre-AP, AP and Advanced Placement program are registered trademarks of the College Entrance Examination Board, which was not involved in the production of and does not endorse these products.

Brief Contents

Contents

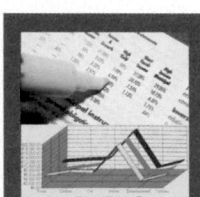

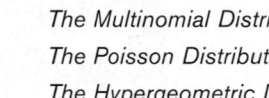

CHAPTER 4

Probability and Counting Rules 171

CHAPTER 5

Discrete Probability Distributions 237

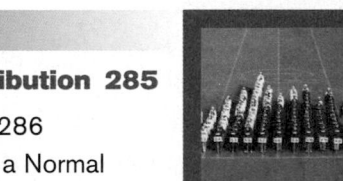

CHAPTER 6

The Normal Distribution 285

CHAPTER 7

Confidence Intervals and Sample Size 347

Preface

Approach

Elementary Statistics: A Step by Step Approach was written to help students in the beginning statistics course whose mathematical background is limited to basic algebra. The book follows a nontheoretical approach without formal proofs, explaining concepts intuitively and supporting them with abundant examples. The applications span a broad range of topics certain to appeal to the interests of students of diverse backgrounds and include problems in business, sports, health, architecture, education, entertainment, political science, psychology, history, criminal justice, the environment, transportation, physical sciences, demographics, eating habits, and travel and leisure.

About This Book

While a number of important changes have been made to the sixth edition, the learning system remains untouched and provides students with a useful framework in which to learn and apply concepts. Some of the retained features include the following:

- **Over 1800** exercises are located at the end of major sections within each chapter.

- **Hypothesis-Testing Summaries** are found at the end of Chapter 9 (z, t, χ^2, and F tests for testing means, proportions, and variances), Chapter 12 (correlation, chi-square, and ANOVA), and Chapter 13 (nonparametric tests) to show students the different types of hypotheses and the types of tests to use.

- A **Data Bank** listing various attributes (educational level, cholesterol level, gender, etc.) for 100 people and 13 additional data sets using real data are included and referenced in various exercises and projects throughout the book, including the projects presented in Data Projects sections.

- A **reference card** containing the formulas and the z, t, χ^2, and PPMC tables is included with this textbook.

- End-of-chapter **Summaries, Important Terms,** and **Important Formulas** give students a concise summary of the chapter topics and provide a good source for quiz or test preparation.

- **Review Exercises** are found at the end of each chapter.

- Special sections called **Data Analysis** require students to work with a data set to perform various statistical tests or procedures and then summarize the results. The data are included in the Data Bank in Appendix D and can be downloaded from the book's website at www.mhhe.com/bluman

- **Chapter Quizzes,** found at the end of each chapter, include multiple-choice, true/false, and completion questions along with exercises to test students' knowledge and comprehension of chapter content.

- The **Appendices** provide students with an essential algebra review, an outline for report writing, Bayes' theorem, extensive reference tables, a glossary of key terms and symbols, and answers to all quiz questions, all odd-numbered exercises, selected even-numbered exercises, and an alternate method for using the standard normal distribution.

Changes in the Sixth Edition

This edition of *Elementary Statistics* is updated and improved for students and instructors in the following ways:

- **Over 300** new exercises have been added, **most using real data,** and many questions now incorporate thought-provoking questions requiring students to interpret their results.

- The text is updated throughout with current data and statistics including **44** new *Unusual Stats* and *Interesting Facts;* **7** new *Speaking of Statistics;* **5** new *Critical Thinking Challenges;* **2** new *Statistics Today* openers; **8** new worked examples; **14** new *Data Analysis Exercises;* and **5** new Data Sets.

- A new feature, *Applying the Concepts,* is added to each section and gives students an opportunity to think about the concepts and to apply them to hypothetical examples and scenarios similar to those found in newspapers, magazines, and news programs.

- The text layout and color palette have been redesigned to increase the readability and ease of use by students and instructors.

Based on user suggestions and reviewer comments on the fifth edition, the following improvements were made:

Chapter 1   Another example of interval-level data has been added. The explanation of random sampling was expanded so students would not have to refer to Chapter 14.

Chapter 2   The explanation of class, frequency, relative frequency, and open-ended frequency distributions was expanded. An explanation was given on how to analyze frequency distributions.

Chapter 3   A greater explanation was given of the mode, including bimodal and multimodal data sets. Also added were the range rule of thumb and an exercise on finding the median for grouped data.

Chapter 4   More detailed explanation was added on the use of the words *and* and *or* in classical probability. A tree diagram was included to help determine the sample space for Exercise 4–40.

Chapter 5   Coverage of discrete variables was expanded.

Chapter 6   An explanation was included on how the area under a continuous curve relates to a probability by using a uniform distribution. More information on the distribution of sample means was given.

Chapter 7   A brief explanation of the sampling distribution of a sample proportion was added.

Chapter 8   The explanation on using the *P*-value is now boxed.

Chapter 10   The concepts of independent and dependent variables and simple and multiple relationships were expanded. The topic of the relationship of the scatter plot to the strength of the correlation coefficient was moved from Section 10–4 to Section 10–3.

Acknowledgments

It is important to acknowledge the many people whose contributions have gone into the Sixth Edition of *Elementary Statistics*. Very special thanks are due to Jackie Miller of The Ohio State University for her provision of the Index of Applications, her exhaustive accuracy check of the page proofs, and her general availability and advice concerning all matters statistical. The Technology Step by Step sections were provided by Gerry Moultine of Northwood University (MINITAB), John Thomas of College of Lake County (Excel), and Michael Keller of St. Johns River Community College (TI-83 Plus and TI-84 Plus). Finally, at McGraw-Hill Higher Education, thanks to Steve Stembridge, Sponsoring Editor; David Dietz, Director of Development; Peter Galuardi, Developmental Editor; Vicki Krug, Senior Project Manager; Jeff Huettman, Lead Media Technology Producer; and Sandra Schnee, Senior Media Project Manager.

Allan G. Bluman

Special thanks for their advice and recommendations for revisions found in the Sixth Edition go to

Rosalie Abraham, *Florida Community College-North*

Anne Albert, *The University of Findlay*

Raid Amin, *University of West Florida*

Trania Aquino, *Del Mar College*

John J. Avioli, *Christopher Newport University*

Rona Axelrod, *Edison Community College*

Mark D. Baker, M.S., *Illinois State University*

Sivanandan Balakumar, *Lincoln University*

Freda Bennett, *Massachusetts College of Liberal Arts*

Matthew Bognar, *University of Iowa*

Andrea Boito, *Pennsylvania State University–Altoona*

Dean Burbank, *Gulf Coast Community College*

Christine Bush, *Palm Beach Community College–Palm Beach Gardens*

Carlos Canas, *Florida Memorial College*

Gregory Daubenmire, *Las Positas College*

Joseph Glaz, *University of Connecticut*

Rebekah A. Griffith, *McNeese State University*

Renu A. Gupta, *Louisiana State University–Alexandria*

Harold S. Hayford, *Pennsylvania State University–Altoona*

Helene Humphrey, *San Joaquin Delta College*

Anand Katiyar, *McNeese State University*

Brother Donald Kelly, *Marist College*

Dr. Susan Kelly, *University of Wisconsin–La Crosse*

Michael Kent, *Borough of Manhattan Community College*

B. M. Golam Kibria, *Florida International University–Miami*

Jong Sung Kim, *Portland State University*

Joseph Kunicki, *University of Findlay*

Marie Langston, *Palm Beach Community College–Lakeworth*

Susan S. Lenker, *Central Michigan University*

Judith McCrory, *University of Findlay*

Charles J. Miller, Jr., *Camden County College*

Carla A. Monticelli, *Camden County College*

Dr. Christina Anne Morian, *Lincoln University*

Ken Mulzet, *Florida Community College–Jacksonville*

Irene Palacios, *Grossmont College*

Elaine Paris, *Mercy College*

Samuel Park, *Long Island University–Brooklyn*

Chester Piascik, *Bryant University*

Leela Rakesh, *Central Michigan University*

Don R. Robinson, *Illinois State University*

Kathy Rogotzke, *North Iowa Area Community College–Mason City*

Dr. J. N. Singh, *Barry University*

George Smeltzer, *Pennsylvania State University–Abington*

Diana Staats, *Dutchess Community College*

Richard Stockbridge, *University of Wisconsin–Milwaukee*

Linda Sturges, *SUNY Maritime College*

Klement Teixeira, *Borough of Manhattan Community College*

Christina Vertullo, *Marist College*

Cassandra L. Vincent, *Plattsburgh State University*

Cheng Wang, *Nova Southeastern University*

Glenn Weber, *Christopher Newport University*

Also, special thanks for their help with the Fifth Edition go to

Naveen K. Bansal, *Marquette University*

James Condor, *Manatee Community College–Bradenton*

Diane Cope, *Washington & Jefferson College*

Melody E. Eldred, *State University College–Oneonta*

Abdul Elfessi, *University of Wisconsin–LaCrosse*

Gholamhosse Gharehgozlo Hamedani, *Marquette University*

Liliana Gonzalez, *University of Rhode Island–Kingston*

Shahryar Heydari, *Piedmont College*

Patricia Humphrey, *Georgia Southern University*

Charles W. Johnson, *Collin County Community College–Plano*

Jeffery C. Jones, *County College of Morris*

Anand S. Katiyar, *McNeese State University*

Hyun-Joo Kim, *Truman State University*

Benny Lo, *DeVry University*

Chip Mason, *Belhaven College*

Judith McCrory, *Findlay University*

Lynnette Meslinsky, *Erie Community College*

Lindsay Packer, *College of Charleston*

Fernando Rincón, *Piedmont Technical College*

Deb Rumsey, *The Ohio State University*

Salvatore Sciandra, Jr., *Niagara County Community College–Sandborn*

Carolyn Shealy, *Piedmont Technical College*

Jeganathan Sriskandarajah, *Madison Area Technical College*

Richard Stevens, *University of Alaska–Fairbanks*

Sherry Taylor, *Piedmont Technical College*

Diane Van Deusen, *Napa Valley College*

David Wallach, *Findlay University*

Guided Tour: Features and Supplements

Each chapter begins with an **outline** and a list of **learning objectives.** The objectives are repeated at the beginning of each section to help students focus on the concepts presented within that section.

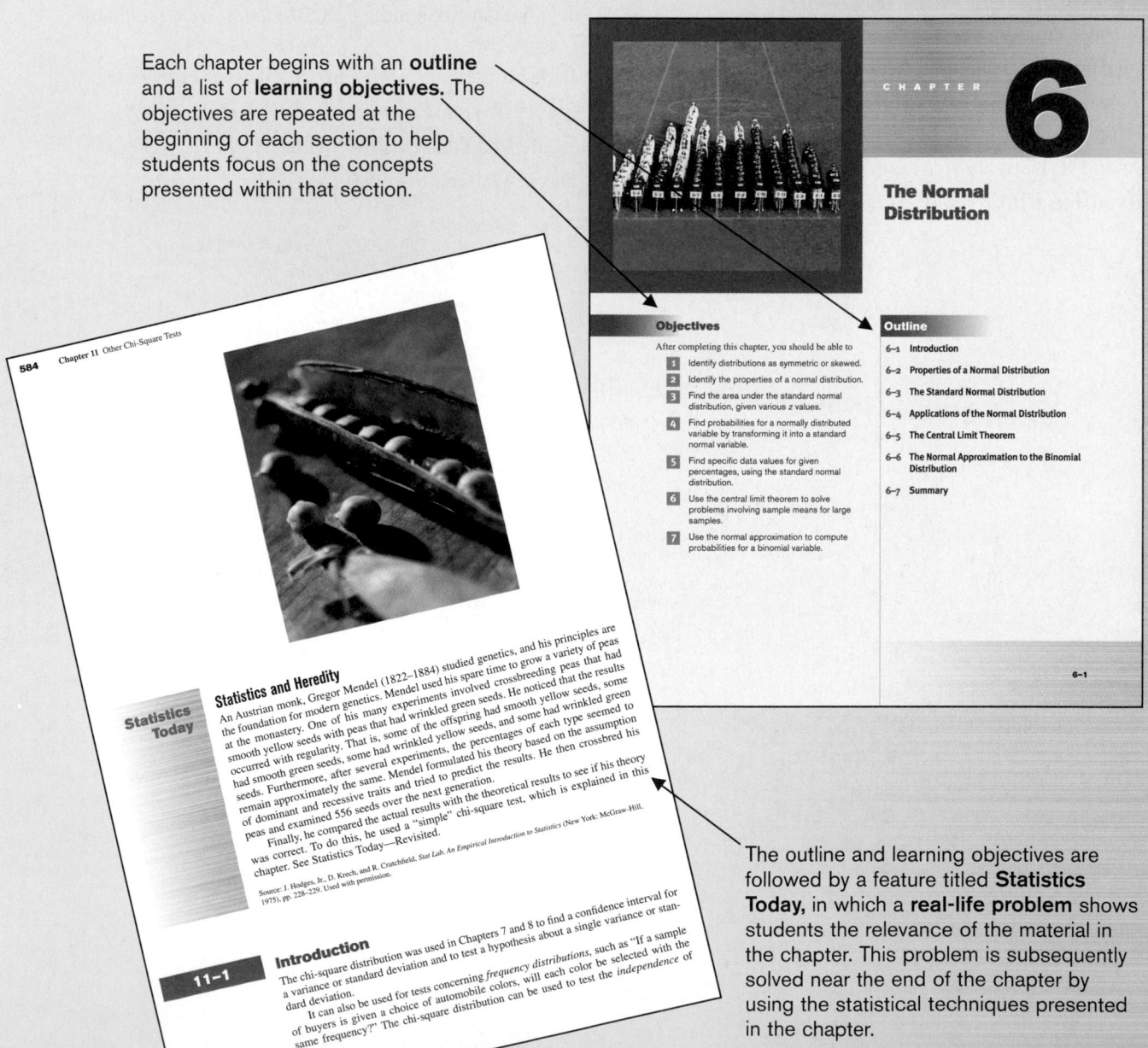

CHAPTER 6

The Normal Distribution

Objectives

After completing this chapter, you should be able to

1 Identify distributions as symmetric or skewed.

2 Identify the properties of a normal distribution.

3 Find the area under the standard normal distribution, given various z values.

4 Find probabilities for a normally distributed variable by transforming it into a standard normal variable.

5 Find specific data values for given percentages, using the standard normal distribution.

6 Use the central limit theorem to solve problems involving sample means for large samples.

7 Use the normal approximation to compute probabilities for a binomial variable.

Outline

6–1 Introduction

6–2 Properties of a Normal Distribution

6–3 The Standard Normal Distribution

6–4 Applications of the Normal Distribution

6–5 The Central Limit Theorem

6–6 The Normal Approximation to the Binomial Distribution

6–7 Summary

6–1

584   Chapter 11   Other Chi-Square Tests

Statistics Today

Statistics and Heredity

An Austrian monk, Gregor Mendel (1822–1884) studied genetics, and his principles are the foundation for modern genetics. Mendel used his spare time to grow a variety of peas at the monastery. One of his many experiments involved crossbreeding peas that had smooth yellow seeds with peas that had wrinkled green seeds. He noticed that the results occurred with regularity. That is, some of the offspring had smooth yellow seeds, some had smooth green seeds, some had wrinkled yellow seeds, and some had wrinkled green seeds. Furthermore, after several experiments, the percentages of each type seemed to remain approximately the same. Mendel formulated his theory based on the assumption of dominant and recessive traits and tried to predict the results.

Finally, he compared the actual results with the theoretical results to see if his theory was correct. To do this, he used a "simple" chi-square test, which is explained in this chapter. He then crossbred his peas and examined 556 seeds over the next generation. See Statistics Today—Revisited.

Source: J. Hodges, Jr., D. Krech, and R. Crutchfield, *Stat Lab, An Empirical Introduction to Statistics* (New York: McGraw-Hill, 1975), pp. 228–229. Used with permission.

11–1 **Introduction**

The chi-square distribution was used in Chapters 7 and 8 to find a confidence interval for a variance or standard deviation and to test a hypothesis about a single variance or standard deviation.

It can also be used for tests concerning *frequency distributions*, such as "If a sample of buyers is given a choice of automobile colors, will each color be selected with the same frequency?" The chi-square distribution can be used to test the *independence* of

11–2

The outline and learning objectives are followed by a feature titled **Statistics Today,** in which a **real-life problem** shows students the relevance of the material in the chapter. This problem is subsequently solved near the end of the chapter by using the statistical techniques presented in the chapter.

Over 300 **examples** with detailed solutions serve as models to help students solve problems on their own. Examples are solved by using a step-by-step explanation, and illustrations provide a clear display of results for students.

Numerous examples and exercises use **real data.** The icon shown here indicates that the data set for the exercise is available in a variety of file formats on the text's Online Learning Center and CD-ROM.

---

36 Chapter 2 Frequency Distributions and Graphs

Categorical Frequency Distributions

The **categorical frequency distribution** is used for data that can be placed in specific categories, such as nominal- or ordinal-level data. For example, data such as political affiliation, religious affiliation, or major field of study would use categorical frequency distributions.

Example 2–1

Twenty-five army inductees were given a blood test to determine their blood type. The data set is

A	B	B	AB	O
O	B	B	AB	B
B	O	B	O	
A	B	O	A	O
AB	A	O	O	AB
				A
			B	

Construct a frequency distribution for the data.

Solution

Since the data are categorical, discrete classes can be used. There are four blood types: A, B, O, and AB. These types will be used as the classes for the distribution.

The procedure for constructing a frequency distribution for categorical data is given next.

Step 1 Make a table as shown.

A Class	B Tally	C Frequency	D Percent
A			
B			
O			
AB			

Step 2 Tally the data and place the results in column B.

Step 3 Count the tallies and place the results in column C.

Step 4 Find the percentage of values in each class by using the formula

$$\% = \frac{f}{n} \cdot 100\%$$

where f = frequency of the class and n = total number of values. For example, in the class of type A blood, the percentage is

$$\% = \frac{5}{25} \cdot 100\% = 20\%$$

Percentages are not normally part of a frequency distribution, but they can be added since they are used in certain types of graphs such as pie graphs. Also, the decimal equivalent of a percent is called a *relative frequency*.

Step 5 Find the totals for columns C (frequency) and D (percent). The completed table is shown.

---

414 Chapter 8 Hypothesis Testing

33	42	125	62	134	73
39	69	23	94	73	24
51	55	26	66	41	67
15	53	56	91	20	78
70	25	62	115	17	36
58	56	33	75	20	16

Source: Based on information from the National Insurance Crime Bureau.

Using this information, answer these questions.

1. What are the hypotheses that you would use?
2. Is the sample considered small or large?
3. What assumption must be met before the hypothesis test can be conducted?
4. Which probability distribution would you use?
5. Would you select a one- or two-tailed test? Why?
6. What critical value(s) would you use?
7. Conduct a hypothesis test.
8. What is your decision?
9. What is your conclusion?
10. Write a brief statement summarizing your conclusion.
11. If you lived in a city whose population was about 50,000, how many automobile thefts per year would you expect to occur?

See page 460 for the answers.

Exercises 8–3

For Exercises 1 through 13, perform each of the following steps.

a. State the hypotheses and identify the claim.
b. Find the critical value(s).
c. Compute the test value.
d. Make the decision.
e. Summarize the results.

Use diagrams to show the critical region (or regions), and use the traditional method of hypothesis testing unless otherwise specified.

1. A survey claims that the average cost of a hotel room in Atlanta is $69.21. To test the claim, a researcher selects a sample of 30 hotel rooms and finds that the average cost is $68.43. The standard deviation of the population is $3.72. At $\alpha = 0.05$, is there enough evidence to reject the claim?
Source: *USA TODAY.*

2. It has been reported that the average credit card debt for college seniors is $3262. The student senate at a large university feels that their seniors have a debt much less than this, so it conducts a study of 50 randomly selected seniors and finds that the average debt is $2995 with a sample standard deviation of $1100. With $\alpha = 0.05$, is the student senate correct?
Source: *USA TODAY.*

3. A researcher estimates that the average revenue of the largest businesses in the United States is greater than $24 billion. A sample of 50 companies is selected, and the revenues (in billions of dollars) are shown. At $\alpha = 0.05$, is there enough evidence to support the researcher's claim?

178	122	91	44	35
61	56	46	20	32
30	28	28	20	27
29	16	16	19	15
41	38	36	15	25
31	30	19	19	19
24	16	15	15	19
25	25	18	14	15
24	23	17	17	22
22	21	20	17	20

Source: *N.Y. Times Almanac.*

4. Full-time Ph.D. students receive an average salary of $12,837 according to the U.S. Department of Education. The dean of graduate studies at a large state university feels that Ph.D. students in his state earn more than this. He surveys 44 randomly selected students and finds their average salary is $14,445 with a standard deviation of $1500. With $\alpha = 0.05$, is the dean correct?
Source: U.S. Department of Education/*Chronicle of Higher Education.*

5. A report in *USA TODAY* stated that the average age of commercial jets in the United States is 14 years. An

At the end of appropriate sections, **Technology Step by Step** boxes show students how to use MINITAB, the TI-83 Plus and TI-84 Plus graphing calculators, and Excel to solve the types of problems covered in the section. Instructions are presented in numbered steps, usually in the context of examples—including examples from the main part of the section. Numerous computer or calculator screens are displayed, showing intermediate steps as well as the final answer.

---

38. An instructor gives a 100-point examination in which the grades are normally distributed. The mean is 60 and the standard deviation is 10. If there are 5% A's and 5% F's, 15% B's and 15% D's, and 60% C's, find the scores that divide the distribution into those categories.

39. The data shown represent the number of outdoor drive-in movies in the United States for a 14-year period. Check for normality.

| 2084 | 1497 | 1014 | 910 | 899 | 870 | 837 | 859 |
| 848 | 826 | 815 | 750 | 637 | 737 | | |

Source: National Association of Theater Owners.

40. The data shown represent the cigarette tax (in cents) for 30 randomly selected states. Check for normality.

3	58	5	65	17	48	52	75	21	76	58	36
100	111	34	41	23	44	33	50	13	18	7	12
20		24	66	28	28	31					

Source: Commerce Clearing House.

41. The data shown represent the box office total revenue (in millions of dollars) for a randomly selected sample of the top-grossing films in 2001. Check for normality.

| 294 | 241 | 130 | 144 | 113 | 70 | 97 | 94 | 91 | 202 | 74 | 79 |
| 71 | 67 | 67 | 56 | 180 | 199 | 165 | 114 | 60 | 56 | 53 | 51 |

Source: USA TODAY.

42. The data shown represent the number of runs made each year during Bill Mazeroski's career. Check for normality.

| 30 | 59 | 69 | 50 | 58 | 71 | 55 | 43 | 66 | 52 | 56 | 62 |
| 36 | 13 | 29 | 17 | 3 | | | | | | | |

Source: Greensburg Tribune Review.

Technology Step by Step

MINITAB
Step by Step

Determining Normality

There are several ways in which statisticians test a data set for normality. Four are shown here.

Construct a Histogram

Inspect the histogram for shape.

1. Enter the data for Example 6–19 in the first column of a new worksheet. Name the column Inventory.
2. Use **Stat>Basic Statistics>Graphical Summary** presented in Section 3–4 to create the histogram. Is it symmetric? Is there a single peak?

Applying the Concepts 10–5

Interpreting Simple Linear Regression

Answer the questions about the following computer-generated information.

Linear correlation coefficient $r = 0.794556$
Coefficient of determination $= 0.631319$
Standard error of estimate $= 12.9668$
Explained variation $= 5182.41$
Unexplained variation $= 3026.49$
Total variation $= 8208.90$
Equation of regression line $y' = 0.725983X + 16.5523$
Level of significance $= 0.1$
Test statistic $= 0.794556$
Critical value $= 0.378419$

1. Are both variables moving in the same direction?
2. Which number measures the distances from the prediction line to the actual values?
3. Which number is the slope of the regression line?
4. Which number is the y intercept of the regression line?
5. Which number can be found in a table?
6. Which number is the allowable risk of making a type I error?
7. Which number measures the variation explained by the regression?
8. Which number measures the scatter of points about the regression line?
9. What is the null hypothesis?
10. Which number is compared to the critical value to see if the null hypothesis should be rejected?
11. Should the null hypothesis be rejected?

See page 581 for the answers.

A new feature called **Applying the Concepts** has been added to the Sixth Edition. These exercises are found at the end of each section, and their purpose is to reinforce the concepts explained in the section. They give the student an opportunity to think about the concepts and apply them to hypothetical examples similar to real-life ones found in newspapers, magazines, and professional journals. Most contain open-ended questions—questions that require interpretation and may have more than one correct answer. These exercises can also be used as classroom discussion topics for instructors who like to use this type of teaching technique. The majority of these exercises were written and class-tested by Dr. James A. Condor and were previously published in *Critical Thinking Workbook*. The rest were written by the author.

Data Projects

Use MINITAB, the TI-83 Plus, the TI-84 Plus, or a computer program of your choice to complete these exercises.

1. Select several variables, such as the number of points a football team scored in each game of a specific season, the number of passes completed, or the number of yards gained. Using confidence intervals for the mean, determine the 90, 95, and 99% confidence intervals. (Use z or t, whichever is relevant.) Decide which you think is more appropriate. When this is completed, write a summary of your findings by answering the following questions.

   a. What was the purpose of the study?
   b. What was the population?
   c. How was the sample selected?

   d. What were the results obtained by using confidence intervals?
   e. Did you use z or t? Why?

2. Using the same data or different data, construct a confidence interval for a proportion. For example, you might want to find the proportion of passes completed by the quarterback or the proportion of passes that were intercepted. Write a short paragraph summarizing the results.

You may use the following websites to obtain raw data:

Visit the data sets at the book's website found at **http://www.mhhe.com/math/stat/bluman**
Click on the 6th edition.
http://lib.stat.cmu.edu/DASL
http://www.statcan.ca

7–42

Data Projects further challenge students' understanding and application of the material presented in the chapter. Many of these require the student to gather, analyze, and report on real data. These projects, which appear at the end of each chapter, may include a World Wide Web icon , indicating that websites are listed as possible sources of data.

Multimedia Supplements

MathZone—www.mathzone.com

McGraw-Hill's **MathZone 3.0** is a complete **web-based tutorial and course management system** for mathematics and statistics, designed for greater ease of use than any other system available. Free upon adoption of a McGraw-Hill textbook, the system enables instructors to **create and share courses and assignments** with colleagues with only a few mouse clicks. All **assignments, exercises, e-Professor multimedia tutorials, video lectures, and NetTutor® live tutors** follow the textbook's learning objectives and problem-solving style and notation. Using MathZone's **assignment builder,** instructors can **edit questions and algorithms, import their own content,** and **create announcements and due dates** for homework and quizzes. MathZone's **automated grading function** reports the results of easy-to-assign algorithmically generated homework, quizzes, and tests. All student activity within MathZone is recorded and available through a **fully integrated gradebook** that can be downloaded to Microsoft Excel®. MathZone also is available on CD-ROM. (See "Supplements for the Student" for descriptions of the elements of MathZone.) High-School adopters, please contact your McGraw-Hill representative regarding student pass codes.

ALEKS

ALEKS (**A**ssessment and **LE**arning in **K**nowledge **S**paces) is an artificial intelligence-based system for mathematics learning, available over the web 24/7. Using unique adaptive questioning, ALEKS accurately assesses what topics each student knows and then determines exactly what each student is ready to learn next. ALEKS interacts with the students much as a skilled human tutor would, moving between explanation and practice as needed, correcting and analyzing errors, defining terms and changing topics on request, and helping them master the course content more quickly and easily. Moreover, the new ALEKS 3.0 now links to text-specific videos, multimedia tutorials, and text book pages in PDF format. ALEKS also offers a robust classroom management system that allows instructors to monitor and direct student progress toward mastery of curricular goals. See www.highed.aleks.com

Instructor's Testing and Resource CD-ROM (instructors only)

The computerized test bank contains a variety of questions, including true/false, multiple-choice, short answer, and short problems requiring analysis and written answers. The testing material is coded by type of question and level of difficulty. The Brownstone Diploma® system enables you to efficiently select, add, and organize questions, such as by type of question or level of difficulty. It also allows for printing tests along with answer keys as well as editing the original questions, and it is available for Windows and Macintosh systems. The CD-ROM also contains PowerPoint® slides, printable tests, and a print version of the test bank.

Text-Specific Videos

Available with this edition are text-specific DVDs that demonstrate key concepts and worked-out exercises from the text plus tutorials in using the TI-83 Plus and TI-84 Plus calculators, Excel, and MINITAB, in a dynamic, engaging format.

NetTutor

NetTutor is a revolutionary system that enables students to interact with a live tutor over the Web by using NetTutor's Web-based, graphical chat capabilities. Students can also submit questions and receive answers, browse previously answered questions, and view previous live chat sessions. NetTutor can be accessed through MathZone.

SPSS Student Version 13 for Windows

A student version of SPSS statistical software is available with copies of this text. Consult your McGraw-Hill representative for details.

Visual Statistics

Visual Statistics is an easy-to-use interactive multimedia tool that is used to teach and learn statistical concepts graphically. It provides complete and thorough coverage of major statistical concepts, giving both student and instructor a visually oriented teaching and learning package to complement his or her text. It's available in two formats: CD with Student Workbook, ISBN-13: 978–0–07–240094–6 (ISBN-10: 0–07–240094–3); CD only, ISBN-13: 978–0–07–240012–0 (ISBN-10: 0–07–240012–9). And remember, too, that the CD actually contains a printable, pdf-formatted version of the entire workbook!

Additional Videos Series (instructors only)

Against All Odds and *Decisions through Data* are video series available to qualified adopters. Please contact your local sales representative for more information about these programs.

Print Supplements

Annotated Instructors Edition (instructors only)

The Annotated Instructor's Edition contains answers to all exercises and tests. The answers to most questions are printed in red next to each problem. Answers not appearing on the page can be found in the Answer Appendix at the end of the book.

Instructor's Solutions Manual (instructors only)

By Sally Robinson of South Plains College, this manual includes worked-out solutions to all the exercises in the text and answers to all quiz questions.

Student Study Guide

By Pat Foard of South Plains College, this study guide will assist students in understanding and reviewing key concepts and preparing for exams. It emphasizes all important concepts contained in each chapter, includes explanations, and provides opportunities for students to test their understanding by completing related exercises and problems.

Student Solutions Manual

By Sally Robinson of South Plains College, this manual contains detailed solutions to all odd-numbered text problems and answers to all quiz questions.

MINITAB 14 Manual

This manual provides the student with how-to information on data and file management, conducting various statistical analyses, and creating presentation-style graphics while following each text chapter.

TI-83 Plus and TI-84 Plus Graphing Calculator Manual

This friendly, practical manual teaches students to learn about statistics and solve problems by using these calculators while following each text chapter.

Excel Manual

This workbook, specially designed to accompany the text, provides additional practice in applying the chapter concepts while using Excel.

Index of Applications

Elementary Statistics: A Step by Step Approach contains a large number of applications—in the text's Examples, Exercises, and Critical Thinking Challenges—to illuminate students' understanding of **how statistical concepts are practiced and incorporated into many diverse personal, professional, and academic fields.** You will find these applications on the pages listed.

CHAPTER **7**

Confidence Intervals and Sample Size

CHAPTER **8**

Hypothesis Testing

CHAPTER **9**

Testing the Difference Between Two Means, Two Variables, and Two Proportions

CHAPTER 10

Correlation and Regression

CHAPTER 11

Other Chi-Square Tests

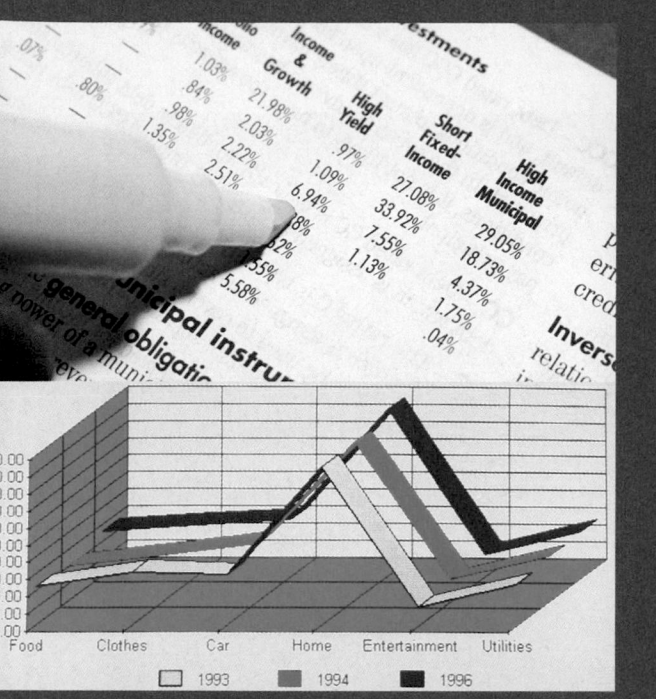

The Nature of Probability and Statistics

Objectives

After completing this chapter, you should be able to

1 Demonstrate knowledge of statistical terms.

2 Differentiate between the two branches of statistics.

3 Identify types of data.

4 Identify the measurement level for each variable.

5 Identify the four basic sampling techniques.

6 Explain the difference between an observational and an experimental study.

7 Explain how statistics can be used and misused.

8 Explain the importance of computers and calculators in statistics.

Outline

Are We Improving Our Diet?

It has been determined that diets rich in fruits and vegetables are associated with a lower risk of chronic diseases such as cancer. Nutritionists recommend that Americans consume five or more servings of fruits and vegetables each day. Several researchers from the Division of Nutrition, the National Center for Chronic Disease Control and Prevention, the National Cancer Institute, and the National Institutes of Health decided to use statistical procedures to see how much progress is being made toward this goal.

The procedures they used and the results of the study will be explained in this chapter. See Statistics Today—Revisited at the end of this chapter.

1–1

Introduction

Most people become familiar with probability and statistics through radio, television, newspapers, and magazines. For example, the following statements were found in newspapers.

- Nearly one in seven U.S. families are struggling with bills from medical expenses even though they have health insurance. Source: *Psychology Today,* October 2004.

- Eating 10 grams (g) of fiber a day reduces the risk of heart attack by 14%. Source: Archives of Internal Medicine, *Reader's Digest,* May 2004.

- Thirty minutes (of exercise) two or three times each week can raise HDLs 10 to 15%. Source: *Prevention,* July 2004.

- The average credit card debt per household in 2003 was $9205. Source: www.cardweb.com.

- About 15% of men in the United States are left-handed and 9% of women are left-handed. Source: Scripps Survey Research Center.

- The median age of couples who watch Jay Leno is 48.1 years. Source: Nielsen Media Research 2003–2004.

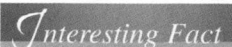

Statistics is used in almost all fields of human endeavor. In sports, for example, a statistician may keep records of the number of yards a running back gains during a football game, or the number of hits a baseball player gets in a season. In other areas, such as public health, an administrator might be concerned with the number of residents who contract a new strain of flu virus during a certain year. In education, a researcher might want to know if new methods of teaching are better than old ones. These are only a few examples of how statistics can be used in various occupations.

Furthermore, statistics is used to analyze the results of surveys and as a tool in scientific research to make decisions based on controlled experiments. Other uses of statistics include operations research, quality control, estimation, and prediction.

> **Statistics** is the science of conducting studies to collect, organize, summarize, analyze, and draw conclusions from data.

Students study statistics for several reasons:

1. Students, like professional people, must be able to read and understand the various statistical studies performed in their fields. To have this understanding, they must be knowledgeable about the vocabulary, symbols, concepts, and statistical procedures used in these studies.

2. Students and professional people may be called on to conduct research in their fields, since statistical procedures are basic to research. To accomplish this, they must be able to design experiments; collect, organize, analyze, and summarize data; and possibly make reliable predictions or forecasts for future use. They must also be able to communicate the results of the study in their own words.

3. Students and professional people can also use the knowledge gained from studying statistics to become better consumers and citizens. For example, they can make intelligent decisions about what products to purchase based on consumer studies, about government spending based on utilization studies, and so on.

These reasons can be considered the goals for students and professionals who study statistics.

It is the purpose of this chapter to introduce the student to the basic concepts of *probability* and *statistics* by answering questions such as the following:

What are the branches of statistics?

What are data?

How are samples selected?

1–2   Descriptive and Inferential Statistics

To gain knowledge about seemingly haphazard events, statisticians collect information for *variables,* which describe the event.

Objective 1

Demonstrate knowledge of statistical terms.

> A **variable** is a characteristic or attribute that can assume different values.

Data are the values (measurements or observations) that the variables can assume. Variables whose values are determined by chance are called **random variables.**

Suppose that an insurance company studies its records over the past several years and determines that, on average, 3 out of every 100 automobiles the company insured were involved in accidents during a 1-year period. Although there is no way to predict

Objective 2

Differentiate between the two branches of statistics.

Historical Note

The origin of descriptive statistics can be traced to data collection methods used in censuses taken by the Babylonians and Egyptians between 4500 and 3000 B.C. In addition, the Roman Emperor Augustus (27 B.C.–A.D. 17) conducted surveys on births and deaths of the citizens of the empire, as well as the number of livestock each owned and the crops each citizen harvested yearly.

the specific automobiles that will be involved in an accident (random occurrence), the company can adjust its rates accordingly, since the company knows the general pattern over the long run. (That is, on average, 3% of the insured automobiles will be involved in an accident each year.)

A collection of data values forms a **data set.** Each value in the data set is called a **data value** or a **datum.**

Data can be used in different ways. The body of knowledge called statistics is sometimes divided into two main areas, depending on how data are used. The two areas are

1. Descriptive statistics

2. Inferential statistics

In *descriptive statistics* the statistician tries to describe a situation. Consider the national census conducted by the U.S. government every 10 years. Results of this census give the average age, income, and other characteristics of the U.S. population. To obtain this information, the Census Bureau must have some means to collect relevant data. Once data are collected, the bureau must organize and summarize them. Finally, the bureau needs a means of presenting the data in some meaningful form, such as charts, graphs, or tables.

Descriptive statistics consists of the collection, organization, summarization, and presentation of data.

The second area of statistics is called *inferential statistics.* Here, the statistician tries to make inferences from *samples* to *populations.* Inferential statistics uses **probability,** i.e., the chance of an event occurring. Many people are familiar with the concepts of probability through various forms of gambling. People who play cards, dice, bingo, and lotteries win or lose according to the laws of probability. Probability theory is also used in the insurance industry and other areas.

It is important to distinguish between a sample and a population.

A **population** consists of all subjects (human or otherwise) that are being studied.

Most of the time, due to the expense, time, size of population, medical concerns, etc., it is not possible to use the entire population for a statistical study; therefore, researchers use samples.

A **sample** is a group of subjects selected from a population.

Unusual Stat

Twenty-nine percent of Americans want their boss's job.

If the subjects of a sample are properly selected, most of the time they should possess the same or similar characteristics as the subjects in the population. The techniques used to properly select a sample will be explained in Section 1–4.

An area of inferential statistics called **hypothesis testing** is a decision-making process for evaluating claims about a population, based on information obtained from samples. For example, a researcher may wish to know if a new drug will reduce the number of heart attacks in men over 70 years of age. For this study, two groups of men over 70 would be selected. One group would be given the drug, and the other would be given a placebo (a substance with no medical benefits or harm). Later, the number of heart attacks occurring in each group of men would be counted, a statistical test would be run, and a decision would be made about the effectiveness of the drug.

Statisticians also use statistics to determine *relationships* among variables. For example, relationships were the focus of the most noted study in the past few decades,

This *USA TODAY* Snapshot shows the percentages of adults who say these activities are extremely important. Based on the information shown in the Snapshot, do you think people take better care of their automobiles than they do of themselves?

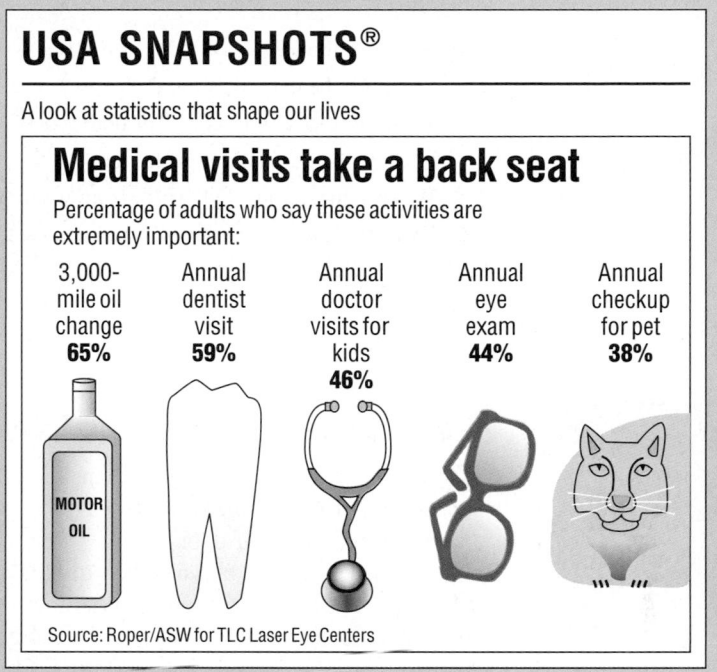

Source: Copyright 2002, *USA TODAY*. Reprinted with permission.

"Smoking and Health," published by the Surgeon General of the United States in 1964. He stated that after reviewing and evaluating the data, his group found a definite relationship between smoking and lung cancer. He did not say that cigarette smoking actually causes lung cancer, but that there is a relationship between smoking and lung cancer. This conclusion was based on a study done in 1958 by Hammond and Horn. In this study, 187,783 men were observed over a period of 45 months. The death rate from lung cancer in this group of volunteers was 10 times as great for smokers as for nonsmokers.

Finally, by studying past and present data and conditions, statisticians try to make predictions based on this information. For example, a car dealer may look at past sales records for a specific month to decide what types of automobiles and how many of each type to order for that month next year.

Inferential statistics consists of generalizing from samples to populations, performing estimations and hypothesis tests, determining relationships among variables, and making predictions.

Applying the Concepts 1–2

Attendance and Grades

Read the following on attendance and grades, and answer the questions.

A study conducted at Manatee Community College revealed that students who attended class 95 to 100% of the time usually received an A in the class. Students who attended class 80 to 90% of the time usually received a B or C in the class. Students who attended class less than 80% of the time usually received a D or an F or eventually withdrew from the class.

Based on this information, attendance and grades are related. The more you attend class, the more likely you will receive a higher grade. If you improve your attendance, your grades

Only one-third of crimes committed are reported to the police.

will probably improve. Many factors affect your grade in a course. One factor that you have considerable control over is attendance. You can increase your opportunities for learning by attending class more often.

1. What are the variables under study?
2. What are the data in the study?
3. Are descriptive, inferential, or both types of statistics used?
4. What is the population under study?
5. Was a sample collected? If so, from where?
6. From the information given, comment on the relationship between the variables.

See page 32 for the answers.

1–3   Variables and Types of Data

Objective 3

Identify types of data.

As stated in Section 1–2, statisticians gain information about a particular situation by collecting data for random variables. This section will explore in greater detail the nature of variables and types of data.

Variables can be classified as qualitative or quantitative. **Qualitative variables** are variables that can be placed into distinct categories, according to some characteristic or attribute. For example, if subjects are classified according to gender (male or female), then the variable *gender* is qualitative. Other examples of qualitative variables are religious preference and geographic locations.

Quantitative variables are numerical and can be ordered or ranked. For example, the variable *age* is numerical, and people can be ranked in order according to the value of their ages. Other examples of quantitative variables are heights, weights, and body temperatures.

Quantitative variables can be further classified into two groups: discrete and continuous. *Discrete variables* can be assigned values such as 0, 1, 2, 3 and are said to be countable. Examples of discrete variables are the number of children in a family, the number of students in a classroom, and the number of calls received by a switchboard operator each day for a month.

Discrete variables assume values that can be counted.

Continuous variables, by comparison, can assume an infinite number of values in an interval between any two specific values. Temperature, for example, is a continuous variable, since the variable can assume an infinite number of values between any two given temperatures.

Continuous variables can assume an infinite number of values between any two specific values. They are obtained by measuring. They often include fractions and decimals.

The classification of variables can be summarized as follows:

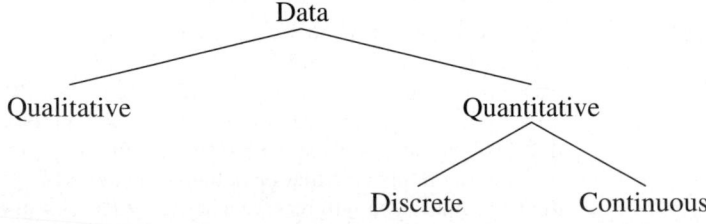

Since continuous data must be measured, answers must be rounded because of the limits of the measuring device. Usually, answers are rounded to the nearest given unit. For example, heights might be rounded to the nearest inch, weights to the nearest ounce, etc. Hence, a recorded height of 73 inches could mean any measure from 72.5 inches up to but not including 73.5 inches. Thus, the boundary of this measure is given as 72.5–73.5 inches. *Boundaries are written for convenience as 72.5–73.5 but are understood to mean all values up to but not including 73.5.* Actual data values of 73.5 would be rounded to 74 and would be included in a class with boundaries of 73.5 up to but not including 74.5, written as 73.5–74.5. As another example, if a recorded weight is 86 pounds, the exact boundaries are 85.5 up to but not including 86.5, written as 85.5–86.5 pounds. Table 1–1 helps to clarify this concept. The boundaries of a continuous variable are given in one additional decimal place and always end with the digit 5.

Table 1-1	Recorded Values and Boundaries	
Variable	**Recorded value**	**Boundaries**
Length	15 centimeters (cm)	14.5–15.5 cm
Temperature	86 degrees Fahrenheit (°F)	85.5–86.5°F
Time	0.43 second (sec)	0.425–0.435 sec
Mass	1.6 grams (g)	1.55–1.65 g

In addition to being classified as qualitative or quantitative, variables can be classified by how they are categorized, counted, or measured. For example, can the data be organized into specific categories, such as area of residence (rural, suburban, or urban)? Can the data values be ranked, such as first place, second place, etc.? Or are the values obtained from measurement, such as heights, IQs, or temperature? This type of classification—i.e., how variables are categorized, counted, or measured—uses **measurement scales,** and four common types of scales are used: nominal, ordinal, interval, and ratio.

The first level of measurement is called the *nominal level* of measurement. A sample of college instructors classified according to subject taught (e.g., English, history, psychology, or mathematics) is an example of nominal-level measurement. Classifying survey subjects as male or female is another example of nominal-level measurement. No ranking or order can be placed on the data. Classifying residents according to zip codes is also an example of the nominal level of measurement. Even though numbers are assigned as zip codes, there is no meaningful order or ranking. Other examples of nominal-level data are political party (Democratic, Republican, Independent, etc.), religion (Christianity, Judaism, Islam, etc.), and marital status (single, married, divorced, widowed, separated).

Objective 4

Identify the measurement level for each variable.

The **nominal level of measurement** classifies data into mutually exclusive (nonoverlapping), exhausting categories in which no order or ranking can be imposed on the data.

The next level of measurement is called the *ordinal level.* Data measured at this level can be placed into categories, and these categories can be ordered, or ranked. For example, from student evaluations, guest speakers might be ranked as superior, average, or poor. Floats in a homecoming parade might be ranked as first place, second place, etc. *Note that precise measurement of differences in the ordinal level of measurement* does not *exist.* For instance, when people are classified according to their build (small, medium, or large), a large variation exists among the individuals in each class.

Other examples of ordinal data are letter grades (A, B, C, D, F).

> The **ordinal level of measurement** classifies data into categories that can be ranked; however, precise differences between the ranks do not exist.

The third level of measurement is called the *interval level.* This level differs from the ordinal level in that precise differences do exist between units. For example, many standardized psychological tests yield values measured on an interval scale. IQ is an example of such a variable. There is a meaningful difference of 1 point between an IQ of 109 and an IQ of 110. Temperature is another example of interval measurement, since there is a meaningful difference of 1°F between each unit, such as 72°F and 73°F. *One property is lacking in the interval scale: There is no true zero.* For example, IQ tests do not measure people who have no intelligence. For temperature, 0°F does not mean no heat at all.

> The **interval level of measurement** ranks data, and precise differences between units of measure do exist; however, there is no meaningful zero.

The final level of measurement is called the *ratio level.* Examples of ratio scales are those used to measure height, weight, area, and number of phone calls received. Ratio scales have differences between units (1 inch, 1 pound, etc.) and a true zero. In addition, the ratio scale contains a true ratio between values. For example, if one person can lift 200 pounds and another can lift 100 pounds, then the ratio between them is 2 to 1. Put another way, the first person can lift twice as much as the second person.

> The **ratio level of measurement** possesses all the characteristics of interval measurement, and there exists a true zero. In addition, true ratios exist when the same variable is measured on two different members of the population.

There is not complete agreement among statisticians about the classification of data into one of the four categories. For example, some researchers classify IQ data as ratio data rather than interval. Also, data can be altered so that they fit into a different category. For instance, if the incomes of all professors of a college are classified into the three categories of low, average, and high, then a ratio variable becomes an ordinal variable. Table 1–2 gives some examples of each type of data.

Table 1–2	Examples of Measurement Scales		
Nominal-level data	**Ordinal-level data**	**Interval-level data**	**Ratio-level data**
Zip code	Grade (A, B, C, D, F)	SAT score	Height
Gender (male, female)		IQ	Weight
Eye color (blue, brown, green, hazel)	Judging (first place, second place, etc.)	Temperature	Time
Political affiliation	Rating scale (poor, good, excellent)		Salary
Religious affiliation			Age
Major field (mathematics, computers, etc.)	Ranking of tennis players		
Nationality			

Applying the Concepts **1–3**

Safe Travel

Read the following information about the transportation industry and answer the questions.

Transportation Safety

The chart shows the number of job-related injuries for each of the transportation industries for 1998.

Industry	Number of injuries
Railroad	4520
Intercity bus	5100
Subway	6850
Trucking	7144
Airline	9950

1. What are the variables under study?

2. Categorize each variable as quantitative or qualitative.

3. Categorize each quantitative variable as discrete or continuous.

4. Identify the level of measurement for each variable.

5. The railroad is shown as the safest transportation industry. Does that mean railroads have fewer accidents than the other industries? Explain.

6. What factors other than safety influence a person's choice of transportation?

7. From the information given, comment on the relationship between the variables.

See page 32 for the answers.

1–4

Objective 5

Identify the four basic sampling techniques.

Data Collection and Sampling Techniques

In research, statisticians use data in many different ways. As stated previously, data can be used to describe situations or events. For example, a manufacturer might want to know something about the consumers who will be purchasing his product so he can plan an effective marketing strategy. In another situation, the management of a company might survey its employees to assess their needs in order to negotiate a new contract with the employees' union. Data can be used to determine whether the educational goals of a school district are being met. Finally, trends in various areas, such as the stock market, can be analyzed, enabling prospective buyers to make more intelligent decisions concerning what stocks to purchase. These examples illustrate a few situations where collecting data will help people make better decisions on courses of action.

Data can be collected in a variety of ways. One of the most common methods is through the use of surveys. Surveys can be done by using a variety of methods. Three of the most common methods are the telephone survey, the mailed questionnaire, and the personal interview.

Telephone surveys have an advantage over personal interview surveys in that they are less costly. Also, people may be more candid in their opinions since there is no face-to-face contact. A major drawback to the telephone survey is that some people in the population will not have phones or will not answer when the calls are made; hence, not all people have a chance of being surveyed. Also, many people now have unlisted numbers and cell phones, so they cannot be surveyed. Finally, even the tone of the voice of the interviewer might influence the response of the person who is being interviewed.

Mailed questionnaire surveys can be used to cover a wider geographic area than telephone surveys or personal interviews since mailed questionnaire surveys are less expensive to conduct. Also, respondents can remain anonymous if they desire. Disadvantages

of mailed questionnaire surveys include a low number of responses and inappropriate answers to questions. Another drawback is that some people may have difficulty reading or understanding the questions.

Personal interview surveys have the advantage of obtaining in-depth responses to questions from the person being interviewed. One disadvantage is that interviewers must be trained in asking questions and recording responses, which makes the personal interview survey more costly than the other two survey methods. Another disadvantage is that the interviewer may be biased in his or her selection of respondents.

Data can also be collected in other ways, such as *surveying records* or *direct observation* of situations.

As stated in Section 1–2, researchers use samples to collect data and information about a particular variable from a large population. Using samples saves time and money and, in some cases, enables the researcher to get more detailed information about a particular subject. Samples cannot be selected in haphazard ways because the information obtained might be biased. For example, interviewing people on a street corner during the day would not include responses from people working in offices at that time or from people attending school; hence, not all subjects in a particular population would have a chance of being selected.

To obtain samples that are unbiased—i.e., give each subject in the population an equally likely chance of being selected—statisticians use four basic methods of sampling: random, systematic, stratified, and cluster sampling.

Random Sampling

Random samples are selected by using chance methods or random numbers. One such method is to number each subject in the population. Then place numbered cards in a bowl, mix them thoroughly, and select as many cards as needed. The subjects whose numbers are selected constitute the sample. Since it is difficult to mix the cards thoroughly, there is a chance of obtaining a biased sample. For this reason, statisticians use another method of obtaining numbers. They generate random numbers with a computer or calculator. Before the invention of computers, random numbers were obtained from tables.

Some two-digit random numbers are shown in Table 1–3. To select a random sample of, say, 15 subjects out of 85 subjects, it is necessary to number each subject from 1 to 85.

This study of Internet search sites was conducted by Nielsen/NetRatings. What type of sample do you think they used?

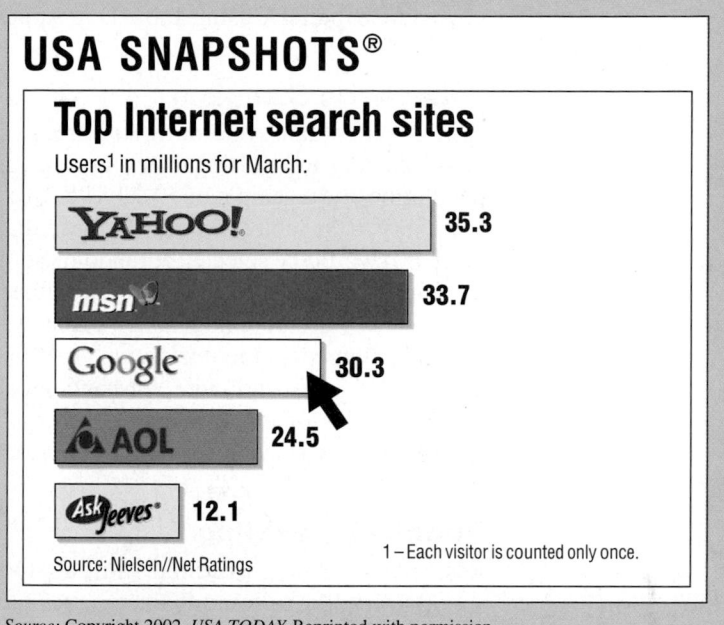

Source: Copyright 2002, *USA TODAY.* Reprinted with permission.

Table 1–3	Random Numbers											
79	41	71	93	60	35	04	67	96	04	79	10	86
26	52	53	13	43	50	92	09	87	21	83	75	17
18	13	41	30	56	20	37	74	49	56	45	46	83
19	82	02	69	34	27	77	34	24	93	16	77	00
14	57	44	30	93	76	32	13	55	29	49	30	77
29	12	18	50	06	33	15	79	50	28	50	45	45
01	27	92	67	93	31	97	55	29	21	64	27	29
55	75	65	68	65	73	07	95	66	43	43	92	16
84	95	95	96	62	30	91	64	74	83	47	89	71
62	62	21	37	82	62	19	44	08	64	34	50	11
66	57	28	69	13	99	74	31	58	19	47	66	89
48	13	69	97	29	01	75	58	05	40	40	18	29
94	31	73	19	75	76	33	18	05	53	04	51	41
00	06	53	98	01	55	08	38	49	42	10	44	38
46	16	44	27	80	15	28	01	64	27	89	03	27
77	49	85	95	62	93	25	39	63	74	54	82	85
81	96	43	27	39	53	85	61	12	90	67	96	02
40	46	15	73	23	75	96	68	13	99	49	64	11

Then select a starting number by closing your eyes and placing your finger on a number in the table. (Although this may sound somewhat unusual, it enables us to find a starting number at random.) In this case suppose your finger landed on the number 12 in the second column. (It is the sixth number down from the top.) Then proceed downward until you have selected 15 different numbers between 01 and 85. When you reach the bottom of the column, go to the top of the next column. If you select a number greater than 85 or the number 00 or a duplicate number, just omit it. In our example, we will use the subjects numbered 12, 27, 75, 62, 57, 13, 31, 06, 16, 49, 46, 71, 53, 41, and 02. A more detailed procedure for selecting a random sample using a table of random numbers is given in Chapter 14, using Table D in Appendix C.

Systematic Sampling

Researchers obtain **systematic samples** by numbering each subject of the population and then selecting every kth subject. For example, suppose there were 2000 subjects in the population and a sample of 50 subjects were needed. Since $2000 \div 50 = 40$, then $k = 40$, and every 40th subject would be selected; however, the first subject (numbered between 1 and 40) would be selected at random. Suppose subject 12 were the first subject selected; then the sample would consist of the subjects whose numbers were 12, 52, 92, etc., until 50 subjects were obtained. When using systematic sampling, one must be careful about how the subjects in the population are numbered. If subjects were arranged in a manner such as wife, husband, wife, husband, and every 40th subject were selected, the sample would consist of all husbands. Numbering is not always necessary. For example, a researcher may select every tenth item from an assembly line to test for defects.

Stratified Sampling

Researchers obtain **stratified samples** by dividing the population into groups (called strata) according to some characteristic that is important to the study, then sampling from each group. Samples within the strata should be randomly selected. For example, suppose the president of a two-year college wants to learn how students feel about a certain issue. Furthermore, the president wishes to see if the opinions of the first-year students differ from those of the second-year students. The president will select students from each group to use in the sample.

Cluster Sampling

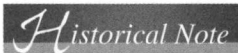

In 1936, the *Literary Digest,* on the basis of a biased sample of its subscribers, predicted that Alf Landon would defeat Franklin D. Roosevelt in the upcoming presidential election. Roosevelt won by a landslide. The magazine ceased publication the following year.

Researchers also use **cluster samples.** Here the population is divided into groups called clusters by some means such as geographic area or schools in a large school district, etc. Then the researcher randomly selects some of these clusters and uses all members of the selected clusters as the subjects of the samples. Suppose a researcher wishes to survey apartment dwellers in a large city. If there are 10 apartment buildings in the city, the researcher can select at random 2 buildings from the 10 and interview all the residents of these buildings. Cluster sampling is used when the population is large or when it involves subjects residing in a large geographic area. For example, if one wanted to do a study involving the patients in the hospitals in New York City, it would be very costly and time-consuming to try to obtain a random sample of patients since they would be spread over a large area. Instead, a few hospitals could be selected at random, and the patients in these hospitals would be interviewed in a cluster.

The four basic sampling methods are summarized in Table 1–4.

Table 1–4	Summary of Sampling Methods
Random	Subjects are selected by random numbers.
Systematic	Subjects are selected by using every kth number after the first subject is randomly selected from 1 through k.
Stratified	Subjects are selected by dividing up the population into groups (strata), and subjects within groups are randomly selected.
Cluster	Subjects are selected by using an intact group that is representative of the population.

Interesting Facts

Older Americans are less likely to sacrifice happiness for a higher-paying job. According to one survey, 38% of those aged 18–29 said they would choose more money over happiness, while only 3% of those over 65 would.

Other Sampling Methods

In addition to the four basic sampling methods, researchers use other methods to obtain samples. One such method is called a **convenience sample.** Here a researcher uses subjects that are convenient. For example, the researcher may interview subjects entering a local mall to determine the nature of their visit or perhaps what stores they will be patronizing. This sample is probably not representative of the general customers for several reasons. For one thing, it was probably taken at a specific time of day, so not all customers entering the mall have an equal chance of being selected since they were not there when the survey was being conducted. But convenience samples can be representative of the population. If the researcher investigates the characteristics of the population and determines that the sample is representative, then it can be used.

Other sampling techniques, such as *sequential sampling, double sampling,* and *multistage sampling,* are explained in Chapter 14, along with a more detailed explanation of the four basic sampling techniques.

Applying the Concepts **1–4**

American Culture and Drug Abuse

Assume you are a member of the Family Research Council and have become increasingly concerned about the drug use by professional sports players. You set up a plan and conduct a survey on how people believe the American culture (television, movies, magazines, and popular music) influences illegal drug use. Your survey consists of 2250 adults and adolescents from around the country. A consumer group petitions you for more information about your survey. Answer the following questions about your survey.

1. What type of survey did you use (phone, mail, or interview)?
2. What are the advantages and disadvantages of the surveying methods you did not use?
3. What type of scores did you use? Why?
4. Did you use a random method for deciding who would be in your sample?
5. Which of the methods (stratified, systematic, cluster, or convenience) did you use?
6. Why was that method more appropriate for this type of data collection?
7. If a convenience sample were obtained, consisting of only adolescents, how would the results of the study be affected?

See page 32 for the answers.

1–5

Objective 6

Explain the difference between an observational and an experimental study.

Observational and Experimental Studies

There are several different ways to classify statistical studies. This section explains two types of studies: *observational studies* and *experimental studies.*

In an **observational study,** the researcher merely observes what is happening or what has happened in the past and tries to draw conclusions based on these observations.

For example, data from the Motorcycle Industry Council (*USA TODAY*) stated that "Motorcycle owners are getting older and richer." Data were collected on the ages and incomes of motorcycle owners for the years 1980 and 1998 and then compared. The findings showed considerable differences in the ages and incomes of motorcycle owners for the two years.

In this study, the researcher merely observed what had happened to the motorcycle owners over a period of time. There was no type of research intervention.

> In an **experimental study,** the researcher manipulates one of the variables and tries to determine how the manipulation influences other variables.

For example, a study conducted at Virginia Polytechnic Institute and presented in *Psychology Today* divided female undergraduate students into two groups and had the students perform as many sit-ups as possible in 90 sec. The first group was told only to "Do your best," while the second group was told to try to increase the actual number of sit-ups done each day by 10%. After four days, the subjects in the group who were given the vague instructions to "Do your best" averaged 43 sit-ups, while the group that was given the more specific instructions to increase the number of sit-ups by 10% averaged 56 sit-ups by the last day's session. The conclusion then was that athletes who were given specific goals performed better than those who were not given specific goals.

This study is an example of a statistical experiment since the researchers intervened in the study by manipulating one of the variables, namely, the type of instructions given to each group.

In a true experimental study, the subjects should be assigned to groups randomly. Also, the treatments should be assigned to the groups at random. In the sit-up study, the article did not mention whether the subjects were randomly assigned to the groups.

Sometimes when random assignment is not possible, researchers use intact groups. These types of studies are done quite often in education where already intact groups are available in the form of existing classrooms. When these groups are used, the study is said to be a **quasi-experimental study.** The treatments, though, should be assigned at random. Most articles do not state whether random assignment of subjects was used.

Statistical studies usually include one or more *independent variables* and one *dependent variable.*

> The **independent variable** in an experimental study is the one that is being manipulated by the researcher. The independent variable is also called the **explanatory variable.** The resultant variable is called the **dependent variable** or the **outcome variable.**

The outcome variable is the variable that is studied to see if it has changed significantly due to the manipulation of the independent variable. For example, in the sit-up study, the researchers gave the groups two different types of instructions, general and specific. Hence, the independent variable is the type of instruction. The dependent variable, then, is the resultant variable, that is, the number of sit-ups each group was able to perform after four days of exercise. If the differences in the dependent or outcome variable are large and other factors are equal, these differences can be attributed to the manipulation of the independent variable. In this case, specific instructions were shown to increase athletic performance.

In the sit-up study, there were two groups. The group that received the special instruction is called the **treatment group** while the other is called the **control group.** The treatment group receives a specific treatment (in this case, instructions for improvement) while the control group does not.

Both types of statistical studies have advantages and disadvantages. Experimental studies have the advantage that the researcher can decide how to select subjects and how to assign them to specific groups. The researcher can also control or manipulate the independent variable. For example, in studies that require the subjects to consume a certain amount of medicine each day, the researcher can determine the precise dosages and, if necessary, vary the dosage for the groups.

There are several disadvantages to experimental studies. First, they may occur in unnatural settings, such as laboratories and special classrooms. This can lead to several problems. One such problem is that the results might not apply to the natural setting. The age-old question then is, "This mouthwash may kill 10,000 germs in a test tube, but how many germs will it kill in my mouth?"

Another disadvantage with an experimental study is the **Hawthorne effect.** This effect was discovered in 1924 in a study of workers at the Hawthorne plant of the Western Electric Company. In this study, researchers found that the subjects who knew they were participating in an experiment actually changed their behavior in ways that affected the results of the study.

Another problem is called *confounding of variables.*

A **confounding variable** is one that influences the dependent or outcome variable but cannot be separated from the independent variable.

Researchers try to control most variables in a study, but this is not possible in some studies. For example, subjects who are put on an exercise program might also improve their diet unbeknownst to the researcher and perhaps improve their health in other ways not due to exercise alone. Then diet becomes a confounding variable.

Observational studies also have advantages and disadvantages. One advantage of an observational study is that it usually occurs in a natural setting. For example, researchers can observe people's driving patterns on streets and highways in large cities. Another advantage of an observational study is that it can be done in situations where it would be unethical or downright dangerous to conduct an experiment. Using observational studies, researchers can study suicides, rapes, murders, etc. In addition, observational studies can be done using variables that cannot be manipulated by the researcher, such as drug users versus nondrug users and right-handedness versus left-handedness.

Observational studies have disadvantages, too. As mentioned previously, since the variables are not controlled by the researcher, a definite cause-and-effect situation cannot be shown since other factors may have had an effect on the results. Observational studies can be expensive and time-consuming. For example, if one wanted to study the habitat of lions in Africa, one would need a lot of time and money, and there would be a certain amount of danger involved. Finally, since the researcher may not be using his or her own measurements, the results could be subject to the inaccuracies of those who collected the data. For example, if the researchers were doing a study of events that occurred in the 1800s, they would have to rely on information and records obtained by others from a previous era. There is no way to ensure the accuracy of these records.

When you read the results of statistical studies, decide if the study was observational or experimental. Then see if the conclusion follows logically, based on the nature of these studies.

No matter what type of study is conducted, two studies on the same subject sometimes have conflicting conclusions. Why might this occur? An article entitled "Bottom Line: Is It Good for You?" (*USA TODAY Weekend*) states that in the 1960s studies suggested that margarine was better for the heart than butter since margarine contains less saturated fat and users had lower cholesterol levels. In a 1980 study, researchers found that butter was better than margarine since margarine contained trans-fatty acids, which are worse for the heart than butter's saturated fat. Then in a 1998 study, researchers found that margarine was better for a person's health. Now, what is to be believed? Should one use butter or margarine?

The answer here is to take a closer look at these studies. Actually, it is not a choice between butter or margarine that counts, but the type of margarine used. In the 1980s, studies showed that solid margarine contains trans-fatty acids, and scientists believe that they are worse for the heart than butter's saturated fat. In the 1998 study, liquid margarine was used. It is very low in trans-fatty acids, and hence it is more healthful than butter because trans-fatty acids have been shown to raise cholesterol. Hence, the conclusion is to use liquid margarine instead of solid margarine or butter.

Before decisions based on research studies are made, it is important to get all the facts and examine them in light of the particular situation.

Applying the Concepts **1–5**

Just a Pinch Between Your Cheek and Gum

As the evidence on the adverse effects of cigarette smoke grew, people tried many different ways to quit smoking. Some people tried chewing tobacco or, as it was called, smokeless tobacco. A small amount of tobacco was placed between the cheek and gum. Certain chemicals from the tobacco were absorbed into the bloodstream and gave the sensation of smoking cigarettes. This prompted studies on the adverse effects of smokeless tobacco. One study in particular used 40 university students as subjects. Twenty were given smokeless tobacco to chew, and twenty given a substance that looked and tasted like smokeless tobacco, but did not contain any of the harmful substances. The students were randomly assigned to one of the groups. The students' blood pressure and heart rate were measured before they started chewing and 20 minutes after they had been chewing. A significant increase in heart rate occurred in the group that chewed the smokeless tobacco. Answer the following questions.

1. What type of study was this (observational, quasi-experimental, or experimental)?
2. What are the independent and dependent variables?
3. Which was the treatment group?
4. Could the students' blood pressures be affected by knowing that they are part of a study?
5. List some possible confounding variables.
6. Do you think this is a good way to study the effect of smokeless tobacco?

See page 32 for the answers.

1–6

Uses and Misuses of Statistics

Objective 7

Explain how statistics can be used and misused.

As explained previously, statistical techniques can be used to describe data, compare two or more data sets, determine if a relationship exists between variables, test hypotheses, and make estimates about population characteristics. However, there is another aspect of statistics, and that is the misuse of statistical techniques to sell products that don't work properly, to attempt to prove something true that is really not true, or to get our attention by using statistics to evoke fear, shock, and outrage.

There are two sayings that have been around for a long time that illustrate this point:

"There are three types of lies—lies, damn lies, and statistics."

"Figures don't lie, but liars figure."

Just because we read or hear the results of a research study or an opinion poll in the media, this does not mean that these results are reliable or that they can be applied to any and all situations. For example, reporters sometimes leave out critical details such as the size of the sample used or how the research subjects were selected. Without this information, one cannot properly evaluate the research and properly interpret the conclusions of the study or survey.

It is the purpose of this section to show some ways that statistics can be misused. One should not infer that all research studies and surveys are suspect, but that there are many factors to consider when making decisions based on the results of research studies and surveys. Here are some ways that statistics can be misrepresented.

Suspect Samples

The first thing to consider is the sample that was used in the research study. Sometimes researchers use very small samples to obtain information. Several years ago, advertisements contained such statements as "Three out of four doctors surveyed recommend brand such and such." If only 4 doctors were surveyed, the results could have been obtained by chance alone; however, if 100 doctors were surveyed, the results might be quite different.

Not only is it important to have a sample size that is large enough, but also it is necessary to see how the subjects in the sample were selected. Studies using volunteers sometimes have a built-in bias. Volunteers generally do not represent the population at large. Sometimes they are recruited from a particular socioeconomic background, and sometimes unemployed people volunteer for research studies to get a stipend. Studies that require the subjects to spend several days or weeks in an environment other than their home or workplace automatically exclude people who are employed and cannot take time away from work. Sometimes only college students or retirees are used in studies. In the past, many studies have used only men, but have attempted to generalize the results to both men and women. Opinion polls that require a person to phone or mail in a response most often are not representative of the population in general, since only those with strong feelings for or against the issue usually call or respond by mail.

Another type of sample that may not be representative is the convenience sample. Educational studies sometimes use students in intact classrooms since it is convenient. Quite often, the students in these classrooms do not represent the student population of the entire school district.

When results are interpreted from studies using small samples, convenience samples, or volunteer samples, care should be used in generalizing the results to the entire population.

Ambiguous Averages

In Chapter 3, you will learn that there are four commonly used measures that are loosely called *averages*. They are the *mean, median, mode,* and *midrange.* For the same data set, these averages can differ markedly. People who know this can, without lying, select the one measure of average which lends the most evidence to support their position. This fact is explained on page 105.

Changing the Subject

Another type of statistical distortion can occur when different values are used to represent the same data. For example, one political candidate who is running for reelection might say, "During my administration, expenditures increased a mere 3%." His opponent, who is trying to unseat him, might say, "During my opponent's administration, expenditures have increased a whopping $6,000,000." Here both figures are correct; however, expressing a 3% increase as $6,000,000 makes it sound like a very large increase. Here again, ask yourself, Which measure best represents the data?

Detached Statistics

A claim that uses a detached statistic is one in which no comparison is made. For example, you may hear a claim such as "Our brand of crackers has one-third fewer calories." Here, no comparison is made. One-third fewer calories than what? Another example is a claim that uses a detached statistic such as "Brand A aspirin works four times faster." Four times faster than what? When you see statements such as this, always ask yourself, Compared to what?

Implied Connections

Many claims attempt to imply connections between variables that may not actually exist. For example, consider the following statement: "Eating fish may help to reduce your cholesterol." Notice the words *may help*. There is no guarantee that eating fish will definitely help you reduce your cholesterol.

"Studies suggest that using our exercise machine will reduce your weight." Here the word *suggest* is used; and again, there is no guarantee that you will lose weight by using the exercise machine advertised.

Another claim might say, "Taking calcium will lower blood pressure in some people." Note the word *some* is used. You may not be included in the group of "some" people. Be careful when you draw conclusions from claims that use words such as *may, in some people,* and *might help.*

Misleading Graphs

Statistical graphs give a visual representation of data that enables viewers to analyze and interpret data more easily than by simply looking at numbers. In Chapter 2, you will see how some graphs are used to represent data. However, if graphs are drawn inappropriately, they can misrepresent the data and lead the reader to false conclusions. The misuse of graphs is also explained in Chapter 2, on pages 70–73.

Faulty Survey Questions

When analyzing the results of a survey using questionnaires, you should be sure that the questions are properly written since the way questions are phrased can often influence the way people answer them. For example, the responses to a question such as "Do you feel that the North Huntingdon School District should build a new football stadium?" might be answered differently than a question such as "Do you favor increasing school taxes so that the North Huntingdon School District can build a new football stadium?" Each question asks something a little different, and the responses could be radically different. When you read and interpret the results obtained from questionnaire surveys, watch out for some of these common mistakes made in the writing of the survey questions.

In Chapter 14, you will find some common ways that survey questions could be misinterpreted by those responding and could therefore result in incorrect conclusions.

To restate the premise of this section, statistics, when used properly, can be beneficial in obtaining much information, but when used improperly, can lead to much misinformation. It is like your automobile. If you use your automobile to get to school or work or to go on a vacation, that's good. But if you use it to run over your neighbor's dog because it barks all night long and tears up your flower garden, that's not so good!

1–7   Computers and Calculators

Objective 8

Explain the importance of computers and calculators in statistics.

In the past, statistical calculations were done with pencil and paper. However, with the advent of calculators, numerical computations became much easier. Computers do all the numerical calculation. All one does is to enter the data into the computer and use the appropriate command; the computer will print the answer or display it on the screen. Now the TI-83 Plus or TI-84 Plus graphing calculator accomplishes the same thing.

There are many statistical packages available; this book uses MINITAB and Microsoft Excel. Instructions for using MINITAB, the TI-83 Plus or TI-84 Plus graphing calculator, and Excel have been placed at the end of each relevant section, in subsections entitled Technology Step by Step.

Students should realize that the computer and calculator merely give numerical answers and save the time and effort of doing calculations by hand. The student is still responsible for understanding and interpreting each statistical concept. In addition, students should realize that the results come from the data and do not appear magically on the computer. Doing calculations using the procedure tables will help reinforce this idea.

The author has left it up to instructors to choose how much technology they will incorporate into the course.

Technology *Step by Step*

MINITAB
Step by Step

General Information

MINITAB statistical software provides a wide range of statistical analysis and graphing capabilities.

Take Note

In this text you will see captured screen images from computers running MINITAB Release 14. If you are using an earlier release of MINITAB, the screens you see on your computer may bear slight visual differences from the screens pictured in this text. **But don't be alarmed!** All the Step by Step operations described in this text, including the commands, the menu options, and the functionality, will work just fine on your computer.

Start the Program

1. Click the Windows XP Start Menu, then **All Programs**.
2. Click the **MINITAB 14** folder and then click 🔲 MINITAB 14 , the program icon. The program screen will look similar to the one shown here. You will see the Session Window, the Worksheet Window, and perhaps the Project Manager Window.
3. Click the Project Manager icon on the toolbar to bring the project manager to the front.

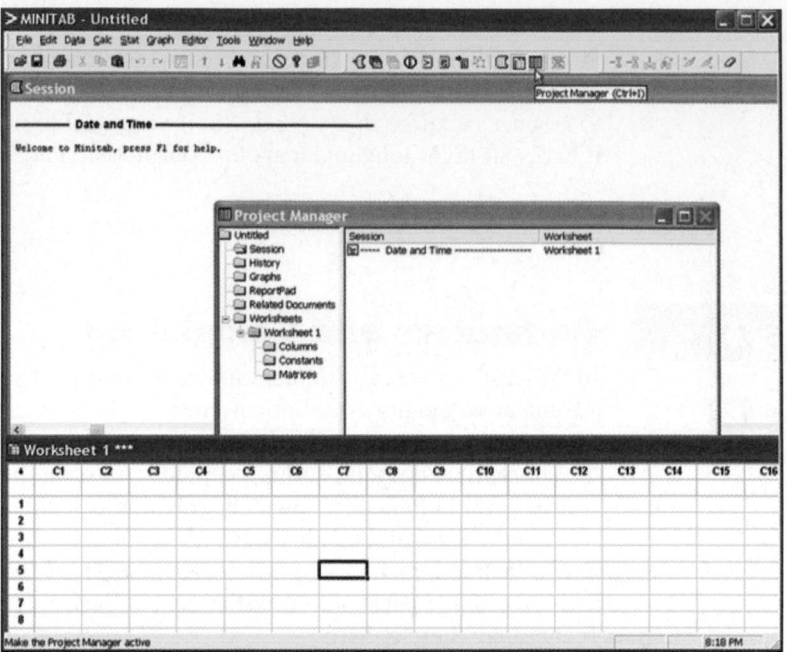

To use the program, data must be entered from the keyboard or from a file.

Entering Data in MINITAB

In MINITAB, all the data for one variable are stored in a column. Step-by-step instructions for entering these data follow.

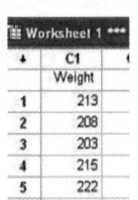

Data

213   208   203   215   222

1. Click in row one of Worksheet 1***. This makes the worksheet the active window and puts the cursor in the first cell. The small data entry arrow in the upper left-hand corner of the worksheet should be pointing down. If it is not, click it to change the direction in which the cursor will move when you press the [Enter] key.

2. Type in each number, pressing [Enter] after each entry, including the last number typed.

3. *Optional:* Click in the space above row 1 to type in **Weight,** the column label.

Save a Worksheet File

4. Click on the **File Menu.** *Note:* This is *not* the same as clicking the disk icon 💾 .

5. Click **Save Current Worksheet As . . .**

6. In the dialog box you will need to verify three items:

 a) Save in: Click on or type in the disk drive and directory where you will store your data. For a floppy disk this would be **A:.**

 b) File Name: Type in the name of the file, such as **MyData.**

 c) Save as Type: The default here is MINITAB. An extension of mtw is added to the name.

 Click [Save]. The name of the worksheet will change from Worksheet 1*** to MyData.MTW.

Open the Databank File

The raw data are shown in Appendix D. There is a row for each person's data and a column for each variable. MINITAB data files comprised of data sets used in this book, including the

Databank, are available on the MathZone CD-ROM or at the book's MathZone site (www.mhhe.com/bluman). Here is how to get the data from a file into a worksheet.

1. Click **File>Open Worksheet.** A sequence of menu instructions will be shown this way.

   Note: This is *not* the same as clicking the file icon 📂 . If the dialog box says Open Project instead of Open Worksheet, click [Cancel] and use the correct menu item. The Open Worksheet dialog box will be displayed.

2. You must check three items in this dialog box.

   a) The Look In: dialog box should show the directory where the file is located.

   b) Make sure the Files of Type: shows the correct type, MINITAB [*.mtw].

   c) Double-click the file name in the list box Databank.mtw. A dialog box may inform you that a copy of this file is about to be added to the project. Click on the checkbox if you do not want to see this warning again.

3. Click the [OK] button. The data will be copied into a second worksheet. Part of the worksheet is shown here.

	C1	C2	C3	C4	C5	C6	C7	C8	C9	C10	C11-T	C12-T
	ID-NUMBER	AGE	ED-LEVEL	SMOKING	EXERCISE	WEIGHT	SERUM	SYSTOLIC	IQ	SODIUM	GENDER	MARITAL-ST
1	1	27	2	1	1	120	193	126	118	136	F	M
2	2	18	1	0	1	145	210	120	105	137	M	S
3	3	32	2	0	0	118	196	128	115	135	F	M

   a) You may maximize the window and scroll if desired.

   b) C12-T Marital Status has a T appended to the label to indicate alphanumeric data. MyData.MTW is not erased or overwritten. Multiple worksheets can be available; however, only the active worksheet is available for analysis.

4. To switch between the worksheets, select **Window>MyData.MTW.**

5. Select **File>Exit** to quit. To save the project, click [Yes].

6. Type in the name of the file, **Chapter01.** The Data Window, the Session Window, and settings are all in one file called a project. Projects have an extension of mpj instead of mtw.

   Clicking the disk icon 💾 on the menu bar is the same as selecting **File>Save Project.**

   Clicking the file icon 📂 is the same as selecting **File>Open Project.**

7. Click [Save]. The mpj extension will be added to the name. The computer will return to the Windows desktop. The two worksheets, the Session Window results, and settings are saved in this project file. When a project file is opened, the program will start up right where you left off.

TI-83 Plus or TI-84 Plus
Step by Step

The TI-83 Plus or TI-84 Plus graphing calculator can be used for a variety of statistical graphs and tests.

General Information

To turn calculator on:
Press **ON** key.
To turn calculator off:
Press **2nd [OFF].**

To reset defaults only:

1. Press **2nd,** then **[MEM].**

2. Select **7,** then **2,** then **2.**

(Optional). To reset settings on calculator and clear memory: (*Note:* This will clear all settings and programs in the calculator's memory.)
Press **2nd,** then **[MEM].** Then press **7,** then **1,** then **2.**
(Also, the contrast may need to be adjusted after this.)

To adjust contrast (if necessary):

Press **2nd.** Then press and hold ▲ to darken or ▼ to lighten contrast.

To clear screen:

Press **CLEAR.**

(*Note:* This will return you to the screen you were using.)

To display a menu:

Press appropriate menu key. Example: **STAT.**

To return to home screen:

Press **2nd,** then **[QUIT].**

To move around on the screens:

Use the arrow keys.

To select items on the menu:

Press the corresponding number or move the cursor to the item, using the arrow keys. Then press **ENTER.**

(*Note:* In some cases, you do not have to press **ENTER,** and in other cases you may need to press **ENTER** twice.)

Entering Data

To enter single-variable data (if necessary, clear the old list):

1. Press **STAT** to display the Edit menu.

2. Press **ENTER** to select 1:Edit.

3. Enter the data in L_1 and press **ENTER** after each value.

4. After all data values are entered, press **STAT** to get back to the Edit menu or **2nd [QUIT]** to end.

Example TI1–1

Enter the following data values in L_1: **213, 208, 203, 215, 222.**

To enter multiple-variable data:

The TI-83 Plus or TI-84 Plus will take up to six lists designated L_1, L_2, L_3, L_4, L_5, and L_6.

Output

L1	L2	L3	1
213	------	------	
208			
203			
215			
222			

L1(6)=			

1. To enter more than one set of data values, complete the preceding steps. Then move the cursor to L_2 by pressing the ▶ key.

2. Repeat the steps in the preceding part.

Editing Data

To correct a data value before pressing **ENTER,** use ◀ and retype the value and press **ENTER.**

To correct a data value in a list after pressing **ENTER,** move cursor to incorrect value in list and type in the correct value. Then press **ENTER.**

To delete a data value in a list:

Move cursor to value and press **DEL.**

To insert a data value in a list:

1. Move cursor to position where data value is to be inserted, then press **2nd [INS].**

2. Type data value; then press **ENTER.**

To clear a list:

1. Press **STAT,** then **4.**

2. Enter list to be cleared. Example: To clear L_1, press **2nd [L₁].** Then press **ENTER.**

   (*Note:* To clear several lists, follow step 1, but enter each list to be cleared, separating them with commas.)

Sorting Data

To sort the data in a list:

1. Enter the data in L_1.
2. Press **STAT 2** to get SortA to sort the list in ascending order.
3. Then press **2nd [L₁] ENTER.**

The calculator will display Done.

4. Press **STAT ENTER** to display sorted list.

   (*Note:* The SortD or **3** sorts the list in descending order.)

Example TI1–2

Sort in ascending order the data values entered in Example TI1–1.

Output

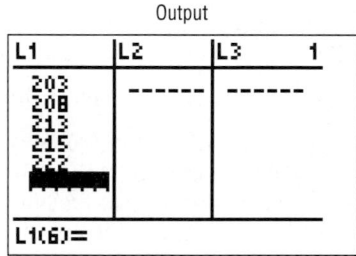

Excel
Step by Step

Excel's Analysis
ToolPak Add-In

General Information

Microsoft Excel has two different ways to solve statistical problems. First, there are built-in functions, such as STDEV and CHITEST, available from the standard toolbar by clicking the f_x icon. For most of the problems in this textbook, however, it is easier to use the packaged tests in the Data Analysis Add-in.

To activate the Data Analysis Add-in:

1. Selcct **Tools** on the Worksheet menu bar.
2. Select Add-Ins from the **Tools** menu.
3. Check the box for Analysis ToolPak.

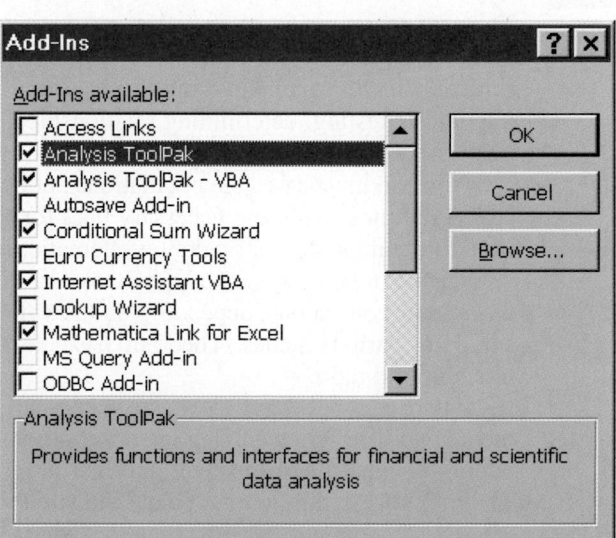

MegaStat

Later in this text you will encounter a few Excel Technology Step by Step operations that would not be feasible without another Excel add-in: MegaStat. MegaStat is provided on the CD that came with your textbook, as well as on your text's Online Learning Center at www.mhhe.com/bluman.

1. Locate the MegaStat installation program by double-clicking "My Computer" on your desktop and looking in the proper directory. (If you downloaded MegaStat from the OLC, it will be in the location to which you saved it. If you are accessing it from your CD, simply double-click on your CD drive.)

2. Follow the instructions to install MegaStat on your computer.

3. Activate the MegaStat add-in in the same way you activated the Data Analysis add-in: select **Tools>Add-ins,** and check the box for MegaStat.

Entering Data

1. Click the cell at the top of the column where you want to enter data. When working with values for a single variable, you usually will want to enter the values into a column.

2. Type each data value and press **[Enter]** or **[Tab].**

You can also enter data into Excel by opening an Excel workbook file. You can access and download data sets used in this book by using the accompanying CD-ROM or the Online Learning Center (www.mhhe.com/bluman). Excel data are in files called Workbooks. Follow the steps in Example XL1–1 to open a workbook file.

Example XL1–1

1. Click **File** on the menu bar, then click Open . . . (i.e., select **File>Open . . .**). The Open dialog box will be displayed.

2. In the Look in box, click the folder where the workbook file is located.

3. In the folder list, double-click folders until you open the folder containing the file you want.

4. Double-click the file name in the list box. The data will be copied into Excel.

1–8   Summary

The two major areas of statistics are descriptive and inferential. Descriptive statistics includes the collection, organization, summarization, and presentation of data. Inferential statistics includes making inferences from samples to populations, estimations and hypothesis testing, determining relationships, and making predictions. Inferential statistics is based on *probability theory.*

Since in most cases the populations under study are large, statisticians use subgroups called samples to get the necessary data for their studies. There are four basic methods used to obtain samples: random, systematic, stratified, and cluster.

Data can be classified as qualitative or quantitative. Quantitative data can be either discrete or continuous, depending on the values they can assume. Data can also be measured by various scales. The four basic levels of measurement are nominal, ordinal, interval, and ratio.

There are two basic types of statistical studies: observational studies and experimental studies. When conducting observational studies, researchers observe what is happening or what has happened and then draw conclusions based on these observations. They do not attempt to manipulate the variables in any way.

When conducting an experimental study, researchers manipulate one or more of the independent or explanatory variables and see how this manipulation influences the dependent or outcome variable.

Finally, the applications of statistics are many and varied. People encounter them in everyday life, such as in reading newspapers or magazines, listening to the radio, or watching television. Since statistics is used in almost every field of endeavor, the educated individual should be knowledgeable about the vocabulary, concepts, and procedures of statistics.

Today, computers and calculators are used extensively in statistics to facilitate the computations.

Source: © 1993 King Features Syndicate, Inc. World Rights reserved. Reprinted with special permission of King Features Syndicate.

Important Terms

cluster sample   12

confounding variable   15

continuous variables   6

control group   14

convenience sample   13

data   3

data set   4

data value or datum   4

dependent variable   14

descriptive statistics   4

discrete variables   6

experimental study   14

explanatory variable   14

Hawthorne effect   15

hypothesis testing   4

independent variable   14

inferential statistics   5

interval level of measurement   8

measurement scales   7

nominal level of measurement   7

observational study   13

ordinal level of measurement   8

outcome variable   14

population   4

probability   4

qualitative variables   6

quantitative variables   6

quasi-experimental study   14

random sample   10

random variable   3

ratio level of measurement   8

sample   4

statistics   3

stratified sample   12

systematic sample   12

treatment group   14

variable   3

Review Exercises

Note: **All odd-numbered problems and even-numbered problems marked with "ans" are included in the answer section at the end of this book.**

1. Name and define the two areas of statistics.

2. What is probability? Name two areas where probability is used.

3. Suggest some ways statistics can be used in everyday life.

4. Explain the differences between a sample and a population.

5. Why are samples used in statistics?

6. **(ans)** In each of these statements, tell whether descriptive or inferential statistics have been used.

   a. In the year 2010, 148 million Americans will be enrolled in an HMO (Source: *USA TODAY*).
   b. Nine out of ten on-the-job fatalities are men (Source: *USA TODAY Weekend*).
   c. Expenditures for the cable industry were $5.66 billion in 1996 (Source: *USA TODAY*).
   d. The median household income for people aged 25–34 is $35,888 (Source: *USA TODAY*).
   e. Allergy therapy makes bees go away (Source: *Prevention*).
   f. Drinking decaffeinated coffee can raise cholesterol levels by 7% (Source: American Heart Association).
   g. The national average annual medicine expenditure per person is $1052 (Source: *The Greensburg Tribune Review*).
   h. Experts say that mortgage rates may soon hit bottom (Source: *USA TODAY*).

7. Classify each as nominal-level, ordinal-level, interval-level, or ratio-level measurement.

   a. Pages in the city of Cleveland telephone book.
   b. Rankings of tennis players.
   c. Weights of air conditioners.
   d. Temperatures inside 10 refrigerators.
   e. Salaries of the top five CEOs in the United States.
   f. Ratings of eight local plays (poor, fair, good, excellent).
   g. Times required for mechanics to do a tune-up.
   h. Ages of students in a classroom.
   i. Marital status of patients in a physician's office.
   j. Horsepower of tractor engines.

8. **(ans)** Classify each variable as qualitative or quantitative.

   a. Number of bicycles sold in 1 year by a large sporting goods store.
   b. Colors of baseball caps in a store.
   c. Times it takes to cut a lawn.
   d. Capacity in cubic feet of six truck beds.
   e. Classification of children in a day-care center (infant, toddler, preschool).
   f. Weights of fish caught in Lake George.
   g. Marital status of faculty members in a large university.

9. Classify each variable as discrete or continuous.

   a. Number of doughnuts sold each day by Doughnut Heaven.
   b. Water temperatures of six swimming pools in Pittsburgh on a given day.
   c. Weights of cats in a pet shelter.
   d. Lifetime (in hours) of 12 flashlight batteries.
   e. Number of cheeseburgers sold each day by a hamburger stand on a college campus.
   f. Number of DVDs rented each day by a video store.
   g. Capacity (in gallons) of six reservoirs in Jefferson County.

10. Give the boundaries of each value.

    a. 42.8 miles.
    b. 1.6 milliliters.
    c. 5.36 ounces.
    d. 18 tons.
    e. 93.8 ounces.
    f. 40 inches.

11. Name and define the four basic sampling methods.

12. **(ans)** Classify each sample as random, systematic, stratified, or cluster.

    a. In a large school district, all teachers from two buildings are interviewed to determine whether they believe the students have less homework to do now than in previous years.
    b. Every seventh customer entering a shopping mall is asked to select her or his favorite store.
    c. Nursing supervisors are selected using random numbers in order to determine annual salaries.
    d. Every 100th hamburger manufactured is checked to determine its fat content.
    e. Mail carriers of a large city are divided into four groups according to gender (male or female) and according to whether they walk or ride on their

routes. Then 10 are selected from each group and interviewed to determine whether they have been bitten by a dog in the last year.

13. Give three examples each of nominal, ordinal, interval, and ratio data.

14. For each of these statements, define a population and state how a sample might be obtained.

    a. The average cost of an airline meal is $4.55 (Source: *Everything Has Its Price,* Richard E. Donley, Simon and Schuster).

    b. More than 1 in 4 United States children have cholesterol levels of 180 milligrams or higher (Source: The American Health Foundation).

    c. Every 10 minutes, 2 people die in car crashes and 170 are injured (Source: National Safety Council estimates).

    d. When older people with mild to moderate hypertension were given mineral salt for 6 months, the average blood pressure reading dropped by 8 points systolic and 3 points diastolic (Source: *Prevention*).

    e. The average amount spent per gift for Mom on Mother's Day is $25.95 (Source: The Gallup Organization).

15. Select a newspaper or magazine article that involves a statistical study, and write a paper answering these questions.

    a. Is this study descriptive or inferential? Explain your answer.

    b. What are the variables used in the study? In your opinion, what level of measurement was used to obtain the data from the variables?

    c. Does the article define the population? If so, how is it defined? If not, how could it be defined?

    d. Does the article state the sample size and how the sample was obtained? If so, determine the size of the sample and explain how it was selected. If not, suggest a way it could have been obtained.

    e. Explain *in your own words* what procedure (survey, comparison of groups, etc.) might have been used to determine the study's conclusions.

    f. Do you agree or disagree with the conclusions? State your reasons.

16. Information from research studies is sometimes taken out of context. Explain why the claims of these studies might be suspect.

    a. The average salary of the graduates of the class of 1980 is $32,500.

    b. It is estimated that in Podunk there are 27,256 cats.

    c. Only 3% of the men surveyed read *Cosmopolitan* magazine.

    d. Based on a recent mail survey, 85% of the respondents favored gun control.

    e. A recent study showed that high school dropouts drink more coffee than students who graduated; therefore, coffee dulls the brain.

    f. Since most automobile accidents occur within 15 miles of a person's residence, it is safer to make long trips.

17. Identify each study as being either observational or experimental.

    a. Subjects were randomly assigned to two groups, and one group was given an herb and the other group a placebo. After 6 months, the numbers of respiratory tract infections each group had were compared.

    b. A researcher stood at a busy intersection to see if the color of the automobile that a person drives is related to running red lights.

    c. A researcher finds that people who are more hostile have higher total cholesterol levels than those who are less hostile.

    d. Subjects are randomly assigned to four groups. Each group is placed on one of four special diets—a low-fat diet, a high-fish diet, a combination of low-fat diet and high-fish diet, and a regular diet. After 6 months, the blood pressures of the groups are compared to see if diet has any effect on blood pressure.

18. Identify the independent variable(s) and the dependent variable for each of the studies in Exercise 17.

19. For each of the studies in Exercise 17, suggest possible confounding variables.

20. According to a pilot study of 20 people conducted at the University of Minnesota, daily doses of a compound called arabinogalactan over a period of 6 months resulted in a significant increase in the beneficial lactobacillus species of bacteria. Why can't it be concluded that the compound is beneficial for the majority of people?

21. Comment on the following statement, taken from a magazine advertisement: "In a recent clinical study, Brand ABC [actual brand will not be named] was proved to be 1950% better than creatine!"

22. In an ad for women, the following statement was made: "For every 100 women, 91 have taken the road less traveled." Comment on this statement.

23. In many ads for weight loss products, under the product claims and in small print, the following statement is made: "These results are not typical." What does this say about the product being advertised?

24. In an ad for moisturizing lotion, the following claim is made: ". . . it's the #1 dermatologist-recommended brand." What is misleading about this claim?

25. An ad for an exercise product stated: "Using this product will burn 74% more calories." What is misleading about this statement?

26. "Vitamin E is a proven antioxidant and may help in fighting cancer and heart disease." Is there anything ambiguous about this claim? Explain.

27. "Just 1 capsule of Brand X can provide 24 hours of acid control." (Actual brand will not be named.) What needs to be more clearly defined in this statement?

28. ". . . Male children born to women who smoke during pregnancy run a risk of violent and criminal behavior that lasts well into adulthood." Can we infer that smoking during pregnancy is responsible for criminal behavior in people?

29. In the 1980s, a study linked coffee to a higher risk of heart disease and pancreatic cancer. In the early 1990s, studies showed that drinking coffee posed minimal health threats. However, in 1994, a study showed that pregnant women who drank 3 or more cups of tea daily may be at risk for spontaneous abortion. In 1998, a study claimed that women who drank more than a half-cup of caffeinated tea every day may actually increase their fertility. In 1998, a study showed that over a lifetime, a few extra cups of coffee a day can raise blood pressure, heart rate, and stress (Source: "Bottom Line: Is It Good for You? Or Bad?" by Monika Guttman, *USA TODAY Weekend*). Suggest some reasons why these studies appear to be conflicting.

Extending the Concepts

30. Find an article that describes a statistical study, and identify the study as observational or experimental.

31. For the article that you used in Exercise 30, identify the independent variable(s) and dependent variable for the study.

32. For the article that you selected in Exercise 30, suggest some confounding variables that may have an effect on the results of the study.

Statistics Today

Are We Improving Our Diet?–Revisited

Researchers selected a *sample* of 23,699 adults in the United States, using phone numbers selected at *random,* and conducted a *telephone survey.* All respondents were asked six questions:

1. How often do you drink juices such as orange, grapefruit, or tomato?
2. Not counting juice, how often do you eat fruit?
3. How often do you eat green salad?
4. How often do you eat potatoes (not including french fries, fried potatoes, or potato chips)?
5. How often do you eat carrots?
6. Not counting carrots, potatoes, or salad, how many servings of vegetables do you usually eat?

Researchers found that men consumed fewer servings of fruits and vegetables per day (3.3) than women (3.7). Only 20% of the population consumed the recommended 5 or more daily servings. In addition, they found that youths and less-educated people consumed an even lower amount than the average.

Based on this study, they recommend that greater educational efforts are needed to improve fruit and vegetable consumption by Americans and to provide environmental and institutional support to encourage increased consumption.

Source: Mary K. Serdula, M.D., et al., "Fruit and Vegetable Intake Among Adults in 16 States: Results of a Brief Telephone Survey," *American Journal of Public Health* 85, no. 2. Copyright by the American Public Health Association.

Chapter Quiz

Determine whether each statement is true or false. If the statement is false, explain why.

1. Probability is used as a basis for inferential statistics.

2. The height of President Lincoln is an example of a variable.

3. The highest level of measurement is the interval level.

4. When the population of college professors is divided into groups according to their rank (instructor, assistant professor, etc.) and then several are selected from each group to make up a sample, the sample is called a cluster sample.

5. The variable *age* is an example of a qualitative variable.

6. The weight of pumpkins is considered to be a continuous variable.

7. The boundary of a value such as 6 inches would be 5.9–6.1 inches.

Select the best answer.

8. The number of absences per year that a worker has is an example of what type of data?
   a. Nominal
   b. Qualitative
   c. Discrete
   d. Continuous

9. What are the boundaries of 25.6 ounces?
   a. 25–26 ounces
   b. 25.55–25.65 ounces
   c. 25.5–25.7 ounces
   d. 20–39 ounces

10. A researcher divided subjects into two groups according to gender and then selected members from each group for her sample. What sampling method was the researcher using?
    a. Cluster
    b. Random
    c. Systematic
    d. Stratified

11. Data that can be classified according to color are measured on what scale?
    a. Nominal
    b. Ratio
    c. Ordinal
    d. Interval

12. A study that involves no researcher intervention is called
    a. An experimental study.
    b. A noninvolvement study.
    c. An observational study.
    d. A quasi-experimental study.

13. A variable that interferes with other variables in the study is called
    a. A confounding variable.
    b. An explanatory variable.
    c. An outcome variable.
    d. An interfering variable.

Use the best answer to complete these statements.

14. Two major branches of statistics are _____ and _____.

15. Two uses of probability are _____ and _____.

16. The group of all subjects under study is called a(n) _____.

17. A group of subjects selected from the group of all subjects under study is called a(n) _____.

18. Three reasons why samples are used in statistics are
    a. _____ b. _____ c. _____.

19. The four basic sampling methods are
    a. _____ b. _____ c. _____ d. _____.

20. A study that uses intact groups when it is not possible to randomly assign participants to the groups is called a(n) _____ study.

21. In a research study, participants should be assigned to groups using _____ methods, if possible.

22. For each statement, decide whether descriptive or inferential statistics is used.
    a. The average life expectancy in New Zealand is 78.49 years. Source: *World Factbook 2004.*
    b. A diet high in fruits and vegetables will lower blood pressure. Source: Institute of Medicine.
    c. The total amount of estimated losses from hurricane Hugo was $4.2 billion. Source: Insurance Service Office.

   d. Researchers stated that the shape of a person's ears is related to the person's aggression. Source: *American Journal of Human Biology.*

   e. In 2013, the number of high school graduates will be 3.2 million students. Source: National Center for Education.

23. Classify each as nominal-level, ordinal-level, interval-level, or ratio-level measurement.

   a. Rating of movies as G, PG, and R.
   b. Number of candy bars sold on a fund drive.
   c. Classification of automobiles as subcompact, compact, standard, and luxury.
   d. Temperatures of hair dryers.
   e. Weights of suitcases on a commercial airline.

24. Classify each variable as discrete or continuous.

   a. Ages of people working in a large factory.

   b. Number of cups of coffee served at a restaurant.
   c. The amount of a drug injected into a guinea pig.
   d. The time it takes a student to drive to school.
   e. The number of gallons of milk sold each day at a grocery store.

25. Give the boundaries of each.

   a. 48 seconds
   b. 0.56 centimeter
   c. 9.1 quarts
   d. 13.7 pounds
   e. 7 feet

Critical Thinking Challenges

1. A study of the world's busiest airports was conducted by *Airports Council International.* Describe three variables that one could use to determine which airports are the busiest. What *units* would one use to measure these variables? Are these variables categorical, discrete, or continuous?

2. The results of a study published in *Archives of General Psychiatry* stated that male children born to women who smoke during pregnancy run a risk of violent and criminal behavior that lasts into adulthood. The results of this study were challenged by some people in the media. Give several reasons why the results of this study would be challenged.

3. The results of a study published in *Neurological Research* stated that second-graders who took piano lessons and played a computer math game more readily grasped math problems in fractions and proportions than a similar group who took an English class and played the same math game. What type of inferential study was this? Give several reasons why the piano lessons could improve a student's math ability.

4. A study of 2958 collegiate soccer players showed that in 46 anterior cruciate ligament (ACL) tears, 36 were in women. Calculate the percentages of tears for each gender.

   a. Can it be concluded that female athletes tear their knees more often than male athletes?
   b. Comment on how this study's conclusion might have been reached.

5. Read the article entitled "Anger Can Cause Snap Judgments" and answer the following questions.

   a. Is the study experimental or observational?
   b. What is the independent variable?
   c. What is the dependent variable?
   d. Do you think the sample sizes are large enough to merit the conclusion?
   e. Based on the results of the study, what changes would you recommend to persons to help them reduce their anger?

6. Read the article entitled "Hostile Children Fight Unemployment" and answer the following questions.

   a. Is the study experimental or observational?
   b. What is the independent variable?
   c. What is the dependent variable?
   d. Suggest some confounding variables that may have influenced the results of the study.
   e. Identify the three groups of subjects used in the study.

ANGER CAN CAUSE SNAP JUDGMENTS

Anger can make a normally unbiased person act with prejudice, according to a forthcoming study in the journal *Psychological Science*.

Assistant psychology professors David DeSteno at Northeastern University in Boston and Nilanjana Dasgupta at the University of Massachusetts, Amherst, randomly divided 81 study participants into two groups and assigned them a writing task designed to induce angry, sad or neutral feelings. In a subsequent test to uncover nonconscious associations, angry subjects were quicker to connect negatively charged words —like war, death and vomit—with members of the opposite group—even though the groupings were completely arbitrary.

"These automatic responses guide our behavior when we're not paying attention," says DeSteno, and they can lead to discriminatory acts when there is pressure to make a quick decision. "If you're aware that your emotions might be coloring these gut reactions," he says, "you should take time to consider that possibility and adjust your actions accordingly."

—*Eric Strand*

Source: Reprinted with permission from *Psychology Today,* Copyright © (2004) Sussex Publishers, Inc.

UNEMPLOYMENT

Hostile Children Fight Unemployment

Aggressive children may be destined for later long-term unemployment. In a study that began in 1968, researchers at the University of Jyvaskyla in Finland examined about 300 participants at ages 8, 14, 27, and 36. They looked for aggressive behaviors like hurting other children, kicking objects when angry, or attacking others without reason.

Their results, published recently in the *International Journal of Behavioral Development*, suggest that children with low self-control of emotion —especially aggression—were significantly more prone to long-term unemployment. Children with behavioral inhibitions—such as passive and anxious behaviors—were also indirectly linked to unemployment as they lacked the preliminary initiative needed for school success. And while unemployment rates were high in Finland during the last data collection, jobless participants who were aggressive as children were less likely to have a job two years later than their nonaggressive counterparts.

Ongoing unemployment can have serious psychological consequences, including depression, anxiety and stress. But lead researcher Lea Pulkkinen, Ph.D., a Jyvaskyla psychology professor, does have encouraging news for parents: Aggressive children with good social skills and child-centered parents were significantly less likely to be unemployed for more than two years as adults.

—*Tanya Zimbardo*

Source: Reprinted with permission from *Psychology Today,* Copyright © (2001) Sussex Publishers, Inc.

Answers to Applying the Concepts

Section 1–2 Attendance and Grades

1. The variables are grades and attendance.

2. The data consist of specific grades and attendance numbers.

3. These are descriptive statistics.

4. The population under study is students at Manatee Community College (MCC).

5. While not specified, we probably have data from a sample of MCC students.

6. Based on the data, it appears that in general, the better your attendance the higher your grade.

Section 1–3 Safe Travel

1. The variables are industry and number of job-related injuries.

2. The type of industry is a qualitative variable, while the number of job-related injuries is quantitative.

3. The number of job-related injuries is discrete.

4. Type of industry is nominal, and the number of job-related injuries is ratio.

5. The railroads do show fewer job-related injuries; however, there may be other things to consider. For example, railroads employ fewer people than the other transportation industries in the study.

6. A person's choice of transportation might also be affected by convenience issues, cost, service, etc.

7. Answers will vary. One possible answer is that the railroads have the fewest job-related injuries, while the airline industry has the most job-related industries (more than twice those of the railroad industry). The numbers of job-related industries in the subway and trucking industries are fairly comparable.

Section 1–4 American Culture and Drug Abuse

Answers will vary, so this is one possible answer.

1. I used a telephone survey. The advantage to my survey method is that this was a relatively inexpensive survey method (although more expensive than using the mail) that could get a fairly sizable response. The disadvantage to my survey method is that I have not included anyone without a telephone. (*Note:* My survey used a random dialing method to include unlisted numbers and cell phone exchanges.)

2. A mail survey also would have been fairly inexpensive, but my response rate may have been much lower than what I got with my telephone survey. Interviewing would have allowed me to use follow-up questions and to clarify any questions of the respondents at the time of the interview. However, interviewing is very labor- and cost-intensive.

3. I used ordinal data on a scale of 1 to 5. The scores were 1 = strongly disagree, 2 = disagree, 3 = neutral, 4 = agree, 5 = strongly agree.

4. The random method that I used was a random dialing method.

5. To include people from each state, I used a stratified random sample, collecting data randomly from each of the area codes and telephone exchanges available.

6. This method allowed me to make sure that I had representation from each area of the United States.

7. Convenience samples may not be representative of the population, and a convenience sample of adolescents would probably differ greatly from the general population with regard to the influence of American culture on illegal drug use.

Section 1–5 Just a Pinch Between Your Cheek and Gum

1. This was an experiment, since the researchers imposed a treatment on each of the two groups involved in the study.

2. The independent variable is whether the participant chewed tobacco or not. The dependent variables are the students' blood pressures and heart rates.

3. The treatment group is the tobacco group—the other group was used as a control.

4. A student's blood pressure might not be affected by knowing that he or she was part of a study. However, if the student's blood pressure were affected by this knowledge, all the students (in both groups) would be affected similarly. This might be an example of the placebo effect.

5. Answers will vary. One possible answer is that confounding variables might include the way that the students chewed the tobacco, whether or not the students smoked (although this would hopefully have been evened out with the randomization), and that all the participants were university students.

6. Answers will vary. One possible answer is that the study design was fine, but that it cannot be generalized beyond the population of university students (or people around that age).

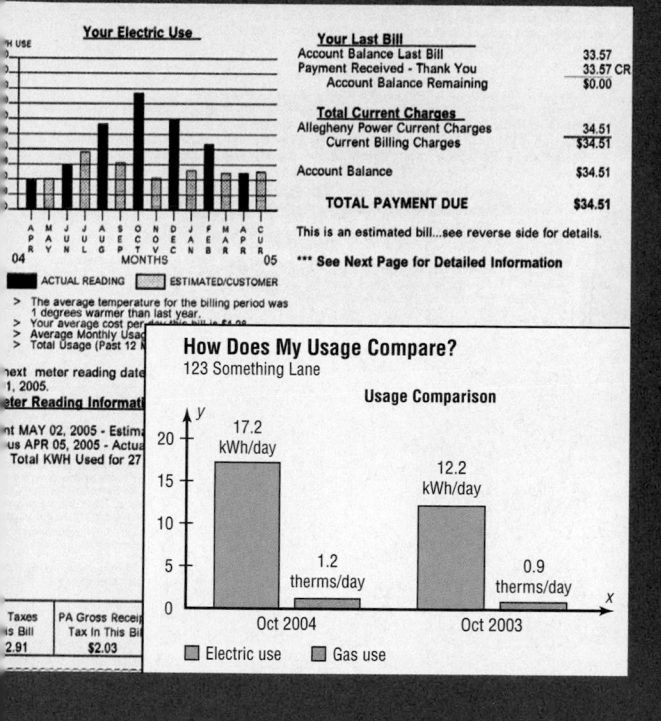

Frequency Distributions and Graphs

Objectives

After completing this chapter, you should be able to

1 Organize data using frequency distributions.

2 Represent data in frequency distributions graphically using histograms, frequency polygons, and ogives.

3 Represent data using Pareto charts, time series graphs, and pie graphs.

4 Draw and interpret a stem and leaf plot.

Outline

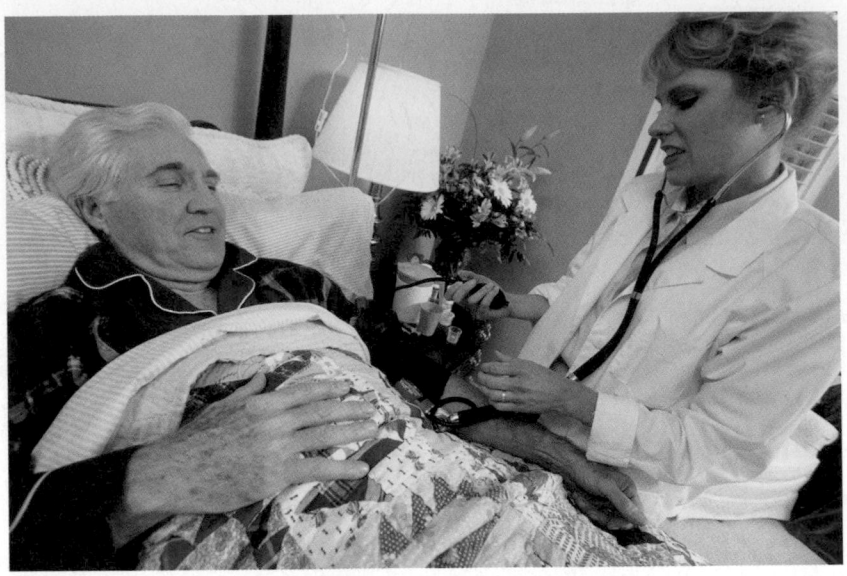

Statistics Today

How Serious Are Hospital Infections?

According to an article in the *Pittsburgh Tribune Review,* hospital infections occur in nearly 2 million patients every year. Just how serious a problem is this? It is very serious since the article further reports that one out of every six patients who develop an infection while in the hospital dies. In the first 3 months of 2004, hospitals in Pennsylvania reported that there were 2253 hospital-acquired infections, and 388 deaths resulted from these infections. That is about 17%. The type and number of infections are shown in the following table.

Type of infection	Infections reported	Number of deaths	Death rate
Urinary tract	931	99	10.6%
Surgical site	229	6	2.6
Pneumonia	291	100	34.4
Bloodstream	410	107	26.1
Other	392	76	Varies
	2253	388	

Looking at the numbers presented in a table does not have the same impact as presenting numbers in a well-drawn chart or graph. The article did not include any graphs. This chapter will show you how to construct appropriate graphs to represent data and help you to get your point across to your audience.

See Statistics Today—Revisited at the end of the chapter for some suggestions on how to represent the data graphically.

2–1

Introduction

When conducting a statistical study, the researcher must gather data for the particular variable under study. For example, if a researcher wishes to study the number of people who were bitten by poisonous snakes in a specific geographic area over the past several years, he or she has to gather the data from various doctors, hospitals, or health departments.

To describe situations, draw conclusions, or make inferences about events, the researcher must organize the data in some meaningful way. The most convenient method of organizing data is to construct a *frequency distribution.*

After organizing the data, the researcher must present them so they can be understood by those who will benefit from reading the study. The most useful method of presenting the data is by constructing *statistical charts* and *graphs.* There are many different types of charts and graphs, and each one has a specific purpose.

This chapter explains how to organize data by constructing frequency distributions and how to present the data by constructing charts and graphs. The charts and graphs illustrated here are histograms, frequency polygons, ogives, pie graphs, Pareto charts, and time series graphs. A graph that combines the characteristics of a frequency distribution and a histogram, called a stem and leaf plot, is also explained.

2–2

Objective 1

Organize data using frequency distributions.

Organizing Data

Suppose a researcher wished to do a study on the number of miles that the employees of a large department store traveled to work each day. The researcher first would have to collect the data by asking each employee the approximate distance the store is from his or her home. When data are collected in original form, they are called **raw data.** In this case, the data are

1	2	6	7	12	13	2	6	9	5
18	7	3	15	15	4	17	1	14	5
4	16	4	5	8	6	5	18	5	2
9	11	12	1	9	2	10	11	4	10
9	18	8	8	4	14	7	3	2	6

Since little information can be obtained from looking at raw data, the researcher organizes the data into what is called a *frequency distribution.* A frequency distribution consists of *classes* and their corresponding *frequencies.* Each raw data value is placed into a quantitative or qualitative category called a **class.** The **frequency** of a class then is the number of data values contained in a specific class. A frequency distribution is shown for the data set above.

Class limits (in miles)	Tally	Frequency
1–3	〣〣	10
4–6	〣〣 ////	14
7–9	〣〣	10
10–12	〣/	6
13–15	〣	5
16–18	〣	5
		Total 50

Now some general observations can be made from looking at the data in the form of a frequency distribution. For example, the majority of employees live within 9 miles of the store.

A **frequency distribution** is the organization of raw data in table form, using classes and frequencies.

Unusual Stat

Of Americans 50 years old and over, 23% think their greatest achievements are still ahead of them.

The classes in this distribution are 1–3, 4–6, etc. These values are called *class limits.* The data values 1, 2, 3 can be tallied in the first class; 4, 5, 6 in the second class; and so on.

Two types of frequency distributions that are most often used are the *categorical frequency distribution* and the *grouped frequency distribution.* The procedures for constructing these distributions are shown now.

Categorical Frequency Distributions

The **categorical frequency distribution** is used for data that can be placed in specific categories, such as nominal- or ordinal-level data. For example, data such as political affiliation, religious affiliation, or major field of study would use categorical frequency distributions.

Example 2–1

Twenty-five army inductees were given a blood test to determine their blood type. The data set is

A	B	B	AB	O
O	O	B	AB	B
B	B	O	A	O
A	O	O	O	AB
AB	A	O	B	A

Construct a frequency distribution for the data.

Solution

Since the data are categorical, discrete classes can be used. There are four blood types: A, B, O, and AB. These types will be used as the classes for the distribution.

The procedure for constructing a frequency distribution for categorical data is given next.

Step 1 Make a table as shown.

A	B	C	D
Class	Tally	Frequency	Percent
A			
B			
O			
AB			

Step 2 Tally the data and place the results in column B.

Step 3 Count the tallies and place the results in column C.

Step 4 Find the percentage of values in each class by using the formula

$$\% = \frac{f}{n} \cdot 100\%$$

where f = frequency of the class and n = total number of values. For example, in the class of type A blood, the percentage is

$$\% = \frac{5}{25} \cdot 100\% = 20\%$$

Percentages are not normally part of a frequency distribution, but they can be added since they are used in certain types of graphs such as pie graphs. Also, the decimal equivalent of a percent is called a *relative frequency.*

Step 5 Find the totals for columns C (frequency) and D (percent). The completed table is shown.

A Class	B Tally	C Frequency	D Percent
A	卌	5	20
B	卌 //	7	28
O	卌 ////	9	36
AB	////	4	16
		Total 25	100

For the sample, more people have type O blood than any other type.

Grouped Frequency Distributions

When the range of the data is large, the data must be grouped into classes that are more than one unit in width, in what is called a **grouped frequency distribution.** For example, a distribution of the number of hours that boat batteries lasted is the following.

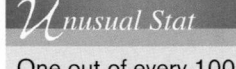
Class limits	Class boundaries	Tally	Frequency	Cumulative frequency
24–30	23.5–30.5	///	3	3
31–37	30.5–37.5	/	1	4
38–44	37.5–44.5	卌	5	9
45–51	44.5–51.5	卌 ////	9	18
52–58	51.5–58.5	卌 /	6	24
59–65	58.5–65.5	/	1	25
			25	

The procedure for constructing the preceding frequency distribution is given in Example 2–2; however, several things should be noted. In this distribution, the values 24 and 30 of the first class are called *class limits*. The **lower class limit** is 24; it represents the smallest data value that can be included in the class. The **upper class limit** is 30; it represents the largest data value that can be included in the class. The numbers in the second column are called **class boundaries.** These numbers are used to separate the classes so that there are no gaps in the frequency distribution. The gaps are due to the limits; for example, there is a gap between 30 and 31.

Students sometimes have difficulty finding class boundaries when given the class limits. The basic rule of thumb is that *the class limits should have the same decimal place value as the data, but the class boundaries should have one additional place value and end in a 5*. For example, if the values in the data set are whole numbers, such as 24, 32, 18, the limits for a class might be 31–37, and the boundaries are 30.5–37.5. Find the boundaries by subtracting 0.5 from 31 (the lower class limit) and adding 0.5 to 37 (the upper class limit).

$$\text{Lower limit} - 0.5 = 31 - 0.5 = 30.5 = \text{lower boundary}$$
$$\text{Upper limit} + 0.5 = 37 + 0.5 = 37.5 = \text{upper boundary}$$

If the data are in tenths, such as 6.2, 7.8, and 12.6, the limits for a class hypothetically might be 7.8–8.8, and the boundaries for that class would be 7.75–8.85. Find these values by subtracting 0.05 from 7.8 and adding 0.05 to 8.8.

Finally, the **class width** for a class in a frequency distribution is found by subtracting the lower (or upper) class limit of one class from the lower (or upper) class limit of the next class. For example, the class width in the preceding distribution on the duration of boat batteries is 7, found from $31 - 24 = 7$.

The class width can also be found by subtracting the lower boundary from the upper boundary for any given class. In this case, $30.5 - 23.5 = 7$.

Note: Do not subtract the limits of a single class. It will result in an incorrect answer.

The researcher must decide how many classes to use and the width of each class. To construct a frequency distribution, follow these rules:

1. *There should be between 5 and 20 classes.* Although there is no hard-and-fast rule for the number of classes contained in a frequency distribution, it is of the utmost importance to have enough classes to present a clear description of the collected data.

2. *It is preferable but not absolutely necessary that the class width be an odd number.* This ensures that the midpoint of each class has the same place value as the data. The **class midpoint** X_m is obtained by adding the lower and upper boundaries and dividing by 2, or adding the lower and upper limits and dividing by 2:

$$X_m = \frac{\text{lower boundary} + \text{upper boundary}}{2}$$

or

$$X_m = \frac{\text{lower limit} + \text{upper limit}}{2}$$

For example, the midpoint of the first class in the example with boat batteries is

$$\frac{24 + 30}{2} = 27 \qquad \text{or} \qquad \frac{23.5 + 30.5}{2} = 27$$

The midpoint is the numeric location of the center of the class. Midpoints are necessary for graphing (see Section 2–3). If the class width is an even number, the midpoint is in tenths. For example, if the class width is 6 and the boundaries are 5.5 and 11.5, the midpoint is

$$\frac{5.5 + 11.5}{2} = \frac{17}{2} = 8.5$$

Rule 2 is only a suggestion, and it is not rigorously followed, especially when a computer is used to group data.

3. *The classes must be mutually exclusive.* Mutually exclusive classes have nonoverlapping class limits so that data cannot be placed into two classes. Many times, frequency distributions such as

Age
10–20
20–30
30–40
40–50

are found in the literature or in surveys. If a person is 40 years old, into which class should she or he be placed? A better way to construct a frequency distribution is to use classes such as

Age
10–20
21–31
32–42
43–53

4. *The classes must be continuous.* Even if there are no values in a class, the class must be included in the frequency distribution. There should be no gaps in a

frequency distribution. The only exception occurs when the class with a zero frequency is the first or last class. A class with a zero frequency at either end can be omitted without affecting the distribution.

5. *The classes must be exhaustive.* There should be enough classes to accommodate all the data.

6. *The classes must be equal in width.* This avoids a distorted view of the data.

One exception occurs when a distribution has a class that is open-ended. That is, the class has no specific beginning value or no specific ending value. A frequency distribution with an open-ended class is called an **open-ended distribution.** Here are two examples of distributions with open-ended classes.

Age	Frequency		Minutes	Frequency
10–20	3		Below 110	16
21–31	6		110–114	24
32–42	4		115–119	38
43–53	10		120–124	14
54 and above	8		125–129	5

The frequency distribution for age is open-ended for the last class, which means that anybody who is 54 years or older will be tallied in the last class. The distribution for minutes is open-ended for the first class, meaning that any minute values below 110 will be tallied in that class.

Example 2–2 shows the procedure for constructing a grouped frequency distribution, i.e., when the classes contain more than one data value.

Example 2–2

These data represent the record high temperatures in °F for each of the 50 states. Construct a grouped frequency distribution for the data using 7 classes.

112	100	127	120	134	118	105	110	109	112
110	118	117	116	118	122	114	114	105	109
107	112	114	115	118	117	118	122	106	110
116	108	110	121	113	120	119	111	104	111
120	113	120	117	105	110	118	112	114	114

Source: *The World Almanac and Book of Facts.*

Solution

The procedure for constructing a grouped frequency distribution for numerical data follows.

Step 1   Determine the classes.

Find the highest value and lowest value: $H = 134$ and $L = 100$.

Find the range: R = highest value − lowest value = $H - L$, so

$R = 134 - 100 = 34$

Select the number of classes desired (usually between 5 and 20). In this case, 7 is arbitrarily chosen.

Find the class width by dividing the range by the number of classes.

$$\text{Width} = \frac{R}{\text{number of classes}} = \frac{34}{7} = 4.9$$

Round the answer up to the nearest whole number if there is a remainder: $4.9 \approx 5$. (Rounding *up* is different from rounding *off*. A number is rounded up if there is any decimal remainder when dividing. For example, $85 \div 6 = 14.167$ and is rounded up to 15. Also, $53 \div 4 = 13.25$ and is rounded up to 14. Also, after dividing, if there is no remainder, you will need to add an extra class to accommodate all the data.)

Select a starting point for the lowest class limit. This can be the smallest data value or any convenient number less than the smallest data value. In this case, 100 is used. Add the width to the lowest score taken as the starting point to get the lower limit of the next class. Keep adding until there are 7 classes, as shown, 100, 105, 110, etc.

Subtract one unit from the lower limit of the second class to get the upper limit of the first class. Then add the width to each upper limit to get all the upper limits.

$$105 - 1 = 104$$

The first class is 100–104, the second class is 105–109, etc.

Find the class boundaries by subtracting 0.5 from each lower class limit and adding 0.5 to each upper class limit:

99.5–104.5, 104.5–109.5, etc.

Step 2   Tally the data.

Step 3   Find the numerical frequencies from the tallies.

Step 4   Find the cumulative frequencies.

A cumulative frequency (cf) column can be added to the distribution by adding the frequency in each class to the total of the frequencies of the classes preceding that class, such as $0 + 2 = 2$, $2 + 8 = 10$, $10 + 18 = 28$, and $28 + 13 = 41$.

The completed frequency distribution is

Class limits	Class boundaries	Tally	Frequency	Cumulative frequency
100–104	99.5–104.5	//	2	2
105–109	104.5–109.5	⫶⫶⫶ ///	8	10
110–114	109.5–114.5	⫶⫶⫶ ⫶⫶⫶ ⫶⫶⫶ ///	18	28
115–119	114.5–119.5	⫶⫶⫶ ⫶⫶⫶ ///	13	41
120–124	119.5–124.5	⫶⫶⫶ //	7	48
125–129	124.5–129.5	/	1	49
130–134	129.5–134.5	/	1	50

$$n = \Sigma f = 50$$

The frequency distribution shows that the class 109.5–114.5 contains the largest number of temperatures (18) followed by the class 114.5–119.5 with 13 temperatures. Hence, most of the temperatures (31) fall between 109.5°F and 119.5°F.

Cumulative frequencies are used to show how many data values are accumulated up to and including a specific class. In Example 2–2, 28 of the total record high temperatures are less than or equal to 114°F. Forty-eight of the total record high temperatures are less than or equal to 124°F.

After the raw data have been organized into a frequency distribution, it will be analyzed by looking for peaks and extreme values. The peaks show which class or classes have the most data values compared to the other classes. Extreme values, called outliers, show large or small data values that are relative to other data values.

When the range of the data values is relatively small, a frequency distribution can be constructed using single data values for each class. This type of distribution is called an **ungrouped frequency distribution** and is shown next.

Example 2–3	The data shown here represent the number of miles per gallon that 30 selected four-wheel-drive sports utility vehicles obtained in city driving. Construct a frequency distribution, and analyze the distribution.

12	17	12	14	16	18
16	18	12	16	17	15
15	16	12	15	16	16
12	14	15	12	15	15
19	13	16	18	16	14

Source: *Model Year 1999 Fuel Economy Guide.* United States Environmental Protection Agency, October 1998.

Solution

Step 1 Determine the classes. Since the range of the data set is small ($19 - 12 = 7$), classes consisting of a single data value can be used. They are 12, 13, 14, 15, 16, 17, 18, 19.

Note: If the data are continuous, class boundaries can be used. Subtract 0.5 from each class value to get the lower class boundary, and add 0.5 to each class value to get the upper class boundary.

Step 2 Tally the data.

Step 3 Find the numerical frequencies from the tallies.

Step 4 Find the cumulative frequencies.

The completed ungrouped frequency distribution is

Class limits	Class boundaries	Tally	Frequency	Cumulative frequency
12	11.5–12.5	////// /	6	6
13	12.5–13.5	/	1	7
14	13.5–14.5	///	3	10
15	14.5–15.5	////// /	6	16
16	15.5–16.5	////// ///	8	24
17	16.5–17.5	//	2	26
18	17.5–18.5	///	3	29
19	18.5–19.5	/	1	30

In this case, almost one-half (14) of the vehicles get 15 or 16 miles per gallon.

The steps for constructing a grouped frequency distribution are summarized in the following Procedure Table.

Procedure Table

Constructing a Grouped Frequency Distribution

Step 1   Determine the classes.

Find the highest and lowest value.

Find the range.

Select the number of classes desired.

Find the width by dividing the range by the number of classes and rounding up.

Select a starting point (usually the lowest value or any convenient number less than the lowest value); add the width to get the lower limits.

Find the upper class limits.

Find the boundaries.

Step 2   Tally the data.

Step 3   Find the numerical frequencies from the tallies.

Step 4   Find the cumulative frequencies.

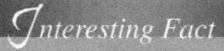

Male dogs bite children more often than female dogs do; however, female cats bite children more often than male cats do.

When one is constructing a frequency distribution, the guidelines presented in this section should be followed. However, one can construct several different but correct frequency distributions for the same data by using a different class width, a different number of classes, or a different starting point.

Furthermore, the method shown here for constructing a frequency distribution is not unique, and there are other ways of constructing one. Slight variations exist, especially in computer packages. But regardless of what methods are used, classes should be mutually exclusive, continuous, exhaustive, and of equal width.

In summary, the different types of frequency distributions were shown in this section. The first type, shown in Example 2–1, is used when the data are categorical (nominal), such as blood type or political affiliation. This type is called a categorical frequency distribution. The second type of distribution is used when the range is large and classes several units in width are needed. This type is called a grouped frequency distribution and is shown in Example 2–2. Another type of distribution is used for numerical data and when the range of data is small, as shown in Example 2–3. Since each class is only one unit, this distribution is called an ungrouped frequency distribution.

All the different types of distributions are used in statistics and are helpful when one is organizing and presenting data.

The reasons for constructing a frequency distribution are as follows:

1. To organize the data in a meaningful, intelligible way.

2. To enable the reader to determine the nature or shape of the distribution.

3. To facilitate computational procedures for measures of average and spread (shown in Sections 3–2 and 3–3).

4. To enable the researcher to draw charts and graphs for the presentation of data (shown in Section 2–3).

5. To enable the reader to make comparisons among different data sets.

The factors used to analyze a frequency distribution are essentially the same as those used to analyze histograms and frequency polygons, which are shown in Section 2–3.

Applying the Concepts **2–2**

Ages of Presidents at Inauguration

The data represent the ages of our presidents at the time they were first inaugurated.

57	61	57	57	58	57	61	54	68
51	49	64	50	48	65	52	56	46
54	49	50	47	55	55	54	42	51
56	55	54	51	60	62	43	55	56
61	52	69	64	46	54			

1. Were the data obtained from a population or a sample? Explain your answer.

2. What was the age of the oldest president?

3. What was the age of the youngest president?

4. Construct a frequency distribution for the data. (Use your own judgment as to the number of classes and class size.)

5. Are there any peaks in the distribution?

6. Identify any possible outliers.

7. Write a brief summary of the nature of the data as shown in the frequency distribution.

See page 93 for the answers.

Exercises 2–2

1. List five reasons for organizing data into a frequency distribution.

2. Name the three types of frequency distributions, and explain when each should be used.

3. Find the class boundaries, midpoints, and widths for each class.

   a. 12–18

   b. 56–74

   c. 695–705

   d. 13.6–14.7

   e. 2.15–3.93

4. How many classes should frequency distributions have? Why should the class width be an odd number?

5. Shown here are four frequency distributions. Each is incorrectly constructed. State the reason why.

   a.
Class	Frequency
27–32	1
33–38	0
39–44	6
45–49	4
50–55	2

   b.
Class	Frequency
5–9	1
9–13	2
13–17	5
17–20	6
20–24	3

   c.
Class	Frequency
123–127	3
128–132	7
138–142	2
143–147	19

   d.
Class	Frequency
9–13	1
14–19	6
20–25	2
26–28	5
29–32	9

6. What are open-ended frequency distributions? Why are they necessary?

7. A survey was taken on how much trust people place in the information they read on the Internet. Construct a categorical frequency distribution for the data. A = trust

in everything they read, M = trust in most of what they read, H = trust in about one-half of what they read, S = trust in a small portion of what they read. (Based on information from the *UCLA Internet Report.*)

```
M    M    M    A    H    M    S    M    H    M
S    M    M    M    M    A    M    M    A    M
M    M    H    M    M    M    H    M    H    M
A    M    M    M    H    M    M    M    M    M
```

8. The heights in inches of commonly grown herbs are shown. Organize the data into a frequency distribution with six classes, and think of a way in which these results would be useful.

```
18    20    18    18    24    10    15
12    20    36    14    20    18    24
18    16    16    20     7
```

Source: *The Old Farmer's Almanac.*

9. The following data are the measured speeds in miles per hour of 30 charging elephants. Construct a grouped frequency distribution for the data. From the distribution, estimate an approximate average speed of a charging elephant. Use 5 classes. (Based on data in the *World Almanac and Book of Facts.*)

```
25    24    25    24    25
23    25    19    32    23
22    24    26    25    23
28    25    25    26    27
22    28    24    23    24
21    25    22    29    23
```

10. The total energy consumption in trillions of BTU for each of the 50 states in the United States is shown. Construct a frequency distribution using 10 classes, and analyze the nature of the data.

```
 1,215    2,706    1,400    4,417    1,868
11,588    1,799    1,199      627    1,099
 1,688    1,083    2,501      561    4,001
 1,035      863      594    2,303      583
   329      620    1,722      744    1,143
   264      417      365      302      250
 8,518    4,779    4,620    3,943    3,121
 1,659      511      246    1,520    1,977
 1,079    2,777    2,769    1,477      632
 3,965    2,173    2,025      718      164
```

Source: *Energy Information Administration.*

11. The average quantitative GRE scores for the top 30 graduate schools of engineering are listed. Construct a frequency distribution with 6 classes.

```
767    770    761    760    771    768    776    771    756    770
763    760    747    766    754    771    771    778    766    762
780    750    746    764    769    759    757    753    758    746
```

Source: *U.S. News & World Report Best Graduate Schools.*

12. The number of unhealthy days in selected U.S. metropolitan areas is shown. Construct a frequency distribution with 7 classes. (The data in this exercise will be used in Exercise 22 in Section 3–2.)

```
61    88    40     5    12    12    18    23     1    15
 6    81    50    21     0    27     5    13     0    24
 5     1    32    12    23    93    38    29    16     0
 1    22    36
```

Source: *N.Y. Times Almanac.*

13. The ages of the signers of the Declaration of Independence are shown. (Age is approximate since only the birth year appeared in the source, and one has been omitted since his birth year is unknown.) Construct a frequency distribution for the data using 7 classes. (The data for this exercise will be used for Exercise 5 in Section 2–3 and Exercise 23 in Section 3–2.)

```
41    54    47    40    39    35    50    37    49    42    70    32
44    52    39    50    40    30    34    69    39    45    33    42
44    63    60    27    42    34    50    42    52    38    36    45
35    43    48    46    31    27    55    63    46    33    60    62
35    46    45    34    53    50    50
```

Source: *The Universal Almanac.*

14. The number of automobile fatalities in 27 states where the speed limits were raised in 1996 is shown here. Construct a frequency distribution using 8 classes. (The data for this exercise will be used for Exercise 6 in Section 2–3 and Exercise 24 in Section 3–2.)

```
1100     460      85
 970     480    1430
4040     405      70
 620     690     180
 125    1160    3630
2805     205     325
1555     300     875
 260     350     705
1430     485     145
```

Source: *USA TODAY.*

15. The following data represent the ages of 47 of the wealthiest people in the United States. Construct a grouped frequency distribution for the data using 7 classes. Analyze the results in terms of peaks, extreme values, etc. (The information in this exercise will be used for Exercise 9 in Section 2–3 and Exercise 25 in Section 3–2.)

```
48    48    74    74    84    51    71
56    55    76    85    68    42    79
73    58    73    81    51    81    55
65    66    87    60    74    62    64
39    60    60    37    90    68    67
61    40    72    61    71    74    31
62    63    67    31    40
```

Source: *Forbes.*

16. The acreage of the 39 U.S. National Parks under 900,000 acres (in thousands of acres) is shown here. Construct a frequency distribution for the data using 8 classes. (The data in this exercise will be used in Exercise 11 in Section 2–3.)

41	66	233	775	169
36	338	233	236	64
183	61	13	308	77
520	77	27	217	5
650	462	106	52	52
505	94	75	265	402
196	70	132	28	220
760	143	46	539	

Source: *The Universal Almanac.*

17. The heights (in feet above sea level) of the major active volcanoes in Alaska are given here. Construct a frequency distribution for the data using 10 classes. (The data in this exercise will be used in Exercise 9 in Section 3–2 and Exercise 17 in Section 3–3.)

4,265	3,545	4,025	7,050	11,413
3,490	5,370	4,885	5,030	6,830
4,450	5,775	3,945	7,545	8,450
3,995	10,140	6,050	10,265	6,965
150	8,185	7,295	2,015	5,055
5,315	2,945	6,720	3,465	1,980
2,560	4,450	2,759	9,430	
7,985	7,540	3,540	11,070	
5,710	885	8,960	7,015	

Source: *The Universal Almanac.*

18. During the 1998 baseball season, Mark McGwire and Sammy Sosa both broke Roger Maris's home run record of 61. The distances (in feet) for each home run follow. Construct a frequency distribution for each player, using 8 classes. (The information in this exercise will be used for Exercise 12 in Section 2–3, Exercise 10 in Section 3–2, and Exercise 14 in Section 3–3.)

McGwire				Sosa			
306	370	370	430	371	350	430	420
420	340	460	410	430	434	370	420
440	410	380	360	440	410	420	460
350	527	380	550	400	430	410	370
478	420	390	420	370	410	380	340
425	370	480	390	350	420	410	415
430	388	423	410	430	380	380	366
360	410	450	350	500	380	390	400
450	430	461	430	364	430	450	440
470	440	400	390	365	420	350	420
510	430	450	452	400	380	380	400
420	380	470	398	370	420	360	368
409	385	369	460	430	433	388	440
390	510	500	450	414	482	364	370
470	430	458	380	400	405	433	390
430	341	385	410	480	480	434	344
420	380	400	440	410	420		
377	370						

Source: *USA TODAY.*

Extending the Concepts

19. A researcher conducted a survey asking people if they believed more than one person was involved in the assassination of John F. Kennedy. The results were as follows: 73% said yes, 19% said no, and 9% had no opinion. Is there anything suspicious about the results?

Technology *Step by Step*

MINITAB
Step by Step

Make a Categorical Frequency Table (Qualitative or Discrete Data)

1. Type in all the blood types from Example 2–1 down C1 of the worksheet.

A B B AB O O O B AB B B B O A O A O O O AB AB A O B A

2. Click above row 1 and name the column **BloodType.**

3. Select **Stat>Tables>Tally Individual Values.**

The cursor should be blinking in the Variables dialog box. If not, click inside the dialog box.

4. Double-click C1 in the Variables list.

5. Check the boxes for the statistics: Counts, Percents, and Cumulative percents.

6. Click [OK]. The results will be displayed in the Session Window as shown.

Tally for Discrete Variables: BloodType

BloodType	Count	Percent	CumPct
A	5	20.00	20.00
AB	4	16.00	36.00
B	7	28.00	64.00
O	9	36.00	100.00
N=	25		

Make a Grouped Frequency Distribution (Quantitative Variable)

1. Select **File>New>New Worksheet.** A new worksheet will be added to the project.

2. Type the data used in Example 2–2 into C1. Name the column **TEMPERATURES.**

3. Use the instructions in the textbook to determine the class limits.

   In the next step you will create a new column of data, converting the numeric variable to text categories that can be tallied.

4. Select **Data>Code>Numeric to Text.**

   a) The cursor should be blinking in Code data from columns. If not, click inside the box, then double-click C1 Temperatures in the list. Only quantitative variables will be shown in this list.

   b) Click in the Into columns: then type the name of the new column, **TempCodes.**

   c) Press [Tab] to move to the next dialog box.

   d) Type in the first interval **100:104.**

      Use a colon to indicate the interval from 100 to 104 with no spaces before or after the colon.

   e) Press [Tab] to move to the New: column, and type the text category **100–104.**

   f) Continue to tab to each dialog box, typing the interval and then the category until the last category has been entered.

   The dialog box should look like the one shown.

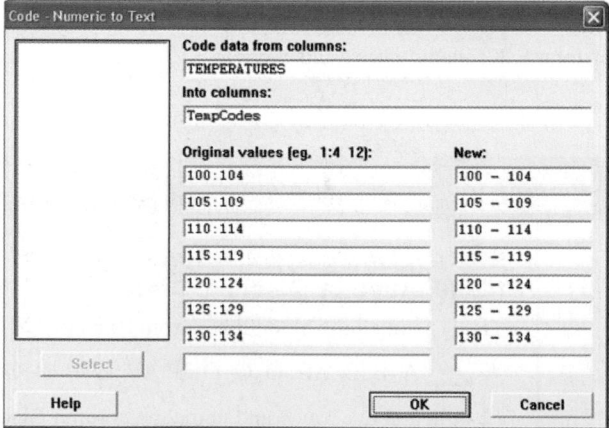

5. Click [OK]. In the worksheet, a new column of data will be created in the first empty column, C2. This new variable will contain the category for each value in C1. The column C2-T contains alphanumeric data.

6. Click **Stat>Tables>Tally Individual Values,** then double-click TempCodes in the Variables list.

   a) Check the boxes for the desired statistics, such as Counts, Percents, and Cumulative percents.

   b) Click [OK].

   The table will be displayed in the Session Window. Eighteen states have high temperatures between 110°F and 114°F. Eighty-two percent of the states have record high temperatures less than or equal to 119°F.

Tally for Discrete Variables: TempCodes

TempCodes	Count	Percent	CumPct
100–104	2	4.00	4.00
105–109	8	16.00	20.00
110–114	18	36.00	56.00
115–119	13	26.00	82.00
120–124	7	14.00	96.00
125–129	1	2.00	98.00
130–134	1	2.00	100.00
N=	50		

7. Click **File>Save Project As** . . . , and type the name of the project file, **Ch2-2.** This will save the two worksheets and the Session Window.

Excel
Step by Step

Categorical Frequency Table (Qualitative or Discrete Variable)

1. Select cell A1 and type in all the blood types from Example 2–1 down column A of the worksheet.

2. Type in the name **BloodType** in cell B1.

3. Select cell B2 and type in the four different blood types down the column.

4. Type in the name **Count** in cell C1.

5. Select cell C2. From the toolbar, select the paste function (f_x) option. Select Statistical from the Function category list. Select COUNTIF from the function name list.

6. In the dialog box, type in **A1:A25** in the Range. Type in the blood type corresponding to the corresponding value from column B.

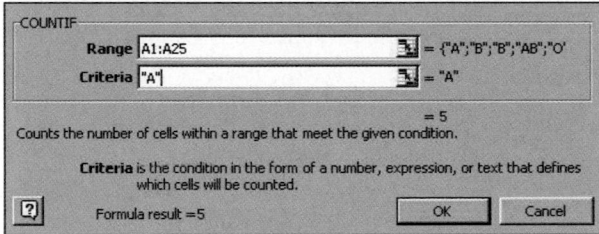

7. After all the data have been counted, select cell C6 from the worksheet.

8. From the toolbar, select the sum (Σ) function. Then type in **C2:C5** and click [Enter].

Making a Grouped Frequency Distribution

1. Press **[Ctrl]-N** for a new worksheet.

2. Enter the data from Examples 2–2 and 2–4 in column A, one number per cell.

3. Select **Tools>Data Analysis.**

4. In Data Analysis, select Histogram and click the [OK] button.

5. In the Histogram dialog box, type **A1:A50** as the Input Range.

6. Select New Worksheet Ply, and check the Cumulative Percentage option. Click [OK].

By leaving the Chart output unchecked, the new worksheet will display the table only. It decides "bins" for the histogram itself (here it picked a bin size of 7 units), but you can also define your own bin range on the data worksheet.

2–3   Histograms, Frequency Polygons, and Ogives

Objective **2**

Represent data in frequency distributions graphically using histograms, frequency polygons, and ogives.

After the data have been organized into a frequency distribution, they can be presented in graphical form. The purpose of graphs in statistics is to convey the data to the viewers in pictorial form. It is easier for most people to comprehend the meaning of data presented graphically than data presented numerically in tables or frequency distributions. This is especially true if the users have little or no statistical knowledge.

Statistical graphs can be used to describe the data set or to analyze it. Graphs are also useful in getting the audience's attention in a publication or a speaking presentation. They can be used to discuss an issue, reinforce a critical point, or summarize a data set. They can also be used to discover a trend or pattern in a situation over a period of time.

The three most commonly used graphs in research are as follows:

1. The histogram.

2. The frequency polygon.

3. The cumulative frequency graph, or ogive (pronounced o-jive).

An example of each type of graph is shown in Figure 2–1. The data for each graph are the distribution of the miles that 20 randomly selected runners ran during a given week.

The Histogram

The **histogram** is a graph that displays the data by using contiguous vertical bars (unless the frequency of a class is 0) of various heights to represent the frequencies of the classes.

Example 2–4

Construct a histogram to represent the data shown for the record high temperatures for each of the 50 states (see Example 2–2).

Class boundaries	Frequency
99.5–104.5	2
104.5–109.5	8
109.5–114.5	18
114.5–119.5	13
119.5–124.5	7
124.5–129.5	1
129.5–134.5	1

Solution

Step 1   Draw and label the x and y axes. The x axis is always the horizontal axis, and the y axis is always the vertical axis.

Figure 2–1

**Examples of
Commonly Used
Graphs**

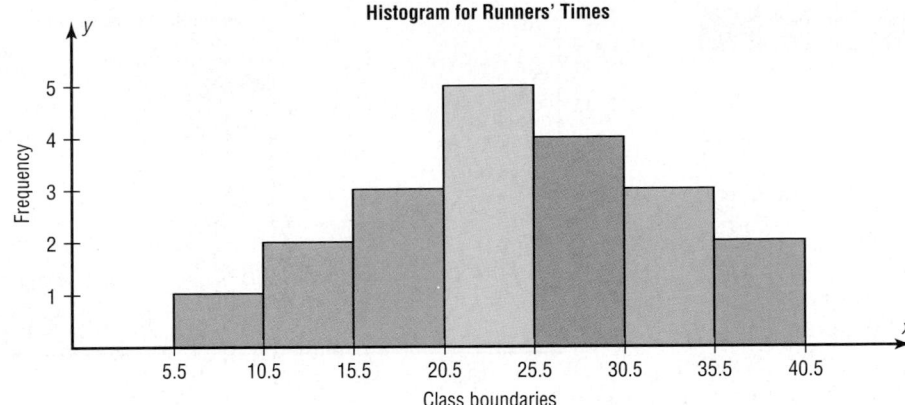

(a) Histogram

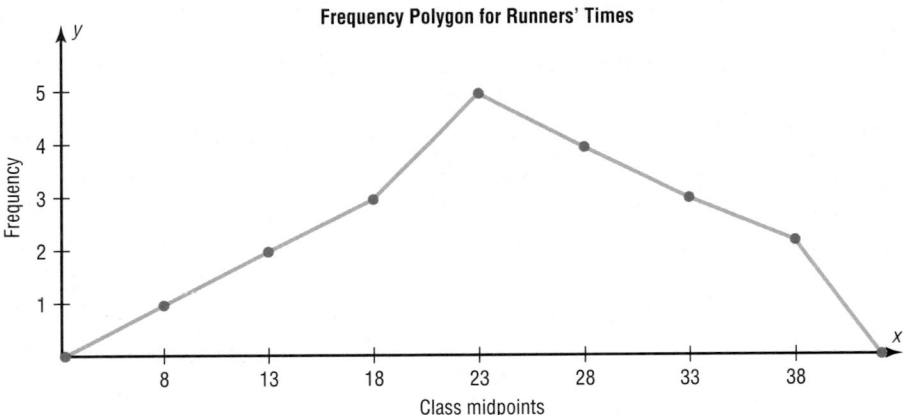

(b) Frequency polygon

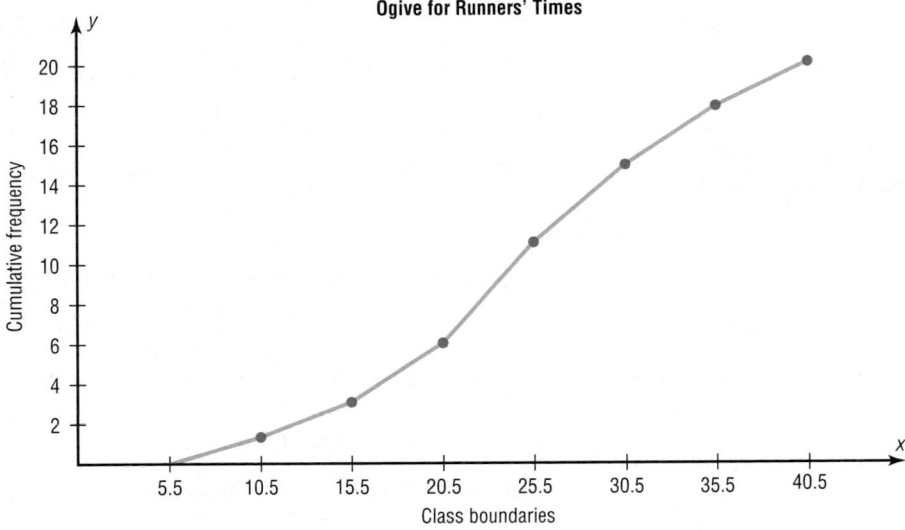

(c) Cumulative frequency graph

Figure 2–2

Histogram for Example 2–4

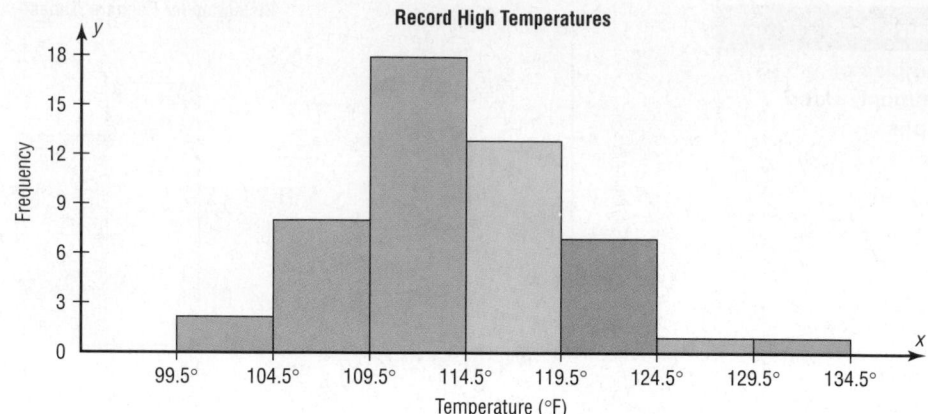

Record High Temperatures

Step 2 Represent the frequency on the *y* axis and the class boundaries on the *x* axis.

Step 3 Using the frequencies as the heights, draw vertical bars for each class. See Figure 2–2.

As the histogram shows, the class with the greatest number of data values (18) is 109.5–114.5, followed by 13 for 114.5–119.5. The graph also has one peak with the data clustering around it.

The Frequency Polygon

Another way to represent the same data set is by using a frequency polygon.

The **frequency polygon** is a graph that displays the data by using lines that connect points plotted for the frequencies at the midpoints of the classes. The frequencies are represented by the heights of the points.

Example 2–5 shows the procedure for constructing a frequency polygon.

Example 2–5

Using the frequency distribution given in Example 2–4, construct a frequency polygon.

Solution

Step 1 Find the midpoints of each class. Recall that midpoints are found by adding the upper and lower boundaries and dividing by 2:

$$\frac{99.5 + 104.5}{2} = 102 \qquad \frac{104.5 + 109.5}{2} = 107$$

and so on. The midpoints are

Class boundaries	Midpoints	Frequency
99.5–104.5	102	2
104.5–109.5	107	8
109.5–114.5	112	18
114.5–119.5	117	13
119.5–124.5	122	7
124.5–129.5	127	1
129.5–134.5	132	1

Figure 2–3

Frequency Polygon for
Example 2–5

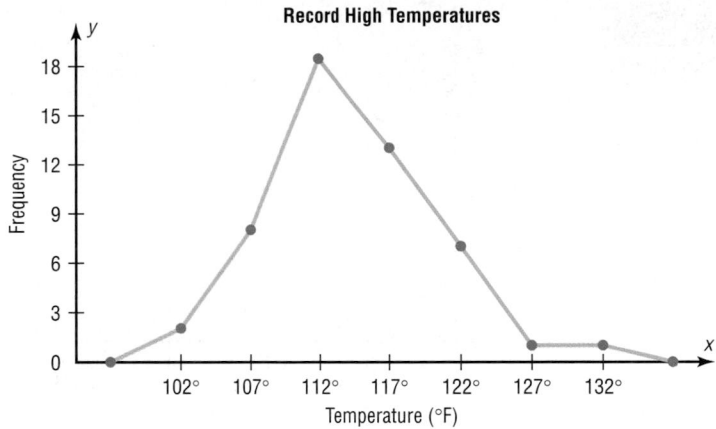

Step 2   Draw the *x* and *y* axes. Label the *x* axis with the midpoint of each class, and then use a suitable scale on the *y* axis for the frequencies.

Step 3   Using the midpoints for the *x* values and the frequencies as the *y* values, plot the points.

Step 4   Connect adjacent points with line segments. Draw a line back to the *x* axis at the beginning and end of the graph, at the same distance that the previous and next midpoints would be located, as shown in Figure 2–3.

The frequency polygon and the histogram are two different ways to represent the same data set. The choice of which one to use is left to the discretion of the researcher.

The Ogive

The third type of graph that can be used represents the cumulative frequencies for the classes. This type of graph is called the *cumulative frequency graph* or *ogive*. The **cumulative frequency** is the sum of the frequencies accumulated up to the upper boundary of a class in the distribution.

The **ogive** is a graph that represents the cumulative frequencies for the classes in a frequency distribution.

Example 2–6 shows the procedure for constructing an ogive.

Example 2–6   Construct an ogive for the frequency distribution described in Example 2–4.

Solution

Step 1   Find the cumulative frequency for each class.

Class boundaries	Cumulative frequency
99.5–104.5	2
104.5–109.5	10
109.5–114.5	28
114.5–119.5	41
119.5–124.5	48
124.5–129.5	49
129.5–134.5	50

Figure 2–4

Plotting the Cumulative
Frequency for
Example 2–6

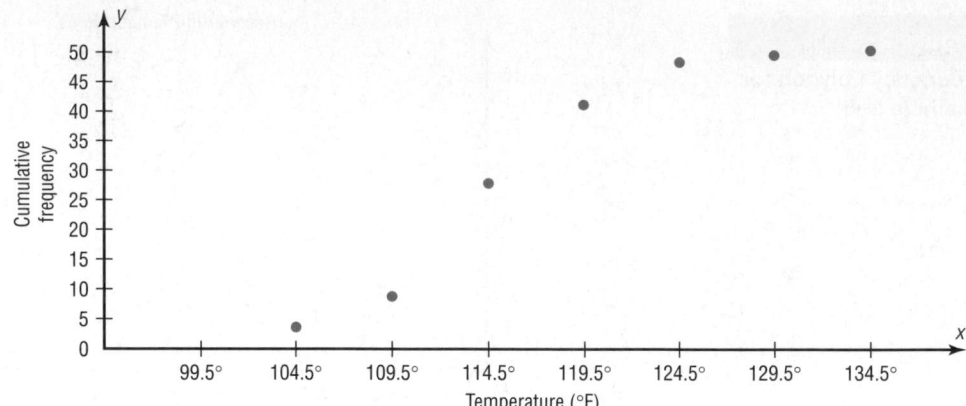

Figure 2–5

Ogive for Example 2–6

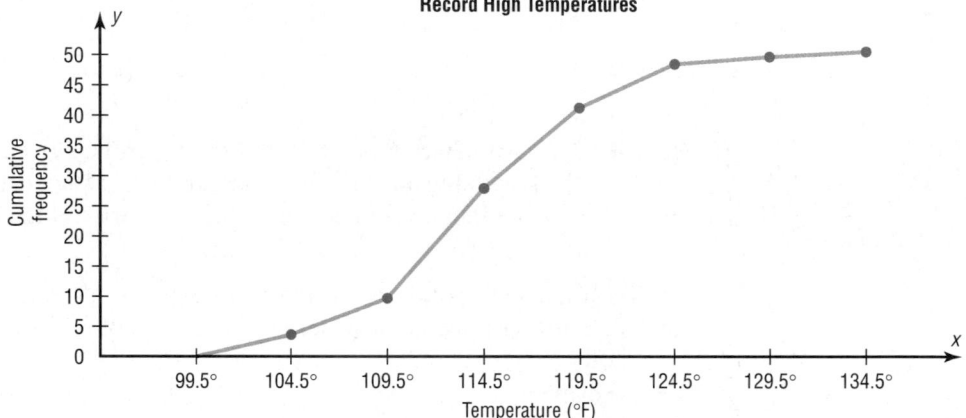

Step 2   Draw the *x* and *y* axes. Label the *x* axis with the class boundaries. Use an
appropriate scale for the *y* axis to represent the cumulative frequencies.
(Depending on the numbers in the cumulative frequency columns, scales such
as 0, 1, 2, 3, . . . , or 5, 10, 15, 20, . . . , or 1000, 2000, 3000, . . . can be used.
Do *not* label the *y* axis with the numbers in the cumulative frequency
column.) In this example, a scale of 0, 5, 10, 15, . . . will be used.

Step 3   Plot the cumulative frequency at each upper class boundary, as shown in
Figure 2–4. Upper boundaries are used since the cumulative frequencies
represent the number of data values accumulated up to the upper boundary
of each class.

Step 4   Starting with the first upper class boundary, 104.5, connect adjacent points
with line segments, as shown in Figure 2–5. Then extend the graph to the first
lower class boundary, 99.5, on the *x* axis.

Cumulative frequency graphs are used to visually represent how many values are
below a certain upper class boundary. For example, to find out how many record high
temperatures are less than 114.5°F, locate 114.5°F on the *x* axis, draw a vertical line up
until it intersects the graph, and then draw a horizontal line at that point to the *y* axis. The
y axis value is 28, as shown in Figure 2–6.

Figure 2–6

Finding a Specific Cumulative Frequency

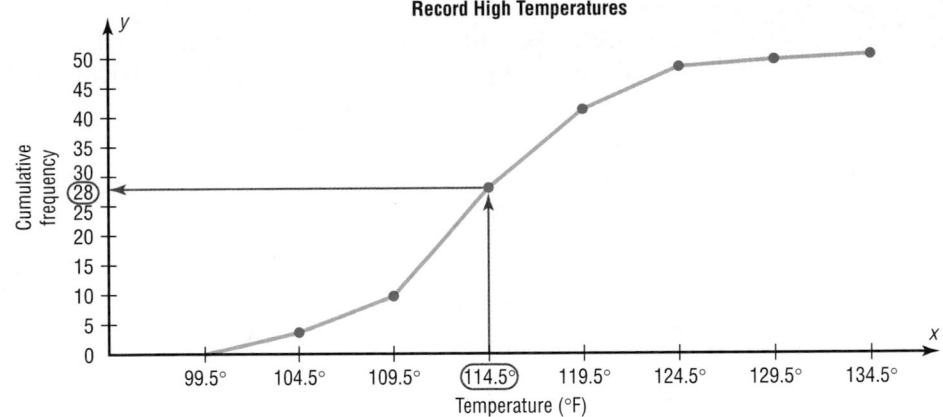

The steps for drawing these three types of graphs are shown in the following Procedure Table.

Procedure Table

Constructing Statistical Graphs

Step 1   Draw and label the x and y axes.

Step 2   Choose a suitable scale for the frequencies or cumulative frequencies, and label it on the y axis.

Step 3   Represent the class boundaries for the histogram or ogive, or the midpoint for the frequency polygon, on the x axis.

Step 4   Plot the points and then draw the bars or lines.

Relative Frequency Graphs

The histogram, the frequency polygon, and the ogive shown previously were constructed by using frequencies in terms of the raw data. These distributions can be converted to distributions using *proportions* instead of raw data as frequencies. These types of graphs are called **relative frequency graphs.**

Graphs of relative frequencies instead of frequencies are used when the proportion of data values that fall into a given class is more important than the actual number of data values that fall into that class. For example, if one wanted to compare the age distribution of adults in Philadelphia, Pennsylvania, with the age distribution of adults of Erie, Pennsylvania, one would use relative frequency distributions. The reason is that since the population of Philadelphia is 1,478,002 and the population of Erie is 105,270, the bars using the actual data values for Philadelphia would be much taller than those for the same classes for Erie.

To convert a frequency into a proportion or relative frequency, divide the frequency for each class by the total of the frequencies. The sum of the relative frequencies will always be 1. These graphs are similar to the ones that use raw data as frequencies, but the values on the y axis are in terms of proportions. Example 2–7 shows the three types of relative frequency graphs.

| Example 2–7 | Construct a histogram, frequency polygon, and ogive using relative frequencies for the distribution (shown here) of the miles that 20 randomly selected runners ran during a given week. |

Class boundaries	Frequency	Cumulative frequency
5.5–10.5	1	1
10.5–15.5	2	3
15.5–20.5	3	6
20.5–25.5	5	11
25.5–30.5	4	15
30.5–35.5	3	18
35.5–40.5	2	20
	20	

Solution

Step 1 Convert each frequency to a proportion or relative frequency by dividing the frequency for each class by the total number of observations.

For class 5.5–10.5, the relative frequency is $\frac{1}{20} = 0.05$; for class 10.5–15.5, the relative frequency is $\frac{2}{20} = 0.10$; for class 15.5–20.5, the relative frequency is $\frac{3}{20} = 0.15$; and so on.

Place these values in the column labeled Relative frequency.

Step 2 Find the cumulative relative frequencies. To do this, add the frequency in each class to the total frequency of the preceding class. In this case, $0 + 0.05 = 0.05, 0.05 + 0.10 = 0.15, 0.15 + 0.15 = 0.30, 0.30 + 0.25 = 0.55$, etc. Place these values in the column labeled Cumulative relative frequency.

Using the same procedure, find the relative frequencies for the Cumulative frequency column. The relative frequencies are shown here.

Class boundaries	Midpoints	Relative frequency	Cumulative relative frequency
5.5–10.5	8	0.05	0.05
10.5–15.5	13	0.10	0.15
15.5–20.5	18	0.15	0.30
20.5–25.5	23	0.25	0.55
25.5–30.5	28	0.20	0.75
30.5–35.5	33	0.15	0.90
35.5–40.5	38	0.10	1.00
		1.00	

Step 3 Draw each graph as shown in Figure 2–7. For the histogram and ogive, use the class boundaries along the x axis. For the frequency polygon, use the midpoints on the x axis. The scale on the y axis uses proportions.

Figure 2–7

Graphs for
Example 2–7

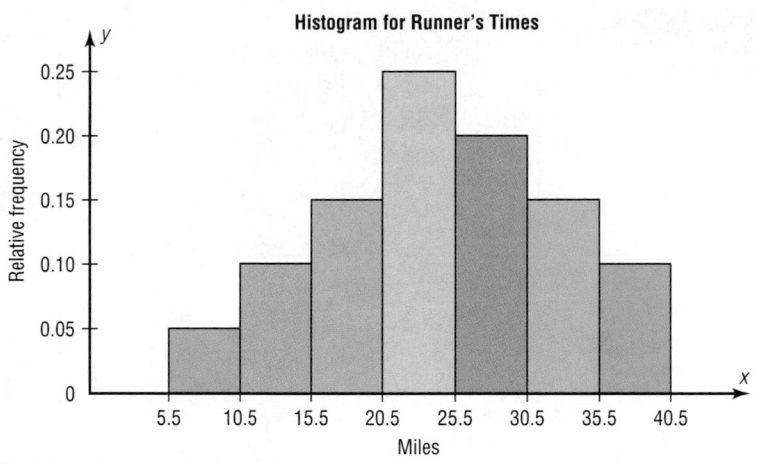

(a) Histogram

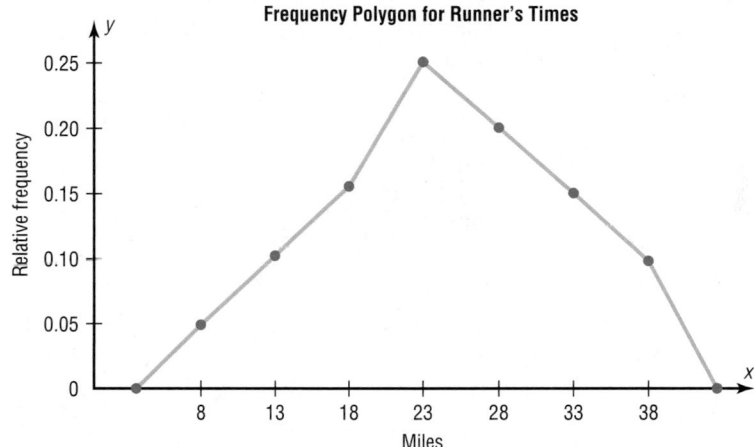

(b) Frequency polygon

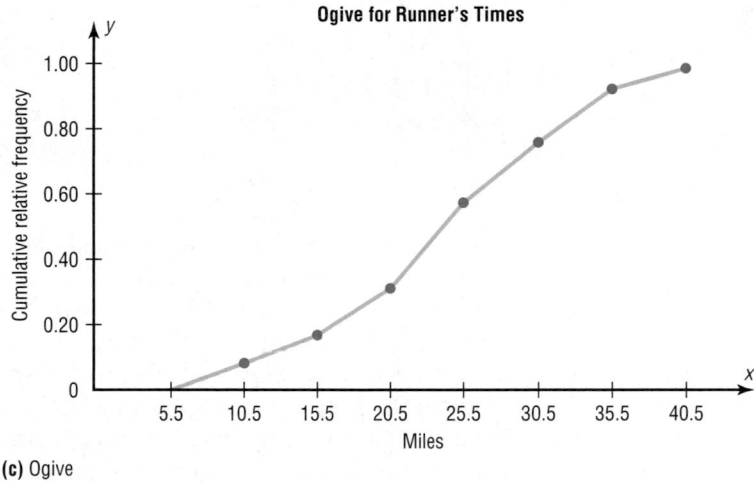

(c) Ogive

Distribution Shapes

When one is describing data, it is important to be able to recognize the shapes of the distribution values. In later chapters you will see that the shape of a distribution also determines the appropriate statistical methods used to analyze the data.

Figure 2–8

Distribution Shapes

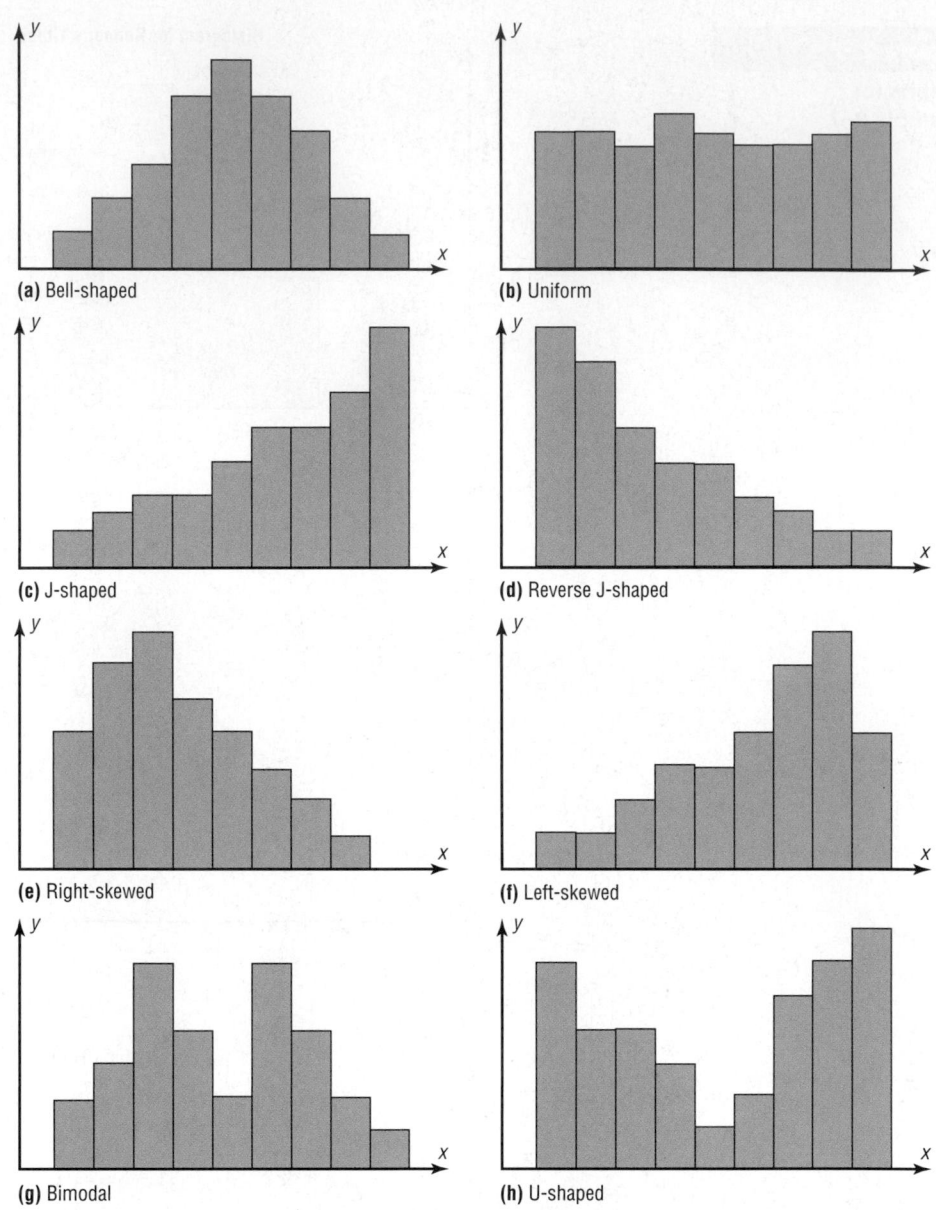

A distribution can have many shapes, and one method of analyzing a distribution is to draw a histogram or frequency polygon for the distribution. Several of the most common shapes are shown in Figure 2–8: *the bell-shaped or mound-shaped, the uniform-shaped, the J-shaped, the reverse J-shaped, the positively or right-skewed shaped, the negatively or left-skewed shaped, the bimodal-shaped, and the U-shaped.*

Distributions are most often not perfectly shaped, so it is not necessary to have an exact shape but rather to identify an overall pattern.

A *bell-shaped distribution* shown in Figure 2–8(a) has a single peak and tapers off at either end. It is approximately symmetric; i.e., it is roughly the same on both sides of a line running through the center.

A *uniform distribution* is basically flat or rectangular. See Figure 2–8(b).

A *J-shaped distribution* is shown in Figure 2–8(c), and it has a few data values on the left side and increases as one moves to the right. A *reverse J-shaped distribution* is the opposite of the J-shaped distribution. See Figure 2–8(d).

When the peak of a distribution is to the left and the data values taper off to the right, a distribution is said to be *positively or right-skewed*. See Figure 2–8(e). When the data values are clustered to the right and taper off to the left, a distribution is said to be *negatively or left-skewed*. See Figure 2–8(f). Skewness will be explained in detail in Chapter 3, pages 108–109. Distributions with one peak, such as those shown in Figure 2–8(a), (e), and (f), are said to be *unimodal*. (The highest peak of a distribution indicates where the mode of the data values is. The mode is the data value that occurs more often than any other data value. Modes are explained in Chapter 3.) When a distribution has two peaks of the same height, it is said to be *bimodal*. See Figure 2–8(g). Finally, the graph shown in Figure 2–8(h) is a *U-shaped* distribution.

Distributions can have other shapes in addition to the ones shown here; however, these are some of the more common ones that you will encounter in analyzing data.

When you are analyzing histograms and frequency polygons, look at the shape of the curve. For example, does it have one peak or two peaks? Is it relatively flat, or is it U-shaped? Are the data values spread out on the graph, or are they clustered around the center? Are there data values in the extreme ends? These may be *outliers*. (See Section 3–4 for an explanation of outliers.) Are there any gaps in the histogram, or does the frequency polygon touch the x axis somewhere other than the ends? Finally, are the data clustered at one end or the other, indicating a *skewed distribution*?

For example, the histogram for the record high temperatures shown in Figure 2–2 shows a single peaked distribution, with the class 109.5–114.5 containing the largest number of temperatures. The distribution has no gaps, and there are fewer temperatures in the highest class than in the lowest class.

Applying the Concepts **2–3**

Selling Real Estate

Assume you are a realtor in Bradenton, Florida. You have recently obtained a listing of the selling prices of the homes that have sold in that area in the last 6 months. You wish to organize that data so you will be able to provide potential buyers with useful information. Use the following data to create a histogram, frequency polygon, and cumulative frequency polygon.

142,000	127,000	99,600	162,000	89,000	93,000	99,500
73,800	135,000	119,500	67,900	156,300	104,500	108,650
123,000	91,000	205,000	110,000	156,300	104,000	133,900
179,000	112,000	147,000	321,550	87,900	88,400	180,000
159,400	205,300	144,400	163,000	96,000	81,000	131,000
114,000	119,600	93,000	123,000	187,000	96,000	80,000
231,000	189,500	177,600	83,400	77,000	132,300	166,000

1. What questions could be answered more easily by looking at the histogram rather than the listing of home prices?

2. What different questions could be answered more easily by looking at the frequency polygon rather than the listing of home prices?

3. What different questions could be answered more easily by looking at the cumulative frequency polygon rather than the listing of home prices?

4. Are there any extremely large or extremely small data values compared to the other data values?

5. Which graph displays these extremes the best?

6. Is the distribution skewed?

See page 93 for the answers.

Exercises 2–3

1. For 108 randomly selected college applicants, the following frequency distribution for entrance exam scores was obtained. Construct a histogram, frequency polygon, and ogive for the data. (The data for this exercise will be used for Exercise 13 in this section.)

Class limits	Frequency
90–98	6
99–107	22
108–116	43
117–125	28
126–134	9

Applicants who score above 107 need not enroll in a summer developmental program. In this group, how many students do not have to enroll in the developmental program?

2. For 75 employees of a large department store, the following distribution for years of service was obtained. Construct a histogram, frequency polygon, and ogive for the data. (The data for this exercise will be used for Exercise 14 in this section.)

Class limits	Frequency
1–5	21
6–10	25
11–15	15
16–20	0
21–25	8
26–30	6

A majority of the employees have worked for how many years or less?

3. The scores for the 2002 LPGA—Giant Eagle are shown.

Score	Frequency
202–204	2
205–207	7
208–210	16
211–213	26
214–216	18
217–219	4

Source: LPGA.com.

Construct a histogram, frequency polygon, and ogive for the distribution. Comment on the skewness of the distribution.

4. The salaries (in millions of dollars) for 31 NFL teams for a specific season are given in this frequency distribution.

Class limits	Frequency
39.9–42.8	2
42.9–45.8	2
45.9–48.8	5
48.9–51.8	5
51.9–54.8	12
54.9–57.8	5

Source: NFL.com.

Construct a histogram, frequency polygon, and ogive for the data; and comment on the shape of the distribution.

5. Thirty automobiles were tested for fuel efficiency, in miles per gallon (mpg). The following frequency distribution was obtained. Construct a histogram, frequency polygon, and ogive for the data.

Class boundaries	Frequency
7.5–12.5	3
12.5–17.5	5
17.5–22.5	15
22.5–27.5	5
27.5–32.5	2

6. Construct a histogram, frequency polygon, and ogive for the data in Exercise 14 in Section 2–2, and analyze the results.

7. The air quality measured for selected cities in the United States for 1993 and 2002 is shown. The data are the number of days per year that the cities failed to meet acceptable standards. Construct a histogram for both years and see if there are any notable changes. If so, explain. (The data in this exercise will be used for Exercise 17 in this section.)

1993		2002	
Class	**Frequency**	**Class**	**Frequency**
0–27	20	0–27	19
28–55	4	28–55	6
56–83	3	56–83	2
84–111	1	84–111	0
112–139	1	112–139	0
140–167	0	140–167	3
168–195	1	168–195	0

Source: *World Almanac and Book of Facts.*

8. In a study of reaction times of dogs to a specific stimulus, an animal trainer obtained the following data, given in seconds. Construct a histogram, frequency polygon, and ogive for the data, and analyze the results.

(The histogram in this exercise will be used for Exercise 18 in this section, Exercise 16 in Section 3–2, and Exercise 26 in Section 3–3.)

Class limits	Frequency
2.3–2.9	10
3.0–3.6	12
3.7–4.3	6
4.4–5.0	8
5.1–5.7	4
5.8–6.4	2

9. Construct a histogram, frequency polygon, and ogive for the data in Exercise 15 of Section 2–2, and analyze the results.

10. The frequency distributions shown indicate the percentages of public school students in fourth-grade reading and mathematics who performed at or above the required proficiency levels for the 50 states in the United States. Draw histograms for each and decide if there is any difference in the performance of the students in the subjects.

Class	Reading Frequency	Math Frequency
17.5–22.5	7	5
22.5–27.5	6	9
27.5–32.5	14	11
32.5–37.5	19	16
37.5–42.5	3	8
42.5–47.5	1	1

Source: *National Center for Educational Statistics.*

11. Construct a histogram, frequency polygon, and ogive for the data in Exercise 16 in Section 2–2, and analyze the results.

12. For the data in Exercise 18 in Section 2–2, construct a histogram for the home run distances for each player and compare them. Are they basically the same, or are there any noticeable differences? Explain your answer.

13. For the data in Exercise 1 in this section, construct a histogram, frequency polygon, and ogive, using relative frequencies. What proportion of the applicants need to enroll in the summer developmental program?

14. For the data in Exercise 2 in this section, construct a histogram, frequency polygon, and ogive, using relative frequencies. What proportion of the employees have been with the store for more than 20 years?

 15. The number of calories per serving for selected ready-to-eat cereals is listed here. Construct a frequency distribution using 7 classes. Draw a histogram, frequency polygon, and ogive for the data, using relative frequencies. Describe the shape of the histogram.

130	190	140	80	100	120	220	220	110	100
210	130	100	90	210	120	200	120	180	120
190	210	120	200	130	180	260	270	100	160
190	240	80	120	90	190	200	210	190	180
115	210	110	225	190	130				

Source: *The Doctor's Pocket Calorie, Fat, and Carbohydrate Counter.*

 16. The amount of protein (in grams) for a variety of fast-food sandwiches is reported here. Construct a frequency distribution using 6 classes. Draw a histogram, frequency polygon, and ogive for the data, using relative frequencies. Describe the shape of the histogram.

23	30	20	27	44	26	35	20	29	29
25	15	18	27	19	22	12	26	34	15
27	35	26	43	35	14	24	12	23	31
40	35	38	57	22	42	24	21	27	33

Source: *The Doctor's Pocket Calorie, Fat, and Carbohydrate Counter.*

17. For the data for year 2002 in Exercise 7 in this section, construct a histogram, frequency polygon, and ogive, using relative frequencies.

18. The animal trainer in Exercise 8 in this section selected another group of dogs who were much older than the first group and measured their reaction times to the same stimulus. Construct a histogram, frequency polygon, and ogive for the data.

Class limits	Frequency
2.3–2.9	1
3.0–3.6	3
3.7–4.3	4
4.4–5.0	16
5.1–5.7	14
5.8–6.4	4

Analyze the results and compare the histogram for this group with the one obtained in Exercise 8 in this section. Are there any differences in the histograms? (The data in this exercise will be used for Exercise 16 in Section 3–2 and Exercise 26 in Section 3–3.)

Extending the Concepts

19. Using the histogram shown here, do the following.

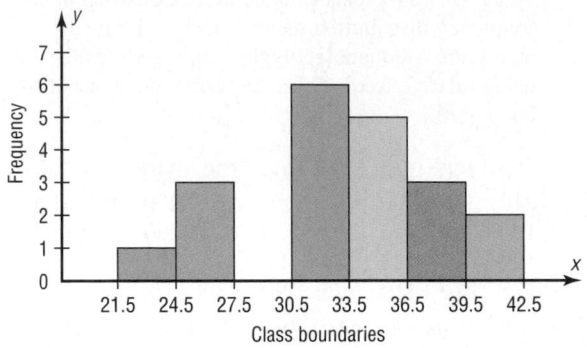

    a. Construct a frequency distribution; include class limits, class frequencies, midpoints, and cumulative frequencies.

    b. Construct a frequency polygon.

    c. Construct an ogive.

20. Using the results from Exercise 19, answer these questions.

    a. How many values are in the class 27.5–30.5?

    b. How many values fall between 24.5 and 36.5?

    c. How many values are below 33.5?

    d. How many values are above 30.5?

Technology *Step by Step*

MINITAB

Step by Step

Construct a Histogram

1. Enter the data from Example 2–2, the high temperatures for the 50 states.

2. Select **Graph>Histogram.**

3. Select [Simple], then click [OK].

4. Click C1 TEMPERATURES in the Graph variables dialog box.

5. Click [Labels]. There are two tabs, Title/Footnote and Data Labels.

    a) Click in the box for Title, and type in Your Name and Course Section.

    b) Click [OK]. The Histogram dialog box is still open.

6. Click [OK]. A new graph window containing the histogram will open.

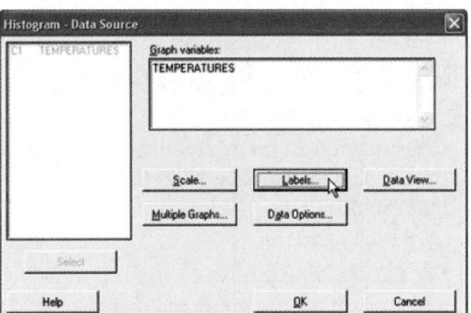

7. Click the **File** menu to print or save the graph.

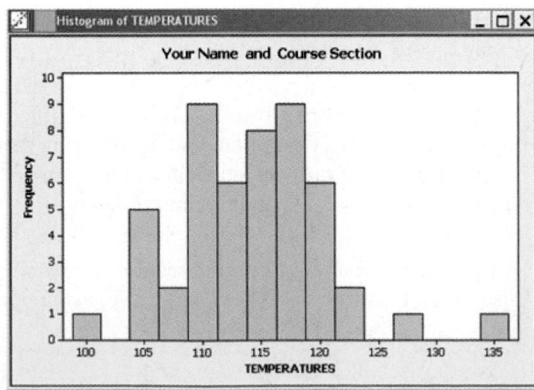

8. Click **File>Exit.**

9. Save the project as **Ch2-3.mpj.**

TI-83 Plus or TI-84 Plus
Step by Step

Constructing a Histogram

To display the graphs on the screen, enter the appropriate values in the calculator, using the WINDOW menu. The default values are $X_{min} = -10$, $X_{max} = +10$, $Y_{min} = -10$, and $Y_{max} = +10$. The X_{scl} changes the distance between the tick marks on the x axis and can be used to change the class width for the histogram.

To change the values in the WINDOW:

1. Press **WINDOW.**

2. Move the cursor to the value that needs to be changed. Then type in the desired value and press **ENTER.**

3. Continue until all values are appropriate.

4. Press **[2nd] [QUIT]** to leave the WINDOW menu.

To plot the histogram from raw data:

1. Enter the data in L_1.

2. Make sure WINDOW values are appropriate for the histogram.

3. Press **[2nd] [STAT PLOT] ENTER.**

4. Press **ENTER** to turn the plot on, if necessary.

5. Move cursor to the Histogram symbol and press **ENTER,** if necessary.

6. Make sure Xlist is L_1.

7. Make sure Freq is 1.

8. Press **GRAPH** to display the histogram.

9. To obtain the number of data values in each class, press the **TRACE** key, followed by ◀ or ▶ keys.

Input

```
WINDOW
 Xmin=100
 Xmax=135
 Xscl=5
 Ymin=-5
 Ymax=20
 Yscl=5
 Xres=1
```

Input

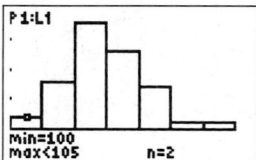

```
Plot1 Plot2 Plot3
On Off
Type: ⊾ ⊿ ⊿
      ⊿ ⊿ ⊿
Xlist:L₁
Freq:1
```

Output

```
P1:L1
Min=100
max<105   n=2
```

Example TI2–1

Plot a histogram for the following data from Examples 2–2 and 2–4.

112	100	127	120	134	118	105	110	109	112
110	118	117	116	118	122	114	114	105	109
107	112	114	115	118	117	118	122	106	110
116	108	110	121	113	120	119	111	104	111
120	113	120	117	105	110	118	112	114	114

Press **TRACE** and use the arrow keys to determine the number of values in each group.

To graph a histogram from grouped data:

1. Enter the midpoints into L_1.

2. Enter the frequencies into L_2.

3. Make sure WINDOW values are appropriate for the histogram.

4. Press **[2nd] [STAT PLOT] ENTER.**

5. Press **ENTER** to turn the plot on, if necessary.

6. Move cursor to the histogram symbol, and press **ENTER,** if necessary.

7. Make sure Xlist is L_1.

8. Make sure Freq is L_2.

9. Press **GRAPH** to display the histogram.

Example TI2–2

Plot a histogram for the data from Examples 2–4 and 2–5.

Class boundaries	Midpoints	Frequency
99.5–104.5	102	2
104.5–109.5	107	8
109.5–114.5	112	18
114.5–119.5	117	13
119.5–124.5	122	7
124.5–129.5	127	1
129.5–134.5	132	1

Input

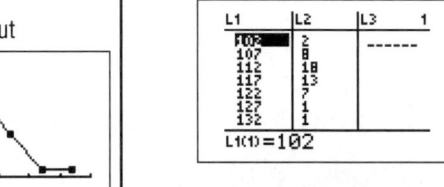

Input

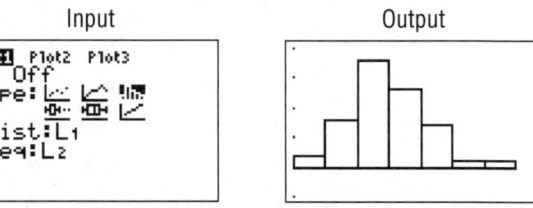

Output

Output

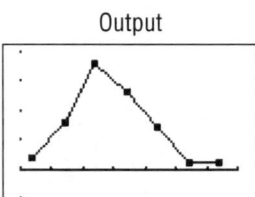

To graph a frequency polygon from grouped data, follow the same steps as for the histogram except change the graph type from histogram (third graph) to a line graph (second graph).

To graph an ogive from grouped data, modify the procedure for the histogram as follows:

1. Enter the upper class boundaries into L_1.

2. Enter the cumulative frequencies into L_2.

3. Change the graph type from histogram (third graph) to line (second graph).

4. Change the Y_{max} from the WINDOW menu to the sample size.

Excel
Step by Step

Constructing a Histogram

1. Press **[Ctrl]-N** for a new worksheet.

2. Enter the data from Examples 2–2 and 2–4 in column A, one number per cell.

3. Select **Tools>Data Analysis.**

4. In Data Analysis, select Histogram and click the [OK] button.

5. In the Histogram dialog box, type **A1:A50** as the Input Range.

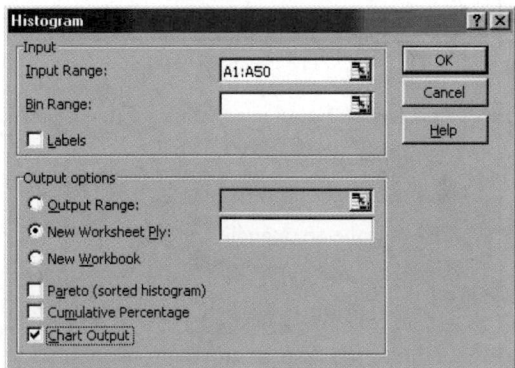

6. Select New Worksheet Ply and Chart Output. Click [OK].

Excel presents both a table and a chart on the new worksheet ply. It decides "bins" for the histogram itself (here it picked a bin size of 7 units), but you can also define your own bin range on the data worksheet.

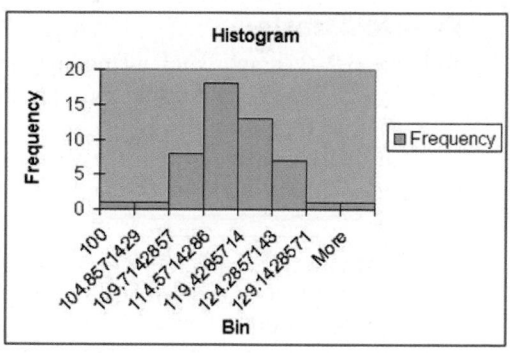

The vertical bars on the histogram can be made contiguous by right-clicking on one of the bars and selecting **Format Data Series.** Select the Options tab, then enter **0** in the Gap Width box.

2–4 Other Types of Graphs

In addition to the histogram, the frequency polygon, and the ogive, several other types of graphs are often used in statistics. They are the Pareto chart, the time series graph, and the pie graph. Figure 2–9 shows an example of each type of graph.

Figure 2–9

Other Types of Graphs Used in Statistics

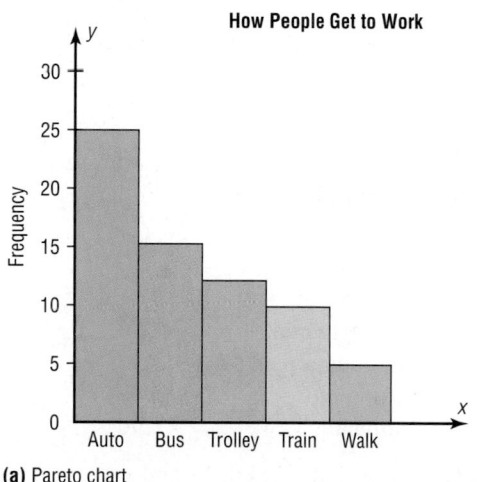

(a) Pareto chart

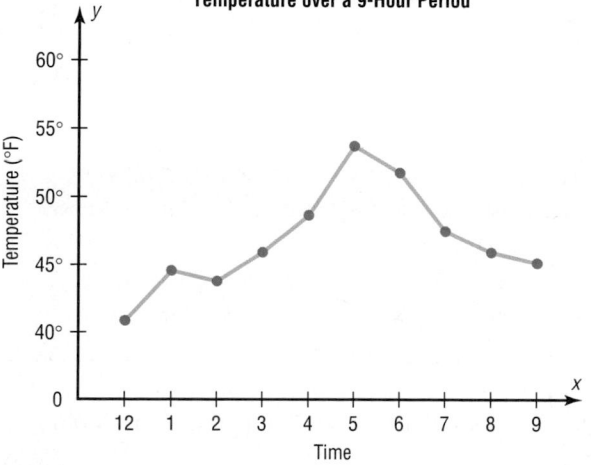

(b) Time series graph

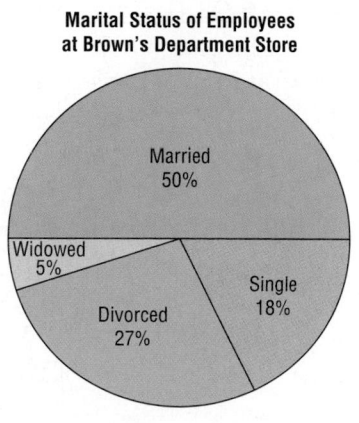

(c) Pie graph

2–31

Objective 3

Represent data using Pareto charts, time series graphs, and pie graphs.

Pareto Charts

In Section 2–3, graphs such as the histogram, frequency polygon, and ogive showed how data can be represented when the variable displayed on the horizontal axis is quantitative, such as heights and weights.

On the other hand, when the variable displayed on the horizontal axis is qualitative or categorical, a *Pareto chart* can be used.

> A **Pareto chart** is used to represent a frequency distribution for a categorical variable, and the frequencies are displayed by the heights of vertical bars, which are arranged in order from highest to lowest.

Example 2–8

The table shown here is the average cost per mile for passenger vehicles on state turnpikes. Construct and analyze a Pareto chart for the data.

State	Number
Indiana	2.9¢
Oklahoma	4.3¢
Florida	6.0¢
Maine	3.8¢
Pennsylvania	5.8¢

Source: *Pittsburgh Tribune Review.*

Historical Note

Vilfredo Pareto (1848–1923) was an Italian scholar who developed theories in economics, statistics, and the social sciences. His contributions to statistics include the development of a mathematical function used in economics. This function has many statistical applications and is called the Pareto distribution. In addition, he researched income distribution, and his findings became known as Pareto's law.

Solution

Step 1   Arrange the data from the largest to smallest according to frequency.

State	Number
Florida	6.0¢
Pennsylvania	5.8¢
Oklahoma	4.3¢
Maine	3.8¢
Indiana	2.9¢

Step 2   Draw and label the x and y axes.

Step 3   Draw the bars corresponding to the frequencies. See Figure 2–10. The Pareto chart shows that Florida has the highest cost per mile. The cost is more than twice as high as the cost for Indiana.

> **Suggestions for Drawing Pareto Charts**
>
> 1. Make the bars the same width.
> 2. Arrange the data from largest to smallest according to frequency.
> 3. Make the units that are used for the frequency equal in size.

When you analyze a Pareto chart, make comparisons by looking at the heights of the bars.

Figure 2–10

Pareto Chart for Example 2–8

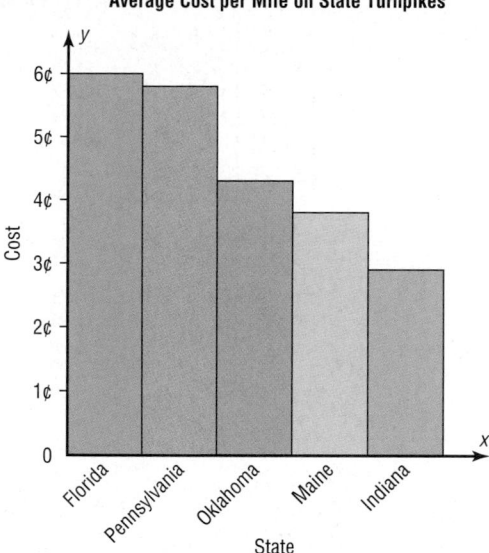

Average Cost per Mile on State Turnpikes

The Time Series Graph

When data are collected over a period of time, they can be represented by a time series graph.

A **time series graph** represents data that occur over a specific period of time.

Example 2–9 shows the procedure for constructing a time series graph.

Example 2–9

The number (in millions) of vehicles, both passenger and commercial, that used the Pennsylvania Turnpike for the years 1999 through 2003 is shown. Construct and analyze a time series graph for the data.

Year	Number
1999	156.2
2000	160.1
2001	162.3
2002	172.8
2003	179.4

Source: *Tribune Review.*

Historical Note

Time series graphs are over 1000 years old. The first ones were used to chart the movements of the planets and the sun.

Solution

Step 1 Draw and label the *x* and *y* axes.

Step 2 Label the *x* axis for years and the *y* axis for the number of vehicles.

Step 3 Plot each point according to the table.

Step 4 Draw line segments connecting adjacent points. Do not try to fit a smooth curve through the data points. See Figure 2–11. The graph shows a steady increase over the 5-year period.

Figure 2–11

Time Series Graph for Example 2–9

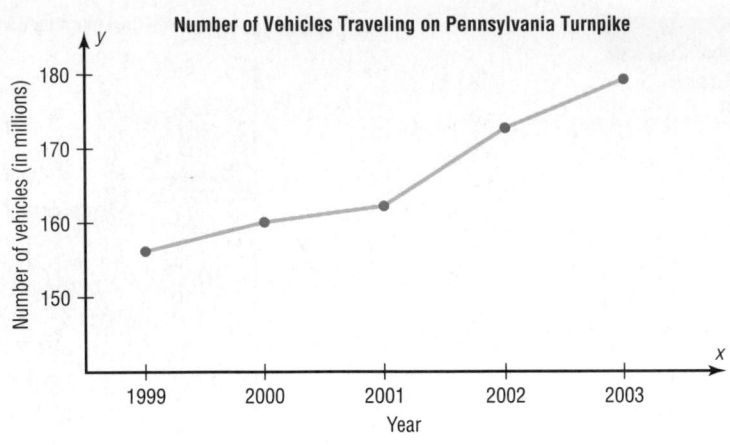

Figure 2–12

Two Time Series Graphs for Comparison

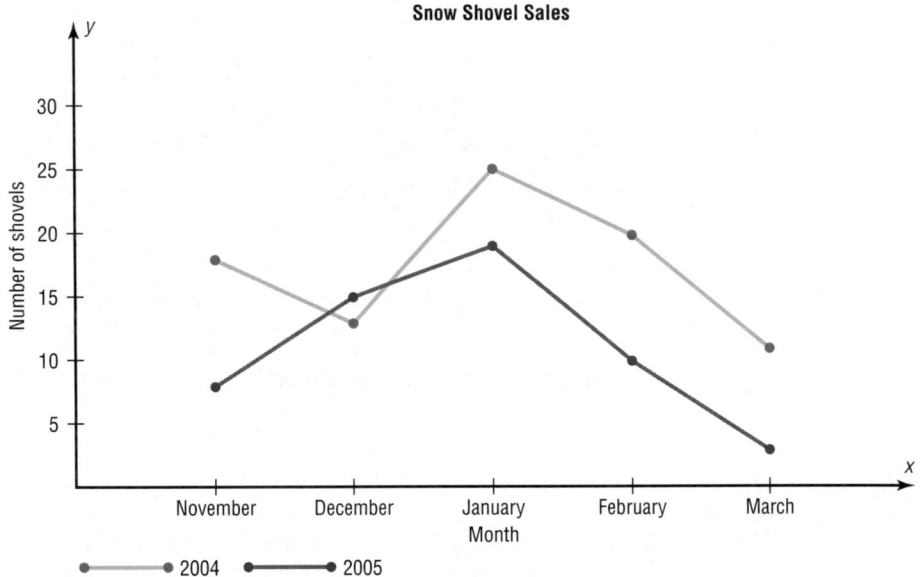

When you analyze a time series graph, look for a trend or pattern that occurs over the time period. For example, is the line ascending (indicating an increase over time) or descending (indicating a decrease over time)? Another thing to look for is the slope, or steepness, of the line. A line that is steep over a specific time period indicates a rapid increase or decrease over that period.

Two data sets can be compared on the same graph (called a *compound time series graph*) if two lines are used, as shown in Figure 2–12. This graph shows the number of snow shovels sold at a store for two seasons.

The Pie Graph

Pie graphs are used extensively in statistics. The purpose of the pie graph is to show the relationship of the parts to the whole by visually comparing the sizes of the sections. Percentages or proportions can be used. The variable is nominal or categorical.

A **pie graph** is a circle that is divided into sections or wedges according to the percentage of frequencies in each category of the distribution.

Example 2–10 shows the procedure for constructing a pie graph.

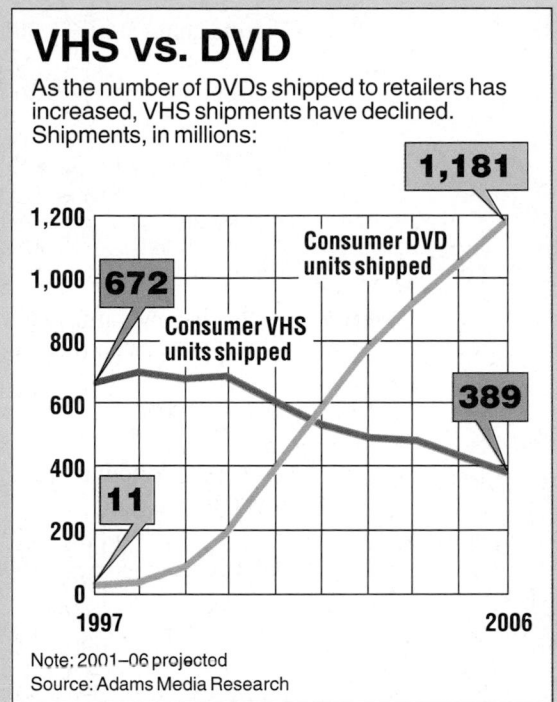

Speaking of
Statistics

This time series graph compares the number of DVD units and the number of VHS units shipped to retailers, using a compound time series graph. Explain in your own words the information that is presented in the graph.

Source: Copyright 2001, *USA TODAY.* Reprinted with permission.

Example 2–10

This frequency distribution shows the number of pounds of each snack food eaten during the Super Bowl. Construct a pie graph for the data.

Snack	Pounds (frequency)
Potato chips	11.2 million
Tortilla chips	8.2 million
Pretzels	4.3 million
Popcorn	3.8 million
Snack nuts	2.5 million
Total $n =$	30.0 million

Source: *USA TODAY Weekend.*

Solution

Step 1 Since there are 360° in a circle, the frequency for each class must be converted into a proportional part of the circle. This conversion is done by using the formula

$$\text{Degrees} = \frac{f}{n} \cdot 360°$$

where f = frequency for each class and n = sum of the frequencies. Hence, the following conversions are obtained. The degrees should sum to 360°.*

*Note: The degrees column does not always sum to 360° due to rounding.

Potato chips $\dfrac{11.2}{30} \cdot 360° = 134°$

Tortilla chips $\dfrac{8.2}{30} \cdot 360° = 98°$

Pretzels $\dfrac{4.3}{30} \cdot 360° = 52°$

Popcorn $\dfrac{3.8}{30} \cdot 360° = 46°$

Snack nuts $\dfrac{2.5}{30} \cdot 360° = 30°$

Total $\overline{360°}$

Step 2 Each frequency must also be converted to a percentage. Recall from Example 2–1 that this conversion is done by using the formula

$$\% = \frac{f}{n} \cdot 100\%$$

Hence, the following percentages are obtained. The percentages should sum to 100%.*

Potato chips $\dfrac{11.2}{30} \cdot 100\% = 37.3\%$

Tortilla chips $\dfrac{8.2}{30} \cdot 100\% = 27.3\%$

Pretzels $\dfrac{4.3}{30} \cdot 100\% = 14.3\%$

Popcorn $\dfrac{3.8}{30} \cdot 100\% = 12.7\%$

Snack nuts $\dfrac{2.5}{30} \cdot 100\% = \ \ 8.3\%$

Total $\overline{99.9\%}$

Step 3 Next, using a protractor and a compass, draw the graph using the appropriate degree measures found in step 1, and label each section with the name and percentages, as shown in Figure 2–13.

Figure 2–13

Pie Graph for Example 2–10

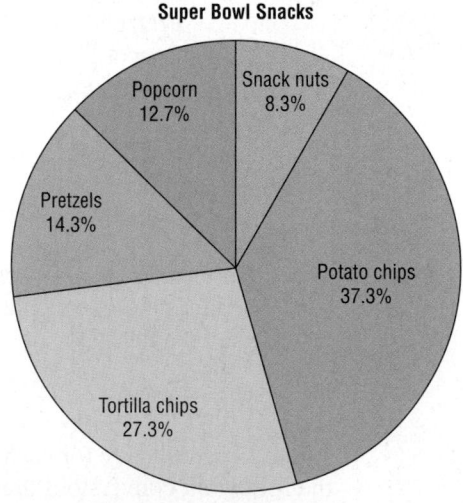

Super Bowl Snacks

Note: The percent column does not always sum to 100% due to rounding.

| Example 2–11 | Construct a pie graph showing the blood types of the army inductees described in Example 2–1. The frequency distribution is repeated here. |

Class	Frequency	Percent
A	5	20
B	7	28
O	9	36
AB	4	16
	25	100

Solution

Step 1 Find the number of degrees for each class, using the formula

$$\text{Degrees} = \frac{f}{n} \cdot 360°$$

For each class, then, the following results are obtained.

A $\frac{5}{25} \cdot 360° = 72°$

B $\frac{7}{25} \cdot 360° = 100.8°$

O $\frac{9}{25} \cdot 360° = 129.6°$

AB $\frac{4}{25} \cdot 360° = 57.6°$

Step 2 Find the percentages. (This was already done in Example 2–1.)

Step 3 Using a protractor, graph each section and write its name and corresponding percentage, as shown in Figure 2–14.

Figure 2–14	
Pie Graph for Example 2–11	

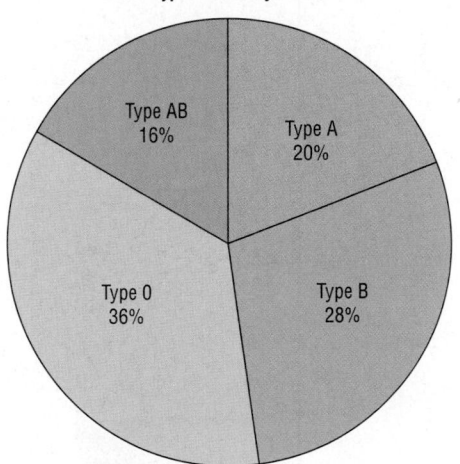

Blood Types for Army Inductees

The graph in Figure 2–14 shows that in this case the most common blood type is type O.

To analyze the nature of the data shown in the pie graph, compare the sections. For example, are any sections relatively large compared to the rest?

Figure 2–14 shows that among the inductees, type O blood is more prevalent than any other type. People who have type AB blood are in the minority. More than twice as many people have type O blood as type AB.

Misleading Graphs

Graphs give a visual representation that enables readers to analyze and interpret data more easily than they could simply by looking at numbers. However, inappropriately drawn graphs can misrepresent the data and lead the reader to false conclusions. For example, a car manufacturer's ad stated that 98% of the vehicles it had sold in the past 10 years were still on the road. The ad then showed a graph similar to the one in Figure 2–15. The graph shows the percentage of the manufacturer's automobiles still on the road and the percentage of its competitors' automobiles still on the road. Is there a large difference? Not necessarily.

Notice the scale on the vertical axis in Figure 2–15. It has been cut off (or truncated) and starts at 95%. When the graph is redrawn using a scale that goes from 0 to 100%, as in Figure 2–16, there is hardly a noticeable difference in the percentages. Thus, changing the units at the starting point on the *y* axis can convey a very different visual representation of the data.

It is not wrong to truncate an axis of the graph; many times it is necessary to do so (see Example 2–9). However, the reader should be aware of this fact and interpret the graph accordingly. Do not be misled if an inappropriate impression is given.

Let's consider another example. The percentage of the world's total motor vehicles produced by manufacturers in the United States declined from 24% in 1998 to 21.5% in 2000, as shown by the data on the next page.

Figure 2–15

Graph of Automaker's Claim Using a Scale from 95 to 100%

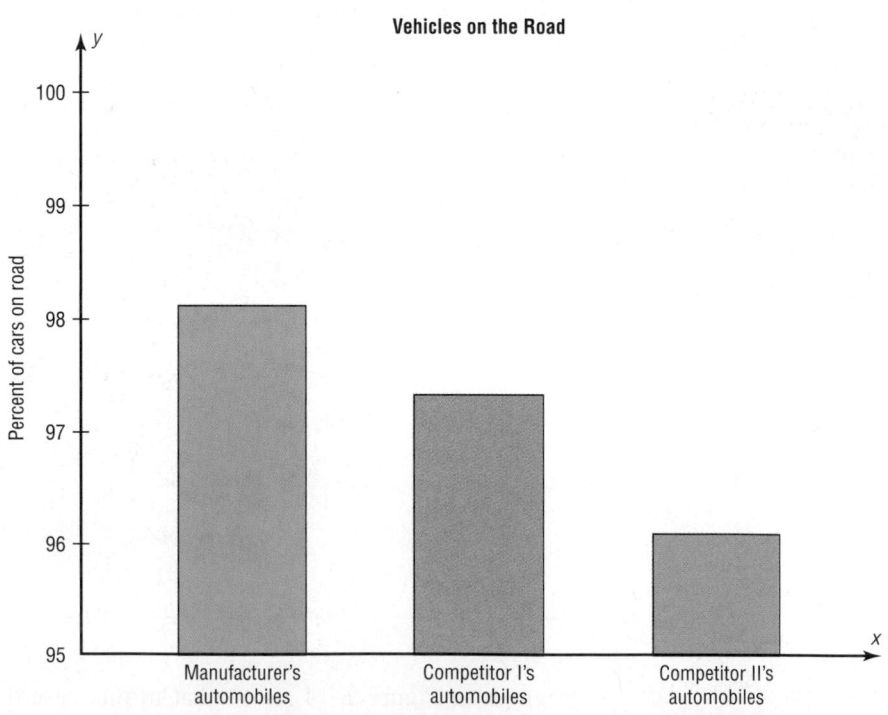

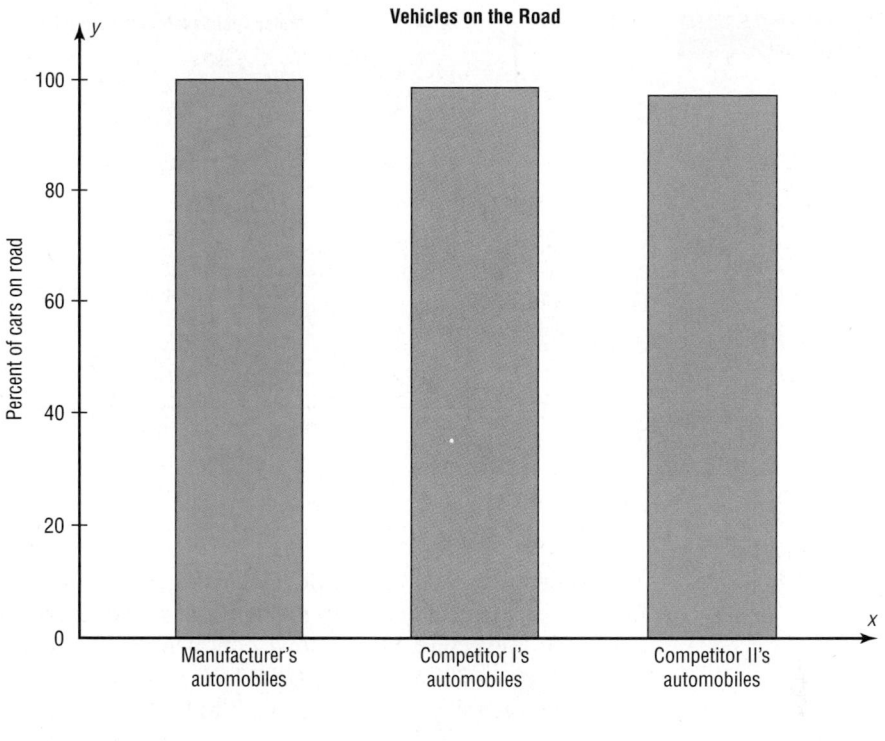

Figure 2–16

Graph in Figure 2–15 Redrawn Using a Scale from 0 to 100%

Interesting Fact

The most popular flavor of ice cream is vanilla, and about one-fourth of the ice cream sold is vanilla.

Year	1995	1996	1997	1998	1999	2000
Percent produced in United States	24.0	23.0	22.7	22.4	22.7	21.5

Source: *The World Almanac and Book of Facts.*

When one draws the graph, as shown in Figure 2–17(a), a scale ranging from 0 to 100% shows a slight decrease. However, this decrease can be emphasized by using a scale that ranges from 15 to 25%, as shown in Figure 2–17(b). Again, by changing the units or the starting point on the *y* axis, one can change the visual message.

Figure 2–17

Percent of World's Motor Vehicles Produced by Manufacturers in the United States

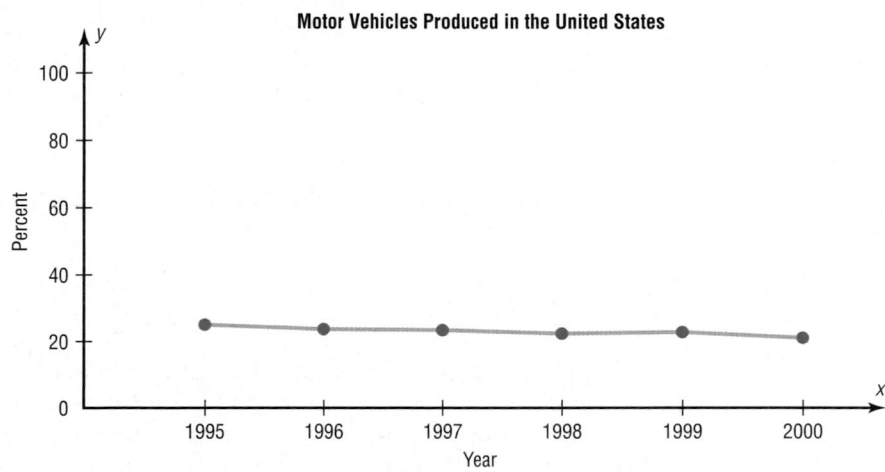

(a) Using a scale from 0% to 100%

Figure 2–17

(*continued*)

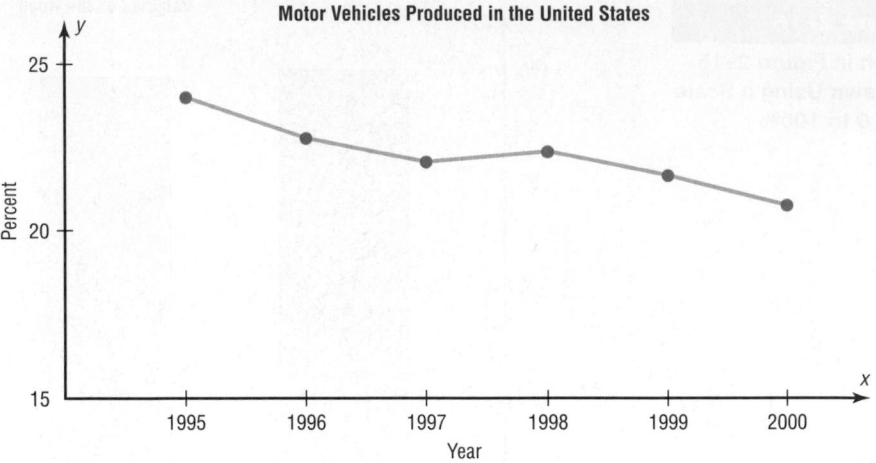

(**b**) Using a scale from 15% to 25%

Another misleading graphing technique sometimes used involves exaggerating a one-dimensional increase by showing it in two dimensions. For example, the average cost of a 30-second Super Bowl commercial has increased from $42,000 in 1967 to $1.9 million in 2002 (Source: *USA TODAY*).

The increase shown by the graph in Figure 2–18(a) represents the change by a comparison of the heights of the two bars in one dimension. The same data are shown two-dimensionally with circles in Figure 2–18(b). Notice that the difference seems much larger because the eye is comparing the areas of the circles rather than the lengths of the diameters.

Note that it is not wrong to use the graphing techniques of truncating the scales or representing data by two-dimensional pictures. But when these techniques are used, the reader should be cautious of the conclusion drawn on the basis of the graphs.

Figure 2–18

Comparison of Costs for a 30-Second Super Bowl Commercial

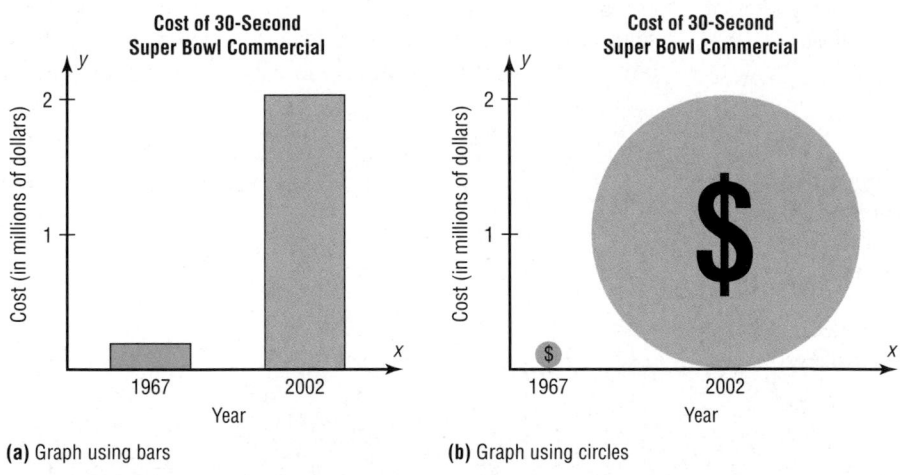

(**a**) Graph using bars   (**b**) Graph using circles

Another way to misrepresent data on a graph is by omitting labels or units on the axes of the graph. The graph shown in Figure 2–19 compares the cost of living, economic growth, population growth, etc., of four main geographic areas in the United States. However, since there are no numbers on the *y* axis, very little information can be gained from this graph, except a crude ranking of each factor. There is no way to decide the actual magnitude of the differences.

Figure 2–19

A Graph with No Units on the *y* Axis

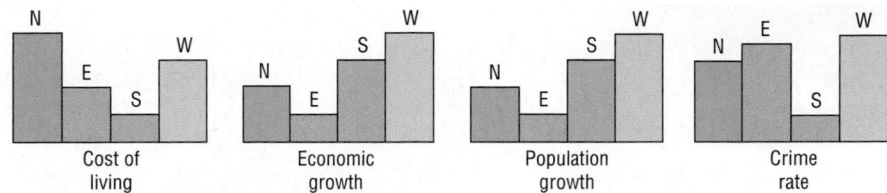

Finally, all graphs should contain a source for the information presented. The inclusion of a source for the data will enable you to check the reliability of the organization presenting the data. A summary of the types of graphs and their uses is shown in Figure 2–20.

Figure 2–20

Summary of Graphs and Uses of Each

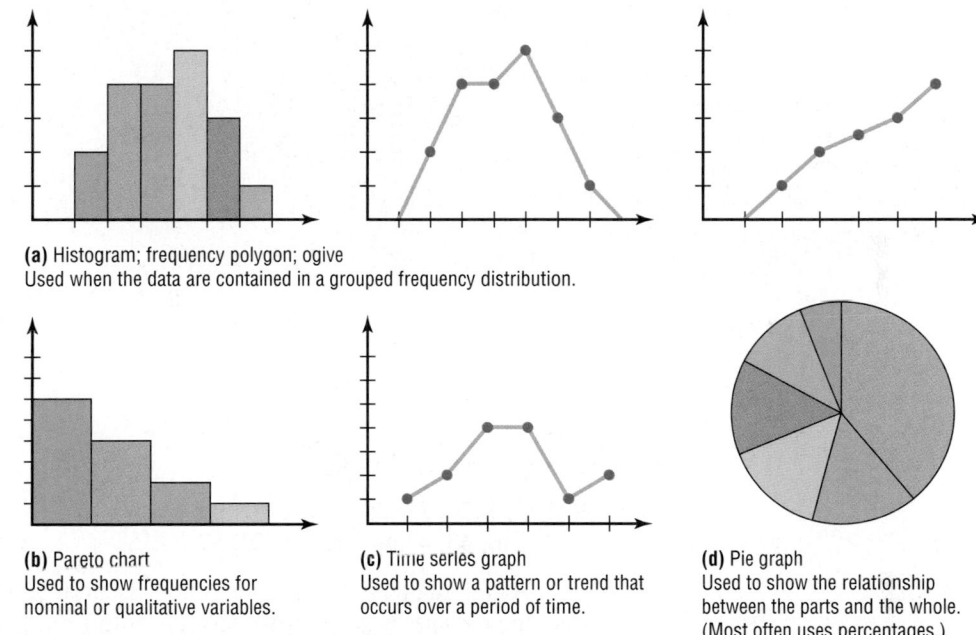

(a) Histogram; frequency polygon; ogive
Used when the data are contained in a grouped frequency distribution.

(b) Pareto chart
Used to show frequencies for nominal or qualitative variables.

(c) Time series graph
Used to show a pattern or trend that occurs over a period of time.

(d) Pie graph
Used to show the relationship between the parts and the whole. (Most often uses percentages.)

Stem and Leaf Plots

The stem and leaf plot is a method of organizing data and is a combination of sorting and graphing. It has the advantage over a grouped frequency distribution of retaining the actual data while showing them in graphical form.

Objective 4

Draw and interpret a stem and leaf plot.

A **stem and leaf plot** is a data plot that uses part of the data value as the stem and part of the data value as the leaf to form groups or classes.

Example 2–12 shows the procedure for constructing a stem and leaf plot.

Example 2–12

 At an outpatient testing center, the number of cardiograms performed each day for 20 days is shown. Construct a stem and leaf plot for the data.

25	31	20	32	13
14	43	02	57	23
36	32	33	32	44
32	52	44	51	45

How Much Paper Money Is in Circulation Today?

The Federal Reserve estimated that during a recent year, there were 22 billion bills in circulation. About 35% of them were $1 bills, 3% were $2 bills, 8% were $5 bills, 7% were $10 bills, 23% were $20 bills, 5% were $50 bills, and 19% were $100 bills. It costs about 3¢ to print each $1 bill.

  The average life of a $1 bill is 22 months, a $10 bill 3 years, a $20 bill 4 years, a $50 bill 9 years, and a $100 bill 9 years. What type of graph would you use to represent the average lifetimes of the bills?

Solution

Step 1   Arrange the data in order:

02, 13, 14, 20, 23, 25, 31, 32, 32, 32,
32, 33, 36, 43, 44, 44, 45, 51, 52, 57

Note: Arranging the data in order is not essential and can be cumbersome when the data set is large; however, it is helpful in constructing a stem and leaf plot. The leaves in the final stem and leaf plot should be arranged in order.

Step 2   Separate the data according to the first digit, as shown.

02   13, 14   20, 23, 25   31, 32, 32, 32, 32, 33, 36
43, 44, 44, 45   51, 52, 57

Step 3   A display can be made by using the leading digit as the *stem* and the trailing digit as the *leaf.* For example, for the value 32, the leading digit, 3, is the stem and the trailing digit, 2, is the leaf. For the value 14, the 1 is the stem and the 4 is the leaf. Now a plot can be constructed as shown in Figure 2–21.

Figure 2–21

Stem and Leaf Plot for Example 2–12

0	2
1	3 4
2	0 3 5
3	1 2 2 2 2 3 6
4	3 4 4 5
5	1 2 7

Leading digit (stem)	Trailing digit (leaf)
0	2
1	3 4
2	0 3 5
3	1 2 2 2 2 3 6
4	3 4 4 5
5	1 2 7

Figure 2–21 shows that the distribution peaks in the center and that there are no gaps in the data. For 7 of the 20 days, the number of patients receiving cardiograms was between 31 and 36. The plot also shows that the testing center treated from a minimum of 2 patients to a maximum of 57 patients in any one day.

If there are no data values in a class, you should write the stem number and leave the leaf row blank. Do not put a zero in the leaf row.

Example 2–13	An insurance company researcher conducted a survey on the number of car thefts in a large city for a period of 30 days last summer. The raw data are shown. Construct a stem and leaf plot by using classes 50–54, 55–59, 60–64, 65–69, 70–74, and 75–79.

52	62	51	50	69
58	77	66	53	57
75	56	55	67	73
79	59	68	65	72
57	51	63	69	75
65	53	78	66	55

Solution

Step 1 Arrange the data in order.

50, 51, 51, 52, 53, 53, 55, 55, 56, 57, 57, 58, 59, 62, 63, 65, 65, 66, 66, 67, 68, 69, 69, 72, 73, 75, 75, 77, 78, 79

Step 2 Separate the data according to the classes.

50, 51, 51, 52, 53, 53   55, 55, 56, 57, 57, 58, 59
62, 63   65, 65, 66, 66, 67, 68, 69, 69   72, 73
75, 75, 77, 78, 79

Figure 2–22

Stem and Leaf Plot for Example 2–13

Step 3 Plot the data as shown here.

5	0 1 1 2 3 3
5	5 5 6 7 7 8 9
6	2 3
6	5 5 6 6 7 8 9 9
7	2 3
7	5 5 7 8 9

Leading digit (stem)	Trailing digit (leaf)
5	0 1 1 2 3 3
5	5 5 6 7 7 8 9
6	2 3
6	5 5 6 6 7 8 9 9
7	2 3
7	5 5 7 8 9

The graph for this plot is shown in Figure 2–22.

When the data values are in the hundreds, such as 325, the stem is 32 and the leaf is 5. For example, the stem and leaf plot for the data values 325, 327, 330, 332, 335, 341, 345, and 347 looks like this.

32	5 7
33	0 2 5
34	1 5 7

When you analyze a stem and leaf plot, look for peaks and gaps in the distribution. See if the distribution is symmetric or skewed. Check the variability of the data by looking at the spread.

Related distributions can be compared by using a back-to-back stem and leaf plot. The back-to-back stem and leaf plot uses the same digits for the stems of both distributions, but the digits that are used for the leaves are arranged in order out from the stems on both sides. Example 2–14 shows a back-to-back stem and leaf plot.

Example 2–14

The number of stories in two selected samples of tall buildings in Atlanta and Philadelphia are shown. Construct a back-to-back stem and leaf plot, and compare the distributions.

Atlanta					Philadelphia				
55	70	44	36	40	61	40	38	32	30
63	40	44	34	38	58	40	40	25	30
60	47	52	32	32	54	40	36	30	30
50	53	32	28	31	53	39	36	34	33
52	32	34	32	50	50	38	36	39	32
26	29								

Source: *The World Almanac and Book of Facts.*

Solution

Step 1 Arrange the data for both data sets in order.

Step 2 Construct a stem and leaf plot using the same digits as stems. Place the digits for the leaves for Atlanta on the left side of the stem and the digits for the leaves for Philadelphia on the right side, as shown. See Figure 2–23.

Figure 2–23

Back-to-Back Stem and Leaf Plot for Example 2–14

Atlanta		Philadelphia
9 8 6	2	5
8 6 4 4 2 2 2 2 2 1	3	0 0 0 0 2 2 3 4 6 6 6 8 8 9 9
7 4 4 0 0	4	0 0 0 0
5 3 2 2 0 0	5	0 3 4 8
3 0	6	1
0	7	

Step 3 Compare the distributions. The buildings in Atlanta have a large variation in the number of stories per building. Although both distributions are peaked in the 30- to 39-story class, Philadelphia has more buildings in this class. Atlanta has more buildings that have 40 or more stories than Philadelphia does.

Stem and leaf plots are part of the techniques called *exploratory data analysis.* More information on this topic is presented in Chapter 3.

Applying the Concepts **2–4**

Leading Cause of Death

The following shows approximations of the leading causes of death among men ages 25–44 years. The rates are per 100,000 men. Answer the following questions about the graph.

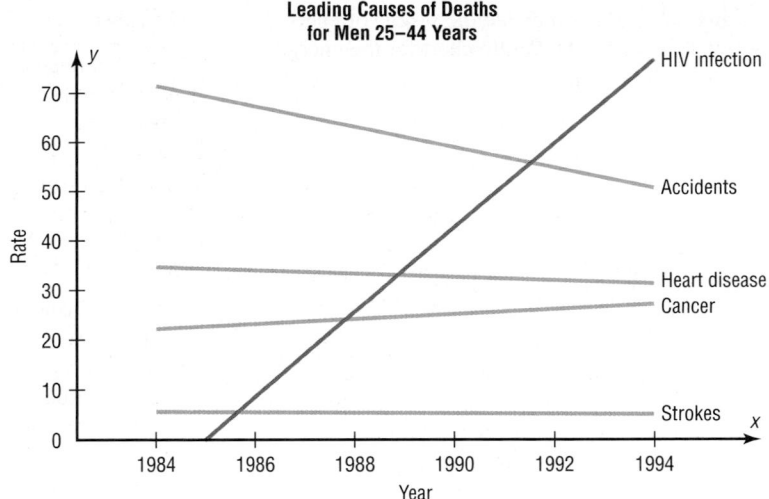

**Leading Causes of Deaths
for Men 25–44 Years**

1. What are the variables in the graph?
2. Are the variables qualitative or quantitative?
3. Are the variables discrete or continuous?
4. What type of graph was used to display the data?
5. Could a Pareto chart be used to display the data?
6. Could a pie chart be used to display the data?
7. List some typical uses for the Pareto chart.
8. List some typical uses for the time series chart.

See page 93 for the answers.

Exercises 2–4

1. The population of federal prisons, according to the most serious offenses, consists of the following. Make a Pareto chart of the population. Based on the Pareto chart, where should most of the money for rehabilitation be spent?

Violent offenses	12.6%
Property offenses	8.5
Drug offenses	60.2
Public order offenses	
Weapons	8.2
Immigration	4.9
Other	5.6

Source: *N.Y. Times Almanac.*

2. Construct a Pareto chart for the number of homicides (rate per 100,000 population) reported for the following states.

State	Number of homicides
Connecticut	4.1
Maine	2.0
New Jersey	4.0
Pennsylvania	5.3
New York	5.1

Source: *FBI Uniform Crime Report.*

3. The following data represent the estimated number (in millions) of computers connected to the Internet worldwide. Construct a Pareto chart for the data. Based on the data, suggest the best place to market appropriate Internet products.

Location	Number of computers
Homes	240
Small companies	102
Large companies	148
Government agencies	33
Schools	47

Source: IDC.

4. The World Roller Coaster Census Report lists the following number of roller coasters on each continent. Represent the data graphically, using a Pareto chart.

Africa	17
Asia	315
Australia	22
Europe	413
North America	643
South America	45

Source: www.rcdb.com.

5. The following percentages indicate the source of energy used worldwide. Construct a Pareto chart for the energy used.

Petroleum	39.8%
Coal	23.2
Dry natural gas	22.4
Hydroelectric	7.0
Nuclear	6.4
Other (wind, solar, etc.)	1.2

Source: *N.Y. Times Almanac.*

6. Draw a time series graph to represent the data for the number of airline departures (in millions) for the given years. Over the years, is the number of departures increasing, decreasing, or about the same?

Year	1996	1997	1998	1999	2000	2001	2002
Number of departures	7.9	9.9	10.5	10.9	11.0	9.8	10.1

Source: *The World Almanac and Book of Facts.*

7. The data represent the personal consumption (in billions of dollars) for tobacco in the United States. Draw a time series graph for the data and explain the trend.

Year	1995	1996	1997	1998	1999	2000	2001	2002
Amount	8.5	8.7	9.0	9.3	9.6	9.9	10.2	10.4

Source: *The World Almanac and Book of Facts.*

8. Draw a time series graph for the data shown and comment on the trend. The data represent the number of active nuclear reactors.

Year	1992	1994	1996	1998	2000	2002
Number	109	109	109	104	104	104

Source: *The World Almanac and Book of Facts.*

9. The percentages of voters voting in 10 presidential elections are shown here. Construct a time series graph and analyze the results.

1964	95.83%	1984	74.63%
1968	89.65	1988	72.48
1972	79.85	1992	78.01
1976	77.64	1996	65.97
1980	76.53	2000	67.50

Source: *N.Y. Times Almanac.*

10. The following data are based on a survey from American Travel Survey on why people travel. Construct a pie graph for the data and analyze the results.

Purpose	Number
Personal business	146
Visit friends or relatives	330
Work-related	225
Leisure	299

Source: *USA TODAY.*

11. The assets of the richest 1% of Americans are distributed as follows. Make a pie graph for the percentages.

Principal residence	7.8%
Liquid assets	5.0
Pension accounts	6.9
Stock, mutual funds, and personal trusts	31.6
Businesses and other real estate	46.9
Miscellaneous	1.8

Source: *The New York Times.*

12. The following elements comprise the earth's crust, the outermost solid layer. Illustrate the composition of the earth's crust with a pie graph.

Oxygen	45.6%
Silicon	27.3
Aluminum	8.4
Iron	6.2
Calcium	4.7
Other	7.8

Source: *N.Y. Times Almanac.*

13. In a recent survey, 3 in 10 people indicated that they are likely to leave their jobs when the economy improves. Of those surveyed, 34% indicated that they would make a career change, 29% want a new job in the same industry, 21% are going to start a business, and 16% are going to retire. Make a pie chart and a Pareto chart for the data. Which chart do you think better represents the data?

Source: *National Survey Institute.*

14. State which graph (Pareto chart, time series graph, or pie graph) would most appropriately represent the given situation.

a. The number of students enrolled at a local college for each year during the last 5 years.

b. The budget for the student activities department at a certain college for each year during the last 5 years.

c. The means of transportation the students use to get to school.

d. The percentage of votes each of the four candidates received in the last election.

e. The record temperatures of a city for the last 30 years.

f. The frequency of each type of crime committed in a city during the year.

15. The age at inauguration for each U.S. President is shown. Construct a stem and leaf plot and analyze the data.

57	54	52	55	51	56
61	68	56	55	54	61
57	51	46	54	51	52
57	49	54	42	60	69
58	64	49	51	62	64
57	48	50	56	43	46
61	65	47	55	55	54

Source: *N.Y. Times Almanac.*

16. The National Insurance Crime Bureau reported that these data represent the number of registered vehicles per car stolen for 35 selected cities in the United States. For example, in Miami, 1 automobile is stolen for every 38 registered vehicles in the city. Construct a stem and leaf plot for the data and analyze the distribution. (The data have been rounded to the nearest whole number.)

38	53	53	56	69	89	94
41	58	68	66	69	89	52
50	70	83	81	80	90	74
50	70	83	59	75	78	73
92	84	87	84	85	84	89

Source: *USA TODAY.*

17. The growth (in centimeters) of two varieties of plant after 20 days is shown in this table. Construct a back-to-back stem and leaf plot for the data, and compare the distributions.

Variety 1				Variety 2			
20	12	39	38	18	45	62	59
41	43	51	52	53	25	13	57
59	55	53	59	42	55	56	38
50	58	35	38	41	36	50	62
23	32	43	53	45	55		

18. The data shown represent the percentage of unemployed males and females in 1995 for a sample of countries of the world. Using the whole numbers as stems and the decimals as leaves, construct a back-to-back stem and leaf plot and compare the distributions of the two groups.

Females					Males				
8.0	3.7	8.6	5.0	7.0	8.8	1.9	5.6	4.6	1.5
3.3	8.6	3.2	8.8	6.8	2.2	5.6	3.1	5.9	6.6
9.2	5.9	7.2	4.6	5.6	9.8	8.7	6.0	5.2	5.6
5.3	7.7	8.0	8.7	0.5	4.4	9.6	6.6	6.0	0.3
6.5	3.4	3.0	9.4		4.6	3.1	4.1	7.7	

Source: *N.Y. Times Almanac.*

19. These data represent the numbers of cities served on nonstop flights by Southwest Airlines's largest airports. Construct a stem and leaf plot.

38	41	25	32	13
19	18	28	14	29

Source: *Southwest Airlines.*

Extending the Concepts

20. The number of successful space launches by the United States and Japan for the years 1993–1997 is shown here. Construct a compound time series graph for the data. What comparison can be made regarding the launches?

Year	1993	1994	1995	1996	1997
United States	29	27	24	32	37
Japan	1	4	2	1	2

Source: *The World Almanac and Book of Facts.*

21. Meat production for veal and lamb for the years 1960–2000 is shown here. (Data are in millions of pounds.) Construct a compound time series graph for the data. What comparison can be made regarding meat production?

Year	1960	1970	1980	1990	2000
Veal	1109	588	400	327	225
Lamb	769	551	318	358	234

Source: *The World Almanac and Book of Facts.*

22. The top 10 airlines with the most aircraft are listed. Represent these data with an appropriate graph.

American	714	Continental	364
United	603	Southwest	327
Delta	600	British Airways	268
Northwest	424	American Eagle	245
U.S. Airways	384	Lufthansa (Ger.)	233

Source: *Top 10 of Everything.*

23. The top prize-winning countries for Nobel Prizes in Physiology or Medicine are listed here. Represent the data with an appropriate graph.

United States	80	Denmark	5
United Kingdom	24	Austria	4
Germany	16	Belgium	4
Sweden	8	Italy	3
France	7	Australia	3
Switzerland	6		

Source: *Top 10 of Everything.*

Source: Cartoon by Bradford Veley, Marquette, Michigan. Used with permission.

24. The graph shows the increase in the price of a quart of milk. Why might the increase appear to be larger than it really is?

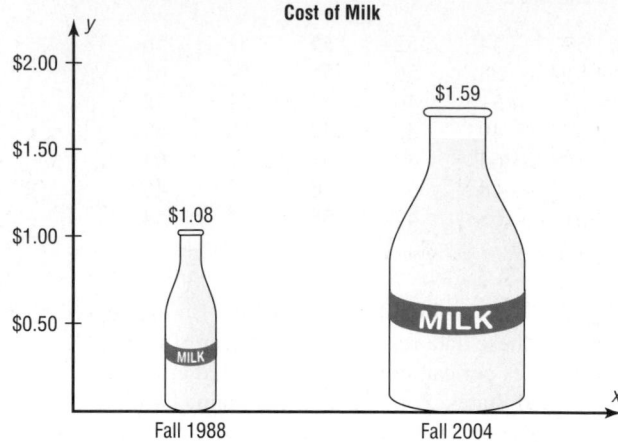

25. The graph shows the projected boom (in millions) in the number of births. Cite several reasons why the graph might be misleading.

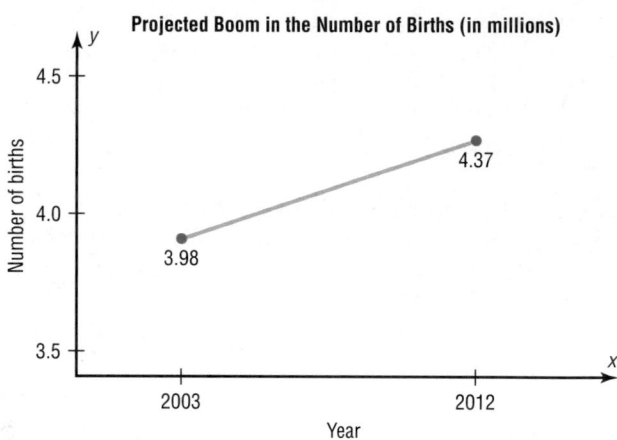

Technology *Step by Step*

MINITAB
Step by Step

Construct a Pie Chart

1. Enter the summary data for snack foods and frequencies from Example 2–10 into C1 and C2.

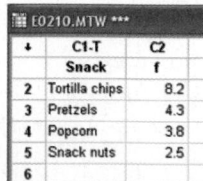

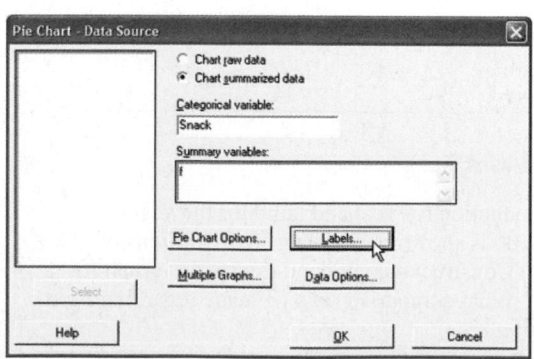

2. Name them **Snack** and **f.**

3. Select **Graph>Pie Chart.**

   a) Click the option for Chart summarized data.

   b) Press [Tab] to move to Categorical variable, then double-click C1 to select it.

   c) Press [Tab] to move to Summary variables, and select the column with the frequencies f.

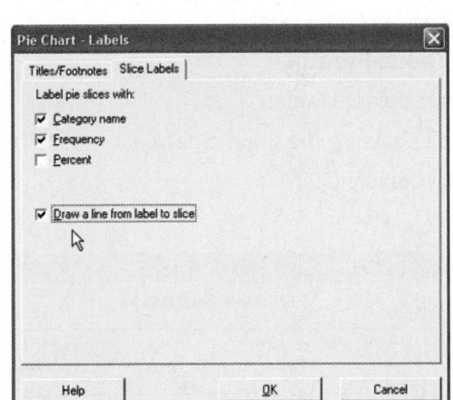
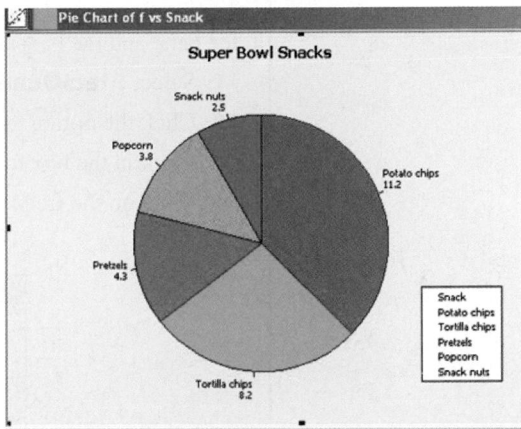

4. Click the [Labels] tab, then Titles/Footnotes.

   a) Type in the title: **Super Bowl Snacks.**

   b) Click the Slice Labels tab, then the options for Category name and Frequency.

   c) Click the option to Draw a line from label to slice.

   d) Click [OK] twice to create the chart.

Construct a Bar Chart

The procedure for constructing a bar chart is similar to that for the pie chart.

1. Select **Graph>Bar Chart.**

   a) Click on the drop-down list in Bars Represent: then select values from a table.

   b) Click on the Simple chart, then click [OK]. The dialog box will be similar to the Pie Chart Dialog Box.

2. Select the frequency column C2 f for Graph variables: and Snack for the Categorical variable.

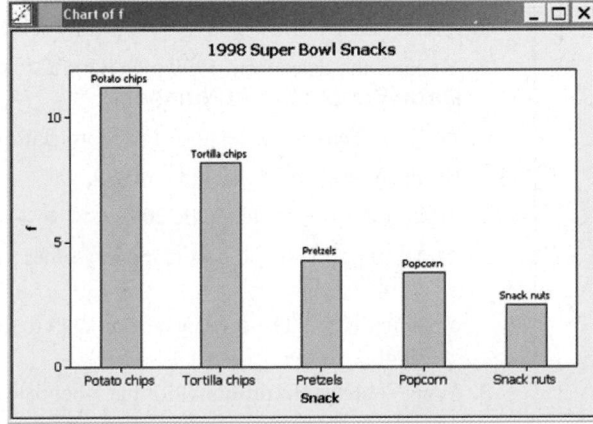

3. Click on [Labels], then type the title in the Titles/Footnote tab: **1998 Super Bowl Snacks.**

4. Click the tab for Data Labels, then click the option to Use labels from column: and select C1 Snacks.

5. Click [OK] twice.

Construct a Pareto Chart

Pareto charts are a quality control tool. They are similar to a bar chart with no gaps between the bars, and the bars are arranged by frequency.

1. Select **Stat>Quality Tools>Pareto.**

2. Click the option to Chart defects table.

3. Click in the box for the Labels in: and select Snack.

4. Click on the frequencies column C2 f.

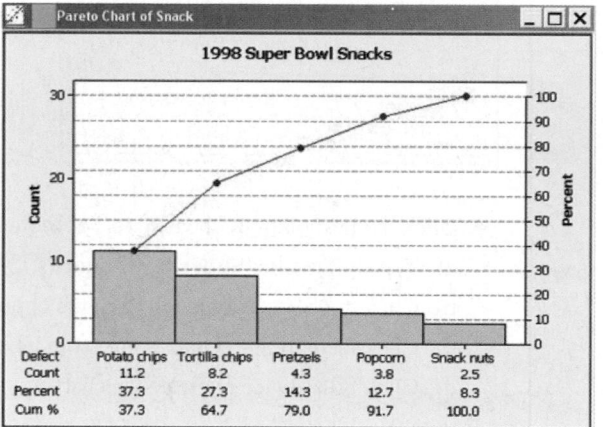

5. Click on [Options].

   a) Check the box for Cumulative percents.

   b) Type in the title, **1998 Super Bowl Snacks.**

6. Click [OK] twice. The chart is completed.

Construct a Time Series Plot

The data used are from Example 2–9, the number of vehicles that used the Pennsylvania Turnpike.

1. Add a blank worksheet to the project by selecting **File>New>New Worksheet.**

2. To enter the dates from 1999 to 2003 in C1, select **Calc>Make Patterned Data>Simple Set of Numbers.**

   a) Type **Year** in the text box for Store patterned data in.

   b) From first value: should be **1999.**

   c) To Last value: should be **2003.**

   d) In steps of should be **1** (for every other year). The last two boxes should be 1, the default value.

   e) Click [OK]. The sequence from 1999 to 2003 will be entered in C1 whose label will be Year.

3. Type **Vehicles (in millions)** for the label row above row 1 in C2.

4. Type **156.2** for the first number, then press [Enter]. Never enter the commas for large numbers!

5. Continue entering the value in each row of C2.

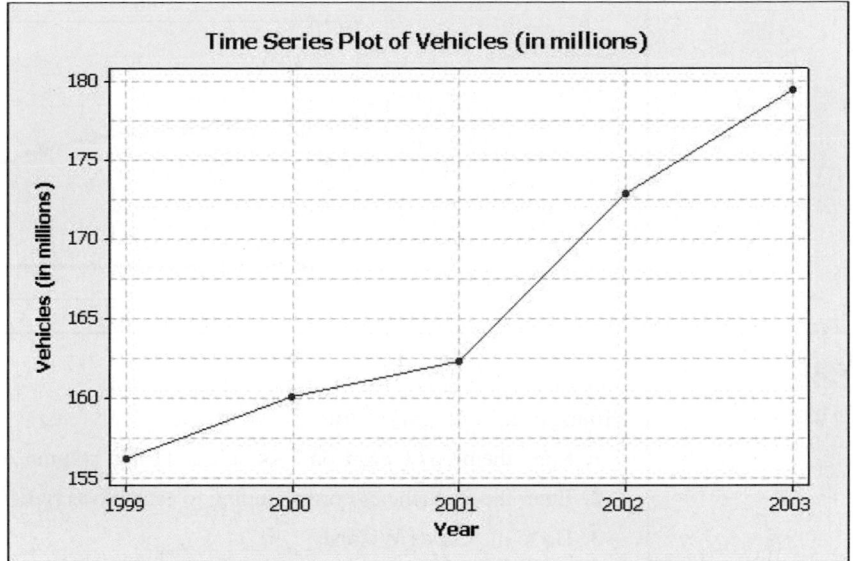

6. To make the graph, select **Graph>Time series plot,** then Simple, and press [OK].

   a) For Series select Vehicles (in millions), then click [Time/scale].

   b) Click the Stamp option and select Year for the Stamp column.

   c) Click the Gridlines tab and select all three boxes, Y major, Y minor, and X major.

   d) Click [OK] twice. A new window will open that contains the graph.

   e) To change the title, double-click the title in the graph window. A dialog box will open, allowing you to edit the text.

Construct a Stem and Leaf Plot

1. Type in the data for Example 2–13. Label the column **CarThefts.**

2. Select **STAT>EDA>Stem-and-Leaf.** This is the same as **Graph>Stem-and-Leaf.**

3. Double-click on C1 CarThefts in the column list.

4. Click in the Increment text box, and enter the class width of **5.**

5. Click [OK]. This character graph will be displayed in the session window.

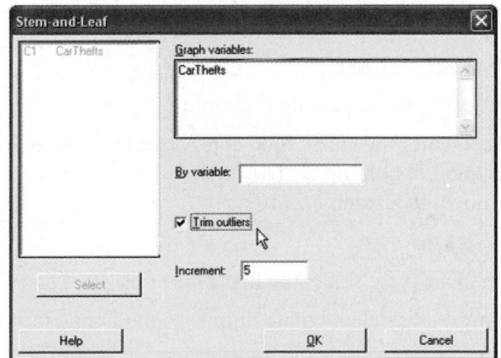

Stem-and-Leaf Display: CarThefts
Stem-and-leaf of CarThefts   N = 30
Leaf Unit = 1.0

6	5	011233
13	5	5567789
15	6	23
15	6	55667899
7	7	23
5	7	55789

TI-83 Plus or TI-84 Plus
Step by Step

To graph a time series, follow the procedure for a frequency polygon from Section 2–3, using the data from Example 2–9.

Output

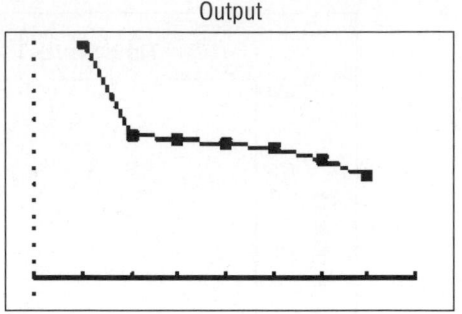

Excel
Step by Step

Constructing a Pie Chart

To make a pie (or bar) chart:

1. Enter the blood types from Example 2–11 into column A of a new worksheet.
2. Enter the frequencies corresponding to each blood type in column B.
3. Go to the **Chart Wizard.**
4. Select Pie (or Bar), and select the first subtype.
5. Click the Data Range tab. Enter both columns as the Data Range.
6. Check column for the Series in option, then click Next.
7. Create a title for your chart, such as **Blood Types of Army Inductees.**
8. Click the Data Labels tab and check the Show percent option.
9. Place chart as a new sheet or as an embedded sheet in the active worksheet, then select Finish.

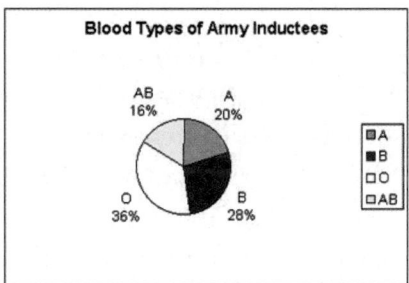

Constructing a Pareto Chart

To make a Pareto chart using the data from Example 2–10:

1. On a new worksheet, enter the snack food categories in column A and the corresponding frequencies (in pounds) in column B. The order is important here. Make sure to enter the data in the order shown in Example 2–10.
2. Go to the **Chart Wizard.**
3. Select the Column chart type, select the first chart subtype, and then click Next.
4. Click the Data Range tab. Select both columns as the Data Range.

5. Check column for the Series in option.

6. Click the Series tab. Select B1:B5 for Values, select A1:A5 for the Category (X) axis labels, and then click Next.

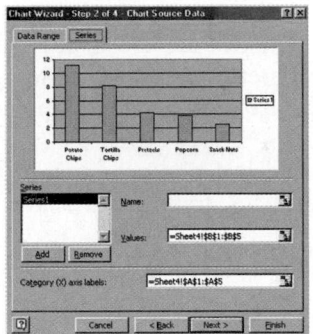
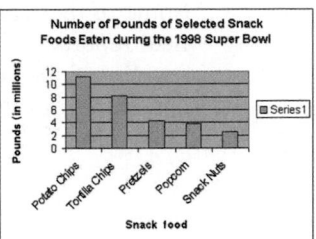

7. Create a title for your chart, such as **Number of Pounds of Selected Snack Foods Eaten During the 1998 Super Bowl.** Enter **Snack Food** as the Category (X) axis and **Pounds in Millions** as the Value (Y) axis. Click Finish.

Constructing a Time Series Plot

To make a time series plot using the data from Example 2–9:

1. On a new worksheet, enter the Years in column A and the corresponding Number (of vehicles) in column B.

2. Go to the **Chart Wizard.**

3. Select the Line chart type, select the fourth subtype, and then click Next.

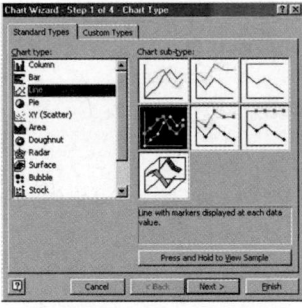
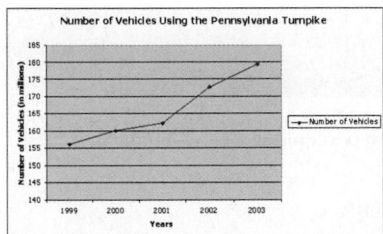

4. Click the Data Range tab. Select B1:B6 as the Data Range.

5. Check column for the Series in option.

6. Click the Series tab. Select A2:A6 for the Category (X) axis labels, and then click Next.

7. Create a title for your chart, such as **Number of Vehicles Using the Pennsylvania Turnpike Between 1999 and 2003.** Enter **Year** as the Category (X) axis and **Number of Vehicles (in millions)** as the Value (Y) axis. Click Finish.

<table>
<tr><td>**2-5**</td><td></td></tr>
</table>

Summary

When data are collected, they are called raw data. Since very little knowledge can be obtained from raw data, they must be organized in some meaningful way. A frequency distribution using classes is the solution. Once a frequency distribution is constructed, the representation of the data by graphs is a simple task. The most commonly used graphs in research statistics are the histogram, frequency polygon, and ogive. Other graphs, such as the Pareto chart, time series graph, and pie graph, can also be used. Some of these graphs are seen frequently in newspapers, magazines, and various statistical reports.

Finally, a stem and leaf plot uses part of the data values as stems and part of the data values as leaves. This graph has the advantages of a frequency distribution and a histogram.

Important Terms

categorical frequency distribution 36

class 35

class boundaries 37

class midpoint 38

class width 37

cumulative frequency 51

frequency 35

frequency distribution 35

frequency polygon 50

grouped frequency distribution 37

histogram 48

lower class limit 37

ogive 51

open-ended distribution 39

Pareto chart 64

pie graph 66

raw data 35

relative frequency graph 53

stem and leaf plot 73

time series graph 65

ungrouped frequency distribution 41

upper class limit 37

Important Formulas

Formula for the percentage of values in each class:

$$\% = \frac{f}{n} \cdot 100\%$$

where

f = **frequency of the class**

n = **total number of values**

Formula for the range:

R = **highest value − lowest value**

Formula for the class width:

Class width = upper boundary − lower boundary

Formula for the class midpoint:

$$X_m = \frac{\text{lower boundary + upper boundary}}{2}$$

or

$$X_m = \frac{\text{lower limit + upper limit}}{2}$$

Formula for the degrees for each section of a pie graph:

$$\textbf{Degrees} = \frac{f}{n} \cdot 360°$$

Review Exercises

1. The Brunswick Research Organization surveyed 50 randomly selected individuals and asked them the primary way they received the daily news. Their choices were via newspaper (N), television (T), radio (R), or Internet (I). Construct a categorical frequency distribution for the data and interpret the results. The data in this exercise will be used for Exercise 2 in this section.

N	N	T	T	T	I	R	R	I	T
I	N	R	R	I	N	N	I	T	N
I	R	T	T	T	T	N	R	R	I
R	R	I	N	T	R	T	I	I	T
T	I	N	T	T	I	R	N	R	T

2. Construct a pie graph for the data in Exercise 1, and analyze the results.

3. A sporting goods store kept a record of sales of five items for one randomly selected hour during a recent sale. Construct a frequency distribution for the data (B = baseballs, G = golf balls, T = tennis balls, S = soccer balls, F = footballs). (The data for this exercise will be used for Exercise 4 in this section.)

F	B	B	B	G	T	F
G	G	F	S	G	T	
F	T	T	T	S	T	
F	S	S	G	S	B	

4. Draw a pie graph for the data in Exercise 3 showing the sales of each item, and analyze the results.

5. The blood urea nitrogen (BUN) count of 20 randomly selected patients is given here in milligrams per deciliter (mg/dl). Construct an ungrouped frequency distribution for the data. (The data for this exercise will be used for Exercise 6.)

17	18	13	14
12	17	11	20
13	18	19	17
14	16	17	12
16	15	19	22

6. Construct a histogram, frequency polygon, and ogive for the data in Exercise 5 in this section, and analyze the results.

7. The data show the estimated added cost per vehicle use due to bad roads. Construct a frequency distribution using 6 classes. (The data for this exercise will be used for Exercises 8 and 11 in this section.)

165	186	122	172	140	153	208	169
156	114	113	135	131	125	177	136
136	127	112	188	171	179	152	155
116	90	187	136	159	97	141	85
91	170	111	147	165	163	159	150

Source: Federal Highway Administration.

8. Construct a histogram, frequency polygon, and ogive for the data in Exercise 7 in this section, and analyze the results.

9. The data shown (in millions of dollars) are the values of the 30 National Football League franchises. Construct a frequency distribution for the data using 8 classes. (The data for this exercise will be used for Exercises 10 and 12 in this section.)

170	191	171	235	173	187	181	191
200	218	243	200	182	320	184	239
186	199	186	210	209	240	204	193
211	186	197	204	188	242		

Source: Pittsburgh Post-Gazette.

10. Construct a histogram, frequency polygon, and ogive for the data in Exercise 9 in this section, and analyze the results.

11. Construct a histogram, frequency polygon, and ogive by using relative frequencies for the data in Exercise 7 in this section.

12. Construct a histogram, frequency polygon, and ogive by using relative frequencies for the data in Exercise 9 in this section.

13. Construct a Pareto chart for the number of homicides reported for the following cities.

City	Number of homicides
New Orleans	363
Washington, D.C.	352
Chicago	824
Baltimore	323
Atlanta	184

Source: USA TODAY.

14. Construct a Pareto chart for the number of trial-ready civil action and equity cases decided in less than 6 months for the selected counties in southwestern Pennsylvania.

County	Number of cases
Westmoreland	427
Washington	298
Green	151
Fayette	106
Somerset	87

Source: Pittsburgh Tribune-Review.

15. The given data represent the federal minimum hourly wage in the years shown. Draw a time series graph to represent the data and analyze the results.

Year	Wage
1960	$1.00
1965	1.25
1970	1.60
1975	2.10
1980	3.10
1985	3.35
1990	3.80
1995	4.25
2000	5.15
2005	5.15

Source: The World Almanac and Book of Facts.

16. The number of bank failures in the United States during the years 1989–2000 is shown. Draw a time series graph to represent the data and analyze the results.

Year	Number of failures
1989	207
1990	169
1991	127
1992	122
1993	41
1994	13
1995	6
1996	5
1997	1
1998	3
1999	8
2000	7
2001	4
2002	11

Source: The World Almanac and Book of Facts.

17. The data show the number (in millions) of viewers who watched the first and second presidential debates. Construct two time series graphs and compare the results.

Year	1992	1996	2000	2004
First debate	62.4	36.1	46.6	62.5
Second debate	69.9	36.3	37.6	46.7

Source: Nielson Media Research.

18. In a study of 100 women, the numbers shown here indicate the major reason why each woman surveyed worked outside the home. Construct a pie graph for the data and analyze the results.

Reason	Number of women
To support self/family	62
For extra money	18
For something different to do	12
Other	8

19. A survey asked if people would like to spend the rest of their careers with their present employers. The results are shown. Construct a pie graph for the data and analyze the results.

Answer	Number of people
Yes	660
No	260
Undecided	80

20. The number of visitors to the Railroad Museum during 24 randomly selected hours is shown here. Construct a stem and leaf plot for the data.

67	62	38	73	34	43	72	35
53	55	58	63	47	42	51	62
32	29	47	62	29	38	36	41

21. The data set shown here represents the number of hours that 25 part-time employees worked at the Sea Side Amusement Park during a randomly selected week in June. Construct a stem and leaf plot for the data and summarize the results.

16	25	18	39	25	17	29	14	37
22	18	12	23	32	35	24	26	
20	19	25	26	38	38	33	29	

22. A special aptitude test is given to job applicants. The data shown here represent the scores of 30 applicants. Construct a stem and leaf plot for the data and summarize the results.

204	210	227	218	254
256	238	242	253	227
251	243	233	251	241
237	247	211	222	231
218	212	217	227	209
260	230	228	242	200

Data Analysis

A Data Bank is found in Appendix D, or on the World Wide Web by following links from
www.mhhe.com/math/stat/bluman

1. From the Data Bank located in Appendix D, choose one of the following variables: age, weight, cholesterol level, systolic pressure, IQ, or sodium level. Select at least 30 values. For these values, construct a grouped frequency distribution. Draw a histogram, frequency polygon, and ogive for the distribution. Describe briefly the shape of the distribution.

2. From the Data Bank, choose one of the following variables: educational level, smoking status, or exercise.

Statistics Today

How Serious Are Hospital Infections?—Revisited

The data presented in numerical form do not convey an easy-to-interpret conclusion; however, when data are presented in graphical form, readers can see the visual impact of the numbers. In this case, a Pareto graph shows that infections from pneumonia are the most common, followed by bloodstream infections. The pie graph also shows that urinary tract infections occur with the highest frequency.

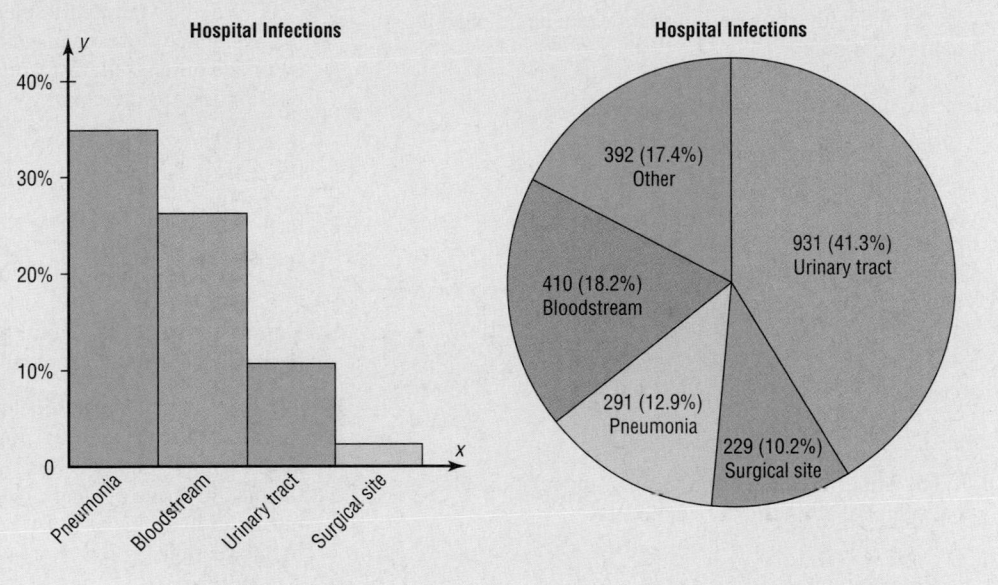

Select at least 20 values. Construct an ungrouped frequency distribution for the data. For the distribution, draw a Pareto chart and describe briefly the nature of the chart.

3. From the Data Bank, select at least 30 subjects and construct a categorical distribution for their marital status. Draw a pie graph and describe briefly the findings.

4. Using the data from Data Set IV in Appendix D, construct a frequency distribution and draw a histogram. Describe briefly the shape of the distribution of the tallest buildings in New York City.

5. Using the data from Data Set XI in Appendix D, construct a frequency distribution and draw a frequency polygon. Describe briefly the shape of the distribution for the number of pages in statistics books.

6. Using the data from Data Set IX in Appendix D, divide the United States into four regions, as follows:

Northeast	CT ME MA NH NJ NY PA RI VT
Midwest	IL IN IA KS MI MN MS NE ND OH SD WI
South	AL AR DE DC FL GA KY LA MD NC OK SC TN TX VA WV
West	AK AZ CA CO HI ID MT NV NM OR UT WA WY

Find the total population for each region, and draw a Pareto chart and a pie graph for the data. Analyze the results. Explain which chart might be a better representation for the data.

7. Using the data from Data Set I in Appendix D, make a stem and leaf plot for the record low temperatures in the United States. Describe the nature of the plot.

Chapter Quiz

Determine whether each statement is true or false. If the statement is false, explain why.

1. In the construction of a frequency distribution, it is a good idea to have overlapping class limits, such as 10–20, 20–30, 30–40.

2. Histograms can be drawn by using vertical or horizontal bars.

3. It is not important to keep the width of each class the same in a frequency distribution.

4. Frequency distributions can aid the researcher in drawing charts and graphs.

5. The type of graph used to represent data is determined by the type of data collected and by the researcher's purpose.

6. In construction of a frequency polygon, the class limits are used for the x axis.

7. Data collected over a period of time can be graphed by using a pie graph.

Select the best answer.

8. What is another name for the ogive?

   a. Histogram
   b. Frequency polygon
   c. Cumulative frequency graph
   d. Pareto chart

9. What are the boundaries for 8.6–8.8?

   a. 8–9
   b. 8.5–8.9
   c. 8.55–8.85
   d. 8.65–8.75

10. What graph should be used to show the relationship between the parts and the whole?

    a. Histogram
    b. Pie graph
    c. Pareto chart
    d. Ogive

11. Except for rounding errors, relative frequencies should add up to what sum?

    a. 0
    b. 1
    c. 50
    d. 100

Complete these statements with the best answers.

12. The three types of frequency distributions are _____, _____, and _____.

13. In a frequency distribution, the number of classes should be between _____ and _____.

14. Data such as blood types (A, B, AB, O) can be organized into a(n) _____ frequency distribution.

15. Data collected over a period of time can be graphed using a(n) _____ graph.

16. A statistical device used in exploratory data analysis that is a combination of a frequency distribution and a histogram is called a(n) _____.

17. On a Pareto chart, the frequencies should be represented on the _____ axis.

18. A questionnaire on housing arrangements showed this information obtained from 25 respondents. Construct a frequency distribution for the data (H = house, A = apartment, M = mobile home, C = condominium).

H	C	H	M	H	A	C	A	M
C	M	C	A	M	A	C	C	M
C	C	H	A	H	H	M		

19. Construct a pie graph for the data in Problem 18.

20. When 30 randomly selected customers left a convenience store, each was asked the number of items he or she purchased. Construct an ungrouped frequency distribution for the data.

2	9	4	3	6
6	2	8	6	5
7	5	3	8	6
6	2	3	2	4
6	9	9	8	9
4	2	1	7	4

21. Construct a histogram, a frequency polygon, and an ogive for the data in Problem 20.

22. For a recent year, the number of murders in 25 selected cities is shown. Construct a frequency distribution using 9 classes, and analyze the nature of the data in terms of shape, extreme values, etc. (The information in this exercise will be used for Exercise 23 in this section.

248	348	74	514	597
270	71	226	41	39
366	73	241	46	34
149	68	73	63	65
109	598	278	69	27

Source: *Pittsburgh Tribune Review.*

23. Construct a histogram, frequency polygon, and ogive for the data in Problem 22. Analyze the histogram.

24. Construct a Pareto chart for the number of tons (in millions) of trash recycled per year by Americans based on an Environmental Protection Agency study.

Type	Amount
Paper	320.0
Iron/steel	292.0
Aluminum	276.0
Yard waste	242.4
Glass	196.0
Plastics	41.6

Source: *USA TODAY.*

25. The data show the number of fatal trespasser casualties on railroad property in the United States. Draw a time series graph and explain any trend.

Year	1998	1999	2000	2001
Number	536	463	511	540

Source: *Federal Railroad Administration.*

26. The number of visitors to the Historic Museum for 25 randomly selected hours is shown. Construct a stem and leaf plot for the data.

15	53	48	19	38
86	63	98	79	38
62	89	67	39	26
28	35	54	88	76
31	47	53	41	68

Critical Thinking Challenges

1. The *USA TODAY* Snapshot shows pumpkin production. Can you see anything misleading about the way the graph is drawn?

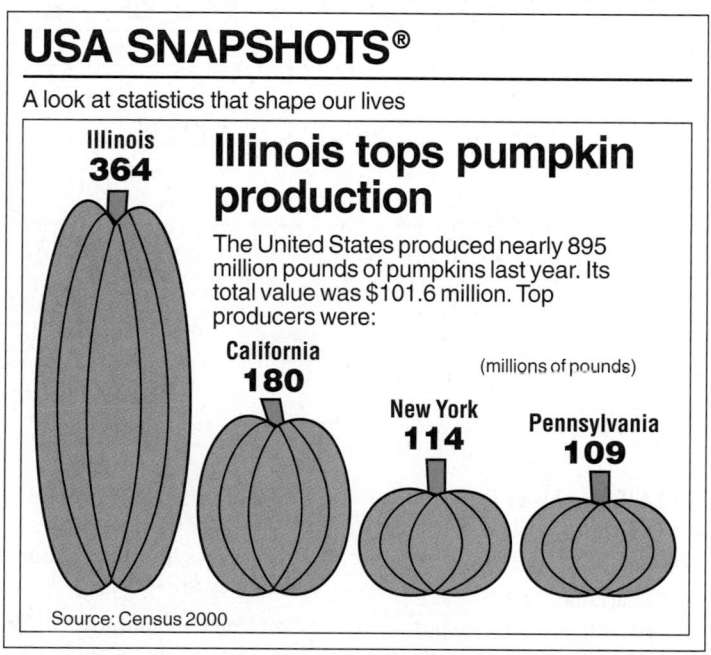

Source: Copyright 2001, *USA TODAY.* Reprinted with permission.

2. Shown are various statistics about the Great Lakes. Using appropriate graphs (your choice) and summary statements, write a report analyzing the data.

	Superior	Michigan	Huron	Erie	Ontario
Length (miles)	350	307	206	241	193
Breadth (miles)	160	118	183	57	53
Depth (feet)	1,330	923	750	210	802
Volume (cubic miles)	2,900	1,180	850	116	393
Area (square miles)	31,700	22,300	23,000	9,910	7,550
Shoreline (U.S., miles)	863	1,400	580	431	300

Source: *The World Almanac and Book of Facts.*

3. A compound time series graph is shown for the percentage of sales of CDs, cassettes, and vinyl records. Using the graph, answer these questions.

a. In what year did cassette sales match CD sales?

b. In what year did cassette sales account for 50% of the sales?

c. What was the percentage difference in sales between CDs and cassettes for the year 1996?

d. In what year did CDs account for 30% of the sales?

e. What was the change in percentage of sales for cassettes from 1987 to 2000?

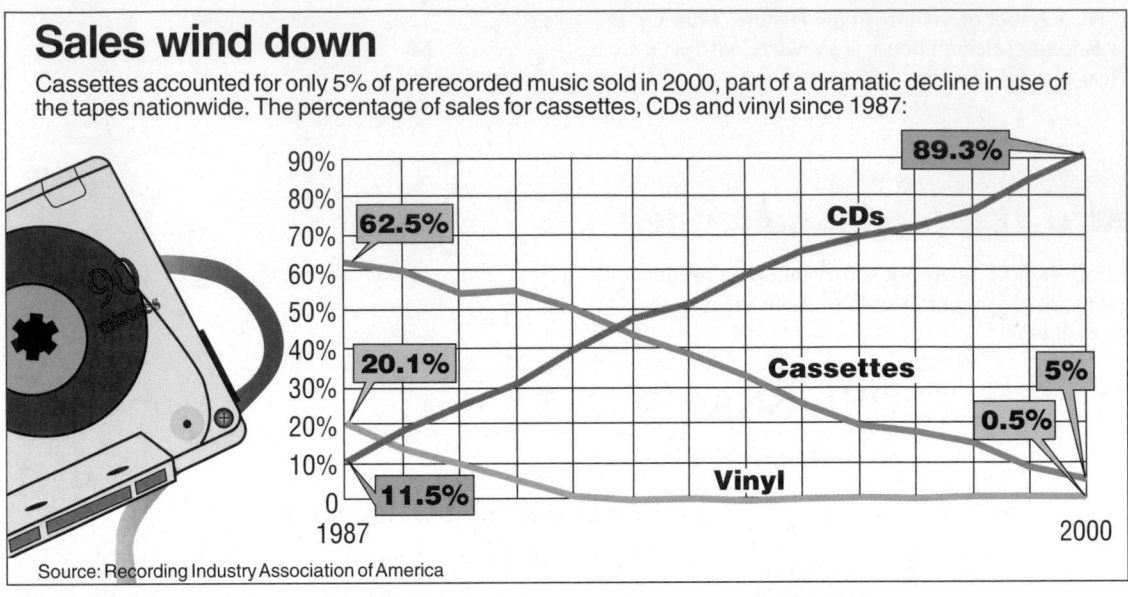

Sales wind down

Cassettes accounted for only 5% of prerecorded music sold in 2000, part of a dramatic decline in use of the tapes nationwide. The percentage of sales for cassettes, CDs and vinyl since 1987:

Source: Recording Industry Association of America

Source: Copyright 2001, *USA TODAY.* Reprinted with permission.

Data Projects

Where appropriate, use MINITAB, the TI-83 Plus, the TI-84 Plus, Excel, or a computer program of your choice to complete the following exercises.

1. Select a categorical (nominal) variable, such as the colors of cars in the school's parking lot or the major fields of the students in statistics class, and collect data on this variable.

   a. State the purpose of the project.
   b. Define the population.
   c. State how the sample was selected.
   d. Show the raw data.
   e. Construct a frequency distribution for the variable.
   f. Draw some appropriate graphs (pie, Pareto, etc.) for the data.
   g. Analyze the results.

2. Using an almanac, select a variable that varies over a period of several years (e.g., silver production), and draw a time series graph for the data. Write a short paragraph interpreting the findings.

3. Select a variable (interval or ratio) and collect at least 30 values. For example, you may ask the students in your class how many hours they study per week or how old they are.

   a. State the purpose of the project.
   b. Define the population.
   c. State how the sample was selected.
   d. Show the raw data.
   e. Construct a frequency distribution for the data.
   f. Draw a histogram, frequency polygon, and ogive for the data.
   g. Analyze the results.

You may use these websites to obtain raw data:

Visit the data sets at the book's website found at
http://www.mhhe.com/math/stat/bluman
Click on the 6th edition.
http://lib.stat.cmu.edu/DASL
http://www.statcan.ca

Answers to Applying the Concepts

Section 2–2   Ages of Presidents at Inauguration

1. The data were obtained from the population of all presidents at the time this text was written.

2. The oldest inauguration age was 69 years old.

3. The youngest inauguration age was 42 years old.

4. Answers will vary. One possible answer is

Age at inauguration	Frequency
42–45	2
46–49	6
50–53	7
54–57	16
58–61	5
62–65	4
66–69	2

5. Answers will vary. For the frequency distribution given in Exercise 4, there is a peak for the 54–57 bin.

6. Answers will vary. This frequency distribution shows no outliers. However, if we had split our frequency into 14 bins instead of 7, then the ages 42, 43, 68, and 69 might appear as outliers.

7. Answers will vary. The data appear to be unimodal and fairly symmetric, centering around 55 years of age.

Section 2–3   Selling Real Estate

1. A histogram of the data gives price ranges and the counts of homes in each price range. We can also talk about how the data are distributed by looking at a histogram.

2. A frequency polygon shows increases or decreases in the number of home prices around values.

3. A cumulative frequency polygon shows the number of homes sold at or below a given price.

4. The house that sold for $321,550 is an extreme value in this data set.

5. Answers will vary. One possible answer is that the histogram displays the outlier well since there is a gap in the prices of the homes sold.

6. The distribution of the data is skewed to the right.

Section 2–4   Leading Cause of Death

1. The variables in the graph are the year, cause of death, and rate of death per 100,000 men.

2. The cause of death is qualitative, while the year and death rates are quantitative.

3. Year is a discrete variable, and death rate is continuous. Since cause of death is qualitative, it is neither discrete nor continuous.

4. A line graph was used to display the data.

5. No, a Pareto chart could not be used to display the data, since we can only have one quantitative variable and one categorical variable in a Pareto chart.

6. We cannot use a pie chart for the same reasons as given for the Pareto chart.

7. A Pareto chart is typically used to show a categorical variable listed from the highest-frequency category to the category with the lowest frequency.

8. A time series chart is used to see trends in the data. It can also be used for forecasting and predicting.

Data Description

Objectives

After completing this chapter, you should be able to

1. Summarize data using measures of central tendency, such as the mean, median, mode, and midrange.

2. Describe data using measures of variation, such as the range, variance, and standard deviation.

3. Identify the position of a data value in a data set, using various measures of position, such as percentiles, deciles, and quartiles.

4. Use the techniques of exploratory data analysis, including boxplots and five-number summaries, to discover various aspects of data.

Outline

How Long Are You Delayed by Road Congestion?

No matter where you live, at one time or another, you have been stuck in traffic. To see whether there are more traffic delays in some cities than in others, statisticians make comparisons using descriptive statistics. A statistical study by the Texas Transportation Institute found that a driver is delayed by road congestion an average of 36 hours per year. To see how selected cities compare to this average, see Statistics Today—Revisited at the end of the chapter.

This chapter will show you how to obtain and interpret descriptive statistics such as measures of average, measures of variation, and measures of position.

3–1

Introduction

Chapter 2 showed how one can gain useful information from raw data by organizing them into a frequency distribution and then presenting the data by using various graphs. This chapter shows the statistical methods that can be used to summarize data. The most familiar of these methods is the finding of averages.

For example, one may read that the average speed of a car crossing midtown Manhattan during the day is 5.3 miles per hour or that the average number of minutes an American father of a 4-year-old spends alone with his child each day is 42.[1]

[1]"Harper's Index," *Harper's* magazine.

In the book *American Averages* by Mike Feinsilber and William B. Meed, the authors state:

"Average" when you stop to think of it is a funny concept. Although it describes all of us it describes none of us. . . . While none of us wants to be the average American, we all want to know about him or her.

The authors go on to give examples of averages:

The average American man is five feet, nine inches tall; the average woman is five feet, 3.6 inches.
The average American is sick in bed seven days a year missing five days of work.
On the average day, 24 million people receive animal bites.
By his or her 70th birthday, the average American will have eaten 14 steers, 1050 chickens, 3.5 lambs, and 25.2 hogs.[2]

In these examples, the word *average* is ambiguous, since several different methods can be used to obtain an average. Loosely stated, the average means the center of the distribution or the most typical case. Measures of average are also called *measures of central tendency* and include the *mean, median, mode,* and *midrange.*

Knowing the average of a data set is not enough to describe the data set entirely. Even though a shoe store owner knows that the average size of a man's shoe is size 10, she would not be in business very long if she ordered only size 10 shoes.

As this example shows, in addition to knowing the average, one must know how the data values are dispersed. That is, do the data values cluster around the mean, or are they spread more evenly throughout the distribution? The measures that determine the spread of the data values are called *measures of variation* or *measures of dispersion.* These measures include the *range, variance,* and *standard deviation.*

Finally, another set of measures is necessary to describe data. These measures are called *measures of position.* They tell where a specific data value falls within the data set or its relative position in comparison with other data values. The most common position measures are *percentiles, deciles,* and *quartiles.* These measures are used extensively in psychology and education. Sometimes they are referred to as *norms.*

The measures of central tendency, variation, and position explained in this chapter are part of what is called *traditional statistics.*

Section 3–5 shows the techniques of what is called *exploratory data analysis.* These techniques include the *boxplot* and the *five-number summary.* They can be used to explore data to see what they show (as opposed to the traditional techniques, which are used to confirm conjectures about the data).

Interesting Fact

A person has on average 1460 dreams in 1 year.

3–2 Measures of Central Tendency

Objective 1

Summarize data using measures of central tendency, such as the mean, median, mode, and midrange.

Chapter 1 stated that statisticians use samples taken from populations; however, when populations are small, it is not necessary to use samples since the entire population can be used to gain information. For example, suppose an insurance manager wanted to know the average weekly sales of all the company's representatives. If the company employed a large number of salespeople, say, nationwide, he would have to use a sample and make an inference to the entire sales force. But if the company had only a few salespeople, say, only 87 agents, he would be able to use all representatives' sales for a randomly chosen week and thus use the entire population.

[2]Mike Feinsilber and William B. Meed, *American Averages* (New York: Bantam Doubleday Dell).

Measures found by using all the data values in the population are called *parameters.* Measures obtained by using the data values from samples are called *statistics;* hence, the average of the sales from a sample of representatives is called a *statistic,* and the average of sales obtained from the entire population is called a *parameter.*

A **statistic** is a characteristic or measure obtained by using the data values from a sample.

A **parameter** is a characteristic or measure obtained by using all the data values from a specific population.

These concepts as well as the symbols used to represent them will be explained in detail in this chapter.

General Rounding Rule In statistics the basic rounding rule is that when computations are done in the calculation, rounding should not be done until the final answer is calculated. When rounding is done in the intermediate steps, it tends to increase the difference between that answer and the exact one. But in the textbook and solutions manual, it is not practical to show long decimals in the intermediate calculations; hence, the values in the examples are carried out to enough places (usually three or four) to obtain the same answer that a calculator would give after rounding on the last step.

The Mean

The *mean,* also known as the *arithmetic average,* is found by adding the values of the data and dividing by the total number of values. For example, the mean of 3, 2, 6, 5, and 4 is found by adding $3 + 2 + 6 + 5 + 4 = 20$ and dividing by 5; hence, the mean of the data is $20 \div 5 = 4$. The values of the data are represented by X's. In this data set, $X_1 = 3$, $X_2 = 2$, $X_3 = 6$, $X_4 = 5$, and $X_5 = 4$. To show a sum of the total X values, the symbol Σ (the capital Greek letter sigma) is used, and ΣX means to find the sum of the X values in the data set. The summation notation is explained in Appendix A.

The **mean** is the sum of the values, divided by the total number of values. The symbol $\overline{X}$ represents the sample mean.

$$\overline{X} = \frac{X_1 + X_2 + X_3 + \cdots + X_n}{n} = \frac{\Sigma X}{n}$$

where n represents the total number of values in the sample.
For a population, the Greek letter μ (mu) is used for the mean.

$$\mu = \frac{X_1 + X_2 + X_3 + \cdots + X_N}{N} = \frac{\Sigma X}{N}$$

where N represents the total number of values in the population.

In statistics, Greek letters are used to denote parameters, and Roman letters are used to denote statistics. Assume that the data are obtained from samples unless otherwise specified.

Example 3–1

The data represent the number of days off per year for a sample of individuals selected from nine different countries. Find the mean.

20, 26, 40, 36, 23, 42, 35, 24, 30

Source: World Tourism Organization.

Solution

$$\bar{X} = \frac{\Sigma X}{n} = \frac{20 + 26 + 40 + 36 + 23 + 42 + 35 + 24 + 30}{9} = \frac{276}{9} = 30.7 \text{ days}$$

Hence, the mean of the number of days off is 30.7 days.

Example 3–2

 The data represent the annual chocolate sales (in billions of dollars) for a sample of seven countries in the world. Find the mean.

2.0, 4.9, 6.5, 2.1, 5.1, 3.2, 16.6

Source: Euromonitor.

Solution

$$\bar{X} = \frac{\Sigma X}{n} = \frac{2.0 + 4.9 + 6.5 + 2.1 + 5.1 + 3.2 + 16.6}{7} = \frac{40.4}{7} = \$5.77 \text{ billion}$$

The mean for the sample is \$5.77 billion.

The mean, in most cases, is not an actual data value.

Rounding Rule for the Mean The mean should be rounded to one more decimal place than occurs in the raw data. For example, if the raw data are given in whole numbers, the mean should be rounded to the nearest tenth. If the data are given in tenths, the mean should be rounded to the nearest hundredth, and so on.

The procedure for finding the mean for grouped data uses the midpoints of the classes. This procedure is shown next.

Example 3–3

Using the frequency distribution for Example 2–7, find the mean. The data represent the number of miles run during one week for a sample of 20 runners.

Solution

The procedure for finding the mean for grouped data is given here.

Step 1 Make a table as shown.

A Class	B Frequency (f)	C Midpoint (X_m)	D $f \cdot X_m$
5.5–10.5	1		
10.5–15.5	2		
15.5–20.5	3		
20.5–25.5	5		
25.5–30.5	4		
30.5–35.5	3		
35.5–40.5	2		
	$n = 20$		

Step 2 Find the midpoints of each class and enter them in column C.

$$X_m = \frac{5.5 + 10.5}{2} = 8 \qquad \frac{10.5 + 15.5}{2} = 13 \qquad \text{etc.}$$

Step 3   For each class, multiply the frequency by the midpoint, as shown, and place the product in column D.

$$1 \cdot 8 = 8 \qquad 2 \cdot 13 = 26 \qquad \text{etc.}$$

The completed table is shown here.

A Class	B Frequency (f)	C Midpoint (X_m)	D $f \cdot X_m$
5.5–10.5	1	8	8
10.5–15.5	2	13	26
15.5–20.5	3	18	54
20.5–25.5	5	23	115
25.5–30.5	4	28	112
30.5–35.5	3	33	99
35.5–40.5	2	38	76
	$n = 20$		$\Sigma f \cdot X_m = 490$

Step 4   Find the sum of column D.

Step 5   Divide the sum by n to get the mean.

$$\overline{X} = \frac{\Sigma f \cdot X_m}{n} = \frac{490}{20} = 24.5 \text{ miles}$$

Unusual Stat

A person looks, on average, at about 14 homes before he or she buys one.

The procedure for finding the mean for grouped data assumes that the mean of all the raw data values in each class is equal to the midpoint of the class. In reality, this is not true, since the average of the raw data values in each class usually will not be exactly equal to the midpoint. However, using this procedure will give an acceptable approximation of the mean, since some values fall above the midpoint and other values fall below the midpoint for each class, and the midpoint represents an estimate of all values in the class.

The steps for finding the mean for grouped data are summarized in the next Procedure Table.

Procedure Table

Finding the Mean for Grouped Data

Step 1   Make a table as shown.

A Class	B Frequency (f)	C Midpoint (X_m)	D $f \cdot X_m$

Step 2   Find the midpoints of each class and place them in column C.

Step 3   Multiply the frequency by the midpoint for each class, and place the product in column D.

Step 4   Find the sum of column D.

Step 5   Divide the sum obtained in column D by the sum of the frequencies obtained in column B.

The formula for the mean is

$$\overline{X} = \frac{\Sigma f \cdot X_m}{n}$$

(Note: The symbols $\Sigma f \cdot X_m$ mean to find the sum of the product of the frequency (f) and the midpoint (X_m) for each class.)

How long do you wait for fast-food service? This Snapshot shows the average times for large fast-food chains. Which type of average (mean, median, or mode) do you think was used?

USA SNAPSHOTS®

Fastest drive-through service

National average **3 min. 3 sec.**

Wendy's **2 min. 22 sec.**

Checkers **2 min. 34 sec.**

Rally's **2 min. 38 sec.**

McDonald's **2 min. 50 sec.**

Hardee's **2 min. 53 sec.**

Source: 2000 QSR Drive-Thru Time Study

Source: Copyright 2001, *USA TODAY.* Reprinted with permission.

The Median

An article recently reported that the median income for college professors was $43,250. This measure of central tendency means that one-half of all the professors surveyed earned more than $43,250, and one-half earned less than $43,250.

The *median* is the halfway point in a data set. Before one can find this point, the data must be arranged in order. When the data set is ordered, it is called a **data array.** The median either will be a specific value in the data set or will fall between two values, as shown in Examples 3–4 through 3–8.

The **median** is the midpoint of the data array. The symbol for the median is MD.

Steps in computing the median of a data array

Step 1 Arrange the data in order.

Step 2 Select the middle point.

Example 3–4

The number of rooms in the seven hotels in downtown Pittsburgh is 713, 300, 618, 595, 311, 401, and 292. Find the median.

Source: Interstate Hotels Corporation.

Solution

Step 1 Arrange the data in order.

292, 300, 311, 401, 595, 618, 713

> **Step 2** Select the middle value.
>
> 292, 300, 311, 401, 595, 618, 713
> ↑
> Median
>
> Hence, the median is 401 rooms.

Example 3–5

 Find the median for the ages of seven preschool children. The ages are 1, 3, 4, 2, 3, 5, and 1.

Solution

1, 1, 2, ③, 3, 4, 5
↑
Median

Hence, the median age is 3 years.

Examples 3–4 and 3–5 each had an odd number of values in the data set; hence, the median was an actual data value. When there are an even number of values in the data set, the median will fall between two given values, as illustrated in Examples 3–6, 3–7, and 3–8.

Example 3–6

 The number of tornadoes that have occurred in the United States over an 8-year period follows. Find the median.

684, 764, 656, 702, 856, 1133, 1132, 1303

Source: The Universal Almanac.

Solution

656, 684, 702, 764, 856, 1132, 1133, 1303
↑
Median

Since the middle point falls halfway between 764 and 856, find the median MD by adding the two values and dividing by 2.

$$MD = \frac{764 + 856}{2} = \frac{1620}{2} = 810$$

The median number of tornadoes is 810.

Example 3–7

 The number of cloudy days for the top 10 cloudiest cities is shown. Find the median.

209, 223, 211, 227, 213, 240, 240, 211, 229, 212

Source: National Climatic Data Center.

Solution

Arrange the data in order.

209, 211, 211, 212, 213, 223, 227, 229, 240, 240
$\uparrow$
Median

$$MD = \frac{213 + 223}{2} = 218$$

Hence, the median is 218 days.

Example 3–8

 Six customers purchased these numbers of magazines: 1, 7, 3, 2, 3, 4. Find the median.

Solution

1, 2, 3, 3, 4, 7 $MD = \dfrac{3 + 3}{2} = 3$
$\uparrow$
Median

Hence, the median number of magazines purchased is 3.

The Mode

The third measure of average is called the *mode*. The mode is the value that occurs most often in the data set. It is sometimes said to be the most typical case.

> The value that occurs most often in a data set is called the **mode.**

A data set that has only one value that occurs with the greatest frequency is said to be **unimodal.**

If a data set has two values that occur with the same greatest frequency, both values are considered to be the mode and the data set is said to be **bimodal.** If a data set has more than two values that occur with the same greatest frequency, each value is used as the mode, and the data set is said to be **multimodal.** When no data value occurs more than once, the data set is said to have *no mode.* A data set can have more than one mode or no mode at all. These situations will be shown in some of the examples that follow.

Example 3–9

 The following data represent the duration (in days) of U.S. Space Shuttle voyages for the years 1992–1994. Find the mode.

8, 9, 9, 14, 8, 8, 10, 7, 6, 9, 7, 8, 10, 14, 11, 8, 14, 11

Source: The Universal Almanac.

Solution

It is helpful to arrange the data in order, although it is not necessary.

6, 7, 7, 8, 8, 8, 8, 8, 9, 9, 9, 10, 10, 11, 11, 14, 14, 14

Since 8-day voyages occurred 5 times—a frequency larger than any other number—the mode for the data set is 8.

Example 3–10

Find the mode for the number of coal employees per county for 10 selected counties in southwestern Pennsylvania.

110, 731, 1031, 84, 20, 118, 1162, 1977, 103, 752

Source: Pittsburgh Tribune-Review.

Solution

Since each value occurs only once, there is no mode.
   Note: Do not say that the mode is zero. That would be incorrect, because in some data, such as temperature, zero can be an actual value.

Example 3–11

The data show the number of licensed nuclear reactors in the United States for a recent 15-year period. Find the mode.

Source: The World Almanac and Book of Facts.

104	104	104	104	104
107	109	109	109	110
109	111	112	111	109

Solution

Since the values 104 and 109 both occur 5 times, the modes are 104 and 109. The data set is said to be bimodal.

The mode for grouped data is the modal class. The **modal class** is the class with the largest frequency.

Example 3–12

Find the modal class for the frequency distribution of miles that 20 runners ran in one week, used in Example 2–7.

Class	Frequency	
5.5–10.5	1	
10.5–15.5	2	
15.5–20.5	3	
20.5–25.5	5	← Modal class
25.5–30.5	4	
30.5–35.5	3	
35.5–40.5	2	

Solution

The modal class is 20.5–25.5, since it has the largest frequency. Sometimes the midpoint of the class is used rather than the boundaries; hence, the mode could also be given as 23 miles per week.

The mode is the only measure of central tendency that can be used in finding the most typical case when the data are nominal or categorical.

Example 3–13

A survey showed this distribution for the number of students enrolled in each field. Find the mode.

Business	1425
Liberal arts	878
Computer science	632
Education	471
General studies	95

Solution

Since the category with the highest frequency is business, the most typical case is a business major.

An extremely high or extremely low data value in a data set can have a striking effect on the mean of the data set. These extreme values are called *outliers*. This is one reason why when analyzing a frequency distribution, you should be aware of any of these values. For the data set shown in Example 3–14, the mean, median, and mode can be quite different because of extreme values. A method for identifying outliers is given in Section 3–4.

Example 3–14

A small company consists of the owner, the manager, the salesperson, and two technicians, all of whose annual salaries are listed here. (Assume that this is the entire population.)

Staff	**Salary**
Owner	$50,000
Manager	20,000
Salesperson	12,000
Technician	9,000
Technician	9,000

Find the mean, median, and mode.

Solution

$$\mu = \frac{\Sigma X}{N} = \frac{50,000 + 20,000 + 12,000 + 9000 + 9000}{5} = \$20,000$$

Hence, the mean is $20,000, the median is $12,000, and the mode is $9000.

In Example 3–14, the mean is much higher than the median or the mode. This is so because the extremely high salary of the owner tends to raise the value of the mean. In this and similar situations, the median should be used as the measure of central tendency.

The Midrange

The *midrange* is a rough estimate of the middle. It is found by adding the lowest and highest values in the data set and dividing by 2. It is a very rough estimate of the average and can be affected by one extremely high or low value.

The **midrange** is defined as the sum of the lowest and highest values in the data set, divided by 2. The symbol MR is used for the midrange.

$$MR = \frac{\text{lowest value} + \text{highest value}}{2}$$

Example 3–15

In the last two winter seasons, the city of Brownsville, Minnesota, reported these numbers of water-line breaks per month. Find the midrange.

2, 3, 6, 8, 4, 1

Solution

$$MR = \frac{1 + 8}{2} = \frac{9}{2} = 4.5$$

Hence, the midrange is 4.5.

If the data set contains one extremely large value or one extremely small value, a higher or lower midrange value will result and may not be a typical description of the middle.

Example 3–16

Suppose the number of water-line breaks was as follows: 2, 3, 6, 16, 4, and 1. Find the midrange.

Solution

$$MR = \frac{1 + 16}{2} = \frac{17}{2} = 8.5$$

Hence, the midrange is 8.5. The value 8.5 is not typical of the average monthly number of breaks, since an excessively high number of breaks, 16, occurred in one month.

In statistics, several measures can be used for an average. The most common measures are the mean, median, mode, and midrange. Each has its own specific purpose and use. Exercises 39 through 41 show examples of other averages, such as the harmonic mean, the geometric mean, and the quadratic mean. Their applications are limited to specific areas, as shown in the exercises.

The Weighted Mean

Sometimes, one must find the mean of a data set in which not all values are equally represented. Consider the case of finding the average cost of a gallon of gasoline for three taxis. Suppose the drivers buy gasoline at three different service stations at a cost of $2.22, $2.53, and $2.63 per gallon. One might try to find the average by using the formula

$$\overline{X} = \frac{\Sigma X}{n}$$

$$= \frac{2.22 + 2.53 + 2.63}{3} = \frac{7.38}{3} = \$2.46$$

But not all drivers purchased the same number of gallons. Hence, to find the true average cost per gallon, one must take into consideration the number of gallons each driver purchased.

The type of mean that considers an additional factor is called the *weighted mean,* and it is used when the values are not all equally represented.

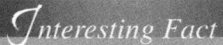

Find the **weighted mean** of a variable X by multiplying each value by its corresponding weight and dividing the sum of the products by the sum of the weights.

$$\overline{X} = \frac{w_1 X_1 + w_2 X_2 + \cdots + w_n X_n}{w_1 + w_2 + \cdots + w_n} = \frac{\Sigma w X}{\Sigma w}$$

where $w_1, w_2, \ldots, w_n$ are the weights and $X_1, X_2, \ldots, X_n$ are the values.

Example 3–17 shows how the weighted mean is used to compute a grade point average. Since courses vary in their credit value, the number of credits must be used as weights.

Example 3–17

A student received an A in English Composition I (3 credits), a C in Introduction to Psychology (3 credits), a B in Biology I (4 credits), and a D in Physical Education (2 credits). Assuming A = 4 grade points, B = 3 grade points, C = 2 grade points, D = 1 grade point, and F = 0 grade points, find the student's grade point average.

Solution

Course	Credits (w)	Grade (X)
English Composition I	3	A (4 points)
Introduction to Psychology	3	C (2 points)
Biology I	4	B (3 points)
Physical Education	2	D (1 point)

$$\overline{X} = \frac{\Sigma w X}{\Sigma w} = \frac{3 \cdot 4 + 3 \cdot 2 + 4 \cdot 3 + 2 \cdot 1}{3 + 3 + 4 + 2} = \frac{32}{12} = 2.7$$

The grade point average is 2.7.

Table 3–1 summarizes the measures of central tendency.

Table 3–1	Summary of Measures of Central Tendency	
Measure	**Definition**	**Symbol(s)**
Mean	Sum of values, divided by total number of values	$\mu, \overline{X}$
Median	Middle point in data set that has been ordered	MD
Mode	Most frequent data value	None
Midrange	Lowest value plus highest value, divided by 2	MR

Researchers and statisticians must know which measure of central tendency is being used and when to use each measure of central tendency. The properties and uses of the four measures of central tendency are summarized next.

Properties and Uses of Central Tendency

The Mean

1. One computes the mean by using all the values of the data.
2. The mean varies less than the median or mode when samples are taken from the same population and all three measures are computed for these samples.
3. The mean is used in computing other statistics, such as the variance.
4. The mean for the data set is unique and not necessarily one of the data values.
5. The mean cannot be computed for the data in a frequency distribution that has an open-ended class.
6. The mean is affected by extremely high or low values, called outliers, and may not be the appropriate average to use in these situations.

The Median

1. The median is used when one must find the center or middle value of a data set.
2. The median is used when one must determine whether the data values fall into the upper half or lower half of the distribution.
3. The median is used for an open-ended distribution.
4. The median is affected less than the mean by extremely high or extremely low values.

The Mode

1. The mode is used when the most typical case is desired.
2. The mode is the easiest average to compute.
3. The mode can be used when the data are nominal, such as religious preference, gender, or political affiliation.
4. The mode is not always unique. A data set can have more than one mode, or the mode may not exist for a data set.

The Midrange

1. The midrange is easy to compute.
2. The midrange gives the midpoint.
3. The midrange is affected by extremely high or low values in a data set.

Distribution Shapes

Frequency distributions can assume many shapes. The three most important shapes are positively skewed, symmetric, and negatively skewed. Figure 3–1 shows histograms of each.

In a **positively skewed** or **right-skewed distribution,** the majority of the data values fall to the left of the mean and cluster at the lower end of the distribution; the "tail" is to the right. Also, the mean is to the right of the median, and the mode is to the left of the median.

For example, if an instructor gave an examination and most of the students did poorly, their scores would tend to cluster on the left side of the distribution. A few high scores would constitute the tail of the distribution, which would be on the right side. Another example of a positively skewed distribution is the incomes of the population of the United States. Most of the incomes cluster about the low end of the distribution; those with high incomes are in the minority and are in the tail at the right of the distribution.

In a **symmetric distribution,** the data values are evenly distributed on both sides of the mean. In addition, when the distribution is unimodal, the mean, median, and mode are the same and are at the center of the distribution. Examples of symmetric distributions are IQ scores and heights of adult males.

Figure 3–1

Types of Distributions

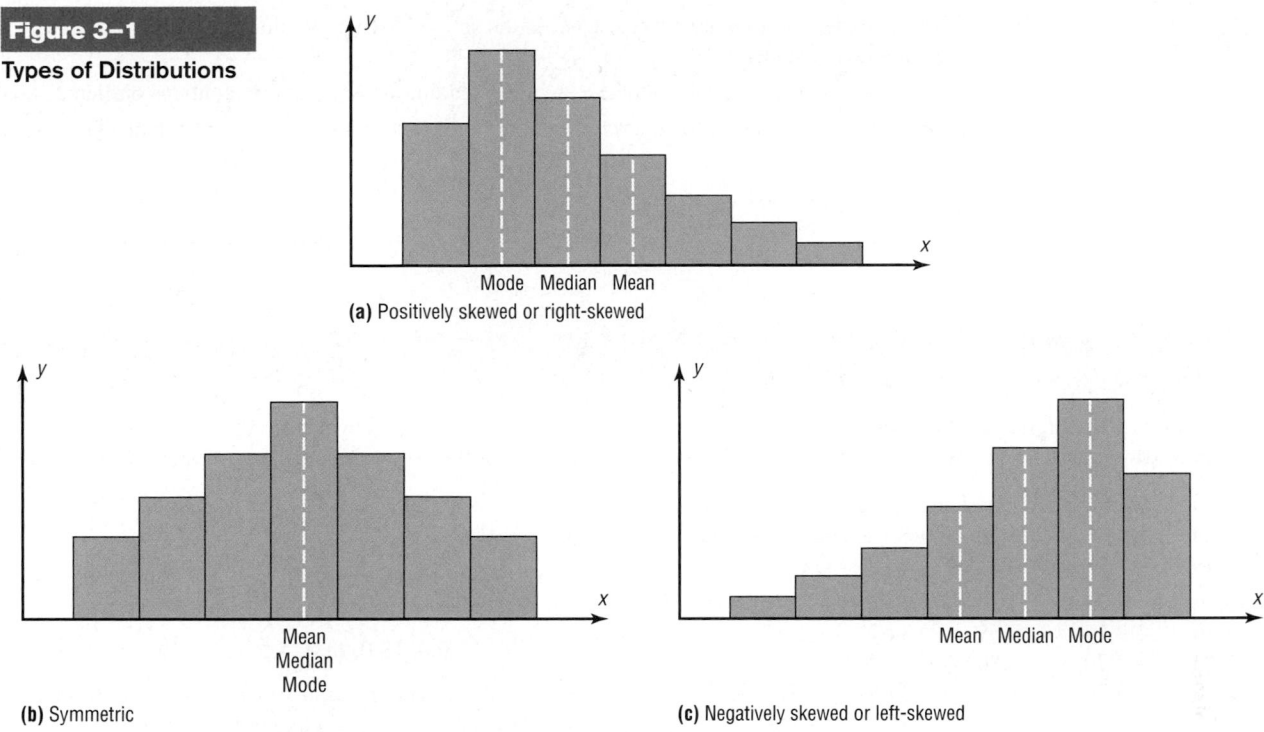

(a) Positively skewed or right-skewed

(b) Symmetric

(c) Negatively skewed or left-skewed

When the majority of the data values fall to the right of the mean and cluster at the upper end of the distribution, with the tail to the left, the distribution is said to be **negatively skewed** or **left-skewed.** Also, the mean is to the left of the median, and the mode is to the right of the median. As an example, a negatively skewed distribution results if the majority of students score very high on an instructor's examination. These scores will tend to cluster to the right of the distribution.

When a distribution is extremely skewed, the value of the mean will be pulled toward the tail, but the majority of the data values will be greater than the mean or less than the mean (depending on which way the data are skewed); hence, the median rather than the mean is a more appropriate measure of central tendency. An extremely skewed distribution can also affect other statistics.

A measure of skewness for a distribution is discussed in Exercise 48 in Section 3–3.

Applying the Concepts **3–2**

Teacher Salaries

The following data represent salaries (in dollars) from a school district in Greenwood, South Carolina.

10,000	11,000	11,000	12,500	14,300	17,500
18,000	16,600	19,200	21,560	16,400	107,000

1. First, assume you work for the school board in Greenwood and do not wish to raise taxes to increase salaries. Compute the mean, median, and mode and decide which one would best support your position to not raise salaries.

2. Second, assume you work for the teachers' union and want a raise for the teachers. Use the best measure of central tendency to support your position.

3. Explain how outliers can be used to support one or the other position.

4. If the salaries represented every teacher in the school district, would the averages be parameters or statistics?

5. Which measure of central tendency can be misleading when a data set contains outliers?

6. When one is comparing the measures of central tendency, does the distribution display any skewness? Explain.

See page 170 for the answers.

Exercises 3–2

For Exercises 1 through 8, find (*a*) the mean, (*b*) the median, (*c*) the mode, and (*d*) the midrange.

 1. The average undergraduate grade point average (GPA) for the 25 top-ranked medical schools is listed below.

3.80	3.77	3.70	3.74	3.70
3.86	3.76	3.68	3.67	3.57
3.83	3.70	3.80	3.74	3.67
3.78	3.74	3.73	3.65	3.66
3.75	3.64	3.78	3.73	3.64

Source: *U.S. News & World Report Best Graduate Schools.*

 2. The heights (in feet) of the 20 highest waterfalls in the world are shown here. (*Note:* The height of Niagara Falls is 182 feet!)

3212 2800 2625 2540 2499 2425 2307 2151 2123 2000
1904 1841 1650 1612 1536 1388 1215 1198 1182 1170

Source: *N.Y. Times Almanac.*

 3. The following data are the number of burglaries reported for a specific year for nine western Pennsylvania universities. Which measure of average might be the best in this case? Explain your answer.

61, 11, 1, 3, 2, 30, 18, 3, 7

Source: *Pittsburgh Post Gazette.*

4. For a recent year, the number of suspensions in a sample of public schools in Beaver County was 67, 12, 11, 92, and 13. The number of suspensions in a sample of public schools in Butler County was 56, 12, 18, and 21. Do you think that there is a difference in the averages?

Source: U.S. Department of Education.

5. A researcher claims that each year, there are an average of 300 victims of identity theft in major cities. Twelve were randomly selected, and the number of victims of identity theft in each city is shown. Can you conclude that the researcher was correct?

574	229	663	372	102	88
117	239	465	136	189	75

6. For a recent year, the worth (in billions of dollars) of a sample of the 10 wealthiest people under the age of 60 is shown:

14, 12, 48, 20, 18, 18, 12.6, 10.4, 7.3, 5.3

The worth of a sample of 10 of the wealthiest people age 60 and over is shown:

41.0, 13.7, 10.0, 18.0, 11.3, 7.6, 7, 18, 20, 60

Based on the averages, can you conclude that those 60 years old and over have a higher net worth?

Source: *Forbes* magazine.

 7. Twelve major earthquakes had Richter magnitudes shown here.

7.0, 6.2, 7.7, 8.0, 6.4, 6.2,
7.2, 5.4, 6.4, 6.5, 7.2, 5.4

Which would you consider the best measure of average?

Source: *The Universal Almanac.*

8. The data shown are the total compensation (in millions of dollars) for the 50 top-paid CEOs for a recent year. Compare the averages, and state which one you think is the best measure.

17.5	18.0	36.8	31.7	31.7
17.3	24.3	47.7	38.5	17.0
23.7	16.5	25.1	17.4	18.0
37.6	19.7	21.4	28.6	21.6
19.3	20.0	16.9	25.2	19.8
25.0	17.2	20.4	20.1	29.1
19.1	25.2	23.2	25.9	24.0
41.7	24.0	16.8	26.8	31.4
16.9	17.2	24.1	35.2	19.1
22.9	18.2	25.4	35.4	25.5

Source: *USA TODAY.*

9. Find the (*a*) mean, (*b*) median, (*c*) mode, and (*d*) midrange for the data in Exercise 17 in Section 2–2. Is the distribution symmetric or skewed? Use the individual data values.

10. Find the (a) mean, (b) median, (c) mode, and (d) midrange for the distances of the home runs for McGwire and Sosa, using the data in Exercise 18 in Section 2–2.

    Compare the means. Decide if the means are approximately equal or if one of the players is hitting longer home runs. Use the individual data values.

11. These data represent the number of traffic fatalities for two specific years for 27 selected states. Find the (a) mean, (b) median, (c) mode, and (d) midrange for each data set. Are the four measures of average for fatalities for year 1 the same as those for year 2? (The data in this exercise will be used in Exercise 15 in Section 3–3.)

Year 1			Year 2		
1113	1488	868	1100	260	205
1031	262	1109	970	1430	300
4192	1586	215	4040	460	350
645	527	254	620	480	485
121	442	313	125	405	85
2805	444	485	2805	690	1430
900	653	170	1555	1160	70
74	1480	69	180	3360	325
158	3181	326	875	705	145

Source: *USA TODAY.*

For Exercises 12 through 21, find the (a) mean and (b) modal class.

12. For 108 randomly selected college students, this exam score frequency distribution was obtained. (The data in this exercise will be used in Exercise 18 in Section 3–3.)

Class limits	Frequency
90–98	6
99–107	22
108–116	43
117–125	28
126–134	9

13. The scores for the LPGA—Giant Eagle were

Score	Frequency
202–204	2
205–207	7
208–210	16
211–213	26
214–216	18
217–219	4

Source: www.LPGA.com

14. Thirty automobiles were tested for fuel efficiency (in miles per gallon). This frequency distribution was obtained. (The data in this exercise will be used in Exercise 20 in Section 3–3.)

Class boundaries	Frequency
7.5–12.5	3
12.5–17.5	5
17.5–22.5	15
22.5–27.5	5
27.5–32.5	2

15. The data show the number of murders in 25 selected cities in a given state.

Number	Frequency
34–96	13
97–159	2
160–222	0
223–285	5
286–348	1
349–411	1
412–474	0
475–537	1
538–600	2

Do you think that the mean is the best measure of average for these data? Explain your answer. (The information in the exercise will be used in Exercise 21 in Section 3–3.)

16. Find the mean and modal class for the two frequency distributions in Exercises 8 and 18 in Section 2–3. Are the "average" reactions the same? Explain your answer.

17. Eighty randomly selected lightbulbs were tested to determine their lifetimes (in hours). This frequency distribution was obtained. (The data in this exercise will be used in Exercise 23 in Section 3–3.)

Class boundaries	Frequency
52.5–63.5	6
63.5–74.5	12
74.5–85.5	25
85.5–96.5	18
96.5–107.5	14
107.5–118.5	5

18. These data represent the net worth (in millions of dollars) of 45 national corporations.

Class limits	Frequency
10–20	2
21–31	8
32–42	15
43–53	7
54–64	10
65–75	3

year of service, the average percentage raise per year is not 15 but 14.89%, as shown.

$$GM = \sqrt{(1.2)(1.1)} = 1.1489$$

or

$$GM = \sqrt{(120)(110)} = 114.89\%$$

His salary is 120% at the end of the first year and 110% at the end of the second year. This is equivalent to an average of 14.89%, since 114.89% − 100% = 14.89%.

This answer can also be shown by assuming that the person makes $10,000 to start and receives two raises of 20% and 10%.

$$Raise\ 1 = 10,000 \cdot 20\% = \$2000$$
$$Raise\ 2 = 12,000 \cdot 10\% = \$1200$$

His total salary raise is $3200. This total is equivalent to

$$\$10,000 \cdot 14.89\% = \$1489.00$$
$$\underline{\$11,489 \cdot 14.89\% = \quad 1710.71}$$
$$\$3199.71 \approx \$3200$$

Find the geometric mean of each of these.

a. The growth rates of the Living Life Insurance Corporation for the past 3 years were 35, 24, and 18%.

b. A person received these percentage raises in salary over a 4-year period: 8, 6, 4, and 5%.

c. A stock increased each year for 5 years at these percentages: 10, 8, 12, 9, and 3%.

d. The price increases, in percentages, for the cost of food in a specific geographic region for the past 3 years were 1, 3, and 5.5%.

41. A useful mean in the physical sciences (such as voltage) is the *quadratic mean* (QM), which is found by taking the square root of the average of the squares of each value. The formula is

$$QM = \sqrt{\frac{\Sigma X^2}{n}}$$

The quadratic mean of 3, 5, 6, and 10 is

$$QM = \sqrt{\frac{3^2 + 5^2 + 6^2 + 10^2}{4}}$$
$$= \sqrt{42.5} = 6.52$$

Find the quadratic mean of 8, 6, 3, 5, and 4.

42. An approximate median can be found for data that have been grouped into a frequency distribution. First it is necessary to find the median class. This is the class that contains the median value. That is the $n/2$ data value. Then it is assumed that the data values are evenly distributed throughout the median class. The formula is

$$MD = \frac{(n/2) - cf}{f}(w) + L_m$$

where

n = sum of frequencies
cf = cumulative frequency of class immediately preceding the median class
w = width of median class
f = frequency of median class
L_m = lower boundary of median class

Using this formula, find the median for data in the frequency distribution of Exercise 15.

Technology *Step by Step*

Excel
Step by Step

Finding the Central Tendency

Example XL3–1

To find the mean, mode, and median of a data set:

1. Enter the numbers in a range of cells (here shown as the numbers in cells A2 to A16). We use the data from Example 3–11 on licensed nuclear reactors:

   104, 104, 104, 104, 104, 107, 109, 109, 109, 110, 109, 111, 112, 111, 109

2. For the mean, enter =AVERAGE (A2:A16) in a blank cell.

3. For the mode, enter =MODE (A2:A16) in a blank cell.

4. For the median, enter =MEDIAN (A2:A16) in a blank cell.

These three functions are available from the standard toolbar by clicking the f_x icon and scrolling down the list of statistical functions. *Note:* For distributions that are *bimodal,* like this one, the Excel MODE function reports the first mode only. A better practice is to use the Histogram routine from the Data Analysis Add-in, which reports actual counts in a table.

	A	B	C	D
1	**Stopping distance**			
2	104		107.7333	**mean**
3	104		104	**mode**
4	104		109	**median**
5	104			
6	104			
7	107			
8	109			
9	109			
10	109			
11	110			
12	109			
13	111			
14	112			
15	111			
16	109			

3–3 Measures of Variation

In statistics, to describe the data set accurately, statisticians must know more than the measures of central tendency. Consider Example 3–18.

Example 3–18

Objective 2

Describe data using measures of variation, such as the range, variance, and standard deviation.

A testing lab wishes to test two experimental brands of outdoor paint to see how long each will last before fading. The testing lab makes 6 gallons of each paint to test. Since different chemical agents are added to each group and only six cans are involved, these two groups constitute two small populations. The results (in months) are shown. Find the mean of each group.

Brand A	Brand B
10	35
60	45
50	30
30	35
40	40
20	25

Solution

The mean for brand A is

$$\mu = \frac{\Sigma X}{N} = \frac{210}{6} = 35 \text{ months}$$

The mean for brand B is

$$\mu = \frac{\Sigma X}{N} = \frac{210}{6} = 35 \text{ months}$$

Figure 3–2

Examining Data Sets Graphically

Variation of paint (in months)

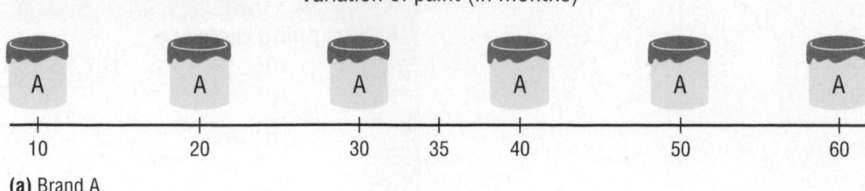

(a) Brand A

Variation of paint (in months)

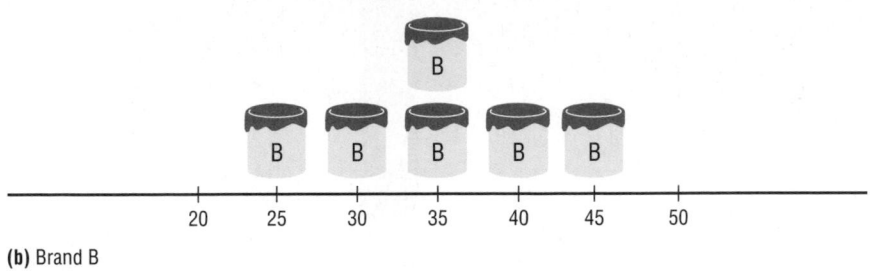

(b) Brand B

Since the means are equal in Example 3–18, one might conclude that both brands of paint last equally well. However, when the data sets are examined graphically, a somewhat different conclusion might be drawn. See Figure 3–2.

As Figure 3–2 shows, even though the means are the same for both brands, the spread, or variation, is quite different. Figure 3–2 shows that brand B performs more consistently; it is less variable. For the spread or variability of a data set, three measures are commonly used: *range, variance,* and *standard deviation.* Each measure will be discussed in this section.

Range

The range is the simplest of the three measures and is defined now.

The **range** is the highest value minus the lowest value. The symbol R is used for the range.

$$R = \text{highest value} - \text{lowest value}$$

Example 3–19

Find the ranges for the paints in Example 3–18.

Solution

For brand A, the range is

$$R = 60 - 10 = 50 \text{ months}$$

For brand B, the range is

$$R = 45 - 25 = 20 \text{ months}$$

Make sure the range is given as a single number.

The range for brand A shows that 50 months separate the largest data value from the smallest data value. For brand B, 20 months separate the largest data value from the smallest data value, which is less than one-half of brand A's range.

One extremely high or one extremely low data value can affect the range markedly, as shown in Example 3–20.

Example 3–20

The salaries for the staff of the XYZ Manufacturing Co. are shown here. Find the range.

Staff	Salary
Owner	$100,000
Manager	40,000
Sales representative	30,000
Workers	25,000
	15,000
	18,000

Solution

The range is $R = \$100,000 - \$15,000 = \$85,000$.

Since the owner's salary is included in the data for Example 3–20, the range is a large number. To have a more meaningful statistic to measure the variability, statisticians use measures called the *variance* and *standard deviation.*

Population Variance and Standard Deviation

Before the variance and standard deviation are defined formally, the computational procedure will be shown, since the definition is derived from the procedure.

Rounding Rule for the Standard Deviation The rounding rule for the standard deviation is the same as that for the mean. The final answer should be rounded to one more decimal place than that of the original data.

Example 3–21

Find the variance and standard deviation for the data set for brand A paint in Example 3–18.

10, 60, 50, 30, 40, 20

Solution

Step 1   Find the mean for the data.

$$\mu = \frac{\Sigma X}{N} = \frac{10 + 60 + 50 + 30 + 40 + 20}{6} = \frac{210}{6} = 35$$

Step 2   Subtract the mean from each data value.

$$10 - 35 = -25 \qquad 50 - 35 = +15 \qquad 40 - 35 = +5$$
$$60 - 35 = +25 \qquad 30 - 35 = -5 \qquad 20 - 35 = -15$$

Step 3   Square each result.

$$(-25)^2 = 625 \qquad (+15)^2 = 225 \qquad (+5)^2 = 25$$
$$(+25)^2 = 625 \qquad (-5)^2 = 25 \qquad (-15)^2 = 225$$

Step 4   Find the sum of the squares.

$$625 + 625 + 225 + 25 + 25 + 225 = 1750$$

Step 5   Divide the sum by N to get the variance.

Variance $= 1750 \div 6 = 291.7$

Step 6   Take the square root of the variance to get the standard deviation. Hence, the standard deviation equals $\sqrt{291.7}$, or 17.1. It is helpful to make a table.

A Values (X)	B $X - \mu$	C $(X - \mu)^2$
10	-25	625
60	$+25$	625
50	$+15$	225
30	-5	25
40	$+5$	25
20	-15	225
		1750

Column A contains the raw data X. Column B contains the differences $X - \mu$ obtained in step 2. Column C contains the squares of the differences obtained in step 3.

The preceding computational procedure reveals several things. First, the square root of the variance gives the standard deviation; and vice versa, squaring the standard deviation gives the variance. Second, the variance is actually the average of the square of the distance that each value is from the mean. Therefore, if the values are near the mean, the variance will be small. In contrast, if the values are far from the mean, the variance will be large.

One might wonder why the squared distances are used instead of the actual distances. One reason is that the sum of the distances will always be zero. To verify this result for a specific case, add the values in column B of the table in Example 3–21. When each value is squared, the negative signs are eliminated.

Finally, why is it necessary to take the square root? The reason is that since the distances were squared, the units of the resultant numbers are the squares of the units of the original raw data. Finding the square root of the variance puts the standard deviation in the same units as the raw data.

When you are finding the square root, always use its positive or principal value, since the variance and standard deviation of a data set can never be negative.

The **variance** is the average of the squares of the distance each value is from the mean. The symbol for the population variance is σ^2 (σ is the Greek lowercase letter sigma). The formula for the population variance is

$$\sigma^2 = \frac{\Sigma(X - \mu)^2}{N}$$

where
   X = individual value
   μ = population mean
   N = population size

The **standard deviation** is the square root of the variance. The symbol for the population standard deviation is σ.
   The corresponding formula for the population standard deviation is

$$\sigma = \sqrt{\sigma^2} = \sqrt{\frac{\Sigma(X - \mu)^2}{N}}$$

Example 3–22

Find the variance and standard deviation for brand B paint data in Example 3–18. The months were

35, 45, 30, 35, 40, 25

Solution

Step 1   Find the mean.

$$\mu = \frac{\Sigma X}{N} = \frac{35 + 45 + 30 + 35 + 40 + 25}{6} = \frac{210}{6} = 35$$

Step 2   Subtract the mean from each value, and place the result in column B of the table.

Step 3   Square each result and place the squares in column C of the table.

A	B	C
X	$X - \mu$	$(X - \mu)^2$
35	0	0
45	10	100
30	−5	25
35	0	0
40	5	25
25	−10	100

Step 4   Find the sum of the squares in column C.

$$\Sigma(X - \mu)^2 = 0 + 100 + 25 + 0 + 25 + 100 = 250$$

Step 5   Divide the sum by N to get the variance.

$$\sigma^2 = \frac{\Sigma(X - \mu)^2}{N} = \frac{250}{6} = 41.7$$

Step 6   Take the square root to get the standard deviation.

$$\sigma = \sqrt{\frac{\Sigma(X - \mu)^2}{N}} = \sqrt{41.7} = 6.5$$

Hence, the standard deviation is 6.5.

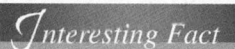

Interesting Fact

Each person receives on average 598 pieces of mail per year.

Since the standard deviation of brand A is 17.1 (see Example 3–21) and the standard deviation of brand B is 6.5, the data are more variable for brand A. *In summary, when the means are equal, the larger the variance or standard deviation is, the more variable the data are.*

Sample Variance and Standard Deviation

When computing the variance for a sample, one might expect the following expression to be used:

$$\frac{\Sigma(X - \overline{X})^2}{n}$$

where $\overline{X}$ is the sample mean and n is the sample size. *This formula is not usually used, however, since in most cases the purpose of calculating the statistic is to estimate the corresponding parameter.* For example, the sample mean $\overline{X}$ is used to estimate the population mean μ. The expression

$$\frac{\Sigma(X - \overline{X})^2}{n}$$

does not give the best estimate of the population variance because when the population is large and the sample is small (usually less than 30), the variance computed by this formula usually underestimates the population variance. Therefore, instead of dividing by n, find the variance of the sample by dividing by $n - 1$, giving a slightly larger value and an *unbiased* estimate of the population variance.

The formula for the sample variance, denoted by s^2, is

$$s^2 = \frac{\Sigma(X - \overline{X})^2}{n - 1}$$

where
$\overline{X}$ = sample mean
n = sample size

To find the standard deviation of a sample, one must take the square root of the sample variance, which was found by using the preceding formula.

Formula for the Sample Standard Deviation

The standard deviation of a sample (denoted by s) is

$$s = \sqrt{s^2} = \sqrt{\frac{\Sigma(X - \overline{X})^2}{n - 1}}$$

where
X = individual value
$\overline{X}$ = sample mean
n = sample size

Shortcut formulas for computing the variance and standard deviation are presented next and will be used in the remainder of the chapter and in the exercises. These formulas are mathematically equivalent to the preceding formulas and do not involve using the mean. They save time when repeated subtracting and squaring occur in the original formulas. They are also more accurate when the mean has been rounded.

Shortcut or Computational Formulas for s^2 and s

The shortcut formulas for computing the variance and standard deviation for data obtained from samples are as follows.

Variance	**Standard deviation**
$s^2 = \dfrac{\Sigma X^2 - [(\Sigma X)^2/n]}{n - 1}$	$s = \sqrt{\dfrac{\Sigma X^2 - [(\Sigma X)^2/n]}{n - 1}}$

Examples 3–23 and 3–24 explain how to use the shortcut formulas.

Example 3–23

Find the sample variance and standard deviation for the amount of European auto sales for a sample of 6 years shown. The data are in millions of dollars.

11.2, 11.9, 12.0, 12.8, 13.4, 14.3

Source: USA TODAY.

Solution

Step 1 Find the sum of the values.

$\Sigma X = 11.2 + 11.9 + 12.0 + 12.8 + 13.4 + 14.3 = 75.6$

Step 2 Square each value and find the sum.

$\Sigma X^2 = 11.2^2 + 11.9^2 + 12.0^2 + 12.8^2 + 13.4^2 + 14.3^2 = 958.94$

Step 3 Substitute in the formulas and solve.

$$s^2 = \frac{\Sigma X^2 - [(\Sigma X)^2/n]}{n-1} = \frac{958.94 - [(75.6)^2/6]}{5}$$

$$= 1.28$$

The variance of the sample is 1.28.

$$s = \sqrt{1.28} = 1.13$$

Hence, the sample standard deviation is 1.13.

Note that ΣX^2 is not the same as $(\Sigma X)^2$. The notation ΣX^2 means to square the values first, then sum; $(\Sigma X)^2$ means to sum the values first, then square the sum.

Variance and Standard Deviation for Grouped Data

The procedure for finding the variance and standard deviation for grouped data is similar to that for finding the mean for grouped data, and it uses the midpoints of each class.

Example 3–24

Find the variance and the standard deviation for the frequency distribution of the data in Example 2–7. The data represent the number of miles that 20 runners ran during one week.

Class	Frequency	Midpoint
5.5–10.5	1	8
10.5–15.5	2	13
15.5–20.5	3	18
20.5–25.5	5	23
25.5–30.5	4	28
30.5–35.5	3	33
35.5–40.5	2	38

Solution

Step 1   Make a table as shown, and find the midpoint of each class.

A Class	B Frequency (f)	C Midpoint (X_m)	D $f \cdot X_m$	E $f \cdot X_m^2$
5.5–10.5	1	8		
10.5–15.5	2	13		
15.5–20.5	3	18		
20.5–25.5	5	23		
25.5–30.5	4	28		
30.5–35.5	3	33		
35.5–40.5	2	38		

Step 2   Multiply the frequency by the midpoint for each class, and place the products in column D.

$$1 \cdot 8 = 8 \qquad 2 \cdot 13 = 26 \qquad \ldots \qquad 2 \cdot 38 = 76$$

Step 3   Multiply the frequency by the square of the midpoint, and place the products in column E.

$$1 \cdot 8^2 = 64 \qquad 2 \cdot 13^2 = 338 \qquad \ldots \qquad 2 \cdot 38^2 = 2888$$

Step 4   Find the sums of columns B, D, and E. The sum of column B is n, the sum of column D is $\Sigma f \cdot X_m$, and the sum of column E is $\Sigma f \cdot X_m^2$. The completed table is shown.

A Class	B Frequency	C Midpoint	D $f \cdot X_m$	E $f \cdot X_m^2$
5.5–10.5	1	8	8	64
10.5–15.5	2	13	26	338
15.5–20.5	3	18	54	972
20.5–25.5	5	23	115	2,645
25.5–30.5	4	28	112	3,136
30.5–35.5	3	33	99	3,267
35.5–40.5	2	38	76	2,888
	$n = 20$		$\Sigma f \cdot X_m = 490$	$\Sigma f \cdot X_m^2 = 13{,}310$

Step 5   Substitute in the formula and solve for s^2 to get the variance.

$$s^2 = \frac{\Sigma f \cdot X_m^2 - [(\Sigma f \cdot X_m)^2/n]}{n-1}$$

$$= \frac{13{,}310 - [(490)^2/20]}{20-1} = 68.7$$

Step 6   Take the square root to get the standard deviation.

$$s = \sqrt{68.7} = 8.3$$

Unusual Stat

At birth men outnumber women by 2%. By age 25, the number of men living is about equal to the number of women living. By age 65, there are 14% more women living than men.

Be sure to use the number found in the sum of column B (i.e., the sum of the frequencies) for n. Do not use the number of classes.

The steps for finding the variance and standard deviation for grouped data are summarized in this Procedure Table.

Procedure Table

Finding the Sample Variance and Standard Deviation for Grouped Data

Step 1   Make a table as shown, and find the midpoint of each class.

A	B	C	D	E
Class	**Frequency**	**Midpoint**	$f \cdot X_m$	$f \cdot X_m^2$

Step 2   Multiply the frequency by the midpoint for each class, and place the products in column D.

Step 3   Multiply the frequency by the square of the midpoint, and place the products in column E.

Step 4   Find the sums of columns B, D, and E. (The sum of column B is n. The sum of column D is $\Sigma f \cdot X_m$. The sum of column E is $\Sigma f \cdot X_m^2$.)

Step 5   Substitute in the formula and solve to get the variance.

$$s^2 = \frac{\Sigma f \cdot X_m^2 - [(\Sigma f \cdot X_m)^2/n]}{n-1}$$

Step 6   Take the square root to get the standard deviation.

The three measures of variation are summarized in Table 3–2.

Table 3–2	Summary of Measures of Variation	
Measure	**Definition**	**Symbol(s)**
Range	Distance between highest value and lowest value	R
Variance	Average of the squares of the distance that each value is from the mean	σ^2, s^2
Standard deviation	Square root of the variance	σ, s

Uses of the Variance and Standard Deviation

1. As previously stated, variances and standard deviations can be used to determine the spread of the data. If the variance or standard deviation is large, the data are more dispersed. This information is useful in comparing two (or more) data sets to determine which is more (most) variable.

2. The measures of variance and standard deviation are used to determine the consistency of a variable. For example, in the manufacture of fittings, such as nuts and bolts, the variation in the diameters must be small, or the parts will not fit together.

3. The variance and standard deviation are used to determine the number of data values that fall within a specified interval in a distribution. For example, Chebyshev's theorem (explained later) shows that, for any distribution, at least 75% of the data values will fall within 2 standard deviations of the mean.

4. Finally, the variance and standard deviation are used quite often in inferential statistics. These uses will be shown in later chapters of this textbook.

Coefficient of Variation

Whenever two samples have the same units of measure, the variance and standard deviation for each can be compared directly. For example, suppose an automobile dealer wanted to compare the standard deviation of miles driven for the cars she received as trade-ins on new cars. She found that for a specific year, the standard deviation for Buicks was 422 miles and the standard deviation for Cadillacs was 350 miles. She could say that the variation in mileage was greater in the Buicks. But what if a manager wanted to compare the standard deviations of two different variables, such as the number of sales per salesperson over a 3-month period and the commissions made by these salespeople?

A statistic that allows one to compare standard deviations when the units are different, as in this example, is called the *coefficient of variation*.

The **coefficient of variation,** denoted by CVar, is the standard deviation divided by the mean. The result is expressed as a percentage.

For samples,

$$\text{CVar} = \frac{s}{\overline{X}} \cdot 100\%$$

For populations,

$$\text{CVar} = \frac{\sigma}{\mu} \cdot 100\%$$

Example 3–25

The mean of the number of sales of cars over a 3-month period is 87, and the standard deviation is 5. The mean of the commissions is $5225, and the standard deviation is $773. Compare the variations of the two.

Solution

The coefficients of variation are

$$\text{CVar} = \frac{s}{\overline{X}} = \frac{5}{87} \cdot 100\% = 5.7\% \qquad \text{sales}$$

$$\text{CVar} = \frac{773}{5225} \cdot 100\% = 14.8\% \qquad \text{commissions}$$

Since the coefficient of variation is larger for commissions, the commissions are more variable than the sales.

Example 3–26

The mean for the number of pages of a sample of women's fitness magazines is 132, with a variance of 23; the mean for the number of advertisements of a sample of women's fitness magazines is 182, with a variance of 62. Compare the variations.

Solution

The coefficients of variation are

$$\text{CVar} = \frac{\sqrt{23}}{132} \cdot 100\% = 3.6\% \qquad \text{pages}$$

$$\text{CVar} = \frac{\sqrt{62}}{182} \cdot 100\% = 4.3\% \qquad \text{advertisements}$$

The number of advertisements is more variable than the number of pages since the coefficient of variation is larger for advertisements.

Range Rule of Thumb

The range can be used to approximate the standard deviation. The approximation is called the **range rule of thumb.**

The Range Rule of Thumb
A rough estimate of the standard deviation is $$s \approx \frac{\text{range}}{4}$$

In other words, if the range is divided by 4, an approximate value for the standard deviation is obtained. For example, the standard deviation for the data set 5, 8, 8, 9, 10, 12, and 13 is 2.7, and the range is $13 - 5 = 8$. The range rule of thumb is $s \approx 2$. The range rule of thumb in this case underestimates the standard deviation somewhat; however, it is in the ballpark.

A note of caution should be mentioned here. The range rule of thumb is only an *approximation* and should be used when the distribution of data values is unimodal and roughly symmetric.

The range rule of thumb can be used to estimate the largest and smallest data values of a data set. The smallest data value will be approximately 2 standard deviations below the mean, and the largest data value will be approximately 2 standard deviations above the mean of the data set. The mean for the previous data set is 9.3; hence,

$$\text{Smallest data value} = \overline{X} - 2s = 9.3 - 2(2.8) = 3.7$$

$$\text{Largest data value} = \overline{X} + 2s = 9.3 + 2(2.8) = 14.9$$

Notice that the smallest data value was 5, and the largest data value was 13. Again, these are rough approximations. For many data sets, almost all data values will fall within 2 standard deviations of the mean. Better approximations can be obtained by using Chebyshev's theorem and the empirical rule. These are explained next.

Chebyshev's Theorem

As stated previously, the variance and standard deviation of a variable can be used to determine the spread, or dispersion, of a variable. That is, the larger the variance or standard deviation, the more the data values are dispersed. For example, if two variables measured in the same units have the same mean, say, 70, and variable 1 has a standard deviation of 1.5 while variable 2 has a standard deviation of 10, then the data for variable 2 will be more spread out than the data for variable 1. *Chebyshev's theorem,* developed by the Russian mathematician Chebyshev (1821–1894), specifies the proportions of the spread in terms of the standard deviation.

Chebyshev's theorem   The proportion of values from a data set that will fall within k standard deviations of the mean will be at least $1 - 1/k^2$, where k is a number greater than 1 (k is not necessarily an integer).

This theorem states that at least three-fourths, or 75%, of the data values will fall within 2 standard deviations of the mean of the data set. This result is found by substituting $k = 2$ in the expression.

$$1 - \frac{1}{k^2} \qquad \text{or} \qquad 1 - \frac{1}{2^2} = 1 - \frac{1}{4} = \frac{3}{4} = 75\%$$

For the example in which variable 1 has a mean of 70 and a standard deviation of 1.5, at least three-fourths, or 75%, of the data values fall between 67 and 73. These values are found by adding 2 standard deviations to the mean and subtracting 2 standard deviations from the mean, as shown:

$$70 + 2(1.5) = 70 + 3 = 73$$

and

$$70 - 2(1.5) = 70 - 3 = 67$$

For variable 2, at least three-fourths, or 75%, of the data values fall between 50 and 90. Again, these values are found by adding and subtracting, respectively, 2 standard deviations to and from the mean.

$$70 + 2(10) = 70 + 20 = 90$$

and

$$70 - 2(10) = 70 - 20 = 50$$

Furthermore, the theorem states that at least eight-ninths, or 88.89%, of the data values will fall within 3 standard deviations of the mean. This result is found by letting $k = 3$ and substituting in the expression.

$$1 - \frac{1}{k^2} \quad \text{or} \quad 1 - \frac{1}{3^2} = 1 - \frac{1}{9} = \frac{8}{9} = 88.89\%$$

For variable 1, at least eight-ninths, or 88.89%, of the data values fall between 65.5 and 74.5, since

$$70 + 3(1.5) = 70 + 4.5 = 74.5$$

and

$$70 - 3(1.5) = 70 - 4.5 = 65.5$$

For variable 2, at least eight-ninths, or 88.89%, of the data values fall between 40 and 100.

This theorem can be applied to any distribution regardless of its shape (see Figure 3–3).

Examples 3–27 and 3–28 illustrate the application of Chebyshev's theorem.

Figure 3–3

Chebyshev's Theorem

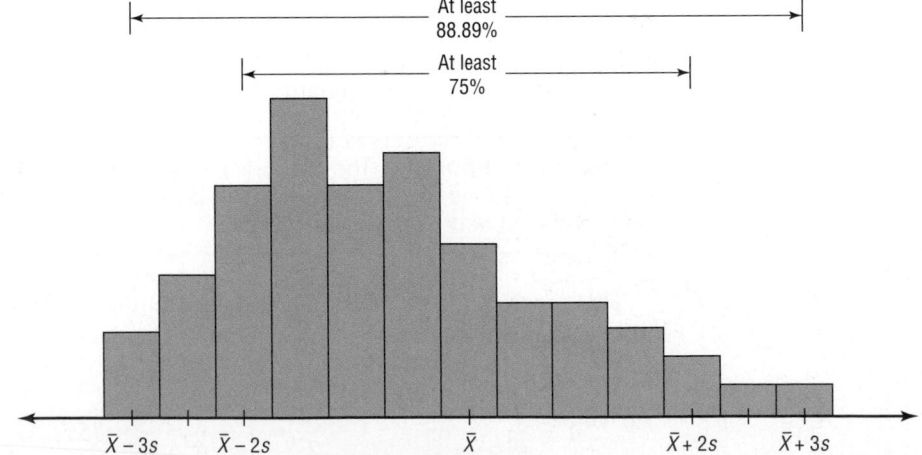

Example 3–27	The mean price of houses in a certain neighborhood is $50,000, and the standard deviation is $10,000. Find the price range for which at least 75% of the houses will sell.

Solution

Chebyshev's theorem states that three-fourths, or 75%, of the data values will fall within 2 standard deviations of the mean. Thus,

$$\$50{,}000 + 2(\$10{,}000) = \$50{,}000 + \$20{,}000 = \$70{,}000$$

and

$$\$50{,}000 - 2(\$10{,}000) = \$50{,}000 - \$20{,}000 = \$30{,}000$$

Hence, at least 75% of all homes sold in the area will have a price range from $30,000 to $70,000.

Chebyshev's theorem can be used to find the minimum percentage of data values that will fall between any two given values. The procedure is shown in Example 3–28.

Example 3–28	A survey of local companies found that the mean amount of travel allowance for executives was $0.25 per mile. The standard deviation was $0.02. Using Chebyshev's theorem, find the minimum percentage of the data values that will fall between $0.20 and $0.30.

Solution

Step 1   Subtract the mean from the larger value.

$$\$0.30 - \$0.25 = \$0.05$$

Step 2   Divide the difference by the standard deviation to get k.

$$k = \frac{0.05}{0.02} = 2.5$$

Step 3   Use Chebyshev's theorem to find the percentage.

$$1 - \frac{1}{k^2} = 1 - \frac{1}{2.5^2} = 1 - \frac{1}{6.25} = 1 - 0.16 = 0.84 \qquad \text{or} \qquad 84\%$$

Hence, at least 84% of the data values will fall between $0.20 and $0.30.

The Empirical (Normal) Rule

Chebyshev's theorem applies to any distribution regardless of its shape. However, when a distribution is *bell-shaped* (or what is called *normal*), the following statements, which make up the **empirical rule,** are true.

Approximately 68% of the data values will fall within 1 standard deviation of the mean.

Approximately 95% of the data values will fall within 2 standard deviations of the mean.

Approximately 99.7% of the data values will fall within 3 standard deviations of the mean.

Figure 3–4

The Empirical Rule

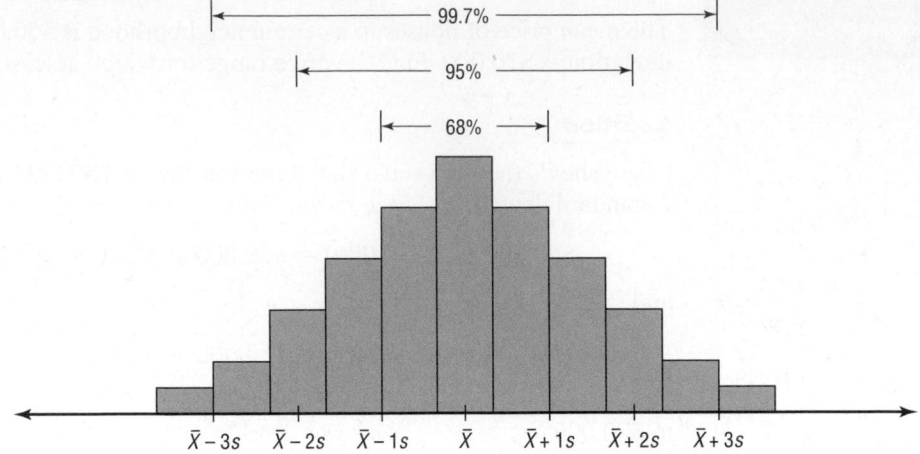

For example, suppose that the scores on a national achievement exam have a mean of 480 and a standard deviation of 90. If these scores are normally distributed, then approximately 68% will fall between 390 and 570 (480 + 90 = 570 and 480 − 90 = 390). Approximately 95% of the scores will fall between 300 and 660 (480 + 2 · 90 = 660 and 480 − 2 · 90 = 300). Approximately 99.7% will fall between 210 and 750 (480 + 3 · 90 = 750 and 480 − 3 · 90 = 210). See Figure 3–4. (The empirical rule is explained in greater detail in Chapter 7.)

Applying the Concepts **3–3**

Blood Pressure

The table lists means and standard deviations. The mean is the number before the plus/minus, and the standard deviation is the number after the plus/minus. The results are from a study attempting to find the average blood pressure of older adults. Use the results to answer the questions.

	Normotensive		Hypertensive	
	Men ($n = 1200$)	**Women** ($n = 1400$)	**Men** ($n = 1100$)	**Women** ($n = 1300$)
Age	55 ± 10	55 ± 10	60 ± 10	64 ± 10
Blood pressure (mm Hg)				
Systolic	123 ± 9	121 ± 11	153 ± 17	156 ± 20
Diastolic	78 ± 7	76 ± 7	91 ± 10	88 ± 10

1. Apply Chebyshev's theorem to the systolic blood pressure of normotensive men. At least how many of the men in the study fall within 1 standard deviation of the mean?

2. At least how many of those men in the study fall within 2 standard deviations of the mean?

Assume that blood pressure is normally distributed among older adults. Answer the following questions, using the empirical rule instead of Chebyshev's theorem.

3. Give ranges for the diastolic blood pressure (normotensive and hypertensive) of older women.

4. Do the normotensive, male, systolic blood pressure ranges overlap with the hypertensive, male, systolic, blood pressure ranges?

See page 170 for the answers.

Exercises 3–3

1. What is the relationship between the variance and the standard deviation?

2. Why might the range *not* be the best estimate of variability?

3. What are the symbols used to represent the population variance and standard deviation?

4. What are the symbols used to represent the sample variance and standard deviation?

5. Why is the unbiased estimator of variance used?

 6. The three data sets have the same mean and range, but is the variation the same? Prove your answer by computing the standard deviation. Assume the data were obtained from samples.

   a. 5, 7, 9, 11, 13, 15, 17
   b. 5, 6, 7, 11, 15, 16, 17
   c. 5, 5, 5, 11, 17, 17, 17

For Exercises 7–13, find the range, variance, and standard deviation. Assume the data represent samples, and use the shortcut formula for the unbiased estimator to compute the variance and standard deviation.

7. The number of incidents where police were needed for a sample of 10 schools in Allegheny County is 7, 37, 3, 8, 48, 11, 6, 0, 10, 3. Are the data consistent or do they vary? Explain your answer.

   Source: U.S. Department of Education.

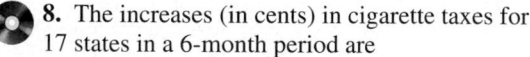 8. The increases (in cents) in cigarette taxes for 17 states in a 6-month period are

   60, 20, 40, 40, 45, 12, 34, 51, 30, 70, 42, 31, 69, 32, 8, 18, 50

   Use the range rule of thumb to estimate the standard deviation. Compare the estimate to the actual standard deviation.

   Source: Federation of Tax Administrators.

9. The normal daily high temperatures (in degrees Fahrenheit) in January for 10 selected cities are as follows.

   50, 37, 29, 54, 30, 61, 47, 38, 34, 61

   The normal monthly precipitation (in inches) for these same 10 cities is listed here.

   4.8, 2.6, 1.5, 1.8, 1.8, 3.3, 5.1, 1.1, 1.8, 2.5

   Which set is more variable?

   Source: N.Y. Times Almanac.

 10. The total surface area (in square miles) for each of six selected Eastern states is listed here.

28,995	PA	37,534	FL
31,361	NY	27,087	VA
20,966	ME	37,741	GA

   The total surface area for each of six selected Western states is listed (in square miles).

72,964	AZ	70,763	NV
101,510	CA	62,161	OR
66,625	CO	54,339	UT

   Which set is more variable?

   Source: N.Y. Times Almanac.

 11. Shown here are the numbers of stories in the 11 tallest buildings in St. Paul, Minnesota.

   32, 36, 46, 20, 32, 18, 16, 34, 26, 27, 26

   Shown here are the numbers of stories in the 11 tallest buildings in Chicago, Illinois.

   100, 100, 83, 60, 64, 65, 66, 74, 60, 67, 57

   Which data set is more variable?

   Source: The World Almanac and Book of Facts.

 12. The following data are the prices of 1 gallon of premium gasoline in U.S. dollars in seven foreign countries.

   3.80, 3.80, 3.20, 3.57, 3.62, 3.74, 3.69

   Do you think the standard deviation of these data is representative of the population standard deviation of gasoline prices in all foreign countries? Explain your answer.

   Source: Pittsburgh Post Gazette.

13. The number of weeks on *The New York Times Best Sellers* list for hardcover fiction is

   1, 4, 2, 2, 3, 18, 5, 5, 10, 4, 3, 6, 2, 2, 22

   Use the range rule of thumb to estimate the standard deviation. Compare the estimate to the actual standard deviation.

   Source: The New York Times Book Review.

14. Find the range, variance, and standard deviation for the distances of the home runs for McGwire and Sosa, using the data in Exercise 18 in Section 2–2. Compare the ranges and standard deviations. Decide which is more variable or if the variability is about the same. (Use individual data.)

15. Find the range, variance, and standard deviation for each data set in Exercise 11 of Section 3–2. Based on the results, which data set is more variable?

16. The Federal Highway Administration reported the number of deficient bridges in each state. Find the range, variance, and standard deviation.

15,458	1,055	5,008	3,598	8,984
1,337	4,132	10,618	17,361	6,081
6,482	25,090	12,681	16,286	18,832
12,470	17,842	16,601	4,587	47,196
23,205	25,213	23,017	27,768	2,686
7,768	25,825	4,962	22,704	2,694
4,131	13,144	15,582	7,279	12,613
810	13,350	1,208	22,242	7,477
10,902	2,343	2,333	2,979	6,578
14,318	4,773	6,252	734	13,220

Source: *USA TODAY.*

17. Find the range, variance, and standard deviation for the data in Exercise 17 of Section 2–2.

For Exercises 18 through 27, find the variance and standard deviation.

18. For 108 randomly selected college students, this exam score frequency distribution was obtained.

Class limits	Frequency
90–98	6
99–107	22
108–116	43
117–125	28
126–134	9

19. The costs per load (in cents) of 35 laundry detergents tested by a consumer organization are shown here.

Class limits	Frequency
13–19	2
20–26	7
27–33	12
34–40	5
41–47	6
48–54	1
55–61	0
62–68	2

20. Thirty automobiles were tested for fuel efficiency (in miles per gallon). This frequency distribution was obtained.

Class boundaries	Frequency
7.5–12.5	3
12.5–17.5	5
17.5–22.5	15
22.5–27.5	5
27.5–32.5	2

21. The data show the number of murders in 25 selected cities.

Class limits	Frequency
34–96	13
97–159	2
160–222	0
223–285	5
286–348	1
349–411	1
412–474	0
475–537	1
538–600	2

22. In a study of reaction times to a specific stimulus, a psychologist recorded these data (in seconds).

Class limits	Frequency
2.1–2.7	12
2.8–3.4	13
3.5–4.1	7
4.2–4.8	5
4.9–5.5	2
5.6–6.2	1

23. Eighty randomly selected lightbulbs were tested to determine their lifetimes (in hours). This frequency distribution was obtained.

Class boundaries	Frequency
52.5–63.5	6
63.5–74.5	12
74.5–85.5	25
85.5–96.5	18
96.5–107.5	14
107.5–118.5	5

24. The data represent the murder rate per 100,000 individuals in a sample of selected cities in the United States.

Class	Frequency
5–11	8
12–18	5
19–25	7
26–32	1
33–39	1
40–46	3

Source: FBI and U.S. Census Bureau.

25. Eighty randomly selected batteries were tested to determine their lifetimes (in hours). The following frequency distribution was obtained.

Class boundaries	Frequency
62.5–73.5	5
73.5–84.5	14
84.5–95.5	18
95.5–106.5	25
106.5–117.5	12
117.5–128.5	6

Can it be concluded that the lifetimes of these brands of batteries are consistent?

26. Find the variance and standard deviation for the two distributions in Exercise 8 in Section 2–3 and Exercise 18 in Section 2–3. Compare the variation of the data sets. Decide if one data set is more variable than the other.

27. This frequency distribution represents the data obtained from a sample of word processor repairers. The values are the days between service calls on 80 machines.

Class boundaries	Frequency
25.5–28.5	5
28.5–31.5	9
31.5–34.5	32
34.5–37.5	20
37.5–40.5	12
40.5–43.5	2

28. The average score of the students in one calculus class is 110, with a standard deviation of 5; the average score of students in a statistics class is 106, with a standard deviation of 4. Which class is more variable in terms of scores?

29. The data show the lengths (in feet) of suspension bridges in the eastern part of North America and the western part of North America. Compare the variability of the two samples, using the coefficient of variation.

East: 4260, 3500, 2300, 2150, 2000, 1750
West: 4200, 2800, 2310, 1550, 1500, 1207

Source: *World Almanac and Book of Facts.*

30. The average score on an English final examination was 85, with a standard deviation of 5; the average score on a history final exam was 110, with a standard deviation of 8. Which class was more variable?

31. The average age of the accountants at Three Rivers Corp. is 26 years, with a standard deviation of 6 years; the average salary of the accountants is $31,000, with a standard deviation of $4000. Compare the variations of age and income.

32. Using Chebyshev's theorem, solve these problems for a distribution with a mean of 80 and a standard deviation of 10.

a. At least what percentage of values will fall between 60 and 100?

b. At least what percentage of values will fall between 65 and 95?

33. The mean of a distribution is 20 and the standard deviation is 2. Use Chebyshev's theorem.

a. At least what percentage of the values will fall between 10 and 30?

b. At least what percentage of the values will fall between 12 and 28?

34. In a distribution of 200 values, the mean is 50 and the standard deviation is 5. Use Chebyshev's theorem.

a. At least how many values will fall between 30 and 70?

b. At most how many values will be less than 40 or more than 60?

35. A sample of the hourly wages of employees who work in restaurants in a large city has a mean of $5.02 and a standard deviation of $0.09. Using Chebyshev's theorem, find the range in which at least 75% of the data values will fall.

36. A sample of the labor costs per hour to assemble a certain product has a mean of $2.60 and a standard deviation of $0.15. Using Chebyshev's theorem, find the range in which at least 88.89% of the data will lie.

37. A survey of a number of the leading brands of cereal shows that the mean content of potassium per serving is 95 milligrams, and the standard deviation is 2 milligrams. Find the range in which at least 88.89% of the data will fall. Use Chebyshev's theorem.

38. The average score on a special test of knowledge of wood refinishing has a mean of 53 and a standard deviation of 6. Using Chebyshev's theorem, find the range of values in which at least 75% of the scores will lie.

39. The average of the number of trials it took a sample of mice to learn to traverse a maze was 12. The standard deviation was 3. Using Chebyshev's theorem, find the minimum percentage of data values that will fall in the range of 4 to 20 trials.

40. The average cost of a certain type of grass seed is $4.00 per box. The standard deviation is $0.10. Using Chebyshev's theorem, find the minimum percentage of data values that will fall in the range of $3.82 to $4.18.

41. The average U.S. yearly per capita consumption of citrus fruit is 26.8 pounds. Suppose that the distribution of fruit amounts consumed is bell-shaped with a standard deviation equal to 4.2 pounds. What percentage of Americans would you expect to consume more than 31 pounds of citrus fruit per year?

Source: USDA/Economic Research Service.

42. The average full-time faculty member in a post-secondary degree-granting institution works an average of 53 hours per week.

a. If we assume the standard deviation is 2.8 hours, what percentage of faculty members work more than 58.6 hours a week?

b. If we assume a bell-shaped distribution, what percentage of faculty members work more than 58.6 hours a week?

Source: National Center for Education Statistics.

Extending the Concepts

43. For this data set, find the mean and standard deviation of the variable. The data represent the serum cholesterol levels of 30 individuals. Count the number of data values that fall within 2 standard deviations of the mean. Compare this with the number obtained from Chebyshev's theorem. Comment on the answer.

211	240	255	219	204
200	212	193	187	205
256	203	210	221	249
231	212	236	204	187
201	247	206	187	200
237	227	221	192	196

44. For this data set, find the mean and standard deviation of the variable. The data represent the ages of 30 customers who ordered a product advertised on television. Count the number of data values that fall within 2 standard deviations of the mean. Compare this with the number obtained from Chebyshev's theorem. Comment on the answer.

42	44	62	35	20
30	56	20	23	41
55	22	31	27	66
21	18	24	42	25
32	50	31	26	36
39	40	18	36	22

45. Using Chebyshev's theorem, complete the table to find the minimum percentage of data values that fall within k standard deviations of the mean.

k	1.5	2	2.5	3	3.5
Percent					

46. Use this data set: 10, 20, 30, 40, 50.

a. Find the standard deviation.

b. Add 5 to each value, and then find the standard deviation.

c. Subtract 5 from each value and find the standard deviation.

d. Multiply each value by 5 and find the standard deviation.

e. Divide each value by 5 and find the standard deviation.

f. Generalize the results of parts b through e.

g. Compare these results with those in Exercise 38.

47. The mean deviation is found by using this formula:

$$\text{Mean deviation} = \frac{\Sigma|X - \bar{X}|}{n}$$

where

X = value
$\bar{X}$ = mean
n = number of values
$|\ |$ = absolute value

Find the mean deviation for these data.

5, 9, 10, 11, 11, 12, 15, 18, 20, 22

48. A measure to determine the skewness of a distribution is called the *Pearson coefficient of skewness*. The formula is

$$\text{Skewness} = \frac{3(\bar{X} - \text{MD})}{s}$$

The values of the coefficient usually range from -3 to $+3$. When the distribution is symmetric, the coefficient is zero; when the distribution is positively skewed, it is positive; and when the distribution is negatively skewed, it is negative.

Using the formula, find the coefficient of skewness for each distribution, and describe the shape of the distribution.

a. Mean = 10, median = 8, standard deviation = 3.

b. Mean = 42, median = 45, standard deviation = 4.

c. Mean = 18.6, median = 18.6, standard deviation = 1.5.

d. Mean = 98, median = 97.6, standard deviation = 4.

49. All values of a data set must be within $s\sqrt{n-1}$ of the mean. If a person collected 25 data values that had a mean of 50 and a standard deviation of 3 and you saw that one data value was 67, what would you conclude?

Technology *Step by Step*

Excel
Step by Step

Finding Measures of Variation

Example XL3–2

To find values that estimate the spread of a distribution of numbers:

1. Enter the numbers in a range (here **A1:A6**). We use the data from Example 3–23 on European automobile sales.

2. For the sample variance, enter **=VAR(A1:A6)** in a blank cell.

3. For the sample standard deviation, enter **=STDEV(A1:A6)** in a blank cell.

4. For the range, you can compute the value **=MAX(A1:A6) − MIN(A1:A6).**

	A	B	C	D	E
1	11.2		*European auto sales in millions*		
2	11.9				
3	12		1.129602	standard deviation	
4	12.8		1.276	variance	
5	13.4				
6	14.3				
7					

There are also functions STDEVP for population standard deviation and VARP for population variances.

3–4

Measures of Position

In addition to measures of central tendency and measures of variation, there are measures of position or location. These measures include standard scores, percentiles, deciles, and quartiles. They are used to locate the relative position of a data value in the data set. For example, if a value is located at the 80th percentile, it means that 80% of the values fall below it in the distribution and 20% of the values fall above it. The *median* is the value that corresponds to the 50th percentile, since one-half of the values fall below it and one-half of the values fall above it. This section discusses these measures of position.

Objective 3

Identify the position of a data value in a data set, using various measures of position, such as percentiles, deciles, and quartiles.

Standard Scores

There is an old saying, "You can't compare apples and oranges." But with the use of statistics, it can be done to some extent. Suppose that a student scored 90 on a music test and 45 on an English exam. Direct comparison of raw scores is impossible, since the exams might not be equivalent in terms of number of questions, value of each question, and so on. However, a comparison of a relative standard similar to both can be made. This comparison uses the mean and standard deviation and is called a standard score or *z* score. (We also use *z* scores in later chapters.)

A **z score** or **standard score** for a value is obtained by subtracting the mean from the value and dividing the result by the standard deviation. The symbol for a standard score is *z*. The formula is

$$z = \frac{\text{value} - \text{mean}}{\text{standard deviation}}$$

For samples, the formula is

$$z = \frac{X - \overline{X}}{s}$$

For populations, the formula is

$$z = \frac{X - \mu}{\sigma}$$

The z score represents the number of standard deviations that a data value falls above or below the mean.

For the purpose of this book, it will be assumed that when we find z scores, the data were obtained from samples.

Example 3–29

A student scored 65 on a calculus test that had a mean of 50 and a standard deviation of 10; she scored 30 on a history test with a mean of 25 and a standard deviation of 5. Compare her relative positions on the two tests.

Solution

First, find the z scores. For calculus the z score is

$$z = \frac{X - \overline{X}}{s} = \frac{65 - 50}{10} = 1.5$$

For history the z score is

$$z = \frac{30 - 25}{5} = 1.0$$

Since the z score for calculus is larger, her relative position in the calculus class is higher than her relative position in the history class.

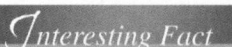
Interesting Fact

The average number of faces that a person learns to recognize and remember during his or her lifetime is 10,000.

Note that if the z score is positive, the score is above the mean. If the z score is 0, the score is the same as the mean. And if the z score is negative, the score is below the mean.

Example 3–30

Find the z score for each test, and state which is higher.

Test A	$X = 38$	$\overline{X} = 40$	$s = 5$
Test B	$X = 94$	$\overline{X} = 100$	$s = 10$

Solution

For test A,

$$z = \frac{X - \overline{X}}{s} = \frac{38 - 40}{5} = -0.4$$

For test B,

$$z = \frac{94 - 100}{10} = -0.6$$

The score for test A is relatively higher than the score for test B.

When all data for a variable are transformed into z scores, the resulting distribution will have a mean of 0 and a standard deviation of 1. A z score, then, is actually the number of standard deviations each value is from the mean for a specific distribution. In Example 3–29, the calculus score of 65 was actually 1.5 standard deviations above the mean of 50. This will be explained in greater detail in Chapter 7.

Percentiles

Percentiles are position measures used in educational and health-related fields to indicate the position of an individual in a group.

Percentiles divide the data set into 100 equal groups.

In many situations, the graphs and tables showing the percentiles for various measures such as test scores, heights, or weights have already been completed. Table 3–3 shows the percentile ranks for scaled scores on the Test of English as a Foreign Language. If a student had a scaled score of 58 for section 1 (listening and comprehension), that student would have a percentile rank of 81. Hence, that student did better than 81% of the students who took section 1 of the exam.

Table 3–3 Percentile Ranks and Scaled Scores on the Test of English as a Foreign Language*

Scaled score	Section 1: Listening comprehension	Section 2: Structure and written expression	Section 3: Vocabulary and reading comprehension	Total scaled score	Percentile rank
68	99	98			
66	98	96	98	660	99
64	96	94	96	640	97
62	92	90	93	620	94
60	87	84	88	600	89
→58	81	76	81	580	82
56	73	68	72	560	73
54	64	58	61	540	62
52	54	48	50	520	50
50	42	38	40	500	39
48	32	29	30	480	29
46	22	21	23	460	20
44	14	15	16	440	13
42	9	10	11	420	9
40	5	7	8	400	5
38	3	4	5	380	3
36	2	3	3	360	1
34	1	2	2	340	1
32		1	1	320	
30		1	1	300	
Mean	51.5	52.2	51.4	517	Mean
S.D.	7.1	7.9	7.5	68	S.D.

*Based on the total group of 1,178,193 examinees tested from July 1989 through June 1991.

Source: Reprinted by permission of Educational Testing Service, the copyright owner.

Figure 3–5

Weights of Girls by Age and Percentile Rankings

Source: Distributed by Mead Johnson Nutritional Division. Reprinted with permission.

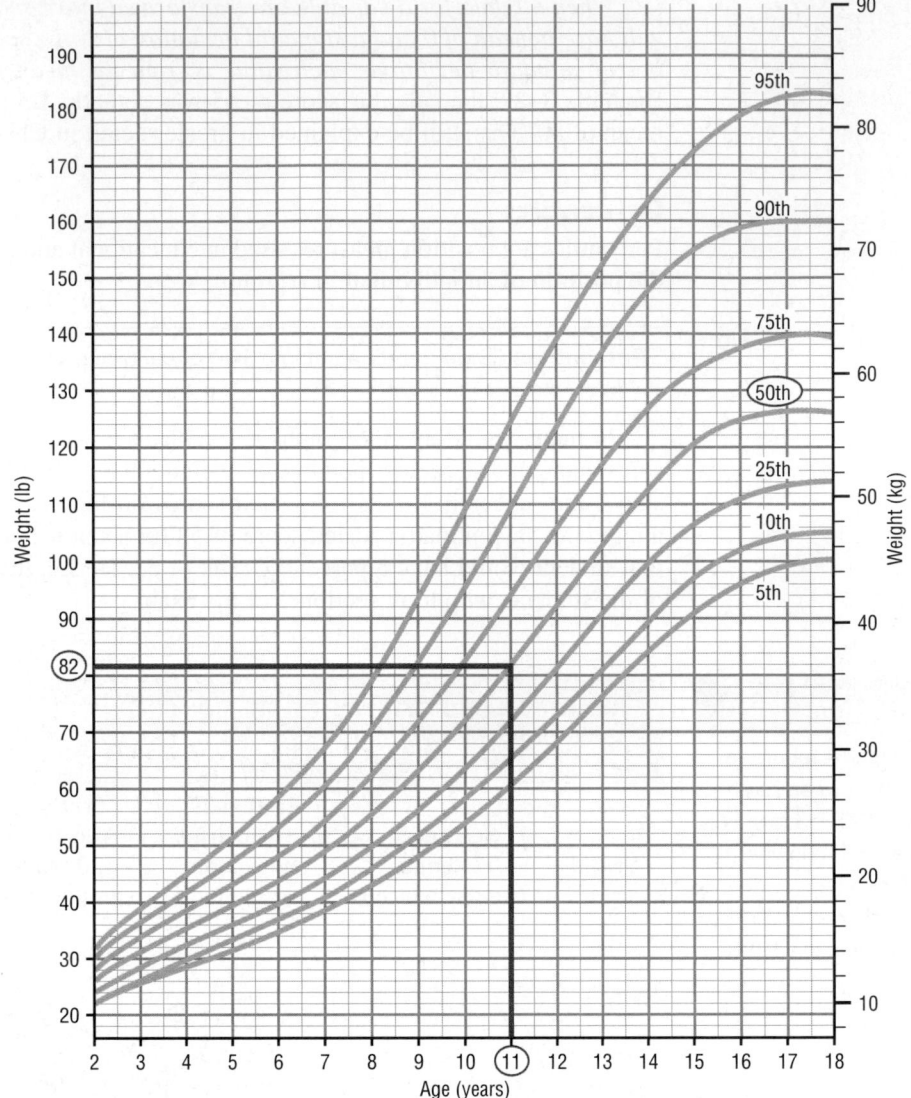

Figure 3–5 shows percentiles in graphical form of weights of girls from ages 2 to 18. To find the percentile rank of an 11-year-old who weighs 82 pounds, start at the 82-pound weight on the left axis and move horizontally to the right. Find 11 on the horizontal axis and move up vertically. The two lines meet at the 50th percentile curved line; hence, an 11-year-old girl who weighs 82 pounds is in the 50th percentile for her age group. If the lines do not meet exactly on one of the curved percentile lines, then the percentile rank must be approximated.

Percentiles are also used to compare an individual's test score with the national norm. For example, tests such as the National Educational Development Test (NEDT) are taken by students in ninth or tenth grade. A student's scores are compared with those of other students locally and nationally by using percentile ranks. A similar test for elementary school students is called the California Achievement Test.

Percentiles are not the same as percentages. That is, if a student gets 72 correct answers out of a possible 100, she obtains a percentage score of 72. There is no indication of her position with respect to the rest of the class. She could have scored the highest, the lowest, or somewhere in between. On the other hand, if a raw score of 72 corresponds to the 64th percentile, then she did better than 64% of the students in her class.

Percentiles are symbolized by

$$P_1, P_2, P_3, \ldots, P_{99}$$

and divide the distribution into 100 groups.

Percentile graphs can be constructed as shown in Example 3–31. Percentile graphs use the same values as the cumulative relative frequency graphs described in Section 2–3, except that the proportions have been converted to percents.

Example 3–31

The frequency distribution for the systolic blood pressure readings (in millimeters of mercury, mm Hg) of 200 randomly selected college students is shown here. Construct a percentile graph.

A Class boundaries	B Frequency	C Cumulative frequency	D Cumulative percent
89.5–104.5	24		
104.5–119.5	62		
119.5–134.5	72		
134.5–149.5	26		
149.5–164.5	12		
164.5–179.5	4		
	200		

Solution

Step 1 Find the cumulative frequencies and place them in column C.

Step 2 Find the cumulative percentages and place them in column D. To do this step, use the formula

$$\text{Cumulative \%} = \frac{\text{cumulative frequency}}{n} \cdot 100\%$$

For the first class,

$$\text{Cumulative \%} = \frac{24}{200} \cdot 100\% = 12\%$$

The completed table is shown here.

A Class boundaries	B Frequency	C Cumulative frequency	D Cumulative percent
89.5–104.5	24	24	12
104.5–119.5	62	86	43
119.5–134.5	72	158	79
134.5–149.5	26	184	92
149.5–164.5	12	196	98
164.5–179.5	4	200	100
	200		

Figure 3–6

Percentile Graph for Example 3–31

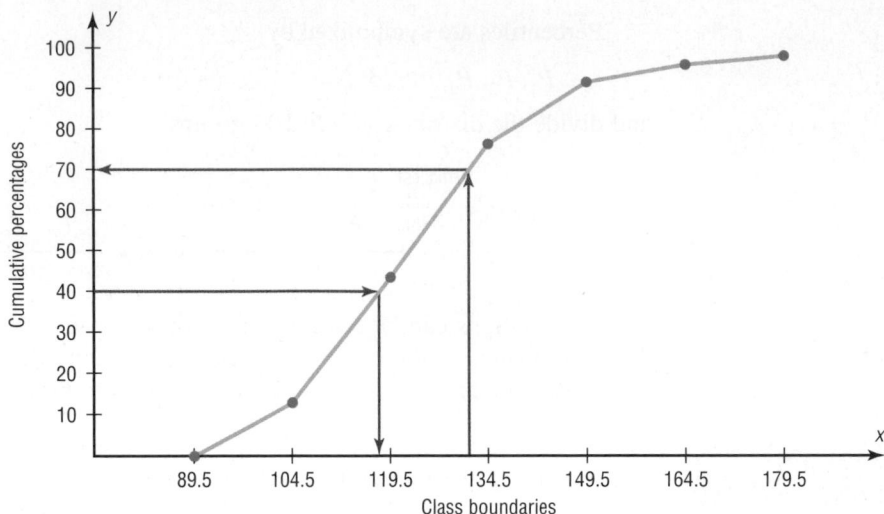

Step 3   Graph the data, using class boundaries for the x axis and the percentages for the y axis, as shown in Figure 3–6.

Once a percentile graph has been constructed, one can find the approximate corresponding percentile ranks for given blood pressure values and find approximate blood pressure values for given percentile ranks.

For example, to find the percentile rank of a blood pressure reading of 130, find 130 on the x axis of Figure 3–6, and draw a vertical line to the graph. Then move horizontally to the value on the y axis. Note that a blood pressure of 130 corresponds to approximately the 70th percentile.

If the value that corresponds to the 40th percentile is desired, start on the y axis at 40 and draw a horizontal line to the graph. Then draw a vertical line to the x axis and read the value. In Figure 3–6, the 40th percentile corresponds to a value of approximately 118. Thus, if a person has a blood pressure of 118, he or she is at the 40th percentile.

Finding values and the corresponding percentile ranks by using a graph yields only approximate answers. Several mathematical methods exist for computing percentiles for data. These methods can be used to find the approximate percentile rank of a data value or to find a data value corresponding to a given percentile. When the data set is large (100 or more), these methods yield better results. Examples 3–32 through 3–35 show these methods.

Percentile Formula

The percentile corresponding to a given value X is computed by using the following formula:

$$\text{Percentile} = \frac{(\text{number of values below } X) + 0.5}{\text{total number of values}} \cdot 100\%$$

Example 3–32

A teacher gives a 20-point test to 10 students. The scores are shown here. Find the percentile rank of a score of 12.

18, 15, 12, 6, 8, 2, 3, 5, 20, 10

Solution

Arrange the data in order from lowest to highest.

$$2, 3, 5, 6, 8, 10, 12, 15, 18, 20$$

Then substitute into the formula.

$$\text{Percentile} = \frac{(\text{number of values below } X) + 0.5}{\text{total number of values}} \cdot 100\%$$

Since there are six values below a score of 12, the solution is

$$\text{Percentile} = \frac{6 + 0.5}{10} \cdot 100\% = 65\text{th percentile}$$

Thus, a student whose score was 12 did better than 65% of the class.

Note: One assumes that a score of 12 in Example 3–32, for instance, means theoretically any value between 11.5 and 12.5.

Example 3–33

Using the data in Example 3–32, find the percentile rank for a score of 6.

Solution

There are three values below 6. Thus

$$\text{Percentile} = \frac{3 + 0.5}{10} \cdot 100\% = 35\text{th percentile}$$

A student who scored 6 did better than 35% of the class.

Examples 3–34 amd 3–35 show a procedure for finding a value corresponding to a given percentile.

Example 3–34

Using the scores in Example 3–32, find the value corresponding to the 25th percentile.

Solution

Step 1 Arrange the data in order from lowest to highest.

$$2, 3, 5, 6, 8, 10, 12, 15, 18, 20$$

Step 2 Compute

$$c = \frac{n \cdot p}{100}$$

where
n = total number of values
p = percentile

Thus,

$$c = \frac{10 \cdot 25}{100} = 2.5$$

Step 3 If c is not a whole number, round it up to the next whole number; in this case, $c = 3$. (If c is a whole number, see Example 3–35.) Start at the lowest value and count over to the third value, which is 5. Hence, the value 5 corresponds to the 25th percentile.

Example 3–35 Using the data set in Example 3–32, find the value that corresponds to the 60th percentile.

Solution

Step 1 Arrange the data in order from smallest to largest.

2, 3, 5, 6, 8, 10, 12, 15, 18, 20

Step 2 Substitute in the formula.

$$c = \frac{n \cdot p}{100} = \frac{10 \cdot 60}{100} = 6$$

Step 3 If c is a whole number, use the value halfway between the c and $c + 1$ values when counting up from the lowest value—in this case, the 6th and 7th values.

2, 3, 5, 6, 8, 10, 12, 15, 18, 20

6th value   7th value

The value halfway between 10 and 12 is 11. Find it by adding the two values and dividing by 2.

$$\frac{10 + 12}{2} = 11$$

Hence, 11 corresponds to the 60th percentile. Anyone scoring 11 would have done better than 60% of the class.

The steps for finding a value corresponding to a given percentile are summarized in this Procedure Table.

Procedure Table

Finding a Data Value Corresponding to a Given Percentile

Step 1 Arrange the data in order from lowest to highest.

Step 2 Substitute into the formula

$$c = \frac{n \cdot p}{100}$$

where

n = total number of values
p = percentile

Step 3A If c is not a whole number, round up to the next whole number. Starting at the lowest value, count over to the number that corresponds to the rounded-up value.

Step 3B If c is a whole number, use the value halfway between the cth and $(c + 1)$st values when counting up from the lowest value.

Quartiles and Deciles

Quartiles divide the distribution into four groups, separated by Q_1, Q_2, Q_3.

Note that Q_1 is the same as the 25th percentile; Q_2 is the same as the 50th percentile, or the median; Q_3 corresponds to the 75th percentile, as shown:

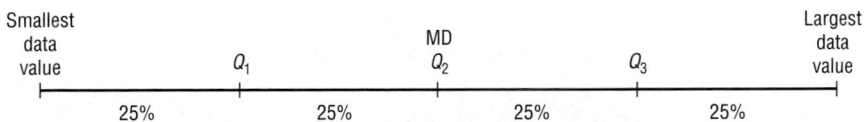

Quartiles can be computed by using the formula given for computing percentiles on page 139. For Q_1 use $p = 25$. For Q_2 use $p = 50$. For Q_3 use $p = 75$. However, an easier method for finding quartiles is found in this Procedure Table.

Procedure Table

Finding Data Values Corresponding to Q_1, Q_2, and Q_3

Step 1 Arrange the data in order from lowest to highest.

Step 2 Find the median of the data values. This is the value for Q_2.

Step 3 Find the median of the data values that fall below Q_2. This is the value for Q_1.

Step 4 Find the median of the data values that fall above Q_2. This is the value for Q_3

Example 3–36 shows how to find the values of Q_1, Q_2, and Q_3.

Example 3–36

 Find Q_1, Q_2, and Q_3 for the data set 15, 13, 6, 5, 12, 50, 22, 18.

Solution

Step 1 Arrange the data in order.

5, 6, 12, 13, 15, 18, 22, 50

Step 2 Find the median (Q_2).

5, 6, 12, 13, 15, 18, 22, 50
↑
MD

$$\text{MD} = \frac{13 + 15}{2} = 14$$

Step 3 Find the median of the data values less than 14.

5, 6, 12, 13
↑
Q_1

$$Q_1 = \frac{6 + 12}{2} = 9$$

So Q_1 is 9.

Step 4   Find the median of the data values greater than 14.

$$15, 18, 22, 50$$
$$\uparrow$$
$$Q_3$$

$$Q_3 = \frac{18 + 22}{2} = 20$$

Here Q_3 is 20. Hence, $Q_1 = 9$, $Q_2 = 14$, and $Q_3 = 20$.

In addition to dividing the data set into four groups, quartiles can be used as a rough measurement of variability. The **interquartile range (IQR)** is defined as the difference between Q_1 and Q_3 and is the range of the middle 50% of the data.

The interquartile range is used to identify outliers, and it is also used as a measure of variability in exploratory data analysis, as shown in Section 3–5.

Deciles divide the distribution into 10 groups, as shown. They are denoted by D_1, D_2, etc.

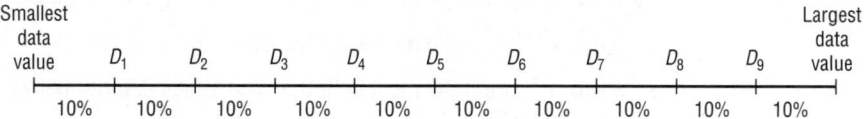

Note that D_1 corresponds to P_{10}; D_2 corresponds to P_{20}; etc. Deciles can be found by using the formulas given for percentiles. Taken altogether then, these are the relationships among percentiles, deciles, and quartiles.

Deciles are denoted by $D_1, D_2, D_3, \ldots, D_9$, and they correspond to $P_{10}, P_{20}, P_{30}, \ldots, P_{90}$.

Quartiles are denoted by Q_1, Q_2, Q_3 and they correspond to P_{25}, P_{50}, P_{75}.

The median is the same as P_{50} or Q_2 or D_5.

The position measures are summarized in Table 3–4.

Table 3–4	Summary of Position Measures	
Measure	**Definition**	**Symbol(s)**
Standard score or z score	Number of standard deviations that a data value is above or below the mean	z
Percentile	Position in hundredths that a data value holds in the distribution	P_n
Decile	Position in tenths that a data value holds in the distribution	D_n
Quartile	Position in fourths that a data value holds in the distribution	Q_n

Outliers

A data set should be checked for extremely high or extremely low values. These values are called *outliers*.

An **outlier** is an extremely high or an extremely low data value when compared with the rest of the data values.

An outlier can strongly affect the mean and standard deviation of a variable. For example, suppose a researcher mistakenly recorded an extremely high data value. This value would then make the mean and standard deviation of the variable much larger than they really were. Outliers can have an effect on other statistics as well.

There are several ways to check a data set for outliers. One method is shown in this Procedure Table.

Procedure Table

Procedure for Identifying Outliers

Step 1   Arrange the data in order and find Q_1 and Q_3.

Step 2   Find the interquartile range: IQR = $Q_3 - Q_1$.

Step 3   Multiply the IQR by 1.5.

Step 4   Subtract the value obtained in step 3 from Q_1 and add the value to Q_3.

Step 5   Check the data set for any data value that is smaller than $Q_1 - 1.5(\text{IQR})$ or larger than $Q_3 + 1.5(\text{IQR})$.

This procedure is shown in Example 3–37.

Example 3–37

 Check the following data set for outliers.

5, 6, 12, 13, 15, 18, 22, 50

Solution

The data value 50 is extremely suspect. These are the steps in checking for an outlier.

Step 1   Find Q_1 and Q_3. This was done in Example 3–36; Q_1 is 9 and Q_3 is 20.

Step 2   Find the interquartile range (IQR), which is $Q_3 - Q_1$.

$$\text{IQR} = Q_3 - Q_1 = 20 - 9 = 11$$

Step 3   Multiply this value by 1.5.

$$1.5(11) = 16.5$$

Step 4   Subtract the value obtained in step 3 from Q_1, and add the value obtained in step 3 to Q_3.

$$9 - 16.5 = -7.5 \qquad \text{and} \qquad 20 + 16.5 = 36.5$$

Step 5   Check the data set for any data values that fall outside the interval from −7.5 to 36.5. The value 50 is outside this interval; hence, it can be considered an outlier.

There are several reasons why outliers may occur. First, the data value may have resulted from a measurement or observational error. Perhaps the researcher measured the variable incorrectly. Second, the data value may have resulted from a recording error. That is, it may have been written or typed incorrectly. Third, the data value may have been obtained from a subject that is not in the defined population. For example, suppose test scores were obtained from a seventh-grade class, but a student in that class was actually in the sixth grade and had special permission to attend the class. This student might have scored extremely low on that particular exam on that day. Fourth, the data value might be a legitimate value that occurred by chance (although the probability is extremely small).

There are no hard-and-fast rules on what to do with outliers, nor is there complete agreement among statisticians on ways to identify them. Obviously, if they occurred as a result of an error, an attempt should be made to correct the error or else the data value should be omitted entirely. When they occur naturally by chance, the statistician must make a decision about whether to include them in the data set.

When a distribution is normal or bell-shaped, data values that are beyond 3 standard deviations of the mean can be considered suspected outliers.

Applying the Concepts **3–4**

Determining Dosages

In an attempt to determine necessary dosages of a new drug (HDL) used to control sepsis, assume you administer varying amounts of HDL to 40 mice. You create four groups and label them *low dosage, moderate dosage, large dosage,* and *very large dosage.* The dosages also vary within each group. After the mice are injected with the HDL and the sepsis bacteria, the time until the onset of sepsis is recorded. Your job as a statistician is to effectively communicate the results of the study.

1. Which measures of position could be used to help describe the data results?
2. If 40% of the rats in the top quartile survived after the injection, how many mice would that be?
3. What information can be given from using percentiles?
4. What information can be given from using quartiles?
5. What information can be given from using standard scores?

See page 170 for the answers.

Exercises 3–4

1. What is a z score?

2. Define *percentile rank.*

3. What is the difference between a percentage and a percentile?

4. Define *quartile.*

5. What is the relationship between quartiles and percentiles?

6. What is a decile?

7. How are deciles related to percentiles?

8. To which percentile, quartile, and decile does the median correspond?

9. If the mean value of major league teams is $127 million and the standard deviation is $9 million, find the corresponding z score for each team's value.

a. 136   *d.* 113.5
b. 109   *e.* 133
c. 104.5

10. The reaction time to a stimulus for a certain test has a mean of 2.5 seconds and a standard deviation of 0.3 second. Find the corresponding z score for each reaction time.

a. 2.7
b. 3.9
c. 2.8
d. 3.1
e. 2.2

11. A final examination for a psychology course has a mean of 84 and a standard deviation of 4. Find the corresponding z score for each raw score.

a. 87   d. 76
b. 79   e. 82
c. 93

12. An aptitude test has a mean of 220 and a standard deviation of 10. Find the corresponding z score for each exam score.

a. 200   d. 212
b. 232   e. 225
c. 218

13. Which of the following exam scores has a better relative position?

a. A score of 42 on an exam with $\overline{X} = 39$ and $s = 4$.
b. A score of 76 on an exam with $\overline{X} = 71$ and $s = 3$.

14. A student scores 60 on a mathematics test that has a mean of 54 and a standard deviation of 3, and she scores 80 on a history test with a mean of 75 and a standard deviation of 2. On which test did she perform better?

15. Which score indicates the highest relative position?

a. A score of 3.2 on a test with $\overline{X} = 4.6$ and $s = 1.5$.
b. A score of 630 on a test with $\overline{X} = 800$ and $s = 200$.
c. A score of 43 on a test with $\overline{X} = 50$ and $s = 5$.

16. This distribution represents the data for weights of fifth-grade boys. Find the approximate weights corresponding to each percentile given by constructing a percentile graph.

Weight (pounds)	Frequency
52.5–55.5	9
55.5–58.5	12
58.5–61.5	17
61.5–64.5	22
64.5–67.5	15

a. 25th   c. 80th
b. 60th   d. 95th

17. For the data in Exercise 16, find the approximate percentile ranks of the following weights.

a. 57 pounds
b. 62 pounds
c. 64 pounds
d. 59 pounds

18. (ans) The data shown represent the scores on a national achievement test for a group of tenth-grade students. Find the approximate percentile ranks of these scores by constructing a percentile graph.

a. 220   d. 280
b. 245   e. 300
c. 276

Score	Frequency
196.5–217.5	5
217.5–238.5	17
238.5–259.5	22
259.5–280.5	48
280.5–301.5	22
301.5–322.5	6

19. For the data in Exercise 18, find the approximate scores that correspond to these percentiles.

a. 15th   d. 65th
b. 29th   e. 80th
c. 43rd

20. (ans) The airborne speeds in miles per hour of 21 planes are shown. Find the approximate values that correspond to the given percentiles by constructing a percentile graph.

Class	Frequency
366–386	4
387–407	2
408–428	3
429–449	2
450–470	1
471–491	2
492–512	3
513–533	4
	21

a. 9th   d. 60th
b. 20th   e. 75th
c. 45th

Source: *The World Almanac and Book of Facts.*

21. Using the data in Exercise 20, find the approximate percentile ranks of the following miles per hour (mph).

a. 380 mph   d. 505 mph
b. 425 mph   e. 525 mph
c. 455 mph

22. Find the percentile ranks of each weight in the data set. The weights are in pounds.

78, 82, 86, 88, 92, 97

23. In Exercise 22, what value corresponds to the 30th percentile?

24. Find the percentile rank for each test score in the data set.

12, 28, 35, 42, 47, 49, 50

25. In Exercise 24, what value corresponds to the 60th percentile?

26. Find the percentile rank for each value in the data set. The data represent the values in billions of dollars of the damage of 10 hurricanes.

1.1, 1.7, 1.9, 2.1, 2.2, 2.5, 3.3, 6.2, 6.8, 20.3

Source: Insurance Services Office.

27. What value in Exercise 26 corresponds to the 40th percentile?

 28. Find the percentile rank for each test score in the data set.

5, 12, 15, 16, 20, 21

29. What test score in Exercise 28 corresponds to the 33rd percentile?

 30. Using the procedure shown in Example 3–37, check each data set for outliers.

a. 16, 18, 22, 19, 3, 21, 17, 20
b. 24, 32, 54, 31, 16, 18, 19, 14, 17, 20
c. 321, 343, 350, 327, 200
d. 88, 72, 97, 84, 86, 85, 100
e. 145, 119, 122, 118, 125, 116
f. 14, 16, 27, 18, 13, 19, 36, 15, 20

 31. Another measure of average is called the *midquartile;* it is the numerical value halfway between Q_1 and Q_3, and the formula is

$$\text{Midquartile} = \frac{Q_1 + Q_3}{2}$$

Using this formula and other formulas, find Q_1, Q_2, Q_3, the midquartile, and the interquartile range for each data set.

a. 5, 12, 16, 25, 32, 38
b. 53, 62, 78, 94, 96, 99, 103

Technology *Step by Step*

MINITAB
Step by Step

Calculate Descriptive Statistics from Data

Example MT3–1

1. Enter the data from Example 3–23 into C1 of MINITAB. Name the column **AutoSales.**
2. Select **Stat>Basic Statistics>Display Descriptive Statistics.**
3. The cursor will be blinking in the Variables text box. Double-click C1 AutoSales.
4. Click [Statistics] to view the statistics that can be calculated with this command.

   a) Check the boxes for Mean, Standard deviation, Variance, Coefficient of variation, Median, Minimum, Maximum, and N nonmissing.

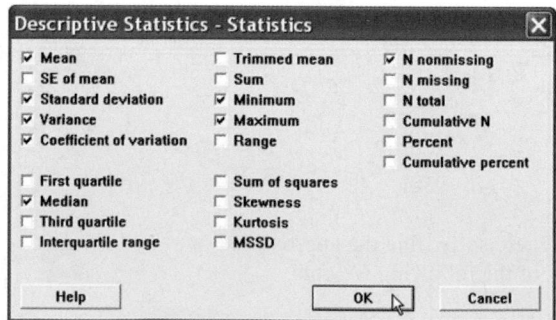

   b) Remove the checks from other options.

5. Click [OK] twice. The results will be displayed in the session window as shown.

Descriptive Statistics: AutoSales

Variable	N	Mean	Median	StDev	Variance	CoefVar	Minimum	Maximum
AutoSales	6	12.6	12.4	1.12960	1.276	8.96509	11.2	14.3

Session window results are in text format. A high-resolution graphical window displays the descriptive statistics, a histogram, and a boxplot.

6. Select **Stat>Basic Statistics>Graphical Summary.**

7. Double-click C1 AutoSales.

8. Click [OK].

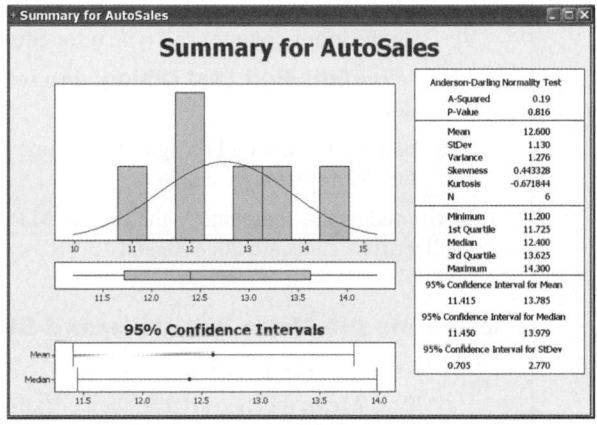

The graphical summary will be displayed in a separate window as shown.

Calculate Descriptive Statistics from a Frequency Distribution

Multiple menu selections must be used to calculate the statistics from a table. We will use data given in Example 3–24.

Enter Midpoints and Frequencies

1. Select **File>New>New Worksheet** to open an empty worksheet.

2. To enter the midpoints into C1, select **Calc>Make Patterned Data>Simple Set of Numbers.**

   a) Type **X** to name the column.

   b) Type in **8** for the First value, **38** for the Last value, and **5** for Steps.

   c) Click [OK].

3. Enter the frequencies in C2. Name the column **f.**

Calculate Columns for f·X and f·X²

4. Select **Calc>Calculator.**

   a) Type in **fX** for the variable and **f*X** in the Expression dialog box. Click [OK].

   b) Select **Edit>Edit Last Dialog** and type in **fX2** for the variable and **f*X**2** for the expression.

   c) Click [OK]. There are now four columns in the worksheet.

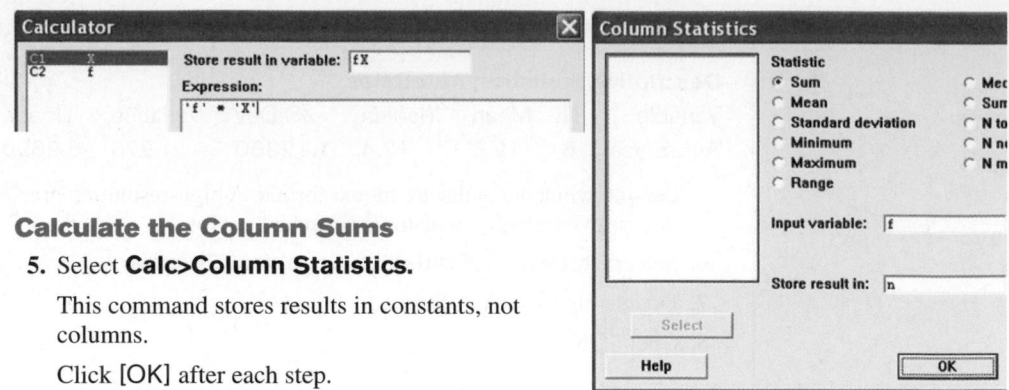

Calculate the Column Sums

5. Select **Calc>Column Statistics.**

    This command stores results in constants, not columns.

    Click [OK] after each step.

    a) Click the option for Sum; then select C2 f for the Input column, and type **n** for Store result in.

    b) Select **Edit>Edit Last Dialog;** then select C3 fX for the column and type **sumX** for storage.

    c) Edit the last dialog box again. This time select C4 fX2 for the column, then type **sumX2** for storage.

To verify the results, navigate to the Project Manager window, then the constants folder of the worksheet. The sums are 20, 490, and 13,310.

Calculate the Mean, Variance, and Standard Deviation

6. Select **Calc>Calculator.**

    a) Type **Mean** for the variable, then click in the box for the Expression and type **sumX/n.** Click [OK]. If you double-click the constants instead of typing them, single quotes will surround the names. The quotes are not required unless the column name has spaces.

    b) Click the **EditLast Dialog** icon and type **Variance** for the variable.

    c) In the expression box type in

    $$(\text{sumX2-sumX}**2/\text{n})/(\text{n-1})$$

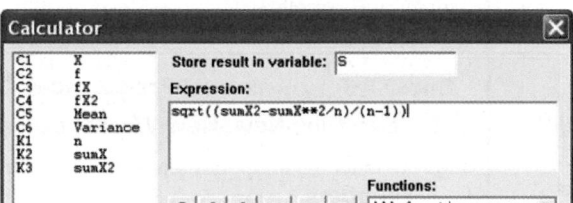

    d) Edit the last dialog box and type **S** for the variable. In the expression box, drag the mouse over the previous expression to highlight it.

    e) Click the button in the keypad for parentheses. Type **SQRT** at the beginning of the line, upper- or lowercase will work. The expression should be SQRT((sumX2-sumX**2/n)/(n-1)).

    f) Click [OK].

Display Results

    g) Select **Data>Display Data,** then highlight all columns and constants in the list.

    h) Click [Select] then [OK].

The session window will display all our work! Create the histogram with instructions from Chapter 2.

Data Display

n   20.0000
sumX   490.000
sumX2 13310.0

Row	X	f	fX	fX2	Mean	Variance	S
1	8	1	8	64	24.5	68.6842	8.28759
2	13	2	26	338			
3	18	3	54	972			
4	23	5	115	2645			
5	28	4	112	3136			
6	33	3	99	3267			
7	38	2	76	2888			

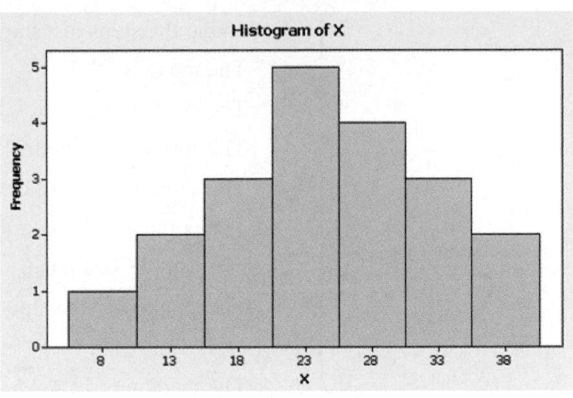

TI-83 Plus or TI-84 Plus
Step by Step

Calculating Descriptive Statistics

To calculate various descriptive statistics:

1. Enter data into L_1.
2. Press **STAT** to get the menu.
3. Press ▶ to move cursor to CALC; then press **1** for 1-Var Stats.
4. Press **2nd [L₁]**, then **ENTER**.

The calculator will display

$\bar{x}$ sample mean

Σx sum of the data values

Σx^2 sum of the squares of the data values

S_x sample standard deviation

σ_x population standard deviation

n number of data values

minX   smallest data value

Q_1 lower quartile

Med   median

Q_3 upper quartile

maxX   largest data value

Example TI3–1

Find the various descriptive statistics for the auto sales data from Example 3–23:

11.2, 11.9, 12.0, 12.8, 13.4, 14.3

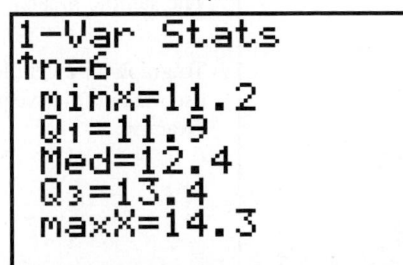

Following the steps just shown, we obtain these results, as shown on the screen:

The mean is 12.6.

The sum of x is 75.6.

The sum of x^2 is 958.94.

The sample standard deviation S_x is 1.1296017.

The population standard deviation σ_x is 1.031180553.

The sample size n is 6.

The smallest data value is 11.2.

Q_1 is 11.9.

The median is 12.4.

Q_3 is 13.4.

The largest data value is 14.3.

To calculate the mean and standard deviation from grouped data:

1. Enter the midpoints into L_1.
2. Enter the frequencies into L_2.
3. Press **STAT** to get the menu.
4. Use the arrow keys to move the cursor to CALC; then press **1** for 1-Var Stats.
5. Press **2nd [L1], 2nd [L2],** then **ENTER.**

Example TI3–2

Calculate the mean and standard deviation for the data given in Examples 3–3 and 3–24.

Class	Frequency	Midpoint
5.5–10.5	1	8
10.5–15.5	2	13
15.5–20.5	3	18
20.5–25.5	5	23
25.5–30.5	4	28
30.5–35.5	3	33
35.5–40.5	2	38

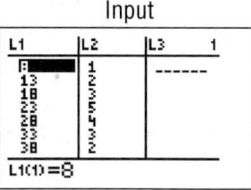

Input

Input

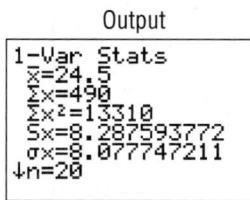

Output

The sample mean is 24.5, and the sample standard deviation is 8.287593772.

To graph a percentile graph, follow the procedure for an ogive but use the cumulative percent in L_2, 100 for Y_{max}, and the data from Example 3–31.

Output

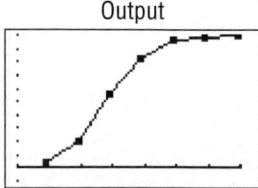

Excel
Step by Step

Descriptive Statistics in Excel

Example XL3–3

 Excel's Data Analysis options include an item called Descriptive Statistics that reports all the standard measures of a data set.

1. Enter the data set shown (nine numbers) in column A of a new worksheet.

> 12 17 15 16 16 14 18 13 10

2. Select **Tools>Data Analysis.**

3. Use these data (A1:A9) as the Input Range in the Descriptive Statistics dialog box.

4. Check the Summary statistics option and click [OK].

Descriptive Statistics
Dialog Box

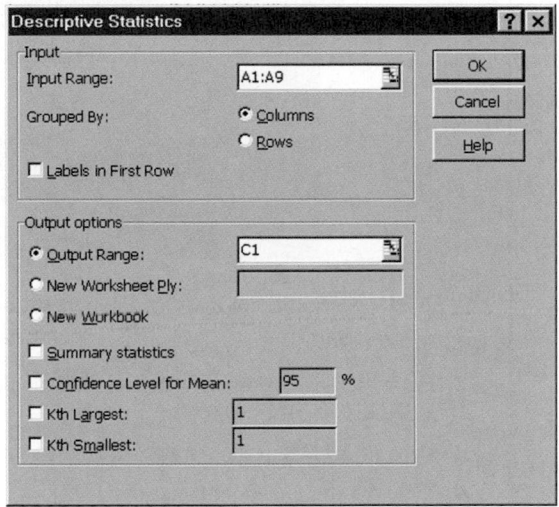

Here's the summary output for this data set. Note that this one operation reports most of the statistics used in this chapter.

Column1	
Mean	14.55555556
Standard Error	0.85165054
Median	15
Mode	16
Standard Deviation	2.554951619
Sample Variance	6.527777778
Kurtosis	-0.3943866
Skewness	-0.51631073
Range	8
Minimum	10
Maximum	18
Sum	131
Count	9
Confidence Level(95.0%)	1.963910937

Measures of Position

Enter the data from Example 3–23 in column A.

To find the z score for a value in a set of data:

1. Select cell **B1** on the worksheet.
2. From the paste function (f_x) icon, select **Statistical** from the function category. Then select the **STANDARDIZE** function.
3. Type in **A1** in the X box.
4. Type in **average(A1:A6)** in the Mean box.
5. Type in **stdev(A1:A6)** in the Standard_dev box. Then click [OK].
6. Repeat this procedure for each data value in column A.

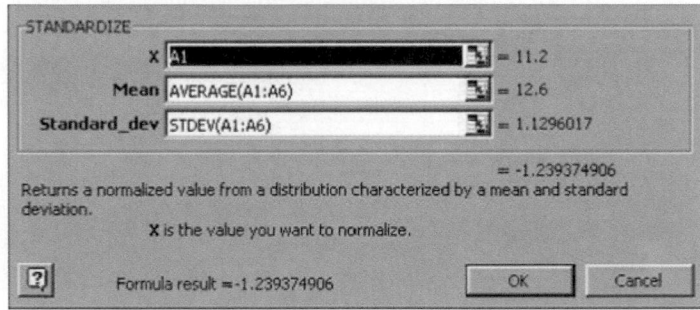

To find the percentile rank for a value in a set of data:

1. Select cell **C1** on the worksheet.
2. From the paste function icon, select **Statistical** from the function category. Then select the **PERCENTRANK** function.
3. Type in **A1:A6** in the Array box.
4. Type **A1** in the X box, then click [OK].

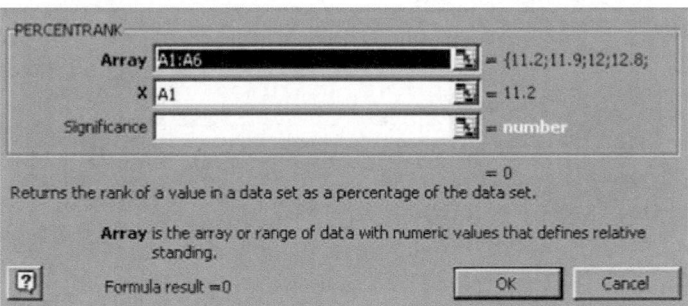

<table>
<tr><td>**3–5**</td><td></td></tr>
</table>

Exploratory Data Analysis

Objective 4

Use the techniques of exploratory data analysis, including boxplots and five-number summaries, to discover various aspects of data.

In traditional statistics, data are organized by using a frequency distribution. From this distribution various graphs such as the histogram, frequency polygon, and ogive can be constructed to determine the shape or nature of the distribution. In addition, various statistics such as the mean and standard deviation can be computed to summarize the data.

The purpose of traditional analysis is to confirm various conjectures about the nature of the data. For example, from a carefully designed study, a researcher might want to know if the proportion of Americans who are exercising today has increased from 10 years ago. This study would contain various assumptions about the population, various definitions such as of exercise, and so on.

In **exploratory data analysis (EDA),** data can be organized using a *stem and leaf plot.* (See Chapter 2.) The measure of central tendency used in EDA is the *median.* The measure of variation used in EDA is the *interquartile range* ($Q_3 - Q_1$). In EDA the data are represented graphically using a **boxplot** (sometimes called a box-and-whisker plot). The purpose of exploratory data analysis is to examine data to find out what information can be discovered about the data such as the center and the spread. Exploratory data analysis was developed by John Tukey and presented in his book *Exploratory Data Analysis* (Addison-Wesley, 1977).

The Five-Number Summary and Boxplots

A **boxplot** can be used to graphically represent the data set. These plots involve five specific values:

1. The lowest value of the data set (i.e., minimum)
2. Q_1
3. The median
4. Q_3
5. The highest value of the data set (i.e., maximum)

These values are called a **five-number summary** of the data set.

A **boxplot** is a graph of a data set obtained by drawing a horizontal line from the minimum data value to Q_1, drawing a horizontal line from Q_3 to the maximum data value, and drawing a box whose vertical sides pass through Q_1 and Q_3 with a vertical line inside the box passing through the median or Q_2.

Example 3–38

A stockbroker recorded the number of clients she saw each day over an 11-day period. The data are shown. Construct a boxplot for the data.

33, 38, 43, 30, 29, 40, 51, 27, 42, 23, 31

Solution

Step 1   Arrange the data in order.

23, 27, 29, 30, 31, 33, 38, 40, 42, 43, 51

Step 2   Find the median.

23, 27, 29, 30, 31, 33, 38, 40, 42, 43, 51

↑

Median

Step 3   Find Q_1.

23, 27, 29, 30, 31

↑

29

Step 4   Find Q_3.

38, 40, 42, 43, 51

↑

42

Figure 3–7

Boxplot for
Example 3–38

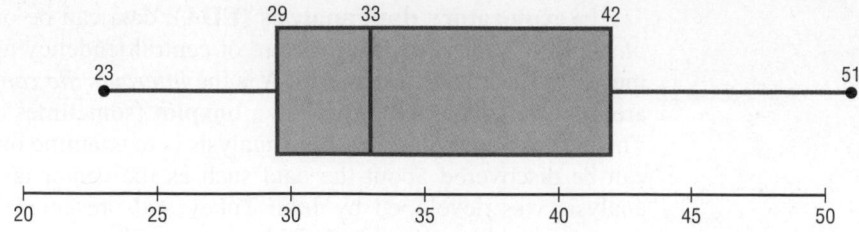

Step 5   Draw a scale for the data on the x axis.

Step 6   Locate the lowest value, Q_1, the median, Q_3, and the highest value on the scale.

Step 7   Draw a box around Q_1 and Q_3, draw a vertical line through the median, and connect the upper and lower values, as shown in Figure 3–7.

The box in Figure 3–7 represents the middle 50% of the data, and the lines represent the lower and upper ends of the data.

Information Obtained from a Boxplot

1. *a.* If the median is near the center of the box, the distribution is approximately symmetric.
   b. If the median falls to the left of the center of the box, the distribution is positively skewed.
   c. If the median falls to the right of the center, the distribution is negatively skewed.
2. *a.* If the lines are about the same length, the distribution is approximately symmetric.
   b. If the right line is larger than the left line, the distribution is positively skewed.
   c. If the left line is larger than the right line, the distribution is negatively skewed.

The boxplot in Figure 3–7 indicates that the distribution is slightly positively skewed.

If the boxplots for two or more data sets are graphed on the same axis, the distributions can be compared. To compare the averages, use the location of the medians. To compare the variability, use the interquartile range, i.e., the length of the boxes. Example 3–39 shows this procedure.

Example 3–39

A dietitian is interested in comparing the sodium content of real cheese with the sodium content of a cheese substitute. The data for two random samples are shown. Compare the distributions, using boxplots.

Real cheese				Cheese substitute			
310	420	45	40	270	180	250	290
220	240	180	90	130	260	340	310

Source: The Complete Book of Food Counts.

Solution

Step 1 Find Q_1, MD, and Q_3 for the real cheese data.

$$40 \quad 45 \quad 90 \quad 180 \quad 220 \quad 240 \quad 310 \quad 420$$

$$\qquad\quad \uparrow \qquad\qquad \uparrow \qquad\qquad \uparrow$$

$$\qquad\quad Q_1 \qquad\qquad \text{MD} \qquad\qquad Q_3$$

$$Q_1 = \frac{45 + 90}{2} = 67.5 \qquad \text{MD} = \frac{180 + 220}{2} = 200$$

$$Q_3 = \frac{240 + 310}{2} = 275$$

Step 2 Find Q_1, MD, and Q_3 for the cheese substitute data.

$$130 \quad 180 \quad 250 \quad 260 \quad 270 \quad 290 \quad 310 \quad 340$$

$$\qquad\quad \uparrow \qquad\qquad \uparrow \qquad\qquad \uparrow$$

$$\qquad\quad Q_1 \qquad\qquad \text{MD} \qquad\qquad Q_3$$

$$Q_1 = \frac{180 + 250}{2} = 215 \qquad \text{MD} = \frac{260 + 270}{2} = 265$$

$$Q_3 = \frac{290 + 310}{2} = 300$$

Step 3 Draw the boxplots for each distribution on the same graph. See Figure 3–8.

Step 4 Compare the plots. It is quite apparent that the distribution for the cheese substitute data has a higher median than the median for the distribution for the real cheese data. The variation or spread for the distribution of the real cheese data is larger than the variation for the distribution of the cheese substitute data.

Figure 3–8

Boxplots for Example 3–39

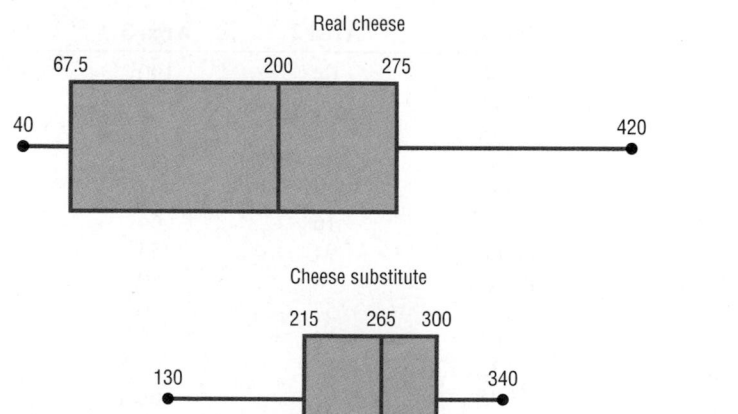

In exploratory data analysis, *hinges* are used instead of quartiles to construct boxplots. When the data set consists of an even number of values, hinges are the same as quartiles. Hinges for a data set with an odd number of values differ somewhat from quartiles. However, since most calculators and computer programs use quartiles, they will be used in this textbook.

Another important point to remember is that the summary statistics (median and interquartile range) used in exploratory data analysis are said to be *resistant statistics*. A **resistant statistic** is relatively less affected by outliers than a *nonresistant statistic*. The mean and standard deviation are nonresistant statistics. Sometimes when a distribution is skewed or contains outliers, the median and interquartile range may more accurately summarize the data than the mean and standard deviation, since the mean and standard deviation are more affected in this case.

Table 3–5 compares the traditional versus the exploratory data analysis approach.

Table 3–5	Traditional versus EDA Techniques	
	Traditional	**Exploratory data analysis**
	Frequency distribution	Stem and leaf plot
	Histogram	Boxplot
	Mean	Median
	Standard deviation	Interquartile range

Applying the Concepts 3–5

The Noisy Workplace

Assume you work for OSHA (Occupational Safety and Health Administration) and have complaints about noise levels from some of the workers at a state power plant. You charge the power plant with taking decibel readings at six different areas of the plant at different times of the day and week. The results of the data collection are listed. Use boxplots to initially explore the data and make recommendations about which plant areas workers must be provided with protective ear wear. The safe hearing level is at approximately 120 decibels.

Area 1	Area 2	Area 3	Area 4	Area 5	Area 6
30	64	100	25	59	67
12	99	59	15	63	80
35	87	78	30	81	99
65	59	97	20	110	49
24	23	84	61	65	67
59	16	64	56	112	56
68	94	53	34	132	80
57	78	59	22	145	125
100	57	89	24	163	100
61	32	88	21	120	93
32	52	94	32	84	56
45	78	66	52	99	45
92	59	57	14	105	80
56	55	62	10	68	34
44	55	64	33	75	21

See page 170 for the answers.

Exercises 3–5

For Exercises 1–6, identify the five-number summary and find the interquartile range.

1. 8, 12, 32, 6, 27, 19, 54

2. 19, 16, 48, 22, 7

3. 362, 589, 437, 316, 192, 188

4. 147, 243, 156, 632, 543, 303

5. 14.6, 19.8, 16.3, 15.5, 18.2

6. 9.7, 4.6, 2.2, 3.7, 6.2, 9.4, 3.8

For Exercises 7–10, use each boxplot to identify the **maximum value, minimum value, median, first quartile, third quartile, and interquartile range.**

7.

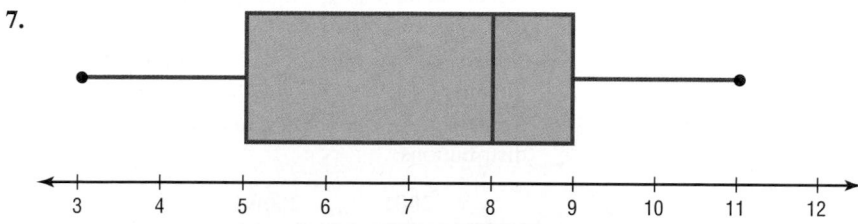

8.

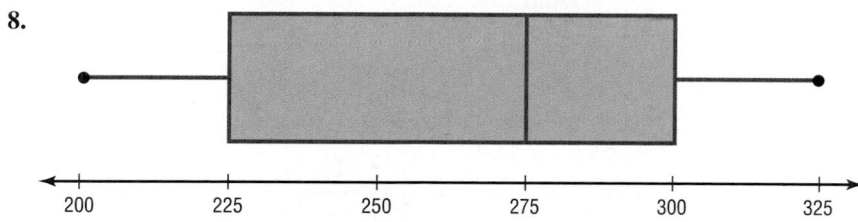

9.

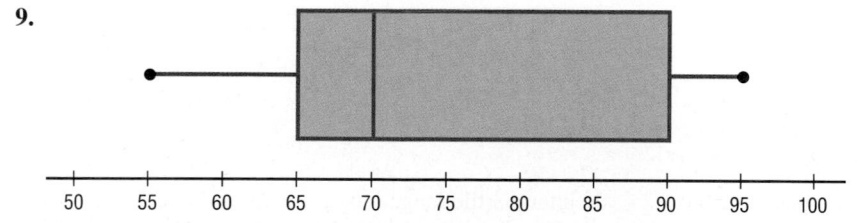

10.

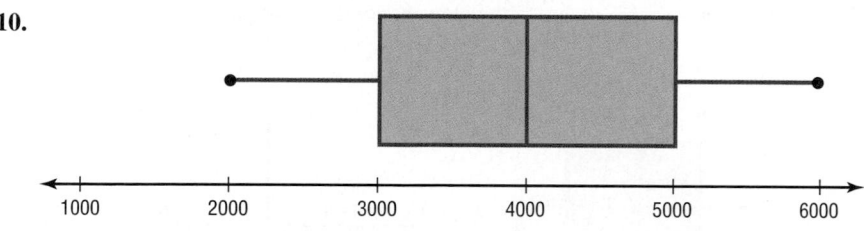

11. Shown next are the sizes of the police forces in the 10 largest cities in the United States in 1993 (the numbers represent hundreds). Construct a boxplot for the data and comment on the shape of the distribution.

29.3, 7.6, 12.1, 4.7, 6.2, 1.9, 3.9, 2.8, 2.0, 1.7

Source: *USA TODAY.*

12. Construct a boxplot for the number of bills enacted by Congress during the last several years. Comment on the shape of the distribution.

88, 245, 153, 241, 170, 410, 136, 241, 198

Source: *USA TODAY.*

13. Construct a boxplot for the following average number of vacation days in selected countries.

42	37	35
34	28	26
13	25	25

Source: World Tourism Organization.

14. Shown here is the number of new theater productions that appeared on Broadway for the past several years. Construct a boxplot for the data and comment on the shape of the distribution.

30	28	33	29	37	39
35	37	37	38	34	

Source: The League of American Theaters and Producers Inc.

15. These data are the number of inches of snow reported in randomly selected U.S. cities for September 1 through January 10. Construct a boxplot and comment on the skewness of the data.

9.8	8.0	13.9	4.4	3.9	21.7
3.2	11.7	24.8	34.1	17.6	15.9

Source: USA TODAY.

16. These data represent the volumes in cubic yards of the largest dams in the United States and in South America. Construct a boxplot of the data for each region and compare the distributions.

United States	South America
125,628	311,539
92,000	274,026
78,008	105,944
77,700	102,014
66,500	56,242
62,850	46,563
52,435	
50,000	

Source: N.Y. Times Almanac.

17. A 4-month record for the number of tornadoes in 1999–2001 is given here.

a. Which month has the highest mean number of tornadoes for this 3-year period?

b. Which year has the highest mean number of tornadoes for this 4-month period?

c. Construct three boxplots and compare the distributions.

	2001	2000	1999
April	135	136	177
May	241	241	311
June	248	135	289
July	120	148	102

Source: National Weather Service, Storm Prediction Center.

Extending the Concepts

18. A *modified boxplot* can be drawn by placing a box around Q_1 and Q_3 and then extending the whiskers to the largest and/or smallest values within 1.5 times the interquartile range (i.e., $Q_3 - Q_1$). *Mild outliers* are values between 1.5 (IQR) and 3 (IQR). *Extreme outliers* are data values beyond 3 (IQR).

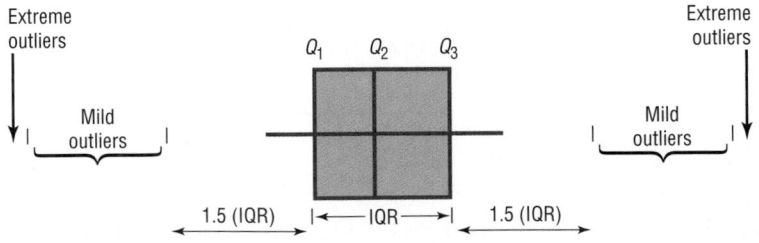

For the data shown here, draw a modified boxplot and identify any mild or extreme outliers. The data represent the number of unhealthful smog days for a specific year for the highest 10 locations.

97	39	43	66	91
43	54	42	53	39

Source: U.S. Public Interest Research Group and Clean Air Network.

Technology *Step by Step*

MINITAB
Step by Step

Construct a Boxplot

1. Type in the data for Example 3–38 (the number of clients seen daily by a stockbroker). Label the column **Clients.**

2. Select **Stat>EDA>Boxplot.**

3. Double-click Clients to select it for the Y variable.

4. Click on [Labels].

   a) In the Title 1: of the Title/Footnotes folder, type **Number of Clients.**

   b) Press the [Tab] key and type **Your Name** in the text box for Subtitle 1:.

5. Click [OK] twice. The graph will be displayed in a graph window.

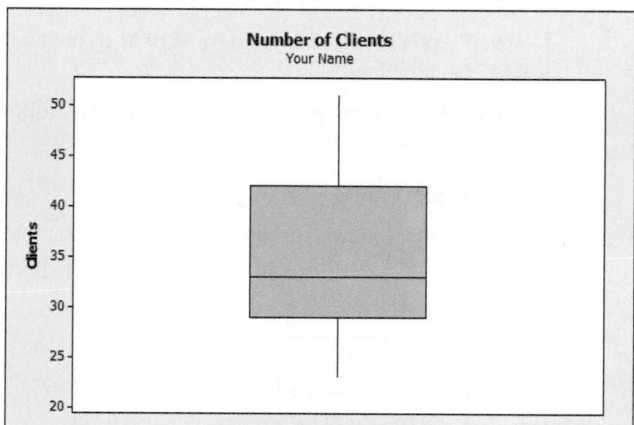

Example MT3–2

1. Enter the data for Example 2–13 in Section 2–4. Label the column **CARS-THEFT.**

2. Select **Stat>EDA>Boxplot.**

3. Double-click CARS-THEFT to select it for the Y variable.

4. Click on the drop-down arrow for Annotation.

5. Click on Title, then enter an appropriate title such as **Car Thefts for Large City, U.S.A.**

6. Click [OK] twice.

A high-resolution graph will be displayed in a graph window.

Boxplot Dialog Box and Boxplot

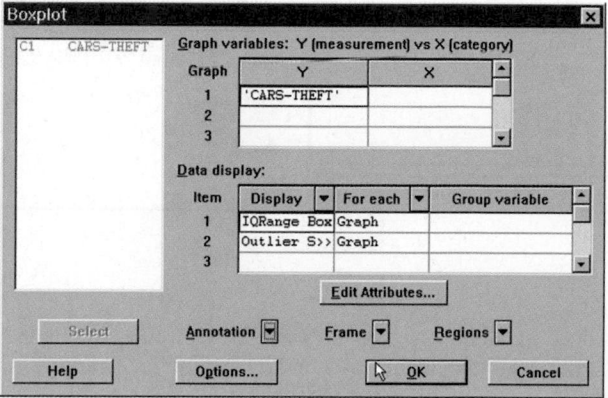

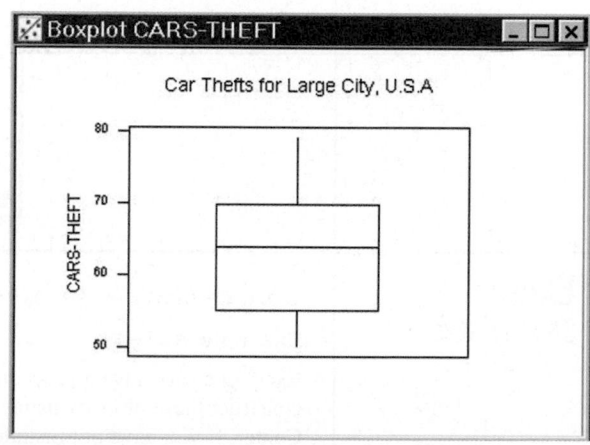

TI-83 Plus or TI-84 Plus
Step by Step

Constructing a Boxplot

To draw a boxplot:

1. Enter data into L_1.
2. Change values in WINDOW menu, if necessary. (*Note:* Make X_{min} somewhat smaller than the smallest data value and X_{max} somewhat larger than the largest data value.) Change Y_{min} to 0 and Y_{max} to 1.
3. Press **[2nd] [STAT PLOT]**, then **1** for Plot 1.
4. Press **ENTER** to turn Plot 1 on.
5. Move cursor to Boxplot symbol (fifth graph) on the Type: line, then press **ENTER.**
6. Make sure Xlist is L_1.
7. Make sure Freq is 1.
8. Press **GRAPH** to display the boxplot.
9. Press **TRACE** followed by ◄ or ► to obtain the values from the five-number summary on the boxplot.

To display two boxplots on the same display, follow the above steps and use the 2: Plot 2 and L_2 symbols.

Example TI3–3

Construct a boxplot for the data values in Example 3–38:

33, 38, 43, 30, 29, 40, 51, 27, 42, 23, 31

Input

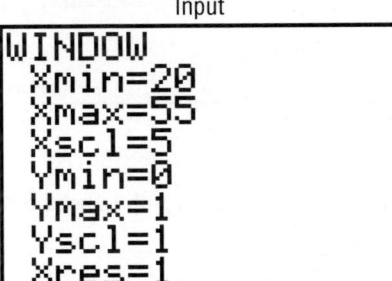

Input

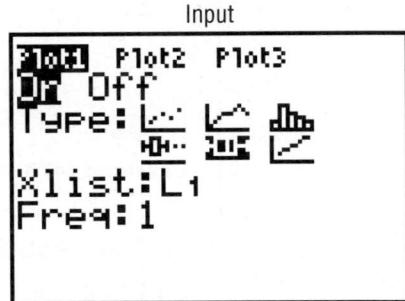

Using the **TRACE** key along with the ◄ and ► keys, we obtain the five-number summary. The minimum value is 23; Q_1 is 29; the median is 33; Q_3 is 42; the maximum value is 51.

Output

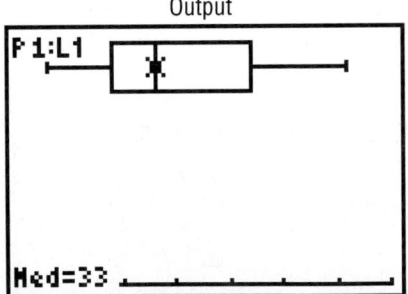

Excel
Step by Step

Constructing a Boxplot

Example XL3–4

Excel does not have a procedure to produce stem and leaf plots or boxplots. However, you may construct these plots by using the MegaStat Add-in available on your CD and Online Learning Center. If you have not installed this add-in, do so by following the instructions on page 24.

To obtain a boxplot and stem and leaf plot:

1. Enter the data from Example 3–38 into column A of a new worksheet.

2. Select **MegaStat>Descriptive Statistics.**

3. Enter the cell range **A1:A50** in the Input range.

4. Check the Boxplot and Stem and Leaf plot options. Click [OK].

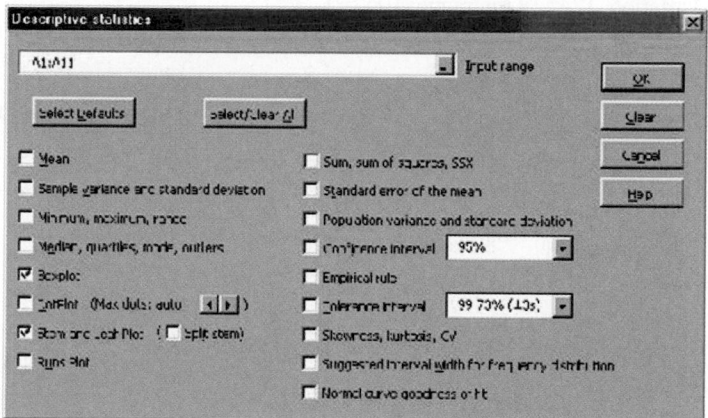

The stem and leaf plot and boxplot that are obtained in the output are featured.

Stem and Leaf plot# 1
    stem unit=10
    leaf unit=1

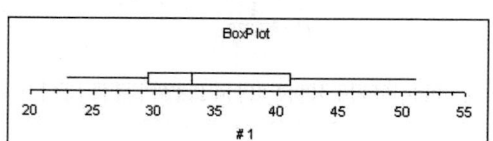

Frequency	Stem	Leaf
3	2	3 7 9
4	3	0 1 3 8
3	4	0 2 3
1	5	1
11		

3–6   Summary

This chapter explains the basic ways to summarize data. These include measures of central tendency, measures of variation or dispersion, and measures of position. The three most commonly used measures of central tendency are the mean, median, and mode. The midrange is also used occasionally to represent an average. The three most commonly used measurements of variation are the range, variance, and standard deviation.

The most common measures of position are percentiles, quartiles, and deciles. This chapter explains how data values are distributed according to Chebyshev's theorem and the empirical rule. The coefficient of variation is used to describe the standard deviation in relationship to the mean. These methods are commonly called traditional statistical methods and are primarily used to confirm various conjectures about the nature of the data.

Other methods, such as the boxplot and five-number summaries, are part of exploratory data analysis; they are used to examine data to see what they reveal.

After learning the techniques presented in Chapter 2 and this chapter, you will have a substantial knowledge of descriptive statistics. That is, you will be able to collect, organize, summarize, and present data.

Important Terms

bimodal 103

boxplot 153

Chebyshev's theorem 125

coefficient of variation 124

data array 101

decile 142

empirical rule 127

exploratory data analysis 153

five-number summary 153

interquartile range 142

mean 98

median 101

midrange 106

modal class 104

mode 103

multimodal 103

negatively skewed or left-skewed distribution 109

outlier 142

parameter 98

percentile 135

positively skewed or right-skewed distribution 108

quartile 141

range 116

range rule of thumb 125

resistant statistic 156

standard deviation 118

statistic 98

symmetric distribution 108

unimodal 103

variance 118

weighted mean 107

z score or standard score 133

Important Formulas

Formula for the mean for individual data:

$$\overline{X} = \frac{\Sigma X}{n} \qquad \mu = \frac{\Sigma X}{N}$$

Formula for the mean for grouped data:

$$\overline{X} = \frac{\Sigma f \cdot X_m}{n}$$

Formula for the weighted mean:

$$\overline{X} = \frac{\Sigma wX}{\Sigma w}$$

Formula for the midrange:

$$MR = \frac{\textbf{lowest value} + \textbf{highest value}}{2}$$

Formula for the range:

$$R = \textbf{highest value} - \textbf{lowest value}$$

Formula for the variance for population data:

$$\sigma^2 = \frac{\Sigma(X - \mu)^2}{N}$$

Formula for the variance for sample data (shortcut formula for the unbiased estimator):

$$s^2 = \frac{\Sigma X^2 - [(\Sigma X)^2/n]}{n - 1}$$

Formula for the variance for grouped data:

$$s^2 = \frac{\Sigma f \cdot X_m^2 - [(\Sigma f \cdot X_m)^2/n]}{n - 1}$$

Formula for the standard deviation for population data:

$$\sigma = \sqrt{\frac{\Sigma(X - \mu)^2}{N}}$$

Formula for the standard deviation for sample data (shortcut formula):

$$s = \sqrt{\frac{\Sigma X^2 - [(\Sigma X)^2/n]}{n - 1}}$$

Formula for the standard deviation for grouped data:

$$s = \sqrt{\frac{\Sigma f \cdot X_m^2 - [(\Sigma f \cdot X_m)^2/n]}{n - 1}}$$

Formula for the coefficient of variation:

$$CVar = \frac{s}{\overline{X}} \cdot 100\% \qquad \text{or} \qquad CVar = \frac{\sigma}{\mu} \cdot 100\%$$

Range rule of thumb:

$$s \approx \frac{\textbf{range}}{4}$$

Expression for Chebyshev's theorem: The proportion of values from a data set that will fall within k standard deviations of the mean will be at least

$$1 - \frac{1}{k^2}$$

where k is a number greater than 1.

Formula for the z score (standard score):

$$z = \frac{X - \mu}{\sigma} \qquad \text{or} \qquad z = \frac{X - \overline{X}}{s}$$

Formula for the cumulative percentage:

$$\text{Cumulative \%} = \frac{\text{cumulative frequency}}{n} \cdot 100\%$$

Formula for the percentile rank of a value X:

$$\text{Percentile} = \frac{(\text{number of values below } X) + 0.5}{\text{total number of values}} \cdot 100\%$$

Formula for finding a value corresponding to a given percentile:

$$c = \frac{n \cdot p}{100}$$

Formula for interquartile range:

$$\text{IQR} = Q_3 - Q_1$$

Review Exercises

 1. The following data represent the number of listeners (in thousands) of 15 radio stations in the 6:00 to 9:00 A.M. time slot in Pittsburgh.

229, 182, 129, 112, 122, 93, 97, 114, 95, 114, 60, 89, 75, 70, 68

Source: Arbitron Inc.

Find each of these.

a. Mean   *e.* Range
b. Median   *f.* Variance
c. Mode   *g.* Standard deviation
d. Midrange

 2. These data represent the area in square miles of major islands in the Caribbean Sea and the Mediterranean Sea.

Caribbean Sea			Mediterranean Sea	
108	926	436	1,927	1,411
75	100	3,339	229	95
5,382	171	116	3,189	540
2,300	290	1,864	3,572	9,301
166	687	59	86	9,926
42,804	4,244	134		
29,389				

Source: The World Almanac and Book of Facts.

Find each of these.

a. Mean   *e.* Range
b. Median   *f.* Variance
c. Mode   *g.* Standard deviation
d. Midrange

Are the averages and variations of the areas approximately equal?

3. Twelve batteries were tested to see how many hours they would last. The frequency distribution is shown here.

Hours	Frequency
1–3	1
4–6	4
7–9	5
10–12	1
13–15	1

Find each of these.

a. Mean   *c.* Variance
b. Modal class   *d.* Standard deviation

4. The following data represent the number of seconds it took 20 students to find information from the Internet on a personal computer.

Class	Frequency
34–38	4
39–43	6
44–48	3
49–53	4
54–58	3

Find each of these.

a. Mean   *c.* Variance
b. Modal class   *d.* Standard deviation

5. Shown here is a frequency distribution for the rise in tides at 30 selected locations in the United States.

Rise in tides (inches)	Frequency
12.5–27.5	6
27.5–42.5	3
42.5–57.5	5
57.5–72.5	8
72.5–87.5	6
87.5–102.5	2

Find each of these.

a. Mean   *c.* Variance
b. Modal class   *d.* Standard deviation

6. The fuel capacity in gallons of 50 randomly selected 1995 cars is shown here.

Class	Frequency
10–12	6
13–15	4
16–18	14
19–21	15
22–24	8
25–27	2
28–30	1
	50

Find each of these.

a. Mean   *c.* Variance
b. Modal class   *d.* Standard deviation

7. In a dental survey of third-grade students, this distribution was obtained for the number of cavities found. Find the average number of cavities for the class. Use the weighted mean.

Number of students	Number of cavities
12	0
8	1
5	2
5	3

8. An investor calculated these percentages of each of three stock investments with payoffs as shown. Find the average payoff. Use the weighted mean.

Stock	Percent	Payoff
A	30	$10,000
B	50	3,000
C	20	1,000

9. In an advertisement, a transmission service center stated that the average years of service of its employees were 13. The distribution is shown here. Using the weighted mean, calculate the correct average.

Number of employees	Years of service
8	3
1	6
1	30

10. If the average number of textbooks in professors' offices is 16, the standard deviation is 5, and the average age of the professors is 43, with a standard deviation of 8, which data set is more variable?

11. A survey of bookstores showed that the average number of magazines carried is 56, with a standard deviation of 12. The same survey showed that the average length of time each store had been in business was 6 years, with a standard deviation of 2.5 years.

Which is more variable, the number of magazines or the number of years?

12. The number of previous jobs held by each of six applicants is shown here.

2, 4, 5, 6, 8, 9

a. Find the percentile for each value.
b. What value corresponds to the 30th percentile?
c. Construct a boxplot and comment on the nature of the distribution.

13. The salaries (in millions of dollars) for 29 NFL teams for the 1999–2000 season are given in this frequency distribution.

Class limits	Frequency
39.9–42.8	2
42.9–45.8	2
45.9–48.8	5
48.9–51.8	5
51.9–54.8	12
54.9–57.8	3

Source: www.NFL.com

a. Construct a percentile graph.
b. Find the values that correspond to the 35th, 65th, and 85th percentiles.
c. Find the percentile of values 44, 48, and 54.

 14. Check each data set for outliers.

a. 506, 511, 517, 514, 400, 521
b. 3, 7, 9, 6, 8, 10, 14, 16, 20, 12
c. 14, 18, 27, 26, 19, 13, 5, 25
d. 112, 157, 192, 116, 153, 129, 131

15. A survey of car rental agencies shows that the average cost of a car rental is $0.32 per mile. The standard deviation is $0.03. Using Chebyshev's theorem, find the range in which at least 75% of the data values will fall.

16. The average cost of a certain type of seed per acre is $42. The standard deviation is $3. Using Chebyshev's theorem, find the range in which at least 88.89% of the data values will fall.

17. The average labor charge for automobile mechanics is $54 per hour. The standard deviation is $4. Find the minimum percentage of data values that will fall within the range of $48 to $60. Use Chebyshev's theorem.

18. For a certain type of job, it costs a company an average of $231 to train an employee to perform the task. The standard deviation is $5. Find the minimum percentage of data values that will fall in the range of $219 to $243. Use Chebyshev's theorem.

19. The average delivery charge for a refrigerator is $32. The standard deviation is $4. Find the minimum percentage of data values that will fall in the range of $20 to $44. Use Chebyshev's theorem.

20. Which of these exam grades has a better relative position?

a. A grade of 82 on a test with $\overline{X} = 85$ and $s = 6$.

b. A grade of 56 on a test with $\overline{X} = 60$ and $s = 5$.

21. The data shown here represent the number of hours that 12 part-time employees at a toy store worked during the weeks before and after Christmas. Construct two boxplots and compare the distributions.

Before	38	16	18	24	12	30	35	32	31	30	24	35
After	26	15	12	18	24	32	14	18	16	18	22	12

22. The mean of the times it takes a commuter to get to work in Baltimore is 29.7 minutes. If the standard deviation is 6 minutes, within what limits would you expect approximately 68% of the times to fall? Assume the distribution is approximately bell-shaped.

Statistics Today

How Long Are You Delayed by Road Congestion?—Revisited

The average number of hours per year that a driver is delayed by road congestion is listed here.

Los Angeles	56
Atlanta	53
Seattle	53
Houston	50
Dallas	46
Washington	46
Austin	45
Denver	45
St. Louis	44
Orlando	42
U.S. average	36

Source: Texas Transportation Institute.

By making comparisons using averages, you can see that drivers in these 10 cities are delayed by road congestion more than the national average.

Data Analysis

A Data Bank is found in Appendix D, or on the World Wide Web by following links from www.mhhe.com/math/stat/bluman/

1. From the Data Bank, choose one of the following variables: age, weight, cholesterol level, systolic pressure, IQ, or sodium level. Select at least 30 values, and find the mean, median, mode, and midrange. State which measurement of central tendency best describes the average and why.

2. Find the range, variance, and standard deviation for the data selected in Exercise 1.

3. From the Data Bank, choose 10 values from any variable, construct a boxplot, and interpret the results.

4. Randomly select 10 values from the number of suspensions in the local school districts in southwestern Pennsylvania in Data Set V in Appendix D. Find the mean, median, mode, range, variance, and standard deviation of the number of suspensions by using the Pearson coefficient of skewness.

5. Using the data from Data Set VII in Appendix D, find the mean, median, mode, range, variance, and standard deviation of the acreage owned by the municipalities. Comment on the skewness of the data, using the Pearson coefficient of skewness.

Chapter Quiz

Determine whether each statement is true or false. If the statement is false, explain why.

1. When the mean is computed for individual data, all values in the data set are used.

2. The mean cannot be found for grouped data when there is an open class.

3. A single, extremely large value can affect the median more than the mean.

4. One-half of all the data values will fall above the mode, and one-half will fall below the mode.

5. In a data set, the mode will always be unique.

6. The range and midrange are both measures of variation.

7. One disadvantage of the median is that it is not unique.

8. The mode and midrange are both measures of variation.

9. If a person's score on an exam corresponds to the 75th percentile, then that person obtained 75 correct answers out of 100 questions.

Select the best answer.

10. What is the value of the mode when all values in the data set are different?
    a. 0
    b. 1
    c. There is no mode.
    d. It cannot be determined unless the data values are given.

11. When data are categorized as, for example, places of residence (rural, suburban, urban), the most appropriate measure of central tendency is the
    a. Mean   c. Mode
    b. Median   d. Midrange

12. P_{50} corresponds to
    a. Q_2
    b. D_5
    c. IQR
    d. Midrange

13. Which is not part of the five-number summary?
    a. Q_1 and Q_3
    b. The mean
    c. The median
    d. The smallest and the largest data values

14. A statistic that tells the number of standard deviations a data value is above or below the mean is called
    a. A quartile
    b. A percentile
    c. A coefficient of variation
    d. A z score

15. When a distribution is bell-shaped, approximately what percentage of data values will fall within 1 standard deviation of the mean?
    a. 50%
    b. 68%
    c. 95%
    d. 99.7%

Complete these statements with the best answer.

16. A measure obtained from sample data is called a(n) _____.

17. Generally, Greek letters are used to represent _____, and Roman letters are used to represent _____.

18. The positive square root of the variance is called the _____.

19. The symbol for the population standard deviation is _____.

20. When the sum of the lowest data value and the highest data value is divided by 2, the measure is called _____.

21. If the mode is to the left of the median and the mean is to the right of the median, then the distribution is _____ skewed.

22. An extremely high or extremely low data value is called a(n) _____.

23. The number of highway miles per gallon of the 10 worst vehicles is shown.

    12   15   13   14   15   16   17   16   17   18

    Source: Pittsburgh Post Gazette.

    Find each of these.
    a. Mean
    b. Median
    c. Mode
    d. Midrange
    e. Range
    f. Variance
    g. Standard deviation

24. The distribution of the number of errors that 10 students made on a typing test is shown.

Errors	Frequency
0–2	1
3–5	3
6–8	4
9–11	1
12–14	1

Find each of these.

a. Mean   c. Variance
b. Modal class   d. Standard deviation

25. Shown here is a frequency distribution for the number of inches of rain received in 1 year in 25 selected cities in the United States.

Number of inches	Frequency
5.5–20.5	2
20.5–35.5	3
35.5–50.5	8
50.5–65.5	6
65.5–80.5	3
80.5–95.5	3

Find each of these.

a. Mean
b. Modal class
c. Variance
d. Standard deviation

26. A survey of 36 selected recording companies showed these numbers of days that it took to receive a shipment from the day it was ordered.

Days	Frequency
1–3	6
4–6	8
7–9	10
10–12	7
13–15	0
16–18	5

Find each of these.

a. Mean
b. Modal class
c. Variance
d. Standard deviation

27. In a survey of third-grade students, this distribution was obtained for the number of "best friends" each had.

Number of students	Number of best friends
8	1
6	2
5	3
3	0

Find the average number of best friends for the class. Use the weighted mean.

28. In an advertisement, a retail store stated that its employees averaged 9 years of service. The distribution is shown here.

Number of employees	Years of service
8	2
2	6
3	10

Using the weighted mean, calculate the correct average.

29. The average number of newspapers for sale in an airport newsstand is 12, and the standard deviation is 4. The average age of the pilots is 37 years, with a standard deviation of 6 years. Which data set is more variable?

30. A survey of grocery stores showed that the average number of brands of toothpaste carried was 16, with a standard deviation of 5. The same survey showed the average length of time each store was in business was 7 years, with a standard deviation of 1.6 years. Which is more variable, the number of brands or the number of years?

31. A student scored 76 on a general science test where the class mean and standard deviation were 82 and 8, respectively; he also scored 53 on a psychology test where the class mean and standard deviation were 58 and 3, respectively. In which class was his relative position higher?

32. Which score has the highest relative position?

a. $X = 12$ $\bar{X} = 10$ $s = 4$
b. $X = 170$ $\bar{X} = 120$ $s = 32$
c. $X = 180$ $\bar{X} = 60$ $s = 8$

33. The number of square feet (in millions) of 8 of the largest malls in southwestern Pennsylvania is shown.

1	0.9	1.3	0.8
1.4	0.77	0.7	1.2

Source: International Council of Shopping Centers.

a. Find the percentile for each value.
b. What value corresponds to the 40th percentile?
c. Construct a boxplot and comment on the nature of the distribution.

34. On a philosophy comprehensive exam, this distribution was obtained from 25 students.

Score	Frequency
40.5–45.5	3
45.5–50.5	8
50.5–55.5	10
55.5–60.5	3
60.5–65.5	1

a. Construct a percentile graph.
b. Find the values that correspond to the 22nd, 78th, and 99th percentiles.
c. Find the percentiles of the values 52, 43, and 64.

35. The first column of these data represents the prebuy gas price of a rental car, and the second column represents the price charged if the car is returned without refilling the gas tank for a selected car rental company. Draw two boxplots for the data and compare the distributions.

Prebuy cost	No prebuy cost
$1.55	$3.80
1.54	3.99
1.62	3.99
1.65	3.85
1.72	3.99
1.63	3.95
1.65	3.94
1.72	4.19
1.45	3.84
1.52	3.94

Source: USA TODAY.

36. The average national SAT score is 1019. If we assume a bell-shaped distribution and a standard deviation equal to 110, what percentage of scores would you expect to fall above 1129? Above 799?

Source: N.Y. Times Almanac, 2002.

Critical Thinking Challenges

1. Averages give us information to help us to see where we stand and enable us to make comparisons. Here is a study on the average cost of a wedding. What type of average—mean, median, mode, or midrange—might have been used for each category?

OTHER PEOPLE'S MONEY

Question: What is the hottest wedding month. **Answer:** It's a tie. September now ranks as high as June in U.S. nuptials. The average attendence is 186 guests. And what kind of tabs are people running up for these affairs? Well, the next time a bride is throwing a bouquet, single women might want to . . . duck!

Reception	**$7246**
Rings	**4042**
Photos/videography.	**1263**
Bridal gown	**790**
Flowers	**775**
Music	**745**
Invitations	**374**
Mother of the bride's dress	**198**
Other (veil, limo, fees, etc.)	**3441**
Average cost of a wedding	**$18,874**

Stats: Bride's 2000 State of the Union Report

Source: Reprinted with permission from the September 2001 Reader's Digest. Copyright © 2001 by The Reader's Digest Assn., Inc.

2. This article states that the average yearly cost of smoking a pack of cigarettes a day is $1190. Find the average cost of a pack of cigarettes in your area, and compute the cost per day for 1 year. Compare your answer with the one in the article.

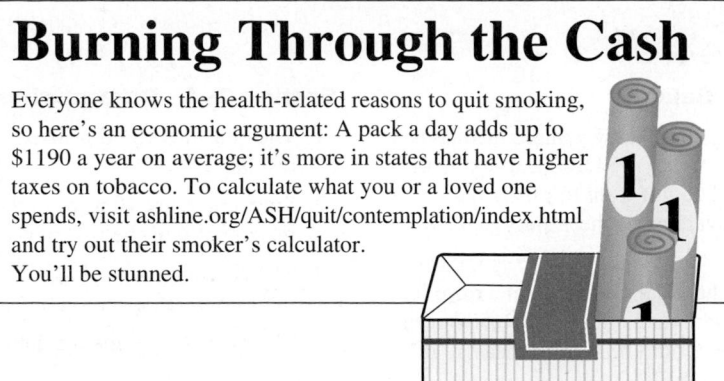

Burning Through the Cash

Everyone knows the health-related reasons to quit smoking, so here's an economic argument: A pack a day adds up to $1190 a year on average; it's more in states that have higher taxes on tobacco. To calculate what you or a loved one spends, visit ashline.org/ASH/quit/contemplation/index.html and try out their smoker's calculator. You'll be stunned.

Source: Reprinted with permission from the April 2002 Reader's Digest. Copyright © 2002 by The Reader's Digest Assn., Inc.

3. The table shows the median ages of residents for the 10 oldest states and the 10 youngest states of the United States including Washington, D.C. Explain why the median is used instead of the mean.

10 Oldest			10 Youngest		
Rank	**State**	**Median age**	**Rank**	**State**	**Median age**
1	West Virginia	38.9	51	Utah	27.1
2	Florida	38.7	50	Texas	32.3
3	Maine	38.6	49	Alaska	32.4
4	Pennsylvania	38.0	48	Idaho	33.2
5	Vermont	37.7	47	California	33.3
6	Montana	37.5	46	Georgia	33.4
7	Connecticut	37.4	45	Mississippi	33.8
8	New Hampshire	37.1	44	Louisiana	34.0
9	New Jersey	36.7	43	Arizona	34.2
10	Rhode Island	36.7	42	Colorado	34.3

Source: U.S. Census Bureau.

Data Projects

Where appropriate, use MINITAB, the TI-83 Plus, the TI-84 Plus, or a computer program of your choice to complete the following exercises.

1. Select a variable and collect about 10 values for two groups. (For example, you may want to ask 10 men how many cups of coffee they drink per day and 10 women the same question.)

   a. Define the variable.
   b. Define the populations.
   c. Describe how the samples were selected.
   d. Write a paragraph describing the similarities and differences between the two groups, using appropriate descriptive statistics such as means, standard deviations, and so on.

2. Collect data consisting of at least 30 values.

   a. State the purpose of the project.
   b. Define the population.
   c. State how the sample was selected.
   d. Using appropriate descriptive statistics, write a paragraph summarizing the data.

You may use the following websites to obtain raw data:

Visit the data sets at the book's website found at http://www.mhhe.com/math/stat/bluman. Click on the 6th edition.
http://lib.stat.cmu.edu/DASL
http://www.statcan.ca

Answers to Applying the Concepts

Section 3–2 Teacher Salaries

1. The sample mean is $22,921.67, the sample median is $16,500, and the sample mode is $11,000. If you work for the school board and do not want to raise salaries, you could say that the average teacher salary is $22,921.67.

2. If you work for the teachers' union and want a raise for the teachers, either the sample median of $16,500 or the sample mode of $11,000 would be a good measure of center to report.

3. The outlier is $107,000. With the outlier removed, the sample mean is $15,278.18, the sample median is $16,400, and the sample mode is still $11,000. The mean is greatly affected by the outlier and allows the school board to report an average teacher salary that is not representative of a "typical" teacher salary.

4. If the salaries represented every teacher in the school district, the averages would be parameters, since we have data from the entire population.

5. The mean can be misleading in the presence of outliers, since it is greatly affected by these extreme values.

6. Since the mean is greater than both the median and the mode, the distribution is skewed to the right (positively skewed).

Section 3–3 Blood Pressure

1. Chebyshev's theorem does not work for scores within 1 standard deviation of the mean.

2. At least 75% (900) of the normotensive men will fall in the interval 105–141 mm Hg.

3. About 95% (1330) of the normotensive women have diastolic blood pressures between 62 and 90 mm Hg. About 95% (1235) of the hypertensive women have diastolic blood pressures between 68 and 108 mm Hg.

4. About 95% (1140) of the normotensive men have systolic blood pressures between 105 and 141 mm Hg. About 95% (1045) of the hypertensive men have systolic blood pressures between 119 and 187 mm Hg. These two ranges do overlap.

Section 3–4 Determining Dosages

1. The quartiles could be used to describe the data results.

2. Since there are 10 mice in the upper quartile, this would mean that 4 of them survived.

3. The percentiles would give us the position of a single mouse with respect to all other mice.

4. The quartiles divide the data into four groups of equal size.

5. Standard scores would give us the position of a single mouse with respect to the mean time until the onset of sepsis.

Section 3–5 The Noisy Workplace

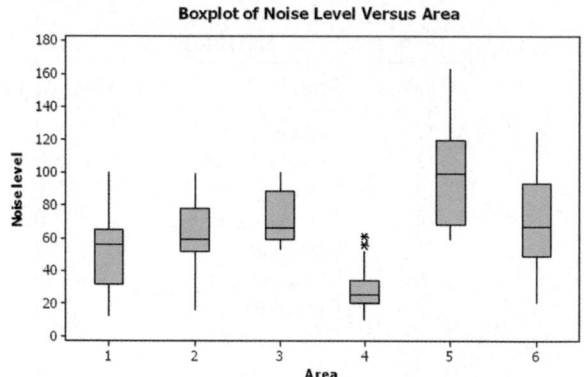

From this boxplot, we see that about 25% of the readings in area 5 are above the safe hearing level of 120 decibels. Those workers in area 5 should definitely have protective ear wear. One of the readings in area 6 is above the safe hearing level. It might be a good idea to provide protective ear wear to those workers in area 6 as well. Areas 1–4 appear to be "safe" with respect to hearing level, with area 4 being the safest.

C H A P T E R

Probability and Counting Rules

Objectives

After completing this chapter, you should be able to

1 Determine sample spaces and find the probability of an event, using classical probability or empirical probability.

2 Find the probability of compound events, using the addition rules.

3 Find the probability of compound events, using the multiplication rules.

4 Find the conditional probability of an event.

5 Find the total number of outcomes in a sequence of events, using the fundamental counting rule.

6 Find the number of ways that r objects can be selected from n objects, using the permutation rule.

7 Find the number of ways that r objects can be selected from n objects without regard to order, using the combination rule.

8 Find the probability of an event, using the counting rules.

Outline

Statistics Today

Would You Bet Your Life?

Humans not only bet money when they gamble, but also bet their lives by engaging in unhealthy activities such as smoking, drinking, using drugs, and exceeding the speed limit when driving. Many people don't care about the risks involved in these activities since they do not understand the concepts of probability. On the other hand, people may fear activities that involve little risk to health or life because these activities have been sensationalized by the press and media.

In his book *Probabilities in Everyday Life* (Ivy Books, p. 191), John D. McGervey states

> *When people have been asked to estimate the frequency of death from various causes, the most overestimated categories are those involving pregnancy, tornadoes, floods, fire, and homicide. The most underestimated categories include deaths from diseases such as diabetes, strokes, tuberculosis, asthma, and stomach cancer (although cancer in general is overestimated).*

The question then is, Would you feel safer if you flew across the United States on a commercial airline or if you drove? How much greater is the risk of one way to travel over the other? See Statistics Today—Revisited at the end of the chapter for the answer.

In this chapter, you will learn about probability—its meaning, how it is computed, and how to evaluate it in terms of the likelihood of an event actually happening.

4-1   Introduction

A cynical person once said, "The only two sure things are death and taxes." This philosophy no doubt arose because so much in people's lives is affected by chance. From the time a person awakes until he or she goes to bed, that person makes decisions regarding the possible events that are governed at least in part by chance. For example, should I carry an umbrella to work today? Will my car battery last until spring? Should I accept that new job?

Probability as a general concept can be defined as the chance of an event occurring. Many people are familiar with probability from observing or playing games of chance, such as card games, slot machines, or lotteries. In addition to being used in games of chance, probability theory is used in the fields of insurance, investments, and weather forecasting and in various other areas. Finally, as stated in Chapter 1, probability is the basis of inferential statistics. For example, predictions are based on probability, and hypotheses are tested by using probability.

The basic concepts of probability are explained in this chapter. These concepts include *probability experiments, sample spaces,* the *addition* and *multiplication rules,* and the *probabilities of complementary events.* Also in this chapter, you will learn the rule for counting, the differences between permutations and combinations, and how to figure out how many different combinations for specific situations exist. Finally, Section 4–6 explains how the counting rules and the probability rules can be used together to solve a wide variety of problems.

<table>
<tr><td>**4–2**</td><td></td></tr>
</table>

Sample Spaces and Probability

The theory of probability grew out of the study of various games of chance using coins, dice, and cards. Since these devices lend themselves well to the application of concepts of probability, they will be used in this chapter as examples. This section begins by explaining some basic concepts of probability. Then the types of probability and probability rules are discussed.

Basic Concepts

Processes such as flipping a coin, rolling a die, or drawing a card from a deck are called *probability experiments.*

Objective 1

Determine sample spaces and find the probability of an event, using classical probability or empirical probability.

A **probability experiment** is a chance process that leads to well-defined results called outcomes.

An **outcome** is the result of a single trial of a probability experiment.

A trial means flipping a coin once, rolling one die once, or the like. When a coin is tossed, there are two possible outcomes: head or tail. (*Note:* We exclude the possibility of a coin landing on its edge.) In the roll of a single die, there are six possible outcomes: 1, 2, 3, 4, 5, or 6. In any experiment, the set of all possible outcomes is called the *sample space.*

A **sample space** is the set of all possible outcomes of a probability experiment.

Some sample spaces for various probability experiments are shown here.

Experiment	Sample space
Toss one coin	Head, tail
Roll a die	1, 2, 3, 4, 5, 6
Answer a true/false question	True, false
Toss two coins	Head-head, tail-tail, head-tail, tail-head

It is important to realize that when two coins are tossed, there are *four* possible outcomes, as shown in the fourth experiment above. Both coins could fall heads up. Both coins could fall tails up. Coin 1 could fall heads up and coin 2 tails up. Or coin 1 could fall tails up and coin 2 heads up. Heads and tails will be abbreviated as H and T throughout this chapter.

Example 4–1	Find the sample space for rolling two dice.

Solution

Since each die can land in six different ways, and two dice are rolled, the sample space can be presented by a rectangular array, as shown in Figure 4–1. The sample space is the list of pairs of numbers in the chart.

Figure 4–1

Sample Space for
Rolling Two Dice
(Example 4–1)

Die 1	Die 2					
	1	2	3	4	5	6
1	(1, 1)	(1, 2)	(1, 3)	(1, 4)	(1, 5)	(1, 6)
2	(2, 1)	(2, 2)	(2, 3)	(2, 4)	(2, 5)	(2, 6)
3	(3, 1)	(3, 2)	(3, 3)	(3, 4)	(3, 5)	(3, 6)
4	(4, 1)	(4, 2)	(4, 3)	(4, 4)	(4, 5)	(4, 6)
5	(5, 1)	(5, 2)	(5, 3)	(5, 4)	(5, 5)	(5, 6)
6	(6, 1)	(6, 2)	(6, 3)	(6, 4)	(6, 5)	(6, 6)

Example 4–2	Find the sample space for drawing one card from an ordinary deck of cards.

Solution

Since there are 4 suits (hearts, clubs, diamonds, and spades) and 13 cards for each suit (ace through king), there are 52 outcomes in the sample space. See Figure 4–2.

Figure 4–2

Sample Space for
Drawing a Card
(Example 4–2)

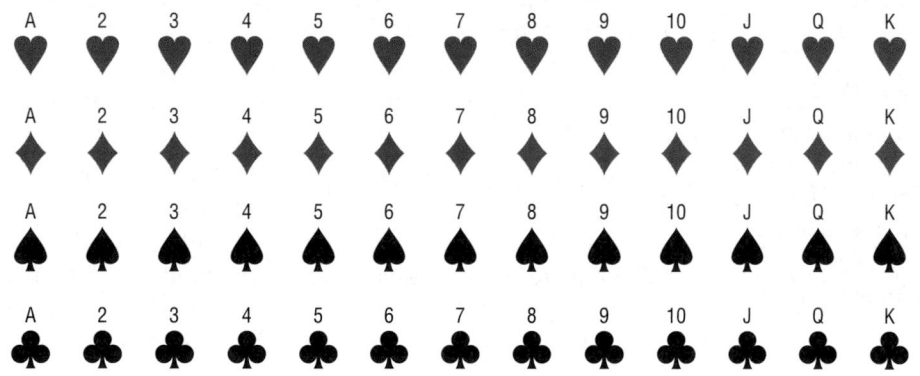

Example 4–3	Find the sample space for the gender of the children if a family has three children. Use B for boy and G for girl.

Solution

There are two genders, male and female, and each child could be either gender. Hence, there are eight possibilities, as shown here.

BBB   BBG   BGB   GBB   GGG   GGB   GBG   BGG

In Examples 4–1 through 4–3, the sample spaces were found by observation and reasoning; however, another way to find all possible outcomes of a probability experiment is to use a *tree diagram*.

A **tree diagram** is a device consisting of line segments emanating from a starting point and also from the outcome point. It is used to determine all possible outcomes of a probability experiment.

| **Example 4–4** | Use a tree diagram to find the sample space for the gender of three children in a family, as in Example 4–3. |

Solution

Since there are two possibilities (boy or girl) for the first child, draw two branches from a starting point and label one B and the other G. Then if the first child is a boy, there are two possibilities for the second child (boy or girl), so draw two branches from B and label one B and the other G. Do the same if the first child is a girl. Follow the same procedure for the third child. The completed tree diagram is shown in Figure 4–3. To find the outcomes for the sample space, trace through all the possible branches, beginning at the starting point for each one.

Figure 4–3

Tree Diagram for Example 4–4

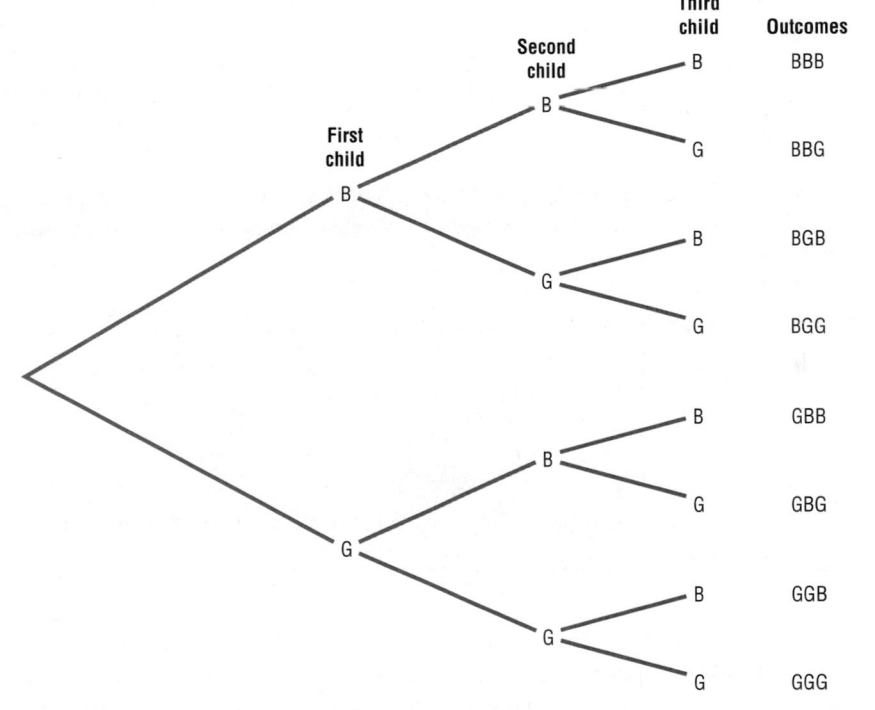

An outcome was defined previously as the result of a single trial of a probability experiment. In many problems, one must find the probability of two or more outcomes. For this reason, it is necessary to distinguish between an outcome and an event.

An **event** consists of a set of outcomes of a probability experiment.

An event can be one outcome or more than one outcome. For example, if a die is rolled and a 6 shows, this result is called an *outcome,* since it is a result of a single trial. An event with one outcome is called a **simple event.** The event of getting an odd number

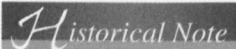

During the mid-1600s, a professional gambler named Chevalier de Méré made a considerable amount of money on a gambling game. He would bet unsuspecting patrons that in four rolls of a die, he could get at least one 6. He was so successful at the game that some people refused to play. He decided that a new game was necessary to continue his winnings. By reasoning, he figured he could roll at least one double 6 in 24 rolls of two dice, but his reasoning was incorrect and he lost systematically. Unable to figure out why, he contacted a mathematician named Blaise Pascal (1623–1662) to find out why.

Pascal became interested and began studying probability theory. He corresponded with a French government official, Pierre de Fermat (1601–1665), whose hobby was mathematics. Together the two formulated the beginnings of probability theory.

when a die is rolled is called a **compound event,** since it consists of three outcomes or three simple events. In general, a compound event consists of two or more outcomes or simple events.

There are three basic interpretations of probability:

1. Classical probability
2. Empirical or relative frequency probability
3. Subjective probability

Classical Probability

Classical probability uses sample spaces to determine the numerical probability that an event will happen. One does not actually have to perform the experiment to determine that probability. Classical probability is so named because it was the first type of probability studied formally by mathematicians in the 17th and 18th centuries.

Classical probability assumes that all outcomes in the sample space are equally likely to occur. For example, when a single die is rolled, each outcome has the same probability of occurring. Since there are six outcomes, each outcome has a probability of $\frac{1}{6}$. When a card is selected from an ordinary deck of 52 cards, one assumes that the deck has been shuffled, and each card has the same probability of being selected. In this case, it is $\frac{1}{52}$.

Equally likely events are events that have the same probability of occurring.

Formula for Classical Probability

The probability of any event E is

$$\frac{\text{Number of outcomes in } E}{\text{Total number of outcomes in the sample space}}$$

This probability is denoted by

$$P(E) = \frac{n(E)}{n(S)}$$

This probability is called *classical probability,* and it uses the sample space S.

Probabilities can be expressed as fractions, decimals, or—where appropriate—percentages. If one asks, "What is the probability of getting a head when a coin is tossed?" typical responses can be any of the following three.

"One-half."

"Point five."

"Fifty percent."[1]

These answers are all equivalent. In most cases, the answers to examples and exercises given in this chapter are expressed as fractions or decimals, but percentages are used where appropriate.

---

[1]Strictly speaking, a percent is not a probability. However, in everyday language, probabilities are often expressed as percents (i.e., there is a 60% chance of rain tomorrow). For this reason, some probabilities will be expressed as percents throughout this book.

Rounding Rule for Probabilities Probabilities should be expressed as reduced fractions or rounded to two or three decimal places. When the probability of an event is an extremely small decimal, it is permissible to round the decimal to the first nonzero digit after the point. For example, 0.0000587 would be 0.00006. When obtaining probabilities from one of the tables in Appendix C, use the number of decimal places given in the table. If decimals are converted to percentages to express probabilities, move the decimal point two places to the right and add a percent sign.

Example 4–5

For a card drawn from an ordinary deck, find the probability of getting a queen.

Solution

Since there are 52 cards in a deck and there are 4 queens, $P(\text{queen}) = \frac{4}{52} = \frac{1}{13}$.

Example 4–6

If a family has three children, find the probability that all the children are girls.

Solution

The sample space for the gender of children for a family that has three children is BBB, BBG, BGB, GBB, GGG, GGB, GBG, and BGG (see Examples 4–3 and 4–4). Since there is one way in eight possibilities for all three children to be girls,

$$P(\text{GGG}) = \frac{1}{8}$$

In probability theory, it is important to understand the meaning of the words *and* and *or*. For example, if you were asked to find the probability of getting a queen *and* a heart when you are drawing a single card from a deck, you would be looking for the queen of hearts. Here the word *and* means "at the same time." The word *or* has two meanings. For example, if you were asked to find the probability of selecting a queen *or* a heart when one card is selected from a deck, you would be looking for one of the 4 queens or one of the 13 hearts. In this case, the queen of hearts would be included in both cases and counted twice. So there would be $4 + 13 - 1 = 16$ possibilities.

On the other hand, if you were asked to find the probability of getting a queen *or* a king, you would be looking for one of the 4 queens or one of the 4 kings. In this case, there would be $4 + 4 = 8$ possibilities. In the first case, both events can occur at the same time; we say that this is an example of the *inclusive or*. In the second case, both events cannot occur at the same time, and we say that this is an example of the *exclusive or*.

Example 4–7

A card is drawn from an ordinary deck. Find these probabilities.

  a. Of getting a jack
  b. Of getting the 6 of clubs (i.e., a 6 and a club)
  c. Of getting a 3 or a diamond
  d. Of getting a 3 or a 6

Solution

  a. Refer to the sample space in Figure 4–2. There are 4 jacks so there are 4 outcomes in event E and 52 possible outcomes in the sample space. Hence,

  $$P(\text{jack}) = \frac{4}{52} = \frac{1}{13}$$

b. Since there is only one 6 of clubs in event E, the probability of getting a 6 of clubs is

$$P(6 \text{ of clubs}) = \tfrac{1}{52}$$

c. There are four 3s and 13 diamonds, but the 3 of diamonds is counted twice in this listing. Hence, there are 16 possibilities of drawing a 3 or a diamond, so

$$P(3 \text{ or diamond}) = \tfrac{16}{52} = \tfrac{4}{13}$$

This is an example of the inclusive or.

d. Since there are four 3s and four 6s,

$$P(3 \text{ or } 6) = \tfrac{8}{52} = \tfrac{2}{13}$$

This is an example of the exclusive or.

There are four basic probability rules. These rules are helpful in solving probability problems, in understanding the nature of probability, and in deciding if your answers to the problems are correct.

Historical Note

Paintings in tombs excavated in Egypt show that the Egyptians played games of chance. One game called *Hounds and Jackals* played in 1800 B.C. is similar to the present-day game of *Snakes and Ladders.*

Probability Rule 1

The probability of any event E is a number (either a fraction or decimal) between and including 0 and 1. This is denoted by $0 \le P(E) \le 1$.

Rule 1 states that probabilities cannot be negative or greater than 1.

Probability Rule 2

If an event E cannot occur (i.e., the event contains no members in the sample space), its probability is 0.

Example 4–8

When a single die is rolled, find the probability of getting a 9.

Solution

Since the sample space is 1, 2, 3, 4, 5, and 6, it is impossible to get a 9. Hence, the probability is $P(9) = \tfrac{0}{6} = 0$.

Probability Rule 3

If an event E is certain, then the probability of E is 1.

In other words, if $P(E) = 1$, then the event E is certain to occur. This rule is illustrated in Example 4–9.

| Example 4–9 | When a single die is rolled, what is the probability of getting a number less than 7? |

Solution

Since all outcomes—1, 2, 3, 4, 5, and 6—are less than 7, the probability is

$$P(\text{number less than 7}) = \tfrac{6}{6} = 1$$

The event of getting a number less than 7 is certain.

In other words, probability values range from 0 to 1. When the probability of an event is close to 0, its occurrence is highly unlikely. When the probability of an event is near 0.5, there is about a 50-50 chance that the event will occur; and when the probability of an event is close to 1, the event is highly likely to occur.

Probability Rule 4

The sum of the probabilities of all the outcomes in the sample space is 1.

For example, in the roll of a fair die, each outcome in the sample space has a probability of $\tfrac{1}{6}$. Hence, the sum of the probabilities of the outcomes is as shown.

Outcome	1	2	3	4	5	6
Probability	$\tfrac{1}{6}$	$\tfrac{1}{6}$	$\tfrac{1}{6}$	$\tfrac{1}{6}$	$\tfrac{1}{6}$	$\tfrac{1}{6}$
Sum	$\tfrac{1}{6}$ +	$\tfrac{1}{6}$ +	$\tfrac{1}{6}$ +	$\tfrac{1}{6}$ +	$\tfrac{1}{6}$ +	$\tfrac{1}{6}$ = $\tfrac{6}{6}$ = 1

Complementary Events

Another important concept in probability theory is that of *complementary events*. When a die is rolled, for instance, the sample space consists of the outcomes 1, 2, 3, 4, 5, and 6. The event E of getting odd numbers consists of the outcomes 1, 3, and 5. The event of not getting an odd number is called the *complement* of event E, and it consists of the outcomes 2, 4, and 6.

The **complement of an event** E is the set of outcomes in the sample space that are not included in the outcomes of event E. The complement of E is denoted by $\overline{E}$ (read "E bar").

Example 4–10 further illustrates the concept of complementary events.

| Example 4–10 | Find the complement of each event. |

    a. Rolling a die and getting a 4

    b. Selecting a letter of the alphabet and getting a vowel

    c. Selecting a month and getting a month that begins with a J

    d. Selecting a day of the week and getting a weekday

Solution

    a. Getting a 1, 2, 3, 5, or 6

    b. Getting a consonant (assume *y* is a consonant)

    c. Getting February, March, April, May, August, September, October, November, or December

    d. Getting Saturday or Sunday

The outcomes of an event and the outcomes of the complement make up the entire sample space. For example, if two coins are tossed, the sample space is HH, HT, TH, and TT. The complement of "getting all heads" is not "getting all tails," since the event "all heads" is HH, and the complement of HH is HT, TH, and TT. Hence, the complement of the event "all heads" is the event "getting at least one tail."

Since the event and its complement make up the entire sample space, it follows that the sum of the probability of the event and the probability of its complement will equal 1. That is, $P(E) + P(\overline{E}) = 1$. In Example 4–10, let E = all heads, or HH, and let $\overline{E}$ = at least one tail, or HT, TH, TT. Then $P(E) = \frac{1}{4}$ and $P(\overline{E}) = \frac{3}{4}$; hence, $P(E) + P(\overline{E}) = \frac{1}{4} + \frac{3}{4} = 1$.

The rule for complementary events can be stated algebraically in three ways.

Rule for Complementary Events

$$P(\overline{E}) = 1 - P(E) \qquad \text{or} \qquad P(E) = 1 - P(\overline{E}) \qquad \text{or} \qquad P(E) + P(\overline{E}) = 1$$

Stated in words, the rule is: *If the probability of an event or the probability of its complement is known, then the other can be found by subtracting the probability from 1.* This rule is important in probability theory because at times the best solution to a problem is to find the probability of the complement of an event and then subtract from 1 to get the probability of the event itself.

Example 4–11

If the probability that a person lives in an industrialized country of the world is $\frac{1}{5}$, find the probability that a person does not live in an industrialized country.

Source: Harper's Index.

Solution

P(not living in an industrialized country) = $1 - P$(living in an industrialized country)

$$= 1 - \frac{1}{5} = \frac{4}{5}$$

Probabilities can be represented pictorially by **Venn diagrams.** Figure 4–4(a) shows the probability of a simple event *E*. The area inside the circle represents the probability of event *E*, that is, $P(E)$. The area inside the rectangle represents the probability of all the events in the sample space $P(S)$.

The Venn diagram that represents the probability of the complement of an event $P(\overline{E})$ is shown in Figure 4–4(b). In this case, $P(\overline{E}) = 1 - P(E)$, which is the area inside the rectangle but outside the circle representing $P(E)$. Recall that $P(S) = 1$ and $P(E) = 1 - P(\overline{E})$. The reasoning is that $P(E)$ is represented by the area of the circle and $P(\overline{E})$ is the probability of the events that are outside the circle.

Figure 4–4

Venn Diagram for the Probability and Complement

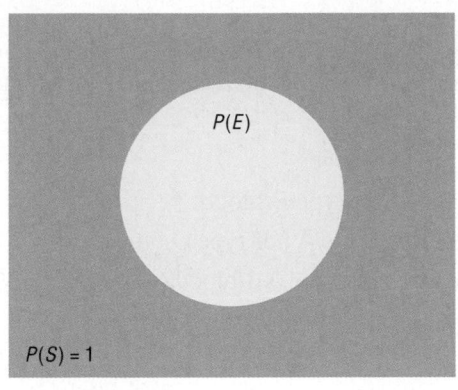

$P(E)$

$P(S) = 1$

(a) Simple probability

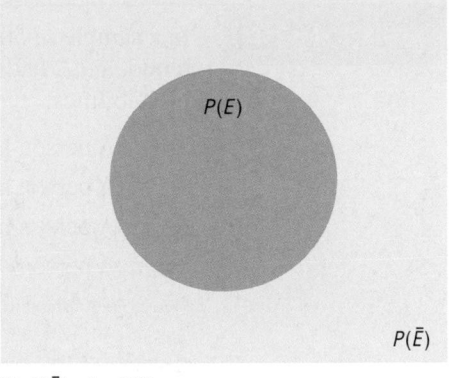

$P(E)$

$P(\bar{E})$

(b) $P(\bar{E}) = 1 - P(E)$

Empirical Probability

The difference between classical and **empirical probability** is that classical probability assumes that certain outcomes are equally likely (such as the outcomes when a die is rolled), while empirical probability relies on actual experience to determine the likelihood of outcomes. In empirical probability, one might actually roll a given die 6000 times, observe the various frequencies, and use these frequencies to determine the probability of an outcome. Suppose, for example, that a researcher asked 25 people if they liked the taste of a new soft drink. The responses were classified as "yes," "no," or "undecided." The results were categorized in a frequency distribution, as shown.

Response	Frequency
Yes	15
No	8
Undecided	2
Total	25

Probabilities now can be compared for various categories. For example, the probability of selecting a person who liked the taste is $\frac{15}{25}$, or $\frac{3}{5}$, since 15 out of 25 people in the survey answered yes.

Formula for Empirical Probability

Given a frequency distribution, the probability of an event being in a given class is

$$P(E) = \frac{\text{frequency for the class}}{\text{total frequencies in the distribution}} = \frac{f}{n}$$

This probability is called *empirical probability* and is based on observation.

Example 4–12

In the soft drink survey just described, find the probability that a person responded no.

Solution

$$P(E) = \frac{f}{n} = \frac{8}{25}$$

Note: This is the same relative frequency explained in Chapter 2.

| Example 4–13 | In a sample of 50 people, 21 had type O blood, 22 had type A blood, 5 had type B blood, and 2 had type AB blood. Set up a frequency distribution and find the following probabilities. |

    a. A person has type O blood.

    b. A person has type A or type B blood.

    c. A person has neither type A nor type O blood.

    d. A person does not have type AB blood.

Source: The American Red Cross.

Solution

Type	Frequency
A	22
B	5
AB	2
O	21
Total	50

a. $P(O) = \dfrac{f}{n} = \dfrac{21}{50}$

b. $P(A \text{ or } B) = \dfrac{22}{50} + \dfrac{5}{50} = \dfrac{27}{50}$

(Add the frequencies of the two classes.)

c. $P(\text{neither A nor O}) = \dfrac{5}{50} + \dfrac{2}{50} = \dfrac{7}{50}$

(Neither A nor O means that a person has either type B or type AB blood.)

d. $P(\text{not AB}) = 1 - P(AB) = 1 - \dfrac{2}{50} = \dfrac{48}{50} = \dfrac{24}{25}$

(Find the probability of not AB by subtracting the probability of type AB from 1.)

| Example 4–14 | Hospital records indicated that maternity patients stayed in the hospital for the number of days shown in the distribution. |

Number of days stayed	Frequency
3	15
4	32
5	56
6	19
7	5
	127

Find these probabilities.

    a. A patient stayed exactly 5 days.   *c.* A patient stayed at most 4 days.

    b. A patient stayed less than 6 days.   *d.* A patient stayed at least 5 days.

Solution

   a. $P(5) = \dfrac{56}{127}$

   b. $P(\text{less than 6 days}) = \dfrac{15}{127} + \dfrac{32}{127} + \dfrac{56}{127} = \dfrac{103}{127}$

      (Less than 6 days means 3, 4, or 5 days.)

   c. $P(\text{at most 4 days}) = \dfrac{15}{127} + \dfrac{32}{127} = \dfrac{47}{127}$

      (At most 4 days means 3 or 4 days.)

   d. $P(\text{at least 5 days}) = \dfrac{56}{127} + \dfrac{19}{127} + \dfrac{5}{127} = \dfrac{80}{127}$

      (At least 5 days means 5, 6, or 7 days.)

Empirical probabilities can also be found by using a relative frequency distribution, as shown in Section 2–3.

For example, the relative frequency distribution of the soft drink data shown before is

Response	Frequency	Relative frequency
Yes	15	0.60
No	8	0.32
Undecided	2	0.08
	25	1.00

Hence, the probability that a person responded no is 0.32, which is equal to $\frac{8}{25}$.

Law of Large Numbers

When a coin is tossed one time, it is common knowledge that the probability of getting a head is $\frac{1}{2}$. But what happens when the coin is tossed 50 times? Will it come up heads 25 times? Not all the time. One should expect about 25 heads if the coin is fair. But due to chance variation, 25 heads will not occur most of the time.

If the empirical probability of getting a head is computed by using a small number of trials, it is usually not exactly $\frac{1}{2}$. However, as the number of trials increases, the empirical probability of getting a head will approach the theoretical probability of $\frac{1}{2}$, if in fact the coin is fair (i.e., balanced). This phenomenon is an example of the **law of large numbers.**

One should be careful to not think that the number of heads and number of tails tend to "even out." As the number of trials increases, the proportion of heads to the total number of trials will approach $\frac{1}{2}$. This law holds for any type of gambling game—tossing dice, playing roulette, and so on.

It should be pointed out that the probabilities that the proportions steadily approach may or may not agree with those theorized in the classical model. If not, it can have important implications, such as "the die is not fair." Pit bosses in Las Vegas watch for empirical trends that do not agree with classical theories, and they will sometimes take a set of dice out of play if observed frequencies are too far out of line with classical expected frequencies.

Subjective Probability

The third type of probability is called *subjective probability*. **Subjective probability** uses a probability value based on an educated guess or estimate, employing opinions and inexact information.

In subjective probability, a person or group makes an educated guess at the chance that an event will occur. This guess is based on the person's experience and evaluation of a solution. For example, a sportswriter may say that there is a 70% probability that the Pirates will win the pennant next year. A physician might say that, on the basis of her diagnosis, there is a 30% chance the patient will need an operation. A seismologist might say there is an 80% probability that an earthquake will occur in a certain area. These are only a few examples of how subjective probability is used in everyday life.

All three types of probability (classical, empirical, and subjective) are used to solve a variety of problems in business, engineering, and other fields.

Probability and Risk Taking

An area in which people fail to understand probability is risk taking. Actually, people fear situations or events that have a relatively small probability of happening rather than those events that have a greater likelihood of occurring. For example, many people think that the crime rate is increasing every year. However, in his book entitled *How Risk Affects Your Everyday Life,* author James Walsh states: "Despite widespread concern about the number of crimes committed in the United States, FBI and Justice Department statistics show that the national crime rate has remained fairly level for 20 years. It even dropped slightly in the early 1990s."

He further states, "Today most media coverage of risk to health and well-being focuses on shock and outrage." Shock and outrage make good stories and can scare us about the wrong dangers. For example, the author states that if a person is 20% overweight, the loss of life expectancy is 900 days (about 3 years), but loss of life expectancy from exposure to radiation emitted by nuclear power plants is 0.02 day. As you can see, being overweight is much more of a threat than being exposed to radioactive emission.

Many people gamble daily with their lives, for example, by using tobacco, drinking and driving, and riding motorcycles. When people are asked to estimate the probabilities or frequencies of death from various causes, they tend to overestimate causes such as accidents, fires, and floods and to underestimate the probabilities of death from diseases (other than cancer), strokes, etc. For example, most people think that their chances of dying of a heart attack are 1 in 20, when in fact they are almost 1 in 3; the chances of dying by pesticide poisoning are 1 in 200,000 (*True Odds* by James Walsh). The reason people think this way is that the news media sensationalize deaths resulting from catastrophic events and rarely mention deaths from disease.

When you are dealing with life-threatening catastrophes such as hurricanes, floods, automobile accidents, or smoking, it is important to get the facts. That is, get the actual numbers from accredited statistical agencies or reliable statistical studies, and then compute the probabilities and make decisions based on your knowledge of probability and statistics.

In summary, then, when you make a decision or plan a course of action based on probability, make sure that you understand the true probability of the event occurring. Also, find out how the information was obtained (i.e., from a reliable source). Weigh the cost of the action and decide if it is worth it. Finally, look for other alternatives or courses of action with less risk involved.

Applying the Concepts **4–2**

Tossing a Coin

Assume you are at a carnival and decide to play one of the games. You spot a table where a person is flipping a coin, and since you have an understanding of basic probability, you believe that the odds of winning are in your favor. When you get to the table, you find out that all you have to do is guess which side of the coin will be facing up after it is tossed. You are assured that the coin is fair, meaning that each of the two sides has an equally likely chance of occurring. You think back about what you learned in your statistics class about probability before you decide what to bet on. Answer the following questions about the coin-tossing game.

1. What is the sample space?
2. What are the possible outcomes?
3. What does the classical approach to probability say about computing probabilities for this type of problem?

You decide to bet on heads, believing that it has a 50% chance of coming up. A friend of yours, who had been playing the game for awhile before you got there, tells you that heads has come up the last 9 times in a row. You remember the law of large numbers.

4. What is the law of large numbers, and does it change your thoughts about what will occur on the next toss?
5. What does the empirical approach to probability say about this problem, and could you use it to solve this problem?
6. Can subjective probabilities be used to help solve this problem? Explain.
7. Assume you could win $1 million if you could guess what the results of the next toss will be. What would you bet on? Why?

See page 234 for the answers.

Exercises 4–2

1. What is a probability experiment?

2. Define *sample space*.

3. What is the difference between an outcome and an event?

4. What are equally likely events?

5. What is the range of the values of the probability of an event?

6. When an event is certain to occur, what is its probability?

7. If an event cannot happen, what value is assigned to its probability?

8. What is the sum of the probabilities of all the outcomes in a sample space?

9. If the probability that it will rain tomorrow is 0.20, what is the probability that it won't rain tomorrow? Would you recommend taking an umbrella?

10. A probability experiment is conducted. Which of these cannot be considered a probability of an outcome?

a. $\frac{1}{3}$ *d.* -0.59 *g.* 1
b. $-\frac{1}{5}$ *e.* 0   *h.* 33%
c. 0.80   *f.* 1.45   *i.* 112%

11. Classify each statement as an example of classical probability, empirical probability, or subjective probability.

a. The probability that a person will watch the 6 o'clock evening news is 0.15.
b. The probability of winning at a Chuck-a-Luck game is $\frac{5}{36}$.
c. The probability that a bus will be in an accident on a specific run is about 6%.
d. The probability of getting a royal flush when five cards are selected at random is $\frac{1}{649,740}$.
e. The probability that a student will get a C or better in a statistics course is about 70%.
f. The probability that a new fast-food restaurant will be a success in Chicago is 35%.
g. The probability that interest rates will rise in the next 6 months is 0.50.

12. **(ans)** If a die is rolled one time, find these probabilities.

    a. Of getting a 4

    b. Of getting an even number

    c. Of getting a number greater than 4

    d. Of getting a number less than 7

    e. Of getting a number greater than 0

    f. Of getting a number greater than 3 or an odd number

    g. Of getting a number greater than 3 and an odd number

13. If two dice are rolled one time, find the probability of getting these results.

    a. A sum of 6

    b. Doubles

    c. A sum of 7 or 11

    d. A sum greater than 9

    e. A sum less than or equal to 4

14. **(ans)** If one card is drawn from a deck, find the probability of getting these results.

    a. An ace

    b. A diamond

    c. An ace of diamonds

    d. A 4 or a 6

    e. A 4 or a club

    f. A 6 or a spade

    g. A heart or a club

    h. A red queen

    i. A red card or a 7

    j. A black card and a 10

15. A shopping mall has set up a promotion as follows. With any mall purchase, the customer gets to spin the wheel shown here. If the number 1 comes up, the customer wins $10. If the number 2 comes up, the customer wins $5, and if the number 3 or 4 comes up, the customer wins a discount coupon. Find the following probabilities.

a. The customer wins $10.

b. The customer wins money.

c. The customer wins a coupon.

16. Choose one of the 50 states at random.

    a. What is the probability that it begins with M?

    b. What is the probability that it doesn't begin with a vowel?

17. In a college class of 250 graduating seniors, 50 have jobs waiting, 10 are going to medical school, 20 are going to law school, and 80 are going to various other kinds of graduate schools. Select one graduate at random.

    a. What is the probability that the student is going to graduate school?

    b. What is the probability that the student is going to medical school?

    c. What is the probability that the student will have to start paying back her deferred student loans after 6 months (i.e., does not continue in school)?

18. Sixty-nine percent of adults favor gun licensing in general. Choose one adult at random. What is the probability that the selected adult doesn't believe in gun licensing?

    Source: *Time* magazine.

19. For a recent year, 51% of the families in the United States had no children under the age of 18; 20% had one child; 19% had two children; 7% had three children; and 3% had four or more children. If a family is selected at random, find the probability that the family has

    a. Two or three children

    b. More than one child

    c. Less than three children

    d. Based on the answers to parts *a*, *b*, and *c*, which is most likely to occur? Explain why.

    Source: U.S. Census Bureau.

20. In a recent year, of the 1,184,000 bachelor's degrees conferred, 233,000 were in the field of business, 125,000 were in the social sciences, and 106,000 were in education. If one degree is selected at random, find the following probabilities.

    a. The degree was awarded in education.

    b. The degree was not awarded in business.

    Source: National Center for Education Statistics.

21. A couple has three children. Find each probability.

    a. All boys

    b. All girls or all boys

    c. Exactly two boys or two girls

    d. At least one child of each gender

22. In the game of craps using two dice, a person wins on the first roll if a 7 or an 11 is rolled. Find the probability of winning on the first roll.

23. In a game of craps, a player loses on the roll if a 2, 3, or 12 is tossed on the first roll. Find the probability of losing on the first roll.

24. For a specific year a total of 2541 postal workers were bitten by dogs. The top six cities for crunching canines were as follows.

Houston	49	Chicago	37
Miami	35	Los Angeles	32
Brooklyn	22	Cleveland	20

If one bitten postal worker is selected at random, what is the probability that he was bitten in Houston, Chicago, or Los Angeles? What is the probability that he was bitten in some other city?

Source: *N.Y. Times Almanac.*

25. A roulette wheel has 38 spaces numbered 1 through 36, 0, and 00. Find the probability of getting these results.

   a. An odd number (Do not count 0 or 00.)
   b. A number greater than 27
   c. A number that contains the digit 0
   d. Based on the answers to parts *a*, *b*, and *c*, which is most likely to occur? Explain why.

26. Thirty-nine of fifty states are currently under court order to alleviate overcrowding and poor conditions in one or more of their prisons. If a state is selected at random, find the probability that it is currently under such a court order.

Source: *Harper's Index.*

27. A CBS News/*New York Times* poll found that of 764 adults surveyed nationwide, 34% felt that we are spending too much on space exploration, 19% felt that we are spending too little, 35% felt that we are spending the right amount, and the rest said "don't know" or had no answer. If one of the respondents is selected at random, what is the probability that the person felt that we are spending the right amount or too little?

Source: www.pollingreport.com.

28. In a survey, 16 percent of American children said they use flattery to get their parents to buy them things. If a child is selected at random, find the probability that the child said he or she does not use parental flattery.

Source: *Harper's Index.*

29. Roll two dice and multiply the numbers together.

   a. Write out the sample space.
   b. What is the probability that the product is a multiple of 6?
   c. What is the probability that the product is less than 10?

30. The source of federal government revenue for a specific year is

   50% from individual income taxes
   32% from social insurance payroll taxes
   10% from corporate income taxes
    3% from excise taxes
    5% other

If a revenue source is selected at random, what is the probability that it comes from individual or corporate income taxes?

Source: *N.Y. Times Almanac.*

31. A box contains a $1 bill, a $5 bill, a $10 bill, and a $20 bill. A bill is selected at random, and it is not replaced; then a second bill is selected at random. Draw a tree diagram and determine the sample space.

32. Draw a tree diagram and determine the sample space for tossing four coins.

33. Four balls numbered 1 through 4 are placed in a box. A ball is selected at random, and its number is noted; then it is replaced. A second ball is selected at random, and its number is noted. Draw a tree diagram and determine the sample space.

34. Kimberly decides to have a computer custom-made. She can select one option from each category:

Megabytes	**Monitor**	**Color**
128	15 inches	Tan
256	17 inches	Ivory
512		

Draw a tree diagram for all possible types of computers she can select.

35. Betty and Claire play a tennis tournament consisting of three games. Draw a tree diagram for all possible outcomes of the tournament.

36. A coin is tossed; if it falls heads up, it is tossed again. If it falls tails up, a die is rolled. Draw a tree diagram and determine the outcomes.

Extending the Concepts

37. The distribution of ages of CEOs is as follows:

Age	Frequency
21–30	1
31–40	8
41–50	27
51–60	29
61–70	24
71–up	11

Source: Information based on
USA TODAY Snapshot.

If a CEO is selected at random, find the probability that his or her age is

a. Between 31 and 40

b. Under 31

c. Over 30 and under 51

d. Under 31 or over 60

38. A person flipped a coin 100 times and obtained 73 heads. Can the person conclude that the coin was unbalanced?

39. A medical doctor stated that with a certain treatment, a patient has a 50% chance of recovering without surgery. That is, "Either he will get well or he won't get well." Comment on this statement.

40. The wheel spinner shown here is spun twice. Find the sample space, and then determine the probability of the following events.

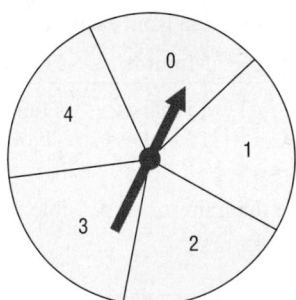

a. An odd number on the first spin and an even number on the second spin (*Note:* 0 is considered even.)

b. A sum greater than 4

c. Even numbers on both spins

d. A sum that is odd

e. The same number on both spins

41. Toss three coins 128 times and record the number of heads (0, 1, 2, or 3); then record your results with the theoretical probabilities. Compute the empirical probabilities of each.

42. Toss two coins 100 times and record the number of heads (0, 1, 2). Compute the probabilities of each outcome, and compare these probabilities with the theoretical results.

43. Odds are used in gambling games to make them fair. For example, if a person rolled a die and won every time he or she rolled a 6, then the person would win on average once every 6 times. So that the game is fair, the odds of 5 to 1 are given. This means that if the person bet $1 and won, he or she could win $5. On average, the player would win $5 once in 6 rolls and lose $1 on the other 5 rolls—hence the term *fair game.*

In most gambling games, the odds given are not fair. For example, if the odds of winning are really 20 to 1, the house might offer 15 to 1 in order to make a profit.

Odds can be expressed as a fraction or as a ratio, such as $\frac{5}{1}$, 5:1, or 5 to 1. Odds are computed in favor of the event or against the event. The formulas for odds are

$$\text{Odds in favor} = \frac{P(E)}{1 - P(E)}$$

$$\text{Odds against} = \frac{P(\overline{E})}{1 - P(\overline{E})}$$

In the die example,

$$\text{Odds in favor of a 6} = \frac{\frac{1}{6}}{\frac{5}{6}} = \frac{1}{5} \text{ or 1:5}$$

$$\text{Odds against a 6} = \frac{\frac{5}{6}}{\frac{1}{6}} = \frac{5}{1} \text{ or 5:1}$$

Find the odds in favor of and against each event.

a. Rolling a die and getting a 2

b. Rolling a die and getting an even number

c. Drawing a card from a deck and getting a spade

d. Drawing a card and getting a red card

e. Drawing a card and getting a queen

f. Tossing two coins and getting two tails

g. Tossing two coins and getting one tail

4–3

Objective [2]

Find the probability of compound events, using the addition rules.

The Addition Rules for Probability

Many problems involve finding the probability of two or more events. For example, at a large political gathering, one might wish to know, for a person selected at random, the probability that the person is a female or is a Republican. In this case, there are three possibilities to consider:

1. The person is a female.
2. The person is a Republican.
3. The person is both a female and a Republican.

Consider another example. At the same gathering there are Republicans, Democrats, and Independents. If a person is selected at random, what is the probability that the person is a Democrat or an Independent? In this case, there are only two possibilities:

1. The person is a Democrat.
2. The person is an Independent.

The difference between the two examples is that in the first case, the person selected can be a female and a Republican at the same time. In the second case, the person selected cannot be both a Democrat and an Independent at the same time. In the second case, the two events are said to be *mutually exclusive;* in the first case, they are not mutually exclusive.

Two events are **mutually exclusive events** if they cannot occur at the same time (i.e., they have no outcomes in common).

In another situation, the events of getting a 4 and getting a 6 when a single card is drawn from a deck are mutually exclusive events, since a single card cannot be both a 4 and a 6. On the other hand, the events of getting a 4 and getting a heart on a single draw are not mutually exclusive, since one can select the 4 of hearts when drawing a single card from an ordinary deck.

Example 4–15

Determine which events are mutually exclusive and which are not, when a single die is rolled.

    a. Getting an odd number and getting an even number

    b. Getting a 3 and getting an odd number

    c. Getting an odd number and getting a number less than 4

    d. Getting a number greater than 4 and getting a number less than 4

Solution

    a. The events are mutually exclusive, since the first event can be 1, 3, or 5 and the second event can be 2, 4, or 6.

    b. The events are not mutually exclusive, since the first event is a 3 and the second can be 1, 3, or 5. Hence, 3 is contained in both events.

    c. The events are not mutually exclusive, since the first event can be 1, 3, or 5 and the second can be 1, 2, or 3. Hence, 1 and 3 are contained in both events.

    d. The events are mutually exclusive, since the first event can be 5 or 6 and the second event can be 1, 2, or 3.

Example 4–16   Determine which events are mutually exclusive and which are not, when a single card is drawn from a deck.

        a. Getting a 7 and getting a jack

        b. Getting a club and getting a king

        c. Getting a face card and getting an ace

        d. Getting a face card and getting a spade

Solution

Only the events in parts *a* and *c* are mutually exclusive.

The probability of two or more events can be determined by the *addition rules*. The first addition rule is used when the events are mutually exclusive.

> ### Addition Rule 1
>
> When two events *A* and *B* are mutually exclusive, the probability that *A* or *B* will occur is
>
> $$P(A \text{ or } B) = P(A) + P(B)$$

Example 4–17   A box contains 3 glazed doughnuts, 4 jelly doughnuts, and 5 chocolate doughnuts. If a person selects a doughnut at random, find the probability that it is either a glazed doughnut or a chocolate doughnut.

Solution

Since the box contains 3 glazed doughnuts, 5 chocolate doughnuts, and a total of 12 doughnuts, $P(\text{glazed or chocolate}) = P(\text{glazed}) + P(\text{chocolate}) = \frac{3}{12} + \frac{5}{12} = \frac{8}{12} = \frac{2}{3}$. The events are mutually exclusive.

Example 4–18   At a political rally, there are 20 Republicans, 13 Democrats, and 6 Independents. If a person is selected at random, find the probability that he or she is either a Democrat or an Independent.

Solution

$$P(\text{Democrat or Independent}) = P(\text{Democrat}) + P(\text{Independent})$$
$$= \frac{13}{39} + \frac{6}{39} = \frac{19}{39}$$

Example 4–19   A day of the week is selected at random. Find the probability that it is a weekend day.

Solution

$$P(\text{Saturday or Sunday}) = P(\text{Saturday}) + P(\text{Sunday}) = \frac{1}{7} + \frac{1}{7} = \frac{2}{7}$$

When two events are not mutually exclusive, we must subtract one of the two probabilities of the outcomes that are common to both events, since they have been counted twice. This technique is illustrated in Example 4–20.

Example 4–20

A single card is drawn from a deck. Find the probability that it is a king or a club.

Solution

Since the king of clubs means a king and a club, it has been counted twice—once as a king and once as a club; therefore, one of the outcomes must be subtracted, as shown.

$$P(\text{king or club}) = P(\text{king}) + P(\text{club}) - P(\text{king of clubs})$$
$$= \tfrac{4}{52} + \tfrac{13}{52} - \tfrac{1}{52} = \tfrac{16}{52} = \tfrac{4}{13}$$

When events are not mutually exclusive, addition rule 2 can be used to find the probability of the events.

Addition Rule 2

If A and B are *not* mutually exclusive, then

$$P(A \text{ or } B) = P(A) + P(B) - P(A \text{ and } B)$$

Note: This rule can also be used when the events are mutually exclusive, since $P(A \text{ and } B)$ will always equal 0. However, it is important to make a distinction between the two situations.

Example 4–21

In a hospital unit there are 8 nurses and 5 physicians; 7 nurses and 3 physicians are females. If a staff person is selected, find the probability that the subject is a nurse or a male.

Solution

The sample space is shown here.

Staff	Females	Males	Total
Nurses	7	1	8
Physicians	3	2	5
Total	10	3	13

The probability is

$$P(\text{nurse or male}) = P(\text{nurse}) + P(\text{male}) - P(\text{male nurse})$$
$$= \tfrac{8}{13} + \tfrac{3}{13} - \tfrac{1}{13} = \tfrac{10}{13}$$

Example 4–22

On New Year's Eve, the probability of a person driving while intoxicated is 0.32, the probability of a person having a driving accident is 0.09, and the probability of a person having a driving accident while intoxicated is 0.06. What is the probability of a person driving while intoxicated or having a driving accident?

Solution

$$P(\text{intoxicated or accident}) = P(\text{intoxicated}) + P(\text{accident})$$
$$- P(\text{intoxicated and accident})$$
$$= 0.32 + 0.09 - 0.06 = 0.35$$

In summary, then, when the two events are mutually exclusive, use addition rule 1. When the events are not mutually exclusive, use addition rule 2.

The probability rules can be extended to three or more events. For three mutually exclusive events *A*, *B*, and *C*,

$$P(A \text{ or } B \text{ or } C) = P(A) + P(B) + P(C)$$

For three events that are *not* mutually exclusive,

$$P(A \text{ or } B \text{ or } C) = P(A) + P(B) + P(C) - P(A \text{ and } B) - P(A \text{ and } C)$$
$$- P(B \text{ and } C) + P(A \text{ and } B \text{ and } C)$$

See Exercises 23, 24, and 25 in this section.

Figure 4–5(a) shows a Venn diagram that represents two mutually exclusive events *A* and *B*. In this case, $P(A \text{ or } B) = P(A) + P(B)$, since these events are mutually exclusive and do not overlap. In other words, the probability of occurrence of event *A* or event *B* is the sum of the areas of the two circles.

Figure 4–5(b) represents the probability of two events that are *not* mutually exclusive. In this case, $P(A \text{ or } B) = P(A) + P(B) - P(A \text{ and } B)$. The area in the intersection or overlapping part of both circles corresponds to $P(A \text{ and } B)$; and when the area of circle *A* is added to the area of circle *B*, the overlapping part is counted twice. It must therefore be subtracted once to get the correct area or probability.

Note: Venn diagrams were developed by mathematician John Venn (1834–1923) and are used in set theory and symbolic logic. They have been adapted to probability theory also. In set theory, the symbol ∪ represents the *union* of two sets, and $A \cup B$ corresponds to *A* or *B*. The symbol ∩ represents the *intersection* of two sets, and $A \cap B$ corresponds to *A* and *B*. Venn diagrams show only a general picture of the probability rules and do not portray all situations, such as $P(A) = 0$, accurately.

Figure 4–5

Venn Diagrams for the Addition Rules

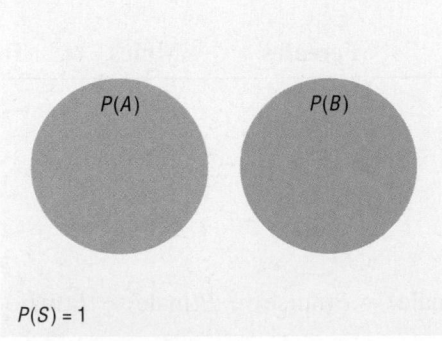

$P(S) = 1$

(a) Mutually exclusive events
$P(A \text{ or } B) = P(A) + P(B)$

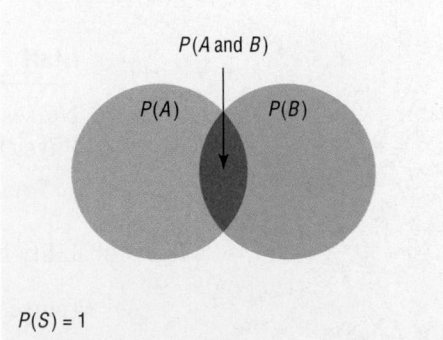

$P(S) = 1$

(b) Non-mutually exclusive events
$P(A \text{ or } B) = P(A) + P(B) - P(A \text{ and } B)$

Applying the Concepts **4–3**

Which Pain Reliever Is Best?

Assume that following an injury you received from playing your favorite sport, you obtain and read information on new pain medications. In that information you read of a study that was conducted to test the side effects of two new pain medications. Use the following table to answer the questions and decide which, if any, of the two new pain medications you will use.

	Number of side effects in 12-week clinical trial		
Side effect	Placebo $n = 192$	Drug A $n = 186$	Drug B $n = 188$
Upper respiratory congestion	10	32	19
Sinus headache	11	25	32
Stomach ache	2	46	12
Neurological headache	34	55	72
Cough	22	18	31
Lower respiratory congestion	2	5	1

1. How many subjects were in the study?

2. How long was the study?

3. What were the variables under study?

4. What type of variables are they, and what level of measurement are they on?

5. Are the numbers in the table exact figures?

6. What is the probability that a randomly selected person was receiving a placebo?

7. What is the probability that a person was receiving a placebo or drug A? Are these mutually exclusive events? What is the complement to this event?

8. What is the probability that a randomly selected person was receiving a placebo or experienced a neurological headache?

9. What is the probability that a randomly selected person was not receiving a placebo or experienced a sinus headache?

See page 234 for the answers.

Exercises 4–3

1. Define mutually exclusive events, and give an example of two events that are mutually exclusive and two events that are not mutually exclusive.

2. Determine whether these events are mutually exclusive.

   a. Roll a die: Get an even number, and get a number less than 3.

   b. Roll a die: Get a prime number (2, 3, 5), and get an odd number.

   c. Roll a die: Get a number greater than 3, and get a number less than 3.

   d. Select a student in your class: The student has blond hair, and the student has blue eyes.

   e. Select a student in your college: The student is a sophomore, and the student is a business major.

   f. Select any course: It is a calculus course, and it is an English course.

   g. Select a registered voter: The voter is a Republican, and the voter is a Democrat.

3. An automobile dealer decides to select a month for its annual sale. Find the probability that it will be September or October. Assume all months have an equal probability of being selected. Compute the probability of selecting September or October, using days, and compare the answers.

4. At a community swimming pool there are 2 managers, 8 lifeguards, 3 concession stand clerks, and 2 maintenance people. If a person is selected at random, find the probability that the person is either a lifeguard or a manager.

5. At a convention there are 7 mathematics instructors, 5 computer science instructors, 3 statistics instructors, and 4 science instructors. If an instructor is selected, find the probability of getting a science instructor or a math instructor.

6. A media rental store rented the following number of movie titles in each of these categories: 170 horror, 230 drama, 120 mystery, 310 romance, and 150 comedy. If a person selects a movie to rent, find the probability that it is a romance or a comedy. Is this event likely or unlikely to occur? Explain your answer.

7. A recent study of 200 nurses found that of 125 female nurses, 56 had bachelor's degrees; and of 75 male nurses, 34 had bachelor's degrees. If a nurse is selected at random, find the probability that the nurse is

 a. A female nurse with a bachelor's degree
 b. A male nurse
 c. A male nurse with a bachelor's degree
 d. Based on your answers to parts *a*, *b*, and *c*, explain which is most likely to occur. Explain why.

8. The probability that a student owns a car is 0.65, and the probability that a student owns a computer is 0.82. If the probability that a student owns both is 0.55, what is the probability that a given student owns neither a car nor a computer?

9. At a particular school with 200 male students, 58 play football, 40 play basketball, and 8 play both. What is the probability that a randomly selected male student plays neither sport?

10. A single card is drawn from a deck. Find the probability of selecting the following.

 a. A 4 or a diamond
 b. A club or a diamond
 c. A jack or a black card

11. In a statistics class there are 18 juniors and 10 seniors; 6 of the seniors are females, and 12 of the juniors are males. If a student is selected at random, find the probability of selecting the following.

 a. A junior or a female
 b. A senior or a female
 c. A junior or a senior

12. At a used-book sale, 100 books are adult books and 160 are children's books. Of the adult books, 70 are nonfiction while 60 of the children's books are nonfiction. If a book is selected at random, find the probability that it is

 a. Fiction
 b. Not a children's nonfiction book
 c. An adult book or a children's nonfiction book

13. The Bargain Auto Mall has these cars in stock.

	SUV	Compact	Mid-sized
Foreign	20	50	20
Domestic	65	100	45

If a car is selected at random, find the probability that it is

 a. Domestic
 b. Foreign and mid-sized
 c. Domestic or an SUV

14. The numbers of endangered species for several groups are listed here.

	Mammals	Birds	Reptiles	Amphibians
United States	63	78	14	10
Foreign	251	175	64	8

If one endangered species is selected at random, find the probability that it is

 a. Found in the United States and is a bird
 b. Foreign or a mammal
 c. Warm-blooded

Source: *N.Y. Times Almanac.*

15. A grocery store employs cashiers, stock clerks, and deli personnel. The distribution of employees according to marital status is shown here.

Marital status	Cashiers	Stock clerks	Deli personnel
Married	8	12	3
Not married	5	15	2

If an employee is selected at random, find these probabilities.

 a. The employee is a stock clerk or married.
 b. The employee is not married.
 c. The employee is a cashier or is not married.

16. In a certain geographic region, newspapers are classified as being published daily morning, daily evening, and weekly. Some have a comics section and others do not. The distribution is shown here.

Have comics section	Morning	Evening	Weekly
Yes	2	3	1
No	3	4	2

If a newspaper is selected at random, find these probabilities.

a. The newspaper is a weekly publication.

b. The newspaper is a daily morning publication or has comics.

c. The newspaper is published weekly or does not have comics.

17. Three cable channels (6, 8, and 10) have quiz shows, comedies, and dramas. The number of each is shown here.

Type of show	Channel 6	Channel 8	Channel 10
Quiz show	5	2	1
Comedy	3	2	8
Drama	4	4	2

If a show is selected at random, find these probabilities.

a. The show is a quiz show, or it is shown on channel 8.

b. The show is a drama or a comedy.

c. The show is shown on channel 10, or it is a drama.

18. A local postal carrier distributes first-class letters, advertisements, and magazines. For a certain day, she distributed the following numbers of each type of item.

Delivered to	First-class letters	Ads	Magazines
Home	325	406	203
Business	732	1021	97

If an item of mail is selected at random, find these probabilities.

a. The item went to a home.

b. The item was an ad, or it went to a business.

c. The item was a first-class letter, or it went to a home.

19. The frequency distribution shown here illustrates the number of medical tests conducted on 30 randomly selected emergency patients.

Number of tests performed	Number of patients
0	12
1	8
2	2
3	3
4 or more	5

If a patient is selected at random, find these probabilities.

a. The patient has had exactly 2 tests done.

b. The patient has had at least 2 tests done.

c. The patient has had at most 3 tests done.

d. The patient has had 3 or fewer tests done.

e. The patient has had 1 or 2 tests done.

20. This distribution represents the length of time a patient spends in a hospital.

Days	Frequency
0–3	2
4–7	15
8–11	8
12–15	6
16+	9

If a patient is selected, find these probabilities.

a. The patient spends 3 days or fewer in the hospital.

b. The patient spends fewer than 8 days in the hospital.

c. The patient spends 16 or more days in the hospital.

d. The patient spends a maximum of 11 days in the hospital.

21. A sales representative who visits customers at home finds she sells 0, 1, 2, 3, or 4 items according to the following frequency distribution.

Items sold	Frequency
0	8
1	10
2	3
3	2
4	1

Find the probability that she sells the following.

a. Exactly 1 item

b. More than 2 items

c. At least 1 item

d. At most 3 items

22. A recent study of 300 patients found that of 100 alcoholic patients, 87 had elevated cholesterol levels, and of 200 nonalcoholic patients, 43 had elevated cholesterol levels. If a patient is selected at random, find the probability that the patient is the following.

a. An alcoholic with elevated cholesterol level

b. A nonalcoholic

c. A nonalcoholic with nonelevated cholesterol level

23. If one card is drawn from an ordinary deck of cards, find the probability of getting the following.

a. A king or a queen or a jack

b. A club or a heart or a spade

c. A king or a queen or a diamond

d. An ace or a diamond or a heart

e. A 9 or a 10 or a spade or a club

24. Two dice are rolled. Find the probability of getting

    a. A sum of 5, 6, or 7

    b. Doubles or a sum of 6 or 8

    c. A sum greater than 8 or less than 3

    d. Based on the answers to parts *a*, *b*, and *c*, which is least likely to occur? Explain why.

25. An urn contains 6 red balls, 2 green balls, 1 blue ball, and 1 white ball. If a ball is drawn, find the probability of getting a red or a white ball.

26. Three dice are rolled. Find the probability of getting

    a. Triples   b. A sum of 5

Extending the Concepts

27. The probability that a customer selects a pizza with mushrooms or pepperoni is 0.43, and the probability that the customer selects only mushrooms is 0.32. If the probability that he or she selects only pepperoni is 0.17, find the probability of the customer selecting both items.

28. In building new homes, a contractor finds that the probability of a home buyer selecting a two-car garage is 0.70 and of selecting a one-car garage is 0.20. Find the probability that the buyer will select no garage. The builder does not build houses with three-car or more garages.

29. In Exercise 28, find the probability that the buyer will not want a two-car garage.

30. Suppose that $P(A) = 0.42$, $P(B) = 0.38$, and $P(A \cup B) = 0.70$. Are A and B mutually exclusive? Explain.

"I know you haven't had an accident in thirteen years. We're raising your rates because you're about due one."

Source: Reprinted with special permission of King Features Syndicate.

Technology *Step by Step*

MINITAB
Step by Step

Calculate Relative Frequency Probabilities

The random variable X represents the number of days patients stayed in the hospital from Example 4–14.

1. In C1 of a worksheet, type in the values of X. Name the column **X.**

2. In C2 enter the frequencies. Name the column **f.**

3. To calculate the relative frequencies and store them in a new column named Px:

   a) Select **Calc>Calculator.**

   b) Type **Px** in the box for Store result in variable:.

   c) Click in the Expression box, then double-click C2 f.

   d) Type or click the division operator.

   e) Scroll down the function list to Sum, then click [Select].

   f) Double-click C2 f to select it.

   g) Click [OK].

The dialog box and completed worksheet are shown.

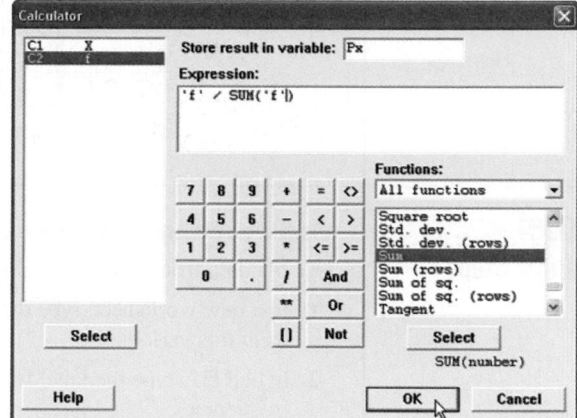

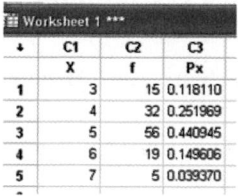

If the original data, rather than the table, are in a worksheet, use **Stat>Tables>Tally** to make the tables with percents (Section 2–2).

MINITAB can also make a two-way classification table.

Construct a Contingency Table

1. Select **File>Open Worksheet** to open the Databank.mtw file.

2. Select **Stat>Tables>Crosstabulation . . .**

   a) Double-click C4 SMOKING STATUS to select it For rows:.

   b) Select C11 GENDER for the For Columns: Field.

   c) Click on option for Counts and then [OK].

The session window and completed dialog box are shown.

Tabulated statistics: SMOKING STATUS, GENDER

Rows: SMOKING STATUS   Columns: GENDER

	F	M	All
0	25	22	47
1	18	19	37
2	7	9	16
All	50	50	100

Cell Contents: Count

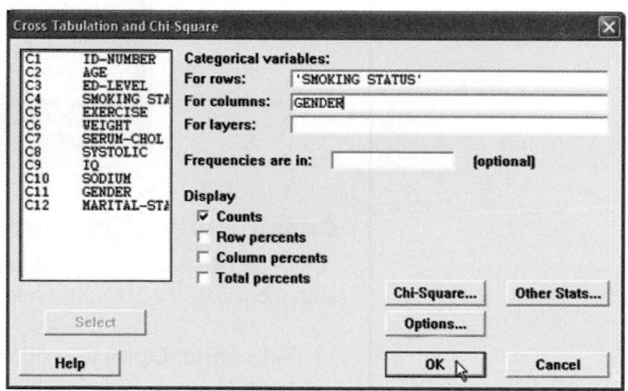

In this sample of 100 there are 25 females who do not smoke compared to 22 men. Sixteen individuals smoke 1 pack or more per day.

TI-83 Plus or TI-84 Plus
Step by Step

To construct a relative frequency table:

1. Enter the data values in L_1 and the frequencies in L_2.

2. Move the cursor to the top of the L_3 column so that L_3 is highlighted.

3. Type L_2 divided by the sample size, then press **ENTER**.

Use the data from Example 4–14.

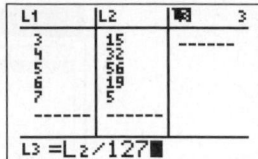

Excel
Step by Step

Constructing a Relative Frequency Distribution

Use the data from Example 4–14.

1. In a new worksheet, type the label **DAYS** in cell A1. Beginning in cell A2, type in the data from this variable.

2. In cell B1, type the label for the frequency, **COUNT.** Beginning in cell B2, type in the frequencies.

3. In the cell below the frequencies, compute the sum of the frequencies by selecting the sum icon (Σ) from the toolbar. Highlight the frequencies in column B and **Enter.**

4. In cell C1, type a label for the relative frequencies, **Rf.** Select cell C2 and type =(B2)/(B7) and **Enter.** Repeat for each frequency.

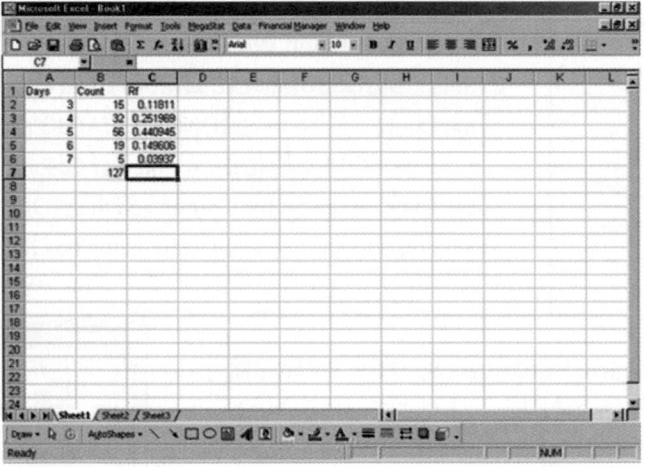

Constructing a Contingency Table

Although Excel was not specifically designed to construct a contingency table, you may do so with the use of the MegaStat Add-in available on your CD and Online Learning Center.

1. Select **File>Open** to open the Databank.xls file as directed in Example XL1–1, on page 24.

2. Highlight the column labeled SMOKING STATUS and select **Edit>Copy.**

3. On the toolbar, select **File>New** to open a new worksheet. Paste the contents of the copied column in column A of the new worksheet by selecting cell A1, then **Edit>Paste.**

4. Return to the previous file Databank.xls by selecting **File>DataBank.xls.**

5. Highlight the column labeled GENDER. Copy and paste these data in column B of the worksheet with the SMOKING STATUS data.

6. In an open cell on the new worksheet, type **M** for male and in the cell directly below in the same column, type **F** for female. In an open column, type in the categories for the SMOKING STATUS data **0, 1,** and **2** in separate cells.

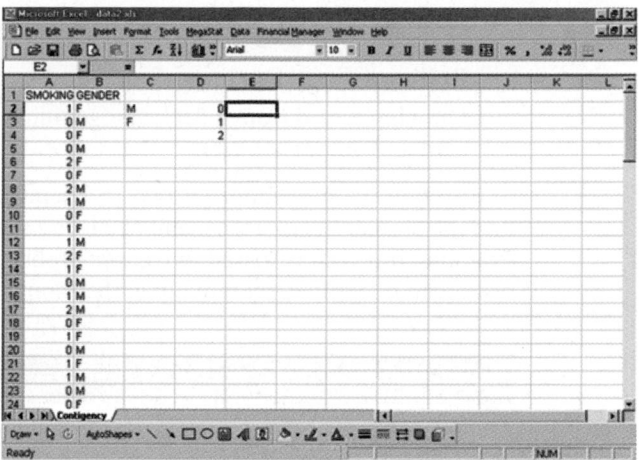

7. Select **MegaStat** from the toolbar. Select **Chi-Square/Crosstab>Crosstabulation.**

8. In the first Data range box, type **A1:A101.** In the Specification range box, type in the range of cells containing the labels for the values of the SMOKING STATUS variable.

9. In the second Data range box, type **B1:B101.** In the Specification range box, type in the range of cells containing the labels for the values of the GENDER variable.

10. Remove any checks from the Output options in the Crosstabulation dialog box. Then click [OK].

This table is obtained in a new sheet labeled Output.

Crosstabulation

		GENDER		
		M	F	Total
SMOKING STATUS	0	22	25	47
	1	19	18	37
	2	9	7	16
	Total	50	50	100

4–4   The Multiplication Rules and Conditional Probability

Section 4–3 showed that the addition rules are used to compute probabilities for mutually exclusive and non-mutually exclusive events. This section introduces the multiplication rules.

Objective 3

Find the probability of compound events, using the multiplication rules.

The Multiplication Rules

The *multiplication rules* can be used to find the probability of two or more events that occur in sequence. For example, if a coin is tossed and then a die is rolled, one can find the probability of getting a head on the coin *and* a 4 on the die. These two events are said to be *independent* since the outcome of the first event (tossing a coin) does not affect the probability outcome of the second event (rolling a die).

Two events *A* and *B* are **independent events** if the fact that *A* occurs does not affect the probability of *B* occurring.

Here are other examples of independent events:

Rolling a die and getting a 6, and then rolling a second die and getting a 3.

Drawing a card from a deck and getting a queen, replacing it, and drawing a second card and getting a queen.

To find the probability of two independent events that occur in sequence, one must find the probability of each event occurring separately and then multiply the answers. For example, if a coin is tossed twice, the probability of getting two heads is $\frac{1}{2} \cdot \frac{1}{2} = \frac{1}{4}$. This result can be verified by looking at the sample space HH, HT, TH, TT. Then $P(\text{HH}) = \frac{1}{4}$.

Multiplication Rule 1

When two events are independent, the probability of both occurring is

$$P(A \text{ and } B) = P(A) \cdot P(B)$$

Example 4–23   A coin is flipped and a die is rolled. Find the probability of getting a head on the coin and a 4 on the die.

Solution

$$P(\text{head and 4}) = P(\text{head}) \cdot P(4) = \frac{1}{2} \cdot \frac{1}{6} = \frac{1}{12}$$

Note that the sample space for the coin is H, T; and for the die it is 1, 2, 3, 4, 5, 6.

The problem in Example 4–23 can also be solved by using the sample space

H1   H2   H3   H4   H5   H6   T1   T2   T3   T4   T5   T6

The solution is $\frac{1}{12}$, since there is only one way to get the head-4 outcome.

Example 4–24   A card is drawn from a deck and replaced; then a second card is drawn. Find the probability of getting a queen and then an ace.

Solution

The probability of getting a queen is $\frac{4}{52}$, and since the card is replaced, the probability of getting an ace is $\frac{4}{52}$. Hence, the probability of getting a queen and an ace is

$$P(\text{queen and ace}) = P(\text{queen}) \cdot P(\text{ace}) = \frac{4}{52} \cdot \frac{4}{52} = \frac{16}{2704} = \frac{1}{169}$$

Example 4–25   An urn contains 3 red balls, 2 blue balls, and 5 white balls. A ball is selected and its color noted. Then it is replaced. A second ball is selected and its color noted. Find the probability of each of these.

    a. Selecting 2 blue balls

    b. Selecting 1 blue ball and then 1 white ball

    c. Selecting 1 red ball and then 1 blue ball

Solution

a. $P(\text{blue and blue}) = P(\text{blue}) \cdot P(\text{blue}) = \dfrac{2}{10} \cdot \dfrac{2}{10} = \dfrac{4}{100} = \dfrac{1}{25}$

b. $P(\text{blue and white}) = P(\text{blue}) \cdot P(\text{white}) = \dfrac{2}{10} \cdot \dfrac{5}{10} = \dfrac{10}{100} = \dfrac{1}{10}$

c. $P(\text{red and blue}) = P(\text{red}) \cdot P(\text{blue}) = \dfrac{3}{10} \cdot \dfrac{2}{10} = \dfrac{6}{100} = \dfrac{3}{50}$

Multiplication rule 1 can be extended to three or more independent events by using the formula

$$P(A \text{ and } B \text{ and } C \text{ and } \ldots \text{ and } K) = P(A) \cdot P(B) \cdot P(C) \cdots P(K)$$

When a small sample is selected from a large population and the subjects are not replaced, the probability of the event occurring changes so slightly that for the most part, it is considered to remain the same. Examples 4–26 and 4–27 illustrate this concept.

Example 4–26

A Harris poll found that 46% of Americans say they suffer great stress at least once a week. If three people are selected at random, find the probability that all three will say that they suffer great stress at least once a week.

Source: 100% American.

Solution

Let S denote stress. Then

$$P(S \text{ and } S \text{ and } S) = P(S) \cdot P(S) \cdot P(S)$$
$$= (0.46)(0.46)(0.46) \approx 0.097$$

Example 4–27

Approximately 9% of men have a type of color blindness that prevents them from distinguishing between red and green. If 3 men are selected at random, find the probability that all of them will have this type of red-green color blindness.

Source: USA TODAY.

Solution

Let C denote red-green color blindness. Then

$$P(C \text{ and } C \text{ and } C) = P(C) \cdot P(C) \cdot P(C)$$
$$= (0.09)(0.09)(0.09)$$
$$= 0.000729$$

Hence, the rounded probability is 0.0007.

In Examples 4–23 through 4–27, the events were independent of one another, since the occurrence of the first event in no way affected the outcome of the second event. On the other hand, when the occurrence of the first event changes the probability of the occurrence of the second event, the two events are said to be *dependent*. For example, suppose a card is drawn from a deck and *not* replaced, and then a second card is drawn. What is the probability of selecting an ace on the first card and a king on the second card?

Before an answer to the question can be given, one must realize that the events are dependent. The probability of selecting an ace on the first draw is $\frac{4}{52}$. If that card is *not* replaced, the probability of selecting a king on the second card is $\frac{4}{51}$, since there are 4 kings and 51 cards remaining. The outcome of the first draw has affected the outcome of the second draw.

Dependent events are formally defined now.

> When the outcome or occurrence of the first event affects the outcome or occurrence of the second event in such a way that the probability is changed, the events are said to be **dependent events.**

Here are some examples of dependent events:

Drawing a card from a deck, not replacing it, and then drawing a second card.

Selecting a ball from an urn, not replacing it, and then selecting a second ball.

Being a lifeguard and getting a suntan.

Having high grades and getting a scholarship.

Parking in a no-parking zone and getting a parking ticket.

To find probabilities when events are dependent, use the multiplication rule with a modification in notation. For the problem just discussed, the probability of getting an ace on the first draw is $\frac{4}{52}$, and the probability of getting a king on the second draw is $\frac{4}{51}$. By the multiplication rule, the probability of both events occurring is

$$\frac{4}{52} \cdot \frac{4}{51} = \frac{16}{2652} = \frac{4}{663}$$

The event of getting a king on the second draw *given* that an ace was drawn the first time is called a *conditional probability*.

The **conditional probability** of an event B in relationship to an event A is the probability that event B occurs after event A has already occurred. The notation for conditional probability is $P(B|A)$. This notation does not mean that B is divided by A; rather, it means the probability that event B occurs given that event A has already occurred. In the card example, $P(B|A)$ is the probability that the second card is a king given that the first card is an ace, and it is equal to $\frac{4}{51}$ since the first card was *not* replaced.

> **Multiplication Rule 2**
>
> When two events are dependent, the probability of both occurring is
>
> $$P(A \text{ and } B) = P(A) \cdot P(B|A)$$

Example 4–28

A person owns a collection of 30 CDs, of which 5 are country music. If 2 CDs are selected at random, find the probability that both are country music.

Solution

Since the events are dependent,

$$P(C_1 \text{ and } C_2) = P(C_1) \cdot P(C_2|C_1) = \tfrac{5}{30} \cdot \tfrac{4}{29} = \tfrac{20}{870} = \tfrac{2}{87}$$

Example 4–29

The World Wide Insurance Company found that 53% of the residents of a city had homeowner's insurance (H) with the company. Of these clients, 27% also had automobile insurance (A) with the company. If a resident is selected at random, find the probability that the resident has both homeowner's and automobile insurance with the World Wide Insurance Company.

Solution

$$P(\text{H and A}) = P(\text{H}) \cdot P(\text{A}|\text{H}) = (0.53)(0.27) = 0.1431$$

This multiplication rule can be extended to three or more events, as shown in Example 4–30.

Example 4–30

Three cards are drawn from an ordinary deck and not replaced. Find the probability of these.

    a. Getting 3 jacks

    b. Getting an ace, a king, and a queen in order

    c. Getting a club, a spade, and a heart in order

    d. Getting 3 clubs

Solution

    a. $P(3 \text{ jacks}) = \dfrac{4}{52} \cdot \dfrac{3}{51} \cdot \dfrac{2}{50} = \dfrac{24}{132,600} = \dfrac{1}{5525}$

    b. $P(\text{ace and king and queen}) = \dfrac{4}{52} \cdot \dfrac{4}{51} \cdot \dfrac{4}{50} = \dfrac{64}{132,600} = \dfrac{8}{16,575}$

    c. $P(\text{club and spade and heart}) = \dfrac{13}{52} \cdot \dfrac{13}{51} \cdot \dfrac{13}{50} = \dfrac{2197}{132,600} = \dfrac{169}{10,200}$

    d. $P(3 \text{ clubs}) = \dfrac{13}{52} \cdot \dfrac{12}{51} \cdot \dfrac{11}{50} = \dfrac{1716}{132,600} = \dfrac{11}{850}$

Tree diagrams can be used as an aid to finding the solution to probability problems when the events are sequential. Example 4–31 illustrates the use of tree diagrams.

Example 4–31

Box 1 contains 2 red balls and 1 blue ball. Box 2 contains 3 blue balls and 1 red ball. A coin is tossed. If it falls heads up, box 1 is selected and a ball is drawn. If it falls tails up, box 2 is selected and a ball is drawn. Find the probability of selecting a red ball.

Solution

With the use of a tree diagram, the sample space can be determined as shown in Figure 4–6. First, assign probabilities to each branch. Next, using the multiplication rule, multiply the probabilities for each branch.

    Finally, use the addition rule, since a red ball can be obtained from box 1 or box 2.

$$P(\text{red}) = \tfrac{2}{6} + \tfrac{1}{8} = \tfrac{8}{24} + \tfrac{3}{24} = \tfrac{11}{24}$$

(*Note:* The sum of all final probabilities will always be equal to 1.)

Figure 4–6

Tree Diagram for
Example 4–31

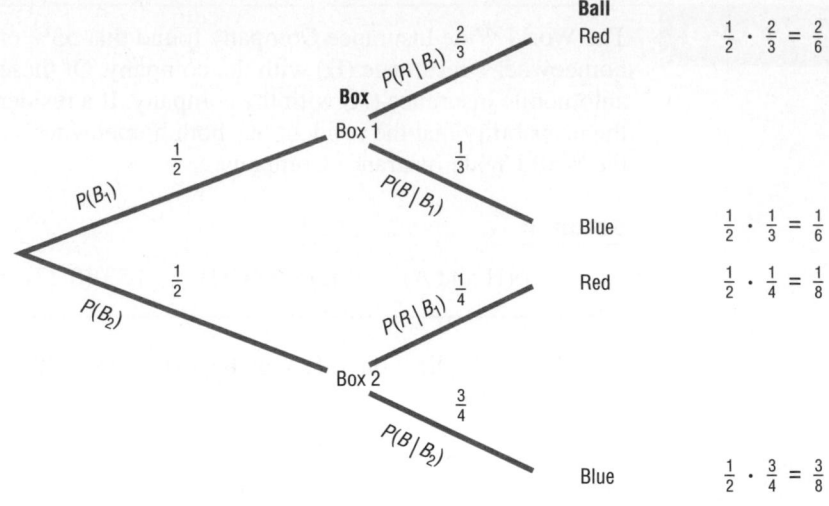

Tree diagrams can be used when the events are independent or dependent, and they can also be used for sequences of three or more events.

Conditional Probability

Objective 4

Find the conditional
probability of an event.

The conditional probability of an event B in relationship to an event A was defined as the probability that event B occurs after event A has already occurred.

The conditional probability of an event can be found by dividing both sides of the equation for multiplication rule 2 by $P(A)$, as shown:

$$P(A \text{ and } B) = P(A) \cdot P(B|A)$$

$$\frac{P(A \text{ and } B)}{P(A)} = \frac{\cancel{P(A)} \cdot P(B|A)}{\cancel{P(A)}}$$

$$\frac{P(A \text{ and } B)}{P(A)} = P(B|A)$$

Formula for Conditional Probability

The probability that the second event B occurs given that the first event A has occurred can be found by dividing the probability that both events occurred by the probability that the first event has occurred. The formula is

$$P(B|A) = \frac{P(A \text{ and } B)}{P(A)}$$

Examples 4–32, 4–33, and 4–34 illustrate the use of this rule.

Example 4–32

A box contains black chips and white chips. A person selects two chips without replacement. If the probability of selecting a black chip *and* a white chip is $\frac{15}{56}$, and the probability of selecting a black chip on the first draw is $\frac{3}{8}$, find the probability of selecting the white chip on the second draw, *given* that the first chip selected was a black chip.

Solution

Let

$$B = \text{selecting a black chip} \qquad W = \text{selecting a white chip}$$

Then

$$P(W|B) = \frac{P(B \text{ and } W)}{P(B)} = \frac{\frac{15}{56}}{\frac{3}{8}}$$

$$= \frac{15}{56} \div \frac{3}{8} = \frac{15}{56} \cdot \frac{8}{3} = \frac{\overset{5}{\cancel{15}}}{\underset{7}{\cancel{56}}} \cdot \frac{\overset{1}{\cancel{8}}}{\underset{1}{\cancel{3}}} = \frac{5}{7}$$

Hence, the probability of selecting a white chip on the second draw given that the first chip selected was black is $\frac{5}{7}$.

Example 4–33

The probability that Sam parks in a no-parking zone *and* gets a parking ticket is 0.06, and the probability that Sam cannot find a legal parking space and has to park in the no-parking zone is 0.20. On Tuesday, Sam arrives at school and has to park in a no-parking zone. Find the probability that he will get a parking ticket.

Solution

Let

$$N = \text{parking in a no-parking zone} \qquad T = \text{getting a ticket}$$

Then

$$P(T|N) = \frac{P(N \text{ and } T)}{P(N)} = \frac{0.06}{0.20} = 0.30$$

Hence, Sam has a 0.30 probability of getting a parking ticket, given that he parked in a no-parking zone.

The conditional probability of events occurring can also be computed when the data are given in table form, as shown in Example 4–34.

Example 4–34

A recent survey asked 100 people if they thought women in the armed forces should be permitted to participate in combat. The results of the survey are shown.

Gender	Yes	No	Total
Male	32	18	50
Female	8	42	50
Total	40	60	100

Find these probabilities.

a. The respondent answered yes, given that the respondent was a female.

b. The respondent was a male, given that the respondent answered no.

Solution

Let

$$M = \text{respondent was a male} \qquad Y = \text{respondent answered yes}$$
$$F = \text{respondent was a female} \qquad N = \text{respondent answered no}$$

a. The problem is to find $P(Y|F)$. The rule states

$$P(Y|F) = \frac{P(F \text{ and } Y)}{P(F)}$$

The probability $P(F \text{ and } Y)$ is the number of females who responded yes, divided by the total number of respondents:

$$P(F \text{ and } Y) = \frac{8}{100}$$

The probability $P(F)$ is the probability of selecting a female:

$$P(F) = \frac{50}{100}$$

Then

$$P(Y|F) = \frac{P(F \text{ and } Y)}{P(F)} = \frac{8/100}{50/100}$$

$$= \frac{8}{100} \div \frac{50}{100} = \frac{\overset{4}{\cancel{8}}}{\underset{1}{\cancel{100}}} \cdot \frac{\overset{1}{\cancel{100}}}{\underset{25}{\cancel{50}}} = \frac{4}{25}$$

b. The problem is to find $P(M|N)$.

$$P(M|N) = \frac{P(N \text{ and } M)}{P(N)} = \frac{18/100}{60/100}$$

$$= \frac{18}{100} \div \frac{60}{100} = \frac{\overset{3}{\cancel{18}}}{\underset{1}{\cancel{100}}} \cdot \frac{\overset{1}{\cancel{100}}}{\underset{10}{\cancel{60}}} = \frac{3}{10}$$

The Venn diagram for conditional probability is shown in Figure 4–7. In this case,

$$P(B|A) = \frac{P(A \text{ and } B)}{P(A)}$$

Figure 4–7

Venn Diagram for Conditional Probability

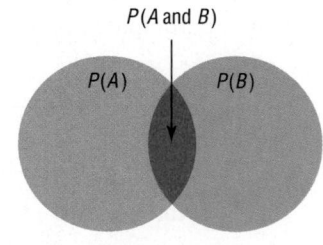

$P(A \text{ and } B)$

$P(A)$ $P(B)$

$P(S)$

$$P(B\,|\,A) = \frac{P(A \text{ and } B)}{P(A)}$$

which is represented by the area in the intersection or overlapping part of the circles A and B, divided by the area of circle A. The reasoning here is that if one assumes A has occurred, then A becomes the sample space for the next calculation and is the denominator of the probability fraction $\dfrac{P(A \text{ and } B)}{P(A)}$. The numerator $P(A \text{ and } B)$ represents the probability of the part of B that is contained in A. Hence, $P(A \text{ and } B)$ becomes the numerator of the probability fraction $\dfrac{P(A \text{ and } B)}{P(A)}$. Imposing a condition reduces the sample space.

Probabilities for "At Least"

The multiplication rules can be used with the complementary event rule (Section 4–2) to simplify solving probability problems involving "at least." Examples 4–35, 4–36, and 4–37 illustrate how this is done.

Example 4–35

A game is played by drawing four cards from an ordinary deck and replacing each card after it is drawn. Find the probability of winning if at least one ace is drawn.

Solution

It is much easier to find the probability that no aces are drawn (i.e., losing) and then subtract that value from 1 than to find the solution directly, because that would involve finding the probability of getting one ace, two aces, three aces, and four aces and then adding the results.

Let E = at least one ace is drawn and $\overline{E}$ = no aces drawn. Then

$$P(\overline{E}) = \frac{48}{52} \cdot \frac{48}{52} \cdot \frac{48}{52} \cdot \frac{48}{52}$$

$$= \frac{12}{13} \cdot \frac{12}{13} \cdot \frac{12}{13} \cdot \frac{12}{13} = \frac{20{,}736}{28{,}561}$$

Hence,

$$P(E) = 1 - P(\overline{E})$$

$$P(\text{winning}) = 1 - P(\text{losing}) = 1 - \frac{20{,}736}{28{,}561} = \frac{7{,}825}{28{,}561} \approx 0.27$$

or a hand with at least one ace will win about 27% of the time.

Example 4–36

A coin is tossed 5 times. Find the probability of getting at least one tail.

Solution

It is easier to find the probability of the complement of the event, which is "all heads," and then subtract the probability from 1 to get the probability of at least one tail.

$$P(E) = 1 - P(\overline{E})$$

$$P(\text{at least 1 tail}) = 1 - P(\text{all heads})$$

$$P(\text{all heads}) = (\tfrac{1}{2})^5 = \tfrac{1}{32}$$

Hence,

$$P(\text{at least 1 tail}) = 1 - \tfrac{1}{32} = \tfrac{31}{32}$$

Example 4–37	The Neckware Association of America reported that 3% of ties sold in the United States are bow ties. If 4 customers who purchased a tie are randomly selected, find the probability that at least one purchased a bow tie.

Solution

Let E = at least one bow tie is purchased and $\overline{E}$ = no bow ties are purchased. Then

$$P(E) = 0.03 \quad \text{and} \quad P(\overline{E}) = 1 - 0.03 = 0.97$$

$P(\text{no bow ties are purchased}) = (0.97)(0.97)(0.97)(0.97) \approx 0.885$; hence,
$P(\text{at least one bow tie is purchased}) = 1 - 0.885 = 0.115$.

Similar methods can be used for problems involving "at most."

Applying the Concepts 4–4

Guilty or Innocent?

In July 1964, an elderly woman was mugged in Costa Mesa, California. In the vicinity of the crime a tall, bearded man sat waiting in a yellow car. Shortly after the crime was committed, a young, tall woman, wearing her blond hair in a ponytail, was seen running from the scene of the crime and getting into the car, which sped off. The police broadcast a description of the suspected muggers. Soon afterward, a couple fitting the description was arrested and convicted of the crime. Although the evidence in the case was largely circumstantial, the two people arrested were nonetheless convicted of the crime. The prosecutor based his entire case on basic probability theory, showing the unlikeness of another couple being in that area while having all the same characteristics that the elderly woman described. The following probabilities were used.

Characteristic	Assumed probability
Drives yellow car	1 out of 12
Man over 6 feet tall	1 out of 10
Man wearing tennis shoes	1 out of 4
Man with beard	1 out of 11
Woman with blond hair	1 out of 3
Woman with hair in a ponytail	1 out of 13
Woman over 6 feet tall	1 out of 100

1. Compute the probability of another couple being in that area with the same characteristics.
2. Would you use the addition or multiplication rule? Why?
3. Are the characteristics independent or dependent?
4. How are the computations affected by the assumption of independence or dependence?
5. Should any court case be based solely on probabilities?
6. Would you convict the couple who was arrested even if there were no eyewitnesses?
7. Comment on why in today's justice system no person can be convicted solely on the results of probabilities.
8. In actuality, aren't most court cases based on uncalculated probabilities?

See page 235 for the answers.

Exercises 4–4

1. State which events are independent and which are dependent.

   a. Tossing a coin and drawing a card from a deck

   b. Drawing a ball from an urn, not replacing it, and then drawing a second ball

   c. Getting a raise in salary and purchasing a new car

   d. Driving on ice and having an accident

   e. Having a large shoe size and having a high IQ

   f. A father being left-handed and a daughter being left-handed

   g. Smoking excessively and having lung cancer

   h. Eating an excessive amount of ice cream and smoking an excessive amount of cigarettes

2. If 37% of high school students said that they exercise regularly, find the probability that 5 randomly selected high school students will say that they exercise regularly. Would you consider this event likely or unlikely to occur? Explain your answer.

3. If 84% of all people who do aerobics are women, find the probability that if 2 people who do aerobics are randomly selected, both are women. Would you consider this event likely or unlikely to occur? Explain your answer.

4. The Gallup Poll reported that 52% of Americans used a seat belt the last time they got into a car. If four people are selected at random, find the probability that they all used a seat belt the last time they got into a car.

Source: *100% American.*

5. If 28% of U.S. medical degrees are conferred to women, find the probability that 3 randomly selected medical school graduates are men. Would you consider this event likely or unlikely to occur? Explain your answer.

6. If 25% of U.S. federal prison inmates are not U.S. citizens, find the probability that 2 randomly selected federal prison inmates will not be U.S. citizens.

Source: *Harper's Index.*

7. At a local university 54.3% of incoming first-year students have computers. If 3 students are selected at random, find the following probabilities.

   a. None have computers.

   b. At least one has a computer.

   c. All have computers.

8. If 2 cards are selected from a standard deck of 52 cards without replacement, find these probabilities.

   a. Both are spades.

   b. Both are the same suit.

   c. Both are kings.

9. Of the 216 players on major league soccer rosters, 80.1% are U.S. citizens. If 3 players are selected at random for an exhibition, what is the probability that all are U.S. citizens?

Source: *USA TODAY.*

10. If one-half of Americans believe that the federal government should take "primary responsibility" for eliminating poverty, find the probability that 3 randomly selected Americans will agree that it is the federal government's responsibility to eliminate poverty.

Source: *Harper's Index.*

11. Of fans who own sports league–licensed apparel, 31% have NFL apparel. If 3 of these fans are selected at random, what is the probability that all have NFL apparel?

Source: *ESPN Chilton Sports Poll.*

12. A flashlight has 6 batteries, 2 of which are defective. If 2 are selected at random without replacement, find the probability that both are defective.

13. Eighty-eight percent of U.S. children are covered by some type of health insurance. If 4 children are selected at random, what is the probability that none are covered?

Source: *Federal Interagency Forum on Child and Family Statistics.*

14. The U.S. Department of Justice reported that 6% of all U.S. murders are committed without a weapon. If 3 murder cases are selected at random, find the probability that a weapon was not used in any one of them.

Source: *100% American.*

15. In a department store there are 120 customers, 90 of whom will buy at least one item. If 5 customers are selected at random, one by one, find the probability that all will buy at least one item.

16. Three cards are drawn from a deck *without* replacement. Find these probabilities.

   a. All are jacks.

   b. All are clubs.

   c. All are red cards.

17. In a scientific study there are 8 guinea pigs, 5 of which are pregnant. If 3 are selected at random without replacement, find the probability that all are pregnant.

18. In Exercise 17, find the probability that none are pregnant.

19. In a civic organization, there are 38 members; 15 are men and 23 are women. If 3 members are selected to plan the July 4th parade, find the probability that all 3 are women. Would you consider this event likely or unlikely to occur? Explain your answer.

20. In Exercise 19, find the probability that all 3 members are men.

21. A manufacturer makes two models of an item: model I, which accounts for 80% of unit sales, and model II, which accounts for 20% of unit sales. Because of defects, the manufacturer has to replace (or exchange) 10% of its model I and 18% of its model II. If a model is selected at random, find the probability that it will be defective.

22. An automobile manufacturer has three factories, A, B, and C. They produce 50, 30, and 20%, respectively, of a specific model of car. Thirty percent of the cars produced in factory A are white, 40% of those produced in factory B are white, and 25% produced in factory C are white. If an automobile produced by the company is selected at random, find the probability that it is white.

23. An insurance company classifies drivers as low-risk, medium-risk, and high-risk. Of those insured, 60% are low-risk, 30% are medium-risk, and 10% are high-risk. After a study, the company finds that during a 1-year period, 1% of the low-risk drivers had an accident, 5% of the medium-risk drivers had an accident, and 9% of the high-risk drivers had an accident. If a driver is selected at random, find the probability that the driver will have had an accident during the year.

24. In a certain geographic location, 25% of the wage earners have a college degree and 75% do not. Of those who have a college degree, 5% earn more than $100,000 a year. Of those who do not have a college degree, 2% earn more than $100,000 a year. If a wage earner is selected at random, find the probability that she or he earns more than $100,000 a year.

25. Urn 1 contains 5 red balls and 3 black balls. Urn 2 contains 3 red balls and 1 black ball. Urn 3 contains 4 red balls and 2 black balls. If an urn is selected at random and a ball is drawn, find the probability it will be red.

26. For a recent year, 0.99 of the incarcerated population is adults and 0.07 is female. If an incarcerated person is selected at random, find the probability that the person is a female given that she is an adult.

Source: Bureau of Justice.

27. In a certain city, the probability that an automobile will be stolen and found within one week is 0.0009. The probability that an automobile will be stolen is 0.0015. Find the probability that a stolen automobile will be found within one week.

28. A circuit to run a model railroad has 8 switches. Two are defective. If a person selects 2 switches at random and tests them, find the probability that the second one is defective, given that the first one is defective.

29. At the Avonlea Country Club, 73% of the members play bridge and swim, and 82% play bridge. If a member is selected at random, find the probability that the member swims, given that the member plays bridge.

30. At a large university, the probability that a student takes calculus and is on the dean's list is 0.042. The probability that a student is on the dean's list is 0.21. Find the probability that the student is taking calculus, given that he or she is on the dean's list.

31. In Rolling Acres Housing Plan, 42% of the houses have a deck and a garage; 60% have a deck. Find the probability that a home has a garage, given that it has a deck.

32. In a pizza restaurant, 95% of the customers order pizza. If 65% of the customers order pizza and a salad, find the probability that a customer who orders pizza will also order a salad.

33. At an exclusive country club, 68% of the members play bridge and drink champagne, and 83% play bridge. If a member is selected at random, find the probability that the member drinks champagne, given that he or she plays bridge.

34. Eighty students in a school cafeteria were asked if they favored a ban on smoking in the cafeteria. The results of the survey are shown in the table.

Class	Favor	Oppose	No opinion
Freshman	15	27	8
Sophomore	23	5	2

If a student is selected at random, find these probabilities.

a. Given that the student is a freshman, he or she opposes the ban.

b. Given that the student favors the ban, the student is a sophomore.

35. Consider this table concerning utility patents granted for a specific year.

	Corporation	Government	Individual
United States	70,894	921	6129
Foreign	63,182	104	6267

Select one patent at random.

a. What is the probability that it is a foreign patent, given that it was issued to a corporation?

b. What is the probability that it was issued to an individual, given that it was a U.S. patent?

Source: *N.Y. Times Almanac.*

36. The medal distribution from the 2000 Summer Olympic Games is shown in the table.

	Gold	Silver	Bronze
United States	39	25	33
Russia	32	28	28
China	28	16	15
Australia	16	25	17
Others	186	205	235

Choose one medal winner at random.

a. Find the probability that the winner won the gold medal, given that the winner was from the United States.

b. Find the probability that the winner was from the United States, given that she or he won a gold medal.

c. Are the events "medal winner is from United States" and "gold medal was won" independent? Explain.

Source: *N.Y. Times Almanac.*

37. According to the *Statistical Abstract of the United States,* 70.3% of females ages 20 to 24 have never been married. Choose 5 young women in this age category at random. Find the probability that

a. None have ever been married.

b. At least one has been married.

Source: *N.Y. Times Almanac.*

38. The American Automobile Association (AAA) reports that of the fatal car and truck accidents, 54% are caused by car driver error. If 3 accidents are chosen at random, find the probability that

a. All are caused by car driver error.

b. None are caused by car driver error.

c. At least 1 is caused by car driver error.

Source: *AAA quoted on CNN.*

39. Seventy-six percent of toddlers get their recommended immunizations. Suppose that 6 toddlers are selected at random. What is the probability that at least one has not received the recommended immunizations?

Source: *Federal Interagency Forum on Child and Family Statistics.*

40. A lot of portable radios contains 15 good radios and 3 defective ones. If 2 are selected and tested, find the probability that at least one will be defective.

41. Fifty-eight percent of American children (ages 3 to 5) are read to every day by someone at home. Suppose 5 children are randomly selected. What is the probability that at least one is read to every day by someone at home?

Source: *Federal Interagency Forum on Child and Family Statistics.*

42. Of Ph.D. students, 60% have paid assistantships. If 3 students are selected at random, find the probabilities.

a. All have assistantships.

b. None have assistantships.

c. At least one has an assistantship.

Source: *U.S. Department of Education/Chronicle of Higher Education.*

43. If 4 cards are drawn from a deck of 52 and not replaced, find the probability of getting at least one club.

44. At a local clinic there are 8 men, 5 women, and 3 children in the waiting room. If 3 patients are randomly selected, find the probability that there is at least one child among them.

45. It has been found that 6% of all automobiles on the road have defective brakes. If 5 automobiles are stopped and checked by the state police, find the probability that at least one will have defective brakes.

46. A medication is 75% effective against a bacterial infection. Find the probability that if 12 people take the medication, at least one person's infection will not improve.

47. A coin is tossed 5 times; find the probability of getting at least one tail. Would you consider this event likely to happen? Explain your answer.

48. If 3 letters of the alphabet are selected at random, find the probability of getting at least one letter x. Letters can be used more than once. Would you consider this event likely to happen? Explain your answer.

49. A die is rolled 7 times. Find the probability of getting at least one 3. Would you consider this event likely to occur? Explain your answer.

50. At a teachers' conference, there were 4 English teachers, 3 mathematics teachers, and 5 science teachers. If 4 teachers are selected for a committee, find the probability that at least one is a science teacher.

51. If a die is rolled 3 times, find the probability of getting at least one even number.

52. In a large vase, there are 8 roses, 5 daisies, 12 lilies, and 9 orchids. If 4 flowers are selected at random, find the probability that at least one of the flowers is a rose. Would you consider this event likely to occur? Explain your answer.

Extending the Concepts

53. Let A and B be two mutually exclusive events. Are A and B independent events? Explain your answer.

54. The Bargain Auto Mall has the following cars in stock.

	SUV	Compact	Mid-sized
Foreign	20	50	20
Domestic	65	100	45

Are the events "compact" and "domestic" independent? Explain.

55. An admissions director knows that the probability a student will enroll after a campus visit is 0.55, or $P(E) = 0.55$. While students are on campus visits, interviews with professors are arranged. The admissions director computes these conditional probabilities for students enrolling after visiting three professors, DW, LP, and MH.

$P(E|\text{DW}) = 0.95$ $P(E|\text{LP}) = 0.55$ $P(E|\text{MH}) = 0.15$

Is there something wrong with the numbers? Explain.

56. Event A is the event that a person remembers a certain product commercial. Event B is the event that a person buys the product. If $P(B) = 0.35$, comment on each of these conditional probabilities if you were vice president for sales.

a. $P(B|A) = 0.20$
b. $P(B|A) = 0.35$
c. $P(B|A) = 0.55$

4–5 Counting Rules

Many times one wishes to know the number of all possible outcomes for a sequence of events. To determine this number, three rules can be used: the *fundamental counting rule,* the *permutation rule,* and the *combination rule.* These rules are explained here, and they will be used in Section 4–6 to find probabilities of events.

The first rule is called the **fundamental counting rule.**

The Fundamental Counting Rule

Objective 5

Find the total number of outcomes in a sequence of events, using the fundamental counting rule.

Fundamental Counting Rule

In a sequence of n events in which the first one has k_1 possibilities and the second event has k_2 and the third has k_3, and so forth, the total number of possibilities of the sequence will be

$$k_1 \cdot k_2 \cdot k_3 \cdots k_n$$

Note: In this case *and* means to multiply.

Examples 4–38 through 4–41 illustrate the fundamental counting rule.

Example 4–38   A coin is tossed and a die is rolled. Find the number of outcomes for the sequence of events.

Figure 4–8

Complete Tree Diagram for Example 4–38

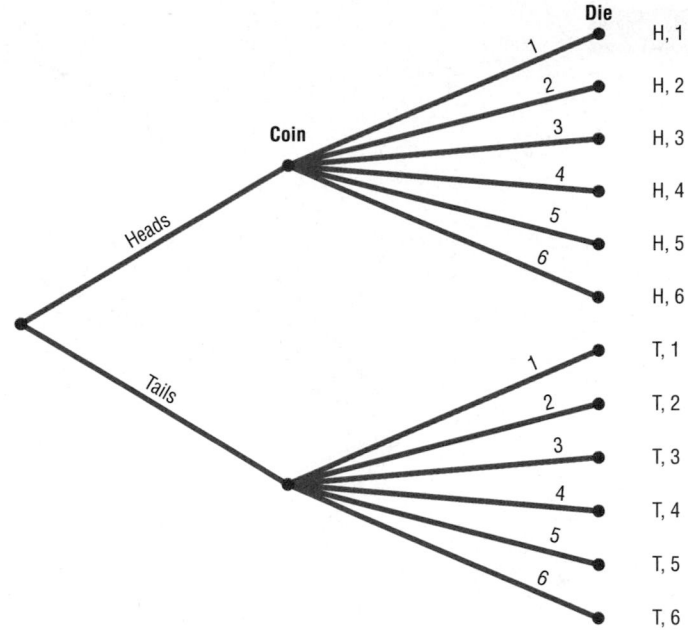

Solution

Since the coin can land either heads up or tails up and since the die can land with any one of six numbers showing face up, there are $2 \cdot 6 = 12$ possibilities. A tree diagram can also be drawn for the sequence of events. See Figure 4–8.

Example 4–39

A paint manufacturer wishes to manufacture several different paints. The categories include

Color	Red, blue, white, black, green, brown, yellow
Type	Latex, oil
Texture	Flat, semigloss, high gloss
Use	Outdoor, indoor

How many different kinds of paint can be made if a person can select one color, one type, one texture, and one use?

Solution

A person can choose one color and one type and one texture and one use. Since there are 7 color choices, 2 type choices, 3 texture choices, and 2 use choices, the total number of possible different paints is

Color		**Type**		**Texture**		**Use**	
7	$\cdot$	2	$\cdot$	3	$\cdot$	2	$= 84$

Example 4–40

There are four blood types, A, B, AB, and O. Blood can also be Rh+ and Rh−. Finally, a blood donor can be classified as either male or female. How many different ways can a donor have his or her blood labeled?

Figure 4–9

Complete Tree Diagram for Example 4–40

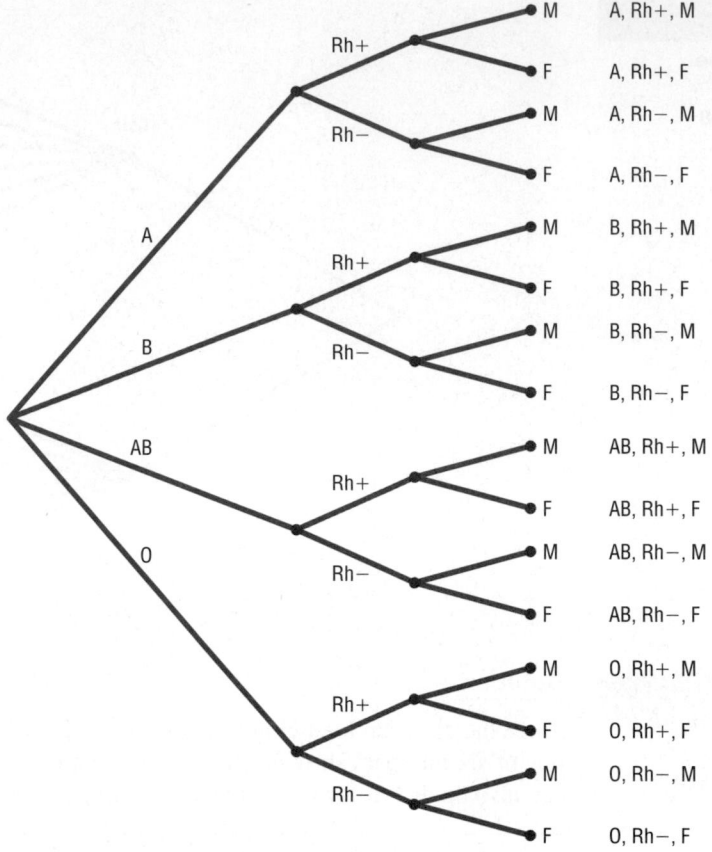

		M	A, Rh+, M
	Rh+	F	A, Rh+, F
A	Rh−	M	A, Rh−, M
		F	A, Rh−, F
	Rh+	M	B, Rh+, M
		F	B, Rh+, F
B	Rh−	M	B, Rh−, M
		F	B, Rh−, F
AB	Rh+	M	AB, Rh+, M
		F	AB, Rh+, F
	Rh−	M	AB, Rh−, M
		F	AB, Rh−, F
	Rh+	M	0, Rh+, M
		F	0, Rh+, F
0	Rh−	M	0, Rh−, M
		F	0, Rh−, F

Solution

Since there are 4 possibilities for blood type, 2 possibilities for Rh factor, and 2 possibilities for the gender of the donor, there are $4 \cdot 2 \cdot 2$, or 16, different classification categories, as shown.

Blood type		**Rh**		**Gender**	
4	·	2	·	2	= 16

A tree diagram for the events is shown in Figure 4–9.

When determining the number of different possibilities of a sequence of events, one must know whether repetitions are permissible.

Example 4–41

The digits 0, 1, 2, 3, and 4 are to be used in a four-digit ID card. How many different cards are possible if repetitions are permitted?

Solution

Since there are 4 spaces to fill and 5 choices for each space, the solution is

$$5 \cdot 5 \cdot 5 \cdot 5 = 5^4 = 625$$

Now, what if repetitions are not permitted? For Example 4–41, the first digit can be chosen in 5 ways. But the second digit can be chosen in only 4 ways, since there are only four digits left, etc. Thus, the solution is

$$5 \cdot 4 \cdot 3 \cdot 2 = 120$$

The same situation occurs when one is drawing balls from an urn or cards from a deck. If the ball or card is replaced before the next one is selected, then repetitions are permitted, since the same one can be selected again. But if the selected ball or card is not replaced, then repetitions are not permitted, since the same ball or card cannot be selected the second time.

These examples illustrate the fundamental counting rule. In summary: *If repetitions are permitted, then the numbers stay the same going from left to right. If repetitions are not permitted, then the numbers decrease by 1 for each place left to right.*

Two other rules that can be used to determine the total number of possibilities of a sequence of events are the permutation rule and the combination rule.

Factorial Notation

In 1808 Christian Kramp first used the factorial notation.

These rules use *factorial notation.* The factorial notation uses the exclamation point.

$$5! = 5 \cdot 4 \cdot 3 \cdot 2 \cdot 1$$
$$9! = 9 \cdot 8 \cdot 7 \cdot 6 \cdot 5 \cdot 4 \cdot 3 \cdot 2 \cdot 1$$

To use the formulas in the permutation and combination rules, a special definition of $0!$ is needed. $0! = 1$.

Factorial Formulas

For any counting n

$$n! = n(n-1)(n-2) \cdots 1$$
$$0! = 1$$

Permutations

A **permutation** is an arrangement of n objects in a specific order.

Examples 4–42 and 4–43 illustrate permutations.

Example 4–42

Suppose a business owner has a choice of five locations in which to establish her business. She decides to rank each location according to certain criteria, such as price of the store and parking facilities. How many different ways can she rank the five locations?

Solution

There are

$$5! = 5 \cdot 4 \cdot 3 \cdot 2 \cdot 1 = 120$$

different possible rankings. The reason is that she has 5 choices for the first location, 4 choices for the second location, 3 choices for the third location, etc.

In Example 4–42 all objects were used up. But what happens when not all objects are used up? The answer to this question is given in Example 4–43.

Example 4–43

Suppose the business owner in Example 4–42 wishes to rank only the top three of the five locations. How many different ways can she rank them?

Solution

Using the fundamental counting rule, she can select any one of the five for first choice, then any one of the remaining four locations for her second choice, and finally, any one of the remaining locations for her third choice, as shown.

First choice		Second choice		Third choice	
5	·	4	·	3	= 60

The solutions in Examples 4–42 and 4–43 are permutations.

Objective 6

Find the number of ways that *r* objects can be selected from *n* objects, using the permutation rule.

Permutation Rule

The arrangement of *n* objects in a specific order using *r* objects at a time is called a *permutation of n objects taking r objects at a time*. It is written as $_nP_r$, and the formula is

$$_nP_r = \frac{n!}{(n-r)!}$$

The notation $_nP_r$ is used for permutations.

$$_6P_4 \text{ means } \frac{6!}{(6-4)!} \quad \text{or} \quad \frac{6!}{2!} = \frac{6 \cdot 5 \cdot 4 \cdot 3 \cdot \cancel{2} \cdot \cancel{1}}{\cancel{2} \cdot \cancel{1}} = 360$$

Although Examples 4–42 and 4–43 were solved by the multiplication rule, they can now be solved by the permutation rule.

In Example 4–42, five locations were taken and then arranged in order; hence,

$$_5P_5 = \frac{5!}{(5-5)!} = \frac{5!}{0!} = \frac{5 \cdot 4 \cdot 3 \cdot 2 \cdot 1}{1} = 120$$

(Recall that 0! = 1.)

In Example 4–43, three locations were selected from five locations, so $n = 5$ and $r = 3$; hence

$$_5P_3 = \frac{5!}{(5-3)!} = \frac{5!}{2!} = \frac{5 \cdot 4 \cdot 3 \cdot \cancel{2} \cdot \cancel{1}}{\cancel{2} \cdot \cancel{1}} = 60$$

Examples 4–44 and 4–45 illustrate the permutation rule.

Example 4–44

A television news director wishes to use three news stories on an evening show. One story will be the lead story, one will be the second story, and the last will be a closing story. If the director has a total of eight stories to choose from, how many possible ways can the program be set up?

Solution

Since order is important, the solution is

$$_8P_3 = \frac{8!}{(8-3)!} = \frac{8!}{5!} = 336$$

Hence, there would be 336 ways to set up the program.

Example 4–45

How many different ways can a chairperson and an assistant chairperson be selected for a research project if there are seven scientists available?

Solution

$$_7P_2 = \frac{7!}{(7-2)!} = \frac{7!}{5!} = 42$$

Combinations

Suppose a dress designer wishes to select two colors of material to design a new dress, and she has on hand four colors. How many different possibilities can there be in this situation?

Objective 7

Find the number of ways that r objects can be selected from n objects without regard to order, using the combination rule.

   This type of problem differs from previous ones in that the order of selection is not important. That is, if the designer selects yellow and red, this selection is the same as the selection red and yellow. This type of selection is called a *combination*. The difference between a permutation and a combination is that in a combination, the order or arrangement of the objects is not important; by contrast, order *is* important in a permutation. Example 4–46 illustrates this difference.

A selection of distinct objects without regard to order is called a **combination**.

Example 4–46

Given the letters A, B, C, and D, list the permutations and combinations for selecting two letters.

Solution

The permutations are

AB	BA	CA	DA
AC	BC	CB	DB
AD	BD	CD	DC

In permutations, AB is different from BA. But in combinations, AB is the same as BA since the order of the objects does not matter in combinations. Therefore, if duplicates are removed from a list of permutations, what is left is a list of combinations, as shown.

AB	B̶A̶	C̶A̶	D̶A̶
AC	BC	C̶B̶	D̶B̶
AD	BD	CD	D̶C̶

Hence the combinations of A, B, C, and D are AB, AC, AD, BC, BD, and CD. (Alternatively, BA could be listed and AB crossed out, etc.) The combinations have been listed alphabetically for convenience, but this is not a requirement.

Combinations are used when the order or arrangement is not important, as in the selecting process. Suppose a committee of 5 students is to be selected from 25 students. The five selected students represent a combination, since it does not matter who is selected first, second, etc.

Combination Rule

The number of combinations of r objects selected from n objects is denoted by ${}_nC_r$ and is given by the formula

$$_nC_r = \frac{n!}{(n-r)!r!}$$

Example 4–47

How many combinations of 4 objects are there, taken 2 at a time?

Solution

Since this is a combination problem, the answer is

$$_4C_2 = \frac{4!}{(4-2)!2!} = \frac{4!}{2!2!} = \frac{\overset{2}{\cancel{4}} \cdot 3 \cdot \cancel{2!}}{\cancel{2} \cdot 1 \cdot \cancel{2!}} = 6$$

This is the same result shown in Example 4–46.

Notice that the expression for ${}_nC_r$ is

$$\frac{n!}{(n-r)!r!}$$

which is the formula for permutations with $r!$ in the denominator. In other words,

$$_nC_r = \frac{_nP_r}{r!}$$

This $r!$ divides out the duplicates from the number of permutations, as shown in Example 4–46. For each two letters, there are two permutations but only one combination. Hence, dividing the number of permutations by $r!$ eliminates the duplicates. This result can be verified for other values of n and r. *Note:* $_nC_n = 1$.

Example 4–48

A bicycle shop owner has 12 mountain bicycles in the showroom. The owner wishes to select 5 of them to display at a bicycle show. How many different ways can a group of 5 be selected?

Solution

$$_{12}C_5 = \frac{12!}{(12-5)!5!} = \frac{12!}{7!5!} = \frac{12 \cdot 11 \cdot \cancel{10}^{2} \cdot \cancel{9}^{3} \cdot 8 \cdot \cancel{7}^{2}}{\cancel{7} \cdot \cancel{5} \cdot \cancel{4} \cdot \cancel{3} \cdot \cancel{2} \cdot 1} = 792$$

Example 4–49

In a club there are 7 women and 5 men. A committee of 3 women and 2 men is to be chosen. How many different possibilities are there?

Solution

Here, one must select 3 women from 7 women, which can be done in $_7C_3$, or 35, ways. Next, 2 men must be selected from 5 men, which can be done in $_5C_2$, or 10, ways. Finally, by the fundamental counting rule, the total number of different ways is $35 \cdot 10 = 350$, since one is choosing both men and women. Using the formula gives

$$_7C_3 \cdot _5C_2 = \frac{7!}{(7-3)!3!} \cdot \frac{5!}{(5-2)!2!} = 350$$

Table 4–1 summarizes the counting rules.

Table 4–1	Summary of Counting Rules	
Rule	**Definition**	**Formula**
Fundamental counting rule	The number of ways a sequence of n events can occur if the first event can occur in k_1 ways, the second event can occur in k_2 ways, etc.	$k_1 \cdot k_2 \cdot k_3 \cdots k_n$
Permutation rule	The number of permutations of n objects taking r objects at a time (order is important)	$_nP_r = \frac{n!}{(n-r)!}$
Combination rule	The number of combinations of r objects taken from n objects (order is not important)	$_nC_r = \frac{n!}{(n-r)!r!}$

Applying the Concepts **4–5**

Garage Door Openers

Garage door openers originally had a series of four on/off switches so that homeowners could personalize the frequencies that opened their garage doors. If all garage door openers were set at the same frequency, anyone with a garage door opener could open anyone else's garage door.

1. Use a tree diagram to show how many different positions four consecutive on/off switches could be in.

After garage door openers became more popular, another set of four on/off switches was added to the systems.

2. Find a pattern of how many different positions are possible with the addition of each on/off switch.

3. How many different positions are possible with eight consecutive on/off switches?

4. Is it reasonable to assume, if you owned a garage door opener with eight switches, that someone could use his/her garage door opener to open your garage door by trying all the different possible positions?

In 1989 it was reported that the ignition keys for 1988 Dodge Caravans were made from a single blank that had five cuts on it. Each cut was made at one out of five possible levels. In 1988, assume there was 420,000 Dodge Caravans sold in the United States.

5. How many different possible keys can be made from the same key blank?

6. How many different 1988 Dodge Caravans could any one key start?

Look at the ignition key for your car and count the number of cuts on it. Assume that the cuts are made at one of any five possible levels. Most car companies use one key blank for all their makes and models of cars.

7. Conjecture how many cars your car company sold over recent years, and then figure out how many other cars your car key could start. What would you do to decrease the odds of someone being able to open another vehicle with his or her key?

See page 235 for the answers.

Exercises 4–5

1. How many 5-digit zip codes are possible if digits can be repeated? If there cannot be repetitions?

2. How many ways can a baseball manager arrange a batting order of 9 players?

3. How many different ways can 7 different video game cartridges be arranged on a shelf?

4. How many different ways can 6 radio commercials be played during a 1-hour radio program?

5. A store manager wishes to display 8 different brands of shampoo in a row. How many ways can this be done?

6. There are 8 different statistics books, 6 different geometry books, and 3 different trigonometry books.

A student must select one book of each type. How many different ways can this be done?

7. At a local cheerleaders' camp, 5 routines must be practiced. A routine may not be repeated. In how many different orders can these 5 routines be presented?

8. The call letters of a radio station must have 4 letters. The first letter must be a K or a W. How many different station call letters can be made if repetitions are not allowed? If repetitions are allowed?

9. How many different 3-digit identification tags can be made if the digits can be used more than once? If the first digit must be a 5 and repetitions are not permitted?

10. How many different ways can 9 trophies be arranged on a shelf?

11. If a baseball manager has 5 pitchers and 2 catchers, how many different possible pitcher-catcher combinations can he field?

12. There are 2 major roads from city X to city Y and 4 major roads from city Y to city Z. How many different trips can be made from city X to city Z passing through city Y?

13. Evaluate each of these.

   a. 8!   e. $_7P_5$ i. $_5P_5$
   b. 10!   f. $_{12}P_4$ j. $_6P_2$
   c. 0!   g. $_5P_3$
   d. 1!   h. $_6P_0$

14. The County Assessment Bureau decides to reassess homes in 8 different areas. How many different ways can this be accomplished?

15. How many different 4-color code stripes can be made on a sports car if each code consists of the colors green, red, blue, and white? All colors are used only once.

16. An inspector must select 3 tests to perform in a certain order on a manufactured part. He has a choice of 7 tests. How many ways can he perform 3 different tests?

17. Anderson Research Co. decides to test-market a product in 6 areas. How many different ways can 3 areas be selected in a certain order for the first test?

18. How many different ways can a city health department inspector visit 5 restaurants in a city with 10 restaurants?

19. How many different 4-letter permutations can be formed from the letters in the word *decagon*?

20. In a board of directors composed of 8 people, how many ways can 1 chief executive officer, 1 director, and 1 treasurer be selected?

21. How many different ID cards can be made if there are 6 digits on a card and no digit can be used more than once?

22. How many different ways can 5 Public Service announcements be run during 1 hour of time?

23. How many different ways can 4 tickets be selected from 50 tickets if each ticket wins a different prize?

24. How many different ways can a researcher select 5 rats from 20 rats and assign each to a different test?

25. How many different signals can be made by using at least 3 distinct flags if there are 5 different flags from which to select?

26. An investigative agency has 7 cases and 5 agents. How many different ways can the cases be assigned if only 1 case is assigned to each agent?

27. **(ans)** Evaluate each expression.

   a. $_5C_2$ d. $_6C_2$ g. $_3C_3$ j. $_4C_3$
   b. $_8C_3$ e. $_6C_4$ h. $_9C_7$
   c. $_7C_4$ f. $_3C_0$ i. $_{12}C_2$

28. How many ways can 3 cards be selected from a standard deck of 52 cards, disregarding the order of selection?

29. How many ways are there to select 3 bracelets from a box of 10 bracelets, disregarding the order of selection?

30. How many ways can 4 baseball players and 3 basketball players be selected from 12 baseball players and 9 basketball players?

31. How many ways can a committee of 4 people be selected from a group of 10 people?

32. If a person can select 3 presents from 10 presents under a Christmas tree, how many different combinations are there?

33. How many different tests can be made from a test bank of 20 questions if the test consists of 5 questions?

34. The general manager of a fast-food restaurant chain must select 6 restaurants from 11 for a promotional program. How many different possible ways can this selection be done?

35. How many different ways can a theatrical group select 2 musicals and 3 dramas from 11 musicals and 8 dramas to be presented during the year?

36. In a train yard there are 4 tank cars, 12 boxcars, and 7 flatcars. How many ways can a train be made up consisting of 2 tank cars, 5 boxcars, and 3 flatcars? (In this case, order is not important.)

37. There are 7 women and 5 men in a department. How many ways can a committee of 4 people be selected? How many ways can this committee be selected if there must be 2 men and 2 women on the committee? How many ways can this committee be selected if there must be at least 2 women on the committee?

38. Wake Up cereal comes in 2 types, crispy and crunchy. If a researcher has 10 boxes of each, how many ways can she select 3 boxes of each for a quality control test?

39. How many ways can a dinner patron select 3 appetizers and 2 vegetables if there are 6 appetizers and 5 vegetables on the menu?

40. How many ways can a jury of 6 women and 6 men be selected from 10 women and 12 men?

41. How many ways can a foursome of 2 men and 2 women be selected from 10 men and 12 women in a golf club?

42. The state narcotics bureau must form a 5-member investigative team. If it has 25 agents from which to choose, how many different possible teams can be formed?

43. How many different ways can an instructor select 2 textbooks from a possible 17?

44. The Environmental Protection Agency must investigate 9 mills for complaints of air pollution. How many different ways can a representative select 5 of these to investigate this week?

45. How many ways can a person select 7 television commercials from 11 television commercials?

46. How many ways can a person select 8 videotapes from 10 tapes?

47. A buyer decides to stock 8 different posters. How many ways can she select these 8 if there are 20 from which to choose?

48. An advertising manager decides to have an ad campaign in which 8 special calculators will be hidden at various locations in a shopping mall. If she has 17 locations from which to pick, how many different possible combinations can she choose?

Extending the Concepts

49. How many different ways can a person select one or more coins if she has 2 nickels, 1 dime, and 1 half-dollar?

50. In a barnyard there is an assortment of chickens and cows. Counting heads, one gets 15; counting legs, one gets 46. How many of each are there?

51. How many different ways can five people—A, B, C, D, and E—sit in a row at a movie theater if (*a*) A and B

must sit together; (*b*) C must sit to the right of, but not necessarily next to, B; (*c*) D and E will not sit next to each other?

52. Using combinations, calculate the number of each poker hand in a deck of cards. (A poker hand consists of 5 cards dealt in any order.)

a. Royal flush   *c.* Four of a kind
b. Straight flush   *d.* Full house

Technology *Step by Step*

TI-83 Plus or TI-84 Plus
Step by Step

Factorials, Permutations, and Combinations

Factorials *n*!

1. Type the value of *n*.
2. Press **MATH** and move the cursor to PRB, then press **4** for !.
3. Press **ENTER**.

Permutations $_nP_r$

1. Type the value of *n*.
2. Press **MATH** and move the cursor to PRB, then press **2** for $_nP_r$.
3. Type the value of *r*.
4. Press **ENTER**.

Combinations $_nC_r$

1. Type the value of *n*.
2. Press **MATH** and move the cursor to PRB, then press **3** for $_nC_r$.
3. Type the value of *r*.
4. Press **ENTER**.

Calculate 5!, $_8P_3$, and $_{12}C_5$ (Examples 4–42, 4–44, and 4–48 from the text).

```
5!
                120
8 nPr 3
                336
12 nCr 5
                792
```

Excel
Step by Step

Permutations, Combinations, and Factorials

To find a value of a permutation, for example, $_5P_3$:

1. Select an open cell in an Excel workbook.
2. Select the paste function (f_x) icon from the toolbar.
3. Select Function category Statistical. Then select the PERMUT function.
4. Type **5** in the Number box.
5. Type **3** in the Number_chosen box and click [OK].

The selected cell will display the answer: 60.

To find a value of a combination, for example, $_5C_3$:

1. Select an open cell in an Excel workbook.
2. Select the paste function icon from the toolbar.
3. Select Function category Math & Trig. Then select the COMBIN function.
4. Type **5** in the Number box.
5. Type **3** in the Number_chosen box and click [OK].

The selected cell will display the answer: 10.

To find a factorial of a number, for example, 7!:

1. Select an open cell in an Excel workbook.
2. Select the paste function icon from the toolbar.
3. Select Function category Math & Trig. Then select the FACT function.
4. Type **7** in the Number box and click [OK].

The selected cell will display the answer: 5040.

4–6 | Probability and Counting Rules

Objective **8**

Find the probability of an event, using the counting rules.

The counting rules can be combined with the probability rules in this chapter to solve many types of probability problems. By using the fundamental counting rule, the permutation rules, and the combination rule, one can compute the probability of outcomes of many experiments, such as getting a full house when 5 cards are dealt or selecting a committee of 3 women and 2 men from a club consisting of 10 women and 10 men.

| **Example 4–50** | Find the probability of getting 4 aces when 5 cards are drawn from an ordinary deck of cards. |

Solution

There are $_{52}C_5$ ways to draw 5 cards from a deck. There is only one way to get 4 aces (i.e., $_4C_4$), but there are 48 possibilities to get the fifth card. Therefore, there are 48 ways to get 4 aces and 1 other card. Hence,

$$P(4 \text{ aces}) = \frac{_4C_4 \cdot 48}{_{52}C_5} = \frac{1 \cdot 48}{2,598,960} = \frac{48}{2,598,960} = \frac{1}{54,145}$$

| **Example 4–51** | A box contains 24 transistors, 4 of which are defective. If 4 are sold at random, find the following probabilities. |

    a. Exactly 2 are defective.   *c.* All are defective.

    b. None is defective.   *d.* At least 1 is defective.

Solution

There are $_{24}C_4$ ways to sell 4 transistors, so the denominator in each case will be 10,626.

    a. Two defective transistors can be selected as $_4C_2$ and 2 nondefective ones as $_{20}C_2$. Hence,

$$P(\text{exactly 2 defectives}) = \frac{_4C_2 \cdot _{20}C_2}{_{24}C_4} = \frac{1140}{10,626} = \frac{190}{1771}$$

    b. The number of ways to choose no defectives is $_{20}C_4$. Hence,

$$P(\text{no defectives}) = \frac{_{20}C_4}{_{24}C_4} = \frac{4845}{10,626} = \frac{1615}{3542}$$

    c. The number of ways to choose 4 defectives from 4 is $_4C_4$, or 1. Hence,

$$P(\text{all defective}) = \frac{1}{_{24}C_4} = \frac{1}{10,626}$$

    d. To find the probability of at least 1 defective transistor, find the probability that there are no defective transistors, and then subtract that probability from 1.

$$P(\text{at least 1 defective}) = 1 - P(\text{no defectives})$$

$$= 1 - \frac{_{20}C_4}{_{24}C_4} = 1 - \frac{1615}{3542} = \frac{1927}{3542}$$

| **Example 4–52** | A store has 6 *TV Graphic* magazines and 8 *Newstime* magazines on the counter. If two customers purchased a magazine, find the probability that one of each magazine was purchased. |

Solution

$$P(1 \text{ } TV \text{ } Graphic \text{ and } 1 \text{ } Newstime) = \frac{_6C_1 \cdot _8C_1}{_{14}C_2} = \frac{6 \cdot 8}{91} = \frac{48}{91}$$

Example 4–53

A combination lock consists of the 26 letters of the alphabet. If a 3-letter combination is needed, find the probability that the combination will consist of the letters ABC in that order. The same letter can be used more than once. (*Note:* A combination lock is really a permutation lock.)

Solution

Since repetitions are permitted, there are $26 \cdot 26 \cdot 26 = 17{,}576$ different possible combinations. And since there is only one ABC combination, the probability is $P(\text{ABC}) = 1/26^3 = 1/17{,}576$.

Example 4–54

There are 8 married couples in a tennis club. If 1 man and 1 woman are selected at random to plan the summer tournament, find the probability that they are married to each other.

Solution

Since there are 8 ways to select the man and 8 ways to select the woman, there are $8 \cdot 8$, or 64, ways to select 1 man and 1 woman. Since there are 8 married couples, the solution is $\frac{8}{64} = \frac{1}{8}$.

As indicated at the beginning of this section, the counting rules and the probability rules can be used to solve a large variety of probability problems found in business, gambling, economics, biology, and other fields.

Applying the Concepts 4–6

Counting Rules and Probability

One of the biggest problems for students when doing probability problems is to decide which formula or formulas to use. Another problem is to decide whether two events are independent or dependent. Use the following problem to help develop a better understanding of these concepts.

Assume you are given a 5-question multiple-choice quiz. Each question has 5 possible answers: A, B, C, D, and E.

1. How many events are there?
2. Are the events independent or dependent?
3. If you guess at each question, what is the probability that you get all of them correct?
4. What is the probability that a person would guess answer A for each question?

Assume that you are given a test in which you are to match the correct answers in the right column with the questions in the left column. You can use each answer only once.

5. How many events are there?
6. Are the events independent or dependent?
7. What is the probability of getting them all correct if you are guessing?
8. What is the difference between the two problems?

See page 235 for the answers.

Speaking of
Statistics

Probabilities can be used to assess the danger of a situation. For example, the chances of dying from a shark attack are 1 in 100 million while the chances of dying from a heart attack are 1 in 400. Hence, we are at greater risk from dying of heart disease. Using the probabilities in the table, explain why we should be more afraid of smoking or a bad diet than of walking across a street.

How much danger are we really in?

You are 71,500 times more likely to die in a car crash this year than to be killed by anthrax. Indeed, you are 8,300 times more likely to get killed walking across the street, says risk expert Fred Kilbourne, a member of the board of Conference of Consulting Actuaries.[1] He calculates an American's chance of dying this year from any cause—from anthrax to drowning—is 1 in 130.

Based on deaths in the USA, these are each American's risk of dying each year from:

Motor vehicle accidents	1 in 7,000
Being shot by a gun	1 in 10,000
Falling down	1 in 20,000
Poison	1 in 40,000
Walking across the street	1 in 60,000
Drowning	1 in 75,000
House fire	1 in 100,000
Bicycle accident	1 in 500,000
Commercial plane crash	1 in 1 million
Lightning strike	1 in 3 million
Shark attack	1 in 100 million
Roller coaster accident	1 in 300 million
Anthrax	**1 in 500 million**

Kilbourne says the major killers are not accidents but illness and disease.

The risk of death per person per year in the USA for:

Heart disease	1 in 400
Cancer	1 in 600
Stroke	1 in 2,000
Flu and pneumonia	1 in 3,000

1 – Kilbourne cautions that actuarial assessments, the predictive numbers used by the insurance industry, are based on "the assumption that the future will look reasonably like the past. We may not be able to make that assumption on every issue right now."

Source: Copyright 2001, *USA TODAY.* Reprinted with permission.

Exercises 4–6

1. Find the probability of getting 2 face cards (king, queen, or jack) when 2 cards are drawn from a deck without replacement.

2. A parent-teacher committee consisting of 4 people is to be formed from 20 parents and 5 teachers. Find the probability that the committee will consist of these people. (Assume that the selection will be random.)

   a. All teachers
   b. 2 teachers and 2 parents
   c. All parents
   d. 1 teacher and 3 parents

3. In a company there are 7 executives: 4 women and 3 men. Three are selected to attend a management seminar. Find these probabilities.

   a. All 3 selected will be women.
   b. All 3 selected will be men.
   c. 2 men and 1 woman will be selected.
   d. 1 man and 2 women will be selected.

4. The composition of the Senate of the 107th Congress is

49 Republicans   1 Independent   50 Democrats

A new committee is being formed to study ways to benefit the arts in education. If 3 Senators are selected at random to head the committee, what is the probability that they will all be Republicans? What is the probability that they will all be Democrats? What is the probability that there will be 1 from each party, including the Independent?

Source: *N.Y. Times Almanac.*

5. The signers of the Declaration of Independence came from the Thirteen Colonies as shown.

Massachusetts	5	New York	4
New Hampshire	3	Georgia	3
Virginia	7	North Carolina	3
Maryland	4	South Carolina	4
New Jersey	5	Connecticut	4
Pennsylvania	9	Delaware	3
Rhode Island	2		

Suppose that 4 are chosen at random to be the subject of a documentary. Find the probability that

a. All 4 come from Pennsylvania

b. 2 come from Pennsylvania and 2 from Virginia

Source: *N.Y. Times Almanac.*

6. A package contains 12 resistors, 3 of which are defective. If 4 are selected, find the probability of getting

a. No defective resistors

b. 1 defective resistor

c. 3 defective resistors

7. If 50 tickets are sold and 2 prizes are to be awarded, find the probability that one person will win 2 prizes if that person buys 2 tickets.

8. Find the probability of getting a full house (3 cards of one denomination and 2 of another) when 5 cards are dealt from an ordinary deck.

9. A committee of 4 people is to be formed from 6 doctors and 8 dentists. Find the probability that the committee will consist of

a. All dentists

b. 2 dentists and 2 doctors

c. All doctors

d. 3 doctors and 1 dentist

e. 1 doctor and 3 dentists

10. An insurance sales representative selects 3 policies to review. The group of policies she can select from contains 8 life policies, 5 automobile policies, and 2 homeowner policies. Find the probability of selecting

a. All life policies

b. Both homeowner policies

c. All automobile policies

d. 1 of each policy

e. 2 life policies and 1 automobile policy

11. A drawer contains 11 identical red socks and 8 identical black socks. Suppose that you choose 2 socks at random in the dark

a. What is the probability that you get a pair of red socks?

b. What is the probability that you get a pair of black socks?

c. What is the probability that you get 2 unmatched socks?

d. Where did the other red sock go?

12. Find the probability of selecting 3 science books and 4 math books from 8 science books and 9 math books. The books are selected at random.

13. When 3 dice are rolled, find the probability of getting a sum of 7.

14. Find the probability of randomly selecting 2 mathematics books and 3 physics books from a box containing 4 mathematics books and 8 physics books.

15. Find the probability that if 5 different-sized washers are arranged in a row, they will be arranged in order of size.

16. Using the information in Exercise 52 in Section 4–5, find the probability of each poker hand.

a. Royal flush

b. Straight flush

c. 4 of a kind

4–7	# Summary

In this chapter, the basic concepts and rules of probability are explained. The three types of probability are classical, empirical, and subjective. Classical probability uses sample spaces. Empirical probability uses frequency distributions and is based on observation. In subjective probability, the researcher makes an educated guess about the chance of an event occurring.

A probability event consists of one or more outcomes of a probability experiment. Two events are said to be mutually exclusive if they cannot occur at the same time. Events can also be classified as independent or dependent. If events are independent, whether or not the first event occurs does not affect the probability of the next event occurring. If the probability of the second event occurring is changed by the occurrence of the first event, then the events are dependent. The complement of an event is the set of outcomes in the sample space that are not included in the outcomes of the event itself. Complementary events are mutually exclusive.

Probability problems can be solved by using the addition rules, the multiplication rules, and the complementary event rules.

Finally, the fundamental counting rule, the permutation rule, and the combination rule can be used to determine the number of outcomes of events; then these numbers can be used to determine the probabilities of events.

Important Terms

classical probability 176

combination 217

complement of an
event 179

compound event 176

conditional probability 202

dependent events 202

empirical probability 181

equally likely events 176

event 175

fundamental counting
rule 212

independent events 199

law of large numbers 183

mutually exclusive
events 189

outcome 173

permutation 215

probability 173

probability experiment 173

sample space 173

simple event 175

subjective probability 183

tree diagram 175

Venn diagrams 180

Important Formulas

Formula for classical probability:

$$P(E) = \frac{\text{number of outcomes in } E}{\text{total number of outcomes in sample space}} = \frac{n(E)}{n(S)}$$

Formula for empirical probability:

$$P(E) = \frac{\text{frequency for class}}{\text{total frequencies in distribution}} = \frac{f}{n}$$

Addition rule 1, for two mutually exclusive events:

$$P(A \text{ or } B) = P(A) + P(B)$$

Addition rule 2, for events that are not mutually exclusive:

$$P(A \text{ or } B) = P(A) + P(B) - P(A \text{ and } B)$$

Multiplication rule 1, for independent events:

$$P(A \text{ and } B) = P(A) \cdot P(B)$$

Multiplication rule 2, for dependent events:

$$P(A \text{ and } B) = P(A) \cdot P(B|A)$$

Formula for conditional probability:

$$P(B|A) = \frac{P(A \text{ and } B)}{P(A)}$$

Formula for complementary events:

$$P(\overline{E}) = 1 - P(E) \qquad \text{or} \qquad P(E) = 1 - P(\overline{E})$$
$$\text{or} \qquad P(E) + P(\overline{E}) = 1$$

Fundamental counting rule: In a sequence of n events in which the first one has k_1 possibilities, the second event has k_2 possibilities, the third has k_3 possibilities, etc., the total number possibilities of the sequence will be

$$k_1 \cdot k_2 \cdot k_3 \cdots k_n$$

Permutation rule: The number of permutations of n objects taking r objects at a time when order is important is

$$_nP_r = \frac{n!}{(n-r)!}$$

Combination rule: The number of combinations of r objects selected from n objects when order is not important is

$$_nC_r = \frac{n!}{(n-r)!r!}$$

Review Exercises

1. When a die is rolled, find the probability of getting a
   a. 5
   b. 6
   c. Number less than 5

2. When a card is selected from a deck, find the probability of getting
   a. A club
   b. A face card or a heart
   c. A 6 and a spade
   d. A king
   e. A red card

3. In a survey conducted at a local restaurant during breakfast hours, 20 people preferred orange juice, 16 preferred grapefruit juice, and 9 preferred apple juice with breakfast. If a person is selected at random, find the probability that she or he prefers grapefruit juice.

4. If a die is rolled one time, find these probabilities.
   a. Getting a 5
   b. Getting an odd number
   c. Getting a number less than 3

5. A recent survey indicated that in a town of 1500 households, 850 had cordless telephones. If a household is randomly selected, find the probability that it has a cordless telephone.

6. During a sale at a men's store, 16 white sweaters, 3 red sweaters, 9 blue sweaters, and 7 yellow sweaters were purchased. If a customer is selected at random, find the probability that he bought
   a. A blue sweater
   b. A yellow or a white sweater
   c. A red, a blue, or a yellow sweater
   d. A sweater that was not white

7. At a swimwear store, the managers found that 16 women bought white bathing suits, 4 bought red suits, 3 bought blue suits, and 7 bought yellow suits. If a customer is selected at random, find the probability that she bought
   a. A blue suit
   b. A yellow or a red suit
   c. A white or a yellow or a blue suit
   d. A suit that was not red

8. When two dice are rolled, find the probability of getting
   a. A sum of 5 or 6
   b. A sum greater than 9
   c. A sum less than 4 or greater than 9
   d. A sum that is divisible by 4
   e. A sum of 14
   f. A sum less than 13

9. The probability that a person owns a car is 0.80, that a person owns a boat is 0.30, and that a person owns both a car and a boat is 0.12. Find the probability that a person owns either a boat or a car.

10. There is a 0.39 probability that John will purchase a new car, a 0.73 probability that Mary will purchase a new car, and a 0.36 probability that both will purchase a new car. Find the probability that neither will purchase a new car.

11. A Gallup Poll found that 78% of Americans worry about the quality and healthfulness of their diet. If 5 people are selected at random, find the probability that all 5 worry about the quality and healthfulness of their diet.
    Source: The Book of Odds.

12. Of Americans using library services, 67% borrow books. If 5 patrons are chosen at random, what is the probability that all borrowed books? That none borrowed books?
    Source: American Library Association.

13. Three cards are drawn from an ordinary deck *without* replacement. Find the probability of getting
    a. All black cards
    b. All spades
    c. All queens

14. A coin is tossed and a card is drawn from a deck. Find the probability of getting
    a. A head and a 6
    b. A tail and a red card
    c. A head and a club

15. A box of candy contains 6 chocolate-covered cherries, 3 peppermints, 2 caramels, and 2 strawberry creams. If a piece of candy is selected, find the probability of getting a caramel or a peppermint.

16. A manufacturing company has three factories: X, Y, and Z. The daily output of each is shown here.

Product	Factory X	Factory Y	Factory Z
TVs	18	32	15
Stereos	6	20	13

    If one item is selected at random, find these probabilities.
    a. It was manufactured at factory X or is a stereo.
    b. It was manufactured at factory Y or factory Z.
    c. It is a TV or was manufactured at factory Z.

17. A vaccine has a 90% probability of being effective in preventing a certain disease. The probability of getting the disease if a person is not vaccinated is 50%. In a certain geographic region, 25% of the people get vaccinated. If a person is selected at random, find the probability that he or she will contract the disease.

18. A manufacturer makes three models of a television set, models A, B, and C. A store sells 40% of model A sets, 40% of model B sets, and 20% of model C sets. Of

model A sets, 3% have stereo sound; of model B sets, 7% have stereo sound; and of model C sets, 9% have stereo sound. If a set is sold at random, find the probability that it has stereo sound.

19. The probability that Sue will live on campus and buy a new car is 0.37. If the probability that she will live on campus is 0.73, find the probability that she will buy a new car, given that she lives on campus.

20. The probability that a customer will buy a television set and buy an extended warranty is 0.03. If the probability that a customer will purchase a television set is 0.11, find the probability that the customer will also purchase the extended warranty.

21. Of the members of the Blue River Health Club, 43% have a lifetime membership and exercise regularly (three or more times a week). If 75% of the club members exercise regularly, find the probability that a randomly selected member is a life member, given that he or she exercises regularly.

22. The probability that it snows and the bus arrives late is 0.023. José hears the weather forecast, and there is a 40% chance of snow tomorrow. Find the probability that the bus will be late, given that it snows.

23. At a large factory, the employees were surveyed and classified according to their level of education and whether they smoked. The data are shown in the table.

| | Educational level | | |
Smoking habit	Not high school graduate	High school graduate	College graduate
Smoke	6	14	19
Do not smoke	18	7	25

If an employee is selected at random, find these probabilities.

a. The employee smokes, given that he or she graduated from college.
b. Given that the employee did not graduate from high school, he or she is a smoker.

24. A survey found that 77% of bike riders sometimes ride without a helmet. If 4 bike riders are randomly selected, find the probability that at least one of the riders does not wear a helmet all the time.

Source: USA TODAY.

25. A coin is tossed 5 times. Find the probability of getting at least one tail.

26. The U.S. Department of Health and Human Services reports that 15% of Americans have chronic sinusitis. If 5 people are selected at random, find the probability that at least one has chronic sinusitis.

Source: 100% American.

27. An automobile license plate consists of 3 letters followed by 4 digits. How many different plates can be made if repetitions are allowed? If repetitions are not allowed?

If repetitions are allowed in the letters but not in the digits?

28. How many different arrangements of the letters in the word *bread* can be made?

29. How many ways can 3 outfielders and 4 infielders be chosen from 5 outfielders and 7 infielders?

30. How many different ways can 8 computer operators be seated in a row?

31. How many ways can a student select 2 electives from a possible choice of 10 electives?

32. There are 6 Republican, 5 Democrat, and 4 Independent candidates. How many different ways can a committee of 3 Republicans, 2 Democrats, and 1 Independent be selected?

33. How many different computer passwords are possible if each consists of 4 symbols and if the first one must be a letter and the other 3 must be digits?

34. A new employee has a choice of 5 health care plans, 3 retirement plans, and 2 different expense accounts. If a person selects one of each option, how many different options does he or she have?

35. There are 12 students who wish to enroll in a particular course. There are only 4 seats left in the classroom. How many different ways can 4 students be selected to attend the class?

36. A candy store allows customers to select 3 different candies to be packaged and mailed. If there are 13 varieties available, how many possible selections can be made?

37. If a student can select 5 novels from a reading list of 20 for a course in literature, how many different possible ways can this selection be done?

38. If a student can select one of 3 language courses, one of 5 mathematics courses, and one of 4 history courses, how many different schedules can be made?

39. License plates are to be issued with 3 letters followed by 4 single digits. How many such license plates are possible? If the plates are issued at random, what is the probability that the license plate says USA followed by a number that is divisible by 5?

40. A newspaper advertises 5 different movies, 3 plays, and 2 baseball games for the weekend. If a couple selects 3 activities, find the probability that they attend 2 plays and 1 movie.

41. In an office there are 3 secretaries, 4 accountants, and 2 receptionists. If a committee of 3 is to be formed, find the probability that one of each will be selected.

42. For a survey, a subject can be classified as follows:
Gender: male or female
Marital status: single, married, widowed, divorced
Occupation: administration, faculty, staff

Draw a tree diagram for the different ways a person can be classified.

Would You Bet Your Life?–Revisited

In his book *Probabilities in Everyday Life,* John D. McGervey states that the chance of being killed on any given commercial airline flight is almost 1 in 1 million and that the chance of being killed during a transcontinental auto trip is about 1 in 8000. The corresponding probabilities are $1/1,000,000 = 0.000001$ as compared to $1/8000 = 0.000125$. Since the second number is 125 times greater than the first number, you have a much higher risk driving than flying across the United States.

Chapter Quiz

Determine whether each statement is true or false. If the statement is false, explain why.

1. Subjective probability has little use in the real world.

2. Classical probability uses a frequency distribution to compute probabilities.

3. In classical probability, all outcomes in the sample space are equally likely.

4. When two events are not mutually exclusive, $P(A \text{ or } B) = P(A) + P(B)$.

5. If two events are dependent, they must have the same probability of occurring.

6. An event and its complement can occur at the same time.

7. The arrangement ABC is the same as BAC for combinations.

8. When objects are arranged in a specific order, the arrangement is called a combination.

Select the best answer.

9. The probability that an event happens is 0.42. What is the probability that the event won't happen?
   a. −0.42 c. 0
   b. 0.58   d. 1

10. When a meteorologist says that there is a 30% chance of showers, what type of probability is the person using?
   a. Classical c. Relative
   b. Empirical d. Subjective

11. The sample space for tossing 3 coins consists of how many outcomes?
   a. 2 c. 6
   b. 4 d. 8

12. The complement of guessing 5 correct answers on a 5-question true/false exam is
   a. Guessing 5 incorrect answers
   b. Guessing at least 1 incorrect answer

   c. Guessing at least 1 correct answer
   d. Guessing no incorrect answers

13. When two dice are rolled, the sample space consists of how many events?
   a. 6   c. 36
   b. 12 d. 54

14. What is $_nP_0$?
   a. 0 c. n
   b. 1 d. It cannot be determined.

15. What is the number of permutations of 6 different objects taken all together?
   a. 0 c. 36
   b. 1 d. 720

16. What is 0!?
   a. 0 c. Undefined
   b. 1 d. 10

17. What is $_nC_n$?
   a. 0 c. n
   b. 1 d. It cannot be determined.

Complete the following statements with the best answer.

18. The set of all possible outcomes of a probability experiment is called the _____.

19. The probability of an event can be any number between and including _____ and _____.

20. If an event cannot occur, its probability is _____.

21. The sum of the probabilities of the events in the sample space is _____.

22. When two events cannot occur at the same time, they are said to be _____.

23. When a card is drawn, find the probability of getting
   a. A jack
   b. A 4
   c. A card less than 6 (an ace is considered above 6)

24. When a card is drawn from a deck, find the probability of getting

    a. A diamond
    b. A 5 or a heart
    c. A 5 and a heart
    d. A king
    e. A red card

25. At a men's clothing store, 12 men purchased blue golf sweaters, 8 purchased green sweaters, 4 purchased gray sweaters, and 7 bought black sweaters. If a customer is selected at random, find the probability that he purchased

    a. A blue sweater
    b. A green or gray sweater
    c. A green or black or blue sweater
    d. A sweater that was not black

26. When 2 dice are rolled, find the probability of getting

    a. A sum of 6 or 7
    b. A sum greater than 8
    c. A sum less than 3 or greater than 8
    d. A sum that is divisible by 3
    e. A sum of 16
    f. A sum less than 11

27. The probability that a person owns a microwave oven is 0.75, that a person owns a compact disk player is 0.25, and that a person owns both a microwave and a CD player is 0.16. Find the probability that a person owns either a microwave or a CD player, but not both.

28. Of the physics graduates of a university, 30% received a starting salary of $30,000 or more. If 5 of the graduates are selected at random, find the probability that all had a starting salary of $30,000 or more.

29. Five cards are drawn from an ordinary deck *without* replacement. Find the probability of getting

    a. All red cards
    b. All diamonds
    c. All aces

30. The probability that Samantha will be accepted by the college of her choice and obtain a scholarship is 0.35. If the probability that she is accepted by the college is 0.65, find the probability that she will obtain a scholarship given that she is accepted by the college.

31. The probability that a customer will buy a car and an extended warranty is 0.16. If the probability that a customer will purchase a car is 0.30, find the probability that the customer will also purchase the extended warranty.

32. Of the members of the Spring Lake Bowling Lanes, 57% have a lifetime membership and bowl regularly (three or more times a week). If 70% of the club members bowl regularly, find the probability that a randomly selected member is a lifetime member, given that he or she bowls regularly.

33. The probability that Mike has to work overtime and it rains is 0.028. Mike hears the weather forecast, and there is a 50% chance of rain. Find the probability that he will have to work overtime, given that it rains.

34. At a large factory, the employees were surveyed and classified according to their level of education and whether they attend a sports event at least once a month. The data are shown in the table.

| | Educational level | | |
| | High school graduate | Two-year college degree | Four-year college degree |
Sports event			
Attend	16	20	24
Do not attend	12	19	25

If an employee is selected at random, find the probability that

    a. The employee attends sports events regularly, given that he or she graduated from college (2- or 4-year degree)
    b. Given that the employee is a high school graduate, he or she does not attend sports events regularly

35. In a certain high-risk group, the chances of a person having suffered a heart attack are 55%. If 6 people are chosen, find the probability that at least 1 will have had a heart attack.

36. A single die is rolled 4 times. Find the probability of getting at least one 5.

37. If 85% of all people have brown eyes and 6 people are selected at random, find the probability that at least 1 of them has brown eyes.

38. How many ways can 5 sopranos and 4 altos be selected from 7 sopranos and 9 altos?

39. How many different ways can 8 speakers be seated on a stage?

40. A soda machine servicer must restock and collect money from 15 machines, each one at a different location. How many ways can she select 4 machines to service in 1 day?

41. One company's ID cards consist of 5 letters followed by 2 digits. How many cards can be made if repetitions are allowed? If repetitions are not allowed?

42. How many different arrangements of the letters in the word *number* can be made?

43. A physics test consists of 25 true/false questions. How many different possible answer keys can be made?

44. How many different ways can 5 cellular telephones be selected from 8 cellular phones?

45. On a lunch counter, there are 3 oranges, 5 apples, and 2 bananas. If 3 pieces of fruit are selected, find the probability that 1 orange, 1 apple, and 1 banana are selected.

46. A cruise director schedules 4 different movies, 2 bridge games, and 3 tennis games for a 2-day period. If a couple selects 3 activities, find the probability that they attend 2 movies and 1 tennis game.

47. At a sorority meeting, there are 6 seniors, 4 juniors, and 2 sophomores. If a committee of 3 is to be formed, find the probability that 1 of each will be selected.

48. For a banquet, a committee can select beef, pork, chicken, or veal; baked potatoes or mashed potatoes; and peas or green beans for a vegetable. Draw a tree diagram for all possible choices of a meat, a potato, and a vegetable.

Critical Thinking Challenges

1. Consider this problem: A con man has 3 coins. One coin has been specially made and has a head on each side. A second coin has been specially made, and on each side it has a tail. Finally, a third coin has a head and a tail on it. All coins are of the same denomination. The con man places the 3 coins in his pocket, selects one, and shows you one side. It is heads. He is willing to bet you even money that it is the two-headed coin. His reasoning is that it can't be the two-tailed coin since a head is showing; therefore, there is a 50-50 chance of it being the two-headed coin. Would you take the bet? (*Hint:* See Exercise 1 in Data Projects.)

2. Chevalier de Méré won money when he bet unsuspecting patrons that in 4 rolls of 1 die, he could get at least one 6, but he lost money when he bet that in 24 rolls of 2 dice, he could get at least a double 6. Using the probability rules, find the probability of each event and explain why he won the majority of the time on the first game but lost the majority of the time when playing the second game. (*Hint:* Find the probabilities of losing each game and subtract from 1.)

3. How many people do you think need to be in a room so that 2 people will have the same birthday (month and day)? You might think it is 366. This would, of course, guarantee it (excluding leap year), but how many people would need to be in a room so that there would be a 90% probability that 2 people would be born on the same day? What about a 50% probability?

Actually, the number is much smaller than you may think. For example, if you have 50 people in a room, the probability that 2 people will have the same birthday is 97%. If you have 23 people in a room, there is a 50% probability that 2 people were born on the same day!

The problem can be solved by using the probability rules. It must be assumed that all birthdays are equally likely, but this assumption will have little effect on the answers. The way to find the answer is by using the complementary event rule as P (2 people having the same birthday) $= 1 - P$ (all have different birthdays).

For example, suppose there were 3 people in the room. The probability that each had a different birthday would be

$$\frac{365}{365} \cdot \frac{364}{365} \cdot \frac{363}{365} = \frac{_{365}P_3}{365^3} = 0.992$$

Hence, the probability that at least 2 of the 3 people will have the same birthday will be

$$1 - 0.992 = 0.008$$

Hence, for k people, the formula is

P(at least 2 people have the same birthday)

$$= 1 - \frac{_{365}P_k}{365^k}$$

Using your calculator, complete the table and verify that for at least a 50% chance of 2 people having the same birthday, 23 or more people will be needed.

Number of people	Probability that at least 2 have the same birthday
1	0.000
2	0.003
5	0.027
10	
15	
20	
21	
22	
23	

4. We know that if the probability of an event happening is 100%, then the event is a certainty. Can it be concluded that if there is a 50% chance of contracting a communicable disease through contact with an infected person, there would be a 100% chance of contracting the disease if 2 contacts were made with the infected person? Explain your answer.

 **Data Projects**

1. Make a set of 3 cards—one with a red star on both sides, one with a black star on both sides, and one with a black star on one side and a red star on the other side. With a partner, play the game described in the first Critical Thinking challenge on page 233 one hundred times, and record the results of how many times you win and how many times your partner wins.
   (*Note:* Do not change options during the 100 trials.)

   a. Do you think the game is fair (i.e., does one person win approximately 50% of the time)?
   b. If you think the game is unfair, explain what the probabilities might be and why.

2. Take a coin and tape a small weight (e.g., part of a paper clip) to one side. Flip the coin 100 times and record the results. Do you think you have changed the probabilities of the results of flipping the coin? Explain.

3. This game is called *Diet Fractions*. Roll 2 dice and use the numbers to make a fraction less than or equal to 1. Player A wins if the fraction cannot be reduced; otherwise, player B wins.

   a. Play the game 100 times and record the results.
   b. Decide if the game is fair or not. Explain why or why not.
   c. Using the sample space for 2 dice, compute the probabilities of winning for player A and for player B. Do these agree with the results obtained in part *a*?

Source: George W. Bright, John G. Harvey, and Margariete Montague Wheeler, "Fair Games, Unfair Games." Chapter 8, *Teaching Statistics and Probability.* NCTM 1981 Yearbook. Reston, Va.: The National Council of Teachers of Mathematics, Inc., 1981, p. 49. Used with permission.

4. Often when playing gambling games or collecting items in cereal boxes, one wonders how long will it be before one achieves a success. For example, suppose there are 6 different types of toys with 1 toy packaged at random in a cereal box. If a person wanted a certain toy, about how many boxes would that person have to buy on average before obtaining that particular toy? Of course, there is a possibility that the particular toy would be in the first box opened or that the person might never obtain the particular toy. These are the extremes.

   a. To find out, simulate the experiment using dice. Start rolling dice until a particular number, say, 3, is obtained, and keep track of how many rolls are necessary. Repeat 100 times. Then find the average.
   b. You may decide to use another number, such as 10 different items. In this case, use 10 playing cards (ace through 10 of diamonds), select a particular card (say, an ace), shuffle the deck each time, deal the cards, and count how many cards are turned over before the ace is obtained. Repeat 100 times, then find the average.
   c. Summarize the findings for both experiments.

Answers to Applying the Concepts

Section 4–2 Tossing a Coin

1. The sample space is the listing of all possible outcomes of the coin toss.

2. The possible outcomes are heads or tails.

3. Classical probability says that a fair coin has a 50-50 chance of coming up heads or tails.

4. The law of large numbers says that as you increase the number of trials, the overall results will approach the theoretical probability. However, since the coin has no "memory," it still has a 50-50 chance of coming up heads or tails on the next toss. Knowing what has already happened should not change your opinion on what will happen on the next toss.

5. The empirical approach to probability is based on running an experiment and looking at the results. You cannot do that at this time.

6. Subjective probabilities could be used if you believe the coin is biased.

7. Answers will vary; however, they should address that a fair coin has a 50-50 chance of coming up heads or tails on the next flip.

Section 4–3 Which Pain Reliever Is Best?

1. There were $192 + 186 + 188 = 566$ subjects in the study.

2. The study lasted for 12 weeks.

3. The variables are the type of pain reliever and the side effects.

4. Both variables are qualitative and nominal.

5. The numbers in the table are exact figures.

6. The probability that a randomly selected person was receiving a placebo is $192/566 = 0.3392$ (about 34%).

7. The probability that a randomly selected person was receiving a placebo or drug A is $(192 + 186)/566 = 378/566 = 0.6678$ (about 67%). These are mutually exclusive events. The complement is that a randomly selected person was receiving drug B.

8. The probability that a randomly selected person was receiving a placebo or experienced a neurological headache is $(192 + 55 + 72)/566 = 319/566 = 0.5636$ (about 56%).

9. The probability that a randomly selected person was not receiving a placebo or experienced a sinus headache is $(186 + 188)/566 + 11/566 = 385/566 = 0.6802$ (about 68%).

Section 4–4   Guilty or Innocent?

1. The probability of another couple with the same characteristics being in that area is
$\frac{1}{12} \cdot \frac{1}{10} \cdot \frac{1}{4} \cdot \frac{1}{11} \cdot \frac{1}{3} \cdot \frac{1}{13} \cdot \frac{1}{100} = \frac{1}{20,592,000}$, assuming the characteristics are independent of one another.

2. You would use the multiplication rule, since we are looking for the probability of multiple events happening together.

3. We do not know if the characteristics are dependent or independent, but we assumed independence for the calculation in question 1.

4. The probabilities would change if there were dependence among two or more events.

5. Answers will vary. One possible answer is that probabilities can be used to explain how unlikely it is to have a set of events occur at the same time (in this case, how unlikely it is to have another couple with the same characteristics in that area).

6. Answers will vary. One possible answer is that if the only eyewitness was the woman who was mugged and the probabilities are accurate, it seems very unlikely that a couple matching these characteristics would be in that area at that time. This might cause you to convict the couple.

7. Answers will vary. One possible answer is that our probabilities are theoretical and serve a purpose when appropriate, but that court cases are based on much more than impersonal chance.

8. Answers will vary. One possible answer is that juries decide whether or not to convict a defendant if they find evidence "beyond a reasonable doubt" that the person is guilty. In probability terms, this means that if the defendant was actually innocent, then the chance of seeing the events that occurred are so unlikely as to have occurred by chance. Therefore, the jury concludes that the defendant is guilty.

Section 4–5   Garage Door Openers

1. Four on/off switches lead to 16 different settings.

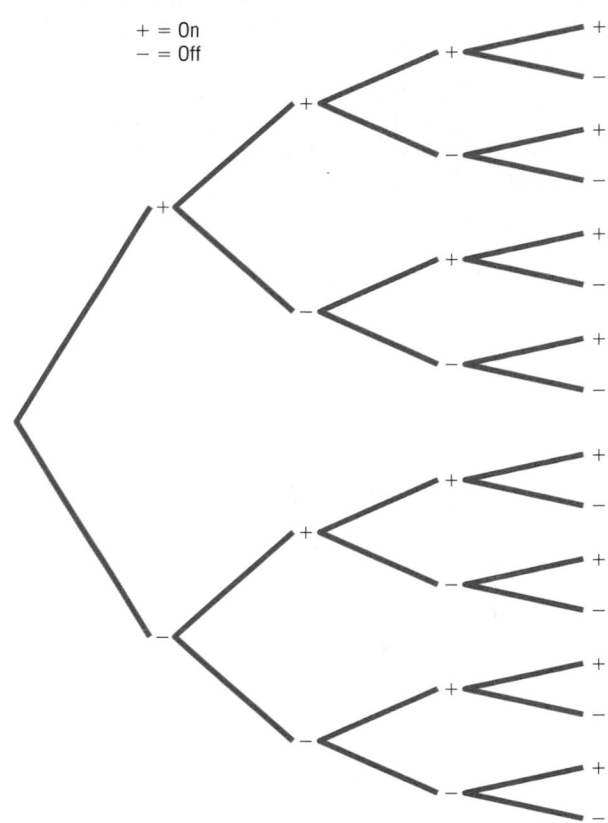

+ = On
− = Off

2. With 5 on/off switches, there are $2^5 = 32$ different settings. With 6 on/off switches, there are $2^6 = 64$ different settings. In general, if there are k on/off switches, there are 2^k different settings.

3. With 8 consecutive on/off switches, there are $2^8 = 256$ different settings.

4. It is less likely for someone to be able to open your garage door if you have 8 on/off settings (probability about 0.4%) than if you have 4 on/off switches (probability about 6.0%). Having 8 on/off switches in the opener seems pretty safe.

5. Each key blank could be made into $5^5 = 3125$ possible keys.

6. If there were 420,000 Dodge Caravans sold in the United States, then any one key could start about $420,000/3125 = 134.4$, or about 134, different Caravans.

7. Answers will vary.

Section 4–6   Counting Rules and Probability

1. There are five different events: each multiple-choice question is an event.

2. These events are independent.

3. If you guess on 1 question, the probability of getting it correct is 0.20. Thus, if you guess on all 5 questions, the probability of getting all of them correct is $(0.20)^5 = 0.00032$.

4. The probability that a person would guess answer A for a question is 0.20, so the probability that a person would guess answer A for each question is $(0.20)^5 = 0.00032$.

5. There are five different events: each matching question is an event.

6. These are dependent events.

7. The probability of getting them all correct if you are guessing is $\frac{1}{5} \cdot \frac{1}{4} \cdot \frac{1}{3} \cdot \frac{1}{2} \cdot \frac{1}{1} = \frac{1}{120} = 0.0083$.

8. The difference between the two problems is that we are sampling without replacement in the second problem, so the denominator changes in the event probabilities.

CHAPTER 5

Discrete Probability Distributions

Objectives

After completing this chapter, you should be able to

1 Construct a probability distribution for a random variable.

2 Find the mean, variance, standard deviation, and expected value for a discrete random variable.

3 Find the exact probability for *X* successes in *n* trials of a binomial experiment.

4 Find the mean, variance, and standard deviation for the variable of a binomial distribution.

5 Find probabilities for outcomes of variables, using the Poisson, hypergeometric, and multinomial distributions.

Outline

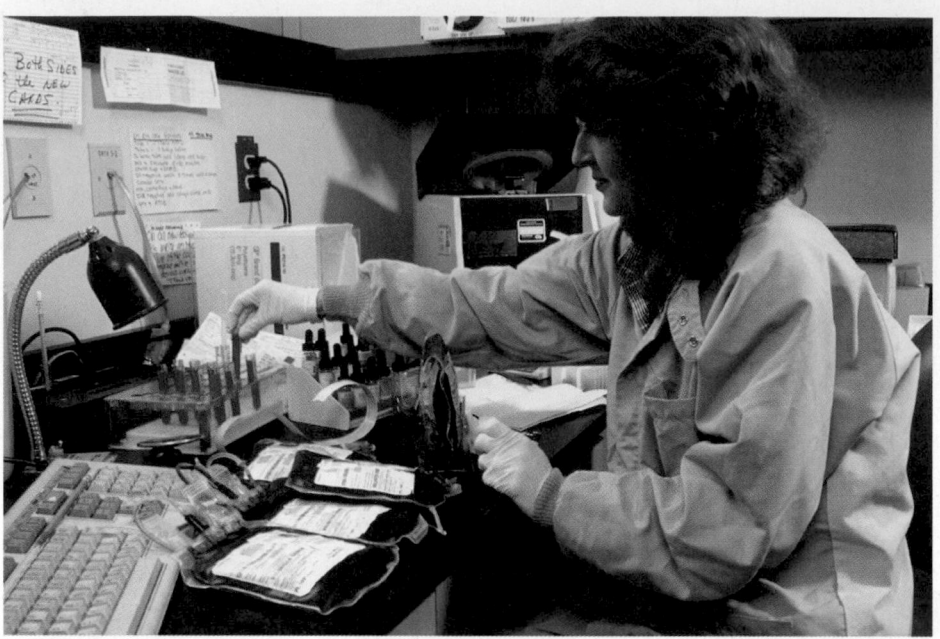

Is Pooling Worthwhile?

Blood samples are used to screen people for certain diseases. When the disease is rare, health care workers sometimes combine or pool the blood samples of a group of individuals into one batch and then test it. If the test result of the batch is negative, no further testing is needed since none of the individuals in the group has the disease. However, if the test result of the batch is positive, each individual in the group must be tested.

Consider this hypothetical example: Suppose the probability of a person having the disease is 0.05, and a pooled sample of 15 individuals is tested. What is the probability that no further testing will be needed for the individuals in the sample? The answer to this question can be found by using what is called the *binomial distribution*. See Statistics Today—Revisited at the end of the chapter.

This chapter explains probability distributions in general and a specific, often used distribution called the binomial distribution. The Poisson, hypergeometric, and multinomial distributions are also explained.

Introduction

Many decisions in business, insurance, and other real-life situations are made by assigning probabilities to all possible outcomes pertaining to the situation and then evaluating the results. For example, a saleswoman can compute the probability that she will make 0, 1, 2, or 3 or more sales in a single day. An insurance company might be able to assign probabilities to the number of vehicles a family owns. A self-employed speaker might be able to compute the probabilities for giving 0, 1, 2, 3, or 4 or more speeches each week. Once these probabilities are assigned, statistics such as the mean, variance, and standard deviation can be computed for these events. With these statistics, various decisions can be made. The saleswoman will be able to compute the average number of sales she makes per week, and if she is working on commission, she will be able to approximate her weekly income over a period of time, say, monthly. The public speaker will be able to plan ahead and approximate his average income and expenses. The insurance company can use its information to design special computer forms and programs to accommodate its customers' future needs.

This chapter explains the concepts and applications of what is called a *probability distribution.* In addition, special probability distributions, such as the *binomial, multinomial, Poisson,* and *hypergeometric* distributions, are explained.

<table>
<tr><td>**5–2**</td></tr>
</table>

Probability Distributions

Objective 1

Construct a probability distribution for a random variable.

Before probability distribution is defined formally, the definition of a variable is reviewed. In Chapter 1, a *variable* was defined as a characteristic or attribute that can assume different values. Various letters of the alphabet, such as X, Y, or Z, are used to represent variables. Since the variables in this chapter are associated with probability, they are called *random variables.*

For example, if a die is rolled, a letter such as X can be used to represent the outcomes. Then the value that X can assume is 1, 2, 3, 4, 5, or 6, corresponding to the outcomes of rolling a single die. If two coins are tossed, a letter, say Y, can be used to represent the number of heads, in this case 0, 1, or 2. As another example, if the temperature at 8:00 A.M. is 43° and at noon it is 53°, then the values T that the temperature assumes are said to be random, since they are due to various atmospheric conditions at the time the temperature was taken.

A **random variable** is a variable whose values are determined by chance.

Also recall from Chapter 1 that one can classify variables as discrete or continuous by observing the values the variable can assume. If a variable can assume only a specific number of values, such as the outcomes for the roll of a die or the outcomes for the toss of a coin, then the variable is called a *discrete variable.*

Discrete variables have a finite number of possible values or an infinite number of values that can be counted. The word *counted* means that they can be enumerated using the numbers 1, 2, 3, etc. For example, the number of joggers in Riverview Park each day and the number of phone calls received after a TV commercial airs are examples of discrete variables, since they can be counted.

Variables that can assume all values in the interval between any two given values are called *continuous variables.* For example, if the temperature goes from 62 to 78° in a 24-hour period, it has passed through every possible number from 62 to 78. *Continuous random variables are obtained from data that can be measured rather than counted.* Continuous random variables can assume an infinite number of values and can be decimal and fractional values. On a continuous scale, a person's weight might be exactly 183.426 pounds if a scale could measure weight to the thousandths place; however, on a digital scale that measures only to tenths of pounds, the weight would be 183.4 pounds. Examples of continuous variables are heights, weights, temperatures, and time. In this chapter only discrete random variables are used; Chapter 6 explains continuous random variables.

The procedure shown here for constructing a probability distribution for a discrete random variable uses the probability experiment of tossing three coins. Recall that when three coins are tossed, the sample space is represented as TTT, TTH, THT, HTT, HHT, HTH, THH, HHH; and if X is the random variable for the number of heads, then X assumes the value 0, 1, 2, or 3.

Probabilities for the values of X can be determined as follows:

No heads	One head			Two heads			Three heads
TTT	TTH	THT	HTT	HHT	HTH	THH	HHH
$\frac{1}{8}$	$\frac{1}{8}$	$\frac{1}{8}$	$\frac{1}{8}$	$\frac{1}{8}$	$\frac{1}{8}$	$\frac{1}{8}$	$\frac{1}{8}$
$\frac{1}{8}$		$\frac{3}{8}$			$\frac{3}{8}$		$\frac{1}{8}$

Hence, the probability of getting no heads is $\frac{1}{8}$, one head is $\frac{3}{8}$, two heads is $\frac{3}{8}$, and three heads is $\frac{1}{8}$. From these values, a probability distribution can be constructed by listing the outcomes and assigning the probability of each outcome, as shown here.

Number of heads X	0	1	2	3
Probability $P(X)$	$\frac{1}{8}$	$\frac{3}{8}$	$\frac{3}{8}$	$\frac{1}{8}$

A **discrete probability distribution** consists of the values a random variable can assume and the corresponding probabilities of the values. The probabilities are determined theoretically or by observation.

Discrete probability distributions can be shown by using a graph or a table. Probability distributions can also be represented by a formula. See Exercises 31–36 at the end of this section for examples.

Example 5–1

Construct a probability distribution for rolling a single die.

Solution

Since the sample space is 1, 2, 3, 4, 5, 6 and each outcome has a probability of $\frac{1}{6}$, the distribution is as shown.

Outcome X	1	2	3	4	5	6
Probability $P(X)$	$\frac{1}{6}$	$\frac{1}{6}$	$\frac{1}{6}$	$\frac{1}{6}$	$\frac{1}{6}$	$\frac{1}{6}$

Probability distributions can be shown graphically by representing the values of X on the x axis and the probabilities $P(X)$ on the y axis.

Example 5–2

Represent graphically the probability distribution for the sample space for tossing three coins.

Number of heads X	0	1	2	3
Probability $P(X)$	$\frac{1}{8}$	$\frac{3}{8}$	$\frac{3}{8}$	$\frac{1}{8}$

Solution

The values that X assumes are located on the x axis, and the values for $P(X)$ are located on the y axis. The graph is shown in Figure 5–1.

Note that for visual appearances, it is not necessary to start with 0 at the origin.

Examples 5–1 and 5–2 are illustrations of *theoretical* probability distributions. One did not need to actually perform the experiments to compute the probabilities. In contrast, to construct actual probability distributions, one must observe the variable over a period of time. They are empirical, as shown in Example 5–3.

Figure 5–1

**Probability Distribution
for Example 5–2**

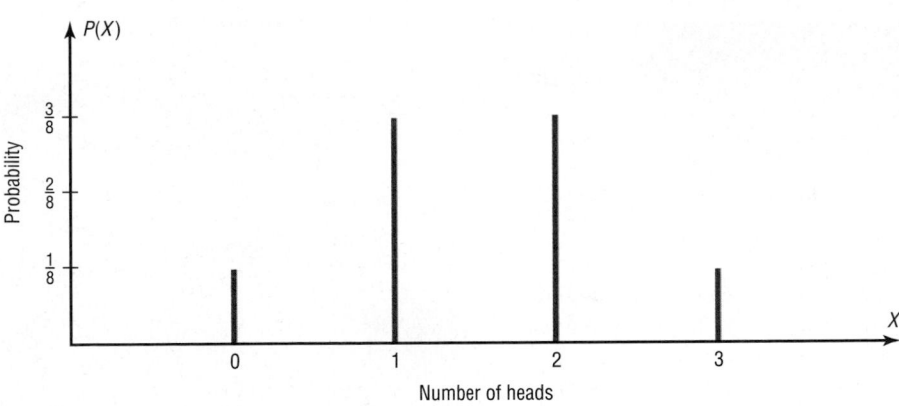

Example 5–3

During the summer months, a rental agency keeps track of the number of chain saws it rents each day during a period of 90 days. The number of saws rented per day is represented by the variable X. The results are shown here. Compute the probability $P(X)$ for each X, and construct a probability distribution and graph for the data.

X	Number of days
0	45
1	30
2	15
Total	90

Solution

The probability $P(X)$ can be computed for each X by dividing the number of days that X saws were rented by total days.

For 0 saws: $\frac{45}{90} = 0.50$

For 1 saw: $\frac{30}{90} = 0.33$

For 2 saws: $\frac{15}{90} = 0.17$

The distribution is shown here.

Number of saws rented X	0	1	2
Probability $P(X)$	0.50	0.33	0.17

The graph is shown in Figure 5–2.

Figure 5–2

**Probability Distribution
for Example 5–3**

Speaking of
Statistics

Coins, Births, and Other Random (?) Events

Examples of random events such as tossing coins are used in almost all books on probability. But is flipping a coin really a random event?

Tossing coins dates back to ancient Roman times when the coins usually consisted of the Emperor's head on one side (i.e., heads) and another icon such as a ship on the other side (i.e., ships). Tossing coins was used in both fortune telling and ancient Roman games.

A Chinese form of divination called the *I-Ching* (pronounced E-Ching) is thought to be at least 4000 years old. It consists of 64 hexagrams made up of six horizontal lines. Each line is either broken or unbroken, representing the yin and the yang. These 64 hexagrams are supposed to represent all possible situations in life. To consult the I-Ching, a question is asked and then three coins are tossed six times. The way the coins fall, either heads up or heads down, determines whether the line is broken (yin) or unbroken (yang). Once the hexagon is determined, its meaning is consulted and interpreted to get the answer to the question. (*Note:* Another method used to determine the hexagon employs yarrow sticks.)

In the 16th century, a mathematician named Abraham DeMoivre used the outcomes of tossing coins to study what later became known as the normal distribution; however, his work at that time was not widely known.

Mathematicians usually consider the outcomes of a coin toss a random event. That is, each probability of getting a head is $\frac{1}{2}$, and the probability of getting a tail is $\frac{1}{2}$. Also, it is not possible to predict with 100% certainty which outcome will occur. But new studies question this theory. During World War II a South African mathematician named John Kerrich tossed a coin 10,000 times while he was interned in a German prison camp. Unfortunately, the results of his experiment were never recorded, so we don't know the number of heads that occurred.

Several studies have shown that when a coin-tossing device is used, the probability that a coin will land on the same side on which it is placed on the coin-tossing device is about 51%. It would take about 10,000 tosses to become aware of this bias. Furthermore, researchers showed that when a coin is spun on its edge, the coin would fall tails up about 80% of the time since there is more metal on the heads side of a coin. This makes the coin slightly heavier on the heads side than on the tails side.

Another assumption commonly made in probability theory is that the number of male births is equal to the number of female births and that the probability of a boy being born is $\frac{1}{2}$ and the probability of a girl being born is $\frac{1}{2}$. We know this is not exactly true.

In the later 1700s, a French mathematician named Pierre Simon Laplace attempted to prove that more males than females are born. He used records from 1745 to 1770 in Paris and showed that the percentage of females born was about 49%. Although these percentages vary somewhat from location to location, further surveys show they are generally true worldwide. Even though there are discrepancies, we generally consider the outcomes to be 50-50 since these discrepancies are relatively small.

Based on this article, would you consider the coin toss at the beginning of a football game fair?

Two Requirements for a Probability Distribution

1. The sum of the probabilities of all the events in the sample space must equal 1; that is, $\Sigma P(X) = 1$.
2. The probability of each event in the sample space must be between or equal to 0 and 1. That is, $0 \leq P(X) \leq 1$.

The first requirement states that the sum of the probabilities of all the events must be equal to 1. This sum cannot be less than 1 or greater than 1 since the sample space includes *all* possible outcomes of the probability experiment. The second requirement states that the probability of any individual event must be a value from 0 to 1. The reason (as stated in Chapter 4) is that the range of the probability of any individual value can be 0, 1, or any value between 0 and 1. A probability cannot be a negative number or greater than 1.

Example 5–4

Determine whether each distribution is a probability distribution.

a.
X	0	5	10	15	20
P(X)	$\frac{1}{5}$	$\frac{1}{5}$	$\frac{1}{5}$	$\frac{1}{5}$	$\frac{1}{5}$

c.
X	1	2	3	4
P(X)	$\frac{1}{4}$	$\frac{1}{8}$	$\frac{1}{16}$	$\frac{9}{16}$

b.
X	0	2	4	6
P(X)	-1.0	1.5	0.3	0.2

d.
X	2	3	7
P(X)	0.5	0.3	0.4

Solution

a. Yes, it is a probability distribution.

b. No, it is not a probability distribution, since $P(X)$ cannot be 1.5 or -1.0.

c. Yes, it is a probability distribution.

d. No, it is not, since $\Sigma P(X) = 1.2$.

Many variables in business, education, engineering, and other areas can be analyzed by using probability distributions. Section 5–3 shows methods for finding the mean and standard deviation for a probability distribution.

Applying the Concepts **5–2**

Dropping College Courses

Use the following table to answer the questions.

Reason for Dropping a College Course	Frequency	Percentage
Too difficult	45	
Illness	40	
Change in work schedule	20	
Change of major	14	
Family-related problems	9	
Money	7	
Miscellaneous	6	
No meaningful reason	3	

1. What is the variable under study? Is it a random variable?

2. How many people were in the study?

3. Complete the table.

4. From the information given, what is the probability that a student will drop a class because of illness? Money? Change of major?

5. Would you consider the information in the table to be a probability distribution?

6. Are the categories mutually exclusive?

7. Are the categories independent?

8. Are the categories exhaustive?

9. Are the two requirements for a discrete probability distribution met?

See page 283 for the answers.

Exercises 5–2

1. Define and give three examples of a random variable.

2. Explain the difference between a discrete and a continuous random variable.

3. Give three examples of a discrete random variable.

4. Give three examples of a continuous random variable.

5. What is a probability distribution? Give an example.

For Exercises 6 through 11, determine whether the distribution represents a probability distribution. If it does not, state why.

6.
X	1	6	11	16	21
P(X)	$\frac{1}{7}$	$\frac{1}{7}$	$\frac{3}{7}$	$\frac{2}{7}$	$\frac{1}{7}$

7.
X	3	6	8	12
P(X)	0.3	0.5	0.7	−0.8

8.
X	3	6	8
P(X)	−0.3	0.6	0.7

9.
X	1	2	3	4	5
P(X)	$\frac{3}{10}$	$\frac{1}{10}$	$\frac{1}{10}$	$\frac{2}{10}$	$\frac{3}{10}$

10.
X	20	30	40	50
P(X)	1.1	0.2	0.9	0.3

11.
X	5	10	15
P(X)	1.2	0.3	0.5

For Exercises 12 through 18, state whether the variable is discrete or continuous.

12. The speed of a jet airplane

13. The number of cheeseburgers a fast-food restaurant serves each day

14. The number of people who play the state lottery each day

15. The weight of a Siberian tiger

16. The time it takes to complete a marathon

17. The number of mathematics majors in your school

18. The blood pressures of all patients admitted to a hospital on a specific day

For Exercises 19 through 26, construct a probability distribution for the data and draw a graph for the distribution.

19. The probabilities that a patient will have 0, 1, 2, or 3 medical tests performed on entering a hospital are $\frac{6}{15}$, $\frac{5}{15}$, $\frac{3}{15}$, and $\frac{1}{15}$, respectively.

20. The probabilities of a return on an investment of $1000, $2000, and $3000 are $\frac{1}{2}$, $\frac{1}{4}$, and $\frac{1}{4}$, respectively.

21. The probabilities of a machine manufacturing 0, 1, 2, 3, 4, or 5 defective parts in one day are 0.75, 0.17, 0.04, 0.025, 0.01, and 0.005, respectively.

22. The probabilities that a customer will purchase 0, 1, 2, or 3 books are 0.45, 0.30, 0.15, and 0.10, respectively.

23. A die is loaded in such a way that the probabilities of getting 1, 2, 3, 4, 5, and 6 are $\frac{1}{2}$, $\frac{1}{6}$, $\frac{1}{12}$, $\frac{1}{12}$, $\frac{1}{12}$, and $\frac{1}{12}$, respectively.

24. The probabilities that a customer selects 1, 2, 3, 4, and 5 items at a convenience store are 0.32, 0.12, 0.23, 0.18, and 0.15, respectively.

25. The probabilities that a surgeon operates on 3, 4, 5, 6, or 7 patients in any one day are 0.15, 0.20, 0.25, 0.20, and 0.20, respectively.

26. Three patients are given a headache relief tablet. The probabilities for 0, 1, 2, or 3 successes are 0.18, 0.52, 0.21, and 0.09, respectively.

27. A box contains two $1 bills, three $5 bills, one $10 bill, and three $20 bills. Construct a probability distribution for the data.

28. Construct a probability distribution for a family of three children. Let X represent the number of boys.

29. Construct a probability distribution for drawing a card from a deck of 40 cards consisting of 10 cards numbered 1, 10 cards numbered 2, 15 cards numbered 3, and 5 cards numbered 4.

30. Using the sample space for tossing two dice, construct a probability distribution for the sums 2 through 12.

Extending the Concepts

A probability distribution can be written in formula notation such as $P(X) = 1/X$, where $X = 2, 3, 6$. The distribution is shown as follows:

X	2	3	6
$P(X)$	$\frac{1}{2}$	$\frac{1}{3}$	$\frac{1}{6}$

For Exercises 31 through 36, write the distribution for the formula and determine whether it is a probability distribution.

31. $P(X) = X/6$ for $X = 1, 2, 3$

32. $P(X) = X$ for $X = 0.2, 0.3, 0.5$

33. $P(X) = X/6$ for $X = 3, 4, 7$

34. $P(X) = X + 0.1$ for $X = 0.1, 0.02, 0.04$

35. $P(X) = X/7$ for $X = 1, 2, 4$

36. $P(X) = X/(X + 2)$ for $X = 0, 1, 2$

5–3 Mean, Variance, Standard Deviation, and Expectation

Objective 2

Find the mean, variance, standard deviation, and expected value for a discrete random variable.

The mean, variance, and standard deviation for a probability distribution are computed differently from the mean, variance, and standard deviation for samples. This section explains how these measures—as well as a new measure called the *expectation*—are calculated for probability distributions.

Mean

In Chapter 3, the mean for a sample or population was computed by adding the values and dividing by the total number of values, as shown in the formulas

$$\overline{X} = \frac{\Sigma X}{n} \qquad \mu = \frac{\Sigma X}{N}$$

But how would one compute the mean of the number of spots that show on top when a die is rolled? One could try rolling the die, say, 10 times, recording the number of spots, and finding the mean; however, this answer would only approximate the true mean. What about 50 rolls or 100 rolls? Actually, the more times the die is rolled, the better the approximation. One might ask, then, How many times must the die be rolled to get the exact answer? *It must be rolled an infinite number of times.* Since this task is impossible, the previous formulas cannot be used because the denominators would be infinity. Hence, a new method of computing the mean is necessary. This method gives the exact theoretical value of the mean as if it were possible to roll the die an infinite number of times.

Before the formula is stated, an example will be used to explain the concept. Suppose two coins are tossed repeatedly, and the number of heads that occurred is recorded. What will be the mean of the number of heads? The sample space is

HH, HT, TH, TT

Historical Note

A professor, Augustin Louis Cauchy (1789–1857), wrote a book on probability. While he was teaching at the Military School of Paris, one of his students was Napoleon Bonaparte.

and each outcome has a probability of $\frac{1}{4}$. Now, in the long run, one would *expect* two heads (HH) to occur approximately $\frac{1}{4}$ of the time, one head to occur approximately $\frac{1}{2}$ of the time (HT or TH), and no heads (TT) to occur approximately $\frac{1}{4}$ of the time. Hence, on average, one would expect the number of heads to be

$$\frac{1}{4} \cdot 2 + \frac{1}{2} \cdot 1 + \frac{1}{4} \cdot 0 = 1$$

That is, if it were possible to toss the coins many times or an infinite number of times, the *average* of the number of heads would be 1.

Hence, to find the mean for a probability distribution, one must multiply each possible outcome by its corresponding probability and find the sum of the products.

Formula for the Mean of a Probability Distribution

The mean of a random variable with a discrete probability distribution is

$$\mu = X_1 \cdot P(X_1) + X_2 \cdot P(X_2) + X_3 \cdot P(X_3) + \cdots + X_n \cdot P(X_n)$$
$$= \Sigma X \cdot P(X)$$

where $X_1, X_2, X_3, \ldots, X_n$ are the outcomes and $P(X_1), P(X_2), P(X_3), \ldots, P(X_n)$ are the corresponding probabilities.

Note: $\Sigma X \cdot P(X)$ means to sum the products.

Rounding Rule for the Mean, Variance, and Standard Deviation for a Probability Distribution The rounding rule for the mean, variance, and standard deviation for variables of a probability distribution is this: The mean, variance, and standard deviation should be rounded to one more decimal place than the outcome X. When fractions are used, they should be reduced to lowest terms.

Examples 5–5 through 5–8 illustrate the use of the formula.

Example 5–5

Find the mean of the number of spots that appear when a die is tossed.

Solution

In the toss of a die, the mean can be computed thus.

Outcome X	1	2	3	4	5	6
Probability $P(X)$	$\frac{1}{6}$	$\frac{1}{6}$	$\frac{1}{6}$	$\frac{1}{6}$	$\frac{1}{6}$	$\frac{1}{6}$

$$\mu = \Sigma X \cdot P(X) = 1 \cdot \frac{1}{6} + 2 \cdot \frac{1}{6} + 3 \cdot \frac{1}{6} + 4 \cdot \frac{1}{6} + 5 \cdot \frac{1}{6} + 6 \cdot \frac{1}{6}$$
$$= \frac{21}{6} = 3\frac{1}{2} \text{ or } 3.5$$

That is, when a die is tossed many times, the theoretical mean will be 3.5. Note that even though the die cannot show a 3.5, the theoretical average is 3.5.

The reason why this formula gives the theoretical mean is that in the long run, each outcome would occur approximately $\frac{1}{6}$ of the time. Hence, multiplying the outcome by its corresponding probability and finding the sum would yield the theoretical mean. In other words, outcome 1 would occur approximately $\frac{1}{6}$ of the time, outcome 2 would occur approximately $\frac{1}{6}$ of the time, etc.

Example 5–6

In a family with two children, find the mean of the number of children who will be girls.

Solution

The probability distribution is as follows:

Number of girls X	0	1	2
Probability $P(X)$	$\frac{1}{4}$	$\frac{1}{2}$	$\frac{1}{4}$

Hence, the mean is

$$\mu = \Sigma X \cdot P(X) = 0 \cdot \tfrac{1}{4} + 1 \cdot \tfrac{1}{2} + 2 \cdot \tfrac{1}{4} = 1$$

Example 5–7

If three coins are tossed, find the mean of the number of heads that occur. (See the table preceding Example 5–1.)

Solution

The probability distribution is

Number of heads X	0	1	2	3
Probability $P(X)$	$\frac{1}{8}$	$\frac{3}{8}$	$\frac{3}{8}$	$\frac{1}{8}$

The mean is

$$\mu = \Sigma X \cdot P(X) = 0 \cdot \tfrac{1}{8} + 1 \cdot \tfrac{3}{8} + 2 \cdot \tfrac{3}{8} + 3 \cdot \tfrac{1}{8} = \tfrac{12}{8} = 1\tfrac{1}{2} \text{ or } 1.5$$

The value 1.5 cannot occur as an outcome. Nevertheless, it is the long-run or theoretical average.

Example 5–8

The probability distribution shown represents the number of trips of five nights or more that American adults take per year. (That is, 6% do not take any trips lasting five nights or more, 70% take one trip lasting five nights or more per year, etc.) Find the mean.

Number of trips X	0	1	2	3	4
Probability $P(X)$	0.06	0.70	0.20	0.03	0.01

Solution

$$\mu = \Sigma X \cdot P(X)$$
$$= (0)(0.06) + (1)(0.70) + (2)(0.20) + (3)(0.03) + (4)(0.01)$$
$$= 0 + 0.70 + 0.40 + 0.09 + 0.04$$
$$= 1.23 \approx 1.2$$

Hence, the mean of the number of trips lasting five nights or more per year taken by American adults is 1.2.

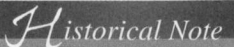

Variance and Standard Deviation

For a probability distribution, the mean of the random variable describes the measure of the so-called long-run or theoretical average, but it does not tell anything about the spread of the distribution. Recall from Chapter 3 that in order to measure this spread or variability, statisticians use the variance and standard deviation. These formulas were used:

$$\sigma^2 = \frac{\Sigma(X - \mu)^2}{N} \qquad \text{or} \qquad \sigma = \sqrt{\frac{\Sigma(X - \mu)^2}{N}}$$

These formulas cannot be used for a random variable of a probability distribution since N is infinite, so the variance and standard deviation must be computed differently.

To find the variance for the random variable of a probability distribution, subtract the theoretical mean of the random variable from each outcome and square the difference. Then multiply each difference by its corresponding probability and add the products. The formula is

$$\sigma^2 = \Sigma[(X - \mu)^2 \cdot P(X)]$$

Finding the variance by using this formula is somewhat tedious. So for simplified computations, a shortcut formula can be used. This formula is algebraically equivalent to the longer one and is used in the examples that follow.

Formula for the Variance of a Probability Distribution

Find the variance of a probability distribution by multiplying the square of each outcome by its corresponding probability, summing those products, and subtracting the square of the mean. The formula for the variance of a probability distribution is

$$\sigma^2 = \Sigma[X^2 \cdot P(X)] - \mu^2$$

The standard deviation of a probability distribution is

$$\sigma = \sqrt{\sigma^2} \qquad \text{or} \qquad \sqrt{\Sigma[X^2 \cdot P(X)] - \mu^2}$$

Remember that the variance and standard deviation cannot be negative.

Example 5–9

Compute the variance and standard deviation for the probability distribution in Example 5–5.

Solution

Recall that the mean is $\mu = 3.5$, as computed in Example 5–5. Square each outcome and multiply by the corresponding probability, sum those products, and then subtract the square of the mean.

$$\sigma^2 = (1^2 \cdot \tfrac{1}{6} + 2^2 \cdot \tfrac{1}{6} + 3^2 \cdot \tfrac{1}{6} + 4^2 \cdot \tfrac{1}{6} + 5^2 \cdot \tfrac{1}{6} + 6^2 \cdot \tfrac{1}{6}) - (3.5)^2 = 2.9$$

To get the standard deviation, find the square root of the variance.

$$\sigma = \sqrt{2.9} = 1.7$$

Example 5–10

Five balls numbered 0, 2, 4, 6, and 8 are placed in a bag. After the balls are mixed, one is selected, its number is noted, and then it is replaced. If this experiment is repeated many times, find the variance and standard deviation of the numbers on the balls.

Historical Note

In 1657 a Dutch mathematician, Huygens, wrote a treatise on the Pascal-Fermat correspondence and introduced the idea of *mathematical expectation*.

Solution

Let X be the number on each ball. The probability distribution is

Number on ball X	0	2	4	6	8
Probability $P(X)$	$\frac{1}{5}$	$\frac{1}{5}$	$\frac{1}{5}$	$\frac{1}{5}$	$\frac{1}{5}$

The mean is

$$\mu = \Sigma X \cdot P(X) = 0 \cdot \tfrac{1}{5} + 2 \cdot \tfrac{1}{5} + 4 \cdot \tfrac{1}{5} + 6 \cdot \tfrac{1}{5} + 8 \cdot \tfrac{1}{5} = 4.0$$

The variance is

$$\sigma^2 = \Sigma[X^2 \cdot P(X)] - \mu^2$$
$$= [0^2 \cdot (\tfrac{1}{5}) + 2^2 \cdot (\tfrac{1}{5}) + 4^2 \cdot (\tfrac{1}{5}) + 6^2 \cdot (\tfrac{1}{5}) + 8^2 \cdot (\tfrac{1}{5})] - 4^2$$
$$= [0 + \tfrac{4}{5} + \tfrac{16}{5} + \tfrac{36}{5} + \tfrac{64}{5}] - 16$$
$$= \tfrac{120}{5} - 16$$
$$= 24 - 16 = 8$$

The standard deviation is $\sigma = \sqrt{8} = 2.8$.

The mean, variance, and standard deviation can also be found by using vertical columns, as shown [0.2 is used for $P(X)$ since $\frac{1}{5} = 0.2$].

X	$P(X)$	$X \cdot P(X)$	$X^2 \cdot P(X)$
0	0.2	0	0
2	0.2	0.4	0.8
4	0.2	0.8	3.2
6	0.2	1.2	7.2
8	0.2	1.6	12.8
		$\Sigma X \cdot P(X) = 4.0$	$\Sigma X^2 \cdot P(X) = 24.0$

Find the mean by summing the $X \cdot P(X)$ column and the variance by summing the $X^2 \cdot P(X)$ column and subtracting the square of the mean:

$$\sigma^2 = 24 - 4^2 = 8 \quad \text{and} \quad \sigma = \sqrt{8} = 2.8$$

Example 5–11

A talk radio station has four telephone lines. If the host is unable to talk (i.e., during a commercial) or is talking to a person, the other callers are placed on hold. When all lines are in use, others who are trying to call in get a busy signal. The probability that 0, 1, 2, 3, or 4 people will get through is shown in the distribution. Find the variance and standard deviation for the distribution.

X	0	1	2	3	4
$P(X)$	0.18	0.34	0.23	0.21	0.04

Should the station have considered getting more phone lines installed?

Solution

The mean is

$$\mu = \Sigma X \cdot P(X)$$
$$= 0 \cdot (0.18) + 1 \cdot (0.34) + 2 \cdot (0.23) + 3 \cdot (0.21) + 4 \cdot (0.04)$$
$$= 1.6$$

The variance is

$$\sigma^2 = \Sigma[X^2 \cdot P(X)] - \mu^2$$
$$= [0^2 \cdot (0.18) + 1^2 \cdot (0.34) + 2^2 \cdot (0.23) + 3^2 \cdot (0.21) + 4^2 \cdot (0.04)] - 1.6^2$$
$$= [0 + 0.34 + 0.92 + 1.89 + 0.64] - 2.56$$
$$= 3.79 - 2.56 = 1.23$$
$$= 1.2 \text{ (rounded)}$$

The standard deviation is $\sigma = \sqrt{\sigma^2}$, or $\sigma = \sqrt{1.2} = 1.1$.

No. The mean number of people calling at any one time is 1.6. Since the standard deviation is 1.1, most callers would be accommodated by having four phone lines because $\mu + 2\sigma$ would be $1.6 + 2(1.1) = 1.6 + 2.2 = 3.8$. Very few callers would get a busy signal since at least 75% of the callers would either get through or be put on hold. (See Chebyshev's theorem in Section 3–3.)

Expectation

Another concept related to the mean for a probability distribution is the concept of expected value or expectation. Expected value is used in various types of games of chance, in insurance, and in other areas, such as decision theory.

The **expected value** of a discrete random variable of a probability distribution is the theoretical average of the variable. The formula is

$$\mu = E(X) = \Sigma X \cdot P(X)$$

The symbol $E(X)$ is used for the expected value.

The formula for the expected value is the same as the formula for the theoretical mean. The expected value, then, is the theoretical mean of the probability distribution. That is, $E(X) = \mu$.

When expected value problems involve money, it is customary to round the answer to the nearest cent.

Example 5–12

One thousand tickets are sold at $1 each for a color television valued at $350. What is the expected value of the gain if a person purchases one ticket?

Solution

The problem can be set up as follows:

	Win	Lose
Gain X	$349	−$1
Probability $P(X)$	$\dfrac{1}{1000}$	$\dfrac{999}{1000}$

Two things should be noted. First, for a win, the net gain is $349, since the person does not get the cost of the ticket ($1) back. Second, for a loss, the gain is represented by a negative number, in this case $-$1. The solution, then, is

$$E(X) = \$349 \cdot \frac{1}{1000} + (-\$1) \cdot \frac{999}{1000} = -\$0.65$$

Expected value problems of this type can also be solved by finding the overall gain (i.e., the value of the prize won or the amount of money won, not considering the cost of the ticket for the prize or the cost to play the game) and subtracting the cost of the tickets or the cost to play the game, as shown:

$$E(X) = \$350 \cdot \frac{1}{1000} - \$1 = -\$0.65$$

Here, the overall gain ($350) must be used.

Note that the expectation is $-$0.65. This does not mean that a person loses $0.65, since the person can only win a television set valued at $350 or lose $1 on the ticket. What this expectation means is that the average of the losses is $0.65 for each of the 1000 ticket holders. Here is another way of looking at this situation: If a person purchased one ticket each week over a long time, the average loss would be $0.65 per ticket, since theoretically, on average, that person would win the set once for each 1000 tickets purchased.

Example 5–13	One thousand tickets are sold at $1 each for four prizes of $100, $50, $25, and $10. After each prize drawing, the winning ticket is then returned to the pool of tickets. What is the expected value if a person purchases two tickets?

Gain X	$98	$48	$23	$8	$-$2
Probability $P(X)$	$\frac{2}{1000}$	$\frac{2}{1000}$	$\frac{2}{1000}$	$\frac{2}{1000}$	$\frac{992}{1000}$

Solution

$$E(X) = \$98 \cdot \frac{2}{1000} + \$48 \cdot \frac{2}{1000} + \$23 \cdot \frac{2}{1000} + \$8 \cdot \frac{2}{1000} + (-\$2) \cdot \frac{992}{1000}$$

$$= -\$1.63$$

An alternate solution is

$$E(X) = \$100 \cdot \frac{2}{1000} + \$50 \cdot \frac{2}{1000} + \$25 \cdot \frac{2}{1000} + \$10 \cdot \frac{2}{1000} - \$2$$

$$= -\$1.63$$

Example 5–14	A financial adviser suggests that his client select one of two types of bonds in which to invest $5000. Bond X pays a return of 4% and has a default rate of 2%. Bond Y has a $2\frac{1}{2}$% return and a default rate of 1%. Find the expected rate of return and decide which bond would be a better investment. When the bond defaults, the investor loses all the investment.

Solution

The return on bond X is $5000 \cdot 4\% = \$200$. The expected return then is

$$E(X) = \$200(0.98) - \$5000(0.02) = \$96$$

The return on bond Y is $5000 \cdot 2\frac{1}{2}\% = \$125$. The expected return then is

$$E(X) = \$125(0.99) - \$5000(0.01) = \$73.75$$

Hence, bond X would be a better investment since the expected return is higher.

In gambling games, if the expected value of the game is zero, the game is said to be fair. If the expected value of a game is positive, then the game is in favor of the player. That is, the player has a better-than-even chance of winning. If the expected value of the game is negative, then the game is said to be in favor of the house. That is, in the long run, the players will lose money.

In his book *Probabilities in Everyday Life* (Ivy Books, 1986), author John D. McGervy gives the expectations for various casino games. For keno, the house wins $0.27 on every $1.00 bet. For Chuck-a-Luck, the house wins about $0.52 on every $1.00 bet. For roulette, the house wins about $0.90 on every $1.00 bet. For craps, the house wins about $0.88 on every $1.00 bet. The bottom line here is that if you gamble long enough, sooner or later you will end up losing money.

Applying the Concepts **5–3**

Expected Value

On March 28, 1979, the nuclear generating facility at Three Mile Island, Pennsylvania, began discharging radiation into the atmosphere. People exposed to even low levels of radiation can experience health problems ranging from very mild to severe, even causing death. A local newspaper reported that 11 babies were born with kidney problems in the three-county area surrounding the Three Mile Island nuclear power plant. The expected value for that problem in infants in that area was 3. Answer the following questions.

1. What does *expected value* mean?
2. Would you expect the exact value of 3 all the time?
3. If a news reporter stated that the number of cases of kidney problems in newborns was nearly four times as much as was usually expected, do you think pregnant mothers living in that area would be overly concerned?
4. Is it unlikely that 11 occurred by chance?
5. Are there any other statistics that could better inform the public?
6. Assume that 3 out of 2500 babies were born with kidney problems in that three-county area the year before the accident. Also assume that 11 out of 2500 babies were born with kidney problems in that three-county area the year after the accident. What is the real percent of increase in that abnormality?
7. Do you think that pregnant mothers living in that area should be overly concerned after looking at the results in terms of rates?

See page 283 for the answers.

Exercises 5–3

1. From past experience, a company has found that in cartons of transistors, 92% contain no defective transistors, 3% contain one defective transistor, 3% contain two defective transistors, and 2% contain three defective transistors. Find the mean, variance, and standard deviation for the defective transistors.

    About how many extra transistors per day would the company need to replace the defective ones if it used 10 cartons per day?

2. The number of suits sold per day at a retail store is shown in the table, with the corresponding probabilities. Find the mean, variance, and standard deviation of the distribution.

Number of suits sold X	19	20	21	22	23
Probability $P(X)$	0.2	0.2	0.3	0.2	0.1

    If the manager of the retail store wants to be sure that he has enough suits for the next 5 days, how many should the manager purchase?

3. A bank vice president feels that each savings account customer has, on average, three credit cards. The following distribution represents the number of credit cards people own. Find the mean, variance, and standard deviation. Is the vice president correct?

Number of cards X	0	1	2	3	4
Probability $P(X)$	0.18	0.44	0.27	0.08	0.03

4. The number of refrigerators sold per day at a local appliance store is shown in the table, along with the corresponding probabilities. Find the mean, variance, and standard deviation.

Number of refrigerators sold X	0	1	2	3	4
Probability $P(X)$	0.1	0.2	0.3	0.2	0.2

5. A public speaker computes the probabilities for the number of speeches she gives each week. Compute the mean, variance, and standard deviation of the distribution shown.

Number of speeches X	0	1	2	3	4	5
Probability $P(X)$	0.06	0.42	0.22	0.12	0.15	0.03

    If she receives $100 per speech, about how much will she earn per week?

6. A recent survey by an insurance company showed the following probabilities for the number of bedrooms in each insured home. Find the mean, variance, and standard deviation for the distribution.

Number of bedrooms X	2	3	4	5
Probability $P(X)$	0.3	0.4	0.2	0.1

7. A concerned parents group determined the number of commercials shown in each of five children's programs over a period of time. Find the mean, variance, and standard deviation for the distribution shown.

Number of commercials X	5	6	7	8	9
Probability $P(X)$	0.2	0.25	0.38	0.10	0.07

8. A study conducted by a TV station showed the number of televisions per household and the corresponding probabilities for each. Find the mean, variance, and standard deviation.

Number of televisions X	1	2	3	4
Probability $P(X)$	0.32	0.51	0.12	0.05

    If you were taking a survey on the programs that were watched on television, how many program diaries would you send to each household in the survey?

9. The following distribution shows the number of students enrolled in CPR classes offered by the local fire department. Find the mean, variance, and standard deviation for the distribution.

Number of students X	12	13	14	15	16
Probability $P(X)$	0.15	0.20	0.38	0.18	0.09

10. A pizza shop owner determines the number of pizzas that are delivered each day. Find the mean, variance, and standard deviation for the distribution shown. If the manager stated that 45 pizzas were delivered on one day, do you think that this is a believable claim?

Number of deliveries X	35	36	37	38	39
Probability $P(X)$	0.1	0.2	0.3	0.3	0.1

11. An insurance company insures a person's antique coin collection worth $20,000 for an annual premium of $300. If the company figures that the probability of the collection being stolen is 0.002, what will be the company's expected profit?

12. A landscape contractor bids on jobs where he can make $3000 profit. The probabilities of getting one, two, three, or four jobs per month are shown.

Number of jobs	1	2	3	4
Probability	0.2	0.3	0.4	0.1

Find the contractor's expected profit per month.

13. If a person rolls doubles when he tosses two dice, he wins $5. For the game to be fair, how much should the person pay to play the game?

14. If a player rolls two dice and gets a sum of 2 or 12, she wins $20. If the person gets a 7, she wins $5. The cost to play the game is $3. Find the expectation of the game.

15. A lottery offers one $1000 prize, one $500 prize, and five $100 prizes. One thousand tickets are sold at $3 each. Find the expectation if a person buys one ticket.

16. In Exercise 15, find the expectation if a person buys two tickets. Assume that the player's ticket is replaced after each draw and that the same ticket can win more than one prize.

17. For a daily lottery, a person selects a three-digit number. If the person plays for $1, she can win $500. Find the expectation. In the same daily lottery, if a person boxes a number, she will win $80. Find the expectation if the

number 123 is played for $1 and boxed. (When a number is "boxed," it can win when the digits occur in any order.)

18. A 35-year-old woman purchases a $100,000 term life insurance policy for an annual payment of $360. Based on a period life table for the U.S. government, the probability that she will survive the year is 0.999057. Find the expected value of the policy for the insurance company.

19. A person decides to invest $50,000 in a gas well. Based on history, the probabilities of the outcomes are as follows.

Outcome	*P(X)*
$80,000 (Highly successful)	0.2
$40,000 (Moderately successful)	0.7
−$50,000 (Dry well)	0.1

Find the expected value of the investment. Would you consider this a good investment?

Extending the Concepts

20. Construct a probability distribution for the sum shown on the faces when two dice are rolled. Find the mean, variance, and standard deviation of the distribution.

21. When one die is rolled, the expected value of the number of spots is 3.5. In Exercise 20, the mean number of spots was found for rolling two dice. What is the mean number of spots if three dice are rolled?

22. The formula for finding the variance for a probability distribution is

$$\sigma^2 = \Sigma[(X - \mu)^2 \cdot P(X)]$$

Verify algebraically that this formula gives the same result as the shortcut formula shown in this section.

23. Roll a die 100 times. Compute the mean and standard deviation. How does the result compare with the theoretical results of Example 5–5?

24. Roll two dice 100 times and find the mean, variance, and standard deviation of the sum of the spots. Compare the result with the theoretical results obtained in Exercise 20.

25. Conduct a survey of the number of extracurricular activities your classmates are enrolled in. Construct a probability distribution and find the mean, variance, and standard deviation.

26. In a recent promotional campaign, a company offered these prizes and the corresponding probabilities. Find the expected value of winning. The tickets are free.

Number of prizes	Amount	Probability
1	$100,000	$\dfrac{1}{1,000,000}$
2	10,000	$\dfrac{1}{50,000}$
5	1,000	$\dfrac{1}{10,000}$
10	100	$\dfrac{1}{1000}$

If the winner has to mail in the winning ticket to claim the prize, what will be the expectation if the cost of the stamp is considered? Use the current cost of a stamp for a first-class letter.

This study shows that a part of the brain reacts to the impact of losing, and it might explain why people tend to increase their bets after losing when gambling. Explain how this type of split decision making may influence fighter pilots, firefighters, or police officers, as the article states.

THE GAMBLER'S FALLACY
WHY WE EXPECT TO STRIKE IT RICH AFTER A LOSING STREAK

A GAMBLER USUALLY WAGERS more after taking a loss, in the misguided belief that a run of bad luck increases the probability of a win. We tend to cling to the misconception that past events can skew future odds. "On some level, you're thinking, 'If I just lost, it's going to even out.' The extent to which you're disturbed by a loss seems to go along with risky behavior," says University of Michigan psychologist William Gehring, Ph.D., co-author of a new study linking dicey decision-making to neurological activity originating in the medial frontal cortex, long thought to be an area of the brain used in error detection.

Because people are so driven to up the ante after a loss, Gehring believes that the medial frontal cortex unconsciously influences future decisions based on the impact of the loss, in addition to registering the loss itself.

Gehring drew this conclusion by asking 12 subjects fitted with electrode caps to choose either the number 5 or 25, with the larger number representing the riskier bet.

On any given round, both numbers could amount to a loss, both could amount to a gain or the results could split, one number signifying a loss, the other a gain.

The medial frontal cortex responded to the outcome of a gamble within a quarter of a second, registering sharp electrical impulses only after a loss. Gehring points out that if the medial frontal cortex simply detected errors it would have reacted after participants chose the lesser of two possible gains. In other words, choosing "5" during a round in which both numbers paid off and betting on "25" would have yielded a larger profit.

After the study appeared in *Science*, Gehring received several e-mails from stock traders likening the "gambler's fallacy" to impulsive trading decisions made directly after off-loading a losing security. Researchers speculate that such risky, split-second decision-making could extend to fighter pilots, firemen and policemen—professions in which rapid-fire decisions are crucial and frequent.

—Dan Schulman

Source: Psychology Today, August 2002, p. 22. Used with permission.

Technology *Step by Step*

TI-83 Plus or TI-84 Plus
Step by Step

To calculate the mean and variance for a discrete random variable by using the formulas:

1. Enter the x values into L_1 and the probabilities into L_2.
2. Move the cursor to the top of the L_3 column so that L_3 is highlighted.
3. Type L_1 multiplied by L_2, then press **ENTER**.
4. Move the cursor to the top of the L_4 column so that L_4 is highlighted.
5. Type L_1 followed by the x^2 key multiplied by L_2, then press **ENTER**.
6. Type **2nd QUIT** to return to the home screen.
7. Type **2nd LIST,** move the cursor to MATH, type **5** for sum(, then type L_3) then press **ENTER**.
8. Type **2nd ENTER,** move the cursor to L_3, type L_4, then press **ENTER**.

Using the data from Example 5–10 gives the following:

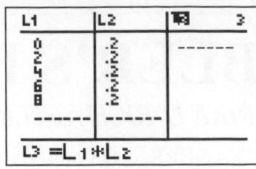

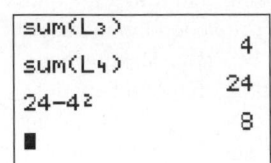

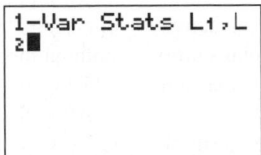

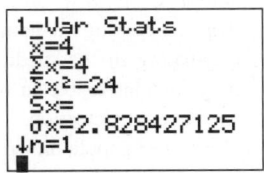

To calculate the mean and standard deviation for a discrete random variable without using the formulas, modify the procedure to calculate the mean and standard deviation from grouped data (Chapter 3) by entering the x values into L_1 and the probabilities into L_2.

5–4   The Binomial Distribution

Many types of probability problems have only two outcomes or can be reduced to two outcomes. For example, when a coin is tossed, it can land heads or tails. When a baby is born, it will be either male or female. In a basketball game, a team either wins or loses. A true/false item can be answered in only two ways, true or false. Other situations can be

Objective 3

Find the exact probability for *X* successes in *n* trials of a binomial experiment.

Historical Note

In 1653, Blaise Pascal created a triangle of numbers called *Pascal's triangle* that can be used in the binomial distribution.

reduced to two outcomes. For example, a medical treatment can be classified as effective or ineffective, depending on the results. A person can be classified as having normal or abnormal blood pressure, depending on the measure of the blood pressure gauge. A multiple-choice question, even though there are four or five answer choices, can be classified as correct or incorrect. Situations like these are called *binomial experiments.*

A **binomial experiment** is a probability experiment that satisfies the following four requirements:

1. There must be a fixed number of trials.
2. Each trial can have only two outcomes or outcomes that can be reduced to two outcomes. These outcomes can be considered as either success or failure.
3. The outcomes of each trial must be independent of each other.
4. The probability of a success must remain the same for each trial.

A binomial experiment and its results give rise to a special probability distribution called the *binomial distribution.*

The outcomes of a binomial experiment and the corresponding probabilities of these outcomes are called a **binomial distribution.**

In binomial experiments, the outcomes are usually classified as successes or failures. For example, the correct answer to a multiple-choice item can be classified as a success, but any of the other choices would be incorrect and hence classified as a failure. The notation that is commonly used for binomial experiments and the binomial distribution is defined now.

Notation for the Binomial Distribution

$P(S)$	The symbol for the probability of success
$P(F)$	The symbol for the probability of failure
p	The numerical probability of a success
q	The numerical probability of a failure

$$P(S) = p \qquad \text{and} \qquad P(F) = 1 - p = q$$

n	The number of trials
X	The number of successes in n trials

Note that $0 \leq X \leq n$ and $X = 0, 1, 2, 3, \ldots, n$

The probability of a success in a binomial experiment can be computed with this formula.

Binomial Probability Formula

In a binomial experiment, the probability of exactly X successes in n trials is

$$P(X) = \frac{n!}{(n - X)!X!} \cdot p^X \cdot q^{n-X}$$

An explanation of why the formula works is given following Example 5–15.

Example 5–15

A coin is tossed 3 times. Find the probability of getting exactly two heads.

Solution

This problem can be solved by looking at the sample space. There are three ways to get two heads.

HHH, <u>HHT, HTH, THH,</u> TTH, THT, HTT, TTT

The answer is $\frac{3}{8}$, or 0.375.

Looking at the problem in Example 5–15 from the standpoint of a binomial experiment, one can show that it meets the four requirements.

1. There are a fixed number of trials (three).
2. There are only two outcomes for each trial, heads or tails.
3. The outcomes are independent of one another (the outcome of one toss in no way affects the outcome of another toss).
4. The probability of a success (heads) is $\frac{1}{2}$ in each case.

In this case, $n = 3$, $X = 2$, $p = \frac{1}{2}$, and $q = \frac{1}{2}$. Hence, substituting in the formula gives

$$P(2 \text{ heads}) = \frac{3!}{(3-2)!2!} \cdot \left(\frac{1}{2}\right)^2\left(\frac{1}{2}\right)^1 = \frac{3}{8} = 0.375$$

which is the same answer obtained by using the sample space.

The same example can be used to explain the formula. First, note that there are three ways to get exactly two heads and one tail from a possible eight ways. They are HHT, HTH, and THH. In this case, then, the number of ways of obtaining two heads from three coin tosses is $_3C_2$, or 3, as shown in Chapter 4. In general, the number of ways to get X successes from n trials without regard to order is

$$_nC_X = \frac{n!}{(n-X)!X!}$$

This is the first part of the binomial formula. (Some calculators can be used for this.)

Next, each success has a probability of $\frac{1}{2}$ and can occur twice. Likewise, each failure has a probability of $\frac{1}{2}$ and can occur once, giving the $(\frac{1}{2})^2(\frac{1}{2})^1$ part of the formula. To generalize, then, each success has a probability of p and can occur X times, and each failure has a probability of q and can occur $n - X$ times. Putting it all together yields the binomial probability formula.

Example 5–16

A survey found that one out of five Americans say he or she has visited a doctor in any given month. If 10 people are selected at random, find the probability that exactly 3 will have visited a doctor last month.

Source: Reader's Digest.

Solution

In this case, $n = 10$, $X = 3$, $p = \frac{1}{5}$, and $q = \frac{4}{5}$. Hence,

$$P(3) = \frac{10!}{(10-3)!3!}\left(\frac{1}{5}\right)^3\left(\frac{4}{5}\right)^7 = 0.201$$

Example 5–17

A survey from Teenage Research Unlimited (Northbrook, Illinois) found that 30% of teenage consumers receive their spending money from part-time jobs. If 5 teenagers are selected at random, find the probability that at least 3 of them will have part-time jobs.

Solution

To find the probability that at least 3 have part-time jobs, it is necessary to find the individual probabilities for 3, or 4, or 5, and then add them to get the total probability.

$$P(3) = \frac{5!}{(5-3)!3!}(0.3)^3(0.7)^2 = 0.132$$

$$P(4) = \frac{5!}{(5-4)!4!}(0.3)^4(0.7)^1 = 0.028$$

$$P(5) = \frac{5!}{(5-5)!5!}(0.3)^5(0.7)^0 = 0.002$$

Hence,

P(at least three teenagers have part-time jobs)
$= 0.132 + 0.028 + 0.002 = 0.162$

Computing probabilities by using the binomial probability formula can be quite tedious at times, so tables have been developed for selected values of n and p. Table B in Appendix C gives the probabilities for individual events. Example 5–18 shows how to use Table B to compute probabilities for binomial experiments.

Example 5–18

Solve the problem in Example 5–15 by using Table B.

Solution

Since $n = 3$, $X = 2$, and $p = 0.5$, the value 0.375 is found as shown in Figure 5–3.

Figure 5–3

Using Table B for Example 5–18

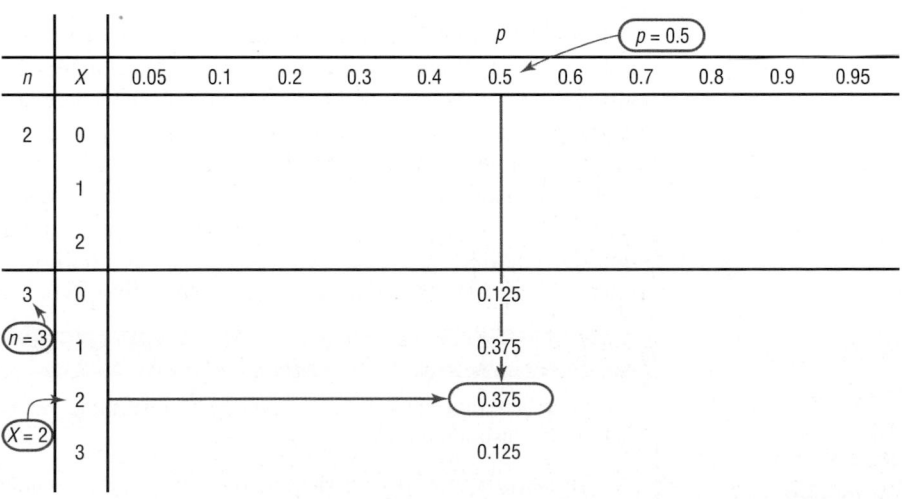

Example 5-19

Public Opinion reported that 5% of Americans are afraid of being alone in a house at night. If a random sample of 20 Americans is selected, find these probabilities by using the binomial table.

a. There are exactly 5 people in the sample who are afraid of being alone at night.

b. There are at most 3 people in the sample who are afraid of being alone at night.

c. There are at least 3 people in the sample who are afraid of being alone at night.

Source: 100% American by Daniel Evan Weiss.

Solution

a. $n = 20$, $p = 0.05$, and $X = 5$. From the table, one gets 0.002.

b. $n = 20$ and $p = 0.05$. "At most 3 people" means 0, or 1, or 2, or 3.
   Hence, the solution is

$$P(0) + P(1) + P(2) + P(3) = 0.358 + 0.377 + 0.189 + 0.060$$
$$= 0.984$$

c. $n = 20$ and $p = 0.05$. "At least 3 people" means 3, 4, 5, . . . , 20. This problem can best be solved by finding $P(0) + P(1) + P(2)$ and subtracting from 1.

$$P(0) + P(1) + P(2) = 0.358 + 0.377 + 0.189 = 0.924$$
$$1 - 0.924 = 0.076$$

Example 5-20

A report from the Secretary of Health and Human Services stated that 70% of single-vehicle traffic fatalities that occur at night on weekends involve an intoxicated driver. If a sample of 15 single-vehicle traffic fatalities that occur at night on a weekend is selected, find the probability that exactly 12 involve a driver who is intoxicated.

Source: 100% American by Daniel Evan Weiss.

Solution

Now, $n = 15$, $p = 0.70$, and $X = 12$. From Table B, $P(12) = 0.170$. Hence, the probability is 0.17.

Remember that in the use of the binomial distribution, the outcomes must be independent. For example, in the selection of components from a batch to be tested, each component must be replaced before the next one is selected. Otherwise, the outcomes are not independent. However, a dilemma arises because there is a chance that the same component could be selected again. This situation can be avoided by not replacing the component and using a distribution called the hypergeometric distribution to calculate the probabilities. The hypergeometric distribution is presented later in this chapter. Note that when the population is large and the sample is small, the binomial probabilities can be shown to be nearly the same as the corresponding hypergeometric probabilities.

Objective 4

Find the mean, variance, and standard deviation for the variable of a binomial distribution.

Mean, Variance, and Standard Deviation for the Binomial Distribution

The mean, variance, and standard deviation of a variable that has the *binomial distribution* can be found by using the following formulas.

Mean $\mu = n \cdot p$ Variance $\sigma^2 = n \cdot p \cdot q$ Standard deviation $\sigma = \sqrt{n \cdot p \cdot q}$

These formulas are algebraically equivalent to the formulas for the mean, variance, and standard deviation of the variables for probability distributions, but because they are for variables of the binomial distribution, they have been simplified by using algebra. The algebraic derivation is omitted here, but their equivalence is shown in Example 5–21.

Example 5–21

A coin is tossed 4 times. Find the mean, variance, and standard deviation of the number of heads that will be obtained.

Solution

With the formulas for the binomial distribution and $n = 4$, $p = \frac{1}{2}$, and $q = \frac{1}{2}$, the results are

$$\mu = n \cdot p = 4 \cdot \tfrac{1}{2} = 2$$
$$\sigma^2 = n \cdot p \cdot q = 4 \cdot \tfrac{1}{2} \cdot \tfrac{1}{2} = 1$$
$$\sigma = \sqrt{1} = 1$$

From Example 5–21, when four coins are tossed many, many times, the average of the number of heads that appear is 2, and the standard deviation of the number of heads is 1. Note that these are theoretical values.

As stated previously, this problem can be solved by using the formulas for expected value. The distribution is shown.

No. of heads X	0	1	2	3	4
Probability $P(X)$	$\frac{1}{16}$	$\frac{4}{16}$	$\frac{6}{16}$	$\frac{4}{16}$	$\frac{1}{16}$

$$\mu = E(X) = \Sigma X \cdot P(X) = 0 \cdot \tfrac{1}{16} + 1 \cdot \tfrac{4}{16} + 2 \cdot \tfrac{6}{16} + 3 \cdot \tfrac{4}{16} + 4 \cdot \tfrac{1}{16} = \tfrac{32}{16} = 2$$
$$\sigma^2 = \Sigma X^2 \cdot P(X) - \mu^2$$
$$= 0^2 \cdot \tfrac{1}{16} + 1^2 \cdot \tfrac{4}{16} + 2^2 \cdot \tfrac{6}{16} + 3^2 \cdot \tfrac{4}{16} + 4^2 \cdot \tfrac{1}{16} - 2^2 = \tfrac{80}{16} - 4 = 1$$
$$\sigma = \sqrt{1} = 1$$

Hence, the simplified binomial formulas give the same results.

Example 5–22

A die is rolled 480 times. Find the mean, variance, and standard deviation of the number of 2s that will be rolled.

Solution

This is a binomial situation, where getting a 2 is a success and not getting a 2 is a failure; hence, $n = 480$, $p = \frac{1}{6}$, and $q = \frac{5}{6}$.

$$\mu = n \cdot p = 480 \cdot \tfrac{1}{6} = 80$$
$$\sigma^2 = n \cdot p \cdot q = 480 \cdot (\tfrac{1}{6})(\tfrac{5}{6}) = 66.7$$
$$\sigma = \sqrt{n \cdot p \cdot q} = \sqrt{66.7} = 8.2$$

On average, there will be eighty 2s. The standard deviation is 8.2.

Example 5–23

The *Statistical Bulletin* published by Metropolitan Life Insurance Co. reported that 2% of all American births result in twins. If a random sample of 8000 births is taken, find the mean, variance, and standard deviation of the number of births that would result in twins.

Source: 100% American by Daniel Evan Weiss.

Solution

This is a binomial situation, since a birth can result in either twins or not twins (i.e., two outcomes).

$$\mu = n \cdot p = (8000)(0.02) = 160$$
$$\sigma^2 = n \cdot p \cdot q = (8000)(0.02)(0.98) = 156.8$$
$$\sigma = \sqrt{n \cdot p \cdot q} = \sqrt{156.8} = 12.5$$

For the sample, the average number of births that would result in twins is 160, the variance is 156.8, or 157, and the standard deviation is 12.5, or 13 if rounded.

Applying the Concepts 5–4

Unsanitary Restaurants

Health officials routinely check sanitary conditions of restaurants. Assume you visit a popular tourist spot and read in the newspaper that in 3 out of every 7 restaurants checked, there were unsatisfactory health conditions found. Assuming you are planning to eat out 10 times while you are there on vacation, answer the following questions.

1. How likely is it that you will eat at three restaurants with unsanitary conditions?
2. How likely is it that you will eat at four or five restaurants with unsanitary conditions?
3. Explain how you would compute the probability of eating in at least one restaurant with unsanitary conditions. Could you use the complement to solve this problem?
4. What is the most likely number to occur in this experiment?
5. How variable will the data be around the most likely number?
6. Is this a binomial distribution?
7. If it is a binomial distribution, does that mean that the likelihood of a success is always 50% since there are only two possible outcomes?

Check your answers by using the following computer-generated table.

Mean = 4.3   Std. dev. = 1.56557

X	P(X)	Cum. Prob.
0	0.00362	0.00362
1	0.02731	0.03093
2	0.09272	0.12365
3	0.18651	0.31016
4	0.24623	0.55639
5	0.22291	0.77930
6	0.14013	0.91943
7	0.06041	0.97983
8	0.01709	0.99692
9	0.00286	0.99979
10	0.00022	1.00000

See page 283 for the answers.

Exercises 5–4

1. Which of the following are binomial experiments or can be reduced to binomial experiments?

   a. Surveying 100 people to determine if they like Sudsy Soap

   b. Tossing a coin 100 times to see how many heads occur

   c. Drawing a card with replacement from a deck and getting a heart

   d. Asking 1000 people which brand of cigarettes they smoke

   e. Testing four different brands of aspirin to see which brands are effective

   f. Testing one brand of aspirin by using 10 people to determine whether it is effective

   g. Asking 100 people if they smoke

   h. Checking 1000 applicants to see whether they were admitted to White Oak College

   i. Surveying 300 prisoners to see how many different crimes they were convicted of

   j. Surveying 300 prisoners to see whether this is their first offense

2. **(ans)** Compute the probability of X successes, using Table B in Appendix C.

   a. $n = 2, p = 0.30, X = 1$

   b. $n = 4, p = 0.60, X = 3$

   c. $n = 5, p = 0.10, X = 0$

   d. $n = 10, p = 0.40, X = 4$

   e. $n = 12, p = 0.90, X = 2$

   f. $n = 15, p = 0.80, X = 12$

   g. $n = 17, p = 0.05, X = 0$

   h. $n = 20, p = 0.50, X = 10$

   i. $n = 16, p = 0.20, X = 3$

3. Compute the probability of X successes, using the binomial formula.

   a. $n = 6, X = 3, p = 0.03$

   b. $n = 4, X = 2, p = 0.18$

   c. $n = 5, X = 3, p = 0.63$

   d. $n = 9, X = 0, p - 0.42$

   e. $n = 10, X = 5, p = 0.37$

For Exercises 4 through 13, assume all variables are binomial. (*Note:* If values are not found in Table B of Appendix C, use the binomial formula.)

4. A burglar alarm system has six fail-safe components. The probability of each failing is 0.05. Find these probabilities.

   a. Exactly three will fail.

   b. Fewer than two will fail.

   c. None will fail.

   d. Compare the answers for parts *a*, *b*, and *c*, and explain why the results are reasonable.

5. A student takes a 20-question, true/false exam and guesses on each question. Find the probability of passing if the lowest passing grade is 15 correct out of 20. Would you consider this event likely to occur? Explain your answer.

6. A student takes a 20-question, multiple-choice exam with five choices for each question and guesses on each question. Find the probability of guessing at least 15 out of 20 correctly. Would you consider this event likely or unlikely to occur? Explain your answer.

7. In a survey, 30% of the people interviewed said that they bought most of their books during the last 3 months of the year (October, November, December). If nine people are selected at random, find the probability that exactly three of these people bought most of their books during October, November, and December.

   Source: USA Snapshot, *USA TODAY.*

8. In a Gallup Survey, 90% of the people interviewed were unaware that maintaining a healthy weight could reduce the risk of stroke. If 15 people are selected at random, find the probability that at least 9 are unaware that maintaining a proper weight could reduce the risk of stroke.

   Source: USA Snapshot, *USA TODAY.*

9. In a survey, three of four students said the courts show "too much concern" for criminals. Find the probability that at most three out of seven randomly selected students will agree with this statement.

   Source: *Harper's Index.*

10. It was found that 60% of American victims of health care fraud are senior citizens. If 10 victims are randomly selected, find the probability that exactly 3 are senior citizens.

    Source: *100% American* by Daniel Evan Weiss.

11. R. H. Bruskin Associates Market Research found that 40% of Americans do not think that having a college education is important to succeed in the business world. If a random sample of five Americans is selected, find these probabilities.

    a. Exactly two people will agree with that statement.

    b. At most three people will agree with that statement.

    c. At least two people will agree with that statement.

    d. Fewer than three people will agree with that statement.

    Source: *100% American* by Daniel Evans Weiss.

12. Find these probabilities for a sample of 20 teenagers if 70% of them had compact disk players by the age of 16.

    a. At least 14 had CD players

    b. Exactly 9 had CD players

    c. More than 17 had CD players

    d. Which event, a, b, or c, is most likely to occur? Explain why.

13. If 80% of the people in a community have Internet access from their homes, find these probabilities for a sample of 10 people.

    a. At most 6 have Internet access.

    b. Exactly 6 have Internet access.

    c. At least 6 have Internet access.

    d. Which event, a, b, or c, is most likely to occur? Explain why.

14. (ans) Find the mean, variance, and standard deviation for each of the values of n and p when the conditions for the binomial distribution are met.

    a. $n = 100, p = 0.75$

    b. $n = 300, p = 0.3$

    c. $n = 20, p = 0.5$

    d. $n = 10, p = 0.8$

    e. $n = 1000, p = 0.1$

    f. $n = 500, p = 0.25$

    g. $n = 50, p = \frac{2}{5}$

    h. $n = 36, p = \frac{1}{6}$

15. A study found that 1% of Social Security recipients are too young to vote. If 800 Social Security recipients are randomly selected, find the mean, variance, and standard deviation of the number of recipients who are too young to vote.

    Source: *Harper's Index.*

16. Find the mean, variance, and standard deviation for the number of heads when 20 coins are tossed.

17. If 3% of calculators are defective, find the mean, variance, and standard deviation of a lot of 300 calculators.

18. It has been reported that 83% of federal government employees use e-mail. If a sample of 200 federal government employees is selected, find the mean, variance, and standard deviation of the number who use e-mail.

    Source: *USA TODAY.*

19. A survey found that 21% of Americans watch fireworks on television on July 4. Find the mean, variance, and standard deviation of the number of individuals who watch fireworks on television on July 4 if a random sample of 1000 Americans is selected.

    Source: USA Snapshot, *USA TODAY.*

20. In a restaurant, a study found that 42% of all patrons smoked. If the seating capacity of the restaurant is 80 people, find the mean, variance, and standard deviation of the number of smokers. About how many seats should be available for smoking customers?

21. A survey found that 25% of pet owners had their pets bathed professionally rather than do it themselves. If 18 pet owners are randomly selected, find the probability that exactly five people have their pets bathed professionally.

    Source: USA Snapshot, *USA TODAY.*

22. In a survey, 63% of Americans said they own an answering machine. If 14 Americans are selected at random, find the probability that exactly 9 own an answering machine.

    Source: USA Snapshot, *USA TODAY.*

23. One out of every three Americans believes that the U.S. government should take "primary responsibility" for eliminating poverty in the United States. If 10 Americans are selected, find the probability that at most 3 will believe that the U.S. government should take primary responsibility for eliminating poverty.

    Source: *Harper's Index.*

24. In a survey, 58% of American adults said they had never heard of the Internet. If 20 American adults are selected at random, find the probability that exactly 12 will say they have never heard of the Internet.

    Source: *Harper's Index.*

25. In the past year, 13% of businesses have eliminated jobs. If five businesses are selected at random, find the probability that at least three have eliminated jobs during the last year.

    Source: *USA TODAY.*

26. Of graduating high school seniors, 14% said that their generation will be remembered for their social concerns. If seven graduating seniors are selected at random, find the probability that either two or three will agree with that statement.

    Source: *USA TODAY.*

27. A survey found that 86% of Americans have never been a victim of violent crime. If a sample of 12 Americans is selected at random, find the probability that 10 or more have never been victims of violent crime. Does it seem reasonable that 10 or more have never been victims of violent crime?

    Source: *Harper's Index.*

Extending the Concepts

28. The graph shown here represents the probability distribution for the number of girls in a family of three children. From this graph, construct a probability distribution.

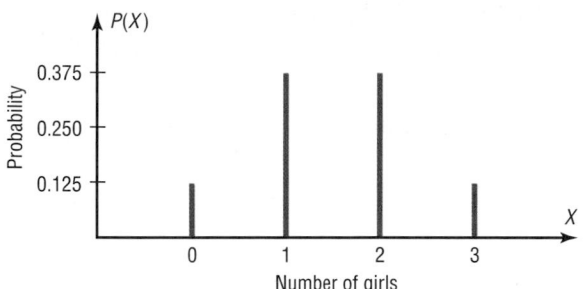

29. Construct a binomial distribution graph for the number of defective computer chips in a lot of four if $p = 0.3$.

Technology *Step by Step*

MINITAB
Step by Step

The Binomial Distribution

Calculate a Binomial Probability

From Example 5–19, it is known that 5% of the population is afraid of being alone at night. If a random sample of 20 Americans is selected, what is the probability that exactly 5 of them are afraid?

$$n = 20 \qquad p = 0.05\,(5\%) \qquad \text{and} \qquad X = 5\,(5 \text{ out of } 20)$$

No data need to be entered in the worksheet.

1. Select **Calc>Probability Distributions>Binomial.**

2. Click the option for Probability.

3. Click in the text box for Number of trials:.

4. Type in **20,** then Tab to Probability of success, then type **.05.**

5. Click the option for Input constant, then type in **5.** Leave the text box for Optional storage empty. If the name of a constant such as K1 is entered here, the results are stored but not displayed in the session window.

6. Click [OK]. The results are visible in the session window.

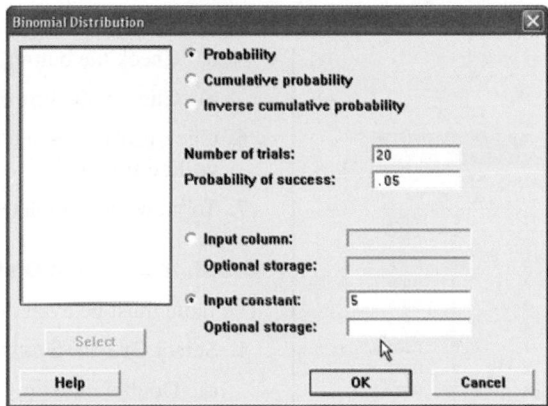

Probability Density Function
Binomial with n = 20 and p = 0.05
x   f(x)
5   0.0022446

Construct a Binomial Distribution

These instructions will use $n = 20$ and $p = 0.05$.

1. Select **Calc>Make Patterned Data>Simple Set of Numbers.**

2. You must enter three items:

   a) Enter **X** in the box for Store patterned data in:. MINITAB will use the first empty column of the active worksheet and name it X.

   b) Press Tab. Enter the value of **0** for the first value. Press Tab.

   c) Enter **20** for the last value. This value should be n. In steps of:, the value should be 1.

3. Click [OK].

4. Select **Calc>Probability Distributions>Binomial.**

5. In the dialog box you must enter five items.

   a) Click the button for Probability.

   b) In the box for Number of trials enter **20.**

   c) Enter **.05** in the Probability of success.

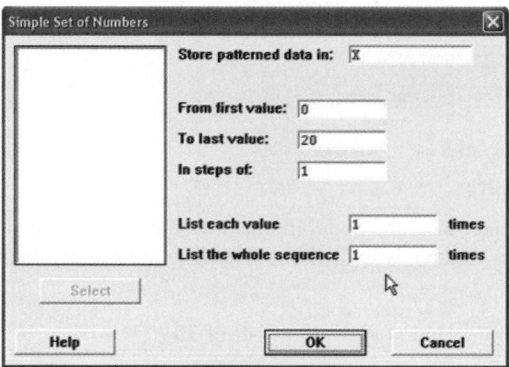
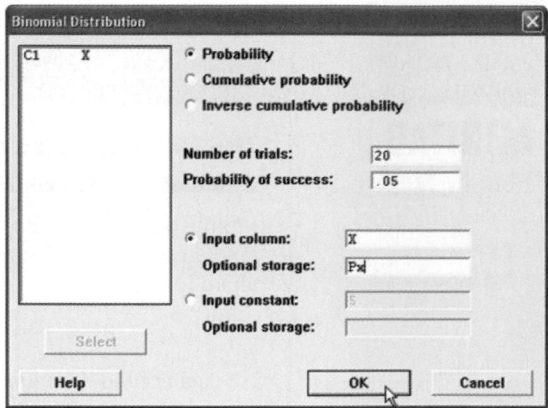

   d) Check the button for Input columns then type the column name, **X,** in the text box.

   e) Click in the box for Optional storage, then type **Px.**

6. Click [OK]. The first available column will be named Px, and the calculated probabilities will be stored in it.

7. To view the completed table, click the worksheet icon on the toolbar.

Graph a Binomial Distribution

The table must be available in the worksheet.

1. Select **Graph>Scatterplot,** then Simple.

   a) Double-click on C2 Px for the Y variable and C1 X for the X variable.

   b) Click [Data view], then Project lines, then [OK]. Deselect any other type of display that may be selected in this list.

   c) Click on [Labels], then Title/Footnotes.

   d) Type an appropriate title, such as **Binomial Distribution n = 20, p = .05.**

   e) Press Tab to the Subtitle 1, then type in Your Name.

   f) Optional: Click [Scales] then [Gridlines] then check the box for Y major ticks.

   g) Click [OK] twice.

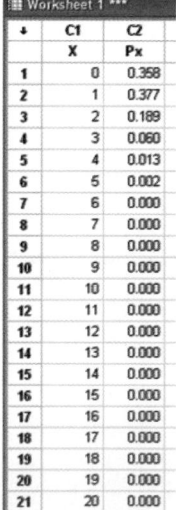

	C1	C2
	X	Px
1	0	0.358
2	1	0.377
3	2	0.189
4	3	0.060
5	4	0.013
6	5	0.002
7	6	0.000
8	7	0.000
9	8	0.000
10	9	0.000
11	10	0.000
12	11	0.000
13	12	0.000
14	13	0.000
15	14	0.000
16	15	0.000
17	16	0.000
18	17	0.000
19	18	0.000
20	19	0.000
21	20	0.000

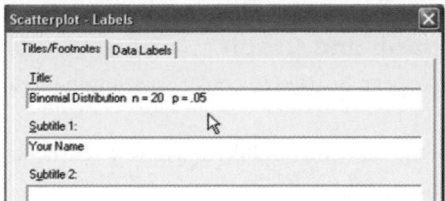

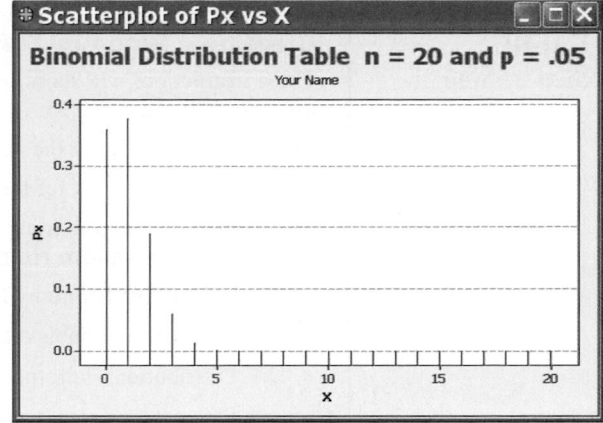

The graph will be displayed in a window. Right-click the control box to save, print, or close the graph.

TI-83 Plus or TI-84 Plus
Step by Step

Binomial Random Variables

To find the probability for a binomial variable:
Press **2nd [DISTR]** then **0** for binomial pdf(
The form is binompdf(n,p,X).

Example: $n = 20$, $X = 5$, $p = .05$. (Example 5–19a from the text) binompdf(20,.05,5)

Example: $n = 20$, $X = 0, 1, 2, 3$, $p = .05$. (Example 5–19b from the text)
binompdf(20,.05,{0,1,2,3})
The calculator will display the probabilities in a list. Use the arrow keys to view entire display.

To find the cumulative probability for a binomial random variable:
Press **2nd [DISTR]** then **A (ALPHA MATH)** for binomcdf(
The form is binomcdf(n,p,X). This will calculate the cumulative probability for values from 0 to X.

Example: $n = 20$, $X = 0, 1, 2, 3$, $p = .05$ (Example 5–19b from the text)
binomcdf(20,.05,3)

```
binompdf(20,.05,
5)
       .002244646
binompdf(20,.05,
{0,1,2,3})
{.3584859224 .3…
```

```
binomcdf(20,.05,
3)
       .984098474
```

To construct a binomial probability table:
1. Enter the X values 0 through n into L$_1$.
2. Move the cursor to the top of the L$_2$ column so that L$_2$ is highlighted.
3. Type the command binompdf(n,p,L$_1$), then press **ENTER.**

Example: $n = 20$, $p = .05$ (Example 5–19 from the text)

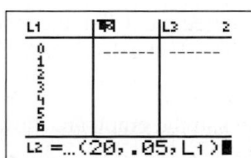
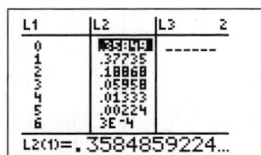

Excel
Step by Step

Creating a Binomial Distribution and Graph

These instructions will show how Excel can be used to construct a binomial distribution table, using $n = 20$ and $p = 0.35$.

1. Label cell A1 **X,** for the value of the random variable in a new worksheet.

2. Label cell B1 **P(X),** for the corresponding probabilities.

3. To enter the integers from 0 to 20 in column A starting at cell A2, select **Tools>Data Analysis>Random Number Generation,** then click OK.

4. In the Random Number Generation dialog box, enter the following:

   a) Number of Variables: **1**

   b) Distribution: Patterned

   c) Parameters: From **0** to **20** in steps of **1,** repeating each number: **1** times and repeating each sequence: **1** times

   d) Output range: **A2:A21**

5. Then click [OK].

Random Number Generation Dialog Box

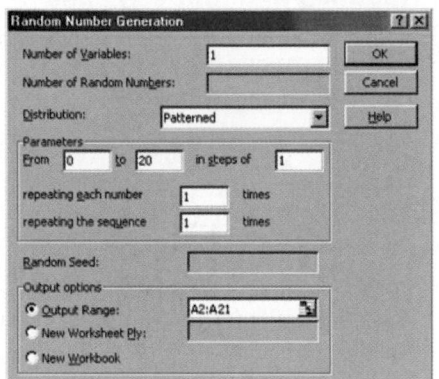

6. To create the probability corresponding to the first value of the binomial random variable, select cell B2 and type: **=BINOMDIST(0,20,.35,FALSE).** This will give the probability of obtaining 0 successes in 20 trials of a binomial experiment in which the probability of success is 0.35.

7. Repeat this procedure for each of the values of the random variable from column A.

Note: If you wish to obtain a column of cumulative probabilities for each outcome, you can type: **=BINOMDIST(0,20,.35,TRUE)** and repeat for each outcome. You can label the cell at the top of the column of cumulative probabilities, cell C1, as **P(X<=x).**

To create the graph:

1. Select the Chart Wizard from the Toolbar. Select the Column type and the first Chart sub-type. Click Next.

2. Click the icon next to the Data Range box to minimize the dialog box. Highlight the cells B1:B21, and then click the icon to return to the dialog box.

3. Select the Series tab. Click the icon next to the Category (X) axis labels box. From the worksheet, highlight cells A1:A21. Click Next. Label as necessary, then click Next and Finish.

The distribution and the graph are shown.

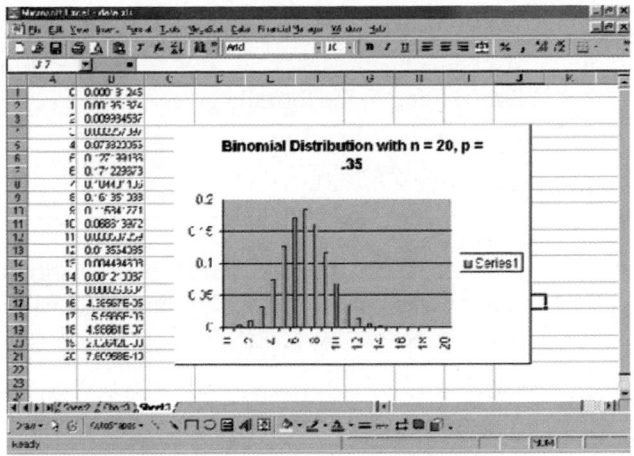

5–5   Other Types of Distributions (Optional)

In addition to the binomial distribution, other types of distributions are used in statistics. Three of the most commonly used distributions are the multinomial distribution, the Poisson distribution, and the hypergeometric distribution. They are described next.

The Multinomial Distribution

Objective 5

Find probabilities for outcomes of variables, using the Poisson, hypergeometric, and multinomial distributions.

Recall that in order for an experiment to be binomial, two outcomes are required for each trial. But if each trial in an experiment has more than two outcomes, a distribution called the **multinomial distribution** must be used. For example, a survey might require the responses of "approve," "disapprove," or "no opinion." In another situation, a person may have a choice of one of five activities for Friday night, such as a movie, dinner, baseball game, play, or party. Since these situations have more than two possible outcomes for each trial, the binomial distribution cannot be used to compute probabilities.

The multinomial distribution can be used for such situations if the probabilities for each trial remain constant and the outcomes are independent for a fixed number of trials. The events must also be mutually exclusive.

Formula for the Multinomial Distribution

If X consists of events $E_1, E_2, E_3, \ldots, E_k$, which have corresponding probabilities $p_1, p_2, p_3, \ldots, p_k$ of occurring, and X_1 is the number of times E_1 will occur, X_2 is the number of times E_2 will occur, X_3 is the number of times E_3 will occur, etc., then the probability that X will occur is

$$P(X) = \frac{n!}{X_1! \cdot X_2! \cdot X_3! \cdots X_k!} \cdot p_1^{X_1} \cdot p_2^{X_2} \cdots p_k^{X_k}$$

where $X_1 + X_2 + X_3 + \cdots + X_k = n$ and $p_1 + p_2 + p_3 + \cdots + p_k = 1$.

Example 5–24

In a large city, 50% of the people choose a movie, 30% choose dinner and a play, and 20% choose shopping as a leisure activity. If a sample of five people is randomly selected, find the probability that three are planning to go to a movie, one to a play, and one to a shopping mall.

Solution

We know that $n = 5$, $X_1 = 3$, $X_2 = 1$, $X_3 = 1$, $p_1 = 0.50$, $p_2 = 0.30$, and $p_3 = 0.20$. Substituting in the formula gives

$$P(X) = \frac{5!}{3! \cdot 1! \cdot 1!} \cdot (0.50)^3 (0.30)^1 (0.20)^1 = 0.15$$

Again, note that the multinomial distribution can be used even though replacement is not done, provided that the sample is small in comparison with the population.

Example 5–25

In a music store, a manager found that the probabilities that a person buys zero, one, or two or more CDs are 0.3, 0.6, and 0.1, respectively. If six customers enter the store, find the probability that one won't buy any CDs, three will buy one CD, and two will buy two or more CDs.

Solution

It is given that $n = 6$, $X_1 = 1$, $X_2 = 3$, $X_3 = 2$, $p_1 = 0.3$, $p_2 = 0.6$, and $p_3 = 0.1$. Then

$$P(X) = \frac{6!}{1!3!2!} \cdot (0.3)^1 (0.6)^3 (0.1)^2$$

$$= 60 \cdot (0.3)(0.216)(0.01) = 0.03888$$

Example 5–26

A box contains four white balls, three red balls, and three blue balls. A ball is selected at random, and its color is written down. It is replaced each time. Find the probability that if five balls are selected, two are white, two are red, and one is blue.

Solution

We know that $n = 5$, $X_1 = 2$, $X_2 = 2$, $X_3 = 1$; $p_1 = \frac{4}{10}$, $p_2 = \frac{3}{10}$, and $p_3 = \frac{3}{10}$; hence,

$$P(X) = \frac{5!}{2!2!1!} \cdot \left(\frac{4}{10}\right)^2 \left(\frac{3}{10}\right)^2 \left(\frac{3}{10}\right)^1 = \frac{81}{625}$$

Historical Notes

Simeon D. Poisson (1781–1840) formulated the distribution that bears his name. It appears only once in his writings and is only one page long. Mathematicians paid little attention to it until 1907, when a statistician named W. S. Gosset found real applications for it.

Thus, the multinomial distribution is similar to the binomial distribution but has the advantage of allowing one to compute probabilities when there are more than two outcomes for each trial in the experiment. That is, the multinomial distribution is a general distribution, and the binomial distribution is a special case of the multinomial distribution.

The Poisson Distribution

A discrete probability distribution that is useful when n is large and p is small and when the independent variables occur over a period of time is called the **Poisson distribution.** In addition to being used for the stated conditions (i.e., n is large, p is small, and the variables occur over a period of time), the Poisson distribution can be used when a density of items is distributed over a given area or volume, such as the number of plants growing per acre or the number of defects in a given length of videotape.

Formula for the Poisson Distribution

The probability of X occurrences in an interval of time, volume, area, etc., for a variable where λ (Greek letter lambda) is the mean number of occurrences per unit (time, volume, area, etc.) is

$$P(X; \lambda) = \frac{e^{-\lambda}\lambda^X}{X!} \quad \text{where } X = 0, 1, 2, \ldots$$

The letter e is a constant approximately equal to 2.7183.

Round the answers to four decimal places.

Example 5–27

If there are 200 typographical errors randomly distributed in a 500-page manuscript, find the probability that a given page contains exactly three errors.

Solution

First, find the mean number λ of errors. Since there are 200 errors distributed over 500 pages, each page has an average of

$$\lambda = \frac{200}{500} = \frac{2}{5} = 0.4$$

or 0.4 error per page. Since $X = 3$, substituting into the formula yields

$$P(X; \lambda) = \frac{e^{-\lambda}\lambda^X}{X!} = \frac{(2.7183)^{-0.4}(0.4)^3}{3!} = 0.0072$$

Thus, there is less than a 1% probability that any given page will contain exactly three errors.

Since the mathematics involved in computing Poisson probabilities is somewhat complicated, tables have been compiled for these probabilities. Table C in Appendix C gives P for various values for λ and X.

In Example 5–27, where X is 3 and λ is 0.4, the table gives the value 0.0072 for the probability. See Figure 5–4.

Figure 5–4

Using Table C

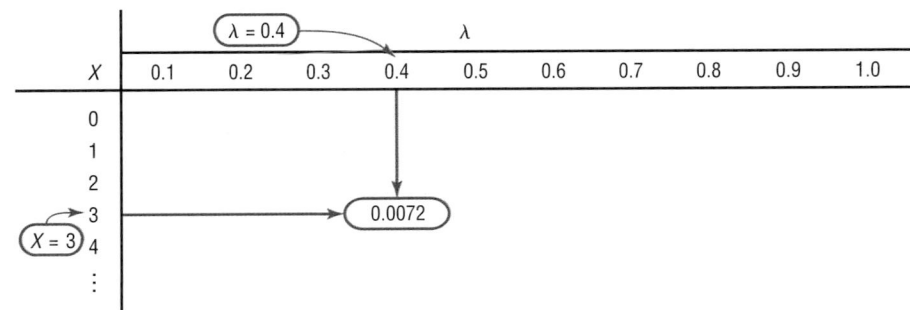

Example 5–28

A sales firm receives, on the average, three calls per hour on its toll-free number. For any given hour, find the probability that it will receive the following.

a. At most three calls   *b.* At least three calls   *c.* Five or more calls

Example 5–31

A recent study found that four out of nine houses were underinsured. If five houses are selected from the nine houses, find the probability that exactly two are underinsured.

Solution

In this problem

$$a = 4 \qquad b = 5 \qquad n = 5 \qquad X = 2 \qquad n - X = 3$$

Then

$$P(X) = \frac{{}_4C_2 \cdot {}_5C_3}{{}_9C_5} = \frac{60}{126} = \frac{10}{21}$$

In many situations where objects are manufactured and shipped to a company, the company selects a few items and tests them to see whether they are satisfactory or defective. If a certain percentage is defective, the company then can refuse the whole shipment. This procedure saves the time and cost of testing every single item. To make the judgment about whether to accept or reject the whole shipment based on a small sample of tests, the company must know the probability of getting a specific number of defective items. To calculate the probability, the company uses the hypergeometric distribution.

Example 5–32

A lot of 12 compressor tanks is checked to see whether there are any defective tanks. Three tanks are checked for leaks. If one or more of the three is defective, the lot is rejected. Find the probability that the lot will be rejected if there are actually three defective tanks in the lot.

Solution

Since the lot is rejected if at least one tank is found to be defective, it is necessary to find the probability that none are defective and subtract this probability from 1.
    Here, $a = 3$, $b = 9$, $n = 3$, $X = 0$; so

$$P(X) = \frac{{}_3C_0 \cdot {}_9C_3}{{}_{12}C_3} = \frac{1 \cdot 84}{220} = 0.38$$

Hence,

$$P(\text{at least one defective}) = 1 - P(\text{no defectives}) = 1 - 0.38 = 0.62$$

There is a 0.62, or 62%, probability that the lot will be rejected when 3 of the 12 tanks are defective.

A summary of the discrete distributions used in this chapter is shown in Table 5–1.

Interesting Fact

An IBM supercomputer set a world record in 2004 by performing 36.01 trillion calculations in 1 second.

Table 5–1	Summary of Discrete Distributions

1. Binomial distribution

$$P(X) = \frac{n!}{(n - X)!X!} \cdot p^X \cdot q^{n-X}$$

$$\mu = n \cdot p \qquad \sigma = \sqrt{n \cdot p \cdot q}$$

Used when there are only two independent outcomes for a fixed number of independent trials and the probability for each success remains the same for each trial.

2. Multinomial distribution

$$P(X) = \frac{n!}{X_1! \cdot X_2! \cdot X_3! \cdots X_k!} \cdot p_1^{X_1} \cdot p_2^{X_2} \cdots p_k^{X_k}$$

where

$$X_1 + X_2 + X_3 + \cdots + X_k = n \qquad \text{and} \qquad p_1 + p_2 + p_3 + \cdots + p_k = 1$$

Used when the distribution has more than two outcomes, the probabilities for each trial remain constant, outcomes are independent, and there are a fixed number of trials.

3. Poisson distribution

$$P(X; \lambda) = \frac{e^{-\lambda}\lambda^X}{X!} \qquad \text{where } X = 0, 1, 2, \ldots$$

Used when n is large and p is small, the independent variable occurs over a period of time, or a density of items is distributed over a given area or volume.

4. Hypergeometric distribution

$$P(X) = \frac{{}_aC_X \cdot {}_bC_{n-X}}{{}_{a+b}C_n}$$

Used when there are two outcomes and sampling is done without replacement.

Applying the Concepts **5–5**

Rockets and Targets

During the latter days of World War II, the Germans developed flying rocket bombs. These bombs were used to attack London. Allied military intelligence didn't know whether these bombs were fired at random or had a sophisticated aiming device. To determine the answer, they used the Poisson distribution.

To assess the accuracy of these bombs, London was divided into 576 square regions. Each region was $\frac{1}{4}$ square kilometer in area. They then compared the number of actual hits with the theoretical number of hits by using the Poisson distribution. If the values in both distributions were close, then they would conclude that the rockets were fired at random. The actual distribution is as follows:

Hits	0	1	2	3	4	5
Regions	229	211	93	35	7	1

1. Using the Poisson distribution, find the theoretical values for each number of hits. In this case, the number of bombs was 535, and the number of regions was 576. So

$$\mu = \frac{535}{576} = 0.929$$

For three hits,

$$P(X) = \frac{\mu^X \cdot e^{-\mu}}{X!}$$

$$= \frac{(0.929)^3(2.7183)^{-0.929}}{3!} = 0.0528$$

Hence the number of hits is $(0.0528)(576) = 30.4128$.
Complete the table for the other number of hits.

Hits	0	1	2	3	4	5
Regions				30.4		

2. Write a brief statement comparing the two distributions.

3. Based on your answer to question 2, can you conclude that the rockets were fired at random?

See page 284 for the answer.

Exercises 5–5

1. Use the multinomial formula and find the probabilities for each.

   a. $n = 6, X_1 = 3, X_2 = 2, X_3 = 1, p_1 = 0.5, p_2 = 0.3, p_3 = 0.2$

   b. $n = 5, X_1 = 1, X_2 = 2, X_3 = 2, p_1 = 0.3, p_2 = 0.6, p_3 = 0.1$

   c. $n = 4, X_1 = 1, X_2 = 1, X_3 = 2, p_1 = 0.8, p_2 = 0.1, p_3 = 0.1$

   d. $n = 3, X_1 = 1, X_2 = 1, X_3 = 1, p_1 = 0.5, p_2 = 0.3, p_3 = 0.2$

   e. $n = 5, X_1 = 1, X_2 = 3, X_3 = 1, p_1 = 0.7, p_2 = 0.2, p_3 = 0.1$

2. The probabilities that a textbook page will have 0, 1, 2, or 3 typographical errors are 0.79, 0.12, 0.07, and 0.02, respectively. If eight pages are randomly selected, find the probability that four will contain no errors, two will contain 1 error, one will contain 2 errors, and one will contain 3 errors.

3. The probabilities are 0.25, 0.40, and 0.35 that an 18-wheel truck will have 0 violations, 1 violation, or 2 or more violations when it is given a safety inspection. If eight trucks are inspected, find the probability that three will have 0 violations, two will have 1 violation, and three will have 2 or more violations.

4. When a customer enters a pharmacy, the probabilities that he or she will have 0, 1, 2, or 3 prescriptions filled are 0.60, 0.25, 0.10, and 0.05, respectively. For a sample of six people who enter the pharmacy, find the probability that two will have 0 prescriptions, two will have 1 prescription, one will have 2 prescriptions, and one will have 3 prescriptions.

5. A die is rolled 4 times. Find the probability of two 1s, one 2, and one 3.

6. According to Mendel's theory, if tall and colorful plants are crossed with short and colorless plants, the corresponding probabilities are $\frac{9}{16}, \frac{3}{16}, \frac{3}{16}$, and $\frac{1}{16}$ for tall and colorful, tall and colorless, short and colorful, and short and colorless, respectively. If eight plants are selected, find the probability that one will be tall and colorful, three will be tall and colorless, three will be short and colorful, and one will be short and colorless.

7. Find each probability $P(X; \lambda)$, using Table C in Appendix C.

   a. $P(5; 4)$

   b. $P(2; 4)$

   c. $P(6; 3)$

   d. $P(10; 7)$

   e. $P(9; 8)$

8. If 2% of the batteries manufactured by a company are defective, find the probability that in a case of 144 batteries, there are 3 defective ones.

9. A recent study of robberies for a certain geographic region showed an average of one robbery per 20,000 people. In a city of 80,000 people, find the probability of the following.

   a. No robberies

   b. One robbery

   c. Two robberies

   d. Three or more robberies

10. In a 400-page manuscript, there are 200 randomly distributed misprints. If a page is selected, find the probability that it has one misprint.

11. A telephone soliciting company obtains an average of five orders per 1000 solicitations. If the company reaches 250 potential customers, find the probability of obtaining at least two orders.

12. A mail-order company receives an average of five orders per 500 solicitations. If it sends out 100 advertisements, find the probability of receiving at least two orders.

13. A videotape has an average of one defect every 1000 feet. Find the probability of at least one defect in 3000 feet.

14. If 3% of all cars fail the emissions inspection, find the probability that in a sample of 90 cars, 3 will fail. Use the Poisson approximation.

15. The average number of phone inquiries per day at the poison control center is four. Find the probability it will receive five calls on a given day. Use the Poisson approximation.

16. In a batch of 2000 calculators, there are, on average, eight defective ones. If a random sample of

150 is selected, find the probability of five defective ones.

17. In a camping club of 18 members, nine prefer hoods and nine prefer hats and earmuffs. On a recent winter outing attended by six members, find the probability that exactly three members wore earmuffs and hats.

18. A bookstore owner examines 5 books from each lot of 25 to check for missing pages. If he finds at least two books with missing pages, the entire lot is returned. If, indeed, there are five books with missing pages, find the probability that the lot will be returned.

19. Shirts are packed at random in two sizes, regular and extra large. Four shirts are selected from a box of 24 and checked for size. If there are 15 regular shirts in the box, find the probability that all 4 will be regular size.

20. A shipment of 24 computer keyboards is rejected if 4 are checked for defects and at least 1 is found to be defective. Find the probability that the shipment will be returned if there are actually 6 defective keyboards.

21. A shipment of 24 electric typewriters is rejected if 3 are checked for defects and at least 1 is found to be defective. Find the probability that the shipment will be returned if there are actually 6 typewriters that are defective.

Technology *Step by Step*

TI-83 Plus or TI-84 Plus
Step by Step

Poisson Random Variables

To find the probability for a Poisson random variable:
Press **2nd [DISTR]** then **B (ALPHA APPS)** for poissonpdf(
The form is poissonpdf(λ,X).

Example: $\lambda = 0.4$, $X = 3$ (Example 5–27 from the text)
poissonpdf(.4,3)

Example: $\lambda = 3$, $X = 0, 1, 2, 3$ (Example 5–28a from the text)
poissonpdf(3,{0,1,2,3})
The calculator will display the probabilities in a list. Use the arrow keys to view the entire display.

To find the cumulative probability for a Poisson random variable:
Press **2nd [DISTR]** then **C (ALPHA PRGM)** for poissoncdf(
The form is poissoncdf(λ,X). This will calculate the cumulative probability for values from 0 to X.

Example: $\lambda = 3$, $X = 0, 1, 2, 3$ (Example 5–28a from the text)
poissoncdf(3,3)

```
poissonpdf(.4,3)
         .0071500805
poissonpdf(3,{0,
1,2,3})
{.0497870684 .1…
■
```

```
poissoncdf(3,3)
         .6472318893
```

To construct a Poisson probability table:

1. Enter the X values 0 through a large possible value of X into L_1.

2. Move the cursor to the top of the L_2 column so that L_2 is highlighted.

3. Enter the command poissonpdf(λ,L_1) then press **ENTER.**

Example: $\lambda = 3$, $X = 0, 1, 2, 3, \ldots, 10$ (Example 5–28 from the text)

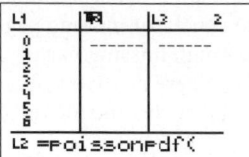

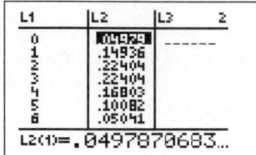

5–6 Summary

Many variables have special probability distributions. This chapter presented several of the most common probability distributions, including the binomial distribution, the multinomial distribution, the Poisson distribution, and the hypergeometric distribution.

The binomial distribution is used when there are only two outcomes for an experiment, there are a fixed number of trials, the probability is the same for each trial, and the outcomes are independent of one another. The multinomial distribution is an extension of the binomial distribution and is used when there are three or more outcomes for an experiment. The hypergeometric distribution is used when sampling is done without replacement. Finally, the Poisson distribution is used in special cases when independent events occur over a period of time, area, or volume.

A probability distribution can be graphed, and the mean, variance, and standard deviation can be found. The mathematical expectation can also be calculated for a probability distribution. Expectation is used in insurance and games of chance.

Important Terms

binomial distribution 257	discrete probability distribution 240	hypergeometric distribution 273	Poisson distribution 270
binomial experiment 257	expected value 250	multinomial distribution 269	random variable 239

Important Formulas

Formula for the mean of a probability distribution:

$$\mu = \Sigma X \cdot P(X)$$

Formulas for the variance and standard deviation of a probability distribution:

$$\sigma^2 = \Sigma[X^2 \cdot P(X)] - \mu^2$$
$$\sigma = \sqrt{\Sigma[X^2 \cdot P(X)] - \mu^2}$$

Formula for expected value:

$$E(X) = \Sigma X \cdot P(X)$$

Binomial probability formula:

$$P(X) = \frac{n!}{(n - X)!X!} \cdot p^X \cdot q^{n-X}$$

Formula for the mean of the binomial distribution:

$$\mu = n \cdot p$$

Formulas for the variance and standard deviation of the binomial distribution:

$$\sigma^2 = n \cdot p \cdot q \qquad \sigma = \sqrt{n \cdot p \cdot q}$$

Formula for the multinomial distribution:

$$P(X) = \frac{n!}{X_1! \cdot X_2! \cdot X_3! \cdots X_k!} \cdot p_1^{X_1} \cdot p_2^{X_2} \cdots p_k^{X_k}$$

Formula for the Poisson distribution:

$$P(X; \lambda) = \frac{e^{-\lambda}\lambda^X}{X!} \qquad \text{where } X = 0, 1, 2, \ldots$$

Formula for the hypergeometric distribution:

$$P(X) = \frac{{}_aC_X \cdot {}_bC_{n-X}}{{}_{a+b}C_n}$$

Review Exercises

For Exercises 1 through 3, determine whether the distribution represents a probability distribution. If it does not, state why.

1.

X	1	2	3	4	5
P(X)	$\frac{1}{10}$	$\frac{3}{10}$	$\frac{1}{10}$	$\frac{2}{10}$	$\frac{3}{10}$

2.

X	10	20	30
P(X)	0.1	0.4	0.3

3.

X	8	12	16	20
P(X)	$\frac{5}{6}$	$\frac{1}{12}$	$\frac{1}{12}$	$\frac{1}{12}$

4. The number of emergency calls a local police department receives per 24-hour period is distributed as shown here. Construct a graph for the data.

Number of calls X	10	11	12	13	14
Probability P(X)	0.02	0.12	0.40	0.31	0.15

5. A study was conducted to determine the number of radios each household has. The data are shown here. Draw a graph for the data.

Number of radios	0	1	2	3	4
Probability P(X)	0.05	0.30	0.45	0.12	0.08

6. A box contains five pennies, three dimes, one quarter, and one half-dollar. Construct a probability distribution and draw a graph for the data.

7. At Tyler's Tie Shop, Tyler found the probabilities that a customer will buy 0, 1, 2, 3, or 4 ties, as shown. Construct a graph for the distribution.

Number of ties X	0	1	2	3	4
Probability P(X)	0.30	0.50	0.10	0.08	0.02

8. A bank has a drive-through service. The number of customers arriving during a 15-minute period is distributed as shown. Find the mean, variance, and standard deviation for the distribution.

Number of customers X	0	1	2	3	4
Probability P(X)	0.12	0.20	0.31	0.25	0.12

9. At a small community museum, the number of visitors per hour during the day has the distribution shown here. Find the mean, variance, and standard deviation for the data.

Number of visitors X	13	14	15	16	17
Probability P(X)	0.12	0.15	0.29	0.25	0.19

10. During a recent paint sale at Corner Hardware, the number of cans of paint purchased was distributed as shown. Find the mean, variance, and standard deviation of the distribution.

Number of cans X	1	2	3	4	5
Probability P(X)	0.42	0.27	0.15	0.10	0.06

11. The number of inquiries received per day for a college catalog is distributed as shown. Find the mean, variance, and standard deviation for the data.

Number of inquiries X	22	23	24	25	26	27
Probability P(X)	0.08	0.19	0.36	0.25	0.07	0.05

12. A producer plans an outdoor regatta for May 3. The cost of the regatta is $8000. This includes advertising, security, printing tickets, entertainment, etc. The producer plans to make $15,000 profit if all goes well. However, if it rains, the regatta would have to be canceled. According to the weather report, the

probability of rain is 0.3. Find the producer's expected profit.

13. A person selects a card from a deck. If it is a red card, he wins $1. If it is a black card between or including 2 and 10, he wins $5. If it is a black face card, he wins $10; and if it is a black ace, he wins $100. Find the expectation of the game. How much should a person bet if the game is to be fair?

14. If 30% of all commuters ride the train to work, find the probability that if 10 workers are selected, 5 will ride the train.

15. If 90% of all people between the ages of 30 and 50 drive a car, find these probabilities for a sample of 20 people in that age group.

    a. Exactly 20 drive a car.
    b. At least 15 drive a car.
    c. At most 15 drive a car.

16. If 10% of the people who are given a certain drug experience dizziness, find these probabilities for a sample of 15 people who take the drug.

    a. At least two people will become dizzy.
    b. Exactly three people will become dizzy.
    c. At most four people will become dizzy.

17. If 75% of nursing students are able to pass a drug calculation test, find the mean, variance, and standard deviation of the number of students who pass the test in a sample of 180 nursing students.

18. A club has 225 members. If there is a 70% attendance rate per meeting, find the mean, variance, and standard deviation of the number of people who will be present at each meeting.

19. The chance that a U.S. police chief believes the death penalty "significantly reduces the number of homicides" is 1 in 4. If a random sample of eight police chiefs is selected, find the probability that at most three believe that the death penalty significantly reduces the number of homicides.

    Source: *Harper's Index.*

20. *American Energy Review* reported that 27% of American households burn wood. If a random sample of 500 American households is selected, find the mean, variance, and standard deviation of the number of households that burn wood.

    Source: *100% American* by Daniel Evan Weiss.

21. Three out of four American adults under age 35 have eaten pizza for breakfast. If a random sample of 20 adults under age 35 is selected, find the probability that exactly 16 have eaten pizza for breakfast.

    Source: *Harper's Index.*

22. One out of four Americans over age 55 has eaten pizza for breakfast. If a sample of 10 Americans over age 55 is selected at random, find the probability that at most 3 have eaten pizza for breakfast.

    Source: *Harper's Index.*

23. **(Opt.)** The probabilities that a person will make 0, 1, 2, and 3 errors on an insurance claim are 0.70, 0.20, 0.08, and 0.02, respectively. If 20 claims are selected, find the probability that 12 will contain no errors, 4 will contain 1 error, 3 will contain 2 errors, and 1 will contain 3 errors.

24. **(Opt.)** Before a VCR leaves the factory, it is given a quality control check. The probabilities that a VCR contains 0, 1, or 2 defects are 0.90, 0.06, and 0.04, respectively. In a sample of 12 recorders, find the probability that 8 have 0 defects, 3 have 1 defect, and 1 has 2 defects.

25. **(Opt.)** In a Christmas display, the probability that all lights are the same color is 0.50; that 2 colors are used is 0.40; and that 3 or more colors are used is 0.10. If a sample of 10 displays is selected, find the probability that 5 have only 1 color of light, 3 have 2 colors, and 2 have 3 or more colors.

26. **(Opt.)** If 4% of the population carries a certain genetic trait, find the probability that in a sample of 100 people, there are exactly 8 people who have the trait. Assume the distribution is approximately Poisson.

27. **(Opt.)** Computer Help Hot Line receives, on the average, six calls per hour asking for assistance. The distribution is Poisson. For any randomly selected hour, find the probability that the company will receive

    a. At least six calls.
    b. Four or more calls.
    c. At most five calls.

28. **(Opt.)** The number of boating accidents on Lake Emilie follows a Poisson distribution. The probability of an accident is 0.003. If there are 1000 boats on the lake during a summer month, find the probability that there will be 6 accidents.

29. **(Opt.)** If five cards are drawn from a deck, find the probability that two will be hearts.

30. **(Opt.)** Of the 50 automobiles in a used-car lot, 10 are white. If five automobiles are selected to be sold at an auction, find the probability that exactly two will be white.

31. **(Opt.)** A board of directors consists of seven men and five women. If a slate of three officers is selected, find these probabilities.

    a. Exactly two are men.
    b. All three are women.
    c. Exactly two are women.

Chapter Quiz

Determine whether each statement is true or false. If the statement is false, explain why.

1. The expected value of a random variable can be thought of as a long-run average.

2. The number of courses a student is taking this semester is an example of a continuous random variable.

3. When the multinomial distribution is used, the outcomes must be dependent.

4. A binomial experiment has a fixed number of trials.

Complete these statements with the best answer.

5. Random variable values are determined by _____.

6. The mean for a binomial variable can be found by using the formula _____.

7. One requirement for a probability distribution is that the sum of all the events in the sample space must equal _____.

Select the best answer.

8. What is the sum of the probabilities of all outcomes in a probability distribution?

   a. 0   c. 1
   b. $\frac{1}{2}$ d. It cannot be determined.

9. How many outcomes are there in a binomial experiment?

   a. 0   c. 2
   b. 1   d. It varies.

10. The number of plants growing in a specific area can be approximated by what distribution?

    a. Binomial
    b. Multinomial
    c. Hypergeometric
    d. Poisson

For questions 11 through 14, determine if the distribution represents a probability distribution. If not, state why.

11.

X	1	2	3	4	5
P(X)	$\frac{1}{7}$	$\frac{2}{7}$	$\frac{2}{7}$	$\frac{3}{7}$	$\frac{2}{7}$

12.

X	3	6	9	12	15
P(X)	0.3	0.5	0.1	0.08	0.02

13.

X	50	75	100
P(X)	0.5	0.2	0.3

14.

X	4	8	12	16
P(X)	$\frac{1}{6}$	$\frac{3}{12}$	$\frac{1}{2}$	$\frac{1}{12}$

15. The number of fire calls the Conestoga Valley Fire Company receives per day is distributed as follows:

Number X	5	6	7	8	9
Probability P(X)	0.28	0.32	0.09	0.21	0.10

Construct a graph for the data.

16. A study was conducted to determine the number of telephones each household has. The data are shown here.

Number of telephones	0	1	2	3	4
Frequency	2	30	48	13	7

Construct a probability distribution and draw a graph for the data.

17. During a recent CD sale at Matt's Music Store, the number of CDs customers purchased was distributed as follows:

Number X	0	1	2	3	4
Probability P(X)	0.10	0.23	0.31	0.27	0.09

Find the mean, variance, and standard deviation of the distribution.

18. The number of calls received per day at a crisis hot line is distributed as follows:

Number X	30	31	32	33	34
Probability $P(X)$	0.05	0.21	0.38	0.25	0.11

Find the mean, variance, and standard deviation of the distribution.

19. There are six playing cards placed face down in a box. They are the 4 of diamonds, the 5 of hearts, the 2 of clubs, the 10 of spades, the 3 of diamonds, and the 7 of hearts. A person selects a card. Find the expected value of the draw.

20. A person selects a card from an ordinary deck of cards. If it is a black card, she wins $2. If it is a red card between or including 3 and 7, she wins $10. If it is a red face card, she wins $25; and if it is a black jack, she wins an extra $100. Find the expectation of the game.

21. If 40% of all commuters ride to work in carpools, find the probability that if eight workers are selected, five will ride in carpools.

22. If 60% of all women are employed outside the home, find the probability that in a sample of 20 women,

   a. Exactly 15 are employed.
   b. At least 10 are employed.
   c. At most five are not employed outside the home.

23. If 80% of the applicants are able to pass a driver's proficiency road test, find the mean, variance, and standard deviation of the number of people who pass the test in a sample of 300 applicants.

24. A history class has 75 members. If there is a 12% absentee rate per class meeting, find the mean, variance, and standard deviation of the number of students who will be absent from each class.

25. The probability that a person will make zero, one, two, or three errors on his or her income tax return is 0.50,

0.30, 0.15, and 0.05, respectively. If 30 claims are selected, find the probability that 15 will contain 0 errors, 8 will contain one error, 5 will contain two errors, and 2 will contain three errors.

26. Before a television set leaves the factory, it is given a quality control check. The probability that a television contains zero, one, or two defects is 0.88, 0.08, and 0.04, respectively. In a sample of 16 televisions, find the probability that 9 will have no defects, 4 will have one defect, and 3 will have two defects.

27. Among the teams in a bowling league, the probability that the uniforms are all one color is 0.45, that two colors are used is 0.35, and that three or more colors are used is 0.20. If a sample of 12 uniforms is selected, find the probability that 5 contain only one color, 4 contain two colors, and 3 contain three or more colors.

28. If 8% of the population of trees are elm trees, find the probability that in a sample of 100 trees, there are exactly 6 elm trees. Assume the distribution is approximately Poisson.

29. Sports Scores Hot Line receives, on the average, eight calls per hour requesting the latest sports scores. The distribution is Poisson in nature. For any randomly selected hour, find the probability that the company will receive

   a. At least eight calls.
   b. Three or more calls.
   c. At most seven calls.

30. There are 48 raincoats for sale at a local men's clothing store. Twelve are black. If six raincoats are selected to be marked down, find the probability that exactly three will be black.

31. A youth group has eight boys and six girls. If a slate of four officers is selected, find the probability that exactly

   a. Three are girls.
   b. Two are girls.
   c. Four are boys.

Critical Thinking Challenges

1. Pennsylvania has a lottery entitled "Big 4." To win, a player must correctly match four digits from a daily lottery in which four digits are selected. Find the probability of winning.

2. In the Big 4 lottery, for a bet of $100, the payoff is $5000. What is the expected value of winning? Is it worth it?

3. If you played the same four-digit number every day (or any four-digit number for that matter) in the Big 4, how often (in years) would you win, assuming you have average luck?

4. In the game Chuck-a-Luck, three dice are rolled. A player bets a certain amount (say $1.00) on a number from 1 to 6. If the number appears on one die, the person wins $1.00. If it appears on two dice, the person wins $2.00, and if it appears on all three dice, the person wins $3.00. What are the chances of winning $1.00? $2.00? $3.00?

5. What is the expected value of the game of Chuck-a-Luck if a player bets $1.00 on one number?

 **Data Projects**

Probability Distributions   Roll three dice 100 times, recording the sum of the spots on the faces as you roll. Then find the average of the spots. How close is this to the theoretical average? Refer to Exercise 21 on page 254.

Answers to Applying the Concepts

Section 5–2   Dropping College Courses

1. The random variable under study is the reason for dropping a college course.

2. There were a total of 144 people in the study.

3. The complete table is as follows:

Reason for dropping a college course	Frequency	Percentage
Too difficult	45	31.25
Illness	40	27.78
Change in work schedule	20	13.89
Change of major	14	9.72
Family-related problems	9	6.25
Money	7	4.86
Miscellaneous	6	4.17
No meaningful reason	3	2.08

4. The probability that a student will drop a class because of illness is about 28%. The probability that a student will drop a class because of money is about 5%. The probability that a student will drop a class because of a change of major is about 10%.

5. The information is not itself a probability distribution, but it can be used as one.

6. The categories are not necessarily mutually exclusive, but we treated them as such in computing the probabilities.

7. The categories are not independent.

8. The categories are exhaustive.

9. Since all the probabilities are between 0 and 1, inclusive, and the probabilities sum to 1, the requirements for a discrete probability distribution are met.

Section 5–3   Expected Value

1. The expected value is the mean in a discrete probability distribution.

2. We would expect variation from the expected value of 3.

3. Answers will vary. One possible answer is that pregnant mothers in that area might be overly concerned upon

hearing that the number of cases of kidney problems in newborns was nearly 4 times what was usually expected. Other mothers (particularly those who had taken a statistics course!) might ask for more information about the claim.

4. Answers will vary. One possible answer is that it does seem unlikely to have 11 newborns with kidney problems when we expect only 3 newborns to have kidney problems.

5. The public might better be informed by percentages or rates (e.g., rate per 1000 newborns).

6. The increase of eight babies born with kidney problems represents a 0.32% increase (less than $\frac{1}{2}$ of 1%).

7. Answers will vary. One possible answer is that the percentage increase does not seem to be something to be overly concerned about.

Section 5–4   Unsanitary Restaurants

1. The probability of eating at 3 restaurants with unsanitary conditions out of the 10 restaurants is 0.18651.

2. The probability of eating at 4 or 5 restaurants with unsanitary conditions out of the 10 restaurants is $0.24623 + 0.22291 = 0.46914$.

3. To find this probability, you could add the probabilities for eating at 1, 2, . . . , 10 unsanitary restaurants. An easier way to compute the probability is to subtract the probability of eating at no unsanitary restaurants from 1 (using the complement rule).

4. The highest probability for this distribution is 4, but the expected number of unsanitary restaurants that you would eat at is $10 \cdot \frac{3}{7} = 4.3$.

5. The standard deviation for this distribution is $\sqrt{10(\frac{3}{7})(\frac{4}{7})} = 1.56$.

6. This is a binomial distribution. We have two possible outcomes: "success" is eating in an unsanitary restaurant; "failure" is eating in a sanitary restaurant. The probability that one restaurant is unsanitary is independent of the probability that any other restaurant

is unsanitary. The probability that a restaurant is unsanitary remains constant at $\frac{3}{7}$. And we are looking at the number of unsanitary restaurants that we eat at out of 10 "trials."

7. The likelihood of success will vary from situation to situation. Just because we have two possible outcomes, this does not mean that each outcome occurs with probability 0.50.

Section 5–5   Rockets and Targets

1. The theoretical values for the number of hits are

Hits	0	1	2	3	4	5
Regions	227.5	211.3	98.2	30.4	7.1	1.3

2. The actual values are very close to the theoretical values.

3. Since the actual values are close to the theoretical values, it does appear that the rockets were fired at random.

The Normal Distribution

Objectives

After completing this chapter, you should be able to

1 Identify distributions as symmetric or skewed.

2 Identify the properties of a normal distribution.

3 Find the area under the standard normal distribution, given various z values.

4 Find probabilities for a normally distributed variable by transforming it into a standard normal variable.

5 Find specific data values for given percentages, using the standard normal distribution.

6 Use the central limit theorem to solve problems involving sample means for large samples.

7 Use the normal approximation to compute probabilities for a binomial variable.

Outline

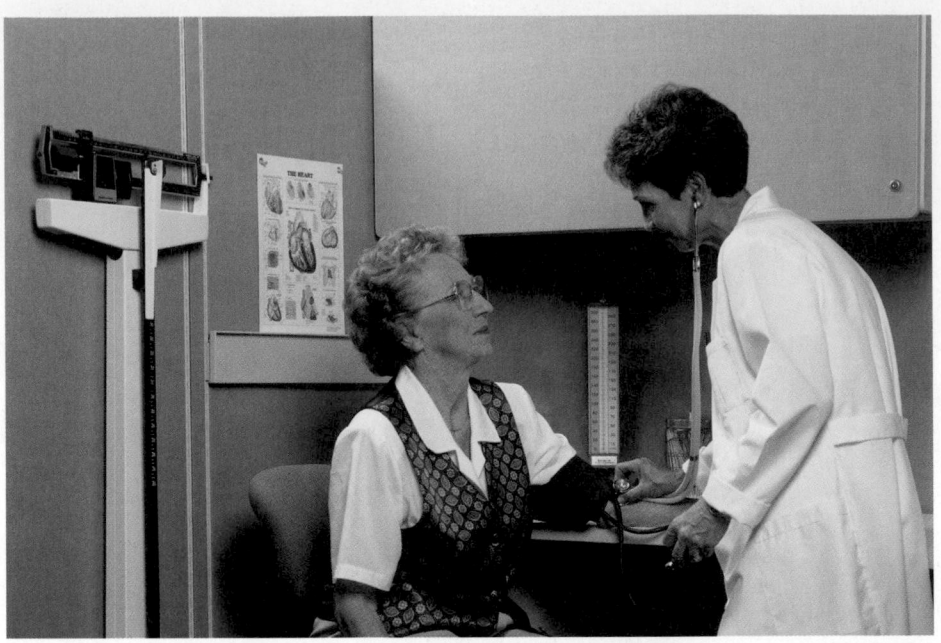

What Is Normal?

Medical researchers have determined so-called normal intervals for a person's blood pressure, cholesterol, triglycerides, and the like. For example, the normal range of systolic blood pressure is 110 to 140. The normal interval for a person's triglycerides is from 30 to 200 milligrams per deciliter (mg/dl). By measuring these variables, a physician can determine if a patient's vital statistics are within the normal interval or if some type of treatment is needed to correct a condition and avoid future illnesses. The question then is, How does one determine the so-called normal intervals? See Statistics Today—Revisited at the end of the chapter.

In this chapter, you will learn how researchers determine normal intervals for specific medical tests by using a normal distribution. You will see how the same methods are used to determine the lifetimes of batteries, the strength of ropes, and many other traits.

6–1

Introduction

Random variables can be either discrete or continuous. Discrete variables and their distributions were explained in Chapter 5. Recall that a discrete variable cannot assume all values between any two given values of the variables. On the other hand, a continuous variable can assume all values between any two given values of the variables. Examples of continuous variables are the heights of adult men, body temperatures of rats, and cholesterol levels of adults. Many continuous variables, such as the examples just mentioned, have distributions that are bell-shaped, and these are called *approximately normally distributed variables.* For example, if a researcher selects a random sample of 100 adult women, measures their heights, and constructs a histogram, the researcher gets a graph similar to the one shown in Figure 6–1(a). Now, if the researcher increases the sample size and decreases the width of the classes, the histograms will look like the ones shown in Figure 6–1(b) and (c). Finally, if it were possible to measure exactly the heights of all adult females in the United States and plot them, the histogram would approach what is called a *normal distribution,* shown in Figure 6–1(d). This distribution is also known as

Figure 6–1

Histograms for the Distribution of Heights of Adult Women

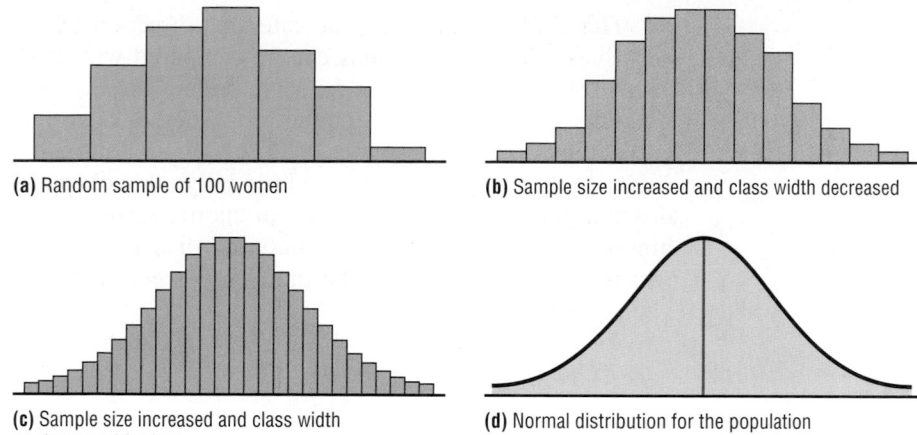

(a) Random sample of 100 women

(b) Sample size increased and class width decreased

(c) Sample size increased and class width decreased further

(d) Normal distribution for the population

Figure 6–2

Normal and Skewed Distributions

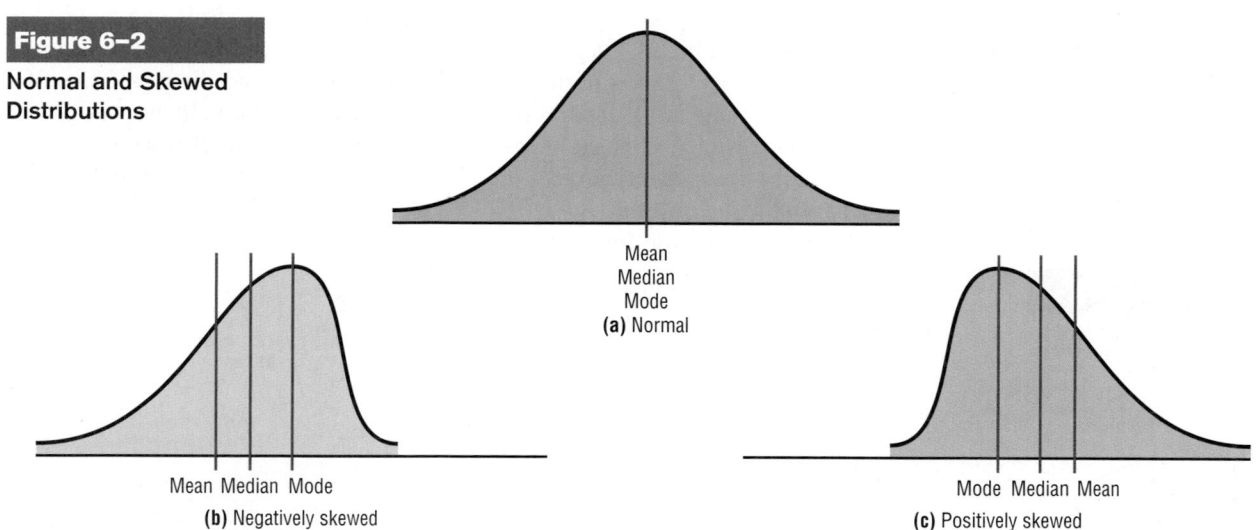

Mean
Median
Mode
(a) Normal

Mean Median Mode
(b) Negatively skewed

Mode Median Mean
(c) Positively skewed

a *bell curve* or a *Gaussian distribution,* named for the German mathematician Carl Friedrich Gauss (1777–1855), who derived its equation.

No variable fits a normal distribution perfectly, since a normal distribution is a theoretical distribution. However, a normal distribution can be used to describe many variables, because the deviations from a normal distribution are very small. This concept will be explained further in Section 6–2.

Objective 1

Identify distributions as symmetric or skewed.

When the data values are evenly distributed about the mean, a distribution is said to be a **symmetric distribution.** (A normal distribution is symmetric.) Figure 6–2(a) shows a symmetric distribution. When the majority of the data values fall to the left or right of the mean, the distribution is said to be *skewed.* When the majority of the data values fall to the right of the mean, the distribution is said to be a **negatively or left-skewed distribution.** The mean is to the left of the median, and the mean and the median are to the left of the mode. See Figure 6–2(b). When the majority of the data values fall to the left of the mean, a distribution is said to be a **positively or right-skewed distribution.** The mean falls to the right of the median, and both the mean and the median fall to the right of the mode. See Figure 6–2(c).

The "tail" of the curve indicates the direction of skewness (right is positive, left negative). These distributions can be compared with the ones shown in Figure 3–1 on page 109. Both types follow the same principles.

This chapter will present the properties of a normal distribution and discuss its applications. Then a very important fact about a normal distribution called the *central limit theorem* will be explained. Finally, the chapter will explain how a normal distribution curve can be used as an approximation to other distributions, such as the binomial distribution. Since a binomial distribution is a discrete distribution, a correction for continuity may be employed when a normal distribution is used for its approximation.

6–2   Properties of a Normal Distribution

Objective 2

Identify the properties of a normal distribution.

In mathematics, curves can be represented by equations. For example, the equation of the circle shown in Figure 6–3 is $x^2 + y^2 = r^2$, where r is the radius. A circle can be used to represent many physical objects, such as a wheel or a gear. Even though it is not possible to manufacture a wheel that is perfectly round, the equation and the properties of a circle can be used to study many aspects of the wheel, such as area, velocity, and acceleration. In a similar manner, the theoretical curve, called a *normal distribution curve,* can be used to study many variables that are not perfectly normally distributed but are nevertheless approximately normal.

The mathematical equation for a normal distribution is

$$y = \frac{e^{-(X-\mu)^2/(2\sigma^2)}}{\sigma\sqrt{2\pi}}$$

where

$e \approx 2.718$ ($\approx$ means "is approximately equal to")

$\pi \approx 3.14$

μ = population mean

σ = population standard deviation

Figure 6–3

Graph of a Circle and an Application

Circle

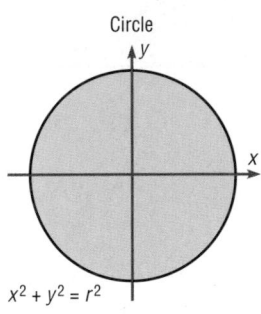

$x^2 + y^2 = r^2$

Wheel

This equation may look formidable, but in applied statistics, tables or technology is used for specific problems instead of the equation.

Another important consideration in applied statistics is that the area under a normal distribution curve is used more often than the values on the y axis. Therefore, when a normal distribution is pictured, the y axis is sometimes omitted.

Circles can be different sizes, depending on their diameters (or radii), and can be used to represent wheels of different sizes. Likewise, normal curves have different shapes and can be used to represent different variables.

The shape and position of a normal distribution curve depend on two parameters, the *mean* and the *standard deviation.* Each normally distributed variable has its own normal distribution curve, which depends on the values of the variable's mean and standard deviation. Figure 6–4(a) shows two normal distributions with the same mean values but different standard deviations. The larger the standard deviation, the more dispersed, or spread out, the distribution is. Figure 6–4(b) shows two normal distributions with the same standard deviation but with different means. These curves have the same shapes but are located at different positions on the x axis. Figure 6–4(c) shows two normal distributions with different means and different standard deviations.

Figure 6–4

Shapes of Normal Distributions

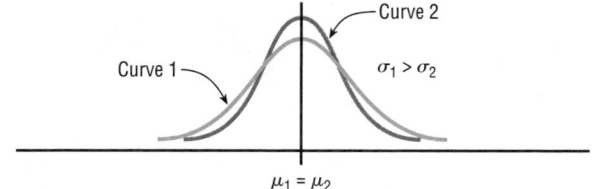

(a) Same means but different standard deviations

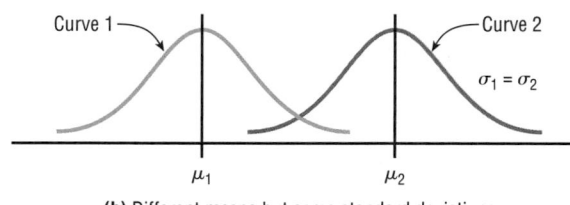

(b) Different means but same standard deviations

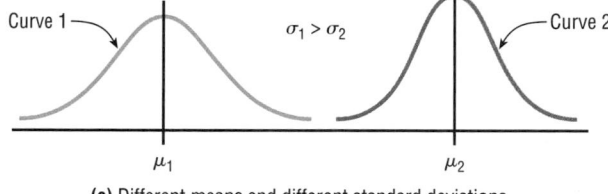

(c) Different means and different standard deviations

Historical Notes

The discovery of the equation for a normal distribution can be traced to three mathematicians. In 1733, the French mathematician Abraham DeMoivre derived an equation for a normal distribution based on the random variation of the number of heads appearing when a large number of coins were tossed. Not realizing any connection with the naturally occurring variables, he showed this formula to only a few friends. About 100 years later, two mathematicians, Pierre Laplace in France and Carl Gauss in Germany, derived the equation of the normal curve independently and without any knowledge of DeMoivre's work. In 1924, Karl Pearson found that DeMoivre had discovered the formula before Laplace or Gauss.

A **normal distribution** is a continuous, symmetric, bell-shaped distribution of a variable.

The properties of a normal distribution, including those mentioned in the definition, are explained next.

Summary of the Properties of the Theoretical Normal Distribution

1. A normal distribution curve is bell-shaped.
2. The mean, median, and mode are equal and are located at the center of the distribution.
3. A normal distribution curve is unimodal (i.e., it has only one mode).
4. The curve is symmetric about the mean, which is equivalent to saying that its shape is the same on both sides of a vertical line passing through the center.
5. The curve is continuous, that is, there are no gaps or holes. For each value of X, there is a corresponding value of Y.
6. The curve never touches the x axis. Theoretically, no matter how far in either direction the curve extends, it never meets the x axis—but it gets increasingly closer.
7. The total area under a normal distribution curve is equal to 1.00, or 100%. This fact may seem unusual, since the curve never touches the x axis, but one can prove it mathematically by using calculus. (The proof is beyond the scope of this textbook.)
8. The area under the part of a normal curve that lies within 1 standard deviation of the mean is approximately 0.68, or 68%; within 2 standard deviations, about 0.95, or 95%; and within 3 standard deviations, about 0.997, or 99.7%. See Figure 6–5, which also shows the area in each region.

The values given in number 8 of the summary follow the *empirical rule* for data given in Section 3–3.

One must know these properties in order to solve problems involving distributions that are approximately normal.

Figure 6–5

Areas Under a Normal
Distribution Curve

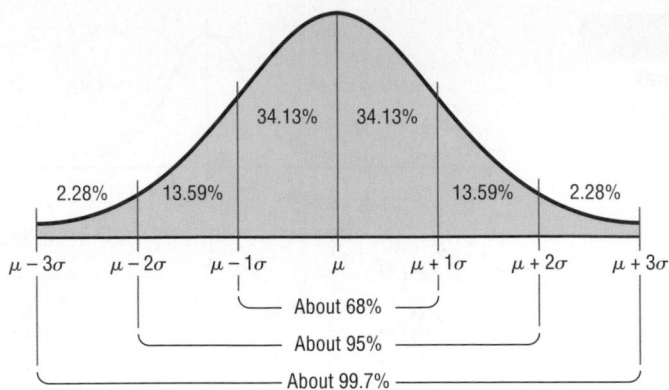

The Standard Normal Distribution

6–3

Since each normally distributed variable has its own mean and standard deviation, as stated earlier, the shape and location of these curves will vary. In practical applications, then, one would have to have a table of areas under the curve for each variable. To simplify this situation, statisticians use what is called the *standard normal distribution.*

Objective **3**

Find the area under the standard normal distribution, given various *z* values.

The **standard normal distribution** is a normal distribution with a mean of 0 and a standard deviation of 1.

The standard normal distribution is shown in Figure 6–6.

The values under the curve indicate the proportion of area in each section. For example, the area between the mean and 1 standard deviation above or below the mean is about 0.3413, or 34.13%.

The formula for the standard normal distribution is

$$y = \frac{e^{-z^2/2}}{\sqrt{2\pi}}$$

All normally distributed variables can be transformed into the standard normally distributed variable by using the formula for the standard score:

$$z = \frac{\text{value} - \text{mean}}{\text{standard deviation}} \qquad \text{or} \qquad z = \frac{X - \mu}{\sigma}$$

This is the same formula used in Section 3–4. The use of this formula will be explained in Section 6–4.

As stated earlier, the area under a normal distribution curve is used to solve practical application problems, such as finding the percentage of adult women whose height is between 5 feet 4 inches and 5 feet 7 inches, or finding the probability that a new battery will last longer than 4 years. Hence, the major emphasis of this section will be to show the procedure for finding the area under the standard normal distribution curve for any *z* value. The applications will be shown in Section 6–4. Once the *X* values are transformed by using the preceding formula, they are called *z* values. The **z value** is actually the number of standard deviations that a particular *X* value is away from the mean. Table E in Appendix C gives the area (to four decimal places) under the standard normal curve for any *z* value from 0 to 3.09.

Figure 6–6

Standard Normal
Distribution

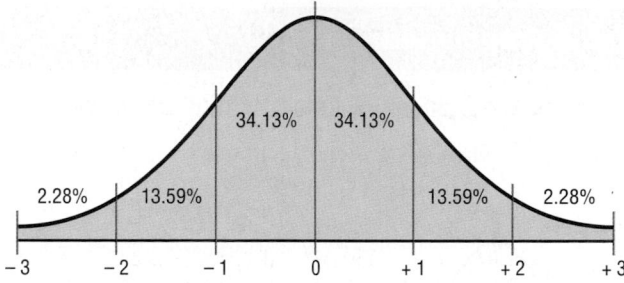

Finding Areas Under the Standard Normal Distribution Curve

For the solution of problems using the standard normal distribution, a four-step procedure is recommended with the use of the Procedure Table shown.

Step 1   Draw a picture.

Step 2   Shade the area desired.

Step 3   Find the correct figure in the following Procedure Table (the figure that is similar to the one you've drawn).

Step 4   Follow the directions given in the appropriate block of the Procedure Table to get the desired area.

There are seven basic types of problems and all seven are summarized in the Procedure Table. Note that this table is presented as an aid in understanding how to use the standard normal distribution table and in visualizing the problems. After learning the procedures, one should *not* find it necessary to refer to the procedure table for every problem.

Interesting Fact

Bell-shaped distributions occurred quite often in early coin-tossing and die-rolling experiments.

Procedure Table

Finding the Area Under the Standard Normal Distribution Curve

1. Between 0 and any *z* value:
    Look up the *z* value in the table to get the area.

2. In any tail:
    a. Look up the *z* value to get the area.
    b. Subtract the area from 0.5000.

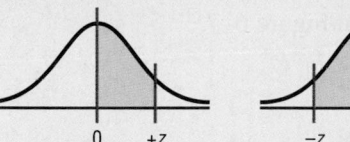

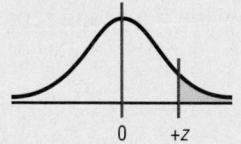

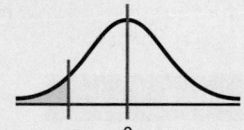

3. Between two *z* values on the same side of the mean:
    a. Look up both *z* values to get the areas.
    b. Subtract the smaller area from the larger area.

4. Between two *z* values on opposite sides of the mean:
    a. Look up both *z* values to get the areas.
    b. Add the areas.

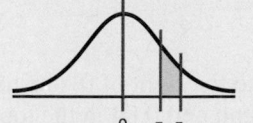

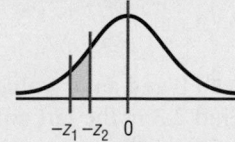

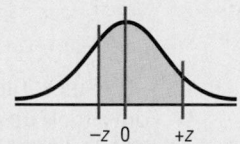

Procedure Table (concluded)

Finding the Area Under the Standard Normal Distribution Curve

5. To the left of any z value, where z is greater than the mean:

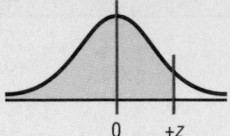

    a. Look up the z value to get the area.
    b. Add 0.5000 to the area.

6. To the right of any z value, where z is less than the mean:

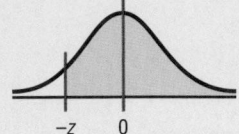

    a. Look up the z value in the table to get the area.
    b. Add 0.5000 to the area.

7. In any two tails:

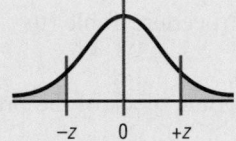

    a. Look up the z values in the table to get the areas.
    b. Subtract both areas from 0.5000.
    c. Add the answers.

Procedure
1. Draw the picture.
2. Shade the area desired.
3. Find the correct figure.
4. Follow the directions.

Note: Table E gives the area between 0 and any z value to the right of 0, and all areas are positive.

Situation 1   Find the area under the standard normal curve between 0 and any z value.

Example 6–1

Find the area under the standard normal distribution curve between $z = 0$ and $z = 2.34$.

Solution

Draw the figure and represent the area as shown in Figure 6–7.

Figure 6–7

Area Under the Standard Normal Distribution Curve for Example 6–1

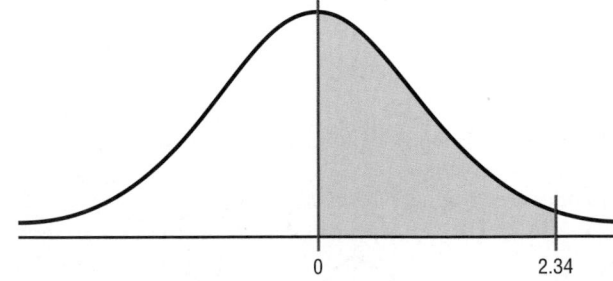

    Since Table E gives the area between 0 and any z value to the right of 0, one need only look up the z value in the table. Find 2.3 in the left column and 0.04 in the top row. The value where the column and row meet in the table is the answer, 0.4904. See Figure 6–8. Hence, the area is 0.4904, or 49.04%.

Figure 6–8

Using Table E in the Appendix for Example 6–1

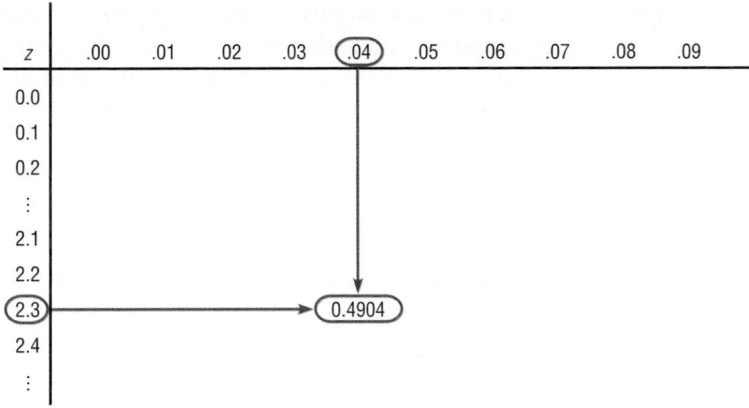

Example 6–2

Find the area between $z = 0$ and $z = 1.8$.

Solution

Draw the figure and represent the area as shown in Figure 6–9.

Figure 6–9

Area Under the Standard Normal Curve for Example 6–2

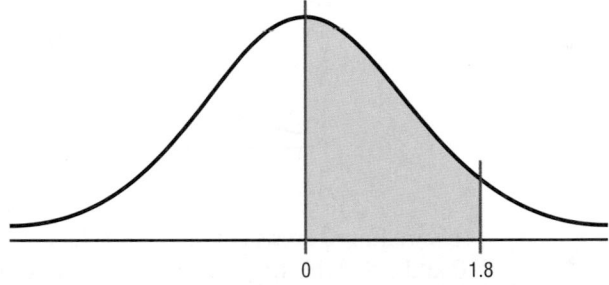

Find the area in Table E by finding 1.8 in the left column and 0.00 in the top row. The area is 0.4641, or 46.41%.

   Next, one must be able to find the areas for values that are not in Table E. This is done by using the properties of the normal distribution described in Section 6–2.

Example 6–3

Find the area between $z = 0$ and $z = -1.75$.

Solution

Represent the area as shown in Figure 6–10.

Figure 6–10

Area Under the Standard Normal Curve for Example 6–3

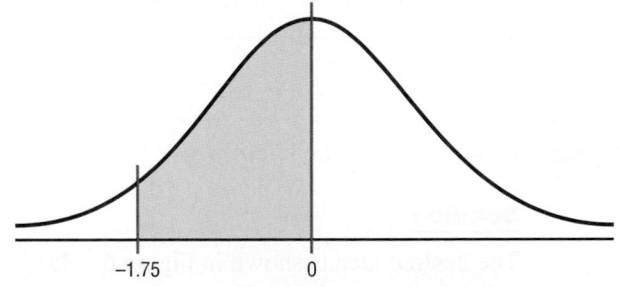

Table E does not give the areas for negative values of z. But since the normal distribution is symmetric about the mean, the area to the left of the mean (in this case, the mean is 0) is the same as the area to the right of the mean. Hence one need only look up the area for $z = +1.75$, which is 0.4599, or 45.99%. This solution is summarized in block 1 in the Procedure Table.

Remember that area is always a positive number, even if the z value is negative.

Situation 2 Find the area under the standard normal distribution curve in either tail.

Example 6–4	Find the area to the right of $z = 1.11$.

Solution

Draw the figure and represent the area as shown in Figure 6–11.

Figure 6–11

Area Under the Standard Normal Distribution Curve for Example 6–4

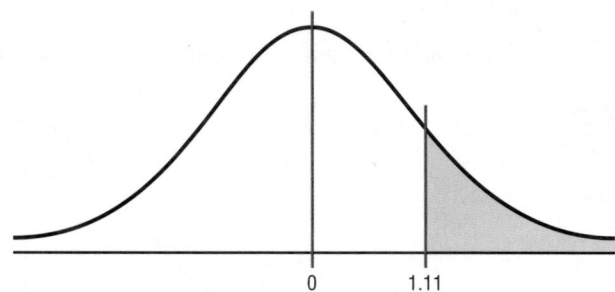

The required area is in the tail of the curve. Since Table E gives the area between $z = 0$ and $z = 1.11$, first find that area. Then subtract this value from 0.5000, since one-half of the area under the curve is to the right of $z = 0$. See Figure 6–12.

Figure 6–12

Finding the Area in the Tail of the Standard Normal Distribution Curve (Example 6–4)

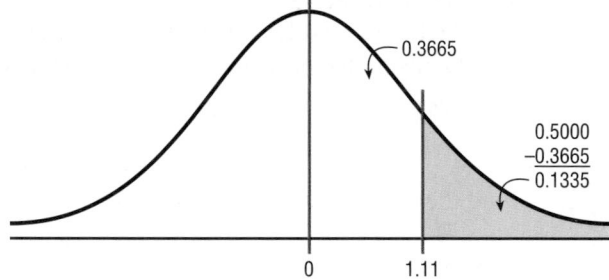

The area between $z = 0$ and $z = 1.11$ is 0.3665, and the area to the right of $z = 1.11$ is 0.1335, or 13.35%, obtained by subtracting 0.3665 from 0.5000.

Example 6–5	Find the area to the left of $z = -1.93$.

Solution

The desired area is shown in Figure 6–13.

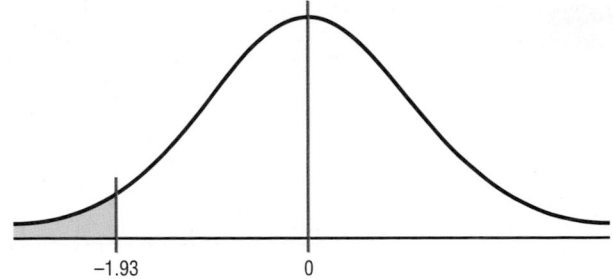

Figure 6–13

Area Under the
Standard Normal
Distribution Curve for
Example 6–5

Again, Table E gives the area for positive z values. But from the symmetric property of the normal distribution, the area to the left of -1.93 is the same as the area to the right of $z = +1.93$, as shown in Figure 6–14.

Now find the area between 0 and $+1.93$ and subtract it from 0.5000, as shown:

$$\begin{array}{r} 0.5000 \\ -0.4732 \\ \hline 0.0268, \text{ or } 2.68\% \end{array}$$

Figure 6–14

Comparison of Areas
to the Right of $+1.93$
and to the Left of
-1.93 (Example 6–5)

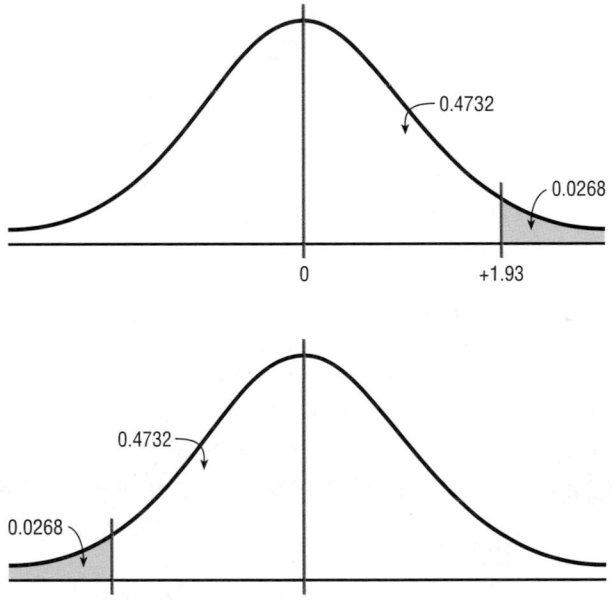

This procedure was summarized in block 2 of the Procedure Table.

Situation 3   Find the area under the standard normal distribution curve between any two z values on the same side of the mean.

Example 6–6   Find the area between $z = 2.00$ and $z = 2.47$.

Solution

The desired area is shown in Figure 6–15.

Figure 6–15

Area Under the Standard Normal Distribution Curve for Example 6–6

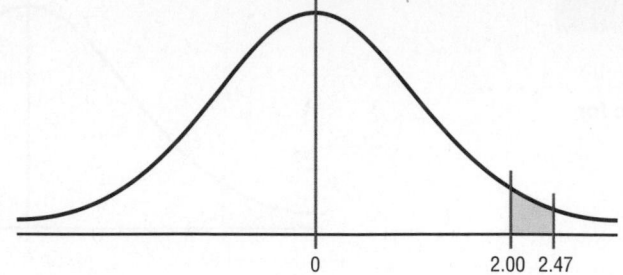

For this situation, look up the area from $z = 0$ to $z = 2.47$ and the area from $z = 0$ to $z = 2.00$. Then subtract the two areas, as shown in Figure 6–16.

Figure 6–16

Finding the Area Under the Standard Normal Distribution Curve for Example 6–6

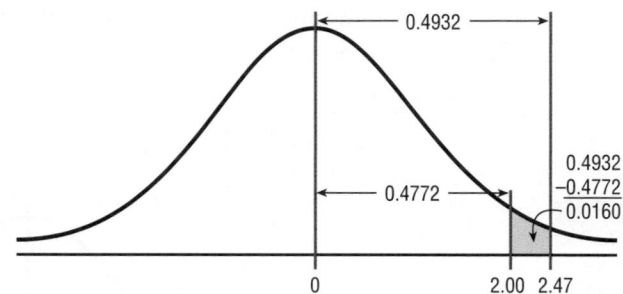

The area between $z = 0$ and $z = 2.47$ is 0.4932. The area between $z = 0$ and $z = 2.00$ is 0.4772. Hence, the desired area is $0.4932 - 0.4772 = 0.0160$, or 1.60%. This procedure is summarized in block 3 of the Procedure Table.

Two things should be noted here. First, the *areas,* not the z values, are subtracted. Subtracting the z values will yield an incorrect answer. Second, the procedure in Example 6–6 is used when both z values are on the same side of the mean.

Example 6–7

Find the area between $z = -2.48$ and $z = -0.83$.

Solution

The desired area is shown in Figure 6–17.

Figure 6–17

Area Under the Standard Normal Distribution Curve for Example 6–7

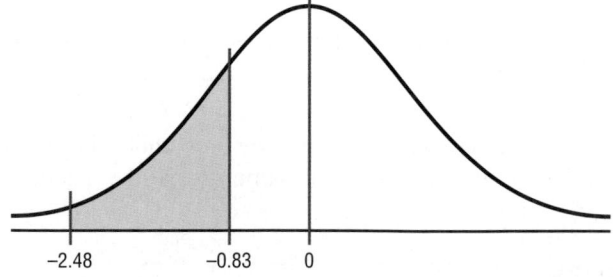

The area between $z = 0$ and $z = -2.48$ is 0.4934. The area between $z = 0$ and $z = -0.83$ is 0.2967. Subtracting yields $0.4934 - 0.2967 = 0.1967$, or 19.67%. This solution is summarized in block 3 of the Procedure Table.

Situation 4   Find the area under the standard normal distribution curve between any two z values on opposite sides of the mean.

Example 6–8

Find the area between $z = +1.68$ and $z = -1.37$.

Solution

The desired area is shown in Figure 6–18.

Figure 6–18

Area Under the
Standard Normal
Distribution Curve for
Example 6–8

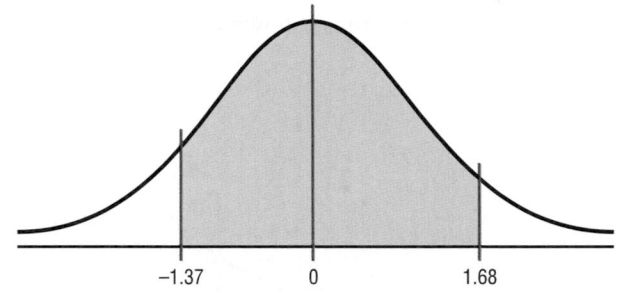

-1.37   0   1.68

    Now, since the two areas are on opposite sides of $z = 0$, one must find both areas and add them. The area between $z = 0$ and $z = 1.68$ is 0.4535. The area between $z = 0$ and $z = -1.37$ is 0.4147. Hence, the total area between $z = -1.37$ and $z = +1.68$ is $0.4535 + 0.4147 = 0.8682$, or 86.82%.

    This type of problem is summarized in block 4 of the Procedure Table.

Situation 5   Find the area under the standard normal distribution curve to the left of any z value, where z is greater than the mean.

Example 6–9

Find the area to the left of $z = 1.99$.

Solution

The desired area is shown in Figure 6–19.

Figure 6–19

Area Under the
Standard Normal
Distribution Curve for
Example 6–9

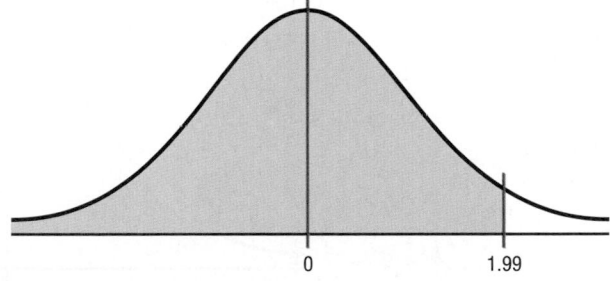

0   1.99

    Since Table E gives only the area between $z = 0$ and $z = 1.99$, one must add 0.5000 to the table area, since 0.5000 (one-half) of the total area lies to the left of $z = 0$. The area between $z = 0$ and $z = 1.99$ is 0.4767, and the total area is $0.4767 + 0.5000 = 0.9767$, or 97.67%.

    This solution is summarized in block 5 of the Procedure Table.

The same procedure is used when the z value is to the left of the mean, as shown in Example 6–10.

Situation 6   Find the area under the standard normal distribution curve to the right of any z value, where z is less than the mean.

| Example 6–10 | Find the area to the right of $z = -1.16$. |

Solution

The desired area is shown in Figure 6–20.

Figure 6–20

Area Under the Standard Normal Distribution Curve for Example 6–10

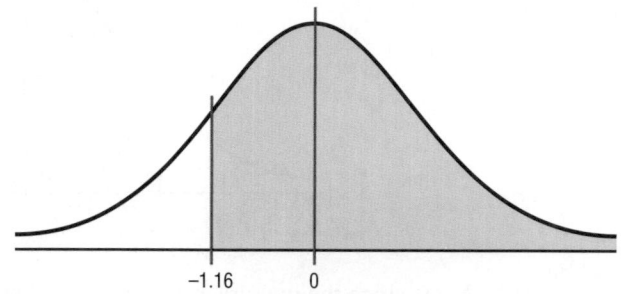

The area between $z = 0$ and $z = -1.16$ is 0.3770. Hence, the total area is $0.3770 + 0.5000 = 0.8770$, or 87.70%.

This type of problem is summarized in block 6 of the Procedure Table.

The final type of problem is that of finding the area in two tails. To solve it, find the area in each tail and add, as shown in Example 6–11.

Situation 7   Find the total area under the standard normal distribution curve in any two tails.

| Example 6–11 | Find the area to the right of $z = +2.43$ and to the left of $z = -3.01$. |

Solution

The desired area is shown in Figure 6–21.

Figure 6–21

Area Under the Standard Normal Distribution Curve for Example 6–11

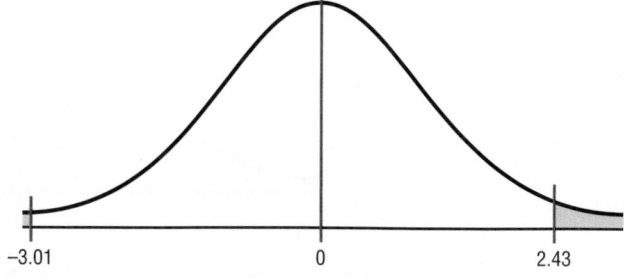

The area to the right of 2.43 is $0.5000 - 0.4925 = 0.0075$. The area to the left of $z = -3.01$ is $0.5000 - 0.4987 = 0.0013$. The total area, then, is $0.0075 + 0.0013 = 0.0088$, or 0.88%.

This solution is summarized in block 7 of the Procedure Table.

A Normal Distribution Curve as a Probability Distribution Curve

A normal distribution curve can be used as a probability distribution curve for normally distributed variables. Recall that a normal distribution is a *continuous distribution,* as opposed to a discrete probability distribution, as explained in Chapter 5. The fact that it is continuous means that there are no gaps in the curve. In other words, for every *z* value on the *x* axis, there is a corresponding height, or frequency, value.

The area under the standard normal distribution curve can also be thought of as a probability. That is, if it were possible to select any *z* value at random, the probability of choosing one, say, between 0 and 2.00 would be the same as the area under the curve between 0 and 2.00. In this case, the area is 0.4772. Therefore, the probability of randomly selecting any *z* value between 0 and 2.00 is 0.4772. The problems involving probability are solved in the same manner as the previous examples involving areas in this section. For example, if the problem is to find the probability of selecting a *z* value between 2.25 and 2.94, solve it by using the method shown in block 3 of the Procedure Table.

For probabilities, a special notation is used. For example, if the problem is to find the probability of any *z* value between 0 and 2.32, this probability is written as $P(0 < z < 2.32)$.

Note: In a continuous distribution, the probability of any exact *z* value is 0 since the area would be represented by a vertical line above the value. But vertical lines in theory have no area. So $P(a \leq z \leq b) = P(a < z < b)$.

Example 6–12	Find the probability for each.

    a. $P(0 < z < 2.32)$

    b. $P(z < 1.65)$

    c. $P(z > 1.91)$

Solution

    a. $P(0 < z < 2.32)$ means to find the area under the standard normal distribution curve between 0 and 2.32. Look up the area in Table E corresponding to $z = 2.32$. It is 0.4898, or 48.98%. The area is shown in Figure 6–22.

Figure 6–22

Area Under the
Standard Normal
Distribution Curve for
Part *a* of Example 6–12

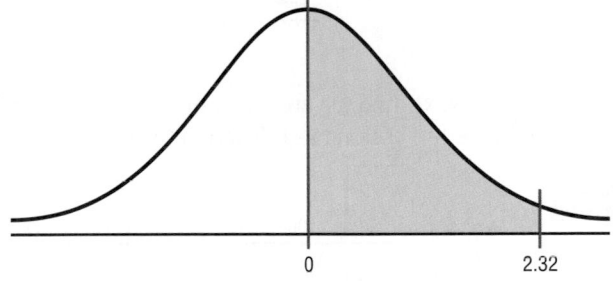

    b. $P(z < 1.65)$ is represented in Figure 6–23.

Figure 6–23

Area Under the
Standard Normal
Distribution Curve for
Part *b* of Example 6–12

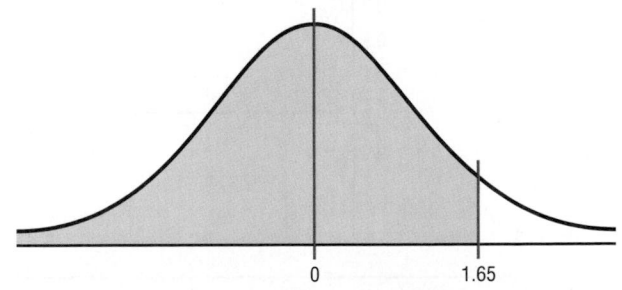

First, find the area between 0 and 1.65 in Table E. Then add it to 0.5000 to get $0.4505 + 0.5000 = 0.9505$, or 95.05%.

c. $P(z > 1.91)$ is shown in Figure 6–24.

Figure 6–24

Area Under the Standard Normal Distribution Curve for Part c of Example 6–12

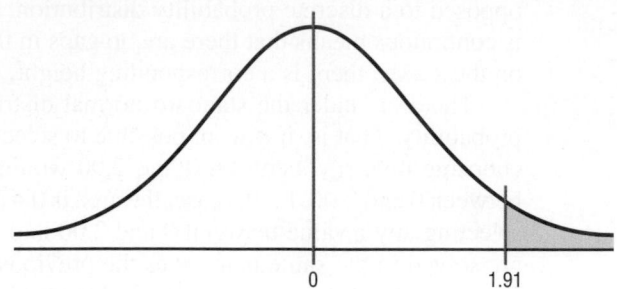

Since this area is a tail area, find the area between 0 and 1.91 and subtract it from 0.5000. Hence, $0.5000 - 0.4719 = 0.0281$, or 2.81%.

Sometimes, one must find a specific z value for a given area under the standard normal distribution curve. The procedure is to work backward, using Table E.

Example 6–13

Find the z value such that the area under the standard normal distribution curve between 0 and the z value is 0.2123.

Solution

Draw the figure. The area is shown in Figure 6–25.

Figure 6–25

Area Under the Standard Normal Distribution Curve for Example 6–13

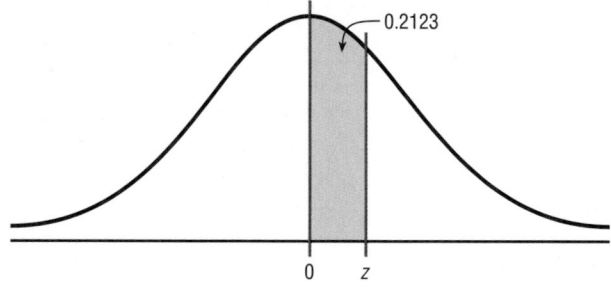

Next, find the area in Table E, as shown in Figure 6–26. Then read the correct z value in the left column as 0.5 and in the top row as 0.06, and add these two values to get 0.56.

Figure 6–26

Finding the z Value from Table E for Example 6–13

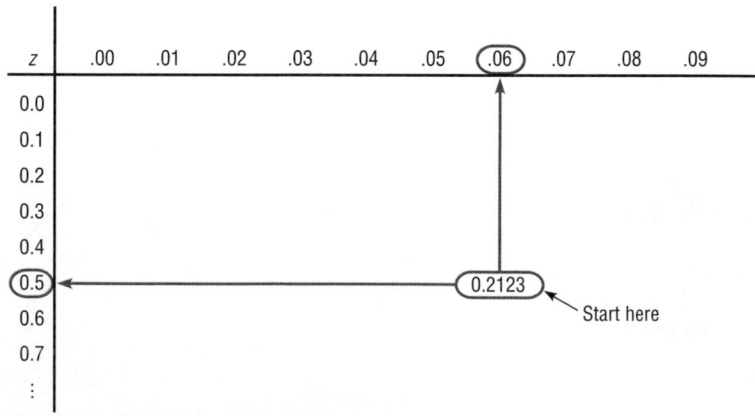

Figure 6–27

The Relationship Between Area and Probability

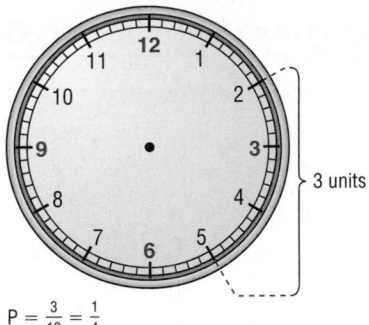

$$P = \frac{3}{12} = \frac{1}{4}$$

(a) Clock

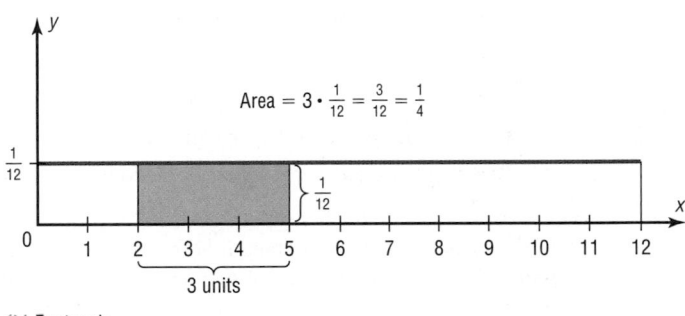

(b) Rectangle

If the exact area cannot be found, use the closest value. For example, if one wanted to find the z value for an area 0.4241, the closest area is 0.4236, which gives a z value of 1.43. See Table E in Appendix C.

The rationale for using an area under a continuous curve to determine a probability can be understood by considering the example of a watch that is powered by a battery. When the battery goes dead, what is the probability that the minute hand will stop somewhere between the numbers 2 and 5 on the face of the watch? In this case, the values of the variable constitute a continuous variable since the hour hand can stop anywhere on the dial's face between 0 and 12 (one revolution of the minute hand). Hence, the sample space can be considered to be 12 units long, and the distance between the numbers 2 and 5 is $5 - 2$, or 3 units. Hence, the probability that the minute hand stops on a number between 2 and 5 is $\frac{3}{12} = \frac{1}{4}$. See Figure 6–27(a).

The problem could also be solved by using a graph of a continuous variable. Let us assume that since the watch can stop anytime at random, the values where the minute hand would land are spread evenly over the range of 0 through 12. The graph would then consist of a *continuous uniform distribution* with a range of 12 units. Now if we require the area under the curve to be 1 (like the area under the standard normal distribution), the height of the rectangle formed by the curve and the x axis would need to be $\frac{1}{12}$. The reason is that the area of a rectangle is equal to the base times the height. If the base is 12 units long, then the height would have to be $\frac{1}{12}$ since $12 \cdot \frac{1}{12} = 1$.

The area of the rectangle with a base from 2 through 5 would be $3 \cdot \frac{1}{12}$, or $\frac{1}{4}$. See Figure 6–27(b). Notice that the area of the small rectangle is the same as the probability found previously. Hence the area of this rectangle corresponds to the probability of this event. The same reasoning can be applied to the standard normal distribution curve shown in Example 6–13.

Finding the area under the standard normal distribution curve is the first step in solving a wide variety of practical applications in which the variables are normally distributed. Some of these applications will be presented in Section 6–4.

Applying the Concepts **6-3**

Assessing Normality

Many times in statistics it is necessary to see if a distribution of data values is approximately normally distributed. There are special techniques that can be used. One technique is to draw a histogram for the data and see if it is approximately bell-shaped. (*Note:* It does not have to be exactly symmetric to be bell-shaped.)

The numbers of branches of the 50 top libraries are shown.

67	84	80	77	97	59	62	37	33	42
36	54	18	12	19	33	49	24	25	22
24	29	9	21	21	24	31	17	15	21
13	19	19	22	22	30	41	22	18	20
26	33	14	14	16	22	26	10	16	24

Source: *The World Almanac and Book of Facts.*

1. Construct a frequency distribution for the data.

2. Construct a histogram for the data.

3. Describe the shape of the histogram.

4. Based on your answer to question 3, do you feel that the distribution is approximately normal?

In addition to the histogram, distributions that are approximately normal have about 68% of the values fall within 1 standard deviation of the mean, about 95% of the data values fall within 2 standard deviations of the mean, and almost 100% of the data values fall within 3 standard deviations of the mean. (See Figure 6–5.)

5. Find the mean and standard deviation for the data.

6. What percent of the data values fall within 1 standard deviation of the mean?

7. What percent of the data values fall within 2 standard deviations of the mean?

8. What percent of data values fall within 3 standard deviations of the mean?

9. How do your answers to questions 6, 7, and 8 compare to 68, 95, and 100%, respectively?

10. Does your answer help support the conclusion you reached in question 4? Explain.

(More techniques for assessing normality are explained in Section 6–4.)
See page 344 for the answers.

Exercises 6-3

1. What are the characteristics of a normal distribution?

2. Why is the standard normal distribution important in statistical analysis?

3. What is the total area under the standard normal distribution curve?

4. What percentage of the area falls below the mean? Above the mean?

5. About what percentage of the area under the normal distribution curve falls within 1 standard deviation above and below the mean? 2 standard deviations? 3 standard deviations?

For Exercises 6 through 25, find the area under the standard normal distribution curve.

6. Between $z = 0$ and $z = 1.66$

7. Between $z = 0$ and $z = 0.75$

8. Between $z = 0$ and $z = -0.35$

9. Between $z = 0$ and $z = -2.07$

10. To the right of $z = 1.10$

11. To the right of $z = 0.23$

12. To the left of $z = -0.48$

13. To the left of $z = -1.43$

14. Between $z = 1.23$ and $z = 1.90$

15. Between $z = 0.79$ and $z = 1.28$

16. Between $z = -0.96$ and $z = -0.36$

17. Between $z = -1.56$ and $z = -1.83$

18. Between $z = 0.24$ and $z = -1.12$

19. Between $z = 2.47$ and $z = -1.03$

20. To the left of $z = 1.31$

21. To the left of $z = 2.11$

22. To the right of $z = -1.92$

23. To the right of $z = -0.15$

24. To the left of $z = -2.15$ and to the right of $z = 1.62$

25. To the right of $z = 1.92$ and to the left of $z = -0.44$

In Exercises 26 through 39, find probabilities for each, using the standard normal distribution.

26. $P(0 < z < 1.69)$

27. $P(0 < z < 0.67)$

28. $P(-1.23 < z < 0)$

29. $P(-1.57 < z < 0)$

30. $P(z > 1.16)$

31. $P(z > 2.83)$

32. $P(z < -1.77)$

33. $P(z < -1.21)$

34. $P(-0.05 < z < 1.10)$

35. $P(-2.46 < z < 1.74)$

36. $P(1.12 < z < 1.43)$

37. $P(1.46 < z < 2.97)$

38. $P(z > -1.39)$

39. $P(z < 1.42)$

For Exercises 40 through 45, find the z value that corresponds to the given area.

40.

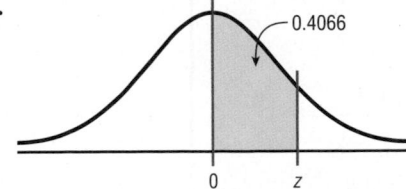

41.

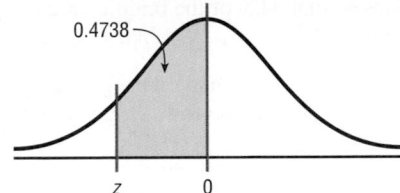

42.

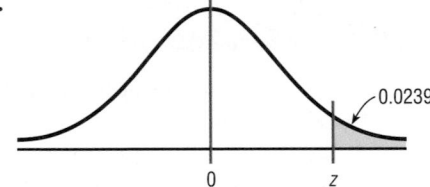

43.

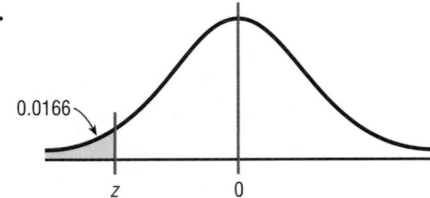

44.

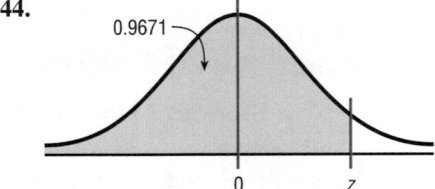

45.

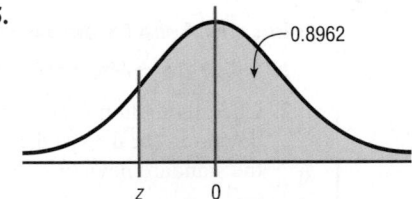

46. Find the z value to the right of the mean so that

   a. 53.98% of the area under the distribution curve lies to the left of it.

   b. 71.90% of the area under the distribution curve lies to the left of it.

   c. 96.78% of the area under the distribution curve lies to the left of it.

47. Find the z value to the left of the mean so that

   a. 98.87% of the area under the distribution curve lies to the right of it.

   b. 82.12% of the area under the distribution curve lies to the right of it.

   c. 60.64% of the area under the distribution curve lies to the right of it.

48. Find two z values so that 44% of the middle area is bounded by them.

49. Find two z values, one positive and one negative, so that the areas in the two tails total the following values.

a. 5%
b. 10%
c. 1%

Extending the Concepts

50. In the standard normal distribution, find the values of z for the 75th, 80th, and 92nd percentiles.

51. Find $P(-1 < z < 1)$, $P(-2 < z < 2)$, and $P(-3 < z < 3)$. How do these values compare with the empirical rule?

52. Find z_0 such that $P(z > z_0) = 0.1234$.

53. Find z_0 such that $P(-1.2 < z < z_0) = 0.8671$.

54. Find z_0 such that $P(z_0 < z < 2.5) = 0.7672$.

55. Find z_0 such that the area between z_0 and $z = -0.5$ is 0.2345 (two answers).

56. Find z_0 such that $P(-z_0 < z < z_0) = 0.86$.

57. Find the equation for the standard normal distribution by substituting 0 for μ and 1 for σ in the equation

$$y = \frac{e^{-(X-\mu)^2/(2\sigma^2)}}{\sigma\sqrt{2\pi}}$$

58. Graph by hand the standard normal distribution by using the formula derived in Exercise 57. Let $\pi \approx 3.14$ and $e \approx 2.718$. Use X values of -2, -1.5, -1, -0.5, 0, 0.5, 1, 1.5, and 2. (Use a calculator to compute the y values.)

Technology *Step by Step*

MINITAB
Step by Step

The Standard Normal Distribution

It is possible to determine the height of the density curve given a value of z, the cumulative area given a value of z, or a z value given a cumulative area. Examples are from Table E in Appendix C.

Find the Area to the Left of $z = 1.39$

1. Select **Calc>Probability Distributions>Normal.** There are three options.

2. Click the button for Cumulative probability. In the center section, the mean and standard deviation for the standard normal distribution are the defaults. The mean should be **0,** and the standard deviation should be **1.**

3. Click the button for Input Constant, then click inside the text box and type in **1.39.** Leave the storage box empty.

4. Click [OK].

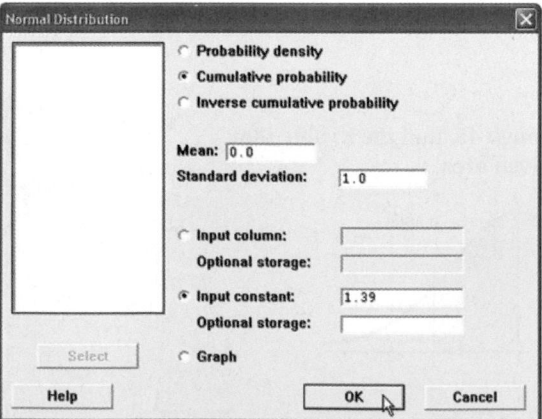

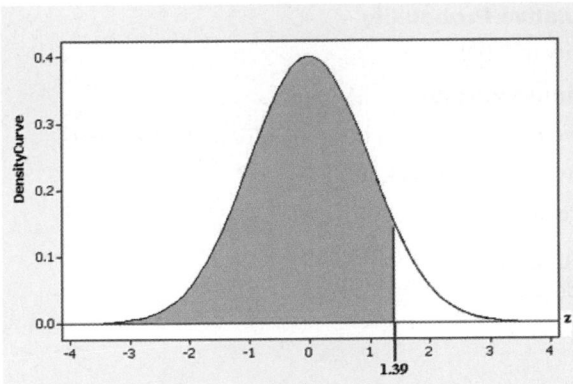

Cumulative Distribution Function

Normal with mean = 0 and standard deviation = 1

$$x \; P(X <= x)$$

1.39   0.917736

The graph is not shown in the output.

The session window displays the result, 0.917736. If you choose the optional storage, type in a variable name such as **K1.** The result will be stored in the constant and will not be in the session window.

Find the Area to the Right of −2.06

1. Select **Calc>Probability Distributions>Normal.**

2. Click the button for Cumulative probability.

3. Click the button for Input Constant, then enter **−2.06** in the text box. Do not forget the minus sign.

4. Click in the text box for Optional storage and type **K1.**

5. Click [OK]. The area to the left of −2.06 is stored in K1 but not displayed in the session window.

   To determine the area to the right of the *z* value, subtract this constant from 1, then display the result.

6. Select **Calc>Calculator.**

   a) Type **K2** in the text box for Store result in:.

   b) Type in the expression **1 − K1,** then click [OK].

7. Select **Data>Display Data.** Drag the mouse over K1 and K2, then click [Select] and [OK].

   The results will be in the session window and stored in the constants.

Data Display

K1   0.0196993

K2   0.980301

8. To see the constants and other information about the worksheet, click the Project Manager icon. In the left pane click on the green worksheet icon, and then click the constants folder. You should see all constants and their values in the right pane of the Project Manager.

9. For the third example calculate the two probabilities and store them in K1 and K2.

10. Use the calculator to subtract K1 from K2 and store in K3.

    The calculator and project manager windows are shown.

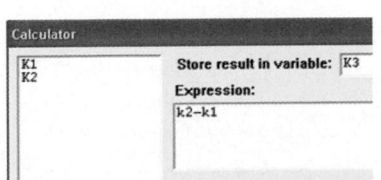

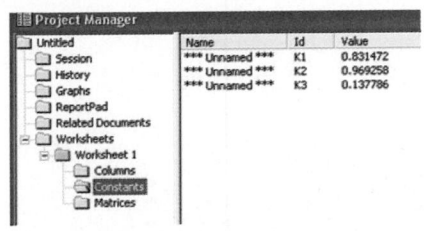

Calculate a *z* Value Given the Cumulative Probability

Find the *z* value for a cumulative probability of 0.025.

1. Select **Calc>Probability Distributions>Normal.**
2. Click the option for Inverse cumulative probability, then the option for Input constant.
3. In the text box type **.025**, the cumulative area, then click [OK].
4. In the dialog box, the *z* value will be returned, -1.960.

Inverse Cumulative Distribution Function
Normal with mean = 0 and standard deviation = 1
P (X <= x)   x
     0.025 -1.95996

In the session window *z* is -1.95996.

TI–83 Plus or TI–84 Plus
Step by Step

```
normalcdf(1.11,E
99)
       .1334995565
normalcdf(-E99,-
1.93)
       .0268033499
∎
```

```
normalcdf(2,2.47
)
       .0159944012
invNorm(.7123)
       .560116461
∎
```

Standard Normal Random Variables

To find the probability for a standard normal random variable:
Press **2nd [DISTR],** then **2** for normalcdf(
The form is normalcdf(lower *z* score, upper *z* score).
Use E99 for ∞ (infinity) and −E99 for −∞ (negative infinity). Press **2nd [EE]** to get E.

Example: Area to the right of $z = 1.11$ (Example 6–4 from the text)
normalcdf(1.11,E99)

Example: Area to the left of $z = -1.93$ (Example 6–5 from the text)
normalcdf(−E99,−1.93)

Example: Area between $z = 2.00$ and $z = 2.47$ (Example 6–6 from the text)
normalcdf(2.00,2.47)

To find the percentile for a standard normal random variable:
Press **2nd [DISTR],** then **3** for the invNorm(
The form is invNorm(area to the left of *z* score)

Example: Find the *z* score such that the area under the standard normal curve to the left of it is 0.7123 (Example 6–13 from the text)
invNorm(.7123)

Excel
Step by Step

The Normal Distribution

To find area under the standard normal curve between two *z* values: $P(-1.23 < z < 2.54)$

1. Open a new worksheet and select a blank cell.
2. Click the f_x icon from the toolbar to call up the function list.
3. Select **NORMSDIST** from the Statistical function category.
4. Enter **−1.23** in the dialog box and click [OK]. This gives the area to the left of -1.23.
5. Select an adjacent blank cell, and repeat steps 1 through 4 for **2.54.**
6. To find the area between -1.23 and 2.54, select another blank cell and subtract the smaller area from the larger area.

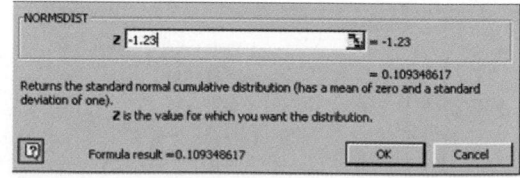

The area between the two values is the answer, 0.885109.

To find a z score corresponding to a cumulative area: $P(Z<=z) = 0.0250$

1. Click the f_x icon and select the **Statistical** function category.

2. Select the **NORMSINV** function and enter **0.0250.**

3. Click [OK].

The z score whose cumulative area is 0.0250 is the answer, -1.96.

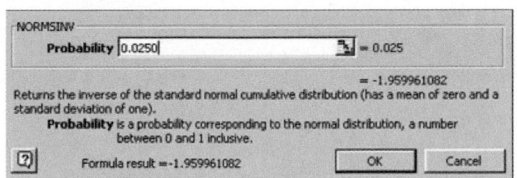

6–4   Applications of the Normal Distribution

Objective 4

Find probabilities for a normally distributed variable by transforming it into a standard normal variable.

The standard normal distribution curve can be used to solve a wide variety of practical problems. The only requirement is that the variable be normally or approximately normally distributed. There are several mathematical tests to determine whether a variable is normally distributed. See the Critical Thinking Challenge on page 342. For all the problems presented in this chapter, one can assume that the variable is normally or approximately normally distributed.

To solve problems by using the standard normal distribution, transform the original variable to a standard normal distribution variable by using the formula

$$z = \frac{\text{value} - \text{mean}}{\text{standard deviation}} \qquad \text{or} \qquad z = \frac{X - \mu}{\sigma}$$

This is the same formula presented in Section 3–4. This formula transforms the values of the variable into standard units or z values. Once the variable is transformed, then the Procedure Table and Table E in Appendix C can be used to solve problems.

For example, suppose that the scores for a standardized test are normally distributed, have a mean of 100, and have a standard deviation of 15. When the scores are transformed to z values, the two distributions coincide, as shown in Figure 6–28. (Recall that the z distribution has a mean of 0 and a standard deviation of 1.)

Figure 6–28

Test Scores and Their Corresponding z Values

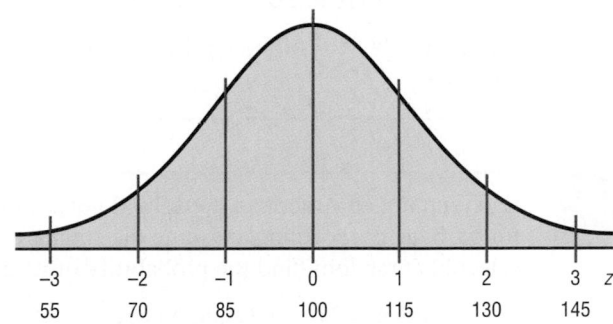

To solve the application problems in this section, transform the values of the variable to z values and then find the areas under the standard normal distribution, as shown in Section 6–3.

Example 6–14

The mean number of hours an American worker spends on the computer is 3.1 hours per workday. Assume the standard deviation is 0.5 hour. Find the percentage of workers who spend less than 3.5 hours on the computer. Assume the variable is normally distributed.

Source: USA TODAY.

Solution

Step 1   Draw the figure and represent the area as shown in Figure 6–29.

Figure 6–29

Area Under a Normal Curve for Example 6–14

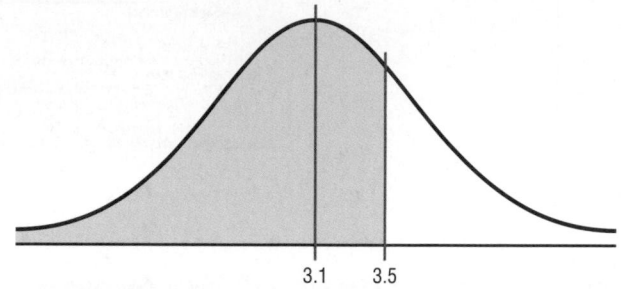

Step 2   Find the z value corresponding to 3.5.

$$z = \frac{X - \mu}{\sigma} = \frac{3.5 - 3.1}{0.5} = 0.80$$

Hence, 3.5 is 0.8 standard deviation above the mean of 3.1, as shown for the z distribution in Figure 6–30.

Figure 6–30

Area and z Values for Example 6–14

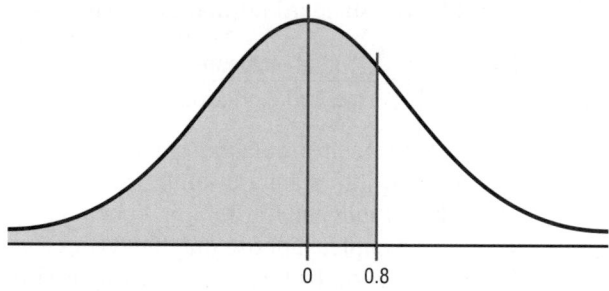

Step 3   Find the area by using Table E. The area between $z = 0$ and $z = 0.8$ is 0.2881. Since the area under the curve to the left of $z = 0.8$ is desired, add 0.5000 to 0.2881 (0.5000 + 0.2881 = 0.7881).

Therefore, 78.81% of the workers spend less than 3.5 hours per workday on the computer.

Example 6–15

Each month, an American household generates an average of 28 pounds of newspaper for garbage or recycling. Assume the standard deviation is 2 pounds. If a household is selected at random, find the probability of its generating

a. Between 27 and 31 pounds per month.

b. More than 30.2 pounds per month.

Assume the variable is approximately normally distributed.

Source: Michael D. Shook and Robert L. Shook, The Book of Odds.

Solution a

Step 1 Draw the figure and represent the area. See Figure 6–31.

Figure 6–31

Area Under a Normal Curve for Part *a* of Example 6–15

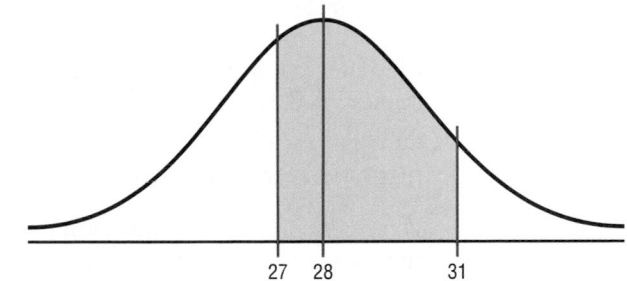

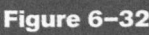

Historical Note

Astronomers in the late 1700s and the 1800s used the principles underlying the normal distribution to correct measurement errors that occurred in charting the positions of the planets.

Step 2 Find the two z values.

$$z_1 = \frac{X - \mu}{\sigma} = \frac{27 - 28}{2} = -\frac{1}{2} = -0.5$$

$$z_2 = \frac{X - \mu}{\sigma} = \frac{31 - 28}{2} = \frac{3}{2} = 1.5$$

Step 3 Find the appropriate area, using Table E. The area between $z = 0$ and $z = -0.5$ is 0.1915. The area between $z = 0$ and $z = 1.5$ is 0.4332. Add 0.1915 and 0.4332 (0.1915 + 0.4332 = 0.6247). Thus, the total area is 62.47%. See Figure 6–32.

Figure 6–32

Area and *z* Values for Part *a* of Example 6–15

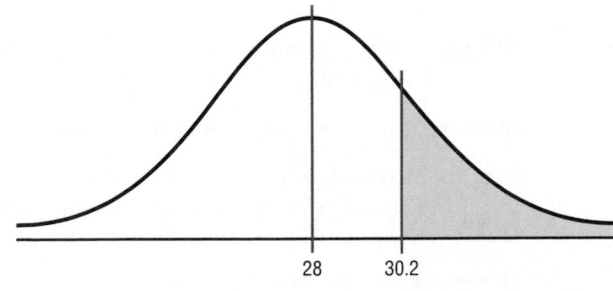

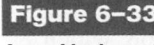

Hence, the probability that a randomly selected household generates between 27 and 31 pounds of newspapers per month is 62.47%.

Solution b

Step 1 Draw the figure and represent the area, as shown in Figure 6–33.

Figure 6–33

Area Under a Normal Curve for Part *b* of Example 6–15

Step 2   Find the z value for 30.2.

$$z = \frac{X - \mu}{\sigma} = \frac{30.2 - 28}{2} = \frac{2.2}{2} = 1.1$$

Step 3   Find the appropriate area. The area between $z = 0$ and $z = 1.1$ obtained from Table E is 0.3643. Since the desired area is in the right tail, subtract 0.3643 from 0.5000.

$$0.5000 - 0.3643 = 0.1357$$

Hence, the probability that a randomly selected household will accumulate more than 30.2 pounds of newspapers is 0.1357, or 13.57%.

A normal distribution can also be used to answer questions of "How many?" This application is shown in Example 6–16.

Example 6–16

The American Automobile Association reports that the average time it takes to respond to an emergency call is 25 minutes. Assume the variable is approximately normally distributed and the standard deviation is 4.5 minutes. If 80 calls are randomly selected, approximately how many will be responded to in less than 15 minutes?

Source: Michael D. Shook and Robert L. Shook, *The Book of Odds.*

Solution

To solve the problem, find the area under a normal distribution curve to the left of 15.

Step 1   Draw a figure and represent the area as shown in Figure 6–34.

Figure 6–34

Area Under a
Normal Curve for
Example 6–16

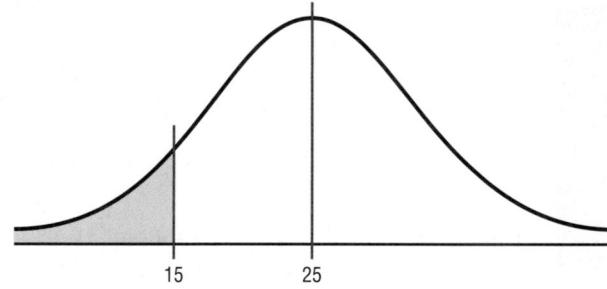

Step 2   Find the z value for 15.

$$z = \frac{X - \mu}{\sigma} = \frac{15 - 25}{4.5} = -2.22$$

Step 3   Find the appropriate area. The area obtained from Table E is 0.4868, which corresponds to the area between $z = 0$ and $z = -2.22$. Use $+2.22$.

Step 4   Subtract 0.4868 from 0.5000 to get 0.0132.

Step 5   To find how many calls will be made in less than 15 minutes, multiply the sample size 80 by 0.0132 to get 1.056. Hence, 1.056, or approximately 1, call will be responded to in under 15 minutes.

Note: For problems using percentages, be sure to change the percentage to a decimal before multiplying. Also, round the answer to the nearest whole number, since it is not possible to have 1.056 calls.

Finding Data Values Given Specific Probabilities

A normal distribution can also be used to find specific data values for given percentages. This application is shown in Example 6–17.

Example 6–17	To qualify for a police academy, candidates must score in the top 10% on a general abilities test. The test has a mean of 200 and a standard deviation of 20. Find the lowest possible score to qualify. Assume the test scores are normally distributed.

Objective **5**

Find specific data values for given percentages, using the standard normal distribution.

Solution

Since the test scores are normally distributed, the test value X that cuts off the upper 10% of the area under a normal distribution curve is desired. This area is shown in Figure 6–35.

Figure 6–35

Area Under a Normal Curve for Example 6–17

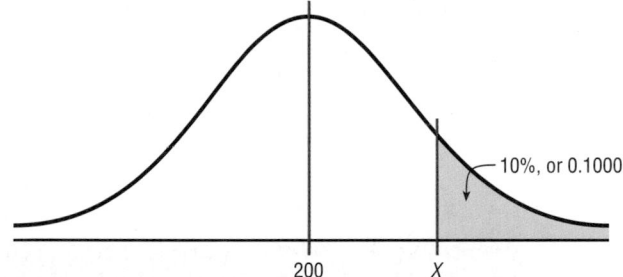

Work backward to solve this problem

Step 1 Subtract 0.1000 from 0.5000 to get the area under the normal distribution between 200 and X: $0.5000 - 0.1000 = 0.4000$.

Step 2 Find the z value that corresponds to an area of 0.4000 by looking up 0.4000 in the area portion of Table E. If the specific value cannot be found, use the closest value—in this case 0.3997, as shown in Figure 6–36. The corresponding z value is 1.28. (If the area falls exactly halfway between two z values, use the larger of the two z values. For example, the area 0.4500 falls halfway between 0.4495 and 0.4505. In this case use 1.65 rather than 1.64 for the z value.)

Figure 6–36

Finding the z Value from Table E (Example 6–17)

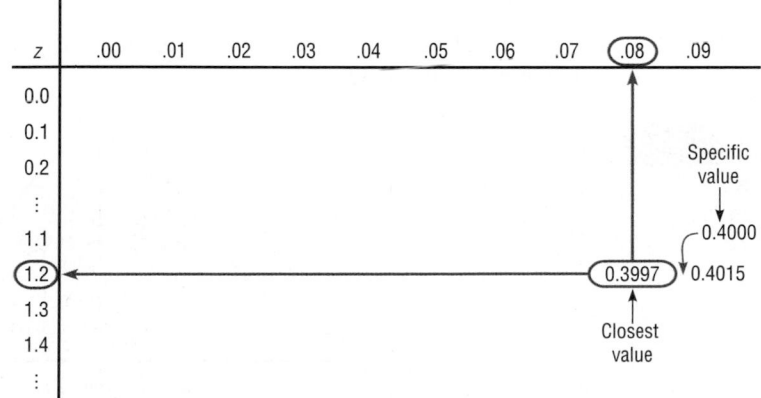

Step 3   Substitute in the formula $z = (X - \mu)/\sigma$ and solve for X.

$$1.28 = \frac{X - 200}{20}$$

$$(1.28)(20) + 200 = X$$

$$25.60 + 200 = X$$

$$225.60 = X$$

$$226 = X$$

A score of 226 should be used as a cutoff. Anybody scoring 226 or higher qualifies.

Instead of using the formula shown in step 3, one can use the formula $X = z \cdot \sigma + \mu$. This is obtained by solving

$$z = \frac{X - \mu}{\sigma}$$

for X as shown.

$$z \cdot \sigma = X - \mu \qquad \text{Multiply both sides by } \sigma.$$

$$z \cdot \sigma + \mu = X \qquad \text{Add } \mu \text{ to both sides.}$$

$$X = z \cdot \sigma + \mu \qquad \text{Exchange both sides of the equation.}$$

Formula for Finding X

When one must find the value of X, the following formula can be used:

$$X = z \cdot \sigma + \mu$$

Example 6–18

For a medical study, a researcher wishes to select people in the middle 60% of the population based on blood pressure. If the mean systolic blood pressure is 120 and the standard deviation is 8, find the upper and lower readings that would qualify people to participate in the study.

Solution

Assume that blood pressure readings are normally distributed; then cutoff points are as shown in Figure 6–37.

Figure 6–37

Area Under a Normal Curve for Example 6–18

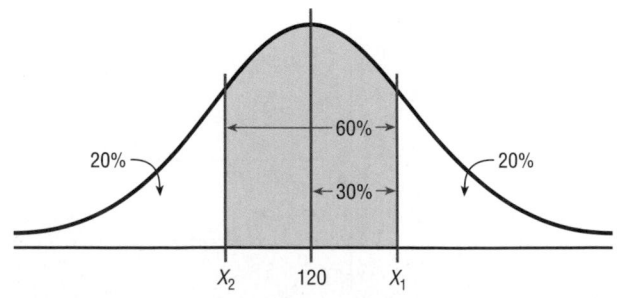

Note that two values are needed, one above the mean and one below the mean. Find the value to the right of the mean first. The closest z value for an area of 0.3000 is 0.84. Substituting in the formula $X = z\sigma + \mu$, one gets

$$X_1 = z\sigma + \mu = (0.84)(8) + 120 = 126.72$$

On the other side, $z = -0.84$; hence,

$$X_2 = (-0.84)(8) + 120 = 113.28$$

Therefore, the middle 60% will have blood pressure readings of $113.28 < X < 126.72$.

As shown in this section, a normal distribution is a useful tool in answering many questions about variables that are normally or approximately normally distributed.

Determining Normality

A normally shaped or bell-shaped distribution is only one of many shapes that a distribution can assume; however, it is very important since many statistical methods require that the distribution of values (shown in subsequent chapters) be normally or approximately normally shaped.

There are several ways statisticians check for normality. The easiest way is to draw a histogram for the data and check its shape. If the histogram is not approximately bell-shaped, then the data are not normally distributed.

Skewness can be checked by using Pearson's index PI of skewness. The formula is

$$PI = \frac{3(\bar{X} - \text{median})}{s}$$

If the index is greater than or equal to $+1$ or less than or equal to -1, it can be concluded that the data are significantly skewed.

In addition, the data should be checked for outliers by using the method shown in Chapter 3, page 143. Even one or two outliers can have a big effect on normality.

Examples 6–19 and 6–20 show how to check for normality.

Example 6–19

 A survey of 18 high-technology firms showed the number of days' inventory they had on hand. Determine if the data are approximately normally distributed.

5	29	34	44	45	63	68	74	74
81	88	91	97	98	113	118	151	158

Source: USA TODAY.

Solution

Step 1 Construct a frequency distribution and draw a histogram for the data, as shown in Figure 6–38.

Class	Frequency
5–29	2
30–54	3
55–79	4
80–104	5
105–129	2
130–154	1
155–179	1

Figure 6–38

Histogram for
Example 6–19

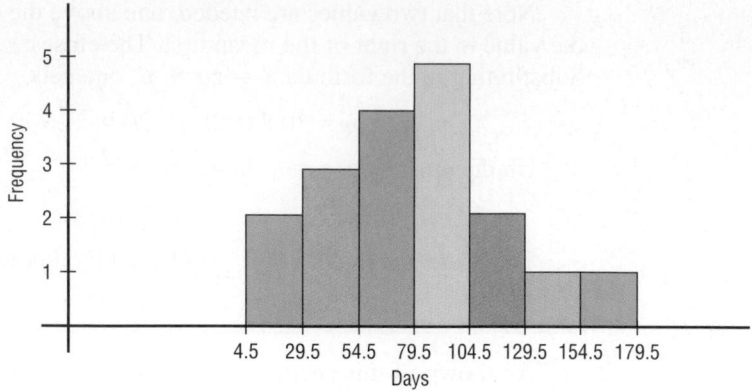

Since the histogram is approximately bell-shaped, one can say that the distribution is approximately normal.

Step 2   Check for skewness. For these data, $\bar{X}$ = 79.5, median = 77.5, and s = 40.5. Using Pearson's index of skewness gives

$$PI = \frac{3(79.5 - 77.5)}{40.5}$$

$$= 0.148$$

In this case, the PI is not greater than $+1$ or less than -1, so it can be concluded that the distribution is not significantly skewed.

Step 3   Check for outliers. Recall that an outlier is a data value that lies more than 1.5 (IQR) units below Q_1 or 1.5 (IQR) units above Q_3. In this case, Q_1 = 45 and Q_3 = 98; hence, IQR = $Q_3 - Q_1$ = 98 − 45 = 53. An outlier would be a data value less than 45 − 1.5(53) = −34.5 or a data value larger than 98 + 1.5(53) = 177.5. In this case, there are no outliers.

Since the histogram is approximately bell-shaped, the data are not significantly skewed, and there are no outliers, it can be concluded that the distribution is approximately normally distributed.

Example 6–20

The data shown consist of the number of games played each year in the career of Baseball Hall of Famer Bill Mazeroski. Determine if the data are approximately normally distributed.

81	148	152	135	151	152
159	142	34	162	130	162
163	143	67	112	70	

Source: Greensburg Tribune Review.

Solution

Step 1   Construct a frequency distribution and draw a histogram for the data. See Figure 6–39.

Figure 6–39

Histogram for
Example 6–20

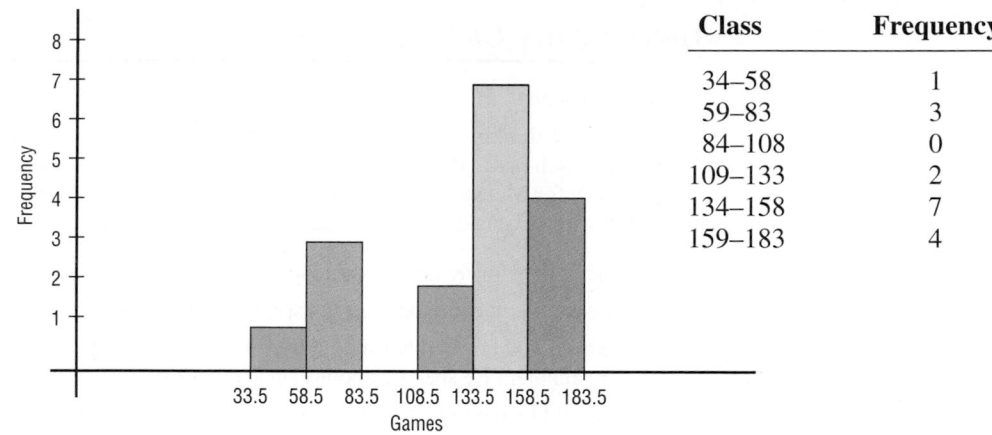

Class	Frequency
34–58	1
59–83	3
84–108	0
109–133	2
134–158	7
159–183	4

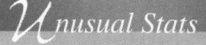

Unusual Stats

The average amount of money stolen by a pickpocket each time is $128.

The histogram shows that the frequency distribution is somewhat negatively skewed.

Step 2 Check for skewness; $\overline{X} = 127.24$, median $= 143$, and $s = 39.87$.

$$PI = \frac{3(\overline{X} - \text{median})}{s}$$

$$= \frac{3(127.24 - 143)}{39.87}$$

$$= -1.19$$

Since the PI is less than -1, it can be concluded that the distribution is significantly skewed to the left.

Step 3 Check for outliers. In this case, $Q_1 = 96.5$ and $Q_3 = 155.5$. IQR $= Q_3 - Q_1 = 155.5 - 96.5 = 59$. Any value less than $96.5 - 1.5(59) = 8$ or above $155.5 + 1.5(59) = 244$ is considered an outlier. There are no outliers.

In summary, the distribution is somewhat negatively skewed.

Another method that is used to check normality is to draw a *normal quantile plot*. *Quantiles,* sometimes called *fractiles,* are values that separate the data set into approximately equal groups. Recall that quartiles separate the data set into four approximately equal groups, and deciles separate the data set into 10 approximately equal groups. A normal quantile plot consists of a graph of points using the data values for the x coordinates and the z values of the quantiles corresponding to the x values for the y coordinates. (*Note:* The calculations of the z values are somewhat complicated, and technology is usually used to draw the graph. The Technology Step by Step section shows how to draw a normal quantile plot.) If the points of the quantile plot do not lie in an approximately straight line, then normality can be rejected.

There are several other methods used to check for normality. A method using normal probability graph paper is shown in the Critical Thinking Challenge section at the end of this chapter, and the chi-square goodness-of-fit test is shown in Chapter 11. Two other tests sometimes used to check normality are the Kolmogorov-Smikirov test and the Lilliefors test. An explanation of these tests can be found in advanced textbooks.

Applying the Concepts **6–4**

Smart People

Assume you are thinking about starting a Mensa chapter in your home town of Visiala, California, which has a population of about 10,000 people. You need to know how many people would qualify for Mensa, which requires an IQ of at least 130. You realize that IQ is normally distributed with a mean of 100 and a standard deviation of 15. Complete the following.

1. Find the approximate number of people in Visiala that are eligible for Mensa.

2. Is it reasonable to continue your quest for a Mensa chapter in Visiala?

3. How would you proceed to find out how many of the eligible people would actually join the new chapter? Be specific about your methods of gathering data.

4. What would be the minimum IQ score needed if you wanted to start an Ultra-Mensa club that included only the top 1% of IQ scores?

See page 344 for the answers.

Exercises 6–4

1. The average admission charge for a movie is $5.39. If the distribution of admission charges is normal with a standard deviation of $0.79, what is the probability that a randomly selected admission charge is less than $3.00?

   Source: *N.Y. Times Almanac.*

2. The average salary for first-year teachers is $27,989. If the distribution is approximately normal with $\sigma = \$3250$, what is the probability that a randomly selected first-year teacher makes these salaries?

   a. Between $20,000 and $30,000 a year

   b. Less than $20,000 a year

   Source: *N.Y. Times Almanac.*

3. The average daily jail population in the United States is 618,319. If the distribution is normal and the standard deviation is 50,200, find the probability that on a randomly selected day the jail population is

   a. Greater than 700,000.

   b. Between 500,000 and 600,000.

   Source: *N.Y. Times Almanac.*

4. The national average SAT score is 1019. If we assume a normal distribution with $\sigma = 90$, what is the 90th percentile score? What is the probability that a randomly selected score exceeds 1200?

   Source: *N.Y. Times Almanac.*

5. The average number of calories in a 1.5-ounce chocolate bar is 225. Suppose that the distribution of calories is approximately normal with $\sigma = 10$. Find the probability that a randomly selected chocolate bar will have

   a. Between 200 and 220 calories.

   b. Less than 200 calories.

   Source: *The Doctor's Pocket Calorie, Fat, and Carbohydrate Counter.*

6. The average age of CEOs is 56 years. Assume the variable is normally distributed. If the standard deviation is 4 years, find the probability that the age of a randomly selected CEO will be in the following range.

   a. Between 53 and 59 years old

   b. Between 58 and 63 years old

   c. Between 50 and 55 years old

   Source: Michael D. Shook and Robert L. Shook, *The Book of Odds.*

7. The average salary for a Queens College full professor is $85,900. If the average salaries are normally distributed with a standard deviation of $11,000, find these probabilities.

   a. The professor makes more than $90,000.

   b. The professor makes more than $75,000.

   Source: AAUP, *Chronicle of Higher Education.*

8. Full-time Ph.D. students receive an average of $12,837 per year. If the average salaries are normally distributed with a standard deviation of $1500, find these probabilities.

   a. The student makes more than $15,000.

   b. The student makes between $13,000 and $14,000.

   Source: U.S. Education Dept., *Chronicle of Higher Education.*

9. A survey found that people keep their microwave ovens an average of 3.2 years. The standard deviation is 0.56 year. If a person decides to buy a new microwave oven, find the probability that he or she has owned the old oven for the following amount of time. Assume the variable is normally distributed.

   a. Less than 1.5 years

   b. Between 2 and 3 years

c. More than 3.2 years

d. What percent of microwave ovens would be replaced if a warranty of 18 months were given?

10. The average commute to work (one way) is 25.5 minutes according to the 2000 Census. If we assume that commuting times are normally distributed with a standard deviation of 6.1 minutes, what is the probability that a randomly selected commuter spends more than 30 minutes a day commuting one way?

Source: *N.Y. Times Almanac.*

11. The average credit card debt for college seniors is $3262. If the debt is normally distributed with a standard deviation of $1100, find these probabilities.

a. That the senior owes at least $1000

b. That the senior owes more than $4000

c. That the senior owes between $3000 and $4000

Source: *USA TODAY.*

12. The average time a person spends at the Barefoot Landing Seaquarium is 96 minutes. The standard deviation is 17 minutes. Assume the variable is normally distributed. If a visitor is selected at random, find the probability that he or she will spend the following time at the seaquarium.

a. At least 120 minutes

b. At most 80 minutes

c. Suggest a time for a bus to return to pick up a group of tourists.

13. The average time for a mail carrier to cover his route is 380 minutes, and the standard deviation is 16 minutes. If one of these trips is selected at random, find the probability that the carrier will have the following route time. Assume the variable is normally distributed.

a. At least 350 minutes

b. At most 395 minutes

c. How might a mail carrier estimate a range for the time he or she will spend en route?

14. During October, the average temperature of Whitman Lake is 53.2° and the standard deviation is 2.3°. Assume the variable is normally distributed. For a randomly selected day in October, find the probability that the temperature will be as follows.

a. Above 54°

b. Below 60°

c. Between 49 and 55°

d. If the lake temperature were above 60°, would you call it very warm?

15. The average waiting time to be seated for dinner at a popular restaurant is 23.5 minutes, with a standard deviation of 3.6 minutes. Assume the variable is normally distributed. When a patron arrives at the restaurant for dinner, find the probability that the patron will have to wait the following time.

a. Between 15 and 22 minutes

b. Less than 18 minutes or more than 25 minutes

c. Is it likely that a person will be seated in less than 15 minutes?

16. A local medical research association proposes to sponsor a footrace. The average time it takes to run the course is 45.8 minutes with a standard deviation of 3.6 minutes. If the association decides to include only the top 25% of the racers, what should be the cutoff time in the tryout run? Assume the variable is normally distributed. Would a person who runs the course in 40 minutes qualify?

17. A marine sales dealer finds that the average price of a previously owned boat is $6492. He decides to sell boats that will appeal to the middle 66% of the market in terms of price. Find the maximum and minimum prices of the boats the dealer will sell. The standard deviation is $1025, and the variable is normally distributed. Would a boat priced at $5550 be sold in this store?

18. The average charitable contribution itemized per income tax return in Pennsylvania is $792. Suppose that the distribution of contributions is normal with a standard deviation of $103. Find the limits for the middle 50% of contributions.

Source: IRS, *Statistics of Income Bulletin.*

19. A contractor decided to build homes that will include the middle 80% of the market. If the average size of homes built is 1810 square feet, find the maximum and minimum sizes of the homes the contractor should build. Assume that the standard deviation is 92 square feet and the variable is normally distributed.

Source: Michael D. Shook and Robert L. Shook, *The Book of Odds.*

20. If the average price of a new home is $145,500, find the maximum and minimum prices of the houses that a contractor will build to include the middle 80% of the market. Assume that the standard deviation of prices is $1500 and the variable is normally distributed.

Source: Michael D. Shook and Robert L. Shook, *The Book of Odds.*

21. The average price of a personal computer (PC) is $949. If the computer prices are approximately normally distributed and $\sigma = \$100$, what is the probability that a randomly selected PC costs more than $1200? The least expensive 10% of personal computers cost less than what amount?

Source: *N.Y. Times Almanac.*

22. To help students improve their reading, a school district decides to implement a reading program. It is to be administered to the bottom 5% of the students in the district, based on the scores on a reading achievement exam. If the average score for the students in the district is 122.6, find the cutoff score that will make a student eligible for the program. The standard deviation is 18. Assume the variable is normally distributed.

23. An automobile dealer finds that the average price of a previously owned vehicle is $8256. He decides to sell cars that will appeal to the middle 60% of the market in terms of price. Find the maximum and minimum prices of the cars the dealer will sell. The standard deviation is $1150, and the variable is normally distributed.

24. The average age of Amtrak passenger train cars is 19.4 years. If the distribution of ages is normal and 20% of the cars are older than 22.8 years, find the standard deviation.

    Source: *N.Y. Times Almanac.*

25. The average length of a hospital stay is 5.9 days. If we assume a normal distribution and a standard deviation of 1.7 days, 15% of hospital stays are less than how many days? Twenty-five percent of hospital stays are longer than how many days?

    Source: *N.Y. Times Almanac.*

26. A mandatory competency test for high school sophomores has a normal distribution with a mean of 400 and a standard deviation of 100.

    a. The top 3% of students receive $500. What is the minimum score you would need to receive this award?

    b. The bottom 1.5% of students must go to summer school. What is the minimum score you would need to stay out of this group?

27. An advertising company plans to market a product to low-income families. A study states that for a particular area, the average income per family is $24,596 and the standard deviation is $6256. If the company plans to target the bottom 18% of the families based on income, find the cutoff income. Assume the variable is normally distributed.

28. If a one-person household spends an average of $40 per week on groceries, find the maximum and minimum dollar amounts spent per week for the middle 50% of one-person households. Assume that the standard deviation is $5 and the variable is normally distributed.

    Source: Michael D. Shook and Robert L. Shook, *The Book of Odds.*

29. The mean lifetime of a wristwatch is 25 months, with a standard deviation of 5 months. If the distribution is normal, for how many months should a guarantee be made if the manufacturer does not want to exchange more than 10% of the watches? Assume the variable is normally distributed.

30. To qualify for security officers' training, recruits are tested for stress tolerance. The scores are normally distributed, with a mean of 62 and a standard deviation of 8. If only the top 15% of recruits are selected, find the cutoff score.

31. In the distributions shown, state the mean and standard deviation for each. *Hint:* See Figures 6–5

and 6–6. Also the vertical lines are 1 standard deviation apart.

a.

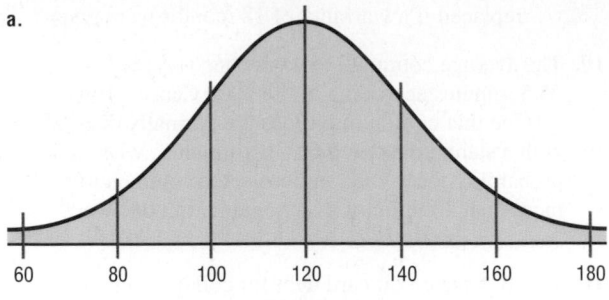

b.

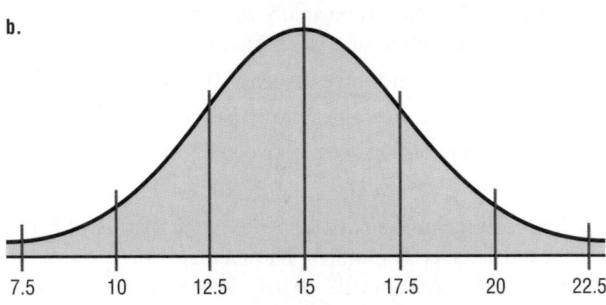

c.

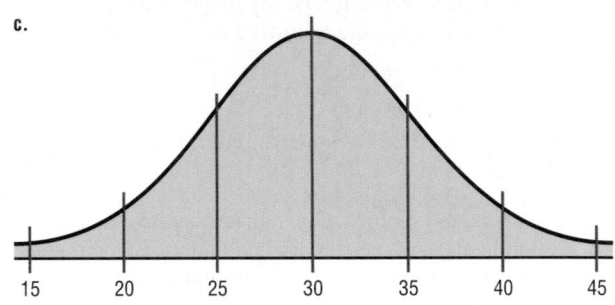

32. Suppose that the mathematics SAT scores for high school seniors for a specific year have a mean of 456 and a standard deviation of 100 and are approximately normally distributed. If a subgroup of these high school seniors, those who are in the National Honor Society, is selected, would you expect the distribution of scores to have the same mean and standard deviation? Explain your answer.

33. Given a data set, how could you decide if the distribution of the data was approximately normal?

34. If a distribution of raw scores were plotted and then the scores were transformed to z scores, would the shape of the distribution change? Explain your answer.

35. In a normal distribution, find σ when $\mu = 110$ and 2.87% of the area lies to the right of 112.

36. In a normal distribution, find μ when σ is 6 and 3.75% of the area lies to the left of 85.

37. In a certain normal distribution, 1.25% of the area lies to the left of 42, and 1.25% of the area lies to the right of 48. Find μ and σ.

38. An instructor gives a 100-point examination in which the grades are normally distributed. The mean is 60 and the standard deviation is 10. If there are 5% A's and 5% F's, 15% B's and 15% D's, and 60% C's, find the scores that divide the distribution into those categories.

39. The data shown represent the number of outdoor drive-in movies in the United States for a 14-year period. Check for normality.

2084	1497	1014	910	899	870	837	859
848	826	815	750	637	737		

Source: National Association of Theater Owners.

40. The data shown represent the cigarette tax (in cents) for 30 randomly selected states. Check for normality.

3	58	5	65	17	48	52	75	21	76	58	36
100	111	34	41	23	44	33	50	13	18	7	12
20	24	66	28	28	31						

Source: Commerce Clearing House.

41. The data shown represent the box office total revenue (in millions of dollars) for a randomly selected sample of the top-grossing films in 2001. Check for normality.

294	241	130	144	113	70	97	94	91	202	74	79
71	67	67	56	180	199	165	114	60	56	53	51

Source: *USA TODAY.*

42. The data shown represent the number of runs made each year during Bill Mazeroski's career. Check for normality.

30	59	69	50	58	71	55	43	66	52	56	62
36	13	29	17	3							

Source: *Greensburg Tribune Review.*

Technology *Step by Step*

MINITAB
Step by Step

Determining Normality

There are several ways in which statisticians test a data set for normality. Four are shown here.

Construct a Histogram

Inspect the histogram for shape.

1. Enter the data for Example 6–19 in the first column of a new worksheet. Name the column Inventory.

2. Use **Stat>Basic Statistics>Graphical Summary** presented in Section 3–4 to create the histogram. Is it symmetric? Is there a single peak?

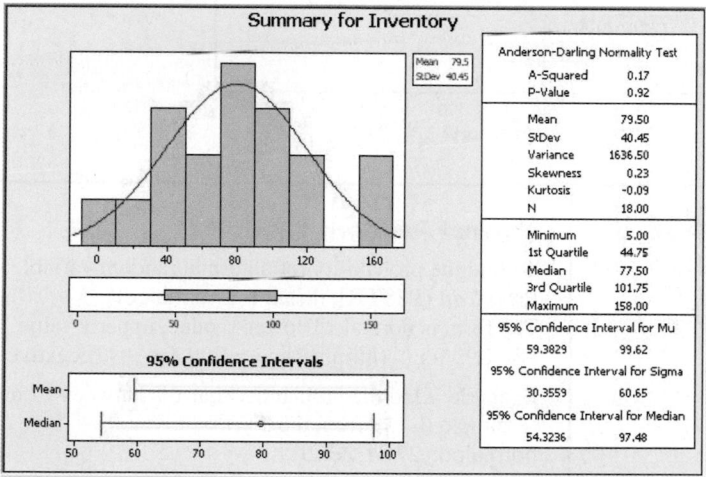

Check for Outliers

Inspect the boxplot for outliers. There are no outliers in this graph. Furthermore, the box is in the middle of the range, and the median is in the middle of the box. Most likely this is not a skewed distribution either.

Calculate Pearson's Index of Skewness

The measure of skewness in the graphical summary is not the same as Pearson's index. Use the calculator and the formula.

$$PI = \frac{3(\overline{X} - \text{median})}{s}$$

3. Select **Calc>Calculator,** then type **PI** in the text box for Store result in:.

4. Enter the expression: **3*(MEAN(C1)−MEDI(C1))/(STDEV(C1)).** Make sure you get all the parentheses in the right place!

5. Click [OK]. The result, 0.148318, will be stored in the first row of **C2** named PI. Since it is smaller than +1, the distribution is not skewed.

Construct a Normal Probability Plot

6. Select **Graph>Probability Plot,** then Single and click [OK].

7. Double-click **C1** Inventory to select the data to be graphed.

8. Click [Distribution] and make sure that Normal is selected. Click [OK].

9. Click [Labels] and enter the title for the graph: **Quantile Plot for Inventory.** You may also put **Your Name** in the subtitle.

10. Click [OK] twice. Inspect the graph to see if the graph of the points is linear.

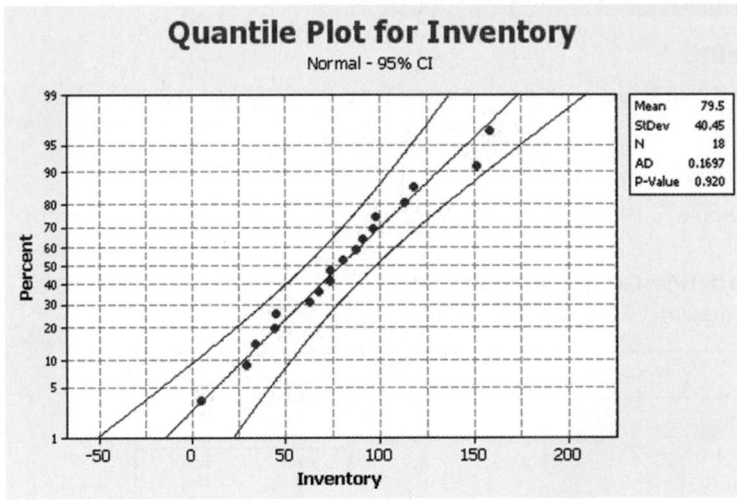

These data are nearly normal.

What do you look for in the plot?

a) An "S curve" indicates a distribution that is too thick in the tails, a uniform distribution, for example.

b) Concave plots indicate a skewed distribution.

c) If one end has a point that is extremely high or low, there may be outliers.

This data set appears to be nearly normal by every one of the four criteria!

TI-83 Plus or TI-84 Plus
Step by Step

Normal Random Variables

To find the probability for a normal random variable:
Press **2nd [DISTR],** then **2** for normalcdf(
The form is normalcdf(lower x value, upper x value, μ, σ)
Use E99 for ∞ (infinity) and −E99 for $-\infty$ (negative infinity). Press **2nd [EE]** to get E.

Example: Find the probability that x is between 27 and 31 when $\mu = 28$ and $\sigma = 2$ (Example 6–15a from the text).
normalcdf(27,31,28,2)

To find the percentile for a normal random variable:
Press **2nd [DISTR],** then **3** for invNorm(
The form is invNorm(area to the left of x value, μ, σ)

Example: Find the 90th percentile when $\mu = 200$ and $\sigma = 20$ (Example 6–17 from text).
invNorm(.9,200,20)

```
normalcdf(27,31,
28,2)
      .6246552391
invNorm(.9,200,2
0)
      225.6310313
```

To construct a normal quantile plot:

1. Enter the data values into L$_1$.

2. Press **2nd [STAT PLOT]** to get the STAT PLOT menu.

3. Press **1** for Plot 1.

4. Turn on the plot by pressing **ENTER** while the cursor is flashing over ON.

5. Move the cursor to the normal quantile plot (6th graph).

6. Make sure L$_1$ is entered for the Data List and X is highlighted for the Data Axis.

7. Press **WINDOW** for the Window menu. Adjust Xmin and Xmax according to the data values. Adjust Ymin and Ymax as well, Ymin $= -3$ and Ymax $= 3$ usually work fine.

8. Press **GRAPH.**

Using the data from Example 6–19 gives

 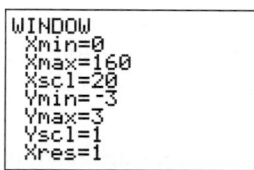 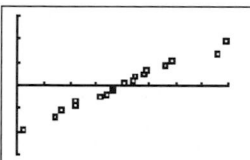

Since the points in the normal quantile plot lie close to a straight line, the distribution is approximately normal.

Excel
Step by Step

Normal Quantile Plot

Excel can be used to construct a normal quantile plot to examine if a set of data is approximately normally distributed.

1. Enter the data from Example 6–19 into column A of a new worksheet. The data should be sorted in ascending order.

2. Since the sample size is 18, each score represents $\frac{1}{18}$, or approximately 5.6%, of the sample. Each data point is assumed to subdivide the data into equal intervals. Each data value corresponds to the midpoint of the particular subinterval.

3. After all the data are entered and sorted in column A, select cell B1. From the function icon, select the NORMSINV command to find the z score corresponding to an area of $\frac{1}{18}$ of the total area under the normal curve. Enter **1/(2*18)** for the Probability.

4. Repeat the procedure from step 3 for each data value in column A. However, for each consecutive z score corresponding to a data value in column A, enter the next odd multiple of $\frac{1}{36}$ in the dialogue box. For example, in cell B2, enter the value **3/(2*18)** in the NORMSINV dialogue box. In cell B3, enter **5/(2*18).** Continue using this procedure to create z scores for each value in column A until all values have corresponding z scores.

5. Highlight the data from columns A and B, and select the Chart Wizard from the toolbar.

6. Select the scatter plot to graph the data from columns A and B as ordered pairs. Click Next.

7. Title and label axes as needed; click [OK].

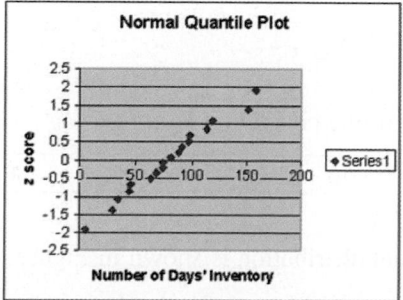

The points appear to lie close to a straight line. Thus, we deduce that the data are approximately normally distributed.

The Central Limit Theorem

In addition to knowing how individual data values vary about the mean for a population, statisticians are interested in knowing how the means of samples of the same size taken from the same population vary about the population mean.

Objective 6

Use the central limit theorem to solve problems involving sample means for large samples.

Distribution of Sample Means

Suppose a researcher selects a sample of 30 adult males and finds the mean of the measure of the triglyceride levels for the sample subjects to be 187 milligrams/deciliter. Then suppose a second sample is selected, and the mean of that sample is found to be 192 milligrams/deciliter. Continue the process for 100 samples. What happens then is that the mean becomes a random variable, and the sample means 187, 192, 184, . . . , 196 constitute a *sampling distribution of sample means.*

> A **sampling distribution of sample means** is a distribution using the means computed from all possible random samples of a specific size taken from a population.

If the samples are randomly selected with replacement, the sample means, for the most part, will be somewhat different from the population mean μ. These differences are caused by sampling error.

> **Sampling error** is the difference between the sample measure and the corresponding population measure due to the fact that the sample is not a perfect representation of the population.

When all possible samples of a specific size are selected with replacement from a population, the distribution of the sample means for a variable has two important properties, which are explained next.

> **Properties of the Distribution of Sample Means**
>
> 1. The mean of the sample means will be the same as the population mean.
> 2. The standard deviation of the sample means will be smaller than the standard deviation of the population, and it will be equal to the population standard deviation divided by the square root of the sample size.

The following example illustrates these two properties. Suppose a professor gave an 8-point quiz to a small class of four students. The results of the quiz were 2, 6, 4, and 8. For the sake of discussion, assume that the four students constitute the population. The mean of the population is

$$\mu = \frac{2 + 6 + 4 + 8}{4} = 5$$

The standard deviation of the population is

$$\sigma = \sqrt{\frac{(2 - 5)^2 + (6 - 5)^2 + (4 - 5)^2 + (8 - 5)^2}{4}} = 2.236$$

The graph of the original distribution is shown in Figure 6–40. This is called a *uniform distribution.*

Figure 6–40

Distribution of
Quiz Scores

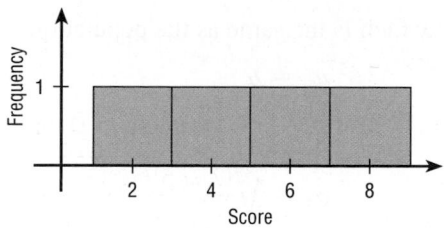

Figure 6–40

Distribution of
Quiz Scores

Now, if all samples of size 2 are taken with replacement and the mean of each sample is found, the distribution is as shown.

Sample	Mean		Sample	Mean
2, 2	2		6, 2	4
2, 4	3		6, 4	5
2, 6	4		6, 6	6
2, 8	5		6, 8	7
4, 2	3		8, 2	5
4, 4	4		8, 4	6
4, 6	5		8, 6	7
4, 8	6		8, 8	8

A frequency distribution of sample means is as follows.

$\overline{X}$	f
2	1
3	2
4	3
5	4
6	3
7	2
8	1

For the data from the example just discussed, Figure 6–41 shows the graph of the sample means. The histogram appears to be approximately normal.

The mean of the sample means, denoted by $\mu_{\overline{X}}$, is

$$\mu_{\overline{X}} = \frac{2 + 3 + \cdots + 8}{16} = \frac{80}{16} = 5$$

Figure 6–41

Distribution of Sample
Means

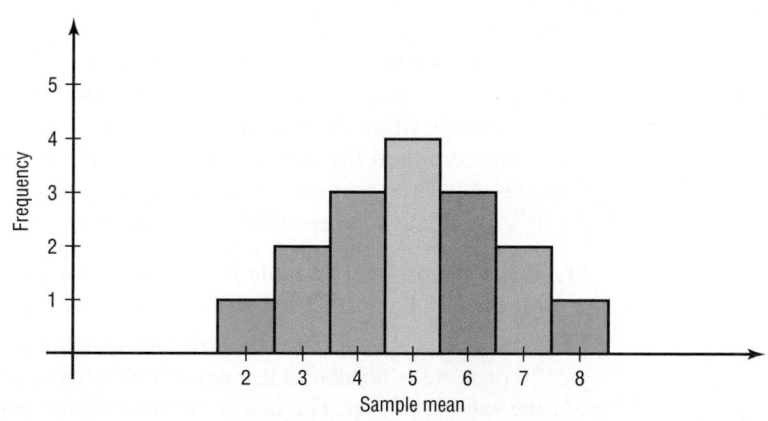

which is the same as the population mean. Hence,

$$\mu_{\bar{X}} = \mu$$

The standard deviation of sample means, denoted by $\sigma_{\bar{X}}$, is

$$\sigma_{\bar{X}} = \sqrt{\frac{(2-5)^2 + (3-5)^2 + \cdots + (8-5)^2}{16}} = 1.581$$

which is the same as the population standard deviation, divided by $\sqrt{2}$:

$$\sigma_{\bar{X}} = \frac{2.236}{\sqrt{2}} = 1.581$$

(*Note:* Rounding rules were not used here in order to show that the answers coincide.)

In summary, if all possible samples of size n are taken with replacement from the same population, the mean of the sample means, denoted by $\mu_{\bar{X}}$, equals the population mean μ; and the standard deviation of the sample means, denoted by $\sigma_{\bar{X}}$, equals $\sigma/\sqrt{n}$. The standard deviation of the sample means is called the **standard error of the mean.** Hence,

$$\sigma_{\bar{X}} = \frac{\sigma}{\sqrt{n}}$$

A third property of the sampling distribution of sample means pertains to the shape of the distribution and is explained by the **central limit theorem.**

The Central Limit Theorem

As the sample size n increases without limit, the shape of the distribution of the sample means taken with replacement from a population with mean μ and standard deviation σ will approach a normal distribution. As previously shown, this distribution will have a mean μ and a standard deviation $\sigma/\sqrt{n}$.

If the sample size is sufficiently large, the central limit theorem can be used to answer questions about sample means in the same manner that a normal distribution can be used to answer questions about individual values. The only difference is that a new formula must be used for the z values. It is

$$z = \frac{\bar{X} - \mu}{\sigma/\sqrt{n}}$$

Notice that $\bar{X}$ is the sample mean, and the denominator must be adjusted since means are being used instead of individual data values. The denominator is the standard deviation of the sample means.

If a large number of samples of a given size are selected from a normally distributed population, or if a large number of samples of a given size that is greater than or equal to 30 are selected from a population that is not normally distributed, and the sample means are computed, then the distribution of sample means will look like the one shown in Figure 6–42. Their percentages indicate the areas of the regions.

It's important to remember two things when you use the central limit theorem:

1. When the original variable is normally distributed, the distribution of the sample means will be normally distributed, for any sample size n.
2. When the distribution of the original variable might not be normal, a sample size of 30 or more is needed to use a normal distribution to approximate the distribution of the sample means. The larger the sample, the better the approximation will be.

Figure 6–42

Distribution of Sample Means for Large Number of Samples

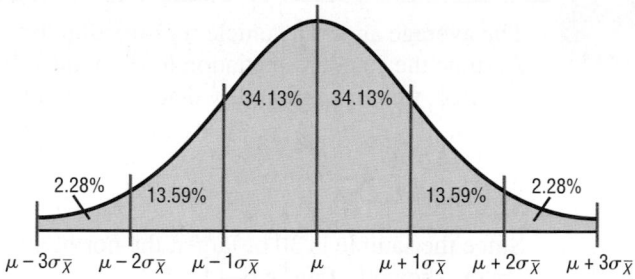

Examples 6–21 through 6–23 show how the standard normal distribution can be used to answer questions about sample means.

Example 6–21

A. C. Neilsen reported that children between the ages of 2 and 5 watch an average of 25 hours of television per week. Assume the variable is normally distributed and the standard deviation is 3 hours. If 20 children between the ages of 2 and 5 are randomly selected, find the probability that the mean of the number of hours they watch television will be greater than 26.3 hours.

Source: Michael D. Shook and Robert L. Shook, *The Book of Odds.*

Solution

Since the variable is approximately normally distributed, the distribution of sample means will be approximately normal, with a mean of 25. The standard deviation of the sample means is

$$\sigma_{\bar{X}} = \frac{\sigma}{\sqrt{n}} = \frac{3}{\sqrt{20}} = 0.671$$

The distribution of the means is shown in Figure 6–43, with the appropriate area shaded.

Figure 6–43

Distribution of the Means for Example 6–21

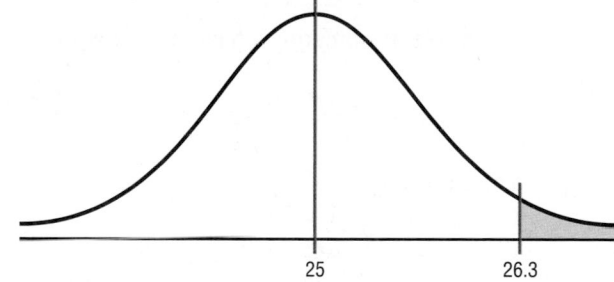

The z value is

$$z = \frac{\bar{X} - \mu}{\sigma/\sqrt{n}} = \frac{26.3 - 25}{3/\sqrt{20}} = \frac{1.3}{0.671} = 1.94$$

The area between 0 and 1.94 is 0.4738. Since the desired area is in the tail, subtract 0.4738 from 0.5000. Hence, 0.5000 − 0.4738 = 0.0262, or 2.62%.

One can conclude that the probability of obtaining a sample mean larger than 26.3 hours is 2.62% [i.e., $P(\bar{X} > 26.3) = 2.62\%$].

Example 6–22

The average age of a vehicle registered in the United States is 8 years, or 96 months. Assume the standard deviation is 16 months. If a random sample of 36 vehicles is selected, find the probability that the mean of their age is between 90 and 100 months.

Source: Harper's Index.

Solution

Since the sample is 30 or larger, the normality assumption is not necessary. The desired area is shown in Figure 6–44.

Figure 6–44

Area Under a
Normal Curve for
Example 6–22

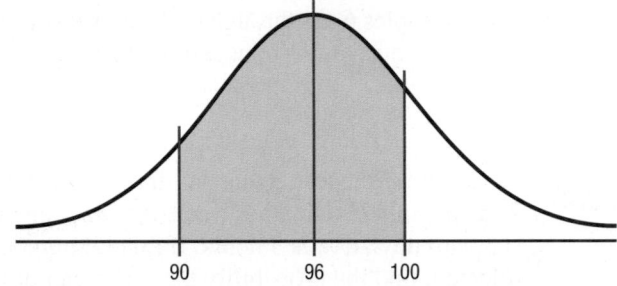

The two z values are

$$z_1 = \frac{90 - 96}{16/\sqrt{36}} = -2.25$$

$$z_2 = \frac{100 - 96}{16/\sqrt{36}} = 1.50$$

The two areas corresponding to the z values of -2.25 and 1.50, respectively, are 0.4878 and 0.4332. Since the z values are on opposite sides of the mean, find the probability by adding the areas: $0.4878 + 0.4332 = 0.921$, or 92.1%.

Hence, the probability of obtaining a sample mean between 90 and 100 months is 92.1%; that is, $P(90 < \bar{X} < 100) = 92.1\%$.

Students sometimes have difficulty deciding whether to use

$$z = \frac{\bar{X} - \mu}{\sigma/\sqrt{n}} \qquad \text{or} \qquad z = \frac{X - \mu}{\sigma}$$

The formula

$$z = \frac{\bar{X} - \mu}{\sigma/\sqrt{n}}$$

should be used to gain information about a sample mean, as shown in this section. The formula

$$z = \frac{X - \mu}{\sigma}$$

is used to gain information about an individual data value obtained from the population. Notice that the first formula contains $\bar{X}$, the symbol for the sample mean, while the second formula contains X, the symbol for an individual data value. Example 6–23 illustrates the uses of the two formulas.

Example 6–23

The average number of pounds of meat that a person consumes a year is 218.4 pounds. Assume that the standard deviation is 25 pounds and the distribution is approximately normal.

Source: Michael D. Shook and Robert L. Shook, *The Book of Odds.*

   a. Find the probability that a person selected at random consumes less than 224 pounds per year.

   b. If a sample of 40 individuals is selected, find the probability that the mean of the sample will be less than 224 pounds per year.

Solution

   a. Since the question asks about an individual person, the formula $z = (X - \mu)/\sigma$ is used. The distribution is shown in Figure 6–45.

Figure 6–45

Area Under a Normal Curve for Part *a* of Example 6–23

218.4 224
Distribution of individual data values for the population

The z value is

$$z = \frac{X - \mu}{\sigma} = \frac{224 - 218.4}{25} = 0.22$$

The area between 0 and 0.22 is 0.0871; this area must be added to 0.5000 to get the total area to the left of $z = 0.22$.

   $0.0871 + 0.5000 = 0.5871$

Hence, the probability of selecting an individual who consumes less than 224 pounds of meat per year is 0.5871, or 58.71% [i.e., $P(X < 224) = 0.5871$].

   b. Since the question concerns the mean of a sample with a size of 40, the formula $z = (\overline{X} - \mu)/(\sigma/\sqrt{n})$ is used. The area is shown in Figure 6–46.

Figure 6–46

Area Under a Normal Curve for Part *b* of Example 6–23

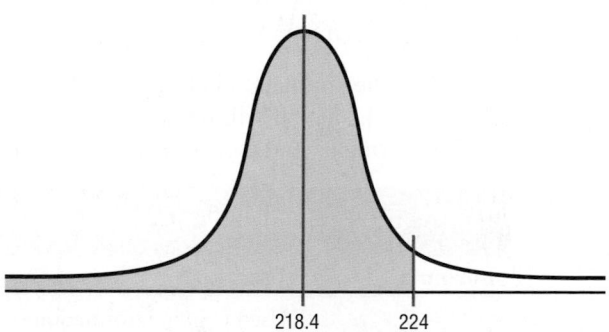

218.4   224
Distribution of means for all samples of size 40 taken from the population

The z value is

$$z = \frac{\overline{X} - \mu}{\sigma/\sqrt{n}} = \frac{224 - 218.4}{25/\sqrt{40}} = 1.42$$

The area between $z = 0$ and $z = 1.42$ is 0.4222; this value must be added to 0.5000 to get the total area.

$$0.4222 + 0.5000 = 0.9222$$

Hence, the probability that the mean of a sample of 40 individuals is less than 224 pounds per year is 0.9222, or 92.22%. That is, $P(\overline{X} < 224) = 0.9222$.

Comparing the two probabilities, one can see that the probability of selecting an individual who consumes less than 224 pounds of meat per year is 58.71%, but the probability of selecting a sample of 40 people with a mean consumption of meat that is less than 224 pounds per year is 92.22%. This rather large difference is due to the fact that the distribution of sample means is much less variable than the distribution of individual data values. (*Note:* An individual person is the equivalent of saying $n = 1$.)

Finite Population Correction Factor (Optional)

The formula for the standard error of the mean $\sigma/\sqrt{n}$ is accurate when the samples are drawn with replacement or are drawn without replacement from a very large or infinite population. Since sampling with replacement is for the most part unrealistic, a *correction factor* is necessary for computing the standard error of the mean for samples drawn without replacement from a finite population. Compute the correction factor by using the expression

$$\sqrt{\frac{N - n}{N - 1}}$$

where N is the population size and n is the sample size.

This correction factor is necessary if relatively large samples are taken from a small population, because the sample mean will then more accurately estimate the population mean and there will be less error in the estimation. Therefore, the standard error of the mean must be multiplied by the correction factor to adjust for large samples taken from a small population. That is,

$$\sigma_{\overline{X}} = \frac{\sigma}{\sqrt{n}} \cdot \sqrt{\frac{N - n}{N - 1}}$$

Finally, the formula for the z value becomes

$$z = \frac{\overline{X} - \mu}{\dfrac{\sigma}{\sqrt{n}} \cdot \sqrt{\dfrac{N - n}{N - 1}}}$$

When the population is large and the sample is small, the correction factor is generally not used, since it will be very close to 1.00.

The formulas and their uses are summarized in Table 6–1.

Table 6–1	Summary of Formulas and Their Uses
Formula	**Use**
1. $z = \dfrac{X - \mu}{\sigma}$	Used to gain information about an individual data value when the variable is normally distributed.
2. $z = \dfrac{\overline{X} - \mu}{\sigma/\sqrt{n}}$	Used to gain information when applying the central limit theorem about a sample mean when the variable is normally distributed or when the sample size is 30 or more.

Applying the Concepts **6–5**

Central Limit Theorem

Twenty students from a statistics class each collected a random sample of times on how long it took students to get to class from their homes. All the sample sizes were 30. The resulting means are listed.

Student	Mean	Std. Dev.	Student	Mean	Std. Dev.
1	22	3.7	11	27	1.4
2	31	4.6	12	24	2.2
3	18	2.4	13	14	3.1
4	27	1.9	14	29	2.4
5	20	3.0	15	37	2.8
6	17	2.8	16	23	2.7
7	26	1.9	17	26	1.8
8	34	4.2	18	21.	2.0
9	23	2.6	19	30	2.2
10	29	2.1	20	29	2.8

1. The students noticed that everyone had different answers. If you randomly sample over and over from any population, with the same sample size, will the results ever be the same?

2. The students wondered whose results were right. How can they find out what the population mean and standard deviation are?

3. Input the means into the computer and check to see if the distribution is normal.

4. Check the mean and standard deviation of the means. How do these values compare to the students' individual scores?

5. Is the distribution of the means a sampling distribution?

6. Check the sampling error for students 3, 7, and 14.

7. Compare the standard deviation of the sample of the 20 means. Is that equal to the standard deviation from student 3 divided by the square of the sample size? How about for student 7, or 14?

See page 344 for the answers.

Exercises 6–5

1. If samples of a specific size are selected from a population and the means are computed, what is this distribution of means called?

2. Why do most of the sample means differ somewhat from the population mean? What is this difference called?

3. What is the mean of the sample means?

4. What is the standard deviation of the sample means called? What is the formula for this standard deviation?

5. What does the central limit theorem say about the shape of the distribution of sample means?

6. What formula is used to gain information about an individual data value when the variable is normally distributed?

7. What formula is used to gain information about a sample mean when the variable is normally distributed or when the sample size is 30 or more?

For Exercises 8 through 25, assume that the sample is taken from a large population and the correction factor can be ignored.

8. A survey found that the American family generates an average of 17.2 pounds of glass garbage each year. Assume the standard deviation of the distribution is 2.5 pounds. Find the probability that the mean of a sample of 55 families will be between 17 and 18 pounds.

Source: Michael D. Shook and Robert L. Shook, *The Book of Odds*.

9. The average yearly cost per household of owning a dog is $186.80. Suppose that we randomly select 50 households that own a dog. What is the probability that

the sample mean for these 50 households is less than $175.00? Assume $\sigma = \$32$.

Source: *N.Y. Times Almanac.*

10. The average teacher's salary in New Jersey (ranked first among states) is $52,174. Suppose that the distribution is normal with standard deviation equal to $7500.

    a. What is the probability that a randomly selected teacher makes less than $50,000 a year?

    b. If we sample 100 teachers' salaries, what is the probability that the sample mean is less than $50,000?

    Source: *N.Y. Times Almanac.*

11. The mean weight of 15-year-old males is 142 pounds, and the standard deviation is 12.3 pounds. If a sample of thirty-six 15-year-old males is selected, find the probability that the mean of the sample will be greater than 144.5 pounds. Assume the variable is normally distributed. Based on your answer, would you consider the group overweight?

12. The average teacher's salary in North Dakota is $29,863. Assume a normal distribution with $\sigma = \$5100$.

    a. What is the probability that a randomly selected teacher's salary is greater than $40,000?

    b. What is the probability that the mean for a sample of 80 teachers' salaries is greater than $30,000?

    Source: *N.Y. Times Almanac.*

13. The average price of a pound of sliced bacon is $2.02. Assume the standard deviation is $0.08. If a random sample of 40 one-pound packages is selected, find the probability that the mean of the sample will be less than $2.00.

    Source: *Statistical Abstract of the United States.*

14. The national average SAT score is 1019. Suppose that nothing is known about the shape of the distribution and that the standard deviation is 100. If a random sample of 200 scores were selected and the sample mean were calculated to be 1050, would you be surprised? Explain.

    Source: *N.Y. Times Almanac.*

15. The average number of milligrams (mg) of sodium in a certain brand of low-salt microwave frozen dinners is 660 mg, and the standard deviation is 35 mg. Assume the variable is normally distributed.

    a. If a single dinner is selected, find the probability that the sodium content will be more than 670 mg.

    b. If a sample of 10 dinners is selected, find the probability that the mean of the sample will be larger than 670 mg.

    c. Why is the probability for part *a* greater than that for part *b*?

16. The average age of chemical engineers is 37 years with a standard deviation of 4 years. If an engineering firm employs 25 chemical engineers, find the probability that the average age of the group is greater than 38.2 years old. If this is the case, would it be safe to assume that the engineers in this group are generally much older than average?

17. The *Old Farmer's Almanac* reports that the average person uses 123 gallons of water daily. If the standard deviation is 21 gallons, find the probability that the mean of a randomly selected sample of 15 people will be between 120 and 126 gallons. Assume the variable is normally distributed.

18. The average public elementary school has 458 students. Assume the standard deviation is 97. If a random sample of 36 public elementary schools is selected, find the probability that the number of students enrolled is between 450 and 465.

19. Procter & Gamble reported that an American family of four washes an average of 1 ton (2000 pounds) of clothes each year. If the standard deviation of the distribution is 187.5 pounds, find the probability that the mean of a randomly selected sample of 50 families of four will be between 1980 and 1990 pounds.

    Source: *The Harper's Index Book.*

20. The average annual salary in Pennsylvania was $24,393 in 1992. Assume that salaries were normally distributed for a certain group of wage earners, and the standard deviation of this group was $4362.

    a. Find the probability that a randomly selected individual earned less than $26,000.

    b. Find the probability that, for a randomly selected sample of 25 individuals, the mean salary was less than $26,000.

    c. Why is the probability for part *b* higher than the probability for part *a*?

    Source: Associated Press.

21. The average time it takes a group of adults to complete a certain achievement test is 46.2 minutes. The standard deviation is 8 minutes. Assume the variable is normally distributed.

    a. Find the probability that a randomly selected adult will complete the test in less than 43 minutes.

    b. Find the probability that, if 50 randomly selected adults take the test, the mean time it takes the group to complete the test will be less than 43 minutes.

    c. Does it seem reasonable that an adult would finish the test in less than 43 minutes? Explain.

    d. Does it seem reasonable that the mean of the 50 adults could be less than 43 minutes?

22. Assume that the mean systolic blood pressure of normal adults is 120 millimeters of mercury (mm Hg) and the standard deviation is 5.6. Assume the variable is normally distributed.

   a. If an individual is selected, find the probability that the individual's pressure will be between 120 and 121.8 mm Hg.

   b. If a sample of 30 adults is randomly selected, find the probability that the sample mean will be between 120 and 121.8 mm Hg.

   c. Why is the answer to part *a* so much smaller than the answer to part *b*?

23. The average cholesterol content of a certain brand of eggs is 215 milligrams, and the standard deviation is 15 milligrams. Assume the variable is normally distributed.

   a. If a single egg is selected, find the probability that the cholesterol content will be greater than 220 milligrams.

   b. If a sample of 25 eggs is selected, find the probability that the mean of the sample will be larger than 220 milligrams.

Source: *Living Fit.*

24. At a large publishing company, the mean age of proofreaders is 36.2 years, and the standard deviation is 3.7 years. Assume the variable is normally distributed.

   a. If a proofreader from the company is randomly selected, find the probability that his or her age will be between 36 and 37.5 years.

   b. If a random sample of 15 proofreaders is selected, find the probability that the mean age of the proofreaders in the sample will be between 36 and 37.5 years.

25. In the United States, one farmworker supplied agricultural products for an average of 106 people. Assume the standard deviation is 16.1. If 35 farmworkers are selected, find the probability that the mean number of people supplied is between 100 and 110.

Extending the Concepts

For Exercises 26 and 27, check to see whether the correction factor should be used. If so, be sure to include it in the calculations.

26. In a study of the life expectancy of 500 people in a certain geographic region, the mean age at death was 72.0 years, and the standard deviation was 5.3 years. If a sample of 50 people from this region is selected, find the probability that the mean life expectancy will be less than 70 years.

27. A study of 800 homeowners in a certain area showed that the average value of the homes was $82,000, and the standard deviation was $5000. If 50 homes are for sale, find the probability that the mean of the values of these homes is greater than $83,500.

28. The average breaking strength of a certain brand of steel cable is 2000 pounds, with a standard deviation of 100 pounds. A sample of 20 cables is selected and tested. Find the sample mean that will cut off the upper 95% of all samples of size 20 taken from the population. Assume the variable is normally distributed.

29. The standard deviation of a variable is 15. If a sample of 100 individuals is selected, compute the standard error of the mean. What size sample is necessary to double the standard error of the mean?

30. In Exercise 29, what size sample is needed to cut the standard error of the mean in half?

6–6

The Normal Approximation to the Binomial Distribution

A normal distribution is often used to solve problems that involve the binomial distribution since, when *n* is large (say, 100), the calculations are too difficult to do by hand using the binomial distribution. Recall from Chapter 5 that a binomial distribution has the following characteristics:

   1. There must be a fixed number of trials.

   2. The outcome of each trial must be independent.

Example 6–24

A magazine reported that 6% of American drivers read the newspaper while driving. If 300 drivers are selected at random, find the probability that exactly 25 say they read the newspaper while driving.

Source: USA Snapshot, *USA TODAY.*

Solution

Here, $p = 0.06$, $q = 0.94$, and $n = 300$.

Step 1 Check to see whether a normal approximation can be used.

$$np = (300)(0.06) = 18 \qquad nq = (300)(0.94) = 282$$

Since $np \geq 5$ and $nq \geq 5$, the normal distribution can be used.

Step 2 Find the mean and standard deviation.

$$\mu = np = (300)(0.06) = 18$$
$$\sigma = \sqrt{npq} = \sqrt{(300)(0.06)(0.94)} = \sqrt{16.92} = 4.11$$

Step 3 Write the problem in probability notation: $P(X = 25)$.

Step 4 Rewrite the problem by using the continuity correction factor. See approximation number 1 in Table 6–2: $P(25 - 0.5 < X < 25 + 0.5) = P(24.5 < X < 25.5)$. Show the corresponding area under the normal distribution curve. See Figure 6–48.

Figure 6–48

Area Under a Normal Curve and *X* Values for Example 6–24

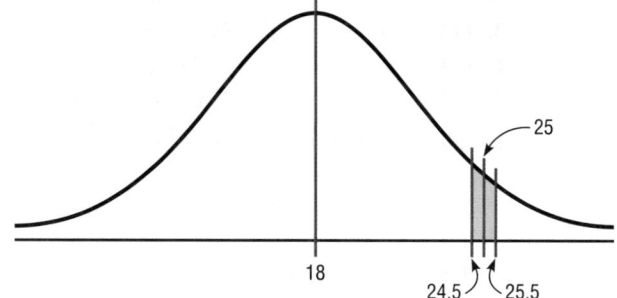

Step 5 Find the corresponding *z* values. Since 25 represents any value between 24.5 and 25.5, find both *z* values.

$$z_1 = \frac{25.5 - 18}{4.11} = 1.82 \qquad z_2 = \frac{24.5 - 18}{4.11} = 1.58$$

Step 6 Find the solution. Find the corresponding areas in the table: The area for $z = 1.82$ is 0.4656, and the area for $z = 1.58$ is 0.4429. Subtract the areas to get the approximate value: $0.4656 - 0.4429 = 0.0227$, or 2.27%.

Hence, the probability that exactly 25 people read the newspaper while driving is 2.27%.

Example 6–25

Of the members of a bowling league, 10% are widowed. If 200 bowling league members are selected at random, find the probability that 10 or more will be widowed.

Solution

Here, $p = 0.10$, $q = 0.90$, and $n = 200$.

Step 1 Since $np = (200)(0.10) = 20$ and $nq = (200)(0.90) = 180$, the normal approximation can be used.

Step 2 $\mu = np = (200)(0.10) = 20$

$\sigma = \sqrt{npq} = \sqrt{(200)(0.10)(0.90)} = \sqrt{18} = 4.24$

Step 3 $P(X \geq 10)$.

Step 4   See approximation number 2 in Table 6–2: $P(X > 10 - 0.5) = P(X > 9.5)$. The desired area is shown in Figure 6–49.

Figure 6–49

Area Under a Normal Curve and X Value for Example 6–25

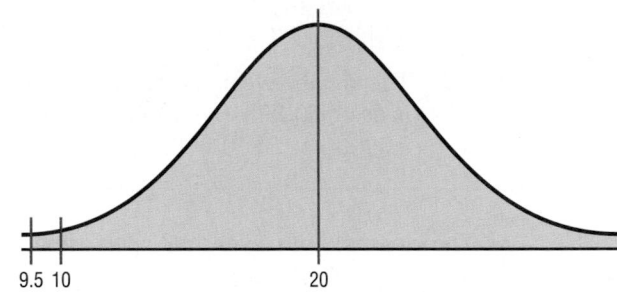

9.5 10   20

Step 5   Since the problem is to find the probability of 10 or more positive responses, a normal distribution graph is as shown in Figure 6–49. Hence, the area between 9.5 and 20 must be added to 0.5000 to get the correct approximation. The z value is

$$z = \frac{9.5 - 20}{4.24} = -2.48$$

Step 6   The area between 9.5 and 20 is 0.4934. Thus, the probability of getting 10 or more responses is $0.4934 + 0.5000 = 0.9934$, or 99.34%.

It can be concluded, then, that the probability of 10 or more widowed people in a random sample of 200 bowling league members is 99.34%.

Example 6–26   If a baseball player's batting average is 0.320 (32%), find the probability that the player will get at most 26 hits in 100 times at bat.

Solution

Here, $p = 0.32$, $q = 0.68$, and $n = 100$.

Step 1   Since $np = (100)(0.320) = 32$ and $nq = (100)(0.680) = 68$, the normal distribution can be used to approximate the binomial distribution.

Step 2 $\mu = np = (100)(0.320) = 32$

$\sigma = \sqrt{npq} = \sqrt{(100)(0.32)(0.68)} = \sqrt{21.76} = 4.66$

Step 3 $P(X \leq 26)$.

Step 4   See approximation number 4 in Table 6–2: $P(X < 26 + 0.5) = P(X < 26.5)$. The desired area is shown in Figure 6–50.

Step 5   The z value is

$$z = \frac{26.5 - 32}{4.66} = -1.18$$

Figure 6–50

Area Under a
Normal Curve for
Example 6–26

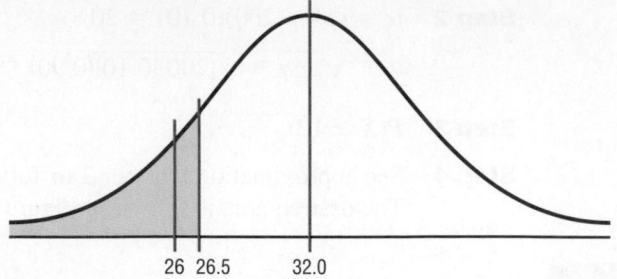

26 26.5   32.0

Step 6 The area between the mean and 26.5 is 0.3810. Since the area in the left tail is desired, 0.3810 must be subtracted from 0.5000. So the probability is $0.5000 - 0.3810 = 0.1190$, or 11.9%.

The closeness of the normal approximation is shown in Example 6–27.

Example 6–27

When $n = 10$ and $p = 0.5$, use the binomial distribution table (Table B in Appendix C) to find the probability that $X = 6$. Then use the normal approximation to find the probability that $X = 6$.

<u>Solution</u>

From Table B, for $n = 10$, $p = 0.5$, and $X = 6$, the probability is 0.205.
    For a normal approximation,

$$\mu = np = (10)(0.5) = 5$$

$$\sigma = \sqrt{npq} = \sqrt{(10)(0.5)(0.5)} = 1.58$$

Now, $X = 6$ is represented by the boundaries 5.5 and 6.5. So the z values are

$$z_1 = \frac{6.5 - 5}{1.58} = 0.95 \qquad z_2 = \frac{5.5 - 5}{1.58} = 0.32$$

The corresponding area for 0.95 is 0.3289, and the corresponding area for 0.32 is 0.1255.
    The solution is $0.3289 - 0.1255 = 0.2034$, which is very close to the binomial table value of 0.205. The desired area is shown in Figure 6–51.

Figure 6–51

Area Under a
Normal Curve for
Example 6–27

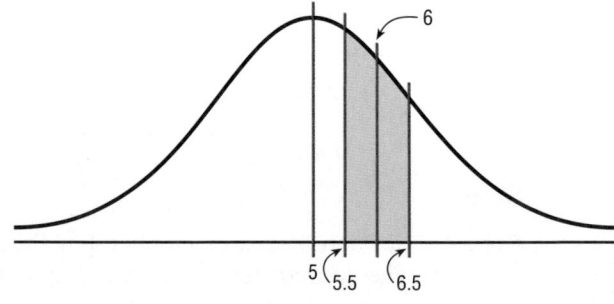

5 5.5   6.5

The normal approximation also can be used to approximate other distributions, such as the Poisson distribution (see Table C in Appendix C).

Applying the Concepts **6–6**

How Safe Are You?

Assume one of your favorite activities is mountain climbing. When you go mountain climbing, you have several safety devices to keep you from falling. You notice that attached to one of your safety hooks is a reliability rating of 97%. You estimate that throughout the next year you will be using this device about 100 times. Answer the following questions.

1. Does a reliability rating of 97% mean that there is a 97% chance that the device will not fail any of the 100 times?
2. What is the probability of at least one failure?
3. What is the complement of this event?
4. Can this be considered a binomial experiment?
5. Can you use the binomial probability formula? Why or why not?
6. Find the probability of at least two failures.
7. Can you use a normal distribution to accurately approximate the binomial distribution? Explain why or why not.
8. Is correction for continuity needed?
9. How much safer would it be to use a second safety hook independently of the first?

See page 345 for the answers.

Exercises 6–6

1. Explain why a normal distribution can be used as an approximation to a binomial distribution. What conditions must be met to use the normal distribution to approximate the binomial distribution? Why is a correction for continuity necessary?

2 (ans) Use the normal approximation to the binomial to find the probabilities for the specific value(s) of X.

   a. $n = 30, p = 0.5, X = 18$
   b. $n = 50, p = 0.8, X = 44$
   c. $n = 100, p = 0.1, X = 12$
   d. $n = 10, p = 0.5, X \geq 7$
   e. $n = 20, p = 0.7, X \leq 12$
   f. $n = 50, p = 0.6, X \leq 40$

3. Check each binomial distribution to see whether it can be approximated by a normal distribution (i.e., are $np \geq 5$ and $nq \geq 5$?).

   a. $n = 20, p = 0.5$ d. $n = 50, p = 0.2$
   b. $n = 10, p = 0.6$ e. $n = 30, p = 0.8$
   c. $n = 40, p = 0.9$ f. $n = 20, p = 0.85$

4. Of all 3- to 5-year-old children, 56% are enrolled in school. If a sample of 500 such children is randomly selected, find the probability that at least 250 will be enrolled in school.

   Source: *Statistical Abstract of the United States.*

5. Two out of five adult smokers acquired the habit by age 14. If 400 smokers are randomly selected, find the probability that 170 or more acquired the habit by age 14.

   Source: *Harper's Index.*

6. A theater owner has found that 5% of patrons do not show up for the performance that they purchased tickets for. If the theater has 100 seats, find the probability that six or more patrons will not show up for the sold-out performance.

7. The percentage of Americans 25 years or older who have at least some college education is 50.9%. In a random sample of 300 Americans 25 years old and older, what is the probability that more than 175 have at least some college education?

   Source: *N.Y. Times Almanac.*

8. According to recent surveys, 53% of households have personal computers. If a random sample of 175 households is selected, what is the probability that more than 75 but fewer than 110 have a personal computer?

   Source: *N.Y. Times Almanac.*

9. The percentage of female Americans 25 years old and older who have completed 4 years of college or more is 23.6%. In a random sample of 180, American women who are at least 25, what is the probability that more than 50 have completed 4 years of college or more?

Source: *N.Y. Times Almanac.*

10. Women make up 24% of the science and engineering workforce. In a random sample of 400 science and engineering employees, what is the probability that more than 120 are women?

Source: *Science and Engineering Indicators,* www.nsf.gov.

11. Women comprise 83.3% of all elementary school teachers. In a random sample of 300 elementary school teachers, what is the probability that more than 50 are men?

Source: *N.Y. Times Almanac.*

12. Seventy-seven percent of U.S. homes have a telephone answering device. In a random sample of 290 homes, what is the probability that more than 50 do not have an answering device?

Source: *N.Y. Times Almanac.*

13. The mayor of a small town estimates that 35% of the residents in his town favor the construction of a municipal parking lot. If there are 350 people at a town meeting, find the probability that at least 100 favor construction of the parking lot. Based on your answer, is it likely that 100 or more people would favor the parking lot?

Extending the Concepts

14. Recall that for use of a normal distribution as an approximation to the binomial distribution, the conditions $np \geq 5$ and $nq \geq 5$ must be met. For each given probability, compute the minimum sample size needed for use of the normal approximation.

   a. $p = 0.1$
   b. $p = 0.3$
   c. $p = 0.5$
   d. $p = 0.8$
   e. $p = 0.9$

6–7   Summary

A normal distribution can be used to describe a variety of variables, such as heights, weights, and temperatures. A normal distribution is bell-shaped, unimodal, symmetric, and continuous; its mean, median, and mode are equal. Since each variable has its own distribution with mean μ and standard deviation σ, mathematicians use the standard normal distribution, which has a mean of 0 and a standard deviation of 1. Other approximately normally distributed variables can be transformed to the standard normal distribution with the formula $z = (X - \mu)/\sigma$.

A normal distribution can also be used to describe a sampling distribution of sample means. These samples must be of the same size and randomly selected with replacement from the population. The means of the samples will differ somewhat from the population mean, since samples are generally not perfect representations of the population from which they came. The mean of the sample means will be equal to the population mean; and the standard deviation of the sample means will be equal to the population standard deviation, divided by the square root of the sample size. The central limit theorem states that as the size of the samples increases, the distribution of sample means will be approximately normal.

A normal distribution can be used to approximate other distributions, such as a binomial distribution. For a normal distribution to be used as an approximation, the conditions $np \geq 5$ and $nq \geq 5$ must be met. Also, a correction for continuity may be used for more accurate results.

Important Terms

central limit theorem 324

correction for continuity 333

negatively or left-skewed distribution 287

normal distribution 289

positively or right-skewed distribution 287

sampling distribution of sample means 322

sampling error 322

standard error of the mean 324

standard normal distribution 290

symmetric distribution 287

z value 290

Important Formulas

Formula for the z value (or standard score):

$$z = \frac{X - \mu}{\sigma}$$

Formula for finding a specific data value:

$$X = z \cdot \sigma + \mu$$

Formula for the mean of the sample means:

$$\mu_{\bar{X}} = \mu$$

Formula for the standard error of the mean:

$$\sigma_{\bar{X}} = \frac{\sigma}{\sqrt{n}}$$

Formula for the z value for the central limit theorem:

$$z = \frac{\bar{X} - \mu}{\sigma / \sqrt{n}}$$

Formulas for the mean and standard deviation for the binomial distribution:

$$\mu = n \cdot p \qquad \sigma = \sqrt{n \cdot p \cdot q}$$

Review Exercises

1. Find the area under the standard normal distribution curve for each.

    a. Between $z = 0$ and $z = 1.95$
    b. Between $z = 0$ and $z = 0.37$
    c. Between $z = 1.32$ and $z = 1.82$
    d. Between $z = -1.05$ and $z = 2.05$
    e. Between $z = -0.03$ and $z = 0.53$
    f. Between $z = +1.10$ and $z = -1.80$
    g. To the right of $z = 1.99$
    h. To the right of $z = -1.36$
    i. To the left of $z = -2.09$
    j. To the left of $z = 1.68$

2. Using the standard normal distribution, find each probability.

    a. $P(0 < z < 2.07)$
    b. $P(-1.83 < z < 0)$
    c. $P(-1.59 < z < +2.01)$
    d. $P(1.33 < z < 1.88)$
    e. $P(-2.56 < z < 0.37)$
    f. $P(z > 1.66)$
    g. $P(z < -2.03)$
    h. $P(z > -1.19)$
    i. $P(z < 1.93)$
    j. $P(z > -1.77)$

3. If the mean salary of auto mechanics in the United States is $27,635 and the standard deviation is $2550, find these probabilities for a randomly selected auto mechanic. Assume the variable is normally distributed.

    a. The mechanic earns more than $27,635.
    b. The mechanic earns less than $25,000.
    c. If you were offered a job at $25,000, how would you compare your salary with the salaries of the population of auto mechanics?

4. The average salary for graduates entering the actuarial field is $40,000. If the salaries are normally distributed with a standard deviation of $5000, find the probability that

    a. An individual graduate will have a salary over $45,000.
    b. A group of nine graduates will have a group average over $45,000.

    Source: www.BeAnActuary.org.

5. The speed limit on Interstate 75 around Findlay, Ohio, is 65 mph. On a clear day with no construction, the mean speed of automobiles was measured at 63 mph with a standard deviation of 8 mph. If the speeds are normally distributed, what percentage of the automobiles are

exceeding the speed limit? If the Highway Patrol decides to ticket only motorists exceeding 72 mph, what percentage of the motorists might they arrest?

6. The national average for a new car loan was 8.28%. If the rate is normally distributed with a standard deviation of 3.5%, find these probabilities.

   a. One can receive a rate less than 9%.
   b. One can receive a rate less than 8%.

   Source: www.bankrate.com/New York Times.

7. For the first 7 months of the year, the average precipitation in Toledo, Ohio, is 19.32 inches. If the average precipitation is normally distributed with a standard deviation of 2.44 inches, find these probabilities.

   a. A randomly selected year will have precipitation greater than 18 inches for the first 7 months.
   b. Five randomly selected years will have an average precipitation greater than 18 inches for the first 7 months.

   Source: Toledo Blade.

8. The average weight of an airline passenger's suitcase is 45 pounds. The standard deviation is 2 pounds. If 15% of the suitcases are overweight, find the maximum weight allowed by the airline. Assume the variable is normally distributed.

9. An educational study to be conducted requires a test score in the middle 40% range. If $\mu = 100$ and $\sigma = 15$, find the highest and lowest acceptable test scores that would enable a candidate to participate in the study. Assume the variable is normally distributed.

10. The average cost of XYZ brand running shoes is $83 per pair, with a standard deviation of $8. If nine pairs of running shoes are selected, find the probability that the mean cost of a pair of shoes will be less than $80. Assume the variable is normally distributed.

11. A recent study of the life span of portable compact disc players found the average to be 3.7 years with a standard deviation of 0.6 year. If a random sample of 32 people who own CD players is selected, find the probability that the mean lifetime of the sample will be less than 3.4 years. If the mean is less than 3.4 years, would you consider that 3.7 years might be incorrect?

12. The probability of winning on a slot machine is 5%. If a person plays the machine 500 times, find the probability of winning 30 times. Use the normal approximation to the binomial distribution.

13. Of the total population of older Americans, 18% live in Florida. For a randomly selected sample of 200 older Americans, find the probability that more than 40 live in Florida.

   Source: Elizabeth Vierck, *Fact Book on Aging.*

14. In a large university, 30% of the incoming freshmen elect to enroll in a personal finance course offered by the university. Find the probability that of 800 randomly selected incoming freshmen, at least 260 have elected to enroll in the course.

15. Of the total population of the United States, 20% live in the northeast. If 200 residents of the United States are selected at random, find the probability that at least 50 live in the northeast.

   Source: Statistical Abstract of the United States.

 16. The heights (in feet above sea level) of a random sample of the world's active volcanoes are shown here. Check for normality.

13,435	5,135	11,339	12,224	7,470
9,482	12,381	7,674	5,223	5,631
3,566	7,113	5,850	5,679	15,584
5,587	8,077	9,550	8,064	2,686
5,250	6,351	4,594	2,621	9,348
6,013	2,398	5,658	2,145	3,038

   Source: N.Y. Times Almanac.

 17. A random sample of enrollments in Pennsylvania's private four-year colleges is listed here. Check for normality.

1350	1886	1743	1290	1767
2067	1118	3980	1773	4605
1445	3883	1486	980	1217
3587				

   Source: N.Y. Times Almanac

 **Statistics Today**

What Is Normal?—Revisited

Many of the variables measured in medical tests—blood pressure, triglyceride level, etc.—are approximately normally distributed for the majority of the population in the United States. Thus, researchers can find the mean and standard deviation of these variables. Then, using these two measures along with the z values, they can find normal intervals for healthy individuals. For example, 95% of the systolic blood pressures of healthy individuals fall within 2 standard deviations of the mean. If an individual's pressure is outside the determined normal range (either above or below), the physician will look for a possible cause and prescribe treatment if necessary.

Chapter Quiz

Determine whether each statement is true or false. If the statement is false, explain why.

1. The total area under a normal distribution is infinite.

2. The standard normal distribution is a continuous distribution.

3. All variables that are approximately normally distributed can be transformed to standard normal variables.

4. The z value corresponding to a number below the mean is always negative.

5. The area under the standard normal distribution to the left of $z = 0$ is negative.

6. The central limit theorem applies to means of samples selected from different populations.

Select the best answer.

7. The mean of the standard normal distribution is

   a. 0   c. 100
   b. 1   d. Variable

8. Approximately what percentage of normally distributed data values will fall within 1 standard deviation above or below the mean?

   a. 68%
   b. 95%
   c. 99.7%
   d. Variable

9. Which is not a property of the standard normal distribution?

   a. It's symmetric about the mean.
   b. It's uniform.
   c. It's bell-shaped.
   d. It's unimodal.

10. When a distribution is positively skewed, the relationship of the mean, median, and mode from left to right will be

    a. Mean, median, mode
    b. Mode, median, mean
    c. Median, mode, mean
    d. Mean, mode, median

11. The standard deviation of all possible sample means equals

    a. The population standard deviation.
    b. The population standard deviation divided by the population mean.
    c. The population standard deviation divided by the square root of the sample size.
    d. The square root of the population standard deviation.

Complete the following statements with the best answer.

12. When one is using the standard normal distribution, $P(z < 0) =$ _____.

13. The difference between a sample mean and a population mean is due to _____.

14. The mean of the sample means equals _____.

15. The standard deviation of all possible sample means is called _____.

16. The normal distribution can be used to approximate the binomial distribution when $n \cdot p$ and $n \cdot q$ are both greater than or equal to _____.

17. The correction factor for the central limit theorem should be used when the sample size is greater than _____ the size of the population.

18. Find the area under the standard normal distribution for each.

    a. Between 0 and 1.50
    b. Between 0 and -1.25
    c. Between 1.56 and 1.96
    d. Between -1.20 and -2.25
    e. Between -0.06 and 0.73
    f. Between 1.10 and -1.80
    g. To the right of $z = 1.75$
    h. To the right of $z = -1.28$
    i. To the left of $z = -2.12$
    j. To the left of $z = 1.36$

19. Using the standard normal distribution, find each probability.

    a. $P(0 < z < 2.16)$
    b. $P(-1.87 < z < 0)$
    c. $P(-1.63 < z < 2.17)$
    d. $P(1.72 < z < 1.98)$
    e. $P(-2.17 < z < 0.71)$
    f. $P(z > 1.77)$
    g. $P(z < -2.37)$
    h. $P(z > -1.73)$
    i. $P(z < 2.03)$
    j. $P(z > -1.02)$

20. The average amount of rain per year in Greenville is 49 inches. The standard deviation is 8 inches. Find the probability that next year Greenville will receive the following amount of rainfall. Assume the variable is normally distributed.

    a. At most 55 inches of rain
    b. At least 62 inches of rain
    c. Between 46 and 54 inches of rain
    d. How many inches of rain would you consider to be an extremely wet year?

21. The average height of a certain age group of people is 53 inches. The standard deviation is 4 inches. If the

variable is normally distributed, find the probability that a selected individual's height will be

a. Greater than 59 inches.
b. Less than 45 inches.
c. Between 50 and 55 inches.
d. Between 58 and 62 inches.

22. The average number of gallons of lemonade consumed by the football team during a game is 20, with a standard deviation of 3 gallons. Assume the variable is normally distributed. When a game is played, find the probability of using

a. Between 20 and 25 gallons.
b. Less than 19 gallons.
c. More than 21 gallons.
d. Between 26 and 28 gallons.

23. The average number of years a person takes to complete a graduate degree program is 3. The standard deviation is 4 months. Assume the variable is normally distributed. If an individual enrolls in the program, find the probability that it will take

a. More than 4 years to complete the program.
b. Less than 3 years to complete the program.
c. Between 3.8 and 4.5 years to complete the program.
d. Between 2.5 and 3.1 years to complete the program.

24. On the daily run of an express bus, the average number of passengers is 48. The standard deviation is 3. Assume the variable is normally distributed. Find the probability that the bus will have

a. Between 36 and 40 passengers.
b. Fewer than 42 passengers.
c. More than 48 passengers.
d. Between 43 and 47 passengers.

25. The average thickness of books on a library shelf is 8.3 centimeters. The standard deviation is 0.6 centimeter. If 20% of the books are oversized, find the minimum thickness of the oversized books on the library shelf. Assume the variable is normally distributed.

26. Membership in an elite organization requires a test score in the upper 30% range. If $\mu = 115$ and $\sigma = 12$, find the lowest acceptable score that would enable a candidate to apply for membership. Assume the variable is normally distributed.

27. The average repair cost of a microwave oven is $55, with a standard deviation of $8. The costs are normally distributed. If 12 ovens are repaired, find the probability that the mean of the repair bills will be greater than $60.

28. The average electric bill in a residential area is $72 for the month of April. The standard deviation is $6. If the amounts of the electric bills are normally distributed, find the probability that the mean of the bill for 15 residents will be less than $75.

29. According to a recent survey, 38% of Americans get 6 hours or less of sleep each night. If 25 people are selected, find the probability that 14 or more people will get 6 hours or less of sleep each night. Does this number seem likely?

Source: Amazing Almanac.

30. If 10% of the people in a certain factory are members of a union, find the probability that, in a sample of 2000, fewer than 180 people are union members.

31. The percentage of U.S. households that have online connections is 44.9%. In a random sample of 420 households, what is the probability that fewer than 200 have online connections?

Source: N.Y. Times Almanac.

32. Fifty-three percent of U.S. households have a personal computer. In a random sample of 250 households, what is the probability that fewer than 120 have a PC?

Source: N.Y. Times Almanac.

 33. The number of calories contained in a selection of fast-food sandwiches is shown here. Check for normality.

390	405	580	300	320
540	225	720	470	560
535	660	530	290	440
390	675	530	1010	450
320	460	290	340	610
430	530			

Source: The Doctor's Pocket Calorie, Fat, and Carbohydrate Counter.

34. The average GMAT scores for the top-30 ranked graduate schools of business are listed here. Check for normality.

718	703	703	703	700	690	695	705	690	688
676	681	689	686	691	669	674	652	680	670
651	651	637	662	641	645	645	642	660	636

Source: U.S. News & World Report Best Graduate Schools.

Critical Thinking Challenges

Sometimes a researcher must decide whether a variable is normally distributed. There are several ways to do this. One simple but very subjective method uses special graph paper,

which is called *normal probability paper.* For the distribution of systolic blood pressure readings given in Chapter 3 of the textbook, the following method can be used:

1. Make a table, as shown.

Boundaries	Frequency	Cumulative frequency	Cumulative percent frequency
89.5–104.5	24		
104.5–119.5	62		
119.5–134.5	72		
134.5–149.5	26		
149.5–164.5	12		
164.5–179.5	4		
	200		

2. Find the cumulative frequencies for each class, and place the results in the third column.

3. Find the cumulative percents for each class by dividing each cumulative frequency by 200 (the total frequencies) and multiplying by 100%. (For the first class, it would be $24/200 \times 100\% = 12\%$.) Place these values in the last column.

4. Using the normal probability paper shown in Table 6–3, label the x axis with the class boundaries as shown and plot the percents.

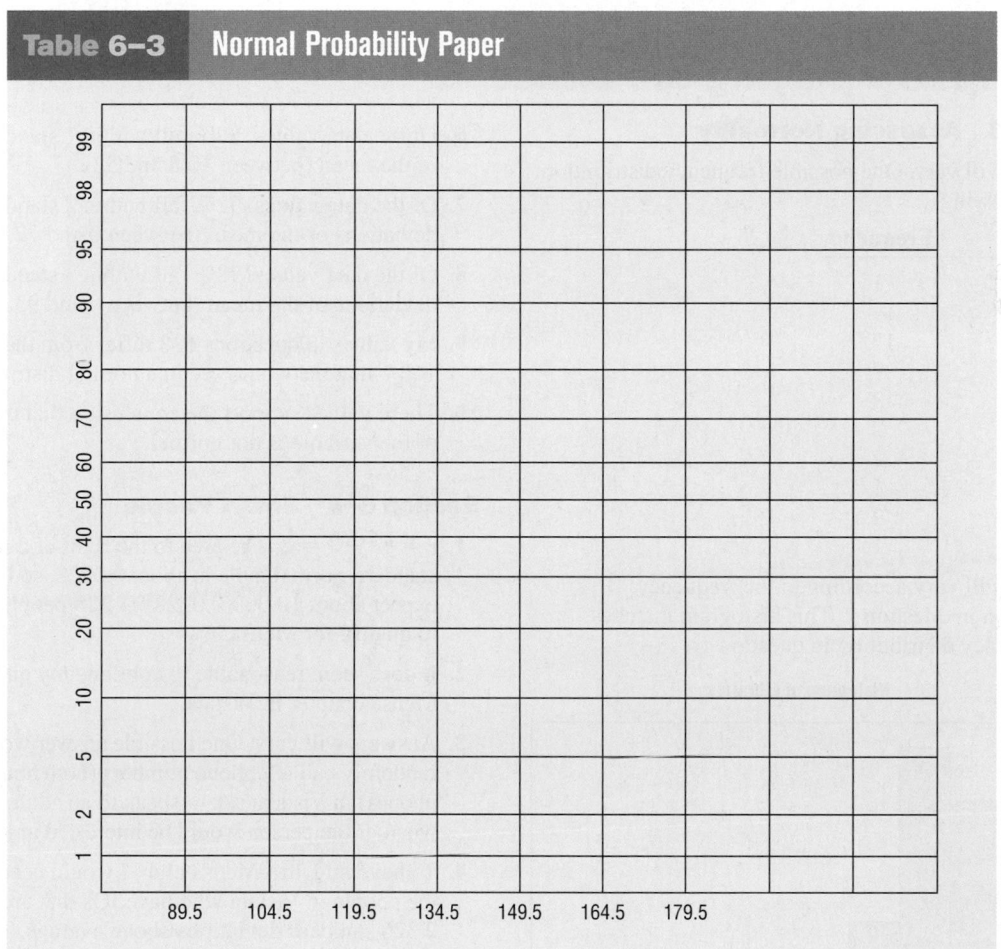

Table 6–3 **Normal Probability Paper**

5. If the points fall approximately in a straight line, it can be concluded that the distribution is normal. Do you feel that this distribution is approximately normal? Explain your answer.

6. To find an approximation of the mean or median, draw a horizontal line from the 50% point on the y axis over to the curve and then a vertical line down to the x axis. Compare this approximation of the mean with the computed mean.

7. To find an approximation of the standard deviation, locate the values on the x axis that correspond to the 16 and 84% values on the y axis. Subtract these two values and divide the result by 2. Compare this approximate standard deviation to the computed standard deviation.

8. Explain why the method used in step 7 works.

Data Projects

1. Select a variable (interval or ratio) and collect 30 data values. Some suggestions might include heights of the students in your class, grade point averages of students, running times of feature length movies, etc.

   a. Construct a frequency distribution for the variable.
   b. Use the procedure described in the Critical Thinking Challenge on page 342 to graph the distribution on normal probability paper.

   c. Can you conclude that the data are approximately normally distributed? Explain your answer.

2. Repeat Exercise 1, using all the values from Data Set II in Appendix D.

3. Repeat Exercise 1, using a random sample of 30 values from Data Set III in Appendix D.

Answers to Applying the Concepts

Section 6–3   Assessing Normality

1. Answers will vary. One possible frequency distribution is the following:

Branches	Frequency
0–9	1
10–19	14
20–29	17
30–39	7
40–49	3
50–59	2
60–69	2
70–79	1
80–89	2
90–99	1

2. Answers will vary according to the frequency distribution in question 1. This histogram matches the frequency distribution in question 1.

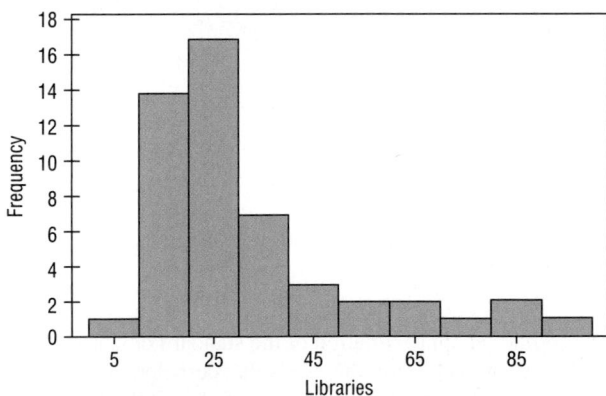

Histogram of Libraries

3. The histogram is unimodal and skewed to the right (positively skewed).

4. The distribution does not appear to be normal.

5. The mean number of branches is $\bar{x} = 31.4$, and the standard deviation is $s = 20.6$.

6. Of the data values, 80% fall within 1 standard deviation of the mean (between 10.8 and 52).

7. Of the data values, 92% fall within 2 standard deviations of the mean (between 0 and 72.6).

8. Of the data values, 98% fall within 3 standard deviations of the mean (between 0 and 93.2).

9. My values in questions 6–8 differ from the 68, 95, and 100% that we would see in a normal distribution.

10. These values support the conclusion that the distribution of the variable is not normal.

Section 6–4   Smart People

1. $z = \frac{130 - 100}{15} = 2$. The area to the right of 2 in the standard normal table is about 0.0228, so I would expect about $10,000(0.0228) = 228$ people in Visiala to qualify for Mensa.

2. It does seem reasonable to continue my quest to start a Mensa chapter in Visiala.

3. Answers will vary. One possible answer would be to randomly call telephone numbers (both home and cell phones) in Visiala, ask to speak to an adult, and ask whether the person would be interested in joining Mensa.

4. To have an Ultra-Mensa club, I would need to find the people in Visiala who have IQs that are at least 2.326 standard deviations above average. This means that I would need to recruit those with IQs that are at least 135:

$$2.326 = \frac{x - 100}{15} \Rightarrow x = 100 + 2.326(15) = 134.89$$

Section 6–5   Central Limit Theorem

1. It is very unlikely that we would ever get the same results for any of our random samples. While it is a remote possibility, it is highly unlikely.

2. A good estimate for the population mean would be to find the average of the students' sample means.

Similarly, a good estimate for the population standard deviation would be to find the average of the students' sample standard deviations.

3. The distribution appears to be somewhat left (negatively) skewed.

Histogram of Central Limit Theorem Means

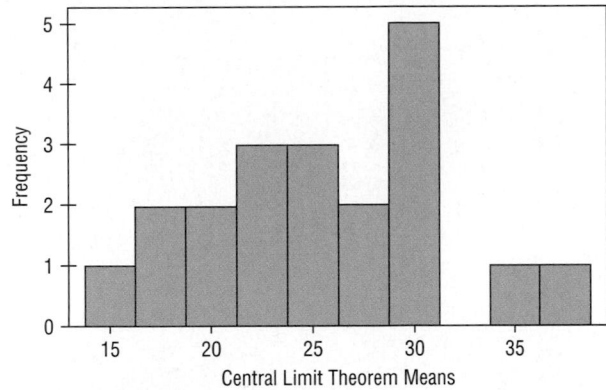

4. The mean of the students' means is 25.4, and the standard deviation is 5.8.

5. The distribution of the means is not a sampling distribution, since it represents just 20 of all possible samples of size 30 from the population.

6. The sampling error for student 3 is $18 - 25.4 = -7.4$; the sampling error for student 7 is $26 - 25.4 = +0.6$; the sampling error for student 14 is $29 - 25.4 = +3.6$.

7. The standard deviation for the sample of the 20 means is greater than the standard deviations for each of the individual students. So it is not equal to the standard deviation divided by the square root of the sample size.

Section 6–6   How Safe Are You?

1. A reliability rating of 97% means that, on average, the device will not fail 97% of the time. We do not know how many times it will fail for any particular set of 100 climbs.

2. The probability of at least 1 failure in 100 climbs is $1 - (0.97)^{100} = 1 - 0.0476 = 0.9524$ (about 95%).

3. The complement of the event in question 2 is the event of "no failures in 100 climbs."

4. This can be considered a binomial experiment. We have two outcomes: success and failure. The probability of the equipment working (success) remains constant at 97%. We have 100 independent climbs. And we are counting the number of times the equipment works in these 100 climbs.

5. We could use the binomial probability formula, but it would be very messy computationally.

6. The probability of at least two failures *cannot* be estimated with the normal distribution (see below). So the probability is $1 - [(0.97)^{100} + 100(0.97)^{99} (0.03)] = 1 - 0.1946 = 0.8054$ (about 80.5%).

7. We *should not* use the normal approximation to the binomial since $nq < 10$.

8. If we had used the normal approximation, we would have needed a correction for continuity, since we would have been approximating a discrete distribution with a continuous distribution.

9. Since a second safety hook will be successful or fail independently of the first safety hook, the probability of failure drops from 3% to $(0.03)(0.03) = 0.0009$, or 0.09%.

CHAPTER

7

Confidence Intervals and Sample Size

Objectives

After completing this chapter, you should be able to

1 Find the confidence interval for the mean when σ is known or $n \geq 30$.

2 Determine the minimum sample size for finding a confidence interval for the mean.

3 Find the confidence interval for the mean when σ is unknown and $n < 30$.

4 Find the confidence interval for a proportion.

5 Determine the minimum sample size for finding a confidence interval for a proportion.

6 Find a confidence interval for a variance and a standard deviation.

Outline

Statistics
Today

Would You Change the Channel?

A survey by the Roper Organization found that 45% of the people who were offended by a television program would change the channel, while 15% would turn off their television sets. The survey further stated that the margin of error is 3 percentage points, and 4000 adults were interviewed.

Several questions arise:

1. How do these estimates compare with the true population percentages?

2. What is meant by a margin of error of 3 percentage points?

3. Is the sample of 4000 large enough to represent the population of all adults who watch television in the United States?

See Statistics Today—Revisited at the end of the chapter for the answers.

After reading this chapter, you will be able to answer these questions, since this chapter explains how statisticians can use statistics to make estimates of parameters.

Source: The Associated Press.

7–1   Introduction

One aspect of inferential statistics is **estimation,** which is the process of estimating the value of a parameter from information obtained from a sample. For example, *The Book of Odds,* by Michael D. Shook and Robert L. Shook (New York: Penguin Putnam, Inc.), contains the following statements:

> *"One out of 4 Americans is currently dieting."* (Calorie Control Council)
> *"Seventy-two percent of Americans have flown on commercial airlines."* (*"The Bristol Meyers Report: Medicine in the Next Century")*
> *"The average kindergarten student has seen more than 5000 hours of television."* (U.S. Department of Education)

"The average school nurse makes $32,786 a year." (National Association of School Nurses)
"The average amount of life insurance is $108,000 per household with life insurance."
(American Council of Life Insurance)

Since the populations from which these values were obtained are large, these values are only *estimates* of the true parameters and are derived from data collected from samples.

The statistical procedures for estimating the population mean, proportion, variance, and standard deviation will be explained in this chapter.

An important question in estimation is that of sample size. How large should the sample be in order to make an accurate estimate? This question is not easy to answer since the size of the sample depends on several factors, such as the accuracy desired and the probability of making a correct estimate. The question of sample size will be explained in this chapter also.

7–2

Confidence Intervals for the Mean (σ Known or $n \geq 30$) and Sample Size

Objective 1

Find the confidence interval for the mean when σ is known or $n \geq 30$.

Suppose a college president wishes to estimate the average age of students attending classes this semester. The president could select a random sample of 100 students and find the average age of these students, say, 22.3 years. From the sample mean, the president could infer that the average age of all the students is 22.3 years. This type of estimate is called a *point estimate*.

A **point estimate** is a specific numerical value estimate of a parameter. The best point estimate of the population mean μ is the sample mean $\overline{X}$.

One might ask why other measures of central tendency, such as the median and mode, are not used to estimate the population mean. The reason is that the means of samples vary less than other statistics (such as medians and modes) when many samples are selected from the same population. Therefore, the sample mean is the best estimate of the population mean.

Sample measures (i.e., statistics) are used to estimate population measures (i.e., parameters). These statistics are called **estimators.** As previously stated, the sample mean is a better estimator of the population mean than the sample median or sample mode.

A good estimator should satisfy the three properties described now.

Three Properties of a Good Estimator

1. The estimator should be an **unbiased estimator.** That is, the expected value or the mean of the estimates obtained from samples of a given size is equal to the parameter being estimated.
2. The estimator should be consistent. For a **consistent estimator,** as sample size increases, the value of the estimator approaches the value of the parameter estimated.
3. The estimator should be a **relatively efficient estimator.** That is, of all the statistics that can be used to estimate a parameter, the relatively efficient estimator has the smallest variance.

Confidence Intervals

As stated in Chapter 6, the sample mean will be, for the most part, somewhat different from the population mean due to sampling error. Therefore, one might ask a second question: How good is a point estimate? The answer is that there is no way of knowing how close a particular point estimate is to the population mean.

This answer places some doubt on the accuracy of point estimates. For this reason, statisticians prefer another type of estimate, called an *interval estimate*.

An **interval estimate** of a parameter is an interval or a range of values used to estimate the parameter. This estimate may or may not contain the value of the parameter being estimated.

In an interval estimate, the parameter is specified as being between two values. For example, an interval estimate for the average age of all students might be $26.9 < \mu < 27.7$, or 27.3 ± 0.4 years.

Either the interval contains the parameter or it does not. A degree of confidence (usually a percent) can be assigned before an interval estimate is made. For instance, one may wish to be 95% confident that the interval contains the true population mean. Another question then arises. Why 95%? Why not 99 or 99.5%?

If one desires to be more confident, such as 99 or 99.5% confident, then the interval must be larger. For example, a 99% confidence interval for the mean age of college students might be $26.7 < \mu < 27.9$, or 27.3 ± 0.6. Hence, a tradeoff occurs. To be more confident that the interval contains the true population mean, one must make the interval wider.

The **confidence level** of an interval estimate of a parameter is the probability that the interval estimate will contain the parameter, assuming that a large number of samples are selected and that the estimation process on the same parameter is repeated.

A **confidence interval** is a specific interval estimate of a parameter determined by using data obtained from a sample and by using the specific confidence level of the estimate.

Intervals constructed in this way are called *confidence intervals*. Three common confidence intervals are used: the 90, the 95, and the 99% confidence intervals.

The algebraic derivation of the formula for determining a confidence interval for a mean will be shown later. A brief intuitive explanation will be given first.

The central limit theorem states that when the sample size is large, approximately 95% of the sample means will fall within ± 1.96 standard errors of the population mean, that is,

$$\mu \pm 1.96\left(\frac{\sigma}{\sqrt{n}}\right)$$

Now, if a specific sample mean is selected, say, $\bar{X}$, there is a 95% probability that it falls within the range of $\mu \pm 1.96(\sigma/\sqrt{n})$. Likewise, there is a 95% probability that the interval specified by

$$\bar{X} \pm 1.96\left(\frac{\sigma}{\sqrt{n}}\right)$$

will contain μ, as will be shown later. Stated another way,

$$\bar{X} - 1.96\left(\frac{\sigma}{\sqrt{n}}\right) < \mu < \bar{X} + 1.96\left(\frac{\sigma}{\sqrt{n}}\right)$$

$\mathcal{H}istorical\ Notes$

Point and interval estimates were known as long ago as the late 1700s. However, it wasn't until 1937 that a mathematician, J. Neyman, formulated practical applications for them.

Hence, one can be 95% confident that the population mean is contained within that interval when the values of the variable are normally distributed in the population.

The value used for the 95% confidence interval, 1.96, is obtained from Table E in Appendix C. For a 99% confidence interval, the value 2.58 is used instead of 1.96 in the formula. This value is also obtained from Table E and is based on the standard normal distribution. Since other confidence intervals are used in statistics, the symbol $z_{\alpha/2}$ (read "zee sub alpha over two") is used in the general formula for confidence intervals. The Greek letter α (alpha) represents the total area in both tails of the standard normal distribution curve, and $\alpha/2$ represents the area in each one of the tails. More will be said after Examples 7–1 and 7–2 about finding other values for $z_{\alpha/2}$.

The relationship between α and the confidence level is that the stated confidence level is the percentage equivalent to the decimal value of $1 - \alpha$, and vice versa. When the 95% confidence interval is to be found, $\alpha = 0.05$, since $1 - 0.05 = 0.95$, or 95%. When $\alpha = 0.01$, then $1 - \alpha = 1 - 0.01 = 0.99$, and the 99% confidence interval is being calculated.

Formula for the Confidence Interval of the Mean for a Specific α

$$\bar{X} - z_{\alpha/2}\left(\frac{\sigma}{\sqrt{n}}\right) < \mu < \bar{X} + z_{\alpha/2}\left(\frac{\sigma}{\sqrt{n}}\right)$$

For a 90% confidence interval, $z_{\alpha/2} = 1.65$; for a 95% confidence interval, $z_{\alpha/2} = 1.96$; and for a 99% confidence interval, $z_{\alpha/2} = 2.58$.

The term $z_{\alpha/2}(\sigma/\sqrt{n})$ is called the *maximum error of estimate*. For a specific value, say, $\alpha = 0.05$, 95% of the sample means will fall within this error value on either side of the population mean, as previously explained. See Figure 7–1.

Figure 7–1

95% Confidence Interval

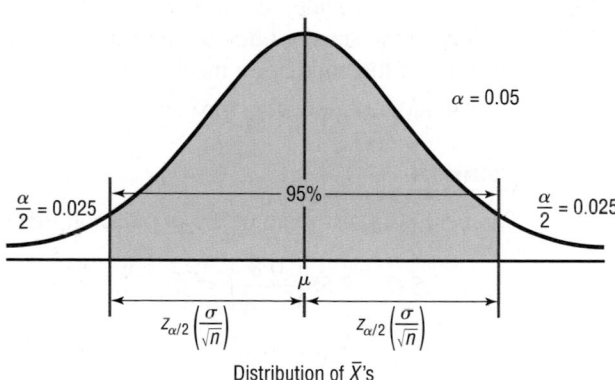

Distribution of $\bar{X}$'s

The **maximum error of estimate** is the maximum likely difference between the point estimate of a parameter and the actual value of the parameter.

A more detailed explanation of the maximum error of estimate follows Examples 7–1 and 7–2, which illustrate the computation of confidence intervals.

Rounding Rule for a Confidence Interval for a Mean

When you are computing a confidence interval for a population mean by using *raw data*, round off to one more decimal place than the number of decimal places in the original data. When you are computing a confidence interval for a population mean by using a sample mean and a standard deviation, round off to the same number of decimal places as given for the mean.

Example 7–1

A researcher wishes to estimate the average amount of money a person spends on lottery tickets each month. A sample of 50 people who play the lottery found the mean to be $19 and the standard deviation to be 6.8. Find the best point estimate of the population mean and the 95% confidence interval of the population mean.

Source: USA TODAY.

Solution

The best point estimate of the mean is $19. For the 95% confidence interval use $z = 1.96$.

$$19 - 1.96\left(\frac{6.8}{\sqrt{50}}\right) < \mu < 19 + 1.96\left(\frac{6.8}{\sqrt{50}}\right)$$

$$19 - 1.9 < \mu < 19 + 1.9$$

$$17.1 < \mu < 20.9$$

or $$19 \pm 1.9$$

Hence, one can say with 95% confidence the true mean of the population is between $17.10 and $20.90, based on a sample of 50 people who play the lottery.

Example 7–2

A survey of 30 adults found that the mean age of a person's primary vehicle is 5.6 years. Assuming the standard deviation of the population is 0.8 year, find the best point estimate of the population mean and the 99% confidence interval of the population mean.

Source: Based on information in *USA TODAY.*

Solution

The best point estimate of the population mean is 5.6 years.

$$5.6 - 2.58\left(\frac{0.8}{\sqrt{30}}\right) < \mu < 5.6 + 2.58\left(\frac{0.8}{\sqrt{30}}\right)$$

$$5.6 - 0.38 < \mu < 5.6 + 0.38$$

$$5.22 < \mu < 5.98$$

or $$5.2 < \mu < 6.0 \text{ (rounded)}$$

Hence, one can be 99% confident that the mean age of all primary vehicles is between 5.2 and 6.0 years, based on 30 vehicles.

Another way of looking at a confidence interval is shown in Figure 7–2. According to the central limit theorem, approximately 95% of the sample means fall within 1.96 standard deviations of the population mean if the sample size is 30 or more or if σ is known when n is less than 30 and the population is normally distributed. If it were possible to build a confidence interval about each sample mean, as was done in Examples 7–1

Figure 7–2

95% Confidence Interval for Sample Means

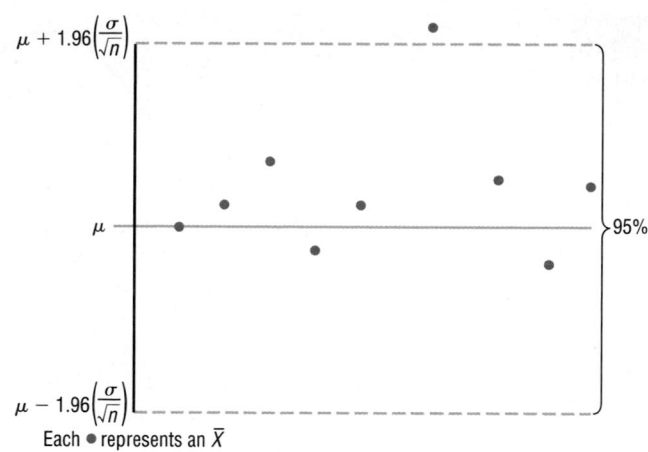

Each ● represents an $\overline{X}$

Figure 7–3

95% Confidence Intervals for Each Sample Mean

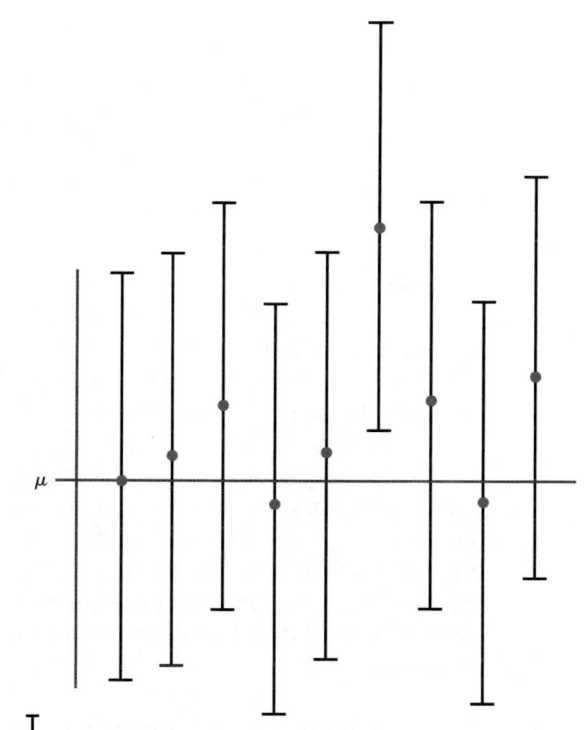

Each I represents an interval about a sample mean

and 7–2 for μ, 95% of these intervals would contain the population mean, as shown in Figure 7–3. Hence, one can be 95% confident that an interval built around a specific sample mean would contain the population mean. If one desires to be 99% confident, the confidence intervals must be enlarged so that 99 out of every 100 intervals contain the population mean.

Since other confidence intervals (besides 90, 95, and 99%) are sometimes used in statistics, an explanation of how to find the values for $z_{\alpha/2}$ is necessary. As stated previously, the Greek letter α represents the total of the areas in both tails of the normal distribution. The value for α is found by subtracting the decimal equivalent for the desired confidence level from 1. For example, if one wanted to find the 98% confidence interval, one would change 98% to 0.98 and find $\alpha = 1 - 0.98$, or 0.02. Then $\alpha/2$ is obtained by dividing α by 2. So $\alpha/2$ is 0.02/2, or 0.01. Finally, $z_{0.01}$ is the z value that will give an area of 0.01 in the right tail of the standard normal distribution curve. See Figure 7–4.

Figure 7–4

Finding $\alpha/2$ for a 98% Confidence Interval

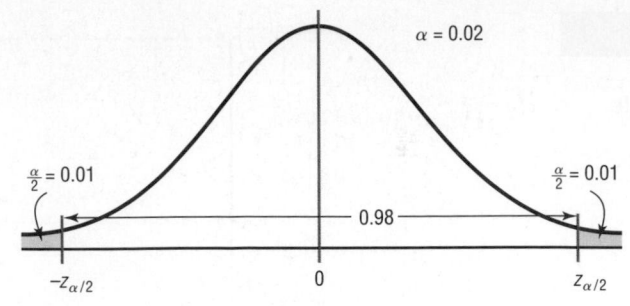

$\alpha = 0.02$

$\frac{\alpha}{2} = 0.01$ $\frac{\alpha}{2} = 0.01$

0.98

$-z_{\alpha/2}$ 0 $z_{\alpha/2}$

Figure 7–5

Finding $z_{\alpha/2}$ for a 98% Confidence Interval

Table E
The Standard Normal Distribution

z	.00	.01	.02	.03	...	.09
0.0						
0.1						
⋮						
2.3				0.4901		

Once $\alpha/2$ is determined, the corresponding $z_{\alpha/2}$ value can be found by using the procedure shown in Chapter 6 (see Example 6–17), which is reviewed here. To get the $z_{\alpha/2}$ value for a 98% confidence interval, subtract 0.01 from 0.5000 to get 0.4900. Next, locate the area that is closest to 0.4900 (in this case, 0.4901) in Table E, and then find the corresponding z value. In this example, it is 2.33. See Figure 7–5.

For confidence intervals, only the positive z value is used in the formula.

When the original variable is normally distributed and α is known, the standard normal distribution can be used to find confidence intervals regardless of the size of the sample. When $n \geq 30$, the distribution of means will be approximately normal even if the original distribution of the variable departs from normality. Also, if $n \geq 30$ (some authors use $n > 30$), s can be substituted for σ in the formula for confidence intervals; and the standard normal distribution can be used to find confidence intervals for means, as shown in Example 7–3.

Example 7–3

The following data represent a sample of the assets (in millions of dollars) of 30 credit unions in southwestern Pennsylvania. Find the 90% confidence interval of the mean.

12.23	16.56	4.39
2.89	1.24	2.17
13.19	9.16	1.42
73.25	1.91	14.64
11.59	6.69	1.06
8.74	3.17	18.13
7.92	4.78	16.85
40.22	2.42	21.58
5.01	1.47	12.24
2.27	12.77	2.76

Source: Pittsburgh Post Gazette.

Solution

Step 1   Find the mean and standard deviation for the data. Use the formulas shown in Chapter 3 or your calculator. The mean $\bar{X} = 11.091$. The standard deviation $s = 14.405$.

Step 2   Find $\alpha/2$. Since the 90% confidence interval is to be used, $\alpha = 1 - 0.90 = 0.10$, and

$$\frac{\alpha}{2} = \frac{0.10}{2} = 0.05$$

Step 3   Find $z_{\alpha/2}$. Subtract 0.05 from 0.5000 to get 0.4500. The corresponding z value obtained from Table E is 1.65. (*Note:* This value is found by using the z value for an area between 0.4495 and 0.4505. A more precise z value obtained mathematically is 1.645 and is sometimes used; however, 1.65 will be used in this textbook.)

Step 4   Substitute in the formula

$$\bar{X} - z_{\alpha/2}\left(\frac{s}{\sqrt{n}}\right) < \mu < \bar{X} + z_{\alpha/2}\left(\frac{s}{\sqrt{n}}\right)$$

(Since $n \geq 30$, s is used in place of σ when σ is unknown.)

$$11.091 - 1.65\left(\frac{14.405}{\sqrt{30}}\right) < \mu < 11.091 + 1.65\left(\frac{14.405}{\sqrt{30}}\right)$$

$$11.091 - 4.339 < \mu < 11.091 + 4.339$$

$$6.752 < \mu < 15.430$$

Hence, one can be 90% confident that the population mean of the assets of all credit unions is between $6.752 million and $15.430 million, based on a sample of 30 credit unions.

Comment to Computer and Statistical Calculator Users

This chapter and subsequent chapters include examples using raw data. If you are using computer or calculator programs to find the solutions, the answers you get may vary somewhat from the ones given in the textbook. This is so because computers and calculators do not round the answers in the intermediate steps and can use 12 or more decimal places for computation. Also, they use more exact values than those given in the tables in the back of this book. These discrepancies are part and parcel of statistics.

Sample Size

Objective 2

Determine the minimum sample size for finding a confidence interval for the mean.

Sample size determination is closely related to statistical estimation. Quite often, one asks, How large a sample is necessary to make an accurate estimate? The answer is not simple, since it depends on three things: the maximum error of estimate, the population standard deviation, and the degree of confidence. For example, how close to the true mean does one want to be (2 units, 5 units, etc.), and how confident does one wish to be (90, 95, 99%, etc.)? For the purpose of this chapter, it will be assumed that the population standard deviation of the variable is known or has been estimated from a previous study.

The formula for sample size is derived from the maximum error of estimate formula

$$E = z_{\alpha/2}\left(\frac{\sigma}{\sqrt{n}}\right)$$

and this formula is solved for n as follows:

$$E\sqrt{n} = z_{\alpha/2}(\sigma)$$

$$\sqrt{n} = \frac{z_{\alpha/2} \cdot \sigma}{E}$$

Hence, $\quad n = \left(\frac{z_{\alpha/2} \cdot \sigma}{E}\right)^2$

Formula for the Minimum Sample Size Needed for an Interval Estimate of the Population Mean

$$n = \left(\frac{z_{\alpha/2} \cdot \sigma}{E}\right)^2$$

where E is the maximum error of estimate. If necessary, round the answer up to obtain a whole number. That is, if there is any fraction or decimal portion in the answer, use the next whole number for sample size n.

Example 7–4

The college president asks the statistics teacher to estimate the average age of the students at their college. How large a sample is necessary? The statistics teacher would like to be 99% confident that the estimate should be accurate within 1 year. From a previous study, the standard deviation of the ages is known to be 3 years.

Solution

Since $\alpha = 0.01$ (or $1 - 0.99$), $z_{\alpha/2} = 2.58$, and $E = 1$, substituting in the formula, one gets

$$n = \left(\frac{z_{\alpha/2} \cdot \sigma}{E}\right)^2 = \left[\frac{(2.58)(3)}{1}\right]^2 = 59.9$$

which is rounded up to 60. Therefore, to be 99% confident that the estimate is within 1 year of the true mean age, the teacher needs a sample size of at least 60 students. (Always round n up to the next whole number. For example, if $n = 59.2$, round it up to 60.)

Notice that when one is finding the sample size, the size of the population is irrelevant when the population is large or infinite or when sampling is done with replacement. In other cases, an adjustment is made in the formula for computing sample size. This adjustment is beyond the scope of this book.

The formula for determining sample size requires the use of the population standard deviation. What happens when σ is unknown? In this case, an attempt is made to estimate σ. One such way is to use the standard deviation s obtained from a sample taken previously as an estimate for σ. The standard deviation can also be estimated by dividing the range by 4.

Sometimes, interval estimates rather than point estimates are reported. For instance, one may read a statement: "On the basis of a sample of 200 families, the survey estimates that an American family of two spends an average of $84 per week for groceries. One

Interesting Fact

It has been estimated that the amount of pizza consumed every day in the United States would cover a farm consisting of 75 acres.

can be 95% confident that this estimate is accurate within $3 of the true mean." This statement means that the 95% confidence interval of the true mean is

$$\$84 - \$3 < \mu < \$84 + \$3$$
$$\$81 < \mu < \$87$$

The algebraic derivation of the formula for a confidence interval is shown next. As explained in Chapter 6, the sampling distribution of the mean is approximately normal when large samples ($n \geq 30$) are taken from a population. Also,

$$z = \frac{\bar{X} - \mu}{\sigma/\sqrt{n}}$$

Furthermore, there is a probability of $1 - \alpha$ that a z will have a value between $-z_{\alpha/2}$ and $+z_{\alpha/2}$. Hence,

$$-z_{\alpha/2} < \frac{\bar{X} - \mu}{\sigma/\sqrt{n}} < z_{\alpha/2}$$

Using algebra, one finds

$$-z_{\alpha/2} \cdot \frac{\sigma}{\sqrt{n}} < \bar{X} - \mu < z_{\alpha/2} \cdot \frac{\sigma}{\sqrt{n}}$$

Subtracting $\bar{X}$ from both sides and from the middle, one gets

$$-\bar{X} - z_{\alpha/2} \cdot \frac{\sigma}{\sqrt{n}} < -\mu < -\bar{X} + z_{\alpha/2} \cdot \frac{\sigma}{\sqrt{n}}$$

Multiplying by -1, one gets

$$\bar{X} + z_{\alpha/2} \cdot \frac{\sigma}{\sqrt{n}} > \mu > \bar{X} - z_{\alpha/2} \cdot \frac{\sigma}{\sqrt{n}}$$

Reversing the inequality, one gets the formula for the confidence interval:

$$\bar{X} - z_{\alpha/2} \cdot \frac{\sigma}{\sqrt{n}} < \mu < \bar{X} + z_{\alpha/2} \cdot \frac{\sigma}{\sqrt{n}}$$

Applying the Concepts **7–2**

Making Decisions with Confidence Intervals

Assume you work for Kimberly Clark Corporation, the makers of Kleenex. The job you are presently working on requires you to decide how many Kleenexes are to be put in the new automobile glove compartment boxes. Complete the following.

1. How will you decide on a reasonable number of Kleenexes to put in the boxes?
2. When do people usually need Kleenexes?
3. What type of data collection technique would you use?
4. Assume you found out that from your sample of 85 people, on average about 57 Kleenexes are used throughout the duration of a cold, with a standard deviation of 15. Use a confidence interval to help you decide how many Kleenexes will go in the boxes.
5. Explain how you decided on how many Kleenexes will go in the boxes.

See page 389 for the answers.

Exercises 7–2

1. What is the difference between a point estimate and an interval estimate of a parameter? Which is better? Why?

2. What information is necessary to calculate a confidence interval?

3. What is the maximum error of estimate?

4. What is meant by the 95% confidence interval of the mean?

5. What are three properties of a good estimator?

6. What statistic best estimates μ?

7. What is necessary to determine the sample size?

8. When one is determining the sample size for a confidence interval, is the size of the population relevant?

9. Find each.

   a. $z_{\alpha/2}$ for the 99% confidence interval
   b. $z_{\alpha/2}$ for the 98% confidence interval
   c. $z_{\alpha/2}$ for the 95% confidence interval
   d. $z_{\alpha/2}$ for the 90% confidence interval
   e. $z_{\alpha/2}$ for the 94% confidence interval

10. Find the 95% confidence interval for the mean paid attendance at the Major League All-Star games. A random sample of the paid attendances is shown.

47,596	68,751	5,838
69,831	28,843	53,107
31,391	48,829	50,706
62,892	55,105	63,974
56,674	38,362	51,549
31,938	31,851	56,088
34,906	38,359	72,086
34,009	50,850	43,801
46,127	49,926	54,960
32,785	48,321	49,671

Source: *Time Almanac.*

11. A sample of the reading scores of 35 fifth-graders has a mean of 82. The standard deviation of the sample is 15.

   a. Find the best point estimate of the mean.
   b. Find the 95% confidence interval of the mean reading scores of all fifth-graders.
   c. Find the 99% confidence interval of the mean reading scores of all fifth-graders.
   d. Which interval is larger? Explain why.

12. Find the 90% confidence interval of the population mean for the incomes of western Pennsylvania credit unions. A random sample of 50 credit unions is shown. The data are in thousands of dollars.

84	14	31	72	26
49	252	104	31	8
3	18	72	23	55
133	16	29	225	138
85	24	391	72	158
4340	346	19	5	846
461	254	125	61	123
60	29	10	366	47
28	254	6	77	21
97	6	17	8	82

Source: *Pittsburgh Post Gazette.*

13. A study of 40 English composition professors showed that they spent, on average, 12.6 minutes correcting a student's term paper.

   a. Find the best point estimate of the mean.
   b. Find the 90% confidence interval of the mean time for all composition papers when $\sigma = 2.5$ minutes.
   c. If a professor stated that he spent, on average, 30 minutes correcting a term paper, what would be your reaction?

14. A study of 35 golfers showed that their average score on a particular course was 92. The standard deviation of the sample is 5.

   a. Find the best point estimate of the mean.
   b. Find the 95% confidence interval of the mean score for all golfers.
   c. Find the 95% confidence interval of the mean score if a sample of 60 golfers is used instead of a sample of 35.
   d. Which interval is smaller? Explain why.

15. A survey of individuals who passed the seven exams and obtained the rank of Fellow in the actuarial field finds the average salary to be $150,000. If the standard deviation for the sample of 35 Fellows was $15,000, construct a 95% confidence interval for all Fellows.

Source: www.BeAnActuary.org.

16. A random sample of the number of farms (in thousands) in various states follows. Estimate the mean number of farms per state with 90% confidence.

47	95	54	33	64	4	8	57	9	80
8	90	3	49	4	44	79	80	48	16
68	7	15	21	52	6	78	109	40	50
29									

Source: *N.Y. Times Almanac.*

17. A study of 415 kindergarten students showed that they have seen on average 5000 hours of television. If the sample standard deviation is 900, find the 95% confidence level of the mean for all students. If a parent claimed that his children watched 4000 hours, would the claim be believable?

Source: U.S. Department of Education.

18. A random sample of 76 four-year-olds attending day-care centers showed that the yearly tuition averaged $3648. The standard deviation of the sample was $630, and the sample size was 50. Find the 90% confidence interval of the true mean. If a day-care center were starting up and wanted to keep tuition low, what would be a reasonable amount to charge?

19. Noise levels at various area urban hospitals were measured in decibels. The mean of the noise levels in 84 corridors was 61.2 decibels, and the standard deviation was 7.9. Find the 95% confidence interval of the true mean.

Source: M. Bayo, A. Garcia, and A. Garcia, "Noise Levels in an Urban Hospital and Workers' Subjective Responses," *Archives of Environmental Health* 50, no. 3, p. 249 (May–June 1995). Reprinted with permission of the Helen Dwight Reid Educational Foundation. Published by Heldref Publications, 1319 Eighteenth St. N.W., Washington, D.C. 20036-1802. Copyright © 1995.

20. The growing seasons for a random sample of 35 U.S. cities were recorded, yielding a sample mean of 190.7 days and a sample standard deviation of 54.2 days. Estimate the true mean population of the growing season with 95% confidence.

Source: *The Old Farmer's Almanac.*

21. A university dean of students wishes to estimate the average number of hours students spend doing homework per week. The standard deviation from a previous study is 6.2 hours. How large a sample must be selected if he wants to be 99% confident of finding whether the true mean differs from the sample mean by 1.5 hours?

22. In the hospital study cited in Exercise 19, the mean noise level in the 171 ward areas was 58.0 decibels, and the standard deviation was 4.8. Find the 90% confidence interval of the true mean.

Source: M. Bayo, A. Garcia, and A. Garcia, "Noise Levels in an Urban Hospital and Workers' Subjective Responses," *Archives of Environmental Health* 50, no. 3, p. 249 (May–June 1995). Reprinted with permission of the Helen Dwight Reid Educational Foundation. Published by Heldref Publications, 1319 Eighteenth St. N.W., Washington, D.C. 20036-1802. Copyright © 1995.

23. An insurance company is trying to estimate the average number of sick days that full-time food service workers use per year. A pilot study found the standard deviation to be 2.5 days. How large a sample must be selected if the company wants to be 95% confident of getting an interval that contains the true mean with a maximum error of 1 day?

24. A pizza shop owner wishes to find the 95% confidence interval of the true mean cost of a large plain pizza. How large should the sample be if she wishes to be accurate to within $0.15? A previous study showed that the standard deviation of the price was $0.26.

25. A researcher is interested in estimating the average monthly salary of sports reporters in a large city. He wants to be 90% confident that his estimate is correct. If the standard deviation is $1100, how large a sample is needed to get the desired information and to be accurate to within $150?

Technology *Step by Step*

MINITAB
Step by Step

Finding a *z* Confidence Interval for the Mean

For Example 7–3, find the 90% confidence interval estimate for the mean amount of assets for credit unions in southwestern Pennsylvania.

1. Maximize the worksheet, then enter the data into C1 of a MINITAB worksheet. If sigma is known, skip to step 3.

2. Calculate the standard deviation for the sample. It will be used as an estimate for sigma.

   a) Select **Calc>Column statistics.**

   b) Click the option for Standard deviation.

   c) Enter **C1 Assets** for the Input variable and **s** for Store in:.

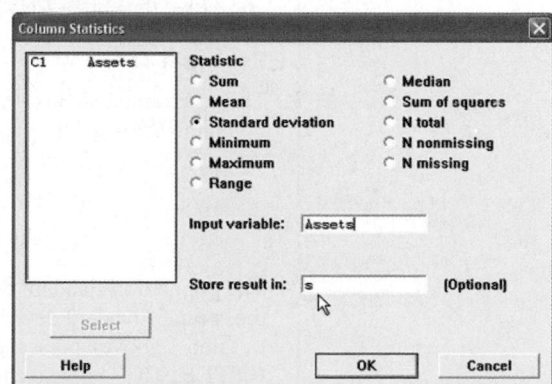

3. Select **Stat>Basic Statistics>1-Sample Z.**

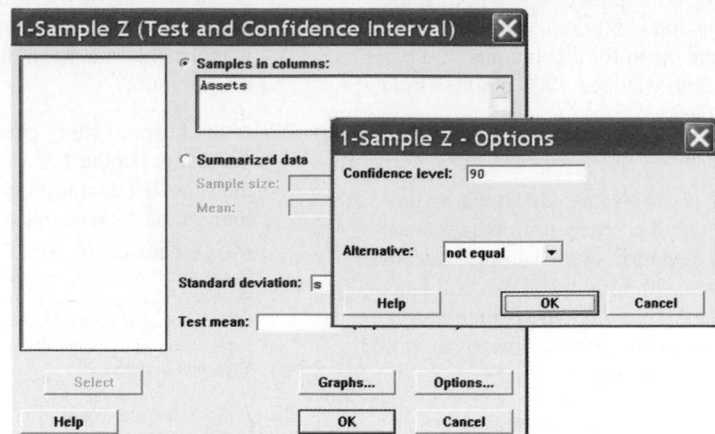

4. Select C1 Assets for the Samples in Columns.

5. Click in the box for Standard Deviation and enter **s.** Leave the box for Test mean empty.

6. Click the [Options] button. In the dialog box make sure the Confidence Level is 90 and the Alternative is not equal.

7. Optional: Click [Graphs], then select Boxplot of data. The boxplot of these data would clearly show the outliers!

8. Click [OK] twice. The results will be displayed in the session window.

One-Sample Z: Assets

```
The assumed sigma = 14.4054
Variable        N        Mean        StDev     SE Mean           90% CI
Assets         30     11.0907      14.4054      2.6301    (6.7646, 15.4167)
```

TI-83 Plus or TI-84 Plus
Step by Step

Finding a z Confidence Interval for the Mean (Data)

1. Enter the data into L_1.

2. Press **STAT** and move the cursor to TESTS.

3. Press **7** for ZInterval.

4. Move the cursor to Data and press **ENTER.**

5. Type in the appropriate values.

6. Move the cursor to Calculate and press **ENTER.**

Example TI7–1

This is Example 7–3 from the text. Find the 90% confidence interval for the population mean, given the data values.

12.23	2.89	13.19	73.25	11.59	8.74	7.92	40.22	5.01	2.27
16.56	1.24	9.16	1.91	6.69	3.17	4.78	2.42	1.47	12.77
4.39	2.17	1.42	14.64	1.06	18.13	16.85	21.58	12.24	2.76

The population standard deviation σ is unknown. Since the sample size is $n = 30$, one can use the sample standard deviation s as an approximation for σ. After the data values are entered in L_1 (step 1 above), press **STAT,** move the cursor to CALC, press **1** for 1-Var Stats, then press **ENTER.** The sample standard deviation of 14.40544747 will be one of the statistics listed. Then continue with step 2. At step 5 on the line for σ, press **VARS** for variables, press **5** for Statistics, press **3** for S_x.

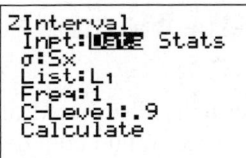   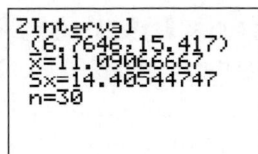

The 90% confidence interval is $6.765 < \mu < 15.417$. The difference between these limits and the ones in Example 7–3 is due to rounding.

Finding a *z* Confidence Interval for the Mean (Statistics)

1. Press **STAT** and move the cursor to TESTS.

2. Press **7** for ZInterval.

3. Move the cursor to Stats and press **ENTER.**

4. Type in the appropriate values.

5. Move the cursor to Calculate and press **ENTER.**

Example TI7–2

This is Example 7–1 from the text. Find the 95% confidence interval for the population mean, given $\sigma = 2$, $\overline{X} = 23.2$, and $n = 50$.

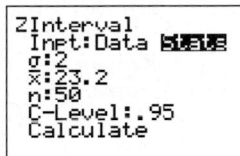   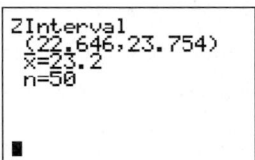

The 95% confidence interval is $22.6 < \mu < 23.8$.

Excel
Step by Step

Finding a *z* Confidence Interval for the Mean

Excel has a procedure to produce the maximum error of the estimate. But it does not produce confidence intervals. However, you may determine confidence intervals for the mean by using the MegaStat Add-in available on your CD and Online Learning Center. If you have not installed this add-in, do so by following the instructions on page 24.

 Find the 95% confidence interval when $\sigma = 11$, using this sample:

43	52	18	20	25	45	43	21	42	32	24	32	19	25	26
44	42	41	41	53	22	25	23	21	27	33	36	47	19	20

1. From the toolbar, select **MegaStat>Confidence Intervals/Sample Size.**

2. In the dialog box, select the Confidence interval mean tab.

3. Enter **32.03** for the mean of the data, and then select *z.* Enter **11** for the standard deviation and **30** for *n,* the sample size.

4. Either type in or scroll to 95% for the confidence level, and then click [OK].

The result of the procedure is shown next.

Confidence interval - mean

```
    95% confidence level
 32.03 mean
    11 std. dev.
    30 n
 1.960 z
 3.936 half-width
35.966 upper confidence limit
28.094 lower confidence limit
```

Objective **3**

Find the confidence interval for the mean when σ is unknown and $n < 30$.

Confidence Intervals for the Mean (σ Unknown and $n < 30$)

When σ is known and the variable is normally distributed or when σ is unknown and $n \geq 30$, the standard normal distribution is used to find confidence intervals for the mean. However, in many situations, the population standard deviation is not known and the sample size is less than 30. In such situations, the standard deviation from the sample can be used in place of the population standard deviation for confidence intervals. But a somewhat different distribution, called the **t distribution,** must be used when the sample size is less than 30 and the variable is normally or approximately normally distributed.

Some important characteristics of the t distribution are described now.

Historical Notes

The t distribution was formulated in 1908 by an Irish brewing employee named W. S. Gosset. Gosset was involved in researching new methods of manufacturing ale. Because brewing employees were not allowed to publish results, Gosset published his finding using the pseudonym *Student;* hence, the t distribution is sometimes called *Student's t distribution.*

Characteristics of the *t* Distribution

The t distribution shares some characteristics of the normal distribution and differs from it in others. The t distribution is similar to the standard normal distribution in these ways.

1. It is bell-shaped.
2. It is symmetric about the mean.
3. The mean, median, and mode are equal to 0 and are located at the center of the distribution.
4. The curve never touches the x axis.

The t distribution differs from the standard normal distribution in the following ways.

1. The variance is greater than 1.
2. The t distribution is actually a family of curves based on the concept of *degrees of freedom,* which is related to sample size.
3. As the sample size increases, the t distribution approaches the standard normal distribution. See Figure 7–6.

Many statistical distributions use the concept of degrees of freedom, and the formulas for finding the degrees of freedom vary for different statistical tests. The **degrees of freedom** are the number of values that are free to vary after a sample statistic has been computed, and they tell the researcher which specific curve to use when a distribution consists of a family of curves.

For example, if the mean of 5 values is 10, then 4 of the 5 values are free to vary. But once 4 values are selected, the fifth value must be a specific number to get a sum of 50, since $50 \div 5 = 10$. Hence, the degrees of freedom are $5 - 1 = 4$, and this value tells the researcher which t curve to use.

The symbol d.f. will be used for degrees of freedom. The degrees of freedom for a confidence interval for the mean are found by subtracting 1 from the sample size. That

Figure 7–6

The *t* Family of Curves

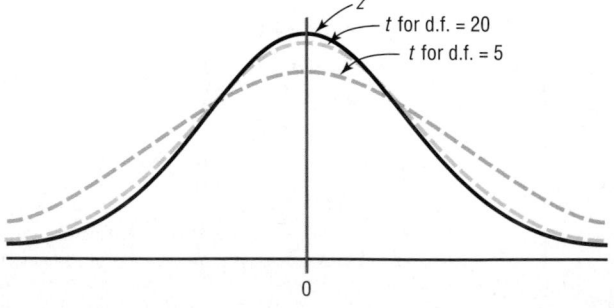

is, d.f. $= n - 1$. *Note:* For some statistical tests used later in this book, the degrees of freedom are not equal to $n - 1$.

The formula for finding a confidence interval about the mean by using the t distribution is given now.

Formula for a Specific Confidence Interval for the Mean When σ Is Unknown and $n < 30$

$$\bar{X} - t_{\alpha/2}\left(\frac{s}{\sqrt{n}}\right) < \mu < \bar{X} + t_{\alpha/2}\left(\frac{s}{\sqrt{n}}\right)$$

The degrees of freedom are $n - 1$.

The values for $t_{\alpha/2}$ are found in Table F in Appendix C. The top row of Table F, labeled Confidence Intervals, is used to get these values. The other two rows, labeled One tail and Two tails, will be explained in Chapter 8 and should not be used here.

Example 7–5 shows how to find the value in Table F for $t_{\alpha/2}$.

Example 7–5

Find the $t_{\alpha/2}$ value for a 95% confidence interval when the sample size is 22.

Solution

The d.f. $= 22 - 1$, or 21. Find 21 in the left column and 95% in the row labeled Confidence Intervals. The intersection where the two meet gives the value for $t_{\alpha/2}$, which is 2.080. See Figure 7–7.

Figure 7–7

Finding $t_{\alpha/2}$ for Example 7–5

Table F							
The t Distribution							
Confidence Intervals	50%	80%	90%	95%	98%	99%
One tail α	0.25	0.10	0.05	0.025	0.01	0.005
Two tails α	0.50	0.20	0.10	0.05	0.02	0.01
1							
2							
3							
21				2.080	2.518	2.831
$z\infty$		0.674	1.282[a]	1.645[b]	1.960	2.326[c]	2.576[d]

Note: At the bottom of Table F where d.f. $= \infty$, the $z_{\alpha/2}$ values can be found for specific confidence intervals. The reason is that as the degrees of freedom increase, the t distribution approaches the standard normal distribution.

Examples 7–6 and 7–7 show how to find the confidence interval when one is using the t distribution.

Example 7–6

Ten randomly selected automobiles were stopped, and the tread depth of the right front tire was measured. The mean was 0.32 inch, and the standard deviation was 0.08 inch. Find the 95% confidence interval of the mean depth. Assume that the variable is approximately normally distributed.

H̶istorical Note

Gosset derived the *t* distribution by selecting small random samples of measurements taken from a population of incarcerated criminals. For the measures he used the lengths of one of their fingers.

Solution

Since σ is unknown and s must replace it, the t distribution (Table F) must be used for 95% confidence interval. Hence, with 9 degrees of freedom, $t_{\alpha/2} = 2.262$.

The 95% confidence interval of the population mean is found by substituting in the formula

$$\bar{X} - t_{\alpha/2}\left(\frac{s}{\sqrt{n}}\right) < \mu < \bar{X} + t_{\alpha/2}\left(\frac{s}{\sqrt{n}}\right)$$

Hence, $0.32 - (2.262)\left(\dfrac{0.08}{\sqrt{10}}\right) < \mu < 0.32 + (2.262)\left(\dfrac{0.08}{\sqrt{10}}\right)$

$$0.32 - 0.057 < \mu < 0.32 + 0.057$$

$$0.26 < \mu < 0.38$$

Therefore, one can be 95% confident that the population mean tread depth of all right front tires is between 0.26 and 0.38 inch based on a sample of 10 tires.

Example 7–7

 The data represent a sample of the number of home fires started by candles for the past several years. (Data are from the National Fire Protection Association.) Find the 99% confidence interval for the mean number of home fires started by candles each year.

5460	5900	6090	6310	7160	8440	9930

Solution

Step 1   Find the mean and standard deviation for the data. Use the formulas in Chapter 3 or your calculator. The mean $\bar{X} = 7041.4$. The standard deviation $s = 1610.3$.

Step 2   Find $t_{\alpha/2}$ in Table F. Use the 99% confidence interval with d.f. = 6. It is 3.707.

Step 3   Substitute in the formula and solve.

$$\bar{X} - t_{\alpha/2}\left(\frac{s}{\sqrt{n}}\right) < \mu < \bar{X} + t_{\alpha/2}\left(\frac{s}{\sqrt{n}}\right)$$

$$7041.4 - 3.707\left(\frac{1610.3}{\sqrt{7}}\right) < \mu < 7041.4 + 3.707\left(\frac{1610.3}{\sqrt{7}}\right)$$

$$7041.4 - 2256.2 < \mu < 7041.4 + 2256.2$$

$$4785.2 < \mu < 9297.6$$

One can be 99% confident that the population mean number of home fires started by candles each year is between 4785.2 and 9297.6, based on a sample of home fires occurring over a period of 7 years.

Students sometimes have difficulty deciding whether to use $z_{\alpha/2}$ or $t_{\alpha/2}$ values when finding confidence intervals for the mean. As stated previously, when σ is known,

Figure 7–8

When to Use the *z* or *t* Distribution

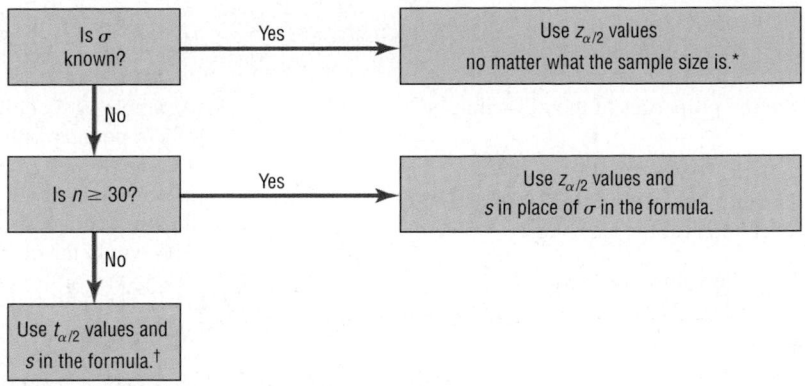

*Variable must be normally distributed when $n < 30$.

†Variable must be approximately normally distributed.

$z_{\alpha/2}$ values can be used *no matter what the sample size is,* as long as the variable is normally distributed or $n \geq 30$. When σ is unknown and $n \geq 30$, then s can be used in the formula and $z_{\alpha/2}$ values can be used. Finally, when σ is unknown and $n < 30$, s is used in the formula and $t_{\alpha/2}$ values are used, as long as the variable is approximately normally distributed. These rules are summarized in Figure 7–8.

It should be pointed out that some statisticians have a different point of view. They use $z_{\alpha/2}$ values when σ is known and $t_{\alpha/2}$ values when σ is unknown. In these circumstances, a *t* table that contains *t* values for sample sizes greater than or equal to 30 would be needed. The procedure shown in Figure 7–8 is the one used throughout this textbook.

Applying the Concepts **7–3**

Sport Drink Decision

Assume you get a new job as a coach for a sports team, and one of your first decisions is to choose the sports drink that the team will use during practices and games. You obtain a *Sports Report* magazine so you can use your statistical background to help you make the best decision. The following table lists the most popular sports drinks and some important information about each of them. Answer the following questions about the table.

Drink	Calories	Sodium	Potassium	Cost
Gatorade	60	110	25	$1.29
Powerade	68	77	32	1.19
All Sport	75	55	55	0.89
10-K	63	55	35	0.79
Exceed	69	50	44	1.59
1st Ade	58	58	25	1.09
Hydra Fuel	85	23	50	1.89

1. Would this be considered a small sample?

2. Compute the mean cost per container, and create a 90% confidence interval about that mean. Do all the costs per container fall inside the confidence interval? If not, which ones do not?

3. Are there any you would consider outliers?

4. How many degrees of freedom are there?

5. If cost is a major factor influencing your decision, would you consider cost per container or cost per serving?

6. List which drink you would recommend and why.

See page 389 for the answers.

Exercises 7–3

1. What are the properties of the *t* distribution?

2. What is meant by *degrees of freedom?*

3. When should the *t* distribution be used to find a confidence interval for the mean?

4. (**ans**) Find the values for each.

   a. $t_{\alpha/2}$ and $n = 18$ for the 99% confidence interval for the mean

   b. $t_{\alpha/2}$ and $n = 23$ for the 95% confidence interval for the mean

   c. $t_{\alpha/2}$ and $n = 15$ for the 98% confidence interval for the mean

   d. $t_{\alpha/2}$ and $n = 10$ for the 90% confidence interval for the mean

   e. $t_{\alpha/2}$ and $n = 20$ for the 95% confidence interval for the mean

For Exercises 5 through 20, assume that all variables are approximately normally distributed.

5. The average hemoglobin reading for a sample of 20 teachers was 16 grams per 100 milliliters, with a sample standard deviation of 2 grams. Find the 99% confidence interval of the true mean.

6. A sample of 17 states had these cigarette taxes (in cents):

112	120	98	55	71	35	99	124	64
150	150	55	100	132	20	70	93	

   Find a 98% confidence interval for the cigarette tax in all 50 states.

   Source: Federation of Tax Administrators.

7. A state representative wishes to estimate the mean number of women representatives per state legislature. A random sample of 17 states is selected, and the number of women representatives is shown. Based on the sample, what is the point estimate of the mean? Find the 90% confidence interval of the mean population. (*Note:* The population mean is actually 31.72, or about 32.) Compare this value to the point estimate and the confidence interval. There is something unusual about the data. Describe it and state how it would affect the confidence interval.

5	33	35	37	24
31	16	45	19	13
18	29	15	39	18
58	132			

8. A random sample of the number of barrels (in millions) of oil produced per day by world oil-producing countries is listed here. Estimate the mean oil production with 95% confidence.

3.56	1.90	7.83	2.83	1.91	5.88	2.91	6.08

   Source: *N.Y. Times Almanac.*

9. A sample of six college wrestlers had an average weight of 276 pounds with a sample standard deviation of 12 pounds. Find the 90% confidence interval of the true mean weight of all college wrestlers. If a coach claimed that the average weight of the wrestlers on his team was 310, would the claim be believable?

10. The daily salaries of substitute teachers for eight local school districts is shown. What is the point estimate for the mean? Find the 90% confidence interval of the mean for the salaries of substitute teachers in the region.

60	56	60	55	70	55	60	55

    Source: *Pittsburgh Tribune Review.*

11. A recent study of 28 employees of XYZ company showed that the mean of the distance they traveled to work was 14.3 miles. The standard deviation of the sample mean was 2 miles. Find the 95% confidence interval of the true mean. If a manager wanted to be sure that most of his employees would not be late, how much time would he suggest they allow for the commute if the average speed were 30 miles per hour?

12. A meteorologist who sampled 13 thunderstorms found that the average speed at which they traveled across a certain state was 15 miles per hour. The standard deviation of the sample was 1.7 miles per hour. Find the 99% confidence interval of the mean. If a meteorologist wanted to use the highest speed to predict the times it would take storms to travel across the state in order to issue warnings, what figure would she likely use?

13. A recent study of 25 students showed that they spent an average of $18.53 for gasoline per week. The standard deviation of the sample was $3.00. Find the 95% confidence interval of the true mean.

14. For a group of 10 men subjected to a stress situation, the mean number of heartbeats per minute was 126, and the standard deviation was 4. Find the 95% confidence interval of the true mean.

15. For the stress test described in Exercise 14, six women had an average heart rate of 115 beats per minute. The standard deviation of the sample was 6 beats. Find the 95% confidence interval of the true mean for the women.

16. For a sample of 24 operating rooms taken in the hospital study mentioned in Exercise 19 in Section 7–2, the mean noise level was 41.6 decibels, and the standard deviation was 7.5. Find the 95% confidence interval of the true mean of the noise levels in the operating rooms.

    Source: M. Bayo, A. Garcia, and A. Garcia, "Noise Levels in an Urban Hospital and Workers' Subjective Responses," *Archives of Environmental Health* 50, no. 3, p. 249 (May–June 1995). Reprinted with permission of the Helen Dwight Reid Educational Foundation. Published by Heldref Publications, 1319 Eighteenth St. N.W., Washington, D.C. 20036-1802. Copyright © 1995.

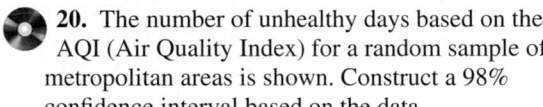

17. The number of grams of carbohydrates in a 12-ounce serving of a regular soft drink is listed here for a random sample of sodas. Estimate the mean number of carbohydrates in all brands of soda with 95% confidence.

48	37	52	40	43	46	41	38
41	45	45	33	35	52	45	41
30	34	46	40				

Source: *The Doctor's Pocket Calorie, Fat, and Carbohydrate Counter.*

18. For a group of 22 college football players, the mean heart rate after a morning workout session was 86 beats per minute, and the standard deviation was 5. Find the 90% confidence interval of the true mean for all college football players after a workout session. If a coach did not want to work his team beyond its capacity, what

would be the maximum value he should use for the mean number of heartbeats per minute?

19. The average yearly income of 28 community college instructors was $56,718. The standard deviation was $650. Find the 95% confidence interval of the true mean. If a faculty member wishes to see if he or she is being paid below average, what salary value should he or she use?

20. The number of unhealthy days based on the AQI (Air Quality Index) for a random sample of metropolitan areas is shown. Construct a 98% confidence interval based on the data.

61	12	6	40	27	38	93	5	13	40

Source: *N.Y. Times Almanac.*

Extending the Concepts

21. A *one-sided confidence* interval can be found for a mean by using

$$\mu > \overline{X} - t_\alpha \frac{s}{\sqrt{n}} \quad \text{or} \quad \mu < \overline{X} + t_\alpha \frac{s}{\sqrt{n}}$$

where t_α is the value found under the row labeled One tail. Find two one-sided 95% confidence intervals of the population mean for the data shown, and interpret the

answers. The data represent the daily revenues in dollars from 20 parking meters in a small municipality.

2.60	1.05	2.45	2.90
1.30	3.10	2.35	2.00
2.40	2.35	2.40	1.95
2.80	2.50	2.10	1.75
1.00	2.75	1.80	1.95

Technology *Step by Step*

MINITAB
Step by Step

Find a *t* Interval for the Mean

For Example 7–7, find the 99% confidence interval for the mean number of home fires started by candles each year.

1. Type the data into C1 of a MINITAB worksheet. Name the column **HomeFires.**

2. Select **Stat>Basic Statistics>1-Sample t.**

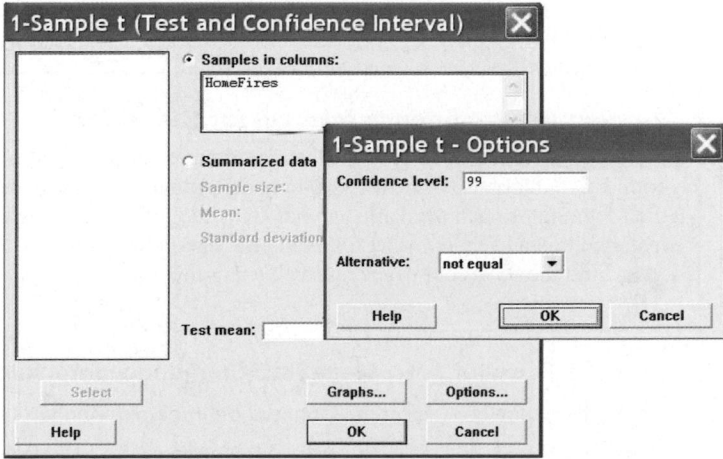

3. Double-click C1 HomeFires for the Samples in Columns.

4. Click on [Options] and be sure the Confidence Level is 99 and the Alternative is not equal.

5. Click [OK] twice.

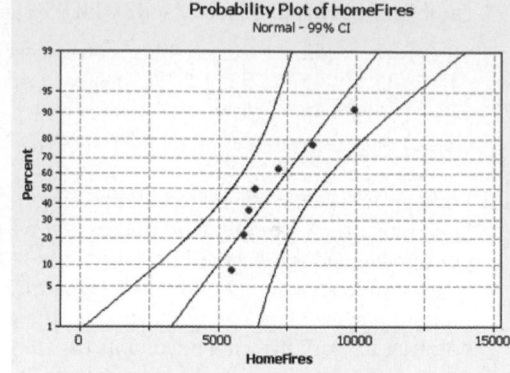

6. Check for normality:

   a) Select **Graph>Probability Plot,** then Single.

   b) Select C1 HomeFires for the variable. The normal plot is concave, a skewed distribution.

In the session window you will see the results. The 99% confidence interval estimate for μ is between 500.4 and 626.0. The sample size, mean, standard deviation, and standard error of the mean are also shown.

However, this small sample appears to have a nonnormal population. The interval is less likely to contain the true mean.

One-Sample T: HomeFires

```
Variable      N     Mean     StDev    SE Mean        99% CI
HomeFires     7   7041.43   1610.27    608.63   (4784.99, 9297.87)
```

TI-83 Plus or TI-84 Plus
Step by Step

Finding a *t* Confidence Interval for the Mean (Data)

1. Enter the data into L₁.

2. Press **STAT** and move the cursor to TESTS.

3. Press **8** for TInterval.

4. Move the cursor to Data and press **ENTER.**

5. Type in the appropriate values.

6. Move the cursor to Calculate and press **ENTER.**

Finding a *t* Confidence Interval for the Mean (Statistics)

1. Press **STAT** and move the cursor to TESTS.

2. Press **8** for TInterval.

3. Move the cursor to Stats and press **ENTER.**

4. Type in the appropriate values.

5. Move the cursor to Calculate and press **ENTER.**

Excel
Step by Step

Finding a *t* Confidence Interval for the Mean

Excel has a procedure to produce the maximum error of the estimate. But it does not produce confidence intervals. However, you may determine confidence intervals for the mean by using the MegaStat Add-in available on your CD and Online Learning Center. If you have not installed this add-in, do so by following the instructions on page 24.

Find the 95% confidence interval, using this sample:

| 625 | 675 | 535 | 406 | 512 | 680 | 483 | 522 | 619 | 575 |

1. From the toolbar, select **MegaStat>Confidence Intervals/Sample Size.**

2. In the dialog box, select the Confidence interval mean tab.

3. Enter **563.2** for the mean of the data, and then select *t*. Enter **87.9** for the standard deviation and 10 for *n*, the sample size.

4. Either type in or scroll to 95% for the confidence level, and then click [OK].

The result of the procedure will show the output in a new chart as indicated.

Confidence interval - mean

```
      95% confidence level
   563.2 mean
    87.9 std. dev.
      10 n
   2.262 t (df = 9)
  62.880 half-width
 626.080 upper confidence limit
 500.320 lower confidence limit
```

7–4

Confidence Intervals and Sample Size for Proportions

Objective 4

Find the confidence interval for a proportion.

A *USA TODAY* Snapshots feature stated that 12% of the pleasure boats in the United States were named *Serenity*. The parameter 12% is called a **proportion.** It means that of all the pleasure boats in the United States, 12 out of every 100 are named *Serenity*. A proportion represents a part of a whole. It can be expressed as a fraction, decimal, or percentage. In this case, $12\% = 0.12 = \frac{12}{100}$ or $\frac{3}{25}$. Proportions can also represent probabilities. In this case, if a pleasure boat is selected at random, the probability that it is called *Serenity* is 0.12.

Proportions can be obtained from samples or populations. The following symbols will be used.

Symbols Used in Proportion Notation

p = population proportion

$\hat{p}$ (read "*p* hat") = sample proportion

For a sample proportion,

$$\hat{p} = \frac{X}{n} \quad \text{and} \quad \hat{q} = \frac{n - X}{n} \quad \text{or} \quad 1 - \hat{p}$$

where X = number of sample units that possess the characteristics of interest and n = sample size.

For example, in a study, 200 people were asked if they were satisfied with their job or profession; 162 said that they were. In this case, $n = 200$, $X = 162$, and $\hat{p} = X/n = 162/200 = 0.81$. It can be said that for this sample, 0.81, or 81%, of those surveyed were satisfied with their job or profession. The sample proportion is $\hat{p} = 0.81$.

The proportion of people who did not respond favorably when asked if they were satisfied with their job or profession constituted $\hat{q}$, where $\hat{q} = (n - X)/n$. For this survey, $\hat{q} = (200 - 162)/200 = 38/200$, or 0.19, or 19%.

When $\hat{p}$ and $\hat{q}$ are given in decimals or fractions, $\hat{p} + \hat{q} = 1$. When $\hat{p}$ and $\hat{q}$ are given in percentages, $\hat{p} + \hat{q} = 100\%$. It follows, then, that $\hat{q} = 1 - \hat{p}$, or $\hat{p} = 1 - \hat{q}$, when $\hat{p}$ and $\hat{q}$ are in decimal or fraction form. For the sample survey on job satisfaction, $\hat{q}$ can also be found by using $\hat{q} = 1 - \hat{p}$, or $1 - 0.81 = 0.19$.

Similar reasoning applies to population proportions; that is, $p = 1 - q$, $q = 1 - p$, and $p + q = 1$, when p and q are expressed in decimal or fraction form. When p and q are expressed as percentages, $p + q = 100\%$, $p = 100\% - q$, and $q = 100\% - p$.

Example 7–8

In a recent survey of 150 households, 54 had central air conditioning. Find $\hat{p}$ and $\hat{q}$, where $\hat{p}$ is the proportion of households that have central air conditioning.

Solution

Since $X = 54$ and $n = 150$,

$$\hat{p} = \frac{X}{n} = \frac{54}{150} = 0.36 = 36\%$$

$$\hat{q} = \frac{n - X}{n} = \frac{150 - 54}{150} = \frac{96}{150} = 0.64 = 64\%$$

One can also find $\hat{q}$ by using the formula $\hat{q} = 1 - \hat{p}$. In this case, $\hat{q} = 1 - 0.36 = 0.64$.

As with means, the statistician, given the sample proportion, tries to estimate the population proportion. Point and interval estimates for a population proportion can be made by using the sample proportion. For a point estimate of p (the population proportion), $\hat{p}$ (the sample proportion) is used. On the basis of the three properties of a good estimator, $\hat{p}$ is unbiased, consistent, and relatively efficient. But as with means, one is not able to decide how good the point estimate of p is. Therefore, statisticians also use an interval estimate for a proportion, and they can assign a probability that the interval will contain the population proportion.

The confidence interval for a particular p is based on the sampling distribution of $\hat{p}$. When the sample size n is no more than 5% of the population size, the sampling distribution of $\hat{p}$ is approximately normally distributed with a mean of p and a standard deviation of $\sqrt{\frac{pq}{n}}$, where $q = 1 - p$.

Confidence Intervals

To construct a confidence interval about a proportion, one must use the maximum error of estimate, which is

$$E = z_{\alpha/2} \sqrt{\frac{\hat{p}\hat{q}}{n}}$$

Confidence intervals about proportions must meet the criteria that $np \geq 5$ and $nq \geq 5$.

Formula for a Specific Confidence Interval for a Proportion

$$\hat{p} - z_{\alpha/2} \sqrt{\frac{\hat{p}\hat{q}}{n}} < p < \hat{p} + z_{\alpha/2} \sqrt{\frac{\hat{p}\hat{q}}{n}}$$

when np and nq are each greater than or equal to 5.

Rounding Rule for a Confidence Interval for a Proportion Round off to three decimal places.

Example 7–9

A sample of 500 nursing applications included 60 from men. Find the 90% confidence interval of the true proportion of men who applied to the nursing program.

Solution

Since $\alpha = 1 - 0.90 = 0.10$ and $z_{\alpha/2} = 1.65$, substituting in the formula

$$\hat{p} - z_{\alpha/2}\sqrt{\frac{\hat{p}\hat{q}}{n}} < p < \hat{p} + z_{\alpha/2}\sqrt{\frac{\hat{p}\hat{q}}{n}}$$

when $\hat{p} = 60/500 = 0.12$ and $\hat{q} = 1 - 0.12 = 0.88$, one gets

$$0.12 - 1.65\sqrt{\frac{(0.12)(0.88)}{500}} < p < 0.12 + 1.65\sqrt{\frac{(0.12)(0.88)}{500}}$$

$$0.12 - 0.024 < p < 0.12 + 0.024$$

$$0.096 < p < 0.144$$

or

$$9.6\% < p < 14.4\%$$

Hence, one can be 90% confident that the percentage of applicants who are men is between 9.6 and 14.4%.

When a specific percentage is given, the percentage becomes $\hat{p}$ when it is changed to a decimal. For example, if the problem states that 12% of the applicants were men, then $\hat{p} = 0.12$.

Example 7–10

A survey of 200,000 boat owners found that 12% of the pleasure boats were named *Serenity*. Find the 95% confidence interval of the true proportion of boats named *Serenity*.

Source: USA TODAY Snapshot.

Solution

From the Snapshot, $\hat{p} = 0.12$ (i.e., 12%), and $n = 200,000$. Since $z_{\alpha/2} = 1.96$, substituting in the formula

$$\hat{p} - z_{\alpha/2}\sqrt{\frac{\hat{p}\hat{q}}{n}} < p < \hat{p} + z_{\alpha/2}\sqrt{\frac{\hat{p}\hat{q}}{n}}$$

yields

$$0.12 - 1.96\sqrt{\frac{(0.12)(0.88)}{200,000}} < p < 0.12 + 1.96\sqrt{\frac{(0.12)(0.88)}{200,000}}$$

$$0.119 < p < 0.121$$

Hence, one can say with 95% confidence that the true percentage of boats named *Serenity* is between 11.9 and 12.1%.

Sample Size for Proportions

To find the sample size needed to determine a confidence interval about a proportion, use this formula:

Objective 5

Determine the minimum sample size for finding a confidence interval for a proportion.

> **Formula for Minimum Sample Size Needed for Interval Estimate of a Population Proportion**
>
> $$n = \hat{p}\hat{q}\left(\frac{z_{\alpha/2}}{E}\right)^2$$
>
> If necessary, round up to obtain a whole number.

This formula can be found by solving the maximum error of estimate value for *n*:

$$E = z_{\alpha/2}\sqrt{\frac{\hat{p}\hat{q}}{n}}$$

There are two situations to consider. First, if some approximation of $\hat{p}$ is known (e.g., from a previous study), that value can be used in the formula.

Second, if no approximation of $\hat{p}$ is known, one should use $\hat{p} = 0.5$. This value will give a sample size sufficiently large to guarantee an accurate prediction, given the confidence interval and the error of estimate. The reason is that when $\hat{p}$ and $\hat{q}$ are each 0.5, the product $\hat{p}\hat{q}$ is at maximum, as shown here.

$\hat{p}$	$\hat{q}$	$\hat{p}\hat{q}$
0.1	0.9	0.09
0.2	0.8	0.16
0.3	0.7	0.21
0.4	0.6	0.24
0.5	**0.5**	**0.25**
0.6	0.4	0.24
0.7	0.3	0.21
0.8	0.2	0.16
0.9	0.1	0.09

Example 7–11

A researcher wishes to estimate, with 95% confidence, the proportion of people who own a home computer. A previous study shows that 40% of those interviewed had a computer at home. The researcher wishes to be accurate within 2% of the true proportion. Find the minimum sample size necessary.

Solution

Since $z_{\alpha/2} = 1.96$, $E = 0.02$, $\hat{p} = 0.40$, and $\hat{q} = 0.60$, then

$$n = \hat{p}\hat{q}\left(\frac{z_{\alpha/2}}{E}\right)^2 = (0.40)(0.60)\left(\frac{1.96}{0.02}\right)^2 = 2304.96$$

which, when rounded up, is 2305 people to interview.

Example 7–12

The same researcher wishes to estimate the proportion of executives who own a car phone. She wants to be 90% confident and be accurate within 5% of the true proportion. Find the minimum sample size necessary.

Solution

Since there is no prior knowledge of $\hat{p}$, statisticians assign the values $\hat{p} = 0.5$ and $\hat{q} = 0.5$. The sample size obtained by using these values will be large enough to ensure the specified degree of confidence. Hence,

$$n = \hat{p}\hat{q}\left(\frac{z_{\alpha/2}}{E}\right)^2 = (0.5)(0.5)\left(\frac{1.65}{0.05}\right)^2 = 272.25$$

which, when rounded up, is 273 executives to ask.

In determining the sample size, the size of the population is irrelevant. Only the degree of confidence and the maximum error are necessary to make the determination.

Does Success Bring Happiness?

W. C. Fields said, "Start every day off with a smile and get it over with."

Do you think people are happy because they are successful, or are they successful because they are just happy people? A recent survey conducted by *Money* magazine showed that 34% of the people surveyed said that they were happy because they were successful; however, 63% said that they were successful because they were happy individuals. The people surveyed had an average household income of $75,000 or more. The margin of error was ±2.5%. Based on the information in this article, what would be the confidence interval for each percent?

Applying the Concepts 7–4

Contracting Influenza

To answer the questions, use the following table describing the percentage of people who reported contracting influenza by gender and race/ethnicity.

	Influenza	
Characteristic	**Percent**	**(95% CI)**
Gender		
Men	48.8	(47.1–50.5%)
Women	51.5	(50.2–52.8%)
Race/ethnicity		
Caucasian	52.2	(51.1–53.3%)
African American	33.1	(29.5–36.7%)
Hispanic	47.6	(40.9–54.3%)
Other	39.7	(30.8–48.5%)
Total	50.4	(49.3–51.5%)

Forty-nine states and the District of Columbia participated in the study. Weighted means were used. The sample size was 19,774. There were 12,774 women and 7000 men.

1. Explain what (95% CI) means.
2. How large is the error for men reporting influenza?
3. What is the sample size?
4. How does sample size affect the size of the confidence interval?
5. Would the confidence intervals be larger or smaller for a 90% CI, using the same data?
6. Where does the 51.5% under influenza for women fit into its associated 95% CI?

See page 389 for the answers.

Exercises 7–4

1. In each case, find $\hat{p}$ and $\hat{q}$.

    a. $n = 80$ and $X = 40$

    b. $n = 200$ and $X = 90$

    c. $n = 130$ and $X = 60$

    d. $n = 60$ and $X = 35$

    e. $n = 95$ and $X = 43$

2. (ans) Find $\hat{p}$ and $\hat{q}$ for each percentage. (Use each percentage for $\hat{p}$.)

    a. 15%

    b. 37%

    c. 71%

    d. 51%

    e. 79%

3. A U.S. Travel Data Center survey conducted for *Better Homes and Gardens* of 1500 adults found that 39% said that they would take more vacations this year than last year. Find the 95% confidence interval for the true proportion of adults who said that they will travel more this year.

    Source: *USA TODAY.*

4. A recent study of 100 people in Miami found 27 were obese. Find the 90% confidence interval of the population proportion of individuals living in Miami who are obese.

    Source: Based on information from the Center for Disease Control and Prevention, *USA TODAY.*

5. The proportion of students in private schools is around 11%. A random sample of 450 students from a wide geographic area indicated that 55 attended private schools. Estimate the true proportion of students attending private schools with 95% confidence. How does your estimate compare to 11%?

    Source: National Center for Education Statistics (www.nces.ed.gov).

6. The Gallup Poll found that 27% of adults surveyed nationwide said they had personally been in a tornado. How many adults should be surveyed to estimate the true proportion of adults who have been in a tornado with a 95% confidence interval 5% wide?

    Source: www.pollingreport.com.

7. A survey found that out of 200 workers, 168 said they were interrupted three or more times an hour by phone messages, faxes, etc. Find the 90% confidence interval of the population proportion of workers who are interrupted three or more times an hour.

    Source: Based on information from *USA TODAY* Snapshot.

8. A CBS News/*New York Times* poll found that 329 out of 763 adults said they would travel to outer space in their lifetime, given the chance. Estimate the true proportion of adults who would like to travel to outer space with 92% confidence.

    Source: www.pollingreport.com.

9. A study by the University of Michigan found that one in five 13- and 14-year-olds is a sometime smoker. To see how the smoking rate of the students at a large school district compared to the national rate, the superintendent surveyed two hundred 13- and 14-year-old students and found that 23% said they were sometime smokers. Find the 99% confidence interval of the true proportion, and compare this with the University of Michigan study.

    Source: *USA TODAY.*

10. A survey of 50 first-time white-water canoers showed that 23 did not want to repeat the experience. Find the 90% confidence interval of the true proportion of canoers who did not wish to canoe the rapids a second time. If a rafting company wants to distribute brochures for repeat trips, what is the minimum number it should print?

11. A survey of 85 families showed that 36 owned at least one DVD player. Find the 99% confidence interval of the true proportion of families who own at least one DVD player. If another survey in a different location found that the proportion of families who owned at least one DVD player was 0.52, would you consider that the proportion of families in this area was larger than in the area where the original survey was done?

12. In a certain countrywide school district, a survey of 350 students showed that 28% carried their lunches to school. Find the 95% confidence interval of the true proportion of students who carried their lunches to school. If the cafeteria manager wanted to be reasonably sure that all the children who didn't bring their lunches could purchase a lunch, how many lunches should she plan to make each day? Explain your answer.

13. In a Gallup Poll of 1005 individuals, 452 thought they were worse off financially than a year ago. Find the 95% confidence interval for the true proportion of individuals that feel they are worse off financially.

    Source: Gallup Poll.

14. In a poll of 1000 likely voters, 560 say that the United States spends too little on fighting hunger at home. Find a 95% confidence interval for the true proportion of voters who feel this way.

    Source: Alliance to End Hunger.

15. A medical researcher wishes to determine the percentage of females who take vitamins. He wishes to be 99% confident that the estimate is within 2 percentage points of the true proportion. A recent study of 180 females showed that 25% took vitamins.

    a. How large should the sample size be?

    b. If no estimate of the sample proportion is available, how large should the sample be?

16. A recent study indicated that 29% of the 100 women over age 55 in the study were widows.

    a. How large a sample must one take to be 90% confident that the estimate is within 0.05 of the true proportion of women over age 55 who are widows?

    b. If no estimate of the sample proportion is available, how large should the sample be?

17. A researcher wishes to estimate the proportion of adult males who are under 5 feet 5 inches tall. She wants to be 90% confident that her estimate is within 5% of the true proportion.

    a. How large a sample should be taken if in a sample of 300 males, 30 were under 5 feet 5 inches tall?

    b. If no estimate of the sample proportion is available, how large should the sample be?

18. Obesity is defined as a *body mass index* (BMI) of 3 kg/m^2 or more. A 95% confidence interval for the percentage of U.S. adults aged 20 years and over who were obese was found to be 22.4% to 23.5%. What was the sample size?

    Source: National Center for Health Statistics (www.cdc.gov/nchs).

19. How large a sample should be surveyed to estimate the true proportion of college students who do laundry once a week within 3% with 95% confidence?

20. A federal report indicated that 27% of children ages 2 to 5 years had a good diet—an increase over previous years. How large a sample is needed to estimate the true proportion of children with good diets within 2% with 95% confidence?

    Source: Federal Interagency Forum on Child and Family Statistics, *Washington Observer-Reporter.*

Extending the Concepts

21. If a sample of 600 people is selected and the researcher decides to have a maximum error of estimate of 4% on the specific proportion who favor gun control, find the degree of confidence. A recent study showed that 50% were in favor of some form of gun control.

22. In a study, 68% of 1015 adults said that they believe the Republicans favor the rich. If the margin of error was 3 percentage points, what was the confidence interval used for the proportion?

    Source: *USA TODAY.*

Technology *Step by Step*

MINITAB
Step by Step

Find a Confidence Interval for a Proportion

MINITAB will calculate a confidence interval, given the statistics from a sample *or* given the raw data. In Example 7–9, in a sample of 500 nursing applications 60 were from men. Find the 90% confidence interval estimate for the true proportion of male applicants.

1. Select **Stat>Basic Statistics>1 Proportion.**

2. Click on the button for Summarized data. No data will be entered in the worksheet.

3. Click in the box for Number of trials and enter **500.**

4. In the Number of events box, enter **60.**

5. Click on [Options].

6. Type **90** for the confidence level.

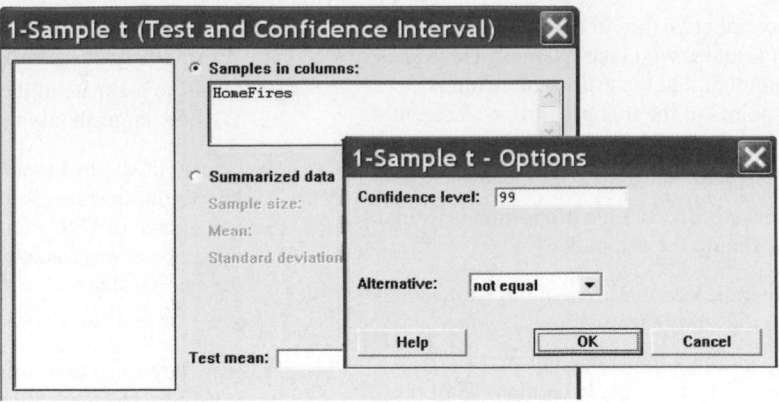

7. Check the box for Use test and interval based on normal distribution.

8. Click [OK] twice.

The results for the confidence interval will be displayed in the session window.

Test and CI for One Proportion

```
Test of p = 0.5 vs p not = 0.
Sample   X     N    Sample p         90% CI          Z-Value   P-Value
1        60    500  0.120000   (0.096096, 0.143904)   -16.99    0.000
```

TI-83 Plus or TI-84 Plus
Step by Step

Finding a Confidence Interval for a Proportion

1. Press **STAT** and move the cursor to TESTS.

2. Press **A (ALPHA, MATH)** for 1-PropZInt.

3. Type in the appropriate values.

4. Move the cursor to Calculate and press **ENTER**.

Input

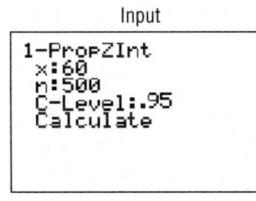

Example TI7–3

Find the 95% confidence interval of p when $X = 60$ and $n = 500$, as in Example 7–9.

The 95% confidence level for p is $0.09152 < p < 0.14848$. Also $\hat{p}$ is given.

Output

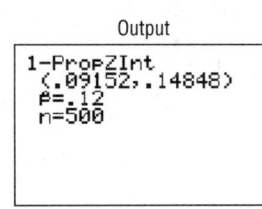

Excel
Step by Step

Finding a Confidence Interval for a Proportion

Excel does not produce confidence intervals for a proportion. However, you may determine confidence intervals for a proportion by using the MegaStat Add-in available on your CD and Online Learning Center. If you have not installed this add-in, do so by following the instructions on page 24.

There were 500 nursing applications in a sample, including 60 from men. Find the 90% confidence interval for the true proportion of male applicants.

1. From the toolbar, select **MegaStat>Confidence Intervals/Sample Size.**

2. In the dialog box, select the Confidence interval–p.

3. Enter **60** in the first box; p will automatically switch to x.

4. Enter **500** in the second box for n.

5. Either type in or scroll to 90% for the confidence level, and then click [OK].

Speaking of
Statistics

Here is a survey about college students' credit card usage. Suggest several ways that the study could have been more meaningful if confidence intervals had been used.

OTHER PEOPLE'S MONEY

Undergrads love their plastic. That means—you guessed it—students are learning to become debtors. According to the Public Interest Research Groups, only half of all students pay off card balances in full each month, 36% sometimes do and 14% never do. Meanwhile, 48% have paid a late fee. Here's how undergrads stack up, according to Nellie Mae, a provider of college loans:

Undergrads with a credit card **78%**

Average number of cards owned . .**3**

Average student card debt **$1236**

Students with 4 or more cards. . . .**32%**

Balances of $3000 to $7000 . . . . .**13%**

Balances over $7000. . . . .**9%**

Reprinted with permission from the January 2002 Reader's Digest.
Copyright © 2002 by The Reader's Digest Assn. Inc.

The result of the procedure will show the output in a new chart, as indicated here.

Confidence interval - proportion

```
   90% confidence level
 0.12 proportion
  500 n
1.645 z
0.024 half-width
0.144 upper confidence limit
0.096 lower confidence limit
```

7–5   Confidence Intervals for Variances and Standard Deviations

Objective 6

Find a confidence interval for a variance and a standard deviation.

In Sections 7–2 through 7–4, confidence intervals were calculated for means and proportions. This section will explain how to find confidence intervals for variances and standard deviations. In statistics, the variance and standard deviation of a variable are as important as the mean. For example, when products that fit together (such as pipes) are manufactured, it is important to keep the variations of the diameters of the products as small as possible; otherwise, they will not fit together properly and will have to be scrapped. In the manufacture of medicines, the variance and standard deviation of the medication in the pills play an important role in making sure patients receive the proper dosage. For these reasons, confidence intervals for variances and standard deviations are necessary.

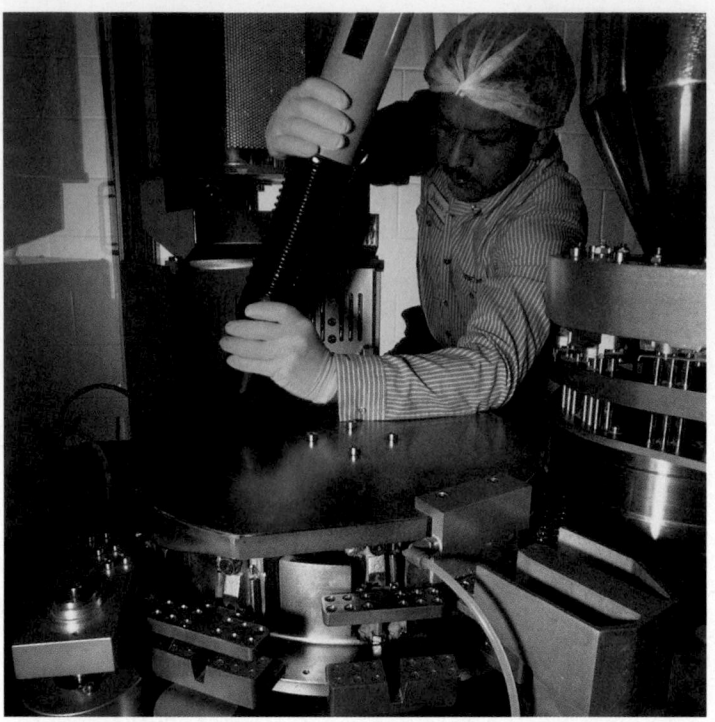

To calculate these confidence intervals, a new statistical distribution is needed. It is called the **chi-square distribution.**

The chi-square variable is similar to the t variable in that its distribution is a family of curves based on the number of degrees of freedom. The symbol for chi-square is χ^2 (Greek letter chi, pronounced "ki"). Several of the distributions are shown in Figure 7–9, along with the corresponding degrees of freedom. The chi-square distribution is obtained from the values of $(n-1)s^2/\sigma^2$ when random samples are selected from a normally distributed population whose variance is σ^2.

A chi-square variable cannot be negative, and the distributions are positively skewed. At about 100 degrees of freedom, the chi-square distribution becomes somewhat symmetric. The area under each chi-square distribution is equal to 1.00, or 100%.

Table G in Appendix C gives the values for the chi-square distribution. These values are used in the denominators of the formulas for confidence intervals. Two different

Figure 7–9

The Chi-Square Family of Curves

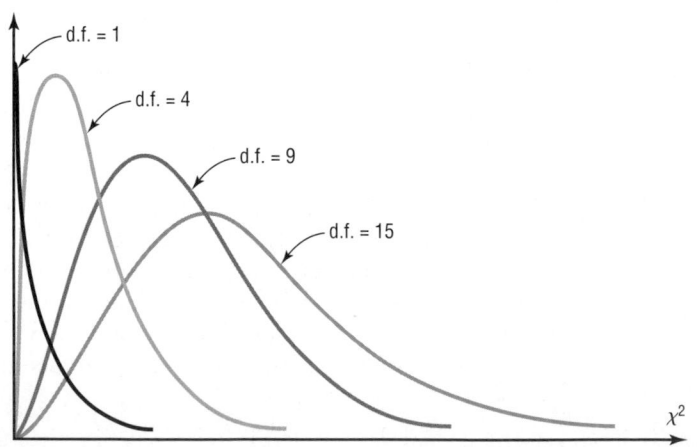

values are used in the formula. One value is found on the left side of the table, and the other is on the right. For example, to find the table values corresponding to the 95% confidence interval, one must first change 95% to a decimal and subtract it from 1 ($1 - 0.95 = 0.05$). Then divide the answer by 2 ($\alpha/2 = 0.05/2 = 0.025$). This is the column on the right side of the table, used to get the values for χ^2_{right}. To get the value for χ^2_{left}, subtract the value of $\alpha/2$ from 1 ($1 - 0.05/2 = 0.975$). Finally, find the appropriate row corresponding to the degrees of freedom $n - 1$. A similar procedure is used to find the values for a 90 or 99% confidence interval.

Example 7–13

Find the values for χ^2_{right} and χ^2_{left} for a 90% confidence interval when $n = 25$.

Solution

To find χ^2_{right}, subtract $1 - 0.90 = 0.10$ and divide by 2 to get 0.05.

To find χ^2_{left}, subtract $1 - 0.05$ to get 0.95. Hence, use the 0.95 and 0.05 columns and the row corresponding to 24 d.f. See Figure 7–10.

Figure 7–10

χ^2 **Table for**
Example 7–13

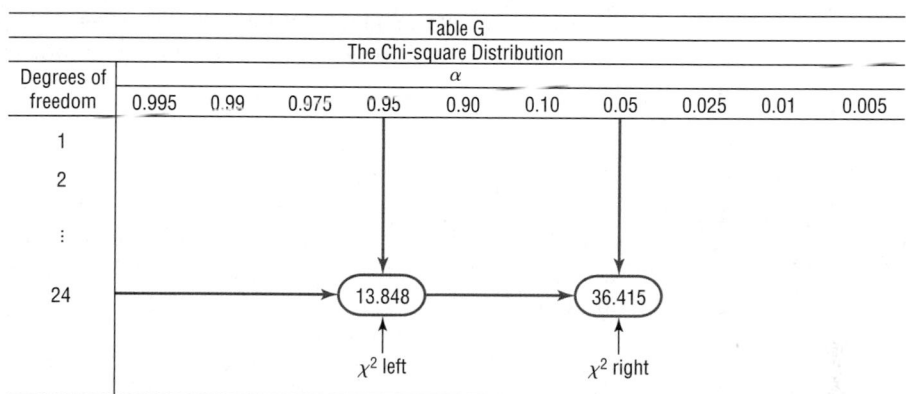

The answers are

$$\chi^2_{\text{right}} = 36.415$$

$$\chi^2_{\text{left}} = 13.848$$

Useful estimates for σ^2 and σ are s^2 and s, respectively.

To find confidence intervals for variances and standard deviations, one must assume that the variable is normally distributed.

The formulas for the confidence intervals are shown here.

Formula for the Confidence Interval for a Variance

$$\frac{(n - 1)s^2}{\chi^2_{\text{right}}} < \sigma^2 < \frac{(n - 1)s^2}{\chi^2_{\text{left}}}$$

d.f. $= n - 1$

> ### Formula for the Confidence Interval for a Standard Deviation
>
> $$\sqrt{\frac{(n-1)s^2}{\chi^2_{\text{right}}}} < \sigma < \sqrt{\frac{(n-1)s^2}{\chi^2_{\text{left}}}}$$
>
> d.f. $= n - 1$

Recall that s^2 is the symbol for the sample variance and s is the symbol for the sample standard deviation. If the problem gives the sample standard deviation s, be sure to *square* it when you are using the formula. But if the problem gives the sample variance s^2, *do not square it* when using the formula, since the variance is already in square units.

Rounding Rule for a Confidence Interval for a Variance or Standard Deviation When you are computing a confidence interval for a population variance or standard deviation by using raw data, round off to one more decimal place than the number of decimal places in the original data.

When you are computing a confidence interval for a population variance or standard deviation by using a sample variance or standard deviation, round off to the same number of decimal places as given for the sample variance or standard deviation.

Example 7–14 shows how to find a confidence interval for a variance and standard deviation.

Example 7–14

Find the 95% confidence interval for the variance and standard deviation of the nicotine content of cigarettes manufactured if a sample of 20 cigarettes has a standard deviation of 1.6 milligrams.

Solution

Since $\alpha = 0.05$, the two critical values, respectively, for the 0.025 and 0.975 levels for 19 degrees of freedom are 32.852 and 8.907. The 95% confidence interval for the variance is found by substituting in the formula.

$$\frac{(n-1)s^2}{\chi^2_{\text{right}}} < \sigma^2 < \frac{(n-1)s^2}{\chi^2_{\text{left}}}$$

$$\frac{(20-1)(1.6)^2}{32.852} < \sigma^2 < \frac{(20-1)(1.6)^2}{8.907}$$

$$1.5 < \sigma^2 < 5.5$$

Hence, one can be 95% confident that the true variance for the nicotine content is between 1.5 and 5.5.

For the standard deviation, the confidence interval is

$$\sqrt{1.5} < \sigma < \sqrt{5.5}$$

$$1.2 < \sigma < 2.3$$

Hence, one can be 95% confident that the true standard deviation for the nicotine content of all cigarettes manufactured is between 1.2 and 2.3 milligrams based on a sample of 20 cigarettes.

Example 7–15

Find the 90% confidence interval for the variance and standard deviation for the price in dollars of an adult single-day ski lift ticket. The data represent a selected sample of nationwide ski resorts. Assume the variable is normally distributed.

59	54	53	52	51
39	49	46	49	48

Source: USA TODAY.

Solution

Step 1 Find the variance for the data. Use the formulas in Chapter 3 or your calculator. The variance $s^2 = 28.2$.

Step 2 Find χ^2_{right} and χ^2_{left} from Table G in Appendix C. Since $\alpha = 0.10$, the two critical values are 3.325 and 16.919, using d.f. = 9 and 0.95 and 0.05.

Step 3 Substitute in the formula and solve.

$$\frac{(n-1)s^2}{\chi^2_{\text{right}}} < \sigma^2 < \frac{(n-1)s^2}{\chi^2_{\text{left}}}$$

$$\frac{(10-1)(28.2)}{16.919} < \sigma^2 < \frac{(10-1)(28.2)}{3.325}$$

$$15.0 < \sigma^2 < 76.3$$

For the standard deviation

$$\sqrt{15} < \sigma < \sqrt{76.3}$$

$$3.87 < \sigma < 8.73$$

Hence one can be 90% confident that the standard deviation for the price of all single-day ski lift tickets of the population is between $3.87 and $8.73 based on a sample of 10 nationwide ski resorts. (Two decimal places are used since the data are in dollars and cents.)

Note: If you are using the standard deviation instead (as in Example 7–14) of the variance, be sure to square the standard deviation when substituting in the formula.

Applying the Concepts **7–5**

Confidence Interval for Standard Deviation

Shown are the ages (in years) of the Presidents at the time of their deaths.

67	90	83	85	73	80	78	79
68	71	53	65	74	64	77	56
66	63	70	49	57	71	67	71
58	60	72	67	57	60	90	63
88	78	46	64	81	93		

1. Do the data represent a population or a sample?
2. Select a random sample of 12 ages and find the variance and standard deviation.
3. Find the 95% confidence interval of the standard deviation.
4. Find the standard deviation of all the data values.
5. Does the confidence interval calculated in question 3 contain the mean?
6. If it does not, give a reason why.
7. What assumption(s) must be considered for constructing the confidence interval in step 3?

See page 389 for the answers.

Exercises 7–5

1. What distribution must be used when computing confidence intervals for variances and standard deviations?

2. What assumption must be made when computing confidence intervals for variances and standard deviations?

3. Using Table G, find the values for χ^2_{left} and χ^2_{right}.

   a. $\alpha = 0.05$, $n = 12$
   b. $\alpha = 0.10$, $n = 20$
   c. $\alpha = 0.05$, $n = 27$
   d. $\alpha = 0.01$, $n = 6$
   e. $\alpha = 0.10$, $n = 41$

4. Find the 90% confidence interval for the variance and standard deviation for the lifetime of disposable camera batteries if a sample of 16 disposable camera batteries has a standard deviation of 2.1 months. Assume the variable is normally distributed. Do you feel that the lifetimes of the batteries are relatively consistent?

5. Find the 95% confidence interval for the variance and standard deviation for the time it takes a customer to place a telephone order with a large catalogue company if a sample of 23 telephone orders has a standard deviation of 3.8 minutes. Assume the variable is normally distributed. Do you think that the times are relatively consistent?

6. Find the 99% confidence interval for the variance and standard deviation of the weights of 25 one-gallon containers of motor oil if a sample of 14 containers has a variance of 3.2. The weights are given in ounces. Assume the variable is normally distributed.

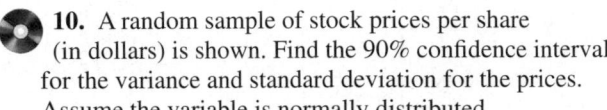 7. The sugar content (in grams) for a random sample of 4-ounce containers of applesauce is shown. Find the 99% confidence interval for the population variance and standard deviation. Assume the variable is normally distributed.

18.6	19.5	20.2	20.4	19.3
21.0	20.3	19.6	20.7	18.9
22.1	19.7	20.8	18.9	20.7
21.6	19.5	20.1	20.3	19.9

8. Find the 90% confidence interval for the variance and standard deviation of the ages of seniors at Oak Park College if a sample of 24 students has a standard deviation of 2.3 years. Assume the variable is normally distributed.

9. The number of calories in a 1-ounce serving of various kinds of regular cheese is shown. Estimate the population variance and standard deviation with 90% confidence.

110	45	100	95	110
110	100	110	95	120
130	100	80	105	105
90	110	70	125	108

Source: *The Doctor's Pocket Calorie, Fat, and Carbohydrate Counter.*

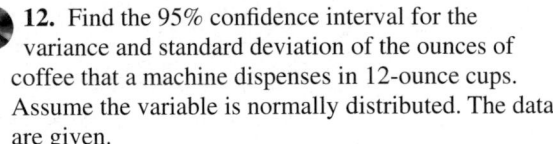 10. A random sample of stock prices per share (in dollars) is shown. Find the 90% confidence interval for the variance and standard deviation for the prices. Assume the variable is normally distributed.

26.69	13.88	28.37	12.00
75.37	7.50	47.50	43.00
3.81	53.81	13.62	45.12
6.94	28.25	28.00	60.50
40.25	10.87	46.12	14.75

Source: *Pittsburgh Tribune Review.*

11. A service station advertises that customers will have to wait no more than 30 minutes for an oil change. A sample of 28 oil changes has a standard deviation of 5.2 minutes. Find the 95% confidence interval of the population standard deviation of the time spent waiting for an oil change.

12. Find the 95% confidence interval for the variance and standard deviation of the ounces of coffee that a machine dispenses in 12-ounce cups. Assume the variable is normally distributed. The data are given.

12.03 12.10 12.02 11.98 12.00 12.05 11.97 11.99

Extending the Concepts

13. A confidence interval for a standard deviation for large samples taken from a normally distributed population can be approximated by

$$s - z_{\alpha/2} \frac{s}{\sqrt{2n}} < \sigma < s + z_{\alpha/2} \frac{s}{\sqrt{2n}}$$

Find the 95% confidence interval for the population standard deviation of calculator batteries. A sample of 200 calculator batteries has a standard deviation of 18 months.

Technology *Step by Step*

TI-83 Plus or TI-84 Plus
Step by Step

The TI-83 Plus and TI-84 Plus do not have a built-in confidence interval for the variance or standard deviation. However, the downloadable program named SDINT is available on your CD and Online Learning Center. Follow the instructions with your CD for downloading the program.

Finding a Confidence Interval for the Variance and Standard Deviation (Data)

1. Enter the data values into L_1.
2. Press **PRGM,** move the cursor to the program named SDINT, and press **ENTER** twice.
3. Press **1** for Data.
4. Type L_1 for the list and press **ENTER.**
5. Type the confidence level and press **ENTER.**
6. Press **ENTER** to clear the screen.

Example TI7–4

This refers to Example 7–15 in the text. Find the 90% confidence interval for the variance and standard deviation for the data:

59 54 53 52 51 39 49 46 49 48

```
LIST ?L₁
```

```
ENTER CONF LEVEL
(0 < CL <1)

CONF LEVEL=.9
```

```
CONF LEVEL .9    ⋮
S= 5.31   n=10

CONF INT FOR σx²
{15.01,76.39}
CONF INT FOR σx
{3.87,8.74}
ENTER TO CLEAR
```

Finding a Confidence Interval for the Variance and Standard Deviation (Statistics)

1. Press **PRGM,** move the cursor to the program named SDINT, and press **ENTER** twice.
2. Press **2** for Stats.
3. Type the sample standard deviation and press **ENTER.**
4. Type the sample size and press **ENTER.**
5. Type the confidence level and press **ENTER.**
6. Press **ENTER** to clear the screen.

Example TI7–5

This refers to Example 7–14 in the text. Find the 95% confidence interval for the variance and standard deviation, given $n = 20$ and $s = 1.6$.

```
S= 1.6
N= 20
```

```
ENTER CONF LEVEL
(0 < CL <1)

CONF LEVEL=.95■
```

```
CONF LEVEL .95   ⋮
S= 1.6   n=20

CONF INT FOR σx²
{1.48,5.46}
CONF INT FOR σx
{1.22,2.34}
ENTER TO CLEAR
```

7–6   Summary

An important aspect of inferential statistics is estimation. Estimations of parameters of populations are accomplished by selecting a random sample from that population and choosing and computing a statistic that is the best estimator of the parameter. A good estimator must be unbiased, consistent, and relatively efficient. The best estimators of μ and p are $\overline{X}$ and $\hat{p}$, respectively. The best estimators of σ^2 and σ are s^2 and s, respectively.

There are two types of estimates of a parameter: point estimates and interval estimates. A point estimate is a specific value. For example, if a researcher wishes to estimate the average length of a certain adult fish, a sample of the fish is selected and measured. The mean of this sample is computed, for example, 3.2 centimeters. From this sample mean, the researcher estimates the population mean to be 3.2 centimeters.

The problem with point estimates is that the accuracy of the estimate cannot be determined. For this reason, statisticians prefer to use the interval estimate. By computing an interval about the sample value, statisticians can be 95 or 99% (or some other percentage) confident that their estimate contains the true parameter. The confidence level is determined by the researcher. The higher the confidence level, the wider the interval of the estimate must be. For example, a 95% confidence interval of the true mean length of a certain species of fish might be

$$3.17 < \mu < 3.23$$

whereas the 99% confidence interval might be

$$3.15 < \mu < 3.25$$

When the confidence interval of the mean is computed, the z or t values are used, depending on whether the population standard deviation is known and depending on the size of the sample. If σ is known or $n \geq 30$, the z values can be used. If σ is not known, the t values must be used when the sample size is less than 30 and the population is normally distributed.

Closely related to computing confidence intervals is the determination of the sample size to make an estimate of the mean. This information is needed to determine the minimum sample size necessary.

1. The degree of confidence must be stated.
2. The population standard deviation must be known or be able to be estimated.
3. The maximum error of estimate must be stated.

Confidence intervals and sample sizes can also be computed for proportions, using the normal distribution; and confidence intervals for variances and standard deviations can be computed, using the chi-square distribution.

Important Terms

chi-square distribution 378	degrees of freedom 362	maximum error of estimate 351	relatively efficient estimator 349
confidence interval 350	estimation 348	point estimate 349	t distribution 362
confidence level 350	estimator 349	proportion 369	unbiased estimator 349
consistent estimator 349	interval estimate 350		

Important Formulas

Formula for the confidence interval of the mean when σ is known (when $n \geq 30$, s can be used if σ is unknown):

$$\overline{X} - z_{\alpha/2}\left(\frac{\sigma}{\sqrt{n}}\right) < \mu < \overline{X} + z_{\alpha/2}\left(\frac{\sigma}{\sqrt{n}}\right)$$

Formula for the sample size for means:

$$n = \left(\frac{z_{\alpha/2} \cdot \sigma}{E}\right)^2$$

where E is the maximum error.

Formula for the confidence interval of the mean when σ is unknown and $n < 30$:

$$\overline{X} - t_{\alpha/2}\left(\frac{s}{\sqrt{n}}\right) < \mu < \overline{X} + t_{\alpha/2}\left(\frac{s}{\sqrt{n}}\right)$$

Formula for the confidence interval for a proportion:

$$\hat{p} - z_{\alpha/2}\sqrt{\frac{\hat{p}\hat{q}}{n}} < p < \hat{p} + z_{\alpha/2}\sqrt{\frac{\hat{p}\hat{q}}{n}}$$

where $\hat{p} = X/n$ and $\hat{q} = 1 - \hat{p}$.

Formula for the sample size for proportions:

$$n = \hat{p}\hat{q}\left(\frac{z_{\alpha/2}}{E}\right)^2$$

Formula for the confidence interval for a variance:

$$\frac{(n-1)s^2}{\chi^2_{\text{right}}} < \sigma^2 < \frac{(n-1)s^2}{\chi^2_{\text{left}}}$$

Formula for confidence interval for a standard deviation:

$$\sqrt{\frac{(n-1)s^2}{\chi^2_{\text{right}}}} < \sigma < \sqrt{\frac{(n-1)s^2}{\chi^2_{\text{left}}}}$$

Review Exercises

1. A study of 36 marathon runners showed that they could run at an average rate of 7.8 miles per hour. The sample standard deviation is 0.6. Find a point estimate of the population mean. Find the 90% confidence interval for the mean of all runners. Based on the results, what minimum speed should a runner obtain to qualify to run in a marathon?

2. In a survey of 1004 individuals, 442 felt that President George W. Bush spent too much time away from Washington. Find a 95% confidence interval for the true population proportion.
   Source: USA TODAY/CNN/Gallup Poll.

3. A U.S. Travel Data Center survey reported that Americans stayed an average of 7.5 nights when they went on vacation. The sample size was 1500. Find a point estimate of the population mean. Find the 95% confidence interval of the true mean. Assume the standard deviation was 0.8 night.
   Source: USA TODAY.

4. A sample of 12 funeral homes in Toledo, Ohio, found these costs for cremation (in dollars).

1320	1052	1090	1285	1570	995	1150
1585	820	1195	1160	590		

   Find a point estimate of the population mean. Find the 95% confidence interval for the population mean.
   Source: Toledo Blade.

5. For a certain urban area, in a sample of 5 months, an average of 28 mail carriers were bitten by dogs each month. The standard deviation of the sample was 3. Find the 90% confidence interval of the true mean number of mail carriers who are bitten by dogs each month. Assume the variable is normally distributed.

6. A researcher is interested in estimating the average salary of teachers in a large urban school district. She wants to be 95% confident that her estimate is correct. If the standard deviation is $1050, how large a sample is needed to be accurate within $200?

7. A researcher wishes to estimate, within $25, the true average amount of postage a community college spends each year. If she wishes to be 90% confident, how large a sample is necessary? The standard deviation is known to be $80.

8. A U.S. Travel Data Center's survey of 1500 adults found that 42% of respondents stated that they favor historical sites as vacations. Find the 95% confidence interval of the true proportion of all adults who favor visiting historical sites as vacations.
   Source: USA TODAY.

9. In a recent study of 75 people, 41 said they were dissatisfied with their community's snow removal service. Find the 95% confidence interval of the true proportion of individuals who are dissatisfied with their

community's snow removal service. Based on the results, should the supervisor consider making improvements in the snow removal service?

10. In a recent study of 100 people, 85 said that they were dissatisfied with their local elected officials. Find the 90% confidence interval of the true proportion of individuals who are dissatisfied with their local elected officials.

11. The federal report in Exercise 10 also stated that 88% of children under age 18 were covered by health insurance in 2000. How large a sample is needed to estimate the true proportion of covered children with 90% confidence with a confidence interval 0.05 wide?

Source: Washington Observer-Reporter.

12. A survey by *Brides* magazine found that 8 out of 10 brides are planning to take the surname of their new husband. How large a sample is needed to estimate the true proportion within 3% with 98% confidence?

Source: Time magazine.

13. The standard deviation of the diameter of 18 baseballs was 0.29 cm. Find the 95% confidence interval of the true standard deviation of the diameters of the baseballs. Do you think the manufacturing process should be checked for inconsistency?

14. A random sample of 22 lawn mowers was selected, and the motors were tested to see how many miles per gallon of gasoline each one obtained. The variance of the measurements was 2.6. Find the 95% confidence interval of the true variance.

15. A random sample of 15 snowmobiles was selected, and the lifetime (in months) of the batteries was measured. The variance of the sample was 8.6. Find the 90% confidence interval of the true variance.

16. The heights of 28 police officers from a large-city police force were measured. The standard deviation of the sample was 1.83 inches. Find the 95% confidence interval of the standard deviation of the heights of the officers.

Statistics Today

Would You Change the Channel?—Revisited

The estimates given in the survey are point estimates. However, since the margin of error is stated to be 3 percentage points, an interval estimate can easily be obtained. For example, if 45% of the people changed the channel, then the confidence interval of the true percentages of people who changed channels would be $42\% < p < 48\%$. The article fails to state whether a 90%, 95%, or some other percentage was used for the confidence interval.

Using the formula given in Section 7–4, a minimum sample size of 1068 would be needed to obtain a 95% confidence interval for p, as shown. Use $\hat{p}$ and $\hat{q}$ as 0.5, since no value is known for $\hat{p}$.

$$n = \hat{p}\hat{q}\left(\frac{z_{\alpha/2}}{E}\right)^2$$

$$= (0.5)(0.5)\left(\frac{1.96}{0.03}\right)^2 = 1067.1$$

$$= 1068$$

Data Analysis

The Data Bank is found in Appendix D, or on the World Wide Web by following links from www.mhhe.com/math/stat/bluman/.

1. From the Data Bank choose a variable, find the mean, and construct the 95 and 99% confidence intervals of the population mean. Use a sample of at least 30 subjects. Find the mean of the population, and determine whether it falls within the confidence interval.

2. Repeat Exercise 1, using a different variable and a sample of 15.

3. Repeat Exercise 1, using a proportion. For example, construct a confidence interval for the proportion of individuals who did not complete high school.

4. From Data Set III in Appendix D, select a sample of 30 values and construct the 95 and 99% confidence intervals of the mean length in miles of major North

American rivers. Find the mean of all the values, and determine if the confidence intervals contain the mean.

5. From Data Set VI in Appendix D, select a sample of 20 values and find the 90% confidence interval of the mean of the number of acres. Find the mean of all the values, and determine if the confidence interval contains the mean.

6. Select a random sample of 20 of the record high temperatures in the United States, found in Data Set I in Appendix D. Find the proportion of temperatures below 110°. Construct a 95% confidence interval for this proportion. Then find the true proportion of temperatures below 110°, using all the data. Is the true proportion contained in the confidence interval? Explain.

Chapter Quiz

Determine whether each statement is true or false. If the statement is false, explain why.

1. Interval estimates are preferred over point estimates since a confidence level can be specified.

2. For a specific confidence interval, the larger the sample size, the smaller the maximum error of estimate will be.

3. An estimator is consistent if, as the sample size decreases, the value of the estimator approaches the value of the parameter estimated.

4. To determine the sample size needed to estimate a parameter, one must know the maximum error of estimate.

Select the best answer.

5. When a 99% confidence interval is calculated instead of a 95% confidence interval with n being the same, the maximum error of estimate will be

   a. Smaller.
   b. Larger.
   c. The same.
   d. It cannot be determined.

6. The best point estimate of the population mean is

   a. The sample mean.
   b. The sample median.
   c. The sample mode.
   d. The sample midrange.

7. When the population standard deviation is unknown and the sample size is less than 30, what table value should be used in computing a confidence interval for a mean?

   a. z
   b. t
   c. Chi-square
   d. None of the above

Complete the following statements with the best answer.

8. A good estimator should be _____, _____, and _____.

9. The maximum difference between the point estimate of a parameter and the actual value of the parameter is called _____.

10. The statement "The average height of an adult male is 5 feet 10 inches" is an example of a(n) _____ estimate.

11. The three confidence intervals used most often are the _____%, _____%, and _____%.

12. A random sample of 49 shoppers showed that they spend an average of $23.45 per visit at the Saturday Mornings Bookstore. The standard deviation of the sample was $2.80. Find a point estimate of the population mean. Find the 90% confidence interval of the true mean.

13. An irate patient complained that the cost of a doctor's visit was too high. She randomly surveyed 20 other patients and found that the mean amount of money they spent on each doctor's visit was $44.80. The standard deviation of the sample was $3.53. Find a point estimate of the population mean. Find the 95% confidence interval of the population mean. Assume the variable is normally distributed.

14. The average weight of 40 randomly selected minivans was 4150 pounds. The standard deviation was 480 pounds. Find a point estimate of the population mean. Find the 99% confidence interval of the true mean weight of the minivans.

15. In a study of 10 insurance sales representatives from a certain large city, the average age of the group was 48.6 years and the standard deviation was 4.1 years. Assume the variable is normally distributed. Find the 95% confidence interval of the population mean age of all insurance sales representatives in that city.

16. In a hospital, a sample of 8 weeks was selected, and it was found that an average of 438 patients were treated in the emergency room each week. The standard deviation was 16. Find the 99% confidence interval of the true mean. Assume the variable is normally distributed.

17. For a certain urban area, it was found that in a sample of 4 months, an average of 31 burglaries occurred each month. The standard deviation was 4. Assume the variable is normally distributed. Find the 90% confidence interval of the true mean number of burglaries each month.

18. A university dean wishes to estimate the average number of hours that freshmen study each week.

The standard deviation from a previous study is 2.6 hours. How large a sample must be selected if he wants to be 99% confident of finding whether the true mean differs from the sample mean by 0.5 hour?

19. A researcher wishes to estimate within $300 the true average amount of money a county spends on road repairs each year. If she wants to be 90% confident, how large a sample is necessary? The standard deviation is known to be $900.

20. A recent study of 75 workers found that 53 people rode the bus to work each day. Find the 95% confidence interval of the proportion of all workers who rode the bus to work.

21. In a study of 150 accidents that required treatment in an emergency room, 36% involved children under 6 years of age. Find the 90% confidence interval of the true proportion of accidents that involve children under the age of 6.

22. A survey of 90 families showed that 40 owned at least one television set. Find the 95% confidence

interval of the true proportion of families who own at least one television set.

23. A nutritionist wishes to determine, within 3%, the true proportion of adults who do not eat any lunch. If he wishes to be 95% confident that his estimate contains the population proportion, how large a sample will be necessary? A previous study found that 15% of the 125 people surveyed said they did not eat lunch.

24. A sample of 25 novels has a standard deviation of 9 pages. Find the 95% confidence interval of the population standard deviation.

25. Find the 90% confidence interval for the variance and standard deviation for the time it takes a state police inspector to check a truck for safety if a sample of 27 trucks has a standard deviation of 6.8 minutes. Assume the variable is normally distributed.

26. A sample of 20 automobiles has a pollution by-product release standard deviation of 2.3 ounces when 1 gallon of gasoline is used. Find the 90% confidence interval of the population standard deviation.

Critical Thinking Challenges

A confidence interval for a median can be found by using these formulas

$$U = \frac{n+1}{2} + \frac{z_{\alpha/2}\sqrt{n}}{2} \qquad \text{(round up)}$$

$$L = n - U + 1$$

to define positions in the set of ordered data values.

Suppose a data set has 30 values, and one wants to find the 95% confidence interval for the median. Substituting in the formulas, one gets

$$U = \frac{30+1}{2} + \frac{1.96\sqrt{30}}{2} = 21 \qquad \text{(rounded up)}$$

$$L = 30 - 21 + 1 = 10$$

when $n = 30$ and $z_{\alpha/2} = 1.96$.

Arrange the data in order from smallest to largest, and then select the 10th and 21st values of the data array; hence, $X_{10} < \text{median} < X_{21}$.

Find the 90% confidence interval for the median for the data in Exercise 12 in Section 7–2.

Data Projects

Use MINITAB, the TI-83 Plus, the TI-84 Plus, or a computer program of your choice to complete these exercises.

1. Select several variables, such as the number of points a football team scored in each game of a specific season, the number of passes completed, or the number of yards gained. Using confidence intervals for the mean, determine the 90, 95, and 99% confidence intervals. (Use z or t, whichever is relevant.) Decide which you think is more appropriate. When this is completed, write a summary of your findings by answering the following questions.

   a. What was the purpose of the study?
   b. What was the population?
   c. How was the sample selected?

   d. What were the results obtained by using confidence intervals?
   e. Did you use z or t? Why?

2. Using the same data or different data, construct a confidence interval for a proportion. For example, you might want to find the proportion of passes completed by the quarterback or the proportion of passes that were intercepted. Write a short paragraph summarizing the results.

You may use the following websites to obtain raw data:

   Visit the data sets at the book's website found at http://www.mhhe.com/math/stat/bluman Click on the 6th edition.
   http://lib.stat.cmu.edu/DASL
   http://www.statcan.ca

Answers to Applying the Concepts

Section 7–2   Making Decisions with Confidence Intervals

1. Answers will vary. One possible answer is to find out the average number of Kleenexes that a group of randomly selected individuals use in a 2-week period.

2. People usually need Kleenexes when they have a cold or when their allergies are acting up.

3. If we want to concentrate on the number of Kleenexes used when people have colds, we select a random sample of people with colds and have them keep a record of how many Kleenexes they use during their colds.

4. Answers may vary. I will use a 95% confidence interval:

$$\bar{x} \pm 1.96 \frac{s}{\sqrt{n}} = 57 \pm 1.96 \frac{15}{\sqrt{85}} = 57 \pm 3.2$$

I am 95% confident that the interval 53.8–60.2 contains the true mean number of Kleenexes used by people when they have colds. It seems reasonable to put 60 Kleenexes in the new automobile glove compartment boxes.

5. Answers will vary. Since I am 95% confident that the interval contains the true average, any number of Kleenexes between 54 and 60 would be reasonable. Sixty seemed to be the most reasonable answer, since it is close to 2 standard deviations above the sample mean.

Section 7–3   Sport Drink Decision

1. Answers will vary. One possible answer is that this is a small sample since we are only looking at seven popular sport drinks.

2. The mean cost per container is $1.25, with standard deviation of $0.39. The 90% confidence interval is

$$\bar{x} \pm t_{\alpha/2} \frac{s}{\sqrt{n}} = 1.25 \pm 1.943 \frac{0.39}{\sqrt{7}} = 1.25 \pm 0.29$$

or $0.96 < \mu < 1.54$.

The 10-K, All Sport, Exceed, and Hydra Fuel all fall outside of the confidence interval.

3. None of the values appear to be outliers.

4. There are $7 - 1 = 6$ degrees of freedom.

5. Cost per serving would impact my decision on purchasing a sport drink, since this would allow me to compare the costs on an equal scale.

6. Answers will vary.

Section 7–4   Contracting Influenza

1. (95% CI) means that these are the 95% confidence intervals constructed from the data.

2. The margin of error for men reporting influenza is $\frac{50.5 - 47.1}{2} = 1.7\%$.

3. The total sample size was 19,774.

4. The larger the sample size, the smaller the margin of error (all other things being held constant).

5. A 90% confidence interval would be narrower (smaller) than a 95% confidence interval, since we need to include fewer values in the interval.

6. The 51.5% is the middle of the confidence interval, since it is the point estimate for the confidence interval.

Section 7–5   Confidence Interval for Standard Deviation

1. The data represent a population, since we have the age at death for all deceased Presidents (at the time of the writing of this book).

2. Answers will vary. One possible sample is 56, 67, 53, 46, 63, 77, 63, 57, 71, 57, 80, 65, which results in a standard deviation of 9.9 years and a variance of 98.0.

3. Answers will vary. The 95% confidence interval for the standard deviation is $\sqrt{\frac{(n-1)s^2}{\chi^2_{\text{right}}}}$ to $\sqrt{\frac{(n-1)s^2}{\chi^2_{\text{left}}}}$. In this case we have $\sqrt{\frac{(12-1)9.9^2}{21.920}} = \sqrt{49.1839} = 7.0$ to $\sqrt{\frac{(12-1)9.9^2}{3.8158}} = \sqrt{282.538} = 16.8$, or 7.0 to 16.8 years.

4. The standard deviation for all the data values is 11.6 years.

5. Answers will vary. Yes, the confidence interval does contain the population standard deviation.

6. Answers will vary.

7. We need to assume that the distribution of ages at death is normal.

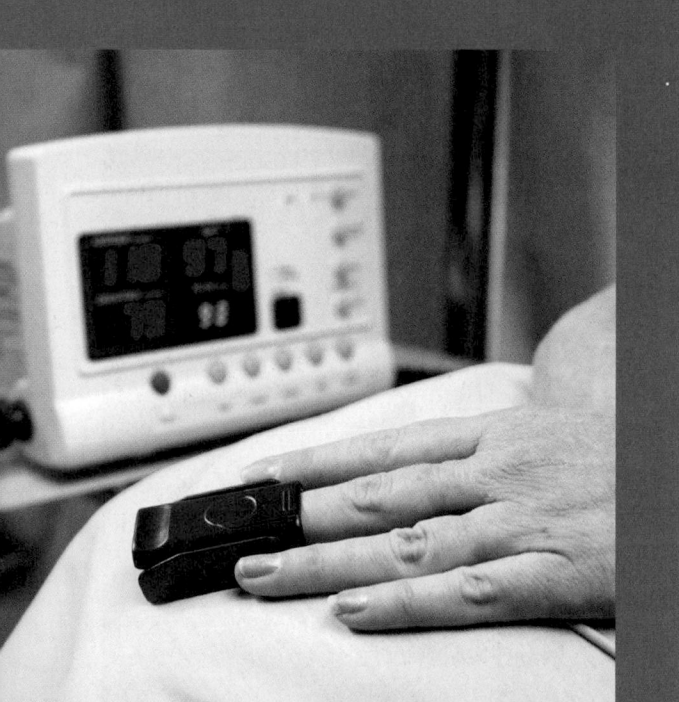

CHAPTER 8

Hypothesis Testing

Objectives

After completing this chapter, you should be able to

1. Understand the definitions used in hypothesis testing.

2. State the null and alternative hypotheses.

3. Find critical values for the z test.

4. State the five steps used in hypothesis testing.

5. Test means for large samples, using the z test.

6. Test means for small samples, using the t test.

7. Test proportions, using the z test.

8. Test variances or standard deviations, using the chi-square test.

9. Test hypotheses, using confidence intervals.

10. Explain the relationship between type I and type II errors and the power of a test.

Outline

How Much Better Is Better?

Suppose a school superintendent reads an article which states that the overall mean score for the SAT is 910. Furthermore, suppose that, for a sample of students, the average of the SAT scores in the superintendent's school district is 960. Can the superintendent conclude that the students in his school district scored higher than average? At first glance, you might be inclined to say yes, since 960 is higher than 910. But recall that the means of samples vary about the population mean when samples are selected from a specific population. So the question arises, Is there a real difference in the means, or is the difference simply due to chance (i.e., sampling error)? In this chapter, you will learn how to answer that question by using statistics that explain hypothesis testing. See Statistics Today—Revisited for the answer. In this chapter, you will learn how to answer many questions of this type by using statistics that are explained in the theory of hypothesis testing.

8–1 Introduction

Researchers are interested in answering many types of questions. For example, a scientist might want to know whether the earth is warming up. A physician might want to know whether a new medication will lower a person's blood pressure. An educator might wish to see whether a new teaching technique is better than a traditional one. A retail merchant might want to know whether the public prefers a certain color in a new line of fashion. Automobile manufacturers are interested in determining whether seat belts will reduce the severity of injuries caused by accidents. These types of questions can be addressed through statistical **hypothesis testing,** which is a decision-making process for evaluating claims about a population. In hypothesis testing, the researcher must define the population under study, state the particular hypotheses that will be investigated, give the significance level, select a sample from the population, collect the data, perform the calculations required for the statistical test, and reach a conclusion.

Hypotheses concerning parameters such as means and proportions can be investigated. There are two specific statistical tests used for hypotheses concerning means: the *z test* and the *t test*. This chapter will explain in detail the hypothesis-testing procedure along with the *z* test and the *t* test. In addition, a hypothesis-testing procedure for testing a single variance or standard deviation using the chi-square distribution is explained in Section 8–6.

The three methods used to test hypotheses are

1. The traditional method
2. The *P*-value method
3. The confidence interval method

The *traditional method* will be explained first. It has been used since the hypothesis-testing method was formulated. A newer method, called the *P-value method,* has become popular with the advent of modern computers and high-powered statistical calculators. It will be explained at the end of Section 8–3. The third method, the *confidence interval method,* is explained in Section 8–7 and illustrates the relationship between hypothesis testing and confidence intervals.

8–2 Steps in Hypothesis Testing–Traditional Method

Every hypothesis-testing situation begins with the statement of a hypothesis.

> A **statistical hypothesis** is a conjecture about a population parameter. This conjecture may or may not be true.

Objective 1

Understand the definitions used in hypothesis testing.

There are two types of statistical hypotheses for each situation: the null hypothesis and the alternative hypothesis.

> The **null hypothesis,** symbolized by H_0, is a statistical hypothesis that states that there is no difference between a parameter and a specific value, or that there is no difference between two parameters.
>
> The **alternative hypothesis,** symbolized by H_1, is a statistical hypothesis that states the existence of a difference between a parameter and a specific value, or states that there is a difference between two parameters.

(*Note:* Although the definitions of null and alternative hypotheses given here use the word *parameter,* these definitions can be extended to include other terms such as *distributions* and *randomness.* This is explained in later chapters.)

As an illustration of how hypotheses should be stated, three different statistical studies will be used as examples.

Situation A   A medical researcher is interested in finding out whether a new medication will have any undesirable side effects. The researcher is particularly concerned with the pulse rate of the patients who take the medication. Will the pulse rate increase, decrease, or remain unchanged after a patient takes the medication?

Objective 2

State the null and alternative hypotheses.

Since the researcher knows that the mean pulse rate for the population under study is 82 beats per minute, the hypotheses for this situation are

$$H_0: \mu = 82 \qquad \text{and} \qquad H_1: \mu \neq 82$$

The null hypothesis specifies that the mean will remain unchanged, and the alternative hypothesis states that it will be different. This test is called a *two-tailed test* (a term that will be formally defined later in this section), since the possible side effects of the medicine could be to raise or lower the pulse rate.

Situation B   A chemist invents an additive to increase the life of an automobile battery. If the mean lifetime of the automobile battery without the additive is 36 months, then her hypotheses are

$$H_0: \mu \leq 36 \qquad \text{and} \qquad H_1: \mu > 36$$

In this situation, the chemist is interested only in increasing the lifetime of the batteries, so her alternative hypothesis is that the mean is greater than 36 months. The null hypothesis is that the mean is less than or equal to 36 months. This test is called *right-tailed,* since the interest is in an increase only.

Situation C   A contractor wishes to lower heating bills by using a special type of insulation in houses. If the average of the monthly heating bills is $78, her hypotheses about heating costs with the use of insulation are

$$H_0: \mu \geq \$78 \qquad \text{and} \qquad H_1: \mu < \$78$$

This test is a *left-tailed test,* since the contractor is interested only in lowering heating costs.

To state hypotheses correctly, researchers must translate the *conjecture* or *claim* from words into mathematical symbols. The basic symbols used are as follows:

Equal to	=	Less than	<
Not equal to	≠	Greater than or equal to	≥
Greater than	>	Less than or equal to	≤

The null and alternative hypotheses are stated together, and the null hypothesis contains the equals sign, as shown (where k represents a specified number).

Two-tailed test	**Right-tailed test**	**Left-tailed test**
$H_0: \mu = k$	$H_0: \mu \leq k$	$H_0: \mu \geq k$
$H_1: \mu \neq k$	$H_1: \mu > k$	$H_1: \mu < k$

The formal definitions of the different types of tests are given later in this section.

Table 8–1 shows some common phrases that are used in hypotheses, conjectures, and the corresponding symbols. This table should be helpful in translating verbal conjectures into mathematical symbols.

Table 8–1	Hypothesis-Testing Common Phrases
>	<
Is greater than	Is less than
Is above	Is below
Is higher than	Is lower than
Is longer than	Is shorter than
Is bigger than	Is smaller than
Is increased	Is decreased or reduced from
≥	≤
Is greater than or equal to	Is less than or equal to
Is at least	Is at most
Is not less than	Is not more than
=	≠
Is equal to	Is not equal to
Is exactly the same as	Is different from
Has not changed from	Has changed from
Is the same as	Is not the same as

| Example 8–1 | State the null and alternative hypotheses for each conjecture. |

a. A researcher thinks that if expectant mothers use vitamin pills, the birth weight of the babies will increase. The average birth weight of the population is 8.6 pounds.

b. An engineer hypothesizes that the mean number of defects can be decreased in a manufacturing process of compact disks by using robots instead of humans for certain tasks. The mean number of defective disks per 1000 is 18.

c. A psychologist feels that playing soft music during a test will change the results of the test. The psychologist is not sure whether the grades will be higher or lower. In the past, the mean of the scores was 73.

Solution

a. $H_0: \mu \le 8.6$ and $H_1: \mu > 8.6$

b. $H_0: \mu \ge 18$ and $H_1: \mu < 18$

c. $H_0: \mu = 73$ and $H_1: \mu \ne 73$

A note of caution: It is now becoming quite common in professional journals and textbooks to state the null hypothesis using only the equals sign. This is so because a single distribution with a parameter equal to the hypothesized value stated in the null hypothesis can be used for the statistical tests. This textbook, however, uses the traditional signs of $\le$ and $\ge$ when a left-tailed or right-tailed hypothesis is being tested. When you are reading statistics in professional journals, then, it is important to read the study carefully and to decide whether a one-tailed or two-tailed hypothesis is being tested.

After stating the hypothesis, the researcher designs the study. The researcher selects the correct *statistical test,* chooses an appropriate *level of significance,* and formulates a plan for conducting the study. In situation A, for instance, the researcher will select a sample of patients who will be given the drug. After allowing a suitable time for the drug to be absorbed, the researcher will measure each person's pulse rate.

Recall that when samples of a specific size are selected from a population, the means of these samples will vary about the population mean, and the distribution of the sample means will be approximately normal when the sample size is 30 or more. (See Section 6–5.) So even if the null hypothesis is true, the mean of the pulse rates of the sample of patients will not, in most cases, be exactly equal to the population mean of 82 beats per minute. There are two possibilities. Either the null hypothesis is true, and the difference between the sample mean and the population mean is due to chance; *or* the null hypothesis is false, and the sample came from a population whose mean is not 82 beats per minute but is some other value that is not known. These situations are shown in Figure 8–1.

The farther away the sample mean is from the population mean, the more evidence there would be for rejecting the null hypothesis. The probability that the sample came from a population whose mean is 82 decreases as the distance or absolute value of the difference between the means increases.

If the mean pulse rate of the sample were, say, 83, the researcher would probably conclude that this difference was due to chance and would not reject the null hypothesis. But if the sample mean were, say, 90, then in all likelihood the researcher would conclude that the medication increased the pulse rate of the users and would reject the null hypothesis. The question is, Where does the researcher draw the line? This decision is not made on feelings or intuition; it is made statistically. That is, the difference must be significant and in all likelihood not due to chance. Here is where the concepts of statistical test and level of significance are used.

Figure 8–1

Situations in
Hypothesis Testing

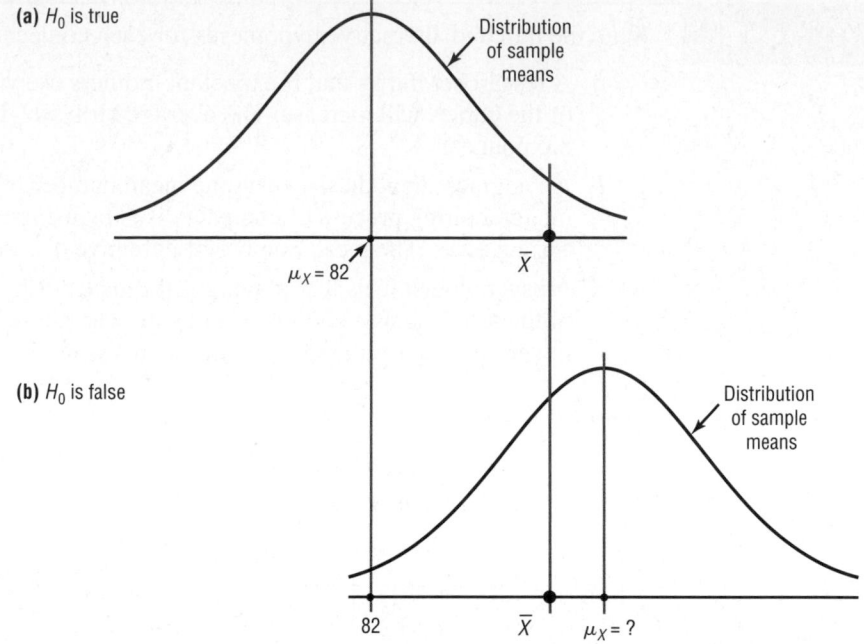

> A **statistical test** uses the data obtained from a sample to make a decision about whether the null hypothesis should be rejected.
>
> The numerical value obtained from a statistical test is called the **test value.**

In this type of statistical test, the mean is computed for the data obtained from the sample and is compared with the population mean. Then a decision is made to reject or not reject the null hypothesis on the basis of the value obtained from the statistical test. If the difference is significant, the null hypothesis is rejected. If it is not, then the null hypothesis is not rejected.

In the hypothesis-testing situation, there are four possible outcomes. In reality, the null hypothesis may or may not be true, and a decision is made to reject or not reject it on the basis of the data obtained from a sample. The four possible outcomes are shown in Figure 8–2. Notice that there are two possibilities for a correct decision and two possibilities for an incorrect decision.

Figure 8–2

Possible Outcomes of
a Hypothesis Test

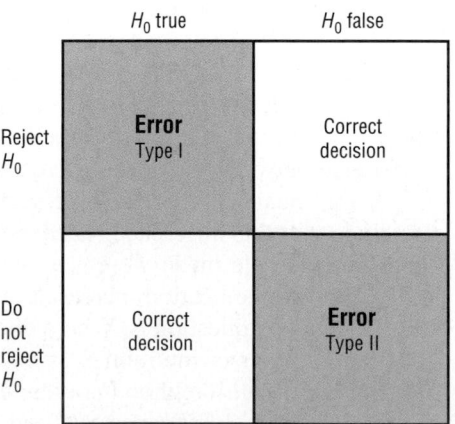

If a null hypothesis is true and it is rejected, then a *type I error* is made. In situation A, for instance, the medication might not significantly change the pulse rate of all the users in the population; but it might change the rate, by chance, of the subjects in the sample. In this case, the researcher will reject the null hypothesis when it is really true, thus committing a type I error.

On the other hand, the medication might not change the pulse rate of the subjects in the sample, but when it is given to the general population, it might cause a significant increase or decrease in the pulse rate of users. The researcher, on the basis of the data obtained from the sample, will not reject the null hypothesis, thus committing a *type II error.*

In situation B, the additive might not significantly increase the lifetimes of automobile batteries in the population, but it might increase the lifetimes of the batteries in the sample. In this case, the null hypothesis would be rejected when it was really true. This would be a type I error. On the other hand, the additive might not work on the batteries selected for the sample, but if it were to be used in the general population of batteries, it might significantly increase their lifetimes. The researcher, on the basis of information obtained from the sample, would not reject the null hypothesis, thus committing a type II error.

A **type I error** occurs if one rejects the null hypothesis when it is true.

A **type II error** occurs if one does not reject the null hypothesis when it is false.

The hypothesis-testing situation can be likened to a jury trial. In a jury trial, there are four possible outcomes. The defendant is either guilty or innocent, and he or she will be convicted or acquitted. See Figure 8–3.

Now the hypotheses are

H_0: The defendant is innocent

H_1: The defendant is not innocent (i.e., guilty)

Next, the evidence is presented in court by the prosecutor, and based on this evidence, the jury decides the verdict, innocent or guilty.

If the defendant is convicted but he or she did not commit the crime, then a type I error has been committed. See block 1 of Figure 8–3. On the other hand, if the defendant is convicted and he or she has committed the crime, then a correct decision has been made. See block 2.

If the defendant is acquitted and he or she did not commit the crime, a correct decision has been made by the jury. See block 3. However, if the defendant is acquitted and he or she did commit the crime, then a type II error has been made. See block 4.

Figure 8–3

Hypothesis Testing and a Jury Trial

H_0: The defendant is innocent.
H_1: The defendant is not innocent.

The results of a trial can be shown as follows:

	H_0 true (innocent)	H_0 false (not innocent)
Reject H_0 (convict)	1. Type I error	2. Correct decision
Do not reject H_0 (acquit)	3. Correct decision	4. Type II error

The decision of the jury does not prove that the defendant did or did not commit the crime. The decision is based on the evidence presented. If the evidence is strong enough, the defendant will be convicted in most cases. If the evidence is weak, the defendant will be acquitted in most cases. Nothing is proved absolutely. Likewise, the decision to reject or not reject the null hypothesis does not prove anything. *The only way to prove anything statistically is to use the entire population,* which, in most cases, is not possible. The decision, then, is made on the basis of probabilities. That is, when there is a large difference between the mean obtained from the sample and the hypothesized mean, the null hypothesis is probably not true. The question is, How large a difference is necessary to reject the null hypothesis? Here is where the level of significance is used.

The **level of significance** is the maximum probability of committing a type I error. This probability is symbolized by α (Greek letter **alpha**). That is, P (type I error) $= \alpha$.

The probability of a type II error is symbolized by β, the Greek letter **beta.** That is, P(type II error) $= \beta$. In most hypothesis-testing situations, β cannot easily be computed; however, α and β are related in that decreasing one increases the other.

Statisticians generally agree on using three arbitrary significance levels: the 0.10, 0.05, and 0.01 levels. That is, if the null hypothesis is rejected, the probability of a type I error will be 10, 5, or 1%, depending on which level of significance is used. Here is another way of putting it: When $\alpha = 0.10$, there is a 10% chance of rejecting a true null hypothesis; when $\alpha = 0.05$, there is a 5% chance of rejecting a true null hypothesis; and when $\alpha = 0.01$, there is a 1% chance of rejecting a true null hypothesis.

In a hypothesis-testing situation, the researcher decides what level of significance to use. It does not have to be the 0.10, 0.05, or 0.01 level. It can be any level, depending on the seriousness of the type I error. After a significance level is chosen, a *critical value* is selected from a table for the appropriate test. If a z test is used, for example, the z table (Table E in Appendix C) is consulted to find the critical value. The critical value determines the critical and noncritical regions.

The **critical value** separates the critical region from the noncritical region. The symbol for critical value is C.V.

The **critical** or **rejection region** is the range of values of the test value that indicates that there is a significant difference and that the null hypothesis should be rejected.

The **noncritical** or **nonrejection region** is the range of values of the test value that indicates that the difference was probably due to chance and that the null hypothesis should not be rejected.

The critical value can be on the right side of the mean or on the left side of the mean for a one-tailed test. Its location depends on the inequality sign of the alternative hypothesis. For example, in situation B, where the chemist is interested in increasing the average lifetime of automobile batteries, the alternative hypothesis is H_1: $\mu > 36$. Since the inequality sign is $>$, the null hypothesis will be rejected only when the sample mean is significantly greater than 36. Hence, the critical value must be on the right side of the mean. Therefore, this test is called a right-tailed test.

A **one-tailed test** indicates that the null hypothesis should be rejected when the test value is in the critical region on one side of the mean. A one-tailed test is either a **right-tailed test** or **left-tailed test,** depending on the direction of the inequality of the alternative hypothesis.

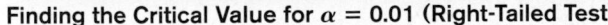

Figure 8-4

Finding the Critical Value for $\alpha = 0.01$ (Right-Tailed Test)

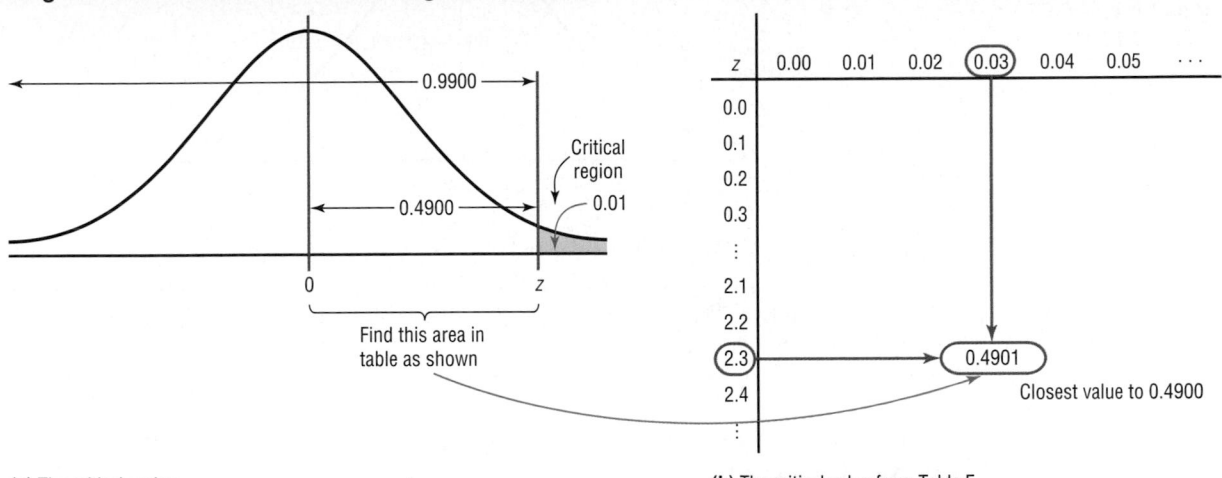

(a) The critical region   **(b)** The critical value from Table E

To obtain the critical value, the researcher must choose an alpha level. In situation B, suppose the researcher chose $\alpha = 0.01$. Then the researcher must find a z value such that 1% of the area falls to the right of the z value and 99% falls to the left of the z value, as shown in Figure 8-4(a).

Objective 3

Find critical values for the z test.

Next, the researcher must find the value in Table E closest to 0.4900. Note that because the table gives the area between 0 and z, 0.5000 must be subtracted from 0.9900 to get 0.4900. The critical z value is 2.33, since that value gives the area closest to 0.4900, as shown in Figure 8-4(b).

The critical and noncritical regions and the critical value are shown in Figure 8-5.

Figure 8-5

Critical and Noncritical Regions for $\alpha = 0.01$ (Right-Tailed Test)

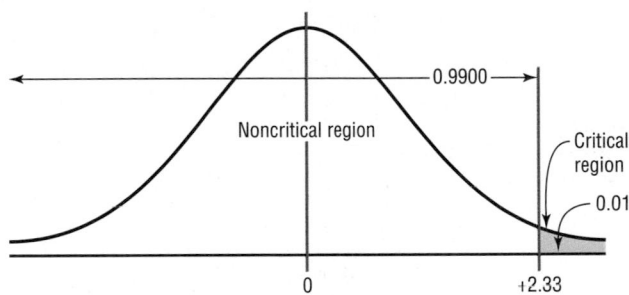

Now, move on to situation C, where the contractor is interested in lowering the heating bills. The alternative hypothesis is H_1: $\mu <$ \$78. Hence, the critical value falls to the left of the mean. This test is thus a left-tailed test. At $\alpha = 0.01$, the critical value is -2.33, as shown in Figure 8-6.

When a researcher conducts a two-tailed test, as in situation A, the null hypothesis can be rejected when there is a significant difference in either direction, above or below the mean.

Figure 8–6

Critical and Noncritical
Regions for $\alpha = 0.01$
(Left-Tailed Test)

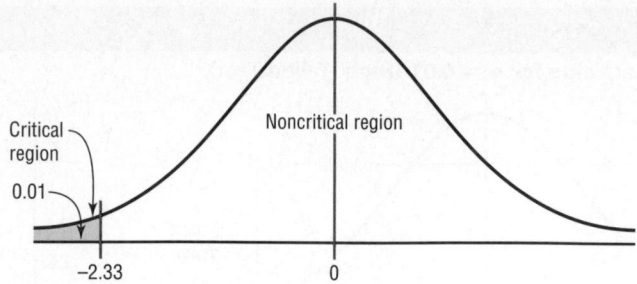

In a **two-tailed test,** the null hypothesis should be rejected when the test value is in either of the two critical regions.

For a two-tailed test, then, the critical region must be split into two equal parts. If $\alpha = 0.01$, then one-half of the area, or 0.005, must be to the right of the mean and one-half must be to the left of the mean, as shown in Figure 8–7.

Figure 8–7

Finding the Critical
Values for $\alpha = 0.01$
(Two-Tailed Test)

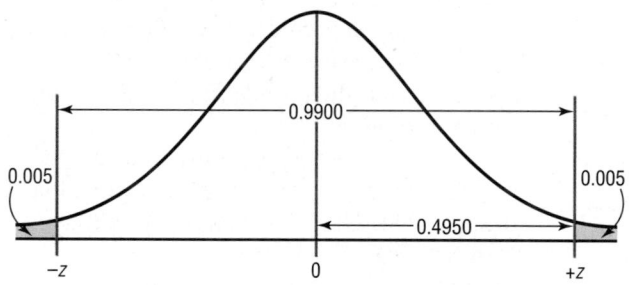

In this case, the area to be found in Table E is 0.4950. The critical values are $+2.58$ and -2.58, as shown in Figure 8–8.

Figure 8–8

Critical and Noncritical
Regions for $\alpha = 0.01$
(Two-Tailed Test)

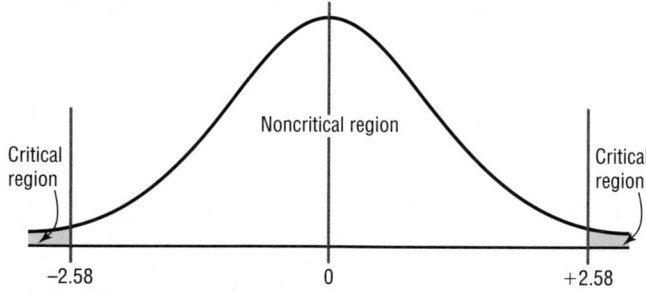

A similar procedure is used for other values of α.

Figure 8–9 with rejection regions shaded shows the critical value (C.V.) for the three situations discussed in this section for values of $\alpha = 0.10$, $\alpha = 0.05$, and $\alpha = 0.01$. The procedure for finding critical values is outlined next (where k is a specified number).

Figure 8–9

Summary of Hypothesis Testing and Critical Values

$H_0: \mu \geq k$
$H_1: \mu < k$ $\begin{cases} \alpha = 0.10, \text{C.V.} = -1.28 \\ \alpha = 0.05, \text{C.V.} = -1.65 \\ \alpha = 0.01, \text{C.V.} = -2.33 \end{cases}$

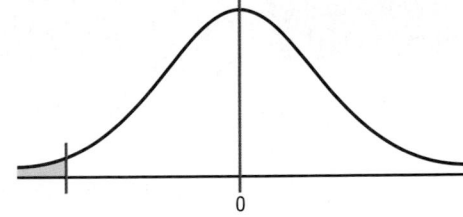

(a) Left-tailed

$H_0: \mu \leq k$
$H_1: \mu > k$ $\begin{cases} \alpha = 0.10, \text{C.V.} = +1.28 \\ \alpha = 0.05, \text{C.V.} = +1.65 \\ \alpha = 0.01, \text{C.V.} = +2.33 \end{cases}$

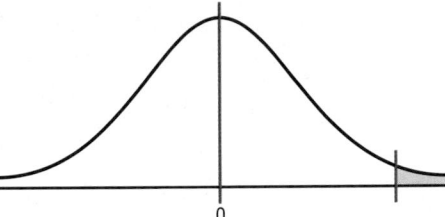

(b) Right-tailed

$H_0: \mu = k$
$H_1: \mu \neq k$ $\begin{cases} \alpha = 0.10, \text{C.V.} = \pm1.65 \\ \alpha = 0.05, \text{C.V.} = \pm1.96 \\ \alpha = 0.01, \text{C.V.} = \pm2.58 \end{cases}$

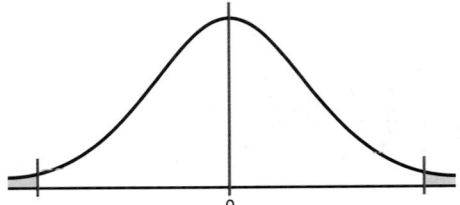

(c) Two-tailed

Procedure Table

Finding the Critical Values for Specific α Values, Using Table E

Step 1   Draw the figure and indicate the appropriate area.

     a. If the test is left-tailed, the critical region, with an area equal to α, will be on the left side of the mean.

     b. If the test is right-tailed, the critical region, with an area equal to α, will be on the right side of the mean.

     c. If the test is two-tailed, α must be divided by 2; one-half of the area will be to the right of the mean, and one-half will be to the left of the mean.

Step 2   For a one-tailed test, subtract the area (equivalent to α) in the critical region from 0.5000, since Table E gives the area under the standard normal distribution curve between 0 and any z to the right of 0. For a two-tailed test, subtract the area (equivalent to $\alpha/2$) from 0.5000.

Step 3   Find the area in Table E corresponding to the value obtained in step 2. If the exact value cannot be found in the table, use the closest value.

Step 4   Find the z value that corresponds to the area. This will be the critical value.

Step 5   Determine the sign of the critical value for a one-tailed test.

     a. If the test is left-tailed, the critical value will be negative.

     b. If the test is right-tailed, the critical value will be positive.

     For a two-tailed test, one value will be positive and the other negative.

| Example 8–2 | Using Table E in Appendix C, find the critical value(s) for each situation and draw the appropriate figure, showing the critical region. |

     a. A left-tailed test with $\alpha = 0.10$.

     b. A two-tailed test with $\alpha = 0.02$.

     c. A right-tailed test with $\alpha = 0.005$.

Solution *a*

Step 1 Draw the figure and indicate the appropriate area. Since this is a left-tailed test, the area of 0.10 is located in the left tail, as shown in Figure 8–10.

Step 2 Subtract 0.10 from 0.5000 to get 0.4000.

Step 3 In Table E, find the area that is closest to 0.4000; in this case, it is 0.3997.

Step 4 Find the z value that corresponds to this area. It is 1.28.

Step 5 Determine the sign of the critical value (i.e., the z value). Since this is a left-tailed test, the sign of the critical value is negative. Hence, the critical value is -1.28. See Figure 8–10.

Figure 8–10

Critical Value and
Critical Region for
part *a* of Example 8–2

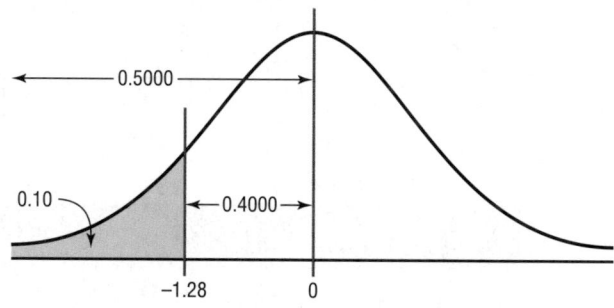

Solution *b*

Step 1 Draw the figure and indicate the appropriate area. In this case, there are two areas equivalent to $\alpha/2$, or $0.02/2 = 0.01$.

Step 2 Subtract 0.01 from 0.5000 to get 0.4900.

Step 3 Find the area in Table E closest to 0.4900. In this case, it is 0.4901.

Step 4 Find the z value that corresponds to this area. It is 2.33.

Step 5 Determine the sign of the critical value. Since this test is a two-tailed test, there are two critical values: one is positive and the other is negative. They are $+2.33$ and -2.33. See Figure 8–11.

Figure 8–11

Critical Values and
Critical Regions for
part *b* of Example 8–2

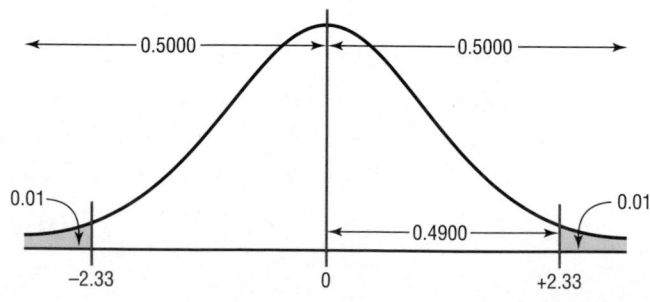

Solution c

Step 1   Draw the figure and indicate the appropriate area. Since this is a right-tailed test, the area 0.005 is located in the right tail, as shown in Figure 8–12.

Figure 8–12

Critical Value and
Critical Region for
part *c* of Example 8–2

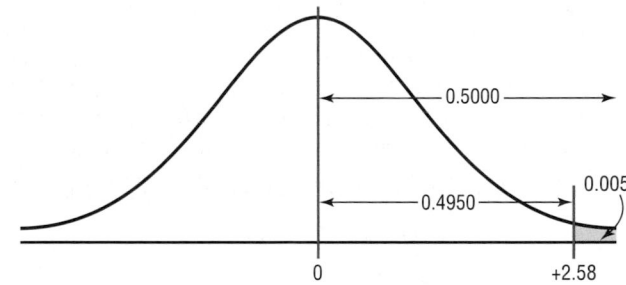

Step 2   Subtract 0.005 from 0.5000 to get 0.4950.

Step 3   Find the area in Table E closest to 0.4950. In this case it is 0.4951.

Step 4   Find the *z* value that corresponds to this area. It is 2.58.

Step 5   Determine the sign of the critical value. Since this is a right-tailed test, the sign is positive; hence, the critical value is +2.58.

Objective 4

State the five steps
used in hypothesis
testing.

In hypothesis testing, the following steps are recommended.

1. State the hypotheses. Be sure to state both the null and the alternative hypotheses.
2. Design the study. This step includes selecting the correct statistical test, choosing a level of significance, and formulating a plan to carry out the study. The plan should include information such as the definition of the population, the way the sample will be selected, and the methods that will be used to collect the data.
3. Conduct the study and collect the data.
4. Evaluate the data. The data should be tabulated in this step, and the statistical test should be conducted. Finally, decide whether to reject or not reject the null hypothesis.
5. Summarize the results.

For the purposes of this chapter, a simplified version of the hypothesis-testing procedure will be used, since designing the study and collecting the data will be omitted. The steps are summarized in the Procedure Table.

Procedure Table

Solving Hypothesis-Testing Problems (Traditional Method)

Step 1   State the hypotheses and identify the claim.

Step 2   Find the critical value(s) from the appropriate table in Appendix C.

Step 3   Compute the test value.

Step 4   Make the decision to reject or not reject the null hypothesis.

Step 5   Summarize the results.

Applying the Concepts **8–2**

Eggs and Your Health

The Incredible Edible Egg company recently found that eating eggs does not increase a person's blood serum cholesterol. Five hundred subjects participated in a study that lasted for 2 years. The participants were randomly assigned to either a no-egg group or a moderate-egg group. The blood serum cholesterol levels were checked at the beginning and at the end of the study. Overall, the groups' levels were not significantly different. The company reminds us that eating eggs is healthy if done in moderation. Many of the previous studies relating eggs and high blood serum cholesterol jumped to improper conclusions.

Using this information, answer these questions.

1. What prompted the study?
2. What is the population under study?
3. Was a sample collected?
4. What was the hypothesis?
5. Was data collected?
6. Were any statistical tests run?
7. What was the conclusion?

See page 460 for the answers.

Exercises 8–2

1. Define *null* and *alternative hypotheses,* and give an example of each.

2. What is meant by a type I error? A type II error? How are they related?

3. What is meant by a statistical test?

4. Explain the difference between a one-tailed and a two-tailed test.

5. What is meant by the critical region? The noncritical region?

6. What symbols are used to represent the null hypothesis and the alternative hypothesis?

7. What symbols are used to represent the probabilities of type I and type II errors?

8. Explain what is meant by a significant difference.

9. When should a one-tailed test be used? A two-tailed test?

10. List the steps in hypothesis testing.

11. In hypothesis testing, why can't the hypothesis be proved true?

12. **(ans)** Using the z table (Table E), find the critical value (or values) for each.

   a. $\alpha = 0.05$, two-tailed test
   b. $\alpha = 0.01$, left-tailed test
   c. $\alpha = 0.005$, right-tailed test
   d. $\alpha = 0.01$, right-tailed test
   e. $\alpha = 0.05$, left-tailed test
   f. $\alpha = 0.02$, left-tailed test
   g. $\alpha = 0.05$, right-tailed test
   h. $\alpha = 0.01$, two-tailed test
   i. $\alpha = 0.04$, left-tailed test
   j. $\alpha = 0.02$, right-tailed test

13. For each conjecture, state the null and alternative hypotheses:

   a. The average age of community college students is 24.6 years.
   b. The average income of accountants is $51,497.
   c. The average age of attorneys is greater than 25.4 years.
   d. The average score of 50 high school basketball games is less than 88.
   e. The average pulse rate of male marathon runners is less than 70 beats per minute.
   f. The average cost of a DVD player is $79.95.
   g. The average weight loss for a sample of people who exercise 30 minutes per day for 6 weeks is 8.2 pounds.

<table>
<tr><td>

8–3

Objective **5**

Test means for large samples, using the *z* test.

</td></tr>
</table>

z Test for a Mean

In this chapter, two statistical tests will be explained: the *z* test, used to test for the mean of a large sample, and the *t* test, used for the mean of a small sample. This section explains the *z* test, and Section 8–4 explains the *t* test.

Many hypotheses are tested using a statistical test based on the following general formula:

$$\text{Test value} = \frac{(\text{observed value}) - (\text{expected value})}{\text{standard error}}$$

The observed value is the statistic (such as the mean) that is computed from the sample data. The expected value is the parameter (such as the mean) that one would expect to obtain if the null hypothesis were true—in other words, the hypothesized value. The denominator is the standard error of the statistic being tested (in this case, the standard error of the mean).

The *z* test is defined formally as follows.

The **z test** is a statistical test for the mean of a population. It can be used when $n \geq 30$, or when the population is normally distributed and σ is known.

The formula for the *z* test is

$$z = \frac{\overline{X} - \mu}{\sigma/\sqrt{n}}$$

where
$\overline{X}$ = sample mean
μ = hypothesized population mean
σ = population standard deviation
n = sample size

For the *z* test, the observed value is the value of the sample mean. The expected value is the value of the population mean, assuming that the null hypothesis is true. The denominator $\sigma/\sqrt{n}$ is the standard error of the mean.

The formula for the *z* test is the same formula shown in Chapter 6 for the situation where one is using a distribution of sample means. Recall that the central limit theorem allows one to use the standard normal distribution to approximate the distribution of sample means when $n \geq 30$. If σ is unknown, *s* can be used when $n \geq 30$.

Note: The student's first encounter with hypothesis testing can be somewhat challenging and confusing, since there are many new concepts being introduced at the same time. *To understand all the concepts, the student must carefully follow each step in the examples and try each exercise that is assigned.* Only after careful study and patience will these concepts become clear.

As stated in Section 8–2, there are five steps for solving *hypothesis-testing* problems:

Step 1 State the hypotheses and identify the claim.

Step 2 Find the critical value(s).

Step 3 Compute the test value.

Step 4 Make the decision to reject or not reject the null hypothesis.

Step 5 Summarize the results.

Example 8–3 illustrates these five steps.

This study found that people who used pedometers reported having increased energy, mood improvement, and weight loss. State possible null and alternative hypotheses for the study. What would be a likely population? What is the sample size? Comment on the sample size.

RD HEALTH

Step to It

I T FITS in your hand, costs less than $30, and will make you feel great. Give up? A pedometer. Brenda Rooney, an epidemiologist at Gundersen Lutheran Medical Center in LaCrosse, Wis., gave 500 people pedometers and asked them to take 10,000 steps—about five miles—a day. (Office workers typically average about 4000 steps a day.) By the end of eight weeks, 56 percent reported having more energy, 47 percent improved their mood and 50 percent lost weight. The subjects reported that seeing their total step-count motivated them to take more.

— JENNIFER BRAUNSCHWEIGER

Source: Reprinted with permission from the April 2002 Reader's Digest. Copyright © 2002 by The Reader's Digest Assn. Inc.

Example 8–3

A researcher reports that the average salary of assistant professors is more than $42,000. A sample of 30 assistant professors has a mean salary of $43,260. At $\alpha = 0.05$, test the claim that assistant professors earn more than $42,000 a year. The standard deviation of the population is $5230.

Solution

Step 1 State the hypotheses and identify the claim.

H_0: $\mu \le \$42,000$ and H_1: $\mu > \$42,000$ (claim)

Step 2 Find the critical value. Since $\alpha = 0.05$ and the test is a right-tailed test, the critical value is $z = +1.65$.

Step 3 Compute the test value.

$$z = \frac{\overline{X} - \mu}{\sigma/\sqrt{n}} = \frac{\$43,260 - \$42,000}{\$5230/\sqrt{30}} = 1.32$$

Step 4 Make the decision. Since the test value, $+1.32$, is less than the critical value, $+1.65$, and is not in the critical region, the decision is not to reject the null hypothesis. This test is summarized in Figure 8–13.

Figure 8–13

Summary of the z Test of Example 8–3

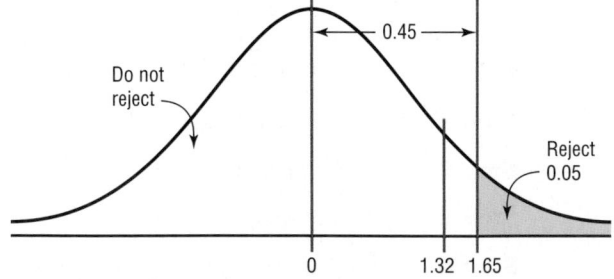

Step 5 Summarize the results. There is not enough evidence to support the claim that assistant professors earn more on average than $42,000 a year.

Comment: Even though in Example 8–3 the sample mean, $43,260, is higher than the hypothesized population mean of $42,000, it is not *significantly* higher. Hence, the difference may be due to chance. When the null hypothesis is not rejected, there is still a probability of a type II error, i.e., of not rejecting the null hypothesis when it is false.

The probability of a type II error is not easily ascertained. Further explanation about the type II error is given in Section 8–7. For now, it is only necessary to realize that the probability of type II error exists when the decision is not to reject the null hypothesis.

Also note that when the null hypothesis is not rejected, it cannot be accepted as true. There is merely not enough evidence to say that it is false. This guideline may sound a little confusing, but the situation is analogous to a jury trial. The verdict is either guilty or not guilty and is based on the evidence presented. If a person is judged not guilty, it does not mean that the person is proved innocent; it only means that there was not enough evidence to reach the guilty verdict.

Example 8–4

A researcher claims that the average cost of men's athletic shoes is less than $80. He selects a random sample of 36 pairs of shoes from a catalog and finds the following costs (in dollars). (The costs have been rounded to the nearest dollar.) Is there enough evidence to support the researcher's claim at $\alpha = 0.10$?

60	70	75	55	80	55
50	40	80	70	50	95
120	90	75	85	80	60
110	65	80	85	85	45
75	60	90	90	60	95
110	85	45	90	70	70

Solution

Step 1 State the hypotheses and identify the claim

H_0: $\mu \geq$ $80 and H_1: $\mu <$ $80 (claim)

Step 2 Find the critical value. Since $\alpha = 0.10$ and the test is a left-tailed test, the critical value is -1.28.

Step 3 Compute the test value. Since the exercise gives raw data, it is necessary to find the mean and standard deviation of the data. Using the formulas in Chapter 3 or your calculator gives $\bar{X} = 75.0$ and $s = 19.2$. Substitute in the formula

$$z = \frac{\bar{X} - \mu}{s/\sqrt{n}} = \frac{75 - 80}{19.2/\sqrt{36}} = -1.56$$

Step 4 Make the decision. Since the test value, -1.56, falls in the critical region, the decision is to reject the null hypothesis. See Figure 8–14.

Figure 8–14

Critical and Test Values for Example 8–4

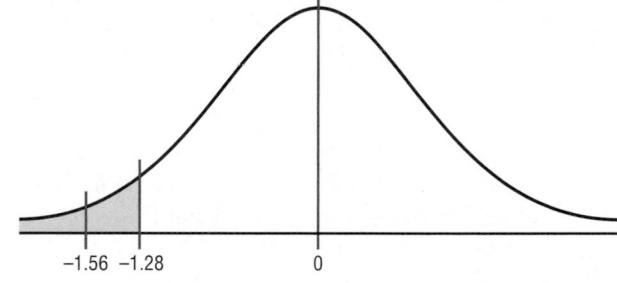

-1.56 -1.28   0

Step 5 Summarize the results. There is enough evidence to support the claim that the average cost of men's athletic shoes is less than $80.

Comment: In Example 8–4, the difference is said to be significant. However, when the null hypothesis is rejected, there is always a chance of a type I error. In this case, the probability of a type I error is at most 0.10, or 10%.

Example 8–5

The Medical Rehabilitation Education Foundation reports that the average cost of rehabilitation for stroke victims is $24,672. To see if the average cost of rehabilitation is different at a particular hospital, a researcher selects a random sample of 35 stroke victims at the hospital and finds that the average cost of their rehabilitation is $25,226. The standard deviation of the population is $3251. At $\alpha = 0.01$, can it be concluded that the average cost of stroke rehabilitation at a particular hospital is different from $24,672?

Source: Snapshot, *USA TODAY.*

Solution

Step 1 State the hypotheses and identify the claim.

$$H_0: \mu = \$24{,}672 \quad \text{and} \quad H_1: \mu \neq \$24{,}672 \text{ (claim)}$$

Step 2 Find the critical values. Since $\alpha = 0.01$ and the test is a two-tailed test, the critical values are $+2.58$ and -2.58.

Step 3 Compute the test value.

$$z = \frac{\bar{X} - \mu}{\sigma/\sqrt{n}} = \frac{25{,}226 - 24{,}672}{3251/\sqrt{35}} = 1.01$$

Step 4 Make the decision. Do not reject the null hypothesis, since the test value falls in the noncritical region, as shown in Figure 8–15.

Figure 8–15

Critical and Test Values for Example 8–5

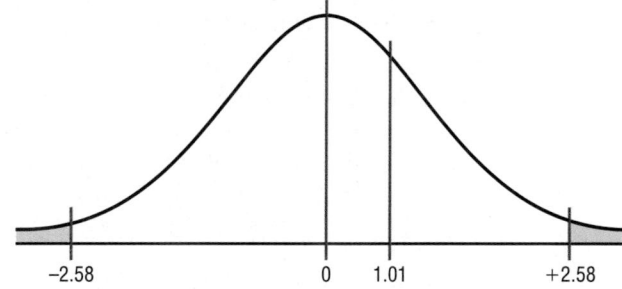

$-2.58 \qquad 0 \quad 1.01 \qquad +2.58$

Step 5 Summarize the results. There is not enough evidence to support the claim that the average cost of rehabilitation at the particular hospital is different from $24,672.

As with confidence intervals, the central limit theorem states that when the population standard deviation σ is unknown, the sample standard deviation s can be used in the formula as long as the sample size is 30 or more. The formula for the z test in this case is

$$z = \frac{\bar{X} - \mu}{s/\sqrt{n}}$$

Figure 8–16

Outcomes of a
Hypothesis-Testing
Situation

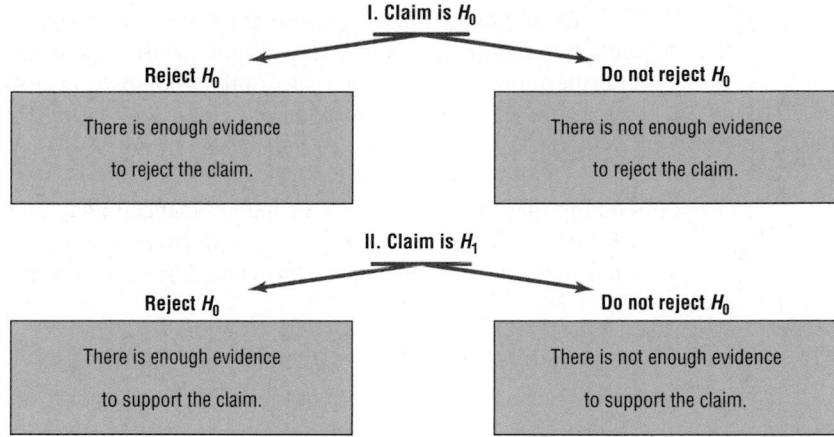

When n is less than 30 and σ is unknown, the t test must be used. The t test will be explained in Section 8–4.

Students sometimes have difficulty summarizing the results of a hypothesis test. Figure 8–16 shows the four possible outcomes and the summary statement for each situation.

First, the claim can be either the null or alternative hypothesis, and one should identify which it is. Second, after the study is completed, the null hypothesis is either rejected or not rejected. From these two facts, the decision can be identified in the appropriate block of Figure 8–16.

For example, suppose a researcher claims that the mean weight of an adult animal of a particular species is 42 pounds. In this case, the claim would be the null hypothesis, H_0: $\mu = 42$, since the researcher is asserting that the parameter is a specific value. If the null hypothesis is rejected, the conclusion would be that there is enough evidence to reject the claim that the mean weight of the adult animal is 42 pounds. See Figure 8–17(a).

Figure 8–17

Outcomes of a
Hypothesis-Testing
Situation for Two
Specific Cases

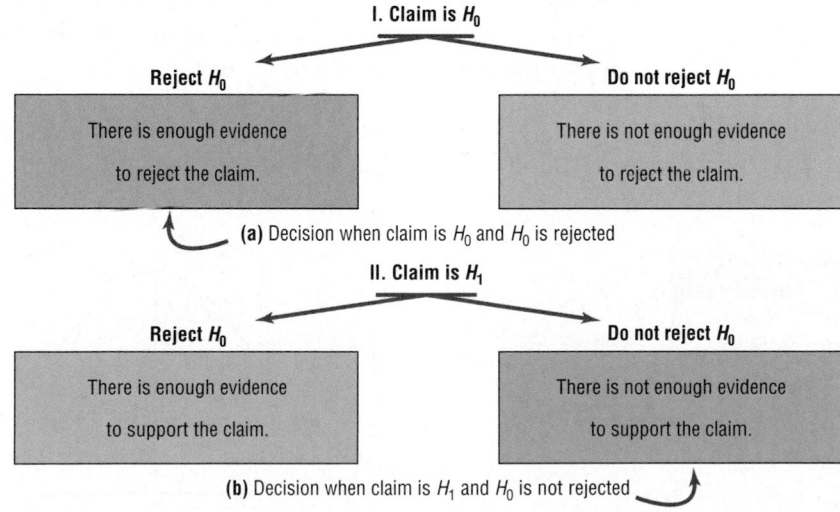

On the other hand, suppose the researcher claims that the mean weight of the adult animals is not 42 pounds. The claim would be the alternative hypothesis $H_1: \mu \neq 42$. Furthermore, suppose that the null hypothesis is not rejected. The conclusion, then, would be that there is not enough evidence to support the claim that the mean weight of the adult animals is not 42 pounds. See Figure 8–17(b).

Again, remember that nothing is being proved true or false. The statistician is only stating that there is or is not enough evidence to say that a claim is *probably* true or false. As noted previously, the only way to prove something would be to use the entire population under study, and usually this cannot be done, especially when the population is large.

P-Value Method for Hypothesis Testing

Statisticians usually test hypotheses at the common α levels of 0.05 or 0.01 and sometimes at 0.10. Recall that the choice of the level depends on the seriousness of the type I error. Besides listing an α value, many computer statistical packages give a P-value for hypothesis tests.

The **P-value** (or probability value) is the probability of getting a sample statistic (such as the mean) or a more extreme sample statistic in the direction of the alternative hypothesis when the null hypothesis is true.

In other words, the P-value is the actual area under the standard normal distribution curve (or other curve, depending on what statistical test is being used) representing the probability of a particular sample statistic or a more extreme sample statistic occurring if the null hypothesis is true.

For example, suppose that a null hypothesis is $H_0: \mu \leq 50$ and the mean of a sample is $\overline{X} = 52$. If the computer printed a P-value of 0.0356 for a statistical test, then the probability of getting a sample mean of 52 or greater is 0.0356 if the true population mean is 50 (for the given sample size and standard deviation). The relationship between the P-value and the α value can be explained in this manner. For $P = 0.0356$, the null hypothesis would be rejected at $\alpha = 0.05$ but not at $\alpha = 0.01$. See Figure 8–18.

When the hypothesis test is two-tailed, the area in one tail must be doubled. For a two-tailed test, if α is 0.05 and the area in one tail is 0.0356, the P-value will be $2(0.0356) = 0.0712$. That is, the null hypothesis should not be rejected at $\alpha = 0.05$, since 0.0712 is greater than 0.05. In summary, then, if the P-value is less than α, reject the null hypothesis. If the P-value is greater than α, do not reject the null hypothesis.

The P-values for the z test can be found by using Table E in Appendix C. First find the area under the standard normal distribution curve corresponding to the z test value; then subtract this area from 0.5000 to get the P-value for a right-tailed or a left-tailed test. To get the P-value for a two-tailed test, double this area after subtracting. This procedure is shown in step 3 of Examples 8–6 and 8–7.

Figure 8–18

Comparison of α Values and P-Values

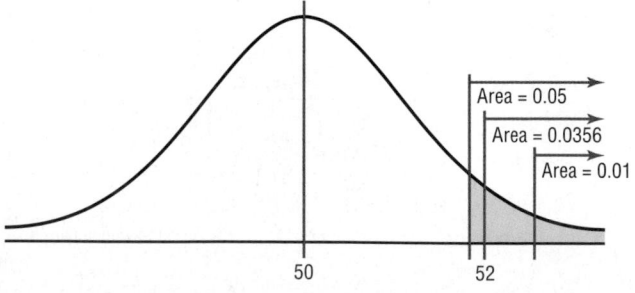

The *P*-value method for testing hypotheses differs from the traditional method somewhat. The steps for the *P*-value method are summarized next.

Procedure Table

Solving Hypothesis-Testing Problems (*P*-Value Method)

Step 1 State the hypotheses and identify the claim.

Step 2 Compute the test value.

Step 3 Find the *P*-value.

Step 4 Make the decision.

Step 5 Summarize the results.

Examples 8–6 and 8–7 show how to use the *P*-value method to test hypotheses.

Example 8–6

A researcher wishes to test the claim that the average age of lifeguards in Ocean City is greater than 24 years. She selects a sample of 36 guards and finds the mean of the sample to be 24.7 years, with a standard deviation of 2 years. Is there evidence to support the claim at $\alpha = 0.05$? Use the *P*-value method.

Solution

Step 1 State the hypotheses and identify the claim.

$H_0: \mu \le 24$ and $H_1: \mu > 24$ (claim)

Step 2 Compute the test value.

$$z = \frac{24.7 - 24}{2/\sqrt{36}} = 2.10$$

Step 3 Find the *P*-value. Using Table E in Appendix C, find the corresponding area under the normal distribution for $z = 2.10$. It is 0.4821. Subtract this value for the area from 0.5000 to find the area in the right tail.

$0.5000 - 0.4821 = 0.0179$

Hence, the *P*-value is 0.0179.

Step 4 Make the decision. Since the *P*-value is less than 0.05, the decision is to reject the null hypothesis. See Figure 8–19.

Figure 8–19

P-Value and α Value for Example 8–6

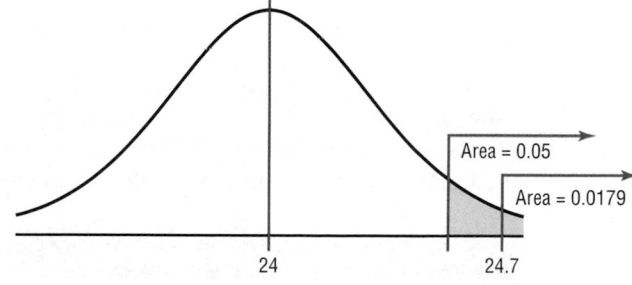

Step 5 Summarize the results. There is enough evidence to support the claim that the average age of lifeguards in Ocean City is greater than 24 years.

Note: Had the researcher chosen $\alpha = 0.01$, the null hypothesis would not have been rejected, since the *P*-value (0.0179) is greater than 0.01.

Example 8–7

A researcher claims that the average wind speed in a certain city is 8 miles per hour. A sample of 32 days has an average wind speed of 8.2 miles per hour. The standard deviation of the sample is 0.6 mile per hour. At $\alpha = 0.05$, is there enough evidence to reject the claim? Use the *P*-value method.

Solution

Step 1   State the hypotheses and identify the claim.

H_0: $\mu = 8$ (claim)   and H_1: $\mu \neq 8$

Step 2   Compute the test value.

$$z = \frac{8.2 - 8}{0.6/\sqrt{32}} = 1.89$$

Step 3   Find the *P*-value. Using Table E, find the corresponding area for $z = 1.89$. It is 0.4706. Subtract the value from 0.5000.

$0.5000 - 0.4706 = 0.0294$

Since this is a two-tailed test, the area 0.0294 must be doubled to get the *P*-value.

$2 (0.0294) = 0.0588$

Step 4   Make the decision. The decision is not to reject the null hypothesis, since the *P*-value is greater than 0.05. See Figure 8–20.

Figure 8–20

P-Values and α Values
for Example 8–7

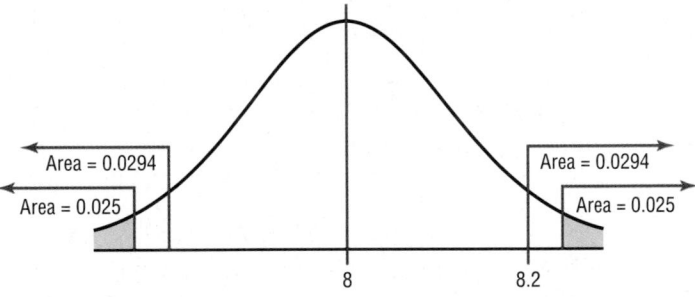

Step 5   Summarize the results. There is not enough evidence to reject the claim that the average wind speed is 8 miles per hour.

In Examples 8–6 and 8–7, the *P*-value and the α value were shown on a normal distribution curve to illustrate the relationship between the two values; however, it is not necessary to draw the normal distribution curve to make the decision whether to reject the null hypothesis. One can use the following rule:

Decision Rule When Using a *P*-Value

If *P*-value $\leq \alpha$, reject the null hypothesis.

If *P*-value $> \alpha$, do not reject the null hypothesis.

In Example 8–6, *P*-value = 0.0179 and $\alpha = 0.05$. Since *P*-value $\leq \alpha$, the null hypothesis was rejected. In Example 8–7, *P*-value = 0.0588 and $\alpha = 0.05$. Since *P*-value $> \alpha$, the null hypothesis was not rejected.

The *P*-values given on calculators and computers are slightly different from those found with Table E. This is due to the fact that *z* values and the values in Table E have been rounded. *Also, most calculators and computers give the exact P-value for two-tailed tests, so it should not be doubled (as it should when the area found in Table E is used).*

A clear distinction between the α value and the *P*-value should be made. The α value is chosen by the researcher before the statistical test is conducted. The *P*-value is computed after the sample mean has been found.

There are two schools of thought on *P*-values. Some researchers do not choose an α value but report the *P*-value and allow the reader to decide whether the null hypothesis should be rejected.

In this case, the following guidelines can be used, but be advised that these guidelines are not written in stone, and some statisticians may have other opinions.

Guidelines for *P*-Values

If *P*-value ≤ 0.01, reject the null hypothesis. The difference is highly significant.

If *P*-value > 0.01 but *P*-value ≤ 0.05, reject the null hypothesis. The difference is significant.

If *P*-value > 0.05 but *P*-value ≤ 0.10, consider the consequences of type I error before rejecting the null hypothesis.

If *P*-value > 0.10, do not reject the null hypothesis. The difference is not significant.

Others decide on the α value in advance and use the *P*-value to make the decision, as shown in Examples 8–6 and 8–7. A note of caution is needed here: If a researcher selects $\alpha = 0.01$ and the *P*-value is 0.03, the researcher may decide to change the α value from 0.01 to 0.05 so that the null hypothesis will be rejected. This, of course, should not be done. If the α level is selected in advance, it should be used in making the decision.

One additional note on hypothesis testing is that the researcher should distinguish between *statistical significance* and *practical significance*. When the null hypothesis is rejected at a specific significance level, it can be concluded that the difference is probably not due to chance and thus is statistically significant. However, the results may not have any practical significance. For example, suppose that a new fuel additive increases the miles per gallon that a car can get by $\frac{1}{4}$ mile for a sample of 1000 automobiles. The results may be statistically significant at the 0.05 level, but it would hardly be worthwhile to market the product for such a small increase. Hence, there is no practical significance to the results. It is up to the researcher to use common sense when interpreting the results of a statistical test.

Applying the Concepts **8–3**

Car Thefts

You recently received a job with a company that manufactures an automobile antitheft device. To conduct an advertising campaign for your product, you need to make a claim about the number of automobile thefts per year. Since the population of various cities in the United States varies, you decide to use rates per 10,000 people. (The rates are based on the number of people living in the cities.) Your boss said that last year the theft rate per 10,000 people was 44 vehicles. You want to see if it has changed. The following are rates per 10,000 people for 36 randomly selected locations in the United States.

55	42	125	62	134	73
39	69	23	94	73	24
51	55	26	66	41	67
15	53	56	91	20	78
70	25	62	115	17	36
58	56	33	75	20	16

Source: Based on information from the National Insurance Crime Bureau.

Using this information, answer these questions.

1. What are the hypotheses that you would use?
2. Is the sample considered small or large?
3. What assumption must be met before the hypothesis test can be conducted?
4. Which probability distribution would you use?
5. Would you select a one- or two-tailed test? Why?
6. What critical value(s) would you use?
7. Conduct a hypothesis test.
8. What is your decision?
9. What is your conclusion?
10. Write a brief statement summarizing your conclusion.
11. If you lived in a city whose population was about 50,000, how many automobile thefts per year would you expect to occur?

See page 460 for the answers.

Exercises 8–3

For Exercises 1 through 13, perform each of the following steps.

   a. State the hypotheses and identify the claim.

   b. Find the critical value(s).

   c. Compute the test value.

   d. Make the decision.

   e. Summarize the results.

Use diagrams to show the critical region (or regions), and use the traditional method of hypothesis testing unless otherwise specified.

1. A survey claims that the average cost of a hotel room in Atlanta is $69.21. To test the claim, a researcher selects a sample of 30 hotel rooms and finds that the average cost is $68.43. The standard deviation of the population is $3.72. At $\alpha = 0.05$, is there enough evidence to reject the claim?
   Source: *USA TODAY.*

2. It has been reported that the average credit card debt for college seniors is $3262. The student senate at a large university feels that their seniors have a debt much less than this, so it conducts a study of 50 randomly selected seniors and finds that the average debt is $2995 with a sample standard deviation of $1100. With $\alpha = 0.05$, is the student senate correct?
   Source: *USA TODAY.*

3. A researcher estimates that the average revenue of the largest businesses in the United States is greater than $24 billion. A sample of 50 companies is selected, and the revenues (in billions of dollars) are shown. At $\alpha = 0.05$, is there enough evidence to support the researcher's claim?

178	122	91	44	35
61	56	46	20	32
30	28	28	20	27
29	16	16	19	15
41	38	36	15	25
31	30	19	19	19
24	16	15	15	19
25	25	18	14	15
24	23	17	17	22
22	21	20	17	20

Source: *N.Y. Times Almanac.*

4. Full-time Ph.D. students receive an average salary of $12,837 according to the U.S. Department of Education. The dean of graduate studies at a large state university feels that Ph.D. students in his state earn more than this. He surveys 44 randomly selected students and finds their average salary is $14,445 with a standard deviation of $1500. With $\alpha = 0.05$, is the dean correct?
   Source: U.S. Department of Education/*Chronicle of Higher Education.*

5. A report in *USA TODAY* stated that the average age of commercial jets in the United States is 14 years. An

executive of a large airline company selects a sample of 36 planes and finds the average age of the planes is 11.8 years. The standard deviation of the sample is 2.7 years. At $\alpha = 0.01$, can it be concluded that the average age of the planes in his company is less than the national average?

Source: *USA TODAY.*

6. The average production of peanuts in the state of Virginia is 3000 pounds per acre. A new plant food has been developed and is tested on 60 individual plots of land. The mean yield with the new plant food is 3120 pounds of peanuts per acre with a standard deviation of 578 pounds. At $\alpha = 0.05$, can one conclude that the average production has increased?

Source: *The Old Farmers' Almanac.*

7. The average 1-year-old (both genders) is 29 inches tall. A random sample of 30 one-year-olds in a large day-care franchise resulted in the following heights. At $\alpha = 0.05$, can it be concluded that the average height differs from 29 inches?

25	32	35	25	30	26.5	26	25.5	29.5	32
30	28.5	30	32	28	31.5	29	29.5	30	34
29	32	27	28	33	28	27	32	29	29.5

Source: www.healthepic.com.

8. At a certain university the mean income of parents of the entering class is reported to be $91,600. The president of another university feels that the parents' income for her entering class is greater than $91,600. She surveys 100 randomly selected families and finds the mean income to be $96,321 with a standard deviation of $9555. With $\alpha = 0.05$, is she correct?

Source: *Chronicle of Higher Education.*

9. Average undergraduate cost for tuition, fees, room, and board for all institutions last year was $19,410. A random sample of costs this year for 40 institutions of higher learning indicated that the sample mean was $22,098, and the sample standard deviation was $6050. At the 0.01 level of significance, is there sufficient evidence to conclude that the cost of attendance has increased?

Source: *N.Y. Times Almanac.*

10. A real estate agent claims that the average price of a home sold in Beaver County, Pennsylvania, is $60,000. A random sample of 36 homes sold in the county is selected, and the prices in dollars are shown. Is there enough evidence to reject the agent's claim at $\alpha = 0.05$?

9,500	54,000	99,000	94,000	80,000
29,000	121,500	184,750	15,000	164,450
6,000	13,000	188,400	121,000	308,000
42,000	7,500	32,900	126,900	25,225
95,000	92,000	38,000	60,000	211,000
15,000	28,000	53,500	27,000	21,000
76,000	85,000	25,225	40,000	97,000
284,000				

Source: *Pittsburgh Tribune-Review.*

11. The average U.S. wedding includes 125 guests. A random sample of 35 weddings during the past year in a particular county had a mean of 110 guests and a standard deviation of 30. Is there sufficient evidence at the 0.01 level of significance that the average number of guests differs from the national average?

Source: www.theknot.com.

12. The average salary for public school teachers for a specific year was reported to be $39,385. A random sample of 50 public school teachers in a particular state had a mean of $41,680 and a standard deviation of $5975. Is there sufficient evidence at the $\alpha = 0.05$ level to conclude that the mean salary differs from $39,385?

Source: *N.Y. Times Almanac.*

13. To see if young men ages 8 through 17 years spend more or less than the national average of $24.44 per shopping trip to a local mall, the manager surveyed 33 young men and found the average amount spent per visit was $22.97. The standard deviation of the sample was $3.70. At $\alpha = 0.02$, can it be concluded that the average amount spent at a local mall is not equal to the national average of $24.44?

Source: *USA TODAY.*

14. What is meant by a *P*-value?

15. State whether the null hypothesis should be rejected on the basis of the given *P*-value.

   a. *P*-value = 0.258, $\alpha = 0.05$, one-tailed test
   b. *P*-value = 0.0684, $\alpha = 0.10$, two-tailed test
   c. *P*-value = 0.0153, $\alpha = 0.01$, one-tailed test
   d. *P*-value = 0.0232, $\alpha = 0.05$, two-tailed test
   e. *P*-value = 0.002, $\alpha = 0.01$, one-tailed test

16. A researcher claims that the yearly consumption of soft drinks per person is 52 gallons. In a sample of 50 randomly selected people, the mean of the yearly consumption was 56.3 gallons. The standard deviation of the sample was 3.5 gallons. Find the *P*-value for the test. On the basis of the *P*-value, is the researcher's claim valid?

Source: U.S. Department of Agriculture.

17. A study found that the average stopping distance of a school bus traveling 50 miles per hour was 264 feet. A group of automotive engineers decided to conduct a study of its school buses and found that for 20 buses, the average stopping distance of buses traveling 50 miles per hour was 262.3 feet. The standard deviation of the population was 3 feet. Test the claim that the average stopping distance of the company's buses is actually less than 264 feet. Find the *P*-value. On the basis of the *P*-value, should the null hypothesis be rejected at $\alpha = 0.01$? Assume that the variable is normally distributed.

Source: *Snapshot, USA TODAY,* March 12, 1992.

18. A store manager hypothesizes that the average number of pages a person copies on the store's copy

machine is less than 40. A sample of 50 customers' orders is selected. At $\alpha = 0.01$, is there enough evidence to support the claim? Use the P-value hypothesis-testing method.

2	2	2	5	32
5	29	8	2	49
21	1	24	72	70
21	85	61	8	42
3	15	27	113	36
37	5	3	58	82
9	2	1	6	9
80	9	51	2	122
21	49	36	43	61
3	17	17	4	1

19. A health researcher read that a 200-pound male can burn an average of 546 calories per hour playing tennis. Thirty-six males were randomly selected and tested. The mean of the number of calories burned per hour was 544.8. Test the claim that the average number of calories burned is actually less than 546, and find the P-value. On the basis of the P-value, should the null hypothesis be rejected at $\alpha = 0.01$? The standard deviation of the sample is 3. Can it be concluded that the average number of calories burned is less than originally thought?

20. A special cable has a breaking strength of 800 pounds. The standard deviation of the population is 12 pounds. A researcher selects a sample of 20 cables and finds that the average breaking strength is 793 pounds. Can one reject the claim that the breaking strength is 800 pounds? Find the P-value. Should the null hypothesis be rejected at $\alpha = 0.01$? Assume that the variable is normally distributed.

21. Several years ago the Department of Agriculture found that the average size of farms in the United States was 47.1 acres. A random sample of 50 farms was selected, and the mean size of the farm was 43.2 acres. The standard deviation of the sample was 8.6 acres. Test the claim at $\alpha = 0.05$ that the average farm size is smaller today, by using the P-value method. Should the Department of Agriculture update its information?

22. Ten years ago, the average acreage of farms in a certain geographic region was 65 acres. The standard deviation of the population was 7 acres. A recent study consisting of 22 farms showed that the average was 63.2 acres per farm. Test the claim, at $\alpha = 0.10$, that the average has not changed by finding the P-value for the test. Assume that σ has not changed and the variable is normally distributed.

23. A car dealer recommends that transmissions be serviced at 30,000 miles. To see whether her customers are adhering to this recommendation, the dealer selects a sample of 40 customers and finds that the average mileage of the automobiles serviced is 30,456. The standard deviation of the sample is 1684 miles. By finding the P-value, determine whether the owners are having their transmissions serviced at 30,000 miles. Use $\alpha = 0.10$. Do you think the α value of 0.10 is an appropriate significance level?

24. A motorist claims that the South Boro Police issue an average of 60 speeding tickets per day. These data show the number of speeding tickets issued each day for a period of one month. Assume σ is 13.42. Is there enough evidence to reject the motorist's claim at $\alpha = 0.05$? Use the P-value method.

72	45	36	68	69	71	57	60
83	26	60	72	58	87	48	59
60	56	64	68	42	57	57	
58	63	49	73	75	42	63	

25. A manager states that in his factory, the average number of days per year missed by the employees due to illness is less than the national average of 10. The following data show the number of days missed by 40 employees last year. Is there sufficient evidence to believe the manager's statement at $\alpha = 0.05$? (Use s to estimate σ.) Use the P-value method.

0	6	12	3	3	5	4	1
3	9	6	0	7	6	3	4
7	4	7	1	0	8	12	3
2	5	10	5	15	3	2	5
3	11	8	2	2	4	1	9

Extending the Concepts

26. Suppose a statistician chose to test a hypothesis at $\alpha = 0.01$. The critical value for a right-tailed test is $+2.33$. If the test value was 1.97, what would the decision be? What would happen if, after seeing the test value, she decided to choose $\alpha = 0.05$? What would the decision be? Explain the contradiction, if there is one.

27. The president of a company states that the average hourly wage of her employees is \$8.65. A sample of 50 employees has the distribution shown. At $\alpha = 0.05$, is the president's statement believable? (Use s to approximate σ.)

Class	Frequency
8.35–8.43	2
8.44–8.52	6
8.53–8.61	12
8.62–8.70	18
8.71–8.79	10
8.80–8.88	2

Technology *Step by Step*

MINITAB
Step by Step

Hypothesis Test for the Mean and the *z* Distribution

MINITAB can be used to calculate the test statistic and its *P*-value. The *P*-value approach does not require a critical value from the table. If the *P*-value is smaller than α, the null hypothesis is rejected. For Example 8–4, test the claim that the mean shoe cost is less than $80.

1. Enter the data into a column of MINITAB. Do not try to type in the dollar signs! Name the column **ShoeCost.**

2. If sigma is known, skip to step 3; otherwise estimate sigma from the sample standard deviation *s*.

Calculate the Standard Deviation in the Sample

    a) Select **Calc>Column Statistics.**

    b) Check the button for Standard deviation.

    c) Select ShoeCost for the Input variable.

    d) Type **s** in the text box for Store the result in:.

    e) Click [OK].

Calculate the Test Statistic and *P*-Value

3. Select **Stat>Basic Statistics>1 Sample Z,** then select ShoeCost in the Variable text box.

4. Click in the text box and enter the value of sigma or type *s*, the sample standard deviation.

5. Click in the text box for Test mean, and enter the hypothesized value of **80.**

6. Click on [Options].

    a) Change the Confidence level to **90.**

    b) Change the Alternative to less than. This setting is crucial for calculating the *P*-value.

7. Click [OK] twice.

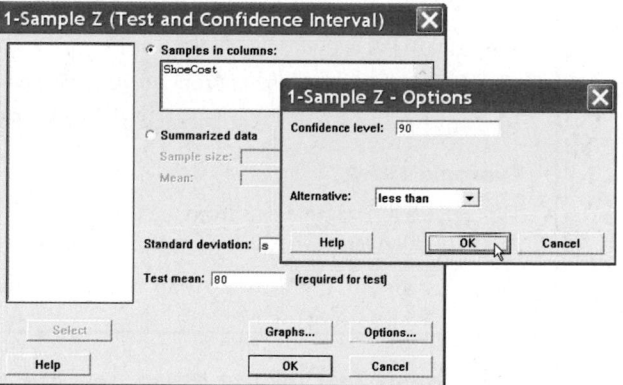

One-Sample Z: ShoeCost
```
Test of mu = 80 vs < 80
The assumed sigma 19.161
```

Variable	N	Mean	StDev	SE Mean	90% Upper Bound	Z	P
ShoeCost	36	75.0000	19.1610	3.1935	79.0926	-1.57	0.059

Since the *P*-value of 0.059 is less than α, reject the null hypothesis. There is enough evidence in the sample to conclude the mean cost is less than $80.

TI-83 Plus or TI-84 Plus
Step by Step

Hypothesis Test for the Mean and the *z* Distribution (Data)

1. Enter the data values into L_1.
2. Press **STAT** and move the cursor to TESTS.
3. Press l for ZTest.
4. Move the cursor to Data and press **ENTER**.
5. Type in the appropriate values.
6. Move the cursor to the appropriate alternative hypothesis and press **ENTER**.
7. Move the cursor to Calculate and press **ENTER**.

Example TI8–1

 This relates to Example 8–4 from the text. At the 10% significance level, test the claim that $\mu < 80$ given the data value.

60	70	75	55	80	55	50	40	80	70	50	95
120	90	75	85	80	60	110	65	80	85	85	45
75	60	90	90	60	95	110	85	45	90	70	70

The population standard deviation σ is unknown. Since the sample size $n = 36 \geq 30$, one can use the sample standard deviation s as an approximation for σ. After the data values are entered in L_1 (step 1), press **STAT,** move the cursor to CALC, press **1** for 1-Var Stats, then press **ENTER**. The sample standard deviation of 19.16097224 will be one of the statistics listed. Then continue with step 2. At step 5 on the line for σ press **VARS** for variables, press **5** for Statistics, press **3** for S_x.

The test statistic is $z = -1.565682556$, and the *P*-value is 0.0587114841.

Hypothesis Test for the Mean and the *z* Distribution (Statistics)

1. Press **STAT** and move the cursor to TESTS.
2. Press **1** for ZTest.
3. Move the cursor to Stats and press **ENTER**.
4. Type in the appropriate values.
5. Move the cursor to the appropriate alternative hypothesis and press **ENTER**.
6. Move the cursor to Calculate and press **ENTER**.

Example TI8–2

This relates to Example 8–3 from the text. At the 5% significance level, test the claim that $\mu > 42,000$ given $\sigma = 5230$, $\overline{X} = 43,260$, and $n = 30$.

The test statistic is $z = 1.319561037$, and the *P*-value is 0.0934908728.

Excel
Step by Step

Hypothesis Test for the Mean and the *z* Distribution

Excel does not have a procedure to conduct a hypothesis test for the mean. However, you may conduct the test of the mean using the MegaStat Add-in available on your CD and Online Learning Center. If you have not installed this add-in, do so by following the instructions on page 24.

This test uses the *P*-value method. Therefore, it is not necessary to enter a significance level.

1. Enter the data from Example 8–4 into column A of a new worksheet.
2. Select **MegaStat>Hypothesis Tests>Mean vs. Hypothesized Value.**
3. Select data input and enter the range of data **A1:A36.**
4. Type **80** for the Hypothesized mean and select the "less than" Alternative.
5. Select *z* test and click [OK].

<table>
<tr><td>**8–4**</td><td></td></tr>
</table>

t Test for a Mean

Objective **6**

Test means for small samples, using the *t* test.

When the population standard deviation is unknown and the sample size is less than 30, the *z* test is inappropriate for testing hypotheses involving means. A different test, called the *t test,* is used. The *t* test is used when σ is unknown, $n < 30$, and the distribution of the variable is approximately normal.

As stated in Chapter 7, the *t* distribution is similar to the standard normal distribution in the following ways.

1. It is bell-shaped.
2. It is symmetric about the mean.
3. The mean, median, and mode are equal to 0 and are located at the center of the distribution.
4. The curve never touches the *x* axis.

The *t* distribution differs from the standard normal distribution in the following ways.

1. The variance is greater than 1.
2. The *t* distribution is a family of curves based on the *degrees of freedom,* which is a number related to sample size. (Recall that the symbol for degrees of freedom is d.f. See Section 7–3 for an explanation of degrees of freedom.)
3. As the sample size increases, the *t* distribution approaches the normal distribution.

The *t* test is defined next.

The **t test** is a statistical test for the mean of a population and is used when the population is normally or approximately normally distributed, σ is unknown, and $n < 30$.
    The formula for the *t* test is

$$t = \frac{\bar{X} - \mu}{s/\sqrt{n}}$$

The degrees of freedom are d.f. $= n - 1$.

The formula for the *t* test is similar to the formula for the *z* test. But since the population standard deviation σ is unknown, the sample standard deviation *s* is used instead.

The critical values for the *t* test are given in Table F in Appendix C. For a one-tailed test, find the α level by looking at the top row of the table and finding the appropriate column. Find the degrees of freedom by looking down the left-hand column. Notice that degrees of freedom are given for values from 1 through 28. When the degrees of freedom are 29 or more, the row with ∞ (infinity) is used. Note that the values in this row are the same as the values for the *z* distribution, since as the sample size increases, the *t* distribution approaches the *z* distribution. When the sample size is 30 or more, statisticians generally agree that the two distributions can be considered identical, since the difference between their values is relatively small.

Example 8–8

Find the critical *t* value for $\alpha = 0.05$ with d.f. $= 16$ for a right-tailed *t* test.

Solution

Find the 0.05 column in the top row and 16 in the left-hand column. Where the row and column meet, the appropriate critical value is found; it is $+1.746$. See Figure 8–21.

Figure 8–21

Finding the Critical Value for the *t* Test in Table F (Example 8–8)

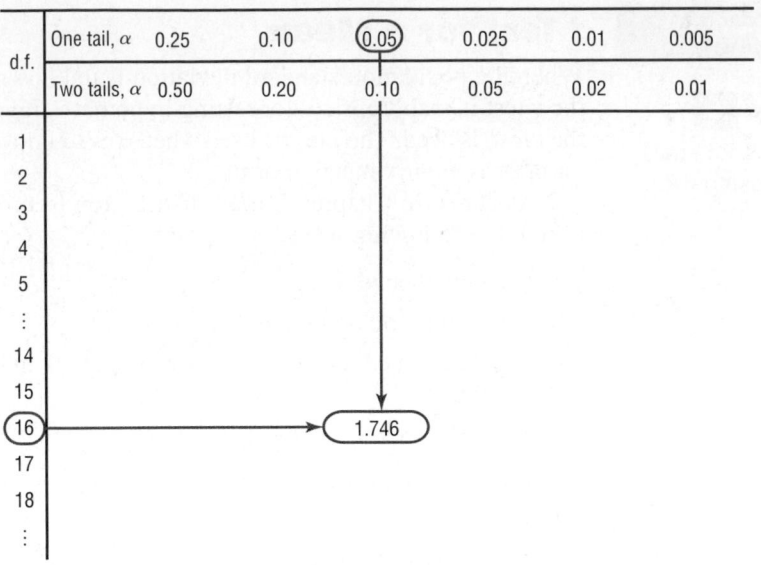

Example 8–9

Find the critical *t* value for $\alpha = 0.01$ with d.f. = 22 for a left-tailed test.

Solution

Find the 0.01 column in the row labeled One tail, and find 22 in the left column. The critical value is -2.508 since the test is a one-tailed left test.

Example 8–10

Find the critical values for $\alpha = 0.10$ with d.f. = 18 for a two-tailed *t* test.

Solution

Find the 0.10 column in the row labeled Two tails, and find 18 in the column labeled d.f. The critical values are $+1.734$ and -1.734.

Example 8–11

Find the critical value for $\alpha = 0.05$ with d.f. = 28 for a right-tailed *t* test.

Solution

Find the 0.05 column in the One tail row and 28 in the left column. The critical value is $+1.701$.

When you test hypotheses by using the *t* test (traditional method), follow the same procedure as for the *z* test, except use Table F.

Step 1   State the hypotheses and identify the claim.

Step 2   Find the critical value(s) from Table F.

Step 3 Compute the test value.

Step 4 Make the decision to reject or not reject the null hypothesis.

Step 5 Summarize the results.

Remember that the t test should be used when the population is approximately normally distributed, the population standard deviation is unknown, and the sample size is less than 30.
    Examples 8–12 through 8–14 illustrate the application of the *t* test.

Example 8–12

A job placement director claims that the average starting salary for nurses is $24,000. A sample of 10 nurses' salaries has a mean of $23,450 and a standard deviation of $400. Is there enough evidence to reject the director's claim at $\alpha = 0.05$?

Solution

Step 1 H_0: $\mu = \$24,000$ (claim) and H_1: $\mu \neq \$24,000$.

Step 2 The critical values are $+2.262$ and -2.262 for $\alpha = 0.05$ and d.f. = 9.

Step 3 The test value is

$$t = \frac{\overline{X} - \mu}{s/\sqrt{n}} = \frac{23,450 - 24,000}{400/\sqrt{10}} = -4.35$$

Step 4 Reject the null hypothesis, since $-4.35 < -2.262$, as shown in Figure 8–22.

Figure 8–22

Summary of the *t* Test of Example 8–12

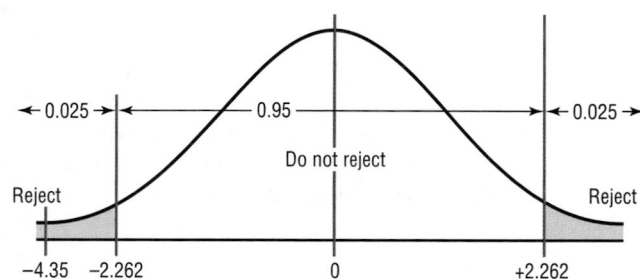

Step 5 There is enough evidence to reject the claim that the starting salary of nurses is $24,000.

Example 8–13

An educator claims that the average salary of substitute teachers in school districts in Allegheny County, Pennsylvania, is less than $60 per day. A random sample of eight school districts is selected, and the daily salaries (in dollars) are shown. Is there enough evidence to support the educator's claim at $\alpha = 0.10$?

    60   56   60   55   70   55   60   55

Source: Pittsburgh Tribune-Review.

Solution

Step 1 H_0: $\mu \geq \$60$ and H_1: $\mu < \$60$ (claim).

Step 2 At $\alpha = 0.10$ and d.f. $= 7$, the critical value is -1.415.

Step 3 To compute the test value, the mean and standard deviation must be found. Using either the formulas in Chapter 3 or your calculator, $\overline{X} = \$58.88$, and $s = 5.08$, you find

$$t = \frac{\overline{X} - \mu}{s/\sqrt{n}} = \frac{58.88 - 60}{5.08/\sqrt{8}} = -0.624$$

Step 4 Do not reject the null hypothesis since -0.624 falls in the noncritical region. See Figure 8–23.

Figure 8–23

Critical Value and
Test Value for
Example 8–13

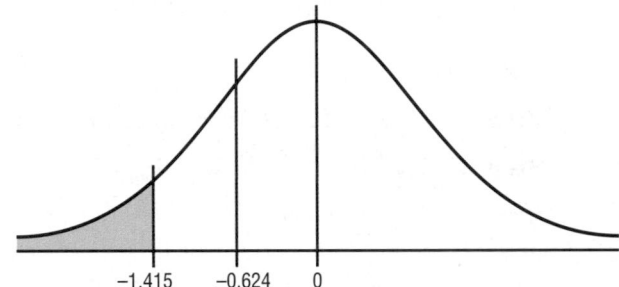

-1.415 -0.624 0

Step 5 There is not enough evidence to support the educator's claim that the average salary of substitute teachers in Allegheny County is less than $60 per day.

The *P*-values for the *t* test can be found by using Table F; however, specific *P*-values for *t* tests cannot be obtained from the table since only selected values of α (for example, 0.01, 0.05) are given. To find specific *P*-values for *t* tests, one would need a table similar to Table E for each degree of freedom. Since this is not practical, only *intervals* can be found for *P*-values. Examples 8–14 to 8–16 show how to use Table F to determine intervals for *P*-values for the *t* test.

Example 8–14

Find the *P*-value when the *t* test value is 2.056, the sample size is 11, and the test is right-tailed.

Solution

To get the *P*-value, look across the row with 10 degrees of freedom (d.f. $= n - 1$) in Table F and find the two values that 2.056 falls between. They are 1.812 and 2.228. Since this is a right-tailed test, look up to the row labeled One tail, α and find the two α values corresponding to 1.812 and 2.228. They are 0.05 and 0.025, respectively. See Figure 8–24. Hence, the *P*-value would be contained in the interval $0.025 < P\text{-value} < 0.05$. This means that the *P*-value is between 0.025 and 0.05. If α were 0.05, one would reject the null hypothesis since the *P*-value is less than 0.05. But if α were 0.01, one would not reject the null hypothesis since the *P*-value is greater than 0.01. (Actually, it is greater than 0.025.)

Figure 8–24

Finding the *P*-Value for Example 8–14

Confidence intervals		50%	80%	90%	95%	98%	99%
One tail, α		0.25	0.10	(0.05)	(0.025)	0.01	0.005
d.f.	Two tails, α	0.50	0.20	0.10	0.05	0.02	0.01
1		1.000	3.078	6.314	12.706	31.821	63.657
2		0.816	1.886	2.920	4.303	6.965	9.925
3		0.765	1.638	2.353	3.182	4.541	5.841
4		0.741	1.533	2.132	2.776	3.747	4.604
5		0.727	1.476	2.015	2.571	3.365	4.032
6		0.718	1.440	1.943	2.447	3.143	3.707
7		0.711	1.415	1.895	2.365	2.998	3.499
8		0.706	1.397	1.860	2.306	2.896	3.355
9		0.703	1.383	1.833	2.262	2.821	3.250
(10)		0.700	1.372	(1.812) *	(2.228)	2.764	3.169
11		0.697	1.363	1.796	2.201	2.718	3.106
12		0.695	1.356	1.782	2.179	2.681	3.055
13		0.694	1.350	1.771	2.160	2.650	3.012
14		0.692	1.345	1.761	2.145	2.624	2.977
15		0.691	1.341	1.753	2.131	2.602	2.947
$\vdots$		$\vdots$	$\vdots$	$\vdots$	$\vdots$	$\vdots$	$\vdots$
(*z*) ∞		0.674	1.282	1.645	1.960	2.326	2.576

*2.056 falls between 1.812 and 2.228.

Example 8–15

Find the *P*-value when the *t* test value is 2.983, the sample size is 6, and the test is two-tailed.

Solution

To get the *P*-value, look across the row with d.f. = 5 and find the two values that 2.983 falls between. They are 2.571 and 3.365. Then look up to the row labeled Two tails, α to find the corresponding α values.

In this case, they are 0.05 and 0.02. Hence the *P*-value is contained in the interval $0.02 < P\text{-value} < 0.05$. This means that the *P*-value is between 0.02 and 0.05. In this case, if $\alpha = 0.05$, the null hypothesis can be rejected since *P*-value < 0.05; but if $\alpha = 0.01$, the null hypothesis cannot be rejected since *P*-value > 0.01 (actually *P*-value > 0.02).

Note: **Since many students will be using calculators or computer programs that give the specific *P*-value for the *t* test and other tests presented later in this textbook, these specific values, in addition to the intervals, will be given for the answers to the examples and exercises.**

The *P*-value obtained from a calculator for Example 8–14 is 0.033. The *P*-value obtained from a calculator for Example 8–15 is 0.031.

To test hypotheses using the *P*-value method, follow the same steps as explained in Section 8–3. These steps are repeated here.

Step 1 State the hypotheses and identify the claim.

Step 2 Compute the test value.

Step 3 Find the *P*-value.

Step 4 Make the decision.

Step 5 Summarize the results.

This method is shown in Example 8–16.

Example 8–16

A physician claims that joggers' maximal volume oxygen uptake is greater than the average of all adults. A sample of 15 joggers has a mean of 40.6 milliliters per kilogram (ml/kg) and a standard deviation of 6 ml/kg. If the average of all adults is 36.7 ml/kg, is there enough evidence to support the physician's claim at $\alpha = 0.05$?

Solution

Step 1 State the hypotheses and identify the claim.

$$H_0\!: \mu \le 36.7 \qquad \text{and} \qquad H_1\!: \mu > 36.7 \text{ (claim)}$$

Step 2 Compute the test value. The test value is

$$t = \frac{\bar{X} - \mu}{s/\sqrt{n}} = \frac{40.6 - 36.7}{6/\sqrt{15}} = 2.517$$

Step 3 Find the *P*-value. Looking across the row with d.f. = 14 in Table F, one sees that 2.517 falls between 2.145 and 2.624, corresponding to $\alpha = 0.025$ and $\alpha = 0.01$ since this is a right-tailed test. Hence, *P*-value > 0.01 and *P*-value < 0.025 or 0.01 < *P*-value < 0.025. That is, the *P*-value is somewhere between 0.01 and 0.025. (The *P*-value obtained from a calculator is 0.012.)

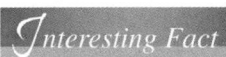
Step 4 Reject the null hypothesis since *P*-value < 0.05 (that is, *P*-value < α).

Step 5 There is enough evidence to support the claim that the joggers' maximal volume oxygen uptake is greater than 36.7 ml/kg.

Students sometimes have difficulty deciding whether to use the *z* test or *t* test. The rules are the same as those pertaining to confidence intervals.

1. If σ is known, use the *z* test. The variable must be normally distributed if $n < 30$.

2. If σ is unknown but $n \ge 30$, use the *z* test and use *s* in place of σ in the formula.

3. If σ is unknown and $n < 30$, use the *t* test. (The population must be approximately normally distributed.)

These rules are summarized in Figure 8–25.

Speaking of Statistics

Can Sunshine Relieve Pain?

A study conducted at the University of Pittsburgh showed that hospital patients in rooms with lots of sunlight required less pain medication the day after surgery and during their total stay in the hospital than patients who were in darker rooms.

Patients in the sunny rooms averaged 3.2 milligrams of pain reliever per hour for their total stay as opposed to 4.1 milligrams per hour for those in darker rooms. This study compared two groups of patients. Although no statistical tests were mentioned in the article, what statistical test do you think the researchers used to compare the groups?

Figure 8–25

Using the *z* or *t* Test

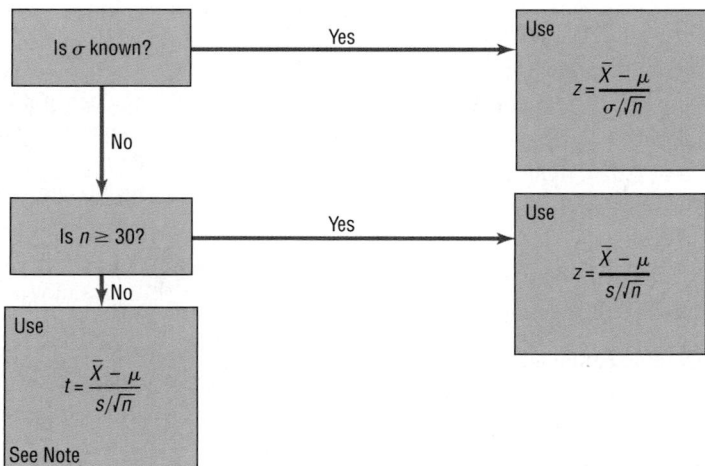

Is σ known? Yes → Use $z = \dfrac{\bar{X} - \mu}{\sigma/\sqrt{n}}$

No

Is $n \geq 30$? Yes → Use $z = \dfrac{\bar{X} - \mu}{s/\sqrt{n}}$

No

Use $t = \dfrac{\bar{X} - \mu}{s/\sqrt{n}}$

See Note

Note: With d.f. = $n - 1$, and the population must be approximately normally distributed.

Applying the Concepts 8–4

How Much Nicotine Is in Those Cigarettes?

A tobacco company claims that its best-selling cigarettes contain at most 40 mg of nicotine. This claim is tested at the 1% significance level by using the results of 15 randomly selected cigarettes. The mean was 42.6 mg and the standard deviation was 3.7 mg. Evidence suggests that nicotine is normally distributed. Information from a computer output of the hypothesis test is listed.

Sample mean = 42.6	*P*-value = 0.008
Sample standard deviation = 3.7	Significance level = 0.01
Sample size = 15	Test statistic *t* = 2.72155
Degrees of freedom = 14	Critical value *t* = 2.62610

1. What are the degrees of freedom?
2. Is this a large or small sample test?
3. Is this a comparison of one or two samples?
4. Is this a right-tailed, left-tailed, or two-tailed test?
5. From observing the *P*-value, what would you conclude?
6. By comparing the test statistic to the critical value, what would you conclude?
7. Is there a conflict in this output? Explain.
8. What has been proved in this study?

See page 460 for the answers.

Exercises 8–4

1. In what ways is the *t* distribution similar to the standard normal distribution? In what ways is the *t* distribution different from the standard normal distribution?

2. What are the degrees of freedom for the *t* test?

3. Find the critical value (or values) for the *t* test for each.

   a. $n = 10$, $\alpha = 0.05$, right-tailed
   b. $n = 18$, $\alpha = 0.10$, two-tailed
   c. $n = 6$, $\alpha = 0.01$, left-tailed
   d. $n = 9$, $\alpha = 0.025$, right-tailed
   e. $n = 15$, $\alpha = 0.05$, two-tailed
   f. $n = 23$, $\alpha = 0.005$, left-tailed
   g. $n = 28$, $\alpha = 0.01$, two-tailed
   h. $n = 17$, $\alpha = 0.02$, two-tailed

4. **(ans)** Using Table F, find the *P*-value interval for each test value.

   a. $t = 2.321$, $n = 15$, right-tailed
   b. $t = 1.945$, $n = 28$, two-tailed
   c. $t = -1.267$, $n = 8$, left-tailed
   d. $t = 1.562$, $n = 17$, two-tailed
   e. $t = 3.025$, $n = 24$, right-tailed
   f. $t = -1.145$, $n = 5$, left-tailed
   g. $t = 2.179$, $n = 13$, two-tailed
   h. $t = 0.665$, $n = 10$, right-tailed

For Exercises 5 through 18, perform each of the following steps.

   a. State the hypotheses and identify the claim.
   b. Find the critical value(s).
   c. Find the test value.
   d. Make the decision.
   e. Summarize the results.

Use the traditional method of hypothesis testing unless otherwise specified.

Assume that the population is approximately normally distributed.

5. The average amount of rainfall during the summer months for the northeastern part of the United States is 11.52 inches. A researcher selects a random sample of 10 cities in the northeast and finds that the average amount of rainfall for 1995 was 7.42 inches. The standard deviation of the sample is 1.3 inches. At $\alpha = 0.05$, can it be concluded that for 1995 the mean rainfall was below 11.52 inches?

   Source: Based on information in *USA TODAY.*

6. A state executive claims that the average number of acres in western Pennsylvania state parks is less than 2000 acres. A random sample of five parks is selected, and the number of acres is shown. At $\alpha = 0.01$, is there enough evidence to support the claim?

959	1187	493	6249	541

   Source: *Pittsburgh Tribune-Review.*

7. The average salary of graduates entering the actuarial field is reported to be $40,000. To test this, a statistics professor surveys 20 graduates and finds their average salary to be $43,228 with a standard deviation of $4000. Using $\alpha = 0.05$, has he shown the reported salary to be incorrect?

   Source: BeAnActuary.org.

8. A survey of 15 large U.S. cities finds that the average commute time one way is 25.4 minutes. A chamber of commerce executive feels that the commute in his city is less and wants to publicize this. He randomly selects 25 commuters and finds the average is 22.1 minutes with a standard deviation of 5.3 minutes. At $\alpha = 0.10$, is he correct?

   Source: *N.Y. Times Almanac.*

9. A researcher estimates that the average height of the buildings of 30 or more stories in a large city is at least 700 feet. A random sample of 10 buildings is selected, and the heights in feet are shown. At $\alpha = 0.025$, is there enough evidence to reject the claim?

485	511	841	725	615
520	535	635	616	582

Source: *Pittsburgh Tribune-Review.*

10. Cushman and Wakefield reported that the average annual rent for office space in Tampa was $17.63 per square foot. A real estate agent selected a random sample of 15 rental properties (offices) and found the mean rent was $18.72 per square foot and the standard deviation was $3.64. At $\alpha = 0.05$, test the claim that there is no difference in the rents.

Source: *USA TODAY.*

11. The average undergraduate cost for tuition, fees, and room and board for two-year institutions last year was $13,252. The following year, a random sample of 20 two-year institutions had a mean of $15,560 and a standard deviation of $3500. Is there sufficient evidence at the $\alpha = 0.01$ level to conclude that the mean cost has increased?

Source: *N.Y. Times Almanac.*

12. A large university reports that the mean salary of parents of an entering class is $91,600. To see how this compares to his university, a president surveys 28 randomly selected families and finds that their average income is $88,500. If the standard deviation is $10,000, can the president conclude that there is a difference? At $\alpha = 0.10$, is he correct?

Source: *Chronicle of Higher Education.*

13. During a recent year the average cost of making a movie was $54.8 million. This year, a random sample of 15 recent action movies had an average production cost of $62.3 million with a variance of $90.25 million. At the 0.05 level of significance, can it be concluded that it costs more than average to produce an action movie?

Source: *N.Y. Times Almanac.*

14. The average 1-ounce chocolate chip cookie contains 110 calories. A random sample of 15 different brands of 1-ounce chocolate chip cookies resulted in the following calorie amounts. At the $\alpha = 0.01$ level, is there sufficient evidence that the average calorie content is greater than 110 calories?

100	125	150	160	185	125	155	145	160
100	150	140	135	120	110			

Source: *The Doctor's Pocket Calorie, Fat, and Carbohydrate Counter.*

15. The average running time for current Broadway shows is 2 hours 12 minutes. A producer in another city claims that the length of time of productions in his city is the same. He samples 8 shows and finds the time to be 2 hours 5 minutes with a standard deviation of 11 minutes. Using $\alpha = 0.05$, is the producer correct?

Source: *New York Times, Arts and Leisure.*

16. The *Old Farmer's Almanac* stated that the average consumption of water per person per day was 123 gallons. To test the hypothesis that this figure may no longer be true, a researcher randomly selected 16 people and found that they used on average 119 gallons per day and $s = 5.3$. At $\alpha = 0.05$, is there enough evidence to say that the *Old Farmer's Almanac* figure might no longer be correct? Use the *P*-value method.

17. A report by the Gallup Poll stated that on average a woman visits her physician 5.8 times a year. A researcher randomly selects 20 women and obtained these data.

3	2	1	3	7	2	9	4	6	6
8	0	5	6	4	2	1	3	4	1

At $\alpha = 0.05$ can it be concluded that the average is still 5.8 visits per year? Use the *P*-value method.

18. The U.S. Bureau of Labor and Statistics reported that a person between the ages of 18 and 34 has had an average of 9.2 jobs. To see if this average is correct, a researcher selected a sample of 8 workers between the ages of 18 and 34 and asked how many different places they had worked. The results were as follows:

8	12	15	6	1	9	13	2

At $\alpha = 0.05$ can it be concluded that the mean is 9.2? Use the *P*-value method. Give one reason why the respondents might not have given the exact number of jobs that they have worked.

19. A random sample of stipends of teaching assistants in economics is listed. Is there sufficient evidence at the $\alpha = 0.05$ level to conclude that the average stipend differs from $15,000? The stipends listed (in dollars) are for the academic year.

14,000	18,000	12,000	14,356	13,185
13,419	14,000	11,981	17,604	12,283
16,338	15,000			

Source: *Chronicle of Higher Education.*

20. The average family size was reported as 3.18. A random sample of families in a particular school district resulted in the following family sizes:

5	4	5	4	4	3	6	4	3	3	5
6	3	3	2	7	4	5	2	2	2	3
5	2									

At $\alpha = 0.05$, does the average family size differ from the national average?

Source: *N.Y. Times Almanac.*

Technology *Step by Step*

MINITAB
Step by Step

Hypothesis Test for the Mean and the *t* Distribution

This relates to Example 8–13. Test the claim that the average salary for substitute teachers is less than $60 per day.

1. Enter the data into C1 of a MINITAB worksheet. Do not use the dollar sign. Name the column **Salary.**
2. Select **Stat>Basic Statistics>1-Sample t.**
3. Choose C1 Salary as the variable.
4. Click inside the text box for Test mean, and enter the hypothesized value of **60.**
5. Click [Options].
6. The Alternative should be less than.
7. Click [OK] twice.

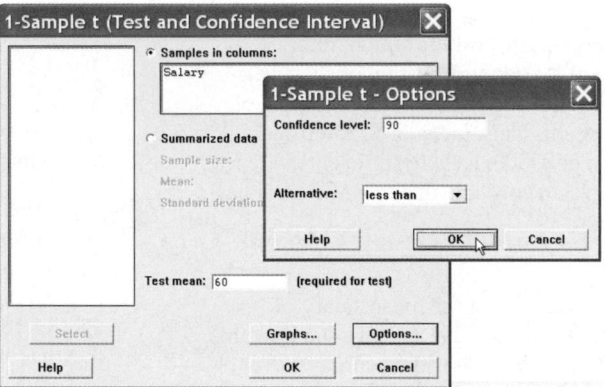

In the session window, the *P*-value for the test is 0.276.

One-Sample T: Salary

```
Test of mu = 60 vs < 60

                                        90% Upper
Variable   N     Mean    StDev   SE Mean    Bound      T       P
Salary     8   58.8750  5.0832   1.7972   61.4179   -0.63   0.276
```

We cannot reject H_0. There is not enough evidence in the sample to conclude the mean salary is less than $60.

TI-83 Plus or TI-84 Plus
Step by Step

Hypothesis Test for the Mean and the *t* Distribution (Data)

1. Enter the data values into L_1.
2. Press **STAT** and move the cursor to TESTS.
3. Press **2** for T-Test.
4. Move the cursor to Data and press **ENTER.**
5. Type in the appropriate values.
6. Move the cursor to the appropriate alternative hypothesis and press **ENTER.**
7. Move the cursor to Calculate and press **ENTER.**

Hypothesis Test for the Mean and the *t* Distribution (Statistics)

1. Press **STAT** and move the cursor to TESTS.
2. Press **2** for T-Test.

3. Move the cursor to Stats and press **ENTER.**

4. Type in the appropriate values.

5. Move the cursor to the appropriate alternative hypothesis and press **ENTER.**

6. Move the cursor to Calculate and press **ENTER.**

Excel
Step by Step

Hypothesis Test for the Mean Using the *t* Distribution

Excel does not have a procedure to conduct a hypothesis test for the mean. However, you may conduct the test of the mean using the MegaStat Add-in available on your CD and Online Learning Center. If you have not installed this add-in, do so by following the instructions on page 24.

This test uses the *P*-value method. Therefore, it is not necessary to enter a significance level.

1. Enter the data from Example 8–13 into column A of a new worksheet.

2. Select **MegaStat>Hypothesis Tests>Mean vs. Hypothesized Value.**

3. Select data input and enter the range of data **A1:A8.**

4. Type **60** for the Hypothesized mean, and select the "less than" Alternative.

5. Select *t* test and click [OK].

8–5

z Test for a Proportion

Objective 7

Test proportions, using the *z* test.

Many hypothesis-testing situations involve proportions. Recall from Chapter 7 that a *proportion* is the same as a percentage of the population.

These data were obtained from *The Book of Odds* by Michael D. Shook and Robert L. Shook (New York: Penguin Putnam, Inc.):

- 59% of consumers purchase gifts for their fathers.
- 85% of people over 21 said they have entered a sweepstakes.
- 51% of Americans buy generic products.
- 35% of Americans go out for dinner once a week.

A hypothesis test involving a population proportion can be considered as a binomial experiment when there are only two outcomes and the probability of a success does not change from trial to trial. Recall from Section 5–4 that the mean is $\mu = np$ and the standard deviation is $\sigma = \sqrt{npq}$ for the binomial distribution.

Since a normal distribution can be used to approximate the binomial distribution when $np \geq 5$ and $nq \geq 5$, the standard normal distribution can be used to test hypotheses for proportions.

Formula for the *z* Test for Proportions

$$z = \frac{\hat{p} - p}{\sqrt{pq/n}}$$

where

$$\hat{p} = \frac{X}{n} \quad \text{(sample proportion)}$$
p = population proportion
n = sample size

The formula is derived from the normal approximation to the binomial and follows the general formula

$$\text{Test value} = \frac{(\text{observed value}) - (\text{expected value})}{\text{standard error}}$$

We obtain $\hat{p}$ from the sample (i.e., observed value), p is the expected value (i.e., hypothesized population proportion), and $\sqrt{pq/n}$ is the standard error.

The formula $z = \dfrac{\hat{p} - p}{\sqrt{pq/n}}$ can be derived from the formula $z = \dfrac{X - \mu}{\sigma}$ by substituting $\mu = np$ and $\sigma = \sqrt{npq}$ and then dividing both numerator and denominator by n. Some algebra is used. See Exercise 23 in this section.

The steps for hypothesis testing are the same as those shown in Section 8–3. Table E is used to find critical values and P-values.

Examples 8–17 to 8–19 show the traditional method of hypothesis testing. Example 8–20 shows the P-value method.

Sometimes it is necessary to find $\hat{p}$, as shown in Examples 8–17, 8–19, and 8–20, and sometimes $\hat{p}$ is given in the exercise. See Example 8–18.

Example 8–17

An educator estimates that the dropout rate for seniors at high schools in Ohio is 15%. Last year, 38 seniors from a random sample of 200 Ohio seniors withdrew. At $\alpha = 0.05$, is there enough evidence to reject the educator's claim?

Solution

Step 1   State the hypotheses and identify the claim.

$H_0: p = 0.15$ (claim)   and $H_1: p \neq 0.15$

Step 2   Find the critical value(s). Since $\alpha = 0.05$ and the test is two-tailed, the critical values are ± 1.96.

Step 3   Compute the test value. First, it is necessary to find $\hat{p}$.

$$\hat{p} = \frac{X}{n} = \frac{38}{200} = 0.19 \quad \text{and} \quad p = 0.15 \quad \text{and} \quad q = 1 - 0.15 = 0.85$$

Substitute in the formula and solve.

$$z = \frac{\hat{p} - p}{\sqrt{pq/n}} = \frac{0.19 - 0.15}{\sqrt{(0.15)(0.85)/200}} = 1.58$$

Step 4   Make the decision. Do not reject the null hypothesis, since the test value falls outside the critical region, as shown in Figure 8–26.

Figure 8–26

Critical and Test Values for Example 8–17

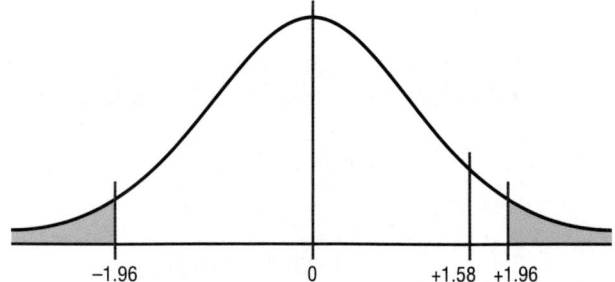

Step 5   Summarize the results. There is not enough evidence to reject the claim that the dropout rate for seniors in high schools in Ohio is 15%.

Example 8–18

A telephone company representative estimates that 40% of its customers have call-waiting service. To test this hypothesis, she selected a sample of 100 customers and found that 37% had call waiting. At $\alpha = 0.01$, is there enough evidence to reject the claim?

Solution

Step 1 State the hypotheses and identify the claim.

$$H_0\colon p = 0.40 \text{ (claim)} \qquad \text{and} \qquad H_1\colon p \neq 0.40$$

Step 2 Find the critical value(s). Since $\alpha = 0.01$ and this test is two-tailed, the critical values are ± 2.58.

Step 3 Compute the test value. It is not necessary to find $\hat{p}$ since it is given in the exercise; $\hat{p} = 0.37$. Substitute in the formula and solve.

$$p = 0.40 \qquad \text{and} \qquad q = 1 - 0.40 = 0.60$$

$$z = \frac{\hat{p} - p}{\sqrt{pq/n}} = \frac{0.37 - 0.40}{\sqrt{(0.40)(0.60)/100}} = -0.612$$

Step 4 Make the decision. Do not reject the null hypothesis, since the test value falls in the noncritical region, as shown in Figure 8–27.

Figure 8–27

Critical and Test Values for Example 8–18

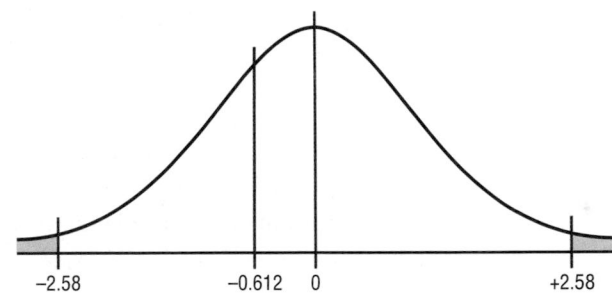

Step 5 Summarize the results. There is not enough evidence to reject the claim that 40% of the telephone company's customers have call waiting.

Example 8–19

A statistician read that at least 77% of the population oppose replacing $1 bills with $1 coins. To see if this claim is valid, the statistician selected a sample of 80 people and found that 55 were opposed to replacing the $1 bills. At $\alpha = 0.01$, test the claim that at least 77% of the population are opposed to the change.

Source: USA TODAY.

Solution

Step 1 State the hypotheses and identify the claim.

$$H_0\colon p \geq 0.77 \text{ (claim)} \qquad \text{and} \qquad H_1\colon p < 0.77$$

Step 2 Find the critical value(s). Since $\alpha = 0.01$ and the test is left-tailed, the critical value is -2.33.

Step 3 Compute the test value.

$$\hat{p} = \frac{X}{n} = \frac{55}{80} = 0.6875$$

$$p = 0.77 \qquad \text{and} \qquad q = 1 - 0.77 = 0.23$$

$$z = \frac{\hat{p} - p}{\sqrt{pq/n}} = \frac{0.6875 - 0.77}{\sqrt{(0.77)(0.23)/80}} = -1.75$$

Step 4 Do not reject the null hypothesis, since the test value does not fall in the critical region, as shown in Figure 8–28.

Figure 8–28

Critical and Test Values for Example 8–19

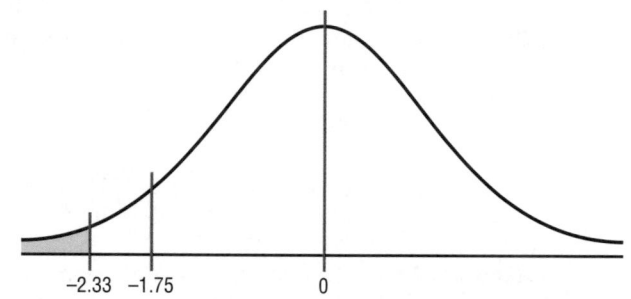

-2.33 -1.75 $\qquad$ 0

Step 5 There is not enough evidence to reject the claim that at least 77% of the population oppose replacing \$1 bills with \$1 coins.

Example 8–20

An attorney claims that more than 25% of all lawyers advertise. A sample of 200 lawyers in a certain city showed that 63 had used some form of advertising. At $\alpha = 0.05$, is there enough evidence to support the attorney's claim? Use the P-value method.

Solution

Step 1 State the hypotheses and identify the claim.

$$H_0\colon p \le 0.25 \qquad \text{and} \qquad H_1\colon p > 0.25 \text{ (claim)}$$

Step 2 Compute the test value

$$\hat{p} = \frac{X}{n} = \frac{63}{200} = 0.315$$

$$p = 0.25 \qquad \text{and} \qquad q = 1 - 0.25 = 0.75$$

$$z = \frac{\hat{p} - p}{\sqrt{pq/n}} = \frac{0.315 - 0.25}{\sqrt{(0.25)(0.75)/200}} = 2.12$$

Step 3 Find the P-value. The area under the curve for $z = 2.12$ is 0.4830. Subtracting the area from 0.5000, one gets $0.5000 - 0.4830 = 0.0170$. The P-value is 0.0170.

Step 4 Reject the null hypothesis, since $0.0170 < 0.05$ (that is, P-value < 0.05). See Figure 8–29.

Figure 8–29
P-Value and α Value for Example 8–20

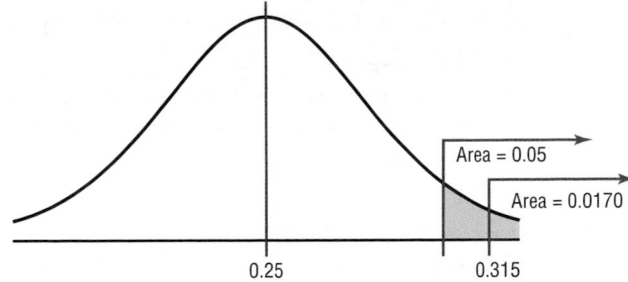

Area = 0.05

Area = 0.0170

0.25   0.315

Step 5 There is enough evidence to support the attorney's claim that more than 25% of the lawyers use some form of advertising.

Applying the Concepts **8–5**

Quitting Smoking

Assume you are part of a research team that compares products designed to help people quit smoking. The Condor Consumer Products Company would like more specific details about the study to be made available to the scientific community. Review the following and then answer the questions about how you would have conducted the study.

New StopSmoke

No method has been proved more effective. StopSmoke provides significant advantages to all other methods. StopSmoke is simpler to use, and it requires no weaning. StopSmoke is also significantly less expensive than the leading brands. StopSmoke's superiority has been proved in two independent studies.

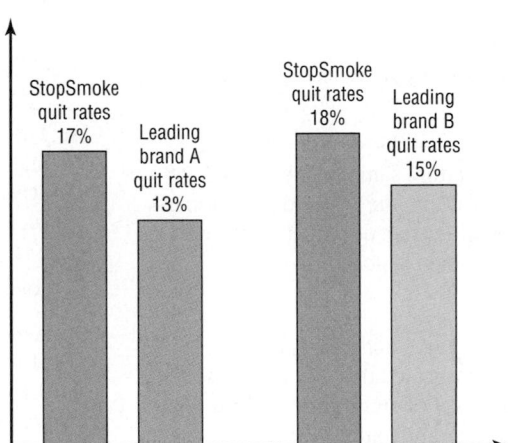

StopSmoke quit rates 17%

Leading brand A quit rates 13%

StopSmoke quit rates 18%

Leading brand B quit rates 15%

1. What were the statistical hypotheses?
2. What were the null hypotheses?
3. What were the alternative hypotheses?
4. Were any statistical tests run?
5. Were one- or two-tailed tests run?
6. What were the levels of significance?
7. If a type I error was committed, explain what it would have been.

8. If a type II error was committed, explain what it would have been.

9. What did the studies prove?

10. Two statements are made about significance. One states that StopSmoke provides significant advantages, and the other states that StopSmoke is significantly less expensive than other leading brands. Are they referring to statistical significance? What other type of significance is there?

See page 461 for the answers.

Exercises 8–5

1. Give three examples of proportions.

2. Why is a proportion considered a binomial variable?

3. When one is testing hypotheses by using proportions, what are the necessary requirements?

4. What are the mean and the standard deviation of a proportion?

For Exercises 5 through 15, perform each of the following steps.

 a. State the hypotheses and identify the claim.

 b. Find the critical value(s).

 c. Compute the test value.

 d. Make the decision.

 e. Summarize the results.

Use the traditional method of hypothesis testing unless otherwise specified.

5. A recent survey found that 64.7% of the population own their homes. In a random sample of 150 heads of households, 92 responded that they owned their homes. At the 0.01 level of significance, does that indicate a difference from the national proportion?

Source: *N.Y. Times Almanac.*

6. Of families 48.8% have stock holdings. A random sample of 250 families indicated that 142 owned some type of stock. At what level of significance would you conclude that this was a significant difference?

Source: *N.Y. Times Almanac.*

7. It has been reported that 40% of the adult population participate in computer hobbies during their leisure time. A random sample of 180 adults found that 65 engaged in computer hobbies. At $\alpha = 0.01$, is there sufficient evidence to conclude that the proportion differs from 40%?

Source: *N.Y. Times Almanac.*

8. The percentage of physicians who are women is 27.9%. In a survey of physicians employed by a large university health system, 45 of 120 randomly selected physicians were women. Is there sufficient evidence at the 0.05 level of significance to conclude that the proportion of women physicians at the university health system exceeds 27.9%?

Source: *N.Y. Times Almanac.*

9. An item in *USA TODAY* reported that 63% of Americans owned an answering machine. A survey of 143 employees at a large school showed that 85 owned an answering machine. At $\alpha = 0.05$, test the claim that the percentage is the same as stated in *USA TODAY.*

Source: *USA TODAY.*

10. The *Statistical Abstract* reported that 17% of adults attended a musical play in the past year. To test this claim, a researcher surveyed 90 people and found that 22 had attended a musical play in the past year. At $\alpha = 0.05$, test the claim that this figure is correct.

Source: *Statistical Abstract of the United States.*

11. The American Automobile Association (AAA) claims that 54% of fatal car/truck accidents are caused by driver error. A researcher studies 30 randomly selected accidents and finds that 14 were caused by driver error. Using $\alpha = 0.05$, can the AAA claim be refuted?

Source: *AAA/CNN.*

12. A survey by *Men's Health* magazine stated that 14% of men said they used exercise to reduce stress. At $\alpha = 0.10$, test the claim that a random sample of 100 men was selected and 10 said that they used exercise to relieve stress. Use the *P*-value method. Could the results be generalized to all adult Americans?

13. In the *Journal of the American Dietetic Association,* it was reported that 54% of kids said that they had a snack after school. Test the claim that a random sample of 60 kids was selected and 36 said that they had a snack after school. Use $\alpha = 0.01$ and the *P*-value method. On the basis of the results, should parents be concerned about their children eating a healthy snack?

14. The Energy Information Administration reported that 51.7% of homes in the United States were heated by

natural gas. A random sample of 200 homes found that 115 were heated by natural gas. Does the evidence support the claim, or has the percentage changed? Use $\alpha = 0.05$ and the *P*-value method. What could be different if the sample were taken in a different geographic area?

15. Researchers suspect that 18% of all high school students smoke at least one pack of cigarettes a day. At Wilson High School, with an enrollment of 300 students, a study found that 50 students smoked at least one pack of cigarettes a day. At $\alpha = 0.05$, test the claim that 18% of all high school students smoke at least one pack of cigarettes a day. Use the *P*-value method.

16. For a certain year a study reports that the percentage of college students using credit cards was 83%. A college dean of student services feels that this is too high for her university, so she randomly selects 50 students and finds that 40 of them use credit cards. At $\alpha = 0.04$, is she correct about her university?

Source: *USA TODAY.*

17. For Americans using library services, the American Library Association (ALA) claims that 67% borrow books. A library director feels that this is not true so he

randomly selects 100 borrowers and finds that 82 borrowed books. Can he show that the ALA claim is incorrect? Use $\alpha = 0.05$.

Source: American Library Association; *USA TODAY.*

18. Nationally, at least 60% of Ph.D. students have paid assistantships. A college dean feels that this is not true in his state, so he randomly selects 50 Ph.D. students and finds that 26 have assistantships. At $\alpha = 0.05$, is the dean correct?

Source: U.S. Department of Education, *Chronicle of Higher Education.*

19. A report by the NCAA states that 57.6% of football injuries occur during practices. A head trainer claims that this is too high for his conference, so he randomly selects 36 injuries and finds that 17 occurred during practices. Is his claim correct, at $\alpha = 0.05$?

Source: *NCAA Sports Medicine Handbook.*

20. For a specific survey, the Gallup Poll reported that 45% of individuals felt they were worse off than 1 year ago. A politician feels that this is too high for her district, so she commissions her own survey and finds that, out of 150 randomly selected citizens, 58 feel they are worse off today than 1 year ago. At $\alpha = 0.05$, is the politician correct about her district?

Source: Gallup Poll, *USA TODAY.*

Extending the Concepts

When *np* or *nq* is not 5 or more, the binomial table (Table B in Appendix C) must be used to find critical values in hypothesis tests involving proportions.

21. A coin is tossed 9 times and 3 heads appear. Can one conclude that the coin is not balanced? Use $\alpha = 0.10$. [*Hint:* Use the binomial table and find $2P(X \leq 3)$ with $p = 0.5$ and $n = 9$.]

22. In the past, 20% of all airline passengers flew first class. In a sample of 15 passengers, 5 flew first class. At $\alpha = 0.10$, can one conclude that the proportions have changed?

23. Show that $z = \dfrac{\hat{p} - p}{\sqrt{pq/n}}$ can be derived from $z = \dfrac{X - \mu}{\sigma}$ by substituting $\mu = np$ and $\sigma = \sqrt{npq}$ and dividing both numerator and denominator by *n*.

Technology *Step by Step*

MINITAB
Step by Step

Hypothesis Test for One Proportion and the *z* Distribution

MINITAB will calculate the test statistic and *P*-value for a test of a proportion, given the statistics from a sample or given the raw data. For Example 8–18, test the claim that 40% of all telephone customers have call-waiting service.

1. Select **Stat>Basic Statistics>1 Proportion.**

2. Click on the button for Summarized data. There are no data to enter in the worksheet.

3. Click in the box for Number of trials and enter **100.**

4. In the Number of events box enter **37.**

5. Click on [Options].

6. Type the complement of α, **99** for the confidence level.

7. Very important! Check the box for Use test and interval based on normal distribution.

8. Click [OK] twice.

The results for the confidence interval will be displayed in the session window. Since the *P*-value of 0.540 is greater than $\alpha = 0.01$, the null hypothesis cannot be rejected.

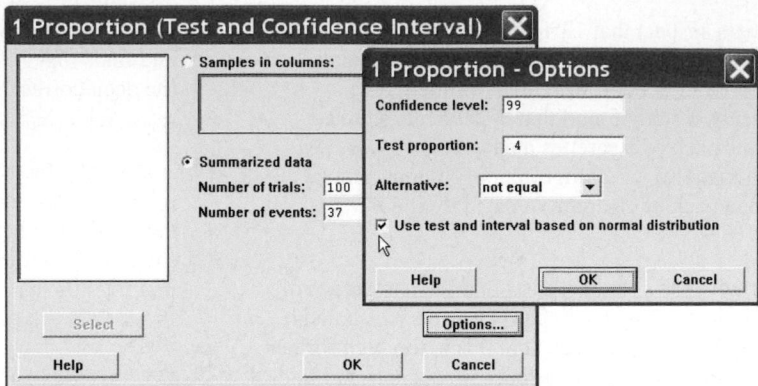

Test and CI for One Proportion

```
Test of p = 0.4 vs p not = 0.4
Sample   X    N   Sample p          99% CI          Z-Value   P-Value
1        37  100  0.370000  (0.245638,  0.494362)    -0.61     0.540
```

There is not enough evidence to conclude that the proportion is different from 40%.

TI-83 Plus or TI-84 Plus
Step by Step

Hypothesis Test for the Proportion

1. Press **STAT** and move the cursor to TESTS.
2. Press **5** for 1-PropZTest.
3. Type in the appropriate values.
4. Move the cursor to the appropriate alternative hypothesis and press **ENTER.**
5. Move the cursor to Calculate and press **ENTER.**

Example TI8–3

This pertains to Example 8–18 in the text. Test the claim that $p = 40\%$, given $n = 100$ and $\hat{p} = 0.37$.

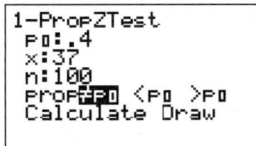   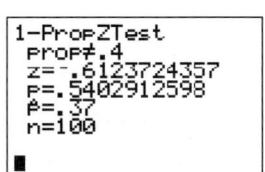

The test statistic is $z = -0.6123724357$, and the *P*-value is 0.5402912598.

Excel
Step by Step

z Test for a Proportion

Excel does not have a procedure to conduct a hypothesis test for a population proportion. However, you may conduct this test by using the MegaStat Add-in available on your CD and Online Learning Center. If you have not installed this add-in, do so by following the instructions on page 24.

This test uses the *P*-value method. Therefore, it is not necessary to enter a significance level.

1. Select **MegaStat>Hypothesis Tests>Proportion vs. Hypothesized Value.**

2. Enter the observed sample proportion **0.37** and the hypothesized proportion **0.40.**

3. Enter the sample size **100** for *n*.

4. Select not equal for the Alternative and click [OK].

8–6 χ^2 **Test for a Variance or Standard Deviation**

Objective 8

Test variances or standard deviations, using the chi-square test.

In Chapter 7, the chi-square distribution was used to construct a confidence interval for a single variance or standard deviation. This distribution is also used to test a claim about a single variance or standard deviation.

To find the area under the chi-square distribution, use Table G in Appendix C. There are three cases to consider:

1. Finding the chi-square critical value for a specific α when the hypothesis test is right-tailed.

2. Finding the chi-square critical value for a specific α when the hypothesis test is left-tailed.

3. Finding the chi-square critical values for a specific α when the hypothesis test is two-tailed.

Example 8–21

Find the critical chi-square value for 15 degrees of freedom when $\alpha = 0.05$ and the test is right-tailed.

Solution

The distribution is shown in Figure 8–30.

Figure 8–30

Chi-Square Distribution for Example 8–21

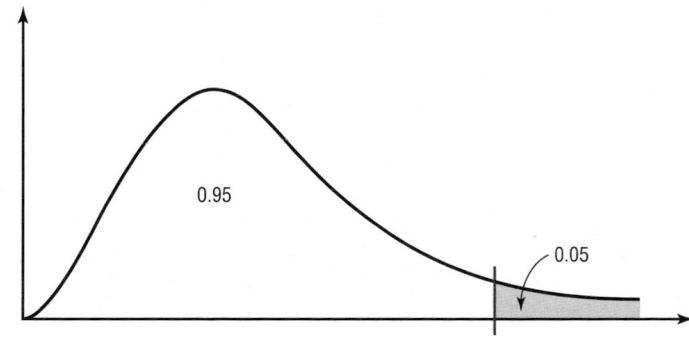

Find the α value at the top of Table G, and find the corresponding degrees of freedom in the left column. The critical value is located where the two columns meet—in this case, 24.996. See Figure 8–31.

Figure 8–31

Locating the Critical Value in Table G for Example 8–21

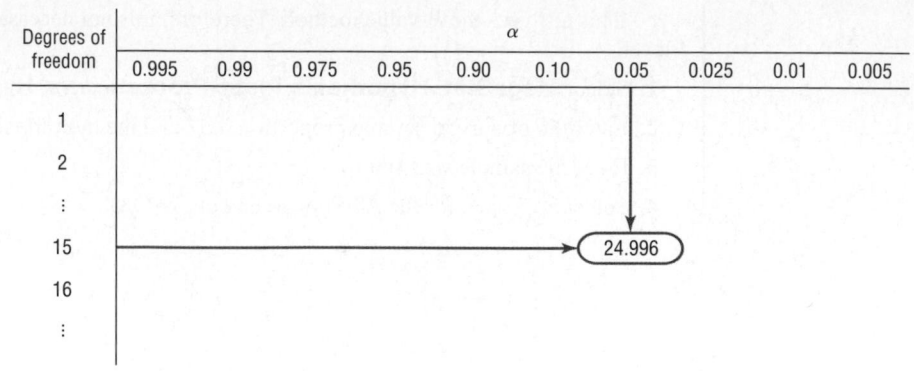

Example 8–22   Find the critical chi-square value for 10 degrees of freedom when $\alpha = 0.05$ and the test is left-tailed.

Solution

This distribution is shown in Figure 8–32.

Figure 8–32

Chi-Square Distribution for Example 8–22

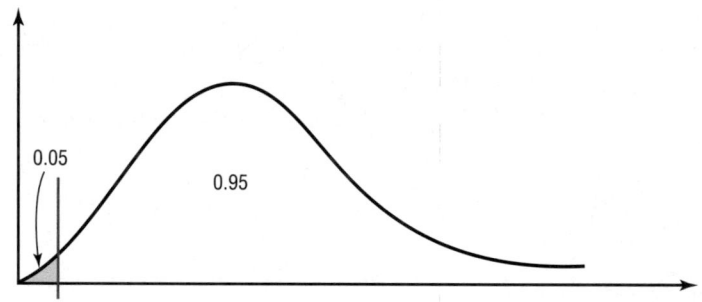

When the test is left-tailed, the α value must be subtracted from 1, that is, $1 - 0.05 = 0.95$. The left side of the table is used, because the chi-square table gives the area to the right of the critical value, and the chi-square statistic cannot be negative. The table is set up so that it gives the values for the area to the right of the critical value. In this case, 95% of the area will be to the right of the value.

For 0.95 and 10 degrees of freedom, the critical value is 3.940. See Figure 8–33.

Figure 8–33

Locating the Critical Value in Table G for Example 8–22

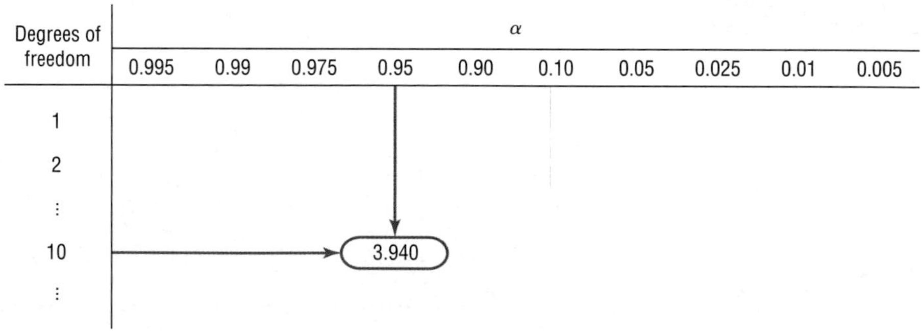

Degrees of freedom	0.995	0.99	0.975	0.95	0.90	0.10	0.05	0.025	0.01	0.005
1										
2										
⋮										
10				3.940						
⋮										

(Header spanning label: α)

Example 8–23   Find the critical chi-square values for 22 degrees of freedom when $\alpha = 0.05$ and a two-tailed test is conducted.

Solution

When a two-tailed test is conducted, the area must be split, as shown in Figure 8–34. Note that the area to the right of the larger value is 0.025 (0.05/2), and the area to the right of the smaller value is 0.975 (1.00 − 0.05/2).

Figure 8–34

Chi-Square Distribution for Example 8–23

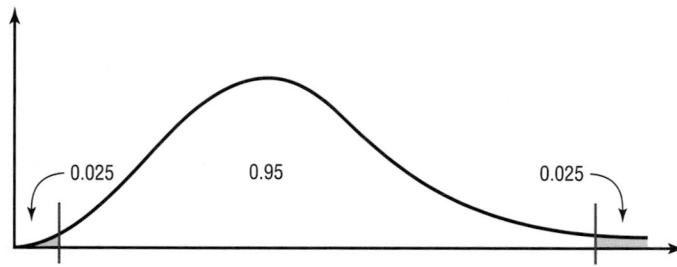

Remember that chi-square values cannot be negative. Hence, one must use α values in the table of 0.025 and 0.975. With 22 degrees of freedom, the critical values are 36.781 and 10.982, respectively.

After the degrees of freedom reach 30, Table G gives values only for multiples of 10 (40, 50, 60, etc.). When the exact degrees of freedom one is seeking are not specified in the table, the closest smaller value should be used. For example, if the given degrees of freedom are 36, use the table value for 30 degrees of freedom. This guideline keeps the type I error equal to or below the α value.

When one is testing a claim about a single variance using the **chi-square test,** there are three possible test situations: right-tailed test, left-tailed test, and two-tailed test.

If a researcher believes the variance of a population to be greater than some specific value, say, 225, then the researcher states the hypotheses as

$$H_0\colon \sigma^2 \le 225 \qquad \text{and} \qquad H_1\colon \sigma^2 > 225$$

and conducts a right-tailed test.

If the researcher believes the variance of a population to be less than 225, then the researcher states the hypotheses as

$$H_0\colon \sigma^2 \ge 225 \qquad \text{and} \qquad H_1\colon \sigma^2 < 225$$

and conducts a left-tailed test.

Finally, if a researcher does not wish to specify a direction, he or she states the hypotheses as

$$H_0\colon \sigma^2 = 225 \qquad \text{and} \qquad H_1\colon \sigma^2 \ne 225$$

and conducts a two-tailed test.

Formula for the Chi-Square Test for a Single Variance

$$\chi^2 = \frac{(n-1)s^2}{\sigma^2}$$

with degrees of freedom equal to $n-1$ and where
n = sample size
s^2 = sample variance
σ^2 = population variance

One might ask, Why is it important to test variances? There are several reasons. First, in any situation where consistency is required, such as in manufacturing, one would like to have the smallest variation possible in the products. For example, when bolts are manufactured, the variation in diameters due to the process must be kept to a minimum, or the nuts will not fit them properly. In education, consistency is required on a test. That is, if the same students take the same test several times, they should get approximately the same grades, and the variance of each of the student's grades should be small. On the other hand, if the test is to be used to judge learning, the overall standard deviation of all the grades should be large so that one can differentiate those who have learned the subject from those who have not learned it.

Three assumptions are made for the chi-square test, as outlined here.

Assumptions for the Chi-Square Test for a Single Variance

1. The sample must be randomly selected from the population.
2. The population must be normally distributed for the variable under study.
3. The observations must be independent of one another.

The traditional method for hypothesis testing follows the same five steps listed earlier. They are repeated here.

Step 1   State the hypotheses and identify the claim.

Step 2   Find the critical value(s).

Step 3   Compute the test value.

Step 4   Make the decision.

Step 5   Summarize the results.

Examples 8–24 through 8–26 illustrate the traditional hypothesis-testing procedure for variances.

Example 8–24

An instructor wishes to see whether the variation in scores of the 23 students in her class is less than the variance of the population. The variance of the class is 198. Is there enough evidence to support the claim that the variation of the students is less than the population variance ($\sigma^2 = 225$) at $\alpha = 0.05$? Assume that the scores are normally distributed.

Solution

Step 1   State the hypotheses and identify the claim.

$$H_0: \sigma^2 \geq 225 \qquad \text{and} \qquad H_1: \sigma^2 < 225 \text{ (claim)}$$

Step 2   Find the critical value. Since this test is left-tailed and $\alpha = 0.05$, use the value $1 - 0.05 = 0.95$. The degrees of freedom are $n - 1 = 23 - 1 = 22$. Hence, the critical value is 12.338. Note that the critical region is on the left, as shown in Figure 8–35.

Figure 8–35

Critical Value for
Example 8–24

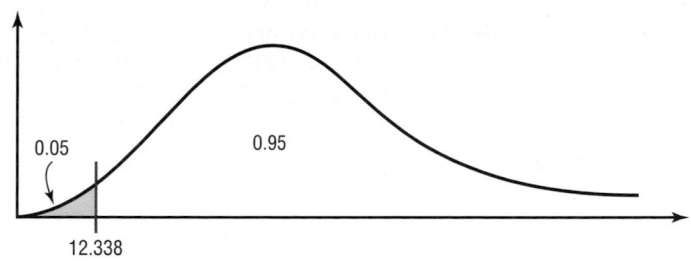

Step 3   Compute the test value.

$$\chi^2 = \frac{(n-1)s^2}{\sigma^2} = \frac{(23-1)(198)}{225} = 19.36$$

Step 4   Make the decision. Since the test value 19.36 falls in the noncritical region, as shown in Figure 8–36, the decision is to not reject the null hypothesis.

Figure 8–36

Critical and Test Values
for Example 8–24

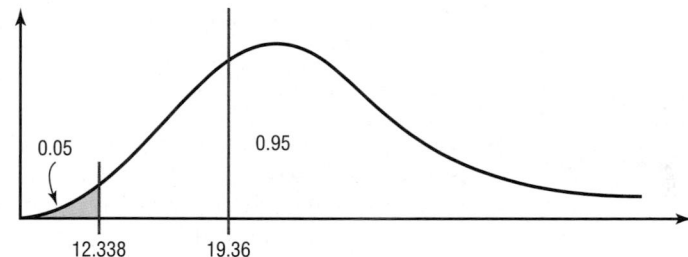

Step 5   Summarize the results. There is not enough evidence to support the claim that the variation in test scores of the instructor's students is less than the variation in scores of the population.

Example 8–25   A hospital administrator believes that the standard deviation of the number of people using outpatient surgery per day is greater than 8. A random sample of 15 days is selected. The data are shown. At $\alpha = 0.10$, is there enough evidence to support the administrator's claim? Assume the variable is normally distributed.

25	30	5	15	18
42	16	9	10	12
12	38	8	14	27

Solution

Step 1   State the hypotheses and identify the claim.

H_0: $\sigma^2 \leq 64$ and H_1: $\sigma^2 > 64$ (claim)

Since the standard deviation is given, it should be squared to get the variance.

Step 2   Find the critical value. Since this test is right-tailed with d.f. of $15 - 1 = 14$ and $\alpha = 0.10$, the critical value is 21.064.

Step 3   Compute the test value. Since raw data are given, the standard deviation of the sample must be found by using the formula in Chapter 3 or your calculator. It is $s = 11.2$.

$$\chi^2 = \frac{(n-1)s^2}{\sigma^2} = \frac{(15-1)(11.2)^2}{64} = 27.44$$

Step 4   Make the decision. The decision is to reject the null hypothesis since the test value, 27.44, is greater than the critical value, 21.064, and falls in the critical region. See Figure 8–37.

Figure 8–37

Critical and Test Value for Example 8–25

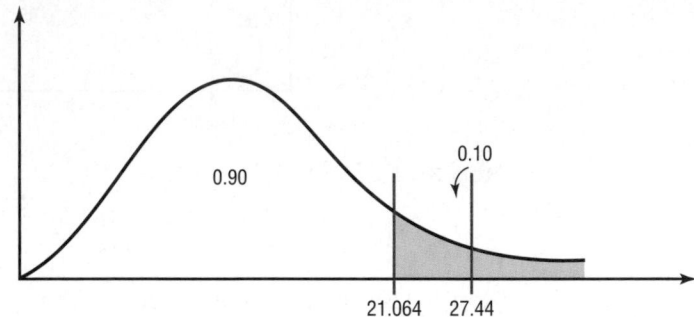

Step 5   Summarize the results. There is enough evidence to support the claim that the standard deviation is greater than 8.

Example 8–26   A cigarette manufacturer wishes to test the claim that the variance of the nicotine content of its cigarettes is 0.644. Nicotine content is measured in milligrams, and assume that it is normally distributed. A sample of 20 cigarettes has a standard deviation of 1.00 milligram. At $\alpha = 0.05$, is there enough evidence to reject the manufacturer's claim?

Solution

Step 1   State the hypotheses and identify the claim.

$$H_0: \sigma^2 = 0.644 \text{ (claim)} \qquad \text{and} \qquad H_1: \sigma^2 \neq 0.644$$

Step 2   Find the critical values. Since this test is a two-tailed test at $\alpha = 0.05$, the critical values for 0.025 and 0.975 must be found. The degrees of freedom are 19; hence, the critical values are 32.852 and 8.907, respectively. The critical or rejection regions are shown in Figure 8–38.

Figure 8–38

Critical Values for Example 8–26

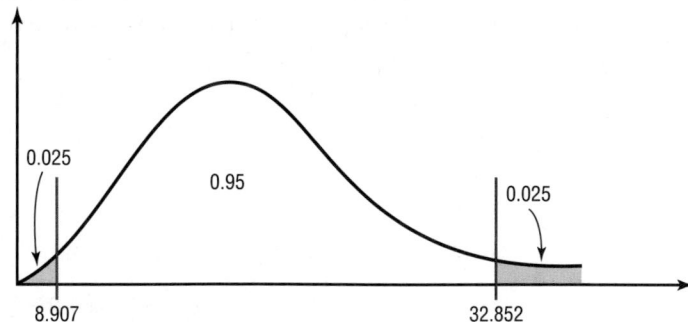

Step 3   Compute the test value.

$$\chi^2 = \frac{(n-1)s^2}{\sigma^2} = \frac{(20-1)(1.0)^2}{0.644} = 29.5$$

Since the standard deviation s is given in the problem, it must be squared for the formula.

Step 4   Make the decision. Do not reject the null hypothesis, since the test value falls between the critical values $(8.907 < 29.5 < 32.852)$ and in the noncritical region, as shown in Figure 8–39.

Figure 8–39

Critical and Test Values for Example 8–26

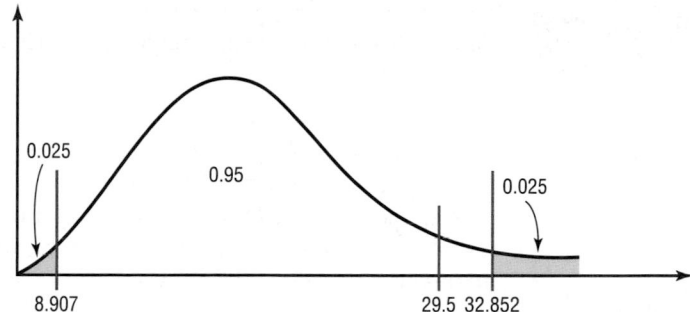

Step 5 Summarize the results. There is not enough evidence to reject the manufacturer's claim that the variance of the nicotine content of the cigarettes is equal to 0.644.

Approximate *P*-values for the chi-square test can be found by using Table G in Appendix C. The procedure is somewhat more complicated than the previous procedures for finding *P*-values for the *z* and *t* tests since the chi-square distribution is not exactly symmetric and χ^2 values cannot be negative. As we did for the *t* test, we will determine an *interval* for the *P*-value based on the table. Examples 8–27 through 8–29 show the procedure.

Example 8–27

Find the *P*-value when $\chi^2 = 19.274$, $n = 8$, and the test is right-tailed.

Solution

To get the *P*-value, look across the row with d.f. = 7 in Table G and find the two values that 19.274 falls between. They are 18.475 and 20.278. Look up to the top row and find the α values corresponding to 18.475 and 20.278. They are 0.01 and 0.005, respectively. See Figure 8–40. Hence the *P*-value is contained in the interval $0.005 < P\text{-value} < 0.01$. (The *P*-value obtained from a calculator is 0.007.)

Figure 8–40

P-Value Interval for Example 8–27

Degrees of freedom	0.995	0.99	0.975	0.95	0.90	0.10	0.05	0.025	0.01	0.005
1	—	—	0.001	0.004	0.016	2.706	3.841	5.024	6.635	7.879
2	0.010	0.020	0.051	0.103	0.211	4.605	5.991	7.378	9.210	10.597
3	0.072	0.115	0.216	0.352	0.584	6.251	7.815	9.348	11.345	12.838
4	0.207	0.297	0.484	0.711	1.064	7.779	9.488	11.143	13.277	14.860
5	0.412	0.554	0.831	1.145	1.610	9.236	11.071	12.833	15.086	16.750
6	0.676	0.872	1.237	1.635	2.204	10.645	12.592	14.449	16.812	18.548
7	0.989	1.239	1.690	2.167	2.833	12.017	11.067	16.013	18.475	20.278
8	1.344	1.646	2.180	2.733	3.490	13.362	15.507	17.535	20.090	21.955
9	1.735	2.088	2.700	3.325	4.168	14.684	16.919	19.023	21.666	23.589
10	2.156	2.558	3.247	3.940	4.865	15.987	18.307	20.483	23.209	25.188
⋮	⋮	⋮	⋮	⋮	⋮	⋮	⋮	⋮	⋮	⋮
100	67.328	70.065	74.222	77.929	82.358	118.498	124.342	129.561	135.807	140.169

The column header for all is α.

*19.274 falls between 18.475 and 20.278

Example 8–28   Find the P-value when $\chi^2 = 3.823$, $n = 13$, and the test is left-tailed.

Solution

To get the P-value, look across the row with d.f. $= 12$ and find the two values that 3.823 falls between. They are 3.571 and 4.404. Look up to the top row and find the values corresponding to 3.571 and 4.404. They are 0.99 and 0.975, respectively. When the χ^2 test value falls on the left side, each of the values must be subtracted from 1 to get the interval that P-value falls between.

$$1 - 0.99 = 0.01 \qquad \text{and} \qquad 1 - 0.975 = 0.025$$

Hence the P-value falls in the interval

$$0.01 < P\text{-value} < 0.025$$

(The P-value obtained from a calculator is 0.014.)

When the χ^2 test is two-tailed, both interval values must be doubled. If a two-tailed test were being used in Example 8–28, then the interval would be $2(0.01) < P\text{-value} < 2(0.025)$, or $0.02 < P\text{-value} < 0.05$.

The P-value method for hypothesis testing for a variance or standard deviation follows the same steps shown in the preceding sections.

Step 1   State the hypotheses and identify the claim.

Step 2   Compute the test value.

Step 3   Find the P-value.

Step 4   Make the decision.

Step 5   Summarize the results.

Example 8–29 shows the P-value method for variances or standard deviations.

Example 8–29   A researcher knows from past studies that the standard deviation of the time it take to inspect a car is 16.8 minutes. A sample of 24 cars is selected and inspected. The standard deviation was 12.5 minutes. At $\alpha = 0.05$, can it be concluded that the standard deviation has changed? Use the P-value method.

Solution

Step 1   State the hypotheses and identify the claim.

$$H_0: \sigma = 16.8 \qquad \text{and} \qquad H_1: \sigma \neq 16.8 \text{ (claim)}$$

Step 2   Compute the test value.

$$\chi^2 = \frac{(n-1)s^2}{\sigma^2} = \frac{(24-1)(12.5)^2}{(16.8)^2} = 12.733$$

Step 3   Find the P-value. Using Table G with d.f. $= 23$, the value 12.733 falls between 11.689 and 13.091, corresponding to 0.975 and 0.95, respectively. Since these values are found on the left side of the distribution, each value must be subtracted from 1. Hence $1 - 0.975 = 0.025$ and $1 - 0.95 = 0.05$. Since this is a two-tailed test, the area must be doubled to obtain the P-value interval. Hence $0.05 < P\text{-value} < 0.10$, or somewhere between 0.05 and 0.10. (The P-value obtained from a calculator is 0.085.)

Step 4 Make the decision. Since $\alpha = 0.05$ and the *P*-value is between 0.05 and 0.10, the decision is to not reject the null hypothesis since *P*-value $> \alpha$.

Step 5 Summarize the results. There is not enough evidence to support the claim that the standard deviation has changed.

Applying the Concepts **8–6**

Testing Gas Mileage Claims

Assume that you are working for the Consumer Protection Agency and have recently been getting complaints about the highway gas mileage of the new Dodge Caravans. Chrysler Corporation agrees to allow you to randomly select 40 of its new Dodge Caravans to test the highway mileage. Chrysler claims that the Caravans get 28 mpg on the highway. Your results show a mean of 26.7 and a standard deviation of 4.2. You support Chrysler's claim.

1. Show why you support Chrysler's claim by listing the *P*-value from your output. After more complaints, you decide to test the variability of the miles per gallon on the highway. From further questioning of Chrysler's quality control engineers, you find they are claiming a standard deviation of 2.1.

2. Test the claim about the standard deviation.

3. Write a short summary of your results and any necessary action that Chrysler must take to remedy customer complaints.

4. State your position about the necessity to perform tests of variability along with tests of the means.

See page 461 for the answers.

Exercises 8–6

1. Using Table G, find the critical value(s) for each, show the critical and noncritical regions, and state the appropriate null and alternative hypotheses. Use $\sigma^2 = 225$.

  a. $\alpha = 0.05, n = 18$, right-tailed

  b. $\alpha = 0.10, n = 23$, left-tailed

  c. $\alpha = 0.05, n = 15$, two-tailed

  d. $\alpha = 0.10, n = 8$, two-tailed

  e. $\alpha = 0.01, n = 17$, right-tailed

  f. $\alpha = 0.025, n = 20$, left-tailed

  g. $\alpha = 0.01, n = 13$, two-tailed

  h. $\alpha = 0.025, n = 29$, left-tailed

2. **(ans)** Using Table G, find the *P*-value interval for each χ^2 test value.

  a. $\chi^2 = 29.321, n = 16$, right-tailed

  b. $\chi^2 = 10.215, n = 25$, left-tailed

  c. $\chi^2 = 24.672, n = 11$, two-tailed

  d. $\chi^2 = 23.722, n = 9$, right-tailed

  e. $\chi^2 = 13.974, n = 28$, two-tailed

  f. $\chi^2 = 10.571, n = 19$, left-tailed

  g. $\chi^2 = 12.144, n = 6$, two-tailed

  h. $\chi^2 = 8.201, n = 23$, two-tailed

For Exercises 3 through 9, assume that the variables are normally or approximately normally distributed. Use the traditional method of hypothesis testing unless otherwise specified.

 3. A nutritionist claims that the standard deviation of the number of calories in 1 tablespoon of the major brands of pancake syrup is 60. A sample of major brands of syrup is selected, and the number of calories is shown. At $\alpha = 0.10$, can the claim be rejected?

53	210	100	200	100	220
210	100	240	200	100	210
100	210	100	210	100	60

Source: Based on information from *The Complete Book of Food Counts* by Corrine T. Netzer, Dell Publishers, New York.

4. A researcher claims that the variance of the number of yearly forest fires in the United States is greater than 140. For a 13-year period, the variance of the number of forest fires in the United States is 146. Test the claim at $\alpha = 0.01$. What factor would influence the variation of the number of forest fires that occurred in the United States?

Source: National Interagency Fire Center.

5. Test the claim that the standard deviation of the number of aircraft stolen each year in the United States is less

than 15 if a sample of 12 years had a standard deviation of 13.6. Use $\alpha = 0.05$.

Source: Aviation Crime Prevention Institute.

6. A random sample of weights (in pounds) of football players at a local college is listed. At $\alpha = 0.05$, is there sufficient evidence that the standard deviation of all football players at the college is less than 10 pounds?

195 185 200 190 210 180 190 185 195 185
175 180 195 185 195 190 180 185

Source: Football Preview, *Washington Observer-Reporter.*

7. The manager of a large company claims that the standard deviation of the time (in minutes) that it takes a telephone call to be transferred to the correct office in her company is 1.2 minutes or less. A sample of 15 calls is selected, and the calls are timed. The standard deviation of the sample is 1.8 minutes. At $\alpha = 0.01$, test the claim that the standard deviation is less than or equal to 1.2 minutes. Use the *P*-value method.

8. A machine fills 12-ounce bottles with soda. For the machine to function properly, the standard deviation of the sample must be less than or equal to 0.03 ounce. A sample of eight bottles is selected, and the number of ounces of soda in each bottle is given. At $\alpha = 0.05$, can we reject the claim that the machine is functioning properly? Use the *P*-value method.

12.03   12.10   12.02   11.98
12.00   12.05   11.97   11.99

9. A random sample of 20 different kinds of doughnuts had the following calorie counts. At $\alpha = 0.01$, is there sufficient evidence to conclude that the standard deviation is greater than 20 calories?

290 320 260 220 300 310 310 270 250 230
270 260 310 200 250 250 270 210 260 300

Source: *The Doctor's Pocket Calorie, Fat, and Carbohydrate Counter.*

10. The high temperatures for a random sample of 7 days in July in southwestern Pennsylvania are listed.

85   86   90   93   77   81   88

At $\alpha = 0.10$, is there sufficient evidence to conclude that the standard deviation for high temperatures is less than 10°?

Source: AccuWeather, *Washington Observer-Reporter.*

11. A researcher claims that the standard deviation of the number of deaths annually from tornadoes in the United States is less than 35. If a sample of 11 randomly selected years had a standard deviation of 32, is the claim believable? Use $\alpha = 0.05$.

Source: National Oceanic and Atmospheric Administration.

12. It has been reported that the standard deviation of the speeds of drivers on Interstate 75 near Findlay, Ohio, is 8 miles per hour for all vehicles. A driver feels from experience that this is very low. A survey is conducted, and for 50 drivers the standard deviation is 10.5 miles per hour. At $\alpha = 0.05$, is the driver correct?

13. A random sample of home run totals for National League Home Run Champions from 1938 to 2001 is shown. At the 0.05 level of significance, is there sufficient evidence to conclude that the variance is greater than 25?

34   47   43   23   36   50   42
44   43   40   39   41   47   45

Source: *N.Y. Times Almanac.*

14. Are some fast-food restaurants healthier than others? The number of grams of fat in a regular serving of French fries for a random sample of restaurants is shown. Is there sufficient evidence to conclude that the standard deviation exceeds 4 grams of fat? Use $\alpha = 0.05$.

15   13   15   22   13   21   10   20
10   17   11   17   15   18   24   18

Source: *The Doctor's Pocket Calorie, Fat, and Carbohydrate Counter.*

Technology *Step by Step*

MINITAB
Step by Step

Hypothesis Test for Variance

For Example 8–25, test the administrator's claim that the standard deviation is greater than 8. There is no menu item to calculate the test statistic and *P*-value directly.

Calculate the Standard Deviation and Sample Size

1. Enter the data into a column of MINITAB. Name the column **OutPatients.**

2. The standard deviation and sample size will be calculated and stored.

   a) Select **Calc>Column Statistics.**

   b) Check the button for Standard deviation. You can only do one of these statistics at a time.

c) Use OutPatients for the Input variable.

d) Store the result in s, then click [OK].

3. Select **Edit>Edit Last Dialog Box,** then do three things:

a) Change the Statistic option from Standard Deviation to N nonmissing.

b) Type **n** in the text box for Store the result.

c) Click [OK].

Calculate the Chi-Square Test Statistic

4. Select **Calc>Calculator.**

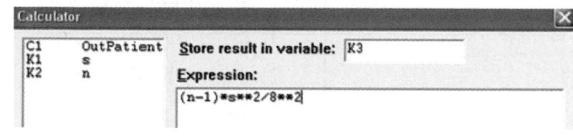

a) In the text box for Store result in variable: type in **K3.** The chi-square value will be stored in a constant so it can be used later.

b) In the expression, type in the formula as shown. The double asterisk is the symbol used for a power.

c) Click [OK]. The chi-square value of 27.44 will be stored in K3.

Calculate the *P*-Value

d) Select **Calc>Probability Distributions>Chi-Square.**

e) Click the button for Cumulative probability.

f) Type in **14** for Degrees of freedom.

g) Click in the text box for Input constant and type **K3.**

h) Type in **K4** for Optional storage.

i) Click [OK]. Now K4 contains the area to the left of the chi-square test statistic.

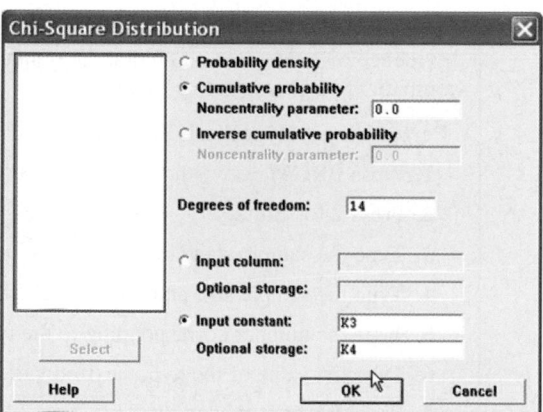

Subtract the cumulative area from 1 to find the area on the right side of the chi-square test statistic. This is the *P*-value for a right-tailed test.

j) Select **Calc>Calculator.**

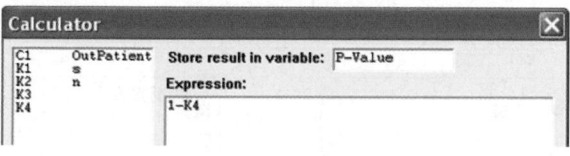

k) In the text box for Store result in variable, type in P-Value.

l) The expression $1 - K4$ calculates the complement of the cumulative area.

m) Click [OK].

The result will be shown in the first row of C2, 0.0168057. Since the P-value is less than α, reject the null hypothesis. The standard deviation in the sample is 11.2, the point estimate for the true standard deviation σ.

TI-83 Plus or TI-84 Plus
Step by Step

The TI-83 Plus and TI-84 Plus do not have a built-in hypothesis test for the variance or standard deviation. However, the downloadable program named SDHYP is available on your CD and Online Learning Center. Follow the instructions with your CD for downloading the program.

Performing a Hypothesis Test for the Variance and Standard Deviation (Data)

1. Enter the values into L_1.
2. Press **PRGM,** move the cursor to the program named SDHYP, and press **ENTER** twice.
3. Press **1** for Data.
4. Type L_1 for the list and press **ENTER.**
5. Type the number corresponding to the type of alternative hypothesis.
6. Type the value of the hypothesized variance and press **ENTER.**
7. Press **ENTER** to clear the screen.

Example TI8–4

This pertains to Example 8–25 in the text. Test the claim that $\sigma > 8$ for these data.

25   30   5   15   18   42   16   9   10   12   12   38   8   14   27

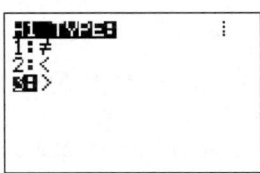 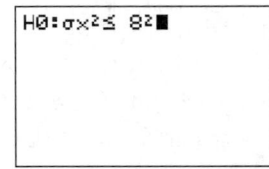 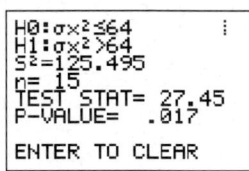

Since P-value $= 0.017 < 0.1$, we reject H_0 and conclude H_1. Therefore, there is enough evidence to support the claim that the standard deviation of the number of people using outpatient surgery is greater than 8.

Performing a Hypothesis Test for the Variance and Standard Deviation (Statistics)

1. Press **PRGM,** move the cursor to the program named SDHYP, and press **ENTER** twice.
2. Press **2** for Stats.
3. Type the sample standard deviation and press **ENTER.**
4. Type the sample size and press **ENTER.**
5. Type the number corresponding to the type of alternative hypothesis.
6. Type the value of the hypothesized variance and press **ENTER.**
7. Press **ENTER** to clear the screen.

Example TI8–5

This pertains to Example 8–26 in the text. Test the claim that $\sigma^2 = 0.644$, given $n = 20$ and $s = 1$.

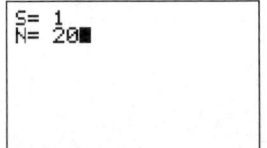 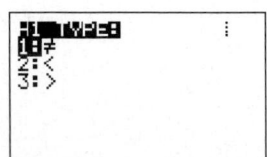 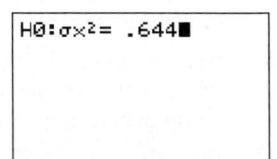 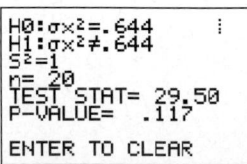

Since *P*-value = 0.117 > 0.05, we do not reject H_0 and do not conclude H_1. Therefore, there is not enough evidence to reject the manufacturer's claim that the variance of the nicotine content of the cigarettes is equal to 0.644.

Excel
Step by Step

Hypothesis Test for the Variance

Excel does not have a procedure to conduct a hypothesis test for the variance. However, you may conduct the test of the variance by using the MegaStat Add-in available on your CD and Online Learning Center. If you have not installed this add-in, do so by following the instructions on page 24.

   This test uses the *P*-value method. Therefore, it is not necessary to enter a significance level.

   We will use the summary information from Example 8–26.

1. Enter a label for the variable: **Nicotine** in cell A1.
2. Enter the observed variance: **1** in cell A2.
3. Enter the sample size: **20** in cell A3.
4. Select **MegaStat>Hypothesis Tests>Chi-Square Variance Test.**
5. Select summary input.
6. Type in **0.644** for the Hypothesized variance and select not equal for Alternative.
7. Click [OK].

8–7   Additional Topics Regarding Hypothesis Testing

In hypothesis testing, there are several other concepts that might be of interest to students in elementary statistics. These topics include the relationship between hypothesis testing and confidence intervals, and some additional information about the type II error.

Objective 9

Test hypotheses, using confidence intervals.

Confidence Intervals and Hypothesis Testing

There is a relationship between confidence intervals and hypothesis testing. When the null hypothesis is rejected in a hypothesis-testing situation, the confidence interval for the mean using the same level of significance *will not* contain the hypothesized mean. Likewise, when the null hypothesis is not rejected, the confidence interval computed using the same level of significance *will* contain the hypothesized mean. Examples 8–30 and 8–31 show this concept for two-tailed tests.

Example 8–30

Sugar is packed in 5-pound bags. An inspector suspects the bags may not contain 5 pounds. A sample of 50 bags produces a mean of 4.6 pounds and a standard deviation of 0.7 pound. Is there enough evidence to conclude that the bags do not contain 5 pounds as stated at $\alpha = 0.05$? Also, find the 95% confidence interval of the true mean.

Solution

Now H_0: $\mu = 5$ and H_1: $\mu \neq 5$ (claim). The critical values are $+1.96$ and -1.96. The test value is

$$z = \frac{\overline{X} - \mu}{s/\sqrt{n}} = \frac{4.6 - 5.0}{0.7/\sqrt{50}} = \frac{-0.4}{0.099} = -4.04$$

Since $-4.04 < -1.96$, the null hypothesis is rejected. There is enough evidence to support the claim that the bags do not weigh 5 pounds.

The 95% confidence for the mean is given by

$$\bar{X} - z_{\alpha/2}\frac{s}{\sqrt{n}} < \mu < \bar{X} + z_{\alpha/2}\frac{s}{\sqrt{n}}$$

$$4.6 - (1.96)\left(\frac{0.7}{\sqrt{50}}\right) < \mu < 4.6 + (1.96)\left(\frac{0.7}{\sqrt{50}}\right)$$

$$4.4 < \mu < 4.8$$

Notice that the 95% confidence interval of μ does *not* contain the hypothesized value $\mu = 5$. Hence, there is agreement between the hypothesis test and the confidence interval.

| **Example 8–31** | A researcher claims that adult hogs fed a special diet will have an average weight of 200 pounds. A sample of 10 hogs has an average weight of 198.2 pounds and a standard deviation of 3.3 pounds. At $\alpha = 0.05$, can the claim be rejected? Also, find the 95% confidence interval of the true mean. |

Solution

Now H_0: $\mu = 200$ pounds (claim) and H_1: $\mu \neq 200$ pounds. The t test must be used since σ is unknown and $n < 30$. The critical values at $\alpha = 0.05$ with 9 degrees of freedom are $+2.262$ and -2.262. The test value is

$$t = \frac{\bar{X} - \mu}{s/\sqrt{n}} = \frac{198.2 - 200}{3.3/\sqrt{10}} = \frac{-1.8}{1.0436} = -1.72$$

Thus, the null hypothesis is not rejected. There is not enough evidence to reject the claim that the weight of the adult hogs is 200 pounds.

The 95% confidence interval of the mean is

$$\bar{X} - t_{\alpha/2}\frac{s}{\sqrt{n}} < \mu < \bar{X} + t_{\alpha/2}\frac{s}{\sqrt{n}}$$

$$198.2 - (2.262)\left(\frac{3.3}{\sqrt{10}}\right) < \mu < 198.2 + (2.262)\left(\frac{3.3}{\sqrt{10}}\right)$$

$$198.2 - 2.361 < \mu < 198.2 + 2.361$$

$$195.8 < \mu < 200.6$$

The 95% confidence interval does contain the hypothesized mean $\mu = 200$. Again there is agreement between the hypothesis test and the confidence interval.

In summary, then, when the null hypothesis is rejected, the confidence interval computed at the same significance level will not contain the value of the mean that is stated in the null hypothesis. On the other hand, when the null hypothesis is not rejected, the confidence interval computed at the same significance level will contain the value of the mean stated in the null hypothesis. These results are true for other hypothesis-testing situations and are not limited to means tests.

The relationship between confidence intervals and hypothesis testing presented here is valid for two-tailed tests. The relationship between one-tailed hypothesis tests and one-sided or one-tailed confidence intervals is also valid; however, this technique is beyond the scope of this textbook.

Type II Error and the Power of a Test

Recall that in hypothesis testing, there are two possibilities: Either the null hypothesis H_0 is true, or it is false. Furthermore, on the basis of the statistical test, the null hypothesis is either rejected or not rejected. These results give rise to four possibilities, as shown in Figure 8–41. This figure is similar to Figure 8–2.

As stated previously, there are two types of errors: type I and type II. A type I error can occur only when the null hypothesis is rejected. By choosing a level of significance, say, 0.05 or 0.01, the researcher can determine the probability of committing a type I error. For example, suppose that the null hypothesis was H_0: $\mu \leq 50$, and it was rejected. At the 0.05 level, the researcher has only a 5% chance of being wrong, i.e., of rejecting a true null hypothesis.

On the other hand, if the null hypothesis is not rejected, then either it is true or a type II error has been committed. A type II error occurs when the null hypothesis is indeed false, but is not rejected. The probability of committing a type II error is denoted as β.

The value of β is not easy to compute. It depends on several things, including the value of α, the size of the sample, the population standard deviation, and the actual difference between the hypothesized value of the parameter being tested and the true parameter. The researcher has control over two of these factors, namely, the selection of α and the size of the sample. The standard deviation of the population is sometimes known or can be estimated. The major problem, then, lies in knowing the actual difference between the hypothesized parameter and the true parameter. If this difference were known, then the value of the parameter would be known; and if the parameter were known, then there would be no need to do any hypothesis testing. Hence, the value of β cannot be computed. But this does not mean that it should be ignored. What the researcher usually does is to try to minimize the size of β or to maximize what is called the **power of a test.**

The power of a statistical test measures the sensitivity of the test to detect a real difference in parameters if one actually exists. The power of a test is a probability and, like all probabilities, can have values ranging from 0 to 1. The higher the power, the more sensitive the test is to detecting a real difference between parameters if there is a difference. In other words, the closer the power of a test is to 1, the better the test is for rejecting the null hypothesis if the null hypothesis is, in fact, false.

The power of a test is equal to $1 - \beta$, that is, 1 minus the probability of committing a type II error. The power of the test is shown in the upper right-hand block of Figure 8–41. If somehow it were known that $\beta = 0.04$, then the power of a test would be $1 - 0.04 = 0.96$, or 96%. In this case, the probability of rejecting the null hypothesis when it is false is 96%.

As stated previously, the power of a test depends on the probability of committing a type II error, and since β is not easily computed, the power of a test cannot be easily computed. (See the Critical Thinking Challenge on page 459.)

Objective 10

Explain the relationship between type I and type II errors and the power of a test.

Figure 8–41

Possibilities in Hypothesis Testing

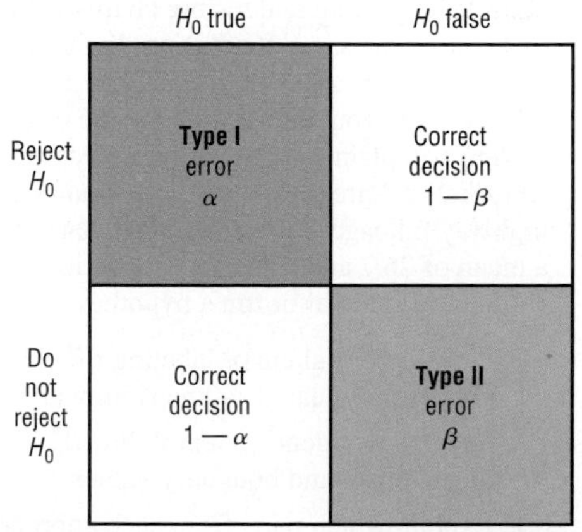

However, there are some guidelines that can be used when you are conducting a statistical study concerning the power of a test. When you are conducting a statistical study, use the test that has the highest power for the data. There are times when the researcher has a choice of two or more statistical tests to test the hypotheses. The tests with the highest power should be used. It is important, however, to remember that statistical tests have assumptions that need to be considered.

If these assumptions cannot be met, then another test with lower power should be used. The power of a test can be increased by increasing the value of α. For example, instead of using $\alpha = 0.01$, use $\alpha = 0.05$. Recall that as α increases, β decreases. So if β is decreased, then $1 - \beta$ will increase, thus increasing the power of the test.

Another way to increase the power of a test is to select a larger sample size. A larger sample size would make the standard error of the mean smaller and consequently reduce β. (The derivation is omitted.)

These two methods should not be used at the whim of the researcher. Before α can be increased, the researcher must consider the consequences of committing a type I error. If these consequences are more serious than the consequences of committing a type II error, then α should not be increased.

Likewise, there are consequences to increasing the sample size. These consequences might include an increase in the amount of money required to do the study and an increase in the time needed to tabulate the data. When these consequences result, increasing the sample size may not be practical.

There are several other methods a researcher can use to increase the power of a statistical test, but these methods are beyond the scope of this book.

One final comment is necessary. When the researcher fails to reject the null hypothesis, this does not mean that there is not enough evidence to support alternative hypotheses. It may be that the null hypothesis is false, but the statistical test has too low a power to detect the real difference; hence, one can conclude only that in this study, there is not enough evidence to reject the null hypothesis.

The relationship among α, β, and the power of a test can be analyzed in greater detail than the explanation given here. However, it is hoped that this explanation will show the student that there is no magic formula or statistical test that can guarantee foolproof results when a decision is made about the validity of H_0. Whether the decision is to reject H_0 or not to reject H_0, there is in either case a chance of being wrong. The goal, then, is to try to keep the probabilities of type I and type II errors as small as possible.

Applying the Concepts **8-7**

Confidence Intervals and Hypothesis Testing

Hypothesis testing and testing claims with confidence intervals are two different approaches that lead to the same conclusion. In the following activities, you will compare and contrast those two approaches.

Assume you are working for the Consumer Protection Agency and have recently been getting complaints about the highway gas mileage of the new Dodge Caravans. Chrysler Corporation agrees to allow you to randomly select 40 of its new Dodge Caravans to test the highway mileage. Chrysler claims that the vans get 28 mpg on the highway. Your results show a mean of 26.7 and a standard deviation of 4.2. You are not certain if you should create a confidence interval or run a hypothesis test. You decide to do both at the same time.

1. Draw a normal curve labeling the critical values, critical regions, test statistic, and population mean. List the significance level and the null and alternative hypotheses.

2. Draw a confidence interval directly below the normal distribution, labeling the sample mean, error, and boundary values.

3. Explain which parts from each approach are the same and which parts are different.

4. Draw a picture of a normal curve and confidence interval where the sample and hypothesized means are equal.

5. Draw a picture of a normal curve and confidence interval where the lower boundary of the confidence interval is equal to the hypothesized mean.

6. Draw a picture of a normal curve and confidence interval where the sample mean falls in the left critical region of the normal curve.

See page 461 for the answers.

Exercises 8–7

1. A ski shop manager claims that the average of the sales for her shop is $1800 a day during the winter months. Ten winter days are selected at random, and the mean of the sales is $1830. The standard deviation of the population is $200. Can one reject the claim at $\alpha = 0.05$? Find the 95% confidence interval of the mean. Does the confidence interval interpretation agree with the hypothesis test results? Explain. Assume that the variable is normally distributed.

2. Charter bus records show that in past years, the buses carried an average of 42 people per trip to Niagara Falls. The standard deviation of the population in the past was found to be 8. This year, the average of 10 trips showed a mean of 48 people booked. Can one reject the claim, at $\alpha = 0.10$, that the average is still the same? Find the 90% confidence interval of the mean. Does the confidence interval interpretation agree with the hypothesis-testing results? Explain. Assume that the variable is normally distributed.

3. The sales manager of a rental agency claims that the monthly maintenance fee for a condominium in the Lakewood region is $86. Past surveys showed that the standard deviation of the population is $6. A sample of 15 owners shows that they pay an average of $84. Test the manager's claim at $\alpha = 0.01$. Find the 99% confidence interval of the mean. Does the confidence interval interpretation agree with the results of the hypothesis test? Explain. Assume that the variable is normally distributed.

4. The average time it takes a person in a one-person canoe to complete a certain river course is 47 minutes. Because of rapid currents in the spring, a group of

10 people traverse the course in 42 minutes. The standard deviation, known from previous trips, is 7 minutes. Test the claim that this group's time was different because of the strong currents. Use $\alpha = 0.10$. Find the 90% confidence level of the true mean. Does the confidence interval interpretation agree with the results of the hypothesis test? Explain. Assume that the variable is normally distributed.

5. From past studies the average time college freshmen spend studying is 22 hours per week. The standard deviation is 4 hours. This year, 60 students were surveyed, and the average time that they spent studying was 20.8 hours. Test the claim that the time students spend studying has changed. Use $\alpha = 0.01$. It is believed that the standard deviation is unchanged. Find the 99% confidence interval of the mean. Do the results agree? Explain.

6. A survey taken several years ago found that the average time a person spent reading the local daily newspaper was 10.8 minutes. The standard deviation of the population was 3 minutes. To see whether the average time had changed since the newspaper's format was revised, the newspaper editor surveyed 36 individuals. The average time that the 36 people spent reading the paper was 12.2 minutes. At $\alpha = 0.02$, is there a change in the average time an individual spends reading the newspaper? Find the 98% confidence interval of the mean. Do the results agree? Explain.

7. What is meant by the power of a test?

8. How is the power of a test related to the type II error?

9. How can the power of a test be increased?

8–8   Summary

This chapter introduces the basic concepts of hypothesis testing. A statistical hypothesis is a conjecture about a population. There are two types of statistical hypotheses: the null and the alternative hypotheses. The null hypothesis states that there is no difference, and the alternative hypothesis specifies a difference. To test the null hypothesis, researchers use a statistical test. Many test values are computed by using

$$\text{Test value} = \frac{(\text{observed value}) - (\text{expected value})}{\text{standard error}}$$

Two common statistical tests are the z test and the t test. The z test is used either when the population standard deviation is known and the variable is normally distributed or when σ is not known and the sample size is greater than or equal to 30. In this case, s is used in place of σ regardless of the shape of the distribution. When the population standard deviation is not known and the variable is normally distributed, the sample standard deviation is used, but a t test should be conducted when the sample size is less than 30. The z test is also used to test proportions when $np \geq 5$ and $nq \geq 5$.

Researchers compute a test value from the sample data in order to decide whether the null hypothesis should be rejected. Statistical tests can be one-tailed or two-tailed, depending on the hypotheses.

The null hypothesis is rejected when the difference between the population parameter and the sample statistic is said to be significant. The difference is significant when the test value falls in the critical region of the distribution. The critical region is determined by α, the level of significance of the test. The level is the probability of committing a type I error. This error occurs when the null hypothesis is rejected when it is true. Three generally agreed upon significance levels are 0.10, 0.05, and 0.01. A second kind of error, the type II error, can occur when the null hypothesis is not rejected when it is false.

Finally, one can test a single variance by using a chi-square test.

All hypothesis-testing situations using the traditional method should include the following steps:

1. State the null and alternative hypotheses and identify the claim.
2. State an alpha level and find the critical value(s).
3. Compute the test value.
4. Make the decision to reject or not reject the null hypothesis.
5. Summarize the results.

All hypothesis-testing situations using the P-value method should include the following steps:

1. State the hypotheses and identify the claim.
2. Compute the test value.
3. Find the P-value.
4. Make the decision.
5. Summarize the results.

Important Terms

α (alpha) 398	chi-square test 439	critical or rejection region 398	hypothesis testing 392
alternative hypothesis 393	β (beta) 398	critical value 398	left-tailed test 398
			level of significance 398

noncritical or nonrejection region 398	power of a test 451	statistical test 396	type I error 397

Important Formulas

Formula for the *z* test for means:

$$z = \frac{\bar{X} - \mu}{\sigma/\sqrt{n}} \quad \text{for any value } n$$

$$z = \frac{\bar{X} - \mu}{s/\sqrt{n}} \quad \text{for } n \geq 30$$

Formula for the *t* test for means:

$$t = \frac{\bar{X} - \mu}{s/\sqrt{n}} \quad \text{for } n < 30$$

Formula for the *z* test for proportions:

$$z = \frac{X - \mu}{\sigma} \quad \text{or} \quad z = \frac{\hat{p} - p}{\sqrt{pq/n}}$$

Formula for the chi-square test for variance or standard deviation:

$$\chi^2 = \frac{(n-1)s^2}{\sigma^2}$$

Review Exercises

For Exercises 1 through 19, perform each of the following steps.

a. State the hypotheses and identify the claim.
b. Find the critical value(s).
c. Compute the test value.
d. Make the decision.
e. Summarize the results.

Use the traditional method of hypothesis testing unless otherwise specified.

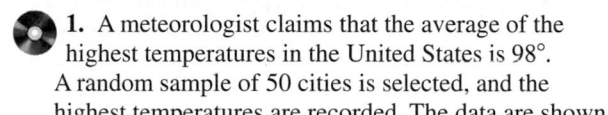 **1.** A meteorologist claims that the average of the highest temperatures in the United States is 98°. A random sample of 50 cities is selected, and the highest temperatures are recorded. The data are shown. At $\alpha = 0.05$, can the claim be rejected?

97	94	96	105	99
96	80	95	101	97
101	87	88	97	94
98	95	88	94	94
99	99	98	96	96
97	98	99	92	97
99	108	97	98	114
91	96	102	99	102
100	93	88	102	99
98	80	95	101	61

Source: The World Almanac & Book of Facts.

2. A stockbroker thought that the average number of shares of stocks traded daily in the stock market was about 500 million. To test the claim, a researcher selected a random sample of 40 days and found the mean number of shares traded each day was 506 million shares. The standard deviation of the sample was 10.3. At $\alpha = 0.05$ is there enough evidence to reject the broker's claim? Based on the results, do you agree or disagree with the broker?

3. Nationwide, graduates entering the actuarial field earn $40,000. A college placement officer feels that this number is too low. She surveys 36 graduates entering the actuarial field and finds the average salary to be $41,000 with a standard deviation of $3000. Can her claim be supported at $\alpha = 0.05$?

Source: BeAnActuary.org.

4. Nationwide, the average salary of actuaries who achieve the rank of Fellow is $150,000. An insurance executive wants to see how this compares with Fellows within his company. He checks the salaries of eight Fellows and finds the average salary to be $155,500 with a standard deviation of $15,000. Can he conclude that Fellows in his company make more than the national average, using $\alpha = 0.05$?

Source: BeAnActuary.org.

5. The average temperature during the summer months for the northeastern part of the United States is 67.0°. A sample of 10 cities had an average temperature of 69.6° for the summer of 1995. The standard deviation of the sample is 1.1°. At $\alpha = 0.10$, can it be concluded that the summer of 1995 was warmer than average?

   Source: USA TODAY.

6. The Tennis Industry Association stated that the average age of a tennis fan is 32 years. To test the claim, a researcher selected a random sample of 18 tennis fans and found that the mean of their ages was 31.3 years and the standard deviation was 2.8 years. At $\alpha = 0.05$ does it appear that the average age is lower than what was stated by the Tennis Industry Association? Use the *P*-value method, and assume the variable is approximately normally distributed.

7. A park manager suggests that the average attendance for the 10 most popular national parks last year was 6 million. The number of visits for the 10 parks this year is shown. At $\alpha = 0.02$, has the average attendance changed? The data are in millions of people.

4.7	17.2	14.0	6.1	6.1
4.9	9.3	9.4	6.4	6.1

   Source: USA TODAY.

8. According to the 2000 Census, 58.5% of women work. A county commissioner feels that more women work in his county, so he conducts a survey of 1000 randomly selected women and finds that 622 work. At $\alpha = 0.05$, is he correct?

   Source: 2000 U.S. Census.

9. Nationally 60.2% of federal prisoners are serving time for drug offenses. A warden feels that in his prison the percentage is even higher. He surveys 400 inmates' records and finds that 260 of the inmates are drug offenders. At $\alpha = 0.05$, is he correct?

   Source: N.Y. Times Almanac.

10. The Tennis Industry Association reported that 41% of tennis fans were female. If, in a sample of 30 tennis fans, 15 were female, do the results suggest that the percentage may have changed? Use $\alpha = 0.02$.

11. A radio manufacturer claims that 65% of teenagers 13 to 16 years old have their own portable radios. A researcher wishes to test the claim and selects a random sample of 80 teenagers. She finds that 57 have their own portable radios. At $\alpha = 0.05$, should the claim be rejected? Use the *P*-value method.

12. A football coach claims that the average weight of all the opposing teams' members is 225 pounds. For a test of the claim, a sample of 50 players is taken from all the opposing teams. The mean is found to be 230 pounds,

and the standard deviation is 15 pounds. At $\alpha = 0.01$, test the coach's claim. Find the *P*-value and make the decision.

13. An advertisement claims that Fasto Stomach Calm will provide relief from indigestion in less than 10 minutes. For a test of the claim, 35 individuals were given the product; the average time until relief was 9.25 minutes. From past studies, the standard deviation is known to be 2 minutes. Can one conclude that the claim is justified? Find the *P*-value and let $\alpha = 0.05$.

14. A film editor feels that the standard deviation for the number of minutes in a video is 3.4 minutes. A sample of 24 videos has a standard deviation of 4.2 minutes. At $\alpha = 0.05$, is the sample standard deviation different from what the editor hypothesized?

15. The standard deviation of the fuel consumption of a certain automobile is hypothesized to be greater than or equal to 4.3 miles per gallon. A sample of 20 automobiles produced a standard deviation of 2.6 miles per gallon. Is the standard deviation really less than previously thought? Use $\alpha = 0.05$ and the *P*-value method.

16. A real estate agent claims that the standard deviation of the rental rates of apartments in a certain county is $95. A random sample of rates in dollars is shown. At $\alpha = 0.02$, can the claim be refuted?

400	345	325	395	400	300
375	435	495	525	290	460
425	250	200	525	375	390

    Source: Pittsburgh Tribune-Review.

17. A manufacturer claims that the standard deviation of the drying time of a certain type of paint is 18 minutes. A sample of five test panels produced a standard deviation of 21 minutes. Test the claim at $\alpha = 0.05$.

18. To see whether people are keeping their car tires inflated to the correct level of 35 pounds per square inch (psi), a tire company manager selects a sample of 36 tires and checks the pressure. The mean of the sample is 33.5 psi, and the standard deviation is 3 psi. Are the tires properly inflated? Use $\alpha = 0.10$. Find the 90% confidence interval of the mean. Do the results agree? Explain.

19. A biologist knows that the average length of a leaf of a certain full-grown plant is 4 inches. The standard deviation of the population is 0.6 inch. A sample of 20 leaves of that type of plant given a new type of plant food had an average length of 4.2 inches. Is there reason to believe that the new food is responsible for a change in the growth of the leaves? Use $\alpha = 0.01$. Find the 99% confidence interval of the mean. Do the results concur? Explain. Assume that the variable is approximately normally distributed.

Statistics Today

How Much Better Is Better?—Revisited

Now that you have learned the techniques of hypothesis testing presented in this chapter, you realize that the difference between the sample mean and the population mean must be *significant* before one can conclude that the students really scored above average. The superintendent should follow the steps in the hypothesis-testing procedure and be able to reject the null hypothesis before announcing that his students scored higher than average.

Data Analysis

The Data Bank is found in Appendix D, or on the World Wide Web by following links from
www.mhhe.com/math/stats/bluman/

1. From the Data Bank, select a random sample of at least 30 individuals, and test one or more of the following hypotheses by using the z test. Use $\alpha = 0.05$.

   a. For serum cholesterol, $H_0: \mu = 220$ milligram percent (mg%).
   b. For systolic pressure, $H_0: \mu = 120$ millimeters of mercury (mm Hg).
   c. For IQ, $H_0: \mu = 100$.
   d. For sodium level, $H_0: \mu = 140$ milliequivalents per liter (mEq/l).

2. Select a random sample of 15 individuals and test one or more of the hypotheses in Exercise 1 by using the t test. Use $\alpha = 0.05$.

3. Select a random sample of at least 30 individuals, and using the z test for proportions, test one or more of the following hypotheses. Use $\alpha = 0.05$.

   a. For educational level, $H_0: p = 0.50$ for level 2.
   b. For smoking status, $H_0: p = 0.20$ for level 1.
   c. For exercise level, $H_0: p = 0.10$ for level 1.
   d. For gender, $H_0: p = 0.50$ for males.

4. Select a sample of 20 individuals and test the hypothesis $H_0: \sigma^2 = 225$ for IQ level. Use $\alpha = 0.05$.

5. Using the data from Data Set XIII, select a sample of 10 hospitals and test $H_0: \mu \geq 250$ for the number of beds. Use $\alpha = 0.05$.

6. Using the data obtained in Exercise 5, test the hypothesis $H_0: \sigma \geq 150$. Use $\alpha = 0.05$.

Chapter Quiz

Determine whether each statement is true or false. If the statement is false, explain why.

1. No error is committed when the null hypothesis is rejected when it is false.

2. When one is conducting the t test, the population must be approximately normally distributed.

3. The test value separates the critical region from the noncritical region.

4. The values of a chi-square test cannot be negative.

5. The chi-square test for variances is always one-tailed.

Select the best answer.

6. When the value of α is increased, the probability of committing a type I error is

   a. Decreased
   b. Increased
   c. The same
   d. None of the above

7. If one wishes to test the claim that the mean of the population is 100, the appropriate null hypothesis is

   a. $\bar{X} = 100$
   b. $\mu \geq 100$
   c. $\mu \leq 100$
   d. $\mu = 100$

8. The degrees of freedom for the chi-square test for variances or standard deviations are

   a. 1
   b. n
   c. $n - 1$
   d. None of the above

9. For the z test, if σ is unknown and $n \geq 30$, one can substitute _____ for σ.

   a. n
   b. s
   c. χ^2
   d. t

Complete the following statements with the best answer.

10. Rejecting the null hypothesis when it is true is called a(n) _____ error.

11. The probability of a type II error is referred to as _____.

12. A conjecture about a population parameter is called a(n) _____.

13. To test the hypothesis H_0: $\mu \leq 87$, one would use a(n) _____-tailed test.

14. The degrees of freedom for the t test are _____.

For the following exercises where applicable:

   a. State the hypotheses and identify the claim.
   b. Find the critical value(s).
   c. Compute the test value.
   d. Make the decision.
   e. Summarize the results.

Use the traditional method of hypothesis testing unless otherwise specified.

15. A sociologist wishes to see if it is true that for a certain group of professional women, the average age at which they have their first child is 28.6 years. A random sample of 36 women is selected, and their ages at the birth of their first children are recorded. At $\alpha = 0.05$, does the evidence refute the sociologist's assertion?

32	28	26	33	35	34
29	24	22	25	26	28
28	34	33	32	30	29
30	27	33	34	28	25
24	33	25	37	35	33
34	36	38	27	29	26

16. A real estate agent believes that the average closing cost of purchasing a new home is $6500 over the purchase price. She selects 40 new home sales at random and finds that the average closing costs are $6600. The standard deviation of the population is $120. Test her belief at $\alpha = 0.05$.

17. A recent study stated that if a person chewed gum, the average number of sticks of gum he or she chewed daily was 8. To test the claim, a researcher selected a random sample of 36 gum chewers and found the mean number of sticks of gum chewed per day was 9. The standard deviation was 1. At $\alpha = 0.05$, is the number of sticks of gum a person chews per day actually greater than 8?

18. A travel agent claims that the average of the number of rooms in hotels in a large city is 500. At $\alpha = 0.01$ is the claim realistic? The data for a sample of six hotels are shown.

713	300	292	311	598	401	618

Give a reason why the claim might be deceptive.

19. In a New York modeling agency, a researcher wishes to see if the average height of female models is really less than 67 inches, as the chief claims. A sample of 20 models has an average height of 65.8 inches. The standard deviation of the sample is 1.7 inches. At $\alpha = 0.05$, is the average height of the models really less than 67 inches? Use the P-value method.

20. A taxi company claims that its drivers have an average of at least 12.4 years' experience. In a study of 15 taxi drivers, the average experience was 11.2 years. The standard deviation was 2. At $\alpha = 0.10$, is the number of years' experience of the taxi drivers really less than the taxi company claimed?

21. A recent study in a small city stated that the average age of robbery victims was 63.5 years. A sample of 20 recent victims had a mean of 63.7 years and a standard deviation of 1.9 years. At $\alpha = 0.05$, is the average age higher than originally believed? Use the P-value method.

22. A magazine article stated that the average age of women who are getting married for the first time is 26 years. A researcher decided to test this hypothesis at $\alpha = 0.02$. She selected a sample of 25 women who were recently married for the first time and found the average was 25.1 years. The standard deviation was 3 years. Should the null hypothesis be rejected on the basis of the sample?

23. A survey in *Men's Health* magazine reported that 39% of cardiologists said that they took vitamin E supplements. To see if this is still true, a researcher randomly selected 100 cardiologists and found that 36 said that they took vitamin E supplements. At $\alpha = 0.05$ test the claim that 39% of the cardiologists took vitamin E supplements. A recent study said that taking too much vitamin E might be harmful. How might this study make the results of the previous study invalid?

24. A dietitian read in a survey that at least 55% of adults do not eat breakfast at least 3 days a week. To verify this, she selected a random sample of 80 adults and asked them how many days a week they skipped breakfast. A total of 50% responded that they skipped breakfast at least 3 days a week. At $\alpha = 0.10$, test the claim.

25. A Harris Poll found that 35% of people said that they drink a caffeinated beverage to combat midday drowsiness. A recent survey found that 19 out of 48 people stated that they drank a caffeinated beverage to combat midday drowsiness. At $\alpha = 0.02$ is the claim of the percentage found in the Harris Poll believable?

26. A magazine claims that 75% of all teenage boys have their own radios. A researcher wished to test the claim and selected a random sample of 60 teenage boys. She found that 54 had their own radios. At $\alpha = 0.01$, should the claim be rejected?

27. Find the *P*-value for the *z* test in Exercise 15.

28. Find the *P*-value for the *z* test in Exercise 16.

29. A copyeditor thinks the standard deviation for the number of pages in a romance novel is greater than 6. A sample of 25 novels has a standard deviation of 9 pages. At $\alpha = 0.05$, is it higher, as the editor hypothesized?

30. It has been hypothesized that the standard deviation of the germination time of radish seeds is 8 days. The standard deviation of a sample of 60 radish plants' germination times was 6 days. At $\alpha = 0.01$, test the claim.

31. The standard deviation of the pollution by-products released in the burning of 1 gallon of gas is 2.3 ounces.

A sample of 20 automobiles tested produced a standard deviation of 1.9 ounces. Is the standard deviation really less than previously thought? Use $\alpha = 0.05$.

32. A manufacturer claims that the standard deviation of the strength of wrapping cord is 9 pounds. A sample of 10 wrapping cords produced a standard deviation of 11 pounds. At $\alpha = 0.05$, test the claim. Use the *P*-value method.

33. Find the 90% confidence interval of the mean in Exercise 15. Is μ contained in the interval?

34. Find the 95% confidence interval for the mean in Exercise 16. Is μ contained in the interval?

Critical Thinking Challenges

The power of a test $(1 - \beta)$ can be calculated when a specific value of the mean is hypothesized in the alternative hypothesis; for example, let H_0: $\mu = 50$ and let H_1: $\mu = 52$. To find the power of a test, it is necessary to find the value of β. This can be done by the following steps:

Step 1 For a specific value of α find the corresponding value of $\bar{X}$, using $z = \dfrac{\bar{X} - \mu}{\sigma/\sqrt{n}}$, where μ is the hypothesized value given in H_0. Use a right-tailed test.

Step 2 Using the value of $\bar{X}$ found in step 1 and the value of μ in the alternative hypothesis, find the area corresponding to z in the formula $z = \dfrac{\bar{X} - \mu}{\sigma/\sqrt{n}}$.

Step 3 Subtract this area from 0.5000. This is the value of β.

Step 4 Subtract the value of β from 1. This will give you the power of a test. See Figure 8–42.

1. Find the power of a test, using the hypotheses given previously and $\alpha = 0.05$, $\sigma = 3$, and $n = 30$.

2. Select several other values for μ in H_1 and compute the power of the test. Generalize the results.

Figure 8–42

Relationship Among α, β, and the Power of a Test

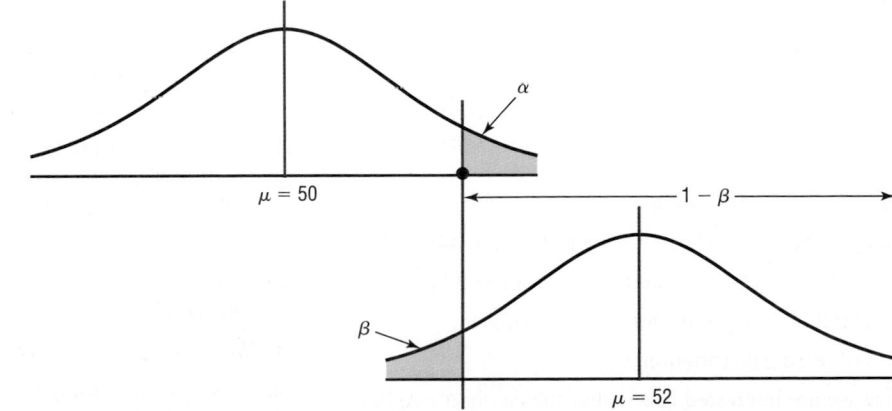

Data Projects

Use MINITAB, the TI-83 Plus, the TI-84 Plus, or a computer program of your choice to complete these exercises.

1. Choose a variable such as the number of miles students live from the college or the number of daily admissions to an ice skating rink for a 1-month period. Before you collect the data, decide what a likely average might be. Then complete the following.

    a. Write a brief statement of the purpose of the study.
    b. Define the population.
    c. State the hypotheses for the study.
    d. Select an α value.
    e. State how the sample was selected.
    f. Show the raw data.
    g. Decide which statistical test is appropriate, and compute the test statistic (z or t). Why is the test appropriate?
    h. Find the critical value(s).
    i. State the decision.
    j. Summarize the results.

2. Decide on a question that could be answered with "yes" or "no" or "not sure (undecided)." For example, "Are you satisfied with the variety of food the cafeteria serves?" Before you collect the data, hypothesize what you expect the proportion of students who will respond yes to be. Then complete the following.

    a. Write a brief statement of the purpose of the study.
    b. Define the population.
    c. State the hypotheses for the study.
    d. Select an α value.
    e. State how the sample was selected.
    f. Show the raw data.
    g. Decide which statistical test is appropriate, and compute the test statistic (z or t). Why is the test appropriate?
    h. Find the critical value(s).
    i. State the decision.
    j. Summarize the results.

Do you think the project supported your initial hypothesis? Explain your answer.

You may use the following websites to obtain raw data:

**Visit the data sets at the book's website found at http://www.mhhe.com/math/stat/bluman Click on the 6th edition.
http://lib.stat.cmu.edu/DASL
http://www.statcan.ca**

Answers to Applying the Concepts

Section 8–2   Eggs and Your Health

1. The study was prompted by claims that linked foods high in cholesterol to high blood serum cholesterol.
2. The population under study is people in general.
3. A sample of 500 subjects was collected.
4. The hypothesis was that eating eggs did not increase blood serum cholesterol.
5. Blood serum cholesterol levels were collected.
6. Most likely but we are not told which test.
7. The conclusion was that eating a moderate amount of eggs will not significantly increase blood serum cholesterol level.

Section 8–3   Car Thefts

1. The hypotheses are H_0: $\mu = 44$ and H_1: $\mu \neq 44$.
2. This sample can be considered large for our purposes.
3. The variable needs to be normally distributed.
4. We will use a z distribution.
5. Since we are interested in whether the car theft rate has changed, we use a two-tailed test.

6. Answers may vary. At the $\alpha = 0.05$ significance level, the critical values are $z = \pm 1.96$.
7. The sample mean is $\overline{X} = 55.97$, and the sample standard deviation is 30.30. Our test statistic is $z = \frac{55.97 - 44}{30.30/\sqrt{36}} = 2.37$.
8. Since $2.37 > 1.96$, we reject the null hypothesis.
9. There is enough evidence to conclude that the car theft rate has changed.
10. Answers will vary. Based on our sample data, it appears that the car theft rate has changed from 44 vehicles per 10,000 people. In fact, the data indicate that the car theft rate has increased.
11. Based on our sample, we would expect 55.97 car thefts per 10,000 people, so we would expect $(55.97)(5) = 279.85$, or about 280, car thefts in the city.

Section 8–4   How Much Nicotine Is in Those Cigarettes?

1. We have $15 - 1 = 14$ degrees of freedom.
2. This is a small-sample test.
3. We are only testing one sample.

4. This is a right-tailed test, since the hypotheses of the tobacco company are $H_0: \mu \le 40$ and $H_1: \mu > 40$.

5. The P-value is 0.008, which is less than the significance level of 0.01. We reject the tobacco company's claim.

6. Since the test statistic (2.72) is greater than the critical value (2.62), we reject the tobacco company's claim.

7. There is no conflict in this output, since the results based on the P-value and on the critical value agree.

8. Answers will vary. It appears that the company's claim is false and that there is more than 40 mg of nicotine in its cigarettes.

Section 8–5   Quitting Smoking

1. The statistical hypotheses were that StopSmoke helps more people quit smoking than the other leading brands.

2. The null hypotheses were that StopSmoke has the same effectiveness as or is not as effective as the other leading brands.

3. The alternative hypotheses were that StopSmoke helps more people quit smoking than the other leading brands. (The alternative hypotheses are the statistical hypotheses.)

4. No statistical tests were run that we know of.

5. Had tests been run, they would have been one-tailed tests.

6. Some possible significance levels are 0.01, 0.05, and 0.10.

7. A type I error would be concluding that StopSmoke is better when it really is not.

8. A type II error would be concluding that StopSmoke is not better when it really is.

9. These studies proved nothing. Had statistical tests been used, we could have tested the effectiveness of StopSmoke.

10. Answers will vary. One possible answer is that more than likely the statements are talking about practical significance and not statistical significance, since we have no indication that any statistical tests were conducted.

Section 8–6   Testing Gas Mileage Claims

1. The hypotheses are $H_0: \mu = 28$ and $H_1: \mu < 28$. The value of our test statistic is $z = -1.96$, and the associated P-value is 0.02514. We would reject Chrysler's claim that the Dodge Caravans are getting 28 mpg.

2. The hypotheses are $H_0: \sigma = 2.1$ and $H_1: \sigma > 2.1$. The value of our test statistic is $\chi^2 = \frac{(n-1)s^2}{\sigma^2} = \frac{(39)4.2^2}{2.1^2} = 156$, and the associated P-value is approximately zero. We would reject Chrysler's claim that the standard deviation is 2.1 mpg.

3. Answers will vary. It is recommended that Chrysler lower its claim about the highway miles per gallon of the Dodge Caravans. Chrysler should also try to reduce variability in miles per gallon and provide confidence intervals for the highway miles per gallon.

4. Answers will vary. There are cases when a mean may be fine, but if there is a lot of variability about the mean, there will be complaints (due to the lack of consistency).

Section 8–7   Confidence Intervals and Hypothesis Testing

1. Answers will vary.

2. Answers will vary.

3. Answers will vary.

4. Answers will vary.

5. Answers will vary.

6. Answers will vary.

CHAPTER

9

Testing the Difference Between Two Means, Two Variances, and Two Proportions

Objectives

After completing this chapter, you should be able to

1 Test the difference between two large sample means, using the *z* test.

2 Test the difference between two variances or standard deviations.

3 Test the difference between two means for small independent samples.

4 Test the difference between two means for small dependent samples.

5 Test the difference between two proportions.

Outline

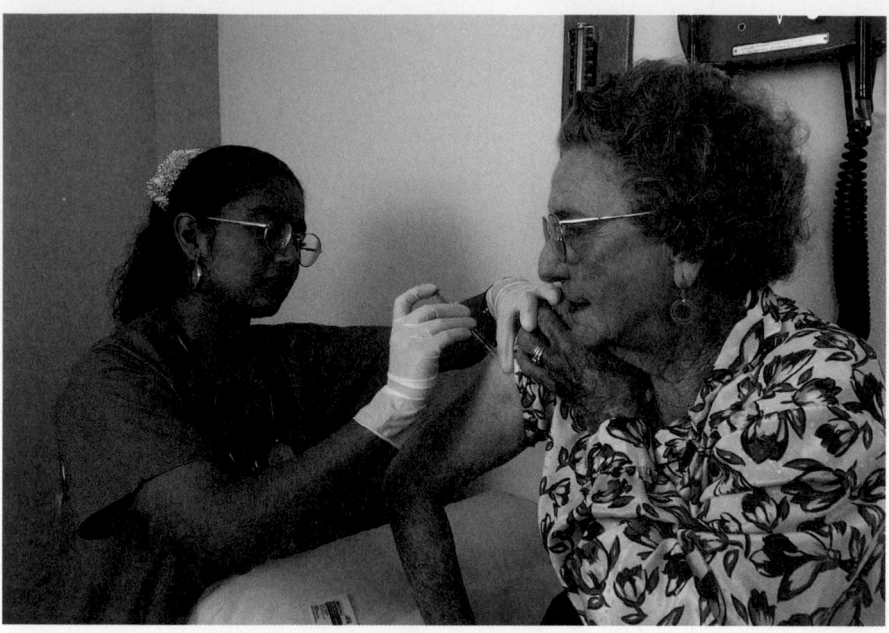

Statistics
Today

To Vaccinate or Not to Vaccinate? Small or Large?

Influenza is a serious disease among the elderly, especially those living in nursing homes. Those residents are more susceptible to influenza than elderly persons living in the community because the former are usually older and more debilitated, and they live in a closed environment where they are exposed more so than community residents to the virus if it is introduced into the home. Three researchers decided to investigate the use of vaccine and its value in determining outbreaks of influenza in small nursing homes.

These researchers surveyed 83 licensed homes in seven counties in Michigan. Part of the study consisted of comparing the number of people being vaccinated in small nursing homes (100 or fewer beds) with the number in larger nursing homes (more than 100 beds). Unlike the statistical methods presented in Chapter 8, these researchers used the techniques explained in this chapter to compare two sample proportions to see if there was a significant difference in the vaccination rates of patients in small nursing homes compared to those in large nursing homes. See Statistics Today—Revisited.

Source: Nancy Arden, Arnold S. Monto, and Suzanne E. Ohmit, "Vaccine Use and the Risk of Outbreaks in a Sample of Nursing Homes During an Influenza Epidemic," *American Journal of Public Health* 85, no. 3 (March 1995), pp. 399–401. Copyright 1995 by the American Public Health Association.

9–1   Introduction

The basic concepts of hypothesis testing were explained in Chapter 8. With the *z, t,* and χ^2 tests, a sample mean, variance, or proportion can be compared to a specific population mean, variance, or proportion to determine whether the null hypothesis should be rejected.

There are, however, many instances when researchers wish to compare two sample means, using experimental and control groups. For example, the average lifetimes of two different brands of bus tires might be compared to see whether there is any difference in tread wear. Two different brands of fertilizer might be tested to see whether one is better than the other for growing plants. Or two brands of cough syrup might be tested to see whether one brand is more effective than the other.

In the comparison of two means, the same basic steps for hypothesis testing shown in Chapter 8 are used, and the z and t tests are also used. When comparing two means by using the t test, the researcher must decide if the two samples are *independent* or *dependent*. The concepts of independent and dependent samples will be explained in Sections 9–4 and 9–5.

Furthermore, when the samples are independent, there are two different formulas that can be used depending on whether or not the variances are equal. To determine if the variances are equal, use the F test shown in Section 9–3. Finally, the z test can be used to compare two proportions, as shown in Section 9–6.

9–2

Testing the Difference Between Two Means: Large Samples

Objective 1

Test the difference between two large sample means, using the z test.

Suppose a researcher wishes to determine whether there is a difference in the average age of nursing students who enroll in a nursing program at a community college and those who enroll in a nursing program at a university. In this case, the researcher is not interested in the average age of all beginning nursing students; instead, he is interested in *comparing* the means of the two groups. His research question is: Does the mean age of nursing students who enroll at a community college differ from the mean age of nursing students who enroll at a university? Here, the hypotheses are

$$H_0: \mu_1 = \mu_2$$
$$H_1: \mu_1 \neq \mu_2$$

where

$\mu_1 =$ mean age of all beginning nursing students at the community college
$\mu_2 =$ mean age of all beginning nursing students at the university

Another way of stating the hypotheses for this situation is

$$H_0: \mu_1 - \mu_2 = 0$$
$$H_1: \mu_1 - \mu_2 \neq 0$$

If there is no difference in population means, subtracting them will give a difference of zero. If they are different, subtracting will give a number other than zero. Both methods of stating hypotheses are correct; however, the first method will be used in this book.

Assumptions for the Test to Determine the Difference Between Two Means

1. The samples must be independent of each other. That is, there can be no relationship between the subjects in each sample.
2. The populations from which the samples were obtained must be normally distributed, and the standard deviations of the variable must be known, or the sample sizes must be greater than or equal to 30.

The theory behind testing the difference between two means is based on selecting pairs of samples and comparing the means of the pairs. The population means need not be known.

All possible pairs of samples are taken from populations. The means for each pair of samples are computed and then subtracted, and the differences are plotted. If both populations have the same mean, then most of the differences will be zero or close to zero.

Figure 9–1

Differences of Means of Pairs of Samples

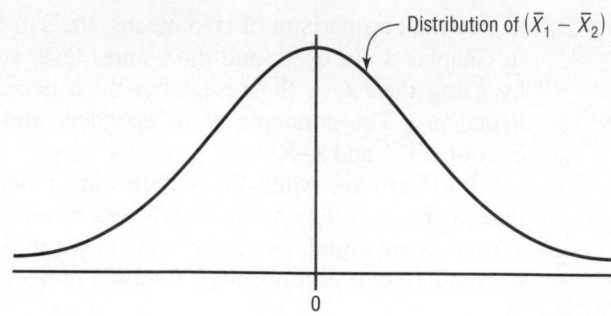

Distribution of $(\overline{X}_1 - \overline{X}_2)$

$Unusual Stats$

Adult children who live with their parents spend more than 2 hours a day doing household chores. According to a study, daughters contribute about 17 hours a week and sons about 14.4 hours.

Occasionally, there will be a few large differences due to chance alone, some positive and others negative. If the differences are plotted, the curve will be shaped like a normal distribution and have a mean of zero, as shown in Figure 9–1.

The variance of the difference $\overline{X}_1 - \overline{X}_2$ is equal to the sum of the individual variances of $\overline{X}_1$ and $\overline{X}_2$. That is,

$$\sigma^2_{\overline{X}_1 - \overline{X}_2} = \sigma^2_{\overline{X}_1} + \sigma^2_{\overline{X}_2}$$

where

$$\sigma^2_{\overline{X}_1} = \frac{\sigma^2_1}{n_1} \qquad \text{and} \qquad \sigma^2_{\overline{X}_2} = \frac{\sigma^2_2}{n_2}$$

So the standard deviation of $\overline{X}_1 - \overline{X}_2$ is

$$\sqrt{\frac{\sigma^2_1}{n_1} + \frac{\sigma^2_2}{n_2}}$$

Formula for the z Test for Comparing Two Means from Independent Populations

$$z = \frac{(\overline{X}_1 - \overline{X}_2) - (\mu_1 - \mu_2)}{\sqrt{\dfrac{\sigma^2_1}{n_1} + \dfrac{\sigma^2_2}{n_2}}}$$

This formula is based on the general format of

$$\text{Test value} = \frac{(\text{observed value}) - (\text{expected value})}{\text{standard error}}$$

where $\overline{X}_1 - \overline{X}_2$ is the observed difference, and the expected difference $\mu_1 - \mu_2$ is zero when the null hypothesis is $\mu_1 = \mu_2$, since that is equivalent to $\mu_1 - \mu_2 = 0$. Finally, the standard error of the difference is

$$\sqrt{\frac{\sigma^2_1}{n_1} + \frac{\sigma^2_2}{n_2}}$$

In the comparison of two sample means, the difference may be due to chance, in which case the null hypothesis will not be rejected, and the researcher can assume that the means of the populations are basically the same. The difference in this case is not significant. See Figure 9–2(a). On the other hand, if the difference is significant, the null hypothesis is rejected and the researcher can conclude that the population means are different. See Figure 9–2(b).

Figure 9–2

Hypothesis-Testing Situations in the Comparison of Means

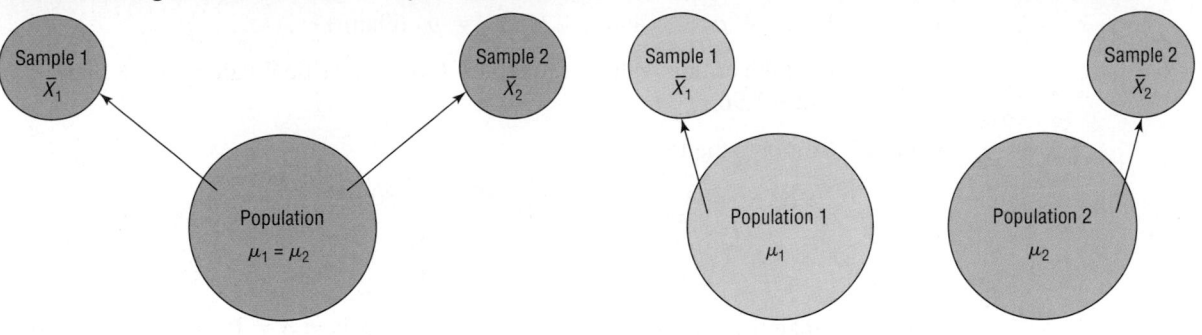

(a) Difference is not significant

Do not reject H_0: $\mu_1 = \mu_2$ since $\bar{X}_1 - \bar{X}_2$ is not significant.

(b) Difference is significant

Reject H_0: $\mu_1 = \mu_2$ since $\bar{X}_1 - \bar{X}_2$ is significant.

These tests can also be one-tailed, using the following hypotheses:

Right-tailed		Left-tailed	
H_0: $\mu_1 \le \mu_2$ H_1: $\mu_1 > \mu_2$	or H_0: $\mu_1 - \mu_2 \le 0$ H_1: $\mu_1 - \mu_2 > 0$	H_0: $\mu_1 \ge \mu_2$ H_1: $\mu_1 < \mu_2$	or H_0: $\mu_1 - \mu_2 \ge 0$ H_1: $\mu_1 - \mu_2 < 0$

The same critical values used in Section 8–3 are used here. They can be obtained from Table E in Appendix C.

If σ_1^2 and σ_2^2 are not known, the researcher can use the variances obtained from each sample s_1^2 and s_2^2, but both sample sizes must be 30 or more. The formula then is

$$z = \frac{(\bar{X}_1 - \bar{X}_2) - (\mu_1 - \mu_2)}{\sqrt{\dfrac{s_1^2}{n_1} + \dfrac{s_2^2}{n_2}}}$$

provided that $n_1 \ge 30$ and $n_2 \ge 30$.

When one or both sample sizes are less than 30 and σ_1 and σ_2 are unknown, the *t* test must be used, as shown in Section 9–4.

The basic format for hypothesis testing using the traditional method is reviewed here.

Step 1   State the hypotheses and identify the claim.

Step 2   Find the critical value(s).

Step 3   Compute the test value.

Step 4   Make the decision.

Step 5   Summarize the results.

Example 9–1

A survey found that the average hotel room rate in New Orleans is $88.42 and the average room rate in Phoenix is $80.61. Assume that the data were obtained from two samples of 50 hotels each and that the standard deviations were $5.62 and $4.83, respectively. At $\alpha = 0.05$, can it be concluded that there is a significant difference in the rates?

Source: USA TODAY.

Solution

Step 1 State the hypotheses and identify the claim.

$$H_0: \mu_1 = \mu_2 \quad \text{and} \quad H_1: \mu_1 \neq \mu_2 \text{ (claim)}$$

Step 2 Find the critical values. Since $\alpha = 0.05$, the critical values are $+1.96$ and -1.96.

Step 3 Compute the test value.

$$z = \frac{(\overline{X}_1 - \overline{X}_2) - (\mu_1 - \mu_2)}{\sqrt{\dfrac{s_1^2}{n_1} + \dfrac{s_2^2}{n_2}}} = \frac{(88.42 - 80.61) - 0}{\sqrt{\dfrac{5.62^2}{50} + \dfrac{4.83^3}{50}}} = 7.45$$

Step 4 Make the decision. Reject the null hypothesis at $\alpha = 0.05$, since $7.45 > 1.96$. See Figure 9–3.

Figure 9–3

Critical and Test Values
for Example 9–1

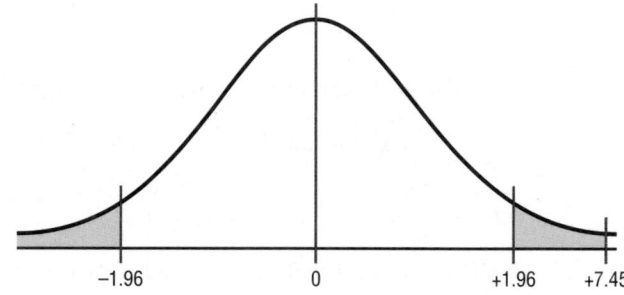

$$-1.96 \qquad 0 \qquad +1.96 \quad +7.45$$

Step 5 Summarize the results. There is enough evidence to support the claim that the means are not equal. Hence, there is a significant difference in the rates.

The *P*-values for this test can be determined by using the same procedure shown in Section 8–3. For example, if the test value for a two-tailed test is 1.40, then the *P*-value obtained from Table E is 0.1616. This value is obtained by looking up the area for $z = 1.40$, which is 0.4192. Then 0.4192 is subtracted from 0.5000 to get 0.0808. Finally, this value is doubled to get 0.1616 since the test is two-tailed. If $\alpha = 0.05$, the decision would be to not reject the null hypothesis, since *P*-value $> \alpha$.

The *P*-value method for hypothesis testing for this chapter also follows the same format as stated in Chapter 8. The steps are reviewed here.

Step 1 State the hypotheses and identify the claim.

Step 2 Compute the test value.

Step 3 Find the *P*-value.

Step 4 Make the decision.

Step 5 Summarize the results.

Example 9–2 illustrates these steps.

Example 9–2

A researcher hypothesizes that the average number of sports that colleges offer for males is greater than the average number of sports that colleges offer for females. A sample of the number of sports offered by colleges is shown. At $\alpha = 0.10$, is there enough evidence to support the claim?

Males					Females				
6	11	11	8	15	6	8	11	13	8
6	14	8	12	18	7	5	13	14	6
6	9	5	6	9	6	5	5	7	6
6	9	18	7	6	10	7	6	5	5
15	6	11	5	5	16	10	7	8	5
9	9	5	5	8	7	5	5	6	5
8	9	6	11	6	9	18	13	7	10
9	5	11	5	8	7	8	5	7	6
7	7	5	10	7	11	4	6	8	7
10	7	10	8	11	14	12	5	8	5

Source: USA TODAY.

Solution

Step 1 State the hypotheses and identify the claim.

$$H_0: \mu_1 \leq \mu_2 \quad \text{and} \quad H_1: \mu_1 > \mu_2 \text{ (claim)}$$

Step 2 Compute the test value. Using a calculator or the formulas in Chapter 3, find the mean and standard deviation for each data set.

For the males $\overline{X}_1 = 8.6$ and $s_1 = 3.3$

For the females $\overline{X}_2 = 7.9$ and $s_2 = 3.3$

Substitute in the formula.

$$z = \frac{(\overline{X}_1 - \overline{X}_2) - (\mu_1 - \mu_2)}{\sqrt{\dfrac{s_1^2}{n_1} + \dfrac{s_2^2}{n_2}}} = \frac{(8.6 - 7.9) - 0}{\sqrt{\dfrac{3.3^2}{50} + \dfrac{3.3^2}{50}}} = 1.06*$$

Step 3 Find the *P*-value. For $z = 1.06$, the area is 0.3554, and $0.5000 - 0.3554 = 0.1446$ or a *P*-value of 0.1446.

Step 4 Make the decision. Since the *P*-value is larger than α (that is, $0.1446 > 0.10$), the decision is to not reject the null hypothesis. See Figure 9–4.

Step 5 Summarize the results. There is not enough evidence to support the claim that colleges offer more sports for males than they do for females.

Figure 9–4

P-Value and α Value for Example 9–2

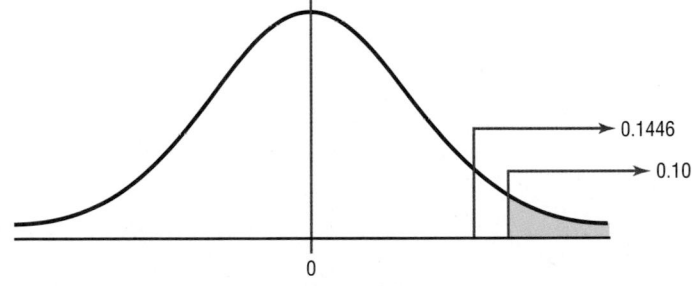

Note: Calculator results may differ due to rounding.

Sometimes, the researcher is interested in testing a specific difference in means other than zero. For example, he or she might hypothesize that the nursing students at a community college are, on the average, 3.2 years older than those at a university. In this case, the hypotheses are

$$H_0: \mu_1 - \mu_2 \le 3.2 \qquad \text{and} \qquad H_1: \mu_1 - \mu_2 > 3.2$$

The formula for the z test is still

$$z = \frac{(\bar{X}_1 - \bar{X}_2) - (\mu_1 - \mu_2)}{\sqrt{\dfrac{\sigma_1^2}{n_1} + \dfrac{\sigma_2^2}{n_2}}}$$

where $\mu_1 - \mu_2$ is the hypothesized difference or expected value. In this case, $\mu_1 - \mu_2 = 3.2$.

Confidence intervals for the difference between two means can also be found. When one is hypothesizing a difference of zero, if the confidence interval contains zero, the null hypothesis is not rejected. If the confidence interval does not contain zero, the null hypothesis is rejected.

Confidence intervals for the difference between two means can be found by using this formula:

Formula for Confidence Interval for Difference Between Two Means: Large Samples

$$(\bar{X}_1 - \bar{X}_2) - z_{\alpha/2}\sqrt{\frac{\sigma_1^2}{n_1} + \frac{\sigma_2^2}{n_2}} < \mu_1 - \mu_2 < (\bar{X}_1 - \bar{X}_2) + z_{\alpha/2}\sqrt{\frac{\sigma_1^2}{n_1} + \frac{\sigma_2^2}{n_2}}$$

When $n_1 \ge 30$ and $n_2 \ge 30$, s_1^2 and s_2^2 can be used in place of σ_1^2 and σ_2^2.

Example 9–3

Find the 95% confidence interval for the difference between the means for the data in Example 9–1.

Solution

Substitute in the formula, using $z_{\alpha/2} = 1.96$.

$$(\bar{X}_1 - \bar{X}_2) - z_{\alpha/2}\sqrt{\frac{\sigma_1^2}{n_1} + \frac{\sigma_2^2}{n_2}} < \mu_1 - \mu_2$$

$$< (\bar{X}_1 - \bar{X}_2) + z_{\alpha/2}\sqrt{\frac{\sigma_1^2}{n_1} + \frac{\sigma_2^2}{n_2}}$$

$$(88.42 - 80.61) - 1.96\sqrt{\frac{5.62^2}{50} + \frac{4.83^2}{50}} < \mu_1 - \mu_2$$

$$< (88.42 - 80.61) + 1.96\sqrt{\frac{5.62^2}{50} + \frac{4.83^2}{50}}$$

$$7.81 - 2.05 < \mu_1 - \mu_2 < 7.81 + 2.05$$

$$5.76 < \mu_1 - \mu_2 < 9.86$$

Since the confidence interval does not contain zero, the decision is to reject the null hypothesis, which agrees with the previous result.

Applying the Concepts **9–2**

Home Runs

For a sports radio talk show, you are asked to research the question whether more home runs are hit by players in the National League or by players in the American League. You decide to use the home run leaders from each league for the last 40 years as your data. The numbers are shown.

National League

47	49	73	50	65	70	49	47	40	43
46	35	38	40	47	39	49	37	37	36
40	37	31	48	48	45	52	38	38	36
44	40	48	45	45	36	39	44	52	47

American League

47	57	52	47	48	56	56	52	50	40
46	43	44	51	36	42	49	49	40	43
39	39	22	41	45	46	39	32	36	32
32	32	37	33	44	49	44	44	49	32

Using the data given, answer the following questions.

1. Define a population.
2. What kind of sample was used?
3. Do you feel that it is representative?
4. What are your hypotheses?
5. What significance level will you use?
6. What statistical test will you use?
7. What are the test results?
8. What is your decision?
9. What can you conclude?
10. Do you feel that using the data given really answers the original question asked?
11. What other data might be used to answer the question?

See page 525 for the answers.

Exercises 9–2

1. Explain the difference between testing a single mean and testing the difference between two means.

2. When a researcher selects all possible pairs of samples from a population in order to find the difference between the means of each pair, what will be the shape of the distribution of the differences when the original distributions are normally distributed? What will be the mean of the distribution? What will be the standard deviation of the distribution?

3. What two assumptions must be met when one is using the z test to test differences between two means? When can the sample standard deviations s_1 and s_2 be used in place of the population standard deviations σ_1 and σ_2?

4. Show two different ways to state that the means of two populations are equal.

For Exercises 5 through 17, perform each of the following steps.

   a. State the hypotheses and identify the claim.
   b. Find the critical value(s).
   c. Compute the test value.
   d. Make the decision.
   e. Summarize the results.

Use the traditional method of hypothesis testing unless otherwise specified.

 5. A researcher wishes to see if the average length of the major rivers in the United States is the same as the

average length of the major rivers in Europe. The data (in miles) of a sample of rivers are shown. At $\alpha = 0.01$, is there enough evidence to reject the claim?

United States			Europe		
729	560	434	481	724	820
329	332	360	532	357	505
450	2315	865	1776	1122	496
330	410	1036	1224	634	230
329	800	447	1420	326	626
600	1310	652	877	580	210
1243	605	360	447	567	252
525	926	722	824	932	600
850	310	430	634	1124	1575
532	375	1979	565	405	2290
710	545	259	675	454	
300	470	425			

Source: *The World Almanac and Book of Facts.*

6. A study was conducted to see if there was a difference between spouses and significant others in coping skills when living with or caring for a person with multiple sclerosis. These skills were measured by questionnaire responses. The results of the two groups are given on one factor, ambivalence. At $\alpha = 0.10$, is there a difference in the means of the two groups?

Spouses	Significant others
$\overline{X}_1 = 2.0$	$\overline{X}_2 = 1.7$
$s_1 = 0.6$	$s_2 = 0.7$
$n_1 = 120$	$n_2 = 34$

Source: Elsie E. Gulick, "Coping Among Spouses or Significant Others of Persons with Multiple Sclerosis," *Nursing Research.*

7. A medical researcher wishes to see whether the pulse rates of smokers are higher than the pulse rates of non-smokers. Samples of 100 smokers and 100 nonsmokers are selected. The results are shown here. Can the researcher conclude, at $\alpha = 0.05$, that smokers have higher pulse rates than nonsmokers?

Smokers	Nonsmokers
$\overline{X}_1 = 90$	$\overline{X}_2 = 88$
$s_1 = 5$	$s_2 = 6$
$n_1 = 100$	$n_2 = 100$

8. At age 9 the average weight (21.3 kg) and the average height (124.5 cm) for both boys and girls are exactly the same. A random sample of 9-year-olds yielded these results. Estimate the mean difference in height between boys and girls with 95% confidence. Does your interval support the given claim?

	Boys	Girls
Sample size	60	50
Mean height, cm	123.5	126.2
Sample variance	98	120

Source: www.healthepic.com.

9. Using data from the "Noise Levels in an Urban Hospital" study cited in Exercise 19 in Exercise set 7–2, test the claim that the noise level in the corridors is higher than that in the clinics. Use $\alpha = 0.02$. The data are shown here.

Corridors	Clinics
$\overline{X}_1 = 61.2$ dBA	$\overline{X}_2 = 59.4$ dBA
$s_1 = 7.9$	$s_2 = 7.9$
$n_1 = 84$	$n_2 = 34$

Source: M. Bayo, A. Garcia, and A. Garcia, "Noise Levels in an Urban Hospital and Workers' Subjective Responses," *Archives of Environmental Health* 50, no. 3.

10. A real estate agent compares the selling prices of homes in two municipalities in southwestern Pennsylvania to see if there is a difference in price. The results of the study are shown. Is there enough evidence to reject the claim that the average cost of a home in both locations is the same? Use $\alpha = 0.01$.

Scott	Ligonier
$\overline{X}_1 = \$93,430*$	$\overline{X}_2 = \$98,043*$
$s_1 = \$5602$	$s_2 = \$4731$
$n_1 = 35$	$n_2 = 40$

*Based on information from RealSTATs.

11. In a study of women science majors, the following data were obtained on two groups, those who left their profession within a few months after graduation (leavers) and those who remained in their profession after they graduated (stayers). Test the claim that those who stayed had a higher science grade point average than those who left. Use $\alpha = 0.05$.

Leavers	Stayers
$\overline{X}_1 = 3.16$	$\overline{X}_2 = 3.28$
$s_1 = 0.52$	$s_2 = 0.46$
$n_1 = 103$	$n_2 = 225$

Source: Paula Rayman and Belle Brett, "Women Science Majors: What Makes a Difference in Persistence after Graduation?" *The Journal of Higher Education.*

12. A survey of 1000 students nationwide showed a mean ACT score of 21.4. A survey of 500 Ohio scores showed a mean of 20.8. If the standard deviation in each case is 3, can we conclude that Ohio is below the national average? Use $\alpha = 0.05$.

Source: Report of WFIN radio.

13. A school administrator hypothesizes that colleges spend more for male sports than they do for female sports. A sample of two different colleges is selected, and the annual expenses (in dollars) per student at each school are shown. At $\alpha = 0.01$, is there enough evidence to support the claim?

Males

7,040	6,576	1,664	12,919	8,605
22,220	3,377	10,128	7,723	2,063
8,033	9,463	7,656	11,456	12,244
6,670	12,371	9,626	5,472	16,175
8,383	623	6,797	10,160	8,725
14,029	13,763	8,811	11,480	9,544
15,048	5,544	10,652	11,267	10,126
8,796	13,351	7,120	9,505	9,571
7,551	5,811	9,119	9,732	5,286
5,254	7,550	11,015	12,403	12,703

Females

10,333	6,407	10,082	5,933	3,991
7,435	8,324	6,989	16,249	5,922
7,654	8,411	11,324	10,248	6,030
9,331	6,869	6,502	11,041	11,597
5,468	7,874	9,277	10,127	13,371
7,055	6,909	8,903	6,925	7,058
12,745	12,016	9,883	14,698	9,907
8,917	9,110	5,232	6,959	5,832
7,054	7,235	11,248	8,478	6,502
7,300	993	6,815	9,959	10,353

Source: *USA TODAY.*

14. Is there a difference in average miles traveled for each of two taxi companies during a randomly selected week? The data are shown. Use $\alpha = 0.05$. Assume that the populations are normally distributed. Use the *P*-value method.

Moonview Cab Company	Starlight Taxi Company
$\bar{X}_1 = 837$	$\bar{X}_2 = 753$
$\sigma_1 = 30$	$\sigma_2 = 40$
$n_1 = 35$	$n_2 = 40$

15. In the study cited in Exercise 11, the researchers collected the data shown here on a self-esteem questionnaire. At $\alpha = 0.05$, can it be concluded that there is a difference in the self-esteem scores of the two groups? Use the *P*-value method.

Leavers	Stayers
$\bar{X}_1 = 3.05$	$\bar{X}_2 = 2.96$
$s_1 = 0.75$	$s_2 = 0.75$
$n_1 = 103$	$n_2 = 225$

Source: Paula Rayman and Belle Brett, "Women Science Majors: What Makes a Difference in Persistence after Graduation?" *The Journal of Higher Education.*

16. The dean of students wants to see whether there is a significant difference in ages of resident students and commuting students. She selects a sample of 50 students from each group. The ages are shown here. At $\alpha = 0.05$, decide if there is enough evidence to reject the claim of no difference in the ages of the two groups. Use the standard deviations from the samples and the *P*-value method.

Resident students

22	25	27	23	26	28	26	24
25	20	26	24	27	26	18	19
18	30	26	18	18	19	32	23
19	19	18	29	19	22	18	22
26	19	19	21	23	18	20	18
22	21	19	21	21	22	18	20
19	23						

Commuter students

18	20	19	18	22	25	24	35
23	18	23	22	28	25	20	24
26	30	22	22	22	21	18	20
19	26	35	19	19	18	19	32
29	23	21	19	36	27	27	20
20	21	18	19	23	20	19	19
20	25						

17. Two groups of students are given a problem-solving test, and the results are compared. Find the 90% confidence interval of the true difference in means.

Mathematics majors	Computer science majors
$\bar{X}_1 = 83.6$	$\bar{X}_2 = 79.2$
$s_1 = 4.3$	$s_2 = 3.8$
$n_1 = 36$	$n_2 = 36$

18. The average credit card debt for a recent year was $9205. Five years earlier the average credit card debt was $6618. Assume sample sizes of 35 were used and the standard deviations of both samples were $1928. Is there enough evidence to believe that the average credit card debt has increased? Use $\alpha = 0.05$. Give a possible reason as to why or why not the debt was increased.

Source: CardWeb.com.

19. Two brands of cigarettes are selected, and their nicotine content is compared. The data are shown here. Find the 99% confidence interval of the true difference in the means.

Brand A	Brand B
$\bar{X}_1 = 28.6$ milligrams	$\bar{X}_2 = 32.9$ milligrams
$\sigma_1 = 5.1$ milligrams	$\sigma_2 = 4.4$ milligrams
$n_1 = 30$	$n_2 = 40$

20. Two brands of batteries are tested, and their voltage is compared. The data follow. Find the 95% confidence interval of the true difference in the means. Assume that both variables are normally distributed.

Brand X	Brand Y
$\bar{X}_1 = 9.2$ volts	$\bar{X}_2 = 8.8$ volts
$\sigma_1 = 0.3$ volt	$\sigma_2 = 0.1$ volt
$n_1 = 27$	$n_2 = 30$

Extending the Concepts

21. A researcher claims that students in a private school have exam scores that are at most 8 points higher than those of students in public schools. Random samples of 60 students from each type of school are selected and given an exam. The results are shown. At $\alpha = 0.05$, test the claim.

Private school	Public school
$\bar{X}_1 = 110$	$\bar{X}_2 = 104$
$s_1 = 15$	$s_2 = 15$
$n_1 = 60$	$n_2 = 60$

Technology Step by Step

MINITAB
Step by Step

Test the Difference Between Two Means: Large Independent Samples*

MINITAB will calculate the test statistic and *P*-value for differences between the means for two populations when the population standard deviations are unknown.

For Example 9–2, is the average number of sports for men higher than the average number for women?

1. Enter the data for Example 9–2 into C1 and C2. Name the columns **MaleS** and **FemaleS**.

2. Select **Stat>Basic Statistics>2-Sample t.**

3. Click the button for Samples in different columns.

There is one sample in each column.

4. Click in the box for First:. Double-click C1 MaleS in the list.

5. Click in the box for Second:, then double-click C2 FemaleS in the list. Do not check the box for Assume equal variances. MINITAB will use the large sample formula. The completed dialog box is shown.

6. Click [Options].

    a) Type in **90** for the Confidence level and **0** for the Test mean.

    b) Select greater than for the Alternative. This option affects the *P*-value. It must be correct.

7. Click [OK] twice. Since the *P*-value is greater than the significance level, $0.172 > 0.1$, do not reject the null hypothesis.

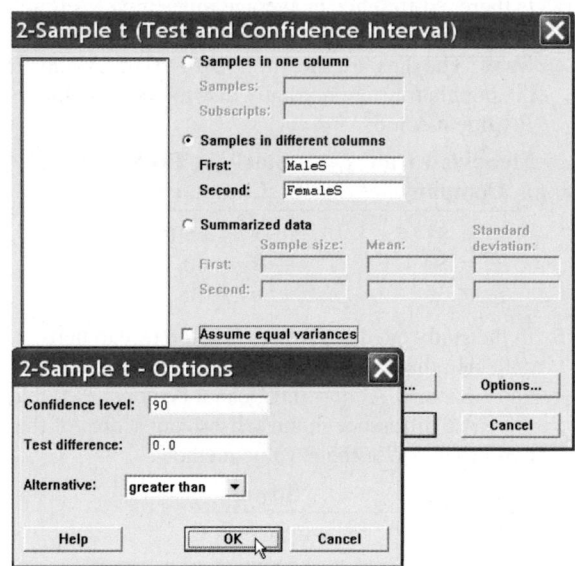

Two-Sample t-Test and CI: MaleS, FemaleS
```
Two-sample t for MaleS vs FemaleS
          N    Mean   StDev   SE Mean
MaleS    50    8.56   3.26    0.46
FemaleS  50    7.94   3.27    0.46
Difference = mu (MaleS) - mu (FemaleS)
Estimate for difference: 0.620000
90% lower bound for difference: -0.221962
t-Test of difference = 0 (vs >): t-Value = 0.95 P-Value = 0.172 DF = 97
```

*MINITAB does not calculate a *z* test statistic. This statistic can be used instead.

TI-83 Plus or TI-84 Plus	**Hypothesis Test for the Difference Between Two Means and *z* Distribution (Data)**

Step by Step

Hypothesis Test for the Difference Between Two Means and *z* Distribution (Data)

1. Enter the data values into L_1 and L_2.
2. Press **STAT** and move the cursor to TESTS.
3. Press **3** for 2-SampZTest.
4. Move the cursor to Data and press **ENTER**.
5. Type in the appropriate values.
6. Move the cursor to the appropriate alternative hypothesis and press **ENTER**.
7. Move the cursor to Calculate and press **ENTER**.

Hypothesis Test for the Difference Between Two Means and *z* Distribution (Statistics)

1. Press **STAT** and move the cursor to TESTS.
2. Press **3** for 2-SampZTest.
3. Move the cursor to Stats and press **ENTER**.
4. Type in the appropriate values.
5. Move the cursor to the appropriate alternative hypothesis and press **ENTER**.
6. Move the cursor to Calculate and press **ENTER**.

Confidence Interval for the Difference Between Two Means and *z* Distribution (Data)

1. Enter the data values into L_1 and L_2.
2. Press **STAT** and move the cursor to TESTS.
3. Press **9** for 2-SampZInt.
4. Move the cursor to Data and press **ENTER**.
5. Type in the appropriate values.
6. Move the cursor to Calculate and press **ENTER**.

Confidence Interval for the Difference Between Two Means and *z* Distribution (Statistics)

1. Press **STAT** and move the cursor to TESTS.
2. Press **9** for 2-SampZInt.
3. Move the cursor to Stats and press **ENTER**.
4. Type in the appropriate values.
5. Move the cursor to Calculate and press **ENTER**.

Excel

Step by Step

z Test for the Difference Between Two Means

Excel has a two-sample *z* test in its Data Analysis tools. To perform a *z* test for the difference between the means of two populations, given two independent samples, do this:

1. Enter the first sample data set in column A.
2. Enter the second sample data set in column B.
3. If the population variances are not known but $n \geq 30$ for both samples, use the formulas =VAR(A1:A*n*) and =VAR(B1:B*n*), where A*n* and B*n* are the last cells with data in each column, to find the variances of the sample data sets.

4. Select **Tools>Data Analysis** and choose z-Test: Two Sample for Means.

5. Enter the ranges for the data in columns A and B and enter **0** for Hypothesized Mean Difference.

6. If the population variances are known, enter them for Variable 1 and Variable 2. Otherwise, use the sample variances obtained in step 3.

7. Specify the confidence level Alpha.

8. Specify a location for output, and click [OK].

Example XL9–1

Test the claim that the two population means are equal, using the sample data provided here, at $\alpha = 0.05$. Assume the population variances are $\sigma_A^2 = 10.067$ and $\sigma_B^2 = 7.067$.

Set A	10	2	15	18	13	15	16	14	18	12	15	15	14	18	16
Set B	5	8	10	9	9	11	12	16	8	8	9	10	11	7	6

The two-sample z test dialog box is shown (before the variances are entered); the results appear in the table that Excel generates. Note that the *P*-value and critical *z* value are provided for both the one-tailed test and the two-tailed test. The *P*-values here are expressed in scientific notation: $7.09045E\text{-}06 = 7.09045 \times 10^{-6} = 0.00000709045$. Because this value is less than 0.05, we reject the null hypothesis and conclude that the population means are not equal.

Two-Sample *z* Test Dialog Box

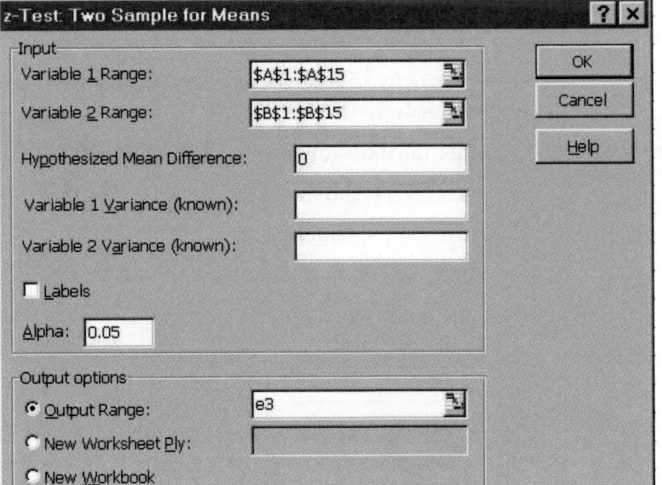

z-Test: Two Sample for Means		
	Variable 1	*Variable 2*
Mean	14.06666667	9.266666667
Known Variance	10.067	7.067
Observations	15	15
Hypothesized Mean Difference	0	
z	4.491149228	
P(Z<=z) one-tail	3.54522E-06	
z Critical one-tail	1.644853	
P(Z<=z) two-tail	7.09045E-06	
z Critical two-tail	1.959961082	

9-3	# Testing the Difference Between Two Variances

In addition to comparing two means, statisticians are interested in comparing two variances or standard deviations. For example, is the variation in the temperatures for a certain month for two cities different?

In another situation, a researcher may be interested in comparing the variance of the cholesterol of men with the variance of the cholesterol of women. For the comparison of two variances or standard deviations, an **F test** is used. The F test should not be confused with the chi-square test, which compares a single sample variance to a specific population variance, as shown in Chapter 8.

Objective 2

Test the difference between two variances or standard deviations.

If two independent samples are selected from two normally distributed populations in which the variances are equal ($\sigma_1^2 = \sigma_2^2$) and if the variances s_1^2 and s_2^2 are compared as $\dfrac{s_1^2}{s_2^2}$, the sampling distribution of the variances is called the **F distribution.**

Characteristics of the *F* Distribution

1. The values of F cannot be negative, because variances are always positive or zero.
2. The distribution is positively skewed.
3. The mean value of F is approximately equal to 1.
4. The F distribution is a family of curves based on the degrees of freedom of the variance of the numerator and the degrees of freedom of the variance of the denominator.

Figure 9-5 shows the shapes of several curves for the F distribution.

Figure 9-5

The *F* Family of Curves

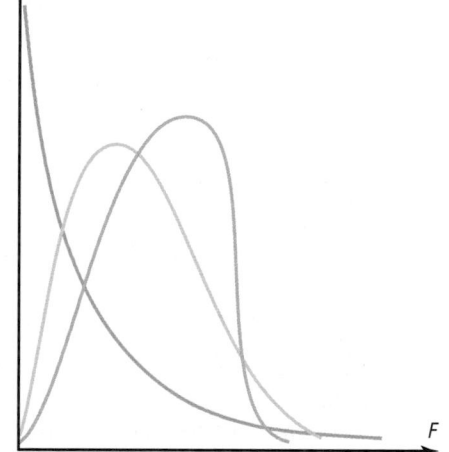

Formula for the *F* Test

$$F = \frac{s_1^2}{s_2^2}$$

where the larger of the two variances is placed in the numerator regardless of the subscripts. (See note on page 482.)

The F test has two terms for the degrees of freedom: that of the numerator, $n_1 - 1$, and that of the denominator, $n_2 - 1$, where n_1 is the sample size from which the larger variance was obtained.

When one is finding the F test value, *the larger of the variances is placed in the numerator of the F formula;* this is not necessarily the variance of the larger of the two sample sizes.

Table H in Appendix C gives the F critical values for $\alpha = 0.005, 0.01, 0.025, 0.05$, and 0.10 (each α value involves a separate table in Table H). These are one-tailed values; if a two-tailed test is being conducted, then the $\alpha/2$ value must be used. For example, if a two-tailed test with $\alpha = 0.05$ is being conducted, then the $0.05/2 = 0.025$ table of Table H should be used.

Example 9–4

Find the critical value for a right-tailed F test when $\alpha = 0.05$, the degrees of freedom for the numerator (abbreviated d.f.N.) are 15, and the degrees of freedom for the denominator (d.f.D.) are 21.

Solution

Since this test is right-tailed with $\alpha = 0.05$, use the 0.05 table. The d.f.N. is listed across the top, and the d.f.D. is listed in the left column. The critical value is found where the row and column intersect in the table. In this case, it is 2.18. See Figure 9–6.

Figure 9–6

Finding the Critical Value in Table H for Example 9–4

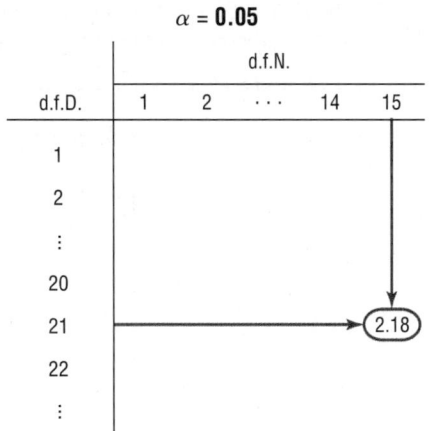

As noted previously, when the F test is used, the larger variance is always placed in the numerator of the formula. When one is conducting a two-tailed test, α is split; and even though there are two values, only the right tail is used. The reason is that the F test value is always greater than or equal to 1.

Example 9–5

Find the critical value for a two-tailed F test with $\alpha = 0.05$ when the sample size from which the variance for the numerator was obtained was 21 and the sample size from which the variance for the denominator was obtained was 12.

Solution

Since this is a two-tailed test with $\alpha = 0.05$, the $0.05/2 = 0.025$ table must be used. Here, d.f.N. $= 21 - 1 = 20$, and d.f.D. $= 12 - 1 = 11$; hence, the critical value is 3.23. See Figure 9–7.

Figure 9–7

Finding the Critical Value in Table H for Example 9–5

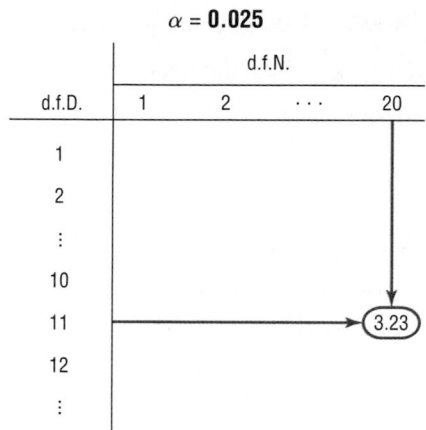

When the degree of freedom values cannot be found in the table, the closest value on the smaller side should be used. For example, if d.f.N. = 14, this value is between the given table values of 12 and 15; therefore, 12 should be used, to be on the safe side.

When one is testing the equality of two variances, these hypotheses are used:

Right-tailed	**Left-tailed**	**Two-tailed**
$H_0: \sigma_1^2 \leq \sigma_2^2$	$H_0: \sigma_1^2 \geq \sigma_2^2$	$H_0: \sigma_1^2 = \sigma_2^2$
$H_1: \sigma_1^2 > \sigma_2^2$	$H_1: \sigma_1^2 < \sigma_2^2$	$H_1: \sigma_1^2 \neq \sigma_2^2$

There are four key points to keep in mind when one is using the *F* test.

Notes for the Use of the *F* Test

1. The larger variance should always be placed in the numerator of the formula regardless of the subscripts. (See note on page 482.)

$$F = \frac{s_1^2}{s_2^2}$$

2. For a two-tailed test, the α value must be divided by 2 and the critical value placed on the right side of the *F* curve.

3. If the standard deviations instead of the variances are given in the problem, they must be squared for the formula for the *F* test.

4. When the degrees of freedom cannot be found in Table H, the closest value on the smaller side should be used.

Assumptions for Testing the Difference Between Two Variances

1. The populations from which the samples were obtained must be normally distributed. (*Note:* The test should not be used when the distributions depart from normality.)

2. The samples must be independent of each other.

Remember also that in tests of hypotheses using the traditional method, these five steps should be taken:

Step 1 State the hypotheses and identify the claim.

Step 2 Find the critical value.

Unusual Stat

Of all U.S. births, 2% are twins.

Step 3   Compute the test value.

Step 4   Make the decision.

Step 5   Summarize the results.

| Example 9–6 | A medical researcher wishes to see whether the variance of the heart rates (in beats per minute) of smokers is different from the variance of heart rates of people who do not smoke. Two samples are selected, and the data are as shown. Using $\alpha = 0.05$, is there enough evidence to support the claim? |

Smokers	Nonsmokers
$n_1 = 26$	$n_2 = 18$
$s_1^2 = 36$	$s_2^2 = 10$

Solution

Step 1   State the hypotheses and identify the claim.

$$H_0: \sigma_1^2 = \sigma_2^2 \qquad \text{and} \qquad H_1: \sigma_1^2 \neq \sigma_2^2 \text{ (claim)}$$

Step 2   Find the critical value. Use the 0.025 table in Table H since $\alpha = 0.05$ and this is a two-tailed test. Here, d.f.N. $= 26 - 1 = 25$, and d.f.D. $= 18 - 1 = 17$. The critical value is 2.56 (d.f.N. $= 24$ was used). See Figure 9–8.

Figure 9–8

Critical Value for
Example 9–6

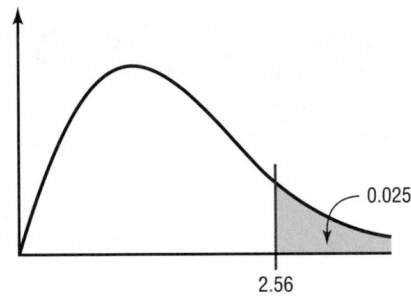

Step 3   Compute the test value.

$$F = \frac{s_1^2}{s_2^2} = \frac{36}{10} = 3.6$$

Step 4   Make the decision. Reject the null hypothesis, since $3.6 > 2.56$.

Step 5   Summarize the results. There is enough evidence to support the claim that the variance of the heart rates of smokers and nonsmokers is different.

| Example 9–7 | The standard deviation of the average waiting time to see a doctor for non-life-threatening problems in the emergency room at an urban hospital is 32 minutes. At a second hospital, the standard deviation is 28 minutes. If a sample of 16 patients was used in the first case and 18 in the second case, is there enough evidence to conclude that the standard deviation of the waiting times in the first hospital is greater than the standard deviation of the waiting times in the second hospital? |

Solution

Step 1   State the hypotheses and identify the claim.

$$H_0: \sigma_1^2 \leq \sigma_2^2 \qquad \text{and} \qquad H_1: \sigma_1^2 > \sigma_2^2 \text{ (claim)}$$

Step 2   Find the critical value. Here, d.f.N. $= 16 - 1 = 15$, and d.f.D. $= 18 - 1 = 17$. From the 0.01 table, the critical value is 3.31.

Step 3   Compute the test value.

$$F = \frac{s_1^2}{s_2^2} = \frac{32^2}{28^2} = 1.31$$

Step 4   Do not reject the null hypothesis since $1.31 < 3.31$.

Step 5   Summarize the results. There is not enough evidence to support the claim that the standard deviation of the waiting times of the first hospital is greater than the standard deviation of the waiting times of the second hospital.

Finding P-values for the F test statistic is somewhat more complicated since it requires looking through all the F tables (Table H in Appendix C) using the specific d.f.N. and d.f.D. values. For example, suppose that a certain test has $F = 3.58$, d.f.N. $= 5$, and d.f.D. $= 10$. To find the P-value interval for $F = 3.58$, one must first find the corresponding F values for d.f.N. $= 5$ and d.f.D. $= 10$ for α equal to 0.005 on page 773, 0.01 on page 774, 0.025 on page 775, 0.05 on page 776, and 0.10 on page 777 in Table H. Then make a table as shown.

α		0.10	0.05	0.025	0.01	0.005
F		2.52	3.33	4.24	5.64	6.87
Reference page		777	776	775	774	773

Now locate the two F values that the test value 3.58 falls between. In this case, 3.58 falls between 3.33 and 4.24, corresponding to 0.05 and 0.025. Hence, the P-value for a right-tailed test for $F = 3.58$ falls between 0.025 and 0.05 (that is, $0.025 < P\text{-value} < 0.05$). For a right-tailed test, then, one would reject the null hypothesis at $\alpha = 0.05$ but not at $\alpha = 0.01$. The P-value obtained from a calculator is 0.0408. Remember that for a two-tailed test the values found in Table H for α must be doubled. In this case, $0.05 < P\text{-value} < 0.10$ for $F = 3.58$.

Once you understand the concept, you can dispense with making a table as shown and find the P-value directly from Table H.

Example 9–8

The CEO of an airport hypothesizes that the variance in the number of passengers for American airports is greater than the variance in the number of passengers for foreign airports. At $\alpha = 0.10$, is there enough evidence to support the hypothesis? The data in millions of passengers per year are shown for selected airports. Use the P-value method. Assume the variable is normally distributed.

American airports		**Foreign airports**	
36.8	73.5	60.7	51.2
72.4	61.2	42.7	38.6
60.5	40.1		

Source: Airports Council International.

Solution

Step 1 State the hypotheses and identify the claim.

$$H_0: \sigma_1^2 \leq \sigma_2^2 \qquad \text{and} \qquad H_1: \sigma_1^2 > \sigma_2^2 \text{ (claim)}$$

Step 2 Compute the test value. Using the formula in Chapter 3 or a calculator, find the variance for each group.

$$s_1^2 = 246.38 \qquad \text{and} \qquad s_2^2 = 95.87$$

Substitute in the formula and solve.

$$F = \frac{s_1^2}{s_2^2} = \frac{246.38}{95.87} = 2.57$$

Step 3 Find the P-value in Table H, using d.f.N. = 5 and d.f.D. = 3.

α	0.10	0.05	0.025	0.01	0.005
F	5.31	9.01	14.88	28.24	45.39

Since 2.57 is less than 5.31, the P-value is greater than 0.10. (The P-value obtained from a calculator is 0.234.)

Step 4 Make the decision. The decision is to not reject the null hypothesis since P-value > 0.10.

Step 5 Summarize the results. There is not enough evidence to support the claim that the variance in the number of passengers for American airports is greater than the variance in the number of passengers for foreign airports.

If the exact degrees of freedom are not specified in Table H, the closest smaller value should be used. For example, if $\alpha = 0.05$ (right-tailed test), d.f.N. = 18, and d.f.D. = 20, use the column d.f.N. = 15 and the row d.f.D. = 20 to get $F = 2.20$.

Note: It is not absolutely necessary to place the larger variance in the numerator when one is performing the F test. Critical values for left-tailed hypotheses tests can be found by interchanging the degrees of freedom and taking the reciprocal of the value found in Table H.

Also, one should use caution when performing the F test since the data can run contrary to the hypotheses on rare occasions. For example, if the hypotheses are $H_0: \sigma_1^2 \leq \sigma_2^2$ and $H_1: \sigma_1^2 > \sigma_2^2$, but if $s_1^2 < s_2^2$, then the F test should not be performed and one would not reject the null hypothesis.

Applying the Concepts **9–3**

Variability and Automatic Transmissions

Assume the following data values are from the June 1996 issue of *Automotive Magazine.* An article compared various parameters of U.S.- and Japanese-made sports cars. This report centers on the price of an optional automatic transmission. Which country has the greater variability in the price of automatic transmissions? Input the data and answer the following questions.

Japanese cars		U.S. cars	
Nissan 300ZX	$1940	Dodge Stealth	$2363
Mazda RX7	1810	Saturn	1230
Mazda MX6	1871	Mercury Cougar	1332
Nissan NX	1822	Ford Probe	932
Mazda Miata	1920	Eagle Talon	1790
Honda Prelude	1730	Chevy Lumina	1833

1. What is the null hypothesis?
2. What test statistic is used to test for any significant differences in the variances?
3. Is there a significant difference in the variability in the prices between the two car companies?
4. What effect does a small sample size have on the standard deviations?
5. What degrees of freedom are used for the statistical test?
6. Could two sets of data have significantly different variances without having significantly different means?

See page 526 for the answers.

Exercises 9–3

1. When one is computing the F test value, what condition is placed on the variance that is in the numerator?

2. Why is the critical region always on the right side in the use of the F test?

3. What are the two different degrees of freedom associated with the F distribution?

4. What are the characteristics of the F distribution?

5. Using Table H, find the critical value for each.
   a. Sample 1: $s_1^2 = 128$, $n_1 = 23$
      Sample 2: $s_2^2 = 162$, $n_2 = 16$
      Two-tailed, $\alpha = 0.01$
   b. Sample 1: $s_1^2 = 37$, $n_1 = 14$
      Sample 2: $s_2^2 = 89$, $n_2 = 25$
      Right-tailed, $\alpha = 0.01$
   c. Sample 1: $s_1^2 = 232$, $n_1 = 30$
      Sample 2: $s_2^2 = 387$, $n_2 = 46$
      Two-tailed, $\alpha = 0.05$
   d. Sample 1: $s_1^2 = 164$, $n_1 = 21$
      Sample 2: $s_2^2 = 53$, $n_2 = 17$
      Two-tailed, $\alpha = 0.10$
   e. Sample 1: $s_1^2 = 92.8$, $n_1 = 11$
      Sample 2: $s_2^2 = 43.6$, $n_2 = 11$
      Right-tailed, $\alpha = 0.05$

6. (ans) Using Table H, find the P-value interval for each F test value.
   a. $F = 2.97$, d.f.N. = 9, d.f.D. = 14, right-tailed
   b. $F = 3.32$, d.f.N. = 6, d.f.D. = 12, two-tailed
   c. $F = 2.28$, d.f.N. = 12, d.f.D. = 20, right-tailed
   d. $F = 3.51$, d.f.N. = 12, d.f.D. = 21, right-tailed
   e. $F = 4.07$, d.f.N. = 6, d.f.D. = 10, two-tailed
   f. $F = 1.65$, d.f.N. = 19, d.f.D. = 28, right-tailed
   g. $F = 1.77$, d.f.N. = 28, d.f.D. = 28, right-tailed
   h. $F = 7.29$, d.f.N. = 5, d.f.D. = 8, two-tailed

For Exercises 7 through 20, perform the following steps. Assume that all variables are normally distributed.
   a. State the hypotheses and identify the claim.
   b. Find the critical value.
   c. Compute the test value.
   d. Make the decision.
   e. Summarize the results.

Use the traditional method of hypothesis testing unless otherwise specified.

7. The standard deviation for the number of weeks 15 *New York Times* hardcover fiction books spent on their bestseller list is 6.17 weeks. The standard deviation for the 15 *New York Times* hardcover nonfiction list is 13.12 weeks. At $\alpha = 0.10$, can we conclude that there is a difference in the variances?
   Source: *The New York Times.*

8. A researcher claims that the standard deviation of the ages of cats is smaller than the standard deviation of the ages of dogs who are owned by families in a large city. A randomly selected sample of 29 cats has a standard deviation of 2.7 years, and a random sample of 16 dogs has a standard deviation of 3.5 years. Is the researcher correct? Use $\alpha = 0.05$. If there is a difference, suggest a reason for the difference.

9. A tax collector wishes to see if the variances of the values of the tax-exempt properties are different for two large cities. The values of the tax-exempt properties for two samples are shown. The data are given in millions of dollars. At $\alpha = 0.05$, is there enough evidence to support the tax collector's claim that the variances are different?

City A				City B			
113	22	14	8	82	11	5	15
25	23	23	30	295	50	12	9
44	11	19	7	12	68	81	2
31	19	5	2	20	16	4	5

10. In the hospital study cited in Exercise 19 in Exercise set 7–2, it was found that the standard deviation of the sound levels from 20 areas designated as "casualty doors" was 4.1 dBA and the standard deviation of 24 areas designated as operating theaters was 7.5 dBA. At $\alpha = 0.05$, can one substantiate the claim that there is a difference in the standard deviations?

Source: M. Bayo, A. Garcia, and A. Garcia, "Noise Levels in an Urban Hospital and Workers' Subjective Responses," *Archives of Environmental Health*.

11. The numbers of calories contained in $\frac{1}{2}$-cup servings of randomly selected flavors of ice cream from two national brands are listed here. At the 0.05 level of significance, is there sufficient evidence to conclude that the variance in the number of calories differs between the two brands?

Brand A		Brand B	
330	300	280	310
310	350	300	370
270	380	250	300
310	300	290	310

Source: *The Doctor's Pocket Calorie, Fat and Carbohydrate Counter.*

12. A researcher wishes to see if the variance in the number of vehicles passing through the tollbooths during a fiscal year on the Pennsylvania Turnpike is different from the variance in the number of vehicles passing through the tollbooths on the expressways in Pennsylvania during the same year. The data are shown. At $\alpha = 0.05$, can it be concluded that the variances are different?

PA Turnpike	PA expressways
3,694,560	2,774,251
719,934	204,369
3,768,285	456,123
1,838,271	1,068,107
3,358,175	3,534,092
6,718,905	235,752
3,469,431	499,043
4,420,553	2,016,046
920,264	253,956
1,005,469	826,710
1,112,722	133,619
2,855,109	3,453,745

Source: *Pittsburgh Post-Gazette.*

13. The standard deviation of the ages of a sample of people who were playing the slot machines is 6.8 years. The standard deviation of the ages of a sample of people who were playing roulette is 3.2 years. If each sample contained 25 people, can it be concluded that the standard deviations of the ages are different? Use $\alpha = 0.05$. If there is a difference, suggest a reason for the difference.

14. The number of grams of carbohydrates contained in 1-ounce servings of randomly selected chocolate and nonchocolate candy is listed here. Is there sufficient

evidence to conclude that the variance in carbohydrate content varies between chocolate and nonchocolate candy? Use $\alpha = 0.10$.

Chocolate:	29	25	17	36	41	25	32	29
	38	34	24	27	29			
Nonchocolate:	41	41	37	29	30	38	39	10
	29	55	29					

Source: *The Doctor's Pocket Calorie, Fat and Carbohydrate Counter.*

15. The yearly tuition costs in dollars for random samples of medical schools that specialize in research and in primary care are listed. At $\alpha = 0.05$, can it be concluded that a difference between the variances of the two groups exists?

Research			Primary care		
30,897	34,280	31,943	26,068	21,044	30,897
34,294	31,275	29,590	34,208	20,877	29,691
20,618	20,500	29,310	33,783	33,065	35,000
21,274			27,297		

Source: *U.S. News & World Report Best Graduate Schools.*

16. A researcher wishes to see if the variance of the areas in square miles for counties in Indiana is less than the variance of the areas for counties in Iowa. A random sample of counties is selected, and the data are shown. At $\alpha = 0.01$, can it be concluded that the variance of the areas for counties in Indiana is less than the variance of the areas for counties in Iowa?

Indiana				Iowa			
406	393	396	485	640	580	431	416
431	430	369	408	443	569	779	381
305	215	489	293	717	568	714	731
373	148	306	509	571	577	503	501
560	384	320	407	568	434	615	402

Source: *The World Almanac and Book of Facts.*

17. Test the claim that the variance of heights of tall buildings in Denver is equal to the variance in heights of tall buildings in Detroit at $\alpha = 0.10$. The data are given in feet.

Denver			Detroit		
714	698	544	620	472	430
504	438	408	562	448	420
404			534	436	

Source: *The World Almanac and Book of Facts.*

18. A researcher claims that the variation in the salaries of elementary school teachers is greater than the variation in the salaries of secondary school teachers. A sample of the salaries of 30 elementary school teachers has a variance of $8324, and a sample of the salaries of 30 secondary school teachers has a variance of $2862. At $\alpha = 0.05$, can the researcher conclude that the variation in the elementary school teachers' salaries is greater than the variation in the secondary teachers' salaries? Use the *P*-value method.

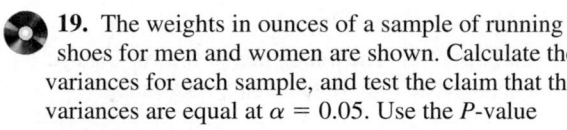 **19.** The weights in ounces of a sample of running shoes for men and women are shown. Calculate the variances for each sample, and test the claim that the variances are equal at $\alpha = 0.05$. Use the *P*-value method.

Men			Women		
11.9	10.4	12.6	10.6	10.2	8.8
12.3	11.1	14.7	9.6	9.5	9.5
9.2	10.8	12.9	10.1	11.2	9.3
11.2	11.7	13.3	9.4	10.3	9.5
13.8	12.8	14.5	9.8	10.3	11.0

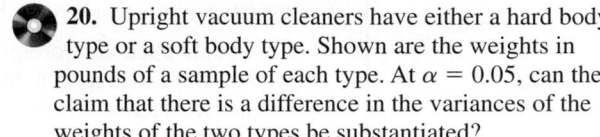 **20.** Upright vacuum cleaners have either a hard body type or a soft body type. Shown are the weights in pounds of a sample of each type. At $\alpha = 0.05$, can the claim that there is a difference in the variances of the weights of the two types be substantiated?

Hard body types				Soft body types			
21	17	17	20	24	13	11	13
16	17	15	20	12	15		
23	16	17	17				
13	15	16	18				
18							

Technology *Step by Step*

MINITAB
Step by Step

Test for the Difference Between Two Variances

For Example 9–8, test the hypothesis that the variance in the number of passengers for American and foreign airports is different. Use the *P*-value approach.

American airports	Foreign airports
36.8	60.7
72.4	42.7
60.5	51.2
73.5	38.6
61.2	
40.1	

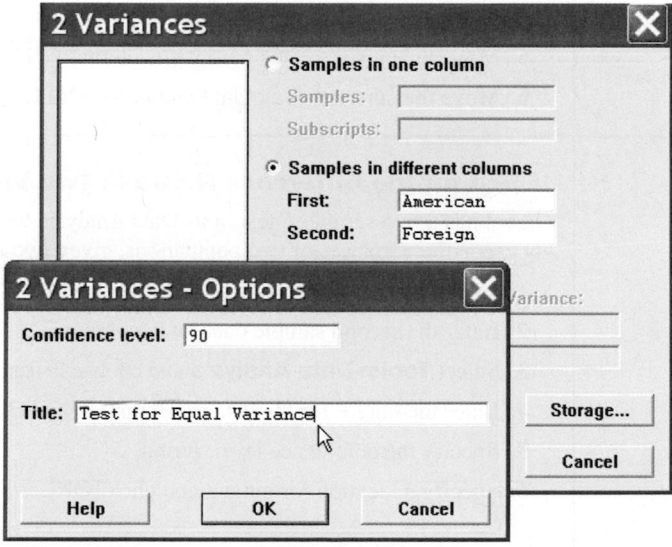

1. Enter the data into two columns of MINITAB.

2. Name the columns American and Foreign.

   a) Select **Stat>Basic Statistics>2-Variances.**

   b) Click the button for Samples in different columns.

c) Click in the text box for First, then double-click C1 American.

d) Double-click C2 Foreign, then click on [Options]. The dialog box is shown. Change the confidence level to **90** and type an appropriate title. In this dialog, we cannot specify a left- or right-tailed test.

3. Click [OK] twice. A graph window will open that includes a small window that says $F = 2.57$ and the P-value is 0.437. Divide this two-tailed P-value by 2 for a one-tailed test.

There is not enough evidence in the sample to conclude there is greater variance in the number of passengers in American airports compared to foreign airports.

TI-83 Plus or TI-84 Plus
Step by Step

Hypothesis Test for the Difference Between Two Variances (Data)

1. Enter the data values into L_1 and L_2.
2. Press **STAT** and move the cursor to TESTS.
3. Press **D (ALPHA X^{-1})** for 2-SampFTest.
4. Move the cursor to Data and press **ENTER**.
5. Type in the appropriate values.
6. Move the cursor to the appropriate alternative hypothesis and press **ENTER**.
7. Move the cursor to Calculate and press **ENTER**.

Hypothesis Test for the Difference Between Two Variances (Statistics)

1. Press **STAT** and move the cursor to TESTS.
2. Press **D (ALPHA X^{-1})** for 2-SampFTest.
3. Move the cursor to Stats and press **ENTER**.
4. Type in the appropriate values.
5. Move the cursor to the appropriate alternative hypothesis and press **ENTER**.
6. Move the cursor to Calculate and press **ENTER**.

Excel
Step by Step

F Test for the Difference Between Two Variances

Excel has a two-sample F test in its Data Analysis tools. To perform an F test for the difference between the variances of two populations, given two independent samples:

1. Enter the first sample data set in column A.
2. Enter the second sample data set in column B.
3. Select **Tools>Data Analysis** and choose F-Test Two-Sample for Variances.
4. Enter the ranges for the data in columns A and B.
5. Specify the confidence level, Alpha.
6. Specify a location for output, and click [OK].

Example XL9–2

 At $\alpha = 0.05$, test the hypothesis that the two population variances are equal, using the sample data provided here.

Set A	63	73	80	60	86	83	70	72	82
Set B	86	93	64	82	81	75	88	63	63

The results appear in the table that Excel generates, shown here. For this example, the output shows that the null hypothesis cannot be rejected at an α level of 0.05.

F-Test Two-Sample for Variances		
	Variable 1	Variable 2
Mean	74.33333333	77.22222222
Variance	82.75	132.9444444
Observations	9	9
df	8	8
F	0.622440451	
P(F<=f) one-tail	0.258814151	
F Critical one-tail	0.290858004	

9–4

Testing the Difference Between Two Means: Small Independent Samples

In Section 9–2, the z test was used to test the difference between two means when the population standard deviations were known and the variables were normally or approximately normally distributed, or when both sample sizes were greater than or equal to 30. In many situations, however, these conditions cannot be met—that is, the population standard deviations are not known, and one or both sample sizes are less than 30. In these cases, a t test is used to test the difference between means when the two samples are independent and when the samples are taken from two normally or approximately normally distributed populations. Samples are **independent samples** when they are not related.

Objective 3

Test the difference between two means for small independent samples.

There are actually two different options for the use of t tests. *One option is used when the variances of the populations are not equal, and the other option is used when the variances are equal.* To determine whether two sample variances are equal, the researcher can use an F test, as shown in Section 9–3.

Note, however, that not all statisticians are in agreement about using the F test before using the t test. Some believe that conducting the F and t tests at the same level of significance will change the overall level of significance of the t test. Their reasons are beyond the scope of this textbook.

Formulas for the t Tests—For Testing the Difference Between Two Means—Small Independent Samples

Variances are assumed to be unequal:

$$t = \frac{(\bar{X}_1 - \bar{X}_2) - (\mu_1 - \mu_2)}{\sqrt{\dfrac{s_1^2}{n_1} + \dfrac{s_2^2}{n_2}}}$$

where the degrees of freedom are equal to the smaller of $n_1 - 1$ or $n_2 - 1$.

Variances are assumed to be equal:

$$t = \frac{(\bar{X}_1 - \bar{X}_2) - (\mu_1 - \mu_2)}{\sqrt{\dfrac{(n_1 - 1)s_1^2 + (n_2 - 1)s_2^2}{n_1 + n_2 - 2}} \sqrt{\dfrac{1}{n_1} + \dfrac{1}{n_2}}}$$

where the degrees of freedom are equal to $n_1 + n_2 - 2$.

When the variances are unequal, the first formula

$$t = \frac{(\bar{X}_1 - \bar{X}_2) - (\mu_1 - \mu_2)}{\sqrt{\dfrac{s_1^2}{n_1} + \dfrac{s_2^2}{n_2}}}$$

follows the format of

$$\text{Test value} = \frac{(\text{observed value}) - (\text{expected value})}{\text{standard error}}$$

where $\bar{X}_1 - \bar{X}_2$ is the observed difference between sample means and where the expected value $\mu_1 - \mu_2$ is equal to zero when no difference between population means is hypothesized. The denominator $\sqrt{s_1^2/n_1 + s_2^2/n_2}$ is the standard error of the difference between two means. Since mathematical derivation of the standard error is somewhat complicated, it will be omitted here.

When the variances are assumed to be equal, the second formula

$$t = \frac{(\bar{X}_1 - \bar{X}_2) - (\mu_1 - \mu_2)}{\sqrt{\dfrac{(n_1 - 1)s_1^2 + (n_2 - 1)s_2^2}{n_1 + n_2 - 2}} \sqrt{\dfrac{1}{n_1} + \dfrac{1}{n_2}}}$$

also follows the format of

$$\text{Test value} = \frac{\cdot(\text{observed value}) - (\text{expected value})}{\text{standard error}}$$

For the numerator, the terms are the same as in the first formula. However, a note of explanation is needed for the denominator of the second test statistic. Since both populations are assumed to have the same variance, the standard error is computed with what is called a pooled estimate of the variance. A **pooled estimate of the variance** is a weighted average of the variance using the two sample variances and the *degrees of freedom* of each variance as the weights. Again, since the algebraic derivation of the standard error is somewhat complicated, it is omitted.

In summary, then, to use the t test, first use the F test to determine whether the variances are equal. Then use the appropriate t test formula. This procedure involves two five-step processes.

Example 9–9

The average size of a farm in Indiana County, Pennsylvania, is 191 acres. The average size of a farm in Greene County, Pennsylvania, is 199 acres. Assume the data were obtained from two samples with standard deviations of 38 and 12 acres, respectively, and sample sizes of 8 and 10, respectively. Can it be concluded at $\alpha = 0.05$ that the average size of the farms in the two counties is different? Assume the populations are normally distributed.

Source: Pittsburgh Tribune-Review.

Solution

Here we will use the F test to determine whether the variances are equal. The null hypothesis is that the variances are equal.

Step 1 State the hypotheses and identify the claim.

$$H_0: \sigma_1^2 = \sigma_2^2 \text{ (claim)} \qquad \text{and} \qquad H_1: \sigma_1^2 \neq \sigma_2^2$$

Step 2   Find the critical value. The critical value for the F test found in Table H (Appendix C) for $\alpha = 0.05$ is 4.20, since there are 7 and 9 degrees of freedom. (*Note:* Use the 0.025 table.)

Step 3   Compute the test value.

$$F = \frac{s_1^2}{s_2^2} = \frac{38^2}{12^2} = 10.03$$

Step 4   Make the decision. Reject the null hypothesis since 10.03 falls in the critical region. See Figure 9–9.

Figure 9–9

Critical and F Test Values for Example 9–9

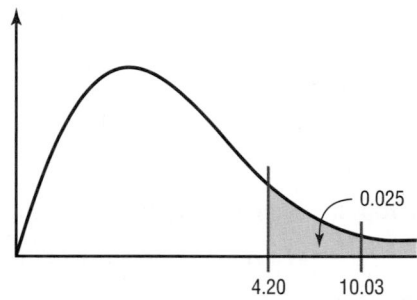

0.025

4.20   10.03

Step 5   Summarize the results. It can be concluded that the variances are not equal.
Since the variances are not equal, the first formula will be used to test the equality of the means.

Step 1   State the hypotheses and identify the claim for the means.

$$H_0: \mu_1 = \mu_2 \qquad \text{and} \qquad H_1: \mu_1 \neq \mu_2 \text{ (claim)}$$

Step 2   Find the critical values. Since the test is two-tailed, since $\alpha = 0.05$, and since the variances are unequal, the degrees of freedom are the smaller of $n_1 - 1$ or $n_2 - 1$. In this case, the degrees of freedom are $8 - 1 = 7$. Hence, from Table F, the critical values are $+2.365$ and -2.365.

Step 3   Compute the test value. Since the variances are unequal, use the first formula.

$$t = \frac{(\bar{X}_1 - \bar{X}_2) - (\mu_1 - \mu_2)}{\sqrt{\dfrac{s_1^2}{n_1} + \dfrac{s_2^2}{n_2}}} = \frac{(191 - 199) - 0}{\sqrt{\dfrac{38^2}{8} + \dfrac{12^2}{10}}} = -0.57$$

Step 4   Make the decision. Do not reject the null hypothesis, since $-0.57 > -2.365$. See Figure 9–10.

Figure 9–10

Critical and Test Values for Example 9–9

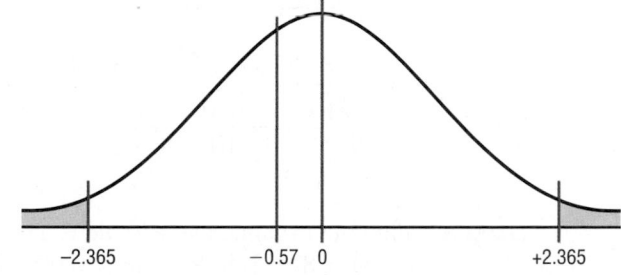

−2.365   −0.57 0   +2.365

Step 5   Summarize the results. There is not enough evidence to support the claim that the average size of the farms is different.

Example 9–10

A researcher wishes to determine whether the salaries of professional nurses employed by private hospitals are higher than those of nurses employed by government-owned hospitals. She selects a sample of nurses from each type of hospital and calculates the means and standard deviations of their salaries. At $\alpha = 0.01$, can she conclude that the private hospitals pay more than the government hospitals? Assume that the populations are approximately normally distributed. Use the *P*-value method.

Private	Government
$\bar{X}_1 = \$26{,}800$	$\bar{X}_2 = \$25{,}400$
$s_1 = \$600$	$s_2 = \$450$
$n_1 = 10$	$n_2 = 8$

Solution

The *F* test will be used to determine whether the variances are equal. The null hypothesis is that the variances are equal.

Step 1 State the hypotheses and identify the claim.

$$H_0: \sigma_1^2 = \sigma_2^2 \text{ (claim)} \qquad \text{and} \qquad H_1: \sigma_1^2 \neq \sigma_2^2$$

Step 2 Compute the test value.

$$F = \frac{s_1^2}{s_2^2} = \frac{600^2}{450^2} = 1.78$$

Step 3 Find the *P*-value in Table H, using d.f.N. = 9 and d.f.D. = 7. Since 1.78 < 2.72, *P*-value > 0.20. (The *P*-value obtained from a calculator is 0.460.)

Step 4 Make the decision. Do not reject the null hypothesis since *P*-value > 0.01 (the α value).

Step 5 Summarize the results. There is not enough evidence to reject the claim that the variances are equal; therefore, the second formula is used to test the difference between the two means, as shown next.

Step 1 State the hypotheses and identify the claim.

$$H_0: \mu_1 \leq \mu_2 \qquad \text{and} \qquad H_1: \mu_1 > \mu_2 \text{ (claim)}$$

Step 2 Compute the test value. Use the second formula, since the variances are assumed to be equal.

$$t = \frac{(\bar{X}_1 - \bar{X}_2) - (\mu_1 - \mu_2)}{\sqrt{\dfrac{(n_1 - 1)s_1^2 + (n_2 - 1)s_2^2}{n_1 + n_2 - 2}}\sqrt{\dfrac{1}{n_1} + \dfrac{1}{n_2}}}$$

$$= \frac{(26{,}800 - 25{,}400) - 0}{\sqrt{\dfrac{(10 - 1)(600)^2 + (8 - 1)(450)^2}{10 + 8 - 2}}\sqrt{\dfrac{1}{10} + \dfrac{1}{8}}}$$

$$= 5.47$$

Step 3 Find the *P*-value, using Table F. The *P*-value for $t = 5.47$ with d.f. = 16 (that is, 10 + 8 − 2) is *P*-value < 0.005. (The *P*-value obtained from a calculator is 0.00002.)

Step 4 Make the decision. Since *P*-value < 0.01 (the α value), the decision is to reject the null hypothesis.

Unusual Stats

The average walking speed of men is 2.8 miles per hour and of women is 2.9 miles per hour.

Step 5   Summarize the results. There is enough evidence to support the claim that the salaries paid to nurses employed by private hospitals are higher than those paid to nurses employed by government-owned hospitals.

When raw data are given in the exercises, use your calculator or the formulas in Chapter 3 to find the means and variances for the data sets. Then follow the procedures shown in this section to test the hypotheses.

Confidence intervals can also be found for the difference between two means with these formulas:

Confidence Intervals for the Difference of Two Means: Small Independent Samples

Variances unequal

$$(\bar{X}_1 - \bar{X}_2) - t_{\alpha/2}\sqrt{\frac{s_1^2}{n_1} + \frac{s_2^2}{n_2}} < \mu_1 - \mu_2 < (\bar{X}_1 - \bar{X}_2) + t_{\alpha/2}\sqrt{\frac{s_1^2}{n_1} + \frac{s_2^2}{n_2}}$$

d.f. = smaller value of $n_1 - 1$ or $n_2 - 1$

Variances equal

$$(\bar{X}_1 - \bar{X}_2) - t_{\alpha/2}\sqrt{\frac{(n_1 - 1)s_1^2 + (n_2 - 1)s_2^2}{n_1 + n_2 - 2}} \cdot \sqrt{\frac{1}{n_1} + \frac{1}{n_2}}$$

$$< \mu_1 - \mu_2 < (\bar{X}_1 - \bar{X}_2) + t_{\alpha/2}\sqrt{\frac{(n_1 - 1)s_1^2 + (n_2 - 1)s_2^2}{n_1 + n_2 - 2}} \cdot \sqrt{\frac{1}{n_1} + \frac{1}{n_2}}$$

d.f. = $n_1 + n_2 - 2$

Remember that when one is testing the difference between two means from independent samples, two different statistical test formulas can be used. One formula is used when the variances are equal, the other when the variances are not equal. As shown in Section 9–3, some statisticians use an F test to determine whether the two variances are equal.

Example 9–11   Find the 95% confidence interval for the data in Example 9–9.

Solution

Substitute in the formula.

$$(\bar{X}_1 - \bar{X}_2) - t_{\alpha/2}\sqrt{\frac{s_1^2}{n_1} + \frac{s_2^2}{n_2}} < \mu_1 - \mu_2$$

$$< (\bar{X}_1 - \bar{X}_2) + t_{\alpha/2}\sqrt{\frac{s_1^2}{n_1} + \frac{s_2^2}{n_2}}$$

$$(191 - 199) - 2.365\sqrt{\frac{38^2}{8} + \frac{12^2}{10}} < \mu_1 - \mu_2$$

$$< (191 - 199) + 2.365\sqrt{\frac{38^2}{8} + \frac{12^2}{10}}$$

$$-41.02 < \mu_1 - \mu_2 < 25.02$$

Since 0 is contained in the interval, the decision is do not reject the null hypothesis H_0: $\mu_1 = \mu_2$.

Applying the Concepts 9–4

Too Long on the Telephone

A company collects data on the lengths of telephone calls made by employees in two different divisions. The mean and standard deviation for the sales division are 10.26 and 8.56, respectively. The mean and standard deviation for the shipping and receiving division are 6.93 and 4.93, respectively. A hypothesis test was run, and the computer output follows.

Test statistic $F = 3.07849$
P-value $= 0.01071$
Significance level $= 0.01$
Degrees of freedom $= 56$
Confidence interval limits $= -0.18979, 6.84979$
Test statistic $t = 1.89566$
Critical value $t = -2.0037, 2.0037$
P-value $= 0.06317$
Significance level $= 0.05$

1. Are the samples independent or dependent?
2. Why was the F test done?
3. Were the results of the F test significant?
4. How many were in the study?
5. Which number from the output is compared to the significance level to check if the null hypothesis should be rejected?
6. Which number from the output gives the probability of a type I error that is calculated from the sample data?
7. Which number from the output is the result of dividing the two sample variances?
8. Was a right-, left-, or two-tailed test done? Why?
9. What are your conclusions?
10. What would your conclusions be if the level of significance were initially set at 0.10?

See page 526 for the answers.

Exercises 9–4

For Exercises 1 through 11, perform each of these steps. Assume that all variables are normally or approximately normally distributed. Be sure to test for equality of variance first.

   a. State the hypotheses and identify the claim.
   b. Find the critical value(s).
   c. Compute the test value.
   d. Make the decision.
   e. Summarize the results.

Use the traditional method of hypothesis testing unless otherwise specified.

1. A real estate agent wishes to determine whether tax assessors and real estate appraisers agree on the values of homes. A random sample of the two groups appraised 10 homes. The data are shown here. Is there a significant difference in the values of the homes for each group? Let $\alpha = 0.05$. Find the 95% confidence interval for the difference of the means.

Real estate appraisers	Tax assessors
$\overline{X}_1 = \$83,256$	$\overline{X}_2 = \$88,354$
$s_1 = \$3256$	$s_2 = \$2341$
$n_1 = 10$	$n_2 = 10$

2. A researcher suggests that male nurses earn more than female nurses. A survey of 16 male nurses and 20 female nurses reports these data. Is there enough evidence to support the claim that male nurses earn more than female nurses? Use $\alpha = 0.05$.

Female	Male
$\overline{X}_2 = \$23,750$	$\overline{X}_1 = \$23,800$
$s_2 = \$250$	$s_1 = \$300$
$n_2 = 20$	$n_1 = 16$

3. An agent claims that there is no difference between the pay of safeties and linebackers in the NFL. A survey of 15 safeties found an average salary of $501,580, and a survey of 15 linebackers found an average salary of $513,360. If the standard deviation in each case was $20,000, is the agent correct? Use $\alpha = 0.05$.

Source: *NFL Players Assn./USA TODAY.*

4. The data show the number of students attending cyber charter schools in Allegheny County and the number of students attending cyber schools in counties surrounding Allegheny County. At $\alpha = 0.01$ is there enough evidence to support the claim that the average number of students in school districts in Allegheny County who attend cyber schools is greater than those who attend cyber schools in school districts outside Allegheny County? Give a factor that should be considered in interpreting this answer.

Allegheny County	Outside Allegheny County
25 75 38 41 27 32	57 25 38 14 10 29

Source: *Pittsburgh Tribune-Review.*

5. A health-care worker wishes to see if the average number of family day-care homes per county is greater than the average number of day-care centers per county. The number of centers for a selected sample of counties is shown. At $\alpha = 0.01$, can it be concluded that the average number of family day-care homes is greater than the average number of day-care centers?

Number of family day-care homes			Number of day-care centers		
25	57	34	5	28	37
42	21	44	16	16	48

Source: *Pittsburgh Tribune-Review.*

6. A researcher wishes to test the claim that, on average, more juveniles than adults are classified as missing persons. Records for the last 5 years are shown. At $\alpha = 0.10$, is there enough evidence to support the claim?

Juveniles	65,513	65,934	64,213	61,954	59,167
Adults	31,364	34,478	36,937	35,946	38,209

Source: *USA TODAY.*

7. The local branch of the Internal Revenue Service spent an average of 21 minutes helping each of 10 people prepare their tax returns. The standard deviation was 5.6 minutes. A volunteer tax preparer spent an average of 27 minutes helping 14 people prepare their taxes. The standard deviation was 4.3 minutes. At $\alpha = 0.02$, is there a difference in the average time spent by the two services? Find the 98% confidence interval for the two means.

8. Females and males alike from the general adult population volunteer an average of 4.2 hours per week. A random sample of 20 female college students and 18 male college students indicated these results concerning the amount of time spent in volunteer service per week. At the 0.01 level of significance, is there sufficient evidence to conclude that a difference exists between the mean number of volunteer hours per week for male and female college students?

	Male	Female
Sample mean	2.5	3.8
Sample variance	2.2	3.5
Sample size	18	20

Source: *N.Y. Times Almanac.*

9. The average cost of a movie ticket in London is $19.63 while the average cost of a movie ticket in New York City is $10.25. Assume the standard deviations are $3.20 and $2.57, respectively, and both samples consisted of 25 theaters. At $\alpha = 0.10$ can it be concluded that movie tickets cost more in London than they do in New York City? Suggest a reason for the difference, if one exists.

Source: *USA TODAY.*

10. Health Care Knowledge Systems reported that an insured woman spends on average 2.3 days in the hospital for a routine childbirth, while an uninsured woman spends on average 1.9 days. Assume two samples of 16 women each were used and the standard deviations are both equal to 0.6 day. At $\alpha = 0.01$, test the claim that the means are equal. Find the 99% confidence interval for the differences of the means. Use the *P*-value method.

Source: Michael D. Shook and Robert L. Shook, *The Book of Odds.*

11. The times (in minutes) it took six white mice to learn to run a simple maze and the times it took six brown mice to learn to run the same maze are given here. At $\alpha = 0.05$, does the color of the mice make a difference in their learning rate? Find the 95% confidence interval for the difference of the means. Use the *P*-value method.

White mice	18	24	20	13	15	12
Brown mice	25	16	19	14	16	10

12. A random sample of enrollments from medical schools that specialize in research and from those that are noted for primary care is listed. Find the 90% confidence interval for the difference in the means.

Research				Primary care			
474	577	605	663	783	605	427	728
783	467	670	414	546	474	371	107
813	443	565	696	442	587	293	277
692	694	277	419	662	555	527	320
884							

Source: *U.S. News & World Report Best Graduate Schools.*

13. The out-of-state tuitions (in dollars) for random samples of both public and private four-year colleges in a New England state are listed. Find the 95% confidence interval for the difference in the means.

Private		Public	
13,600	13,495	7,050	9,000
16,590	17,300	6,450	9,758
23,400	12,500	7,050	7,871
		16,100	

Source: *N.Y. Times Almanac.*

Technology *Step by Step*

MINITAB
Step by Step

Test the Difference Between Two Means: Small Independent Samples with Equal Variance

In Section 9–3 we determined that the variance in the number of passengers at American and foreign airports was not different. To test the hypothesis that the mean number of passengers is the same, we continue using the data from Example 9–8.

American airports	Foreign airports
36.8	60.7
72.4	42.7
60.5	51.2
73.5	38.6
61.2	
40.1	

1. Enter the data into two columns of a MINITAB worksheet.

2. Name the columns **American** and **Foreign.**

3. Select **Stat>Basic Statistics>2-Sample t.**

4. Click on Samples in different columns.

5. Click in the box for First, then double-click C1 American.

6. Double-click C2 Foreign for the Second.

7. Check the box for Assume equal variances. The pooled standard deviation formula from Section 9–4 will be used to calculate the test statistic and *P*-value.

8. Click [OK]. The session window is shown. The *P*-value for the difference is 0.335. Do not reject the null hypothesis. There is no significant difference in the mean number of passengers at American airports compared to foreign airports.

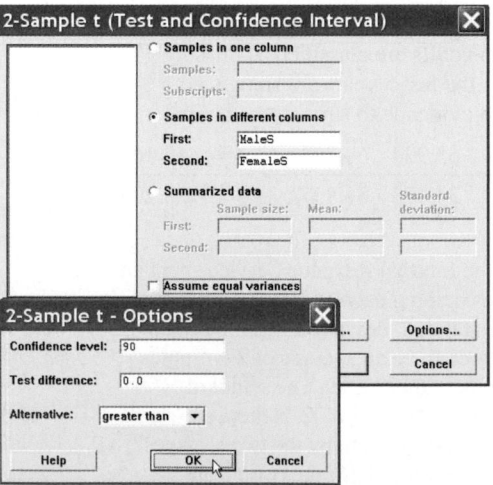

Two-Sample t-Test and CI: American, Foreign

```
Two-sample t for American vs Foreign

            N    Mean    StDev    SE Mean
American    6    57.4    15.7         6.4
Foreign     4    48.30   9.79        4.9

Difference = mu (American) - mu (Foreign)
Estimate for difference:  9.11667
90% CI for difference:  (-7.42622,  25.65955)

t-Test of difference = 0 (vs not =) : t-Value = 1.02 P-Value = 0.335 DF = 8
      Both use Pooled StDev = 13.7819
```

TI-83 Plus or TI-84 Plus
Step by Step

Hypothesis Test for the Difference Between Two Means and *t* Distribution (Data)

1. Enter the data values into L_1 and L_2.
2. Press **STAT** and move the cursor to TESTS.
3. Press **4** for 2-SampTTest.
4. Move the cursor to Data and press **ENTER**.
5. Type in the appropriate values.
6. Move the cursor to the appropriate alternative hypothesis and press **ENTER**.
7. On the line for Pooled, move the cursor to No (standard deviations are assumed not equal) or Yes (standard deviations are assumed equal) and press **ENTER**.
8. Move the cursor to Calculate and press **ENTER**.

Hypothesis Test for the Difference Between Two Means and *t* Distribution (Statistics)

1. Press **STAT** and move the cursor to TESTS.
2. Press **4** for 2-SampTTest.
3. Move the cursor to Stats and press **ENTER**.
4. Type in the appropriate values.
5. Move the cursor to the appropriate alternative hypothesis and press **ENTER**.
6. On the line for Pooled, move the cursor to No (standard deviations are assumed not equal) or Yes (standard deviations are assumed equal) and press **ENTER**.
7. Move the cursor to Calculate and press **ENTER**.

Confidence Interval for the Difference Between Two Means and *t* Distribution (Data)

1. Enter the data values into L_1 and L_2.
2. Press **STAT** and move the cursor to TESTS.
3. Press **0** for 2-SampTInt.
4. Move the cursor to Data and press **ENTER**.
5. Type in the appropriate values.
6. On the line for Pooled, move the cursor to No (standard deviations are assumed not equal) or Yes (standard deviations are assumed equal) and press **ENTER**.
7. Move the cursor to Calculate and press **ENTER**.

Confidence Interval for the Difference Between Two Means and *t* Distribution (Statistics)

1. Press **STAT** and move the cursor to TESTS.
2. Press **0** for 2-SampTInt.
3. Move the cursor to Stats and press **ENTER.**
4. Type in the appropriate values.
5. On the line for Pooled, move the cursor to No (standard deviations are assumed not equal) or Yes (standard deviations are assumed equal) and press **ENTER.**
6. Move the cursor to Calculate and press **ENTER.**

Excel
Step by Step

Testing the Difference Between Two Means: Small Independent Samples

Excel has a two-sample *t* test in its Data Analysis tools. To perform the *t* test for the difference between means, see Example XL9–3.

Example XL9–3

Test the hypothesis that there is no difference between population means based on these sample data. Assume the population variances are not equal. Use $\alpha = 0.05$.

Set A	32	38	37	36	36	34	39	36	37	42
Set B	30	36	35	36	31	34	37	33	32	

1. Enter the 10-number data set A in column A.
2. Enter the 9-number data set B in column B.
3. Select **Tools>Data Analysis** and choose t-Test: Two-Sample Assuming Unequal Variances.
4. Enter the data ranges, hypothesized mean difference (here, 0), and α.
5. Select a location for output and click [OK].

Two-Sample *t* Test in Excel

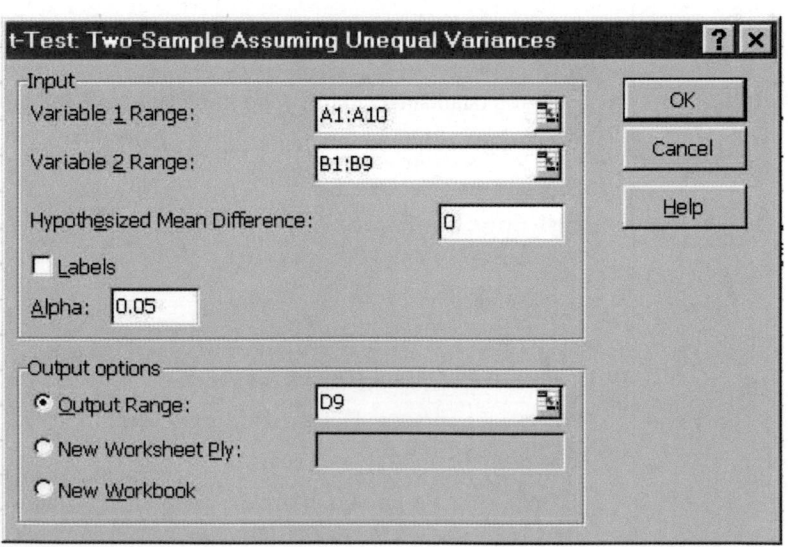

t-Test: Two-Sample Assuming Unequal Variance	*Variable 1*	*Variable 2*
Mean	36.7	33.77777778
Variance	7.344444444	5.944444444
Observations	10	9
Hypothesized Mean Difference	0	
df	17	
t Stat	2.474205364	
P(T<=t) one-tail	0.012095	
t Critical one-tail	1.739606432	
P(T<=t) two-tail	0.024189999	
t Critical two-tail	2.109818524	

The output reports both one- and two-tailed *P*-values.

If the variances are equal, use the two-sample *t* test, assuming equal-variances procedure.

9–5

Testing the Difference Between Two Means: Small Dependent Samples

Objective 4

Test the difference between two means for small dependent samples.

In Section 9–4, the *t* test was used to compare two sample means when the samples were independent. In this section, a different version of the *t* test is explained. This version is used when the samples are dependent. Samples are considered to be **dependent samples** when the subjects are paired or matched in some way.

For example, suppose a medical researcher wants to see whether a drug will affect the reaction time of its users. To test this hypothesis, the researcher must pretest the subjects in the sample first. That is, they are given a test to ascertain their normal reaction times. Then after taking the drug, the subjects are tested again, using a posttest. Finally, the means of the two tests are compared to see whether there is a difference. Since the same subjects are used in both cases, the samples are *related;* subjects scoring high on the pretest will generally score high on the posttest, even after consuming the drug. Likewise, those scoring lower on the pretest will tend to score lower on the posttest. To take this effect into account, the researcher employs a *t* test, using the differences between the pretest values and the posttest values. Thus only the gain or loss in values is compared.

Here are some other examples of dependent samples. A researcher may want to design an SAT preparation course to help students raise their test scores the second time they take the SAT. Hence, the differences between the two exams are compared. A medical specialist may want to see whether a new counseling program will help subjects lose weight. Therefore, the preweights of the subjects will be compared with the postweights.

Besides samples in which the same subjects are used in a pre-post situation, there are other cases where the samples are considered dependent. For example, students might be matched or paired according to some variable that is pertinent to the study; then one student is assigned to one group, and the other student is assigned to a second group. For instance, in a study involving learning, students can be selected and paired according to their IQs. That is, two students with the same IQ will be paired. Then one will be assigned to one sample group (which might receive instruction by computers), and the other student will be assigned to another sample group (which might receive instruction by the lecture discussion method). These assignments will be done randomly. Since a student's IQ is important to learning, it is a variable that should be controlled. By matching subjects on IQ, the researcher can eliminate the variable's influence, for the most part. Matching, then, helps to reduce type II error by eliminating extraneous variables.

In this study, two groups of children were compared. One group dined with their families daily, and the other did not. Suggest a hypothesis for the study. Comment on what variables were used for comparison.

Do Your Kids Have Dinner With You?

If so, they probably eat better than those who don't dine with their folks. In a recent Harvard Medical School study of 16,000 children ages 9 to 14, 24% of those who dined daily with their family got the recommended five servings of fruits and vegetables, compared with 13% of those who rarely or never shared meals at home. They also ate less fried food, drank less soda, and consumed more calcium, fiber, iron and vitamins C and E. Says Matthew W. Gillman, M.D., lead investigator of the study at Harvard, "There are two possible explanations. When kids eat with their parents, there may be more nutritious food on the table. Or maybe there's a discussion of healthful eating."

— JEANNIE RALSTON *in*
Ladies' Home Journal

Source: Reprinted with permission from the September 2000 *Reader's Digest.* Copyright © 2000 by the Reader's Digest Assn. Inc.

Two notes of caution should be mentioned. First, when subjects are matched according to one variable, the matching process does not eliminate the influence of other variables. Matching students according to IQ does not account for their mathematical ability or their familiarity with computers. Since not all variables influencing a study can be controlled, it is up to the researcher to determine which variables should be used in matching. Second, when the same subjects are used for a pre-post study, sometimes the knowledge that they are participating in a study can influence the results. For example, if people are placed in a special program, they may be more highly motivated to succeed simply because they have been selected to participate; the program itself may have little effect on their success.

When the samples are dependent, a special *t* test for dependent means is used. This test employs the difference in values of the matched pairs. The hypotheses are as follows:

Two-tailed	**Left-tailed**	**Right-tailed**
$H_0: \mu_D = 0$	$H_0: \mu_D \geq 0$	$H_0: \mu_D \leq 0$
$H_1: \mu_D \neq 0$	$H_1: \mu_D < 0$	$H_1: \mu_D > 0$

where μ_D is the symbol for the expected mean of the difference of the matched pairs. The general procedure for finding the test value involves several steps.

First, find the differences of the values of the pairs of data.

$$D = X_1 - X_2$$

Second, find the mean $\overline{D}$ of the differences, using the formula

$$\overline{D} = \frac{\Sigma D}{n}$$

where n is the number of data pairs. Third, find the standard deviation s_D of the differences, using the formula

$$s_D = \sqrt{\frac{\Sigma D^2 - \frac{(\Sigma D)^2}{n}}{n - 1}}$$

Fourth, find the estimated standard error $s_{\overline{D}}$ of the differences, which is

$$s_{\overline{D}} = \frac{s_D}{\sqrt{n}}$$

Finally, find the test value, using the formula

$$t = \frac{\overline{D} - \mu_D}{s_D/\sqrt{n}} \quad \text{with d.f.} = n - 1$$

The formula in the final step follows the basic format of

$$\text{Test value} = \frac{(\text{observed value}) - (\text{expected value})}{\text{standard error}}$$

where the observed value is the mean of the differences. The expected value μ_D is zero if the hypothesis is $\mu_D = 0$. The standard error of the difference is the standard deviation of the difference, divided by the square root of the sample size. Both populations must be normally or approximately normally distributed. Example 9–12 illustrates the hypothesis-testing procedure in detail.

Example 9–12

A physical education director claims by taking a special vitamin, a weight lifter can increase his strength. Eight athletes are selected and given a test of strength, using the standard bench press. After 2 weeks of regular training, supplemented with the vitamin, they are tested again. Test the effectiveness of the vitamin regimen at $\alpha = 0.05$. Each value in these data represents the maximum number of pounds the athlete can bench-press. Assume that the variable is approximately normally distributed.

Athlete	1	2	3	4	5	6	7	8
Before (X_1)	210	230	182	205	262	253	219	216
After (X_2)	219	236	179	204	270	250	222	216

Solution

Step 1 State the hypotheses and identify the claim. For the vitamin to be effective, the before weights must be significantly less than the after weights; hence, the mean of the differences must be less than zero.

$$H_0: \mu_D \geq 0 \quad \text{and} \quad H_1: \mu_D < 0 \text{ (claim)}$$

Step 2 Find the critical value. The degrees of freedom are $n - 1$. In this case, d.f. = $8 - 1 = 7$. The critical value for a left-tailed test with $\alpha = 0.05$ is -1.895.

Step 3 Compute the test value.

a. Make a table.

Before (X_1)	After (X_2)	A $D = X_1 - X_2$	B $D^2 = (X_1 - X_2)^2$
210	219		
230	236		
182	179		
205	204		
262	270		
253	250		
219	222		
216	216		

Unusual Stat

Americans eat on average 65 pounds of sugar each year.

b. Find the differences and place the results in column A.

$$
\begin{aligned}
210 - 219 &= -9 \\
230 - 236 &= -6 \\
182 - 179 &= +3 \\
205 - 204 &= +1 \\
262 - 270 &= -8 \\
253 - 250 &= +3 \\
219 - 222 &= -3 \\
216 - 216 &= 0 \\
\Sigma D &= -19
\end{aligned}
$$

c. Find the mean of the differences.

$$\overline{D} = \frac{\Sigma D}{n} = \frac{-19}{8} = -2.375$$

d. Square the differences and place the results in column B.

$$
\begin{aligned}
(-9)^2 &= 81 \\
(-6)^2 &= 36 \\
(+3)^2 &= 9 \\
(+1)^2 &= 1 \\
(-8)^2 &= 64 \\
(+3)^2 &= 9 \\
(-3)^2 &= 9 \\
0^2 &= 0 \\
\Sigma D^2 &= 209
\end{aligned}
$$

The completed table is shown next.

Before (X_1)	After (X_2)	A $D = X_1 - X_2$	B $D^2 = (X_1 - X_2)^2$
210	219	−9	81
230	236	−6	36
182	179	+3	9
205	204	+1	1
262	270	−8	64
253	250	+3	9
219	222	−3	9
216	216	0	0
		$\Sigma D = -19$	$\Sigma D^2 = 209$

    e. Find the standard deviation of the differences.

$$s_D = \sqrt{\dfrac{\Sigma D^2 - \dfrac{(\Sigma D)^2}{n}}{n-1}} = \sqrt{\dfrac{209 - \dfrac{(-19)^2}{8}}{8-1}} = 4.84$$

    f. Find the test value.

$$t = \dfrac{\overline{D} - \mu_D}{s_D/\sqrt{n}} = \dfrac{-2.375 - 0}{4.84/\sqrt{8}} = -1.388$$

Step 4   Make the decision. The decision is not to reject the null hypothesis at $\alpha = 0.05$, since $-1.388 > -1.895$, as shown in Figure 9–11.

Figure 9–11

Critical and Test Values for Example 9–12

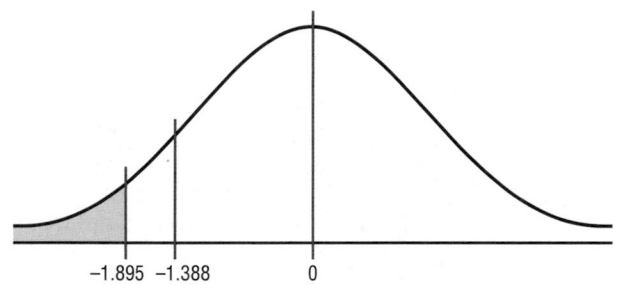

-1.895 -1.388 0

Step 5   Summarize the results. There is not enough evidence to support the claim that the vitamin increases the strength of weight lifters.

The formulas for this *t* test are summarized next.

Formulas for the *t* Test for Dependent Samples

$$t = \dfrac{\overline{D} - \mu_D}{s_D/\sqrt{n}}$$

with d.f. $= n - 1$ and where

$$\overline{D} = \dfrac{\Sigma D}{n} \quad \text{and} \quad s_D = \sqrt{\dfrac{\Sigma D^2 - \dfrac{(\Sigma D)^2}{n}}{n-1}}$$

Example 9–13

A dietitian wishes to see if a person's cholesterol level will change if the diet is supplemented by a certain mineral. Six subjects were pretested, and then they took the mineral supplement for a 6-week period. The results are shown in the table. (Cholesterol level is measured in milligrams per deciliter.) Can it be concluded that the cholesterol level has been changed at $\alpha = 0.10$? Assume the variable is approximately normally distributed.

Subject	1	2	3	4	5	6
Before (X_1)	210	235	208	190	172	244
After (X_2)	190	170	210	188	173	228

Solution

Step 1 State the hypotheses and identify the claim. If the diet is effective, the before cholesterol levels should be different from the after levels.

$$H_0: \mu_D = 0 \quad \text{and} \quad H_1: \mu_D \neq 0 \text{ (claim)}$$

Step 2 Find the critical value. The degrees of freedom are 5. At $\alpha = 0.10$, the critical values are ± 2.015.

Step 3 Compute the test value.

a. Make a table.

Before (X_1)	After (X_2)	A $D = X_1 - X_2$	B $D^2 = (X_1 - X_2)^2$
210	190		
235	170		
208	210		
190	188		
172	173		
244	228		

b. Find the differences and place the results in column A.

$$210 - 190 = \quad 20$$
$$235 - 170 = \quad 65$$
$$208 - 210 = \quad -2$$
$$190 - 188 = \quad 2$$
$$172 - 173 = \quad -1$$
$$244 - 228 = \quad 16$$
$$\Sigma D = 100$$

c. Find the mean of the differences.

$$\overline{D} = \frac{\Sigma D}{n} = \frac{100}{6} = 16.7$$

d. Square the differences and place the results in column B.

$$(20)^2 = \quad 400$$
$$(65)^2 = 4225$$
$$(-2)^2 = \quad 4$$
$$(2)^2 = \quad 4$$
$$(-1)^2 = \quad 1$$
$$(16)^2 = \quad 256$$
$$\Sigma D^2 = 4890$$

Then complete the table as shown.

Before (X_1)	After (X_2)	A $D = X_1 - X_2$	B $D^2 = (X_1 - X_2)^2$
210	190	20	400
235	170	65	4225
208	210	−2	4
190	188	2	4
172	173	−1	1
244	228	16	256
		$\Sigma D = 100$	$\Sigma D^2 = 4890$

e. Find the standard deviation of the differences.

$$s_D = \sqrt{\frac{\Sigma D^2 - \frac{(\Sigma D)^2}{n}}{n-1}} = \sqrt{\frac{4890 - \frac{(100)^2}{6}}{5}} = 25.4$$

f. Find the test value.

$$t = \frac{\overline{D} - \mu_D}{s_D/\sqrt{n}} = \frac{16.7 - 0}{25.4/\sqrt{6}} = 1.610$$

Step 4   Make the decision. The decision is not to reject the null hypothesis, since the test value 1.610 is in the noncritical region, as shown in Figure 9–12.

Figure 9–12

Critical and Test Values for Example 9–13

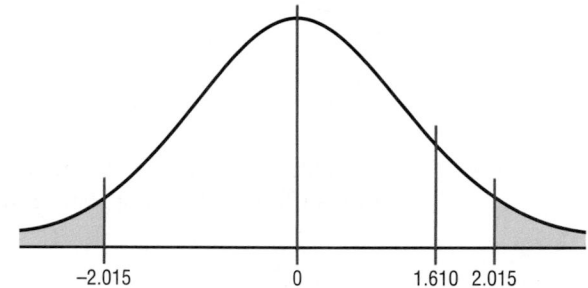

-2.015 0 1.610 2.015

Step 5   Summarize the results. There is not enough evidence to support the claim that the mineral changes a person's cholesterol level.

The steps for this *t* test are summarized in the Procedure Table.

Procedure Table

Testing the Difference Between Means for Dependent Samples

Step 1   State the hypotheses and identify the claim.

Step 2   Find the critical value(s).

Step 3   Compute the test value.

a. Make a table, as shown.

X_1	X_2	**A** $D = X_1 - X_2$	**B** $D^2 = (X_1 - X_2)^2$
$\vdots$	$\vdots$		
		$\Sigma D = $ _____	$\Sigma D^2 = $ _____

b. Find the differences and place the results in column A.

$$D = X_1 - X_2$$

c. Find the mean of the differences.

$$\overline{D} = \frac{\Sigma D}{n}$$

d. Square the differences and place the results in column B. Complete the table.

$$D^2 = (X_1 - X_2)^2$$

Procedure Table (*continued*)

e. Find the standard deviation of the differences.

$$s_D = \sqrt{\frac{\Sigma D^2 - \dfrac{(\Sigma D)^2}{n}}{n - 1}}$$

f. Find the test value.

$$t = \frac{\overline{D} - \mu_D}{s_D/\sqrt{n}} \qquad \text{with d.f.} = n - 1$$

Step 4 Make the decision.

Step 5 Summarize the results.

The *P*-values for the *t* test are found in Table F. For a two-tailed test with d.f. = 5 and $t = 1.610$, the *P*-value is found between 1.476 and 2.015; hence, $0.10 < P\text{-value} < 0.20$. Thus, the null hypothesis cannot be rejected at $\alpha = 0.10$.

If a specific difference is hypothesized, this formula should be used

$$t = \frac{\overline{D} - \mu_D}{s_D/\sqrt{n}}$$

where μ_D is the hypothesized difference.

For example, if a dietitian claims that people on a specific diet will lose an average of 3 pounds in a week, the hypotheses are

$$H_0\text{: } \mu_D = 3 \qquad \text{and} \qquad H_1\text{: } \mu_D \neq 3$$

The value 3 will be substituted in the test statistic formula for μ_D.

Confidence intervals can be found for the mean differences with this formula.

Confidence Interval for the Mean Difference

$$\overline{D} - t_{\alpha/2}\frac{s_D}{\sqrt{n}} < \mu_D < \overline{D} + t_{\alpha/2}\frac{s_D}{\sqrt{n}}$$

$$\text{d.f.} = n - 1$$

Example 9–14

Find the 90% confidence interval for the data in Example 9–13.

Solution

Substitute in the formula.

$$\overline{D} - t_{\alpha/2}\frac{s_D}{\sqrt{n}} < \mu_D < \overline{D} + t_{\alpha/2}\frac{s_D}{\sqrt{n}}$$

$$16.7 - 2.015 \cdot \frac{25.4}{\sqrt{6}} < \mu_D < 16.7 + 2.015 \cdot \frac{25.4}{\sqrt{6}}$$

$$16.7 - 20.89 < \mu_D < 16.7 + 20.89$$

$$-4.19 < \mu_D < 37.59$$

Since 0 is contained in the interval, the decision is do not reject the null hypothesis $H_0\text{: } \mu_D = 0$.

Speaking of
Statistics

Can Video Games Save Lives?

Can playing video games help doctors perform surgery? The answer is yes. A study showed that surgeons who played video games for a least 3 hours each week made about 37% fewer mistakes and finished operations 27% faster than those who did not play video games.

The type of surgery that they performed is called *laparoscopic* surgery, where the surgeon inserts a tiny video camera into the body and uses a joystick to maneuver the surgical instruments while watching the results on a television monitor. This study compares two groups and uses proportions. What statistical test do you think was used to compare the percentages? (See Section 9–6)

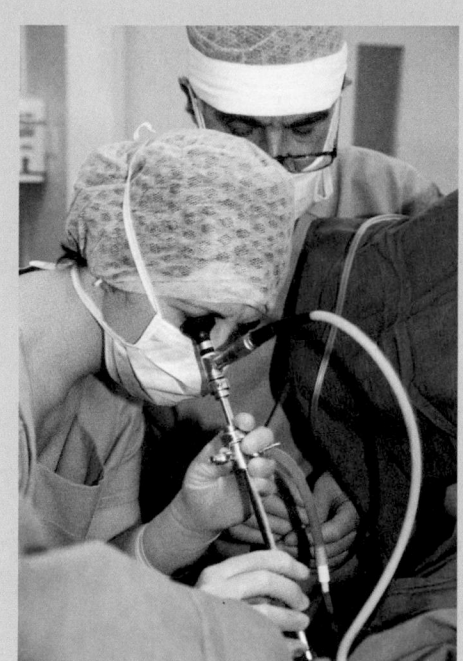

Applying the Concepts 9–5

Air Quality

As a researcher for the EPA, you have been asked to determine if the air quality in the United States has changed over the past 2 years. You select a random sample of 10 metropolitan areas and find the number of days each year that the areas failed to meet acceptable air quality standards. The data are shown.

Year 1	18	125	9	22	138	29	1	19	17	31
Year 2	24	152	13	21	152	23	6	31	34	20

Source: *The World Almanac and Book of Facts.*

Based on the data, answer the following questions.

1. What is the purpose of the study?
2. Are the samples independent or dependent?
3. What are the hypotheses that you would use?
4. What is(are) the critical value(s) that you would use?
5. What statistical test would you use?
6. How many degrees of freedom are there?
7. What is your conclusion?
8. Could an independent means test have been used?
9. Do you think this was a good way to answer the original question?

See page 526 for the answers.

Exercises 9–5

1. Classify each as independent or dependent samples.

   a. Heights of identical twins

   b. Test scores of the same students in English and psychology

   c. The effectiveness of two different brands of aspirin

   d. Effects of a drug on reaction time, measured by a before and an after test

   e. The effectiveness of two different diets on two different groups of individuals

For Exercises 2 through 10, perform each of these steps. Assume that all variables are normally or approximately normally distributed.

   a. State the hypotheses and identify the claim.

   b. Find the critical value(s).

   c. Compute the test value.

   d. Make the decision.

   e. Summarize the results.

Use the traditional method of hypothesis testing unless otherwise specified.

2. A program for reducing the number of days missed by food handlers in a certain restaurant chain was conducted. The owners hypothesized that after the program the workers would miss fewer days of work due to illness. The table shows the number of days 10 workers missed per month before and after completing the program. Is there enough evidence to support the claim, at $\alpha = 0.05$, that the food handlers missed fewer days after the program?

Before	2	3	6	7	4	5	3	1	0	0
After	1	4	3	8	3	3	1	0	1	0

3. As an aid for improving students' study habits, nine students were randomly selected to attend a seminar on the importance of education in life. The table shows the number of hours each student studied per week before and after the seminar. At $\alpha = 0.10$, did attending the seminar increase the number of hours the students studied per week?

Before	9	12	6	15	3	18	10	13	7
After	9	17	9	20	2	21	15	22	6

4. A doctor is interested in determining whether a film about exercise will change 10 persons' attitudes about exercise. The results of his questionnaire are shown. A higher numerical value shows a more favorable attitude toward exercise. Is there enough evidence to support the claim, at $\alpha = 0.05$, that there

was a change in attitude? Find the 95% confidence interval for the difference of the two means.

Before	12	11	14	9	8	6	8	5	4	7
After	13	12	10	9	8	8	7	6	5	5

5. Students in a statistics class were asked to report the number of hours they slept on weeknights and on weekends. At $\alpha = 0.05$, is there sufficient evidence that there is a difference in the mean number of hours slept?

Student	1	2	3	4	5	6	7	8
Hours, Sun.–Thurs.	8	5.5	7.5	8	7	6	6	8
Hours, Fri.–Sat.	4	7	10.5	12	11	9	6	9

6. A sample of municipalities' legal costs (in thousands of dollars) for two recent consecutive years is as shown. At $\alpha = 0.05$ is there a difference in the costs? Suggest a reason for the difference, if one exists.

Year 1	61	26	9	16	61	71	14	86	17	24
Year 2	62	40	10	23	38	118	18	67	21	20

Source: *Pittsburgh Tribune-Review.*

7. A composition teacher wishes to see whether a new grammar program will reduce the number of grammatical errors her students make when writing a two-page essay. The data are shown here. At $\alpha = 0.025$, can it be concluded that the number of errors has been reduced?

Student	1	2	3	4	5	6
Errors before	12	9	0	5	4	3
Errors after	9	6	1	3	2	3

8. A sample of legal costs (in thousands of dollars) for school districts for two recent consecutive years is shown. At $\alpha = 0.05$ is there a difference in the costs? Suggest a reason for the difference, if one exists.

Year 1	108	36	65	108	87	94	10	40
Year 2	138	28	67	181	97	126	18	67

Source: *Pittsburgh Tribune-Review.*

9. A researcher wanted to compare the pulse rates of identical twins to see whether there was any difference. Eight sets of twins were selected. The rates are given in the table as number of beats per minute. At $\alpha = 0.01$, is there a significant difference in the average pulse rates of twins? Find the 99% confidence interval for the difference of the two. Use the *P*-value method.

Twin A	87	92	78	83	88	90	84	93
Twin B	83	95	79	83	86	93	80	86

10. A reporter hypothesizes that the average assessed values of land in a large city have changed during a 5-year period. A random sample of wards is selected, and the data (in millions of dollars) are shown. At $\alpha = 0.05$, can it be concluded that the average taxable assessed values have changed? Use the *P*-value method.

Ward	A	B	C	D	E	F	G	H	I	J	K	L	M	N	O	P
1994	184	414	22	99	116	49	24	50	282	25	141	45	12	37	9	17
1999	161	382	22	109	120	52	28	50	297	40	148	56	20	38	9	19

Source: *Pittsburgh Tribune-Review.*

Extending the Concepts

11. Instead of finding the mean of the differences between X_1 and X_2 by subtracting $X_1 - X_2$, one can find it by finding the means of X_1 and X_2 and then subtracting the means. Show that these two procedures will yield the same results.

Technology *Step by Step*

MINITAB
Step by Step

Test the Difference Between Two Means: Small Dependent Samples

For Example 9–12, test the effectiveness of the vitamin regimen. Is there a difference in the strength of the athletes after the treatment?

1. Enter the data into C1 and C2. Name the columns **Before** and **After.**

2. Select **Stat>Basic Statistics>Paired t.**

3. Double-click C1 Before for First sample.

4. Double-click C2 After for Second sample. The second sample will be subtracted from the first. The differences are not stored or displayed.

5. Click [Options].

6. Change the Alternative to less than.

7. Click [OK] twice.

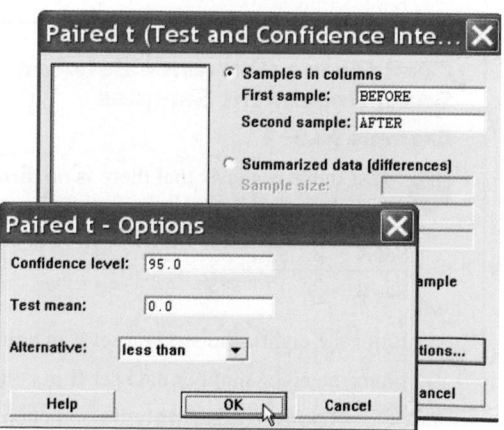

Paired t-Test and CI: BEFORE, AFTER

```
Paired t for BEFORE - AFTER
                N      Mean      StDev     SE Mean
BEFORE          8    222.125    25.920      9.164
AFTER           8    224.500    27.908      9.867
Difference      8    -2.37500    4.83846    1.71065

95% upper bound for mean difference: 0.86597
t-Test of mean difference = 0 (vs < 0) : t-Value = -1.39  P-Value = 0.104.
```

Since the *P*-value is 0.104, do not reject the null hypothesis. The sample difference of -2.38 in the strength measurement is not statistically significant.

TI-83 Plus or TI-84 Plus
Step by Step

Hypothesis Test for the Difference Between Two Means: Dependent Samples

1. Enter the data values into L_1 and L_2.
2. Move the cursor to the top of the L_3 column so that L_3 is highlighted.
3. Type $L_1 - L_2$, then press **ENTER**.
4. Press **STAT** and move the cursor to TESTS.
5. Press **2** for TTest.
6. Move the cursor to Data and press **ENTER**.
7. Type in the appropriate values, using **0** for μ_0 and L_3 for the list.
8. Move the cursor to the appropriate alternative hypothesis and press **ENTER**.
9. Move the cursor to Calculate and press **ENTER**.

Confidence Interval for the Difference Between Two Means: Dependent Samples

1. Enter the data values into L_1 and L_2.
2. Move the cursor to the top of the L_3 column so that L_3 is highlighted.
3. Type $L_1 - L_2$, then press **ENTER**.
4. Press **STAT** and move the cursor to TESTS.
5. Press **8** for TInterval.
6. Move the cursor to Stats and press **ENTER**.
7. Type in the appropriate values, using L_3 for the list.
8. Move the cursor to Calculate and press **ENTER**.

Excel
Step by Step

t Test for the Difference Between Two Means: Small Dependent Samples
Example XL9–4

 Test the hypothesis that there is no difference in population means, based on these sample paired data. Use $\alpha = 0.05$.

Set A	33	35	28	29	32	34	30	34
Set B	27	29	36	34	30	29	28	24

1. Enter the eight-number data set A in column A.
2. Enter the eight-number data set B in column B.
3. Select **Tools>Data Analysis** and choose t-Test: Paired Two Sample for Means.
4. Enter the data ranges and hypothesized mean difference (here, zero), and α.

5. Select a location for output and click [OK].

Dialog Box for Paired-Data
t Test

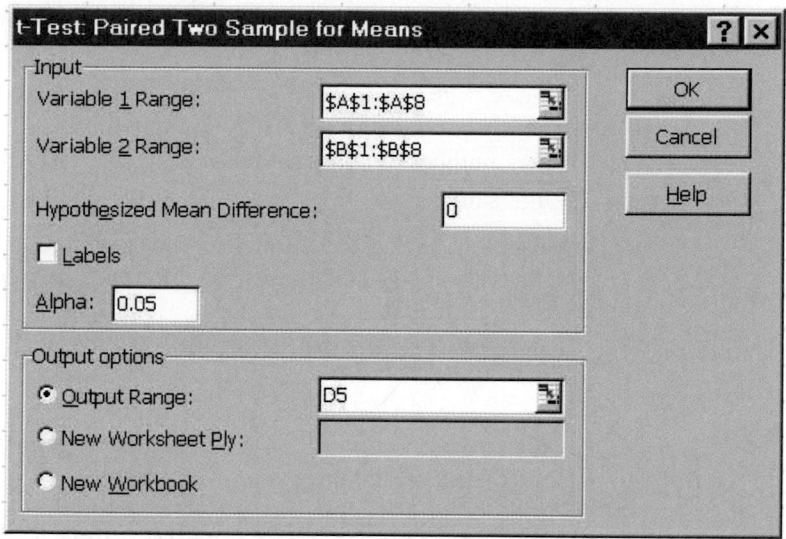

The screen shows a *P*-value of 0.3253988 for the two-tailed case. This is greater than the confidence level 0.05, so we fail to reject the null hypothesis.

t-Test: Paired Two Sample for Means		
	Variable 1	*Variable 2*
Mean	31.875	29.625
Variance	6.696428571	14.55357143
Observations	8	8
Pearson Correlation	-0 757913399	
Hypothesized Mean Difference	0	
df	7	
t Stat	1.057517468	
P(T<=t) one-tail	0.1626994	
t Critical one-tail	1.894577508	
P(T<=t) two-tail	0.3253988	
t Critical two-tail	2.36462256	

9–6 Testing the Difference Between Proportions

Objective 5

Test the difference between two proportions.

The *z* test with some modifications can be used to test the equality of two proportions. For example, a researcher might ask, Is the proportion of men who exercise regularly less than the proportion of women who exercise regularly? Is there a difference in the percentage of students who own a personal computer and the percentage of nonstudents who own one? Is there a difference in the proportion of college graduates who pay cash for purchases and the proportion of non-college graduates who pay cash?

Recall from Chapter 7 that the symbol $\hat{p}$ ("*p* hat") is the sample proportion used to estimate the population proportion, denoted by *p*. For example, if in a sample of

30 college students, 9 are on probation, then the sample proportion is $\hat{p} = \frac{9}{30}$, or 0.3. The population proportion p is the number of all students who are on probation, divided by the number of students who attend the college. The formula for $\hat{p}$ is

$$\hat{p} = \frac{X}{n}$$

where

   X = number of units that possess the characteristic of interest

   n = sample size

When one is testing the difference between two population proportions p_1 and p_2, the hypotheses can be stated thus, if no difference between the proportions is hypothesized.

$$H_0: p_1 = p_2 \qquad H_0: p_1 - p_2 = 0$$
$$H_1: p_1 \neq p_2 \qquad \text{or} \qquad H_1: p_1 - p_2 \neq 0$$

Similar statements using $\geq$ and $<$ or $\leq$ and $>$ can be formed for one-tailed tests.

For two proportions, $\hat{p}_1 = X_1/n_1$ is used to estimate p_1 and $\hat{p}_2 = X_2/n_2$ is used to estimate p_2. The standard error of the difference is

$$\sigma_{(\hat{p}_1 - \hat{p}_2)} = \sqrt{\sigma_{p_1}^2 + \sigma_{p_2}^2} = \sqrt{\frac{p_1 q_1}{n_1} + \frac{p_2 q_2}{n_2}}$$

where $\sigma_{p_1}^2$ and $\sigma_{p_2}^2$ are the variances of the proportions, $q_1 = 1 - p_1$, $q_2 = 1 - p_2$, and n_1 and n_2 are the respective sample sizes.

Since p_1 and p_2 are unknown, a weighted estimate of p can be computed by using the formula

$$\bar{p} = \frac{n_1 \hat{p}_1 + n_2 \hat{p}_2}{n_1 + n_2}$$

and $\bar{q} = 1 - \bar{p}$. This weighted estimate is based on the hypothesis that $p_1 = p_2$. Hence, $\bar{p}$ is a better estimate than either $\hat{p}_1$ or $\hat{p}_2$, since it is a combined average using both $\hat{p}_1$ and $\hat{p}_2$.

Since $\hat{p}_1 = X_1/n_1$ and $\hat{p}_2 = X_2/n_2$, $\bar{p}$ can be simplified to

$$\bar{p} = \frac{X_1 + X_2}{n_1 + n_2}$$

Finally, the standard error of the difference in terms of the weighted estimate is

$$\sigma_{(\hat{p}_1 - \hat{p}_2)} = \sqrt{\bar{p}\,\bar{q}\left(\frac{1}{n_1} + \frac{1}{n_2}\right)}$$

The formula for the test value is shown next.

Formula for the *z* Test for Comparing Two Proportions

$$z = \frac{(\hat{p}_1 - \hat{p}_2) - (p_1 - p_2)}{\sqrt{\bar{p}\,\bar{q}\left(\dfrac{1}{n_1} + \dfrac{1}{n_2}\right)}}$$

where

$$\bar{p} = \frac{X_1 + X_2}{n_1 + n_2} \qquad \hat{p}_1 = \frac{X_1}{n_1}$$

$$\bar{q} = 1 - \bar{p} \qquad \hat{p}_2 = \frac{X_2}{n_2}$$

This formula follows the format

$$\text{Test value} = \frac{(\text{observed value}) - (\text{expected value})}{\text{standard error}}$$

There are two requirements for use of the z test: (1) The samples must be independent of each other, and (2) n_1p_1 and n_1q_1 must be 5 or more, and n_2p_2 and n_2q_2 must be 5 or more.

Example 9–15

In the nursing home study mentioned in the chapter-opening Statistics Today, the researchers found that 12 out of 34 small nursing homes had a resident vaccination rate of less than 80%, while 17 out of 24 large nursing homes had a vaccination rate of less than 80%. At $\alpha = 0.05$, test the claim that there is no difference in the proportions of the small and large nursing homes with a resident vaccination rate of less than 80%.

Source: Nancy Arden, Arnold S. Monto, and Suzanne E. Ohmit, "Vaccine Use and the Risk of Outbreaks in a Sample of Nursing Homes During an Influenza Epidemic," *American Journal of Public Health.*

Solution

Let $\hat{p}_1$ be the proportion of the small nursing homes with a vaccination rate of less than 80% and $\hat{p}_2$ be the proportion of the large nursing homes with a vaccination rate of less than 80%. Then

$$\hat{p}_1 = \frac{X_1}{n_1} = \frac{12}{34} = 0.35 \qquad \text{and} \qquad \hat{p}_2 = \frac{X_2}{n_2} = \frac{17}{24} = 0.71$$

$$\bar{p} = \frac{X_1 + X_2}{n_1 + n_2} = \frac{12 + 17}{34 + 24} = \frac{29}{58} = 0.5$$

$$\bar{q} = 1 - \bar{p} = 1 - 0.5 = 0.5$$

Now, follow the steps in hypothesis testing.

Step 1 State the hypotheses and identify the claim.

$$H_0\colon p_1 = p_2 \text{ (claim)} \qquad \text{and} \qquad H_1\colon p_1 \neq p_2$$

Step 2 Find the critical values. Since $\alpha = 0.05$, the critical values are $+1.96$ and -1.96.

Step 3 Compute the test value.

$$z = \frac{(\hat{p}_1 - \hat{p}_2) - (p_1 - p_2)}{\sqrt{\bar{p}\,\bar{q}\left(\dfrac{1}{n_1} + \dfrac{1}{n_2}\right)}}$$

$$= \frac{(0.35 - 0.71) - 0}{\sqrt{(0.5)(0.5)\left(\dfrac{1}{34} + \dfrac{1}{24}\right)}} = \frac{-0.36}{0.1333} = -2.7$$

Step 4 Make the decision. Reject the null hypothesis, since $-2.7 < -1.96$. See Figure 9–13.

Figure 9–13

Critical and Test Values for Example 9–15

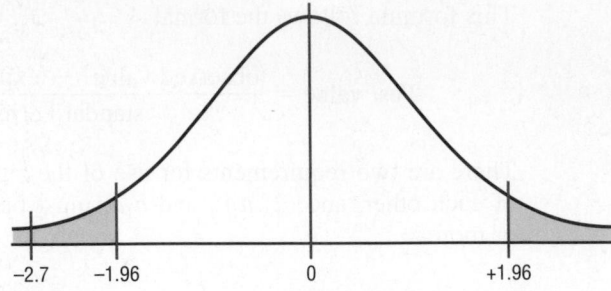

-2.7 -1.96 0 $+1.96$

Step 5   Summarize the results. There is enough evidence to reject the claim that there is no difference in the proportions of small and large nursing homes with a resident vaccination rate of less than 80%.

Example 9–16

In a sample of 200 surgeons, 15% thought the government should control health care. In a sample of 200 general practitioners, 21% felt the same way. At $\alpha = 0.01$, is there a difference in the proportions?

Solution

Since the statistics are given in percentages, $\hat{p}_1 = 15\%$ or 0.15 and $\hat{p}_2 = 21\%$ or 0.21. To compute $\bar{p}$, one must find X_1 and X_2.

$$X_1 = \hat{p}_1 n_1 = 0.15(200) = 30$$

$$X_2 = \hat{p}_2 n_2 = 0.21(200) = 42$$

$$\bar{p} = \frac{X_1 + X_2}{n_1 + n_2} = \frac{30 + 42}{200 + 200} = \frac{72}{400} = 0.18$$

$$\bar{q} = 1 - \bar{p} = 1 - 0.18 = 0.82$$

Step 1   State the hypotheses and identify the claim.

$$H_0: p_1 = p_2 \qquad \text{and} \qquad H_1: p_1 \neq p_2 \text{ (claim)}$$

Step 2   Find the critical values. Since $\alpha = 0.01$, the critical values are $+2.58$ and -2.58.

Step 3   Compute the test value.

$$z = \frac{(\hat{p}_1 - \hat{p}_2) - (p_1 - p_2)}{\sqrt{\bar{p}\,\bar{q}\left(\dfrac{1}{n_1} + \dfrac{1}{n_2}\right)}}$$

$$= \frac{(0.15 - 0.21) - 0}{\sqrt{(0.18)(0.82)\left(\dfrac{1}{200} + \dfrac{1}{200}\right)}} = -1.56$$

Step 4   Make the decision. Do not reject the null hypothesis since $-1.56 > -2.58$. See Figure 9–14.

Figure 9–14

Critical and Test Values for Example 9–16

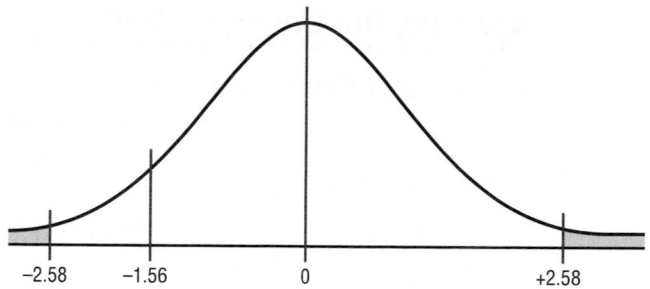

-2.58   -1.56   0   +2.58

Step 5 Summarize the results. There is not enough evidence to support the claim that there is a difference in proportions.

The *P*-value for the difference of proportions can be found from Table E, as shown in Section 8–3. For Example 9–16, 3.99 is beyond 3.09; hence, the null hypothesis can be rejected since the *P*-value is less than 0.001.

The formula for the confidence interval for the difference between two proportions is shown next.

Confidence Interval for the Difference Between Two Proportions

$$(\hat{p}_1 - \hat{p}_2) - z_{\alpha/2}\sqrt{\frac{\hat{p}_1\hat{q}_1}{n_1} + \frac{\hat{p}_2\hat{q}_2}{n_2}} < p_1 - p_2 < (\hat{p}_1 - \hat{p}_2) + z_{\alpha/2}\sqrt{\frac{\hat{p}_1\hat{q}_1}{n_1} + \frac{\hat{p}_2\hat{q}_2}{n_2}}$$

Example 9–17

Find the 95% confidence interval for the difference of proportions for the data in Example 9–15.

Solution

$$\hat{p}_1 = \frac{12}{34} = 0.35 \qquad \hat{q}_1 = 0.65$$

$$\hat{p}_2 = \frac{17}{24} = 0.71 \qquad \hat{q}_2 = 0.29$$

Substitute in the formula.

$$(\hat{p}_1 - \hat{p}_2) - z_{\alpha/2}\sqrt{\frac{\hat{p}_1\hat{q}_1}{n_1} + \frac{\hat{p}_2\hat{q}_2}{n_2}} < p_1 - p_2$$

$$< (\hat{p}_1 - \hat{p}_2) + z_{\alpha/2}\sqrt{\frac{\hat{p}_1\hat{q}_1}{n_1} + \frac{\hat{p}_2\hat{q}_2}{n_2}}$$

$$(0.35 - 0.71) - 1.96\sqrt{\frac{(0.35)(0.65)}{34} + \frac{(0.71)(0.29)}{24}}$$

$$< p_1 - p_2 < (0.35 - 0.71) + 1.96\sqrt{\frac{(0.35)(0.65)}{34} + \frac{(0.71)(0.29)}{24}}$$

$$-0.36 - 0.242 < p_1 - p_2 < -0.36 + 0.242$$

$$-0.602 < p_1 - p_2 < -0.118$$

Since 0 is not contained in the interval, the decision is to reject the null hypothesis $H_0: p_1 = p_2$.

Applying the Concepts 9–6

Smoking and Education

You are researching the hypothesis that there is no difference in the percent of public school students who smoke and the percent of private school students who smoke. You find these results from a recent survey.

School	Percent who smoke
Public	32.3
Private	14.5

Based on these figures, answer the following questions.

1. What hypotheses would you use if you wanted to compare percentages of the public school students who smoke with the private school students who smoke?
2. What critical value(s) would you use?
3. What statistical test would you use to compare the two percentages?
4. What information would you need to complete the statistical test?
5. Suppose you found that 1000 individuals in each group were surveyed. Could you perform the statistical test?
6. If so, complete the test and summarize the results.

See page 526 for the answers.

Exercises 9–6

1a. Find the proportions $\hat{p}$ and $\hat{q}$ for each.

  a. $n = 48, X = 34$
  b. $n = 75, X = 28$
  c. $n = 100, X = 50$
  d. $n = 24, X = 6$
  e. $n = 144, X = 12$

1b. Find each X, given $\hat{p}$.

  a. $\hat{p} = 0.16, n = 100$
  b. $\hat{p} = 0.08, n = 50$
  c. $\hat{p} = 6\%, n = 800$
  d. $\hat{p} = 52\%, n = 200$
  e. $\hat{p} = 20\%, n = 150$

2. Find $\bar{p}$ and $\bar{q}$ for each.

  a. $X_1 = 60, n_1 = 100, X_2 = 40, n_2 = 100$
  b. $X_1 = 22, n_1 = 50, X_2 = 18, n_2 = 30$
  c. $X_1 = 18, n_1 = 60, X_2 = 20, n_2 = 80$
  d. $X_1 = 5, n_1 = 32, X_2 = 12, n_2 = 48$
  e. $X_1 = 12, n_1 = 75, X_2 = 15, n_2 = 50$

For Exercises 3 through 14, perform these steps.

  a. State the hypotheses and identify the claim.
  b. Find the critical value(s).

  c. Compute the test value.
  d. Make the decision.
  e. Summarize the results.

Use the traditional method of hypothesis testing unless otherwise specified.

3. A sample of 150 people from a certain industrial community showed that 80 people suffered from a lung disease. A sample of 100 people from a rural community showed that 30 suffered from the same lung disease. At $\alpha = 0.05$, is there a difference between the proportions of people who suffer from the disease in the two communities?

4. A study is conducted to determine if the percent of women who receive financial aid in undergraduate school is different from the percent of men who receive financial aid in undergraduate school. A random sample of undergraduates revealed these results. At $\alpha = 0.01$, is there significant evidence to reject the null hypothesis?

	Women	Men
Sample size	250	300
Number receiving aid	200	180

Source: U.S. Department of Education, National Center for Education Statistics.

5. Labor statistics indicate that 77% of cashiers and servers are women. A random sample of cashiers and servers in a large metropolitan area found that 112 of 150 cashiers and 150 of 200 servers were women. At the 0.05 level of significance, is there sufficient evidence to conclude that a difference exists between the proportion of servers and the proportion of cashiers who are women?

   Source: *N.Y. Times Almanac.*

6. In Cleveland, a sample of 73 mail carriers showed that 10 had been bitten by an animal during one week. In Philadelphia, in a sample of 80 mail carriers, 16 had received animal bites. Is there a significant difference in the proportions? Use $\alpha = 0.05$. Find the 95% confidence interval for the difference of the two proportions.

7. A survey found that 83% of the men questioned preferred computer-assisted instruction to lecture and 75% of the women preferred computer-assisted instruction to lecture. There were 100 individuals in each sample. At $\alpha = 0.05$, test the claim that there is no difference in the proportion of men and the proportion of women who favor computer-assisted instruction over lecture. Find the 95% confidence interval for the difference of the two proportions.

8. In a sample of 50 men, 44 said that they had less leisure time today than they had 10 years ago. In a sample of 50 women, 48 women said that they had less leisure time than they had 10 years ago. At $\alpha = 0.10$ is there a difference in the proportion? Find the 90% confidence interval for the difference of the two proportions. Does the confidence interval contain 0? Give a reason why this information would be of interest to a researcher.

   Source: Based on statistics from Market Directory.

9. In a sample of 80 Americans, 55% wished that they were rich. In a sample of 90 Europeans, 45% wished that they were rich. At $\alpha = 0.01$, is there a difference in the proportions? Find the 99% confidence interval for the difference of the two proportions.

10. In a sample of 200 men, 130 said they used seat belts. In a sample of 300 women, 63 said they used seat belts. Test the claim that men are more safety conscious than women, at $\alpha = 0.01$. Use the *P*-value method.

11. A survey found that in a sample of 75 families, 26 owned dogs. A survey done 15 years ago found that in a sample of 60 families, 26 owned dogs. At $\alpha = 0.05$ has the proportion of dog owners changed over the 15-year period? Find the 95% confidence interval of the true difference in the proportions. Does the confidence interval contain 0? Why would this fact be important to a researcher?

   Source: Based on statistics from the American Veterinary Medical Association.

12. A recent study showed that in a sample of 100 people, 30% had visited Disneyland. In another sample of 100 people, 24% had visited Disney World. Are the proportions of people who visited each park different? Use $\alpha = 0.02$ and the *P*-value method.

13. A sample of 200 teenagers shows that 50 believe that war is inevitable, and a sample of 300 people over age 60 shows that 93 believe war is inevitable. Is the proportion of teenagers who believe war is inevitable different from the proportion of people over age 60 who do? Use $\alpha = 0.01$. Find the 99% confidence interval for the difference of the two proportions.

14. In a sample of 50 high school seniors, 8 had their own cars. In a sample of 75 college freshmen, 20 had their own cars. At $\alpha = 0.05$, can it be concluded that a higher proportion of college freshmen have their own cars? Use the *P*-value method.

15. Find the 99% confidence interval for the difference in the population proportions for the data of a study in which 80% of the 150 Republicans surveyed favored the bill for a salary increase and 60% of the 200 Democrats surveyed favored the bill for a salary increase.

16. The Miami County commissioners feel that a higher percentage of women work there than in neighboring Greene County. To test this, they randomly select 1000 women in each county and find that in Miami, 622 women work and in Greene, 594 work. Using $\alpha = 0.05$, do you think the Miami County commissioners are correct?

   Source: 2000 U.S. Census/*Dayton Daily News.*

17. In a sample of 100 store customers, 43 used a MasterCard. In another sample of 100, 58 used a Visa card. At $\alpha = 0.05$, is there a difference in the proportion of people who use each type of credit card?

18. Find the 95% confidence interval for the true difference in proportions for the data of a study in which 40% of the 200 males surveyed opposed the death penalty and 56% of the 100 females surveyed opposed the death penalty.

19. The percentages of adults 25 years of age and older who have completed 4 or more years of college are 23.6% for females and 27.8% for males. A random sample of women and men who were 25 years old or older was surveyed with these results. Estimate the true difference in proportions with 95% confidence, and compare your interval with the *Almanac* statistics.

	Women	Men
Sample size	350	400
No. who completed 4 or more years	100	115

   Source: *N.Y. Times Almanac.*

Extending the Concepts

20. If there is a significant difference between p_1 and p_2 and between p_2 and p_3, can one conclude that there is a significant difference between p_1 and p_3?

Technology *Step by Step*

MINITAB
Step by Step

Test the Difference Between Two Proportions

For Example 9–15, test for a difference in the resident vaccination rates between small and large nursing homes.

1. This test does not require data. It doesn't matter what is in the worksheet.
2. Select **Stat>Basic Statistics>2 Proportions.**
3. Click the button for Summarized data.
4. Press **TAB** to move cursor to the first sample box for Trials.
   a) Enter **34, TAB,** then enter **12.**
   b) Press **TAB** or click in the second sample text box for Trials.
   c) Enter **24, TAB,** then enter **17.**
5. Click on [Options]. Check the box for Use pooled estimate of p for test. The Confidence level should be 95%, and the Test difference should be 0.
6. Click [OK] twice. The results are shown in the session window.

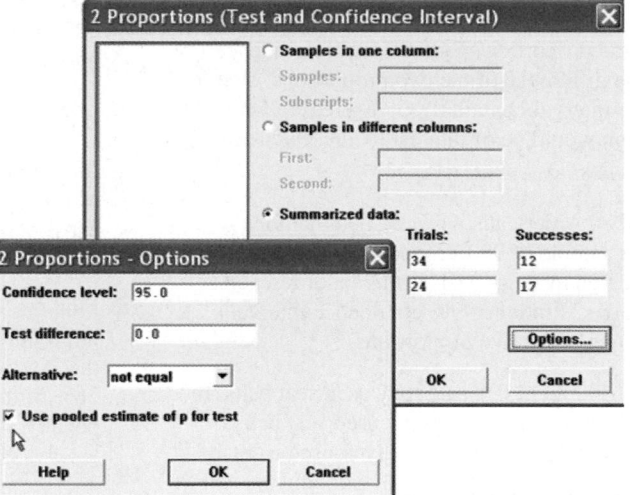

Test and CI for Two Proportions

```
Sample   X    N   Sample p
1       12   34   0.352941
2       17   24   0.708333


Difference = p (1) - p (2)
Estimate for difference:  -0.355392
95% CI for difference:  (-0.598025, -0.112759)
Test for difference = 0 (vs not = 0):  Z = -2.67 P-Value = 0.008
```

The *P*-value of the test is 0.008. Reject the null hypothesis. The difference is statistically significant. Of all small nursing homes 35%, compared to 71% of all large nursing homes, have an immunization rate of 80%. We can't tell why, only that there is a difference.

TI-83 Plus or TI-84 Plus
Step by Step

Hypothesis Test for the Difference Between Two Proportions

1. Press **STAT** and move the cursor to TESTS.

2. Press **6** for 2-PropZTEST.

3. Type in the appropriate values.

4. Move the cursor to the appropriate alternative hypothesis and press **ENTER**.

5. Move the cursor to Calculate and press **ENTER**.

Confidence Interval for the Difference Between Two Proportions

1. Press **STAT** and move the cursor to TESTS.

2. Press **B (ALPHA APPS)** for 2-PropZInt.

3. Type in the appropriate values.

4. Move the cursor to Calculate and press **ENTER**.

Excel
Step by Step

Testing the Difference Between Two Proportions

Excel does not have a procedure to test the difference between two proportions. However, you may test the difference between two proportions by using the MegaStat Add-in available on your CD and Online Learning Center. If you have not installed this add-in, do so by following the instructions on page 24.

We will use the summary information from Example 9–15.

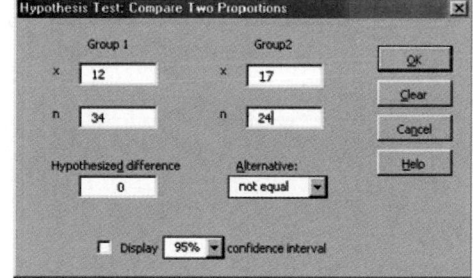

1. Select **MegaStat>Hypothesis tests>Compare Two Independent Proportions.**

2. Enter the information as indicated in the figure and click [OK].

Note: This hypothesis test is conducted using the *P*-value method; thus no significance level is required.

9–7   Summary

Many times researchers are interested in comparing two population parameters, such as means or proportions. This comparison can be accomplished by using special *z* and *t* tests. If the samples are independent and the variances are known, the *z* test is used. The *z* test is also used when the variances are unknown but both sample sizes are 30 or more. If the variances are not known and one or both sample sizes are less than 30, the *t* test must be used. For independent samples, a further requirement is that one must determine whether the variances of the populations are equal. The *F* test is used to determine whether the variances are equal. Different formulas are used in each case. If the samples are dependent, the *t* test for dependent samples is used. Finally, a *z* test is used to compare two proportions.

Important Terms

Important Formulas

Formula for the z test for comparing two means from independent populations:

$$z = \frac{(\overline{X}_1 - \overline{X}_2) - (\mu_1 - \mu_2)}{\sqrt{\dfrac{\sigma_1^2}{n_1} + \dfrac{\sigma_2^2}{n_2}}}$$

Formula for the confidence interval for difference of two means (large samples):

$$(\overline{X}_1 - \overline{X}_2) - z_{\alpha/2}\sqrt{\frac{\sigma_1^2}{n_1} + \frac{\sigma_2^2}{n_2}} < \mu_1 - \mu_2$$

$$< (\overline{X}_1 - \overline{X}_2) + z_{\alpha/2}\sqrt{\frac{\sigma_1^2}{n_1} + \frac{\sigma_2^2}{n_2}}$$

Formula for the F test for comparing two variances:

$$F = \frac{s_1^2}{s_2^2} \qquad \begin{array}{l} \text{d.f.N.} = n_1 - 1 \\ \text{d.f.D.} = n_2 - 1 \end{array}$$

Formula for the t test for comparing two means (small independent samples, variances not equal):

$$t = \frac{(\overline{X}_1 - \overline{X}_2) - (\mu_1 - \mu_2)}{\sqrt{\dfrac{s_1^2}{n_1} + \dfrac{s_2^2}{n_2}}}$$

and d.f. = the smaller of $n_1 - 1$ or $n_2 - 1$.

Formula for the t test for comparing two means (independent samples, variances equal):

$$t = \frac{(\overline{X}_1 - \overline{X}_2) - (\mu_1 - \mu_2)}{\sqrt{\dfrac{(n_1 - 1)s_1^2 + (n_2 - 1)s_2^2}{n_1 + n_2 - 2}}\sqrt{\dfrac{1}{n_1} + \dfrac{1}{n_2}}}$$

and d.f. = $n_1 + n_2 - 2$.

Formula for the confidence interval for the difference of two means (small independent samples, variances unequal):

$$(\overline{X}_1 - \overline{X}_2) - t_{\alpha/2}\sqrt{\frac{s_1^2}{n_1} + \frac{s_2^2}{n_2}} < \mu_1 - \mu_2$$

$$< (\overline{X}_1 - \overline{X}_2) + t_{\alpha/2}\sqrt{\frac{s_1^2}{n_1} + \frac{s_2^2}{n_2}}$$

and d.f. = smaller of $n_1 - 1$ and $n_2 - 2$.

Formula for the confidence interval for the difference of two means (small independent samples, variances equal):

$$(\overline{X}_1 - \overline{X}_2) - t_{\alpha/2}\sqrt{\frac{(n_1 - 1)s_1^2 + (n_2 - 1)s_2^2}{n_1 + n_2 - 2}} \cdot \sqrt{\frac{1}{n_1} + \frac{1}{n_2}}$$

$$< \mu_1 - \mu_2$$

$$< (\overline{X}_1 - \overline{X}_2) + t_{\alpha/2}\sqrt{\frac{(n_1 - 1)s_1^2 + (n_2 - 1)s_2^2}{n_1 + n_2 - 2}} \cdot \sqrt{\frac{1}{n_1} + \frac{1}{n_2}}$$

and d.f. = $n_1 + n_2 - 2$.

Formula for the t test for comparing two means from dependent samples:

$$t = \frac{\overline{D} - \mu_D}{s_D/\sqrt{n}}$$

where $\overline{D}$ is the mean of the differences

$$\overline{D} = \frac{\Sigma D}{n}$$

and s_D is the standard deviation of the differences

$$s_D = \sqrt{\frac{\Sigma D^2 - \dfrac{(\Sigma D)}{n}}{n - 1}}$$

Formula for confidence interval for the mean of the difference for dependent samples:

$$\overline{D} - t_{\alpha/2}\frac{s_D}{\sqrt{n}} < \mu_D < \overline{D} + t_{\alpha/2}\frac{s_D}{\sqrt{n}}$$

and d.f. = $n - 1$.

Formula for the z test for comparing two proportions:

$$z = \frac{(\hat{p}_1 - \hat{p}_2) - (p_1 - p_2)}{\sqrt{\overline{p}\,\overline{q}\left(\dfrac{1}{n_1} + \dfrac{1}{n_2}\right)}}$$

where

$$\overline{p} = \frac{X_1 + X_2}{n_1 + n_2} \qquad \hat{p}_1 = \frac{X_1}{n_1}$$

$$\overline{q} = 1 - \overline{p} \qquad \hat{p}_2 = \frac{X_2}{n_2}$$

Formula for confidence interval for the difference of two proportions:

$$(\hat{p}_1 - \hat{p}_2) - z_{\alpha/2}\sqrt{\frac{\hat{p}_1\hat{q}_1}{n_1} + \frac{\hat{p}_2\hat{q}_2}{n_2}} < p_1 - p_2$$

$$< (\hat{p}_1 - \hat{p}_2) + z_{\alpha/2}\sqrt{\frac{\hat{p}_1\hat{q}_1}{n_1} + \frac{\hat{p}_2\hat{q}_2}{n_2}}$$

Review Exercises

For each exercise, perform these steps. Assume that all variables are normally or approximately normally distributed.

 a. State the hypotheses and identify the claim.
 b. Find the critical value(s).
 c. Compute the test value.
 d. Make the decision.
 e. Summarize the results.

Use the traditional method of hypothesis testing unless otherwise specified.

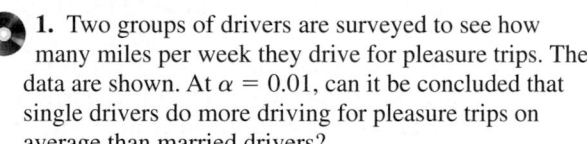 **1.** Two groups of drivers are surveyed to see how many miles per week they drive for pleasure trips. The data are shown. At $\alpha = 0.01$, can it be concluded that single drivers do more driving for pleasure trips on average than married drivers?

Single drivers					Married drivers				
106	110	115	121	132	97	104	138	102	115
119	97	118	122	135	133	120	119	136	96
110	117	116	138	142	139	108	117	145	114
115	114	103	98	99	140	136	113	113	150
108	117	152	147	117	101	114	116	113	135
154	86	115	116	104	115	109	147	106	88
107	133	138	142	140	113	119	99	108	105

2. An educator wishes to compare the variances of the amount of money spent per pupil in two states. The data are given. At $\alpha = 0.05$, is there a significant difference in the variances of the amounts the states spend per pupil?

State 1	State 2
$s_1^2 = \$585$	$s_2^2 = \$261$
$n_1 = 18$	$n_2 = 16$

3. In the hospital study cited in Exercise 19 in Exercise set 7–2, the standard deviation of the noise levels of the 11 intensive care units was 4.1 dBA, and the standard deviation of the noise levels of 24 nonmedical care areas, such as kitchens and machine rooms, was 13.2 dBA. At $\alpha = 0.10$, is there a significant difference between the standard deviations of these two areas?

Source: M. Bayo, A. Garcia, and A. Garcia, "Noise Levels in an Urban Hospital and Workers' Subjective Responses," *Archives of Environmental Health.*

4. A researcher wants to compare the variances of the heights (in inches) of major league baseball players with those of players in the minor leagues. A sample of 25 players from each league is selected, and the variances of the heights for each league are 2.25 and 4.85, respectively. At $\alpha = 0.10$, is there a significant difference between the variances of the heights for the two leagues?

5. A traffic safety commissioner believes the variation in the number of speeding tickets given on Route 19 is greater than the variation in the number of speeding tickets given on Route 22. Ten weeks are randomly selected; the standard deviation of the number of tickets issued for Route 19 is 6.3, and the standard deviation of the number of tickets issued for Route 22 is 2.8. At $\alpha = 0.05$, can the commissioner conclude that the variance of speeding tickets issued on Route 19 is greater than the variance of speeding tickets issued on Route 22? Use the P-value method.

6. The variations in the number of absentees per day in two schools are being compared. A sample of 30 days is selected; the standard deviation of the number of absentees in school A is 4.9, and for school B it is 2.5. At $\alpha = 0.01$, can one conclude that there is a difference in the two standard deviations?

7. A researcher claims the variation in the number of days that factory workers miss per year due to illness is greater than the variation in the number of days that hospital workers miss per year. A sample of 42 workers from a large hospital has a standard deviation of 2.1 days, and a sample of 65 workers from a large factory has a standard deviation of 3.2 days. Test the claim at $\alpha = 0.10$.

8. The average price of 15 cans of tomato soup from different stores is $0.73, and the standard deviation is $0.05. The average price of 24 cans of chicken noodle soup is $0.91, and the standard deviation is $0.03. At $\alpha = 0.01$, is there a significant difference in price?

9. The average temperatures for a 25-day period for Birmingham, Alabama, and Chicago, Illinois, are shown. Based on the samples, at $\alpha = 0.10$, can it be concluded that it is warmer in Birmingham?

Birmingham					Chicago				
78	82	68	67	68	70	74	73	60	77
75	73	75	64	68	71	72	71	74	76
62	73	77	78	79	71	80	65	70	83
74	72	73	78	68	67	76	75	62	65
73	79	82	71	66	66	65	77	66	64

10. A sample of 15 teachers from Rhode Island has an average salary of $35,270, with a standard deviation of $3256. A sample of 30 teachers from New York has an average salary of $29,512, with a standard deviation of $1432. Is there a significant difference in teachers' salaries between the two states? Use $\alpha = 0.02$. Find the 98% confidence interval for the difference of the two means.

11. The data show the amounts (in thousands of dollars) of the contracts for soft drinks in local school districts. At

$\alpha = 0.10$ can it be concluded that there is a difference in the averages? Use the *P*-value method. Give a reason why the result would be of concern to a cafeteria manager.

Pepsi						Coca-Cola		
46	120	80	500	100	59	420	285	57

Source: Local school districts.

 12. In an effort to improve the vocabulary of 10 students, a teacher provides a weekly 1-hour tutoring session for them. A pretest is given before the sessions, and a posttest is given afterward. The results are shown in the table. At $\alpha = 0.01$, can the teacher conclude that the tutoring sessions helped to improve the students' vocabularies?

Before	1	2	3	4	5	6	7	8	9	10
Pretest	83	76	92	64	82	68	70	71	72	63
Posttest	88	82	100	72	81	75	79	68	81	70

 13. In an effort to increase production of an automobile part, the factory manager decides to play music in the manufacturing area. Eight workers are selected, and the number of items each produced for a specific day is recorded. After one week of music, the same workers are monitored again. The data are given in the table. At $\alpha = 0.05$, can the manager conclude that the music has increased production?

Worker	1	2	3	4	5	6	7	8
Before	6	8	10	9	5	12	9	7
After	10	12	9	12	8	13	8	10

14. St. Petersburg, Russia, has 207 foggy days out of 365 days while Stockholm, Sweden, has 166 foggy days out of 365. At $\alpha = 0.02$, can it be concluded that the proportions of foggy days for the two cities are different? Find the 98% confidence interval for the difference of the two proportions.

Source: Jack Williams, *USA TODAY.*

15. A recent survey found that in a study of 120 families, 18 owned a bird. Fifteen years ago, a survey found that in a sample of 100 families, 5 owned a bird. At $\alpha = 0.05$ has the proportion of families who owned a bird changed? Find the confidence interval for the difference of the proportions. Does the confidence interval contain 0? Why would this be important for a researcher to know?

Source: Based on statistics from the American Veterinary Medical Association.

To Vaccinate or Not to Vaccinate? Small or Large?–Revisited

Using a *z* test to compare two proportions, the researchers found that the proportion of residents in smaller nursing homes who were vaccinated (80.8%) was statistically greater than that of residents in large nursing homes who were vaccinated (68.7%). Using statistical methods presented in later chapters, they also found that the larger size of the nursing home and the lower frequency of vaccination were significant predictions of influenza outbreaks in nursing homes.

Data Analysis

The Data Bank is found in Appendix D, or on the World Wide Web by following links from www.mhhe.com/math/stat/bluman/

1. From the Data Bank, select a variable and compare the mean of the variable for a random sample of at least 30 men with the mean of the variable for the random sample of at least 30 women. Use a *z* test.

2. Repeat the experiment in Exercise 1, using a different variable and two samples of size 15. Compare the means by using a *t* test. Assume that the variances are equal.

3. Compare the proportion of men who are smokers with the proportion of women who are smokers. Use the data in the Data Bank. Choose random samples of size 30 or more. Use the *z* test for proportions.

4. Select two samples of 20 values from the data in Data Set IV in Appendix D. Test the hypothesis that the mean heights of the buildings are equal.

5. Using the same data obtained in Exercise 4, test the hypothesis that the variances are equal.

Chapter Quiz

Determine whether each statement is true or false. If the statement is false, explain why.

1. When one is testing the difference between two means for small samples, it is not important to distinguish whether the samples are independent of each other.

2. If the same diet is given to two groups of randomly selected individuals, the samples are considered to be dependent.

3. When computing the F test value, one always places the larger variance in the numerator of the fraction.

4. Tests for variances are always two-tailed.

Select the best answer.

5. To test the equality of two variances, one would use a(n) _____ test.

    a. z *c.* Chi-square
    b. t *d.* F

6. To test the equality of two proportions, one would use a(n) _____ test.

    a. z *c.* Chi-square
    b. t *d.* F

7. The mean value of the F is approximately equal to

    a. 0   *c.* 1
    b. 0.5   *d.* It cannot be determined.

8. What test can be used to test the difference between two small sample means?

    a. z *c.* Chi-square
    b. t *d.* F

Complete these statements with the best answer.

9. If one hypothesizes that there is no difference between means, this is represented as H_0: _____.

10. When one is testing the difference between two means, a(n) _____ estimate of the variances is used when the variances are equal.

11. When the t test is used for testing the equality of two means, the populations must be _____.

12. The values of F cannot be _____.

13. The formula for the F test for variances is _____.

For each of these problems, perform the following steps.

   a. State the hypotheses and identify the claim.
   b. Find the critical value(s).
   c. Compute the test value.
   d. Make the decision.
   e. Summarize the results.

Use the traditional method of hypothesis testing unless otherwise specified.

14. A researcher wishes to see if there is a difference in the cholesterol levels of two groups of men. A random sample of 30 men between the ages of 25 and 40 is selected and tested. The average level is 223. A second sample of 25 men between the ages of 41 and 56 is selected and tested. The average of this group is 229. The population standard deviation for both groups is 6. At $\alpha = 0.01$, is there a difference in the cholesterol levels between the two groups? Find the 99% confidence interval for the difference of the two means.

15. The data shown are the rental fees (in dollars) for two random samples of apartments in a large city. At $\alpha = 0.10$, can it be concluded that the average rental fee for apartments in the East is greater than the average rental fee in the West?

East					West				
495	390	540	445	420	525	400	310	375	750
410	550	499	500	550	390	795	554	450	370
389	350	450	530	350	385	395	425	500	550
375	690	325	350	799	380	400	450	365	425
475	295	350	485	625	375	360	425	400	475
275	450	440	425	675	400	475	430	410	450
625	390	485	550	650	425	450	620	500	400
685	385	450	550	425	295	350	300	360	400

Source: Pittsburgh Post-Gazette.

16. A politician wishes to compare the variances of the amount of money spent for road repair in two different counties. The data are given here. At $\alpha = 0.05$, is there a significant difference in the variances of the amounts spent in the two counties? Use the P-value method.

County A	County B
$s_1 = \$11,596$	$s_2 = \$14,837$
$n_1 = 15$	$n_2 = 18$

17. A researcher wants to compare the variances of the heights (in inches) of four-year college basketball players with those of players in junior colleges. A sample of 30 players from each type of school is selected, and the variances of the heights for each type are 2.43 and 3.15, respectively. At $\alpha = 0.10$, is there a significant difference between the variances of the heights in the two types of schools?

18. The data shown are based on a survey taken in February and July and indicate the number of hours per day of household television usage. At $\alpha = 0.05$, test the claim that there is no difference in the standard deviations of the number of hours that televisions are used.

February			July		
7.6	9.3	8.2	7.4	10.3	9.4
7.4	7.9	6.8	4.6	7.3	7.1
7.5	7.1	6.4	6.8	7.7	8.2
4.3	10.6	9.8	5.4	6.2	7.1

19. The variances in the amount of fat in two different types of ground beef are compared. Eight samples of the first type, Super Lean, have a variance of 18.2 grams; 12 of the second type, Ultimate Lean, have a variance of 9.4 grams. At $\alpha = 0.10$, can it be concluded that there is a difference in the variances of the two types of ground beef?

20. It is hypothesized that the variance in the number of murders committed in selected cities on the East Coast is greater than the variance of the number of murders committed on the West Coast. A sample of six selected cities on the West Coast has a standard deviation of 206. A sample of six cities on the East Coast has a standard deviation of 215. Test the hypothesis at $\alpha = 0.10$. Suggest a factor that may have influenced the results of this study.

Source: Based on FBI statistics.

21. The variations in the number of retail thefts per day in two shopping malls are being compared. A sample of 21 days is selected. The standard deviation of the number of retail thefts in mall A is 6.8, and for mall B it is 5.3. At $\alpha = 0.05$, can it be concluded that there is a difference in the two standard deviations?

22. The average price of a sample of 12 bottles of diet salad dressing taken from different stores is $1.43. The standard deviation is $0.09. The average price of a sample of 16 low-calorie frozen desserts is $1.03. The standard deviation is $0.10. At $\alpha = 0.01$, is there a significant difference in price? Find the 99% confidence interval of the difference in the means.

23. The data shown represent the number of accidents people had when using jet skis and other types of wet bikes. At $\alpha = 0.05$, can it be concluded that the average number of accidents per year has increased from one period to the next?

1987–1991			1992–1996		
376	650	844	1650	2236	3002
1162	1513		4028	4010	

Source: USA TODAY.

24. A sample of 12 chemists from Washington state shows an average salary of $39,420 with a standard deviation of $1659, while a sample of 26 chemists from New Mexico has an average salary of $30,215 with a standard deviation of $4116. Is there a significant difference between the two states in chemists' salaries at $\alpha = 0.02$? Find the 98% confidence interval of the difference in the means.

25. The average income of 15 families who reside in a large metropolitan East Coast city is $62,456. The standard deviation is $9652. The average income of 11 families who reside in a rural area of the Midwest is $60,213, with a standard deviation of $2009. At $\alpha = 0.05$, can it be concluded that the families who live in the cities have a higher income than those who live in the rural areas? Use the *P*-value method.

26. In an effort to improve the mathematical skills of 10 students, a teacher provides a weekly 1-hour tutoring session for the students. A pretest is given before the sessions, and a posttest is given after. The results are shown here. At $\alpha = 0.01$, can it be concluded that the sessions help to improve the students' mathematical skills?

Student	1	2	3	4	5	6	7	8	9	10
Pretest	82	76	91	62	81	67	71	69	80	85
Posttest	88	80	98	80	80	73	74	78	85	93

27. To increase egg production, a farmer decided to increase the amount of time the lights in his hen house were on. Ten hens were selected, and the number of eggs each produced was recorded. After one week of lengthened light time, the same hens were monitored again. The data are given here. At $\alpha = 0.05$, can it be concluded that the increased light time increased egg production?

Hen	1	2	3	4	5	6	7	8	9	10
Before	4	3	8	7	6	4	9	7	6	5
After	6	5	9	7	4	5	10	6	9	6

28. In a sample of 80 workers from a factory in city A, it was found that 5% were unable to read, while in a sample of 50 workers in city B, 8% were unable to read. Can it be concluded that there is a difference in the proportions of nonreaders in the two cities? Use $\alpha = 0.10$. Find the 90% confidence interval for the difference of the two proportions.

29. A recent survey of 200 households showed that 8 had a single male as the head of household. Forty years ago, a survey of 200 households showed that 6 had a single male as the head of household. At $\alpha = 0.05$ can it be concluded that the proportion has changed? Find the 95% confidence interval of the difference of the two proportions. Does the confidence interval contain 0? Why is this important to know?

Source: Based on data from the U.S. Census Bureau.

Critical Thinking Challenges

1. The study cited in the article entitled "Only the Timid Die Young" stated that "Timid rats were 60% more likely to die at any given time than were their outgoing brothers." Based on the results, answer the following questions.

  a. Why were rats used in the study?

  b. What are the variables in the study?

  c. Why were infants included in the article?

  d. What is wrong with extrapolating the results to humans?

  e. Suggest some ways humans might be used in a study of this type.

ONLY THE TIMID DIE YOUNG

DO OVERACTIVE STRESS HORMONES DAMAGE HEALTH?

ABOUT 15 OUT OF 100 CHILDREN ARE BORN SHY, BUT ONLY THREE WILL BE SHY AS ADULTS.

FEARFUL TYPES MAY MEET THEIR maker sooner, at least among rats. Researchers have for the first time connected a personality trait—fear of novelty—to an early death.

Sonia Cavigelli and Martha McClintock, psychologists at the University of Chicago, presented unfamiliar bowls, tunnels and bricks to a group of young male rats. Those hesitant to explore the mystery objects were classified as "neophobic."

The researchers found that the neophobic rats produced high levels of stress hormones, called glucocorticoids—typically involved in the fight-or-flight stress response—when faced with strange situations. Those rats continued to have high levels of the hormones at random times throughout their lives, indicating that timidity is a fixed and stable trait. The team then set out to examine the cumulative effects of this personality trait on the rats' health.

Timid rats were 60 percent more likely to die at any given time than were their outgoing brothers. The causes of death were similar for both groups. "One hypothesis as to why the neophobic rats died earlier is that the stress hormones negatively affected their immune system," Cavigelli says. Neophobes died, on average, three months before their rat brothers, a significant gap, considering that most rats lived only two years.

Shyness—the human equivalent of neophobia—can be detected in infants as young as 14 months. Shy people also produce more stress hormones than "average," or thrill-seeking humans. But introverts don't necessarily stay shy for life, as rats apparently do. Jerome Kagan, a professor of psychology at Harvard University, has found that while 15 out of every 100 children will be born with a shy temperament, only three will appear shy as adults. None, however, will be extroverts.

Extrapolating from the doomed fate of neophobic rats to their human counterparts is difficult. "But it means that something as simple as a personality trait could have physiological consequences," Cavigelli says.

—*Carlin Flora*

Reprinted with permission from *Psychology Today,* Copyright © 2004, Sussex Publishers, Inc.

2. Based on the study presented in the article entitled "Sleeping Brain, Not at Rest," answer these questions.

  a. What were the variables used in the study?

  b. How were they measured?

  c. Suggest a statistical test that might have been used to arrive at the conclusion.

  d. Based on the results, what would you suggest for students preparing for an exam?

SLEEPING BRAIN, NOT AT REST

Regions of the brain that have spent the day learning sleep more heavily at night.

In a study published in the journal *Nature*, Giulio Tononi, a psychiatrist at the University of Wisconsin–Madison, had subjects perform a simple point-and-click task with a computer adjusted so that its cursor didn't track in the right direction. Afterward, the subjects' brain waves were recorded while they slept, then examined for "slow wave" activity, a kind of deep sleep.

Compared with people who'd completed the same task with normal cursors, Tononi's subjects showed elevated slow wave activity in brain areas associated with spatial orientation, indicating that their brains were adjusting to the day's learning by making cellular-level changes. In the morning, Tononi's subjects performed their tasks better than they had before going to sleep.

—*Richard A. Love*

Reprinted with permission from *Psychology Today*, Copyright © 2004, Sussex Publishers, Inc.

 Data Projects

Where appropriate, use MINITAB, the TI-83 Plus, the TI-84 Plus, or a computer program of your choice to complete the following exercises.

1. Choose a variable for which you would like to determine if there is a difference in the averages for two groups. Make sure that the samples are independent. For example, you may wish to see if men see more movies or spend more money on lunch than women. Select a sample of data values (10 to 50) and complete the following:

    a. Write a brief statement of the purpose of the study.
    b. Define the population.
    c. State the hypotheses for the study.
    d. Select an α value.
    e. State how the sample was selected.
    f. Show the raw data.
    g. Decide which statistical test is appropriate and compute the test statistic (z or t). Why is the test appropriate?
    h. Find the critical value(s).
    i. State the decision.
    j. Summarize the results.

2. Choose a variable that will permit the use of dependent samples. For example, you might wish to see if a person's weight has changed after a diet. Select a sample of data (10 to 50) value pairs (e.g., before and after), and then complete the following:

    a. Write a brief statement of the purpose of the study.
    b. Define the population.
    c. State the hypotheses for the study.
    d. Select an α value.
    e. State how the sample was selected.
    f. Show the raw data.
    g. Decide which statistical test is appropriate and compute the test statistic (z or t). Why is the test appropriate?
    h. Find the critical value(s).
    i. State the decision.
    j. Summarize the results.

3. Choose a variable that will enable you to compare proportions of two groups. For example, you might want to see if the proportion of first-year students who buy used books is lower than (or higher than or the same as) the proportion of second-year students who buy used books. After collecting 30 or more responses from the two groups, complete the following:

    a. Write a brief statement of the purpose of the study.
    b. Define the population.
    c. State the hypotheses for the study.
    d. Select an α value.
    e. State how the sample was selected.
    f. Show the raw data.
    g. Decide which statistical test is appropriate and compute the test statistic (z or t). Why is the test appropriate?
    h. Find the critical value(s).
    i. State the decision.
    j. Summarize the results.

You may use the following websites to obtain raw data:

Visit the data sets at the book's website found at http://www.mhhe.com/math/stat/bluman Click on the 6th edition.

http://lib.stat.cmu.edu/DASL

http://www.statcan.ca

Hypothesis-Testing Summary 1

1. Comparison of a sample mean with a specific population mean.

   Example: H_0: $\mu = 100$

   a. Use the z test when σ is known:

   $$z = \frac{\bar{X} - \mu}{\sigma/\sqrt{n}}$$

   b. Use the t test when σ is unknown:

   $$t = \frac{\bar{X} - \mu}{s/\sqrt{n}} \quad \text{with d.f.} = n - 1$$

2. Comparison of a sample variance or standard deviation with a specific population variance or standard deviation.

   Example: H_0: $\sigma^2 = 225$

   Use the chi-square test:

   $$\chi^2 = \frac{(n-1)s^2}{\sigma^2} \quad \text{with d.f.} = n - 1$$

3. Comparison of two sample means.

   Example: H_0: $\mu_1 = \mu_2$

   a. Use the z test when the population variances are known:

   $$z = \frac{(\bar{X}_1 - \bar{X}_2) - (\mu_1 - \mu_2)}{\sqrt{\dfrac{\sigma_1^2}{n_1} + \dfrac{\sigma_2^2}{n_2}}}$$

   b. Use the t test for independent samples when the population variances are unknown and the sample variances are unequal:

   $$t = \frac{(\bar{X}_1 - \bar{X}_2) - (\mu_1 - \mu_2)}{\sqrt{\dfrac{s_1^2}{n_1} + \dfrac{s_2^2}{n_2}}}$$

   with d.f. = the smaller of $n_1 - 1$ or $n_2 - 1$.

   c. Use the t test for independent samples when the population variances are unknown and assumed to be equal:

   $$t = \frac{(\bar{X}_1 - \bar{X}_2) - (\mu_1 - \mu_2)}{\sqrt{\dfrac{(n_1 - 1)s_1^2 + (n_2 - 1)s_2^2}{n_1 + n_2 - 2}}\sqrt{\dfrac{1}{n_1} + \dfrac{1}{n_2}}}$$

   with d.f. = $n_1 + n_2 - 2$.

   d. Use the t test for means for dependent samples:

   Example: H_0: $\mu_D = 0$

   $$t = \frac{\bar{D} - \mu_D}{s_D/\sqrt{n}} \quad \text{with d.f.} = n - 1$$

   where n = number of pairs.

4. Comparison of a sample proportion with a specific population proportion.

   Example: H_0: $p = 0.32$

   Use the z test:

   $$z = \frac{X - \mu}{\sigma} \quad \text{or} \quad z = \frac{\hat{p} - p}{\sqrt{pq/n}}$$

5. Comparison of two sample proportions.

   Example: H_0: $p_1 = p_2$

   Use the z test:

   $$z = \frac{(\hat{p}_1 - \hat{p}_2) - (p_1 - p_2)}{\sqrt{\bar{p}\,\bar{q}\left(\dfrac{1}{n_1} + \dfrac{1}{n_2}\right)}}$$

   where

   $$\bar{p} = \frac{X_1 + X_2}{n_1 + n_2} \qquad \hat{p}_1 = \frac{X_1}{n_1}$$

   $$\bar{q} = 1 - \bar{p} \qquad \hat{p}_2 = \frac{X_2}{n_2}$$

6. Comparison of two sample variances or standard deviations.

   Example: H_0: $\sigma_1^2 = \sigma_2^2$

   Use the F test:

   $$F = \frac{s_1^2}{s_2^2}$$

   where

s_1^2 = larger variance	d.f.N. = $n_1 - 1$
s_2^2 = smaller variance	d.f.D. = $n_2 - 1$

Answers to Applying the Concepts

Section 9–2 Home Runs

1. The population is all home runs hit by major league baseball players.

2. A cluster sample was used.

3. Answers will vary. While this sample is not representative of all major league baseball players per se, it does allow us to compare the leaders in each league.

4. H_0: $\mu_1 = \mu_2$ and H_1: $\mu_1 \neq \mu_2$

5. Answers will vary. Possible answers include the 0.05 and 0.01 significance levels.

6. We will use the z test for the difference in means.

7. Out test statistic is $z = \dfrac{44.75 - 42.88}{\sqrt{\dfrac{8.88^2}{40} + \dfrac{7.82^2}{40}}} = 1.00$, and our

   P-value is 0.3173.

8. We fail to reject the null hypothesis.

9. There is not enough evidence to conclude that there is a difference in the number of home runs hit by National League versus American League baseball players.

10. Answers will vary. One possible answer is that since we do not have a random sample of data from each league, we cannot answer the original question asked.

11. Answers will vary. One possible answer is that we could get a random sample of data from each league from a recent season.

Section 9–3   Variability and Automatic Transmissions

1. The null hypothesis is that the variances are the same: H_0: $\sigma_1^2 = \sigma_2^2$.

2. We will use an F test.

3. The value of the test statistic is $F = \dfrac{s_1^2}{s_2^2} = \dfrac{77.7^2}{514.8^2} = 0.02$,

   and the P-value is 0.0006. There is a significant difference in the variability of the prices between the two countries.

4. Small sample sizes are highly impacted by outliers.

5. The degrees of freedom for the numerator and denominator are both 5.

6. Yes, two sets of data can center on the same mean but have very different standard deviations.

Section 9–4   Too Long on the Telephone

1. These samples are independent.

2. The F test was done to test for any significant difference in the variances.

3. The results of the F test are not significant at the 0.01 level, since the P-value is greater than 0.01.

4. There were $56 + 2 = 58$ people in the study.

5. We compare the P-value of 0.06317 to the significance level to check if the null hypothesis should be rejected.

6. The P-value of 0.06317 also gives the probability of a type I error.

7. The F value of 3.07849 is the result of dividing the two sample variances.

8. Since two critical values are shown, we know that a two-tailed test was done.

9. Since the P-value of 0.06317 is greater than the significance value of 0.05, we fail to reject the null

hypothesis and find that we do not have enough evidence to conclude that there is a difference in the lengths of telephone calls made by employees in the two divisions of the company.

10. If the significance level had been 0.10, we would have rejected the null hypothesis, since the P-value would have been less than the significance level.

Section 9–5   Air Quality

1. The purpose of the study is to determine if the air quality in the United States has changed over the past 2 years.

2. These are dependent samples, since we have two readings from each of 10 metropolitan areas.

3. The hypotheses we will test are H_0: $\mu_D = 0$ and H_1: $\mu_D \neq 0$.

4. We will use the 0.05 significance level and critical values of $t = \pm 2.262$.

5. We will use the t test for dependent samples.

6. There are $10 - 1 = 9$ degrees of freedom.

7. Our test statistic is $t = \dfrac{-6.7 - 0}{11.27/\sqrt{10}} = -1.879$. We fail

   to reject the null hypothesis and find that there is not enough evidence to conclude that the air quality in the United States has changed over the past 2 years.

8. No, we could not use an independent means test since we have two readings from each metropolitan area.

9. Answers will vary. One possible answer is that there are other measures of air quality that we could have examined to answer the question.

Section 9–6   Smoking and Education

1. Our hypotheses are H_0: $p_1 = p_2$ and H_1: $p_1 \neq p_2$.

2. At the 0.05 significance level, our critical values are $z = \pm 1.96$.

3. We will use the z test for the difference between proportions.

4. To complete the statistical test, we would need the sample sizes.

5. Knowing the sample sizes were 1000, we can now complete the test.

6. Our test statistic is $z = \dfrac{0.323 - 0.145}{\sqrt{(0.234)(0.766)\left(\dfrac{1}{1000} + \dfrac{1}{1000}\right)}} =$

   9.40, and our P-value is very close to zero. We reject the null hypothesis and find that there is enough evidence to conclude that there is a difference in the proportions of high school graduates and college graduates who smoke.

10

Correlation and Regression

Objectives

After completing this chapter, you should be able to

1. Draw a scatter plot for a set of ordered pairs.
2. Compute the correlation coefficient.
3. Test the hypothesis H_0: $\rho = 0$.
4. Compute the equation of the regression line.
5. Compute the coefficient of determination.
6. Compute the standard error of the estimate.
7. Find a prediction interval.
8. Be familiar with the concept of multiple regression.

Outline

Outline

10–1 **Introduction**

10–2 **Scatter Plots**

10–3 **Correlation**

10–4 **Regression**

10–5 **Coefficient of Determination and Standard Error of the Estimate**

10–6 **Multiple Regression (Optional)**

10–7 **Summary**

Statistics Today

Do Dust Storms Affect Respiratory Health?

Southeast Washington state has a long history of seasonal dust storms. Several researchers decided to see what effect, if any, these storms had on the respiratory health of the people living in the area. They undertook (among other things) to see if there was a relationship between the amount of dust and sand particles in the air when the storms occur and the number of hospital emergency room visits for respiratory disorders at three community hospitals in southeast Washington. Using methods of correlation and regression, which are explained in this chapter, they were able to determine the effect of these dust storms on local residents. See Statistics Today—Revisited.

Source: B. Hefflin, B. Jalaludin, N. Cobb, C. Johnson, L. Jecha, and R. Etzel, "Surveillance for Dust Storms and Respiratory Diseases in Washington State, 1991," *Archives of Environmental Health* 49, no. 3 (May–June 1994), pp. 170–74. Reprinted with permission of the Helen Dwight Reid Education Foundation. Published by Heldref Publications, 1319 18th St. N.W., Washington, D.C. 20036-1802. Copyright 1994.

10–1   Introduction

In Chapters 7 and 8, two areas of inferential statistics—confidence intervals and hypothesis testing—were explained. Another area of inferential statistics involves determining whether a relationship between two or more numerical or quantitative variables exists. For example, a businessperson may want to know whether the volume of sales for a given month is related to the amount of advertising the firm does that month. Educators are interested in determining whether the number of hours a student studies is related to the student's score on a particular exam. Medical researchers are interested in questions such as, Is caffeine related to heart damage? or Is there a relationship between a person's age and his or her blood pressure? A zoologist may want to know whether the birth weight of a certain animal is related to its life span. These are only a few of the many questions that can be answered by using the techniques of correlation and regression analysis. **Correlation** is a statistical method used to determine whether a relationship between variables exists. **Regression** is a statistical method used to describe the nature of the relationship between variables, that is, positive or negative, linear or nonlinear.

The purpose of this chapter is to answer these questions statistically:

1. Are two or more variables related?
2. If so, what is the strength of the relationship?
3. What type of relationship exists?
4. What kind of predictions can be made from the relationship?

To answer the first two questions, statisticians use a numerical measure to determine whether two or more variables are related and to determine the strength of the relationship between or among the variables. This measure is called a *correlation coefficient.* For example, there are many variables that contribute to heart disease, among them lack of exercise, smoking, heredity, age, stress, and diet. Of these variables, some are more important than others; therefore, a physician who wants to help a patient must know which factors are most important.

To answer the third question, one must ascertain what type of relationship exists. There are two types of relationships: *simple* and *multiple*. In a **simple relationship,** there are two variables—an **independent variable,** also called an explanatory variable or a predictor variable, and a **dependent variable,** also called a response variable. A simple relationship analysis is called *simple regression,* and there is one independent variable that is used to predict the dependent variable. For example, a manager may wish to see whether the number of years the salespeople have been working for the company has anything to do with the amount of sales they make. This type of study involves a simple relationship, since there are only two variables—years of experience and amount of sales.

In a **multiple relationship,** called *multiple regression,* two or more independent variables are used to predict one dependent variable. For example, an educator may wish to investigate the relationship between a student's success in college and factors such as the number of hours devoted to studying, the student's GPA, and the student's high school background. This type of study involves several variables.

Simple relationships can also be positive or negative. A **positive relationship** exists when both variables increase or decrease at the same time. For instance, a person's height and weight are related; and the relationship is positive, since the taller a person is, generally, the more the person weighs. In a **negative relationship,** as one variable increases, the other variable decreases, and vice versa. For example, if one measures the strength of people over 60 years of age, one will find that as age increases, strength generally decreases. The word *generally* is used here because there are exceptions.

Finally, the fourth question asks what type of predictions can be made. Predictions are made in all areas and daily. Examples include weather forecasting, stock market analyses, sales predictions, crop predictions, gasoline predictions, and sports predictions. Some predictions are more accurate than others, due to the strength of the relationship. That is, the stronger the relationship is between variables, the more accurate the prediction is.

10–2 Scatter Plots

Objective 1

Draw a scatter plot for a set of ordered pairs.

In simple correlation and regression studies, the researcher collects data on two numerical or quantitative variables to see whether a relationship exists between the variables. For example, if a researcher wishes to see whether there is a relationship between number of hours of study and test scores on an exam, she must select a random sample of students, determine the hours each studied, and obtain their grades on the exam. A table can be made for the data, as shown here.

Student	Hours of study x	Grade y (%)
A	6	82
B	2	63
C	1	57
D	5	88
E	2	68
F	3	75

As stated previously, the two variables for this study are called the independent variable and the dependent variable. The independent variable is the variable in regression that can be controlled or manipulated. In this case, "number of hours of study" is the independent variable and is designated as the x variable. The dependent variable is the variable in regression that cannot be controlled or manipulated. The grade the student received on the exam is the dependent variable, designated as the y variable. The reason for this distinction between the variables is that one assumes that the grade the student earns *depends* on the number of hours the student studied. Also, one assumes that, to some extent, the student can regulate or *control* the number of hours he or she studies for the exam.

The determination of the x and y variables is not always clear-cut and is sometimes an arbitrary decision. For example, if a researcher studies the effects of age on a person's blood pressure, the researcher can generally assume that age affects blood pressure. Hence, the variable *age* can be called the *independent variable,* and the variable *blood pressure* can be called the *dependent variable*. On the other hand, if a researcher is studying the attitudes of husbands on a certain issue and the attitudes of their wives on the same issue, it is difficult to say which variable is the independent variable and which is the dependent variable. In this study, the researcher can arbitrarily designate the variables as independent and dependent.

The independent and dependent variables can be plotted on a graph called a *scatter plot*. The independent variable x is plotted on the horizontal axis, and the dependent variable y is plotted on the vertical axis.

A **scatter plot** is a graph of the ordered pairs (x, y) of numbers consisting of the independent variable x and the dependent variable y.

The scatter plot is a visual way to describe the nature of the relationship between the independent and dependent variables. The scales of the variables can be different, and the coordinates of the axes are determined by the smallest and largest data values of the variables.

The procedure for drawing a scatter plot is shown in Examples 10–1 through 10–3.

Example 10–1   Construct a scatter plot for the data obtained in a study of age and systolic blood pressure of six randomly selected subjects. The data are shown in the table.

Subject	Age x	Pressure y
A	43	128
B	48	120
C	56	135
D	61	143
E	67	141
F	70	152

Solution

Step 1   Draw and label the *x* and *y* axes.

Step 2   Plot each point on the graph, as shown in Figure 10–1.

Figure 10–1

Scatter Plot for
Example 10–1

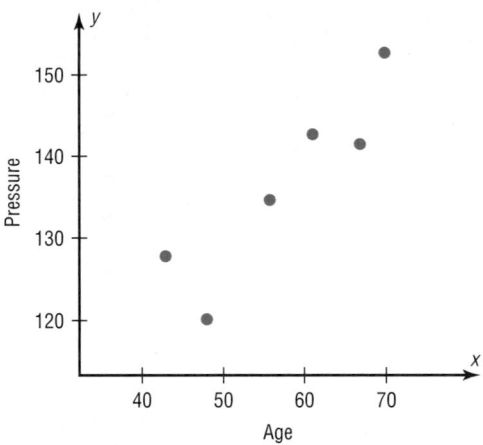

Example 10–2   Construct a scatter plot for the data obtained in a study on the number of absences and the final grades of seven randomly selected students from a statistics class. The data are shown here.

Student	Number of absences *x*	Final grade *y* (%)
A	6	82
B	2	86
C	15	43
D	9	74
E	12	58
F	5	90
G	8	78

Solution

Step 1   Draw and label the *x* and *y* axes.

Step 2   Plot each point on the graph, as shown in Figure 10–2.

Figure 10–2

Scatter Plot for
Example 10–2

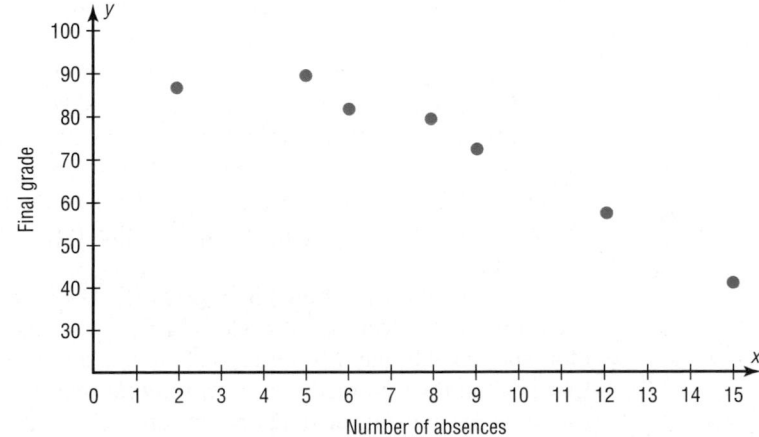

Example 10–3

Construct a scatter plot for the data obtained in a study on the number of hours that nine people exercise each week and the amount of milk (in ounces) each person consumes per week. The data are shown.

Subject	Hours x	Amount y
A	3	48
B	0	8
C	2	32
D	5	64
E	8	10
F	5	32
G	10	56
H	2	72
I	1	48

Solution

Step 1 Draw and label the x and y axes.

Step 2 Plot each point on the graph, as shown in Figure 10–3.

Figure 10–3

Scatter Plot for
Example 10–3

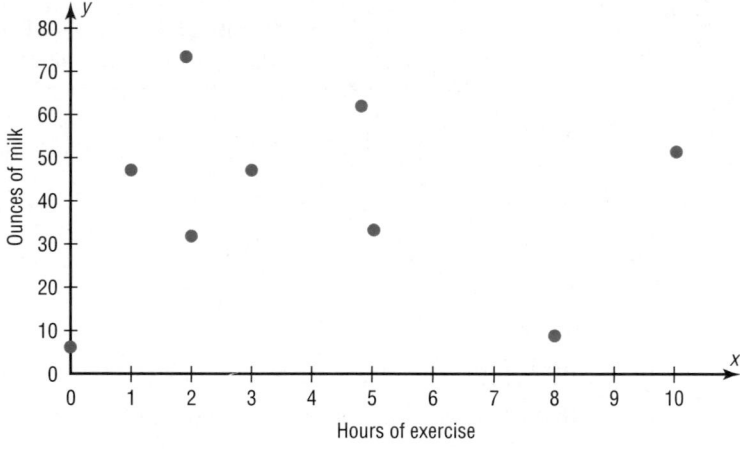

After the plot is drawn, it should be analyzed to determine which type of relationship, if any, exists. For example, the plot shown in Figure 10–1 suggests a positive relationship, since as a person's age increases, blood pressure tends to increase also. The plot of the data shown in Figure 10–2 suggests a negative relationship, since as the number of absences increases, the final grade decreases. Finally, the plot of the data shown in Figure 10–3 shows no specific type of relationship, since no pattern is discernible.

Note that the data shown in Figures 10–1 and 10–2 also suggest a linear relationship, since the points seem to fit a straight line, although not perfectly. Sometimes a scatter plot, such as the one in Figure 10–4, shows a curvilinear relationship between the data. In this situation, the methods shown in this section and in Section 10–3 cannot be used. Methods for curvilinear relationships are beyond the scope of this book.

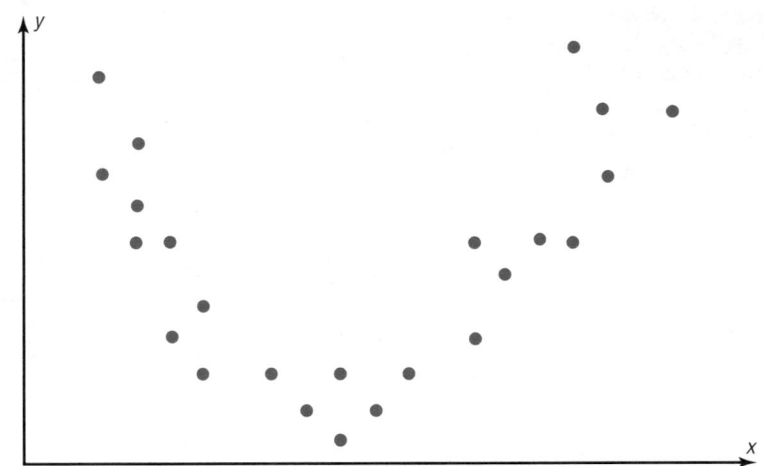

Figure 10–4

Scatter Plot
Suggesting a
Curvilinear
Relationship

10–3

Correlation

Correlation Coefficient

Objective 2

Compute the
correlation coefficient.

As stated in Section 10–1, statisticians use a measure called the *correlation coefficient* to determine the strength of the relationship between two variables. There are several types of correlation coefficients. The one explained in this section is called the **Pearson product moment correlation coefficient** (PPMC), named after statistician Karl Pearson, who pioneered the research in this area.

> The **correlation coefficient** computed from the sample data measures the strength and direction of a linear relationship between two variables. The symbol for the sample correlation coefficient is r. The symbol for the population correlation coefficient is ρ (Greek letter rho).

The *range of the correlation coefficient* is from -1 to $+1$. If there is a *strong positive linear relationship* between the variables, the value of r will be close to $+1$. If there is a *strong negative linear relationship* between the variables, the value of r will be close to -1. When there is no linear relationship between the variables or only a weak relationship, the value of r will be close to 0. See Figure 10–5.

The graphs in Figure 10–6 show the relationship between the correlation coefficients and their corresponding scatter plots. Notice that as the value of the correlation coefficient increases from 0 to $+1$ (parts a, b, and c), data values become closer to an increasingly stronger relationship. As the value of the correlation coefficient decreases from 0 to -1 (parts d, e, and f), the data values also become closer to a straight line. Again this suggests a stronger relationship.

Figure 10–5

Range of Values for the
Correlation Coefficient

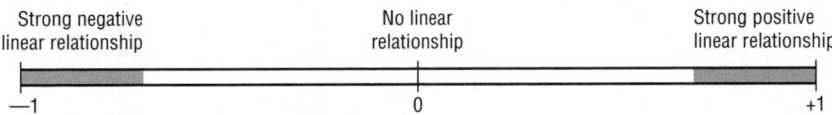

Figure 10–6

Relationship Between the Correlation Coefficient and the Scatter Plot

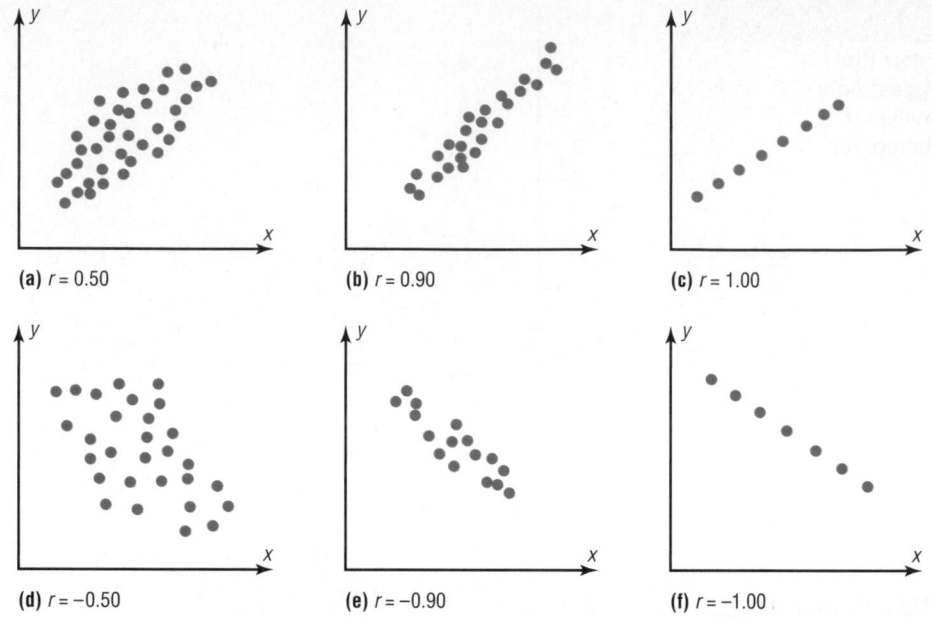

(a) $r = 0.50$ (b) $r = 0.90$ (c) $r = 1.00$

(d) $r = -0.50$ (e) $r = -0.90$ (f) $r = -1.00$

There are several ways to compute the value of the correlation coefficient. One method is to use the formula shown here.

Formula for the Correlation Coefficient r

$$r = \frac{n(\Sigma xy) - (\Sigma x)(\Sigma y)}{\sqrt{[n(\Sigma x^2) - (\Sigma x)^2][n(\Sigma y^2) - (\Sigma y)^2]}}$$

where n is the number of data pairs.

Rounding Rule for the Correlation Coefficient Round the value of r to three decimal places.

The formula looks somewhat complicated, but using a table to compute the values, as shown in Example 10–4, makes it somewhat easier to determine the value of r.

Example 10–4

Compute the value of the correlation coefficient for the data obtained in the study of age and blood pressure given in Example 10–1.

Solution

Step 1 Make a table, as shown here.

Subject	Age x	Pressure y	xy	x^2	y^2
A	43	128			
B	48	120			
C	56	135			
D	61	143			
E	67	141			
F	70	152			

Step 2 Find the values of xy, x^2, and y^2 and place these values in the corresponding columns of the table.

The completed table is shown.

Subject	Age x	Pressure y	xy	x^2	y^2
A	43	128	5,504	1,849	16,384
B	48	120	5,760	2,304	14,400
C	56	135	7,560	3,136	18,225
D	61	143	8,723	3,721	20,449
E	67	141	9,447	4,489	19,881
F	70	152	10,640	4,900	23,104
	$\Sigma x = 345$	$\Sigma y = 819$	$\Sigma xy = 47,634$	$\Sigma x^2 = 20,399$	$\Sigma y^2 = 112,443$

Step 3 Substitute in the formula and solve for r.

$$r = \frac{n(\Sigma xy) - (\Sigma x)(\Sigma y)}{\sqrt{[n(\Sigma x^2) - (\Sigma x)^2][n(\Sigma y^2) - (\Sigma y)^2]}}$$

$$= \frac{(6)(47,634) - (345)(819)}{\sqrt{[(6)(20,399) - (345)^2][(6)(112,443) - (819)^2]}} = 0.897$$

The correlation coefficient suggests a strong positive relationship between age and blood pressure.

Example 10–5

Compute the value of the correlation coefficient for the data obtained in the study of the number of absences and the final grade of the seven students in the statistics class given in Example 10–2.

Solution

Step 1 Make a table.

Step 2 Find the values of xy, x^2, and y^2 and place these values in the corresponding columns of the table.

Student	Number of absences x	Final grade y (%)	xy	x^2	y^2
A	6	82	492	36	6,724
B	2	86	172	4	7,396
C	15	43	645	225	1,849
D	9	74	666	81	5,476
E	12	58	696	144	3,364
F	5	90	450	25	8,100
G	8	78	624	64	6,084
	$\Sigma x = 57$	$\Sigma y = 511$	$\Sigma xy = 3745$	$\Sigma x^2 = 579$	$\Sigma y^2 = 38,993$

Step 3 Substitute in the formula and solve for r.

$$r = \frac{n(\Sigma xy) - (\Sigma x)(\Sigma y)}{\sqrt{[n(\Sigma x^2) - (\Sigma x)^2][n(\Sigma y^2) - (\Sigma y)^2]}}$$

$$= \frac{(7)(3745) - (57)(511)}{\sqrt{[(7)(579) - (57)^2][(7)(38,993) - (511)^2]}} = -0.944$$

The value of r suggests a strong negative relationship between a student's final grade and the number of absences a student has. That is, the more absences a student has, the lower is his or her grade.

Example 10–6

Compute the value of the correlation coefficient for the data given in Example 10–3 for the number of hours a person exercises and the amount of milk a person consumes per week.

Solution

Step 1 Make a table.

Step 2 Find the values of xy, x^2, and y^2 and place these values in the corresponding columns of the table.

Subject	Hours x	Amount y	xy	x^2	y^2
A	3	48	144	9	2,304
B	0	8	0	0	64
C	2	32	64	4	1,024
D	5	64	320	25	4,096
E	8	10	80	64	100
F	5	32	160	25	1,024
G	10	56	560	100	3,136
H	2	72	144	4	5,184
I	1	48	48	1	2,304
	$\Sigma x = 36$	$\Sigma y = 370$	$\Sigma xy = 1{,}520$	$\Sigma x^2 = 232$	$\Sigma y^2 = 19{,}236$

Step 3 Substitute in the formula and solve for r.

$$r = \frac{n(\Sigma xy) - (\Sigma x)(\Sigma y)}{\sqrt{[n(\Sigma x^2) - (\Sigma x)^2][n(\Sigma y^2) - (\Sigma y)^2]}}$$

$$= \frac{(9)(1520) - (36)(370)}{\sqrt{[(9)(232) - (36)^2][(9)(19{,}236) - (370)^2]}} = 0.067$$

The value of r indicates a very weak positive relationship between the variables.

In Example 10–4, the value of r was high (close to 1.00); in Example 10–6, the value of r was much lower (close to 0). This question then arises: When is the value of r due to chance, and when does it suggest a significant linear relationship between the variables? This question will be answered next.

Objective **3**

Test the hypothesis H_0: $\rho = 0$.

The Significance of the Correlation Coefficient

As stated before, the range of the correlation coefficient is between -1 and $+1$. When the value of r is near $+1$ or -1, there is a strong linear relationship. When the value of r is near 0, the linear relationship is weak or nonexistent. Since the value of r is computed from data obtained from samples, there are two possibilities when r is not equal to zero: either the value of r is high enough to conclude that there is a significant linear relationship between the variables, or the value of r is due to chance.

To make this decision, one uses a hypothesis-testing procedure. The traditional method is similar to the one used in previous chapters.

Step 1 State the hypotheses.

Step 2 Find the critical values.

Step 3 Compute the test value.

Step 4 Make the decision.

Step 5 Summarize the results.

The population correlation coefficient is computed from taking all possible (x, y) pairs; it is designated by the Greek letter ρ (rho). The sample correlation coefficient can then be used as an estimator of ρ if the following assumptions are valid.

1. The variables x and y are *linearly* related.
2. The variables are *random* variables.
3. The two variables have a *bivariate normal distribution.*

This means for any given value of x, the y variable is normally distributed.

Formally defined, the **population correlation coefficient** ρ is the correlation computed by using all possible pairs of data values (x, y) taken from a population.

In hypothesis testing, one of these is true:

H_0: $\rho = 0$ This null hypothesis means that there is no correlation between the x and y variables in the population.

H_1: $\rho \neq 0$ This alternative hypothesis means that there is a significant correlation between the variables in the population.

When the null hypothesis is rejected at a specific level, it means that there is a significant difference between the value of r and 0. When the null hypothesis is not rejected, it means that the value of r is not significantly different from 0 (zero) and is probably due to chance.

Several methods can be used to test the significance of the correlation coefficient. Three methods will be shown in this section. The first uses the t test.

Formula for the t Test for the Correlation Coefficient

$$t = r\sqrt{\frac{n - 2}{1 - r^2}}$$

with degrees of freedom equal to $n - 2$.

Although hypothesis tests can be one-tailed, most hypotheses involving the correlation coefficient are two-tailed. Recall that ρ represents the population correlation coefficient. Also, if there is no linear relationship, the value of the correlation coefficient will be 0. Hence, the hypotheses will be

$$H_0: \rho = 0 \qquad \text{and} \qquad H_1: \rho \neq 0$$

One does not have to identify the claim here, since the question will always be whether there is a significant linear relationship between the variables.

The two-tailed critical values are used. These values are found in Table F in Appendix C. Also, when one is testing the significance of a correlation coefficient, both variables x and y must come from normally distributed populations.

Example 10–7

Test the significance of the correlation coefficient found in Example 10–4. Use $\alpha = 0.05$ and $r = 0.897$.

Solution

Step 1 State the hypotheses.

$$H_0: \rho = 0 \quad \text{and} \quad H_1: \rho \neq 0$$

Step 2 Find the critical values. Since $\alpha = 0.05$ and there are $6 - 2 = 4$ degrees of freedom, the critical values obtained from Table F are ± 2.776, as shown in Figure 10–7.

Figure 10–7

Critical Values for Example 10–7

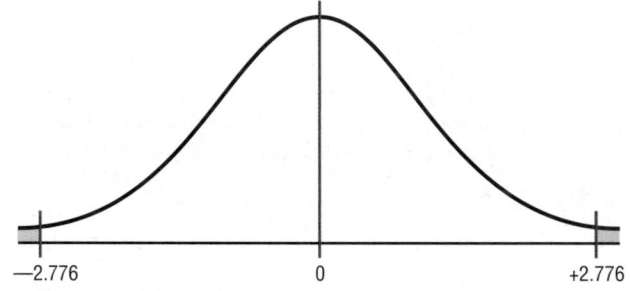

Step 3 Compute the test value.

$$t = r\sqrt{\frac{n-2}{1-r^2}} = (0.897)\sqrt{\frac{6-2}{1-(0.897)^2}} = 4.059$$

Step 4 Make the decision. Reject the null hypothesis, since the test value falls in the critical region, as shown in Figure 10–8.

Figure 10–8

Test Value for Example 10–7

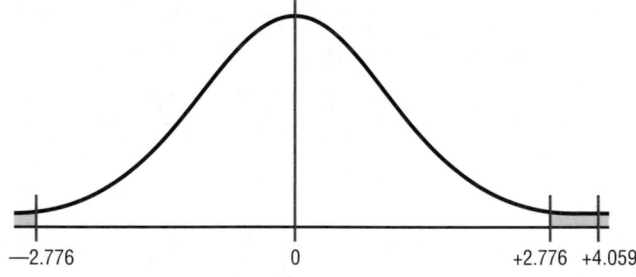

Step 5 Summarize the results. There is a significant relationship between the variables of age and blood pressure.

The second method that can be used to test the significance of r is the P-value method. The method is the same as that shown in Chapters 8 and 9. It uses the following steps.

Step 1 State the hypotheses.

Step 2 Find the test value. (In this case, use the t test.)

Figure 10–9

Finding the Critical Value from Table I

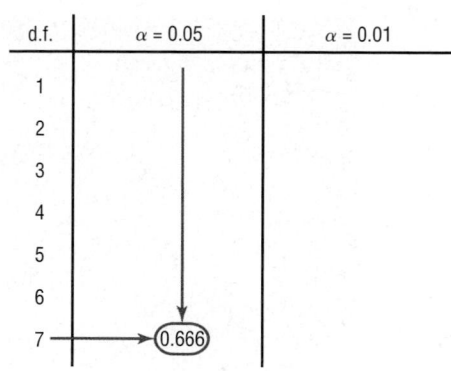

Step 3   Find the *P*-value. (In this case, use Table F.)

Step 4   Make the decision.

Step 5   Summarize the results.

Referring to Example 10–7, we see that the *t* value obtained in step 3 is 4.059 and d.f. = 4. Using Table F with d.f. = 4 and the row Two tails, the value 4.059 falls between 3.747 and 4.604; hence, $0.01 < P\text{-value} < 0.02$. (The *P*-value obtained from a calculator is 0.015.) That is, the *P*-value falls between 0.01 and 0.02. The decision then is to reject the null hypothesis since $P\text{-value} < 0.05$.

The third method of testing the significance of *r* is to use Table I in Appendix C. This table shows the values of the correlation coefficient that are significant for a specific α level and a specific number of degrees of freedom. For example, for 7 degrees of freedom and $\alpha = 0.05$, the table gives a critical value of 0.666. Any value of *r* greater than $+0.666$ or less than -0.666 will be significant, and the null hypothesis will be rejected. See Figure 10–9. When Table I is used, one need not compute the *t* test value. Table I is for two-tailed tests only.

Example 10–8

Using Table I, test the significance of the correlation coefficient $r = 0.067$, obtained in Example 10–6, at $\alpha = 0.01$.

Solution

$$H_0: \rho = 0 \qquad \text{and} \qquad H_1: \rho \neq 0$$

Since the sample size is 9, there are 7 degrees of freedom. When $\alpha = 0.01$ and with 7 degrees of freedom, the value obtained from Table I is 0.798. For a significant relationship, a value of *r* greater than $+0.798$ or less than -0.798 is needed. Since $r = 0.067$, the null hypothesis is not rejected. Hence, there is not enough evidence to say that there is a significant linear relationship between the variables. See Figure 10–10.

Figure 10–10

Rejection and Nonrejection Regions for Example 10–8

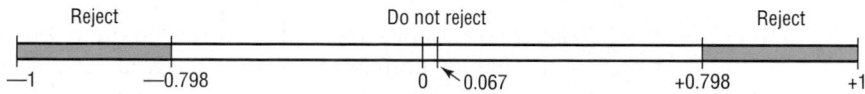

Correlation and Causation

Researchers must understand the nature of the linear relationship between the independent variable x and the dependent variable y. When a hypothesis test indicates that a significant linear relationship exists between the variables, researchers must consider the possibilities outlined next.

Possible Relationships Between Variables

When the null hypothesis has been rejected for a specific α value, any of the following five possibilities can exist.

1. *There is a direct cause-and-effect relationship between the variables.* That is, x causes y. For example, water causes plants to grow, poison causes death, and heat causes ice to melt.

2. *There is a reverse cause-and-effect relationship between the variables.* That is, y causes x. For example, suppose a researcher believes excessive coffee consumption causes nervousness, but the researcher fails to consider that the reverse situation may occur. That is, it may be that an extremely nervous person craves coffee to calm his or her nerves.

3. *The relationship between the variables may be caused by a third variable.* For example, if a statistician correlated the number of deaths due to drowning and the number of cans of soft drink consumed daily during the summer, he or she would probably find a significant relationship. However, the soft drink is not necessarily responsible for the deaths, since both variables may be related to heat and humidity.

4. *There may be a complexity of interrelationships among many variables.* For example, a researcher may find a significant relationship between students' high school grades and college grades. But there probably are many other variables involved, such as IQ, hours of study, influence of parents, motivation, age, and instructors.

5. *The relationship may be coincidental.* For example, a researcher may be able to find a significant relationship between the increase in the number of people who are exercising and the increase in the number of people who are committing crimes. But common sense dictates that any relationship between these two values must be due to coincidence.

When two variables are highly correlated, item 3 in the box states that there exists a possibility that the correlation is due to a third variable. If this is the case and the third variable is unknown to the researcher or not accounted for in the study, it is called a **lurking variable.** An attempt should be made by the researcher to identify such variables and to use methods to control their influence.

Also, one should be cautious when the data for one or both of the variables involve averages rather than individual data. It is not wrong to use averages, but the results cannot be generalized to individuals since averaging tends to smooth out the variability among individual data values. The result could be a higher correlation than actually exists.

Thus, when the null hypothesis is rejected, the researcher must consider all possibilities and select the appropriate one as determined by the study. Remember, correlation does not necessarily imply causation.

Applying the Concepts **10–3**

Stopping Distances

In a study on speed control, it was found that the main reasons for regulations were to make traffic flow more efficient and to minimize the risk of danger. An area that was focused on in the study was the distance required to completely stop a vehicle at various speeds. Use the following table to answer the questions.

MPH	Braking distance (feet)
20	20
30	45
40	81
50	133
60	205
80	411

Assume MPH is going to be used to predict stopping distance.

1. Which of the two variables is the independent variable?
2. Which is the dependent variable?
3. What type of variable is the independent variable?
4. What type of variable is the dependent variable?
5. Construct a scatter plot for the data.
6. Is there a linear relationship between the two variables?
7. Redraw the scatter plot, and change the distances between the independent-variable numbers. Does the relationship look different?
8. Is the relationship positive or negative?
9. Can braking distance be accurately predicted from MPH?
10. List some other variables that affect braking distance.
11. Compute the value of r.
12. Is r significant at $\alpha = 0.05$?

See page 580 for the answers.

Exercises 10–3

1. What is meant by the statement that two variables are related?

2. How is a linear relationship between two variables measured in statistics? Explain.

3. What is the symbol for the sample correlation coefficient? The population correlation coefficient?

4. What is the range of values for the correlation coefficient?

5. What is meant when the relationship between the two variables is positive? Negative?

6. Give examples of two variables that are positively correlated and two that are negatively correlated.

Speaking of
Statistics

In correlation and regression studies, it is difficult to control all variables. This study shows some of the consequences when researchers overlook certain aspects in studies. Suggest ways that the extraneous variables might be controlled in future studies.

Coffee Not Disease Culprit, Study Says

NEW YORK (AP)—Two new studies suggest that coffee drinking, even up to $5\frac{1}{2}$ cups per day, does not increase the risk of heart disease, and other studies that claim to have found increased risks might have missed the true culprits, a researcher says.

"It might not be the coffee cup in one hand, it might be the cigarette or coffee roll in the other," said Dr. Peter W. F. Wilson, the author of one of the new studies.

He noted in a telephone interview Thursday that many coffee drinkers, particularly heavy coffee drinkers, are smokers. And one of the new studies found that coffee drinkers had excess fat in their diets.

The findings of the new studies conflict sharply with a study reported in November 1985 by Johns Hopkins University scientists in Baltimore.

The Hopkins scientists found that coffee drinkers who consumed five or more cups of coffee per day had three times the heart-disease risk of non-coffee drinkers.

The reason for the discrepancy appears to be that many of the coffee drinkers in the Hopkins study also smoked—and it was the smoking that increased their heart-disease risk, said Wilson.

Wilson, director of laboratories for the Framingham Heart Study in Framingham, Mass., said Thursday at a conference sponsored by the American Heart Association in Charleston, S.C., that he had examined the coffee intake of 3,937 participants in the Framingham study during 1956–66 and an additional 2,277 during the years 1972–1982.

In contrast to the subjects in the Hopkins study, most of these coffee drinkers consumed two or three cups per day, Wilson said. Only 10 percent drank six or more cups per day.

He then looked at blood cholesterol levels and heart and blood vessel disease in the two groups. "We ran these analyses for coronary heart disease, heart attack, sudden death and stroke and in absolutely every analysis, we found no link with coffee," Wilson said.

He found that coffee consumption was linked to a significant decrease in total blood cholesterol in men, and to a moderate increase in total cholesterol in women.

Source: Reprinted with permission of the Associated Press.

---

7. Give an example of a correlation study, and identify the independent and dependent variables.

8. What is the diagram of the independent and dependent variables called? Why is drawing this diagram important?

9. What is the name of the correlation coefficient used in this section?

10. What statistical test is used to test the significance of the correlation coefficient?

11. When two variables are correlated, can the researcher be sure that one variable causes the other? Why or why not?

For Exercises 12 through 27, perform the following steps.

   a. Draw the scatter plot for the variables.

   b. Compute the value of the correlation coefficient.

   c. State the hypotheses.

   d. Test the significance of the correlation coefficient at $\alpha = 0.05$, using Table I.

   e. Give a brief explanation of the type of relationship.

12. A medical researcher wishes to see if there is a relationship between prescription drug prices for identical drugs and identical dosages that are prescribed for humans and for animals. The prices are shown. Is there a relationship between the prices?

Prices for humans x	0.67	0.64	1.20	0.51	0.87	0.74	0.50	1.22
Prices for animals y	0.13	0.18	0.42	0.25	0.57	0.57	0.49	1.28

Source: House Committee on Government Reform.

(The information in this exercise will be used for Exercise 12 in Section 10–4.)

13. A researcher wishes to determine if a person's age is related to the number of hours he or she exercises per week. The data for the sample are shown here.

(The information in this exercise will be used for Exercises 13 and 36 in Section 10–4 and Exercise 15 in Section 10–5.)

Age x	18	26	32	38	52	59
Hours y	10	5	2	3	1.5	1

14. An environmentalist wants to determine the relationships between the numbers (in thousands) of forest fires over the year and the number (in hundred thousands) of acres burned. The data for 8 recent years are shown. Describe the relationship.

Number of fires x	72	69	58	47	84	62	57	45
Number of acres burned y	62	42	19	26	51	15	30	15

Source: National Interagency Fire Center.

(The information in this exercise will be used for Exercises 14 and 36 in Section 10–4 and Exercises 16 and 20 in Section 10–5.)

15. The director of an alumni association for a small college wants to determine whether there is any type of relationship between the amount of an alumnus's contribution (in dollars) and the years the alumnus has been out of school. The data follow. (The information in this exercise will be used for Exercises 15, 36, and 37 in Section 10–4 and Exercise 17 in Section 10–5.)

Years x	1	5	3	10	7	6
Contribution y	500	100	300	50	75	80

16. A store manager wishes to find out whether there is a relationship between the age of her employees and the number of sick days they take each year. The data for the sample are shown. (The information in this exercise will be used for Exercises 16 and 37 in Section 10–4 and Exercise 18 in Section 10–5.)

Age x	18	26	39	48	53	58
Days y	16	12	9	5	6	2

17. A criminology student wishes to see if there is a relationship between the number of larceny crimes and the number of vandalism crimes on college campuses in southwestern Pennsylvania. The data are shown. Is there a relationship between the two types of crimes?

Number of larceny crimes x	24	6	16	64	10	25	35
Number of vandalism crimes y	21	3	6	15	21	61	20

(The information in this exercise will be used for Exercise 17 of Section 10–4.)

18. A football fan wishes to see how the number of pass attempts (not completions) relates to the number of yards gained for quarterbacks in past NFL season playoff games. The data are shown for five quarterbacks. Describe the relationships.

Pass attempts x	116	90	82	108	92
Yards gained y	1001	823	851	873	839

(The information in this exercise will be used for Exercises 18 and 38 in Section 10–4.)

19. A meteorologist wants to see if there is a relationship between the number of tornadoes that occur each year and the number of deaths attributed to the tornadoes. The data are shown for 10 recent years. Is there a relationship?

No. of tornadoes x	1133	1132	1297	1173	1082
No. of deaths y	53	39	39	33	69

No. of tornadoes x	1234	1148	1424	1342	898
No. of deaths y	30	67	130	94	40

(The information in this exercise will be used for Exercise 19 of Section 10–4.)

20. An emergency service wishes to see whether a relationship exists between the outside temperature and the number of emergency calls it receives for a 7-hour period. The data are shown. (The information in this exercise will be used for Exercises 20 and 38 in Section 10–4.)

Temperature x	68	74	82	88	93	99	101
No. of calls y	7	4	8	10	11	9	13

21. A random sample of U.S. cities is selected to determine if there is a relationship between the population (in thousands) of people under 5 years of age and the population (in thousands) of those 65 years of age and older. The data for the sample are shown here. (The information in this exercise will be used for Exercises 21 and 36 in Section 10–4.)

Under 5 x	178	27	878	314	322	143
65 and over y	361	72	1496	501	585	207

Source: N.Y. Times Almanac.

22. The results of a survey of the average monthly rents (in dollars) for one-bedroom apartments and two-bedroom apartments in randomly selected metropolitan areas are shown. Determine if there is a relationship between the rents. (The information in this exercise will be used for Exercise 22 in Section 10–4.)

One-bedroom x	782	486	451	529	618	520	845
Two-bedroom y	1223	902	739	954	1055	875	1455

Source: N.Y. Times Almanac.

23. The average normal daily temperature (in degrees Fahrenheit) and the corresponding average monthly precipitation (in inches) for the month of June are shown here for seven randomly selected cities in the United States. Determine if there is a relationship between the two variables. (The information in this exercise will be used for Exercise 23 in Section 10–4.)

Avg. daily temp. x	86	81	83	89	80	74	64
Avg. mo. precip. y	3.4	1.8	3.5	3.6	3.7	1.5	0.2

Source: *N.Y. Times Almanac.*

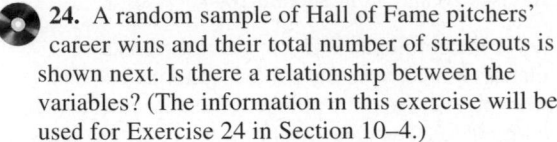 **24.** A random sample of Hall of Fame pitchers' career wins and their total number of strikeouts is shown next. Is there a relationship between the variables? (The information in this exercise will be used for Exercise 24 in Section 10–4.)

Wins x	329	150	236	300	284	207
Strikeouts y	4136	1155	1956	2266	3192	1277

Wins x	247	314	273	324
Strikeouts y	1068	3534	1987	3574

Source: *N.Y. Times Almanac.*

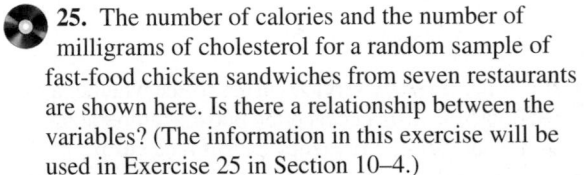 **25.** The number of calories and the number of milligrams of cholesterol for a random sample of fast-food chicken sandwiches from seven restaurants are shown here. Is there a relationship between the variables? (The information in this exercise will be used in Exercise 25 in Section 10–4.)

Calories x	390	535	720	300	430	500	440
Cholesterol y	43	45	80	50	55	52	60

Source: *The Doctor's Pocket Calorie, Fat, and Carbohydrate Counter.*

26. An architect wants to determine the relationship between the heights (in feet) of a building and the number of stories in the building. The data for a sample of 10 buildings in Pittsburgh are shown. Explain the relationship.

Stories x	64	54	40	31	45	38	42	41	37	40
Height y	841	725	635	616	615	582	535	520	511	485

Source: *World Almanac Book of Facts.*

(The information in this exercise will be used for Exercise 26 of Section 10–4.)

27. A hospital administrator wants to see if there is a relationship between the number of licensed beds and the number of staffed beds in local hospitals. The data for a specific day are shown. Describe the relationship.

Licensed beds x	144	32	175	185	208	100	169
Staffed beds y	112	32	162	141	103	80	118

Source: *Pittsburgh Tribune-Review.*

(The information in this exercise will be used for Exercise 28 of this section and Exercise 27 in Section 10–4.)

Extending the Concepts

28. One of the formulas for computing r is

$$r = \frac{\Sigma(x - \bar{x})(y - \bar{y})}{(n - 1)(s_x)(s_y)}$$

Using the data in Exercise 27, compute r with this formula. Compare the results.

 29. Compute r for the data set shown. Explain the reason for this value of r. Now, interchange the values of x and y and compute r again. Compare this value with the previous one. Explain the results of the comparison.

x	1	2	3	4	5
y	3	5	7	9	11

30. Compute r for the following data and test the hypothesis H_0: $\rho = 0$. Draw the scatter plot; then explain the results.

x	−3	−2	−1	0	1	2	3
y	9	4	1	0	1	4	9

10–4 Regression

Objective 4

Compute the equation of the regression line.

In studying relationships between two variables, collect the data and then construct a scatter plot. The purpose of the scatter plot, as indicated previously, is to determine the nature of the relationship. The possibilities include a positive linear relationship, a negative linear relationship, a curvilinear relationship, or no discernible relationship. After the scatter plot is drawn, the next steps are to compute the value of the correlation coefficient and to test the significance of the relationship. If the value of the correlation coefficient is significant, the next step is to determine the equation of the **regression line,** which is the data's line of best fit. (*Note:* Determining the regression line when r is not significant and then making predictions using the regression line are meaningless.) The purpose of

Figure 10–11

Scatter Plot with Three Lines Fit to the Data

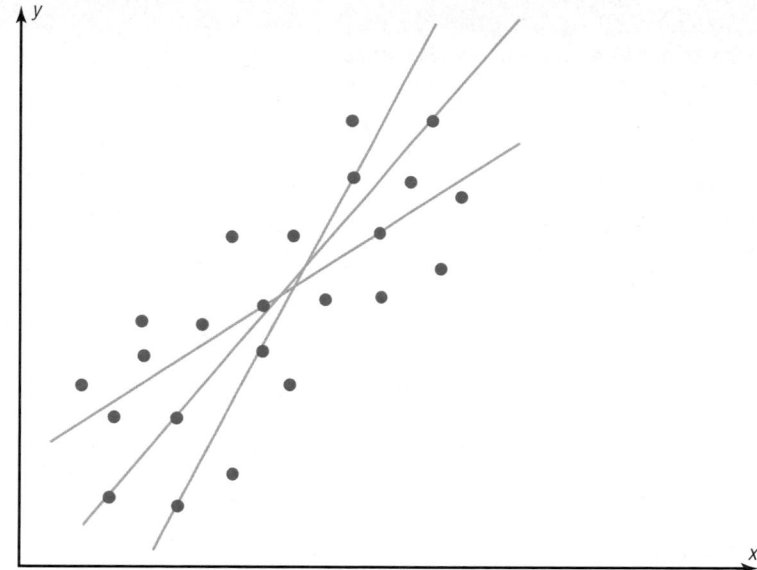

Figure 10–12

Line of Best Fit for a Set of Data Points

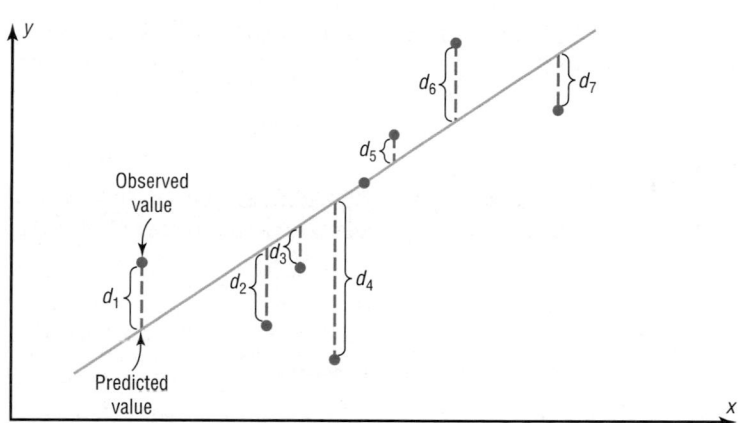

Historical Notes

Francis Galton drew the line of best fit visually. An assistant of Karl Pearson's named G. Yule devised the mathematical solution using the least-squares method, employing a mathematical technique developed by Adrien-Marie Legendre about 100 years earlier.

the regression line is to enable the researcher to see the trend and make predictions on the basis of the data.

Line of Best Fit

Figure 10–11 shows a scatter plot for the data of two variables. It shows that several lines can be drawn on the graph near the points. Given a scatter plot, one must be able to draw the *line of best fit*. *Best fit* means that the sum of the squares of the vertical distances from each point to the line is at a minimum. The reason one needs a line of best fit is that the values of y will be predicted from the values of x; hence, the closer the points are to the line, the better the fit and the prediction will be. See Figure 10–12. When r is positive, the line slopes upward and to the right. When r is negative, the line slopes downward from left to right.

Determination of the Regression Line Equation

In algebra, the equation of a line is usually given as $y = mx + b$, where m is the slope of the line and b is the y intercept. (Students who need an algebraic review of the properties of a line should refer to Appendix A, Section A–3, before studying this section.) In

Figure 10–13

A Line as Represented in Algebra and in Statistics

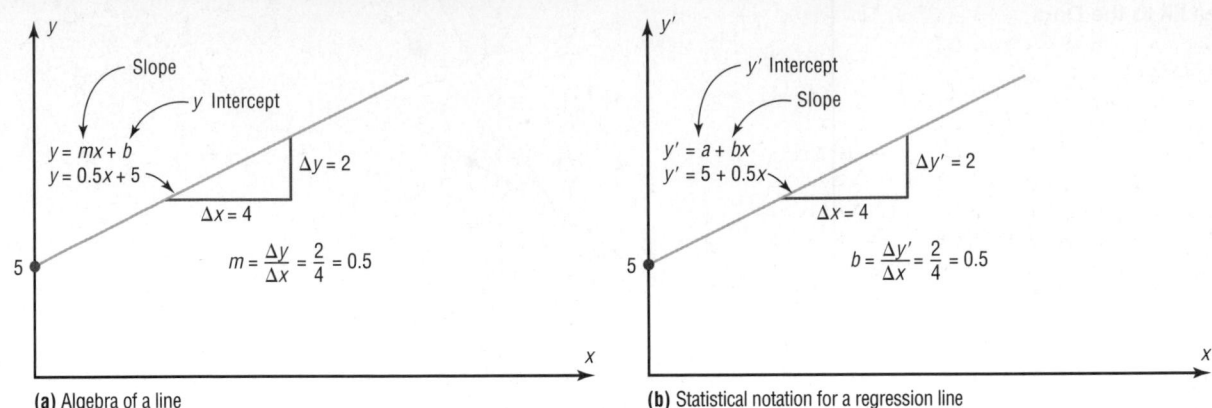

(a) Algebra of a line

(b) Statistical notation for a regression line

statistics, the equation of the regression line is written as $y' = a + bx$, where a is the y' intercept and b is the slope of the line. See Figure 10–13.

There are several methods for finding the equation of the regression line. Two formulas are given here. *These formulas use the same values that are used in computing the value of the correlation coefficient.* The mathematical development of these formulas is beyond the scope of this book.

Formulas for the Regression Line $y' = a + bx$

$$a = \frac{(\Sigma y)(\Sigma x^2) - (\Sigma x)(\Sigma xy)}{n(\Sigma x^2) - (\Sigma x)^2}$$

$$b = \frac{n(\Sigma xy) - (\Sigma x)(\Sigma y)}{n(\Sigma x^2) - (\Sigma x)^2}$$

where a is the y' intercept and b is the slope of the line.

Rounding Rule for the Intercept and Slope Round the values of a and b to three decimal places.

Example 10–9

Find the equation of the regression line for the data in Example 10–4, and graph the line on the scatter plot of the data.

Solution

The values needed for the equation are $n = 6$, $\Sigma x = 345$, $\Sigma y = 819$, $\Sigma xy = 47,634$, and $\Sigma x^2 = 20,399$. Substituting in the formulas, one gets

$$a = \frac{(\Sigma y)(\Sigma x^2) - (\Sigma x)(\Sigma xy)}{n(\Sigma x^2) - (\Sigma x)^2} = \frac{(819)(20,399) - (345)(47,634)}{(6)(20,399) - (345)^2} = 81.048$$

$$b = \frac{n(\Sigma xy) - (\Sigma x)(\Sigma y)}{n(\Sigma x^2) - (\Sigma x)^2} = \frac{(6)(47,634) - (345)(819)}{(6)(20,399) - (345)^2} = 0.964$$

Hence, the equation of the regression line $y' = a + bx$ is

$$y' = 81.048 + 0.964x$$

The graph of the line is shown in Figure 10–14.

Figure 10–14

Regression Line for
Example 10–9

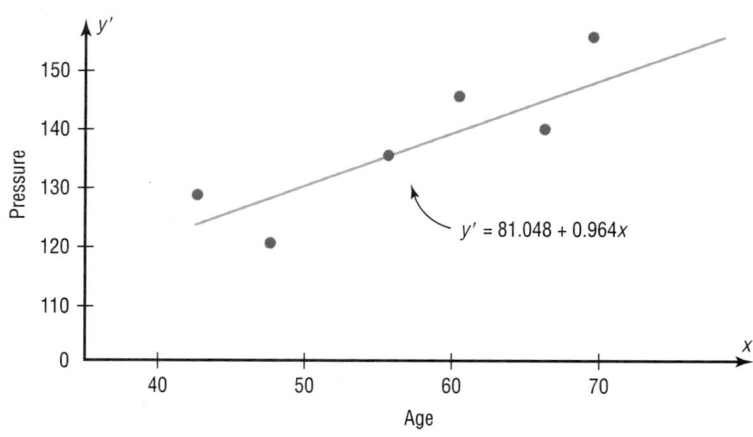

Historical Note

In 1795, Adrien-
Marie Legendre
(1752–1833)
measured the meridian
arc on the earth's
surface from
Barcelona, Spain, to
Dunkirk, England. This
measure was used as
the basis for the
measure of the meter.
Legendre developed
the least-squares
method around the
year 1805.

Note: When one is drawing the scatter plot and the regression line, it is sometimes desirable to *truncate* the graph (see Chapter 2). The motive is to show the line drawn in the range of the independent and dependent variables. For example, the regression line in Figure 10–14 is drawn between the x values of approximately 43 and 82 and the y' values of approximately 120 and 152. The range of the x values in the original data shown in Example 10–4 is $70 - 43 = 27$, and the range of the y' values is $152 - 120 = 32$. Notice that the x axis has been truncated; the distance between 0 and 40 is not shown in the proper scale compared to the distance between 40 and 50, 50 and 60, etc. The y' axis has been similarly truncated.

The important thing to remember is that when the x axis and sometimes the y' axis have been truncated, do not use the y' intercept value a to graph the line. To be on the safe side when graphing the regression line, use a value for x selected from the range of x values.

Example 10–10

Find the equation of the regression line for the data in Example 10–5, and graph the line on the scatter plot.

Solution

The values needed for the equation are $n = 7$, $\Sigma x = 57$, $\Sigma y = 511$, $\Sigma xy = 3745$, and $\Sigma x^2 = 579$. Substituting in the formulas, one gets

$$a = \frac{(\Sigma y)(\Sigma x^2) - (\Sigma x)(\Sigma xy)}{n(\Sigma x^2) - (\Sigma x)^2} = \frac{(511)(579) - (57)(3745)}{(7)(579) - (57)^2} = 102.493$$

$$b = \frac{n(\Sigma xy) - (\Sigma x)(\Sigma y)}{n(\Sigma x^2) - (\Sigma x)^2} = \frac{(7)(3745) - (57)(511)}{(7)(579) - (57)^2} = -3.622$$

Hence, the equation of the regression line $y' = a + bx$ is

$$y' = 102.493 - 3.622x$$

The graph of the line is shown in Figure 10–15.

Figure 10–15

Regression Line for
Example 10–10

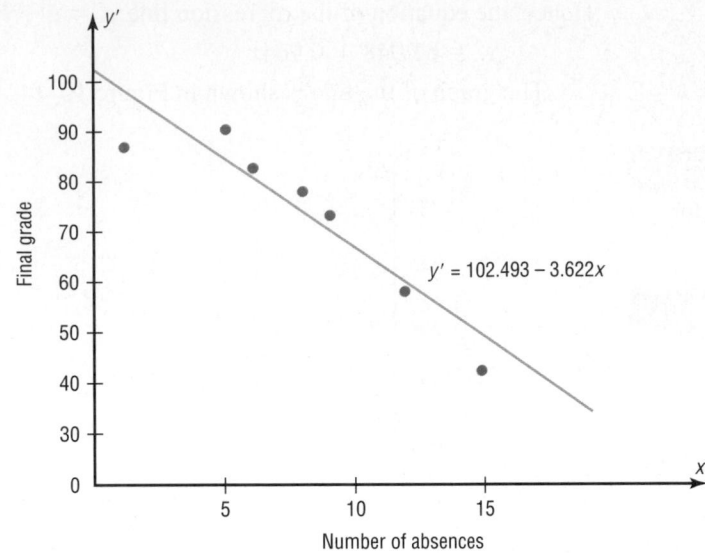

$$y' = 102.493 - 3.622x$$

The sign of the correlation coefficient and the sign of the slope of the regression line will always be the same. That is, if r is positive, then b will be positive; if r is negative, then b will be negative. The reason is that the numerators of the formulas are the same and determine the signs of r and b, and the denominators are always positive. The regression line will always pass through the point whose x coordinate is the mean of the x values and whose y coordinate is the mean of the y values, that is, $(\bar{x}, \bar{y})$.

The regression line can be used to make predictions for the dependent variable. The method for making predictions is shown in Example 10–11.

Example 10–11

Using the equation of the regression line found in Example 10–9, predict the blood pressure for a person who is 50 years old.

Solution

Substituting 50 for x in the regression line $y' = 81.048 + 0.964x$ gives

$$y' = 81.048 + (0.964)(50) = 129.248 \text{ (rounded to 129)}$$

In other words, the predicted systolic blood pressure for a 50-year-old person is 129.

The value obtained in Example 10–11 is a point prediction, and with point predictions, no degree of accuracy or confidence can be determined. More information on prediction is given in Section 10–5.

The magnitude of the change in one variable when the other variable changes exactly 1 unit is called a **marginal change.** The value of slope b of the regression line equation represents the marginal change. For example, the slope of the regression line equation in Example 10–9 is 0.964. This means that for each increase of 1 year, the value of y (systolic blood pressure reading) changes by 0.964 unit on average. In other words, for each year a person ages, her or his blood pressure rises about 1 unit.

When r is not significantly different from 0, the best predictor of y is the mean of the data values of y. For valid predictions, the value of the correlation coefficient must be significant. Also, two other assumptions must be met.

Figure 10–16

Assumptions for Predictions

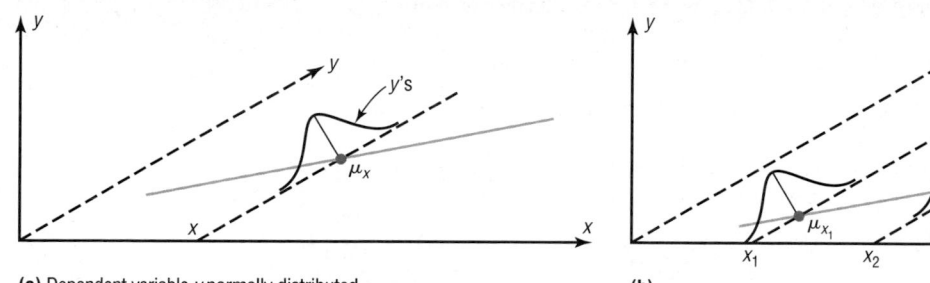

(a) Dependent variable *y* normally distributed

(b) $\sigma_1 = \sigma_2 = \cdots = \sigma_n$

Assumptions for Valid Predictions in Regression

1. For any specific value of the independent variable *x,* the value of the dependent variable *y* must be normally distributed about the regression line. See Figure 10–16(a).
2. The standard deviation of each of the dependent variables must be the same for each value of the independent variable. See Figure 10–16(b).

    Extrapolation, or making predictions beyond the bounds of the data, must be interpreted cautiously. For example, in 1979, some experts predicted that the United States would run out of oil by the year 2003. This prediction was based on the current consumption and on known oil reserves at that time. However, since then, the automobile industry has produced many new fuel-efficient vehicles. Also, there are many as yet undiscovered oil fields. Finally, science may someday discover a way to run a car on something as unlikely but as common as peanut oil. In addition, the price of a gallon of gasoline was predicted to reach $10 a few years later. Fortunately this has not come to pass. *Remember that when predictions are made, they are based on present conditions or on the premise that present trends will continue.* This assumption may or may not prove true in the future.

    The steps for finding the value of the correlation coefficient and the regression line equation are summarized in this Procedure Table:

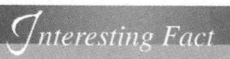

It is estimated that wearing a motorcycle helmet reduces the risk of a fatal accident by 30%.

Procedure Table

Finding the Correlation Coefficient and the Regression Line Equation

Step 1   Make a table, as shown in step 2.

Step 2   Find the values of xy, x^2, and y^2. Place them in the appropriate columns and sum each column.

x	y	xy	x^2	y^2
.	.	.	.	.
.	.	.	.	.
.	.	.	.	.
$\Sigma x =$	$\Sigma y =$	$\Sigma xy =$	$\Sigma x^2 =$	$\Sigma y^2 =$

Procedure Table (*Continued*)

Step 3 Substitute in the formula to find the value of r.

$$r = \frac{n(\Sigma xy) - (\Sigma x)(\Sigma y)}{\sqrt{[n(\Sigma x^2) - (\Sigma x)^2][n(\Sigma y^2) - (\Sigma y)^2]}}$$

Step 4 When r is significant, substitute in the formulas to find the values of a and b for the regression line equation $y' = a + bx$.

$$a = \frac{(\Sigma y)(\Sigma x^2) - (\Sigma x)(\Sigma xy)}{n(\Sigma x^2) - (\Sigma x)^2} \qquad b = \frac{n(\Sigma xy) - (\Sigma x)(\Sigma y)}{n(\Sigma x^2) - (\Sigma x)^2}$$

A scatter plot should be checked for outliers. An outlier is a point that seems out of place when compared with the other points (see Chapter 3). Some of these points can affect the equation of the regression line. When this happens, the points are called **influential points** or **influential observations.**

When a point on the scatter plot appears to be an outlier, it should be checked to see if it is an influential point. An influential point tends to "pull" the regression line toward the point itself. To check for an influential point, the regression line should be graphed with the point included in the data set. Then a second regression line should be graphed that excludes the point from the data set. If the position of the second line is changed considerably, the point is said to be an influential point. Points that are outliers in the x direction tend to be influential points.

Researchers should use their judgment as to whether to include influential observations in the final analysis of the data. If the researcher feels that the observation is not necessary, then it should be excluded so that it does not influence the results of the study. However, if the researcher feels that it is necessary, then he or she may want to obtain additional data values whose x values are near the x value of the influential point and then include them in the study.

"Explain that to me."

Source: Reprinted with special permission of King Features Syndicate.

Applying the Concepts **10–4**

Stopping Distances Revisited

In a study on speed and braking distance, researchers looked for a method to estimate how fast a person was traveling before an accident by measuring the length of their skid marks. An area that was focused on in the study was the distance required to completely stop a vehicle at various speeds. Use the following table to answer the questions.

MPH	Braking distance (feet)
20	20
30	45
40	81
50	133
60	205
80	411

Assume MPH is going to be used to predict stopping distance.

1. Find the linear regression equation.
2. What does the slope tell you about MPH and the braking distance? How about the y intercept?
3. Find the braking distance when MPH $= 45$.
4. Find the braking distance when MPH $= 100$.
5. Comment on predicting beyond the given data values.

See page 580 for the answers.

Exercises 10–4

1. What two things should be done before one performs a regression analysis?

2. What are the assumptions for regression analysis?

3. What is the general form for the regression line used in statistics?

4. What is the symbol for the slope? For the y intercept?

5. What is meant by the *line of best fit?*

6. When all the points fall on the regression line, what is the value of the correlation coefficient?

7. What is the relationship between the sign of the correlation coefficient and the sign of the slope of the regression line?

8. As the value of the correlation coefficient increases from 0 to 1, or decreases from 0 to -1, how do the points of the scatter plot fit the regression line?

9. How is the value of the correlation coefficient related to the accuracy of the predicted value for a specific value of x?

10. If the value of r is not significant, what can be said about the regression line?

11. When the value of r is not significant, what value should be used to predict y?

For Exercises 12 through 27, use the same data as for the corresponding exercises in Section 10–3. For each exercise, find the equation of the regression line and find the y' value for the specified x value. Remember that no regression should be done when r is not significant.

12. Price of drugs for humans and animals

Humans x	0.67	0.64	1.20	0.51	0.87	0.74	0.50	1.22
Animals y	0.13	0.18	0.42	0.25	0.57	0.57	0.49	1.28

Find y' when $x = 0.75$.

13. Ages and exercise

Age x	18	26	32	38	52	59
Hours y	10	5	2	3	1.5	1

Find y' when $x = 35$ years.

14. Number of fires and number of acres burned

Fires x	72	69	58	47	84	62	57	45
Acres y	62	41	19	26	51	15	30	15

Find y' when $x = 60$.

15. Years and contribution

Years x	1	5	3	10	7	6
Contribution y, $	500	100	300	50	75	80

Find y' when $x = 4$ years.

16. Age and sick days

Age x	18	26	39	48	53	58
Days y	16	12	9	5	6	2

Find y' when $x = 47$ years.

17. Larceny crimes and vandalism crimes

Larceny x	24	6	16	64	10	25	35
Vandalism y	21	3	6	15	21	61	20

Find y' when $x = 40$.

18. Pass attempts and yards gained

Attempts x	116	90	82	108	92
Yards y	1001	823	851	873	837

Find y' when $x = 95$.

19. Tornadoes and deaths

Tornadoes x	1113	1132	1297	1173	1082
Deaths y	53	39	39	33	69
Tornadoes x	1234	1148	1424	1342	898
Deaths y	30	67	130	94	40

Find y' when $x = 1000$.

20. Temperature in degrees Fahrenheit and number of emergency calls

Temperature x	68	74	82	88	93	99	101
No. of calls y	7	4	8	10	11	9	13

Find y' when $x = 80°F$.

21. Number (in thousands) of people under 5 years old and people 65 and over living in six randomly selected cities in the United States

Under 5 x	178	27	878	314	322	143
65 and older y	361	72	1496	501	585	207

Find y' when $x = 200$ thousand.

22. Rents for one-bedroom and two-bedroom apartments

One-bedroom x, $	782	486	451	529	618	520	845
Two-bedroom y, $	1223	902	739	954	1055	875	1455

Find y' when $x = \$700$.

23. Temperatures (in degrees Fahrenheit) and precipitation (in inches)

Avg. daily temp. x	86	81	83	89	80	74	64
Avg. mo. precip. y	3.4	1.8	3.5	3.6	3.7	1.5	0.2

Find y' when $x = 70°F$.

24. Wins and strikeouts for Hall of Fame pitchers

Wins x	329	150	236	300	284	207
Strikeouts y	4136	1155	1956	2266	3192	1277
Wins x	247	314	273	324		
Strikeouts y	1068	3534	1987	3574		

Find y' when $x = 260$ wins.

25. Calories and cholesterol

Calories x	390	535	720	300	430	500	440
Cholesterol y	43	45	80	50	55	52	60

Find y' when $x = 600$ calories.

26. Stories and heights of buildings

Stories x	64	54	40	31	45	38	42	41	37	40
Heights y	841	725	635	616	615	582	535	520	511	485

Find y' when $x = 44$.

27. Licensed beds and staffed beds

Licensed beds x	144	32	175	185	208	100	169
Staffed beds y	112	32	162	141	103	80	118

Find y' when $x = 44$.

For Exercises 28 through 33, do a complete regression analysis by performing these steps.

 a. Draw a scatter plot.
 b. Compute the correlation coefficient.
 c. State the hypotheses.
 d. Test the hypotheses at $\alpha = 0.05$. Use Table I.
 e. Determine the regression line equation.
 f. Plot the regression line on the scatter plot.
 g. Summarize the results.

28. These data were obtained for the years 1993 through 1998 and indicate the number of fireworks (in millions) used and the related injuries. Predict the number of injuries if 100 million fireworks are used during a given year.

Fireworks in use x	67.6	87.1	117	115	118	113
Related injuries y	12,100	12,600	12,500	10,900	7800	7000

Source: National Council of Fireworks Safety, American Pyrotechnic Assoc.

29. These data were obtained from a survey of the number of years people smoked and the percentage of lung damage they sustained. Predict the percentage of lung damage for a person who has smoked for 30 years.

Years x	22	14	31	36	9	41	19
Damage y	20	14	54	63	17	71	23

30. A medical researcher wishes to describe the relationship between the prescription cost of a brand

name drug and its generic equivalent. The data (in dollars) are shown. Describe the relationship.

Brand name x	96	93	59	80	44	47	15	56
Generic y	42	31	17	16	8	12	6	22

31. These data were obtained from a sample of counties in southwestern Pennsylvania and indicate the number (in thousands) of tons of bituminous coal produced in each county and the number of employees working in coal production in each county. Predict the number of employees needed to produce 500 thousand tons of coal. The data are given here.

Tons x	227	5410	5328	147	729
No. of employees y	110	731	1031	20	118
Tons x	8095	635	6157		
No. of employees y	1162	103	752		

32. A television executive selects 10 television shows and compares the average number of viewers the show had last year with the average number of viewers this year. The data (in millions) are shown. Describe the relationship.

Viewers last year x	26.6	17.85	20.3	16.8	20.8
Viewers this year y	28.9	19.2	26.4	13.7	20.2
Viewers last year x	16.7	19.1	18.9	16.0	15.8
Viewers this year y	18.8	25.0	21.0	16.8	15.3

Source: Nielson Media Research.

33. An educator wants to see how the number of absences for a student in her class affects the student's final grade. The data obtained from a sample are shown.

No. of absences x	10	12	2	0	8	5
Final grade y	70	65	96	94	75	82

For Exercises 34 and 35, do a complete regression analysis and test the significance of r at $\alpha = 0.05$, using the P-value method.

34. A physician wishes to know whether there is a relationship between a father's weight (in pounds) and his newborn son's weight (in pounds). The data are given here.

Father's weight x	176	160	187	210	196	142	205	215
Son's weight y	6.6	8.2	9.2	7.1	8.8	9.3	7.4	8.6

35. Is a person's age related to his or her net worth? A sample of 10 billionaires is selected, and the person's age and net worth are compared. The data are given here.

Age x	56	39	42	60	84	37	68	66	73	55
Net worth (in billions) y	18	14	12	14	11	10	10	7	7	5

Source: The Associated Press.

Extending the Concepts

36. For Exercises 13, 15, and 21 in Section 10–3, find the mean of the x and y variables. Then substitute the mean of the x variable into the corresponding regression line equations found in Exercises 12, 13, and 14 in this section and find y'. Compare the value of y' with $\bar{y}$ for each exercise. Generalize the results.

37. The y intercept value a can also be found by using the equation

$$a = \bar{y} - b\bar{x}$$

Verify this result by using the data in Exercises 15 and 16 of Sections 10–3 and 10–4.

38. The value of the correlation coefficient can also be found by using the formula

$$r = \frac{bs_x}{s_y}$$

where s_x is the standard deviation of the x value and s_y is the standard deviation of the y values. Verify this result for Exercises 18 and 20 of Section 10–3.

Technology *Step by Step*

MINITAB
Step by Step

Create a Scatter Plot

1. These instructions use Example 10–1. Enter the data into three columns. The subject column is optional (see step 6b).

2. Name the columns **C1 Subject, C2 Age,** and **C3 Pressure.**

3. Select **Graph>Scatterplot,** then select Simple and click [OK].

4. Double-click on C3 Pressure for the [Y] variable and C2 Age for the predictor [X] variable.

5. Click [Data View]. The Data Display should be Symbols. If not, click the option box to select it. Click [OK].

6. Click [Labels].

   a) Type **Pressure vs. Age** in the text box for Titles/Footnotes, then type **Your Name** in the box for Subtitle 1.

   b) *Optional:* Click the tab for Data Labels, then click the option to Use labels from column.

   c) Select C1 Subject.

7. Click [OK] twice.

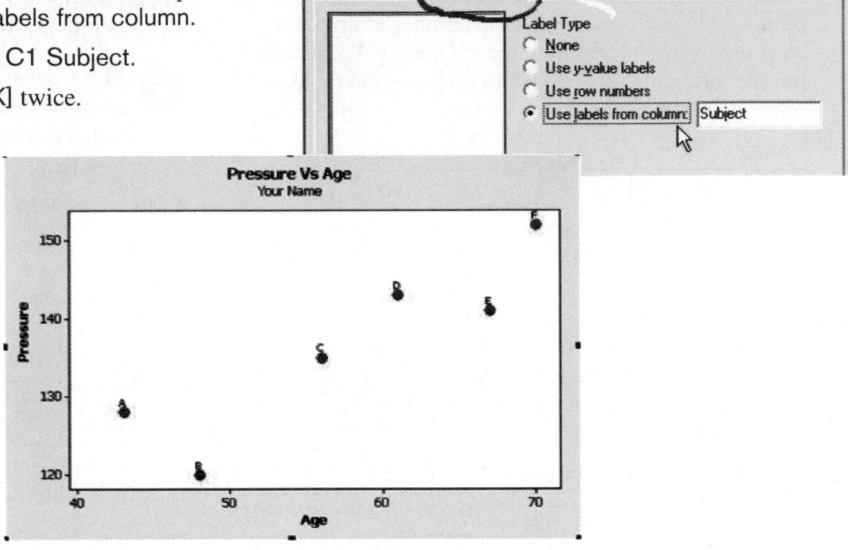

Calculate the Correlation Coefficient

8. Select **Stat>Basic Statistics>Correlation.**

9. Double-click C3 Pressure, then double-click C2 Age. The box for Display p-values should be checked.

10. Click [OK]. The correlation coefficient will be displayed in the session window, $r = +0.897$ with a P-value of 0.015.

Determine the Equation of the Least-Squares Regression Line

11. Select **Stat>Regression>Regression.**

12. Double-click Pressure in the variable list to select it for the Response variable Y.

13. Double-click C2 Age in the variable list to select it for the Predictors variable X.

14. Click on [Storage], then check the boxes for Residuals and Fits.

15. Click [OK] twice.

The session window will contain the regression analysis as shown.

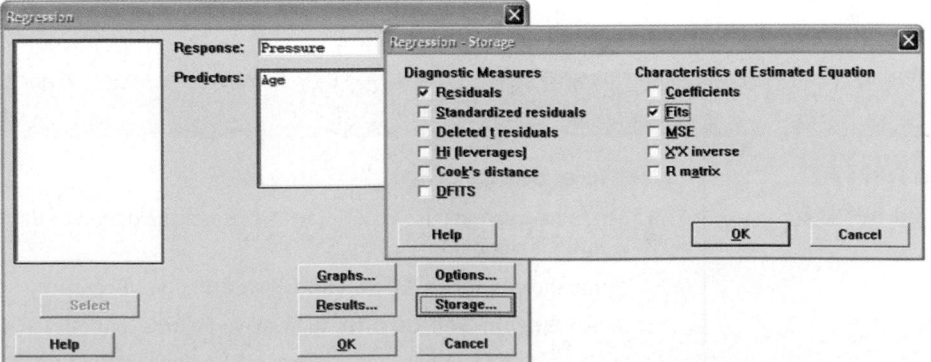

In the worksheet two new columns will be added with the fitted values and residuals. Summary: The scatter plot and correlation coefficient confirm a strong positive linear correlation between pressure and age. The null hypothesis would be rejected at a significance level of 0.015. The equation of the regression equation is pressure = 81.0 + 0.964 (age).

↓	C1-T	C2	C3	C4	C5
	Subject	Age	Pressure	RESI1	FITS1
1	A	43	128	5.48353	122.516
2	B	48	120	-7.33838	127.338
3	C	56	135	-0.05343	135.053
4	D	61	143	3.12467	139.875
5	E	67	141	-4.66162	145.662
6	F	70	152	3.44524	148.555

Regression Analysis: Pressure versus Age

```
The regression equation is
Pressure = 81.0 + 0.964 Age

Predictor          Coef      SE Coef           T         P
Constant          81.05        13.88        5.84     0.004
Age              0.9644       0.2381        4.05     0.015
S = 5.641      R-Sq = 80.4%      R-Sq (adj) = 75.5%

Analysis of Variance
Source             DF          SS          MS         F         P
Regression          1      522.21      522.21     16.41     0.015
Residual Error      4      127.29       31.82
Total               5      649.50
```

TI-83 Plus or TI-84 Plus
Step by Step

Correlation and Regression

To graph a scatter plot:

1. Enter the x values in L_1 and the y values in L_2.

2. Make sure the Window values are appropriate. Select an Xmin slightly less than the smallest x data value and an Xmax slightly larger than the largest x data value. Do the same for Ymin and Ymax. Also, you may need to change the Xscl and Yscl values, depending on the data.

3. Press **2nd [STAT PLOT] 1** for Plot 1. The other y functions should be turned off.

4. Move the cursor to On and press **ENTER** on the Plot 1 menu.

5. Move the cursor to the graphic that looks like a scatter plot next to Type (first graph), and press **ENTER.** Make sure the X list is L_1, and the Y list is L_2.

6. Press **GRAPH.**

Example TI10–1

Draw a scatter plot for the data from Example 10–1.

x	43	48	56	61	67	70
y	128	120	135	143	141	152

The input and output screens are shown.

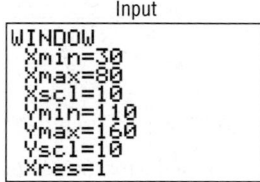
Input

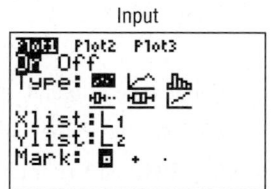
Input

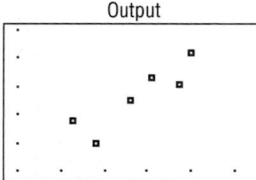
Output

To find the equation of the regression line:

1. Press **STAT** and move the cursor to Calc.

2. Press **8** for LinReg(a+bx) then **ENTER.** The values for a and b will be displayed.

In order to have the calculator compute and display the correlation coefficient and coefficient of determination as well as the equation of the line, you must set the diagnostics display mode to on. Follow these steps:

1. Press **2nd [CATALOG].**
2. Use the arrow keys to scroll down to DiagnosticOn.
3. Press **ENTER** to copy the command to the home screen.
4. Press **ENTER** to execute the command.

You will have to do this only once. Diagnostic display mode will remain on until you perform a similar set of steps to turn it off.

Example TI10–2

Find the equation of the regression line for the data in Example TI10–1, as shown in Example 10–9. The input and output screens are shown.

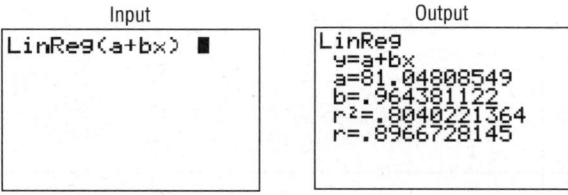

The equation of the regression line is $y' = 81.04808549 + 0.964381122x$.

To plot the regression line on the scatter plot:

1. Press **Y=** and **CLEAR** to clear any previous equations.
2. Press **VARS** and then **5** for Statistics.
3. Move the cursor to EQ and press **1** for RegEQ. The line will be in the Y= screen.
4. Press **GRAPH.**

Example TI10–3

Draw the regression line found in Example TI10–2 on the scatter plot.

The output screens are shown.

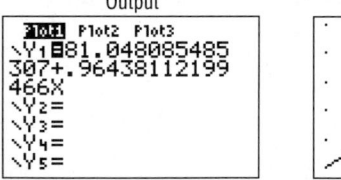
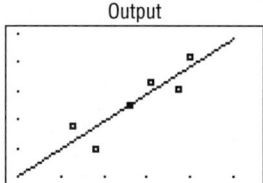

To test the significance of b and ρ:

1. Press **STAT** and move the cursor to TESTS.
2. Press **E (ALPHA SIN)** for LinRegTTest. Make sure the Xlist is L_1, the Ylist is L_2, and the Freq is 1.
3. Select the appropriate alternative hypothesis.
4. Move the cursor to Calculate and press **ENTER.**

Example TI10–4

Test the hypothesis from Examples 10–4 and 10–7, H_0: $\rho = 0$ for the data in Example 10–1. Use $\alpha = 0.05$.

Input	Output	Output
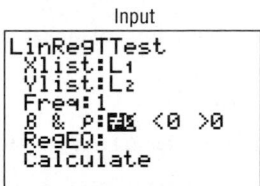	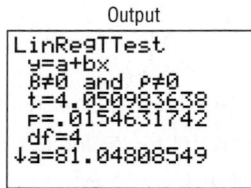	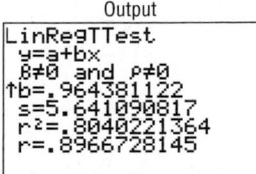

In this case, the t test value is 4.050983638. The P-value is 0.0154631742, which is significant. The decision is to reject the null hypothesis at $\alpha = 0.05$, since $0.0154631742 < 0.05$; $r = 0.8966728145$, $r^2 = 0.8040221364$.

There are two other ways to store the equation for the regression line in Y_1 for graphing.

1. Type Y_1 after the LinReg(a+bx) command.

2. Type Y_1 in the RegEQ: spot in the LinRegTTest.

To get Y_1 do this:

Press **VARS** for variables, move cursor to Y-VARS, press **1** for Function, press **1** for Y_1.

Excel
Step by Step

Scatter Plots

Creating scatter plots in Excel is straightforward when one uses the Chart Wizard.

1. Click on the Chart Wizard icon (it looks like a colorful histogram).

2. Select chart type XY (Scatter) under the Standard Types tab. Click on [Next >].

3. Enter the data range, and specify whether the data for each variable are stored in columns (as we have done in our examples) or rows. Click on [Next >].

4. The next dialog box enables you to set various options for displaying the plot. In most cases, the defaults will be okay. After entering the desired options (note that there are several tabs for this screen), click on [Next >].

5. Use this final dialog box to specify where the chart will be located. Click on [Finish].

Correlation Coefficient

The CORREL function returns the correlation coefficient.

1. Enter the data in columns A and B.

2. Select a blank cell, then click on the f_x button.

3. Under Function category, select Statistical. From the Function name list, select CORREL.

4. Enter the data range (**A1:AN,** where N is the number of sample data pairs) for the first variable in Array1. Enter the data range for the second variable in Array2. The correlation coefficient will be displayed in the selected cell.

Correlation and Regression

This procedure will allow you to calculate the Pearson product moment correlation coefficient without performing a regression analysis.

1. Enter the data from Example 10–2 in a new worksheet. Enter the seven values for the numbers of absences in column A and the corresponding final grades in column B.

2. Select **Tools>Data Analysis>Correlation.** Click the icon next to the Input Range box. This will minimize the dialog box. Use the mouse to highlight the data from columns A and B. Once the data are selected, click the icon to restore the box.

3. Make sure that the data are grouped by columns, and select New Worksheet Ply, then click [OK].

This procedure will allow you to conduct a regression analysis and compute the correlation coefficient.

1. Enter the seven values for the number of absences in column A and the corresponding final grades in column B.

2. Select **Tools>Data Analysis>Regression.**

3. Enter **B1:B7** for the Input Y Range, and then enter **A1:A7** for the Input X Range.

4. Click [OK].

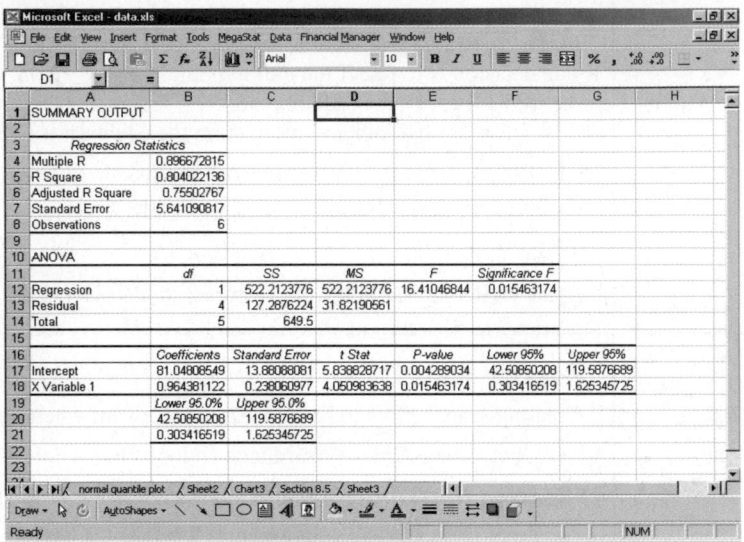

10–5

Coefficient of Determination and Standard Error of the Estimate

The previous sections stated that if the correlation coefficient is significant, the equation of the regression line can be determined. Also, for various values of the independent variable x, the corresponding values of the dependent variable y can be predicted. Several other measures are associated with the correlation and regression techniques. They include the coefficient of determination, the standard error of estimate, and the prediction interval. But before these concepts can be explained, the different types of variation associated with the regression model must be defined.

Types of Variation for the Regression Model

Consider the following hypothetical regression model.

x	1	2	3	4	5
y	10	8	12	16	20

The equation of the regression line is $y' = 4.8 + 2.8x$, and $r = 0.919$. The sample y values are 10, 8, 12, 16, and 20. The predicted values, designated by y', for each x can be found by substituting each x value into the regression equation and finding y'. For example, when $x = 1$,

$$y' = 4.8 + 2.8x = 4.8 + (2.8)(1) = 7.6$$

Now, for each x, there is an observed y value and a predicted y' value, for example, when $x = 1$, $y = 10$, and $y' = 7.6$. Recall that the closer the observed values are to the predicted values, the better the fit is and the closer r is to $+1$ or -1.

The *total variation* $\Sigma(y - \bar{y})^2$ is the sum of the squares of the vertical distances each point is from the mean. The total variation can be divided into two parts: that which is attributed to the relationship of x and y and that which is due to chance. The variation obtained from the relationship (i.e., from the predicted y' values) is $\Sigma(y' - \bar{y})^2$ and is called the *explained variation*. Most of the variations can be explained by the relationship. The closer the value r is to $+1$ or -1, the better the points fit the line and the closer $\Sigma(y' - \bar{y})^2$ is to $\Sigma(y - \bar{y})^2$. In fact, if all points fall on the regression line, $\Sigma(y' - \bar{y})^2$ will equal $\Sigma(y - \bar{y})^2$, since y' would be equal to y in each case.

On the other hand, the variation due to chance, found by $\Sigma(y - y')^2$, is called the *unexplained variation*. This variation cannot be attributed to the relationship. When the unexplained variation is small, the value of r is close to $+1$ or -1. If all points fall on the regression line, the unexplained variation $\Sigma(y - y')^2$ will be 0. Hence, the *total variation* is equal to the sum of the explained variation and the unexplained variation. That is,

$$\Sigma(y - \bar{y})^2 = \Sigma(y' - \bar{y})^2 + \Sigma(y - y')^2$$

These values are shown in Figure 10–17. For a single point, the differences are called *deviations*. For the hypothetical regression model given earlier, for $x = 1$ and $y = 10$, one gets $y' = 7.6$ and $\bar{y} = 13.2$.

The procedure for finding the three types of variation is illustrated next.

Step 1 Find the predicted y' values.

For $x = 1$ $y' = 4.8 + 2.8x = 4.8 + (2.8)(1) = 7.6$

For $x = 2$ $y' = 4.8 + (2.8)(2) = 10.4$

For $x = 3$ $y' = 4.8 + (2.8)(3) = 13.2$

For $x = 4$ $y' = 4.8 + (2.8)(4) = 16.0$

For $x = 5$ $y' = 4.8 + (2.8)(5) = 18.8$

Figure 10–17

Deviations for the Regression Equation

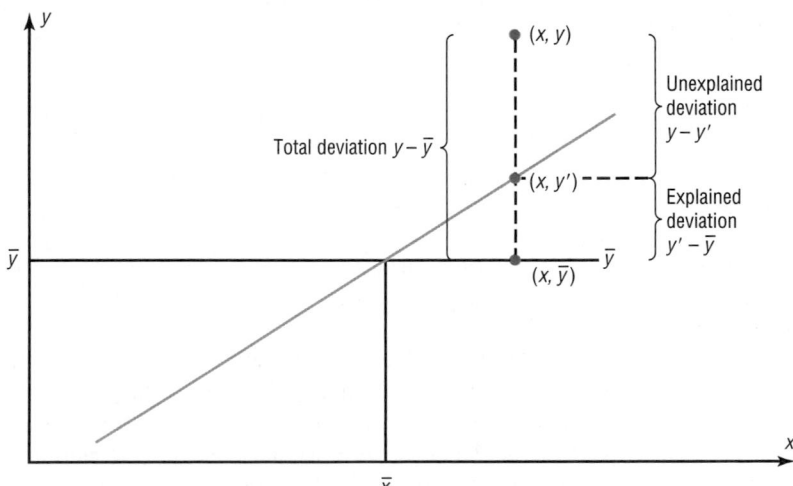

Hence, the values for this example are as follows:

x	y	y′
1	10	7.6
2	8	10.4
3	12	13.2
4	16	16.0
5	20	18.8

Step 2 Find the mean of the y values.

$$\bar{y} = \frac{10 + 8 + 12 + 16 + 20}{5} = 13.2$$

Step 3 Find the total variation $\Sigma(y - \bar{y})^2$.

$$(10 - 13.2)^2 = 10.24$$
$$(8 - 13.2)^2 = 27.04$$
$$(12 - 13.2)^2 = 1.44$$
$$(16 - 13.2)^2 = 7.84$$
$$(20 - 13.2)^2 = \underline{46.24}$$
$$\Sigma(y - \bar{y})^2 = 92.8$$

Step 4 Find the explained variation $\Sigma(y' - \bar{y})^2$.

$$(7.6 - 13.2)^2 = 31.36$$
$$(10.4 - 13.2)^2 = 7.84$$
$$(13.2 - 13.2)^2 = 0.00$$
$$(16 - 13.2)^2 = 7.84$$
$$(18.8 - 13.2)^2 = \underline{31.36}$$
$$\Sigma(y' - \bar{y})^2 = 78.4$$

Step 5 Find the unexplained variation $\Sigma(y - y')^2$.

$$(10 - 7.6)^2 = 5.76$$
$$(8 - 10.4)^2 = 5.76$$
$$(12 - 13.2)^2 = 1.44$$
$$(16 - 16)^2 = 0.00$$
$$(20 - 18.8)^2 = \underline{1.44}$$
$$\Sigma(y - y')^2 = 14.4$$

Notice that

Total variation = Explained variation + Unexplained variation
$$92.8 = \qquad 78.4 \qquad + \qquad 14.4$$

Note: The values $(y - y')$ are called *residuals*. A **residual** is the difference between the actual value of y and the predicted value y' for a given x value. The mean of the residuals is always zero. As stated previously, the regression line determined by the formulas in Section 10–4 is the line that best fits the points of the scatter plot. The sum of the squares of the residuals computed by using the regression line is the smallest possible value. For this reason, a regression line is also called a **least-squares line.**

Objective 5

Compute the coefficient of determination.

Coefficient of Determination

The *coefficient of determination* is the ratio of the explained variation to the total variation and is denoted by r^2. That is,

$$r^2 = \frac{\text{explained variation}}{\text{total variation}}$$

For the example, $r^2 = 78.4/92.8 = 0.845$. The term r^2 is usually expressed as a percentage. So in this case, 84.5% of the total variation is explained by the regression line using the independent variable.

Another way to arrive at the value for r^2 is to square the correlation coefficient. In this case, $r = 0.919$ and $r^2 = 0.845$, which is the same value found by using the variation ratio.

> The **coefficient of determination** is a measure of the variation of the dependent variable that is explained by the regression line and the independent variable. The symbol for the coefficient of determination is r^2.

Of course, it is usually easier to find the coefficient of determination by squaring r and converting it to a percentage. Therefore, if $r = 0.90$, then $r^2 = 0.81$, which is equivalent to 81%. This result means that 81% of the variation in the dependent variable is accounted for by the variations in the independent variable. The rest of the variation, 0.19, or 19%, is unexplained. This value is called the *coefficient of nondetermination* and is found by subtracting the coefficient of determination from 1. As the value of r approaches 0, r^2 decreases more rapidly. For example, if $r = 0.6$, then $r^2 = 0.36$, which means that only 36% of the variation in the dependent variable can be attributed to the variation in the independent variable.

Coefficient of Nondetermination

$$1.00 - r^2$$

Objective 6

Compute the standard error of the estimate.

Standard Error of the Estimate

When a y' value is predicted for a specific x value, the prediction is a point prediction. However, a prediction interval about the y' value can be constructed, just as a confidence interval was constructed for an estimate of the population mean. The prediction interval uses a statistic called the *standard error of the estimate*.

> The **standard error of the estimate,** denoted by s_{est}, is the standard deviation of the observed y values about the predicted y' values. The formula for the standard error of estimate is
>
> $$s_{est} = \sqrt{\frac{\Sigma(y - y')^2}{n - 2}}$$

The standard error of the estimate is similar to the standard deviation, but the mean is not used. As can be seen from the formula, the standard error of the estimate is the square root of the unexplained variation—that is, the variation due to the difference of the observed values and the expected values—divided by $n - 2$. So the closer the observed values are to the predicted values, the smaller the standard error of the estimate will be.

Example 10–12 shows how to compute the standard error of the estimate.

Example 10–12

A researcher collects the following data and determines that there is a significant relationship between the age of a copy machine and its monthly maintenance cost. The regression equation is $y' = 55.57 + 8.13x$. Find the standard error of the estimate.

Machine	Age x (years)	Monthly cost y
A	1	$ 62
B	2	78
C	3	70
D	4	90
E	4	93
F	6	103

Solution

Step 1 Make a table, as shown.

x	y	y'	$y - y'$	$(y - y')^2$
1	62			
2	78			
3	70			
4	90			
4	93			
6	103			

Step 2 Using the regression line equation $y' = 55.57 + 8.13x$, compute the predicted values y' for each x and place the results in the column labeled y'.

$x = 1$ $y' = 55.57 + (8.13)(1) = 63.70$
$x = 2$ $y' = 55.57 + (8.13)(2) = 71.83$
$x = 3$ $y' = 55.57 + (8.13)(3) = 79.96$
$x = 4$ $y' = 55.57 + (8.13)(4) = 88.09$
$x = 6$ $y' = 55.57 + (8.13)(6) = 104.35$

Step 3 For each y, subtract y' and place the answer in the column labeled $y - y'$.

$62 - 63.70 = -1.70$ $90 - 88.09 = 1.91$
$78 - 71.83 = 6.17$ $93 - 88.09 = 4.91$
$70 - 79.96 = -9.96$ $103 - 104.35 = -1.35$

Step 4 Square the numbers found in step 3 and place the squares in the column labeled $(y - y')^2$.

Step 5 Find the sum of the numbers in the last column. The completed table is shown.

x	y	y'	$y - y'$	$(y - y')^2$
1	62	63.70	-1.70	2.89
2	78	71.83	6.17	38.0689
3	70	79.96	-9.96	99.2016
4	90	88.09	1.91	3.6481
4	93	88.09	4.91	24.1081
6	103	104.35	-1.35	1.8225
				169.7392

Step 6 Substitute in the formula and find s_{est}.

$$s_{est} = \sqrt{\frac{\Sigma(y - y')^2}{n - 2}} = \sqrt{\frac{169.7392}{6 - 2}} = 6.51$$

In this case, the standard deviation of observed values about the predicted values is 6.51.

The standard error of the estimate can also be found by using the formula

$$s_{est} = \sqrt{\frac{\Sigma y^2 - a\,\Sigma y - b\,\Sigma xy}{n - 2}}$$

Example 10–13   Find the standard error of the estimate for the data for Example 10–12 by using the preceding formula. The equation of the regression line is $y' = 55.57 + 8.13x$.

Solution

Step 1 Make a table.

Step 2 Find the product of x and y values, and place the results in the third column.

Step 3 Square the y values, and place the results in the fourth column.

Step 4 Find the sums of the second, third, and fourth columns. The completed table is shown here.

x	y	xy	y^2
1	62	62	3,844
2	78	156	6,084
3	70	210	4,900
4	90	360	8,100
4	93	372	8,649
6	103	618	10,609
	$\Sigma y = 496$	$\Sigma xy = 1{,}778$	$\Sigma y^2 = 42{,}186$

Step 5 From the regression equation $y' = 55.57 + 8.13x$, $a = 55.57$ and $b = 8.13$.

Step 6 Substitute in the formula and solve for s_{est}.

$$s_{est} = \sqrt{\frac{\Sigma y^2 - a\,\Sigma y - b\,\Sigma xy}{n - 2}}$$

$$= \sqrt{\frac{42{,}186 - (55.57)(496) - (8.13)(1778)}{6 - 2}} = 6.48$$

This value is close to the value found in Example 10–12. The difference is due to rounding.

Objective 7

Find a prediction interval.

Prediction Interval

The standard error of estimate can be used for constructing a **prediction interval** (similar to a confidence interval) about a y' value.

When a specific value x is substituted into the regression equation, one gets y', which is a point estimate for y. For example, if the regression line equation for the age of a machine and the monthly maintenance cost is $y' = 55.57 + 8.13x$ (Example 10–12), then the predicted maintenance cost for a 3-year-old machine would be $y' = 55.57 + 8.13(3)$, or \$79.96. Since this is a point estimate, one has no idea how accurate it is. But one can construct a prediction interval about the estimate. By selecting an α value, one can achieve a $(1 - \alpha) \cdot 100\%$ confidence that the interval contains the actual mean of the y values that correspond to the given value of x.

The reason is that there are possible sources of prediction errors in finding the regression line equation. One source occurs when finding the standard error of the estimate s_{est}. Two others are errors made in estimating the slope and the y' intercept, since the equation of the regression line will change somewhat if different random samples are used when calculating the equation.

Formula for the Prediction Interval about a Value y'

$$y' - t_{\alpha/2}s_{est}\sqrt{1 + \frac{1}{n} + \frac{n(x - \overline{X})^2}{n\,\Sigma x^2 - (\Sigma x)^2}} < y < y' + t_{\alpha/2}s_{est}\sqrt{1 + \frac{1}{n} + \frac{n(x - \overline{X})^2}{n\,\Sigma x^2 - (\Sigma x)^2}}$$

with d.f. $= n - 2$.

Example 10–14

For the data in Example 10–12, find the 95% prediction interval for the monthly maintenance cost of a machine that is 3 years old.

Solution

Step 1 Find Σx, Σx^2, and $\overline{X}$.

$$\Sigma x = 20 \qquad \Sigma x^2 = 82 \qquad \overline{X} = \frac{20}{6} = 3.3$$

Step 2 Find y' for $x = 3$.

$$y' = 55.57 + 8.13x$$
$$y' = 55.57 + 8.13(3) = 79.96$$

Step 3 Find s_{est}.

$$s_{est} = 6.48$$

as shown in Example 10–13.

Step 4 Substitute in the formula and solve: $t_{\alpha/2} = 2.776$, d.f. $= 6 - 2 = 4$ for 95%.

$$y' - t_{\alpha/2}s_{est}\sqrt{1 + \frac{1}{n} + \frac{n(x - \overline{X})^2}{n\,\Sigma x^2 - (\Sigma x)^2}} < y < y'$$

$$+ t_{\alpha/2}s_{est}\sqrt{1 + \frac{1}{n} + \frac{n(x - \overline{X})^2}{n\,\Sigma x^2 - (\Sigma x)^2}}$$

$$79.96 - (2.776)(6.48)\sqrt{1 + \frac{1}{6} + \frac{6(3 - 3.3)^2}{6(82) - (20)^2}} < y < 79.96$$

$$+ (2.776)(6.48)\sqrt{1 + \frac{1}{6} + \frac{6(3 - 3.3)^2}{6(82) - (20)^2}}$$

$$79.96 - (2.776)(6.48)(1.08) < y < 79.96 + (2.776)(6.48)(1.08)$$
$$79.96 - 19.43 < y < 79.96 + 19.43$$
$$60.53 < y < 99.39$$

Hence, one can be 95% confident that the interval $60.53 < y < 99.39$ contains the actual value of y.

Applying the Concepts **10–5**

Interpreting Simple Linear Regression

Answer the questions about the following computer-generated information.

Linear correlation coefficient $r = 0.794556$
Coefficient of determination $= 0.631319$
Standard error of estimate $= 12.9668$
Explained variation $= 5182.41$
Unexplained variation $= 3026.49$
Total variation $= 8208.90$
Equation of regression line $y' = 0.725983X + 16.5523$
Level of significance $= 0.1$
Test statistic $= 0.794556$
Critical value $= 0.378419$

1. Are both variables moving in the same direction?
2. Which number measures the distances from the prediction line to the actual values?
3. Which number is the slope of the regression line?
4. Which number is the y intercept of the regression line?
5. Which number can be found in a table?
6. Which number is the allowable risk of making a type I error?
7. Which number measures the variation explained by the regression?
8. Which number measures the scatter of points about the regression line?
9. What is the null hypothesis?
10. Which number is compared to the critical value to see if the null hypothesis should be rejected?
11. Should the null hypothesis be rejected?

See page 581 for the answers.

Exercises 10–5

1. What is meant by the *explained variation?* How is it computed?

2. What is meant by the *unexplained variation?* How is it computed?

3. What is meant by the *total variation?* How is it computed?

4. Define the coefficient of determination.

5. How is the coefficient of determination found?

6. Define the coefficient of nondetermination.

7. How is the coefficient of nondetermination found?

For Exercises 8 through 13, find the coefficients of determination and nondetermination and explain the meaning of each.

8. $r = 0.81$

9. $r = 0.70$

10. $r = 0.45$

11. $r = 0.37$

12. $r = 0.15$

13. $r = 0.05$

14. Define the standard error of the estimate for regression. When can the standard error of the estimate be used to construct a prediction interval about a value y'?

15. Compute the standard error of the estimate for Exercise 13 in Section 10–3. The regression line equation was found in Exercise 13 in Section 10–4.

16. Compute the standard error of the estimate for Exercise 14 in Section 10–3. The regression line equation was found in Exercise 14 in Section 10–4.

17. Compute the standard error of the estimate for Exercise 15 in Section 10–3. The regression line equation was found in Exercise 15 in Section 10–4.

18. Compute the standard error of the estimate for Exercise 16 in Section 10–3. The regression line equation was found in Exercise 16 in Section 10–4.

19. For the data in Exercises 13 in Sections 10–3 and 10–4 and 15 in Section 10–5, find the 90% prediction interval when $x = 20$ years.

20. For the data in Exercises 14 in Sections 10–3 and 10–4 and 16 in Section 10–5, find the 95% prediction interval when $x = 60$.

21. For the data in Exercises 15 in Sections 10–3 and 10–4 and 17 in Section 10–5, find the 90% prediction interval when $x = 4$ years.

22. For the data in Exercises 16 in Sections 10–3 and 10–4 and 18 in Section 10–5, find the 98% prediction interval when $x = 47$ years.

10–6 Multiple Regression (Optional)

Objective 8

Be familiar with the concept of multiple regression.

The previous sections explained the concepts of simple linear regression and correlation. In simple linear regression, the regression equation contains one independent variable x and one dependent variable y' and is written as

$$y' = a + bx$$

where a is the y' intercept and b is the slope of the regression line.

In **multiple regression,** there are several independent variables and one dependent variable, and the equation is

$$y' = a + b_1x_1 + b_2x_2 + \cdots + b_kx_k$$

where $x_1, x_2, \ldots, x_k$ are the independent variables.

For example, suppose a nursing instructor wishes to see whether there is a relationship between a student's grade point average, age, and score on the state board nursing examination. The two independent variables are GPA (denoted by x_1) and age (denoted by x_2). The instructor will collect the data for all three variables for a sample of nursing students. Rather than conduct two separate simple regression studies, one using the GPA and state board scores and another using ages and state board scores, the instructor can conduct one study using multiple regression analysis with two independent variables—GPA and ages—and one dependent variable—state board scores.

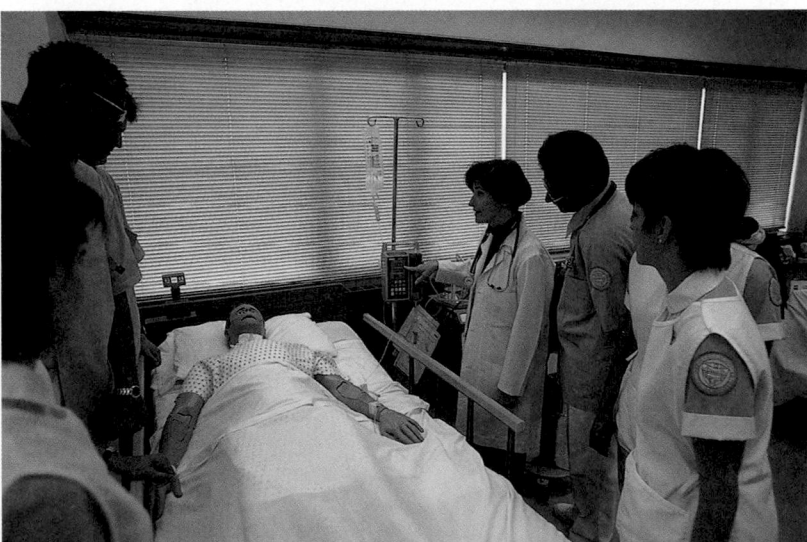

In this study, researchers found a correlation between the cleanliness of the homes children are raised in and the years of schooling completed and earning potential for those children. What interfering variables were controlled? How might these have been controlled? Summarize the conclusions of the study.

SUCCESS

HOME SMART HOME

KIDS WHO GROW UP IN A CLEAN HOUSE FARE BETTER AS ADULTS

Good-bye, GPA. So long, SATs. New research suggests that we may be able to predict children's future success from the level of cleanliness in their homes.

A University of Michigan study presented at the annual meeting of the American Economic Association uncovered a surprising correlation: children raised in clean homes were later found to have completed more school and to have higher earning potential than those raised in dirty homes. The clean homes may indicate a family that values organization and similarly helpful skills at school and work, researchers say.

Cleanliness ratings for about 5,000 households were assessed between 1968 and 1972, and respondents were interviewed 25 years later to determine educational achievement and professional earnings of the young adults who had grown up there, controlling for variables such as race, socio-economic status and level of parental education. The data showed that those raised in homes rated "clean" to "very clean" had completed an average of 1.6 more years of school than those raised in "not very clean" or "dirty" homes. Plus, the first group's annual wages averaged about $3,100 more than the second's.

But don't buy stock in Mr. Clean and Pine Sol just yet. "We're not advocating that everyone go out and clean their homes right this minute," explains Rachel Dunifon, a University of Michigan doctoral candidate and a researcher on the study. Rather, the main implication of the study, Dunifon says, is that there is significant evidence that non-cognitive factors, such as organization and efficiency, play a role in determining academic and financial success.

— Jackie Fisherman

A multiple regression correlation R can also be computed to determine if a significant relationship exists between the independent variables and the dependent variable. Multiple regression analysis is used when a statistician thinks there are several independent variables contributing to the variation of the dependent variable. This analysis then can be used to increase the accuracy of predictions for the dependent variable over one independent variable alone.

Two other examples for multiple regression analysis are when a store manager wants to see whether the amount spent on advertising and the amount of floor space used for a display affect the amount of sales of a product, and when a sociologist wants to see whether the amount of time children spend watching television and playing video games is related to their weight. Multiple regression analysis can also be conducted by using more than two independent variables, denoted by $x_1, x_2, x_3, \ldots, x_m$. Since these computations are quite complicated and for the most part would be done on a computer, this chapter will show the computations for two independent variables only.

For example, the nursing instructor wishes to see whether a student's grade point average and age are related to the student's score on the state board nursing examination. She selects five students and obtains the following data.

Student	GPA x_1	Age x_2	State board score y
A	3.2	22	550
B	2.7	27	570
C	2.5	24	525
D	3.4	28	670
E	2.2	23	490

The multiple regression equation obtained from the data is

$$y' = -44.81 + 87.64x_1 + 14.533x_2$$

If a student has a GPA of 3.0 and is 25 years old, her predicted state board score can be computed by substituting these values in the equation for x_1 and x_2, respectively, as shown.

$$y' = -44.81 + 87.64(3.0) + 14.533(25)$$
$$= 581.44 \text{ or } 581$$

Hence, if a student has a GPA of 3.0 and is 25 years old, the student's predicted state board score is 581.

The Multiple Regression Equation

A multiple regression equation with two independent variables (x_1 and x_2) and one dependent variable would have the form

$$y' = a + b_1x_1 + b_2x_2$$

A multiple regression with three independent variables (x_1, x_2, and x_3) and one dependent variable would have the form

$$y' = a + b_1x_1 + b_2x_2 + b_3x_3$$

General Form of the Multiple Regression Equation

The general form of the multiple regression equation with k independent variables is

$$y' = a + b_1x_1 + b_2x_2 + \cdots + b_kx_k$$

The x's are the independent variables. The value for a is more or less an intercept, although a multiple regression equation with two independent variables constitutes a plane rather than a line. The b's are called *partial regression coefficients*. Each b represents the amount of change in y' for one unit of change in the corresponding x value when the other x values are held constant. In the example just shown, the regression equation was $y' = -44.81 + 87.64x_1 + 14.533x_2$. In this case, for each unit of change in the student's GPA, there is a change of 87.64 units in the state board score with the student's age x_2 being held constant. And for each unit of change in x_2 (the student's age), there is a change of 14.533 units in the state board score with the GPA held constant.

Assumptions for Multiple Regression

The assumptions for multiple regression are similar to those for simple regression.

1. For any specific value of the independent variable, the values of the y variable are normally distributed. (This is called the *normality* assumption.)
2. The variances (or standard deviations) for the y variables are the same for each value of the independent variable. (This is called the *equal-variance* assumption.)
3. There is a linear relationship between the dependent variable and the independent variables. (This is called the *linearity* assumption.)
4. The independent variables are not correlated. (This is called the *nonmulticolinearity* assumption.)
5. The values for the y variables are independent. (This is called the *independence* assumption.)

In multiple regression, as in simple regression, the strength of the relationship between the independent variables and the dependent variable is measured by a correlation coefficient. This **multiple correlation coefficient** is symbolized by R. The value of R can range from 0 to $+1$; R can never be negative. The closer to $+1$, the stronger the relationship; the closer to 0, the weaker the relationship. The value of R takes into account all the independent variables and can be computed by using the values of the individual correlation coefficients. The formula for the multiple correlation coefficient when there are two independent variables is shown next.

Formula for the Multiple Correlation Coefficient

The formula for R is

$$R = \sqrt{\frac{r_{yx_1}^2 + r_{yx_2}^2 - 2r_{yx_1} \cdot r_{yx_2} \cdot r_{x_1x_2}}{1 - r_{x_1x_2}^2}}$$

where r_{yx_1} is the value of the correlation coefficient for variables y and x_1; r_{yx_2} is the value of the correlation coefficient for variables y and x_2; and $r_{x_1x_2}$ is the value of the correlation coefficient for variables x_1 and x_2.

In this case, R is 0.989, as shown in Example 10–15. The multiple correlation coefficient is always higher than the individual correlation coefficients. For this specific example, the multiple correlation coefficient is higher than the two individual correlation coefficients computed by using grade point average and state board scores ($r_{yx_1} = 0.845$) or age and state board scores ($r_{yx_2} = 0.791$). *Note:* $r_{x_1x_2} = 0.371$.

Example 10–15

For the data regarding state board scores, find the value of R.

Solution

The values of the correlation coefficients are

$$r_{yx_1} = 0.845$$
$$r_{yx_2} = 0.791$$
$$r_{x_1x_2} = 0.371$$

Substituting in the formula, one gets

$$R = \sqrt{\frac{r_{yx_1}^2 + r_{yx_2}^2 - 2r_{yx_1} \cdot r_{yx_2} \cdot r_{x_1x_2}}{1 - r_{x_1x_2}^2}}$$

$$= \sqrt{\frac{(0.845)^2 + (0.791)^2 - 2(0.845)(0.791)(0.371)}{1 - 0.371^2}}$$

$$= \sqrt{\frac{0.8437569}{0.862359}} = \sqrt{0.9784288} = 0.989$$

Hence, the correlation between a student's grade point average and age with the student's score on the nursing state board examination is 0.989. In this case, there is a strong relationship among the variables; the value of R is close to 1.00.

As with simple regression, R^2 is the *coefficient of multiple determination,* and it is the amount of variation explained by the regression model. The expression $1 - R^2$ represents the amount of unexplained variation, called the *error* or *residual variation.* Since $R = 0.989$, $R^2 = 0.978$ and $1 - R^2 = 1 - 0.978 = 0.022$.

Testing the Significance of *R*

An F test is used to test the significance of R. The hypotheses are

$$H_0: \rho = 0 \qquad \text{and} \qquad H_1: \rho \neq 0$$

where ρ represents the population correlation coefficient for multiple correlation.

F Test for Significance of *R*

The formula for the F test is

$$F = \frac{R^2/k}{(1 - R^2)/(n - k - 1)}$$

where n is the number of data groups $(x_1, x_2, \ldots, y)$ and k is the number of independent variables.

The degrees of freedom are d.f.N. $= n - k$ and d.f.D. $= n - k - 1$.

Example 10–16

Test the significance of the R obtained in Example 10–15 at $\alpha = 0.05$.

Solution

$$F = \frac{R^2/k}{(1 - R^2)/(n - k - 1)}$$

$$= \frac{0.978/2}{(1 - 0.978)/(5 - 2 - 1)} = \frac{0.489}{0.011} = 44.45$$

The critical value obtained from Table H with $\alpha = 0.05$, d.f.N. $= 3$, and d.f.D. $= 5 - 2 - 1 = 2$ is 19.16. Hence, the decision is to reject the null hypothesis and conclude that there is a significant relationship among the student's GPA, age, and score on the nursing state board examination.

Adjusted R^2

Since the value of R^2 is dependent on n (the number of data pairs) and k (the number of variables), statisticians also calculate what is called an **adjusted R^2,** denoted by R^2_{adj}. This is based on the number of degrees of freedom.

Formula for the Adjusted R^2

The formula for the adjusted R^2 is

$$R^2_{adj} = 1 - \left[\frac{(1 - R^2)(n - 1)}{n - k - 1} \right]$$

The adjusted R^2 is smaller than R^2 and takes into account the fact that when n and k are approximately equal, the value of R may be artificially high, due to sampling error rather than a true relationship among the variables. This occurs because the chance variations of all the variables are used in conjunction with each other to derive the regression equation. Even if the individual correlation coefficients for each independent variable and the dependent variable were all zero, the multiple correlation coefficient due to sampling error could be higher than zero.

Hence, both R^2 and R^2_{adj} are usually reported in a multiple regression analysis.

Example 10–17

Calculate the adjusted R^2 for the data in Example 10–16. The value for R is 0.989.

Solution

$$\begin{aligned}
R^2_{adj} &= 1 - \left[\frac{(1 - R^2)(n - 1)}{n - k - 1} \right] \\
&= 1 - \left[\frac{(1 - 0.989^2)(5 - 1)}{5 - 2 - 1} \right] \\
&= 1 - 0.043758 \\
&= 0.956
\end{aligned}$$

In this case, when the number of data pairs and the number of independent variables are accounted for, the adjusted multiple coefficient of determination is 0.956.

Applying the Concepts **10–6**

More Math Means More Money

In a study to determine a person's yearly income 10 years after high school, it was found that the two biggest predictors are number of math courses taken and number of hours worked per week during a person's senior year of high school. The multiple regression equation generated from a sample of 20 individuals is

$$y' = 6000 + 4540x_1 + 1290x_2$$

Let x_1 represent the number of mathematics courses taken and x_2 represent hours worked. The correlation between income and mathematics courses is 0.63. The correlation between income and hours worked is 0.84, and the correlation between mathematics courses and hours worked is 0.31. Use this information to answer the following questions.

1. What is the dependent variable?
2. What are the independent variables?

3. What are the multiple regression assumptions?

4. Explain what 4540 and 1290 in the equation tell us.

5. What is the predicted income if a person took 8 math classes and worked 20 hours per week during her or his senior year in high school?

6. What does a multiple correlation coefficient of 0.77 mean?

7. Compute R^2.

8. Compute the adjusted R^2.

9. Would the equation be considered a good predictor of income?

10. What are your conclusions about the relationship between courses taken, hours worked, and yearly income?

See page 581 for the answers.

Exercises 10–6

1. Explain the similarities and differences between simple linear regression and multiple regression.

2. What is the general form of the multiple regression equation? What does a represent? What do the b's represent?

3. Why would a researcher prefer to conduct a multiple regression study rather than separate regression studies using one independent variable and the dependent variable?

4. What are the assumptions for multiple regression?

5. How do the values of the individual correlation coefficients compare to the value of the multiple correlation coefficient?

6. A researcher has determined that a significant relationship exists among an employee's age x_1, grade point average x_2, and income y. The multiple regression equation is $y' = -34,127 + 132x_1 + 20,805x_2$. Predict the income of a person who is 32 years old and has a GPA of 3.4.

7. A manufacturer found that a significant relationship exists among the number of hours an assembly line employee works per shift x_1, the total number of items produced x_2, and the number of defective items produced y. The multiple regression equation is $y' = 9.6 + 2.2x_1 - 1.08x_2$. Predict the number of defective items produced by an employee who has worked 9 hours and produced 24 items.

8. A real estate agent found that there is a significant relationship among the number of acres on a farm x_1, the number of rooms in the farmhouse x_2, and the selling price in thousands of dollars y of farms in a specific area. The regression equation is $y' = 44.9 - 0.0266x_1 + 7.56x_2$. Predict the selling price of a farm that has 371 acres and a farmhouse with six rooms.

9. An educator has found a significant relationship among a college graduate's IQ x_1, score on the verbal section of the SAT x_2, and income for the first year following graduation from college y. Predict the income of a college graduate whose IQ is 120 and verbal SAT score is 650. The regression equation is $y' = 5000 + 97x_1 + 35x_2$.

10. A medical researcher found a significant relationship among a person's age x_1, cholesterol level x_2, sodium level of the blood x_3, and systolic blood pressure y. The regression equation is $y' = 97.7 + 0.691x_1 + 219x_2 - 299x_3$. Predict the systolic blood pressure of a person who is 35 years old and has a cholesterol level of 194 milligrams per deciliter (mg/dl) and a sodium blood level of 142 milliequivalents per liter (mEq/l).

11. Explain the meaning of the multiple correlation coefficient R.

12. What is the range of values R can assume?

13. Define R^2 and R^2_{adj}.

14. What are the hypotheses used to test the significance of R?

15. What test is used to test the significance of R?

16. What is the meaning of the adjusted R^2? Why is it computed?

Technology *Step by Step*

MINITAB
Step by Step

Multiple Regression

In Example 10–15, is there a correlation between a student's score and her or his age and grade point average?

1. Enter the data for the example into three columns of MINITAB. Name the columns **GPA, AGE,** and **SCORE.**

2. Click **Stat>Regression> Regression.**

3. Double-click on C3 SCORE, the response variable.

4. Double-click C1 GPA, then C2 AGE.

5. Click on [Storage].

   a) Check the box for Residuals.

   b) Check the box for Fits.

6. Click [OK] twice.

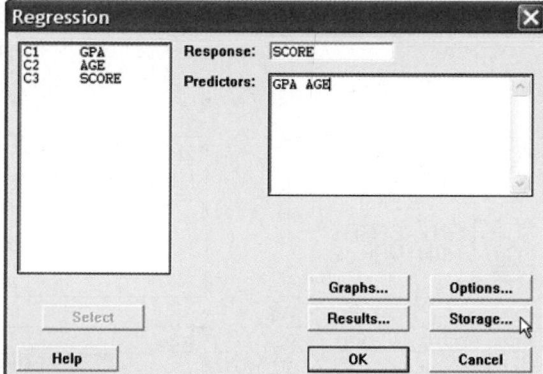

Regression Analysis: SCORE versus GPA, AGE
```
The regression equation is
SCORE = -44.8 + 87.6 GPA + 14.5 Age

Predictor     Coef   SE Coef      T       P
Constant    -44.81     69.25  -0.65   0.584
GPA          87.64     15.24   5.75   0.029
AGE         14.533     2.914   4.99   0.038
S = 14.0091   R-Sq = 97.9%   R-Sq(adj) = 95.7%

Analysis of Variance
Source         DF       SS       MS      F       P
Regression      2  18027.5   9013.7  45.93   0.021
Residual Error  2    392.5    196.3
Total           4  18420.0
```

The test statistic and *P*-value are 45.93 and 0.021, respectively. Since the *P*-value is less than α, reject the null hypothesis. There is enough evidence in the sample to conclude the scores are related to age and grade point average.

TI-83 Plus or TI-84 Plus
Step by Step

The TI-83 Plus and the TI-84 Plus do not have a built-in function for multiple regression. However, the downloadable program named MULREG is available on your CD and Online Learning Center. Follow the instructions with your CD for downloading the program.

Finding a Multiple Regression Equation

1. Enter the sets of data values into L_1, L_2, L_3, etc. Make note of which lists contain the independent variables and which list contains the dependent variable as well as how many data values are in each list.

2. Press **PRGM,** move the cursor to the program named MULREG, and press **ENTER** twice.

3. Type the number of independent variables and press **ENTER.**

4. Type the number of cases for each variable and press **ENTER.**

5. Type the name of the list that contains the data values for the first independent variable and press **ENTER.** Repeat this for all independent variables and the dependent variable.

6. The program will show the regression coefficients.

7. Press **ENTER** to see the values of R^2 and adjusted R^2.

8. Press **ENTER** to see the values of the F test statistics and the P-value.

Find the multiple regression equation for these data used in this section:

Student	GPA x_1	Age x_2	State board score y
A	3.2	22	550
B	2.7	27	570
C	2.5	24	525
D	3.4	28	670
E	2.2	23	490

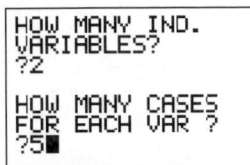
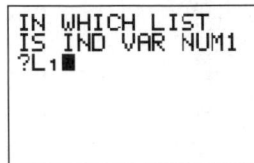
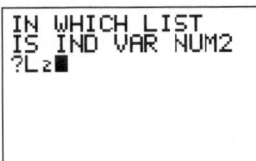

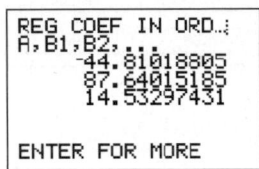
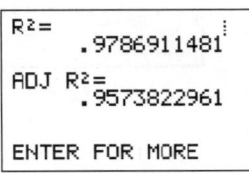
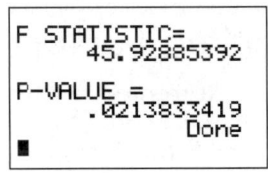

Excel
Step by Step

Multiple Regression

These instructions use data from the nursing examination example discussed at the beginning of Section 10–6.

1. Enter the data from the example into three separate columns of a new worksheet—GPAs in cells A1:A5, ages in cells B1:B5, and scores in cells C1:C5.

2. Select **Tools>Data Analysis>Regression.**

3. Select cells C1:C5 for the Input Y Range.

4. Select cells A1:B5 for the Input X Range.

5. Click [OK].

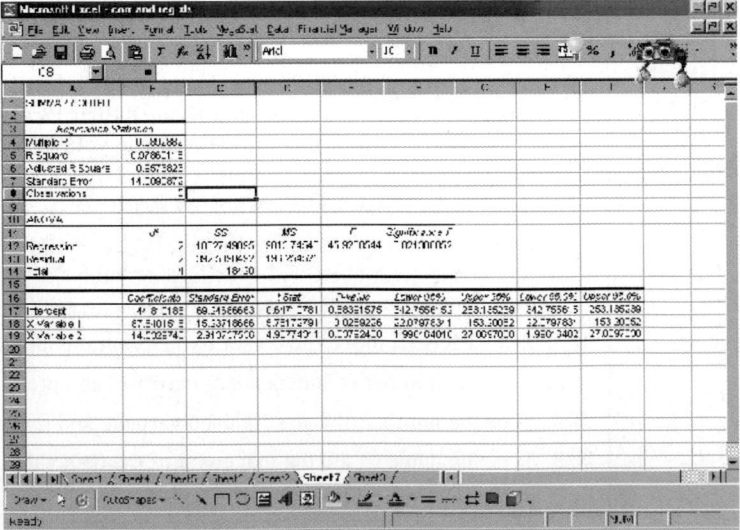

Regression Analysis: SCORE versus GPA, AGE
```
The regression equation is
SCORE = -44.8 + 87.6 GPA + 14.5 AGE

Predictor          Coef      SE Coef              T       P
Constant         -44.81        69.25          -0.65   0.584
GPA               87.64        15.24           5.75   0.029
AGE              14.533        2.914           4.99   0.038
S = 14.01  R-Sq = 97.9%      R-Sq(adj) = 95.7%

Analysis of Variance
Source            DF         SS        MS       F       P
Regression         2    18027.5    9013.7   45.93   0.021
Residual Error     2      392.5     196.3
Total              4    18420.0

Source            DF     Seq SS
GPA                1    13145.2
AGE                1     4882.3
```

The session window shows the correlation coefficient for each pair of variables. The multiple correlation coefficient is significant at 0.021. Ninety-six percent of the variation from the mean is explained by regression. The regression equation is SCORE = −44.8 + 87.6GPA + 14.5AGE.

10–7   Summary

Many relationships among variables exist in the real world. One way to determine whether a relationship exists is to use the statistical techniques known as correlation and regression. The strength and direction of the relationship are measured by the value of the correlation coefficient. It can assume values between and including +1 and −1. The closer the value of the correlation coefficient is to +1 or −1, the stronger the linear relationship is between the variables. A value of +1 or −1 indicates a perfect linear relationship. A positive relationship between two variables means that for small values of the independent variable, the values of the dependent variable will be small, and that for large values of the independent variable, the values of the dependent variable will be large. A negative relationship between two variables means that for small values of the independent variable, the values of the dependent variable will be large, and that for large values of the independent variable, the values of the dependent variable will be small.

Relationships can be linear or curvilinear. To determine the shape, one draws a scatter plot of the variables. If the relationship is linear, the data can be approximated by a straight line, called the *regression line,* or the *line of best fit.* The closer the value of *r* is to +1 or −1, the closer the points will fit the line.

In addition, relationships can be multiple. That is, there can be two or more independent variables and one dependent variable. A coefficient of correlation and a regression equation can be found for multiple relationships, just as they can be found for simple relationships.

"At this point in my report, I'll ask all of you to follow me to the conference room directly below us!"

Source: Cartoon by Bradford Veley, Marquette, Michigan. Reprinted with permission.

The coefficient of determination is a better indicator of the strength of a relationship than the correlation coefficient. It is better because it identifies the percentage of variation of the dependent variable that is directly attributable to the variation of the independent variable. The coefficient of determination is obtained by squaring the correlation coefficient and converting the result to a percentage.

Another statistic used in correlation and regression is the standard error of the estimate, which is an estimate of the standard deviation of the y values about the predicted y' values. The standard error of the estimate can be used to construct a prediction interval about a specific value point estimate y' of the mean of the y values for a given value of x.

Finally, remember that a significant relationship between two variables does not necessarily mean that one variable is a direct cause of the other variable. In some cases this is true, but other possibilities that should be considered include a complex relationship involving other (perhaps unknown) variables, a third variable interacting with both variables, or a relationship due solely to chance.

Important Terms

adjusted R^2 571

coefficient of determination 561

correlation 528

correlation coefficient 533

dependent variable 529

extrapolation 549

independent variable 529

influential point or observation 550

least-squares line 560

lurking variable 540

marginal change 548

multiple correlation coefficient 569

multiple regression 566

multiple relationship 529

negative relationship 529

Pearson product moment correlation coefficient 533

population correlation coefficient 537

positive relationship 529

prediction interval 563

regression 529

regression line 544

residual 560

scatter plot 530

simple relationship 529

standard error of the estimate 561

Important Formulas

Formula for the correlation coefficient:

$$r = \frac{n(\Sigma xy) - (\Sigma x)(\Sigma y)}{\sqrt{[n(\Sigma x^2) - (\Sigma x)^2][n(\Sigma y^2) - (\Sigma y)^2]}}$$

Formula for the t test for the correlation coefficient:

$$t = r\sqrt{\frac{n-2}{1-r^2}} \qquad \text{d.f.} = n - 2$$

The regression line equation: $y' = a + bx$, where

$$a = \frac{(\Sigma y)(\Sigma x^2) - (\Sigma x)(\Sigma xy)}{n(\Sigma x^2) - (\Sigma x)^2}$$

$$b = \frac{n(\Sigma xy) - (\Sigma x)(\Sigma y)}{n(\Sigma x^2) - (\Sigma x)^2}$$

Formula for the standard error of the estimate:

$$s_{\text{est}} = \sqrt{\frac{\Sigma(y - y')^2}{n-2}}$$

or

$$s_{\text{est}} = \sqrt{\frac{\Sigma y^2 - a\,\Sigma y - b\,\Sigma xy}{n-2}}$$

Formula for the prediction interval for a value y':

$$y' - t_{\alpha/2}s_{\text{est}}\sqrt{1 + \frac{1}{n} + \frac{n(x - \bar{X})^2}{n\,\Sigma x^2 - (\Sigma x)^2}} < y$$

$$< y' + t_{\alpha/2}s_{\text{est}}\sqrt{1 + \frac{1}{n} + \frac{n(x - \bar{X})^2}{n\,\Sigma x^2 - (\Sigma x)^2}}$$

$$\text{d.f.} = n - 2$$

Formula for the multiple correlation coefficient:

$$R = \sqrt{\frac{r_{yx_1}^2 + r_{yx_2}^2 - 2r_{yx_1} \cdot r_{yx_2} \cdot r_{x_1 x_2}}{1 - r_{x_1 x_2}^2}}$$

Formula for the F test for the multiple correlation coefficient:

$$F = \frac{R^2/k}{(1 - R^2)/(n - k - 1)}$$

with d.f.N $= n - k$ and d.f.D $= n - k - 1$.

Formula for the adjusted R^2:

$$R_{\text{adj}}^2 = 1 - \left[\frac{(1 - R^2)(n - 1)}{n - k - 1}\right]$$

Review Exercises

For Exercises 1 through 7, do a complete regression analysis by performing the following steps.

   a. Draw the scatter plot.
   b. Compute the value of the correlation coefficient.
   c. Test the significance of the correlation coefficient at $\alpha = 0.01$, using Table I.
   d. Determine the regression line equation.
   e. Plot the regression line on the scatter plot.
   f. Predict y' for a specific value of x.

1. These data represent the number of hits and the number of strikeouts for 15 players on a college baseball team. If there is a significant relationship between the variables, predict the number of strikeouts a baseball player is likely to have if he has 30 hits.

Hits x	54	16	41	43	24	21	6	2
Strikeouts y	12	6	30	33	21	29	10	4

Hits x	54	41	29	39	11	24	1
Strikeouts y	26	20	23	27	10	12	3

Source: University of Findlay baseball statistics.

2. A researcher wishes to determine if there is a relationship between the number of day-care centers and the number of group day-care homes for counties in Pennsylvania. If there is a significant relationship, predict the number of group care homes a county has if the county has 20 day-care centers.

Day-care centers x	5	28	37	16	16	48
Group day-care homes y	2	7	4	10	6	9

Source: State Department of Public Welfare.

3. A study is done to see whether there is a relationship between a mother's age and the number of children she has. The data are shown here. If there is a significant relationship, predict the number of children of a mother whose age is 34.

Mother's age x	18	22	29	20	27	32	33	36
No. of children y	2	1	3	1	2	4	3	5

4. A study is conducted to determine the relationship between a driver's age and the number of accidents he or she has over a 1-year period. The data are shown here. (This information will be used for Exercise 8.) If there is a significant relationship, predict the number of accidents of a driver who is 28.

Driver's age x	16	24	18	17	23	27	32
No. of accidents y	3	2	5	2	0	1	1

5. A researcher desires to know whether the typing speed of a secretary (in words per minute) is related to the time (in hours) that it takes the secretary to learn to use a new word processing program. The data are shown.

Speed x	48	74	52	79	83	56	85	63	88	74	90	92
Time y	7	4	8	3.5	2	6	2.3	5	2.1	4.5	1.9	1.5

If there is a significant relationship, predict the time it will take the average secretary who has a typing speed of 72 words per minute to learn the word processing program. (This information will be used for Exercises 9 and 11.)

6. A study was conducted with vegetarians to see whether the number of grams of protein each ate per day was related to diastolic blood pressure. The data are given here. (This information will be used for Exercises 10 and 12.) If there is a significant relationship, predict the diastolic pressure of a vegetarian who consumes 8 grams of protein per day.

Grams x	4	6.5	5	5.5	8	10	9	8.2	10.5
Pressure y	73	79	83	82	84	92	88	86	95

7. A researcher wishes to determine the relationship between the number of cows (in thousands) in counties in southwestern Pennsylvania and the milk production (in millions of pounds). The data are shown. Describe the relationship.

Cows x	70	3	194	12	46	65
Pounds y	115	5	289	15	72	92

Source: Pittsburgh Tribune-Review.

8. For Exercise 4, find the standard error of the estimate.

9. For Exercise 5, find the standard error of the estimate.

10. For Exercise 6, find the standard error of the estimate.

11. For Exercise 5, find the 90% prediction interval for time when the speed is 72 words per minute.

12. For Exercise 6, find the 95% prediction interval for pressure when the number of grams is 8.

13. (Opt.) A study found a significant relationship among a person's years of experience on a particular job x_1, the number of workdays missed per month x_2, and the person's age y. The regression equation is $y' = 12.8 + 2.09x_1 + 0.423x_2$. Predict a person's age if he or she has been employed for 4 years and has missed 2 workdays a month.

14. (Opt.) Find R when $r_{yx_1} = 0.681$ and $r_{yx_2} = 0.872$ and $r_{x_1x_2} = 0.746$.

15. (Opt.) Find R_{adj}^2 when $R = 0.873$, $n = 10$, and $k = 3$.

Statistics Today

Do Dust Storms Affect Respiratory Health?—Revisited

The researchers correlated the dust pollutant levels in the atmosphere and the number of daily emergency room visits for several respiratory disorders, such as bronchitis, sinusitis, asthma, and pneumonia. Using the Pearson correlation coefficient, they found overall a significant but low correlation, $r = 0.13$, for bronchitis visits only. However, they found a much higher correlation value for sinusitis, $P\text{-value} = 0.08$, when pollutant levels exceeded maximums set by the Environmental Protection Agency (EPA). In addition, they found statistically significant correlation coefficients $r = 0.94$ for sinusitis visits and $r = 0.74$ for upper-respiratory-tract infection visits 2 days after the dust pollutants exceeded the maximum levels set by the EPA.

Data Analysis

The Data Bank is found in Appendix D, or on the World Wide Web by following links from www.mhhe.com/math/stat/bluman/

1. From the Data Bank, choose two variables that might be related: for example, IQ and educational level; age and cholesterol level; exercise and weight; or weight and systolic pressure. Do a complete correlation and regression analysis by performing the following steps. Select a random sample of at least 10 subjects.

   a. Draw a scatter plot.
   b. Compute the correlation coefficient.
   c. Test the hypothesis $H_0: \rho = 0$.
   d. Find the regression line equation.
   e. Summarize the results.

2. Repeat Exercise 1, using samples of values of 10 or more obtained from Data Set V in Appendix D. Let $x =$ the number of suspensions and $y =$ the enrollment size.

3. Repeat Exercise 1, using samples of 10 or more values obtained from Data Set XIII. Let $x =$ the number of beds and $y =$ the number of personnel employed.

Chapter Quiz

Determine whether each statement is true or false. If the statement is false, explain why.

1. A negative relationship between two variables means that for the most part, as the x variable increases, the y variable increases.

2. A correlation coefficient of -1 implies a perfect linear relationship between the variables.

3. Even if the correlation coefficient is high or low, it may not be significant.

4. When the correlation coefficient is significant, one can assume x causes y.

5. It is not possible to have a significant correlation by chance alone.

6. In multiple regression, there are several dependent variables and one independent variable.

Select the best answer.

7. The strength of the relationship between two variables is determined by the value of

   a. r
   b. a
   c. x
   d. s_{est}

8. To test the significance of r, a(n) _____ test is used.

   a. t
   b. F
   c. χ^2
   d. None of the above

9. The test of significance for r has _____ degrees of freedom.

   a. 1
   b. n
   c. $n - 1$
   d. $n - 2$

10. The equation of the regression line used in statistics is

    a. $x = a + by$
    b. $y = bx + a$
    c. $y' = a + bx$
    d. $x = ay + b$

11. The coefficient of determination is

    a. r
    b. r^2
    c. a
    d. b

Complete the following statements with the best answer.

12. A statistical graph of two variables is called a(n) _____.

13. The x variable is called the _____ variable.

14. The range of r is from _____ to _____.

15. The sign of r and _____ will always be the same.

16. The regression line is called the _____.

17. If all the points fall on a straight line, the value of r will be _____ or _____.

For Exercises 18 through 21, do a complete regression analysis.

    a. Draw the scatter plot.

    b. Compute the value of the correlation coefficient.

    c. Test the significance of the correlation coefficient at $\alpha = 0.05$.

    d. Determine the regression line equation.

    e. Plot the regression line on the scatter plot.

    f. Predict y' for a specific value of x.

18. A medical researcher wants to determine the relationship between the price per dose of prescription drugs in the United States and the price of the same dose in Australia. The data are shown. Describe the relationship.

U.S. price x	3.31	3.16	2.27	3.13	2.54	1.98	2.22
Australian price y	1.29	1.75	0.82	0.83	1.32	0.84	0.82

19. A study is conducted to determine the relationship between a driver's age and the number of accidents he or she has over a 1-year period. The data are shown here. If there is a significant relationship, predict the number of accidents of a driver who is 64.

Driver's age x	63	65	60	62	66	67	59
No. of accidents y	2	3	1	0	3	1	4

20. A researcher desires to know if the age of a child is related to the number of cavities he or she has. The data are shown here. If there is a significant

relationship, predict the number of cavities for a child of 11.

Age of child x	6	8	9	10	12	14
No. of cavities y	2	1	3	4	6	5

21. A study is conducted with a group of dieters to see if the number of grams of fat each consumes per day is related to cholesterol level. The data are shown here. If there is a significant relationship, predict the cholesterol level of a dieter who consumes 8.5 grams of fat per day.

Fat grams x	6.8	5.5	8.2	10	8.6	9.1	8.6	10.4
Cholesterol level y	183	201	193	283	222	250	190	218

22. For Exercise 20, find the standard error of the estimate.

23. For Exercise 21, find the standard error of the estimate.

24. For Exercise 20, find the 90% prediction interval of the number of cavities for a 7-year-old.

25. For Exercise 21, find the 95% prediction interval of the cholesterol level of a person who consumes 10 grams of fat.

26. (Opt.) A study was conducted, and a significant relationship was found among the number of hours a teenager watches television per day x_1, the number of hours the teenager talks on the telephone per day x_2, and the teenager's weight y. The regression equation is $y' = 98.7 + 3.82x_1 + 6.51x_2$. Predict a teenager's weight if she averages 3 hours of TV and 1.5 hours on the phone per day.

27. (Opt.) Find R when $r_{yx_1} = 0.561$ and $r_{yx_2} = 0.714$ and $r_{x_1x_2} = 0.625$.

28. (Opt.) Find R^2_{adj} when $R = 0.774$, $n = 8$, and $k = 2$.

Critical Thinking Challenges

When the points in a scatter plot show a curvilinear trend rather than a linear trend, statisticians have methods of fitting curves rather than straight lines to the data, thus obtaining a better fit and a better prediction model. One type of curve that can be used is the logarithmic regression curve. The data shown are the number of items of a new product sold over a period of 15 months at a certain store. Notice that sales rise during the beginning months and then level off later on.

Month x	1	3	6	8	10	12	15
No. of items sold y	10	12	15	19	20	21	21

  1. Draw the scatter plot for the data.

  2. Find the equation of the regression line.

3. Describe how the line fits the data.

4. Using the log key on your calculator, transform the x values into log x values.

5. Using the log x values instead of the x values, find the equation of a and b for the regression line.

6. Next, plot the curve $y = a + b \log x$ on the graph.

7. Compare the line $y = a + bx$ with the curve $y = a + b \log x$ and decide which one fits the data better.

8. Compute r, using the x and y values; then compute r, using the log x and y values. Which is higher?

9. In your opinion, which (the line or the logarithmic curve) would be a better predictor for the data? Why?

Data Projects

Where appropriate, use MINITAB, the TI-83 Plus, the TI-84 Plus, or a computer program of your choice to complete the following exercises.

1. Select two variables that might be related, such as the age of a person and the number of cigarettes the person smokes, or the number of credits a student has and the number of hours the student watches television. Sample at least 10 people.

   a. Write a brief statement as to the purpose of the study.
   b. Define the population.
   c. State how the sample was selected.
   d. Show the raw data.
   e. Draw a scatter plot for the data values.
   f. Write a statement analyzing the scatter plot.
   g. Compute the value of the correlation coefficient.
   h. Test the significance of r. (State the hypotheses, select α, find the critical values, make the decision, and analyze the results.)

   i. Find the equation of the regression line and draw it on the scatter plot. (*Note:* Even if r is not significant, complete this step.)
   j. Summarize the overall results.

2. For the data in Exercise 1, use MINITAB to answer these.

   a. Does a linear correlation exist between x and y?
   b. If so, find the regression equation.
   c. Explain how good a model the regression equation is by finding the coefficient of determination and coefficient of correlation and interpreting the strength of these values.
   d. Find the prediction interval for y. Use the α value that you selected in Exercise 1.

You may use the following websites to obtain raw data:

Visit the data sets at the book's website found at http://www.mhhe.com/math/stat/bluman Click on the 6th edition.
http://lib.stat.cmu.edu/DASL
http://www.statcan.ca

Answers to Applying the Concepts

Section 10–3   Stopping Distances

1. The independent variable is miles per hour (mph).

2. The dependent variable is braking distance (feet).

3. Miles per hour is a continuous quantitative variable.

4. Braking distance is a continuous quantitative variable.

5. A scatter plot of the data is shown.

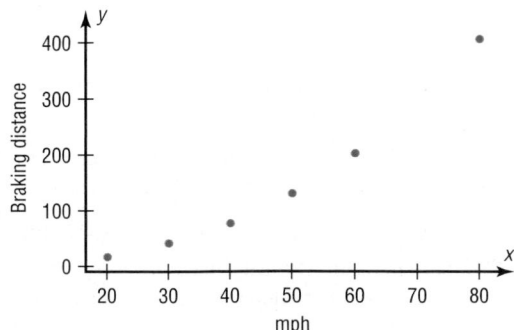

Scatterplot of braking distance vs. mph

6. There might be a linear relationship between the two variables, but there is a bit of a curve in the data.

7. Changing the distances between the mph increments will change the appearance of the relationship.

8. There is a positive relationship between the two variables—higher speeds are associated with longer braking distances.

9. The strong relationship between the two variables suggests that braking distance can be accurately predicted from mph. We might still have some concern about the curve in the data.

10. Answers will vary. Some other variables that might affect braking distance include road conditions, driver response time, and condition of the brakes.

11. The correlation coefficient is $r = 0.966$.

12. The value for $r = 0.966$ is significant at $\alpha = 0.05$. This confirms the strong positive relationship between the variables.

Section 10–4   Stopping Distances Revisited

1. The linear regression equation is
$\overline{y'} = -151.90 + 6.4514x$.

2. The slope says that for each additional mile per hour a car is traveling, we expect the stopping distance to

increase by 6.45 feet, on average. The y intercept is the braking distance we would expect for a car traveling 0 mph—this is meaningless in this context, but is an important part of the model.

3. $y' = -151.90 + 6.4514(45) = 138.4$ The braking distance for a car traveling 45 mph is approximately 138 feet.

4. $y' = -151.90 + 6.4514(100) = 493.2$ The braking distance for a car traveling 100 mph is approximately 493 feet.

5. It is not appropriate to make predictions of braking distance for speeds outside of the given data values (for example, the 100 mph above) because we know nothing about the relationship between the two variables outside of the range of the data.

Section 10–5   Interpreting Simple Linear Regression

1. Both variables are moving in the same direction. In others words, the two variables are positively associated. We know this because the correlation coefficient is positive.

2. The unexplained variation of 3026.49 measures the distances from the prediction line to the actual values.

3. The slope of the regression line is 0.725983.

4. The y intercept is 16.5523.

5. The critical value of 0.378419 can be found in a table.

6. The allowable risk of making a type I error is 0.10, the level of significance.

7. The variation explained by the regression is 0.631319, or about 63.1%.

8. The average scatter of points about the regression line is 12.9668, the standard error of the estimate.

9. The null hypothesis is that there is no correlation, $H_0: \rho = 0$.

10. We compare the test statistic of 0.794556 to the critical value to see if the null hypothesis should be rejected.

11. Since $0.794556 > 0.378419$, we reject the null hypothesis and find that there is enough evidence to conclude that the correlation is not equal to zero.

Section 10–6   More Math Means More Money

1. The dependent variable is yearly income 10 years after high school.

2. The independent variables are number of math courses taken and number of hours worked per week during the senior year of high school.

3. Multiple regression assumes that the independent variables are not highly correlated.

4. We expect a person's yearly income 10 years after high school to be $4540 more, on average, for each additional math course taken, all other variables held constant. We expect a person's yearly income 10 years after high school to be $1290 more, on average, for each additional hour worked per week during the senior year of high school, all other variables held constant.

5. $y' = 6000 + 4540(8) + 1290(20) = 68,120$. The predicted yearly income 10 years after high school is $68,120.

6. The multiple correlation coefficient of 0.77 means that there is a fairly strong positive relationship between the independent variables (number of math courses and hours worked during senior year of high school) and the dependent variable (yearly income 10 years after high school).

7. $R^2 = (0.77)^2 = 0.5929$

8. $\begin{aligned} R^2_{adj} &= 1 - \left[\dfrac{(1 - R^2)(n - 1)}{n - k - 1} \right] \\ &= 1 - \left[\dfrac{(1 - 0.5929)(20 - 1)}{20 - 2 - 1} \right] \\ &= 1 - \left[\dfrac{(0.4071)(19)}{17} \right] = 0.5450 \end{aligned}$

9. The equation appears to be a fairly good predictor of income, since 54.5% of the variation in yearly income 10 years after high school is explained by the regression model.

10. Answers will vary. One possible answer is that yearly income 10 years after high school increases with more math classes and more hours of work during the senior year of high school. The number of math classes has a higher coefficient, so more math does mean more money!

Other Chi-Square Tests

Objectives

After completing this chapter, you should be able to

1 Test a distribution for goodness of fit, using chi-square.

2 Test two variables for independence, using chi-square.

3 Test proportions for homogeneity, using chi-square.

Outline

**Statistics
Today**

Statistics and Heredity

An Austrian monk, Gregor Mendel (1822–1884) studied genetics, and his principles are the foundation for modern genetics. Mendel used his spare time to grow a variety of peas at the monastery. One of his many experiments involved crossbreeding peas that had smooth yellow seeds with peas that had wrinkled green seeds. He noticed that the results occurred with regularity. That is, some of the offspring had smooth yellow seeds, some had smooth green seeds, some had wrinkled yellow seeds, and some had wrinkled green seeds. Furthermore, after several experiments, the percentages of each type seemed to remain approximately the same. Mendel formulated his theory based on the assumption of dominant and recessive traits and tried to predict the results. He then crossbred his peas and examined 556 seeds over the next generation.

Finally, he compared the actual results with the theoretical results to see if his theory was correct. To do this, he used a "simple" chi-square test, which is explained in this chapter. See Statistics Today—Revisited.

Source: J. Hodges, Jr., D. Krech, and R. Crutchfield, *Stat Lab, An Empirical Introduction to Statistics* (New York: McGraw-Hill, 1975), pp. 228–229. Used with permission.

11–1   Introduction

The chi-square distribution was used in Chapters 7 and 8 to find a confidence interval for a variance or standard deviation and to test a hypothesis about a single variance or standard deviation.

It can also be used for tests concerning *frequency distributions,* such as "If a sample of buyers is given a choice of automobile colors, will each color be selected with the same frequency?" The chi-square distribution can be used to test the *independence* of two variables, for example, "Are senators' opinions on gun control independent of party affiliations?" That is, do the Republicans feel one way and the Democrats feel differently, or do they have the same opinion?

Finally, the chi-square distribution can be used to test the *homogeneity of proportions.* For example, is the proportion of high school seniors who attend college immediately after graduating the same for the northern, southern, eastern, and western parts of the United States?

This chapter explains the chi-square distribution and its applications. In addition to the applications mentioned here, chi-square has many other uses in statistics.

Test for Goodness of Fit

In addition to being used to test a single variance, the chi-square statistic can be used to see whether a frequency distribution fits a specific pattern. For example, to meet customer demands, a manufacturer of running shoes may wish to see whether buyers show a preference for a specific style. A traffic engineer may wish to see whether accidents occur more often on some days than on others, so that she can increase police patrols accordingly. An emergency service may want to see whether it receives more calls at certain times of the day than at others, so that it can provide adequate staffing.

When one is testing to see whether a frequency distribution fits a specific pattern, the chi-square **goodness-of-fit test** is used. For example, suppose a market analyst wished to see whether consumers have any preference among five flavors of a new fruit soda. A sample of 100 people provided these data:

Cherry	Strawberry	Orange	Lime	Grape
32	28	16	14	10

If there were no preference, one would expect each flavor to be selected with equal frequency. In this case, the equal frequency is $100/5 = 20$. That is, *approximately* 20 people would select each flavor.

Since the frequencies for each flavor were obtained from a sample, these actual frequencies are called the **observed frequencies.** The frequencies obtained by calculation (as if there were no preference) are called the **expected frequencies.** A completed table for the test is shown.

Frequency	Cherry	Strawberry	Orange	Lime	Grape
Observed	32	28	16	14	10
Expected	20	20	20	20	20

The observed frequencies will almost always differ from the expected frequencies due to sampling error; that is, the values differ from sample to sample. But the question is: Are these differences significant (a preference exists), or are they due to chance? The chi-square goodness-of-fit test will enable the researcher to determine the answer.

Before computing the test value, one must state the hypotheses. The null hypothesis should be a statement indicating that there is no difference or no change. For this example, the hypotheses are as follows:

H_0: Consumers show no preference for flavors of the fruit soda.

H_1: Consumers show a preference.

In the goodness-of-fit test, the degrees of freedom are equal to the number of categories minus 1. For this example, there are five categories (cherry, strawberry, orange, lime, and grape); hence, the degrees of freedom are $5 - 1 = 4$. This is so because the

number of subjects in each of the first four categories is free to vary. But in order for
the sum to be 100—the total number of subjects—the number of subjects in the last category is fixed.

Formula for the Chi-Square Goodness-of-Fit Test

$$\chi^2 = \sum \frac{(O - E)^2}{E}$$

with degrees of freedom equal to the number of categories minus 1, and where
  O = observed frequency
  E = expected frequency

Two assumptions are needed for the goodness-of-fit test. These assumptions are
given next.

Assumptions for the Chi-Square Goodness-of-Fit Test

1. The data are obtained from a random sample.
2. The expected frequency for each category must be 5 or more.

This test is a right-tailed test, since when the $O - E$ values are squared, the answer
will be positive or zero. This formula is explained in Example 11–1.

Example 11–1

Is there enough evidence to reject the claim that there is no preference in the selection
of fruit soda flavors, using the data shown previously? Let $\alpha = 0.05$.

Solution

Step 1 State the hypotheses and identify the claim.
  H_0: Consumers show no preference for flavors (claim).
  H_1: Consumers show a preference.

Step 2 Find the critical value. The degrees of freedom are $5 - 1 = 4$, and $\alpha = 0.05$.
Hence, the critical value from Table G in Appendix C is 9.488.

Step 3 Compute the test value by subtracting the expected value from the
corresponding observed value, squaring the result and dividing by the
expected value, and finding the sum. The expected value for each category
is 20, as shown previously.

$$\chi^2 = \sum \frac{(O - E)^2}{E}$$
$$= \frac{(32 - 20)^2}{20} + \frac{(28 - 20)^2}{20} + \frac{(16 - 20)^2}{20} + \frac{(14 - 20)^2}{20} + \frac{(10 - 20)^2}{20}$$
$$= 18.0$$

Step 4 Make the decision. The decision is to reject the null hypothesis, since
18.0 > 9.488, as shown in Figure 11–1.

Figure 11–1

Critical and Test Values for Example 11–1

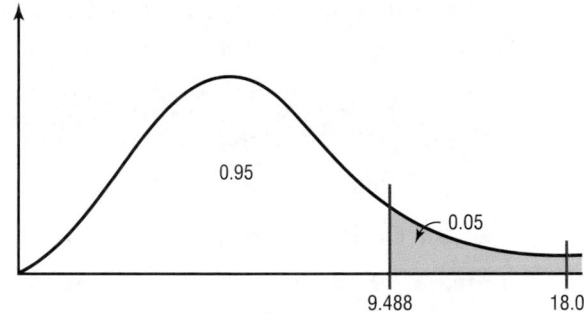

Step 5 Summarize the results. There is enough evidence to reject the claim that consumers show no preference for the flavors.

To get some idea of why this test is called the goodness-of-fit test, examine graphs of the observed values and expected values. See Figure 11–2. From the graphs, one can see whether the observed values and expected values are close together or far apart.

Figure 11–2

Graphs of the Observed and Expected Values for Soda Flavors

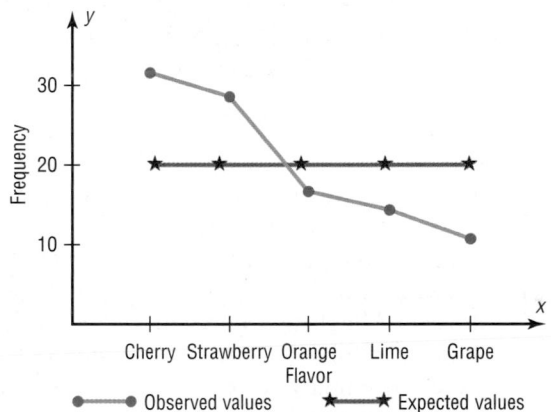

When the observed values and expected values are close together, the chi-square test value will be small. Then the decision will be to not reject the null hypothesis—hence, there is "a good fit." See Figure 11–3(a). When the observed values and the expected values are far apart, the chi-square test value will be large. Then the null hypothesis will be rejected—hence, there is "not a good fit." See Figure 11–3(b).

Figure 11–3

Results of the Goodness-of-Fit Test

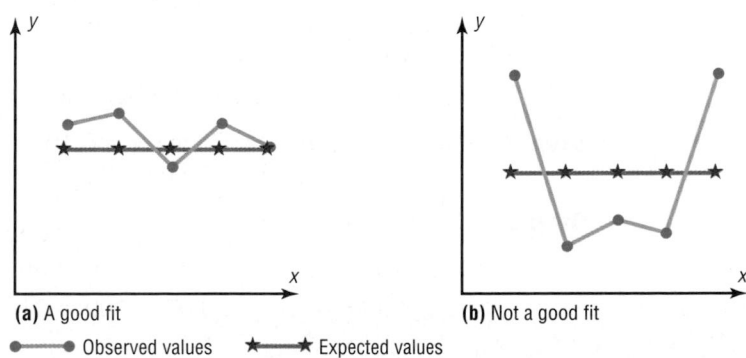

The steps for the chi-square goodness-of-fit test are summarized in this Procedure Table.

Procedure Table

The Chi-Square Goodness-of-Fit Test

Step 1   State the hypotheses and identify the claim.

Step 2   Find the critical value. The test is always right-tailed.

Step 3   Compute the test value.

Find the sum of the $\dfrac{(O - E)^2}{E}$ values.

Step 4   Make the decision.

Step 5   Summarize the results.

When there is perfect agreement between the observed and the expected values, $\chi^2 = 0$. Also, χ^2 can never be negative. Finally, the test is right-tailed because "H_0: Good fit" and "H_1: Not a good fit" mean that χ^2 will be small in the first case and large in the second case.

Example 11–2

The Russel Reynold Association surveyed retired senior executives who had returned to work. They found that after returning to work, 38% were employed by another organization, 32% were self-employed, 23% were either freelancing or consulting, and 7% had formed their own companies. To see if these percentages are consistent with those of Allegheny County residents, a local researcher surveyed 300 retired executives who had returned to work and found that 122 were working for another company, 85 were self-employed, 76 were either freelancing or consulting, and 17 had formed their own companies. At $\alpha = 0.10$, test the claim that the percentages are the same for those people in Allegheny County.

Source: Michael L. Shook and Robert D. Shook, *The Book of Odds.*

Solution

Step 1   State the hypotheses and identify the claim.

H_0: The retired executives who returned to work are distributed as follows: 38% are employed by another organization, 32% are self-employed, 23% are either freelancing or consulting, and 7% have formed their own companies (claim).

H_1: The distribution is not the same as stated in the null hypothesis.

Step 2   Find the critical value. Since $\alpha = 0.10$ and the degrees of freedom are $4 - 1 = 3$, the critical value is 6.251.

Step 3   Compute the test value. The expected values are computed as follows:

$0.38 \times 300 = 114$ $0.23 \times 300 = 69$

$0.32 \times 300 = 96$ $0.07 \times 300 = 21$

$$\chi^2 = \sum \frac{(O - E)^2}{E}$$

$$= \frac{(122 - 114)^2}{114} + \frac{(85 - 96)^2}{96} + \frac{(76 - 69)^2}{69} + \frac{(17 - 21)^2}{21}$$

$$= 3.2939$$

Step 4   Make the decision. Since $3.2939 < 6.251$, the decision is not to reject the null hypothesis. See Figure 11-4.

Figure 11-4

Critical and Test Values for Example 11-2

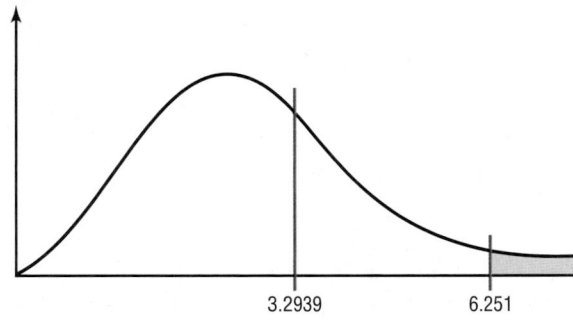

3.2939   6.251

Step 5   Summarize the results. There is not enough evidence to reject the claim. It can be concluded that the percentages are not significantly different from those given in the null hypothesis.

Example 11-3

The adviser of an ecology club at a large college believes that the group consists of 10% freshmen, 20% sophomores, 40% juniors, and 30% seniors. The membership for the club this year consisted of 14 freshmen, 19 sophomores, 51 juniors, and 16 seniors. At $\alpha = 0.10$, test the adviser's conjecture.

Solution

Step 1   State the hypotheses and identify the claim.

H_0: The club consists of 10% freshmen, 20% sophomores, 40% juniors, and 30% seniors (claim).

H_1: The distribution is not the same as stated in the null hypothesis.

Step 2   Find the critical value. Since $\alpha = 0.10$ and the degrees of freedom are $4 - 1 = 3$, the critical value is 6.251.

Step 3   Compute the test value. The expected values are computed as follows:

$0.10 \times 100 = 10$ $0.40 \times 100 = 40$

$0.20 \times 100 = 20$ $0.30 \times 100 = 30$

$$\chi^2 = \sum \frac{(O - E)^2}{E}$$

$$= \frac{(14 - 10)^2}{10} + \frac{(19 - 20)^2}{20} + \frac{(51 - 40)^2}{40} + \frac{(16 - 30)^2}{30}$$

$$= 11.208$$

Step 4   Reject the null hypothesis, since $11.208 > 6.251$, as shown in Figure 11–5.

Figure 11–5

Critical and Test Values
for Example 11–3

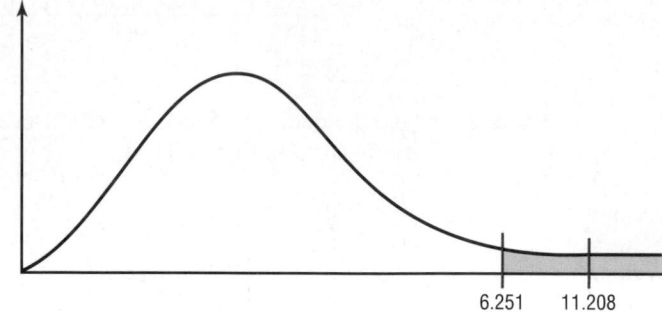

6.251   11.208

Step 5   Summarize the results. There is enough evidence to reject the adviser's claim.

The P-value method of hypothesis testing can also be used for the chi-square tests explained in this chapter. The P-values for chi-square are found in Table G in Appendix C. The method used to find the P-value for a chi-square test value is the same as the method shown in Section 8–6. The P-value for $\chi^2 = 3.2939$ with d.f. $= 3$ (for the data in Example 11–2) is greater than 0.10 since 6.251 is the value in Table G for $\alpha = 0.10$. (The P-value obtained from a calculator is 0.348.) Hence P-value > 0.10. The decision is to not reject the null hypothesis, which is consistent with the decision made in Example 11–2 using the traditional method of hypothesis testing.

For use of the chi-square goodness-of-fit test, statisticians have determined that the expected frequencies should be at least 5, as stated in the assumptions. The reasoning is as follows: The chi-square distribution is continuous, whereas the goodness-of-fit test is discrete. However, the continuous distribution is a good approximation and can be used when the expected value for each class is at least 5. If an expected frequency of a class is less than 5, then that class can be combined with another class so that the expected frequency is 5 or more.

Test of Normality (Optional)

The chi-square goodness-of-fit test can be used to test a variable to see if it is normally distributed. The null hypotheses are

H_0: The variable is normally distributed.

H_1: The variable is not normally distributed.

The procedure is somewhat complicated. It involves finding the expected frequencies for each class of a frequency distribution by using the standard normal distribution. Then the actual frequencies (i.e., observed frequencies) are compared to the expected frequencies, using the chi-square goodness-of-fit test. If the observed frequencies are close in value to the expected frequencies, the chi-square test value will be small, and the null hypothesis cannot be rejected. In this case, it can be concluded that the variable is approximately normally distributed.

On the other hand, if there is a large difference between the observed frequencies and the expected frequencies, the chi-square test value will be larger, and the null hypothesis can be rejected. In this case, it can be concluded that the variable is not normally distributed. Example 11–4 illustrates the procedure for the chi-square test of normality. To find the areas in the examples, you might want to review Section 6–3.

Example 11–4 shows how to do the calculations.

Example 11–4

Use chi-square to determine if the variable shown in the frequency distribution is normally distributed. Use $\alpha = 0.05$.

Boundaries	Frequency
89.5–104.5	24
104.5–119.5	62
119.5–134.5	72
134.5–149.5	26
149.5–164.5	12
164.5–179.5	4
	200

Solution

H_0: The variable is normally distributed.

H_1: The variable is not normally distributed.

First find the mean and standard deviation of the variable. Then find the area under the standard normal distribution, using z values and Table E for each class. Find the expected frequencies for each class by multiplying the area by 200. Finally, find the chi-square test value by using the formula $\chi^2 = \sum \dfrac{(O - E)^2}{E}$.

Boundaries	f	X_m	$f \cdot X_m$	$f \cdot X_m^2$
89.5–104.5	24	97	2,328	225,816
104.5–119.5	62	112	6,944	777,728
119.5–134.5	72	127	9,144	1,161,288
134.5–149.5	26	142	3,692	524,264
149.5–164.5	12	157	1,884	295,788
164.5–179.5	4	172	688	118,336
	200		24,680	3,103,220

$$\bar{X} = \frac{24,680}{200} = 123.4$$

$$s = \sqrt{\frac{3,103,220 - 24,680^2/200}{199}} = \sqrt{290} = 17.03$$

The area to the left of $x = 104.5$ is found as

$$z = \frac{104.5 - 123.4}{17.03} = -1.11$$

The area for $z < -1.11$ is $0.5000 - 0.3665 = 0.1335$.
   The area between 104.5 and 119.5 is found as

$$z = \frac{119.5 - 123.4}{17.03} = -0.23$$

The area for $-1.11 < z < -0.23$ is $0.3665 - 0.0910 = 0.2755$.
   The area between 119.5 and 134.5 is found as

$$z = \frac{134.5 - 123.4}{17.03} = 0.65$$

Unusual Stat

Drinking milk may lower your risk of stroke. A 22-year study of men over 55 found that only 4% of men who drank 16 ounces of milk every day suffered a stroke, compared with 8% of the nonmilk drinkers.

The area for $-0.23 < z < 0.65$ is $0.2422 + 0.0910 = 0.3332$.
The area between 134.5 and 149.5 is found as

$$z = \frac{149.5 - 123.4}{17.03} = 1.53$$

The area for $0.65 < z < 1.53$ is $0.4370 - 0.2422 = 0.1948$.
The area between 149.5 and 164.5 is found as

$$z = \frac{164.5 - 123.4}{17.03} = 2.41$$

The area for $1.53 < z < 2.41$ is $0.4920 - 0.4370 = 0.0550$.
The area to the right of $x = 164.5$ is found as

$$z = \frac{164.5 - 123.4}{17.03} = 2.41$$

The area is $0.5000 - 0.4920 = 0.0080$.
The expected frequencies are found by

$$0.1335 \cdot 200 = 26.7$$
$$0.2755 \cdot 200 = 55.1$$
$$0.3332 \cdot 200 = 66.64$$
$$0.1948 \cdot 200 = 38.96$$
$$0.0550 \cdot 200 = 11.0$$
$$0.0080 \cdot 200 = 1.6$$

Note: Since the expected frequency for the last category is less than 5, it can be combined with the previous category.
The χ^2 is found by

O	24	62	72	26	16
E	26.7	55.1	66.64	38.96	12.6

$$\chi^2 = \frac{(24 - 26.7)^2}{26.7} + \frac{(62 - 55.1)^2}{55.1} + \frac{(72 - 66.64)^2}{66.64} + \frac{(26 - 38.96)^2}{38.96}$$
$$+ \frac{(16 - 12.6)^2}{12.6}$$
$$= 6.797$$

The C.V. with d.f. $= 4$ and $\alpha = 0.05$ is 9.488, so the null hypothesis is not rejected. Hence, the distribution can be considered approximately normal.

Applying the Concepts **11–2**

Never the Same Amounts

M&M/Mars, the makers of Skittles candies, states that the flavor blend is 20% for each flavor. Skittles is a combination of lemon, lime, orange, strawberry, and grape flavored candies. The following data list the results of four randomly selected bags of Skittles and their flavor blends. Use the data to answer the questions.

	Flavor				
Bag	**Green**	**Orange**	**Red**	**Purple**	**Yellow**
1	7	20	10	7	14
2	20	5	5	13	17
3	4	16	13	21	4
4	12	9	16	3	17
Total	43	50	44	44	52

1. Are the variables quantitative or qualitative?

2. What type of test can be used to compare the observed values to the expected values?

3. Perform a chi-square test on the total values.

4. What hypotheses did you use?

5. What were the degrees of freedom for the test? What is the critical value?

6. What is your conclusion?

See page 618 for the answers.

Exercises 11–2

1. How does the goodness-of-fit test differ from the chi-square variance test?

2. How are the degrees of freedom computed for the goodness-of-fit test?

3. How are the expected values computed for the goodness-of-fit test?

4. When the expected frequencies are less than 5 for a specific class, what should be done so that one can use the goodness-of-fit test?

For Exercises 5 through 19, perform these steps.

   a. State the hypotheses and identify the claim.

   b. Find the critical value.

   c. Compute the test value.

   d. Make the decision.

   e. Summarize the results.

Use the traditional method of hypothesis testing unless otherwise specified.

5. A researcher for an automobile manufacturer wishes to see if the ages of automobiles are equally distributed among three categories: less than 3 years old, 3 to 7 years old, and 8 years or older. A sample of 30 adult automobile owners is selected, and the results are shown. At $\alpha = 0.05$ can it be considered that the ages of the automobiles are equally distributed among the three categories? What would the results suggest to a tire manufacturer?

Category	Less than three years	3 to 7 years	8 years or older
Number	8	10	12

Source: Based on information from Goodyear.

6. A researcher wishes to see if the five ways (drinking decaffeinated beverages, taking a nap, going for a walk, eating a sugary snack, other) people use to combat midday drowsiness are equally distributed among office workers. A sample of 60 office workers is selected, and the following data are obtained. At $\alpha = 0.10$ can it be concluded that there is no preference? Why would the results be of interest to an employer?

Method	Beverage	Nap	Walk	Snack	Other
Number	21	16	10	8	5

Source: Based on information from Harris Interactive.

7. In a recent study, the following percentages of U.S. retail car sales based on size were reported: 28.1% small, 47.8% midsize, 7% large, and 17.1% luxury. A recent survey of retail sales in a particular county indicated that of 100 cars sold, 25 were small, 50 were midsize, 10 were large, and 15 were luxury cars. At $\alpha = 0.05$, is there sufficient evidence to conclude that the proportions differ from those stated in the report?

Source: *N.Y. Times Almanac.*

8. According to a recent census report, 68% of families have two parents present, 23% have only a mother present, 5% have only a father present, and 4% have no parent present. A random sample of families from a large school district revealed these results:

Two parents	Mother only	Father only	No parent
120	40	30	10

Is there sufficient evidence to conclude that the proportions of families by type of parent(s) present differ from those reported by the census?

Source: *N.Y. Times Almanac.*

9. An ABC News poll asked adults whether they felt genetically modified food was safe to eat. Thirty-five percent felt it was safe, 52% felt it was not safe, and 13% had no opinion. A random sample of 120 adults was asked the same question at a local county fair. Forty people felt that genetically modified food was safe, 60 felt that it was not safe, and 20 had no opinion. At the 0.01 level of significance, is there sufficient evidence to conclude that the proportions differ from those reported in the survey?

Source: ABCNews.com Poll, www.pollingreport.com.

10. A recent survey asked adults nationwide if they thought that the federal government should continue to fund NASA's efforts to send unstaffed missions to Mars. Fifty-six percent said they should continue, 40% said that they should not continue, and 4% had no opinion. A random sample of 200 college students resulted in these numbers:

Should continue: 126

Should not continue: 65

No opinion: 9

Is there sufficient evidence at $\alpha = 0.05$ to conclude that the opinions of the college students differ from those of the report?

Source: CNN/*USA TODAY*/Gallup Poll, www.pollingreport.com.

11. *USA TODAY* reported that 21% of loans granted by credit unions were for home mortgages, 39% were for automobile purchases, 20% were for credit card and other unsecured loans, 12% were for real estate other than home loans, and 8% were for other miscellaneous needs. In order to see if her credit union customers had similar needs, a manager surveyed a random sample of 100 loans and found that 25 were for home mortgages, 44 for automobile purchases, 19 for credit card and unsecured loans, 8 for real estate other than home loans, and 4 for miscellaneous needs. At $\alpha = 0.05$, is the distribution the same as reported in the newspaper?

Source: *USA TODAY*.

12. Nationwide the shares of carbon emissions for the year 2000 are transportation, 33%; industry, 30%; residential, 20%; and commercial, 17%. A state hazardous materials official wants to see if her state is the same. Her study of 300 emissions sources finds transportation, 36%; industry, 31%; residential, 17%; and commercial, 16%. At $\alpha = 0.05$, can she claim the percentages are the same?

Source: Energy Information Administration/*New York Times*.

13. A *USA TODAY* Snapshot states that 53% of adult shoppers prefer to pay cash for purchases, 30% use checks, 16% use credit cards, and 1% have no preference. The owner of a large store randomly selected 800 shoppers and asked their payment preferences. The results were that 400 paid cash, 210 paid by check, 170 paid with a credit card, and 20 had no preference. At $\alpha = 0.01$, test the claim that the owner's customers have the same preferences as those surveyed.

Source: *USA TODAY*.

14. The population distribution of federal prisons nationwide by serious offenses is the following: violent offenses, 12.6%; property offenses, 8.5%; drug offenses, 60.2%; public order offenses—weapons, 8.2%; immigration, 4.9%; other, 5.6%. A warden wants to see how his prison compares, so he surveys 500 prisoners and finds 64 are violent offenders, 40 are property offenders, 326 are drug offenders, 42 are public order offenders, 25 are immigration offenders, and 3 have other offenses. Can the warden conclude that the percentages are the same for his prison? Use $\alpha = 0.05$.

15. The populations of state prisons nationwide by serious offenses are the following: violent offenses, 29.5%; property offenses, 29%; drug offenses, 30.2%; public order offenses—weapons, 10.6%; other, 0.7%. A state prison official wants to check how this compares to her state. She surveys 1000 inmates and finds 298 are violent offenders, 275 are property offenders, 344 are drug offenders, 80 are public order offenders, and 3 have other offenses. Can she conclude that the percentages for her prison are the same as national statistics? Use $\alpha = 0.05$.

Source: *N.Y. Times Almanac*.

16. A researcher wishes to see if the number of adults who do not have health insurance is equally distributed among three categories (less than 12 years of education, 12 years of education, more than 12 years of education). A sample of 60 adults who do not have health insurance is selected, and the results are shown. At $\alpha = 0.05$ can it be concluded that the frequencies are not equal? Use the P-value method. If the null hypothesis is rejected, give a possible reason for this.

Category	Less than 12 years	12 years	More than 12 years
Frequency	29	20	11

Source: U.S. Census Bureau.

17. A medical researcher wishes to determine if the way people pay for their medical prescriptions is distributed as follows: 60% personal funds, 25% insurance, 15% Medicare. A sample of 50 people found that 32 paid with their own money, 10 paid using insurance, and 8 paid using Medicare. At $\alpha = 0.05$ is the assumption correct? Use the P-value method. What would be an implication of the results?

Source: U.S. Health Care Financing.

Extending the Concepts

18. Three coins are tossed 72 times, and the number of heads is shown. At $\alpha = 0.05$, test the null hypothesis that the coins are balanced and randomly tossed. (*Hint:* Use the binomial distribution.)

No. of heads	0	1	2	3
Frequency	3	10	17	42

19. Select a three-digit state lottery number over a period of 50 days. Count the number of times each digit, 0 through 9, occurs. Test the claim, at $\alpha = 0.05$, that the digits occur at random.

Technology *Step by Step*

MINITAB
Step by Step

Chi-Square Test for Goodness of Fit

For Example 11–1, is there a preference for flavor of soda? There is no menu command to do this directly. Use the calculator.

1. Enter the observed counts into C1 and the expected counts into C2. Name the columns O and E.

2. Select **Calc>Calculator.**

    a) Type **K1** in the Store result in variable.

    b) In the Expression box type the formula **SUM((O-E)**2/E).**

    c) Click [OK]. The chi-square test statistic will be displayed in the constant K1.

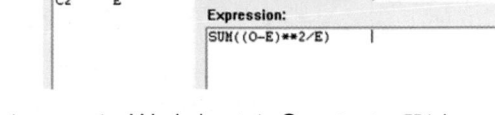

    d) Click the Project Manager icon, then navigate to the Worksheet 1>Constants. K1 is unnamed and equal to 18.

Calculate the *P*-Value

3. Select **Calc>Probability Distributions.**

4. Click on Chi-square.

    a) Click the button for Cumulative probability.

    b) In the box for Degrees of freedom type **4.**

    c) Click the button for Input constant, then click in the text box and select K1 in the variable list.

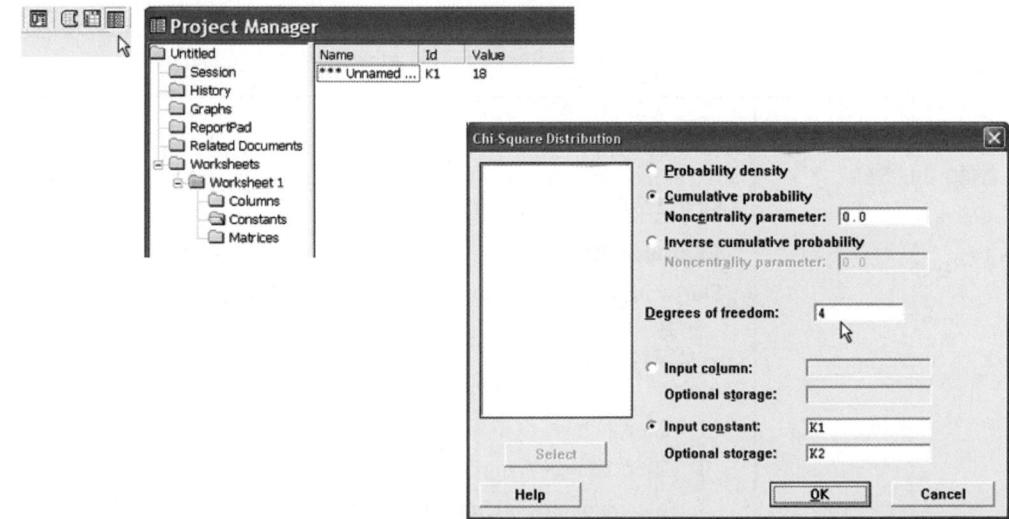

---

d) In the text box, Optional storage type **K2.** This is the area to the left of the test statistic. To calculate the P-value, we need the complement—the area to the right.

e) Select **Calc>Calculator,** then type **K3** for the storage variable and **1 − K2** for the expression.

f) Click [OK]. In the Project Manager you will see K3 = 0.00121341. This is the P-value for the test.

Reject the null hypothesis. There is enough evidence in the sample to conclude there is not 20% of each flavor.

TI-83 Plus or TI-84 Plus
Step by Step

Goodness-of-Fit Test

Example TI11–1

This pertains to Example 11–1 from the text. At the 5% significance level, test the claim that there is no preference in the selection of fruit soda flavors for the data.

Frequency	Cherry	Strawberry	Orange	Lime	Grape
Observed	32	28	16	14	10
Expected	20	20	20	20	20

To calculate the test statistic:

1. Enter the observed frequencies in L_1 and the expected frequencies in L_2.

2. Press **2nd [QUIT]** to return to the home screen.

3. Press **2nd [LIST],** move the cursor to MATH, and press **5** for sum(.

4. Type **$(L_1 − L_2)^2/L_2$)**, then press **ENTER.**

To calculate the P-value:
Press **2nd [DISTR]** then press **7** to get χ^2cdf(.
For this P-value, the χ^2cdf(command has form χ^2cdf(test statistic, ∞, degrees of freedom).
Use E99 for ∞. Type **2nd [EE]** to get the small E.
For this example use χ^2cdf(18, E99,4):

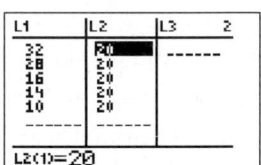 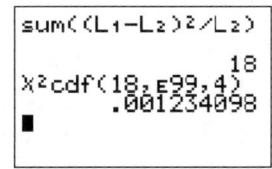

Since P-value = 0.001234098 < 0.05 = significance level, reject H_0 and conclude H_1. Therefore, there is enough evidence to reject the claim that consumers show no preference for soda flavors.

Excel
Step by Step

Chi-Square Test

Excel has several functions for χ^2 calculations in its statistical functions category. We can use CHITEST to solve Example 11–4.

Example XL11–1

Start with the table of observed and expected frequencies:

O	24	62	72	26	16
E	26.7	55.1	66.64	38.96	12.6

1. Enter data set O in row 1.

2. Enter data set E in row 2.

3. Select a blank cell, and enter the formula **=CHITEST(B1:F1,B2:F2).**

CHITEST Dialog Box and
Worksheet

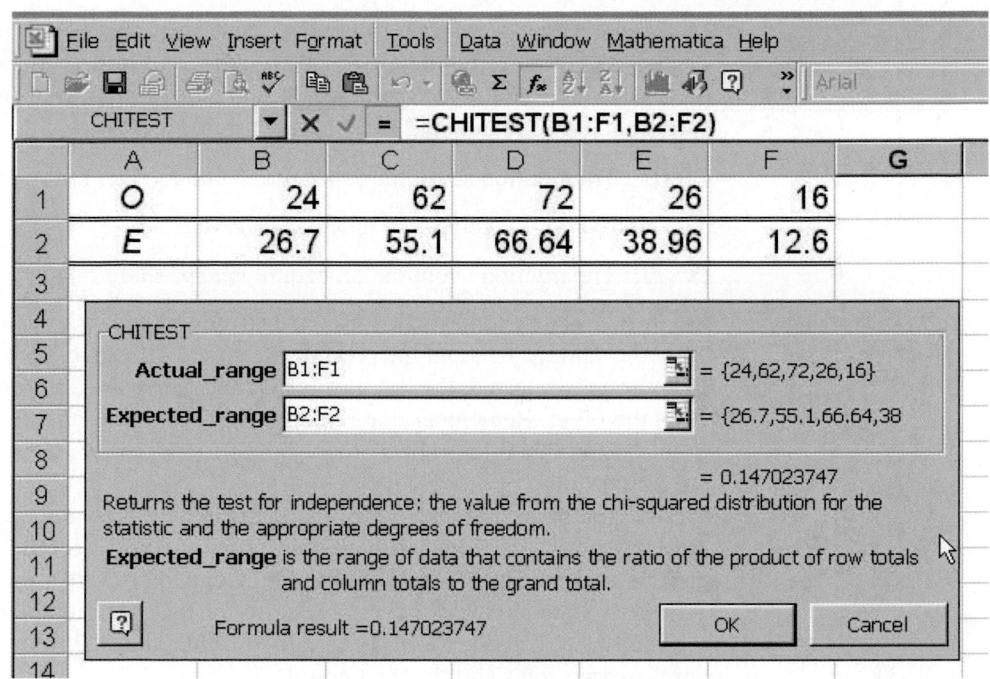

This test gives a *P*-value directly. Here the *P*-value of 0.1470 is greater than the significance
level of 0.05 specified in the example, so we do not reject the null hypothesis.

11–3   Tests Using Contingency Tables

When data can be tabulated in table form in terms of frequencies, several types of
hypotheses can be tested by using the chi-square test.

Two such tests are the independence of variables test and the homogeneity of proportions test. The test of independence of variables is used to determine whether two
variables are independent of or related to each other when a single sample is selected.
The test of homogeneity of proportions is used to determine whether the proportions for
a variable are equal when several samples are selected from different populations. Both
tests use the chi-square distribution and a contingency table, and the test value is found
in the same way. The independence test will be explained first.

Objective 2

Test two variables for
independence, using
chi-square.

Test for Independence

The chi-square **independence test** can be used to test the independence of two variables.
For example, suppose a new postoperative procedure is administered to a number of
patients in a large hospital. One can ask the question, Do the doctors feel differently
about this procedure from the nurses, or do they feel basically the same way? Note that
the question is not whether they prefer the procedure but whether there is a difference of
opinion between the two groups.

To answer this question, a researcher selects a sample of nurses and doctors and tab-
ulates the data in table form, as shown.

Group	Prefer new procedure	Prefer old procedure	No preference
Nurses	100	80	20
Doctors	50	120	30

As the survey indicates, 100 nurses prefer the new procedure, 80 prefer the old procedure, and 20 have no preference; 50 doctors prefer the new procedure, 120 like the old procedure, and 30 have no preference. Since the main question is whether there is a difference in opinion, the null hypothesis is stated as follows:

H_0: The opinion about the procedure is *independent* of the profession.

The alternative hypothesis is stated as follows:

H_1: The opinion about the procedure is *dependent* on the profession.

If the null hypothesis is not rejected, the test means that both professions feel basically the same way about the procedure and the differences are due to chance. If the null hypothesis is rejected, the test means that one group feels differently about the procedure from the other. Remember that rejection does *not* mean that one group favors the procedure and the other does not. Perhaps both groups favor it or both dislike it, but in different proportions.

To test the null hypothesis by using the chi-square independence test, one must compute the expected frequencies, assuming that the null hypothesis is true. These frequencies are computed by using the observed frequencies given in the table.

When data are arranged in table form for the chi-square independence test, the table is called a **contingency table.** The table is made up of R rows and C columns. The table here has two rows and three columns.

Group	Prefer new procedure	Prefer old procedure	No preference
Nurses	100	80	20
Doctors	50	120	30

Note that row and column headings do not count in determining the number of rows and columns.

A contingency table is designated as an $R \times C$ (rows by columns) table. In this case, $R = 2$ and $C = 3$; hence, this table is a 2×3 contingency table. Each block in the table is called a *cell* and is designated by its row and column position. For example, the cell with a frequency of 80 is designated as $C_{1,2}$, or row 1, column 2. The cells are shown below.

	Column 1	Column 2	Column 3
Row 1	$C_{1,1}$	$C_{1,2}$	$C_{1,3}$
Row 2	$C_{2,1}$	$C_{2,2}$	$C_{2,3}$

The degrees of freedom for any contingency table are (rows − 1) times (columns − 1); that is, d.f. = $(R - 1)(C - 1)$. In this case, $(2 - 1)(3 - 1) = (1)(2) = 2$. The reason for this formula for d.f. is that all the expected values except one are free to vary in each row and in each column.

Using the previous table, one can compute the expected frequencies for each block (or cell), as shown next.

1. Find the sum of each row and each column, and find the grand total, as shown.

Group	Prefer new procedure	Prefer old procedure	No preference	Total
Nurses	100	80	20	Row 1 sum 200
Doctors	+50	+120	+30	Row 2 sum 200
Total	150	200	50	400
	Column 1 sum	Column 2 sum	Column 3 sum	Grand total

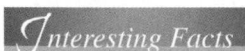

2. For each cell, multiply the corresponding row sum by the column sum and divide by the grand total, to get the expected value:

$$\text{Expected value} = \frac{\text{row sum} \times \text{column sum}}{\text{grand total}}$$

For example, for $C_{1,2}$, the expected value, denoted by $E_{1,2}$, is (refer to the previous tables)

$$E_{1,2} = \frac{(200)(200)}{400} = 100$$

For each cell, the expected values are computed as follows:

$$E_{1,1} = \frac{(200)(150)}{400} = 75 \qquad E_{1,2} = \frac{(200)(200)}{400} = 100 \qquad E_{1,3} = \frac{(200)(50)}{400} = 25$$

$$E_{2,1} = \frac{(200)(150)}{400} = 75 \qquad E_{2,2} = \frac{(200)(200)}{400} = 100 \qquad E_{2,3} = \frac{(200)(50)}{400} = 25$$

The expected values can now be placed in the corresponding cells along with the observed values, as shown.

Group	Prefer new procedure	Prefer old procedure	No preference	Total
Nurses	100 (75)	80 (100)	20 (25)	200
Doctors	50 (75)	120 (100)	30 (25)	200
Total	150	200	50	400

The rationale for the computation of the expected frequencies for a contingency table uses proportions. For $C_{1,1}$ a total of 150 out of 400 people prefer the new procedure. And since there are 200 nurses, one would expect, if the null hypothesis were true, $(150/400)(200)$, or 75, of the nurses to be in favor of the new procedure.

The formula for the test value for the independence test is the same as the one used for the goodness-of-fit test. It is

$$\chi^2 = \sum \frac{(O - E)^2}{E}$$

For the previous example, compute the $(O - E)^2/E$ values for each cell, and then find the sum.

$$\chi^2 = \sum \frac{(O - E)^2}{E}$$

$$= \frac{(100 - 75)^2}{75} + \frac{(80 - 100)^2}{100} + \frac{(20 - 25)^2}{25} + \frac{(50 - 75)^2}{75}$$

$$+ \frac{(120 - 100)^2}{100} + \frac{(30 - 25)^2}{25}$$

$$= 26.67$$

The final steps are to make the decision and summarize the results. This test is always a right-tailed test, and the degrees of freedom are $(R - 1)(C - 1) = (2 - 1)(3 - 1) = 2$. If $\alpha = 0.05$, the critical value from Table G is 5.991. Hence, the decision is to reject the null hypothesis, since $26.67 > 5.991$. See Figure 11–6.

Figure 11–6

Critical and Test Values for the Postoperative Procedures Example

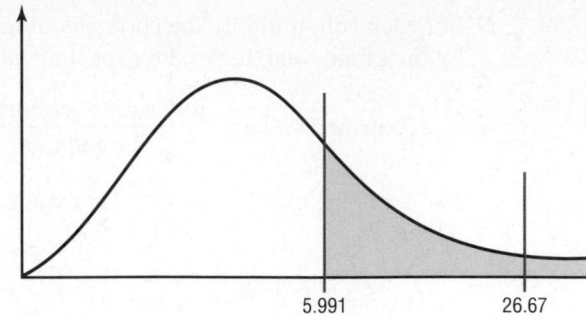

5.991   26.67

The conclusion is that there is enough evidence to support the claim that opinion is related to (dependent on) profession—that is, that the doctors and nurses differ in their opinions about the procedure.

Examples 11–5 and 11–6 illustrate the procedure for the chi-square test of independence.

Example 11–5

A sociologist wishes to see whether the number of years of college a person has completed is related to her or his place of residence. A sample of 88 people is selected and classified as shown.

Location	No college	Four-year degree	Advanced degree	Total
Urban	15	12	8	35
Suburban	8	15	9	32
Rural	6	8	7	21
Total	29	35	24	88

At $\alpha = 0.05$, can the sociologist conclude that a person's location is dependent on the number of years of college?

Solution

Step 1 State the hypotheses and identify the claim.

H_0: A person's place of residence is independent of the number of years of college completed.

H_1: A person's place of residence is dependent on the number of years of college completed (claim).

Step 2 Find the critical value. The critical value is 9.488, since the degrees of freedom are $(3 - 1)(3 - 1) = (2)(2) = 4$.

Step 3 Compute the test value. To compute the test value, one must first compute the expected values.

$$E_{1,1} = \frac{(35)(29)}{88} = 11.53 \quad E_{1,2} = \frac{(35)(35)}{88} = 13.92 \quad E_{1,3} = \frac{(35)(24)}{88} = 9.55$$

$$E_{2,1} = \frac{(32)(29)}{88} = 10.55 \quad E_{2,2} = \frac{(32)(35)}{88} = 12.73 \quad E_{2,3} = \frac{(32)(24)}{88} = 8.73$$

$$E_{3,1} = \frac{(21)(29)}{88} = 6.92 \quad E_{3,2} = \frac{(21)(35)}{88} = 8.35 \quad E_{3,3} = \frac{(21)(24)}{88} = 5.73$$

The completed table is shown.

Location	No college	Four-year degree	Advanced degree	Total
Urban	15 (11.53)	12 (13.92)	8 (9.55)	35
Suburban	8 (10.55)	15 (12.73)	9 (8.73)	32
Rural	6 (6.92)	8 (8.35)	7 (5.73)	21
Total	29	35	24	88

Then the chi-square test value is

$$\chi^2 = \sum \frac{(O - E)^2}{E}$$

$$= \frac{(15 - 11.53)^2}{11.53} + \frac{(12 - 13.92)^2}{13.92} + \frac{(8 - 9.55)^2}{9.55}$$

$$+ \frac{(8 - 10.55)^2}{10.55} + \frac{(15 - 12.73)^2}{12.73} + \frac{(9 - 8.73)^2}{8.73}$$

$$+ \frac{(6 - 6.92)^2}{6.92} + \frac{(8 - 8.35)^2}{8.35} + \frac{(7 - 5.73)^2}{5.73}$$

$$= 3.01$$

Step 4 Make the decision. The decision is not to reject the null hypothesis since $3.01 < 9.488$. See Figure 11–7.

Figure 11–7

Critical and Test Values for Example 11–5

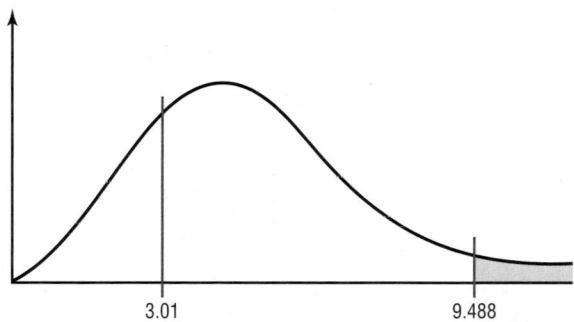

3.01   9.488

Step 5 Summarize the results. There is not enough evidence to support the claim that a person's place of residence is dependent on the number of years of college completed.

Example 11–6

A researcher wishes to determine whether there is a relationship between the gender of an individual and the amount of alcohol consumed. A sample of 68 people is selected, and the following data are obtained.

	Alcohol consumption			
Gender	Low	Moderate	High	Total
Male	10	9	8	27
Female	13	16	12	41
Total	23	25	20	68

At $\alpha = 0.10$, can the researcher conclude that alcohol consumption is related to gender?

Solution

Step 1 State the hypotheses and identify the claim.

H_0: The amount of alcohol that a person consumes is independent of the individual's gender.

H_1: The amount of alcohol that a person consumes is dependent on the individual's gender (claim).

Step 2 Find the critical value. The critical value is 4.605, since the degrees of freedom are $(2 - 1)(3 - 1) = 2$.

Step 3 Compute the test value. First, compute the expected values.

$$E_{1,1} = \frac{(27)(23)}{68} = 9.13 \qquad E_{1,2} = \frac{(27)(25)}{68} = 9.93 \qquad E_{1,3} = \frac{(27)(20)}{68} = 7.94$$

$$E_{2,1} = \frac{(41)(23)}{68} = 13.87 \qquad E_{2,2} = \frac{(41)(25)}{68} = 15.07 \qquad E_{2,3} = \frac{(41)(20)}{68} = 12.06$$

The completed table is shown.

	Alcohol consumption			
Gender	Low	Moderate	High	Total
Male	10 (9.13)	9 (9.93)	8 (7.94)	27
Female	13 (13.87)	16 (15.07)	12 (12.06)	41
Total	23	25	20	68

Then the test value is

$$\chi^2 = \sum \frac{(O - E)^2}{E}$$

$$= \frac{(10 - 9.13)^2}{9.13} + \frac{(9 - 9.93)^2}{9.93} + \frac{(8 - 7.94)^2}{7.94}$$

$$+ \frac{(13 - 13.87)^2}{13.87} + \frac{(16 - 15.07)^2}{15.07} + \frac{(12 - 12.06)^2}{12.06}$$

$$= 0.283$$

Step 4 Make the decision. The decision is not to reject the null hypothesis, since $0.283 < 4.605$. See Figure 11–8.

Figure 11–8

Critical and Test Values for Example 11–6

0.283   4.605

Step 5 Summarize the results. There is not enough evidence to support the claim that the amount of alcohol a person consumes is dependent on the individual's gender.

Objective 3

Test proportions for homogeneity, using chi-square.

Test for Homogeneity of Proportions

The second chi-square test that uses a contingency table is called the **homogeneity of proportions test.** In this situation, samples are selected from several different populations, and the researcher is interested in determining whether the proportions of elements that have a common characteristic are the same for each population. The sample sizes are specified in advance, making either the row totals or column totals in the contingency table known before the samples are selected. For example, a researcher may select a sample of 50 freshmen, 50 sophomores, 50 juniors, and 50 seniors and then find the proportion of students who are smokers in each level. The researcher will then compare the proportions for each group to see if they are equal. The hypotheses in this case would be

H_0: $p_1 = p_2 = p_3 = p_4$
H_1: At least one proportion is different from the others.

If one does not reject the null hypothesis, it can be assumed that the proportions are equal and the differences in them are due to chance. Hence, the proportion of students who smoke is the same for grade levels freshmen through senior. When the null hypothesis is rejected, it can be assumed that the proportions are not all equal. The computational procedure is the same as that for the test of independence shown in Example 11–7.

Example 11–7

Interesting Facts

Water is the most critical nutrient in your body. It is needed for just about everything that happens. Water is lost fast: 2 cups daily is lost just exhaling, 10 cups through normal waste and body cooling, and 1 to 2 quarts per hour running, biking, or working out.

A researcher selected a sample of 50 seniors from each of three area high schools and asked each senior, "Do you drive to school in a car owned by either you or your parents?" The data are shown in the table. At $\alpha = 0.05$, test the claim that the proportion of students who drive their own or their parents' car is the same at all three schools.

	School 1	School 2	School 3	Total
Yes	18	22	16	56
No	32	28	34	94
Total	50	50	50	150

Solution

Step 1 State the hypotheses.

H_0: $p_1 = p_2 = p_3$

H_1: At least one proportion is different from the others.

Step 2 Find the critical value. The formula for the degrees of freedom is the same as before: (rows $-$ 1)(columns $-$ 1) $=$ (2 $-$ 1)(3 $-$ 1) $=$ 1(2) $=$ 2. The critical value is 5.991.

Step 3 Compute the test value. First, compute the expected values.

$$E_{1,1} = \frac{(56)(50)}{150} = 18.67 \qquad E_{2,1} = \frac{(94)(50)}{150} = 31.33$$

$$E_{1,2} = \frac{(56)(50)}{150} = 18.67 \qquad E_{2,2} = \frac{(94)(50)}{150} = 31.33$$

$$E_{1,3} = \frac{(56)(50)}{150} = 18.67 \qquad E_{2,3} = \frac{(94)(50)}{150} = 31.33$$

The completed table is shown here.

	School 1	School 2	School 3	Total
Yes	18 (18.67)	22 (18.67)	16 (18.67)	56
No	32 (31.33)	28 (31.33)	34 (31.33)	94
Total	50	50	50	150

The test value is

$$\chi^2 = \sum \frac{(O - E)^2}{E}$$

$$= \frac{(18 - 18.67)^2}{18.67} + \frac{(22 - 18.67)^2}{18.67} + \frac{(16 - 18.67)^2}{18.67}$$

$$+ \frac{(32 - 31.33)^2}{31.33} + \frac{(28 - 31.33)^2}{31.33} + \frac{(34 - 31.33)^2}{31.33}$$

$$= 1.596$$

Step 4 Make the decision. The decision is not to reject the null hypothesis, since $1.596 < 5.991$.

Step 5 Summarize the results. There is not enough evidence to reject the null hypothesis that the proportions of high school students who drive their own or their parents' car to school are equal for each school.

When the degrees of freedom for a contingency table are equal to 1—that is, the table is a 2×2 table—some statisticians suggest using the *Yates correction for continuity*. The formula for the test is then

$$\chi^2 = \sum \frac{(|O - E| - 0.5)^2}{E}$$

Since the chi-square test is already conservative, most statisticians agree that the Yates correction is not necessary. (See Exercise 33 in Exercises 11–3.)

The steps for the chi-square independence and homogeneity tests are summarized in this Procedure Table.

Procedure Table

The Chi-Square Independence and Homogeneity Tests

Step 1 State the hypotheses and identify the claim.

Step 2 Find the critical value in the right tail. Use Table G.

Step 3 Compute the test value. To compute the test value, first find the expected values. For each cell of the contingency table, use the formula

$$E = \frac{(\text{row sum})(\text{column sum})}{\text{grand total}}$$

to get the expected value. To find the test value, use the formula

$$\chi^2 = \sum \frac{(O - E)^2}{E}$$

Step 4 Make the decision.

Step 5 Summarize the results.

The assumptions for the two chi-square tests are given next.

Assumptions for the Chi-Square Independence and Homogeneity Tests

1. The data are obtained from a random sample.
2. The expected value in each cell must be 5 or more.

If the expected values are not 5 or more, combine categories.

Applying the Concepts **11–3**

Satellite Dishes in Restricted Areas

The Senate is expected to vote on a bill to allow the installation of satellite dishes of any size in deed-restricted areas. The House had passed a similar bill. An opinion poll was taken to see if how a person felt about satellite dish restrictions was related to his or her age. A chi-square test was run, creating the following computer-generated information.

Degrees of freedom d.f. = 6
Test statistic $\chi^2 = 61.25$
Critical value C.V. = 12.6
P-value = 0.00
Significance level = 0.05

	18–29	30–49	50–64	65 and up
For	96 (79.5)	96 (79.5)	90 (79.5)	36 (79.5)
Against	201 (204.75)	189 (204.75)	195 (204.75)	234 (204.75)
Don't know	3 (15.75)	15 (15.75)	15 (15.75)	30 (15.75)

1. Which number from the output is compared to the significance level to check if the null hypothesis should be rejected?

2. Which number from the output gives the probability of a type I error that is calculated from your sample data?

3. Was a right-, left-, or two-tailed test run? Why?

4. Can you tell how many rows and columns there were by looking at the degrees of freedom?

5. Does increasing the sample size change the degrees of freedom?

6. What are your conclusions? Look at the observed and expected frequencies in the table to draw some of your own specific conclusions about response and age.

7. What would your conclusions be if the level of significance were initially set at 0.10?

8. Does chi-square tell you which cell's observed and expected frequencies are significantly different?

See page 618 for the answers.

Exercises 11–3

1. How is the chi-square independence test similar to the goodness-of-fit test? How is it different?

2. How are the degrees of freedom computed for the independence test?

3. Generally, how would the null and alternative hypotheses be stated for the chi-square independence test?

4. What is the name of the table used in the independence test?

5. How are the expected values computed for each cell in the table?

6. Explain how the chi-square independence test differs from the chi-square homogeneity of proportions test.

7. How are the null and alternative hypotheses stated for the test of homogeneity of proportions?

For Exercises 8 through 31, perform the following steps.

   a. State the hypotheses and identify the claim.

   b. Find the critical value.

   c. Compute the test value.

   d. Make the decision.

   e. Summarize the results.

Use the traditional method of hypothesis testing unless otherwise specified.

8. A study is conducted as to whether there is a relationship between joggers and the consumption of nutritional supplements. A random sample of 210 subjects is selected, and they are classified as shown. At $\alpha = 0.05$, test the claim that jogging and the consumption of supplements are not related. Why might supplement manufacturers use the results of this study?

Jogging status	Daily	Weekly	As needed
Joggers	34	52	23
Nonjoggers	18	65	18

9. Is the type of pet owned dependent on annual household income? Use $\alpha = 0.05$.

Income ($)	Type of pet			
	Dog	Cat	Bird	Horse
Under 12,500	127	139	173	95
12,500–24,999	191	197	209	203
25,000–39,999	216	215	220	218
40,000–59,999	215	212	175	231
60,000 and over	254	237	223	254

Source: *N.Y. Times Almanac.*

10. This table lists the numbers of officers and enlisted personnel for women in the military. At $\alpha = 0.05$, is there sufficient evidence to conclude that a relationship exists between rank and branch of the Armed Forces?

	Officers	Enlisted
Army	10,791	62,491
Navy	7,816	42,750
Marine Corps	932	9,525
Air Force	11,819	54,344

Source: *N.Y. Times Almanac.*

11. Is the composition of state legislatures in the House of Representatives related to the specific state? Use $\alpha = 0.05$.

	Democrats	Republicans
Pennsylvania	100	103
Ohio	39	59
West Virginia	75	25
Maryland	106	35

Source: *N.Y. Times Almanac.*

12. Is the size of the population by age related to the state that it's in? Use $\alpha = 0.05$. (Population values are in thousands.)

	Under 5	5–17	18–24	25–44	45–64	65+
Pennsylvania	721	2140	1025	3515	2702	1899
Ohio	740	2104	1065	3359	2487	1501

Source: *N.Y. Times Almanac.*

13. Some 300 men and 210 women were asked about how many ads in all media they think they saw or heard during one day. The results are shown.

	Number				
	1–30	31–50	51–100	101–300	301 or more
Men	45	60	90	54	51
Women	50	50	54	30	26

At $\alpha = 0.01$, is the number of ads people feel that they see or hear related to the gender of the person?

Source: Based on information from *USA TODAY.*

14. An instructor wishes to see if the way people obtain information is independent of their educational background. A survey of 400 high school and college graduates yielded this information. At $\alpha = 0.05$, test the claim that the way people obtain information is independent of their educational background.

	Television	Newspapers	Other sources
High school	159	90	51
College	27	42	31

Source: *USA TODAY.*

15. A study was conducted to see if there was a relationship between the gender of an attorney and the type of practice he or she is engaged in. A sample of 240 attorneys is selected, and the results are shown. At $\alpha = 0.05$, can it be assumed that gender and employment are independent?

Gender	Private practice	Law firm	Government
Male	112	16	12
Female	64	18	18

16. A researcher wants to see if there is a relationship between speeding and states of residence on Interstate 75 near Findlay, Ohio. These data are collected:

	Ohio	Michigan
Speeders (66 mph or more)	18	25
Nonspeeders (65 mph or less)	27	15

Using $\alpha = 0.05$, can it be concluded that the state of residence and speeding are independent?

17. A study is being conducted to determine whether the age of the customer is related to the type of movie he or she rents. A sample of renters gives the data

shown here. At $\alpha = 0.10$, is the type of movie selected related to the customer's age?

Type of movie

Age	Documentary	Comedy	Mystery
12–20	14	9	8
21–29	15	14	9
30–38	9	21	39
39–47	7	22	17
48 and over	6	38	12

18. A researcher wishes to see if the gender of the person who prepares dinner is independent of the time when the person decides what to eat for dinner. A sample of 70 men and women is selected, and the results are shown. At $\alpha = 0.05$, can it be concluded that the variables are independent? Summarize the results in your own words.

Gender	In the afternoon	In the morning	Day or days before
Female	20	9	10
Male	16	12	3

19. A survey at a ballpark shows this selection of snacks purchased. At $\alpha = 0.10$, is the snack chosen independent of the gender of the consumer?

Snack

Gender	Hot dog	Peanuts	Popcorn
Male	12	21	19
Female	13	8	25

20. To test the effectiveness of a new drug, a researcher gives one group of individuals the new drug and another group a placebo. The results of the study are shown here. At $\alpha = 0.10$, can the researcher conclude that the drug is effective? Use the *P*-value method.

Medication	Effective	Not effective
Drug	32	9
Placebo	12	18

21. A book publisher wishes to determine whether there is a difference in the type of book selected by males and females for recreational reading. A random sample provides the data given here. At $\alpha = 0.05$, test the claim that the type of book selected is independent of the gender of the individual. Use the *P*-value method.

Type of book

Gender	Mystery	Romance	Self-help
Male	243	201	191
Female	135	149	202

22. According to a recent survey, 32% of Americans say they are "very likely" to become organ donors. A researcher surveys 50 drivers in each of three neighborhoods to determine the percentage of those

willing to donate their organs. The results are shown here. At $\alpha = 0.01$, test the claim that the proportions of those who will donate their organs are equal in all three neighborhoods.

	Neighbor-hood A	Neighbor-hood B	Neighbor-hood C
Will donate	28	14	21
Will not donate	22	36	29
Total	50	50	50

Source: *The Harper's Index Book.*

23. According to a recent survey, 64% of Americans between the ages of 6 and 17 cannot pass a basic fitness test. A physical education instructor wishes to determine if the percentages of such students in different schools in his school district are the same. He administers a basic fitness test to 120 students in each of four schools. The results are shown here. At $\alpha = 0.05$, test the claim that the proportions who pass the test are equal.

	Southside	West End	East Hills	Jefferson
Passed	49	38	46	34
Failed	71	82	74	86
Total	120	120	120	120

Source: *The Harper's Index Book.*

24. An advertising firm has decided to ask 92 customers at each of three local shopping malls if they are willing to take part in a market research survey. According to previous studies, 38% of Americans refuse to take part in such surveys. The results are shown here. At $\alpha = 0.01$, test the claim that the proportions of those who are willing to participate are equal.

	Mall A	Mall B	Mall C
Will participate	52	45	36
Will not participate	40	47	56
Total	92	92	92

Source: *The Harper's Index Book.*

25. A researcher wishes to see if the proportions of workers for each type of job have changed during the last 10 years. A sample of 100 workers is selected, and the results are shown. At $\alpha = 0.05$, test the claim that the proportions have not changed. Can the results be generalized to the population of the United States?

Types of jobs	Services	Manu-facturing	Government	Other
10 years ago	33	13	11	3
Now	18	12	8	2
Total	51	25	19	5

Source: Pennsylvania Department of Labor and Industry.

26. According to a recent survey, 59% of Americans aged 8 to 17 would prefer that their mother work outside the home, regardless of what she does now. A school district psychologist decided to select three samples of 60 students each in elementary, middle, and high school to see how the students in her district felt about the issue. At $\alpha = 0.10$, test the claim that the proportions of the students who prefer that their mother have a job are equal.

	Elementary	Middle	High
Prefers mother work	29	38	51
Prefers mother not work	31	22	9
Total	60	60	60

Source: Daniel Weiss, *100% American.*

27. A researcher surveyed 100 randomly selected lawyers in each of four areas of the country and asked them if they had performed *pro bono* work for 25 or fewer hours in the last year. The results are shown here. At $\alpha = 0.10$, is there enough evidence to reject the claim that the proportions of those who accepted *pro bono* work for 25 hours or less are the same in each area?

	North	South	East	West
Yes	43	39	22	28
No	57	61	78	72
Total	100	100	100	100

Source: Daniel Weiss, *100% American.*

28. On average, 79% of American fathers are in the delivery room when their children are born. A physician's assistant surveyed 300 first-time fathers to determine if they had been in the delivery room when their children were born. The results are shown here. At $\alpha = 0.05$, is there enough evidence to reject the claim that the proportions of those who were in the delivery room at the time of birth are the same?

	Hos-pital A	Hos-pital B	Hos-pital C	Hos-pital D
Present	66	60	57	56
Not present	9	15	18	19
Total	75	75	75	75

Source: Daniel Weiss, *100% American.*

29. A children's playground equipment manufacturer read in a survey that 55% of all U.S. playground injuries occur on the monkey bars. The manufacturer wishes to investigate playground injuries in four different parts of the country to determine if the proportions of accidents on the monkey bars are equal. The results are shown here. At $\alpha = 0.05$, test the claim that the proportions are equal. Use the *P*-value method.

Accidents	North	South	East	West
On monkey bars	15	18	13	16
Not on monkey bars	15	12	17	14
Total	30	30	30	30

Source: Michael D. Shook and Robert L. Shook, *The Book of Odds.*

30. According to the American Automobile Association, 31 million Americans travel over the Thanksgiving holiday. To determine whether to stay open or not, a national restaurant chain surveyed 125 customers at each of four locations to see if they would be traveling over the holiday. The results are shown here. At $\alpha = 0.10$, test the claim that the proportions of Americans who will travel over the Thanksgiving holiday are equal. Use the *P*-value method.

	Loca-tion A	Loca-tion B	Loca-tion C	Loca-tion D
Will travel	37	52	46	49
Will not travel	88	73	79	76
Total	125	125	125	125

Source: Michael D. Shook and Robert L. Shook, *The Book of Odds.*

31. The vice president of a large supermarket chain wished to determine if her customers made a list before going grocery shopping. She surveyed 288 customers in three stores. The results are shown here. At $\alpha = 0.10$, test the claim that the proportions of the customers in the three stores who made a list before going shopping are equal.

	Store A	Store B	Store C
Made list	77	74	68
No list	19	22	28
Total	96	96	96

Source: Daniel Weiss, *100% American.*

Extending the Concepts

32. For a 2 × 2 table, *a*, *b*, *c*, and *d* are the observed values for each cell, as shown.

a	b
c	d

The chi-square test value can be computed as

$$\chi^2 = \frac{n(ad - bc)^2}{(a + b)(a + c)(c + d)(b + d)}$$

Speaking of Statistics

Does Color Affect Your Appetite?

It has been suggested that color is related to appetite in humans. For example, if the walls in a restaurant are painted certain colors, it is thought that the customer will eat more food. A study was done at the University of Illinois and the University of Pennsylvania. When people were given six varieties of jellybeans mixed in a bowl or separated by color, they ate about twice as many from the bowl with the mixed jellybeans as from the bowls that were separated by color.

It is thought that when the jellybeans were mixed, people felt that it offered a greater variety of choices, and the variety of choices increased their appetites.

In this case one variable—color—is categorical, and the other variable—amount of jellybeans eaten—is numerical. Could a chi-square goodness-of-fit test be used here? If so, suggest how it could be set up.

where $n = a + b + c + d$. Compute the χ^2 test value by using the above formula and the formula $\Sigma(O - E)^2/E$, and compare the results for the following table.

12	15
9	23

33. For the contingency table shown in Exercise 32, compute the chi-square test value by using the Yates correction for continuity.

34. When the chi-square test value is significant and there is a relationship between the variables, the strength of this relationship can be measured by using the *contingency coefficient*. The formula for the contingency coefficient is

$$C = \sqrt{\frac{\chi^2}{\chi^2 + n}}$$

where χ^2 is the test value and n is the sum of frequencies of the cells. The contingency coefficient will always be less than 1. Compute the contingency coefficient for Exercises 8 and 20.

Technology *Step by Step*

MINITAB
Step by Step

Tests Using Contingency Tables
Calculate the Chi-Square Test Statistic and *P*-Value

1. Enter the observed frequencies for Example 11–5 into three columns of MINITAB. Name the columns but not the rows. Exclude totals. The complete worksheet is shown.

↓	C1	C2	C3
	NoCollege	Four-year	Advanced
1	15	12	8
2	8	15	9
3	6	8	7

Chi-Square Test (Table in Worksheet) ✕

C1	NoCollege	Columns containing the table:
C2	Four-year	NoCollege–Advanced
C3	Advanced	

2. Select **Stat>Tables>Chi-Square Test.**

3. Drag the mouse over the three columns in the list.

4. Click [Select]. The three columns will be placed in the Columns box as a sequence, NoCollege through Advanced.

5. Click [OK].

The chi-square test statistic 3.006 has a *P*-value of 0.557. Do not reject the null hypothesis.

There is no relationship between level of education and place of residence.

Chi-Square Test: NoCollege, Four-year, Advanced
```
Expected counts are printed below observed counts
Chi-Square contributions are printed below expected counts
```

	NoCollege	Four-year	Advanced	Total
1	15	12	8	35
	11.53	13.92	9.55	
	1.041	0.265	0.250	
2	8	15	9	32
	10.55	12.73	8.73	
	0.614	0.406	0.009	
3	6	8	7	21
	6.92	8.35	5.73	
	0.122	0.015	0.283	
Total	29	35	24	88

```
Chi-Sq = 3.006, DF = 4, P-Value = 0.557
```

Construct a Contingency Table and Calculate the Chi-Square Test Statistic
In Chapter 4 we learned how to construct a contingency table by using gender and smoking status in the Data Bank file described in Appendix D. Are smoking status and gender related? Who is more likely to smoke, men or women?

1. Use **File>Open Worksheet** to open the Data Bank file. Remember do *not* click the file icon.

2. Select **Stat>Tables>Cross Tabulation and Chi-Square.**

3. Double-click Smoking Status for rows and Gender for columns.

4. The Display option for Counts should be checked.

5. Click [Chi-Square].

   a) Check Chi-Square analysis.

   b) Check Expected cell counts.

6. Click [OK] twice.

In the session window the contingency table and the chi-square analysis will be displayed.

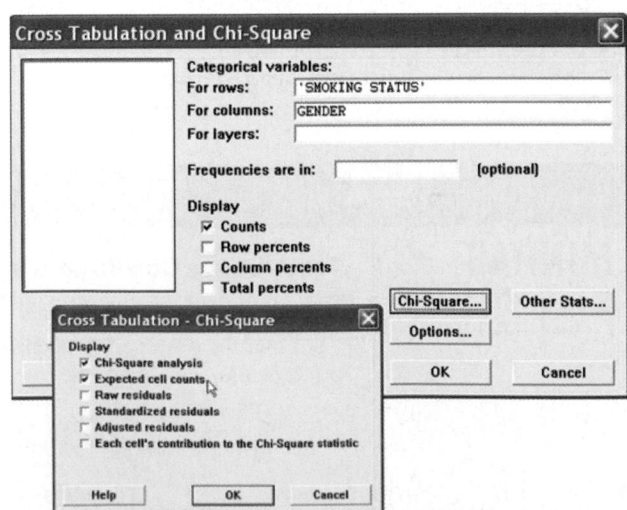

Tabulated statistics: SMOKING STATUS, GENDER

Rows: SMOKING STATUS   Columns: Gender

```
                F        M      All

   0           25       22       47
             23.50    23.50    47.00

   1           18       19       37
             18.50    18.50    37.00

   2            7        9       16
              8.00     8.00    16.00

 All           50       50      100
             50.00    50.00   100.00
Cell Contents:           Count
                         Expected count
```

Pearson Chi-Square = 0.469, DF = 2, P-Value = 0.791

There is not enough evidence to conclude that smoking is related to gender.

TI-83 Plus or TI-84 Plus
Step by Step

Chi-Square Test for Independence

1. Press **2nd [χ^{-1}]** for **MATRIX** and move the cursor to Edit, then press **ENTER.**
2. Enter the number of rows and columns. Then press **ENTER.**
3. Enter the values in the matrix as they appear in the contingency table.
4. Press **STAT** and move the cursor to TESTS. Press **C (ALPHA PRGM)** for χ^2-Test. Make sure the observed matrix is [A] and the expected matrix is [B].
5. Move the cursor to Calculate and press **ENTER.**

Example TI11–2

Using the data shown from Example 11–6, test the claim of independence at $\alpha = 0.10$.

10	9	8
13	16	12

Input	Input	Output
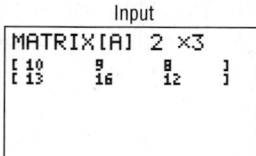	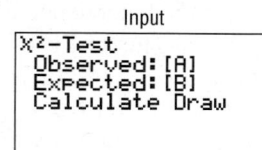	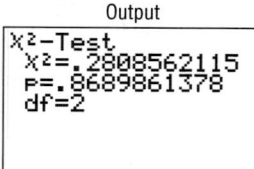

The test value is 0.2808562115. The *P*-value is 0.8689861378. The decision is to not reject the null hypothesis, since this value is greater than 0.10. You can find the expected values by pressing **MATRIX,** moving the cursor to [B], and pressing **ENTER** twice.

Excel
Step by Step

Tests Using Contingency Tables

Excel does not have a procedure to conduct tests using a contingency table without inclusion of the expected values. However, you may conduct such tests by using the MegaStat Add-in available on your CD and Online Learning Center. If you have not installed this add-in, do so by following the instructions on page 24.

These instructions use the data from Example 11–5.

1. Enter the location labels—**Urban, Suburban,** and **Rural**—into Column A of a new worksheet.

2. Enter the categories for the number of years of college in the first row in cells B1, C1, and D1.

3. Enter the data in the appropriate cells.

4. Select **MegaStat>Chi-Square/Crosstab>Contingency Table.**

5. Select cells A1:D4 for the Input range.

6. Check chi-square from the Output Options.

7. Click [OK].

The results will appear in a new sheet labeled Output as shown.

Chi-Square Contingency Table Test for Independence

	None	4-yr	Advanced	Total
Urban	15	12	8	35
Suburban	8	15	9	32
Rural	6	8	7	21
Total	29	35	24	88

3.01 chi-square
4 df
0.5569 p-value

11–4   Summary

Three uses of the chi-square distribution were explained in this chapter. It can be used as a goodness-of-fit test to determine whether the frequencies of a distribution are the same as the hypothesized frequencies. For example, is the number of defective parts produced by a factory the same each day? This test is always a right-tailed test.

The test of independence is used to determine whether two variables are related or are independent. This test uses a contingency table and is always a right-tailed test. An example of its use is a test to determine whether the attitudes of urban residents about the recycling of trash differ from the attitudes of rural residents.

Finally, the homogeneity of proportions test is used to determine if several proportions are all equal when samples are selected from different populations.

The chi-square distribution is also used for other types of statistical hypothesis tests, such as the Kruskal-Wallis test, which is explained in Chapter 13.

Important Terms

contingency table 598

expected frequency 585

goodness-of-fit test 585

homogeneity of proportions test 603

independence test 597

observed frequency 585

Important Formulas

Formula for the chi-square test for goodness of fit:

$$\chi^2 = \sum \frac{(O - E)^2}{E}$$

with degrees of freedom equal to the number of categories minus 1 and where

O = observed frequency

E = expected frequency

Formula for the chi-square independence and homogeneity of proportions tests:

$$\chi^2 = \sum \frac{(O - E)^2}{E}$$

with degrees of freedom equal to (rows $-$ 1) times (columns $-$ 1). Formula for the expected value for each cell:

$$E = \frac{(\textbf{row sum})(\textbf{column sum})}{\textbf{grand total}}$$

Review Exercises

For Exercises 1 through 10, follow these steps.

   a. State the hypotheses and identify the claim.
   b. Find the critical value(s).
   c. Compute the test value.
   d. Make the decision.
   e. Summarize the results.

Use the traditional method of hypothesis testing unless otherwise specified.

1. A storeowner wishes to see if people have a favorite day of the week to shop. A sample of 400 people is selected, and each is asked his or her preference. The data are shown. At $\alpha = 0.05$, test the claim that shoppers have no preference. Give one example of how retail merchants would be able to use the numbers in this study.

Day	Sun.	Mon.	Tues.	Wed.	Thurs.	Fri.	Sat.
Number	28	16	20	26	74	96	140

Source: Based on information from the International Mass Retail Association.

2. A researcher reads that the marital status of U.S. adults is distributed as follows: married, 60%; single, 24%; divorced, 10%; widowed 6%. She selects a sample of 150 people, and the results are shown. At $\alpha = 0.01$, can it be concluded that marital status is distributed as she has read? Give one factor that might influence the results of this study.

Marital status	Married	Single	Divorced	Widowed
Number	72	42	21	15

Source: U.S. Census Bureau.

3. The federal government has proposed labeling tires by fuel efficiency to save fuel and cut emissions. A survey was taken to see who would use these labels. At $\alpha = 0.10$, is the gender of the individual related to whether or not a person would use these labels? The data from a sample are shown here.

Gender	Yes	No	Undecided
Men	114	30	6
Women	136	16	8

Source: USA TODAY.

4. A police investigator read that the reasons why gun sales to applicants were denied were distributed as follows: criminal history of felonies, 75%; domestic violence conviction, 11%; and drug abuse, fugitive, etc., 14%. A sample of applicants in a large study who were refused sales is obtained and is distributed as follows. At $\alpha = 0.10$, can it be concluded that the distribution is as stated? Do you think the results might be different in a rural area?

Reason	Criminal history	Domestic violence	Drug abuse etc.
Number	120	42	38

Source: Based on FBI statistics.

5. A survey was taken on how a lump-sum pension would be invested by 45-year-olds and 65-year-olds. The data are shown here. At $\alpha = 0.05$, is there a relationship between the age of the investor and the way the money would be invested?

	Large company stock funds	Small company stock funds	International stock funds	CDs or money market funds	Bonds
Age 45	20	10	10	15	45
Age 65	42	24	24	6	24

Source: USA TODAY.

6. A car manufacturer wishes to determine whether the type of car purchased is related to the individual's gender. The data obtained from a sample are shown here. At $\alpha = 0.01$, is the gender of the purchaser related to the type of car purchased?

Statistics Today

Statistics and Heredity–Revisited

Using probability, Mendel predicted the following:

	Smooth		Wrinkled	
	Yellow	Green	Yellow	Green
Expected	0.5625	0.1875	0.1875	0.0625

The observed results were these:

	Smooth		Wrinkled	
	Yellow	Green	Yellow	Green
Observed	0.5666	0.1942	0.1816	0.0556

Using chi-square tests on the data, Mendel found that his predictions were accurate in most cases (i.e., a good fit), thus supporting his theory. He reported many highly successful experiments. Mendel's genetic theory is simple but useful in predicting the results of hybridization.

A Fly in the Ointment

Although Mendel's theory is basically correct, an English statistician named R. A. Fisher examined Mendel's data some 50 years later. He found that the observed (actual) results agreed too closely with the expected (theoretical) results and concluded that the data had in some way been falsified. The results were too good to be true. Several explanations have been proposed, ranging from deliberate misinterpretation to an assistant's error, but no one can be sure how this happened.

Gender of purchaser	Type of vehicle purchased			
	Sedan	Compact	Station wagon	SUV
Male	33	27	23	17
Female	21	34	41	18

7. A guidance counselor wishes to determine if the proportions of high school girls in his school district who have jobs are equal to the national average of 36%. He surveys 80 female students, ages 16 through 18, to determine if they work or not. The results are shown. At $\alpha = 0.01$, test the claim that the proportions of girls who work are equal. Use the *P*-value method.

	16-year-olds	17-year-olds	18-year-olds
Work	45	31	38
Don't work	35	49	42
Total	80	80	80

Source: Michael D. Shook and Robert L. Shook, *The Book of Odds.*

8. The risk of injury is higher for males compared to females (57% versus 43%). A hospital emergency room supervisor wishes to determine if the proportions of injuries to males in his hospital are the same for each of four months. He surveys 100 injuries treated in his ER for each month. The results are shown here. At $\alpha = 0.05$, can he reject the claim that the proportions of injuries for males are equal for each of the four months?

	May	June	July	August
Male	51	47	58	63
Female	49	53	42	37
Total	100	100	100	100

Source: Michael D. Shook and Robert L. Shook, *The Book of Odds.*

9. A researcher surveyed 50 randomly selected subjects in four cities and asked if they felt that their anger was the most difficult behavior to control. The results are shown. At $\alpha = 0.10$, is there enough evidence to reject the claim that the proportion of those who felt this way in each city is the same?

	City A	City B	City C	City D
Yes	12	15	10	21
No	38	35	40	29
Total	50	50	50	50

10. A researcher surveyed 50 randomly selected males and 50 randomly selected females to see how they paid their bills. The data are shown. At $\alpha = 0.01$, test the claim that the proportions are not equal. What might be a reason for the difference, if one exists?

Type of payment	Checks	Electronically	In person
Males	27	15	8
Females	22	19	9
Total	49	34	17

Source: Based on information from Gallup.

Data Analysis

The Data Bank is located in Appendix D, or on the World Wide Web by following links from www.mhhe.com/math/stat/bluman

1. Select a sample of 40 individuals from the Data Bank. Use the chi-square goodness-of-fit test to see if the marital status of individuals is equally distributed.

2. Use the chi-square test of independence to test the hypothesis that smoking is independent of gender. Use a sample of at least 75 people.

3. Using the data from Data Set X in Appendix D, classify the data as 1–3, 4–6, 7–9, etc. Use the chi-square goodness-of-fit test to see if the number of times each ball is drawn is equally distributed.

Chapter Quiz

Determine whether each statement is true or false. If the statement is false, explain why.

1. The chi-square test of independence is always two-tailed.

2. The test values for the chi-square goodness-of-fit test and the independence test are computed by using the same formula.

3. When the null hypothesis is rejected in the goodness-of-fit test, it means there is close agreement between the observed and expected frequencies.

Select the best answer.

4. The values of the chi-square variable cannot be

   a. Positive   c. Negative
   b. 0   d. None of the above

5. The null hypothesis for the chi-square test of independence is that the variables are

   a. Dependent   c. Related
   b. Independent   d. Always 0

6. The degrees of freedom for the goodness-of-fit test are

   a. 0   c. Sample size − 1
   b. 1   d. Number of categories − 1

Complete the following statements with the best answer.

7. The degrees of freedom for a 4 × 3 contingency table are _____.

8. An important assumption for the chi-square test is that the observations must be _____.

9. The chi-square goodness-of-fit test is always _____-tailed.

10. In the chi-square independence test, the expected frequency for each class must always be _____.

For Exercises 11 through 19, follow these steps.

   a. State the hypotheses and identify the claim.
   b. Find the critical value.
   c. Compute the test value.
   d. Make the decision.
   e. Summarize the results.

Use the traditional method of hypothesis testing unless otherwise specified.

11. A survey of why people lost their jobs produced the following results. At $\alpha = 0.05$, test the claim that the number of responses is equally distributed. Do you think the results might be different if the study were done 10 years ago?

Reason	Company closing	Position abolished	Insufficient work
Number	26	18	28

Source: Based on information from U.S. Department of Labor.

12. A food service manager read that the place where people consumed takeout food is distributed as follows: home, 53%; car, 19%; work, 14%; other, 14%. A survey of 300 individuals showed the following results. At $\alpha = 0.01$, can it be concluded that the distribution is as stated? Where would a fast-food restaurant want to target its advertisements?

Place	Home	Car	Work	Other
Number	142	57	51	50

Source: Beef Industry Council.

13. A survey found that 62% of the respondents stated that they never watched the home shopping channels on cable television, 23% stated that they watched the channels rarely, 11% stated that they watched them occasionally, and 4% stated that they watched them frequently. A group of 200 college students was surveyed, and 105 stated that they never watched the home shopping channels, 72 stated that they watched them rarely, 13 stated that they watched them occasionally, and 10 stated that they watched them frequently. At $\alpha = 0.05$, can it be concluded that the college students differ in their preference for the home shopping channels?

Source: Based on information obtained from *USA TODAY* Snapshots.

14. The 2000 Census indicated the following percentages for means of commuting to work for workers over 15 years of age.

Alone	75.7
Carpooling	12.2
Public	4.7
Walked	2.9
Other	1.2
Worked at home	3.3

A random sample of workers found that 320 drove alone, 100 carpooled, 30 used public transportation, 20 walked, 10 used other forms of transportation, and 20 worked at home. Is there sufficient evidence to conclude that the proportions of workers using each type of transportation differ from those in the Census report? Use $\alpha = 0.05$.

Source: Census Bureau, *Washington Observer-Reporter.*

15. A survey of women and men asked what their favorite ice cream flavor was. The results are shown. At $\alpha = 0.05$, can it be concluded that the favorite flavor is independent of gender?

	Flavor			
	Vanilla	**Chocolate**	**Strawberry**	**Other**
Women	62	36	10	2
Men	49	37	5	9

16. A pizza shop owner wishes to determine if the type of pizza a person selects is related to the age of the individual. The data obtained from a sample are shown here. At $\alpha = 0.10$, is the age of the purchaser related to the type of pizza ordered? Use the *P*-value method.

	Type of pizza			
Age	**Plain**	**Pepperoni**	**Mushroom**	**Double cheese**
10–19	12	21	39	71
20–29	18	76	52	87
30–39	24	50	40	47
40–49	52	30	12	28

17. A survey at a ballpark shows the following selection of pennants sold to fans. The data are presented here. At $\alpha = 0.10$, is the color of the pennant purchased independent of the gender of the individual?

	Blue	**Yellow**	**Red**
Men	519	659	876
Women	487	702	787

18. In a survey of children ages 8 through 11, these data were obtained as to what their parents should do with the money from a $400 tax credit.

	Keep it for themselves	**Give it to their children**	**Don't know**
Girls	162	132	6
Boys	147	147	6

At $\alpha = 0.10$, is there a relationship between the feelings of the children and the gender of the children?

Source: Based on information from *USA TODAY* Snapshot.

19. A survey of 60 men and 60 women asked if they would be happy spending the rest of their careers with their present employers. The results are shown. At $\alpha = 0.10$, can it be concluded that the proportions are equal? If they are not equal, give a possible reason for the difference.

	Yes	**No**	**Undecided**
Men	40	15	5
Women	36	9	15
Total	76	24	20

Source: Based on information from a Maritz Poll.

Critical Thinking Challenges

1. Use your calculator or the MINITAB random number generator to generate 100 two-digit random numbers. Make a grouped frequency distribution, using the chi-square goodness-of-fit test to see if the distribution is random. To do this, use an expected frequency of 10 for each class. Can it be concluded that the distribution is random? Explain.

2. Simulate the state lottery by using your calculator or MINITAB to generate 100 three-digit random numbers. Group these numbers 100–199, 200–299, etc. Use the chi-square goodness-of-fit test to see if the numbers are random. The expected frequency for each class should be 10. Explain why.

3. Purchase a bag of M & M candy and count the number of pieces of each color. Using the information as your sample, state a hypothesis for the distribution of colors and compare your hypothesis to one obtained from the *USA TODAY* Snapshot. What did you discover?

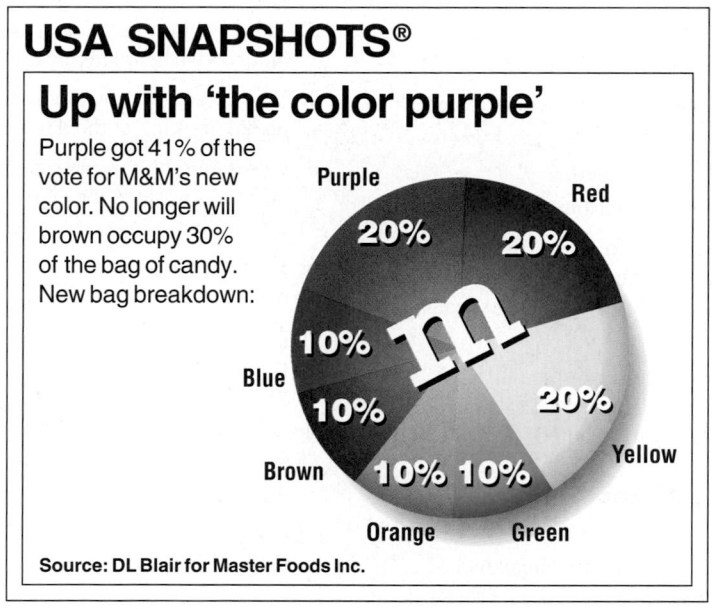

USA SNAPSHOTS®

Up with 'the color purple'

Purple got 41% of the vote for M&M's new color. No longer will brown occupy 30% of the bag of candy. New bag breakdown:

Purple 20%
Red 20%
Blue 10%
Yellow 20%
Brown 10%
Orange 10%
Green 10%

Source: DL Blair for Master Foods Inc.

Source: Copyright 2002. *USA TODAY.* Reprinted with permission.

Data Projects

Where appropriate, use MINITAB, the TI-83 Plus, the TI-84 Plus, or a computer program of your choice to complete the following exercises.

1. Select a variable and collect some data over a period of a week or several months. For example, you may want to record the number of phone calls you received over 7 days, or the number of times you used your credit card each month for the past several months. Using the chi-square goodness-of-fit test, see if the occurrences are equally distributed over the period.

a. State the purpose of the study.
b. Define the population.
c. State how the sample was selected.
d. State the hypotheses.
e. Select an α value.
f. Compute the chi-square test value.
g. Make the decision.
h. Write a paragraph summarizing the results.

2. Collect some data on a variable and construct a frequency distribution.

a. Using the method shown in Section 11–2, decide if the variable you have chosen is approximately normally distributed.

b. Write a short paper describing your findings, and cite some reasons why the variable you selected is or is not approximately normally distributed.

3. Collect some data on a variable that can be divided into groups. For example, you may want to see if there is a difference in the color of cars men own versus the color of cars women own. Using the chi-square independence test, determine if the one variable is independent of the other.

a. State the purpose of the study.
b. Define the population.
c. State how the sample was selected.
d. State the hypotheses for the study.
e. Select an α value.
f. Compute the chi-square test value.
g. Make the decision.
h. Summarize the results.

You may use the following websites to obtain raw data:

**Visit the data sets at the book's website found at http://www.mhhe.com/math/stat/bluman Click on the 6th edition.
http://lib.stat.cmu.edu/DASL
http://www.statcan.ca**

Answers to Applying the Concepts

Section 11–2   Never the Same Amounts

1. The variables are quantitative.

2. We can use a chi-square goodness-of-fit test.

3. There are a total of 233 candies, so we would expect 46.6 of each color. Our test statistic is $\chi^2 = 1.442$.

4. H_0: The colors are equally distributed.
   H_1: The colors are not equally distributed.

5. There are $5 - 1 = 4$ degrees of freedom for the test. The critical value depends on the choice of significance level. At the 0.05 significance level, the critical value is 9.488.

6. Since $1.442 < 9.488$, we fail to reject the null hypothesis. There is not enough evidence to conclude that the colors are not equally distributed.

Section 11–3   Satellite Dishes in Restricted Areas

1. We compare the P-value to the significance level of 0.05 to check if the null hypothesis should be rejected.

2. The P-value gives the probability of a type I error.

3. This is a right-tailed test, since chi-square tests of independence are always right-tailed.

4. You cannot tell how many rows and columns there were just by looking at the degrees of freedom.

5. Increasing the sample size does not increase the degrees of freedom, since the degrees of freedom are based on the number of rows and columns.

6. We will reject the null hypothesis. There are a number of cells where the observed and expected frequencies are quite different.

7. If the significance level were initially set at 0.10, we would still reject the null hypothesis.

8. No, the chi-square value does not tell us which cells have observed and expected frequencies that are very different.

CHAPTER 12

Analysis of Variance

Objectives

After completing this chapter, you should be able to

1. Use the one-way ANOVA technique to determine if there is a significant difference among three or more means.

2. Determine which means differ, using the Scheffé or Tukey test if the null hypothesis is rejected in the ANOVA.

3. Use the two-way ANOVA technique to determine if there is a significant difference in the main effects or interaction.

Outline

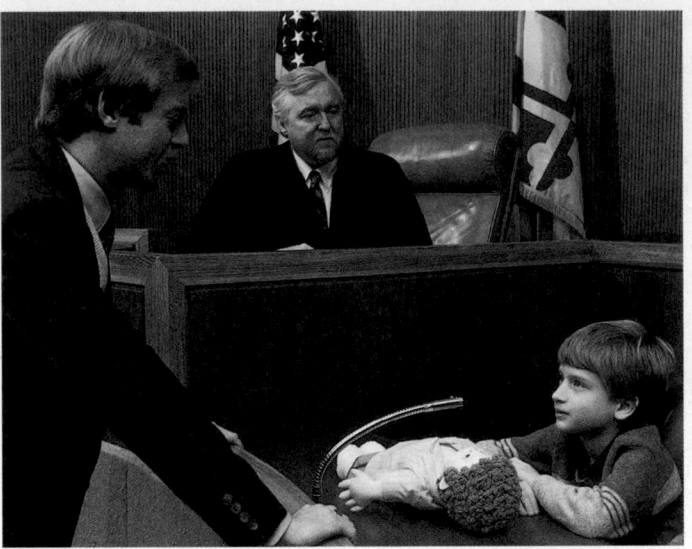

Statistics Today

Is Seeing Really Believing?

Many adults look on the eyewitness testimony of children with skepticism. They believe that young witnesses' testimony is less accurate than the testimony of adults in court cases. Several statistical studies have been done on this subject.

In a preliminary study, three researchers selected fourteen 8-year-olds, fourteen 12-year-olds, and fourteen adults. The researchers showed each group the same video of a crime being committed. The next day, each witness responded to direct and cross-examination questioning. Then the researchers, using statistical methods explained in this chapter, were able to determine if there were differences in the accuracy of the testimony of the three groups on direct examination and on cross-examination. The statistical methods used here differ from the ones explained in Chapter 9 because there are three groups rather than two. See Statistics Today—Revisited.

Source: C. Luus, G. Wells, and J. Turtle, "Child Eyewitnesses: Seeing Is Believing," *Journal of Applied Psychology* 80, no. 2, pp. 317–26.

12–1

Introduction

The F test, used to compare two variances shown in Chapter 9, can also be used to compare three or more means. This technique is called *analysis of variance,* or *ANOVA*. It is used to test claims involving three or more means. (*Note:* The F test can also be used to test the equality of two means. But since it is equivalent to the t test in this case, the t test is usually used instead of the F test when there are only two means.) For example, suppose a researcher wishes to see whether the means of the time it takes three groups of students to solve a computer problem using Fortran, Basic, and Pascal are different. The researcher will use the ANOVA technique for this test. The z and t tests should not be used when three or more means are compared, for reasons given later in this chapter.

For three groups, the F test can only show whether a difference exists among the three means. It cannot reveal where the difference lies—that is, between $\overline{X}_1$ and $\overline{X}_2$, or $\overline{X}_1$ and $\overline{X}_3$, or $\overline{X}_2$ and $\overline{X}_3$. If the F test indicates that there is a difference among the means, other statistical tests are used to find where the difference exists. The most commonly used tests are the Scheffé test and the Tukey test, which are also explained in this chapter.

The analysis of variance that is used to compare three or more means is called a *one-way analysis of variance* since it contains only one variable. In the previous example, the variable is the type of computer language used. The analysis of variance can be extended to studies involving two variables, such as type of computer language used and mathematical background of the students. These studies involve a *two-way analysis of variance*. Section 12–4 explains the two-way analysis of variance.

12–2   One-Way Analysis of Variance

Objective 1

Use the one-way ANOVA technique to determine if there is a significant difference among three or more means.

When an *F* test is used to test a hypothesis concerning the means of three or more populations, the technique is called **analysis of variance** (commonly abbreviated as **ANOVA**). At first glance, one might think that to compare the means of three or more samples, the *t* test can be used, comparing two means at a time. But there are several reasons why the *t* test should not be done.

First, when one is comparing two means at a time, the rest of the means under study are ignored. With the *F* test, all the means are compared simultaneously. Second, when one is comparing two means at a time and making all pairwise comparisons, the probability of rejecting the null hypothesis when it is true is increased, since the more *t* tests that are conducted, the greater is the likelihood of getting significant differences by chance alone. Third, the more means there are to compare, the more *t* tests are needed. For example, for the comparison of 3 means two at a time, 3 *t* tests are required. For the comparison of 5 means two at a time, 10 tests are required. And for the comparison of 10 means two at a time, 45 tests are required.

Assumptions for the *F* Test for Comparing Three or More Means

1. The populations from which the samples were obtained must be normally or approximately normally distributed.
2. The samples must be independent of one another.
3. The variances of the populations must be equal.

Even though one is comparing three or more means in this use of the *F* test, *variances* are used in the test instead of means.

With the *F* test, two different estimates of the population variance are made. The first estimate is called the **between-group variance,** and it involves finding the variance of the means. The second estimate, the **within-group variance,** is made by computing the variance using all the data and is not affected by differences in the means. If there is no difference in the means, the between-group variance estimate will be approximately equal to the within-group variance estimate, and the *F* test value will be approximately equal to 1. The null hypothesis will not be rejected. However, when the means differ significantly, the between-group variance will be much larger than the within-group variance; the *F* test value will be significantly greater than 1; and the null hypothesis will be rejected. Since variances are compared, this procedure is called *analysis of variance* (ANOVA).

For a test of the difference among three or more means, the following hypotheses should be used:

H_0: $\mu_1 = \mu_2 = \cdots = \mu_n$

H_1: At least one mean is different from the others.

As stated previously, a significant test value means that there is a high probability that this difference in means is not due to chance, but it does not indicate where the difference lies.

The degrees of freedom for this F test are d.f.N. $= k - 1$, where k is the number of groups, and d.f.D. $= N - k$, where N is the sum of the sample sizes of the groups $N = n_1 + n_2 + \cdots + n_k$. The sample sizes need not be equal. The F test to compare means is always right-tailed.

Examples 12–1 and 12–2 illustrate the computational procedure for the ANOVA technique for comparing three or more means, and the steps are summarized in the Procedure Table shown after the examples.

Example 12–1	

A researcher wishes to try three different techniques to lower the blood pressure of individuals diagnosed with high blood pressure. The subjects are randomly assigned to three groups; the first group takes medication, the second group exercises, and the third group follows a special diet. After four weeks, the reduction in each person's blood pressure is recorded. At $\alpha = 0.05$, test the claim that there is no difference among the means. The data are shown.

Medication	**Exercise**	**Diet**
10	6	5
12	8	9
9	3	12
15	0	8
13	2	4
$\overline{X}_1 = 11.8$	$\overline{X}_2 = 3.8$	$\overline{X}_3 = 7.6$
$s_1^2 = 5.7$	$s_2^2 = 10.2$	$s_3^2 = 10.3$

Solution

Step 1 State the hypotheses and identify the claim.

H_0: $\mu_1 = \mu_2 = \mu_3$ (claim)

H_1: At least one mean is different from the others.

Step 2 Find the critical value. Since $k = 3$ and $N = 15$,

d.f.N. $= k - 1 = 3 - 1 = 2$

d.f.D. $= N - k = 15 - 3 = 12$

The critical value is 3.89, obtained from Table H in Appendix C with $\alpha = 0.05$.

Step 3 Compute the test value, using the procedure outlined here.

a. Find the mean and variance of each sample (these values are shown below the data).

b. Find the grand mean. The *grand mean,* denoted by $\overline{X}_{GM}$, is the mean of all values in the samples.

$$\overline{X}_{GM} = \frac{\Sigma X}{N} = \frac{10 + 12 + 9 + \cdots + 4}{15} = \frac{116}{15} = 7.73$$

When samples are equal in size, find $\overline{X}_{GM}$ by summing the $\overline{X}$'s and dividing by k, where $k =$ the number of groups.

c. Find the between-group variance, denoted by s_B^2.

$$s_B^2 = \frac{\sum n_i(\bar{X}_i - \bar{X}_{GM})^2}{k - 1}$$

$$= \frac{5(11.8 - 7.73)^2 + 5(3.8 - 7.73)^2 + 5(7.6 - 7.73)^2}{3 - 1}$$

$$= \frac{160.13}{2} = 80.07$$

Note: This formula finds the variance among the means by using the sample sizes as weights and considers the differences in the means.

d. Find the within-group variance, denoted by s_W^2.

$$s_W^2 = \frac{\sum(n_i - 1)s_i^2}{\sum(n_i - 1)}$$

$$= \frac{(5 - 1)(5.7) + (5 - 1)(10.2) + (5 - 1)(10.3)}{(5 - 1) + (5 - 1) + (5 - 1)}$$

$$= \frac{104.80}{12} = 8.73$$

Note: This formula finds an overall variance by calculating a weighted average of the individual variances. It does not involve using differences of the means.

e. Find the F test value.

$$F = \frac{s_B^2}{s_W^2} = \frac{80.07}{8.73} = 9.17$$

Step 4 Make the decision. The decision is to reject the null hypothesis, since $9.17 > 3.89$.

Step 5 Summarize the results. There is enough evidence to reject the claim and conclude that at least one mean is different from the others.

The numerator of the fraction obtained in step 3, part *c*, of the computational procedure is called the **sum of squares between groups,** denoted by SS_B. The numerator of the fraction obtained in step 3, part *d*, of the computational procedure is called the **sum of squares within groups,** denoted by SS_W. This statistic is also called the *sum of squares for the error.* SS_B is divided by d.f.N. to obtain the between-group variance. SS_W is divided by $N - k$ to obtain the within-group or error variance. These two variances are sometimes called **mean squares,** denoted by MS_B and MS_W. These terms are used to summarize the analysis of variance and are placed in a summary table, as shown in Table 12–1.

Table 12–1 **Analysis of Variance Summary Table**

Source	Sum of squares	d.f.	Mean square	F
Between	SS_B	$k - 1$	MS_B	
Within (error)	SS_W	$N - k$	MS_W	
Total				

In the table,

$$SS_B = \text{sum of squares between groups}$$

$$SS_W = \text{sum of squares within groups}$$

$$k = \text{number of groups}$$

$$N = n_1 + n_2 + \cdots + n_k = \text{sum of sample sizes for groups}$$

$$MS_B = \frac{SS_B}{k-1}$$

$$MS_W = \frac{SS_W}{N-k}$$

$$F = \frac{MS_B}{MS_W}$$

The totals are obtained by adding the corresponding columns. For Example 12–1, the **ANOVA summary table** is shown in Table 12–2.

Table 12–2	Analysis of Variance Summary Table for Example 12-1			
Source	Sum of squares	d.f.	Mean square	F
Between	160.13	2	80.07	9.17
Within (error)	104.80	12	8.73	
Total	264.93	14		

Most computer programs will print out an ANOVA summary table.

Example 12–2

A state employee wishes to see if there is a significant difference in the number of employees at the interchanges of three state toll roads. The data are shown. At $\alpha = 0.05$, can it be concluded that there is a significant difference in the average number of employees at each interchange?

Pennsylvania Turnpike	Greensburg Bypass/ Mon-Fayette Expressway	Beaver Valley Expressway
7	10	1
14	1	12
32	1	1
19	0	9
10	11	1
11	1	11
$\bar{X}_1 = 15.5$	$\bar{X}_2 = 4.0$	$\bar{X}_3 = 5.8$
$s_1^2 = 81.9$	$s_2^2 = 25.6$	$s_3^2 = 29.0$

Source: Pennsylvania Turnpike Commission.

Solution

Step 1 State the hypotheses and identify the claim.

$$H_0: \mu_1 = \mu_2 = \mu_3$$

H_1: At least one mean is different from the others (claim)

Step 2 Find the critical value. Since $k = 3$, $N = 18$, and $\alpha = 0.05$,

d.f.N. $= k - 1 = 3 - 1 = 2$

d.f.D. $= N - k = 18 - 3 = 15$

The critical value is 3.68.

Step 3 Compute the test value.

a. Find the mean and variance of each sample (these values are shown below the data columns in the example).

b. Find the grand mean.

$$\bar{X}_{\text{GM}} = \frac{\Sigma X}{N} = \frac{7 + 14 + 32 + \cdots + 11}{18} = \frac{152}{18} = 8.4$$

c. Find the between-group variance.

$$s_B^2 = \frac{\Sigma n_i(\bar{X}_i - \bar{X}_{\text{GM}})^2}{k - 1}$$

$$= \frac{6(15.5 - 8.4)^2 + 6(4 - 8.4)^2 + 6(5.8 - 8.4)^2}{3 - 1}$$

$$= \frac{459.18}{2} = 229.59$$

d. Find the within-group variance.

$$s_W^2 = \frac{\Sigma(n_i - 1)s_i^2}{\Sigma(n_i - 1)}$$

$$= \frac{(6 - 1)(81.9) + (6 - 1)(25.6) + (6 - 1)(29.0)}{(6 - 1) + (6 - 1) + (6 - 1)}$$

$$= \frac{682.5}{15} = 45.5$$

e. Find the F test value.

$$F = \frac{s_B^2}{s_W^2} = \frac{229.59}{45.5} = 5.05$$

Step 4 Make the decision. Since $5.05 > 3.68$, the decision is to reject the null hypothesis.

Step 5 Summarize the results. There is enough evidence to support the claim that there is a difference among the means. The ANOVA summary table for this example is shown in Table 12–3.

Table 12–3	Analysis of Variance Summary Table for Example 12-2			
Source	**Sum of squares**	**d.f.**	**Mean square**	**F**
Between	459.18	2	229.59	5.05
Within	682.5	15	45.5	
Total	1141.68	17		

The steps for computing the F test value for the ANOVA are summarized in this Procedure Table.

Procedure Table

Finding the F Test Value for the Analysis of Variance

Step 1 Find the mean and variance of each sample:

$$(\overline{X}_1, s_1^2), (\overline{X}_2, s_2^2), \ldots, (\overline{X}_k, s_k^2)$$

Step 2 Find the grand mean.

$$\overline{X}_{GM} = \frac{\Sigma X}{N}$$

Step 3 Find the between-group variance.

$$s_B^2 = \frac{\Sigma n_i(\overline{X}_i - \overline{X}_{GM})^2}{k - 1}$$

Step 4 Find the within-group variance.

$$s_W^2 = \frac{\Sigma(n_i - 1)s_i^2}{\Sigma(n_i - 1)}$$

Step 5 Find the F test value.

$$F = \frac{s_B^2}{s_W^2}$$

The degrees of freedom are

d.f.N. $= k - 1$

where k is the number of groups, and

d.f.D. $= N - k$

where N is the sum of the sample sizes of the groups,

$$N = n_1 + n_2 + \cdots + n_k$$

The *P*-values for ANOVA are found by using the procedure shown in Section 9–3. For Example 12–2, find the two α values in the tables for the *F* distribution (Table H), using d.f.N. = 2 and d.f.D. = 15, where $F = 5.05$ falls between. In this case, 5.05 falls between 4.77 and 6.36, corresponding, respectively, to $\alpha = 0.025$ and $\alpha = 0.01$; hence, $0.01 < P\text{-value} < 0.025$. Since the *P*-value is between 0.01 and 0.025 and since *P*-value < 0.05 (the originally chosen value for α), the decision is to reject the null hypothesis. (The *P*-value obtained from a calculator is 0.021.)

Technology *Step by Step*

MINITAB
Step by Step

One-Way Analysis of Variance (ANOVA)

Which treatment is more effective in lowering cholesterol—medication, diet, or exercise?

1. Enter the data for Example 12–1 into columns of MINITAB.
2. Name the columns **Medication, Exercise,** and **Diet.**
3. Select **Stat>ANOVA>One-Way (Unstacked).**
4. Drag the mouse over the three columns in the list box and then click [Select].
5. Click [OK]. In the session window the ANOVA table will be displayed, showing the test statistic $F = 9.17$ whose *P*-value is 0.004.

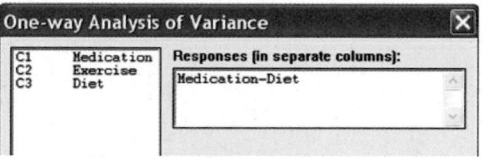

One-Way ANOVA: Medication, Exercise, Diet

```
Source   DF       SS      MS      F      P
Factor    2   160.13   80.07   9.17   0.004
Error    12   104.80    8.73
Total    14   264.93
```

```
                                 Individual 95% CIs For Mean
                                 Based on Pooled StDev
Level        N     Mean   StDev  -------+---------+---------+---------+--
Medication   5   11.800   2.387                        (--------*-------)
Exercise     5    3.800   3.194  (-------*-------)
Diet         5    7.600   3.209           (-------*-------)
                                 -------+---------+---------+---------+--
                                    3.5       7.0      10.5      14.0
Pooled StDev = 2.955
```

Reject the null hypothesis. There is enough evidence to conclude that there is a difference between the treatments. Section 12–3 will explain.

TI-83 Plus or TI-84 Plus
Step by Step

One-Way Analysis of Variance (ANOVA)

1. Enter the data into L_1, L_2, L_3, etc.
2. Press **STAT** and move the cursor to TESTS.
3. Press **F (ALPHA COS)** for ANOVA(.
4. Type each list followed by a comma. End with) and press **ENTER.**

Example TI12–1

Test the claim H_0: $\mu_1 = \mu_2 = \mu_3$ at $\alpha = 0.05$ for these data from Example 12–1:

Medication	Exercise	Diet
10	6	5
12	8	9
9	3	12
15	0	8
13	2	4

Input

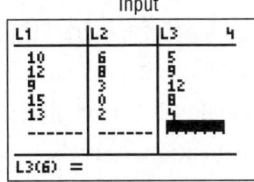

Input

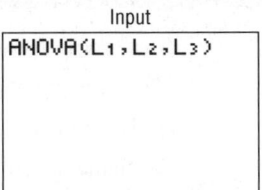

Output

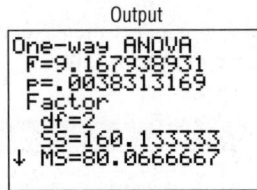

Output

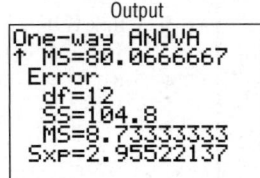

The F test value is 9.167938931. The P-value is 0.0038313169, which is significant at $\alpha = 0.05$. The factor variable has

d.f. = 2

SS = 160.133333

MS = 80.0666667

The error has

d.f. = 12

SS = 104.8

MS = 8.73333333

Excel
Step by Step

One-Way Analysis of Variance (ANOVA)

Example XL12–1

1. Enter this data set in columns A, B, and C.

9	8	12
6	7	15
15	12	18
4	3	9
3	5	10

2. Select **Tools>Data Analysis** and choose Anova: Single Factor.

3. Enter the data range for the three columns.

ANOVA: Single-Factor
Dialog Box and Worksheet

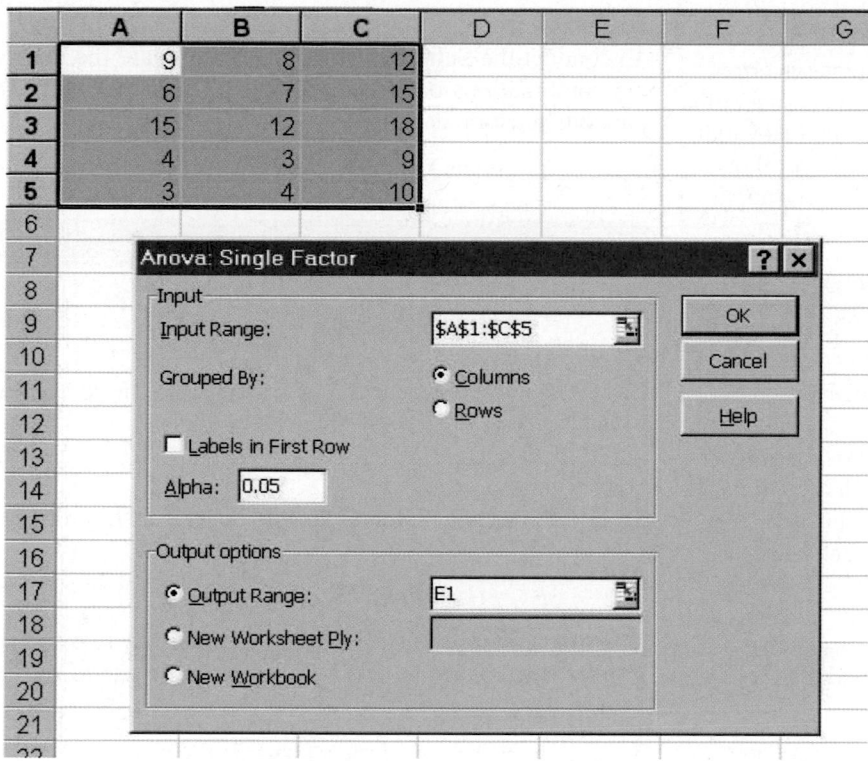

4. Set an alpha level; here we enter **0.05.**

5. Select a location for output and click [OK].

The results are shown.

Anova: Single Factor						
SUMMARY						
Groups	Count	Sum	Average	Variance		
Column 1	5	37	7.4	23.3		
Column 2	5	35	7	11.5		
Column 3	5	64	12.8	13.7		
ANOVA						
Source of Variation	SS	df	MS	F	P-value	F crit
Between Groups	104.9333	2	52.46667	3.245361	0.074708	3.88529
Within Groups	194	12	16.16667			
Total	298.9333	14				

12–3 The Scheffé Test and the Tukey Test

When the null hypothesis is rejected using the F test, the researcher may want to know where the difference among the means is. Several procedures have been developed to determine where the significant differences in the means lie after the ANOVA procedure has been performed. Among the most commonly used tests are the *Scheffé test* and the *Tukey test.*

Objective 2

Determine which means differ, using the Scheffé or Tukey test if the null hypothesis is rejected in the ANOVA.

Scheffé Test

To conduct the **Scheffé test,** one must compare the means two at a time, using all possible combinations of means. For example, if there are three means, the following comparisons must be done:

$$\overline{X}_1 \text{ versus } \overline{X}_2 \qquad \overline{X}_1 \text{ versus } \overline{X}_3 \qquad \overline{X}_2 \text{ versus } \overline{X}_3$$

Formula for the Scheffé Test

$$F_S = \frac{(\overline{X}_i - \overline{X}_j)^2}{s_W^2[(1/n_i) + (1/n_j)]}$$

where $\overline{X}_i$ and $\overline{X}_j$ are the means of the samples being compared, n_i and n_j are the respective sample sizes, and s_W^2 is the within-group variance.

To find the critical value F' for the Scheffé test, multiply the critical value for the F test by $k - 1$:

$$F' = (k - 1)(\text{C.V.})$$

There is a significant difference between the two means being compared when F_S is greater than F'. Example 12–3 illustrates the use of the Scheffé test.

Example 12–3

Using the Scheffé test, test each pair of means in Example 12–1 to see whether a specific difference exists, at $\alpha = 0.05$.

Solution

a. For $\overline{X}_1$ versus $\overline{X}_2$,

$$F_S = \frac{(\overline{X}_1 - \overline{X}_2)^2}{s_W^2[(1/n_1) + (1/n_2)]} = \frac{(11.8 - 3.8)^2}{8.73[(1/5) + (1/5)]} = 18.33$$

b. For $\overline{X}_2$ versus $\overline{X}_3$,

$$F_S = \frac{(\overline{X}_2 - \overline{X}_3)^2}{s_W^2[(1/n_2) + (1/n_3)]} = \frac{(3.8 - 7.6)^2}{8.73[(1/5) + (1/5)]} = 4.14$$

c. For $\overline{X}_1$ versus $\overline{X}_3$,

$$F_S = \frac{(\overline{X}_1 - \overline{X}_3)^2}{s_W^2[(1/n_1) + (1/n_3)]} = \frac{(11.8 - 7.6)^2}{8.73[(1/5) + (1/5)]} = 5.05$$

The critical value for the analysis of variance for Example 12–1 was 3.89, found by using Table H with $\alpha = 0.05$, d.f.N. $= k - 1 = 2$, and d.f.D. $= N - k = 12$. In this case, it is multiplied by $k - 1$ as shown.

The critical value for F' at $\alpha = 0.05$, with d.f.N. $= 2$ and d.f.D. $= 12$, is

$$F' = (k - 1)(\text{C.V.}) = (3 - 1)(3.89) = 7.78$$

Since only the F test value for part a ($\overline{X}_1$ versus $\overline{X}_2$) is greater than the critical value, 7.78, the only significant difference is between $\overline{X}_1$ and $\overline{X}_2$, that is, between medication and exercise.

This study involved three groups. The results showed that patients in all three groups felt better after 2 years. State possible null and alternative hypotheses for this study. Was the null hypothesis rejected? Explain how the statistics could have been used to arrive at the conclusion.

HEALTH

TRICKING KNEE PAIN

You sign up for a clinical trial of arthroscopic surgery used to relieve knee pain caused by arthritis. You're sedated and wake up with tiny incisions. Soon your bum knee feels better. Two years later you find out you had "placebo" surgery. In a study at the Houston VA Medical Center, researchers divided 180 patients into three groups: two groups had damaged cartilage removed, while the third got simulated surgery. Yet an equal number of patients in all groups felt better after two years. Some 650,000 people have the surgery annually, but they're wasting their money, says Dr. Nelda P. Wray, who led the study. And the patients who got fake surgery? "They aren't angry at us," she says. "They still report feeling better."

— STEPHEN P. WILLIAMS

On occasion, when the *F* test value is greater than the critical value, the Scheffé test may not show any significant differences in the pairs of means. This result occurs because the difference may actually lie in the average of two or more means when compared with the other mean. The Scheffé test can be used to make these types of comparisons, but the technique is beyond the scope of this book.

Tukey Test

The **Tukey test** can also be used after the analysis of variance has been completed to make pairwise comparisons between means when the groups have the same sample size. The symbol for the test value in the Tukey test is *q*.

Formula for the Tukey Test

$$q = \frac{\bar{X}_i - \bar{X}_j}{\sqrt{s_W^2/n}}$$

where $\bar{X}_i$ and $\bar{X}_j$ are the means of the samples being compared, *n* is the size of the samples, and s_W^2 is the within-group variance.

When the absolute value of q is greater than the critical value for the Tukey test, there is a significant difference between the two means being compared. The procedures for finding q and the critical value from Table N in Appendix C for the Tukey test are shown in Example 12–4.

Example 12–4

Using the Tukey test, test each pair of means in Example 12–1 to see whether a specific difference exists, at $\alpha = 0.05$.

Solution

a. For $\bar{X}_1$ versus $\bar{X}_2$,

$$q = \frac{\bar{X}_1 - \bar{X}_2}{\sqrt{s_W^2/n}} = \frac{11.8 - 3.8}{\sqrt{8.73/5}} = \frac{8}{1.32} = 6.06$$

b. For $\bar{X}_1$ versus $\bar{X}_3$,

$$q = \frac{\bar{X}_1 - \bar{X}_3}{\sqrt{s_W^2/n}} = \frac{11.8 - 7.6}{\sqrt{8.73/5}} = \frac{4.2}{1.32} = 3.18$$

c. For $\bar{X}_2$ versus $\bar{X}_3$,

$$q = \frac{\bar{X}_2 - \bar{X}_3}{\sqrt{s_W^2/n}} = \frac{3.8 - 7.6}{\sqrt{8.73/5}} = \frac{-3.8}{1.32} = -2.88$$

To find the critical value for the Tukey test, use Table N in Appendix C. The number of means k is found in the row at the top, and the degrees of freedom for s_W^2 are found in the left column (denoted by v). Since $k = 3$, d.f. $= 12$, and $\alpha = 0.05$, the critical value is 3.77. See Figure 12–1. Hence, the only q value that is greater in absolute value than the critical value is the one for the difference between $\bar{X}_1$ and $\bar{X}_2$. The conclusion, then, is that there is a significant difference in means for medication and exercise. These results agree with the Scheffé analysis.

Figure 12–1

Finding the Critical
Value in Table N for
the Tukey Test
(Example 12–4)

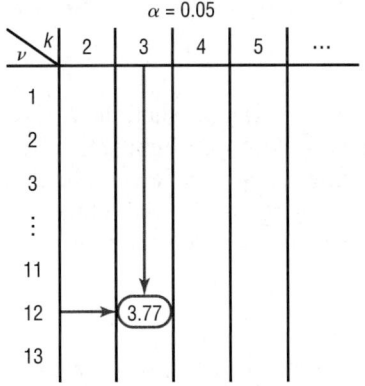

The student might wonder why there are two different tests that can be used after the ANOVA. Actually, there are several other tests that can be used in addition to the Scheffé and Tukey tests. It is up to the researcher to select the most appropriate test. The Scheffé test is the most general, and it can be used when the samples are of different sizes. Furthermore, the Scheffé test can be used to make comparisons such as the average of $\bar{X}_1$ and $\bar{X}_2$ compared with $\bar{X}_3$. However, the Tukey test is more powerful than the Scheffé test for making pairwise comparisons for the means. A rule of thumb for pairwise comparisons is to use the Tukey test when the samples are equal in size and the Scheffé test when the samples differ in size. This rule will be followed in this textbook.

Applying the Concepts **12–3**

Colors That Make You Smarter

The following set of data values was obtained from a study of people's perceptions on whether the color of a person's clothing is related to how intelligent the person looks. The subjects rated the person's intelligence on a scale of 1 to 10. Group 1 subjects were randomly shown people with clothing in shades of blue and gray. Group 2 subjects were randomly shown people with clothing in shades of brown and yellow. Group 3 subjects were randomly shown people with clothing in shades of pink and orange. The results follow.

Group 1	Group 2	Group 3
8	7	4
7	8	9
7	7	6
7	7	7
8	5	9
8	8	8
6	5	5
8	8	8
8	7	7
7	6	5
7	6	4
8	6	5
8	6	4

1. Use ANOVA to test for any significant differences between the means.
2. What is the purpose of this study?
3. Could the Scheffé test be used to test all possible pairwise comparisons?
4. Use the Tukey test to test all possible pairwise comparisons.
5. Are there any contradictions in the results?
6. Explain why separate *t* tests are not accepted in this situation.
7. When would Tukey's test be preferred over the Scheffé method? Explain.

See page 658 for the answers.

Exercises 12–3

1. What test is used to compare three or more means?

2. State three reasons why multiple *t* tests cannot be used to compare three or more means.

3. What are the assumptions for ANOVA?

4. Define between-group variance and within-group variance.

5. What is the *F* test formula for comparing three or more means?

6. State the hypotheses used in the ANOVA test.

7. What two tests are used to compare individual means if the null hypothesis is rejected in using the ANOVA technique?

If the null hypothesis is rejected in Exercises 8 through 19, use the Scheffé test when the sample sizes are unequal to test the differences between the means, and use the Tukey test when the sample sizes are equal. Assume that all variables are normally distributed, that the samples are independent, and that the population variances are equal. Also, for each exercise, perform the following steps.

   a. State the hypotheses and identify the claim.
   b. Find the critical value.
   c. Compute the test value.
   d. Make the decision.
   e. Summarize the results, and explain where the differences in the means are.

Use the traditional method of hypothesis testing unless otherwise specified.

8. The amount of sodium (in milligrams) in one serving for a random sample of three different kinds of foods is listed here. At the 0.05 level of significance, is there sufficient evidence to conclude that a difference in mean sodium amounts exists among condiments, cereals, and desserts?

Condiments	Cereals	Desserts
270	260	100
130	220	180
230	290	250
180	290	250
80	200	300
70	320	360
200	140	300
		160

Source: *The Doctor's Pocket Calorie, Fat, and Carbohydrate Counter.*

9. The number of grams of fiber per serving for a random sample of three different kinds of foods is listed. Is there sufficient evidence at the 0.05 level of significance to conclude that there is a difference in mean fiber content among breakfast cereals, fruits, and vegetables?

Breakfast cereals	Fruits	Vegetables
3	5.5	10
4	2	1.5
6	4.4	3.5
4	1.6	2.7
10	3.8	2.5
5	4.5	6.5
6	2.8	4
8		3
5		

Source: *The Doctor's Pocket Calorie, Fat, and Carbohydrate Counter.*

10. A random sample of enrollments from public institutions of higher learning (with enrollments under 10,000) is shown. At $\alpha = 0.10$, test the claim that the mean enrollments are the same in all parts of the country.

West	Midwest	Northeast	South
3737	5585	9264	3903
3706	8205	4673	4539
2457	4170	7320	2649
4309	5440	5401	6414
4103	3355	6050	2935
5048	4412	4087	7147
2463	5115	1579	3354
	8739		8669
			2431

Source: *N.Y. Times Almanac.*

11. The lengths (in feet) of a random sample of suspension bridges in the United States, Europe, and Asia are shown. At $\alpha = 0.05$, is there sufficient evidence to conclude that there is a difference in mean lengths?

United States	Europe	Asia
4260	5238	6529
3500	4626	4543
2300	4347	3668
2000	3300	3379
1850		2874

Source: *N.Y. Times Almanac.*

12. A researcher wishes to see whether there is any difference in the weight gains of athletes following one of three special diets. Athletes are randomly assigned to three groups and placed on the diet for 6 weeks. The weight gains (in pounds) are shown here. At $\alpha = 0.05$, can the researcher conclude that there is a difference in the diets?

Diet A	Diet B	Diet C
3	10	8
6	12	3
7	11	2
4	14	5
	8	
	6	

A computer printout for this problem is shown. Use the *P*-value method and the information in this printout to test the claim.

Computer Printout for Exercise 12

```
ANALYSIS OF VARIANCE SOURCE TABLE
Source          df    Sum of Squares    Mean Square      F      P-value

Bet Groups      2        101.095          50.548       7.740    0.00797
W/I Groups      11        71.833           6.530

Total           13       172.929

DESCRIPTIVE STATISTICS
Condit          N           Means         St Dev

diet A          4           5.000         1.826
diet B          6          10.167         2.858
diet C          4           4.500         2.646
```

13. The per-pupil costs (in thousands of dollars) for cyber charter school tuition for school districts in three areas of southwestern Pennsylvania are shown. At $\alpha = 0.05$, is there a difference in the means? If so, give a possible reason for the difference.

Area I	Area II	Area III
6.2	7.5	5.8
9.3	8.2	6.4
6.8	8.5	5.6
6.1	8.2	7.1
6.7	7.0	3.0
6.9	9.3	3.5

Source: *Tribune-Review.*

14. Random samples of populations (in thousands) of U.S. cities in three sections of the country are listed. Is there evidence to conclude a difference in means at the 0.01 level of significance?

Eastern third	Middle third	Western third
187	217	276
146	478	110
362	277	177
186	113	122
174	128	134
248	150	137
335	198	338
		255

Source: *N.Y. Times Almanac.*

15. The numbers (in thousands) of farms per state found in three sections of the country are listed next. Test the claim at $\alpha = 0.05$ that the mean number of farms is the same across these three geographic divisions.

Eastern third	Middle third	Western third
48	95	29
57	52	40
24	64	40
10	64	68
38		

Source: *N.Y. Times Almanac.*

16. The expenditures (in dollars) per pupil for states in three sections of the country are listed. Using $\alpha = 0.05$, can one conclude that there is a difference in means?

Eastern third	Middle third	Western third
4946	6149	5282
5953	7451	8605
6202	6000	6528
7243	6479	6911
6113		

Source: *N.Y. Times Almanac.*

17. A research organization tested microwave ovens. At $\alpha = 0.10$, is there a significant difference in the average prices of the three types of oven?

Watts		
1000	**900**	**800**
270	240	180
245	135	155
190	160	200
215	230	120
250	250	140
230	200	180
	200	140
	210	130

A computer printout for this exercise is shown. Use the *P*-value method and the information in this printout to test the claim.

Computer Printout for Exercise 17

```
ANALYSIS OF VARIANCE SOURCE TABLE
Source          df      Sum of Squares      Mean Square        F        P-value

Bet Groups      2        21729.735          10864.867        10.118     0.00102
W/I Groups      19       20402.083          1073.794

Total           21       42131.818

DESCRIPTIVE STATISTICS
Condit          N                 Means             St Dev

  1000          6               233.333            28.23
   900          8               203.125            39.36
   800          8               155.625            28.21
```

18. Three random samples of times (in minutes) that commuters are stuck in traffic are shown. At $\alpha = 0.05$, is there a difference in the mean times among the three cities? What factor might have influenced the results of the study?

Dallas	Boston	Detroit
59	54	53
62	52	56
58	55	54
63	58	49
61	53	52

Source: Based on information from Texas Transportation Institute.

19. The data consist of the number of pupils who were sent to alternative forms of education for schools in four different counties. At $\alpha = 0.01$, is there a difference in the means? Give a few reasons why some people would be enrolled in an alternative type of school.

County A	County B	County C	County D
2	6	4	0
0	0	0	3
8	1	2	0
1	5	3	1
0	3	2	1

12–4   Two-Way Analysis of Variance

Objective 3

Use the two-way ANOVA technique to determine if there is a significant difference in the main effects or interaction.

The analysis of variance technique shown previously is called a **one-way ANOVA** since there is only *one independent variable*. The **two-way ANOVA** is an extension of the one-way analysis of variance; it involves *two independent variables*. The independent variables are also called **factors.**

The two-way analysis of variance is quite complicated, and many aspects of the subject should be considered when one is using a research design involving a two-way ANOVA. For the purposes of this textbook, only a brief introduction to the subject will be given.

In doing a study that involves a two-way analysis of variance, the researcher is able to test the effects of two independent variables or factors on one *dependent variable*. In addition, the interaction effect of the two variables can be tested.

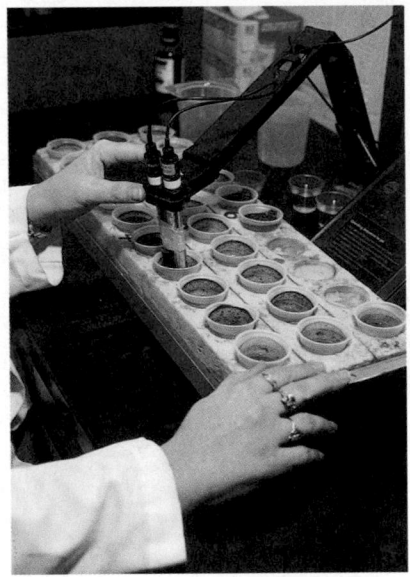

For example, suppose a researcher wishes to test the effects of two different types of plant food and two different types of soil on the growth of certain plants. The two independent variables are the type of plant food and the type of soil, while the dependent variable is the plant growth. Other factors, such as water, temperature, and sunlight, are held constant.

To conduct this experiment, the researcher sets up four groups of plants. See Figure 12–2.

Figure 12–2

Treatment Groups for the Plant Food–Soil Type Experiment

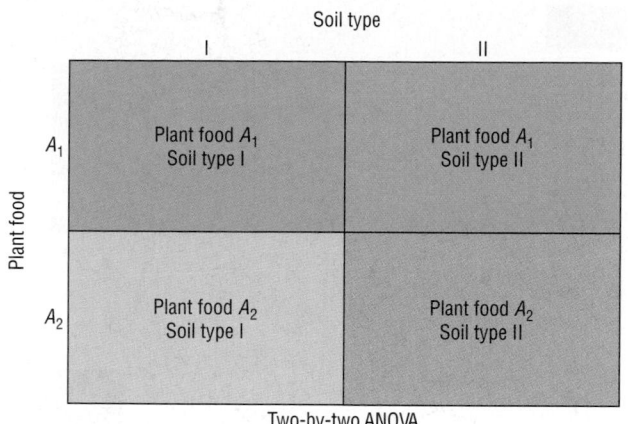

Assume that the plant food type is designated by the letters A_1 and A_2 and the soil type by the Roman numerals I and II. The groups for such a two-way ANOVA are sometimes called **treatment groups.** The four groups are

Group 1	Plant food A_1, soil type I
Group 2	Plant food A_1, soil type II
Group 3	Plant food A_2, soil type I
Group 4	Plant food A_2, soil type II

The plants are assigned to the groups at random. This design is called a 2×2 (read "two-by-two") design, since each variable consists of two **levels,** that is, two different treatments.

The two-way ANOVA enables the researcher to test the effects of the plant food and the soil type in a single experiment rather than in separate experiments involving the plant food alone and the soil type alone. Furthermore, the researcher can test an additional hypothesis about the effect of the *interaction* of the two variables—plant food and soil type—on plant growth. For example, is there a difference between the growth of plants using plant food A_1 and soil type II and the growth of plants using plant food A_2 and soil type I? When a difference of this type occurs, the experiment is said to have a significant **interaction effect.** That is, the types of plant food affect the plant growth differently in different soil types.

There are many different kinds of two-way ANOVA designs, depending on the number of levels of each variable. Figure 12–3 shows a few of these designs. As stated previously, the plant food–soil type experiment uses a 2×2 ANOVA.

The design in Figure 12–3(a) is called a 3×2 design, since the factor in the rows has three levels and the factor in the columns has two levels. Figure 12–3(b) is a 3×3 design, since each factor has three levels. Figure 12–3(c) is a 4×3 design.

The two-way ANOVA design has several null hypotheses. There is one for each independent variable and one for the interaction. In the plant food–soil type problem, the hypotheses are as follows:

1. H_0: There is no interaction effect between type of plant food used and type of soil used on plant growth.

   H_1: There is an interaction effect between food type and soil type on plant growth.

2. H_0: There is no difference in means of heights of plants grown using different foods.

   H_1: There is a difference in means of heights of plants grown using different foods.

Figure 12–3

Some Types of
Two-Way ANOVA
Designs

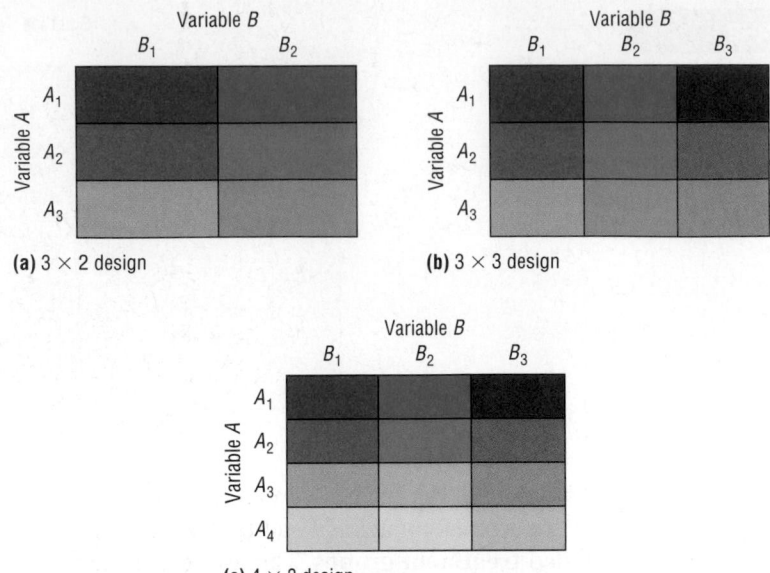

(a) 3 × 2 design

(b) 3 × 3 design

(c) 4 × 3 design

3. H_0: There is no difference in means of heights of plants grown in different soil
   types.

   H_1: There is a difference in means of heights of plants grown in different soil types.

The first set of hypotheses concerns the interaction effect; the second and third sets test
the effects of the independent variables, which are sometimes called the **main effects.**

As with the one-way ANOVA, a between-group variance estimate is calculated, and
a within-group variance estimate is calculated. An F test is then performed for each of
the independent variables and the interaction. The results of the two-way ANOVA are
summarized in a two-way table, as shown in Table 12–4 for the plant experiment.

Table 12–4	ANOVA Summary Table for Plant Food and Soil Type			
Source	**Sum of squares**	**d.f.**	**Mean square**	**F**
Plant food				
Soil type				
Interaction				
Within (error)				
Total				

In general, the two-way **ANOVA summary table** is set up as shown in Table 12–5.

Table 12–5	ANOVA Summary Table			
Source	**Sum of squares**	**d.f.**	**Mean square**	**F**
A	SS_A	$a - 1$	MS_A	F_A
B	SS_B	$b - 1$	MS_B	F_B
$A \times B$	$SS_{A \times B}$	$(a - 1)(b - 1)$	$MS_{A \times B}$	$F_{A \times B}$
Within (error)	SS_W	$ab(n - 1)$	MS_W	
Total				

In the table,

$$SS_A = \text{sum of squares for factor } A$$
$$SS_B = \text{sum of squares for factor } B$$
$$SS_{A \times B} = \text{sum of squares for interaction}$$
$$SS_W = \text{sum of squares for error term (within-group)}$$
$$a = \text{number of levels of factor } A$$
$$b = \text{number of levels of factor } B$$
$$n = \text{number of subjects in each group}$$

$$MS_A = \frac{SS_A}{a - 1}$$

$$MS_B = \frac{SS_B}{b - 1}$$

$$MS_{A \times B} = \frac{SS_{A \times B}}{(a - 1)(b - 1)}$$

$$MS_W = \frac{SS_W}{ab(n - 1)}$$

$$F_A = \frac{MS_A}{MS_W} \qquad \text{with d.f.N.} = a - 1, \text{d.f.D.} = ab(n - 1)$$

$$F_B = \frac{MS_B}{MS_W} \qquad \text{with d.f.N.} = b - 1, \text{d.f.D.} = ab(n - 1)$$

$$F_{A \times B} = \frac{MS_{A \times B}}{MS_W} \qquad \text{with d.f.N.} = (a - 1)(b - 1), \text{d.f.D.} = ab(n - 1)$$

The assumptions for the two-way analysis of variance are basically the same as those for the one-way ANOVA, except for sample size.

Assumptions for the Two-Way ANOVA

1. The populations from which the samples were obtained must be normally or approximately normally distributed.
2. The samples must be independent.
3. The variances of the populations from which the samples were selected must be equal.
4. The groups must be equal in sample size.

The computational procedure for the two-way ANOVA is quite lengthy. For this reason, it will be omitted in Example 12–5, and only the two-way ANOVA summary table will be shown. The table used in Example 12–5 is similar to the one generated by most computer programs. The student should be able to interpret the table and summarize the results.

Example 12–5

A researcher wishes to see whether the type of gasoline used and the type of automobile driven have any effect on gasoline consumption. Two types of gasoline, regular and high-octane, will be used, and two types of automobiles, two-wheel- and four-wheel-drive, will be used in each group. There will be two automobiles in each group, for a total of eight automobiles used. Using a two-way analysis of variance, the researcher will perform the following steps.

Step 1 State the hypotheses.

Step 2 Find the critical value for each F test, using $\alpha = 0.05$.

Step 3 Complete the summary table to get the test value.

Step 4 Make the decision.

Step 5 Summarize the results.

 The data (in miles per gallon) are shown here, and the summary table is given in Table 12–6.

	Type of automobile	
Gas	**Two-wheel-drive**	**Four-wheel-drive**
Regular	26.7	28.6
	25.2	29.3
High-octane	32.3	26.1
	32.8	24.2

Table 12–6	ANOVA Summary Table for Example 12–5				
Source	**SS**	**d.f.**	**MS**	***F***	
Gasoline *A*	3.920				
Automobile *B*	9.680				
Interaction ($A \times B$)	54.080				
Within (error)	3.300				
Total	70.980				

Solution

Step 1 State the hypotheses. The hypotheses for the interaction are these:

H_0: There is no interaction effect between type of gasoline used and type of automobile a person drives on gasoline consumption.

H_1: There is an interaction effect between type of gasoline used and type of automobile a person drives on gasoline consumption.

The hypotheses for the gasoline types are:

H_0: There is no difference between the means of gasoline consumption for two types of gasoline.

H_1: There is a difference between the means of gasoline consumption for two types of gasoline.

The hypotheses for the types of automobile driven are:

H_0: There is no difference between the means of gasoline consumption for two-wheel-drive and four-wheel-drive automobiles.

H_1: There is a difference between the means of gasoline consumption for two-wheel-drive and four-wheel-drive automobiles.

Step 2 Find the critical values for each *F* test. In this case, each independent variable, or factor, has two levels. Hence, a 2×2 ANOVA table is used. Factor *A* is designated as the gasoline type. It has two levels, regular and high-octane; therefore, $a = 2$. Factor *B* is designated as the automobile type. It also has two levels; therefore, $b = 2$. The degrees of freedom for each factor are as follows:

$$\text{Factor } A: \quad \text{d.f.N.} = a - 1 = 2 - 1 = 1$$

$$\text{Factor } B: \quad \text{d.f.N.} = b - 1 = 2 - 1 = 1$$

$$\text{Interaction } (A \times B): \quad \text{d.f.N.} = (a - 1)(b - 1)$$

$$= (2 - 1)(2 - 1) = 1 \cdot 1 = 1$$

$$\text{Within (error)}: \quad \text{d.f.D.} = ab(n - 1)$$

$$= 2 \cdot 2(2 - 1) = 4$$

where n is the number of data values in each group. In this case, $n = 2$.

The critical value for the F_A test is found by using $\alpha = 0.05$, d.f.N. = 1, and d.f.D. = 4. In this case, $F_A = 7.71$. The critical value for the F_B test is found by using $\alpha = 0.05$, d.f.N. = 1, and d.f.D. = 4; F_B is also 7.71. Finally, the critical value for the $F_{A \times B}$ test is found by using d.f.N. = 1 and d.f.D. = 4; it is also 7.71.

Note: If there are different levels of the factors, the critical values will not all be the same. For example, if factor A has three levels and factor b has four levels, and if there are two subjects in each group, then the degrees of freedom are as follows:

$$\text{d.f.N.} = a - 1 = 3 - 1 = 2 \qquad \text{factor } A$$

$$\text{d.f.N.} = b - 1 = 4 - 1 = 3 \qquad \text{factor } B$$

$$\text{d.f.N.} = (a - 1)(b - 1) = (3 - 1)(4 - 1)$$

$$= 2 \cdot 3 = 6 \qquad \text{factor } A \times B$$

$$\text{d.f.N.} = ab(n - 1) = 3 \cdot 4(2 - 1) = 12 \qquad \text{within (error) factor}$$

Step 3   Complete the ANOVA summary table to get the test values. The mean squares are computed first.

$$\text{MS}_A = \frac{\text{SS}_A}{a - 1} = \frac{3.920}{2 - 1} = 3.920$$

$$\text{MS}_B = \frac{\text{SS}_B}{b - 1} = \frac{9.680}{2 - 1} = 9.680$$

$$\text{MS}_{A \times B} = \frac{\text{SS}_{A \times B}}{(a - 1)(b - 1)} = \frac{54.080}{(2 - 1)(2 - 1)} = 54.080$$

$$\text{MS}_W = \frac{\text{SS}_W}{ab(n - 1)} = \frac{3.300}{4} = 0.825$$

The F values are computed next.

$$F_A = \frac{\text{MS}_A}{\text{MS}_W} = \frac{3.920}{0.825} = 4.752 \qquad \text{d.f.N.} = a - 1 = 1 \qquad \text{d.f.D.} = ab(n - 1) = 4$$

$$F_B = \frac{\text{MS}_B}{\text{MS}_W} = \frac{9.680}{0.825} = 11.733 \qquad \text{d.f.N.} = b - 1 = 1 \qquad \text{d.f.D.} = ab(n - 1) = 4$$

$$F_{A \times B} = \frac{\text{MS}_{A \times B}}{\text{MS}_W} = \frac{54.080}{0.825} = 65.552 \qquad \text{d.f.N.} = (a - 1)(b - 1) = 1 \qquad \text{d.f.D.} = ab(n - 1) = 4$$

The completed ANOVA table is shown in Table 12–7.

Table 12–7	Completed ANOVA Summary Table for Example 12–5			
Source	SS	d.f.	MS	F
Gasoline A	3.920	1	3.920	4.752
Automobile B	9.680	1	9.680	11.733
Interaction ($A \times B$)	54.080	1	54.080	65.552
Within (error)	3.300	4	0.825	
Total	70.980	7		

Step 4 Make the decision. Since $F_B = 11.733$ and $F_{A \times B} = 65.552$ are greater than the critical value 7.71, the null hypotheses concerning the type of automobile driven and the interaction effect should be rejected.

Step 5 Summarize the results. Since the null hypothesis for the interaction effect was rejected, it can be concluded that the combination of type of gasoline and type of automobile does affect gasoline consumption.

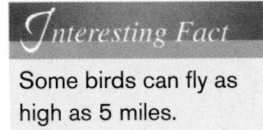

Interesting Fact

Some birds can fly as high as 5 miles.

In the preceding analysis, the effect of the type of gasoline used and the effect of the type of automobile driven are called the *main effects*. If there is no significant interaction effect, the main effects can be interpreted independently. However, if there is a significant interaction effect, the main effects must be interpreted cautiously.

To interpret the results of a two-way analysis of variance, researchers suggest drawing a graph, plotting the means of each group, analyzing the graph, and interpreting the results. In Example 12–5, find the means for each group or cell by adding the data values in each cell and dividing by *n*. The means for each cell are shown in the chart here.

	Type of automobile	
Gas	**Two-wheel-drive**	**Four-wheel-drive**
Regular	$\bar{X} = \dfrac{26.7 + 25.2}{2} = 25.95$	$\bar{X} = \dfrac{28.6 + 29.3}{2} = 28.95$
High-octane	$\bar{X} = \dfrac{32.3 + 32.8}{2} = 32.55$	$\bar{X} = \dfrac{26.1 + 24.2}{2} = 25.15$

The graph of the means for each of the variables is shown in Figure 12–4. In this graph, the lines cross each other. When such an intersection occurs and the interaction is significant, the interaction is said to be a **disordinal interaction.** When there is a disordinal interaction, one should not interpret the main effects without considering the interaction effect.

The other type of interaction that can occur is an *ordinal interaction*. Figure 12–5 shows a graph of means in which an ordinal interaction occurs between two variables. The lines do not cross each other, nor are they parallel. If the *F* test value for the interaction is significant and the lines do not cross each other, then the interaction is said to be an **ordinal interaction** and the main effects can be interpreted independently of each other.

Finally, when there is no significant interaction effect, the lines in the graph will be parallel or approximately parallel. When this situation occurs, the main effects can be interpreted independently of each other because there is no significant interaction.

Figure 12–4

Graph of the Means
of the Variables in
Example 12–5

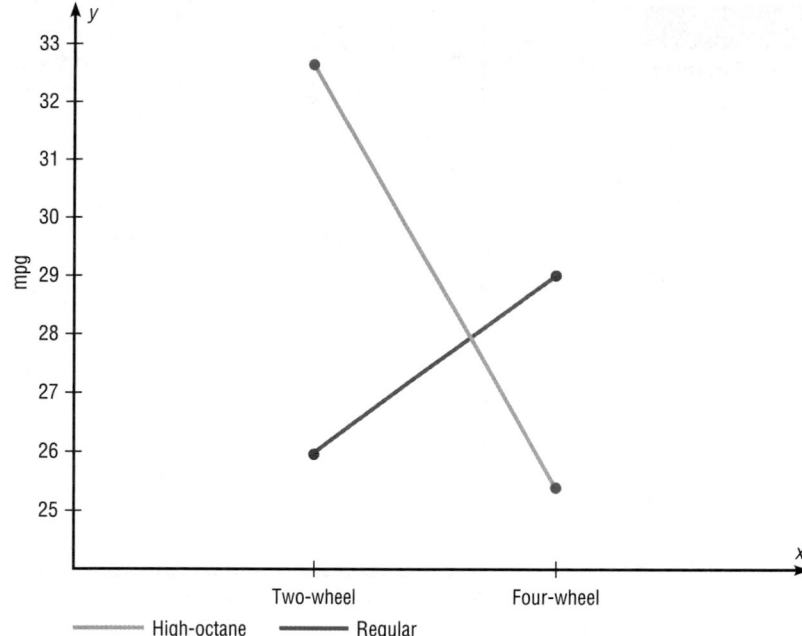

Figure 12–5

Graph of Two Variables
Indicating an Ordinal
Interaction

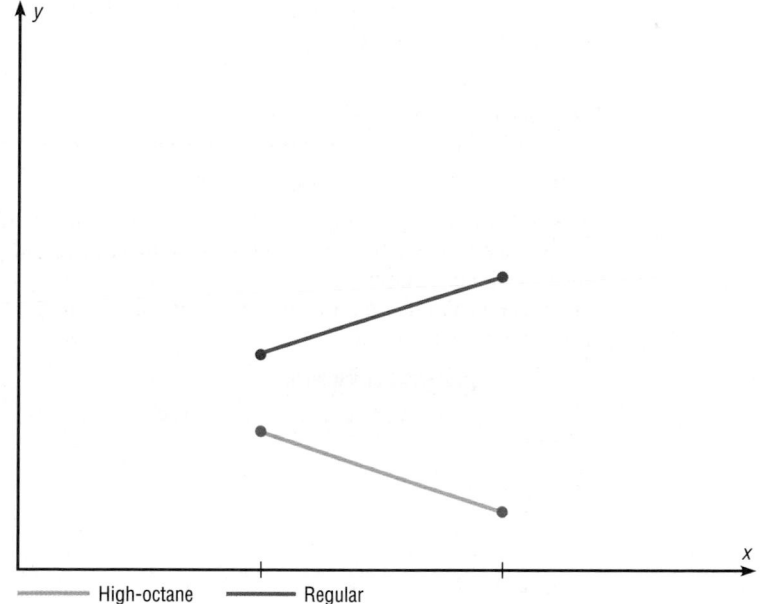

Figure 12–6 shows the graph of two variables when the interaction effect is not significant; the lines are parallel.

Example 12–5 was an example of a 2×2 two-way analysis of variance, since each independent variable had two levels. For other types of variance problems, such as a 3×2 or a 4×3 ANOVA, interpretation of the results can be quite complicated. Procedures using tests such as the Tukey and Scheffé tests for analyzing the cell means exist and are similar to the tests shown for the one-way ANOVA, but they are beyond the scope of this textbook. Many other designs for analysis of variance are available to researchers, such as three-factor designs and repeated-measure designs; they are also beyond the scope of this book.

Figure 12–6

Graph of Two Variables Indicating No Interaction

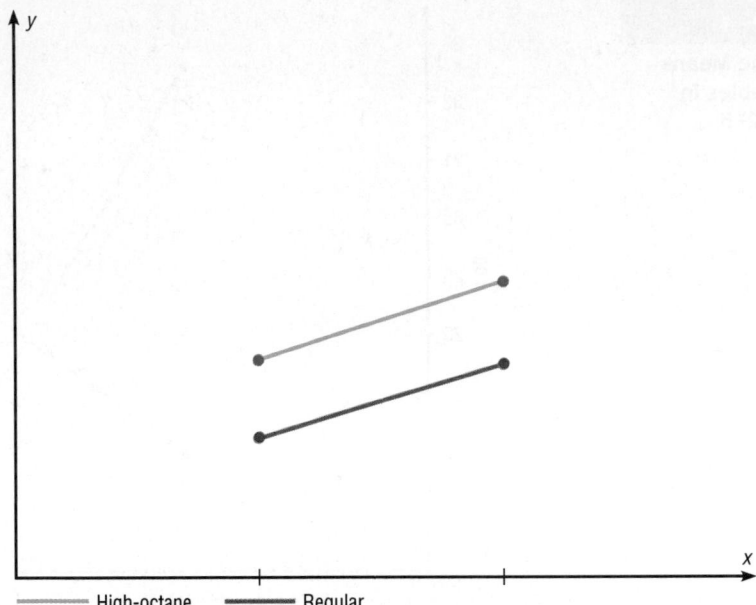

In summary, the two-way ANOVA is an extension of the one-way ANOVA. The former can be used to test the effects of two independent variables and a possible interaction effect on a dependent variable.

Applying the Concepts 12–4

Automobile Sales Techniques

The following outputs are from the result of an analysis of how car sales are affected by the experience of the salesperson and the type of sales technique used. Experience was broken up into four levels, and two different sales techniques were used. Analyze the results and draw conclusions about level of experience with respect to the two different sales techniques and how they affect car sales.

Two-Way Analysis of Variance

```
Analysis of Variance for Sales
Source          DF         SS         MS
Experience       3     3414.0     1138.0
Presentation     1        6.0        6.0
Interaction      3      414.0      138.0
Error           16      838.0       52.4
Total           23     4672.0

                             Individual 95% CI
Experience      Mean     -----+---------+---------+---------+------
1               62.0     (-----*-----)
2               63.0      (-----*-----)
3               78.0                      (-----*-----)
4               91.0                                  (-----*-----)
                         -----+---------+---------+---------+------
                            60.0      70.0      80.0      90.0

                             Individual 95% CI
Presentation    Mean     ------+---------+---------+---------+------
1               74.0          (-----------------*-----------------)
2               73.0     (-----------------*-----------------)
                         ------+---------+---------+---------+------
```

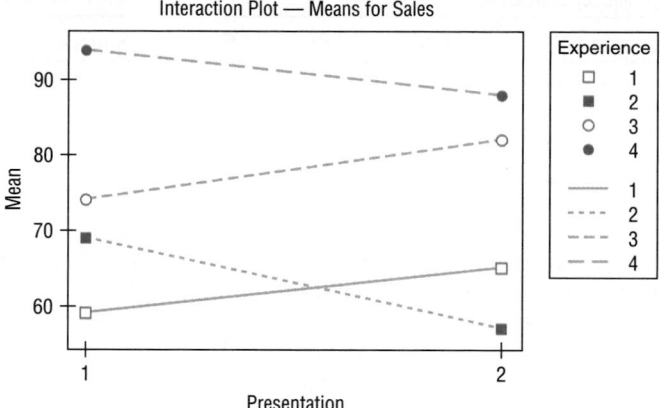

Interaction Plot — Means for Sales

See page 658 for the answers.

Exercises 12–4

1. How does the two-way ANOVA differ from the one-way ANOVA?

2. Explain what is meant by *main effects* and *interaction effect.*

3. How are the values for the mean squares computed?

4. How are the *F* test values computed?

5. In a two-way ANOVA, variable *A* has three levels and variable *B* has two levels. There are five data values in each cell. Find each degrees-of-freedom value.

   a. d.f.N. for factor *A*
   b. d.f.N. for factor *B*
   c. d.f.N. for factor *A* × *B*
   d. d.f.D. for the within (error) factor

6. In a two-way ANOVA, variable *A* has six levels and variable *B* has five levels. There are seven data values in each cell. Find each degrees-of-freedom value.

   a. d.f.N. for factor *A*
   b. d.f.N. for factor *B*
   c. d.f.N. for factor *A* × *B*
   d. d.f.D. for the within (error) factor

7. What are the two types of interactions that can occur in the two-way ANOVA?

8. When can the main effects for the two-way ANOVA be interpreted independently?

9. Describe what the graph of the variables would look like for each situation in a two-way ANOVA experiment.

   a. No interaction effect occurs.
   b. An ordinal interaction effect occurs.
   c. A disordinal interaction effect occurs.

For Exercises 10 through 15, perform these steps. Assume that all variables are normally or approximately normally distributed, that the samples are independent, and that the population variances are equal.

   a. State the hypotheses.
   b. Find the critical value for each *F* test.
   c. Complete the summary table and find the test value.
   d. Make the decision.
   e. Summarize the results. (Draw a graph of the cell means if necessary.)

10. A company wishes to test the effectiveness of its advertising. A product is selected, and two types of ads are written; one is serious and one is humorous. Also the ads are run on both television and radio. Sixteen potential customers are selected and assigned randomly to one of four groups. After seeing or listening to the ad, each customer is asked to rate its effectiveness on a scale of 1 to 20. Various points are assigned for clarity, conciseness, etc. The data are shown here. At $\alpha = 0.01$, analyze the data, using a two-way ANOVA.

Type of ad	Medium	
	Radio	**Television**
Humorous	6, 10, 11, 9	15, 18, 14, 16
Serious	8, 13, 12, 10	19, 20, 13, 17

ANOVA Summary Table for Exercise 10

Source	SS	d.f.	MS	F
Type	10.563			
Medium	175.563			
Interaction	0.063			
Within	66.250			
Total	252.439			

ANOVA Summary Table for Exercise 11

Source	SS	d.f.	MS	F
Time	1800.0			
Diet	242.0			
Interaction	264.5			
Within	279.0			
Total	2585.5			

11. A medical researcher wishes to test the effects of two diets and the time of day on the sodium level in a person's blood. Eight people are randomly selected, and two are randomly assigned to each of the four groups. Analyze the data shown in the tables here, using a two-way ANOVA at $\alpha = 0.05$. The sodium content is measured in milliequivalents per liter.

Time	Diet type I		Diet type II	
8:00 A.M.	135	145	138	141
8:00 P.M.	155	162	171	191

12. A contractor wishes to see whether there is a difference in the time (in days) it takes two subcontractors to build three different types of homes. At $\alpha = 0.05$, analyze the data shown here, using a two-way ANOVA. See above for raw data.

ANOVA Summary Table for Exercise 12

Source	SS	d.f.	MS	F
Subcontractor	1672.553			
Home type	444.867			
Interaction	313.267			
Within	328.800			
Total	2759.487			

Data for Exercise 12

Subcontractor	Home type I	Home type II	Home type III
A	25, 28, 26, 30, 31	30, 32, 35, 29, 31	43, 40, 42, 49, 48
B	15, 18, 22, 21, 17	21, 27, 18, 15, 19	23, 25, 24, 17, 13

13. Two special training programs in outdoor survival are available for army recruits. One lasts one week and the other lasts two weeks. The officer wishes to test the effectiveness of the programs and see whether there are any gender differences. Six subjects are randomly assigned to each of the programs according to gender. After completing the program, each is given a written test on his or her knowledge of survival skills. The test consists of 100 questions. The scores of the groups are shown here. Use $\alpha = 0.10$ and analyze the data, using a two-way ANOVA.

Gender	Duration One week	Duration Two weeks
Female	86, 92, 87, 88, 78, 95	78, 62, 56, 54, 65, 63
Male	52, 67, 53, 42, 68, 71	85, 94, 82, 84, 78, 91

ANOVA Summary Table for Exercise 13

Source	SS	d.f.	MS	F
Gender	57.042			
Duration	7.042			
Interaction	3978.375			
Within	1365.500			
Total	5407.959			

14. Two types of outdoor paint, enamel and latex, were tested to see how long (in months) each lasted before it began to crack, flake, and peel. They were tested in four geographic locations in the United States to study the effects of climate on the paint. At $\alpha = 0.01$, analyze the data shown, using a two-way ANOVA shown on the next page. Each group contained five test panels. See below for raw data.

Data for Exercise 14

Type of paint	Geographic location North	Geographic location East	Geographic location South	Geographic location West
Enamel	60, 53, 58, 62, 57	54, 63, 62, 71, 76	80, 82, 62, 88, 71	62, 76, 55, 48, 61
Latex	36, 41, 54, 65, 53	62, 61, 77, 53, 64	68, 72, 71, 82, 86	63, 65, 72, 71, 63

ANOVA Summary Table for Exercise 14

Source	SS	d.f.	MS	F
Paint type	12.1			
Location	2501.0			
Interaction	268.1			
Within	2326.8			
Total	5108.0			

15. A company sells three items: swimming pools, spas, and saunas. The owner decides to see whether the age of the sales representative and the type of item affect monthly sales. At $\alpha = 0.05$, analyze the data shown,

using a two-way ANOVA. Sales are given in hundreds of dollars for a randomly selected month, and five salespeople were selected for each group.

ANOVA Summary Table for Exercise 15

Source	SS	d.f.	MS	F
Age	168.033			
Product	1,762.067			
Interaction	7,955.267			
Within	2,574.000			
Total	12,459.367			

Data for Exercise 15

Age of salesperson	Product		
	Pool	Spa	Sauna
Over 30	56, 23, 52, 28, 35	43, 25, 16, 27, 32	47, 43, 52, 61, 74
30 or under	16, 14, 18, 27, 31	58, 62, 68, 72, 83	15, 14, 22, 16, 27

Technology *Step by Step*

MINITAB
Step by Step

Two-Way Analysis of Variance

For Example 12–5, how do gasoline type and vehicle type affect gasoline mileage?

1. Enter the data into three columns of a worksheet. The data for this analysis have to be "stacked" as shown.

a) All the gas mileage data are entered in a single column named MPG.

b) The second column contains codes identifying the gasoline type, a 1 for regular or a 2 for high-octane.

c) The third column will contain codes identifying the type of automobile, 1 for two-wheel-drive or 2 for four-wheel-drive.

2. Select **Stat>ANOVA>Two-Way.**

a) Double-click MPG in the list box.

b) Double-click GasCode as Row factor.

c) Double-click TypeCode as Column factor.

d) Check the boxes for Display means, then click [OK].

The session window will contain the results.

↓	C1	C2	C3
	MPG	**GasCode**	**TypeCode**
1	26.7	1	1
2	25.2	1	1
3	32.3	2	1
4	32.8	2	1
5	28.6	1	2
6	29.3	1	2
7	26.1	2	2
8	24.2	2	2

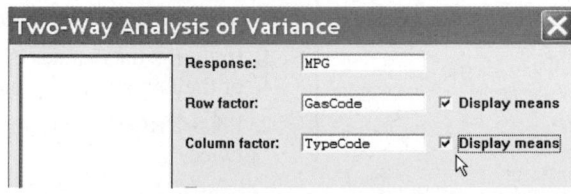

Two-Way ANOVA: MPG versus GasCode, TypeCode

```
Source          DF     SS      MS       F      P
GasCode          1   3.92   3.920    4.75  0.095
TypeCode         1   9.68   9.680   11.73  0.027
Interaction      1  54.08  54.080   65.55  0.001
Error            4   3.30   0.825
Total            7  70.98
```

```
                    Individual 95% CIs For Mean Based on
                    Pooled StDev
GasCode   Mean     --------+--------+--------+--------+-
1         27.45    (------------*------------)
2         28.85               (------------*-------------)
                   --------+--------+--------+--------+-
                       27.0     28.0     29.0     30.0
                    Individual 95% CIs For Mean Based on
                    Pooled StDev
TypeCode  Mean     -----+---------+---------+---------+----
1         29.25                      (----------*---------)
2         27.05    (---------*-----------)
                   -----+---------+---------+---------+----
                       26.4      27.6      28.8      30.0
```

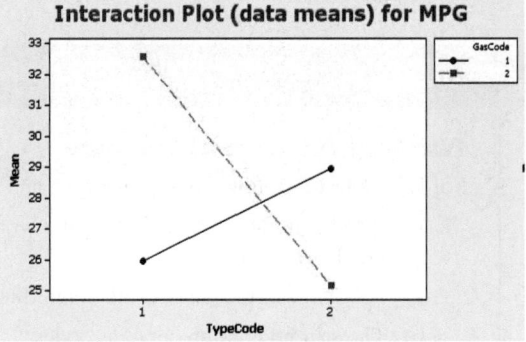

Interaction Plot (data means) for MPG

Plot Interactions

3. Select **Stat>ANOVA>Interactions Plot.**

a) Double-click MPG for the response variable and GasCodes and TypeCodes for the factors.

b) Click [OK].

Intersecting lines indicate a significant interaction of the two independent variables.

TI-83 Plus or TI-84 Plus
Step by Step

The TI-83 Plus and TI-84 Plus do not have a built-in function for two-way analysis of variance. However, the downloadable program named TWOWAY is available on your CD and Online Learning Center. Follow the instructions with your CD for downloading the program.

Performing a Two-Way Analysis of Variance

1. Enter the data values of the dependent variable into L_1 and the coded values for the levels of the factors into L_2 and L_3.

2. Press **PRGM,** move the cursor to the program named TWOWAY, and press **ENTER** twice.

3. Type L_1 for the list that contains the dependent variable and press **ENTER**.

4. Type L_2 for the list that contains the coded values for the first factor and press **ENTER**.

5. Type L_3 for the list that contains the coded values for the second factor and press **ENTER**.

6. The program will show the statistics for the first factor.

7. Press **ENTER** to see the statistics for the second factor.

8. Press **ENTER** to see the statistics for the interaction.

9. Press **ENTER** to see the statistics for the error.

10. Press **ENTER** to clear the screen.

Example TI12–2

Perform a two-way analysis of variance for the gasoline data (Example 12–5 in the text). The gas mileages are the data values for the dependent variable. Factor A is the type of gasoline (1 for regular, 2 for high-octane). Factor B is the type of automobile (1 for two-wheel-drive, 2 for four-wheel-drive).

Gas mileages (L_1)	Type of gasoline (L_2)	Type of automobile (L_3)
26.7	1	1
25.2	1	1
32.3	2	1
32.8	2	1
28.6	1	2
29.3	1	2
26.1	2	2
24.2	2	2

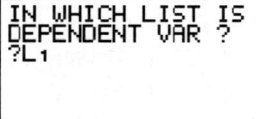 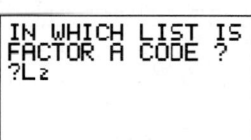 

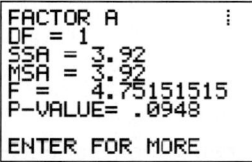 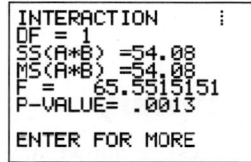 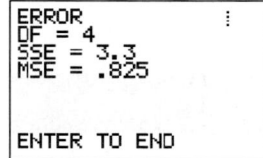

Excel
Step by Step

Two-Way Analysis of Variance (ANOVA)

The Data Analysis tools include "two-factor" analysis of variance, which is used to solve two-way ANOVA problems. Solve Example 12–5 about the effects of type of gasoline and type of automobile on gas mileage.

Example XL12–2

1. Enter the data from Example 12–5 exactly as shown in the figure.

2. Select **Tools>Data Analysis** and choose Anova: Two-Factor With Replication.

3. Enter the data range including labels as shown. Note that there are two rows of data for regular gasoline and two for high-octane.

Worksheet and Dialog
Box for Two-Way ANOVA

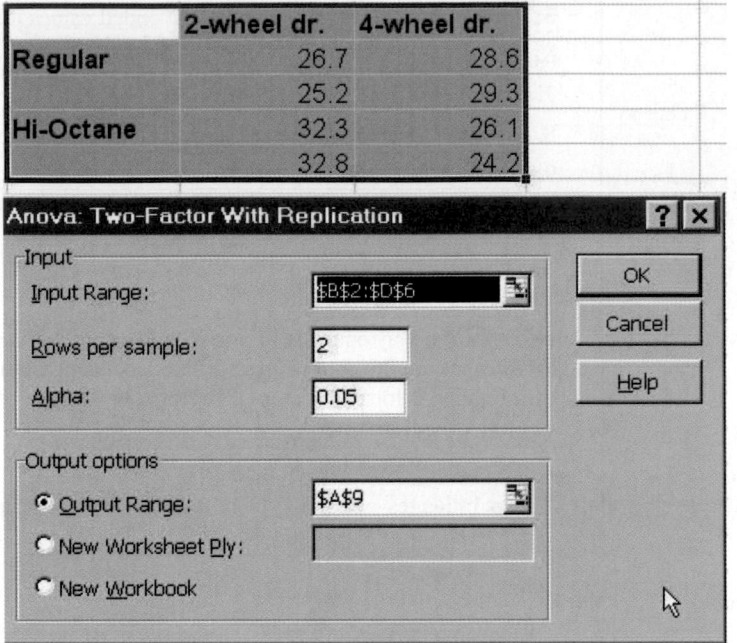

	2-wheel dr.	4-wheel dr.
Regular	26.7	28.6
	25.2	29.3
Hi-Octane	32.3	26.1
	32.8	24.2

Anova: Two-Factor With Replication

Input
Input Range: $B$2:$D$6
Rows per sample: 2
Alpha: 0.05

Output options
◉ Output Range: $A$9
○ New Worksheet Ply:
○ New Workbook

OK
Cancel
Help

4. Set an α level; use **0.05.**

5. Select a location for output and click [OK].

The results are shown. Compare these Excel results with those shown in Table 12–7.

ANOVA						
Source of Variation	*SS*	*df*	*MS*	*F*	*P-value*	*F crit*
Sample	3.92	1	3.92	4.752	0.094766	7.7086
Columns	9.68	1	9.68	11.73	0.026648	7.7086
Interaction	54.08	1	54.08	65.55	0.001265	7.7086
Within	3.3	4	0.825			
Total	70.98	7				

12–5   Summary

The F test, as shown in Chapter 9, can be used to compare two sample variances to determine whether they are equal. It can also be used to compare three or more means. When three or more means are compared, the technique is called analysis of variance (ANOVA). The ANOVA technique uses two estimates of the population variance. The between-group variance is the variance of the sample means; the within-group variance is the overall variance of all the values. When there is no significant difference among the means, the two estimates will be approximately equal and the F test value will be close to 1. If there is a significant difference among the means, the between-group variance estimate will be larger than the within-group variance estimate and a significant test value will result.

    If there is a significant difference among means, the researcher may wish to see where this difference lies. Several statistical tests can be used to compare the sample means after the ANOVA technique has been done. The most common are the Scheffé test and the Tukey test. When the sample sizes are the same, the Tukey test can be used. The Scheffé test is more general and can be used when the sample sizes are equal or not equal.

When there is one independent variable, the analysis of variance is called a one-way ANOVA. When there are two independent variables, the analysis of variance is called a two-way ANOVA. The two-way ANOVA enables the researcher to test the effects of two independent variables and a possible interaction effect on one dependent variable.

Important Terms

analysis of variance (ANOVA) 621

ANOVA summary table 624

between-group variance 621

disordinal interaction 642

factors 636

interaction effect 637

level 637

main effect 638

mean square 623

one-way ANOVA 636

ordinal interaction 642

Scheffé test 630

sum of squares between groups 623

sum of squares within groups 623

treatment groups 637

Tukey test 631

two-way ANOVA 636

within-group variance 621

Important Formulas

Formulas for the ANOVA test:

$$\bar{X}_{\text{GM}} = \frac{\Sigma X}{N}$$

$$F = \frac{s_B^2}{s_W^2}$$

where

$$s_B^2 = \frac{\Sigma n_i(\bar{X}_i - \bar{X}_{\text{GM}})^2}{k - 1} \qquad s_W^2 = \frac{\Sigma(n_i - 1)s_i^2}{\Sigma(n_i - 1)}$$

d.f.N. $= k - 1$ $\qquad\qquad N = n_1 + n_2 + \cdots + n_k$

d.f.D. $= N - k$ $\qquad\qquad k =$ **number of groups**

Formulas for the Scheffé test:

$$F_s = \frac{(\bar{X}_i - \bar{X}_j)^2}{s_W^2\,[(1/n_i) + (1/n_j)]} \qquad \text{and} \qquad F' = (k - 1)(\text{C.V.})$$

Formula for the Tukey test:

$$q = \frac{\bar{X}_i - \bar{X}_j}{\sqrt{s_W^2/n}}$$

d.f.N. $= k$ $\qquad$ and $\qquad$ **d.f.D. = degrees of freedom for** s_W^2

Formulas for the two-way ANOVA:

$$\text{MS}_A = \frac{\text{SS}_A}{a - 1} \qquad\qquad F_A = \frac{\text{MS}_A}{\text{MS}_W} \qquad \begin{array}{l} \textbf{d.f.N.} = a - 1 \\ \textbf{d.f.D.} = ab(n - 1) \end{array}$$

$$\text{MS}_B = \frac{\text{SS}_B}{b - 1} \qquad\qquad F_B = \frac{\text{MS}_B}{\text{MS}_W} \qquad \begin{array}{l} \textbf{d.f.N.} = b - 1 \\ \textbf{d.f.D.} = ab(n - 1) \end{array}$$

$$\text{MS}_{A \times B} = \frac{\text{SS}_{A \times B}}{(a - 1)(b - 1)} \qquad F_{A \times B} = \frac{\text{MS}_{A \times B}}{\text{MS}_W} \qquad \begin{array}{l} \textbf{d.f.N.} = (a - 1)(b - 1) \\ \textbf{d.f.D.} = ab(n - 1) \end{array}$$

$$\text{MS}_W = \frac{\text{SS}_W}{ab(n - 1)}$$

Review Exercises

If the null hypothesis is rejected in Exercises 1 through 9, use the Scheffé test when the sample sizes are unequal to test the differences between the means, and use the Tukey test when the sample sizes are equal. For these exercises, perform these steps.

  a. State the hypotheses and identify the claim.
  b. Find the critical value(s).
  c. Compute the test value.
  d. Make the decision.
  e. Summarize the results, and explain where the differences in means are.

Use the traditional method of hypothesis testing unless otherwise specified.

1. The data represent the lengths in feet of three types of bridges in the United States. At $\alpha = 0.01$, test the claim that there is no significant difference in the means of the lengths of the types of bridges.

Simple truss	Segmented concrete	Continuous plate
745	820	630
716	750	573
700	790	525
650	674	510
647	660	480
625	640	460
608	636	451
598	620	450
550	520	450
545	450	425
534	392	420
528	370	360

Source: World Almanac and Book of Facts.

2. Is there a difference in the price per bottle of three different types of wine? The data show the costs per bottle (in dollars) of three different types of wines. Use $\alpha = 0.01$ to answer the question, Why would some wines be more expensive than others?

Type 1	Type 2	Type 3
11	15	13
10	7	11
20	9	18
10	10	10
17	4	22
9	5	15

3. The number of carbohydrates per serving in randomly selected cereals from three manufacturers is shown. At the 0.05 level of significance, is there sufficient evidence to conclude a difference in the average number of carbohydrates?

Manufacturer 1	Manufacturer 2	Manufacturer 3
25	23	24
26	44	39
24	24	28
26	24	25
26	36	23
41	27	32
26	25	
43		

Source: The Doctor's Pocket Calorie, Fat, and Carbohydrate Counter.

4. The number of grams of fat per serving for three different kinds of pizza from several manufacturers is listed below. At the 0.01 level of significance, is there sufficient evidence that a difference exists in mean fat content?

Cheese	Pepperoni	Supreme/Deluxe
18	20	16
11	17	27
19	15	17
20	18	17
16	23	12
21	23	27
16	21	20

Source: The Doctor's Pocket Calorie, Fat, and Carbohydrate Counter.

5. The iron content in three different types of food is shown. At the 0.10 level of significance, is there sufficient evidence to conclude that a difference in mean iron content exists for meats and fish, breakfast cereals, and nutritional high-protein drinks?

Meats and fish	Breakfast cereals	Nutritional drinks
3.4	8	3.6
2.5	2	3.6
5.5	1.5	4.5
5.3	3.8	5.5
2.5	3.8	2.7
1.3	6.8	3.6
2.7	1.5	6.3
	4.5	

Source: The Doctor's Pocket Calorie, Fat, and Carbohydrate Counter.

6. The data consist of the weights in ounces of three different types of digital camera. Use $\alpha = 0.05$ to see if the means are equal.

2–3 Megapixels	4–5 Megapixels	6–8 Megapixels
6	14	19
8	11	27
7	15	21
11	24	23
4	17	24
8	10	33

Is Seeing Really Believing?—Revisited

To see if there were differences in the testimonies of the witnesses in the three age groups, the witnesses responded to 17 questions, 10 on direct examination and 7 on cross-examination. These were then scored for accuracy. An analysis of variance test with age as the independent variable was used to compare the total number of questions answered correctly by the groups. The results showed no significant differences among the age groups for the direct examination questions. However, there was a significant difference among the groups on the cross-examination questions. Further analysis showed the 8-year-olds were significantly less accurate under cross-examination compared to the other two groups. The 12-year-old and adult eyewitnesses did not differ in the accuracy of their cross-examination responses.

7. A researcher wishes to see if there is a difference in the average number of times local police were called in school incidents. Samples of school districts were selected, and the numbers of incidents for a specific year were reported. At $\alpha = 0.05$, is there a difference in the means? If so, suggest a reason for the difference.

County A	County B	County C	County D
13	16	15	11
11	33	12	31
2	12	19	3
	2	2	
	2		

Source: U.S. Department of Education.

8. A teacher wishes to test the math anxiety level of her students in two classes at the beginning of the semester. The classes are Calculus I and Statistics. Furthermore, she wishes to see whether there is a difference owing to the students' ages. Math anxiety is measured by the score on a 100-point anxiety test. Use $\alpha = 0.10$ and a two-way analysis of variance to see whether there is a difference. Five students are randomly assigned to each group. The data are shown here.

	Class	
Age	**Calculus I**	**Statistics**
Under 20	43, 52, 61, 57, 55	19, 20, 31, 36, 24
20 or over	56, 55, 42, 48, 61	63, 78, 67, 71, 75

ANOVA Summary Table for Exercise 8

Source	SS	d.f.	MS	F
Age	2376.2			
Class	105.8			
Interaction	2645.0			
Within	763.2			
Total	5890.2			

9. A medical researcher wishes to test the effects of two different diets and two different exercise programs on the glucose level in a person's blood. The glucose level is measured in milligrams per deciliter (mg/dl). Three subjects are randomly assigned to each group. Analyze the data shown here, using a two-way ANOVA with $\alpha = 0.05$.

Exercise	Diet	
program	**A**	**B**
I	62, 64, 66	58, 62, 53
II	65, 68, 72	83, 85, 91

ANOVA Summary Table for Exercise 9

Source	SS	d.f.	MS	F
Exercise	816.750			
Diet	102.083			
Interaction	444.083			
Within	108.000			
Total	1470.916			

Data Analysis

The Data Bank is found in Appendix D, or on the World Wide Web by following links from
www.mhhe.com/math/stat/bluman

1. From the Data Bank, select a random sample of subjects, and test the hypothesis that the mean cholesterol levels of the nonsmokers, less-than-one-

pack-a-day smokers, and one-pack-plus smokers are equal. Use an ANOVA test. If the null hypothesis is rejected, conduct the Scheffé test to find where the difference is. Summarize the results.

2. Repeat Exercise 2 for the mean IQs of the various educational levels of the subjects.

3. Using the Data Bank, randomly select 12 subjects and randomly assign them to one of the four groups in the following classifications.

	Smoker	**Nonsmoker**
Male		
Female		

Use one of these variables—weight, cholesterol, or systolic pressure—as the dependent variable, and perform a two-way ANOVA on the data. Use a computer program to generate the ANOVA table.

Chapter Quiz

Determine whether each statement is true or false. If the statement is false, explain why.

1. In analysis of variance, the null hypothesis should be rejected only when there is a significant difference among all pairs of means.

2. The F test does not use the concept of degrees of freedom.

3. When the F test value is close to 1, the null hypothesis should be rejected.

4. The Tukey test is generally more powerful than the Scheffé test for pairwise comparisons.

Select the best answer.

5. Analysis of variance uses the _____ test.

   a. z c. χ^2
   b. t d. F

6. The null hypothesis in ANOVA is that all the means are _____.

   a. Equal   c. Variable
   b. Unequal   d. None of the above

7. When one conducts an F test, _____ estimates of the population variance are compared.

   a. Two   c. Any number of
   b. Three   d. No

8. If the null hypothesis is rejected in ANOVA, one can use the _____ test to see where the difference in the means is found.

   a. z or t c. Scheffé or Tukey
   b. F or χ^2 d. Any of the above

Complete the following statements with the best answer.

9. When three or more means are compared, one uses the _____ technique.

10. If the null hypothesis is rejected in ANOVA, the _____ test should be used when sample sizes are equal.

11. In a two-way ANOVA, one can test _____ main hypotheses and one interactive hypothesis.

For Exercises 12 through 16 use the traditional method of hypothesis testing unless otherwise specified.

12. In a recent Presidential election, a sample of the percentage of voters who voted is shown. At $\alpha = 0.05$, is there a difference in the mean percentage of voters who voted?

Northeast	Southeast	Northwest	Southwest
65.3	54.8	60.5	42.3
59.9	61.8	61.0	61.2
66.9	49.6	74.0	54.7
64.2	58.6	61.4	56.7

Source: Committee for the Study of the American Electorate.

13. A media researcher wishes to see if there is a difference in the ages of viewers of three late-night television talk shows. Three samples of viewers are selected, and the ages of the viewers are shown. At $\alpha = 0.01$, is there a difference in the means of the ages of the viewers? Why is the average age of a viewer important to a television show writer?

David Letterman	Jay Leno	Conan O'Brien
53	48	40
46	51	36
48	57	35
42	46	42
35	38	39

Source: Based on information from Nielsen Media Research.

14. Prices (in dollars) of men's, women's, and children's athletic shoes are shown. At the 0.05 level of significance, can it be concluded that there is a difference in mean price?

Women's	Men's	Children's
59	65	40
36	70	45
44	66	40
49	59	56
48	48	46
50	70	36

15. The birth weights of randomly selected newborns at three area hospitals are shown. Using the 0.10 level

of significance, test the claim that the mean weights are equal.

Hospital A	Hospital B	Hospital C
7 lb 12 oz	9 lb 6 oz	8 lb 6 oz
8 lb 3 oz	5 lb 9 oz	9 lb 5 oz
11 lb 6 oz	6 lb 8 oz	7 lb 13 oz
6 lb 10 oz	8 lb 9 oz	8 lb 2 oz
7 lb 3 oz	10 lb 5 oz	9 lb 2 oz
8 lb 2 oz	7 lb 6 oz	6 lb 5 oz

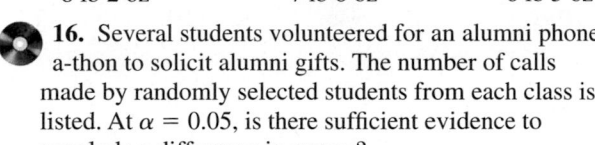 **16.** Several students volunteered for an alumni phone-a-thon to solicit alumni gifts. The number of calls made by randomly selected students from each class is listed. At $\alpha = 0.05$, is there sufficient evidence to conclude a difference in means?

Freshmen	Sophomores	Juniors	Seniors
25	17	20	20
29	25	24	25
32	20	25	26
15	26	30	32
18	30	15	19
26	28	18	20
35			

17. A researcher conducted a study of two different diets and two different exercise programs. Three randomly selected subjects were assigned to each group for one month. The values indicate the amount of weight each lost.

Exercise program	Diet	
	A	B
I	5, 6, 4	8, 10, 15
II	3, 4, 8	12, 16, 11

Answer the following questions for the information in the printout shown below.

a. What procedure is being used?
b. What are the names of the two variables?
c. How many levels does each variable contain?
d. What are the hypotheses for the study?
e. What are the F values for the hypotheses? State which are significant, using the P-values.
f. Based on the answers to part *e*, which hypotheses can be rejected?

Computer Printout for Problem 17

```
Datafile: NONAME.SST    Procedure: Two-way ANOVA

TABLE OF MEANS:
                          DIET
                    A .....       B .....      Row Mean
EX PROG I .....     5.000        11.000        8.000
      II .....      5.000        13.000        9.000
      Col Mean      5.000        12.000
      Tot Mean      8.500

SOURCE TABLE:
       Source      df    Sums of Squares    Mean Square    F Ratio    p-value
         DIET      1         147.000          147.000      21.000     0.00180
       EX PROG     1           3.000            3.000       0.429     0.53106
    DIET X EX P    1           3.000            3.000       0.429     0.53106
       Within      8          56.000            7.000
        Total     11         209.000
```

Critical Thinking Challenges

Shown here are the abstract and two tables from a research study entitled "Adult Children of Alcoholics: Are They at Greater Risk for Negative Health Behaviors?" by Arlene E. Hall. Based on the abstract and the tables, answer these questions.

1. What was the purpose of the study?
2. How many groups were used in the study?
3. By what means were the data collected?
4. What was the sample size?
5. What type of sampling method was used?

6. How might the population be defined?

7. What may have been the hypothesis for the ANOVA part of the study?

8. Why was the one-way ANOVA procedure used, as opposed to another test, such as the *t* test?

9. What part of the ANOVA table did the conclusion "ACOAs had significantly lower wellness scores (WS) than non-ACOAs" come from?

10. What level of significance was used?

11. In the following excerpts from the article, the researcher states that

> ... using the Tukey-HSD procedure revealed a significant difference between ACOAs and non-ACOAs, p = 0.05, but no significant difference was found between ACOAs and Unsures or between non-ACOAs and Unsures.

Using Tables 12–8 and 12–9 and the means, explain why the Tukey test would have enabled the researcher to draw this conclusion.

Abstract *The purpose of the study was to examine and compare the health behaviors of adult children of alcoholics (ACOAs) and their non-ACOA peers within a university population. Subjects were 980 undergraduate students from a major university in the East. Three groups (ACOA, non-ACOA, and Unsure) were identified from subjects' responses to three direct questions regarding parental drinking behaviors. A questionnaire was used to collect data for the study. Included were questions related to demographics, parental drinking behaviors, and the College Wellness Check (WS), a health risk appraisal designed especially for college students (Dewey & Cabral, 1986). Analysis of variance procedures revealed that ACOAs had significantly lower wellness scores (WS) than non-ACOAs. Chi-square analyses of the individual variables revealed that ACOAs and non-ACOAs were significantly different on 15 of the 50 variables of the WS. A discriminant analysis procedure revealed the similarities between Unsure subjects and ACOA subjects. The results provide valuable information regarding ACOAs in a nonclinical setting and contribute to our understanding of the influences related to their health risk behaviors.*

Table 12–8	Means and Standard Deviations for the Wellness Scores (WS) Group by ($N = 945$)		
Group	**N**	**$\overline{X}$**	**S.D.**
ACOAs	143	69.0	13.6
Non-ACOAs	746	73.2	14.5
Unsure	56	70.1	14.0
Total	945	212.3	42.1

Table 12–9	ANOVA of Group Means for the Wellness Scores (WS)			
Source	**d.f.**	**SS**	**MS**	**F**
Between groups	2	2,403.5	1,201.7	5.9*
Within groups	942	193,237.4	205.1	
Total	944	195,640.8		

$*p < 0.01$

Source: Arlene E. Hall, "Adult Children of Alcoholics: Are They at Greater Risk for Negative Health Behaviors?" *Journal of Health Education* 12, no. 4, pp. 232–238.

 Data Projects

Where appropriate, use MINITAB, the TI-83 Plus, the TI-84 Plus, or a computer program of your choice to complete these exercises.

Select a variable and collect data for at least three different groups (samples). For example, you could ask students, faculty, and clerical staff how many cups of coffee they drink per day or how many hours they watch television per day. Compare the means, using the one-way ANOVA technique, and then answer each of these.

 a. What is the purpose of the study?
 b. Define the population.
 c. How were the samples selected?
 d. What α value was used?
 e. State the hypotheses.
 f. What was the *F* test value?
 g. What was the decision?
 h. Summarize the results.

You may use the following websites to obtain raw data:

Visit the data sets at the book's website found at http://www.mhhe.com/math/stat/bluman Click on the 6th edition.
http://lib.stat.cmu.edu/DASL
http://www.statcan.ca

Hypothesis-Testing Summary 2*

7. Test of the significance of the correlation coefficient.

Example: H_0: $\rho = 0$

Use a t test:

$$t = r\sqrt{\frac{n-2}{1-r^2}} \qquad \text{with d.f.} = n - 2$$

8. Formula for the F test for the multiple correlation coefficient.

Example: H_0: $\rho = 0$

$$F = \frac{R^2/k}{(1-R^2)/(n-k-1)}$$

$$\text{d.f.N.} = n - k \qquad \text{d.f.D.} = n - k - 1$$

9. Comparison of a sample distribution with a specific population.

Example: H_0: There is no difference between the two distributions.

Use the chi-square goodness-of-fit test:

$$\chi^2 = \sum \frac{(O - E)^2}{E}$$

$$\text{d.f.} = \text{no. of categories} - 1$$

10. Comparison of the independence of two variables.

Example: H_0: Variable A is independent of variable B.

Use the chi-square independence test:

$$\chi^2 = \sum \frac{(O - E)^2}{E}$$

$$\text{d.f.} = (R - 1)(C - 1)$$

11. Test for homogeneity of proportions.

Example: H_0: $p_1 = p_2 = p_3$

Use the chi-square test:

$$\chi^2 = \sum \frac{(O - E)^2}{E}$$

$$\text{d.f.} = (R - 1)(C - 1)$$

12. Comparison of three or more sample means.

Example: H_0: $\mu_1 = \mu_2 = \mu_3$

Use the analysis of variance test:

$$F = \frac{s_B^2}{s_W^2}$$

where

$$s_B^2 = \frac{\sum n_i(\bar{X}_i - \bar{X}_{\text{GM}})^2}{k - 1}$$

$$s_W^2 = \frac{\sum(n_i - 1)s_i^2}{\sum(n_i - 1)}$$

$$\text{d.f.N.} = k - 1 \qquad N = n_1 + n_2 + \cdots + n_k$$

$$\text{d.f.D.} = N - k \qquad k = \text{number of groups}$$

13. Test when the F value for the ANOVA is significant. Use the Scheffé test to find what pairs of means are significantly different.

$$F_s = \frac{(\bar{X}_i - \bar{X}_j)^2}{s_W^2[(1/n_i) + (1/n_j)]}$$

$$F' = (k - 1)(\text{C.V.})$$

Use the Tukey test to find which pairs of means are significantly different.

$$q = \frac{\bar{X}_i - \bar{X}_j}{\sqrt{s_W^2/n}} \qquad \begin{array}{l} \text{d.f.N.} = k \\ \text{d.f.D.} = \text{degrees of freedom for } s_W^2 \end{array}$$

14. Test for the two-way ANOVA.

Example:

H_0: There is no significant difference for the main effects.

H_1: There is no significant difference for the interaction effect.

$$\text{MS}_A = \frac{\text{SS}_A}{a - 1}$$

$$\text{MS}_B = \frac{\text{SS}_B}{b - 1}$$

$$\text{MS}_{A \times B} = \frac{\text{SS}_{A \times B}}{(a - 1)(b - 1)}$$

$$\text{MS}_W = \frac{\text{SS}_W}{ab(n - 1)}$$

$$F_A = \frac{\text{MS}_A}{\text{MS}_W} \qquad \begin{array}{l} \text{d.f.N.} = a - 1 \\ \text{d.f.D.} = ab(n - 1) \end{array}$$

$$F_B = \frac{\text{MS}_B}{\text{MS}_W} \qquad \begin{array}{l} \text{d.f.N.} = (b - 1) \\ \text{d.f.D.} = ab(n - 1) \end{array}$$

$$F_{A \times B} = \frac{\text{MS}_{A \times B}}{\text{MS}_W} \qquad \begin{array}{l} \text{d.f.N.} = (a - 1)(b - 1) \\ \text{d.f.D.} = ab(n - 1) \end{array}$$

*This summary is a continuation of Hypothesis-Testing Summary 1, at the end of Chapter 9.

Answers to Applying the Concepts

Section 12–3 Colors That Make You Smarter

1. The ANOVA produces a test statistic of $F = 3.06$, with a P-value of 0.059. We would fail to reject the null hypothesis and find that there is not enough evidence to conclude that the color of a person's clothing is related to people's perceptions of how intelligent the person looks.

2. Answers will vary. One possible answer is that the purpose of the study was to determine if the color of a person's clothing is related to people's perceptions of how intelligent the person looks.

3. We could use the Scheffé test to test all possible pairwise comparisons.

4. Tukey's pairwise comparisons

```
Family error rate = 0.0500
Individual error rate = 0.0195

Critical value = 3.46
Intervals for (column level mean) - (row level mean)
               1              2
   2         -0.399
              2.091

   3         -0.014        -0.861
              2.476         1.630
```

5. There are no contradictions in the results. Tukey's pairwise comparisons do not show any differences between any of the means, and the ANOVA failed to reject the null hypothesis of the means being equal.

6. We would have to perform three separate t tests, which would inflate the error rate.

7. We prefer to use Tukey's test over the Scheffé method when the sample sizes are all equal.

Section 12–4 Automobile Sales Techniques

There is no significant difference between levels 1 and 2 of experience. Level 3 and level 4 salespersons did significantly better than those at levels 1 and 2, with level 4 showing the best results, on average. If type of presentation is taken into consideration, the interaction plot shows a significant difference. The best combination seems to be level 4 experience with presentation style 1.

13

Nonparametric Statistics

Objectives

After completing this chapter, you should be able to

1 State the advantages and disadvantages of nonparametric methods.

2 Test hypotheses, using the sign test.

3 Test hypotheses, using the Wilcoxon rank sum test.

4 Test hypotheses, using the signed-rank test.

5 Test hypotheses, using the Kruskal-Wallis test.

6 Compute the Spearman rank correlation coefficient.

7 Test hypotheses, using the runs test.

Outline

**Statistics
Today**

Too Much or Too Little?

Suppose a manufacturer of ketchup wishes to check the bottling machines to see if they are functioning properly. That is, are they dispensing the right amount of ketchup per bottle? A 40-ounce bottle is currently used. Because of the natural variation in the manufacturing process, the amount of ketchup in a bottle will not always be exactly 40 ounces. Some bottles will contain less than 40 ounces, and others will contain more than 40 ounces. To see if the variation is due to chance or to a malfunction in the manufacturing process, a runs test can be used. The runs test is a nonparametric statistical technique. See Statistics Today—Revisited. This chapter explains such techniques, which can be used to help the manufacturer determine the answer to the question.

13–1   ## Introduction

Statistical tests, such as the z, t, and F tests, are called parametric tests. **Parametric tests** are statistical tests for population parameters such as means, variances, and proportions that involve assumptions about the populations from which the samples were selected. One assumption is that these populations are normally distributed. But what if the population in a particular hypothesis-testing situation is *not* normally distributed? Statisticians have developed a branch of statistics known as **nonparametric statistics** or **distribution-free statistics** to use when the population from which the samples are selected is not normally distributed. Nonparametric statistics can also be used to test hypotheses that do not involve specific population parameters, such as μ, σ, or p.

For example, a sportswriter may wish to know whether there is a relationship between the rankings of two judges on the diving abilities of 10 Olympic swimmers. In another situation, a sociologist may wish to determine whether men and women enroll at random for a specific drug rehabilitation program. The statistical tests used in these situations are nonparametric or distribution-free tests. The term *nonparametric* is used for both situations.

The nonparametric tests explained in this chapter are the sign test, the Wilcoxon rank sum test, the Wilcoxon signed-rank test, the Kruskal-Wallis test, and the runs test. In addition, the Spearman rank correlation coefficient, a statistic for determining the relationship between ranks, is explained.

13–2 Advantages and Disadvantages of Nonparametric Methods

As stated previously, nonparametric tests and statistics can be used in place of their parametric counterparts (z, t, and F) when the assumption of normality cannot be met. However, one should not assume that these statistics are a better alternative than the parametric statistics. There are both advantages and disadvantages in the use of nonparametric methods.

Advantages

Objective 1

State the advantages and disadvantages of nonparametric methods.

There are five advantages that nonparametric methods have over parametric methods:

1. They can be used to test population parameters when the variable is not normally distributed.
2. They can be used when the data are nominal or ordinal.
3. They can be used to test hypotheses that do not involve population parameters.
4. In most cases, the computations are easier than those for the parametric counterparts.
5. They are easy to understand.

Disadvantages

There are three disadvantages of nonparametric methods:

1. They are *less sensitive* than their parametric counterparts when the assumptions of the parametric methods are met. Therefore, larger differences are needed before the null hypothesis can be rejected.
2. They tend to use *less information* than the parametric tests. For example, the sign test requires the researcher to determine only whether the data values are above or below the median, not how much above or below the median each value is.
3. They are *less efficient* than their parametric counterparts when the assumptions of the parametric methods are met. That is, larger sample sizes are needed to overcome the loss of information. For example, the nonparametric sign test is about 60% as efficient as its parametric counterpart, the z test. Thus, a sample size of 100 is needed for use of the sign test, compared with a sample size of 60 for use of the z test to obtain the same results.

Since there are both advantages and disadvantages to the nonparametric methods, the researcher should use caution in selecting these methods. If the parametric assumptions can be met, the parametric methods are preferred. However, when parametric assumptions cannot be met, the nonparametric methods are a valuable tool for analyzing the data.

Ranking

Many nonparametric tests involve the **ranking** of data, that is, the positioning of a data value in a data array according to some rating scale. Ranking is an ordinal variable. For example, suppose a judge decides to rate five speakers on an ascending scale of 1 to 10, with 1 being the best and 10 being the worst, for categories such as voice, gestures, logical presentation, and platform personality. The ratings are shown in the chart.

Speaker	A	B	C	D	E
Rating	8	6	10	3	1

The rankings are shown next.

Speaker	E	D	B	A	C
Rating	1	3	6	8	10
Ranking	1	2	3	4	5

Since speaker E received the lowest score, 1 point, he or she is ranked first. Speaker D received the next-lower score, 3 points; he or she is ranked second; and so on.

What happens if two or more speakers receive the same number of points? Suppose the judge awards points as follows:

Speaker	A	B	C	D	E
Rating	8	6	10	6	3

The speakers are then ranked as follows:

Speaker	E	D	B	A	C
Rating	3	6	6	8	10
Ranking	1	Tie for 2nd and 3rd		4	5

When there is a tie for two or more places, the average of the ranks must be used. In this case, each would be ranked as

$$\frac{2 + 3}{2} = \frac{5}{2} = 2.5$$

Hence, the rankings are as follows:

Speaker	E	D	B	A	C
Rating	3	6	6	8	10
Ranking	1	2.5	2.5	4	5

Many times, the data are already ranked, so no additional computations must be done. For example, if the judge does not have to award points but can simply select the speakers who are best, second-best, third-best, and so on, then these ranks can be used directly.

P-values can also be found for nonparametric statistical tests, and the *P*-value method can be used to test hypotheses that use nonparametric tests. For this chapter, the *P*-value method will be limited to some of the nonparametric tests that use the standard normal distribution or the chi-square distribution.

Applying the Concepts **13–2**

Ranking Data

The following table lists the percentages of patients who experienced side effects from a drug used to lower a person's cholesterol level.

Side effect	Percent
Chest pain	4.0
Rash	4.0
Nausea	7.0
Heartburn	5.4
Fatigue	3.8
Headache	7.3
Dizziness	10.0
Chills	7.0
Cough	2.6

Rank each value in the table.

See page 705 for the answer.

Exercises 13–2

1. What is meant by *nonparametric statistics?*

2. When should nonparametric statistics be used?

3. List the advantages and disadvantages of nonparametric statistics.

For Exercises 4 through 10, rank each set of data.

4. 3, 8, 6, 1, 4, 10, 7

5. 22, 66, 32, 43, 65, 43, 71, 34

6. 83, 460, 582, 177, 241

7. 9, 7, 4, 9, 8, 6, 6, 10, 13, 16, 18, 15

8. 22, 25, 28, 28, 18, 32, 37, 41, 41, 43

9. 188, 256, 197, 188, 321, 530, 763

10. 2.8, 6.8, 2.6, 3.1, 1.5, 8.9, 3.15, 2.12, 3.1

13–3 The Sign Test

Single-Sample Sign Test

Objective 2

Test hypotheses, using the sign test.

The simplest nonparametric test, the **sign test** for single samples, is used to test the value of a median for a specific sample. When using the sign test, the researcher hypothesizes the specific value for the median of a population; then he or she selects a sample of data and compares each value with the conjectured median. If the data value is above the conjectured median, it is assigned a plus sign. If it is below the conjectured median, it is assigned a minus sign. And if it is exactly the same as the conjectured median, it is assigned a 0. Then the numbers of plus and minus signs are compared. If the null hypothesis is true, the number of plus signs should be approximately equal to the number of minus signs. If the null hypothesis is not true, there will be a disproportionate number of plus or minus signs.

Test Value for the Sign Test

The test value is the smaller number of plus or minus signs.

For example, if there are 8 positive signs and 3 negative signs, the test value is 3. When the sample size is 25 or less, Table J in Appendix C is used to determine the critical value. For a specific α, if the test value is less than or equal to the critical value obtained from the table, the null hypothesis should be rejected. The values in Table J are obtained from the binomial distribution. The derivation is omitted here.

Example 13–1

A convenience store owner hypothesizes that the median number of snow cones she sells per day is 40. A random sample of 20 days yields the following data for the number of snow cones sold each day.

18	43	40	16	22
30	29	32	37	36
39	34	39	45	28
36	40	34	39	52

At $\alpha = 0.05$, test the owner's hypothesis.

Solution

Step 1   State the hypotheses and identify the claim.

H_0: median $= 40$ (claim)   and H_1: median $\neq 40$

Step 2   Find the critical value. Compare each value of the data with the median. If the value is greater than the median, replace the value with a plus sign. If it is less than the median, replace it with a minus sign. And if it is equal to the median, replace it with a 0. The completed table follows.

−	+	0	−	−
−	−	−	−	−
−	−	−	+	−
−	0	−	−	+

Refer to Table J in Appendix C, using $n = 18$ (the total number of plus and minus signs; omit the zeros) and $\alpha = 0.05$ for a two-tailed test; the critical value is 4. See Figure 13–1.

Figure 13–1

Finding the Critical Value in Table J for Example 13–1.

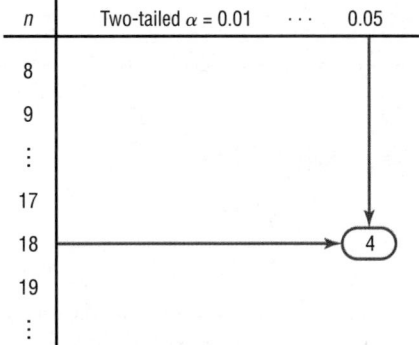

Step 3   Compute the test value. Count the number of plus and minus signs obtained in step 2, and use the smaller value as the test value. Since there are 3 plus signs and 15 minus signs, 3 is the test value.

Step 4   Make the decision. Compare the test value 3 with the critical value 4. If the test value is less than or equal to the critical value, the null hypothesis is rejected. In this case, the null hypothesis is rejected since $3 < 4$.

Step 5   Summarize the results. There is enough evidence to reject the claim that the median number of snow cones sold per day is 40.

When the sample size is 26 or more, the normal approximation can be used to find the test value. The formula is given. The critical value is found in Table E in Appendix C.

Formula for the z Test Value in the Sign Test When $n \geq 26$

$$z = \frac{(X + 0.5) - (n/2)}{\sqrt{n}/2}$$

where

X = smaller number of $+$ or $-$ signs
n = sample size

Example 13–2

Based on past experience, a manufacturer claims that the median (MD) lifetime of a rubber washer is at least 8 years. A sample of 50 washers showed that 21 lasted more than 8 years. At $\alpha = 0.05$, is there enough evidence to reject the manufacturer's claim?

Solution

Step 1 State the hypotheses and identify the claim.

H_0: MD $\geq$ 8 (claim)   and H_1: MD $<$ 8

Step 2 Find the critical value. Since $\alpha = 0.05$ and $n = 50$, and since this is a left-tailed test, the critical value is -1.65, obtained from Table E.

Step 3 Compute the test value.

$$z = \frac{(X + 0.5) - (n/2)}{\sqrt{n}/2} = \frac{(21 + 0.5) - (50/2)}{\sqrt{50}/2} = \frac{-3.5}{3.5355} = -0.99$$

Step 4 Make the decision. Since the test value of -0.99 is greater than -1.65, the decision is do not reject the null hypothesis.

Step 5 Summarize the results. There is not enough evidence to reject the claim that the median lifetime of the washers is at least 8 years.

In Example 13–2, the sample size was 50, and 21 washers lasted more than 8 years, so $50 - 21$, or 29, washers did not last 8 years. The value of X corresponds to the smaller of the two numbers 21 and 29. In this case, $X = 21$ is used in the formula. The reason is that there would be 21 positive signs, since subtracting 8 years from the value in years of a washer that lasted longer than 8 years results in a positive answer. When 8 is subtracted from the value in years of a washer that did not last 8 years, the answer is negative. Assuming that no washer lasted exactly 8 years results in 21 positive answers and 29 negative answers. Since 21 is the smaller of the two numbers, the value of X is 21.

Suppose a researcher hypothesized that the median age of houses in a certain municipality was 40 years. In a random sample of 100 houses, 68 were older than 40 years. Then the value used for X in the formula would be $100 - 68$, or 32, since it is the smaller of the two numbers 68 and 32. When 40 is subtracted from the age of a house older than 40 years, the answer is positive. When 40 is subtracted from the age of a house that is less than 40 years old, the result is negative. There would be 68 positive signs and 32 negative signs (assuming that no house was exactly 40 years old). Hence, 32 would be used for X, since it is the smaller of the two values.

Paired-Sample Sign Test

The sign test can also be used to test sample means in a comparison of two dependent samples, such as a before-and-after test. Recall that when dependent samples are taken from normally distributed populations, the *t* test is used (Section 9–5). When the condition of normality cannot be met, the nonparametric sign test can be used, as shown in Example 13–3.

Example 13–3

A medical researcher believed the number of ear infections in swimmers can be reduced if the swimmers use earplugs. A sample of 10 people was selected, and the number of infections for a four-month period was recorded. During the first two months, the swimmers did not use the earplugs; during the second two months, they did. At the beginning of the second two-month period, each swimmer was examined to make sure that no infections were present. The data are shown here. At $\alpha = 0.05$, can the researcher conclude that using earplugs reduced the number of ear infections?

Number of ear infections

Swimmer	Before, X_B	After, X_A
A	3	2
B	0	1
C	5	4
D	4	0
E	2	1
F	4	3
G	3	1
H	5	3
I	2	2
J	1	3

Solution

Step 1 State the hypotheses and identify the claim.

H_0: The number of ear infections will not be reduced.

H_1: The number of ear infections will be reduced (claim).

Step 2 Find the critical value. Subtract the after values X_A from the before values X_B and indicate the difference by a positive or negative sign or 0, according to the value, as shown in the table.

Swimmer	Before, X_B	After, X_A	Sign of difference
A	3	2	+
B	0	1	−
C	5	4	+
D	4	0	+
E	2	1	+
F	4	3	+
G	3	1	+
H	5	3	+
I	2	2	0
J	1	3	−

From Table J, with $n = 9$ (the total number of positive and negative signs; the 0 is not counted) and $\alpha = 0.05$ (one-tailed), at most 1 negative sign is needed to reject the null hypothesis because 1 is the smallest entry in the $\alpha = 0.05$ column of Table J.

Step 3   Compute the test value. Count the number of positive and negative signs found in step 2, and use the smaller value as the test value. There are 2 negative signs, so the test value is 2.

Step 4   Make the decision. There are 2 negative signs. The decision is to not reject the null hypothesis. The reason is that with $n = 9$, C.V. $= 1$ and $1 < 2$.

Step 5   Summarize the results. There is not enough evidence to support the claim that the use of earplugs reduced the number of ear infections.

When conducting a one-tailed sign test, the researcher must scrutinize the data to determine whether they support the null hypothesis. If the data support the null hypothesis, there is no need to conduct the test. In Example 13–3, the null hypothesis states that the number of ear infections will not be reduced. The data would support the null hypothesis if there were more negative signs than positive signs. The reason is that the before values X_B in most cases would be smaller than the after values X_A, and the $X_B - X_A$ values would be negative more often than positive. This would indicate that there is not enough evidence to reject the null hypothesis. The researcher would stop here, since there is no need to continue the procedure.

On the other hand, if the number of ear infections were reduced, the X_B values, for the most part, would be larger than the X_A values, and the $X_B - X_A$ values would most often be positive, as in Example 13–3. Hence, the researcher would continue the procedure. A word of caution is in order, and a little reasoning is required.

When the sample size is 26 or more, the normal approximation can be used in the same manner as in Example 13–2. The steps for conducting the sign test for single or paired samples are given in the Procedure Table.

Procedure Table

Sign Test for Single and Paired Samples

Step 1   State the hypotheses and identify the claim.

Step 2   Find the critical value(s). For the single-sample test, compare each value with the conjectured median. If the value is larger than the conjectured median, replace it with a positive sign. If it is smaller than the conjectured median, replace it with a negative sign.

For the paired-sample sign test, subtract the after values from the before values, and indicate the difference with a positive or negative sign or 0, according to the value. Use Table J and $n = $ total number of positive and negative signs.

Check the data to see whether they support the null hypothesis. If they do, do not reject the null hypothesis. If not, continue with step 3.

Step 3   Compute the test value. Count the number of positive and negative signs found in step 2, and use the smaller value as the test value.

Step 4   Make the decision. Compare the test value with the critical value in Table J. If the test value is less than or equal to the critical value, reject the null hypothesis.

Step 5   Summarize the results.

Note: If the sample size n is 26 or more, use Table E and the following formula for the test value:

$$z = \frac{(X + 0.5) - (n/2)}{\sqrt{n}/2}$$

where
   $X = $ smaller number of $+$ or $-$ signs
   $n = $ sample size

Applying the Concepts **13–3**

Clean Air

An environmentalist suggests that the median of the number of days per year that a large city failed to meet the EPA acceptable standards for clean air is 11 days per month. A random sample of 20 months shows the number of days per month that the air quality was below the EPA's standards.

15	14	1	9	0	3	3	1	10	8
6	16	21	22	3	19	16	5	23	13

1. What is the claim?
2. What test would you use to test the claim? Why?
3. What would the hypotheses be?
4. Select a value for α and find the corresponding critical value.
5. What is the test value?
6. What is your decision?
7. Summarize the results.
8. Could a parametric test be used?

See page 705 for the answers.

Exercises 13–3

1. Why is the sign test the simplest nonparametric test to use?

2. What population parameter can be tested with the sign test?

3. In the sign test, what is used as the test value when $n < 26$?

4. When $n \geq 26$, what is used in place of Table J for the sign test?

For Exercises 5 through 20, perform these steps.

 a. State the hypotheses and identify the claim.
 b. Find the critical value(s).
 c. Compute the test value.
 d. Make the decision.
 e. Summarize the results.

Use the traditional method of hypothesis testing unless otherwise specified.

 5. An oceanographer believes that the median height of the waves at Ocean City is 2.8 feet. The wave heights (in feet) are measured for a random sample of 20 days. The data are shown here. At $\alpha = 0.05$, is there enough evidence to reject the oceanographer's claim?

3.6	2.1	2.3	2.1	2.7
3.2	3.9	3.4	3.0	2.9
2.0	1.9	3.2	3.5	2.8
1.8	2.3	3.7	3.9	4.2

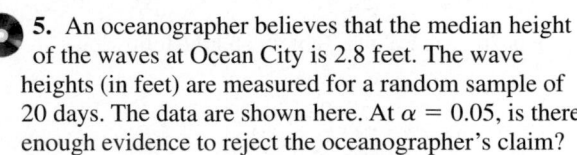 6. An athletic director suggests the median number for the paid attendance at 20 local football games is 3000. The data for a sample are shown. At $\alpha = 0.05$, is there enough evidence to reject the claim? If you were printing the programs for the games, would you use this figure as a guide?

6210	3150	2700	3012	4875
3540	6127	2581	2642	2573
2792	2800	2500	3700	6030
5437	2758	3490	2851	2720

Source: *Pittsburgh Post Gazette.*

 7. An educator hypothesizes that the median of the number of students enrolled in cyber schools in school districts in southwestern Pennsylvania is 25. At $\alpha = 0.05$, is there enough evidence to reject the educator's claim? The data are shown here. What benefit would this information provide to the school board of a local school district?

12	41	26	14	4
38	27	27	9	11
17	11	66	5	14
8	35	16	25	17

Source: *Tribune-Review.*

 8. A government economist estimates that the median cost per pound of beef is $5.00. A sample of 22 livestock buyers shows the following costs (in dollars) per pound of beef. Is there enough evidence to reject the economist's hypothesis at $\alpha = 0.10$?

5.35	5.16	4.97	4.83	5.05	5.19
4.78	4.93	4.86	5.00	4.63	5.06
5.19	5.00	5.05	5.10	5.16	5.25
5.16	5.42	5.13	5.27		

9. For a specific year, the median price of natural gas was $10.86 per 1000 cubic feet. A researcher wishes to see if there is enough evidence to reject the claim. Out of 42 households, 18 paid less that $10.86 per 1000 cubic feet for natural gas. Test the claim at $\alpha = 0.05$. How could a prospective home buyer use this information?

Source: Based on information from the Energy Information Administration.

10. One hundred people were placed on a special exercise program. After one month, 58 lost weight, 12 gained weight, and 30 weighed the same as before. Test the hypothesis that the exercise program is effective at $\alpha = 0.10$. (*Note:* It will be effective if fewer than 50% of the people did not lose weight.)

11. Of 50 students surveyed, 29 favored single-room dormitories. At $\alpha = 0.02$, test the hypothesis that more than 50% of the students favor single-room dormitories. Use the *P*-value method.

12. A researcher read that the median age for viewers of the Carson Daly show is 39. To test the claim, 75 viewers were surveyed, and 27 were under the age of 39. At $\alpha = 0.02$ test the claim. Give one reason why an advertiser might like to know the results of this study.

Source: Nielsen Media Research.

13. One hundred students are asked if they favor increasing the school year by 20 days. The responses are 62 no, 36 yes, and 2 undecided. At $\alpha = 0.10$, test the hypothesis that 50% of the students are against extending the school year. Use the *P*-value method.

14. A meteorologist suggests that the median number of deaths per year from tornadoes in the United States is 60. The number of deaths for a sample of 11 years is shown. At $\alpha = 0.05$ is there enough evidence to reject the claim? If you took proper safety precautions during a tornado, would you feel relatively safe?

| 53 | 39 | 39 | 67 | 69 | 40 |
| 25 | 33 | 30 | 130 | 94 | |

Source: NOAA.

15. A study was conducted to see whether a certain diet medication had an effect on the weights (in pounds) of eight women. Their weights were taken before and six weeks after daily administration of the medication. The data are shown here. At $\alpha = 0.05$, can one conclude that the medication had an effect (increase or decrease) on the weights of the women?

Subject	A	B	C	D	E	F	G	H
Weight before	187	163	201	158	139	143	198	154
Weight after	178	162	188	156	133	150	175	150

16. Two different laboratory machines measure the sodium content (in milligrams) of the same 10 blood samples. The data are shown here. At $\alpha = 0.01$, test the claim that both machines gave the same reading.

Sample	1	2	3	4	5	6	7	8	9	10
Machine 1	138	136	142	151	154	141	140	138	132	136
Machine 2	140	136	141	150	153	144	143	136	131	138

17. An educator designed a reasoning skills course. Nine students were selected and given a pretest to determine their reasoning abilities. After completing the course, the same students were given an equivalent form of the test to see whether their reasoning skills had improved. The data are shown here. At $\alpha = 0.05$, did the course improve their reasoning skills?

Student	1	2	3	4	5	6	7	8	9
Pretest	80	76	74	83	92	78	91	74	88
Posttest	82	78	73	85	95	79	93	78	90

18. A researcher wishes to test the effects of a pill on a person's appetite. Twelve subjects are allowed to eat a meal of their choice, and their caloric intake is measured. The next day, the same subjects take the pill and eat a meal of their choice. The caloric intake of the second meal is measured. The data are shown here. At $\alpha = 0.02$, can the researcher conclude that the pill had an effect on a person's appetite?

Subject	1	2	3	4	5	6	7
Meal 1	856	732	900	1321	843	642	738
Meal 2	843	721	872	1341	805	531	740

Subject	8	9	10	11	12
Meal 1	1005	888	756	911	998
Meal 2	900	805	695	878	914

19. A researcher wishes to determine if the number of viewers for 10 returning television shows has not changed since last year. The data are given in millions of viewers. At $\alpha = 0.01$, test the claim that the number of viewers has not changed. Depending on your answer, would a television executive plan to air these programs for another year?

Show	1	2	3	4	5	6
Last year	28.9	26.4	20.8	25.0	21.0	19.2
This year	26.6	20.5	20.2	19.1	18.9	17.8

Show	7	8	9	10
Last year	13.7	18.8	16.8	15.3
This year	16.8	16.7	16.0	15.8

Source: Based on information from Nielson Media Research.

20. A manufacturer believes that if routine maintenance (cleaning and oiling of machines) is increased to once a day rather than once a week, the

number of defective parts produced by the machines will decrease. Nine machines are selected, and the number of defective parts produced over a 24-hour operating period is counted. Maintenance is then increased to once a day for a week, and the number of defective parts each machine produces is again counted over a 24-hour operating period. The data are shown here. At $\alpha = 0.01$,

can the manufacturer conclude that increased maintenance reduces the number of defective parts manufactured by the machines?

Machine	1	2	3	4	5	6	7	8	9
Before	6	18	5	4	16	13	20	9	3
After	5	16	7	4	18	12	14	7	1

Extending the Concepts

The confidence interval for the median of a set of values less than or equal to 25 in number can be found by ordering the data from smallest to largest, finding the median, and using Table J. For example, to find the 95% confidence interval of the true median for 17, 19, 3, 8, 10, 15, 1, 23, 2, 12, order the data:

1, 2, 3, 8, 10, 12, 15, 17, 19, 23

From Table J, select $n = 10$ and $\alpha = 0.05$, and find the critical value. Use the two-tailed row. In this case, the critical value is 1. Add 1 to this value to get 2. In the ordered list, count from the left two numbers and from the right two numbers, and use these numbers to get the confidence interval, as shown:

1, 2, 3, 8, 10, 12, 15, 17, 19, 23

$2 \leq MD \leq 19$

Always add 1 to the number obtained from the table before counting. For example, if the critical value is 3, then count 4 values from the left and right.

For Exercises 21 through 25, find the confidence interval of the median, indicated in parentheses, for each set of data.

21. 3, 12, 15, 18, 16, 15, 22, 30, 25, 4, 6, 9 (95%)

22. 101, 115, 143, 106, 100, 142, 157, 163, 155, 141, 145, 153, 152, 147, 143, 115, 164, 160, 147, 150 (90%)

23. 8.2, 7.1, 6.3, 5.2, 4.8, 9.3, 7.2, 9.3, 4.5, 9.6, 7.8, 5.6, 4.7, 4.2, 9.5, 5.1 (98%)

24. 1, 8, 2, 6, 10, 15, 24, 33, 56, 41, 58, 54, 5, 3, 42, 31, 15, 65, 21 (99%)

25. 12, 15, 18, 14, 17, 19, 25, 32, 16, 47, 14, 23, 27, 42, 33, 35, 39, 41, 21, 19 (95%)

Technology *Step by Step*

MINITAB

Step by Step

The Sign Test

1. Type the data for Example 13–1 into a column of MINITAB. Name the column SnowCones.

2. Select **Stat>Nonparametrics> 1-Sample Sign Test.**

3. Double-click SnowCones in the list box.

4. Click on Test median, then enter the hypothesized value of **40.**

5. Click [OK]. In the session window the *P*-value is 0.0075.

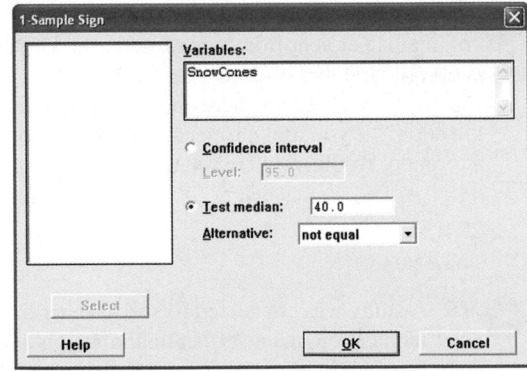

The Paired-Sample Sign Test

1. Enter the data for Example 13–3 into a worksheet; only the Before and After columns are necessary. Calculate a column with the differences to begin the process.

2. Select **Calc>Calculator.**

3. Type **D** in the box for Store result in variable.

4. Move to the Expression box, then click on Before, the subtraction sign, and After. The completed entry is shown.

5. Click [OK].

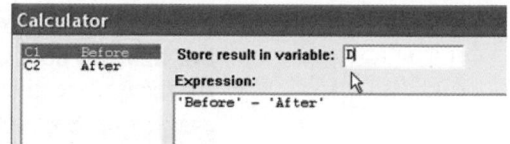

MINITAB will calculate the differences and store them in the first available column with the name "D." Use the instructions for the Sign Test on the differences D with a hypothesized value of zero.

Sign Test for Median: D

```
Sign test of median = 0.00000 versus not = 0.00000
     N  Below  Equal  Above     P    Median
D   10      2      1      7  0.1797   1.000
```

The P-value is 0.1797. Do not reject the null hypothesis.

Excel
Step by Step

The Sign Test

Excel does not have a procedure to conduct the sign test. However, you may conduct this test by using the MegaStat Add-in available on your CD and Online Learning Center. If you have not installed this add-in, do so by following the instructions on page 24.

1. Enter the data from Example 13–1 into column A of a new worksheet.

2. Select **MegaStat>Nonparametric Tests>Sign Test.**

3. Type **A1:A20** in the Input range box.

4. Type **40** in the box for the Hypothesized value, and select not equal for Alternative.

5. Click [OK].

The P-value is 0.0075. Reject the null hypothesis.

13–4 The Wilcoxon Rank Sum Test

Objective 3

Test hypotheses, using the Wilcoxon rank sum test.

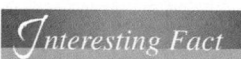

One in four married women now earns more than her husband.

The sign test does not consider the magnitude of the data. For example, whether a value is 1 point or 100 points below the median, it will receive a negative sign. And when one compares values in the pretest/posttest situation, the magnitude of the differences is not considered. The Wilcoxon tests consider differences in magnitudes by using ranks.

The two tests considered in this section and in Section 13–5 are the **Wilcoxon rank sum test,** which is used for independent samples, and the **Wilcoxon signed-rank test,** which is used for dependent samples. Both tests are used to compare distributions. The parametric equivalents are the z and t tests for independent samples (Sections 9–2 and 9–4) and the t test for dependent samples (Section 9–5). For the parametric tests, as stated previously, the samples must be selected from approximately normally distributed populations, but the only assumption for the Wilcoxon signed-rank tests is that the population of differences has a symmetric distribution.

In the Wilcoxon tests, the values of the data for both samples are combined and then ranked. If the null hypothesis is true—meaning that there is no difference in the population distributions—then the values in each sample should be ranked approximately the same. Therefore, when the ranks are summed for each sample, the sums should be approximately equal, and the null hypothesis will not be rejected. If there is a large difference in the sums of the ranks, then the distributions are not identical, and the null hypothesis will be rejected.

The first test to be considered is the Wilcoxon rank sum test for independent samples. For this test, both sample sizes must be greater than or equal to 10. The formulas needed for the test are given next.

Formula for the Wilcoxon Rank Sum Test When Samples Are Independent

$$z = \frac{R - \mu_R}{\sigma_R}$$

where

$$\mu_R = \frac{n_1(n_1 + n_2 + 1)}{2}$$

$$\sigma_R = \sqrt{\frac{n_1 n_2 (n_1 + n_2 + 1)}{12}}$$

R = sum of ranks for smaller sample size (n_1)

n_1 = smaller of sample sizes

n_2 = larger of sample sizes

$n_1 \geq 10$ and $n_2 \geq 10$

Note that if both samples are the same size, either size can be used as n_1.

Example 13–4 illustrates the Wilcoxon rank sum test for independent samples.

Example 13–4

Two independent samples of army and marine recruits are selected, and the time in minutes it takes each recruit to complete an obstacle course is recorded, as shown in the table. At $\alpha = 0.05$, is there a difference in the times it takes the recruits to complete the course?

Army	15	18	16	17	13	22	24	17	19	21	26	28	Mean = 19.67
Marines	14	9	16	19	10	12	11	8	15	18	25		Mean = 14.27

Solution

Step 1 State the hypotheses and identify the claim.

H_0: There is no difference in the times it takes the recruits to complete the obstacle course.

H_1: There is a difference in the times it takes the recruits to complete the obstacle course (claim).

Step 2 Find the critical value. Since $\alpha = 0.05$ and this test is a two-tailed test, use the z values of $+1.96$ and -1.96 from Table E.

Step 3 Compute the test value.

a. Combine the data from the two samples, arrange the combined data in order, and rank each value. Be sure to indicate the group.

Time	8	9	10	11	12	13	14	15	15	16	16	17
Group	M	M	M	M	M	A	M	A	M	A	M	A
Rank	1	2	3	4	5	6	7	8.5	8.5	10.5	10.5	12.5

Time	17	18	18	19	19	21	22	24	25	26	28
Group	A	M	A	A	M	A	A	A	M	A	A
Rank	12.5	14.5	14.5	16.5	16.5	18	19	20	21	22	23

b. Sum the ranks of the group with the smaller sample size. (*Note:* If both groups have the same sample size, either one can be used.) In this case, the sample size for the marines is smaller.

$$R = 1 + 2 + 3 + 4 + 5 + 7 + 8.5 + 10.5 + 14.5 + 16.5 + 21$$
$$= 93$$

c. Substitute in the formulas to find the test value.

$$\mu_R = \frac{n_1(n_1 + n_2 + 1)}{2} = \frac{(11)(11 + 12 + 1)}{2} = 132$$

$$\sigma_R = \sqrt{\frac{n_1 n_2(n_1 + n_2 + 1)}{12}} = \sqrt{\frac{(11)(12)(11 + 12 + 1)}{12}}$$
$$= \sqrt{264} = 16.2$$

$$z = \frac{R - \mu_R}{\sigma_R} = \frac{93 - 132}{16.2} = -2.41$$

Step 4 Make the decision. The decision is to reject the null hypothesis, since $-2.41 < -1.96$.

Step 5 Summarize the results. There is enough evidence to support the claim that there is a difference in the times it takes the recruits to complete the course.

The steps for the Wilcoxon rank sum test are given in the Procedure Table.

Procedure Table

Wilcoxon Rank Sum Test

Step 1 State the hypotheses and identify the claim.

Step 2 Find the critical value(s). Use Table E.

Step 3 Compute the test value.
  a. Combine the data from the two samples, arrange the combined data in order, and rank each value.
  b. Sum the ranks of the group with the smaller sample size. (*Note:* If both groups have the same sample size, either one can be used.)
  c. Use these formulas to find the test value.

$$\mu_R = \frac{n_1(n_1 + n_2 + 1)}{2}$$

$$\sigma_R = \sqrt{\frac{n_1 n_2(n_1 + n_2 + 1)}{12}}$$

$$z = \frac{R - \mu_R}{\sigma_R}$$

where R is the sum of the ranks of the data in the smaller sample and n_1 and n_2 are each greater than or equal to 10.

Step 4 Make the decision.

Step 5 Summarize the results.

Applying the Concepts **13–4**

School Lunch

A nutritionist decided to see if there was a difference in the number of calories served for lunch in elementary and secondary schools. She selected a random sample of eight elementary schools and another random sample of eight secondary schools in Pennsylvania. The data are shown.

Elementary	Secondary
648	694
589	730
625	750
595	810
789	860
727	702
702	657
564	761

1. Are the samples independent or dependent?
2. What are the hypotheses?
3. What nonparametric test would you use to test the claim?
4. What critical value would you use?
5. What is the test value?
6. What is your decision?
7. What is the corresponding parametric test?
8. What assumption would you need to meet to use the parametric test?
9. If this assumption were not met, would the parametric test yield the same results?

See page 705 for the answers.

Exercises 13–4

1. What are the minimum samples sizes for the Wilcoxon rank sum test?

2. What are the parametric equivalent tests for the Wilcoxon rank sum tests?

3. What distribution is used for the Wilcoxon rank sum test?

For Exercises 4 through 11, use the Wilcoxon rank sum test. Assume that the samples are independent. Also perform each of these steps.

   a. State the hypotheses and identify the claim.
   b. Find the critical value(s).
   c. Compute the test value.
   d. Make the decision.
   e. Summarize the results.

Use the traditional method of hypothesis testing unless otherwise specified.

 4. A random sample of men and women in prison was asked to give the length of sentence each received for a

certain type of crime. At $\alpha = 0.05$, test the claim that there is no difference in the sentence received by each gender. The data (in months) are shown here.

Males	8	12	6	14	22	27	32	24	26
Females	7	5	2	3	21	26	30	9	4

Males	19	15	13		
Females	17	23	12	11	16

5. A researcher wishes to see if there is a difference in the number of jetplanes grounded for repairs and the number of jetplanes in storage. A sample of data for the past several years is shown. At $\alpha = 0.05$, test the claim that there is no difference in the number of planes that are stored and the number of planes grounded. Based on these results, should an airline plan for more storage facilities or repair facilities?

Year	1	2	3	4	5	6
In storage	179	149	140	152	191	432
Grounded	647	716	1,480	1,338	1,285	208

Year	7	8	9	10
In storage	727	345	1054	371
Grounded	511	806	615	913

Source: Back Aviation Solutions.

6. To test the claim that there is no difference in the lifetimes of two brands of handheld video games, a researcher selects a sample of 11 video games of each brand. The lifetimes (in months) of each brand are shown here. At $\alpha = 0.01$, can the researcher conclude that there is a difference in the distributions of lifetimes for the two brands?

Brand A	42	34	39	42	22	47	51	34	41	39	28
Brand B	29	39	38	43	45	49	53	38	44	43	32

7. A researcher wishes to see if the stopping distance for midsize automobiles is different from the stopping distance for compact automobiles at a speed of 70 miles per hour. The data are shown. At $\alpha = 0.10$, test the claim that the stopping distances are the same. If one of your safety concerns is stopping distance, would it make a difference which type of automobile you purchase?

Automobile	1	2	3	4	5	6	7	8	9	10
Midsize	188	190	195	192	186	194	188	187	214	203
Compact	200	211	206	297	198	204	218	212	196	193

Source: Based on information from the National Highway Traffic Safety Administration.

8. Two groups of employees were given a questionnaire to ascertain their degree of job satisfaction. The scale ranged from 0 to 100. The groups were divided into those who had under 5 years of work experience and those who had 5 or more years of experience. The data are shown here. At $\alpha = 0.10$, test the claim that there is no difference in the job satisfaction of the two groups, as measured by the questionnaire. Use the *P*-value method.

Under 5	78	98	83	86	75	77	72	68
5 and over	94	79	82	85	73	66	64	59

Under 5	56	93	97	99	93
5 and over	52	58	63	68	88

9. A game commissioner wishes to see if the number of hunting accidents in counties in western Pennsylvania is different from the number of hunting accidents in counties in eastern Pennsylvania. A sample of counties from the two regions is selected, and the numbers of hunting accidents are shown. At $\alpha = 0.05$, is there a difference in the number of accidents in the two areas? If so, give a possible reason for the difference.

Western PA	10	21	11	11	9	17	13	8	15	17
Eastern PA	14	3	7	13	11	2	8	5	5	6

Source: Pennsylvania Game Commission.

10. Supervisors were asked to rate the productivity of employees on their jobs. A researcher wishes to see whether married men receive higher ratings than single men. A rating scale of 1 to 50 yielded the data shown here. At $\alpha = 0.01$, is there evidence to support this claim?

Single men	48	46	42	50	38	36	40	31	28	24	49	34
Married men	44	35	41	37	42	43	29	31	37	32	36	

11. A study was conducted to see whether there is a difference in the time it takes employees of a factory to assemble the product. Samples of high school graduates and nongraduates were timed. At $\alpha = 0.05$, is there a difference in the distributions for the two groups in the times needed to assemble the product? The data (in minutes) are shown here.

Graduates	3.6	3.2	4.4	3.0	5.6	6.3	8.2
Nongraduates	2.7	3.8	5.3	1.6	1.9	2.4	2.9
Graduates	7.1	5.8	7.3	6.4	4.2	4.7	
Nongraduates	1.7	2.6	2.0	3.1	3.4	3.9	

Technology *Step by Step*

MINITAB
Step by Step

Wilcoxon Rank Sum Test (Mann-Whitney)

1. Enter the data for Example 13–4 into two columns of a worksheet.
2. Name the columns **Army** and **Marines**.
3. Select **Stat>Nonparametric>Mann-Whitney.**
4. Double-click Army for the First Sample.
5. Double-click Marines for the Second Sample.
6. Click [OK].

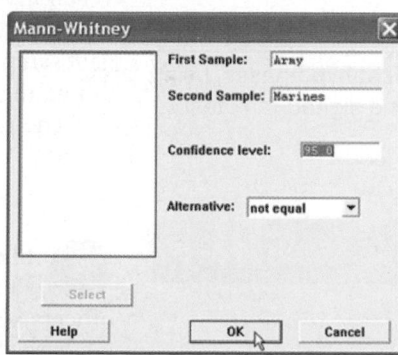

Mann-Whitney Test and CI: Army, Marines

```
          N  Median
Army     12  18.500
Marines  11  14.000
```

```
Point estimate for ETA1-ETA2 is 6.000
95.5 Percent CI for ETA1-ETA2 is (1.003, 9.998)
W = 183.0
Test of ETA1 = ETA2 vs ETA1 not = ETA2 is significant at 0.0178
The test is significant at 0.0177 (adjusted for ties)
```

The P-value for the test is 0.0177. Reject the null hypothesis. There is a significant difference in the times it takes the recruits to complete the course.

Excel
Step by Step

The Wilcoxon Mann-Whitney Test

Excel does not have a procedure to conduct the Mann-Whitney test. However, you may conduct this test by using the MegaStat Add-in available on your CD and Online Learning Center. If you have not installed this add-in, do so by following the instructions on page 24.

1. Enter the data from Example 13–4 into columns A and B of a new worksheet.
2. Select **MegaStat>Nonparametric Tests>Wilcoxon-Mann/Whitney Test.**
3. Type **A1:A12** in the box for Group 1.
4. Type **B1:B11** in the box for Group 2.
5. Check the option labeled Correct for ties, and select not equal for Alternative.
6. Click [OK].

Wilcoxon-Mann/Whitney Test

n	sum of ranks	
12	183	Group 1
11	93	Group 2
23	276	total

```
144.00 expected value
 16.23 standard deviation
  2.37 z, corrected for ties
0.0177 p-value (two-tailed)
```

The P-value is 0.0177. Reject the null hypothesis.

13–5 The Wilcoxon Signed-Rank Test

Objective 4

Test hypotheses, using the signed-rank test.

When the samples are dependent, as they would be in a before-and-after test using the same subjects, the Wilcoxon signed-rank test can be used in place of the t test for dependent samples. Again, this test does not require the condition of normality. Table K is used to find the critical values.

The procedure for this test is shown in Example 13–5.

Example 13–5

In a large department store, the owner wishes to see whether the number of shoplifting incidents per day will change if the number of uniformed security officers is doubled. A sample of 7 days before security is increased and 7 days after the increase shows the number of shoplifting incidents.

Number of shoplifting incidents

Day	Before	After
Monday	7	5
Tuesday	2	3
Wednesday	3	4
Thursday	6	3
Friday	5	1
Saturday	8	6
Sunday	12	4

Is there enough evidence to support the claim, at $\alpha = 0.05$, that there is a difference in the number of shoplifting incidents before and after the increase in security?

Solution

Step 1 State the hypotheses and identify the claim.

H_0: There is no difference in the number of shoplifting incidents before and after the increase in security.

H_1: There is a difference in the number of shoplifting incidents before and after the increase in security (claim).

Step 2 Find the critical value from Table K. Since $n = 7$ and $\alpha = 0.05$ for this two-tailed test, the critical value is 2. See Figure 13–2.

Figure 13–2

Finding the Critical Value in Table K for Example 13–5

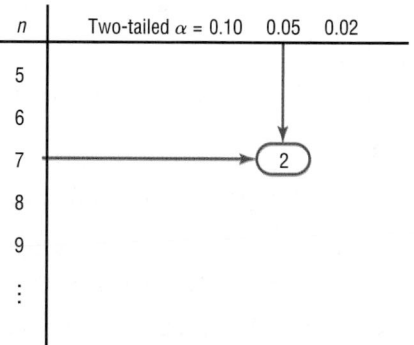

Step 3 Find the test value.

a. Make a table as shown here.

Day	Before, X_B	After, X_A	Difference $D = X_B - X_A$	Absolute value $|D|$	Rank	Signed rank
Mon.	7	5				
Tues.	2	3				
Wed.	3	4				
Thurs.	6	3				
Fri.	5	1				
Sat.	8	6				
Sun.	12	4				

b. Find the differences (before minus after), and place the values in the Difference column.

$7 - 5 = 2$ $6 - 3 = 3$ $8 - 6 = 2$

$2 - 3 = -1$ $5 - 1 = 4$ $12 - 4 = 8$

$3 - 4 = -1$

c. Find the absolute value of each difference, and place the results in the Absolute value column. (*Note:* The absolute value of any number except 0 is the positive value of the number. Any differences of 0 should be ignored.)

$$|2| = 2 \qquad |3| = 3 \qquad |2| = 2$$
$$|-1| = 1 \qquad |4| = 4 \qquad |8| = 8$$
$$|-1| = 1$$

d. Rank each absolute value from lowest to highest, and place the rankings in the Rank column. In the case of a tie, assign the values that rank plus 0.5.

Value	2	1	1	3	4	2	8
Rank	3.5	1.5	1.5	5	6	3.5	7

e. Give each rank a plus or minus sign, according to the sign in the Difference column. The completed table is shown here.

Interesting Fact

Nearly one in three unmarried adults lives with a parent today.

Day	Before, X_B	After, X_A	Difference $D = X_B - X_A$	Absolute value $|D|$	Rank	Signed rank
Mon.	7	5	2	2	3.5	+3.5
Tues.	2	3	−1	1	1.5	−1.5
Wed.	3	4	−1	1	1.5	−1.5
Thurs.	6	3	3	3	5	+5
Fri.	5	1	4	4	6	+6
Sat.	8	6	2	2	3.5	+3.5
Sun.	12	4	8	8	7	+7

f. Find the sum of the positive ranks and the sum of the negative ranks separately.

Positive rank sum $(+3.5) + (+5) + (+6) + (+3.5) + (+7) = +25$
Negative rank sum $(-1.5) + (-1.5) \qquad\qquad\qquad\qquad = -3$

g. Select the smaller of the absolute values of the sums ($|-3|$), and use this absolute value as the test value w_s. In this case, $w_s = |-3| = 3$.

Step 4   Make the decision. Reject the null hypothesis if the test value is less than or equal to the critical value. In this case, $3 > 2$; hence, the decision is not to reject the null hypothesis.

Step 5   Summarize the results. There is not enough evidence to support the claim that there is a difference in the number of shoplifting incidents. Hence, the security increase probably made no difference in the number of shoplifting incidents.

The rationale behind the signed-rank test can be explained by a diet example. If the diet is working, then the majority of the postweights will be smaller than the preweights. When the postweights are subtracted from the preweights, the majority of the signs will be positive, and the absolute value of the sum of the negative ranks will be small. This sum will probably be smaller than the critical value obtained from Table K, and the null hypothesis will be rejected. On the other hand, if the diet does not work, some people will gain weight, other people will lose weight, and still other people will remain about the same weight. In this case, the sum of the positive ranks and the absolute value of the sum of the negative ranks will be approximately equal and will be about one-half of the sum of the absolute value of all the ranks. In this case, the smaller of the absolute values of the two sums will still be larger than the critical value obtained from Table K, and the null hypothesis will not be rejected.

When $n \geq 30$, the normal distribution can be used to approximate the Wilcoxon distribution. The same critical values from Table E used for the z test for specific α values are used. The formula is

$$z = \frac{w_s - \dfrac{n(n+1)}{4}}{\sqrt{\dfrac{n(n+1)(2n+1)}{24}}}$$

where

n = number of pairs where difference is not 0
w_s = smaller sum in absolute value of signed ranks

The steps for the Wilcoxon signed-rank test are given in the Procedure Table.

Procedure Table

Wilcoxon Signed-Rank Test

Step 1 State the hypotheses and identify the claim.

Step 2 Find the critical value from Table K.

Step 3 Compute the test value.

    a. Make a table, as shown.

Before, X_B	After, X_A	Difference $D = X_B - X_A$	Absolute value $|D|$	Rank	Signed rank

    b. Find the differences (before − after), and place the values in the Difference column.

    c. Find the absolute value of each difference, and place the results in the Absolute value column.

    d. Rank each absolute value from lowest to highest, and place the rankings in the Rank column.

    e. Give each rank a positive or negative sign, according to the sign in the Difference column.

    f. Find the sum of the positive ranks and the sum of the negative ranks separately.

    g. Select the smaller of the absolute values of the sums, and use this absolute value as the test value w_s.

Step 4 Make the decision. Reject the null hypothesis if the test value is less than or equal to the critical value.

Step 5 Summarize the results.

    Note: When $n \geq 30$, use Table E and the test value

$$z = \frac{w_s - \dfrac{n(n+1)}{4}}{\sqrt{\dfrac{n(n+1)(2n+1)}{24}}}$$

where

n = number of pairs where difference is not 0
w_s = smaller sum in absolute value of signed ranks

Applying the Concepts **13–5**

Pain Medication

A researcher decides to see how effective a pain medication is. Eight subjects were asked to determine the severity of their pain by using a scale of 1 to 10, with 1 being very minor and 10 being very severe. Then each was given the medication, and after 1 hour, they were asked to rate the severity of their pain, using the same scale.

Subject	1	2	3	4	5	6	7	8
Before	8	6	2	3	4	6	2	7
After	6	5	3	1	2	6	1	6

1. What is the purpose of the study?
2. Are the samples independent or dependent?
3. What are the hypotheses?
4. What nonparametric test could be used to test the claim?
5. What significance level would you use?
6. What is your decision?
7. What parametric test could you use?
8. Would the results be the same?

See page 705 for the answers.

Exercises 13–5

1. What is the parametric equivalent test for the Wilcoxon signed-rank test?

For Exercises 2 and 3, find the sum of the signed ranks. Assume that the samples are dependent. State which sum is used as the test value.

2.
Pretest	18	32	35	37	25	41	52	43	56	62
Posttest	20	21	26	37	29	40	31	37	51	65

3.
Pretest	108	97	115	162	156	105	153
Posttest	110	97	103	168	143	112	141

For Exercises 4 through 8, use Table K to determine whether the null hypothesis should be rejected.

4. $w_s = 62$, $n = 21$, $\alpha = 0.05$, two-tailed test

5. $w_s = 18$, $n = 15$, $\alpha = 0.02$, two-tailed test

6. $w_s = 53$, $n = 25$, $\alpha = 0.05$, one-tailed test

7. $w_s = 142$, $n = 28$, $\alpha = 0.05$, one-tailed test

8. $w_s = 109$, $n = 27$, $\alpha = 0.025$, one-tailed test

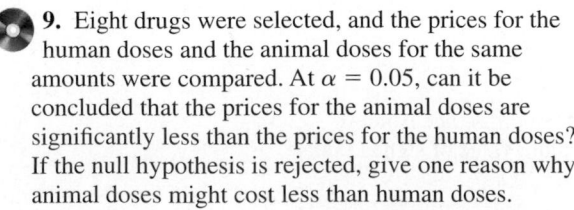 9. Eight drugs were selected, and the prices for the human doses and the animal doses for the same amounts were compared. At $\alpha = 0.05$, can it be concluded that the prices for the animal doses are significantly less than the prices for the human doses? If the null hypothesis is rejected, give one reason why animal doses might cost less than human doses.

Human dose	0.67	0.64	1.20	0.51	0.87	0.74	0.50	1.22
Animal dose	0.13	0.18	0.42	0.25	0.57	0.57	0.49	1.28

Source: House Committee on Government Reform.

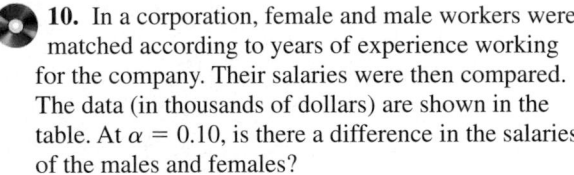 10. In a corporation, female and male workers were matched according to years of experience working for the company. Their salaries were then compared. The data (in thousands of dollars) are shown in the table. At $\alpha = 0.10$, is there a difference in the salaries of the males and females?

Males	18	43	32	27	15	45	21	22
Females	16	38	35	29	15	46	25	28

11. A researcher wished to compare the number of viewers who watched the Presidential debates with the number (in millions) of viewers who watched the Vice Presidential debates for the same years. At $\alpha = 0.05$, can it be concluded that the number of people who watched the Presidential debates is larger than the number of people who watched the Vice Presidential debates? Based on your answer, should the Vice Presidential debates continue? Explain your answer.

Year/Debates	1992	1996	2000	2004
Presidential	62.4	36.1	46.6	62.5
Vice Presidential	51.2	26.6	29.0	43.6

Source: Nielsen Media Research.

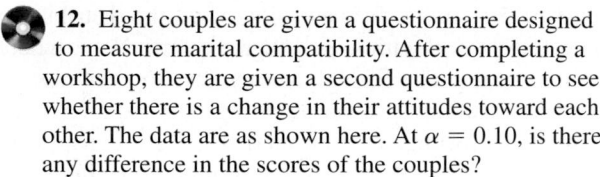

12. Eight couples are given a questionnaire designed to measure marital compatibility. After completing a workshop, they are given a second questionnaire to see whether there is a change in their attitudes toward each other. The data are as shown here. At $\alpha = 0.10$, is there any difference in the scores of the couples?

Before	43	52	37	29	51	62	57	61
After	48	59	36	29	60	68	59	72

13. A researcher wishes to compare the prices for prescription drugs in the United States with those in Canada. The same drugs and dosages were compared in

each country. At $\alpha = 0.05$, can it be concluded that the drugs in Canada are cheaper?

Drug	1	2	3	4	5	6
United States	3.31	2.27	2.54	3.13	23.40	3.16
Canada	1.47	1.07	1.34	1.34	21.44	1.47

Drug	7	8	9	10
United States	1.98	5.27	1.96	1.11
Canada	1.07	3.39	2.22	1.13

Source: IMS Health and other sources.

MINITAB
Step by Step

Wilcoxon Signed-Rank Test

Test the median value for the differences of two dependent samples. Use Example 13–5.

1. Enter the data into two columns of a worksheet. Name the columns **Before** and **After.**

2. Calculate the differences, using **Calc>Calculator.**

3. Type **D** in the box for Store result in variable.

4. In the expression box, type **Before − After.**

5. Click [OK].

6. Select **Stat>Nonparametric> 1-Sample Wilcoxon.**

7. Select C3 for the Variable.

8. Click on Test median. The value should be 0.

9. Click [OK].

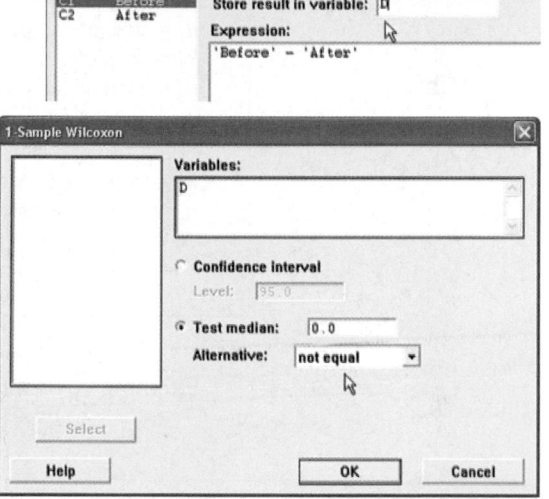

Wilcoxon Signed-Rank Test: D
```
Test of median = 0.000000 versus median not = 0.000000
             N
         for   Wilcoxon              Estimated
    N   Test  Statistic     P         Median
D   7     7        25.0   0.076        2.250
```

The *P*-value of the test is 0.076. Do not reject the null hypothesis.

13–6   The Kruskal-Wallis Test

Objective 5

Test hypotheses, using the Kruskal-Wallis test.

The analysis of variance uses the *F* test to compare the means of three or more populations. The assumptions for the ANOVA test are that the populations are normally distributed and that the population variances are equal. When these assumptions cannot be met, the nonparametric **Kruskal-Wallis test,** sometimes called the *H test,* can be used to compare three or more means.

In this test, each sample size must be 5 or more. In these situations, the distribution can be approximated by the chi-square distribution with $k - 1$ degrees of freedom, where k = number of groups. This test also uses ranks. The formula for the test is given next.

In the Kruskal-Wallis test, one considers all the data values as a group and then ranks them. Next, the ranks are separated and the H formula is computed. This formula approximates the variance of the ranks. If the samples are from different populations, the sums of the ranks will be different and the H value will be large; hence, the null hypothesis will be rejected if the H value is large enough. If the samples are from the same population, the sums of the ranks will be approximately the same and the H value will be small; therefore, the null hypothesis will not be rejected. This test is always a right-tailed test. The chi-square table, Table G, with d.f. = $k - 1$, should be used for critical values.

Formula for the Kruskal-Wallis Test

$$H = \frac{12}{N(N + 1)}\left(\frac{R_1^2}{n_1} + \frac{R_2^2}{n_2} + \cdots + \frac{R_k^2}{n_k}\right) - 3(N + 1)$$

where

R_1 = sum of ranks of sample 1
n_1 = size of sample 1
R_2 = sum of ranks of sample 2
n_2 = size of sample 2
    .
    .
    .
R_k = sum of ranks of sample k
n_k = size of sample k
$N = n_1 + n_2 + \cdots + n_k$
k = number of samples

Example 13–6 illustrates the procedure for conducting the Kruskal-Wallis test.

Example 13-6

A researcher tests three different brands of breakfast drink to see how many milliequivalents of potassium per quart each contains. These data are obtained.

Brand A	Brand B	Brand C
4.7	5.3	6.3
3.2	6.4	8.2
5.1	7.3	6.2
5.2	6.8	7.1
5.0	7.2	6.6

At $\alpha = 0.05$, is there enough evidence to reject the hypothesis that all brands contain the same amount of potassium?

Solution

Step 1   State the hypotheses and identify the claim.

H_0: There is no difference in the amount of potassium contained in the brands (claim).

H_1: There is a difference in the amount of potassium contained in the brands.

Step 2   Find the critical value. Use the chi-square table, Table G, with d.f. = $k - 1$ (k = number of groups). With $\alpha = 0.05$ and d.f. = $3 - 1 = 2$, the critical value is 5.991.

Step 3   Compute the test value.

    a. Arrange all the data from the lowest to highest, and rank each value.

Amount	Brand	Rank
3.2	A	1
4.7	A	2
5.0	A	3
5.1	A	4
5.2	A	5
5.3	B	6
6.2	C	7
6.3	C	8
6.4	B	9
6.6	C	10
6.8	B	11
7.1	C	12
7.2	B	13
7.3	B	14
8.2	C	15

    b. Find the sum of the ranks of each brand.

Brand A $1 + 2 + 3 + 4 + 5 = 15$

Brand B $6 + 9 + 11 + 13 + 14 = 53$

Brand C $7 + 8 + 10 + 12 + 15 = 52$

    c. Substitute in the formula.

$$H = \frac{12}{N(N + 1)}\left(\frac{R_1^2}{n_1} + \frac{R_2^2}{n_2} + \frac{R_3^2}{n_3}\right) - 3(N + 1)$$

where

$$N = 15 \qquad R_1 = 15 \qquad R_2 = 53 \qquad R_3 = 52$$
$$n_1 = n_2 = n_3 = 5$$

Therefore,

$$H = \frac{12}{15(15 + 1)}\left(\frac{15^2}{5} + \frac{53^2}{5} + \frac{52^2}{5}\right) - 3(15 + 1) = 9.38$$

Step 4   Make the decision. Since the test value 9.38 is greater than the critical value 5.991, the decision is to reject the null hypothesis.

Step 5   Summarize the results. There is enough evidence to reject the claim that there is no difference in the amount of potassium contained in the three brands. Hence, not all brands contain the same amount of potassium.

The steps for the Kruskal-Wallis test are given in the Procedure Table.

Interesting Fact

Albert Einstein was born on March 14. This is sometimes called pi day since it is the 14th day of the third month of any year and 3.14 is the first three digits of pi.

Procedure Table

Kruskal-Wallis Test

Step 1 State the hypotheses and identify the claim.

Step 2 Find the critical value. Use the chi-square table, Table G, with d.f. $= k - 1$ ($k =$ number of groups).

Step 3 Compute the test value.
   a. Arrange the data from lowest to highest and rank each value.
   b. Find the sum of the ranks of each group.
   c. Substitute in the formula.

$$H = \frac{12}{N(N + 1)}\left(\frac{R_1^2}{n_1} + \frac{R_2^2}{n_2} + \cdots + \frac{R_k^2}{n_k}\right) - 3(N + 1)$$

   where
   $$N = n_1 + n_2 + \cdots + n_k$$
   $R_k =$ sum of ranks for kth group
   $k =$ number of groups

Step 4 Make the decision.

Step 5 Summarize the results.

Applying the Concepts 13–6

Heights of Waterfalls

You are doing research for an article on the waterfalls on our planet. You want to make a statement about the heights of waterfalls on three continents. Three samples of waterfall heights (in feet) are shown.

North America	Africa	Asia
600	406	330
1200	508	830
182	630	614
620	726	1100
1170	480	885
442	2014	330

1. What questions are you trying to answer?
2. What nonparametric test would you use to find the answer?
3. What are the hypotheses?
4. Select a significance level and run the test. What is the H value?
5. What is your conclusion?
6. What is the corresponding parametric test?
7. What assumptions would you need to make to conduct this test?

See page 705 for the answers.

Exercises 13–6

For Exercises 1 through 11, perform these steps.

a. State the hypotheses and identify the claim.
b. Find the critical value.
c. Compute the test value.
d. Make the decision.
e. Summarize the results.

Use the traditional method of hypothesis testing unless otherwise specified.

1. Samples of four different cereals show the following number of calories for the suggested servings of each brand. At $\alpha = 0.05$, is there a difference in the number of calories for the different brands?

Brand A	Brand B	Brand C	Brand D
112	110	109	106
120	118	116	122
135	123	125	130
125	128	130	117
108	102	128	116
121	101	132	114

2. A test to measure self-esteem is given to three different samples of individuals based on birth order. The scores range from 0 to 50. The data are shown here. At $\alpha = 0.05$, is there a difference in the scores?

Oldest child	Middle child	Youngest child
48	50	47
46	49	45
42	42	46
41	43	30
37	39	32
32	28	41

3. A researcher wishes to compare the prices of three types of lawnmowers. At $\alpha = 0.10$, can it be concluded that there is a difference in the prices? Based on your answer, do you feel that the cost should be a factor in determining which type of lawnmower a person would purchase?

Gas-powered self-propelled	Gas-powered push	Electric
290	320	188
325	360	245
210	200	470
300	229	395
330	160	

4. Three brands of microwave dinners were advertised as low in sodium. Samples of the three different brands show the following milligrams of sodium. At $\alpha = 0.05$, is there a difference in the amount of sodium among the brands?

Brand A	Brand B	Brand C
810	917	893
702	912	790
853	952	603
703	958	744
892	893	623
732		743
713		609
613		

5. A nutritionist wishes to compare the number of carbohydrates in one serving of three low-carbohydrate foods. At $\alpha = 0.01$, is there a difference in the number of calories? Based on your answer, which type of food would you recommend if a person wished to limit carbohydrates?

Pasta	Ice cream	Bread
11	5	43
22	13	62
16	10	71
29	8	49
25	12	50

6. A recent study recorded the number of job offers received by newly graduated chemical engineers at three colleges. The data are shown here. At $\alpha = 0.05$, is there a difference in the average number of job offers received by the graduates at the three colleges?

College A	College B	College C
6	2	10
8	1	12
7	0	9
5	3	13
6	6	4

7. A meteorologist wishes to see if there is a difference in the number of deaths in the United States due to severe weather. The data from the past 6 years are shown here. At $\alpha = 0.10$, is there a difference in the number of deaths from the different weather conditions?

Lightning	Tornado	Flash flood	Blizzard
39	30	46	54
41	39	55	43
73	39	45	39
74	53	109	35
67	50	62	56
68	32	30	48

Source: Jack Williams, *The USA TODAY Weather Almanac.*

8. An electronics store manager wishes to compare the costs (in dollars) of three types of computer printer. The data are shown. At $\alpha = 0.05$, can it be concluded that there is a difference in the prices? Based on your answer, do you think that a certain type of printer generally costs more than the other types?

Inkjet printers	Multifunction printers	Laser printers
149	98	192
199	119	159
249	149	198
239	249	198
99	99	229
79	199	

 9. In a large city, the number of crimes per week in five precincts is recorded for five weeks. The data are shown here. At $\alpha = 0.01$, is there a difference in the number of crimes?

Precinct 1	Precinct 2	Precinct 3	Precinct 4	Precinct 5
105	87	74	56	103
108	86	83	43	98
99	91	78	52	94
97	93	74	58	89
92	82	60	62	88

 10. A recent study examined the number of unemployed people in five cities who are actively

seeking employment. They are listed here according to the education each received. At $\alpha = 0.05$, is there a difference in the number of unemployed based on education received? Use the *P*-value method.

High school diploma	College degree	Postgraduate degree
49	23	7
43	49	38
51	54	23
108	87	52
68	28	26

 11. Three different methods of first aid instruction are given to students. The same final examination is given to each class. The data are shown here. At $\alpha = 0.10$, is there a difference in the final examination scores? Use the *P*-value method.

Method A	Method B	Method C
98	97	99
100	88	94
95	82	96
92	84	89
86	75	81
76	73	72
71	74	

Technology *Step by Step*

MINITAB
Step by Step

Kruskal-Wallis Test

The data for this test must be "stacked." All the numeric data must be in one column, and the second column identifies the brand.

↓	C1	C2-T
	Potassium	Brand
1	4.7	A
2	3.2	A
3	5.1	A
4	5.2	A
5	5.0	A
6	5.3	B
7	6.4	B
8	7.3	B
9	6.8	B
10	7.2	B
11	6.3	C
12	8.2	C
13	6.2	C
14	7.1	C
15	6.6	C

1. Stack the data for Example 13–6 into two columns of a worksheet.

   a) First, enter all the potassium amounts into one column.

   b) Name this column **Potassium.**

   c) Enter code **A, B,** or **C** for the brand into the next column.

   d) Name this column **Brand.**

   The worksheet is shown.

2. Select **Stat>Nonparametric>Kruskal-Wallis.**

3. Double-click C1 Potassium to select it for Response.

This variable must be quantitative so the column for Brand will not be available in the list until the cursor is in the Factor text box.

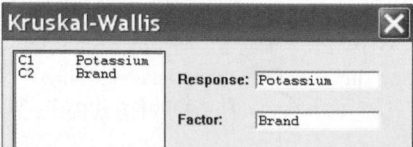

4. Select C2 Brand for Factor.

5. Click [OK].

Kruskal-Wallis Test: Potassium versus Brand
```
Kruskal-Wallis Test on Potassium
Brand    N   Median   Ave Rank      Z
A        5   5.000        3.0    -3.06
B        5   6.800       10.6     1.59
C        5   6.600       10.4     1.47
Overall 15                8.0
H = 9.38   DF = 2   P = 0.009
```

The value $H = 9.38$ has a P-value of 0.009. Reject the null hypothesis.

Excel
Step by Step

The Kruskal-Wallis Test

Excel does not have a procedure to conduct the Kruskal-Wallis test. However, you may conduct this test by using the MegaStat Add-in available on your CD and Online Learning Center. If you have not installed this add-in, do so by following the instructions on page 24.

1. Enter the brand names **A, B,** and **C** in cells in the first row.

2. Enter the data in three separate columns A, B, and C below the labels.

3. Select **MegaStat>Nonparametric Tests>Kruskal-Wallis Test.**

4. Type **A1:C6** in the Input range box.

5. Check the option labeled Correct for ties, and select not equal for Alternative.

6. Click [OK].

Kruskal-Wallis Test

Median	n	Avg. Rank	
5.00	5	3.00	A
6.80	5	10.60	B
6.60	5	10.40	C
6.30	15		Total

```
                        9.380  H
                            2  d.f.
                        .0092  p-value
        multiple comparison values for avg. ranks
          6.77(.05)             8.30(.01)
```

The P-value is 0.0092. Reject the null hypothesis.

13–7

The Spearman Rank Correlation Coefficient and the Runs Test

The techniques of regression and correlation were explained in Chapter 10. To determine whether two variables are linearly related, one uses the Pearson product moment correlation coefficient. Its values range from $+1$ to -1. One assumption for testing the hypothesis that $\rho = 0$ for the Pearson coefficient is that the populations from which the samples are obtained are normally distributed. If this requirement cannot be met, the nonparametric equivalent, called the **Spearman rank correlation coefficient** (denoted by r_s), can be used when the data are ranked.

Rank Correlation Coefficient

The computations for the rank correlation coefficient are simpler than those for the Pearson coefficient and involve ranking each set of data. The difference in ranks is found,

Objective 6

Compute the Spearman rank correlation coefficient.

and r_s is computed by using these differences. If both sets of data have the same ranks, r_s will be $+1$. If the sets of data are ranked in exactly the opposite way, r_s will be -1. If there is no relationship between the rankings, r_s will be near 0.

Formula for Computing the Spearman Rank Correlation Coefficient

$$r_s = 1 - \frac{6\,\Sigma d^2}{n(n^2 - 1)}$$

where
  d = difference in ranks
  n = number of data pairs

This formula is algebraically equivalent to the formula for r given in Chapter 10, except that ranks are used instead of raw data.

The computational procedure is shown in Example 13–7. For a test of the significance of r_s, Table L is used for values of n up to 30. For larger values, the normal distribution can be used. (See Exercises 24 through 28 in the exercise section.)

Example 13–7

Two students were asked to rate eight different textbooks for a specific course on an ascending scale from 0 to 20 points. Points were assigned for each of several categories, such as reading level, use of illustrations, and use of color. At $\alpha = 0.05$, test the hypothesis that there is a significant linear correlation between the two students' ratings. The data are shown in the following table.

Textbook	Student 1's rating	Student 2's rating
A	4	4
B	10	6
C	18	20
D	20	14
E	12	16
F	2	8
G	5	11
H	9	7

Solution

Step 1 State the hypotheses.

  $H_0: \rho = 0$ and $H_1: \rho \neq 0$

Step 2 Find the critical value. Use Table L to find the value for $n = 8$ and $\alpha = 0.05$. It is 0.738. See Figure 13–3.

Figure 13–3

Finding the Critical Value in Table L for Example 13–7

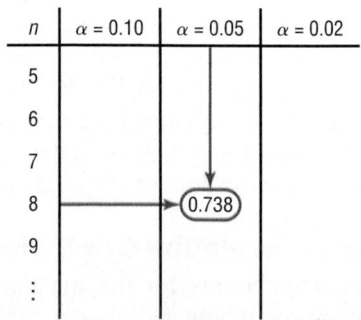

n	$\alpha = 0.10$	$\alpha = 0.05$	$\alpha = 0.02$
5			
6			
7			
8		0.738	
9			
⋮			

Step 3   Find the test value.

   a. Rank each data set, as shown in the table.

Textbook	Student 1	Rank	Student 2	Rank
A	4	7	4	8
B	10	4	6	7
C	18	2	20	1
D	20	1	14	3
E	12	3	16	2
F	2	8	8	5
G	5	6	11	4
H	9	5	7	6

   Let X_1 be the first student's rankings and X_2 be the second student's rankings.

   b. Subtract the rankings $(X_1 - X_2)$.

   $$7 - 8 = -1 \qquad 4 - 7 = -3 \qquad \text{etc.}$$

   c. Square the differences.

   $$(-1)^2 = 1 \qquad (-3)^2 = 9 \qquad \text{etc.}$$

   d. Find the sum of the squares.

   $$1 + 9 + 1 + 4 + 1 + 9 + 4 + 1 = 30$$

   The results can be summarized in a table, as shown here.

X_1	X_2	$d = X_1 - X_2$	d^2
7	8	−1	1
4	7	−3	9
2	1	1	1
1	3	−2	4
3	2	1	1
8	5	3	9
6	4	2	4
5	6	−1	1
			$\Sigma d^2 = 30$

   e. Substitute in the formula to find r_s.

   $$r_s = 1 - \frac{6\,\Sigma d^2}{n(n^2 - 1)}$$

   where n = the number of data pairs. For this problem,

   $$r_s = 1 - \frac{(6)(30)}{8(8^2 - 1)} = 1 - \frac{180}{504} = 0.643$$

Step 4   Make the decision. Do not reject the null hypothesis since $r_s = 0.643$, which is less than the critical value of 0.738.

Step 5   Summarize the results. There is not enough evidence to say that there is a correlation between the rankings of the two students.

Unusual Stat

You are almost twice as likely to be killed while walking with your back to traffic as you are when facing traffic, according to the National Safety Council.

The steps for finding and testing the Spearman rank correlation coefficient are given in the Procedure Table.

Procedure Table

Finding and Testing the Spearman Rank Correlation Coefficient

Step 1 State the hypotheses.

Step 2 Rank each data set.

Step 3 Subtract the rankings $(X_1 - X_2)$.

Step 4 Square the differences.

Step 5 Find the sum of the squares.

Step 6 Substitute in the formula.

$$r_s = 1 - \frac{6 \sum d^2}{n(n^2 - 1)}$$

where
  d = difference in ranks
  n = number of pairs of data

Step 7 Find the critical value.

Step 8 Make the decision.

Step 9 Summarize the results.

Objective 7

Test hypotheses, using the runs test.

The Runs Test

When samples are selected, one assumes that they are selected at random. How does one know if the data obtained from a sample are truly random? Before the answer to this question is given, consider the following situations for a researcher interviewing 20 people for a survey. Let their genders be denoted by M for male and F for female. Suppose the participants were chosen as follows:

Situation 1   M M M M M M M M M M F F F F F F F F F F

It does not look as if the people in this sample were selected at random, since 10 males were selected first, followed by 10 females.
    Consider a different selection:

Situation 2   F M F M F M F M F M F M F M F M F M F M

In this case, it seems as if the researcher selected a female, then a male, etc. This selection is probably not random either.
    Finally, consider the following selection:

Situation 3   F F F M M F M F M M F F M M F F M M M F

This selection of data looks as if it may be random, since there is a mix of males and females and no apparent pattern to their selection.
    Rather than try to guess whether the data of a sample have been selected at random, statisticians have devised a nonparametric test to determine randomness. This test is called the **runs test.**

A **run** is a succession of identical letters preceded or followed by a different letter or no letter at all, such as the beginning or end of the succession.

For example, the first situation presented has two runs:

Run 1: M M M M M M M M M

Run 2: F F F F F F F F F

The second situation has 20 runs. (Each letter constitutes one run.) The third situation has 11 runs.

Run 1:	F F F	Run 5:	F	Run 9:	F F
Run 2:	M M	Run 6:	M M	Run 10:	M M M
Run 3:	F	Run 7:	F F	Run 11:	F
Run 4:	M	Run 8:	M M		

Example 13–8

Determine the number of runs in each sequence.

a. F F F M M M F F F F M

b. H H H T T T T

c. A A B B A A B B A A B B

Solution

a. There are four runs, as shown.

F F F   M M   F F F F   M

  1   2   3   4

b. There are two runs, as shown.

H H H   T T T T

  1   2

c. There are six runs, as shown.

A A   B B   A A   B B   A A   B B

  1   2   3   4   5   6

The test for randomness considers the number of runs rather than the frequency of the letters. For example, for data to be selected at random, there should not be too few or too many runs, as in situations 1 and 2. The runs test does not consider the questions of how many males or females were selected or how many of each are in a specific run.

To determine whether the number of runs is within the random range, use Table M in Appendix C. The values are for a two-tailed test with $\alpha = 0.05$. For a sample of 12 males and 8 females, the table values shown in Figure 13–4 mean that any number of runs from 7 to 15 would be considered random. If the number of runs is 6 or less or 16 or more, the sample is probably not random, and the null hypothesis should be rejected.

Example 13–9 shows the procedure for conducting the runs test by using letters as data. Example 13–10 shows how the runs test can be used for numerical data.

Figure 13–4

Finding the Critical Value in Table M

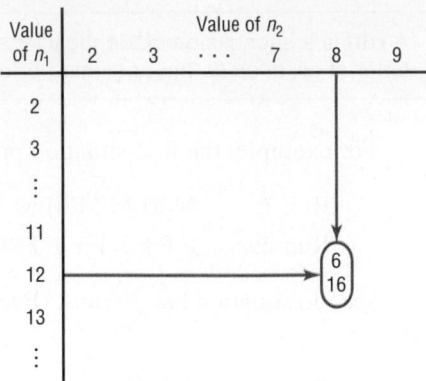

Example 13–9

On a commuter train, the conductor wishes to see whether the passengers enter the train at random. He observes the first 25 people, with the following sequence of males (M) and females (F).

$$\text{F F F M M F F F F M F M M M F F F F M M F F F M M}$$

Test for randomness at $\alpha = 0.05$.

Solution

Step 1 State the hypotheses and identify the claim.

H_0: The passengers board the train at random, according to gender (claim).

H_1: The null hypothesis is not true.

Step 2 Find the number of runs. Arrange the letters according to runs of males and females, as shown.

Run	Gender
1	F F F
2	M M
3	F F F F
4	M
5	F
6	M M M
7	F F F F
8	M M
9	F F F
10	M M

There are 15 females (n_1) and 10 males (n_2).

Step 3 Find the critical value. Find the number of runs in Table M for $n_1 = 15$, $n_2 = 10$, and $\alpha = 0.05$. The values are 7 and 18. *Note:* In this situation the critical value is found after the number of runs is determined.

Step 4 Make the decision. Compare these critical values with the number of runs. Since the number of runs is 10 and 10 is between 7 and 18, do not reject the null hypothesis.

Step 5 Summarize the results. There is not enough evidence to reject the hypothesis that the passengers board the train at random according to gender.

Example 13–10	Twenty people enrolled in a drug abuse program. Test the claim that the ages of the people, according to the order in which they enroll, occur at random, at $\alpha = 0.05$. The data are 18, 36, 19, 22, 25, 44, 23, 27, 27, 35, 19, 43, 37, 32, 28, 43, 46, 19, 20, 22.

Solution

Step 1 State the hypotheses and identify the claim.

H_0: The ages of the people, according to the order in which they enroll in a drug program, occur at random (claim).

H_1: The null hypothesis is not true.

Step 2 Find the number of runs.

a. Find the median of the data. Arrange the data in ascending order.

18 19 19 19 20 22 22 23 25 27 27

28 32 35 36 37 43 43 44 46

The median is 27.

b. Replace each number in the original sequence with an A if it is above the median and with a B if it is below the median. Eliminate any numbers that are equal to the median.

B A B B B A B A B A A A A A A B B B

c. Arrange the letters according to runs.

Run	Letters
1	B
2	A
3	B B B
4	A
5	B
6	A
7	B
8	A A A A A
9	B B B

Step 3 Find the critical value. Table M shows that with $n_1 = 9$, $n_2 = 9$, and $\alpha = 0.05$, the number of runs should be between 5 and 15.

Step 4 Make the decision. Since there are 9 runs and 9 falls between 5 and 15, the null hypothesis is not rejected.

Step 5 Summarize the results. There is not enough evidence to reject the hypothesis that the ages of the people who enroll occur at random.

The steps for the runs test are given in the Procedure Table.

Procedure Table

The Runs Test

Step 1 State the hypotheses and identify the claim.

Step 2 Find the number of runs.

Procedure Table (*continued*)

Note: When the data are numerical, find the median. Then compare each data value with the median and classify it as above or below the median. Other methods such as odd-even can also be used. (Discard any value that is equal to the median.)

Step 3 Find the critical value. Use Table M.

Step 4 Make the decision. Compare the actual number of runs with the critical value.

Step 5 Summarize the results.

Applying the Concepts 13–7

Tall Trees

As a biologist, you wish to see if there is a relationship between the heights of tall trees and their diameters. You find the following data for the diameter (in inches) of the tree at 4.5 feet from the ground and the corresponding heights (in feet).

Diameter (in.)	Height (ft)
1024	261
950	321
451	219
505	281
761	159
644	83
707	191
586	141
442	232
546	108

Source: *The World Almanac and Book of Facts.*

1. What question are you trying to answer?
2. What type of nonparametric analysis could be used to answer the question?
3. What would be the corresponding parametric test that could be used?
4. Which test do you think would be better?
5. Perform both tests and write a short statement comparing the results.

See page 705 for the answer.

Exercises 13–7

For Exercises 1 through 4, find the critical value from Table L for the rank correlation coefficient, given sample size n and α. Assume that the test is two-tailed.

1. $n = 14$, $\alpha = 0.01$
2. $n = 28$, $\alpha = 0.02$
3. $n = 10$, $\alpha = 0.05$
4. $n = 9$, $\alpha = 0.01$

For Exercises 5 through 14, perform these steps.

a. Find the Spearman rank correlation coefficient.
b. State the hypotheses.
c. Find the critical value. Use $\alpha = 0.05$.
d. Make the decision.
e. Summarize the results.

Use the traditional method of hypothesis testing unless otherwise specified.

5. The table shows the total number of tornadoes that occurred in 10 states from 1962 to 1991 and the record high temperatures for the same states. At $\alpha = 0.10$, is there a relationship between the number of tornadoes and the record high temperatures?

State	Tornadoes	Record high temperatures
AL	668	112
CO	781	118
FL	1590	109
IL	798	117
KS	1198	121
NY	169	108
PA	310	111
TN	360	113
VT	21	105
WI	625	114

Source: *The World Almanac and Book of Facts.*

6. Six cities are selected, and the number of daily passenger trips (in thousands) for subways and commuter rail service is obtained. At $\alpha = 0.05$, is there a relationship between the variables? Suggest one reason why the transportation authority might use the results of this study.

City	1	2	3	4	5	6
Subway	845	494	425	313	108	41
Rail	39	291	142	103	33	39

Source: American Public Transportation Association.

7. The table shows the average maximum sentence length for certain crimes and the actual time served in months. At $\alpha = 0.05$, is there a relationship between the two?

Crime	Sentence	Time served
Murder	227	97
Rape	120	45
Robbery	106	40
Burglary	77	26
Drug offenses	60	18
Weapons offenses	49	21

Source: *World Almanac and Book of Facts.*

8. Six different summer theater actors were ranked by male and female patrons on the basis of diction and appearance. The data are shown here (1 is the highest rating). At $\alpha = 0.05$, is there a relationship between the rankings?

Actors	A	B	C	D	E	F
Males	6	3	2	5	1	4
Females	4	5	1	6	3	2

9. Eight music videos were ranked by teenagers and their parents on style and clarity, with 1 being the highest ranking. The data are shown here. At $\alpha = 0.05$, is there a relationship between the rankings?

Music videos	1	2	3	4	5	6	7	8
Teenagers	4	6	2	8	1	7	3	5
Parents	1	7	5	4	3	8	2	6

10. Eight school districts are randomly selected, and the number of employees who retired and the payment for their unused sick days are obtained. At $\alpha = 0.05$, is there a relationship between the data? How can a school board use the results of this study?

No. of employees	10	5	17	8	10	10	11	9
Amount paid in thousands of dollars	83	30	29	35	40	18	90	54

11. Shown is a comparison between the average gasoline prices charged by a gasoline station and a car rental company for 10 cities in the United States. At $\alpha = 0.05$, is there a relationship between the prices? How might a person who travels a lot and rents an automobile use the information obtained from this study?

Car rental agency price	5.12	5.27	5.29	5.18	5.59
Gas station price	2.09	1.96	2.29	1.94	2.20
Car rental agency price	5.30	5.83	5.46	5.12	5.15
Gas station price	2.20	2.40	2.12	2.15	2.11

Source: AAA Oil Price Information Service and car rental agencies.

12. A meteorologist wishes to see if the number of deaths from tornadoes is related to the number of tornadoes per year. At $\alpha = 0.05$, is there a relationship? The data for a randomly selected 11-year period are shown. How could an insurance agency use this information?

Number of deaths	53	39	39	33	69	30
Number of tornadoes	1133	1132	1297	1173	1082	1234
Number of deaths	25	67	130	40	94	
Number of tornadoes	1173	1148	1424	898	1342	

Source: NOAA.

13. Shown are the number of students enrolled in cyber school for five randomly selected school districts and the per-pupil costs for the cyber school education. At $\alpha = 0.10$, is there a relationship between the two variables? How might this information be useful to school administrators?

Number of students	10	6	17	8	11
Per-pupil cost	7200	9393	7385	4500	8203

Source: *Tribune-Review.*

14. Shown are the price for a human dose of several prescription drugs and the price for an equivalent dose for animals. At $\alpha = 0.10$, is there a relationship between the variables?

Humans	0.67	0.64	1.20	0.51	0.87	0.74	0.50	1.22
Animals	0.13	0.18	0.42	0.25	0.57	0.57	0.49	1.28

Source: House Committee on Government Reform.

15. A school dentist wanted to test the claim, at $\alpha = 0.05$, that the number of cavities in fourth-grade students is random. Forty students were checked, and the number of cavities each had is shown here. Test for randomness of the values above or below the median.

0	4	6	0	6	2	5	3	1	5	1
2	2	1	3	7	3	6	0	2	6	0
2	3	1	5	2	1	3	0	2	3	7
3	1	5	1	1	2	2				

16. A drawing was held each day for a month. Categorize the winning numbers as odd or even. The data follow. Test for randomness, at $\alpha = 0.05$.

409	872	235	338	472	481	318	129	229
084	291	991	356	212	457	473	834	304
361	301	051	652	405	458	094	633	809
299	712	802						

17. The winning numbers for the Pennsylvania State Lotto drawing for April are listed here. Classify each as odd or even and test for randomness, at $\alpha = 0.05$. No drawings were held on weekends.

457 605 348 927 463 300 620 261 614 098 467
961 957 870 262 571 633 448 187 462 565
180 050

18. An irate student believes that the answers to his history professor's final true/false examination are not random. Test the claim, at $\alpha = 0.05$. The answers to the questions are shown.

T T T F F T T T F F F F F F T
T T F F F T T T F T F F T T F

19. A machine manufactures audiocassette cases that are either defective (D) or acceptable (A). The sequence is shown here. At $\alpha = 0.05$, test for randomness.

D A A A A A A D D A D D A A A
D D A A A A A A A D D D A A A

20. Twenty shoppers are in a checkout line at a grocery store. At $\alpha = 0.05$, test for randomness of their gender: male (M) or female (F). The data are shown here.

F M M F F M F M M F
F M M M F F F F F M

21. A supervisor records the number of employees absent over a 30-day period. Test for randomness, at $\alpha = 0.05$.

27	6	19	24	18	12	15	17	18	20
0	9	4	12	3	2	7	7	0	5
32	16	38	31	27	15	5	9	4	10

22. A ski lodge manager observes the weather for the month of February. If his customers are able to ski, he records S; if weather conditions do not permit skiing, he records N. Test for randomness, at $\alpha = 0.05$.

S S S S S S N N N N N N N N
N S S S N N S S S S S S S S

23. These data are the scores on an IQ exam in the order that the students finished the test. At $\alpha = 0.05$, test for randomness.

101	98	99	110	119	121	118
106	96	88	91	97	92	106
94	93	100	89	86	95	99

Extending the Concepts

When $n \geq 30$, the formula $r = \dfrac{\pm z}{\sqrt{n-1}}$ can be used to find the critical values for the rank correlation coefficient. For example, if $n = 40$ and $\alpha = 0.05$ for a two-tailed test,

$$r = \frac{\pm 1.96}{\sqrt{40 - 1}} = \pm 0.314$$

Hence, any r_s greater than or equal to $+0.314$ or less than or equal to -0.314 is significant.

For Exercises 24 through 28, find the critical r value for each (assume that the test is two-tailed).

24. $n = 50$, $\alpha = 0.05$

25. $n = 30$, $\alpha = 0.01$

26. $n = 35$, $\alpha = 0.02$

27. $n = 60$, $\alpha = 0.10$

28. $n = 40$, $\alpha = 0.01$

Technology *Step by Step*

MINITAB
Step by Step

Runs Test for Randomness

1. Sequence is important! Enter the data down C1 in the same order they were collected. Do not sort them! Use the data from Example 13–10.

2. Calculate the median and store it as a constant.

   a) Select **Calc>Column Statistics.**

   b) Check the option for Median.

   c) Use C1 Age for the Input Variable.

   d) Type the name of the constant MedianAge in the Store result in text box.

   e) Click [OK].

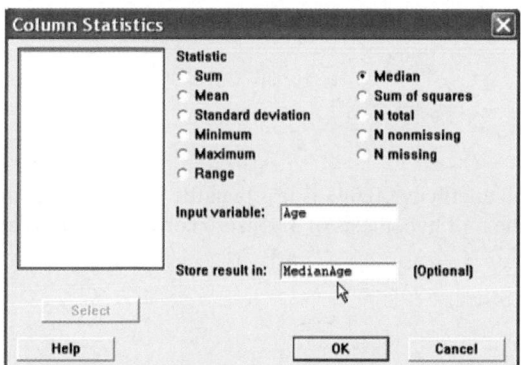 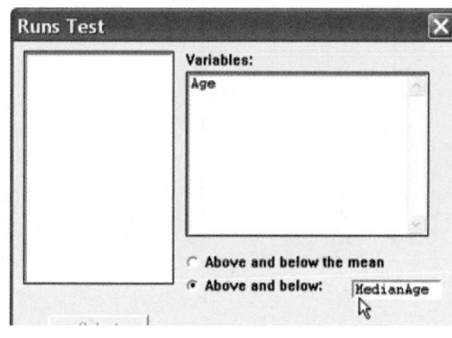

3. Select **Stat>Nonparametric>Runs Test.**

4. Select C1 Age as the variable.

5. Click the button for Above and below, then select MedianAge in the text box.

6. Click [OK]. The results will be displayed in the session window.

Runs Test: Age
```
Runs test for Age
Runs above and below K = 27

The observed number of runs = 9
The expected number of runs = 10.9
9 observations above K, 11 below
* N is small, so the following approximation may be invalid.
P-value = 0.378
```

The *P*-value is 0.378. Do not reject the null hypothesis.

Excel
Step by Step

Spearman Rank Correlation Coefficient

Excel does not have a procedure to compute the Spearman rank correlation coefficient. However, you may compute this statistic by using the MegaStat Add-in available on your CD and Online Learning Center. If you have not installed this add-in, do so by following the instructions on page 24.

1. Enter the rating scores from Example 13–7 in two separate columns of a new worksheet.

2. Select **MegaStat>Nonparametric Tests>Spearman Coefficient of Rank Correlation.**

3. In the dialog box, type **A1:B8** to select the rating scores.

4. Click [OK].

The results, including the final ranks from the resulting output sheet, are shown.

Spearman Coefficient of Rank Correlation

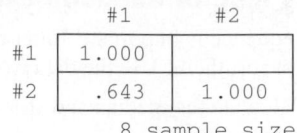

```
           #1        #2
    #1 ┌──────────┬──────────┐
       │ 1.000    │          │
    #2 │  .643    │ 1.000    │
       └──────────┴──────────┘
          8 sample size
```

```
±.707 critical value .05 (two-tail)
±.834 critical value .01 (two-tail)
```

```
Group 1    Group 2
    2          1
    5          2
    7          8
    8          6
    6          7
    1          4
    3          5
    4          3
```

Since the correlation coefficient 0.643 is less than the critical value, there is not enough evidence to reject the null hypothesis of a nonzero correlation between the variables.

13–8 Summary

In many research situations, the assumptions (particularly that of normality) for the use of parametric statistics cannot be met. Also, some statistical studies do not involve parameters such as means, variances, and proportions. For both situations, statisticians have developed nonparametric statistical methods, also called *distribution-free methods*.

There are several advantages to the use of nonparametric methods. The most important one is that no knowledge of the population distributions is required. Other advantages include ease of computation and understanding. The major disadvantage is that they are less efficient than their parametric counterparts when the assumptions for the parametric methods are met. In other words, larger sample sizes are needed to get results as accurate as those given by their parametric counterparts.

This list gives the nonparametric statistical tests presented in this chapter, along with their parametric counterparts.

Nonparametric test	Parametric test	Condition
Single-sample sign test	z or t test	One sample
Paired-sample sign test	z or t test	Two dependent samples
Wilcoxon rank sum test	z or t test	Two independent samples
Wilcoxon signed-rank test	t test	Two dependent samples
Kruskal-Wallis test	ANOVA	Three or more independent samples
Spearman rank correlation coefficient	Pearson's correlation coefficient	Relationships between variables
Runs test	None	Randomness

When the assumptions of the parametric tests can be met, the parametric tests should be used instead of their nonparametric counterparts.

Important Terms

distribution-free statistics 660

Kruskal-Wallis test 681

nonparametric statistics 660

parametric tests 660

ranking 662

run 691

runs test 690

sign test 663

Spearman rank correlation coefficient 687

Wilcoxon rank sum test 671

Wilcoxon signed-rank test 671

Important Formulas

Formula for the z test value in the sign test:

$$z = \frac{(X + 0.5) - (n/2)}{\sqrt{n}/2}$$

where

n = sample size (greater than or equal to 26)

X = smaller number of positive or negative signs

Formula for the Wilcoxon rank sum test:

$$z = \frac{R - \mu_R}{\sigma_R}$$

where

$$\mu_R = \frac{n_1(n_1 + n_2 + 1)}{2}$$

$$\sigma_R = \sqrt{\frac{n_1 n_2(n_1 + n_2 + 1)}{12}}$$

R = sum of ranks for smaller sample size (n_1)

n_1 = smaller of sample sizes

n_2 = larger of sample sizes

$n_1 \geq 10$ and $n_2 \geq 10$

Formula for the Wilcoxon signed-rank test:

$$z = \frac{w_s - \dfrac{n(n + 1)}{4}}{\sqrt{\dfrac{n(n + 1)(2n + 1)}{24}}}$$

where

n = number of pairs where difference is not 0 and $n \geq 30$

w_s = smaller sum in absolute value of signed ranks

Formula for the Kruskal-Wallis test:

$$H = \frac{12}{N(N + 1)}\left(\frac{R_1^2}{n_1} + \frac{R_2^2}{n_2} + \cdots + \frac{R_k^2}{n_k}\right) - 3(N + 1)$$

where

R_1 = sum of ranks of sample 1

n_1 = size of sample 1

R_2 = sum of ranks of sample 2

n_2 = size of sample 2

⋮

R_k = sum of ranks of sample k

n_k = size of sample k

$N = n_1 + n_2 + \cdots + n_k$

k = number of samples

Formula for the Spearman rank correlation coefficient:

$$r_s = 1 - \frac{6\,\Sigma d^2}{n(n^2 - 1)}$$

where

d = difference in ranks

n = number of data pairs

Review Exercises

For Exercises 1 through 13, follow this procedure:

  a. State the hypotheses and identify the claim.

  b. Find the critical value(s).

  c. Compute the test value.

  d. Make the decision.

  e. Summarize the results.

Use the traditional method of hypothesis testing unless otherwise specified.

 1. A researcher believes that the median number of wet days per month in Yosemite National Park is 5. A sample of the number of wet days per month for last year is shown. At $\alpha = 0.05$, is there enough evidence to reject the researcher's claim?

8   6   7   6   4   2   1   1   1   2   6   7

Source: Jack Williams, The USA TODAY Weather Almanac.

2. A tire manufacturer claims that the median lifetime of a certain brand of truck tires is 40,000 miles. A sample of 30 tires shows that 12 lasted longer than 40,000 miles. Is there enough evidence to reject the claim at $\alpha = 0.05$? Use the sign test.

3. A special diet including hormones is fed to adult hogs to see whether they will gain weight. The before and after weights (in pounds) are given here. Use the paired-sample sign test at $\alpha = 0.05$ to determine whether there is a weight gain.

Before	320	432	456	358	371	394	362	359	319
After	333	430	459	362	381	395	367	356	315

4. Shown here are the record high temperatures for Dawson Creek in British Columbia, Canada, and for Whitehorse in Yukon, Canada, for 12 months. Using the Wilcoxon rank sum test at $\alpha = 0.05$, do you find a difference in the record high temperatures? Use the P-value method.

Dawson Creek	52	60	57	71	86	89	94	93	88	80	66	52
White-horse	47	50	51	69	86	89	91	86	80	66	51	47

Source: Jack Williams, *The USA TODAY Weather Almanac.*

5. Samples of students majoring in business and engineering are selected, and the amount (in dollars) each spent on a required textbook for the fall semester is recorded. The data are shown here. For the Wilcoxon rank sum test at $\alpha = 0.10$, is there a difference in the amount spent by each group?

Business		Engineering	
48	36	98	73
52	62	72	78
74	50	63	93
63	46	78	88
51	53	55	86
49	58	58	85
		64	

6. Twelve automobiles were tested to see how many miles per gallon each one obtained. Under similar driving conditions, they were tested again, using a special additive. The data are shown here. At $\alpha = 0.05$, did the additive improve gas mileage? Use the Wilcoxon signed-rank test.

Before		After	
13.6	18.3	22.6	23.7
18.2	19.5	21.9	20.8
16.1	18.2	25.3	25.3
15.3	16.7	28.6	27.2
19.2	21.3	15.2	17.2
18.8	17.2	16.3	18.5

7. The number of sick days taken by seven assembly-line workers was recorded for one year. The owners of the company then installed brighter lighting and

permitted the employees to take a 10-minute break in the morning and afternoon. The number of sick days taken was recorded for another year. The data are shown here. From the Wilcoxon signed-rank test at $\alpha = 0.05$, can one conclude that the number of sick days taken was reduced?

Before	6	15	18	14	27	17	9
After	8	12	16	9	23	14	15

8. Samples of three types of ropes are tested for breaking strength. The data (in pounds) are shown here. At $\alpha = 0.05$, is there a difference in the breaking strength of the ropes? Use the Kruskal-Wallis test.

Cotton	Nylon	Hemp
230	356	506
432	303	527
505	361	581
487	405	497
451	432	459
380	378	507
462	361	562
531	399	571
366	372	499
372	363	475
453	306	505
488	304	561
462	318	532
467	322	501

9. Rats were fed three different diets for one month to see whether diet has any effect on learning. Each rat was then taught to traverse a simple maze. The number of trials it took each rat to learn the correct path is shown in the table. At $\alpha = 0.05$, does diet have any effect on learning? Use the Kruskal-Wallis test.

Diet 1	8	6	12	15	9	7	5
Diet 2	2	3	6	8	7	4	
Diet 3	9	15	17	8	4	13	18	20

10. A statistics instructor wishes to see whether there is a relationship between the number of homework exercises a student completes and her or his exam score. The data are shown here. Using the Spearman rank correlation coefficient, test the hypothesis that there is no relationship at $\alpha = 0.05$.

Homework problems	63	55	58	87	89	52	46	75	105
Exam score	85	71	75	98	93	63	72	89	100

11. Shown are the average number of viewers for 10 television shows for two consecutive years. At $\alpha = 0.05$, is there a relationship between the number of viewers?

Last year	28.9	26.4	20.8	25.0	21.0	19.2
This year	26.6	20.5	20.2	19.1	18.9	17.8
Last year	13.7	18.8	16.8	15.3		
This year	16.8	16.7	16.0	15.8		

Statistics Today

Too Much or Too Little?—Revisited

In this case, the manufacturer would select a sequence of bottles and see how many bottles contained more than 40 ounces, denoted by plus, and how many bottles contained less than 40 ounces, denoted by minus. The sequence could then be analyzed according to the number of runs, as explained in Section 13–7. If the sequence were not random, then the machine would need to be checked to see if it was malfunctioning. Another method that can be used to see if machines are functioning properly is *statistical quality control*. This method is beyond the scope of this book.

12. In a recent survey, 20 college students were asked, as they arrived for class, if they worked during the academic year (W) or did not work (N). At $\alpha = 0.05$, test for randomness. The data are shown here. Assume the probabilities are the same.

W	N	N	N	W	W	W	N	W	N
N	W	W	N	N	W	N	N	W	N

13. An instructor wishes to see whether grades of students who finish an exam occur at random. Shown here are the grades of 30 students in the order that they finished an exam. (Read from left to right across each row, and then proceed to the next row.) Test for randomness, at $\alpha = 0.05$.

87	93	82	77	64	98
100	93	88	65	72	73
56	63	85	92	95	91
88	63	72	79	55	53
65	68	54	71	73	72

Data Analysis

The Data Bank is found in Appendix D, or on the World Wide Web by following links from www.mhhe.com/math/stat/bluman

1. From the Data Bank, choose a sample and use the sign test to test one of the following hypotheses.

    a. For serum cholesterol, test H_0: median = 220 milligram percent (mg%).

    b. For systolic pressure, test H_0: median = 120 millimeters of mercury (mm Hg).

    c. For IQ, test H_0: median = 100.

    d. For sodium level, test H_0: median = 140 mEq/l.

2. From the Data Bank, select a sample of subjects. Use the Kruskal-Wallis test to see if the sodium levels of smokers and nonsmokers are equal.

3. From the Data Bank select a sample of 50 subjects. Use the Wilcoxon rank sum test to see if the means of the sodium levels of the males differ from those of the females.

Chapter Quiz

Determine whether each statement is true or false. If the statement is false, explain why.

1. Nonparametric statistics cannot be used to test the difference between two means.

2. Nonparametric statistics are more sensitive than their parametric counterparts.

3. Nonparametric statistics can be used to test hypotheses about parameters other than means, proportions, and standard deviations.

4. Parametric tests are preferred over their nonparametric counterparts, if the assumptions can be met.

Select the best answer.

5. The _____ test is used to test means when samples are dependent and the normality assumption cannot be met.

    a. Wilcoxon signed-rank   c. Sign

    b. Wilcoxon rank sum   d. Kruskal-Wallis

6. The Kruskal-Wallis test uses the _____ distribution.

   a. z   *c.* Chi-square
   b. t   *d.* F

7. The nonparametric counterpart of ANOVA is the
   _____.

   a. Wilcoxon signed-rank test
   b. Sign test
   c. Runs test
   d. None of the above

8. To see if two rankings are related, one can use the
   _____.

   a. Runs test
   b. Spearman correlation coefficient
   c. Sign test
   d. Kruskal-Wallis test

Complete the following statements with the best answer.

9. When the assumption of normality cannot be met, one
   can use _____ tests.

10. When data are _____ or _____ in nature,
    nonparametric methods are used.

11. To test to see whether a median was equal to a specific
    value, one would use the _____ test.

12. Nonparametric tests are less _____ than their
    parametric counterparts.

For the following exercises, use the traditional method of hypothesis testing unless otherwise specified.

13. The owner of a candy store states that she sells on
    average 300 candy bars per day. A random sample of
    18 days shows the number of candy bars sold each day.
    At $\alpha = 0.10$, is the claim correct? Use the sign test.

271	297	315	282	106	297	268	215
262	305	315	256	311	375	319	297
311	299						

14. A battery manufacturer claims that the median lifetime
    of a certain brand of heavy-duty battery is 1200 hours.
    A sample of 25 batteries shows that 15 lasted longer
    than 1200 hours. Test the claim at $\alpha = 0.05$. Use the
    sign test.

15. A special diet is fed to adult turkeys to see if
    they will gain weight. The before and after weights
    (in pounds) are given here. Use the paired-sample sign
    test at $\alpha = 0.05$ to see if there is weight gain.

Before	28	24	29	30	32	33	25	26	28
After	30	29	31	32	32	35	29	25	31

16. Two groups of alcoholics, one group male and the
    other female, were asked at what age they first drank
    alcohol. The data are shown here. Using the Wilcoxon
    rank sum test at $\alpha = 0.05$, is there a difference in the
    ages of the females and males?

Males	6	12	14	16	17	17	13	12	10	11
Females	8	9	9	12	14	15	12	16	17	19

17. Samples of students majoring in law and nursing
    are selected, and the amount each spent on textbooks
    for the spring semester is recorded here, in dollars.
    Using the Wilcoxon rank sum test at $\alpha = 0.10$, is there
    a difference in the amount spent by each group?

Law	167	158	162	106	98	206	112	121
Nursing	98	198	209	168	157	126	104	122

Law	133	145	151	199
Nursing	111	138	116	201

18. The grade point average of a group of students
    was recorded for one month. During the next nine-
    week grading period, the students attended a workshop
    on study skills. Their GPAs were recorded at the end of
    the grading period, and the data appear here. Using the
    Wilcoxon signed-rank test at $\alpha = 0.05$, can it be
    concluded that the GPA increased?

Before	3.0	2.9	2.7	2.5	2.1	2.6	1.9	2.0
After	3.2	3.4	2.9	2.5	3.0	3.1	2.4	2.8

19. Samples of three different types of wrapping tape
    are tested for breaking strength, in pounds. The data
    are shown here. At $\alpha = 0.05$, is there a difference in the
    breaking strength of the tapes? Use the Kruskal-Wallis
    test.

Type A	225	332	404	387	351	280	362	431	266
Type B	256	203	261	305	232	278	261	299	272
Type C	406	427	481	397	351	409	462	471	399

Type A	353	288	362	367	272
Type B	206	206	218	222	263
Type C	405	461	432	401	375

20. Three different groups of monkeys were fed
    three different medications for one month to see if
    the medication has any effect on reaction time.
    Each monkey was then taught to repeat a series of
    steps to receive a reward. The number of trials it
    took each to receive the reward is shown here.
    At $\alpha = 0.05$, does the medication have an effect on
    reaction time? Use the Kruskal-Wallis test. Use the
    P-value method.

Med. 1	8	7	11	14	8	6	5
Med. 2	3	4	6	7	9	3	4
Med. 3	8	14	13	7	5	9	12

21. Is there a relationship between the prescription
    drug prices in Canada and Great Britain? Use
    $\alpha = 0.10$.

| Canada | 1.47 | 1.07 | 1.34 | 1.34 | 1.47 | 1.07 | 3.39 | 1.11 | 1.13 |
| Great Britain | 1.67 | 1.08 | 1.67 | 0.82 | 1.73 | 0.95 | 2.86 | 0.41 | 1.70 |

Source: USA TODAY.

22. Is there a relationship between the amount of money (in millions of dollars) spent on the Head Start Program by the states and the number of students enrolled (in thousands)? Use $\alpha = 0.10$.

| Funding | 100 | 50 | 22 | 88 | 49 | 219 |
| Enrollment | 16 | 7 | 3 | 14 | 8 | 31 |

Source: Gannet News Service.

23. At the state registry of vital statistics, the birth certificates issued for females (F) and males (M) were tallied. At $\alpha = 0.05$, test for randomness. The data are shown here.

M M F F F F F F F F M M M M F F
M F M F M M M F F F

24. The output in revolutions per minute (rpm) of 10 motors was obtained. The motors were tested again under similar conditions after they had been reconditioned. The data are shown here. At $\alpha = 0.05$, did the reconditioning improve the motors' performance? Use the Wilcoxon signed-rank test.

| Before | 413 | 701 | 397 | 602 | 405 | 512 | 450 | 487 | 388 | 351 |
| After | 433 | 712 | 406 | 650 | 450 | 550 | 450 | 500 | 402 | 415 |

25. A statistician wishes to determine if a state's lottery numbers are selected at random. The winning numbers selected for the month of February are shown here. Test for randomness at $\alpha = 0.05$.

321 909 715 700 487 808 509 606 943 761
200 123 367 012 444 576 409 128 567 908
103 407 890 193 672 867 003 578

Critical Thinking Challenges

1. Two commuters ride to work together in one car. To decide who pays the toll for a bridge on the way to work, they flip a coin and the loser pays. Explain why over a period of one year, one person might have to pay the toll 5 days in a row. There is no toll on the return trip. (*Hint:* You may want to use random numbers.)

2. Shown in the next column are the type and number of medals each country won in the 2000 Summer Olympic Games. You are to rank the countries from highest to lowest. Gold medals are highest, followed by silver, followed by bronze. There are many different ways to rank objects and events. Here are several suggestions.

a. Rank the countries according to the total medals won.

b. List some advantages and disadvantages of this method.

c. Rank each country separately for the number of gold medals won, then for the number of silver medals won, and then for the number of bronze medals won. Then rank the countries according to the sum of the *ranks* for the categories.

d. Are the rankings of the countries the same as those in step *a*? Explain any differences.

e. List some advantages and disadvantages of this method of ranking.

f. A third way to rank the countries is to assign a weight to each medal. In this case, assign 3 points

for each gold medal, 2 points for each silver medal, and 1 point for each bronze medal the country won. Multiply the number of medals by the weights for each medal and find the sum. For example, since Austria won 2 gold medals, 1 silver medal, and 0 bronze medals, its rank sum is $(2 \times 3) + (1 \times 2) + (0 \times 1) = 8$. Rank the countries according to this method.

g. Compare the ranks using this method with those using the other two methods. Are the rankings the same or different? Explain.

h. List some advantages and disadvantages of this method.

i. Select two of the rankings, and run the Spearman rank correlation test to see if they differ significantly.

Summer Olympic Games 2000 Final Medal Standings

Country	Gold	Silver	Bronze
Austria	2	1	0
Canada	3	3	8
Germany	14	17	26
Italy	13	8	13
Norway	4	3	3
Russia	32	28	28
Switzerland	1	6	2
U.S.A.	40	24	33

Source: Reprinted with permission from the *World Almanac and Book of Facts.* World Almanac Education Group Inc.

Data Projects

Where appropriate, use MINITAB, the TI-83 Plus, the TI-84 Plus, or a computer program of your choice to complete these nonparametric exercises.

1. There are many nonparametric statistical tests. Decide on a project that will use one of the nonparametric tests, and collect data from a sample.

    a. Describe the purpose of the study.
    b. Define the population.
    c. State how the sample was obtained.
    d. Select an α value.
    e. State the hypotheses for the study.
    f. Decide which nonparametric test statistic will be used and compute the test value.
    g. Make the decision.
    h. Summarize the results.
    i. Conduct the corresponding parametric test and compare the results.
    j. Write a brief paragraph on which test is more appropriate, and give reasons why.

2. Select a variable on which you can perform a runs test. For example, you might observe the gender of 20 or 30 individuals waiting in the cafeteria, bookstore checkout, or registration line.

    a. Conduct the runs test and decide whether the sequence is random.
    b. Write a brief summary of the results.

You may use the following websites to obtain raw data:

Visit the data sets at the book's website found at
http://www.mhhe.com/math/stat/bluman
Click on the 6th edition.
http://lib.stat.cmu.edu/DASL
http://www.statcan.ca

Hypothesis-Testing Summary 3*

15. Test to see whether the median of a sample is a specific value when $n \geq 26$.

    Example: H_0: median $= 100$

    Use the sign test:

    $$z = \frac{(X + 0.5) - (n/2)}{\sqrt{n}/2}$$

16. Test to see whether two independent samples are obtained from populations that have identical distributions.

    Example: H_0: There is no difference in the ages of the subjects.

    Use the Wilcoxon rank sum test:

    $$z = \frac{R - \mu_R}{\sigma_R}$$

    where

    $$\mu_R = \frac{n_1(n_1 + n_2 + 1)}{2}$$

    $$\sigma_R = \sqrt{\frac{n_1 n_2 (n_1 + n_2 + 1)}{12}}$$

17. Test to see whether two dependent samples have identical distributions.

    Example: H_0: There is no difference in the effects of a tranquilizer on the number of hours a person sleeps at night.

    Use the Wilcoxon signed-rank test:

    $$z = \frac{w_s - \frac{n(n + 1)}{4}}{\sqrt{\frac{n(n + 1)(2n + 1)}{24}}}$$

    when $n \geq 30$.

18. Test to see whether three or more samples come from identical populations.

    Example: H_0: There is no difference in the weights of the three groups.

    Use the Kruskal-Wallis test:

    $$H = \frac{12}{N(N + 1)}\left(\frac{R_1^2}{n_1} + \frac{R_2^2}{n_2} + \cdots + \frac{R_k^2}{n_k}\right) - 3(N + 1)$$

19. Rank correlation coefficient.

    $$r_s = 1 - \frac{6 \, \Sigma d^2}{n(n^2 - 1)}$$

20. Test for randomness: Use the runs test.

*This summary is a continuation of Hypothesis-Testing Summary 2 at the end of Chapter 12.

Answers to Applying the Concepts

Section 13–2   Ranking Data

Percent	2.6	3.8	4.0	4.0	5.4	7.0	7.0	7.3	10.0
Rank	1	2	3.5	3.5	5	6.5	6.5	8	9

Section 13–3   Clean Air

1. The claim is that the median number of days that a large city failed to meet EPA standards is 11 days per month.

2. We will use the sign test, since we do not know anything about the distribution of the variable and we are testing the median.

3. H_0: median = 11 and H_1: median ≠ 11.

4. If $\alpha = 0.05$, then the critical value is 5.

5. The test value is 9.

6. Since $9 > 5$, we reject the null hypothesis. There is enough evidence to conclude that the median is not 11 days per month.

7. We cannot use a parametric test in this situation.

Section 13–4   School Lunch

1. The samples are independent since two different random samples were selected.

2. H_0: There is no difference in the number of calories served for lunch in elementary and secondary schools.

   H_1: There is a difference in the number of calories served for lunch in elementary and secondary schools.

3. We will use the Wilcoxon rank sum test.

4. The critical value is ± 1.96 if we use $\alpha = 0.05$.

5. The test statistic is $z = -2.15$.

6. Since $-2.15 < -1.96$, we reject the null hypothesis and conclude that there is a difference in the number of calories served for lunch in elementary and secondary schools.

7. The corresponding parametric test is the two-sample t test.

8. We would need to know that the samples were normally distributed to use the parametric test.

9. Since t tests are robust against variations from normality, the parametric test would yield the same results.

Section 13–5   Pain Medication

1. The purpose of the study is to see how effective a pain medication is.

2. These are dependent samples, since we have before and after readings on the same subjects.

3. H_0: The severity of pain after is the same as the severity of pain before the medication was administered.

   H_1: The severity of pain after is less than the severity of pain before the medication was administered.

4. We will use the Wilcoxon signed-rank test.

5. We will choose to use a significance level of 0.05.

6. The test statistic is $w_s = 25.5$. The critical value is 4. Since $25.5 > 4$, we reject the null hypothesis. There is enough evidence to conclude that the severity of pain after is less than the severity of pain before the medication was administered.

7. The parametric test that could be used is the t test for small dependent samples.

8. The results for the parametric test would be the same.

Section 13–6   Heights of Waterfalls

1. We are investigating the heights of waterfalls on three continents.

2. We will use the Kruskal-Wallis test.

3. H_0: There is no difference in the heights of waterfalls on the three continents.

   H_1: There is a difference in the heights of waterfalls on the three continents.

4. We will use the 0.05 significance level. The critical value is 5.991. Our test statistic is $H = 0.01$.

5. Since $0.01 < 5.991$, we fail to reject the null hypothesis. There is not enough evidence to conclude that there is a difference in the heights of waterfalls on the three continents.

6. The corresponding parametric test is analysis of variance (ANOVA).

7. To perform an ANOVA, the population must be normally distributed, the samples must be independent of each other, and the variances of the samples must be equal.

Section 13–7   Tall Trees

1. The biologist is trying to see if there is a relationship between the heights and diameters of tall trees.

2. We will use a Spearman rank correlation analysis.

3. The corresponding parametric test is the Pearson product moment correlation analysis.

4. Answers will vary.

5. The Pearson correlation coefficient is $r = 0.329$. The associated P-value is 0.353. We would fail to reject the null hypothesis that the correlation is zero. The Spearman's rank correlation coefficient is $r_s = 0.115$. We would reject the null hypothesis, at the 0.05 significance level, if $r_s > 0.648$. Since $0.115 < 0.648$, we fail to reject the null hypothesis that the correlation is zero. Both the parametric and nonparametric tests find that the correlation is not statistically significantly different from zero—it appears that no linear relationship exists between the heights and diameters of tall trees.

Sampling and Simulation

Objectives

After completing this chapter, you should be able to

1 Demonstrate a knowledge of the four basic sampling methods.

2 Recognize faulty questions on a survey and other factors that can bias responses.

3 Solve problems, using simulation techniques.

The Monty Hall Problem

On the game show *Let's Make A Deal,* host Monty Hall gave a contestant a choice of three doors. A valuable prize was behind one door, and nothing was behind the other two doors. When the contestant selected one door, host Monty Hall opened one of the other doors that the contestant didn't select and that had no prize behind it. (Monty Hall knew in advance which door had the prize.) Then he asked the contestant if he or she wanted to change doors or keep the one that the contestant originally selected. Now the question is, Should the contestant switch doors, or does it really matter? This chapter will show you how you can solve this problem by simulation. For the answer, see Statistics Today—Revisited.

14–1

Introduction

Most people have heard of Gallup, Harris, and Nielsen. These and other pollsters gather information about the habits and opinions of the U.S. people. Such survey firms, and the U.S. Census Bureau, gather information by selecting samples from well-defined populations. Recall from Chapter 1 that the subjects in the sample should be a subgroup of the subjects in the population. Sampling methods often use what are called *random numbers* to select samples.

Since many statistical studies use surveys and questionnaires, some information about these is presented in Section 14–3.

Random numbers are also used in *simulation techniques.* Instead of studying a real-life situation, which may be costly or dangerous, researchers create a similar situation in a laboratory or with a computer. Then, by studying the simulated situation, researchers can gain the necessary information about the real-life situation in a less expensive or safer manner. This chapter will explain some common methods used to obtain samples as well as the techniques used in simulations.

<table>
<tr><td>**14–2**</td></tr>
</table>

Common Sampling Techniques

In Chapter 1, a *population* was defined as all subjects (human or otherwise) under study. Since some populations can be very large, researchers cannot use every single subject, so a sample must be selected. A *sample* is a subgroup of the population. Any subgroup of the population, technically speaking, can be called a sample. However, for researchers to make valid inferences about population characteristics, the sample must be random.

Objective 1

Demonstrate a knowledge of the four basic sampling methods.

For a sample to be a **random sample,** every member of the population must have an equal chance of being selected.

When a sample is chosen at random from a population, it is said to be an **unbiased sample.** That is, the sample, for the most part, is representative of the population. Conversely, if a sample is selected incorrectly, it may be a biased sample. Samples are said to be **biased samples** when some type of systematic error has been made in the selection of the subjects.

A sample is used to get information about a population for several reasons:

1. *It saves the researcher time and money.*

2. *It enables the researcher to get information that he or she might not be able to obtain otherwise.* For example, if a person's blood is to be analyzed for cholesterol, a researcher cannot analyze every single drop of blood without killing the person. Or if the breaking strength of cables is to be determined, a researcher cannot test to destruction every cable manufactured, since the company would not have any cables left to sell.

3. *It enables the researcher to get more detailed information about a particular subject.* If only a few people are surveyed, the researcher can conduct in-depth interviews by spending more time with each person, thus getting more information about the subject. This is not to say that the smaller the sample, the better; in fact, the opposite is true. In general, larger samples—if correct sampling techniques are used—give more reliable information about the population.

It would be ideal if the sample were a perfect miniature of the population in all characteristics. This ideal, however, is impossible to achieve, because there are so many human traits (height, weight, IQ, etc.). The best that can be done is to select a sample that will be representative with respect to *some* characteristics, preferably those pertaining to the study. For example, if one-half of the population subjects are female, then approximately one-half of the sample subjects should be female. Likewise, other characteristics, such as age, socioeconomic status, and IQ, should be represented proportionately. To obtain unbiased samples, statisticians have developed several basic sampling methods. The most common methods are *random, systematic, stratified,* and *cluster sampling.* Each method will be explained in detail in this section.

In addition to the basic methods, there are other methods used to obtain samples. Some of these methods are also explained in this section.

Random Sampling

A random sample is obtained by using methods such as random numbers, which can be generated from calculators, computers, or tables. In *random sampling,* the basic requirement is that for a sample of size n, all possible samples of this size must have an equal chance of being selected from the population. But before the correct method of obtaining a random sample is explained, several incorrect methods commonly used by various researchers and agencies to gain information are discussed.

One incorrect method commonly used is to ask "the person on the street." News reporters use this technique quite often. Selecting people haphazardly on the street does not meet the requirement for simple random sampling, since not all possible samples of a specific size have an equal chance of being selected. Many people will be at home or at work when the interview is being conducted and therefore do not have a chance of being selected.

Another incorrect technique is to ask a question by either radio or television and have the listeners or viewers call the station to give their responses or opinions. Again, this sample is not random, since only those who feel strongly for or against the issue may respond and people may not have heard or seen the program. A third erroneous method is to ask people to respond by mail. Again, only those who are concerned and who have the time are likely to respond.

These methods do not meet the requirement of random sampling, since not all possible samples of a specific size have an equal chance of being selected. To meet this requirement, researchers can use one of two methods. The first method is to number each element of the population and then place the numbers on cards. Place the cards in a hat or fishbowl, mix them, and then select the sample by drawing the cards. When using this procedure, researchers must ensure that the numbers are well mixed. On occasion, when this procedure is used, the numbers are not mixed well, and the numbers chosen for the sample are those that were placed in the bowl last.

The second and preferred way of selecting a random sample is to use random numbers. Figure 14–1 shows a table of two-digit random numbers generated by a computer. A more detailed table of random numbers is found in Table D of Appendix C.

The theory behind random numbers is that each digit, 0 through 9, has an equal probability of occurring. That is, in every sequence of 10 digits, each digit has a probability of $\frac{1}{10}$ of occurring. This does not mean that in every sequence of 10 digits, one will find each digit. Rather, it means that on the average, each digit will occur once. For example, the digit 2 may occur 3 times in a sequence of 10 digits, but in later sequences, it may not occur at all, thus averaging to a probability of $\frac{1}{10}$.

To obtain a sample by using random numbers, number the elements of the population sequentially and then select each person by using random numbers. This process is shown in Example 14–1.

Random samples can be selected with or without replacement. If the same member of the population cannot be used more than once in the study, then the sample is selected without replacement. That is, once a random number is selected, it cannot be used later.

79	41	71	93	60	35	04	67	96	04	79	10	86
26	52	53	13	43	50	92	09	87	21	83	75	17
18	13	41	30	56	20	37	74	49	56	45	46	83
19	82	02	69	34	27	77	34	24	93	16	77	00
14	57	44	30	93	76	32	13	55	29	49	30	77
29	12	18	50	06	33	15	79	50	28	50	45	45
01	27	92	67	93	31	97	55	29	21	64	27	29
55	75	65	68	65	73	07	95	66	43	43	92	16
84	95	95	96	62	30	91	64	74	83	47	89	71
62	62	21	37	82	62	19	44	08	64	34	50	11
66	57	28	69	13	99	74	31	58	19	47	66	89
48	13	69	97	29	01	75	58	05	40	40	18	29
94	31	73	19	75	76	33	18	05	53	04	51	41
00	06	53	98	01	55	08	38	49	42	10	44	38
46	16	44	27	80	15	28	01	64	27	89	03	27
77	49	85	95	62	93	25	39	63	74	54	82	85
81	96	43	27	39	53	85	61	12	90	67	96	02
40	46	15	73	23	75	96	68	13	99	49	64	11

Figure 14–1

Table of Random Numbers

Note: In the explanations and examples of the sampling procedures, a small population will be used, and small samples will be selected from this population. Small populations are used for illustrative purposes only, because the entire population could be included with little difficulty. In real life, however, researchers must usually sample from very large populations, using the procedures shown in this chapter.

Example 14–1	Suppose a researcher wants to produce a television show featuring in-depth interviews with state governors on the subject of capital punishment. Because of time constraints, the 60-minute program will have room for only 10 governors. The researcher wishes to select the governors at random. Select a random sample of 10 states from 50. *Note:* This answer is not unique.

Solution

Step 1   Number each state from 1 to 50, as shown. In this case, they are numbered alphabetically.

01. Alabama	14. Indiana	27. Nebraska	40. South Carolina
02. Alaska	15. Iowa	28. Nevada	41. South Dakota
03. Arizona	16. Kansas	29. New Hampshire	42. Tennessee
04. Arkansas	17. Kentucky	30. New Jersey	43. Texas
05. California	18. Louisiana	31. New Mexico	44. Utah
06. Colorado	19. Maine	32. New York	45. Vermont
07. Connecticut	20. Maryland	33. North Carolina	46. Virginia
08. Delaware	21. Massachusetts	34. North Dakota	47. Washington
09. Florida	22. Michigan	35. Ohio	48. West Virginia
10. Georgia	23. Minnesota	36. Oklahoma	49. Wisconsin
11. Hawaii	24. Mississippi	37. Oregon	50. Wyoming
12. Idaho	25. Missouri	38. Pennsylvania	
13. Illinois	26. Montana	39. Rhode Island	

Step 2   Using the random numbers shown in Figure 14–1, find a starting point. To find a starting point, one generally closes one's eyes and places one's finger anywhere on the table. In this case, the first number selected was 27 in the fourth column. Going down the column and continuing on to the next column, select the first 10 numbers. They are 27, 95, 27, 73, 60, 43, 56, 34, 93, and 06. See Figure 14–2. (Note that 06 represents 6.)

Figure 14–2												
Selecting a Starting Point and 10 Numbers from the Random Number Table												

79	41	71	93	60 ✔	35	04	67	96	04	79	10	86
26	52	53	13	43 ✔	50	92	09	87	21	83	75	17
18	13	41	30	56 ✔	20	37	74	49	56	45	46	83
19	82	02	69	34 ✔	27	77	34	24	93	16	77	00
14	57	44	30	93 ✔	76	32	13	55	29	49	30	77
29	12	18	50	06 ✔	33	15	79	50	28	50	45	45
01	27	92	67	93	31	97	55	29	21	64	27	29
55	75	65	68	65	73	07	95	66	43	43	92	16
84	95	95	96	62	30	91	64	74	83	47	89	71
62	62	21	37	82	62	19	44	08	64	34	50	11
66	57	28	69	13	99	74	31	58	19	47	66	89
48	13	69	97	29	01	75	58	05	40	40	18	29
94	31	73	19	75	76	33	18	05	53	04	51	41
00	06	53	*Start here 01	55	08	38	49	42	10	44	38
46	16	44	㉗ ✔	80	15	28	01	64	27	89	03	27
77	49	85	95 ✔	62	93	25	39	63	74	54	82	85
81	96	43	27 ✔	39	53	85	61	12	90	67	96	02
40	46	15	73 ✔	23	75	96	68	13	99	49	64	11

Now, refer to the list of states and identify the state corresponding to each number. The sample consists of the following states:

27	Nebraska	43	Texas
95		56	
27	Nebraska	34	North Dakota
73		93	
60		06	Colorado

Step 3   Since the numbers 95, 73, 60, 56, and 93 are too large, they are disregarded. And since 27 appears twice, it is also disregarded the second time. Now, one must select six more random numbers between 1 and 50 and omit duplicates, since this sample will be selected without replacement. Make this selection by continuing down the column and moving over to the next column until a total of 10 numbers is selected. The final 10 numbers are 27, 43, 34, 06, 13, 29, 01, 39, 23, and 35. See Figure 14–3.

Figure 14–3

The Final 10 Numbers Selected

79	41	71	93	60	㉟	04	67	96	04	79	10	86
26	52	53	13	㊸	50	92	09	87	21	83	75	17
18	13	41	30	56	20	37	74	49	56	45	46	83
19	82	02	69	㉞	27	77	34	24	93	16	77	00
14	57	44	30	93	76	32	13	55	29	49	30	77
29	12	18	50	⑥	33	15	79	50	28	50	45	45
01	27	92	67	93	31	97	55	29	21	64	27	29
55	75	65	68	65	73	07	95	66	43	43	92	16
84	95	95	96	62	30	91	64	74	83	47	89	71
62	62	21	37	82	62	19	44	08	64	34	50	11
66	57	28	69	⑬	99	74	31	58	19	47	66	89
48	13	69	97	㉙	01	75	58	05	40	40	18	29
94	31	73	19	75	76	33	18	05	53	04	51	41
00	06	53	98	ⓞ1	55	08	38	49	42	10	44	38
46	16	44	㉗	80	15	28	01	64	27	89	03	27
77	49	85	95	62	93	25	39	63	74	54	82	85
81	96	43	27	㊴	53	85	61	12	90	67	96	02
40	46	15	73	㉓	75	96	68	13	99	49	64	11

These numbers correspond to the following states:

27	Nebraska	29	New Hampshire
43	Texas	01	Alabama
34	North Dakota	39	Rhode Island
06	Colorado	23	Minnesota
13	Illinois	35	Ohio

Thus, the governors of these 10 states will constitute the sample.

Random sampling has one limitation. If the population is extremely large, it is time-consuming to number and select the sample elements. Also, notice that the random numbers in the table are two-digit numbers. If three digits are needed, then the first digit from the next column can be used, as shown in Figure 14–4. Table D in Appendix C gives five-digit random numbers.

Speaking of Statistics

Should We Be Afraid of Lightning?

The National Weather Service collects various types of data about the weather. For example, each year in the United States about 400 million lightning strikes occur. On average, 400 people are struck by lightning, and 85% of those struck are men. About 100 of these people die. The cause of most of these deaths is not burns, even though temperatures as high as 54,000°F are reached, but heart attacks. The lightning strike short-circuits the body's autonomic nervous system, causing the heart to stop beating. In some instances, the heart will restart on its own. In other cases, the heart victim will need emergency resuscitation.

The most dangerous places to be during a thunderstorm are open fields, golf courses, under trees, and near water, such as a lake or swimming pool. It's best to be inside a building during a thunderstorm although there's no guarantee that the building won't be struck by lightning. Are these statistics descriptive or inferential? Why do you think more men are struck by lightning than women? Should you be afraid of lightning?

79	41	71	93	60	35	04	67	96	04	79	10	86
26	52	53	13	43	50	92	09	87	21	83	75	17
18	13	41	30	56	20	37	74	49	56	45	46	83
19	82	02	69	34	27	77	34	24	93	16	77	00
14	57	44	30	93	76	32	13	55	29	49	30	77
29	12	18	50	06	33	15	79	50	28	50	45	45
01	27	92	67	93	31	97	55	29	21	64	27	29
55	75	65	68	65	73	07	95	66	43	43	92	16
84	95	95	96	62	30	91	64	74	83	47	89	71
62	62	21	37	82	62	19	44	08	64	34	50	11
66	57	28	69	13	99	74	31	58	19	47	66	89
48	13	69	97	29	01	75	58	05	40	40	18	29
94	31	73	19	75	76	33	18	05	53	04	51	41
00	06	53	98	01	55	08	38	49	42	10	44	38
46	16	44	27	80	15	28	01	64	27	89	03	27
77	49	85	95	62	93	25	39	63	74	54	82	85
81	96	43	27	39	53	85	61	12	90	67	96	02
40	46	15	73	23	75	96	68	13	99	49	64	11

Figure 14–4

Method for Selecting Three-Digit Numbers

Use one column and part of the next column for three digits, that is, 404.

Systematic Sampling

A **systematic sample** is a sample obtained by numbering each element in the population and then selecting every third or fifth or tenth, etc., number from the population to be included in the sample. This is done after the first number is selected at random.

The procedure of systematic sampling is illustrated in Example 14–2.

Example 14–2

Using the population of 50 states in Example 14–1, select a systematic sample of 10 states.

Solution

Step 1 Number the population units as shown in Example 14–1.

Step 2 Since there are 50 states and 10 are to be selected, the rule is to select every fifth state. This rule was determined by dividing 50 by 10, which yields 5.

Step 3 Using the table of random numbers, select the first digit (from 1 to 5) at random. In this case, 4 was selected.

Step 4 Select every fifth number on the list, starting with 4. The numbers include the following:

1 2 3 ④ 5 6 7 8 ⑨ 10 11 12 13 ⑭ . . .

The selected states are as follows:

4	Arkansas	29	New Hampshire
9	Florida	34	North Dakota
14	Indiana	39	Rhode Island
19	Maine	44	Utah
24	Mississippi	49	Wisconsin

The advantage of systematic sampling is the ease of selecting the sample elements. Also, in many cases, a numbered list of the population units may already exist. For example, the manager of a factory may have a list of employees who work for the company, or there may be an in-house telephone directory.

When doing systematic sampling, one must be careful how the items are arranged on the list. For example, if each unit were arranged, say, as

1. Husband
2. Wife
3. Husband
4. Wife

then the selection of the starting number could produce a sample of all males or all females, depending on whether the starting number is even or odd and whether the number to be added is even or odd. As another example, if the list were arranged in order of heights of individuals, one would get a different average from two samples if the first were selected by using a small starting number and the second by using a large starting number.

Stratified Sampling

A **stratified sample** is a sample obtained by dividing the population into subgroups, called *strata,* according to various homogeneous characteristics and then selecting members from each stratum for the sample.

For example, a population may consist of males and females who are smokers or nonsmokers. The researcher will want to include in the sample people from each group— that is, males who smoke, males who do not smoke, females who smoke, and females who do not smoke. To accomplish this selection, the researcher divides the population into four subgroups and then selects a random sample from each subgroup. This method ensures that the sample is representative on the basis of the characteristics of gender and smoking. Of course, it may not be representative on the basis of other characteristics.

Example 14–3

Using the population of 20 students shown in Figure 14–5, select a sample of eight students on the basis of gender (male/female) and grade level (freshman/sophomore) by stratification.

Figure 14–5

Population of Students for Example 14–3

1.	Ald, Peter	M	Fr	11.	Martin, Janice	F	Fr
2.	Brown, Danny	M	So	12.	Meloski, Gary	M	Fr
3.	Bear, Theresa	F	Fr	13.	Oeler, George	M	So
4.	Carson, Susan	F	Fr	14.	Peters, Michele	F	So
5.	Collins, Carolyn	F	Fr	15.	Peterson, John	M	Fr
6.	Davis, William	M	Fr	16.	Smith, Nancy	F	Fr
7.	Hogan, Michael	M	Fr	17.	Thomas, Jeff	M	So
8.	Jones, Lois	F	So	18.	Toms, Debbie	F	So
9.	Lutz, Harry	M	So	19.	Unger, Roberta	F	So
10.	Lyons, Larry	M	So	20.	Zibert, Mary	F	So

Solution

Step 1 Divide the population into two subgroups, consisting of males and females, as shown in Figure 14–6.

Figure 14–6

Population Divided into Subgroups by Gender

Males				**Females**			
1.	Ald, Peter	M	Fr	1.	Bear, Theresa	F	Fr
2.	Brown, Danny	M	So	2.	Carson, Susan	F	Fr
3.	Davis, William	M	Fr	3.	Collins, Carolyn	F	Fr
4.	Hogan, Michael	M	Fr	4.	Jones, Lois	F	So
5.	Lutz, Harry	M	So	5.	Martin, Janice	F	Fr
6.	Lyons, Larry	M	So	6.	Peters, Michele	F	So
7.	Meloski, Gary	M	Fr	7.	Smith, Nancy	F	Fr
8.	Oeler, George	M	So	8.	Toms, Debbie	F	So
9.	Peterson, John	M	Fr	9.	Unger, Roberta	F	So
10.	Thomas, Jeff	M	So	10.	Zibert, Mary	F	So

Step 2 Divide each subgroup further into two groups of freshmen and sophomores, as shown in Figure 14–7.

Figure 14–7

Each Subgroup Divided into Subgroups by Grade Level

Group 1				**Group 2**			
1.	Ald, Peter	M	Fr	1.	Bear, Theresa	F	Fr
2.	Davis, William	M	Fr	2.	Carson, Susan	F	Fr
3.	Hogan, Michael	M	Fr	3.	Collins, Carolyn	F	Fr
4.	Meloski, Gary	M	Fr	4.	Martin, Janice	F	Fr
5.	Peterson, John	M	Fr	5.	Smith, Nancy	F	Fr

Group 3				**Group 4**			
1.	Brown, Danny	M	So	1.	Jones, Lois	F	So
2.	Lutz, Harry	M	So	2.	Peters, Michele	F	So
3.	Lyons, Larry	M	So	3.	Toms, Debbie	F	So
4.	Oeler, George	M	So	4.	Unger, Roberta	F	So
5.	Thomas, Jeff	M	So	5.	Zibert, Mary	F	So

Step 3   Determine how many students need to be selected from each subgroup to have a proportional representation of each subgroup in the sample. There are four groups, and since a total of eight students is needed for the sample, two students must be selected from each subgroup.

Step 4   Select two students from each group by using random numbers. In this case, the random numbers are as follows:

Group 1	Students 5 and 4	Group 2	Students 5 and 2
Group 3	Students 1 and 3	Group 4	Students 3 and 4

The stratified sample then consists of the following people:

Peterson, John	M	Fr	Smith, Nancy	F	Fr
Meloski, Gary	M	Fr	Carson, Susan	F	Fr
Brown, Danny	M	So	Toms, Debbie	F	So
Lyons, Larry	M	So	Unger, Roberta	F	So

The major advantage of stratification is that it ensures representation of all population subgroups that are important to the study. There are two major drawbacks to stratification, however. First, if there are many variables of interest, dividing a large population into representative subgroups requires a great deal of effort. Second, if the variables are somewhat complex or ambiguous (such as beliefs, attitudes, or prejudices), it is difficult to separate individuals into the subgroups according to these variables.

Cluster Sampling

A **cluster sample** is a sample obtained by selecting a preexisting or natural group, called a *cluster,* and using the members in the cluster for the sample.

For example, many studies in education use already existing classes, such as the seventh grade in Wilson Junior High School. The voters of a certain electoral district might be surveyed to determine their preferences for a mayoral candidate in the upcoming election. Or the residents of an entire city block might be polled to ascertain the percentage of households that have two or more incomes. In cluster sampling, researchers may use all units of a cluster if that is feasible, or they may select only part of a cluster to use as a sample. This selection is done by random methods.

There are three advantages to using a cluster sample instead of other types of samples: (1) A cluster sample can reduce costs, (2) it can simplify fieldwork, and (3) it is convenient. For example, in a dental study involving X-raying fourth-grade students' teeth to see how many cavities each child had, it would be a simple matter to select a single classroom and bring the X-ray equipment to the school to conduct the study. If other sampling methods were used, researchers might have to transport the machine to several different schools or transport the pupils to the dental office.

The major disadvantage of cluster sampling is that the elements in a cluster may not have the same variations in characteristics as elements selected individually from a population. The reason is that groups of people may be more homogeneous (alike) in specific clusters such as neighborhoods or clubs. For example, the people who live in a certain neighborhood tend to have similar incomes, drive similar cars, live in similar houses, and, for the most part, have similar habits.

In this study, the researchers found that subjects did better on fill-in-the-blank questions than on multiple-choice questions. Do you agree with the professor's statement, "Trusting your first impulse is your best strategy?" Explain your answer.

TESTS

Is That Your Final Answer?

Beating game shows takes more than smarts: Contestants must also overcome self-doubt and peer pressure. Two new studies suggest today's hottest game shows are particularly challenging because the very mechanisms employed to help contestants actually lead them astray.

Multiple-choice questions are one such offender, as alternative answers seem to make test-takers ignore gut instincts. To learn why, researchers at Southern Methodist University (SMU) gave two identical tests: one using multiple-choice questions and the other fill-in-the-blank. The results, recently published in the *Journal of Educational Psychology*, show that test-takers were incorrect more often when given false alternatives, and that the longer they considered those alternatives, the more credible the answers looked.

"If you sit and stew, you forget that you know the right answer," says Alan Brown, Ph.D., a psychology professor at SMU. "Trusting your first impulse is your best strategy."

Audiences can also be trouble, says Jennifer Butler, Ph.D., a Wittenberg University psychology professor. Her recent study in the *Journal of Personality and Social Psychology* found that contestants who see audience participation as peer pressure slow down to avoid making embarrassing mistakes. But this strategy backfires, as more contemplation produces more wrong answers. Worse, Butler says, if perceived peer pressure grows unbearable, contestants may opt out of answering at all, "thinking that it's better to stop than to have your once supportive audience come to believe you're an idiot."

— *Sarah Smith*

Source: Reprinted with permission from *Psychology Today,* Copyright © 2000 Sussex Publishers, Inc.

Other Types of Sampling Techniques

In addition to the four basic sampling methods, other methods are sometimes used. In **sequence sampling,** which is used in quality control, successive units taken from production lines are sampled to ensure that the products meet certain standards set by the manufacturing company.

In **double sampling,** a very large population is given a questionnaire to determine those who meet the qualifications for a study. After the questionnaires are reviewed, a second, smaller population is defined. Then a sample is selected from this group.

In **multistage sampling,** the researcher uses a combination of sampling methods. For example, suppose a research organization wants to conduct a nationwide survey for a new product being manufactured. A sample can be obtained by using the following combination of methods. First the researchers divide the 50 states into four or five regions (or clusters). Then several states from each region are selected at random. Next the states are divided into various areas by using large cities and small towns. Samples of these areas are then selected. Next, each city and town is divided into districts or wards. Finally, streets in these wards are selected at random, and the families living on these streets are given samples of the product to test and are asked to report the results. This hypothetical example illustrates a typical multistage sampling method.

The steps for conducting a sample survey are given in the Procedure Table.

Procedure Table

Conducting a Sample Survey

Step 1   Decide what information is needed.

Step 2   Determine how the data will be collected (phone interview, mail survey, etc.).

Step 3   Select the information-gathering instrument or design the questionnaire if one is not available.

Step 4   Set up a sampling list, if possible.

Step 5   Select the best method for obtaining the sample (random, systematic, stratified, cluster, or other).

Step 6   Conduct the survey and collect the data.

Step 7   Tabulate the data.

Step 8   Conduct the statistical analysis.

Step 9   Report the results.

Applying the Concepts 14–2

The White or Wheat Bread Debate

Read the following study and answer the questions.

A baking company selected 36 women weighing different amounts and randomly assigned them to four different groups. The four groups were white bread only, brown bread only, low-fat white bread only, and low-fat brown bread only. Each group could eat only the type of bread assigned to the group. The study lasted for eight weeks. No other changes in any of the women's diets were allowed. A trained evaluator was used to check for any differences in the women's diets. The results showed that there were no differences in weight gain between the groups over the eight-week period.

1. Did the researchers use a population or a sample for their study?
2. Based on who conducted this study, would you consider the study to be biased?
3. Which sampling method do you think was used to obtain the original 36 women for the study (random, systematic, stratified, or clustered)?
4. Which sampling method would you use? Why?
5. How would you collect a random sample for this study?
6. Does random assignment help representativeness the same as random selection does? Explain.

See page 736 for the answers.

Exercises 14–2

1. Name the four basic sampling techniques.

2. Why are samples used in statistics?

3. What is the basic requirement for a sample?

4. Why should random numbers be used when one is selecting a random sample?

5. List three incorrect methods that are often used to obtain a sample.

Figure 14–8

Student Survey at Utopia University (for Exercises 11 through 15)

Student number	Gen-der	Class rank	GPA	Miles traveled to school	IQ	Major field	Student number	Gen-der	Class rank	GPA	Miles traveled to school	IQ	Major field
1	M	Fr	1.4	1	104	Bio	26	M	Fr	1.1	8	100	Ed
2	M	Fr	2.3	2	95	Ed	27	F	Jr	2.1	3	101	Bus
3	M	So	2.7	6	108	Psy	28	M	Gr	3.7	5	99	Bio
4	F	So	3.2	7	119	Eng	29	M	Se	2.4	8	105	Eng
5	F	Gr	3.8	12	114	Ed	30	M	So	2.1	15	108	Bus
6	M	Jr	4.0	13	91	Psy	31	M	Gr	3.9	2	112	Ed
7	F	Jr	3.0	2	106	Eng	32	F	Jr	2.4	4	111	Psy
8	M	Jr	3.3	6	100	Bio	33	M	Se	2.7	6	107	Eng
9	F	Se	2.7	9	102	Eng	34	F	So	2.5	1	104	Bio
10	F	So	2.3	5	99	Ed	35	M	Se	3.2	3	96	Bus
11	M	Se	1.6	18	100	Bus	36	F	Fr	3.4	7	98	Bio
12	M	Gr	3.2	7	105	Psy	37	M	Gr	3.6	14	105	Ed
13	F	Gr	3.8	3	103	Bus	38	M	Jr	3.8	4	115	Psy
14	F	Se	3.1	5	97	Eng	39	F	Se	2.2	8	113	Eng
15	F	Jr	2.7	5	106	Bio	40	F	So	2.0	8	103	Psy
16	F	Fr	1.4	4	114	Bus	41	F	Fr	2.3	9	103	Eng
17	M	So	3.6	17	102	Ed	42	F	Se	2.5	10	99	Bus
18	M	Fr	2.2	1	101	Psy	43	M	Gr	3.7	13	114	Ed
19	F	Gr	4.0	7	108	Bus	44	M	Fr	3.0	11	121	Bus
20	M	Jr	2.1	4	97	Ed	45	M	Jr	2.1	10	101	Eng
21	F	Fr	2.0	3	113	Bio	46	F	Jr	3.4	2	104	Ed
22	F	So	3.6	4	104	Bio	47	M	So	3.6	9	105	Psy
23	F	Gr	3.3	16	110	Eng	48	M	Se	2.1	1	97	Psy
24	F	Se	2.5	4	99	Psy	49	F	Gr	3.3	12	111	Bio
25	M	So	3.0	5	96	Psy	50	F	Fr	2.2	11	102	Bio

6. What is the principle behind random numbers?

7. List the advantages and disadvantages of random sampling.

8. List the advantages and disadvantages of systematic sampling.

9. List the advantages and disadvantages of stratified sampling.

10. List the advantages and disadvantages of cluster sampling.

Using the student survey at Utopia University, shown in Figure 14–8, as the population, complete Exercises 11 through 15.

11. Using the table of random numbers in Figure 14–1, select 10 students and find the sample mean (average) of the GPA, IQ, and distance traveled to school. Compare these sample means with the population means.

12. Select a sample of 10 students by the systematic method, and compute the sample means of the GPA, IQ,

and distance traveled to school of this sample. Compare these sample means with the population means.

13. Select a cluster of 10 students, for example, students 9 through 18, and compute the sample means of their GPA, IQ, and distance traveled to school. Compare these sample means with the population means.

14. Divide the 50 students into subgroups according to class rank. Then select a sample of 2 students from each rank and compute the means of these 10 students for the GPA, IQ, and distance traveled to school each day. Compare these sample means with the population means.

15. In your opinion, which sampling method(s) provided the best sample to represent the population?

Figure 14–9 shows the 50 states and the number of electoral votes each state has in the Presidential election. Using this listing as a population, complete Exercises 16 through 19.

16. Select a random sample of 10 states and find the mean number of electoral votes for this sample. Compare this mean with the population mean.

Figure 14–9

States and Number of Electoral Votes for Each (for Exercises 16 through 19)

1. Alabama	9	14. Indiana	12	27. Nebraska	5	40. South Carolina	8
2. Alaska	3	15. Iowa	8	28. Nevada	4	41. South Dakota	3
3. Arizona	7	16. Kansas	7	29. New Hampshire	4	42. Tennessee	11
4. Arkansas	6	17. Kentucky	9	30. New Jersey	16	43. Texas	29
5. California	47	18. Louisiana	10	31. New Mexico	5	44. Utah	5
6. Colorado	8	19. Maine	4	32. New York	36	45. Vermont	3
7. Connecticut	8	20. Maryland	10	33. North Carolina	13	46. Virginia	12
8. Delaware	3	21. Massachusetts	13	34. North Dakota	3	47. Washington	10
9. Florida	21	22. Michigan	20	35. Ohio	23	48. West Virginia	6
10. Georgia	12	23. Minnesota	10	36. Oklahoma	8	49. Wisconsin	11
11. Hawaii	4	24. Mississippi	7	37. Oregon	7	50. Wyoming	3
12. Idaho	4	25. Missouri	11	38. Pennsylvania	25		
13. Illinois	24	26. Montana	4	39. Rhode Island	4		

AMERICAN POLLSTER INC.

"Now think carefully. The answer you give will represent the opinion of millions of Americans."

Source: *The Saturday Evening Post*, BFL&MS, Inc.

17. Select a systematic sample of 10 states and compute the mean number of electoral votes for the sample. Compare this mean with the population mean.

18. Divide the 50 states into five subgroups by geographic location, using a map of the United States. Each subgroup should include 10 states. The subgroups should be northeast, southeast, central, northwest, and southwest. Select two states from each subgroup, and find the mean number of electoral votes for the sample. Compare these means with the population mean.

19. Select a cluster of 10 states and compute the mean number of electoral votes for the sample. Compare this mean with the population mean.

20. Many research studies described in newspapers and magazines do not report the sample size or the sampling method used. Try to find a research article that gives this information; state the sampling method that was used and the sample size.

Technology *Step by Step*

MINITAB
Step by Step

Select a Random Sample with Replacement

A simple random sample selected with replacement allows some values to be used more than once, duplicates. In the first example, a random sample of integers will be selected with replacement.

1. Select **Calc>Random Data>Integer.**
2. Type **10** for rows of data.
3. Type the name of a column, Random1, in the box for Store in column(s).
4. Type **1** for Minimum and **50** for Maximum, then click [OK].

A sample of 10 integers between 1 and 50 will be displayed in the first column of the worksheet. Every list will be different.

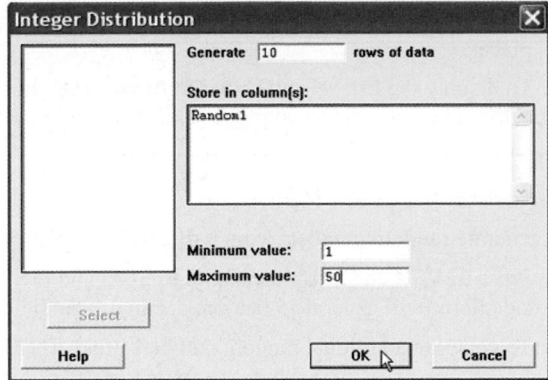

Select a Random Sample Without Replacement

To sample without replacement, make a list of integers and sample from the columns.

1. Select **Calc>Make Patterned Data>Simple Set of Numbers.**

2. Type **Integers** in the text box for Store patterned data in.

3. Type **1** for Minimum and **50** for Maximum. Leave 1 for steps and click [OK]. A list of the integers from 1 to 50 will be created in the worksheet.

4. Select **Calc>Random Data>Sample from columns.**

5. Sample **10** for the number of rows and Integers for the name of the column.

6. Type Random2 as the name of the new column. Be sure to leave the option for Sample with replacement unchecked.

7. Click [OK]. The new sample will be in the worksheet. There will be no duplicates.

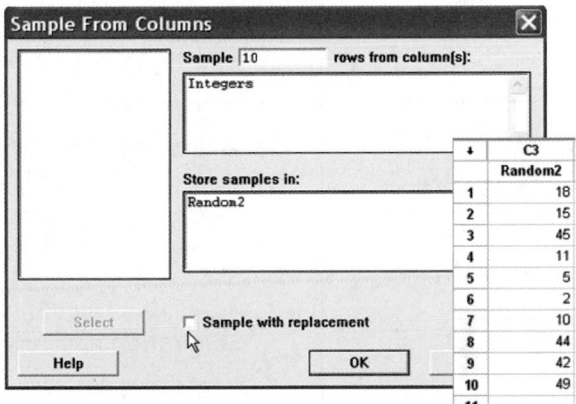

Select a Random Sample from a Normal Distribution

No data are required in the worksheet.

1. Select **Calc>Random Data>Normal . . .**

2. Type **50** for the number of rows.

3. Press TAB or click in the box for Store in columns. Type in RandomNormal.

4. Type in **500** for the Mean and **75** for the Standard deviation.

5. Click [OK]. The random numbers are in a column of the worksheet. The distribution is sampled "with replacement." However, duplicates are not likely since this distribution is continuous. They are displayed to 3 decimal places, but many more places are stored.

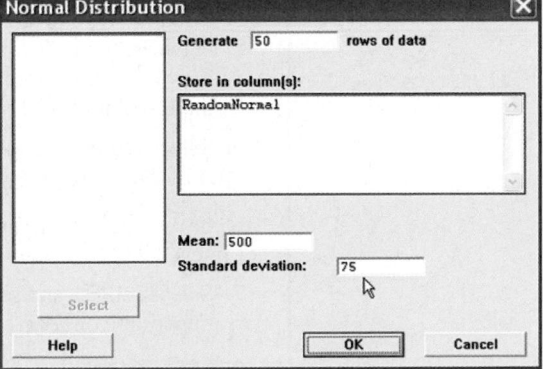

Click in any cell such as row 5 of C4 RandomNormal, and you will see more decimal places.

6. To display the list, select **Data>Display data,** then select C1 RandomNormal and click [OK]. They are displayed in the same order they were selected, but going across not down.

TI-83 Plus or TI-84 Plus
Step by Step

Generate Random Numbers

To generate random numbers from 0 to 1 by using the TI-83 Plus or TI-84 Plus:

1. Press **MATH** and move the cursor to PRB and press **1** for rand, then press **ENTER.** The calculator will generate a random decimal from 0 to 1.

2. To generate additional random numbers press **ENTER.**

To generate a list of random integers between two specific values:

1. Press **MATH** and move the cursor to PRB.

2. Press **5** for randInt(.

3. Enter the lowest value followed by a comma, then the largest value followed by a comma, then the number of random numbers desired followed by).
Press **ENTER.**

Example: Generate five three-digit random numbers.

Enter **0, 999, 5)** at the randInt(as shown.

The calculator will generate five three-digit random numbers. Use the arrow keys to view the entire list.

```
rand
             .9435974025
randInt(0,999,5)
{908 146 514 40…
```

Excel
Step by Step

Generate Random Numbers

The Data Analysis Add-In in Excel has a feature to generate random numbers from a specific probability distribution. For this example, a list of 50 random real numbers will be generated from a uniform distribution. The real numbers will then be rounded to integers between 1 and 50.

1. Open a new worksheet and select **Tool>Data Analysis>Random Number Generation** from Analysis Tools. Click [OK].

2. In the dialog box, type **1** for the Number of Variables. Leave the Number of Random Numbers box blank.

3. For Distribution, select Uniform.

4. In the Parameters box, type **1** for the lower bound and **51** for the upper bound.

5. You may type in an integer value between 1 and 51 for the Random Seed. For this example, type **3** for the Random Seed.

6. Select Output Range and type in **A1:A50.** Click [OK].

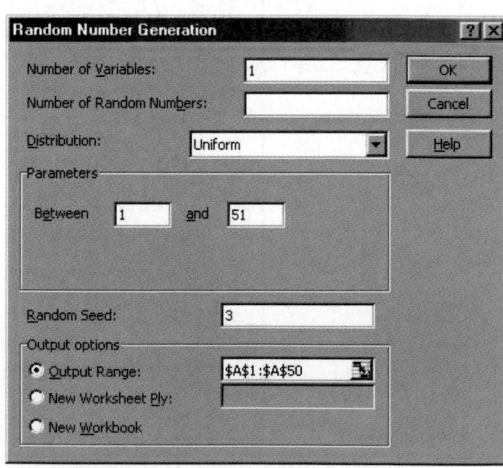

To convert the random numbers to a list of integers:

7. Select cell B1 and select the Paste Function icon from the toolbar.

8. Select the Math & Trig Function category and scroll to the Function name INT to convert the data in column A to integers.

Note that the INT function rounds the argument (input) down to the nearest integer.

9. Type cell **A1** for the Number in the INT dialog box. Click [OK].

10. While cell B1 is selected in the worksheet, move the pointer to the lower right-hand corner of the cell until a thick plus sign appears. Right-click on the mouse and drag the plus down to cell B50; then release the mouse key.

11. The numbers from column A should have been rounded to integers in column B.

Here is a sample of the data produced from the preceding procedures.

```
1.073244      1
11.98056     11
15.18195     15
14.87219     14
11.72878     11
36.97674     36
28.01193     28
36.86383     36
42.53111     42
19.56746     19
```

14–3

Surveys and Questionnaire Design

Objective 2

Recognize faulty questions on a survey and other factors that can bias responses.

Many statistical studies obtain information from *surveys*. A survey is conducted when a sample of individuals is asked to respond to questions about a particular subject. There are two types of surveys: interviewer-administered and self-administered. Interviewer-administered surveys require a person to ask the questions. The interview can be conducted face to face in an office, on a street, or in the mall, or via telephone.

Self-administered surveys can be done by mail or in a group setting such as a classroom.

When analyzing the results of surveys, you should be very careful about the interpretations. The way a question is phrased can influence the way people respond. For example, when a group of people were asked if they favored a waiting period and background check before guns could be sold, 91% of the respondents were in favor of it and 7% were against it. However, when asked if there should be a national gun registration program costing about 20% of all dollars spent on crime control, only 33% of the respondents were in favor of it and 61% were against it.

As you can see, by phrasing questions in different ways, different responses can be obtained, since the purpose of a national gun registry would include a waiting period and a background check.

When you are writing questions for a questionnaire, it is important to avoid these common mistakes.

1. *Asking biased questions.* By asking questions in a certain way, the researcher can lead the respondents to answer in the way he or she wants them to. For example, asking a question such as "Are you going to vote for the candidate Jones even though the latest survey indicates that he will lose the election?" instead of "Are you going to vote for candidate Jones?" may dissuade some people from answering in the affirmative.

2. *Using confusing words.* In this case, the participant misinterprets the meaning of the words and answers the questions in a biased way. For example, the question "Do you think people would live longer if they were on a diet?" could be misinterpreted since there are many different types of diets—weight loss diets, low-salt diets, medically prescribed diets, etc.

3. *Asking double-barreled questions.* Sometimes questions contain compound sentences that require the participant to respond to two questions at the same time. For example, the question "Are you in favor of a special tax to provide national health care for the citizens of the United States?" asks two questions: "Are you in favor of a national health care program?" and "Do you favor a tax to support it?"

4. *Using double negatives in questions.* Questions with double negatives can be confusing to the respondents. For example, the question "Do you feel that it is not appropriate to have areas where people cannot smoke?" is very confusing since *not* is used twice in the sentence.

5. *Ordering questions improperly.* By arranging the questions in a certain order, the researcher can lead the participant to respond in a way that he or she may otherwise not have done. For example, a question might ask the respondent, "At what age should an elderly person not be permitted to drive?" A later question might ask the respondent to list some problems of elderly people. The respondent may indicate that transportation is a problem based on reading the previous question.

Other factors can also bias a survey. For example, the participant may not know anything about the subject of the question but will answer the question anyway to avoid being considered uninformed. For example, many people might respond yes or no to the following question: "Would you be in favor of giving pensions to the widows of unknown soldiers?" In this case, the question makes no sense since if the soldiers were unknown, their widows would also be unknown.

Many people will make responses on the basis of what they think the person asking the questions wants to hear. For example, if a question states, "How often do you lie?" people may *understate* the incidences of their lying.

Participants will, in some cases, respond differently to questions depending on whether their identity is known. This is especially true if the questions concern sensitive issues such as income, sexuality, and abortion. Researchers try to ensure confidentiality (i.e., keeping the respondent's identity secret) rather than anonymity (soliciting unsigned responses); however, many people will be suspicious in either case.

Still other factors that could bias a survey include the time and place of the survey and whether the questions are open-ended or closed-ended. The time and place where a survey is conducted can influence the results. For example, if a survey on airline safety is conducted immediately after a major airline crash, the results may differ from those obtained in a year in which no major airline disasters occurred.

Finally, the type of questions asked influences the responses. In this case, the concern is whether the question is open-ended or closed-ended.

An *open-ended question* would be one such as "List three activities that you plan to spend more time on when you retire." A *closed-ended question* would be one such as "Select three activities that you plan to spend more time on after you retire: traveling; eating out; fishing, hunting; exercising; visiting relatives."

One problem with a closed-ended question is that the respondent is forced to choose the answers that the researcher gives and cannot supply his or her own. But there is also a problem with open-ended questions in that the results may be so varied that attempting to summarize them might be difficult, if not impossible. Hence, you should be aware of what types of questions are being asked before you draw any conclusions from the survey.

There are several other things to consider when you are conducting a study that uses questionnaires. For example, a pilot study should be done to test the design and usage of the questionnaire (i.e., the *validity* of the questionnaire). The pilot study helps the researcher to pretest the questionnaire to determine if it meets the objectives of the study. It also helps the researcher to rewrite any questions that may be misleading, ambiguous, etc.

*U*nusual Stat

Of people who are struck by lightning, 85% are men.

If the questions are being asked by an interviewer, some training should be given to that person. If the survey is being done by mail, a cover letter and clear directions should accompany the questionnaire.

Questionnaires help researchers to gather needed statistical information for their studies; however, much care must be given to proper questionnaire design and usage; otherwise, the results will be unreliable.

Applying the Concepts 14–3

Smoking Bans and Profits

Assume you are a restaurant owner and are concerned about the recent bans on smoking in public places. Will your business lose money if you do not allow smoking in your restaurant? You decide to research this question and find two related articles in regional newspapers. The first article states that randomly selected restaurants in Derry, Pennsylvania, that have completely banned smoking have lost 25% of their business. In that study, a survey was used and the owners were asked how much business they thought they lost. The survey was conducted by an anonymous group. It was reported in the second article that there had been a modest increase in business among restaurants that banned smoking in that same area. Sales receipts were collected and analyzed against last year's profits. The second survey was conducted by the Restaurants Business Association.

1. How has the public smoking ban affected restaurant business in Derry, Pennsylvania?
2. Why do you think the surveys reported conflicting results?
3. Should surveys based on anecdotal responses be allowed to be published?
4. Can the results of a sample be representative of a population and still offer misleading information?
5. How critical is measurement error in survey sampling?

See page 737 for the answers.

Exercises 14–3

Exercises 1 through 8 include questions that contain a flaw. Identify the flaw and rewrite the question, following the guidelines presented in this section.

1. Will you continue to shop at XYZ Department Store even though it does not carry brand names?

2. Would you buy an ABC car even if you knew the manufacturer used imported parts?

3. Should banks charge their checking account customers a fee to balance their checkbooks when customers are not able to do so?

4. Do you feel that it is not appropriate for shopping malls to have activities for children who cannot read?

5. How long have you studied for this examination?

6. Do you think children would watch less television if they read more?

7. If a plane were to crash on the border of New York and New Jersey, where should the survivors be buried?

8. Are you in favor of imposing a tax on tobacco to pay for health care related to diseases caused by smoking?

9. Find a study that uses a questionnaire. Select any questions that you feel are improperly written.

10. Many television and radio stations have a phone vote poll. If there is one in your area, select a specific day and write a brief paragraph stating the question of the day and state if it could be misleading in any way.

<table>
<tr><td>**14–4**</td></tr>
</table>

Simulation Techniques

Many real-life problems can be solved by employing simulation techniques.

A **simulation technique** uses a probability experiment to mimic a real-life situation.

Instead of studying the actual situation, which might be too costly, too dangerous, or too time-consuming, scientists and researchers create a similar situation but one that is less expensive, less dangerous, or less time-consuming. For example, NASA uses space shuttle flight simulators so that its astronauts can practice flying the shuttle. Most video games use the computer to simulate real-life sports such as boxing, wrestling, baseball, and hockey.

Simulation techniques go back to ancient times when the game of chess was invented to simulate warfare. Modern techniques date to the mid-1940s when two physicists, John Von Neumann and Stanislaw Ulam, developed simulation techniques to study the behavior of neutrons in the design of atomic reactors.

Mathematical simulation techniques use probability and random numbers to create conditions similar to those of real-life problems. Computers have played an important role in simulation techniques, since they can generate random numbers, perform experiments, tally the outcomes, and compute the probabilities much faster than human beings. The basic simulation technique is called the *Monte Carlo method*. This topic is discussed next.

<table>
<tr><td>**14–5**</td></tr>
</table>

The Monte Carlo Method

Objective   **3**

Solve problems, using simulation techniques.

The **Monte Carlo method** is a simulation technique using random numbers. Monte Carlo simulation techniques are used in business and industry to solve problems that are extremely difficult or involve a large number of variables. The steps for simulating real-life experiments in the Monte Carlo method are as follows:

1. List all possible outcomes of the experiment.
2. Determine the probability of each outcome.
3. Set up a correspondence between the outcomes of the experiment and the random numbers.
4. Select random numbers from a table and conduct the experiment.
5. Repeat the experiment and tally the outcomes.
6. Compute any statistics and state the conclusions.

Before examples of the complete simulation technique are given, an illustration is needed for step 3 (set up a correspondence between the outcomes of the experiment and the random numbers). Tossing a coin, for instance, can be simulated by using random numbers as follows: Since there are only two outcomes, heads and tails, and since each outcome has a probability of $\frac{1}{2}$, the odd digits (1, 3, 5, 7, and 9) can be used to represent a head, and the even digits (0, 2, 4, 6, and 8) can represent a tail.

Suppose a random number 8631 is selected. This number represents four tosses of a single coin and the results T, T, H, H. Or this number could represent one toss of four coins with the same results.

An experiment of rolling a single die can also be simulated by using random numbers. In this case, the digits 1, 2, 3, 4, 5, and 6 can represent the number of spots that appear on the face of the die. The digits 7, 8, 9, and 0 are ignored, since they cannot be rolled.

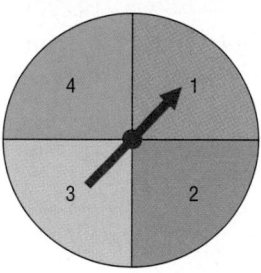

Figure 14–10

Spinner with Four Numbers

When two dice are rolled, two random digits are needed. For example, the number 26 represents a 2 on the first die and a 6 on the second die. The random number 37 represents a 3 on the first die, but the 7 cannot be used, so another digit must be selected. As another example, a three-digit daily lotto number can be simulated by using three-digit random numbers. Finally, a spinner with four numbers, as shown in Figure 14–10, can be simulated by letting the random numbers 1 and 2 represent 1 on the spinner, 3 and 4 represent 2 on the spinner, 5 and 6 represent 3 on the spinner, and 7 and 8 represent 4 on the spinner, since each number has a probability of $\frac{1}{4}$ of being selected. The random numbers 9 and 0 are ignored in this situation.

Many real-life games, such as bowling and baseball, can be simulated by using random numbers, as shown in Figure 14–11.

Figure 14–11

Example of Simulation of a Game

Source: Albert Shuylte, "Simulated Bowling Game," Student Math Notes, March 1986. Published by the National Council of Teachers of Mathematics. Reprinted with permission.

Simulated Bowling Game

Let's use the random digit table to simulate a bowling game. Our game is much simpler than commercial simulation games.

First Ball		Second Ball			
		2-Pin Split		No split	
Digit	Results	Digit	Results	Digit	Results
1–3	Strike	1	Spare	1–3	Spare
4–5	2-pin split	2–8	Leave one pin	4–6	Leave 1 pin
6–7	9 pins down	9–0	Miss both pins	7–8	*Leave 2 pins
8	8 pins down			9	+Leave 3 pins
9	7 pins down			0	Leave all pins
0	6 pins down				

*If there are fewer than 2 pins, result is a spare.
+If there are fewer than 3 pins, those pins are left.

Here's how to score bowling:
1. There are 10 frames to a **game** or **line**.
2. You roll two balls for each frame, unless you knock all the pins down with the first ball (a **strike**).
3. Your score for a frame is the sum of the pins knocked down by the two balls, if you don't knock down all 10.
4. If you knock all 10 pins down with two balls (a **spare**, shown as ▢), your score is 10 pins plus the number knocked down with the next ball.
5. If you knock all 10 pins down with the first ball (a **strike**, shown as ⊠), your score is 10 pins plus the number knocked down by the next **two** balls.
6. A **split** (shown as ⊙) is when there is a big space between the remaining pins. Place in the circle the number of pins remaining after the second ball.
7. A **miss** is shown as —.

Here is how one person simulated a bowling game using the random digits 7 2 7 4 8 2 2 3 6 1 6 0 4 6 1 5 5, chosen in that order from the table.

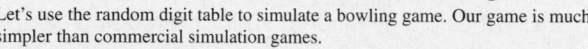

	Frame									
	1	**2**	**3**	**4**	**5**	**6**	**7**	**8**	**9**	**10**
Digit(s)	7/2	7/4	8/2	2	3	6/1	6/0	4/6	1	5/5
Bowling result	9▢	9—	8▢	⊠	⊠	9▢	9—	8⊙	⊠	8⊙
	19	28	48	77	97	116	125	134	153	162	162

Now you try several.

	Frame									
	1	**2**	**3**	**4**	**5**	**6**	**7**	**8**	**9**	**10**
Digit(s)										
Bowling result										

	1	**2**	**3**	**4**	**5**	**6**	**7**	**8**	**9**	**10**
Digit(s)										
Bowling result										

If you wish to, you can change the probabilities in the simulation to better reflect *your* actual bowling ability.

| **Example 14–4** | Using random numbers, simulate the gender of children born. |

Solution

There are only two possibilities, female and male. Since the probability of each outcome is 0.5, the odd digits can be used to represent male births and the even digits to represent female births.

| **Example 14–5** | Using random numbers, simulate the outcomes of a tennis game between Bill and Mike, with the additional condition that Bill is twice as good as Mike. |

Solution

Since Bill is twice as good as Mike, he will win approximately two games for every one Mike wins; hence, the probability that Bill wins will be $\frac{2}{3}$, and the probability that Mike wins will be $\frac{1}{3}$. The random digits 1 through 6 can be used to represent a game Bill wins; the random digits 7, 8, and 9 can be used to represent Mike's wins. The digit 0 is disregarded. Suppose they play five games, and the random number 86314 is selected. This number means that Bill won games 2, 3, 4, and 5 and Mike won the first game. The sequence is

8	**6**	**3**	**1**	**4**
M	B	B	B	B

Unusual Stats

The average 6-year-old laughs 300 times a day; the average adult, just 17.

More complex problems can be solved by using random numbers, as shown in Examples 14–6 to 14–8.

| **Example 14–6** | A die is rolled until a 6 appears. Using simulation, find the average number of rolls needed. Try the experiment 20 times. |

Solution

Step 1 List all possible outcomes. They are 1, 2, 3, 4, 5, 6.

Step 2 Assign the probabilities. Each outcome has a probability of $\frac{1}{6}$.

Step 3 Set up a correspondence between the random numbers and the outcome. Use random numbers 1 through 6. Omit the numbers 7, 8, 9, and 0.

Step 4 Select a block of random numbers, and count each digit 1 through 6 until the first 6 is obtained. For example, the block 857236 means that it takes 4 rolls to get a 6.

8	5	7	2	3	6
	↑		↑	↑	↑
	5		2	3	6

Interesting Fact

A recent survey of more than 300 Californians ranked exercise as the surest way out of a bad mood. Listening to music was a close second.

Step 5 Repeat the experiment 19 more times and tally the data as shown.

Trial	Random number	Number of rolls
1	8 5 7 2 3 6	4
2	2 1 0 4 8 0 1 5 1 1 0 1 5 3 6	11
3	2 3 3 6	4
4	2 4 1 3 0 4 8 3 6	7
5	4 2 1 6	4
6	3 7 5 2 0 3 9 8 7 5 8 1 8 3 7 1 6	9
7	7 7 9 2 1 0 6	3
8	9 9 5 6	2
9	9 6	1
10	8 9 5 7 9 1 4 3 4 2 6	7
11	8 5 4 7 5 3 6	5
12	2 8 9 1 8 6	3
13	6	1
14	0 9 4 2 9 9 3 9 6	4
15	1 0 3 6	3
16	0 7 1 1 9 9 7 3 3 6	5
17	5 1 0 8 5 1 2 7 6	6
18	0 2 3 6	3
19	0 1 0 1 1 5 4 0 9 2 3 3 3 6	10
20	5 2 1 6	4
	Total	96

Step 6 Compute the results and draw a conclusion. In this case, one must find the average.

$$\overline{X} = \frac{\Sigma X}{n} = \frac{96}{20} = 4.8$$

Hence, the average is about 5 rolls.

Note: The theoretical average obtained from the expected value formula is 6. If this experiment is done many times, say 1000 times, the results should be closer to the theoretical results.

Example 14–7

A person selects a key at random from four keys to open a lock. Only one key fits. If the first key does not fit, she tries other keys until one fits. Find the average of the number of keys a person will have to try to open the lock. Try the experiment 25 times.

Solution

Assume that each key is numbered from 1 through 4 and that key 2 fits the lock. Naturally, the person doesn't know this, so she selects the keys at random. For the simulation, select a sequence of random digits, using only 1 through 4, until the digit 2 is reached. The trials are shown here.

Trial	Random digit (key)	Number	Trial	Random digit (key)	Number
1	2	1	14	2	1
2	2	1	15	4 2	2
3	1 2	2	16	1 3 2	3
4	1 4 3 2	4	17	1 2	2
5	3 2	2	18	2	1
6	3 1 4 2	4	19	3 4 2	3
7	4 2	2	20	2	1
8	4 3 2	3	21	2	1
9	4 2	2	22	2	1
10	2	1	23	4 2	2
11	4 2	2	24	4 3 1 2	4
12	3 1 2	3	25	3 1 2	3
13	3 1 2	3		Total	54

Next, find the average:

$$\overline{X} = \frac{\Sigma X}{n} = \frac{1 + 1 + \cdots + 3}{25} = \frac{54}{25} = 2.16$$

The theoretical average is 2.2. Again, only 25 repetitions were used; more repetitions should give a result closer to the theoretical average.

Example 14–8

A box contains five \$1 bills, three \$5 bills, and two \$10 bills. A person selects a bill at random. What is the expected value of the bill? Perform the experiment 25 times.

Solution

Step 1 List all possible outcomes. They are \$1, \$5, and \$10.

Step 2 Assign the probabilities to each outcome:

$$P(\$1) = \tfrac{5}{10} \qquad P(\$5) = \tfrac{3}{10} \qquad P(\$10) = \tfrac{2}{10}$$

Step 3 Set up a correspondence between the random numbers and the outcomes. Use random numbers 1 through 5 to represent a \$1 bill being selected, 6 through 8 to represent a \$5 bill being selected, and 9 and 0 to represent a \$10 bill being selected.

Steps 4 and 5 Select 25 random numbers and tally the results.

Number	Results (\$)
4 5 8 2 9	1, 1, 5, 1, 10
2 5 6 4 6	1, 1, 5, 1, 5
9 1 8 0 3	10, 1, 5, 10, 1
8 4 0 6 0	5, 1, 10, 5, 10
9 6 9 4 3	10, 5, 10, 1, 1

Step 6 Compute the average:

$$\overline{X} = \frac{\Sigma X}{n} = \frac{\$1 + \$1 + \$5 + \cdots + \$1}{25} = \frac{\$116}{25} = \$4.64$$

Hence, the average (expected value) is \$4.64.

Recall that using the expected value formula $E(X) = \Sigma X \cdot P(X)$ gives a theoretical average of

$$E(X) = \Sigma[X \cdot P(X)] = (0.5)(\$1) + (0.3)(\$5) + (0.2)(\$10) = \$4.00$$

Remember that simulation techniques do not give exact results. The more times the experiment is performed, though, the closer the actual results should be to the theoretical results. (Recall the law of large numbers.)

The steps for solving problems using the Monte Carlo method are summarized in the Procedure Table.

Procedure Table

Simulating Experiments Using the Monte Carlo Method

Step 1   List all possible outcomes of the experiment.

Step 2   Determine the probability of each outcome.

Step 3   Set up a correspondence between the outcomes of the experiment and the random numbers.

Step 4   Select random numbers from a table and conduct the experiment.

Step 5   Repeat the experiment and tally the outcomes.

Step 6   Compute any statistics and state the conclusions.

Applying the Concepts 14–4

Simulations

Answer the following questions:

1. Define simulation technique.
2. Have simulation techniques been used for very many years?
3. Is it cost-effective to do simulation testing on some things such as airplanes or automobiles?
4. Why might simulation testing be better than real-life testing? Give examples.
5. When did physicists develop computer simulation techniques to study neutrons?
6. When could simulations be misleading or harmful? Give examples.
7. Could simulations have prevented previous disasters such as the Hindenburg or the Space Shuttle disaster?
8. What discipline is simulation theory based in?

See page 737 for the answers.

Exercises 14–5

1. Define simulation techniques.

2. Give three examples of simulation techniques.

3. Who is responsible for the development of modern simulation techniques?

4. What role does the computer play in simulation?

5. What are the steps in the simulation of an experiment?

6. What purpose do random numbers play in simulation?

7. What happens when the number of repetitions is increased?

For Exercises 8 through 13, explain how each experiment can be simulated by using random numbers.

8. A spinner contains six equal areas.

9. A basketball player makes 70% of her shots.

10. A certain brand of DVD player manufactured has a 10% defective rate.

11. An archer hits a target 80% of the time.

12. Two players match pennies.

13. Three players play odd man out. (Three coins are tossed; if all three match, the game is repeated and no one wins. If two players match, the third person wins all three coins.)

For Exercises 14 through 21, use random numbers to simulate the experiments. The number in parentheses is the number of times the experiment should be repeated.

14. A coin is tossed until four heads are obtained. Find the average number of tosses necessary. (50)

15. A die is rolled until all faces appear at least once. Find the average number of tosses. (30)

16. A caramel corn company gives four different prizes, one in each box. They are placed in the boxes at random. Find the average number of boxes a person needs to buy to get all four prizes. (40)

17. Two teams are evenly matched. They play a tournament in which the first team to win three games wins the tournament. Find the average number of games the tournament will last. (20)

18. To win a certain lotto, a person must spell the word *big*. Sixty percent of the tickets contain the letter *b*, 30% contain the letter *i*, and 10% contain the letter *g*. Find the average number of tickets a person must buy to win the prize. (30)

19. Two shooters shoot clay pigeons. Gail has an 80% accuracy rate and Paul has a 60% accuracy rate. Paul shoots first. The first person who hits the target wins. Find the probability that each wins. (30).

20. In Exercise 19, find the average number of shots fired. (30)

21. A basketball player has a 60% success rate for shooting foul shots. If she gets two shots, find the probability that she will make one or both shots. (50).

22. Select a game such as baseball or football and write a simulation using random numbers.

23. Explain how cards can be used to generate random numbers.

24. Explain how a pair of dice can be used to generate random numbers.

14–6 ## Summary

To obtain information and make inferences about a large population, researchers select a sample. A sample is a subgroup of the population. Using a sample rather than a population, researchers can save time and money, get more detailed information, and get information that otherwise would be impossible to obtain.

The four most common methods researchers use to obtain samples are random, systematic, stratified, and cluster sampling methods. In random sampling, some type of random method (usually random numbers) is used to obtain the sample. In systematic sampling, the researcher selects every *k*th person or item after selecting the first one at random. In stratified sampling, the population is divided into subgroups according to various characteristics, and elements are then selected at random from the subgroups. In cluster sampling, the researcher selects an intact group to use as a sample. When the population is large, multistage sampling (a combination of methods) is used to obtain a subgroup of the population.

Researchers must use caution when conducting surveys and designing questionnaires; otherwise, conclusions obtained from these will be inaccurate. Guidelines were presented in Section 14–3.

Most sampling methods use random numbers, which can also be used to simulate many real-life problems or situations. The basic method of simulation is known as the Monte Carlo method. The purpose of simulation is to duplicate situations that are too dangerous, too costly, or too time-consuming to study in real life. Most simulation techniques can be done on the computer or calculator, since they can rapidly generate random numbers, count the outcomes, and perform the necessary computations.

Sampling and simulation are two techniques that enable researchers to gain information that might otherwise be unobtainable.

Important Terms

biased sample 709	Monte Carlo method 726	sequence sampling 717	systematic sample 713
cluster sample 716	multistage sampling 717	simulation technique 726	unbiased sample 709
double sampling 717	random sample 709	stratified sample 714	

Review Exercises

Use Figure 14–12 for Exercises 1 through 8.

1. Select a random sample of 10 people, and find the mean of the weights of the individuals. Compare this mean with the population mean.

2. Select a systematic sample of 10 people, and compute the mean of their weights. Compare this mean with the population mean.

3. Divide the individuals into subgroups of males and females. Select five individuals from each group, and find the mean of their weights. Compare these means with the population mean.

4. Select a cluster of 10 people, and find the mean of their weights. Compare this mean with the population mean.

Figure 14–12

Population for Exercises 1 through 8

Individual	Gender	Weight	Systolic blood pressure	Individual	Gender	Weight	Systolic blood pressure	Individual	Gender	Weight	Systolic blood pressure
1	F	122	132	18	F	118	125	35	M	172	116
2	F	128	116	19	F	107	138	36	M	175	123
3	M	183	140	20	M	214	121	37	F	101	114
4	M	165	136	21	F	114	127	38	F	123	113
5	M	192	120	22	M	119	125	39	M	186	145
6	F	116	118	23	F	125	114	40	F	100	119
7	M	206	116	24	M	182	137	41	M	202	135
8	F	131	120	25	F	127	127	42	F	117	121
9	M	155	118	26	F	132	130	43	F	120	130
10	F	106	122	27	M	198	114	44	M	193	125
11	F	103	119	28	F	135	119	45	M	200	115
12	M	169	136	29	M	183	137	46	F	118	132
13	M	173	134	30	F	140	123	47	F	121	143
14	M	195	145	31	M	189	135	48	M	189	128
15	F	107	113	32	M	165	121	49	M	114	118
16	M	201	111	33	M	211	117	50	M	174	138
17	F	114	141	34	F	111	127				

5. Repeat Exercise 1 for blood pressure.

6. Repeat Exercise 2 for blood pressure.

7. Repeat Exercise 3 for blood pressure.

8. Repeat Exercise 4 for blood pressure.

For Exercises 9 through 13, explain how to simulate each experiment by using random numbers.

9. A baseball player strikes out 40% of the time.

10. An airline overbooks 15% of the time.

11. Two players roll a die. The higher number wins.

12. Player 1 rolls two dice. Player 2 rolls one die. If the number on the single die matches one number of the player who rolled the two dice, player 2 wins. Otherwise, player 1 wins.

13. Two players play rock, paper, scissors. The rules are as follows: Since paper covers rock, paper wins. Since rock breaks scissors, rock wins. Since scissors cut paper, scissors win. Each person selects rock, paper, or scissors by random numbers and then compares results.

For Exercises 14 through 18, use random numbers to simulate the experiments. The number in parentheses is the number of times the experiment should be repeated.

14. A football is placed on the 10-yard line, and a team has four downs to score a touchdown. The team can move the ball only 0 to 5 yards per play. Find the average number of times the team will score a touchdown. (30)

15. In Exercise 14, find the average number of plays it will take to score a touchdown. Ignore the four-downs rule and keep playing until a touchdown is scored. (30)

16. Four dice are rolled 50 times. Find the average of the sum of the number of spots that will appear. (50)

17. A field goal kicker is successful in 60% of his kicks inside the 35-yard line. Find the probability of kicking three field goals in a row. (50)

18. A sales representative finds that there is a 30% probability of making a sale by visiting the potential customer personally. For every 20 calls, find the probability of making three sales in a row. (50)

For Exercises 19 through 22, explain what is wrong with each question. Rewrite each one following the guidelines in this chapter.

19. How often do you run red lights?

20. Do you think students who are not failing should not be tutored?

21. Do you think all automobiles should have heavy-duty bumpers, even though it will raise the price of the cars by $500?

22. Explain the difference between an open-ended question and a closed-ended question.

Data Analysis

The Data Bank is found in Appendix D.

1. From the Data Bank, choose a variable. Select a random sample of 20 individuals, and find the mean of the data.

2. Select a systematic sample of 20 individuals, and using the same variable as in Exercise 1, find the mean.

3. Select a cluster sample of 20 individuals, and using the same variable as in Exercise 1, find the mean.

4. Stratify the data according to marital status and gender, and sample 20 individuals. Compute the mean of the sample variable selected in Exercise 1 (use four groups of five individuals).

5. Compare all four means and decide which one is most appropriate. (*Hint:* Find the population mean.)

Chapter Quiz

Determine whether each statement is true or false. If the statement is false, explain why.

1. When researchers are sampling from large populations, such as adult citizens living in the United States, they may use a combination of sampling techniques to ensure representativeness.

2. Simulation techniques using random numbers are a substitute for performing the actual statistical experiment.

3. When researchers perform simulation experiments, they do not need to use random numbers since they can make up random numbers.

4. Random samples are said to be unbiased.

The Monty Hall Problem—Revisited

It appears that it does not matter whether the contestant switches doors because he is given a choice of two doors, and the chance of winning the prize is 1 out of 2, or $\frac{1}{2}$. This reasoning, however, is incorrect. Consider the three possibilities for the prize. It could be behind door A, B, or C. Also consider the fact that the contestant has selected door A. Now the three situations look like this:

Case	Door		
	A	**B**	**C**
1	Prize	Empty	Empty
2	Empty	Prize	Empty
3	Empty	Empty	Prize

In case 1, the contestant selected door A, and if the contestant switched after being shown that there was no prize behind either door B or door C, he'd lose. In case 2, the contestant selected door A, and Monty will open door C, so if the contestant would switch, he would win the prize. In case 3, the contestant selected door A, and Monty will open door B, so if the contestant would switch, he would win the prize. Hence, by switching, the probability of winning is $\frac{2}{3}$ and the probability of losing is $\frac{1}{3}$. The same reasoning can be used no matter which door you select.

You can simulate this problem by using three cards, say, an ace (prize) and two other cards. Have a person arrange the cards in a row and let you select a card. After the person turns over one of the cards (a nonace), then switch. Keep track of the number of times you win. You can also play this game on the Internet by going to the website http://www.stat.sc.edu/~west/javahtml/LetsMakeaDeal.html

Select the best answer.

5. When all subjects under study are used, the group is called a _____.

   a. Population   *c.* Sample
   b. Large group   *d.* Study group

6. When a population is divided into subgroups with similar characteristics and then a sample is obtained, this method is called _____ sampling.

   a. Random   *c.* Stratified
   b. Systematic   *d.* Cluster

7. Interviewing selected people at a local supermarket can be considered an example of _____ sampling.

   a. Random   *c.* Convenience
   b. Systematic   *d.* Stratified

Complete the following statements with the best answer.

8. In general, when one conducts sampling, the _____ the sample, the more representative it will be.

9. When samples are not representative, they are said to be _____.

10. When all residents of a street are interviewed for a survey, the sampling method used is _____.

Use Figure 14–12 in the Review Exercises (page 733) for Exercises 11 through 14.

11. Select a random sample of 12 people, and find the mean of the blood pressures of the individuals. Compare this with the population mean.

12. Select a systematic sample of 12 people, and compute the mean of their blood pressures. Compare this with the population mean.

13. Divide the individuals into subgroups of six males and six females. Find the means of their blood pressures. Compare these means with the population mean.

14. Select a cluster of 12 people, and find the mean of their blood pressures. Compare this with the population mean.

For Exercises 15 through 19, explain how each could be simulated by using random numbers.

15. A chess player wins 45% of his games.

16. A travel agency has a 5% cancellation rate.

17. Two players select a card from a deck with no face cards. The player who gets the higher card wins.

18. One player rolls two dice. The other player selects a card from a deck. Face cards count as 11 for a jack, 12 for a queen, and 13 for a king. The player with the higher total points wins.

19. Two players toss two coins. If they match, player 1 wins; otherwise, player 2 wins.

For Exercises 20 through 24, use random numbers to simulate the experiments. The number in parentheses is the number of times the experiment should be done.

20. A telephone solicitor finds that there is a 15% probability of selling her product over the phone. For every 20 calls, find the probability of making two sales in a row. (100)

21. A field goal kicker is successful in 65% of his kicks inside the 40-yard line. Find the probability of his kicking four field goals in a row. (40)

22. Two coins are tossed. Find the average number of times two tails will appear. (40)

23. A single card is drawn from a deck. Find the average number of times it takes to draw an ace. (30)

24. A bowler finds that there is a 30% probability that he will make a strike. For every 15 frames he bowls, find the probability of making two strikes. (30)

Critical Thinking Challenges

1. Explain why two different opinion polls might yield different results on a survey. Also, give an example of an opinion poll and explain how the data may have been collected.

2. Use a computer to generate random numbers to simulate the following real-life problem.

   In a certain geographic region, 40% of the people have type O blood. On a certain day, the blood center needs 4 pints of type O blood. On average, how many donors are needed to obtain 4 pints of type O blood?

 ## Data Projects

Where appropriate, use MINITAB, the TI-83 Plus, the TI-84 Plus, or a computer program of your choice to complete the following exercises.

1. Using the rules given in Figure 14–11 on page 727 of your textbook, play the simulated bowling game at least 10 times. Each game consists of 10 frames.

   a. Analyze the results of the scores by finding the mean, median, mode, range, variance, and standard deviation.
   b. Draw a box plot and explain the nature of the distribution.
   c. Write several paragraphs explaining the results.
   d. Compare this simulation with real bowling. Do you think the game actually simulates bowling? Why or why not?

2. Select a sports game that you like to play or watch on television (e.g., baseball, golf, or hockey). Write a simulated version of the game, using random numbers or dice. Play the game several times and answer these questions.

   a. Does your simulated game represent the real game accurately?
   b. Is your game one of pure chance, or is strategy involved?
   c. What are some shortcomings of your game?
   d. What parts of the real game cannot be simulated in your game?
   e. Is there any way that you could improve your simulated game by changing some rules?

Answers to Applying the Concepts

Section 14–2 The White or Wheat Bread Debate

1. The researchers used a sample for their study.

2. Answers will vary. One possible answer is that we might have doubts about the validity of the study, since the baking company that conducted the experiment has an interest in the outcome of the experiment.

3. The sample was probably a convenience sample.

4. Answers will vary. One possible answer would be to use a simple random sample.

5. Answers will vary. One possible answer is that a list of women's names could be obtained from the city in which the women live. Then a simple random sample could be selected from this list.

6. The random assignment helps to spread variation among the groups. The random selection helps to generalize from the sample back to the population. These are two different issues.

Section 14–3   Smoking Bans and Profits

1. It is uncertain how public smoking bans affected restaurant business in Derry, Pennsylvania, since the survey results were conflicting.

2. Since the data were collected in different ways, the survey results were bound to have different answers. Perceptions of the owners will definitely be different from an analysis of actual sales receipts, particularly if the owners assumed that the public smoking bans would hurt business.

3. Answers will vary. One possible answer is that it would be difficult to not allow surveys based on anecdotal responses to be published. At the same time, it would be good for those publishing such survey results to comment on the limitations of these surveys.

4. We can get results from a representative sample that offer misleading information about the population.

5. Answers will vary. One possible answer is that measurement error is important in survey sampling in order to give ranges for the population parameters that are being investigated.

Section 14–4   Simulations

1. A simulation uses a probability experiment to mimic a real-life situation.

2. Simulation techniques date back to ancient times.

3. It is definitely cost-effective to run simulations for expensive items such as airplanes and automobiles.

4. Simulation testing is safer, faster, and less expensive than many real-life testing situations.

5. Computer simulation techniques were developed in the mid-1940s.

6. Answers will vary. One possible answer is that some simulations are far less harmful than conducting an actual study on the real-life situation of interest.

7. Answers will vary. Simulations could have possibly prevented disasters such as the Hindenburg or the Space Shuttle disaster. For example, data analysis after the Space Shuttle disaster showed that there was a decent chance that something would go wrong on that flight.

8. Simulation theory is based in probability theory.

Appendix A

Algebra Review

A-1 Factorials

A-2 Summation Notation

A-3 The Line

A-1 Factorials

Definition and Properties of Factorials

The notation called factorial notation is used in probability. *Factorial notation* uses the exclamation point and involves multiplication. For example,

$$5! = 5 \cdot 4 \cdot 3 \cdot 2 \cdot 1 = 120$$
$$4! = 4 \cdot 3 \cdot 2 \cdot 1 = 24$$
$$3! = 3 \cdot 2 \cdot 1 = 6$$
$$2! = 2 \cdot 1 = 2$$
$$1! = 1$$

In general, a factorial is evaluated as follows:

$$n! = n(n - 1)(n - 2) \cdots 3 \cdot 2 \cdot 1$$

Note that the factorial is the product of n factors, with the number decreased by 1 for each factor.

One property of factorial notation is that it can be stopped at any point by using the exclamation point. For example,

$$5! = 5 \cdot 4! \qquad \text{since} \quad 4! = 4 \cdot 3 \cdot 2 \cdot 1$$
$$= 5 \cdot 4 \cdot 3! \qquad \text{since} \quad 3! = 3 \cdot 2 \cdot 1$$
$$= 5 \cdot 4 \cdot 3 \cdot 2! \qquad \text{since} \quad 2! = 2 \cdot 1$$
$$= 5 \cdot 4 \cdot 3 \cdot 2 \cdot 1$$

Thus, $\quad n! = n(n - 1)!$
$$= n(n - 1)(n - 2)!$$
$$= n(n - 1)(n - 2)(n - 3)! \qquad \text{etc.}$$

Another property of factorials is

$$0! = 1$$

This fact is needed for formulas.

Operations with Factorials

Factorials cannot be added or subtracted directly. They must be multiplied out. Then the products can be added or subtracted.

Example A-1

Evaluate $3! + 4!$.

Solution

$$3! + 4! = (3 \cdot 2 \cdot 1) + (4 \cdot 3 \cdot 2 \cdot 1)$$
$$= 6 + 24 = 30$$

Note: $3! + 4! \neq 7!$, since $7! = 5040$.

Example A-2

Evaluate $5! - 3!$.

Solution

$$5! - 3! = (5 \cdot 4 \cdot 3 \cdot 2 \cdot 1) - (3 \cdot 2 \cdot 1)$$
$$= 120 - 6 = 114$$

Note: $5! - 3! \neq 2!$, since $2! = 2$.

Factorials cannot be multiplied directly. Again, one must multiply them out and then multiply the products.

Example A-3

Evaluate $3! \cdot 2!$.

Solution

$$3! \cdot 2! = (3 \cdot 2 \cdot 1) \cdot (2 \cdot 1) = 6 \cdot 2 = 12$$

Note: $3! \cdot 2! \neq 6!$, since $6! = 720$.

Finally, factorials cannot be divided directly unless they are equal.

Example A-4

Evaluate $6! \div 3!$.

Solution

$$\frac{6!}{3!} = \frac{6 \cdot 5 \cdot 4 \cdot 3 \cdot 2 \cdot 1}{3 \cdot 2 \cdot 1} = \frac{720}{6} = 120$$

Note: $\quad \dfrac{6!}{3!} \neq 2! \qquad \text{since} \qquad 2! = 2$

But $\quad \dfrac{3!}{3!} = \dfrac{3 \cdot 2 \cdot 1}{3 \cdot 2 \cdot 1} = \dfrac{6}{6} = 1$

In division, one can take some shortcuts, as shown:

$$\frac{6!}{3!} = \frac{6 \cdot 5 \cdot 4 \cdot 3!}{3!} \qquad \text{and} \qquad \frac{3!}{3!} = 1$$
$$= 6 \cdot 5 \cdot 4 = 120$$

$$\frac{8!}{6!} = \frac{8 \cdot 7 \cdot 6!}{6!} \qquad \text{and} \qquad \frac{6!}{6!} = 1$$
$$= 8 \cdot 7 = 56$$

Another shortcut that can be used with factorials is cancellation, after factors have been expanded. For example,

$$\frac{7!}{(4!)(3!)} = \frac{7 \cdot 6 \cdot 5 \cdot 4!}{3 \cdot 2 \cdot 1 \cdot 4!}$$

Now cancel both instances of 4!. Then cancel the $3 \cdot 2$ in the denominator with the 6 in the numerator.

$$\frac{7 \cdot \overset{1}{\cancel{6}} \cdot 5 \cdot \overset{1}{\cancel{4!}}}{\underset{1}{\cancel{3}} \cdot \underset{1}{\cancel{2}} \cdot 1 \cdot \underset{1}{\cancel{4!}}} = 7 \cdot 5 = 35$$

Example A–5

Evaluate $10! \div (6!)(4!)$.

Solution

$$\frac{10!}{(6!)(4!)} = \frac{10 \cdot \overset{3}{\cancel{9}} \cdot \overset{1}{\cancel{8}} \cdot 7 \cdot \overset{1}{\cancel{6!}}}{\underset{1}{\cancel{4}} \cdot \underset{1}{\cancel{3}} \cdot \underset{1}{\cancel{2}} \cdot 1 \cdot \underset{1}{\cancel{6!}}} = 10 \cdot 3 \cdot 7 = 210$$

Exercises

Evaluate each expression.

A–1. $9!$

A–2. $7!$

A–3. $5!$

A–4. $0!$

A–5. $1!$

A–6. $3!$

A–7. $\dfrac{12!}{9!}$

A–8. $\dfrac{10!}{2!}$

A–9. $\dfrac{5!}{3!}$

A–10. $\dfrac{11!}{7!}$

A–11. $\dfrac{9!}{(4!)(5!)}$

A–12. $\dfrac{10!}{(7!)(3!)}$

A–13. $\dfrac{8!}{(4!)(4!)}$

A–14. $\dfrac{15!}{(12!)(3!)}$

A–15. $\dfrac{10!}{(10!)(0!)}$

A–16. $\dfrac{5!}{(3!)(2!)(1!)}$

A–17. $\dfrac{8!}{(3!)(3!)(2!)}$

A–18. $\dfrac{11!}{(7!)(2!)(2!)}$

A–19. $\dfrac{10!}{(3!)(2!)(5!)}$

A–20. $\dfrac{6!}{(2!)(2!)(2!)}$

A–2 Summation Notation

In mathematics, the symbol Σ (Greek capital letter sigma) means to add or find the sum. For example, ΣX means to add the numbers represented by the variable X. Thus, when X represents 5, 8, 2, 4, and 6, then ΣX means $5 + 8 + 2 + 4 + 6 = 25$.

Sometimes, a subscript notation is used, such as

$$\sum_{i=1}^{5} X_i$$

This notation means to find the sum of five numbers represented by X, as shown:

$$\sum_{i=1}^{5} X_i = X_1 + X_2 + X_3 + X_4 + X_5$$

When the number of values is not known, the unknown number can be represented by n, such as

$$\sum_{i=1}^{n} X_i = X_1 + X_2 + X_3 + \cdots + X_n$$

There are several important types of summation used in statistics. The notation ΣX^2 means to square each value before summing. For example, if the values of the X's are 2, 8, 6, 1, and 4, then

$$\Sigma X^2 = 2^2 + 8^2 + 6^2 + 1^2 + 4^2$$
$$= 4 + 64 + 36 + 1 + 16 = 121$$

The notation $(\Sigma X)^2$ means to find the sum of X's and then square the answer. For instance, if the values for X are 2, 8, 6, 1, and 4, then

$$(\Sigma X)^2 = (2 + 8 + 6 + 1 + 4)^2$$
$$= (21)^2 = 441$$

Another important use of summation notation is in finding the mean (shown in Section 3–2). The mean $\overline{X}$ is defined as

$$\overline{X} = \frac{\Sigma X}{n}$$

For example, to find the mean of 12, 8, 7, 3, and 10, use the formula and substitute the values, as shown:

$$\overline{X} = \frac{\Sigma X}{n} = \frac{12 + 8 + 7 + 3 + 10}{5} = \frac{40}{5} = 8$$

The notation $\Sigma(X - \overline{X})^2$ means to perform the following steps.

STEP 1 Find the mean.

STEP 2 Subtract the mean from each value.

STEP 3 Square the answers.

STEP 4 Find the sum.

Example A–6

Find the value of $\Sigma(X - \overline{X})^2$ for the values 12, 8, 7, 3, and 10 of X.

Solution

STEP 1 Find the mean.

$$\overline{X} = \frac{12 + 8 + 7 + 3 + 10}{5} = \frac{40}{5} = 8$$

STEP 2 Subtract the mean from each value.

$$12 - 8 = 4 \qquad 7 - 8 = -1 \qquad 10 - 8 = 2$$
$$8 - 8 = 0 \qquad 3 - 8 = -5$$

STEP 3 Square the answers.

$$4^2 = 16 \qquad (-1)^2 = 1 \qquad 2^2 = 4$$
$$0^2 = 0 \qquad (-5)^2 = 25$$

STEP 4 Find the sum.

$$16 + 0 + 1 + 25 + 4 = 46$$

Example A–7

Find $\Sigma(X - \overline{X})^2$ for the following values of X: 5, 7, 2, 1, 3, 6.

Solution

Find the mean:

$$\overline{X} = \frac{5 + 7 + 2 + 1 + 3 + 6}{6} = \frac{24}{6} = 4$$

Then the steps in Example A–6 can be shortened as follows:

$$\begin{aligned}
\Sigma(X - \overline{X})^2 &= (5 - 4)^2 + (7 - 4)^2 + (2 - 4)^2 \\
&\quad + (1 - 4)^2 + (3 - 4)^2 + (6 - 4)^2 \\
&= 1^2 + 3^2 + (-2)^2 + (-3)^2 \\
&\quad + (-1)^2 + 2^2 \\
&= 1 + 9 + 4 + 9 + 1 + 4 = 28
\end{aligned}$$

Exercises

For each set of values, find ΣX, ΣX^2, $(\Sigma X)^2$, and $\Sigma(X - \overline{X})^2$.

A–21. 9, 17, 32, 16, 8, 2, 9, 7, 3, 18

A–22. 4, 12, 9, 13, 0, 6, 2, 10

A–23. 5, 12, 8, 3, 4

A–24. 6, 2, 18, 30, 31, 42, 16, 5

A–25. 80, 76, 42, 53, 77

A–26. 123, 132, 216, 98, 146, 114

A–27. 53, 72, 81, 42, 63, 71, 73, 85, 98, 55

A–28. 43, 32, 116, 98, 120

A–29. 12, 52, 36, 81, 63, 74

A–30. -9, -12, 18, 0, -2, -15

A–3 The Line

The following figure shows the *rectangular coordinate system* or *Cartesian plane*. This figure consists of two axes: the horizontal axis, called the x axis, and the vertical axis, called the y axis. Each axis has numerical scales. The point of intersection of the axes is called the *origin*.

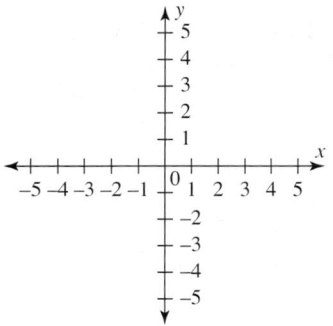

Points can be graphed by using coordinates. For example, the notation for point $P(3, 2)$ means that the x coordinate is 3 and the y coordinate is 2. Hence, P is located at the intersection of $x = 3$ and $y = 2$, as shown.

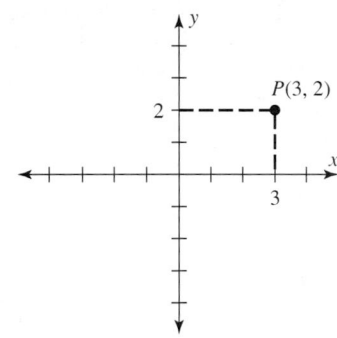

Other points, such as $Q(-5, 2)$, $R(4, 1)$, and $S(-3, -4)$, can be plotted, as shown in the next figure.

When a point lies on the y axis, the x coordinate is 0, as in $(0, 6)(0, -3)$, etc. When a point lies on the x axis, the y coordinate is 0, as in $(6, 0)(-8, 0)$, etc., as shown at the top of the next page.

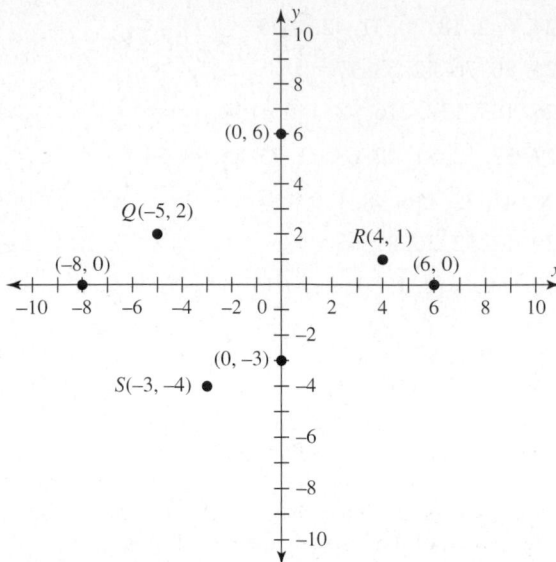

Two points determine a line. There are two properties of a line: its slope and its equation. The *slope m* of a line is determined by the ratio of the rise (called Δy) to the run (Δx).

$$m = \frac{\text{rise}}{\text{run}} = \frac{\Delta y}{\Delta x}$$

For example, the slope of the line shown below is $\frac{3}{2}$, or 1.5, since the height Δy is 3 units and the run Δx is 2 units.

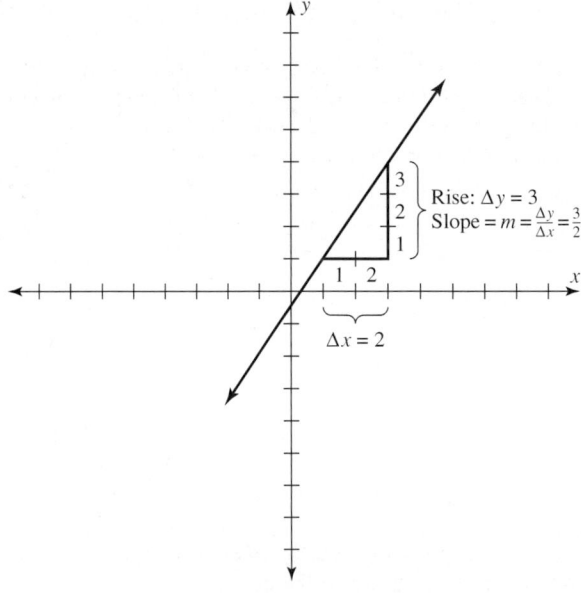

The slopes of lines can be positive, negative, or zero. A line going uphill from left to right has a positive slope. A line going downhill from left to right has a negative slope. And a line that is horizontal has a slope of zero.

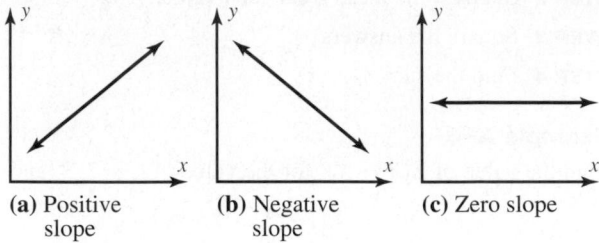

(a) Positive slope

(b) Negative slope

(c) Zero slope

A point b where the line crosses the x axis is called the *x intercept* and has the coordinates $(b, 0)$. A point a where the line crosses the y axis is called the *y intercept* and has the coordinates $(0, a)$.

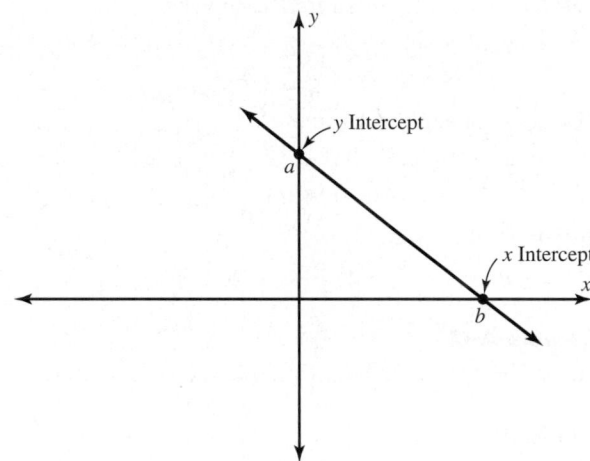

Every line has a unique equation of the form $y = a + bx$. For example, the equations

$$y = 5 + 3x$$
$$y = 8.6 + 3.2x$$
$$y = 5.2 - 6.1x$$

all represent different, unique lines. The number represented by a is the y intercept point; the number represented by b is the slope. The line whose equation is $y = 3 + 2x$ has a y intercept at 3 and a slope of 2, or $\frac{2}{1}$. This line can be shown as in the following graph.

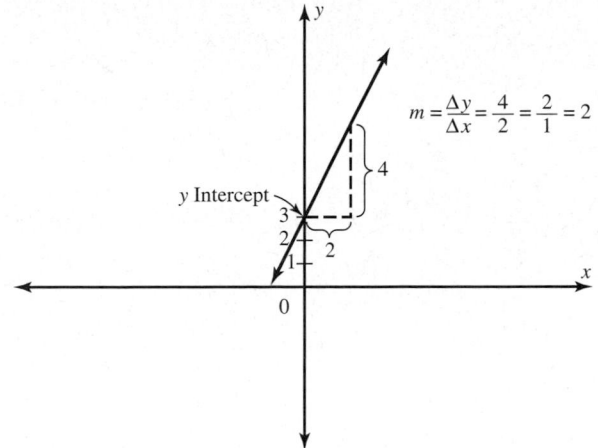

If two points are known, then the graph of the line can be plotted. For example, to find the graph of a line passing through the points $P(2, 1)$ and $Q(3, 5)$, plot the points and connect them as shown below.

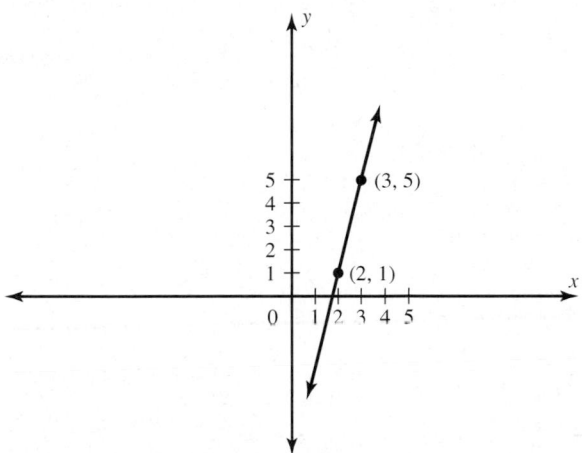

Given the equation of a line, one can graph the line by finding two points and then plotting them.

Example A–8

Plot the graph of the line whose equation is $y = 3 + 2x$.

Solution

Select any number as an x value, and substitute it in the equation to get the corresponding y value. Let $x = 0$.

Then

$$y = 3 + 2x = 3 + 2(0) = 3$$

Hence, when $x = 0$, then $y = 3$, and the line passes through the point $(0, 3)$.

Now select any other value of x, say, $x = 2$.

$$y = 3 + 2x = 3 + 2(2) = 7$$

Hence, a second point is $(2, 7)$. Then plot the points and graph the line.

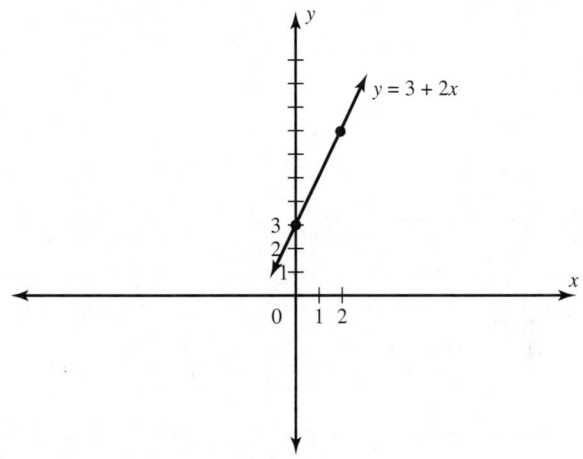

Exercises

Plot the line passing through each set of points.

A–31. $P(3, 2)$, $Q(1, 6)$ **A–34.** $P(-1, -2)$, $Q(-7, 8)$

A–32. $P(0, 5)$, $Q(8, 0)$ **A–35.** $P(6, 3)$, $Q(10, 3)$

A–33. $P(-2, 4)$, $Q(3, 6)$

Find at least two points on each line, and then graph the line containing these points.

A–36. $y = 5 + 2x$ **A–39.** $y = -2 - 2x$

A–37. $y = -1 + x$ **A–40.** $y = 4 - 3x$

A–38. $y = 3 + 4x$

Writing the Research Report

After conducting a statistical study, a researcher must write a final report explaining how the study was conducted and giving the results. The formats of research reports, theses, and dissertations vary from school to school; however, they tend to follow the general format explained here.

Front Materials

The front materials typically include the following items:

Title page
Copyright page
Acknowledgments
Table of contents
Table of appendixes
List of tables
List of figures

Chapter 1: Nature and Background of the Study

This chapter should introduce the reader to the nature of the study and present some discussion on the background. It should contain the following information:

Introduction
Statement of the problem
Background of the problem
Rationale for the study
Research questions and/or hypotheses
Assumptions, limitations, and delimitations
Definitions of terms

Chapter 2: Review of Literature

This chapter should explain what has been done in previous research related to the study. It should contain the following information:

Prior research
Related literature

Chapter 3: Methodology

This chapter should explain how the study was conducted. It should contain the following information:

Development of questionnaires, tests, survey instruments, etc.
Definition of the population
Sampling methods used
How the data were collected
Research design used
Statistical tests that will be used to analyze the data

Chapter 4: Analysis of Data

This chapter should explain the results of the statistical analysis of the data. It should state whether the null hypothesis should be rejected. Any statistical tables used to analyze the data should be included here.

Chapter 5: Summary, Conclusions, and Recommendations

This chapter summarizes the results of the study and explains any conclusions that have resulted from the statistical analysis of the data. The researchers should cite and explain any shortcomings of the study. Recommendations obtained from the study should be included here, and further studies should be suggested.

Bayes' Theorem

Objective B-1

Find the probability of an event, using Bayes' theorem.

Given two dependent events A and B, the previous formulas for conditional probability allow one to find $P(A$ and $B)$, or $P(B|A)$. Related to these formulas is a rule developed by the English Presbyterian minister Thomas Bayes (1702–61). The rule is known as **Bayes' theorem.**

It is possible, given the outcome of the second event in a sequence of two events, to determine the probability of various possibilities for the first event. In Example 4–31, there were two boxes, each containing red balls and blue balls. A box was selected and a ball was drawn. The example asked for the probability that the ball selected was red. Now, a different question can be asked: If the ball is red, what is the probability it came from box 1? In this case, the outcome is known, a red ball was selected, and one is asked to find the probability that it is a result of a previous event, that it came from box 1. Bayes' theorem can enable one to compute this probability and can be explained by using tree diagrams.

The tree diagram for the solution of Example 4–31 is shown in Figure B–1, along with the appropriate notation and the corresponding probabilities. In this case, A_1 is the event of selecting box 1, A_2 is the event of selecting box 2, R is the event of selecting a red ball, and B is the event of selecting a blue ball.

To answer the question "If the ball selected is red, what is the probability that it came from box 1?" the two formulas

$$P(B|A) = \frac{P(A \text{ and } B)}{P(A)} \tag{1}$$

$$P(A \text{ and } B) = P(A) \cdot P(B|A) \tag{2}$$

can be used. The notation that will be used is that of Example 4–31, shown in Figure B–1. Finding the probability that box 1 was selected given that the ball selected was red can be written symbolically as $P(A_1|R)$. By formula 1,

$$P(A_1|R) = \frac{P(R \text{ and } A_1)}{P(R_1)}$$

Note: $P(R \text{ and } A_1) = P(A_1 \text{ and } R)$.

Figure B–1

Tree Diagram for Example 4–31

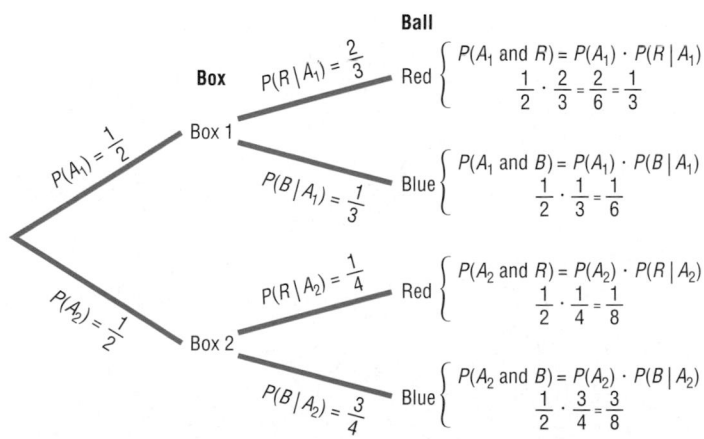

By formula 2,

$$P(A_1 \text{ and } R) = P(A_1) \cdot P(R|A_1)$$

and

$$P(R) = P(A_1 \text{ and } R) + P(A_2 \text{ and } R)$$

as shown in Figure B–1; $P(R)$ was found by adding the products of the probabilities of the branches in which a red ball was selected. Now,

$$P(A_1 \text{ and } R) = P(A_1) \cdot P(R|A_1)$$
$$P(A_2 \text{ and } R) = P(A_2) \cdot P(R|A_2)$$

Substituting these values in the original formula for $P(A_1|R)$, one gets

$$P(A_1|R) = \frac{P(A_1) \cdot P(R|A_1)}{P(A_1) \cdot P(R|A_1) + P(A_2) \cdot P(R|A_2)}$$

Refer to Figure B–1. The numerator of the fraction is the product of the top branch of the tree diagram, which consists of selecting a red ball and selecting box 1. And the denominator is the sum of the products of the two branches of the tree where the red ball was selected.

Using this formula and the probability values shown in Figure B–1, one can find the probability that box 1 was selected given that the ball was red, as shown.

$$P(A_1|R) = \frac{P(A_1) \cdot P(R|A_1)}{P(A_1) \cdot P(R|A_1) + P(A_2) \cdot P(R|A_2)}$$

$$= \frac{\frac{1}{2} \cdot \frac{2}{3}}{\frac{1}{2} \cdot \frac{2}{3} + \frac{1}{2} \cdot \frac{1}{4}} = \frac{\frac{1}{3}}{\frac{1}{3} + \frac{1}{8}} = \frac{\frac{1}{3}}{\frac{8}{24} + \frac{3}{24}} = \frac{\frac{1}{3}}{\frac{11}{24}}$$

$$= \frac{1}{3} \div \frac{11}{24} = \frac{1}{\overset{1}{\cancel{3}}} \cdot \frac{\overset{8}{\cancel{24}}}{11} = \frac{8}{11}$$

This formula is a simplified version of Bayes' theorem.

Before Bayes' theorem is stated, another example is shown.

Example B–1

A shipment of two boxes, each containing six telephones, is received by a store. Box 1 contains one defective phone, and box 2 contains two defective phones. After the boxes are unpacked, a phone is selected and found to be defective. Find the probability that it came from box 2.

Solution

STEP 1 Select the proper notation. Let A_1 represent box 1 and A_2 represent box 2. Let D represent a defective phone and ND represent a phone that is not defective.

STEP 2 Draw a tree diagram and find the corresponding probabilities for each branch. The probability of selecting box 1 is $\frac{1}{2}$, and the probability of selecting box 2 is $\frac{1}{2}$. Since there is one defective phone in box 1, the probability of selecting it is $\frac{1}{6}$. The probability of selecting a nondefective phone from box 1 is $\frac{5}{6}$.

Since there are two defective phones in box 2, the probability of selecting a defective phone from box 2 is $\frac{2}{6}$, or $\frac{1}{3}$; and the probability of selecting a nondefective phone is $\frac{4}{6}$, or $\frac{2}{3}$. The tree diagram is shown in Figure B–2.

STEP 3 Write the corresponding formula. Since the example is asking for the probability that, given a defective phone, it came from box 2, the corresponding formula is as shown.

$$P(A_2|D) = \frac{P(A_2) \cdot P(D|A_2)}{P(A_1) \cdot P(D|A_1) + P(A_2) \cdot P(D|A_2)}$$

$$= \frac{\frac{1}{2} \cdot \frac{2}{6}}{\frac{1}{2} \cdot \frac{1}{6} + \frac{1}{2} \cdot \frac{2}{6}} = \frac{\frac{1}{6}}{\frac{1}{12} + \frac{2}{12}} = \frac{\frac{1}{6}}{\frac{3}{12}}$$

$$= \frac{1}{6} \div \frac{3}{12} = \frac{1}{\overset{1}{\cancel{6}}} \cdot \frac{\overset{2}{\cancel{12}}}{3} = \frac{2}{3}$$

Figure B–2

Tree Diagram for Example B–1

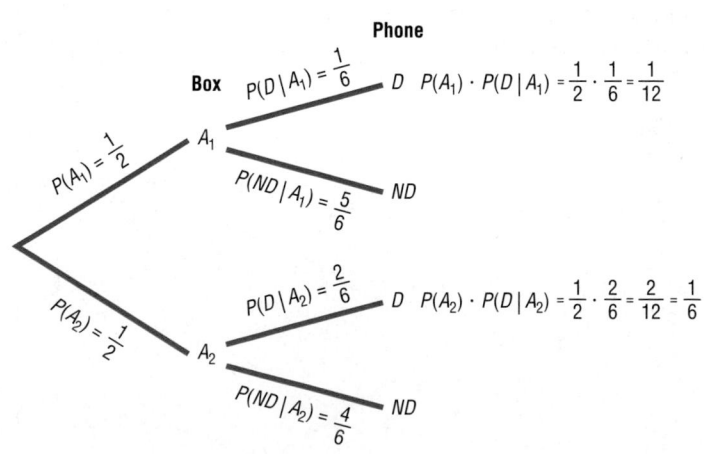

Bayes' theorem can be generalized to events with three or more outcomes and formally stated as in the next box.

Bayes' theorem For two events A and B, where event B follows event A, event A can occur in $A_1, A_2, \ldots, A_n$ mutually exclusive ways, and event B can occur in $B_1, B_2, \ldots, B_m$ mutually exclusive ways,

$$P(A_1|B_1) = \frac{P(A_1) \cdot P(B_1|A_1)}{[P(A_1) \cdot P(B_1|A_1) + P(A_2) \cdot P(B_1|A_2)} + \cdots + P(A_n) \cdot P(B_1|A_n)]$$

for any specific events A_1 and B_1.

The numerator is the product of the probabilities on the branch of the tree that consists of outcomes A_1 and B_1. The denominator is the sum of the products of the probabilities of the branches containing B_1 and A_1, B_1 and A_2, $\ldots$, B_1 and A_n.

Example B–2

On a game show, a contestant can select one of four boxes. Box 1 contains one $100 bill and nine $1 bills. Box 2 contains two $100 bills and eight $1 bills. Box 3 contains three $100 bills and seven $1 bills. Box 4 contains five $100 bills and five $1 bills. The contestant selects a box at random and selects a bill from the box at random. If a $100 bill is selected, find the probability that it came from box 4.

Solution

STEP 1 Select the proper notation. Let B_1, B_2, B_3, and B_4 represent the boxes and 100 and 1 represent the values of the bills in the boxes.

STEP 2 Draw a tree diagram and find the corresponding probabilities. The probability of selecting each box is $\frac{1}{4}$, or 0.25. The probabilities of selecting the $100 bill from each box, respectively, are $\frac{1}{10} = 0.1$, $\frac{2}{10} = 0.2$, $\frac{3}{10} = 0.3$, and $\frac{5}{10} = 0.5$. The tree diagram is shown in Figure B–3.

STEP 3 Using Bayes' theorem, write the corresponding formula. Since the example asks for the probability that box 4 was selected, given that $100 was obtained, the corresponding formula is as follows:

$$P(B_4|100) = \frac{P(B_4) \cdot P(100|B_4)}{[P(B_1) \cdot P(100|B_1) + P(B_2) \cdot P(100|B_2)} + P(B_3) \cdot P(100|B_3) + P(B_4) \cdot P(100|B_4)]$$

$$= \frac{0.125}{0.025 + 0.05 + 0.075 + 0.125}$$

$$= \frac{0.125}{0.275} = 0.455$$

Figure B–3

Tree Diagram for Example B–2

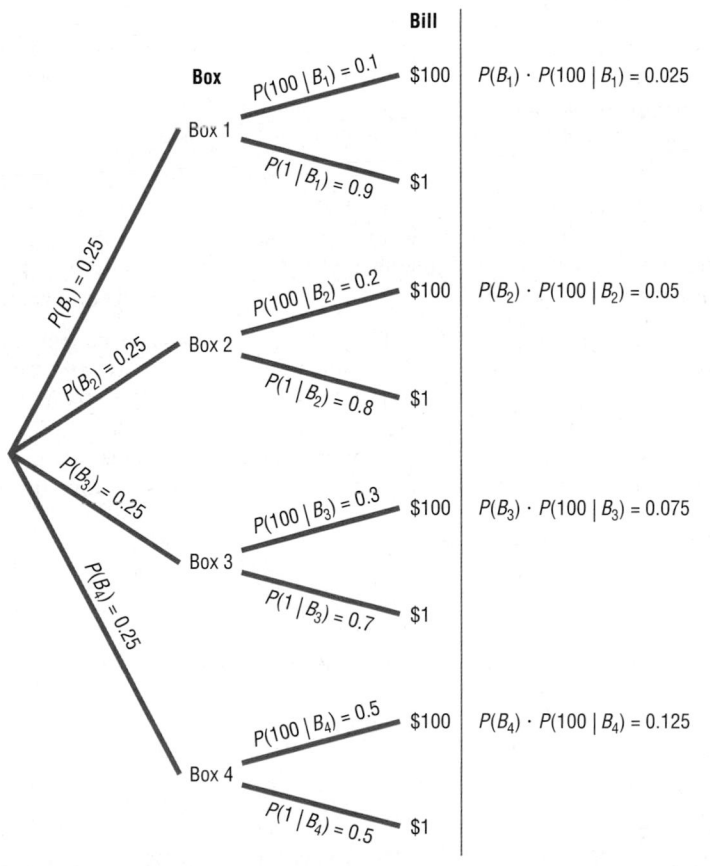

In Example B–2, the original probability of selecting box 4 was 0.25. However, once additional information was obtained—and the condition was considered, that a $100 bill was selected—the revised probability of selecting box 4 became 0.455.

Bayes' theorem can be used to revise probabilities of events once additional information becomes known. Bayes' theorem is used as the basis for a branch of statistics called *Bayesian decision making,* which includes the use of subjective probabilities in making statistical inferences.

Exercises

B–1. An appliance store purchases electric ranges from two companies. From company A, 500 ranges are purchased and 2% are defective. From company B, 850 ranges are purchased and 2% are defective. Given that a range is defective, find the probability that it came from company B.

B–2. Two manufacturers supply blankets to emergency relief organizations. Manufacturer A supplies 3000 blankets, and 4% are irregular in workmanship. Manufacturer B supplies 2400 blankets, and 7% are found to be irregular. Given that a blanket is irregular, find the probability that it came from manufacturer B.

B–3. A test for a certain disease is found to be 95% accurate, meaning that it will correctly diagnose the disease in 95 out of 100 people who have the ailment. For a certain segment of the population, the incidence of the disease is 9%. If a person tests positive, find the probability that the person actually has the disease. The test is also 95% accurate for a negative result.

B–4. Using the test in Exercise B–3, if a person tests negative for the disease, find the probability that the person actually has the disease. Remember, 9% of the population has the disease.

B–5. A corporation has three methods of training employees. Because of time, space, and location, it sends 20% of its employees to location A, 35% to location B, and 45% to location C. Location A has an 80% success rate. That is, 80% of the employees who complete the course will pass the licensing

exam. Location B has a 75% success rate, and location C has a 60% success rate. If a person has passed the exam, find the probability that the person went to location B.

B–6. In Exercise B–5, if a person failed the exam, find the probability that the person went to location C.

B–7. A store purchases baseball hats from three different manufacturers. In manufacturer A's box, there are 12 blue hats, 6 red hats, and 6 green hats. In manufacturer B's box, there are 10 blue hats, 10 red hats, and 4 green hats. In manufacturer C's box, there are 8 blue hats, 8 red hats, and 8 green hats. A box is selected at random, and a hat is selected at random from that box. If the hat is red, find the probability that it came from manufacturer A's box.

B–8. In Exercise B–7, if the hat selected is green, find the probability that it came from manufacturer B's box.

B–9. A driver has three ways to get from one city to another. There is an 80% probability of encountering a traffic jam on route 1, a 60% probability on route 2, and a 30% probability on route 3. Because of other factors, such as distance and speed limits, the driver uses route 1 fifty percent of the time and routes 2 and 3 each 25% of the time. If the driver calls the dispatcher to inform him that she is in a traffic jam, find the probability that she has selected route 1.

B–10. In Exercise B–9, if the driver did not encounter a traffic jam, find the probability that she selected route 3.

B–11. A store owner purchases telephones from two companies. From company A, 350 telephones are purchased and 2% are defective. From company B, 1050 telephones are purchased and 4% are defective. Given that a phone is defective, find the probability that it came from company B.

B–12. Two manufacturers supply food to a large cafeteria. Manufacturer A supplies 2400 cans of soup, and 3% are found to be dented. Manufacturer B supplies 3600 cans, and 1% are found to be dented. Given that a can of soup is dented, find the probability that it came from manufacturer B.

Appendix B–3

Alternate Approach to the Standard Normal Distribution

This approach can be used in lieu of that shown in Section 6–3. This approach uses a cumulative table of values and starts at $z = -3.09$ and ends at $z = +3.09$. For z values less than $z = -3.09$, use 0.0000 for the area; and for z values greater than 3.09, use 0.9999 for the area.

To find the area, follow these steps:

1. Draw the standard normal distribution.
2. Shade the desired area.
3. Use one of these rules:
   a. To find the area to the left of any z value, use the area in the table that corresponds to the z value.
   b. To find the area to the right of any z value, find the area in the table corresponding to the z value and subtract this area from 1.
   c. To find the area between two z values, find the corresponding areas and subtract the smaller value from the larger value.

Examples B–3 through B–5 illustrate the procedures.

Example B–3

Find the area under the standard normal distribution to the left of $z = 1.39$.

Solution

The area is shown as

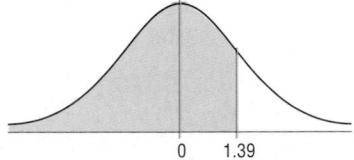

Since this is an example of situation a, look up the corresponding area as shown.

The area is 0.9177.

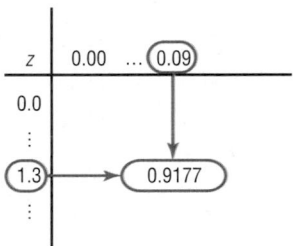

Example B–4

Find the area under the standard normal distribution to the right of $z = -2.06$.

Solution

The area is shown as

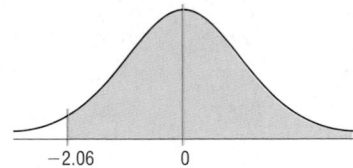

Since this is an example of situation b, look up the area corresponding to -2.06 as shown. The area is 0.0197. Subtract this area from 1.0000.

$$1.0000 - 0.0197 = 0.9803$$

Hence, the area to the right of $z = -2.06$ is 0.9803.

A–13

Example B–5

Find the area under the standard normal distribution between $z = 0.96$ and $z = 1.87$.

The area corresponding to $z = 0.96$ is 0.8315, and the area corresponding to $z = 1.87$ is 0.9693. Hence, the area between $z = 0.96$ and $z = 0.8315$ is $0.9693 - 0.8315 = 0.1378$.

Solution

The area is shown as

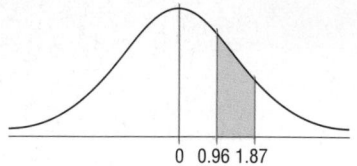

0 0.96 1.87

Cumulative Standard Normal Distribution

z	.00	.01	.02	.03	.04	.05	.06	.07	.08	.09
−3.0	.0013	.0013	.0013	.0012	.0012	.0011	.0011	.0011	.0010	.0010
−2.9	.0019	.0018	.0018	.0017	.0016	.0016	.0015	.0015	.0014	.0014
−2.8	.0026	.0025	.0024	.0023	.0023	.0022	.0021	.0021	.0020	.0019
−2.7	.0035	.0034	.0033	.0032	.0031	.0030	.0029	.0028	.0027	.0026
−2.6	.0047	.0045	.0044	.0043	.0041	.0040	.0039	.0038	.0037	.0036
−2.5	.0062	.0060	.0059	.0057	.0055	.0054	.0052	.0051	.0049	.0048
−2.4	.0082	.0080	.0078	.0075	.0073	.0071	.0069	.0068	.0066	.0064
−2.3	.0107	.0104	.0102	.0099	.0096	.0094	.0091	.0089	.0087	.0084
−2.2	.0139	.0136	.0132	.0129	.0125	.0122	.0119	.0116	.0113	.0110
−2.1	.0179	.0174	.0170	.0166	.0162	.0158	.0154	.0150	.0146	.0143
−2.0	.0228	.0222	.0217	.0212	.0207	.0202	.0197	.0192	.0188	.0183
−1.9	.0287	.0281	.0274	.0268	.0262	.0256	.0250	.0244	.0239	.0233
−1.8	.0359	.0351	.0344	.0336	.0329	.0322	.0314	.0307	.0301	.0294
−1.7	.0446	.0436	.0427	.0418	.0409	.0401	.0392	.0384	.0375	.0367
−1.6	.0548	.0537	.0526	.0516	.0505	.0495	.0485	.0475	.0465	.0455
−1.5	.0668	.0655	.0643	.0630	.0618	.0606	.0594	.0582	.0571	.0559
−1.4	.0808	.0793	.0778	.0764	.0749	.0735	.0721	.0708	.0694	.0681
−1.3	.0968	.0951	.0934	.0918	.0901	.0885	.0869	.0853	.0838	.0823
−1.2	.1151	.1131	.1112	.1093	.1075	.1056	.1038	.1020	.1003	.0985
−1.1	.1357	.1335	.1314	.1292	.1271	.1251	.1230	.1210	.1190	.1170
−1.0	.1587	.1562	.1539	.1515	.1492	.1469	.1446	.1423	.1401	.1379
−0.9	.1841	.1814	.1788	.1762	.1736	.1711	.1685	.1660	.1635	.1611
−0.8	.2119	.2090	.2061	.2033	.2005	.1977	.1949	.1922	.1894	.1867
−0.7	.2420	.2389	.2358	.2327	.2296	.2266	.2236	.2206	.2177	.2148
−0.6	.2743	.2709	.2676	.2643	.2611	.2578	.2546	.2514	.2483	.2451
−0.5	.3085	.3050	.3015	.2981	.2946	.2912	.2877	.2843	.2810	.2776
−0.4	.3446	.3409	.3372	.3336	.3300	.3264	.3228	.3192	.3156	.3121
−0.3	.3821	.3783	.3745	.3707	.3669	.3632	.3594	.3557	.3520	.3483
−0.2	.4207	.4168	.4129	.4090	.4052	.4013	.3974	.3936	.3897	.3859
−0.1	.4602	.4562	.4522	.4483	.4443	.4404	.4364	.4325	.4286	.4247
−0.0	.5000	.4960	.4920	.4880	.4840	.4801	.4761	.4721	.4681	.4641

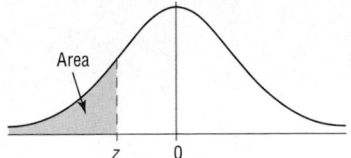

Cumulative Standard Normal Distribution

z	.00	.01	.02	.03	.04	.05	.06	.07	.08	.09
0.0	.5000	.5040	.5080	.5120	.5160	.5199	.5239	.5279	.5319	.5359
0.1	.5398	.5438	.5478	.5517	.5557	.5596	.5636	.5675	.5714	.5753
0.2	.5793	.5832	.5871	.5910	.5948	.5987	.6026	.6064	.6103	.6141
0.3	.6179	.6217	.6255	.6293	.6331	.6368	.6406	.6443	.6480	.6517
0.4	.6554	.6591	.6628	.6664	.6700	.6736	.6772	.6808	.6844	.6879
0.5	.6915	.6950	.6985	.7019	.7054	.7088	.7123	.7157	.7190	.7224
0.6	.7257	.7291	.7324	.7357	.7389	.7422	.7454	.7486	.7517	.7549
0.7	.7580	.7611	.7642	.7673	.7704	.7734	.7764	.7794	.7823	.7852
0.8	.7881	.7910	.7939	.7967	.7995	.8023	.8051	.8078	.8106	.8133
0.9	.8159	.8186	.8212	.8238	.8264	.8289	.8315	.8340	.8365	.8389
1.0	.8413	.8438	.8461	.8485	.8508	.8531	.8554	.8577	.8599	.8621
1.1	.8643	.8665	.8686	.8708	.8729	.8749	.8770	.8790	.8810	.8830
1.2	.8849	.8869	.8888	.8907	.8925	.8944	.8962	.8980	.8997	.9015
1.3	.9032	.9049	.9066	.9082	.9099	.9115	.9131	.9147	.9162	.9177
1.4	.9192	.9207	.9222	.9236	.9251	.9265	.9279	.9292	.9306	.9319
1.5	.9332	.9345	.9357	.9370	.9382	.9394	.9406	.9418	.9429	.9441
1.6	.9452	.9463	.9474	.9484	.9495	.9505	.9515	.9525	.9535	.9545
1.7	.9554	.9564	.9573	.9582	.9591	.9599	.9608	.9616	.9625	.9633
1.8	.9641	.9649	.9656	.9664	.9671	.9678	.9686	.9693	.9699	.9706
1.9	.9713	.9719	.9726	.9732	.9738	.9744	.9750	.9756	.9761	.9767
2.0	.9772	.9778	.9783	.9788	.9793	.9798	.9803	.9808	.9812	.9817
2.1	.9821	.9826	.9830	.9834	.9838	.9842	.9846	.9850	.9854	.9857
2.2	.9861	.9864	.9868	.9871	.9875	.9878	.9881	.9884	.9887	.9890
2.3	.9893	.9896	.9898	.9901	.9904	.9906	.9909	.9911	.9913	.9916
2.4	.9918	.9920	.9922	.9925	.9927	.9929	.9931	.9932	.9934	.9936
2.5	.9938	.9940	.9941	.9943	.9945	.9946	.9948	.9949	.9951	.9952
2.6	.9953	.9955	.9956	.9957	.9959	.9960	.9961	.9962	.9963	.9964
2.7	.9965	.9966	.9967	.9968	.9969	.9970	.9971	.9972	.9973	.9974
2.8	.9974	.9975	.9976	.9977	.9977	.9978	.9979	.9979	.9980	.9981
2.9	.9981	.9982	.9982	.9983	.9984	.9984	.9985	.9985	.9986	.9986
3.0	.9987	.9987	.9987	.9988	.9988	.9989	.9989	.9989	.9990	.9990

Appendix C

Tables

Table A	Factorials
n	$n!$
0	1
1	1
2	2
3	6
4	24
5	120
6	720
7	5,040
8	40,320
9	362,880
10	3,628,800
11	39,916,800
12	479,001,600
13	6,227,020,800
14	87,178,291,200
15	1,307,674,368,000
16	20,922,789,888,000
17	355,687,428,096,000
18	6,402,373,705,728,000
19	121,645,100,408,832,000
20	2,432,902,008,176,640,000

Table B		The Binomial Distribution										

							p					
n	x	0.05	0.1	0.2	0.3	0.4	0.5	0.6	0.7	0.8	0.9	0.95
2	0	0.902	0.810	0.640	0.490	0.360	0.250	0.160	0.090	0.040	0.010	0.002
	1	0.095	0.180	0.320	0.420	0.480	0.500	0.480	0.420	0.320	0.180	0.095
	2	0.002	0.010	0.040	0.090	0.160	0.250	0.360	0.490	0.640	0.810	0.902
3	0	0.857	0.729	0.512	0.343	0.216	0.125	0.064	0.027	0.008	0.001	
	1	0.135	0.243	0.384	0.441	0.432	0.375	0.288	0.189	0.096	0.027	0.007
	2	0.007	0.027	0.096	0.189	0.288	0.375	0.432	0.441	0.384	0.243	0.135
	3		0.001	0.008	0.027	0.064	0.125	0.216	0.343	0.512	0.729	0.857
4	0	0.815	0.656	0.410	0.240	0.130	0.062	0.026	0.008	0.002		
	1	0.171	0.292	0.410	0.412	0.346	0.250	0.154	0.076	0.026	0.004	
	2	0.014	0.049	0.154	0.265	0.346	0.375	0.346	0.265	0.154	0.049	0.014
	3		0.004	0.026	0.076	0.154	0.250	0.346	0.412	0.410	0.292	0.171
	4			0.002	0.008	0.026	0.062	0.130	0.240	0.410	0.656	0.815
5	0	0.774	0.590	0.328	0.168	0.078	0.031	0.010	0.002			
	1	0.204	0.328	0.410	0.360	0.259	0.156	0.077	0.028	0.006		
	2	0.021	0.073	0.205	0.309	0.346	0.312	0.230	0.132	0.051	0.008	0.001
	3	0.001	0.008	0.051	0.132	0.230	0.312	0.346	0.309	0.205	0.073	0.021
	4			0.006	0.028	0.077	0.156	0.259	0.360	0.410	0.328	0.204
	5				0.002	0.010	0.031	0.078	0.168	0.328	0.590	0.774
6	0	0.735	0.531	0.262	0.118	0.047	0.016	0.004	0.001			
	1	0.232	0.354	0.393	0.303	0.187	0.094	0.037	0.010	0.002		
	2	0.031	0.098	0.246	0.324	0.311	0.234	0.138	0.060	0.015	0.001	
	3	0.002	0.015	0.082	0.185	0.276	0.312	0.276	0.185	0.082	0.015	0.002
	4		0.001	0.015	0.060	0.138	0.234	0.311	0.324	0.246	0.098	0.031
	5			0.002	0.010	0.037	0.094	0.187	0.303	0.393	0.354	0.232
	6				0.001	0.004	0.016	0.047	0.118	0.262	0.531	0.735
7	0	0.698	0.478	0.210	0.082	0.028	0.008	0.002				
	1	0.257	0.372	0.367	0.247	0.131	0.055	0.017	0.004			
	2	0.041	0.124	0.275	0.318	0.261	0.164	0.077	0.025	0.004		
	3	0.004	0.023	0.115	0.227	0.290	0.273	0.194	0.097	0.029	0.003	
	4		0.003	0.029	0.097	0.194	0.273	0.290	0.227	0.115	0.023	0.004
	5			0.004	0.025	0.077	0.164	0.261	0.318	0.275	0.124	0.041
	6				0.004	0.017	0.055	0.131	0.247	0.367	0.372	0.257
	7					0.002	0.008	0.028	0.082	0.210	0.478	0.698
8	0	0.663	0.430	0.168	0.058	0.017	0.004	0.001				
	1	0.279	0.383	0.336	0.198	0.090	0.031	0.008	0.001			
	2	0.051	0.149	0.294	0.296	0.209	0.109	0.041	0.010	0.001		
	3	0.005	0.033	0.147	0.254	0.279	0.219	0.124	0.047	0.009		
	4		0.005	0.046	0.136	0.232	0.273	0.232	0.136	0.046	0.005	
	5			0.009	0.047	0.124	0.219	0.279	0.254	0.147	0.033	0.005
	6			0.001	0.010	0.041	0.109	0.209	0.296	0.294	0.149	0.051
	7				0.001	0.008	0.031	0.090	0.198	0.336	0.383	0.279
	8					0.001	0.004	0.017	0.058	0.168	0.430	0.663

Table B	(continued)										
							p					

n	x	0.05	0.1	0.2	0.3	0.4	0.5	0.6	0.7	0.8	0.9	0.95
9	0	0.630	0.387	0.134	0.040	0.010	0.002					
	1	0.299	0.387	0.302	0.156	0.060	0.018	0.004				
	2	0.063	0.172	0.302	0.267	0.161	0.070	0.021	0.004			
	3	0.008	0.045	0.176	0.267	0.251	0.164	0.074	0.021	0.003		
	4	0.001	0.007	0.066	0.172	0.251	0.246	0.167	0.074	0.017	0.001	
	5		0.001	0.017	0.074	0.167	0.246	0.251	0.172	0.066	0.007	0.001
	6			0.003	0.021	0.074	0.164	0.251	0.267	0.176	0.045	0.008
	7				0.004	0.021	0.070	0.161	0.267	0.302	0.172	0.063
	8					0.004	0.018	0.060	0.156	0.302	0.387	0.299
	9						0.002	0.010	0.040	0.134	0.387	0.630
10	0	0.599	0.349	0.107	0.028	0.006	0.001					
	1	0.315	0.387	0.268	0.121	0.040	0.010	0.002				
	2	0.075	0.194	0.302	0.233	0.121	0.044	0.011	0.001			
	3	0.010	0.057	0.201	0.267	0.215	0.117	0.042	0.009	0.001		
	4	0.001	0.011	0.088	0.200	0.251	0.205	0.111	0.037	0.006		
	5		0.001	0.026	0.103	0.201	0.246	0.201	0.103	0.026	0.001	
	6			0.006	0.037	0.111	0.205	0.251	0.200	0.088	0.011	0.001
	7			0.001	0.009	0.042	0.117	0.215	0.267	0.201	0.057	0.010
	8				0.001	0.011	0.044	0.121	0.233	0.302	0.194	0.075
	9					0.002	0.010	0.040	0.121	0.268	0.387	0.315
	10						0.001	0.006	0.028	0.107	0.349	0.599
11	0	0.569	0.314	0.086	0.020	0.004						
	1	0.329	0.384	0.236	0.093	0.027	0.005	0.001				
	2	0.087	0.213	0.295	0.200	0.089	0.027	0.005	0.001			
	3	0.014	0.071	0.221	0.257	0.177	0.081	0.023	0.004			
	4	0.001	0.016	0.111	0.220	0.236	0.161	0.070	0.017	0.002		
	5		0.002	0.039	0.132	0.221	0.226	0.147	0.057	0.010		
	6			0.010	0.057	0.147	0.226	0.221	0.132	0.039	0.002	
	7			0.002	0.017	0.070	0.161	0.236	0.220	0.111	0.016	0.001
	8				0.004	0.023	0.081	0.177	0.257	0.221	0.071	0.014
	9				0.001	0.005	0.027	0.089	0.200	0.295	0.213	0.087
	10					0.001	0.005	0.027	0.093	0.236	0.384	0.329
	11							0.004	0.020	0.086	0.314	0.569
12	0	0.540	0.282	0.069	0.014	0.002						
	1	0.341	0.377	0.206	0.071	0.017	0.003					
	2	0.099	0.230	0.283	0.168	0.064	0.016	0.002				
	3	0.017	0.085	0.236	0.240	0.142	0.054	0.012	0.001			
	4	0.002	0.021	0.133	0.231	0.213	0.121	0.042	0.008	0.001		
	5		0.004	0.053	0.158	0.227	0.193	0.101	0.029	0.003		
	6			0.016	0.079	0.177	0.226	0.177	0.079	0.016		
	7			0.003	0.029	0.101	0.193	0.227	0.158	0.053	0.004	
	8			0.001	0.008	0.042	0.121	0.213	0.231	0.133	0.021	0.002
	9				0.001	0.012	0.054	0.142	0.240	0.236	0.085	0.017
	10					0.002	0.016	0.064	0.168	0.283	0.230	0.099
	11						0.003	0.017	0.071	0.206	0.377	0.341
	12							0.002	0.014	0.069	0.282	0.540

Table B		*(continued)*									

						p						
n	*x*	0.05	0.1	0.2	0.3	0.4	0.5	0.6	0.7	0.8	0.9	0.95
13	0	0.513	0.254	0.055	0.010	0.001						
	1	0.351	0.367	0.179	0.054	0.011	0.002					
	2	0.111	0.245	0.268	0.139	0.045	0.010	0.001				
	3	0.021	0.100	0.246	0.218	0.111	0.035	0.006	0.001			
	4	0.003	0.028	0.154	0.234	0.184	0.087	0.024	0.003			
	5		0.006	0.069	0.180	0.221	0.157	0.066	0.014	0.001		
	6		0.001	0.023	0.103	0.197	0.209	0.131	0.044	0.006		
	7			0.006	0.044	0.131	0.209	0.197	0.103	0.023	0.001	
	8			0.001	0.014	0.066	0.157	0.221	0.180	0.069	0.006	
	9				0.003	0.024	0.087	0.184	0.234	0.154	0.028	0.003
	10				0.001	0.006	0.035	0.111	0.218	0.246	0.100	0.021
	11					0.001	0.010	0.045	0.139	0.268	0.245	0.111
	12						0.002	0.011	0.054	0.179	0.367	0.351
	13							0.001	0.010	0.055	0.254	0.513
14	0	0.488	0.229	0.044	0.007	0.001						
	1	0.359	0.356	0.154	0.041	0.007	0.001					
	2	0.123	0.257	0.250	0.113	0.032	0.006	0.001				
	3	0.026	0.114	0.250	0.194	0.085	0.022	0.003				
	4	0.004	0.035	0.172	0.229	0.155	0.061	0.014	0.001			
	5		0.008	0.086	0.196	0.207	0.122	0.041	0.007			
	6		0.001	0.032	0.126	0.207	0.183	0.092	0.023	0.002		
	7			0.009	0.062	0.157	0.209	0.157	0.062	0.009		
	8			0.002	0.023	0.092	0.183	0.207	0.126	0.032	0.001	
	9				0.007	0.041	0.122	0.207	0.196	0.086	0.008	
	10				0.001	0.014	0.061	0.155	0.229	0.172	0.035	0.004
	11					0.003	0.022	0.085	0.194	0.250	0.114	0.026
	12					0.001	0.006	0.032	0.113	0.250	0.257	0.123
	13						0.001	0.007	0.041	0.154	0.356	0.359
	14							0.001	0.007	0.044	0.229	0.488
15	0	0.463	0.206	0.035	0.005							
	1	0.366	0.343	0.132	0.031	0.005						
	2	0.135	0.267	0.231	0.092	0.022	0.003					
	3	0.031	0.129	0.250	0.170	0.063	0.014	0.002				
	4	0.005	0.043	0.188	0.219	0.127	0.042	0.007	0.001			
	5	0.001	0.010	0.103	0.206	0.186	0.092	0.024	0.003			
	6		0.002	0.043	0.147	0.207	0.153	0.061	0.012	0.001		
	7			0.014	0.081	0.177	0.196	0.118	0.035	0.003		
	8			0.003	0.035	0.118	0.196	0.177	0.081	0.014		
	9			0.001	0.012	0.061	0.153	0.207	0.147	0.043	0.002	
	10				0.003	0.024	0.092	0.186	0.206	0.103	0.010	0.001
	11				0.001	0.007	0.042	0.127	0.219	0.188	0.043	0.005
	12					0.002	0.014	0.063	0.170	0.250	0.129	0.031
	13						0.003	0.022	0.092	0.231	0.267	0.135
	14							0.005	0.031	0.132	0.343	0.366
	15							0.005	0.035	0.206	0.463

Table B		(continued)										
							p					
n	*x*	0.05	0.1	0.2	0.3	0.4	0.5	0.6	0.7	0.8	0.9	0.95
16	0	0.440	0.185	0.028	0.003							
	1	0.371	0.329	0.113	0.023	0.003						
	2	0.146	0.275	0.211	0.073	0.015	0.002					
	3	0.036	0.142	0.246	0.146	0.047	0.009	0.001				
	4	0.006	0.051	0.200	0.204	0.101	0.028	0.004				
	5	0.001	0.014	0.120	0.210	0.162	0.067	0.014	0.001			
	6		0.003	0.055	0.165	0.198	0.122	0.039	0.006			
	7			0.020	0.101	0.189	0.175	0.084	0.019	0.001		
	8			0.006	0.049	0.142	0.196	0.142	0.049	0.006		
	9			0.001	0.019	0.084	0.175	0.189	0.101	0.020		
	10				0.006	0.039	0.122	0.198	0.165	0.055	0.003	
	11				0.001	0.014	0.067	0.162	0.210	0.120	0.014	0.001
	12					0.004	0.028	0.101	0.204	0.200	0.051	0.006
	13					0.001	0.009	0.047	0.146	0.246	0.142	0.036
	14						0.002	0.015	0.073	0.211	0.275	0.146
	15							0.003	0.023	0.113	0.329	0.371
	16								0.003	0.028	0.185	0.440
17	0	0.418	0.167	0.023	0.002							
	1	0.374	0.315	0.096	0.017	0.002						
	2	0.158	0.280	0.191	0.058	0.010	0.001					
	3	0.041	0.156	0.239	0.125	0.034	0.005					
	4	0.008	0.060	0.209	0.187	0.080	0.018	0.002				
	5	0.001	0.017	0.136	0.208	0.138	0.047	0.008	0.001			
	6		0.004	0.068	0.178	0.184	0.094	0.024	0.003			
	7		0.001	0.027	0.120	0.193	0.148	0.057	0.009			
	8			0.008	0.064	0.161	0.185	0.107	0.028	0.002		
	9			0.002	0.028	0.107	0.185	0.161	0.064	0.008		
	10				0.009	0.057	0.148	0.193	0.120	0.027	0.001	
	11				0.003	0.024	0.094	0.184	0.178	0.068	0.004	
	12				0.001	0.008	0.047	0.138	0.208	0.136	0.017	0.001
	13					0.002	0.018	0.080	0.187	0.209	0.060	0.008
	14						0.005	0.034	0.125	0.239	0.156	0.041
	15						0.001	0.010	0.058	0.191	0.280	0.158
	16							0.002	0.017	0.096	0.315	0.374
	17								0.002	0.023	0.167	0.418

Table B (continued)

n	x	0.05	0.1	0.2	0.3	0.4	0.5	0.6	0.7	0.8	0.9	0.95
18	0	0.397	0.150	0.018	0.002							
	1	0.376	0.300	0.081	0.013	0.001						
	2	0.168	0.284	0.172	0.046	0.007	0.001					
	3	0.047	0.168	0.230	0.105	0.025	0.003					
	4	0.009	0.070	0.215	0.168	0.061	0.012	0.001				
	5	0.001	0.022	0.151	0.202	0.115	0.033	0.004				
	6		0.005	0.082	0.187	0.166	0.071	0.015	0.001			
	7		0.001	0.035	0.138	0.189	0.121	0.037	0.005			
	8			0.012	0.081	0.173	0.167	0.077	0.015	0.001		
	9			0.003	0.039	0.128	0.185	0.128	0.039	0.003		
	10			0.001	0.015	0.077	0.167	0.173	0.081	0.012		
	11				0.005	0.037	0.121	0.189	0.138	0.035	0.001	
	12				0.001	0.015	0.071	0.166	0.187	0.082	0.005	
	13					0.004	0.033	0.115	0.202	0.151	0.022	0.001
	14					0.001	0.012	0.061	0.168	0.215	0.070	0.009
	15						0.003	0.025	0.105	0.230	0.168	0.047
	16						0.001	0.007	0.046	0.172	0.284	0.168
	17							0.001	0.013	0.081	0.300	0.376
	18								0.002	0.018	0.150	0.397
19	0	0.377	0.135	0.014	0.001							
	1	0.377	0.285	0.068	0.009	0.001						
	2	0.179	0.285	0.154	0.036	0.005						
	3	0.053	0.180	0.218	0.087	0.017	0.002					
	4	0.011	0.080	0.218	0.149	0.047	0.007	0.001				
	5	0.002	0.027	0.164	0.192	0.093	0.022	0.002				
	6		0.007	0.095	0.192	0.145	0.052	0.008	0.001			
	7		0.001	0.044	0.153	0.180	0.096	0.024	0.002			
	8			0.017	0.098	0.180	0.144	0.053	0.008			
	9			0.005	0.051	0.146	0.176	0.098	0.022	0.001		
	10			0.001	0.022	0.098	0.176	0.146	0.051	0.005		
	11				0.008	0.053	0.144	0.180	0.098	0.071		
	12				0.002	0.024	0.096	0.180	0.153	0.044	0.001	
	13				0.001	0.008	0.052	0.145	0.192	0.095	0.007	
	14					0.002	0.022	0.093	0.192	0.164	0.027	0.002
	15					0.001	0.007	0.047	0.149	0.218	0.080	0.011
	16						0.002	0.017	0.087	0.218	0.180	0.053
	17							0.005	0.036	0.154	0.285	0.179
	18							0.001	0.009	0.068	0.285	0.377
	19								0.001	0.014	0.135	0.377

Table B	(concluded)											
							p					
n	*x*	0.05	0.1	0.2	0.3	0.4	0.5	0.6	0.7	0.8	0.9	0.95
20	0	0.358	0.122	0.012	0.001							
	1	0.377	0.270	0.058	0.007							
	2	0.189	0.285	0.137	0.028	0.003						
	3	0.060	0.190	0.205	0.072	0.012	0.001					
	4	0.013	0.090	0.218	0.130	0.035	0.005					
	5	0.002	0.032	0.175	0.179	0.075	0.015	0.001				
	6		0.009	0.109	0.192	0.124	0.037	0.005				
	7		0.002	0.055	0.164	0.166	0.074	0.015	0.001			
	8			0.022	0.114	0.180	0.120	0.035	0.004			
	9			0.007	0.065	0.160	0.160	0.071	0.012			
	10			0.002	0.031	0.117	0.176	0.117	0.031	0.002		
	11				0.012	0.071	0.160	0.160	0.065	0.007		
	12				0.004	0.035	0.120	0.180	0.114	0.022		
	13				0.001	0.015	0.074	0.166	0.164	0.055	0.002	
	14					0.005	0.037	0.124	0.192	0.109	0.009	
	15					0.001	0.015	0.075	0.179	0.175	0.032	0.002
	16						0.005	0.035	0.130	0.218	0.090	0.013
	17						0.001	0.012	0.072	0.205	0.190	0.060
	18							0.003	0.028	0.137	0.285	0.189
	19								0.007	0.058	0.270	0.377
	20								0.001	0.012	0.122	0.358

Note: All values of 0.0005 or less are omitted.

Source: John E. Freund, *Modern Elementary Statistics,* 8th ed., © 1992. Reprinted by permission of Prentice-Hall, Inc., Upper Saddle River, N.J.

| Table C | The Poisson Distribution |

	λ									
x	0.1	0.2	0.3	0.4	0.5	0.6	0.7	0.8	0.9	1.0
0	.9048	.8187	.7408	.6703	.6065	.5488	.4966	.4493	.4066	.3679
1	.0905	.1637	.2222	.2681	.3033	.3293	.3476	.3595	.3659	.3679
2	.0045	.0164	.0333	.0536	.0758	.0988	.1217	.1438	.1647	.1839
3	.0002	.0011	.0033	.0072	.0126	.0198	.0284	.0383	.0494	.0613
4	.0000	.0001	.0003	.0007	.0016	.0030	.0050	.0077	.0111	.0153
5	.0000	.0000	.0000	.0001	.0002	.0004	.0007	.0012	.0020	.0031
6	.0000	.0000	.0000	.0000	.0000	.0000	.0001	.0002	.0003	.0005
7	.0000	.0000	.0000	.0000	.0000	.0000	.0000	.0000	.0000	.0001

	λ									
x	1.1	1.2	1.3	1.4	1.5	1.6	1.7	1.8	1.9	2.0
0	.3329	.3012	.2725	.2466	.2231	.2019	.1827	.1653	.1496	.1353
1	.3662	.3614	.3543	.3452	.3347	.3230	.3106	.2975	.2842	.2707
2	.2014	.2169	.2303	.2417	.2510	.2584	.2640	.2678	.2700	.2707
3	.0738	.0867	.0998	.1128	.1255	.1378	.1496	.1607	.1710	.1804
4	.0203	.0260	.0324	.0395	.0471	.0551	.0636	.0723	.0812	.0902
5	.0045	.0062	.0084	.0111	.0141	.0176	.0216	.0260	.0309	.0361
6	.0008	.0012	.0018	.0026	.0035	.0047	.0061	.0078	.0098	.0120
7	.0001	.0002	.0003	.0005	.0008	.0011	.0015	.0020	.0027	.0034
8	.0000	.0000	.0001	.0001	.0001	.0002	.0003	.0005	.0006	.0009
9	.0000	.0000	.0000	.0000	.0000	.0000	.0001	.0001	.0001	.0002

	λ									
x	2.1	2.2	2.3	2.4	2.5	2.6	2.7	2.8	2.9	3.0
0	.1225	.1108	.1003	.0907	.0821	.0743	.0672	.0608	.0550	.0498
1	.2572	.2438	.2306	.2177	.2052	.1931	.1815	.1703	.1596	.1494
2	.2700	.2681	.2652	.2613	.2565	.2510	.2450	.2384	.2314	.2240
3	.1890	.1966	.2033	.2090	.2138	.2176	.2205	.2225	.2237	.2240
4	.0992	.1082	.1169	.1254	.1336	.1414	.1488	.1557	.1622	.1680
5	.0417	.0476	.0538	.0602	.0668	.0735	.0804	.0872	.0940	.1008
6	.0146	.0174	.0206	.0241	.0278	.0319	.0362	.0407	.0455	.0504
7	.0044	.0055	.0068	.0083	.0099	.0118	.0139	.0163	.0188	.0216
8	.0011	.0015	.0019	.0025	.0031	.0038	.0047	.0057	.0068	.0081
9	.0003	.0004	.0005	.0007	.0009	.0011	.0014	.0018	.0022	.0027
10	.0001	.0001	.0001	.0002	.0002	.0003	.0004	.0005	.0006	.0008
11	.0000	.0000	.0000	.0000	.0000	.0001	.0001	.0001	.0002	.0002
12	.0000	.0000	.0000	.0000	.0000	.0000	.0000	.0000	.0000	.0001

	λ									
x	3.1	3.2	3.3	3.4	3.5	3.6	3.7	3.8	3.9	4.0
0	.0450	.0408	.0369	.0334	.0302	.0273	.0247	.0224	.0202	.0183
1	.1397	.1304	.1217	.1135	.1057	.0984	.0915	.0850	.0789	.0733
2	.2165	.2087	.2008	.1929	.1850	.1771	.1692	.1615	.1539	.1465
3	.2237	.2226	.2209	.2186	.2158	.2125	.2087	.2046	.2001	.1954
4	.1734	.1781	.1823	.1858	.1888	.1912	.1931	.1944	.1951	.1954

Table C	(continued)									

					λ					
x	**3.1**	**3.2**	**3.3**	**3.4**	**3.5**	**3.6**	**3.7**	**3.8**	**3.9**	**4.0**
5	.1075	.1140	.1203	.1264	.1322	.1377	.1429	.1477	.1522	.1563
6	.0555	.0608	.0662	.0716	.0771	.0826	.0881	.0936	.0989	.1042
7	.0246	.0278	.0312	.0348	.0385	.0425	.0466	.0508	.0551	.0595
8	.0095	.0111	.0129	.0148	.0169	.0191	.0215	.0241	.0269	.0298
9	.0033	.0040	.0047	.0056	.0066	.0076	.0089	.0102	.0116	.0132
10	.0010	.0013	.0016	.0019	.0023	.0028	.0033	.0039	.0045	.0053
11	.0003	.0004	.0005	.0006	.0007	.0009	.0011	.0013	.0016	.0019
12	.0001	.0001	.0001	.0002	.0002	.0003	.0003	.0004	.0005	.0006
13	.0000	.0000	.0000	.0000	.0001	.0001	.0001	.0001	.0002	.0002
14	.0000	.0000	.0000	.0000	.0000	.0000	.0000	.0000	.0000	.0001

					λ					
x	**4.1**	**4.2**	**4.3**	**4.4**	**4.5**	**4.6**	**4.7**	**4.8**	**4.9**	**5.0**
0	.0166	.0150	.0136	.0123	.0111	.0101	.0091	.0082	.0074	.0067
1	.0679	.0630	.0583	.0540	.0500	.0462	.0427	.0395	.0365	.0337
2	.1393	.1323	.1254	.1188	.1125	.1063	.1005	.0948	.0894	.0842
3	.1904	.1852	.1798	.1743	.1687	.1631	.1574	.1517	.1460	.1404
4	.1951	.1944	.1933	.1917	.1898	.1875	.1849	.1820	.1789	.1755
5	.1600	.1633	.1662	.1687	.1708	.1725	.1738	.1747	.1753	.1755
6	.1093	.1143	.1191	.1237	.1281	.1323	.1362	.1398	.1432	.1462
7	.0640	.0686	.0732	.0778	.0824	.0869	.0914	.0959	.1002	.1044
8	.0328	.0360	.0393	.0428	.0463	.0500	.0537	.0575	.0614	.0653
9	.0150	.0168	.0188	.0209	.0232	.0255	.0280	.0307	.0334	.0363
10	.0061	.0071	.0081	.0092	.0104	.0118	.0132	.0147	.0164	.0181
11	.0023	.0027	.0032	.0037	.0043	.0049	.0056	.0064	.0073	.0082
12	.0008	.0009	.0011	.0014	.0016	.0019	.0022	.0026	.0030	.0034
13	.0002	.0003	.0004	.0005	.0006	.0007	.0008	.0009	.0011	.0013
14	.0001	.0001	.0001	.0001	.0002	.0002	.0003	.0003	.0004	.0005
15	.0000	.0000	.0000	.0000	.0001	.0001	.0001	.0001	.0001	.0002

					λ					
x	**5.1**	**5.2**	**5.3**	**5.4**	**5.5**	**5.6**	**5.7**	**5.8**	**5.9**	**6.0**
0	.0061	.0055	.0050	.0045	.0041	.0037	.0033	.0030	.0027	.0025
1	.0311	.0287	.0265	.0244	.0225	.0207	.0191	.0176	.0162	.0149
2	.0793	.0746	.0701	.0659	.0618	.0580	.0544	.0509	.0477	.0446
3	.1348	.1293	.1239	.1185	.1133	.1082	.1033	.0985	.0938	.0892
4	.1719	.1681	.1641	.1600	.1558	.1515	.1472	.1428	.1383	.1339

Table C   *(continued)*

	λ									
x	5.1	5.2	5.3	5.4	5.5	5.6	5.7	5.8	5.9	6.0
5	.1753	.1748	.1740	.1728	.1714	.1697	.1678	.1656	.1632	.1606
6	.1490	.1515	.1537	.1555	.1571	.1584	.1594	.1601	.1605	.1606
7	.1086	.1125	.1163	.1200	.1234	.1267	.1298	.1326	.1353	.1377
8	.0692	.0731	.0771	.0810	.0849	.0887	.0925	.0962	.0998	.1033
9	.0392	.0423	.0454	.0486	.0519	.0552	.0586	.0620	.0654	.0688
10	.0200	.0220	.0241	.0262	.0285	.0309	.0334	.0359	.0386	.0413
11	.0093	.0104	.0116	.0129	.0143	.0157	.0173	.0190	.0207	.0225
12	.0039	.0045	.0051	.0058	.0065	.0073	.0082	.0092	.0102	.0113
13	.0015	.0018	.0021	.0024	.0028	.0032	.0036	.0041	.0046	.0052
14	.0006	.0007	.0008	.0009	.0011	.0013	.0015	.0017	.0019	.0022
15	.0002	.0002	.0003	.0003	.0004	.0005	.0006	.0007	.0008	.0009
16	.0001	.0001	.0001	.0001	.0001	.0002	.0002	.0002	.0003	.0003
17	.0000	.0000	.0000	.0000	.0000	.0000	.0001	.0001	.0001	.0001

	λ									
x	6.1	6.2	6.3	6.4	6.5	6.6	6.7	6.8	6.9	7.0
0	.0022	.0020	.0018	.0017	.0015	.0014	.0012	.0011	.0010	.0009
1	.0137	.0126	.0116	.0106	.0098	.0090	.0082	.0076	.0070	.0064
2	.0417	.0390	.0364	.0340	.0318	.0296	.0276	.0258	.0240	.0223
3	.0848	.0806	.0765	.0726	.0688	.0652	.0617	.0584	.0552	.0521
4	.1294	.1249	.1205	.1162	.1118	.1076	.1034	.0992	.0952	.0912
5	.1579	.1549	.1519	.1487	.1454	.1420	.1385	.1349	.1314	.1277
6	.1605	.1601	.1595	.1586	.1575	.1562	.1546	.1529	.1511	.1490
7	.1399	.1418	.1435	.1450	.1462	.1472	.1480	.1486	.1489	.1490
8	.1066	.1099	.1130	.1160	.1188	.1215	.1240	.1263	.1284	.1304
9	.0723	.0757	.0791	.0825	.0858	.0891	.0923	.0954	.0985	.1014
10	.0441	.0469	.0498	.0528	.0558	.0588	.0618	.0649	.0679	.0710
11	.0245	.0265	.0285	.0307	.0330	.0353	.0377	.0401	.0426	.0452
12	.0124	.0137	.0150	.0164	.0179	.0194	.0210	.0227	.0245	.0264
13	.0058	.0065	.0073	.0081	.0089	.0098	.0108	.0119	.0130	.0142
14	.0025	.0029	.0033	.0037	.0041	.0046	.0052	.0058	.0064	.0071
15	.0010	.0012	.0014	.0016	.0018	.0020	.0023	.0026	.0029	.0033
16	.0004	.0005	.0005	.0006	.0007	.0008	.0010	.0011	.0013	.0014
17	.0001	.0002	.0002	.0002	.0003	.0003	.0004	.0004	.0005	.0006
18	.0000	.0001	.0001	.0001	.0001	.0001	.0001	.0002	.0002	.0002
19	.0000	.0000	.0000	.0000	.0000	.0000	.0000	.0001	.0001	.0001

Table C *(continued)*

					λ					
x	7.1	7.2	7.3	7.4	7.5	7.6	7.7	7.8	7.9	8.0
0	.0008	.0007	.0007	.0006	.0006	.0005	.0005	.0004	.0004	.0003
1	.0059	.0054	.0049	.0045	.0041	.0038	.0035	.0032	.0029	.0027
2	.0208	.0194	.0180	.0167	.0156	.0145	.0134	.0125	.0116	.0107
3	.0492	.0464	.0438	.0413	.0389	.0366	.0345	.0324	.0305	.0286
4	.0874	.0836	.0799	.0764	.0729	.0696	.0663	.0632	.0602	.0573
5	.1241	.1204	.1167	.1130	.1094	.1057	.1021	.0986	.0951	.0916
6	.1468	.1445	.1420	.1394	.1367	.1339	.1311	.1282	.1252	.1221
7	.1489	.1486	.1481	.1474	.1465	.1454	.1442	.1428	.1413	.1396
8	.1321	.1337	.1351	.1363	.1373	.1382	.1388	.1392	.1395	.1396
9	.1042	.1070	.1096	.1121	.1144	.1167	.1187	.1207	.1224	.1241
10	.0740	.0770	.0800	.0829	.0858	.0887	.0914	.0941	.0967	.0993
11	.0478	.0504	.0531	.0558	.0585	.0613	.0640	.0667	.0695	.0722
12	.0283	.0303	.0323	.0344	.0366	.0388	.0411	.0434	.0457	.0481
13	.0154	.0168	.0181	.0196	.0211	.0227	.0243	.0260	.0278	.0296
14	.0078	.0086	.0095	.0104	.0113	.0123	.0134	.0145	.0157	.0169
15	.0037	.0041	.0046	.0051	.0057	.0062	.0069	.0075	.0083	.0090
16	.0016	.0019	.0021	.0024	.0026	.0030	.0033	.0037	.0041	.0045
17	.0007	.0008	.0009	.0010	.0012	.0013	.0015	.0017	.0019	.0021
18	.0003	.0003	.0004	.0004	.0005	.0006	.0006	.0007	.0008	.0009
19	.0001	.0001	.0001	.0002	.0002	.0002	.0003	.0003	.0003	.0004
20	.0000	.0000	.0001	.0001	.0001	.0001	.0001	.0001	.0001	.0002
21	.0000	.0000	.0000	.0000	.0000	.0000	.0000	.0000	.0001	.0001

					λ					
x	8.1	8.2	8.3	8.4	8.5	8.6	8.7	8.8	8.9	9.0
0	.0003	.0003	.0002	.0002	.0002	.0002	.0002	.0002	.0001	.0001
1	.0025	.0023	.0021	.0019	.0017	.0016	.0014	.0013	.0012	.0011
2	.0100	.0092	.0086	.0079	.0074	.0068	.0063	.0058	.0054	.0050
3	.0269	.0252	.0237	.0222	.0208	.0195	.0183	.0171	.0160	.0150
4	.0544	.0517	.0491	.0466	.0443	.0420	.0398	.0377	.0357	.0337
5	.0882	.0849	.0816	.0784	.0752	.0722	.0692	.0663	.0635	.0607
6	.1191	.1160	.1128	.1097	.1066	.1034	.1003	.0972	.0941	.0911
7	.1378	.1358	.1338	.1317	.1294	.1271	.1247	.1222	.1197	.1171
8	.1395	.1392	.1388	.1382	.1375	.1366	.1356	.1344	.1332	.1318
9	.1256	.1269	.1280	.1290	.1299	.1306	.1311	.1315	.1317	.1318

Table C	(continued)

λ

x	8.1	8.2	8.3	8.4	8.5	8.6	8.7	8.8	8.9	9.0
10	.1017	.1040	.1063	.1084	.1104	.1123	.1140	.1157	.1172	.1186
11	.0749	.0776	.0802	.0828	.0853	.0878	.0902	.0925	.0948	.0970
12	.0505	.0530	.0555	.0579	.0604	.0629	.0654	.0679	.0703	.0728
13	.0315	.0334	.0354	.0374	.0395	.0416	.0438	.0459	.0481	.0504
14	.0182	.0196	.0210	.0225	.0240	.0256	.0272	.0289	.0306	.0324
15	.0098	.0107	.0116	.0126	.0136	.0147	.0158	.0169	.0182	.0194
16	.0050	.0055	.0060	.0066	.0072	.0079	.0086	.0093	.0101	.0109
17	.0024	.0026	.0029	.0033	.0036	.0040	.0044	.0048	.0053	.0058
18	.0011	.0012	.0014	.0015	.0017	.0019	.0021	.0024	.0026	.0029
19	.0005	.0005	.0006	.0007	.0008	.0009	.0010	.0011	.0012	.0014
20	.0002	.0002	.0002	.0003	.0003	.0004	.0004	.0005	.0005	.0006
21	.0001	.0001	.0001	.0001	.0001	.0002	.0002	.0002	.0002	.0003
22	.0000	.0000	.0000	.0000	.0001	.0001	.0001	.0001	.0001	.0001

λ

x	9.1	9.2	9.3	9.4	9.5	9.6	9.7	9.8	9.9	10.0
0	.0001	.0001	.0001	.0001	.0001	.0001	.0001	.0001	.0001	.0000
1	.0010	.0009	.0009	.0008	.0007	.0007	.0006	.0005	.0005	.0005
2	.0046	.0043	.0040	.0037	.0034	.0031	.0029	.0027	.0025	.0023
3	.0140	.0131	.0123	.0115	.0107	.0100	.0093	.0087	.0081	.0076
4	.0319	.0302	.0285	.0269	.0254	.0240	.0226	.0213	.0201	.0189
5	.0581	.0555	.0530	.0506	.0483	.0460	.0439	.0418	.0398	.0378
6	.0881	.0851	.0822	.0793	.0764	.0736	.0709	.0682	.0656	.0631
7	.1145	.1118	.1091	.1064	.1037	.1010	.0982	.0955	.0928	.0901
8	.1302	.1286	.1269	.1251	.1232	.1212	.1191	.1170	.1148	.1126
9	.1317	.1315	.1311	.1306	.1300	.1293	.1284	.1274	.1263	.1251
10	.1198	.1210	.1219	.1228	.1235	.1241	.1245	.1249	.1250	.1251
11	.0991	.1012	.1031	.1049	.1067	.1083	.1098	.1112	.1125	.1137
12	.0752	.0776	.0799	.0822	.0844	.0866	.0888	.0908	.0928	.0948
13	.0526	.0549	.0572	.0594	.0617	.0640	.0662	.0685	.0707	.0729
14	.0342	.0361	.0380	.0399	.0419	.0439	.0459	.0479	.0500	.0521
15	.0208	.0221	.0235	.0250	.0265	.0281	.0297	.0313	.0330	.0347
16	.0118	.0127	.0137	.0147	.0157	.0168	.0180	.0192	.0204	.0217
17	.0063	.0069	.0075	.0081	.0088	.0095	.0103	.0111	.0119	.0128
18	.0032	.0035	.0039	.0042	.0046	.0051	.0055	.0060	.0065	.0071
19	.0015	.0017	.0019	.0021	.0023	.0026	.0028	.0031	.0034	.0037

| Table C | (continued) |

					λ					
x	**9.1**	**9.2**	**9.3**	**9.4**	**9.5**	**9.6**	**9.7**	**9.8**	**9.9**	**10.0**
20	.0007	.0008	.0009	.0010	.0011	.0012	.0014	.0015	.0017	.0019
21	.0003	.0003	.0004	.0004	.0005	.0006	.0006	.0007	.0008	.0009
22	.0001	.0001	.0002	.0002	.0002	.0002	.0003	.0003	.0004	.0004
23	.0000	.0001	.0001	.0001	.0001	.0001	.0001	.0001	.0002	.0002
24	.0000	.0000	.0000	.0000	.0000	.0000	.0000	.0001	.0001	.0001

					λ					
x	**11**	**12**	**13**	**14**	**15**	**16**	**17**	**18**	**19**	**20**
0	.0000	.0000	.0000	.0000	.0000	.0000	.0000	.0000	.0000	.0000
1	.0002	.0001	.0000	.0000	.0000	.0000	.0000	.0000	.0000	.0000
2	.0010	.0004	.0002	.0001	.0000	.0000	.0000	.0000	.0000	.0000
3	.0037	.0018	.0008	.0004	.0002	.0001	.0000	.0000	.0000	.0000
4	.0102	.0053	.0027	.0013	.0006	.0003	.0001	.0001	.0000	.0000
5	.0224	.0127	.0070	.0037	.0019	.0010	.0005	.0002	.0001	.0001
6	.0411	.0255	.0152	.0087	.0048	.0026	.0014	.0007	.0004	.0002
7	.0646	.0437	.0281	.0174	.0104	.0060	.0034	.0018	.0010	.0005
8	.0888	.0655	.0457	.0304	.0194	.0120	.0072	.0042	.0024	.0013
9	.1085	.0874	.0661	.0473	.0324	.0213	.0135	.0083	.0050	.0029
10	.1194	.1048	.0859	.0663	.0486	.0341	.0230	.0150	.0095	.0058
11	.1194	.1144	.1015	.0844	.0663	.0496	.0355	.0245	.0164	.0106
12	.1094	.1144	.1099	.0984	.0829	.0661	.0504	.0368	.0259	.0176
13	.0926	.1056	.1099	.1060	.0956	.0814	.0658	.0509	.0378	.0271
14	.0728	.0905	.1021	.1060	.1024	.0930	.0800	.0655	.0514	.0387
15	.0534	.0724	.0885	.0989	.1024	.0992	.0906	.0786	.0650	.0516
16	.0367	.0543	.0719	.0866	.0960	.0992	.0963	.0884	.0772	.0646
17	.0237	.0383	.0550	.0713	.0847	.0934	.0963	.0936	.0863	.0760
18	.0145	.0256	.0397	.0554	.0706	.0830	.0909	.0936	.0911	.0844
19	.0084	.0161	.0272	.0409	.0557	.0699	.0814	.0887	.0911	.0888
20	.0046	.0097	.0177	.0286	.0418	.0559	.0692	.0798	.0866	.0888
21	.0024	.0055	.0109	.0191	.0299	.0426	.0560	.0684	.0783	.0846
22	.0012	.0030	.0065	.0121	.0204	.0310	.0433	.0560	.0676	.0769
23	.0006	.0016	.0037	.0074	.0133	.0216	.0320	.0438	.0559	.0669
24	.0003	.0008	.0020	.0043	.0083	.0144	.0226	.0328	.0442	.0557
25	.0001	.0004	.0010	.0024	.0050	.0092	.0154	.0237	.0336	.0446
26	.0000	.0002	.0005	.0013	.0029	.0057	.0101	.0164	.0246	.0343
27	.0000	.0001	.0002	.0007	.0016	.0034	.0063	.0109	.0173	.0254
28	.0000	.0000	.0001	.0003	.0009	.0019	.0038	.0070	.0117	.0181
29	.0000	.0000	.0001	.0002	.0004	.0011	.0023	.0044	.0077	.0125

Table C	(concluded)								
					λ				
x	11	12	13	14	15	16	17	18	19	20
30	.0000	.0000	.0000	.0001	.0002	.0006	.0013	.0026	.0049	.0083
31	.0000	.0000	.0000	.0000	.0001	.0003	.0007	.0015	.0030	.0054
32	.0000	.0000	.0000	.0000	.0001	.0001	.0004	.0009	.0018	.0034
33	.0000	.0000	.0000	.0000	.0000	.0001	.0002	.0005	.0010	.0020
34	.0000	.0000	.0000	.0000	.0000	.0000	.0001	.0002	.0006	.0012
35	.0000	.0000	.0000	.0000	.0000	.0000	.0000	.0001	.0003	.0007
36	.0000	.0000	.0000	.0000	.0000	.0000	.0000	.0001	.0002	.0004
37	.0000	.0000	.0000	.0000	.0000	.0000	.0000	.0000	.0001	.0002
38	.0000	.0000	.0000	.0000	.0000	.0000	.0000	.0000	.0000	.0001
39	.0000	.0000	.0000	.0000	.0000	.0000	.0000	.0000	.0000	.0001

Reprinted with permission from W. H. Beyer, *Handbook of Tables for Probability and Statistics,* 2nd ed. Copyright CRC Press, Boca Raton, Fla., 1986.

Table D Random Numbers

10480	15011	01536	02011	81647	91646	67179	14194	62590	36207	20969	99570	91291	90700
22368	46573	25595	85393	30995	89198	27982	53402	93965	34095	52666	19174	39615	99505
24130	48360	22527	97265	76393	64809	15179	24830	49340	32081	30680	19655	63348	58629
42167	93093	06243	61680	07856	16376	39440	53537	71341	57004	00849	74917	97758	16379
37570	39975	81837	16656	06121	91782	60468	81305	49684	60672	14110	06927	01263	54613
77921	06907	11008	42751	27756	53498	18602	70659	90655	15053	21916	81825	44394	42880
99562	72905	56420	69994	98872	31016	71194	18738	44013	48840	63213	21069	10634	12952
96301	91977	05463	07972	18876	20922	94595	56869	69014	60045	18425	84903	42508	32307
89579	14342	63661	10281	17453	18103	57740	84378	25331	12566	58678	44947	05584	56941
85475	36857	43342	53988	53060	59533	38867	62300	08158	17983	16439	11458	18593	64952
28918	69578	88231	33276	70997	79936	56865	05859	90106	31595	01547	85590	91610	78188
63553	40961	48235	03427	49626	69445	18663	72695	52180	20847	12234	90511	33703	90322
09429	93969	52636	92737	88974	33488	36320	17617	30015	08272	84115	27156	30613	74952
10365	61129	87529	85689	48237	52267	67689	93394	01511	26358	85104	20285	29975	89868
07119	97336	71048	08178	77233	13916	47564	81056	97735	85977	29372	74461	28551	90707
51085	12765	51821	51259	77452	16308	60756	92144	49442	53900	70960	63990	75601	40719
02368	21382	52404	60268	89368	19885	55322	44819	01188	65255	64835	44919	05944	55157
01011	54092	33362	94904	31273	04146	18594	29852	71585	85030	51132	01915	92747	64951
52162	53916	46369	58586	23216	14513	83149	98736	23495	64350	94738	17752	35156	35749
07056	97628	33787	09998	42698	06691	76988	13602	51851	46104	88916	19509	25625	58104
48663	91245	85828	14346	09172	30168	90229	04734	59193	22178	30421	61666	99904	32812
54164	58492	22421	74103	47070	25306	76468	26384	58151	06646	21524	15227	96909	44592
32639	32363	05597	24200	13363	38005	94342	28728	35806	06912	17012	64161	18296	22851
29334	27001	87637	87308	58731	00256	45834	15398	46557	41135	10367	07684	36188	18510
02488	33062	28834	07351	19731	92420	60952	61280	50001	67658	32586	86679	50720	94953
81525	72295	04839	96423	24878	82651	66566	14778	76797	14780	13300	87074	79666	95725
29676	20591	68086	26432	46901	20849	89768	81536	86645	12659	92259	57102	80428	25280
00742	57392	39064	66432	84673	40027	32832	61362	98947	96067	64760	64584	96096	98253
05366	04213	25669	26422	44407	44048	37937	63904	45766	66134	75470	66520	34693	90449
91921	26418	64117	94305	26766	25940	39972	22209	71500	64568	91402	42416	07844	69618
00582	04711	87917	77341	42206	35126	74087	99547	81817	42607	43808	76655	62028	76630
00725	69884	62797	56170	86324	88072	76222	36086	84637	93161	76038	65855	77919	88006
69011	65797	95876	55293	18988	27354	26575	08625	40801	59920	29841	80150	12777	48501
25976	57948	29888	88604	67917	48708	18912	82271	65424	69774	33611	54262	85963	03547
09763	83473	73577	12908	30883	18317	28290	35797	05998	41688	34952	37888	38917	88050
91567	42595	27958	30134	04024	86385	29880	99730	55536	84855	29080	09250	79656	73211
17955	56349	90999	49127	20044	59931	06115	20542	18059	02008	73708	83517	36103	42791
46503	18584	18845	49618	02304	51038	20655	58727	28168	15475	56942	53389	20562	87338
92157	89634	94824	78171	84610	82834	09922	25417	44137	48413	25555	21246	35509	20468
14577	62765	35605	81263	39667	47358	56873	56307	61607	49518	89656	20103	77490	18062
98427	07523	33362	64270	01638	92477	66969	98420	04880	45585	46565	04102	46880	45709
34914	63976	88720	82765	34476	17032	87589	40836	32427	70002	70663	88863	77775	69348
70060	28277	39475	46473	23219	53416	94970	25832	69975	94884	19661	72828	00102	66794
53976	54914	06990	67245	68350	82948	11398	42878	80287	88267	47363	46634	06541	97809
76072	29515	40980	07391	58745	25774	22987	80059	39911	96189	41151	14222	60697	59583
90725	52210	83974	29992	65831	38857	50490	83765	55657	14361	31720	57375	56228	41546
64364	67412	33339	31926	14883	24413	59744	92351	97473	89286	35931	04110	23726	51900
08962	00358	31662	25388	61642	34072	81249	35648	56891	69352	48373	45578	78547	81788
95012	68379	93526	70765	10593	04542	76463	54328	02349	17247	28865	14777	62730	92277
15664	10493	20492	38391	91132	21999	59516	81652	27195	48223	46751	22923	32261	85653

Reprinted with permission from W. H. Beyer, *Handbook of Tables for Probability and Statistics*, 2nd ed. Copyright CRC Press, Boca Raton, Fla., 1986.

Table E		The Standard Normal Distribution								
z	.00	.01	.02	.03	.04	.05	.06	.07	.08	.09
0.0	.0000	.0040	.0080	.0120	.0160	.0199	.0239	.0279	.0319	.0359
0.1	.0398	.0438	.0478	.0517	.0557	.0596	.0636	.0675	.0714	.0753
0.2	.0793	.0832	.0871	.0910	.0948	.0987	.1026	.1064	.1103	.1141
0.3	.1179	.1217	.1255	.1293	.1331	.1368	.1406	.1443	.1480	.1517
0.4	.1554	.1591	.1628	.1664	.1700	.1736	.1772	.1808	.1844	.1879
0.5	.1915	.1950	.1985	.2019	.2054	.2088	.2123	.2157	.2190	.2224
0.6	.2257	.2291	.2324	.2357	.2389	.2422	.2454	.2486	.2517	.2549
0.7	.2580	.2611	.2642	.2673	.2704	.2734	.2764	.2794	.2823	.2852
0.8	.2881	.2910	.2939	.2967	.2995	.3023	.3051	.3078	.3106	.3133
0.9	.3159	.3186	.3212	.3238	.3264	.3289	.3315	.3340	.3365	.3389
1.0	.3413	.3438	.3461	.3485	.3508	.3531	.3554	.3577	.3599	.3621
1.1	.3643	.3665	.3686	.3708	.3729	.3749	.3770	.3790	.3810	.3830
1.2	.3849	.3869	.3888	.3907	.3925	.3944	.3962	.3980	.3997	.4015
1.3	.4032	.4049	.4066	.4082	.4099	.4115	.4131	.4147	.4162	.4177
1.4	.4192	.4207	.4222	.4236	.4251	.4265	.4279	.4292	.4306	.4319
1.5	.4332	.4345	.4357	.4370	.4382	.4394	.4406	.4418	.4429	.4441
1.6	.4452	.4463	.4474	.4484	.4495	.4505	.4515	.4525	.4535	.4545
1.7	.4554	.4564	.4573	.4582	.4591	.4599	.4608	.4616	.4625	.4633
1.8	.4641	.4649	.4656	.4664	.4671	.4678	.4686	.4693	.4699	.4706
1.9	.4713	.4719	.4726	.4732	.4738	.4744	.4750	.4756	.4761	.4767
2.0	.4772	.4778	.4783	.4788	.4793	.4798	.4803	.4808	.4812	.4817
2.1	.4821	.4826	.4830	.4834	.4838	.4842	.4846	.4850	.4854	.4857
2.2	.4861	.4864	.4868	.4871	.4875	.4878	.4881	.4884	.4887	.4890
2.3	.4893	.4896	.4898	.4901	.4904	.4906	.4909	.4911	.4913	.4916
2.4	.4918	.4920	.4922	.4925	.4927	.4929	.4931	.4932	.4934	.4936
2.5	.4938	.4940	.4941	.4943	.4945	.4946	.4948	.4949	.4951	.4952
2.6	.4953	.4955	.4956	.4957	.4959	.4960	.4961	.4962	.4963	.4964
2.7	.4965	.4966	.4967	.4968	.4969	.4970	.4971	.4972	.4973	.4974
2.8	.4974	.4975	.4976	.4977	.4977	.4978	.4979	.4979	.4980	.4981
2.9	.4981	.4982	.4982	.4983	.4984	.4984	.4985	.4985	.4986	.4986
3.0	.4987	.4987	.4987	.4988	.4988	.4989	.4989	.4989	.4990	.4990

Note: Use 0.4999 for z values above 3.09.

Source: Frederick Mosteller and Robert E. K. Rourke, *Sturdy Statistics,* Table A–1 (Reading, Mass.: Addison-Wesley, 1973). Reprinted with permission of the copyright owners.

Area given in table

Table F	The *t* Distribution						
	Confidence intervals	**50%**	**80%**	**90%**	**95%**	**98%**	**99%**
	One tail, α	0.25	0.10	0.05	0.025	0.01	0.005
d.f.	**Two tails, α**	0.50	0.20	0.10	0.05	0.02	0.01
1		1.000	3.078	6.314	12.706	31.821	63.657
2		.816	1.886	2.920	4.303	6.965	9.925
3		.765	1.638	2.353	3.182	4.541	5.841
4		.741	1.533	2.132	2.776	3.747	4.604
5		.727	1.476	2.015	2.571	3.365	4.032
6		.718	1.440	1.943	2.447	3.143	3.707
7		.711	1.415	1.895	2.365	2.998	3.499
8		.706	1.397	1.860	2.306	2.896	3.355
9		.703	1.383	1.833	2.262	2.821	3.250
10		.700	1.372	1.812	2.228	2.764	3.169
11		.697	1.363	1.796	2.201	2.718	3.106
12		.695	1.356	1.782	2.179	2.681	3.055
13		.694	1.350	1.771	2.160	2.650	3.012
14		.692	1.345	1.761	2.145	2.624	2.977
15		.691	1.341	1.753	2.131	2.602	2.947
16		.690	1.337	1.746	2.120	2.583	2.921
17		.689	1.333	1.740	2.110	2.567	2.898
18		.688	1.330	1.734	2.101	2.552	2.878
19		.688	1.328	1.729	2.093	2.539	2.861
20		.687	1.325	1.725	2.086	2.528	2.845
21		.686	1.323	1.721	2.080	2.518	2.831
22		.686	1.321	1.717	2.074	2.508	2.819
23		.685	1.319	1.714	2.069	2.500	2.807
24		.685	1.318	1.711	2.064	2.492	2.797
25		.684	1.316	1.708	2.060	2.485	2.787
26		.684	1.315	1.706	2.056	2.479	2.779
27		.684	1.314	1.703	2.052	2.473	2.771
28		.683	1.313	1.701	2.048	2.467	2.763
(z) ∞		.674	1.282[a]	1.645[b]	1.960	2.326[c]	2.576[d]

[a]This value has been rounded to 1.28 in the textbook.
[b]This value has been rounded to 1.65 in the textbook.
[c]This value has been rounded to 2.33 in the textbook.
[d]This value has been rounded to 2.58 in the textbook.

Source: Adapted from W. H. Beyer, *Handbook of Tables for Probability and Statistics,* 2nd ed., CRC Press, Boca Raton, Fla., 1986. Reprinted with permission.

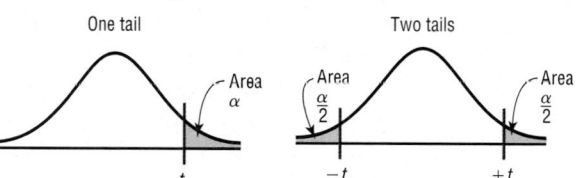

Table G   The Chi-Square Distribution

Degrees of freedom	α									
	0.995	0.99	0.975	0.95	0.90	0.10	0.05	0.025	0.01	0.005
1	—	—	0.001	0.004	0.016	2.706	3.841	5.024	6.635	7.879
2	0.010	0.020	0.051	0.103	0.211	4.605	5.991	7.378	9.210	10.597
3	0.072	0.115	0.216	0.352	0.584	6.251	7.815	9.348	11.345	12.838
4	0.207	0.297	0.484	0.711	1.064	7.779	9.488	11.143	13.277	14.860
5	0.412	0.554	0.831	1.145	1.610	9.236	11.071	12.833	15.086	16.750
6	0.676	0.872	1.237	1.635	2.204	10.645	12.592	14.449	16.812	18.548
7	0.989	1.239	1.690	2.167	2.833	12.017	14.067	16.013	18.475	20.278
8	1.344	1.646	2.180	2.733	3.490	13.362	15.507	17.535	20.090	21.955
9	1.735	2.088	2.700	3.325	4.168	14.684	16.919	19.023	21.666	23.589
10	2.156	2.558	3.247	3.940	4.865	15.987	18.307	20.483	23.209	25.188
11	2.603	3.053	3.816	4.575	5.578	17.275	19.675	21.920	24.725	26.757
12	3.074	3.571	4.404	5.226	6.304	18.549	21.026	23.337	26.217	28.299
13	3.565	4.107	5.009	5.892	7.042	19.812	22.362	24.736	27.688	29.819
14	4.075	4.660	5.629	6.571	7.790	21.064	23.685	26.119	29.141	31.319
15	4.601	5.229	6.262	7.261	8.547	22.307	24.996	27.488	30.578	32.801
16	5.142	5.812	6.908	7.962	9.312	23.542	26.296	28.845	32.000	34.267
17	5.697	6.408	7.564	8.672	10.085	24.769	27.587	30.191	33.409	35.718
18	6.265	7.015	8.231	9.390	10.865	25.989	28.869	31.526	34.805	37.156
19	6.844	7.633	8.907	10.117	11.651	27.204	30.144	32.852	36.191	38.582
20	7.434	8.260	9.591	10.851	12.443	28.412	31.410	34.170	37.566	39.997
21	8.034	8.897	10.283	11.591	13.240	29.615	32.671	35.479	38.932	41.401
22	8.643	9.542	10.982	12.338	14.042	30.813	33.924	36.781	40.289	42.796
23	9.262	10.196	11.689	13.091	14.848	32.007	35.172	38.076	41.638	44.181
24	9.886	10.856	12.401	13.848	15.659	33.196	36.415	39.364	42.980	45.559
25	10.520	11.524	13.120	14.611	16.473	34.382	37.652	40.646	44.314	46.928
26	11.160	12.198	13.844	15.379	17.292	35.563	38.885	41.923	45.642	48.290
27	11.808	12.879	14.573	16.151	18.114	36.741	40.113	43.194	46.963	49.645
28	12.461	13.565	15.308	16.928	18.939	37.916	41.337	44.461	48.278	50.993
29	13.121	14.257	16.047	17.708	19.768	39.087	42.557	45.722	49.588	52.336
30	13.787	14.954	16.791	18.493	20.599	40.256	43.773	46.979	50.892	53.672
40	20.707	22.164	24.433	26.509	29.051	51.805	55.758	59.342	63.691	66.766
50	27.991	29.707	32.357	34.764	37.689	63.167	67.505	71.420	76.154	79.490
60	35.534	37.485	40.482	43.188	46.459	74.397	79.082	83.298	88.379	91.952
70	43.275	45.442	48.758	51.739	55.329	85.527	90.531	95.023	100.425	104.215
80	51.172	53.540	57.153	60.391	64.278	96.578	101.879	106.629	112.329	116.321
90	59.196	61.754	65.647	69.126	73.291	107.565	113.145	118.136	124.116	128.299
100	67.328	70.065	74.222	77.929	82.358	118.498	124.342	129.561	135.807	140.169

Source: Donald B. Owen, *Handbook of Statistics Tables,* The Chi-Square Distribution Table, © 1962 by Addison-Wesley Publishing Company, Inc. Copyright renewal © 1990. Reprinted by permission of Pearson Education, Inc.

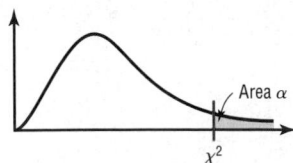

Table H The *F* Distribution

$\alpha = 0.005$

d.f.N.: degrees of freedom, numerator

d.f.D.: degrees of freedom, denominator	1	2	3	4	5	6	7	8	9	10	12	15	20	24	30	40	60	120	∞
1	16,211	20,000	21,615	22,500	23,056	23,437	23,715	23,925	24,091	24,224	24,426	24,630	24,836	24,940	25,044	25,148	25,253	25,359	25,465
2	198.5	199.0	199.2	199.2	199.3	199.3	199.4	199.4	199.4	199.4	199.4	199.4	199.4	199.5	199.5	199.5	199.5	199.5	199.5
3	55.55	49.80	47.47	46.19	45.39	44.84	44.43	44.13	43.88	43.69	43.39	43.08	42.78	42.62	42.47	42.31	42.15	41.99	41.83
4	31.33	26.28	24.26	23.15	22.46	21.97	21.62	21.35	21.14	20.97	20.70	20.44	20.17	20.03	19.89	19.75	19.61	19.47	19.32
5	22.78	18.31	16.53	15.56	14.94	14.51	14.20	13.96	13.77	13.62	13.38	13.15	12.90	12.78	12.66	12.53	12.40	12.27	12.14
6	18.63	14.54	12.92	12.03	11.46	11.07	10.79	10.57	10.39	10.25	10.03	9.81	9.59	9.47	9.36	9.24	9.12	9.00	8.88
7	16.24	12.40	10.88	10.05	9.52	9.16	8.89	8.68	8.51	8.38	8.18	7.97	7.75	7.65	7.53	7.42	7.31	7.19	7.08
8	14.69	11.04	9.60	8.81	8.30	7.95	7.69	7.50	7.34	7.21	7.01	6.81	6.61	6.50	6.40	6.29	6.18	6.06	5.95
9	13.61	10.11	8.72	7.96	7.47	7.13	6.88	6.69	6.54	6.42	6.23	6.03	5.83	5.73	5.62	5.52	5.41	5.30	5.19
10	12.83	9.43	8.08	7.34	6.87	6.54	6.30	6.12	5.97	5.85	5.66	5.47	5.27	5.17	5.07	4.97	4.86	4.75	4.64
11	12.23	8.91	7.60	6.88	6.42	6.10	5.86	5.68	5.54	5.42	5.24	5.05	4.86	4.76	4.65	4.55	4.44	4.34	4.23
12	11.75	8.51	7.23	6.52	6.07	5.76	5.52	5.35	5.20	5.09	4.91	4.72	4.53	4.43	4.33	4.23	4.12	4.01	3.90
13	11.37	8.19	6.93	6.23	5.79	5.48	5.25	5.08	4.94	4.82	4.64	4.46	4.27	4.17	4.07	3.97	3.87	3.76	3.65
14	11.06	7.92	6.68	6.00	5.56	5.26	5.03	4.86	4.72	4.60	4.43	4.25	4.06	3.96	3.86	3.76	3.66	3.55	3.44
15	10.80	7.70	6.48	5.80	5.37	5.07	4.85	4.67	4.54	4.42	4.25	4.07	3.88	3.79	3.69	3.58	3.48	3.37	3.26
16	10.58	7.51	6.30	5.64	5.21	4.91	4.69	4.52	4.38	4.27	4.10	3.92	3.73	3.64	3.54	3.44	3.33	3.22	3.11
17	10.38	7.35	6.16	5.50	5.07	4.78	4.56	4.39	4.25	4.14	3.97	3.79	3.61	3.51	3.41	3.31	3.21	3.10	2.98
18	10.22	7.21	6.03	5.37	4.96	4.66	4.44	4.28	4.14	4.03	3.86	3.68	3.50	3.40	3.30	3.20	3.10	2.99	2.87
19	10.07	7.09	5.92	5.27	4.85	4.56	4.34	4.18	4.04	3.93	3.76	3.59	3.40	3.31	3.21	3.11	3.00	2.89	2.78
20	9.94	6.99	5.82	5.17	4.76	4.47	4.26	4.09	3.96	3.85	3.68	3.50	3.32	3.22	3.12	3.02	2.92	2.81	2.69
21	9.83	6.89	5.73	5.09	4.68	4.39	4.18	4.01	3.88	3.77	3.60	3.43	3.24	3.15	3.05	2.95	2.84	2.73	2.61
22	9.73	6.81	5.65	5.02	4.61	4.32	4.11	3.94	3.81	3.70	3.54	3.36	3.18	3.08	2.98	2.88	2.77	2.66	2.55
23	9.63	6.73	5.58	4.95	4.54	4.26	4.05	3.88	3.75	3.64	3.47	3.30	3.12	3.02	2.92	2.82	2.71	2.60	2.48
24	9.55	6.66	5.52	4.89	4.49	4.20	3.99	3.83	3.69	3.59	3.42	3.25	3.06	2.97	2.87	2.77	2.66	2.55	2.43
25	9.48	6.60	5.46	4.84	4.43	4.15	3.94	3.78	3.64	3.54	3.37	3.20	3.01	2.92	2.82	2.72	2.61	2.50	2.38
26	9.41	6.54	5.41	4.79	4.38	4.10	3.89	3.73	3.60	3.49	3.33	3.15	2.97	2.87	2.77	2.67	2.56	2.45	2.33
27	9.34	6.49	5.36	4.74	4.34	4.06	3.85	3.69	3.56	3.45	3.28	3.11	2.93	2.83	2.73	2.63	2.52	2.41	2.29
28	9.28	6.44	5.32	4.70	4.30	4.02	3.81	3.65	3.52	3.41	3.25	3.07	2.89	2.79	2.69	2.59	2.48	2.37	2.25
29	9.23	6.40	5.28	4.66	4.26	3.98	3.77	3.61	3.48	3.38	3.21	3.04	2.86	2.76	2.66	2.56	2.45	2.33	2.21
30	9.18	6.35	5.24	4.62	4.23	3.95	3.74	3.58	3.45	3.34	3.18	3.01	2.82	2.73	2.63	2.52	2.42	2.30	2.18
40	8.83	6.07	4.98	4.37	3.99	3.71	3.51	3.35	3.22	3.12	2.95	2.78	2.60	2.50	2.40	2.30	2.18	2.06	1.93
60	8.49	5.79	4.73	4.14	3.76	3.49	3.29	3.13	3.01	2.90	2.74	2.57	2.39	2.29	2.19	2.08	1.96	1.83	1.69
120	8.18	5.54	4.50	3.92	3.55	3.28	3.09	2.93	2.81	2.71	2.54	2.37	2.19	2.09	1.98	1.87	1.75	1.61	1.43
∞	7.88	5.30	4.28	3.72	3.35	3.09	2.90	2.74	2.62	2.52	2.36	2.19	2.00	1.90	1.79	1.67	1.53	1.36	1.00

Table H (continued)

$\alpha = 0.01$

d.f.N.: degrees of freedom, numerator

d.f.D.: degrees of freedom, denominator	1	2	3	4	5	6	7	8	9	10	12	15	20	24	30	40	60	120	∞
1	4052	4999.5	5403	5625	5764	5859	5928	5982	6022	6056	6106	6157	6209	6235	6261	6287	6313	6339	6366
2	98.50	99.00	99.17	99.25	99.30	99.33	99.36	99.37	99.39	99.40	99.42	99.43	99.45	99.46	99.47	99.47	99.48	99.49	99.50
3	34.12	30.82	29.46	28.71	28.24	27.91	27.67	27.49	27.35	27.23	27.05	26.87	26.69	26.60	26.50	26.41	26.32	26.22	26.13
4	21.20	18.00	16.69	15.98	15.52	15.21	14.98	14.80	14.66	14.55	14.37	14.20	14.02	13.93	13.84	13.75	13.65	13.56	13.46
5	16.26	13.27	12.06	11.39	10.97	10.67	10.46	10.29	10.16	10.05	9.89	9.72	9.55	9.47	9.38	9.29	9.20	9.11	9.02
6	13.75	10.92	9.78	9.15	8.75	8.47	8.26	8.10	7.98	7.87	7.72	7.56	7.40	7.31	7.23	7.14	7.06	6.97	6.88
7	12.25	9.55	8.45	7.85	7.46	7.19	6.99	6.84	6.72	6.62	6.47	6.31	6.16	6.07	5.99	5.91	5.82	5.74	5.65
8	11.26	8.65	7.59	7.01	6.63	6.37	6.18	6.03	5.91	5.81	5.67	5.52	5.36	5.28	5.20	5.12	5.03	4.95	4.86
9	10.56	8.02	6.99	6.42	6.06	5.80	5.61	5.47	5.35	5.26	5.11	4.96	4.81	4.73	4.65	4.57	4.48	4.40	4.31
10	10.04	7.56	6.55	5.99	5.64	5.39	5.20	5.06	4.94	4.85	4.71	4.56	4.41	4.33	4.25	4.17	4.08	4.00	3.91
11	9.65	7.21	6.22	5.67	5.32	5.07	4.89	4.74	4.63	4.54	4.40	4.25	4.10	4.02	3.94	3.86	3.78	3.69	3.60
12	9.33	6.93	5.95	5.41	5.06	4.82	4.64	4.50	4.39	4.30	4.16	4.01	3.86	3.78	3.70	3.62	3.54	3.45	3.36
13	9.07	6.70	5.74	5.21	4.86	4.62	4.44	4.30	4.19	4.10	3.96	3.82	3.66	3.59	3.51	3.43	3.34	3.25	3.17
14	8.86	6.51	5.56	5.04	4.69	4.46	4.28	4.14	4.03	3.94	3.80	3.66	3.51	3.43	3.35	3.27	3.18	3.09	3.00
15	8.68	6.36	5.42	4.89	4.56	4.32	4.14	4.00	3.89	3.80	3.67	3.52	3.37	3.29	3.21	3.13	3.05	2.96	2.87
16	8.53	6.23	5.29	4.77	4.44	4.20	4.03	3.89	3.78	3.69	3.55	3.41	3.26	3.18	3.10	3.02	2.93	2.84	2.75
17	8.40	6.11	5.18	4.67	4.34	4.10	3.93	3.79	3.68	3.59	3.46	3.31	3.16	3.08	3.00	2.92	2.83	2.75	2.65
18	8.29	6.01	5.09	4.58	4.25	4.01	3.84	3.71	3.60	3.51	3.37	3.23	3.08	3.00	2.92	2.84	2.75	2.66	2.57
19	8.18	5.93	5.01	4.50	4.17	3.94	3.77	3.63	3.52	3.43	3.30	3.15	3.00	2.92	2.84	2.76	2.67	2.58	2.49
20	8.10	5.85	4.94	4.43	4.10	3.87	3.70	3.56	3.46	3.37	3.23	3.09	2.94	2.86	2.78	2.69	2.61	2.52	2.42
21	8.02	5.78	4.87	4.37	4.04	3.81	3.64	3.51	3.40	3.31	3.17	3.03	2.88	2.80	2.72	2.64	2.55	2.46	2.36
22	7.95	5.72	4.82	4.31	3.99	3.76	3.59	3.45	3.35	3.26	3.12	2.98	2.83	2.75	2.67	2.58	2.50	2.40	2.31
23	7.88	5.66	4.76	4.26	3.94	3.71	3.54	3.41	3.30	3.21	3.07	2.93	2.78	2.70	2.62	2.54	2.45	2.35	2.26
24	7.82	5.61	4.72	4.22	3.90	3.67	3.50	3.36	3.26	3.17	3.03	2.89	2.74	2.66	2.58	2.49	2.40	2.31	2.21
25	7.77	5.57	4.68	4.18	3.85	3.63	3.46	3.32	3.22	3.13	2.99	2.85	2.70	2.62	2.54	2.45	2.36	2.27	2.17
26	7.72	5.53	4.64	4.14	3.82	3.59	3.42	3.29	3.18	3.09	2.96	2.81	2.66	2.58	2.50	2.42	2.33	2.23	2.13
27	7.68	5.49	4.60	4.11	3.78	3.56	3.39	3.26	3.15	3.06	2.93	2.78	2.63	2.55	2.47	2.38	2.29	2.20	2.10
28	7.64	5.45	4.57	4.07	3.75	3.53	3.36	3.23	3.12	3.03	2.90	2.75	2.60	2.52	2.44	2.35	2.26	2.17	2.06
29	7.60	5.42	4.54	4.04	3.73	3.50	3.33	3.20	3.09	3.00	2.87	2.73	2.57	2.49	2.41	2.33	2.23	2.14	2.03
30	7.56	5.39	4.51	4.02	3.70	3.47	3.30	3.17	3.07	2.98	2.84	2.70	2.55	2.47	2.39	2.30	2.21	2.11	2.01
40	7.31	5.18	4.31	3.83	3.51	3.29	3.12	2.99	2.89	2.80	2.66	2.52	2.37	2.29	2.20	2.11	2.02	1.92	1.80
60	7.08	4.98	4.13	3.65	3.34	3.12	2.95	2.82	2.72	2.63	2.50	2.35	2.20	2.12	2.03	1.94	1.84	1.73	1.60
120	6.85	4.79	3.95	3.48	3.17	2.96	2.79	2.66	2.56	2.47	2.34	2.19	2.03	1.95	1.86	1.76	1.66	1.53	1.38
∞	6.63	4.61	3.78	3.32	3.02	2.80	2.64	2.51	2.41	2.32	2.18	2.04	1.88	1.79	1.70	1.59	1.47	1.32	1.00

Table H (continued)

α = 0.025

d.f.N.: degrees of freedom, numerator

d.f.D.: degrees of freedom, denominator	1	2	3	4	5	6	7	8	9	10	12	15	20	24	30	40	60	120	∞
1	647.8	799.5	864.2	899.6	921.8	937.1	948.2	956.7	963.3	968.6	976.7	984.9	993.1	997.2	1001	1006	1010	1014	1018
2	38.51	39.00	39.17	39.25	39.30	39.33	39.36	39.37	39.39	39.40	39.41	39.43	39.45	39.46	39.46	39.47	39.48	39.49	39.50
3	17.44	16.04	15.44	15.10	14.88	14.73	14.62	14.54	14.47	14.42	14.34	14.25	14.17	14.12	14.08	14.04	13.99	13.95	13.90
4	12.22	10.65	9.98	9.60	9.36	9.20	9.07	8.98	8.90	8.84	8.75	8.66	8.56	8.51	8.46	8.41	8.36	8.31	8.26
5	10.01	8.43	7.76	7.39	7.15	6.98	6.85	6.76	6.68	6.62	6.52	6.43	6.33	6.28	6.23	6.18	6.12	6.07	6.02
6	8.81	7.26	6.60	6.23	5.99	5.82	5.70	5.60	5.52	5.46	5.37	5.27	5.17	5.12	5.07	5.01	4.96	4.90	4.85
7	8.07	6.54	5.89	5.52	5.29	5.12	4.99	4.90	4.82	4.76	4.67	4.57	4.47	4.42	4.36	4.31	4.25	4.20	4.14
8	7.57	6.06	5.42	5.05	4.82	4.65	4.53	4.43	4.36	4.30	4.20	4.10	4.00	3.95	3.89	3.84	3.78	3.73	3.67
9	7.21	5.71	5.08	4.72	4.48	4.32	4.20	4.10	4.03	3.96	3.87	3.77	3.67	3.61	3.56	3.51	3.45	3.39	3.33
10	6.94	5.46	4.83	4.47	4.24	4.07	3.95	3.85	3.78	3.72	3.62	3.52	3.42	3.37	3.31	3.26	3.20	3.14	3.08
11	6.72	5.26	4.63	4.28	4.04	3.88	3.76	3.66	3.59	3.53	3.43	3.33	3.23	3.17	3.12	3.06	3.00	2.94	2.88
12	6.55	5.10	4.47	4.12	3.89	3.73	3.61	3.51	3.44	3.37	3.28	3.18	3.07	3.02	2.96	2.91	2.85	2.79	2.72
13	6.41	4.97	4.35	4.00	3.77	3.60	3.48	3.39	3.31	3.25	3.15	3.05	2.95	2.89	2.84	2.78	2.72	2.66	2.60
14	6.30	4.86	4.24	3.89	3.66	3.50	3.38	3.29	3.21	3.15	3.05	2.95	2.84	2.79	2.73	2.67	2.61	2.55	2.49
15	6.20	4.77	4.15	3.80	3.58	3.41	3.29	3.20	3.12	3.06	2.96	2.86	2.76	2.70	2.64	2.59	2.52	2.46	2.40
16	6.12	4.69	4.08	3.73	3.50	3.34	3.22	3.12	3.05	2.99	2.89	2.79	2.68	2.63	2.57	2.51	2.45	2.38	2.32
17	6.04	4.62	4.01	3.66	3.44	3.28	3.16	3.06	2.98	2.92	2.82	2.72	2.62	2.56	2.50	2.44	2.38	2.32	2.25
18	5.98	4.56	3.95	3.61	3.38	3.22	3.10	3.01	2.93	2.87	2.77	2.67	2.56	2.50	2.44	2.38	2.32	2.26	2.19
19	5.92	4.51	3.90	3.56	3.33	3.17	3.05	2.96	2.88	2.82	2.72	2.62	2.51	2.45	2.39	2.33	2.27	2.20	2.13
20	5.87	4.46	3.86	3.51	3.29	3.13	3.01	2.91	2.84	2.77	2.68	2.57	2.46	2.41	2.35	2.29	2.22	2.16	2.09
21	5.83	4.42	3.82	3.48	3.25	3.09	2.97	2.87	2.80	2.73	2.64	2.53	2.42	2.37	2.31	2.25	2.18	2.11	2.04
22	5.79	4.38	3.78	3.44	3.22	3.05	2.93	2.84	2.76	2.70	2.60	2.50	2.39	2.33	2.27	2.21	2.14	2.08	2.00
23	5.75	4.35	3.75	3.41	3.18	3.02	2.90	2.81	2.73	2.67	2.57	2.47	2.36	2.30	2.24	2.18	2.11	2.04	1.97
24	5.72	4.32	3.72	3.38	3.15	2.99	2.87	2.78	2.70	2.64	2.54	2.44	2.33	2.27	2.21	2.15	2.08	2.01	1.94
25	5.69	4.29	3.69	3.35	3.13	2.97	2.85	2.75	2.68	2.61	2.51	2.41	2.30	2.24	2.18	2.12	2.05	1.98	1.91
26	5.66	4.27	3.67	3.33	3.10	2.94	2.82	2.73	2.65	2.59	2.49	2.39	2.28	2.22	2.16	2.09	2.03	1.95	1.88
27	5.63	4.24	3.65	3.31	3.08	2.92	2.80	2.71	2.63	2.57	2.47	2.36	2.25	2.19	2.13	2.07	2.00	1.93	1.85
28	5.61	4.22	3.63	3.29	3.06	2.90	2.78	2.69	2.61	2.55	2.45	2.34	2.23	2.17	2.11	2.05	1.98	1.91	1.83
29	5.59	4.20	3.61	3.27	3.04	2.88	2.76	2.67	2.59	2.53	2.43	2.32	2.21	2.15	2.09	2.03	1.96	1.89	1.81
30	5.57	4.18	3.59	3.25	3.03	2.87	2.75	2.65	2.57	2.51	2.41	2.31	2.20	2.14	2.07	2.01	1.94	1.87	1.79
40	5.42	4.05	3.46	3.13	2.90	2.74	2.62	2.53	2.45	2.39	2.29	2.18	2.07	2.01	1.94	1.88	1.80	1.72	1.64
60	5.29	3.93	3.34	3.01	2.79	2.63	2.51	2.41	2.33	2.27	2.17	2.06	1.94	1.88	1.82	1.74	1.67	1.58	1.48
120	5.15	3.80	3.23	2.89	2.67	2.52	2.39	2.30	2.22	2.16	2.05	1.94	1.82	1.76	1.69	1.61	1.53	1.43	1.31
∞	5.02	3.69	3.12	2.79	2.57	2.41	2.29	2.19	2.11	2.05	1.94	1.83	1.71	1.64	1.57	1.48	1.39	1.27	1.00

Table H (continued)

$\alpha = 0.05$

| | d.f.N.: degrees of freedom, numerator | | | | | | | | | | | | | | | | | | |
d.f.D.: degrees of freedom, denominator	1	2	3	4	5	6	7	8	9	10	12	15	20	24	30	40	60	120	∞
1	161.4	199.5	215.7	224.6	230.2	234.0	236.8	238.9	240.5	241.9	243.9	245.9	248.0	249.1	250.1	251.1	252.2	253.3	254.3
2	18.51	19.00	19.16	19.25	19.30	19.33	19.35	19.37	19.38	19.40	19.41	19.43	19.45	19.45	19.46	19.47	19.48	19.49	19.50
3	10.13	9.55	9.28	9.12	9.01	8.94	8.89	8.85	8.81	8.79	8.74	8.70	8.66	8.64	8.62	8.59	8.57	8.55	8.53
4	7.71	6.94	6.59	6.39	6.26	6.16	6.09	6.04	6.00	5.96	5.91	5.86	5.80	5.77	5.75	5.72	5.69	5.66	5.63
5	6.61	5.79	5.41	5.19	5.05	4.95	4.88	4.82	4.77	4.74	4.68	4.62	4.56	4.53	4.50	4.46	4.43	4.40	4.36
6	5.99	5.14	4.76	4.53	4.39	4.28	4.21	4.15	4.10	4.06	4.00	3.94	3.87	3.84	3.81	3.77	3.74	3.70	3.67
7	5.59	4.74	4.35	4.12	3.97	3.87	3.79	3.73	3.68	3.64	3.57	3.51	3.44	3.41	3.38	3.34	3.30	3.27	3.23
8	5.32	4.46	4.07	3.84	3.69	3.58	3.50	3.44	3.39	3.35	3.28	3.22	3.15	3.12	3.08	3.04	3.01	2.97	2.93
9	5.12	4.26	3.86	3.63	3.48	3.37	3.29	3.23	3.18	3.14	3.07	3.01	2.94	2.90	2.86	2.83	2.79	2.75	2.71
10	4.96	4.10	3.71	3.48	3.33	3.22	3.14	3.07	3.02	2.98	2.91	2.85	2.77	2.74	2.70	2.66	2.62	2.58	2.54
11	4.84	3.98	3.59	3.36	3.20	3.09	3.01	2.95	2.90	2.85	2.79	2.72	2.65	2.61	2.57	2.53	2.49	2.45	2.40
12	4.75	3.89	3.49	3.26	3.11	3.00	2.91	2.85	2.80	2.75	2.69	2.62	2.54	2.51	2.47	2.43	2.38	2.34	2.30
13	4.67	3.81	3.41	3.18	3.03	2.92	2.83	2.77	2.71	2.67	2.60	2.53	2.46	2.42	2.38	2.34	2.30	2.25	2.21
14	4.60	3.74	3.34	3.11	2.96	2.85	2.76	2.70	2.65	2.60	2.53	2.46	2.39	2.35	2.31	2.27	2.22	2.18	2.13
15	4.54	3.68	3.29	3.06	2.90	2.79	2.71	2.64	2.59	2.54	2.48	2.40	2.33	2.29	2.25	2.20	2.16	2.11	2.07
16	4.49	3.63	3.24	3.01	2.85	2.74	2.66	2.59	2.54	2.49	2.42	2.35	2.28	2.24	2.19	2.15	2.11	2.06	2.01
17	4.45	3.59	3.20	2.96	2.81	2.70	2.61	2.55	2.49	2.45	2.38	2.31	2.23	2.19	2.15	2.10	2.06	2.01	1.96
18	4.41	3.55	3.16	2.93	2.77	2.66	2.58	2.51	2.46	2.41	2.34	2.27	2.19	2.15	2.11	2.06	2.02	1.97	1.92
19	4.38	3.52	3.13	2.90	2.74	2.63	2.54	2.48	2.42	2.38	2.31	2.23	2.16	2.11	2.07	2.03	1.98	1.93	1.88
20	4.35	3.49	3.10	2.87	2.71	2.60	2.51	2.45	2.39	2.35	2.28	2.20	2.12	2.08	2.04	1.99	1.95	1.90	1.84
21	4.32	3.47	3.07	2.84	2.68	2.57	2.49	2.42	2.37	2.32	2.25	2.18	2.10	2.05	2.01	1.96	1.92	1.87	1.81
22	4.30	3.44	3.05	2.82	2.66	2.55	2.46	2.40	2.34	2.30	2.23	2.15	2.07	2.03	1.98	1.94	1.89	1.84	1.78
23	4.28	3.42	3.03	2.80	2.64	2.53	2.44	2.37	2.32	2.27	2.20	2.13	2.05	2.01	1.96	1.91	1.86	1.81	1.76
24	4.26	3.40	3.01	2.78	2.62	2.51	2.42	2.36	2.30	2.25	2.18	2.11	2.03	1.98	1.94	1.89	1.84	1.79	1.73
25	4.24	3.39	2.99	2.76	2.60	2.49	2.40	2.34	2.28	2.24	2.16	2.09	2.01	1.96	1.92	1.87	1.82	1.77	1.71
26	4.23	3.37	2.98	2.74	2.59	2.47	2.39	2.32	2.27	2.22	2.15	2.07	1.99	1.95	1.90	1.85	1.80	1.75	1.69
27	4.21	3.35	2.96	2.73	2.57	2.46	2.37	2.31	2.25	2.20	2.13	2.06	1.97	1.93	1.88	1.84	1.79	1.73	1.67
28	4.20	3.34	2.95	2.71	2.56	2.45	2.36	2.29	2.24	2.19	2.12	2.04	1.96	1.91	1.87	1.82	1.77	1.71	1.65
29	4.18	3.33	2.93	2.70	2.55	2.43	2.35	2.28	2.22	2.18	2.10	2.03	1.94	1.90	1.85	1.81	1.75	1.70	1.64
30	4.17	3.32	2.92	2.69	2.53	2.42	2.33	2.27	2.21	2.16	2.09	2.01	1.93	1.89	1.84	1.79	1.74	1.68	1.62
40	4.08	3.23	2.84	2.61	2.45	2.34	2.25	2.18	2.12	2.08	2.00	1.92	1.84	1.79	1.74	1.69	1.64	1.58	1.51
60	4.00	3.15	2.76	2.53	2.37	2.25	2.17	2.10	2.04	1.99	1.92	1.84	1.75	1.70	1.65	1.59	1.53	1.47	1.39
120	3.92	3.07	2.68	2.45	2.29	2.17	2.09	2.02	1.96	1.91	1.83	1.75	1.66	1.61	1.55	1.50	1.43	1.35	1.25
∞	3.84	3.00	2.60	2.37	2.21	2.10	2.01	1.94	1.88	1.83	1.75	1.67	1.57	1.52	1.46	1.39	1.32	1.22	1.00

Table H *(concluded)*

α = 0.10

d.f.N.: degrees of freedom, numerator

d.f.D.: degrees of freedom, denominator	1	2	3	4	5	6	7	8	9	10	12	15	20	24	30	40	60	120	∞
1	39.86	49.50	53.59	55.83	57.24	58.20	58.91	59.44	59.86	60.19	60.71	61.22	61.74	62.00	62.26	62.53	62.79	63.06	63.33
2	8.53	9.00	9.16	9.24	9.29	9.33	9.35	9.37	9.38	9.39	9.41	9.42	9.44	9.45	9.46	9.47	9.47	9.48	9.49
3	5.54	5.46	5.39	5.34	5.31	5.28	5.27	5.25	5.24	5.23	5.22	5.20	5.18	5.18	5.17	5.16	5.15	5.14	5.13
4	4.54	4.32	4.19	4.11	4.05	4.01	3.98	3.95	3.94	3.92	3.90	3.87	3.84	3.83	3.82	3.80	3.79	3.78	3.76
5	4.06	3.78	3.62	3.52	3.45	3.40	3.37	3.34	3.32	3.30	3.27	3.24	3.21	3.19	3.17	3.16	3.14	3.12	3.10
6	3.78	3.46	3.29	3.18	3.11	3.05	3.01	2.98	2.96	2.94	2.90	2.87	2.84	2.82	2.80	2.78	2.76	2.74	2.72
7	3.59	3.26	3.07	2.96	2.88	2.83	2.78	2.75	2.72	2.70	2.67	2.63	2.59	2.58	2.56	2.54	2.51	2.49	2.47
8	3.46	3.11	2.92	2.81	2.73	2.67	2.62	2.59	2.56	2.54	2.50	2.46	2.42	2.40	2.38	2.36	2.34	2.32	2.29
9	3.36	3.01	2.81	2.69	2.61	2.55	2.51	2.47	2.44	2.42	2.38	2.34	2.30	2.28	2.25	2.23	2.21	2.18	2.16
10	3.29	2.92	2.73	2.61	2.52	2.46	2.41	2.38	2.35	2.32	2.28	2.24	2.20	2.18	2.16	2.13	2.11	2.08	2.06
11	3.23	2.86	2.66	2.54	2.45	2.39	2.34	2.30	2.27	2.25	2.21	2.17	2.12	2.10	2.08	2.05	2.03	2.00	1.97
12	3.18	2.81	2.61	2.48	2.39	2.33	2.28	2.24	2.21	2.19	2.15	2.10	2.06	2.04	2.01	1.99	1.96	1.93	1.90
13	3.14	2.76	2.56	2.43	2.35	2.28	2.23	2.20	2.16	2.14	2.10	2.05	2.01	1.98	1.96	1.93	1.90	1.88	1.85
14	3.10	2.73	2.52	2.39	2.31	2.24	2.19	2.15	2.12	2.10	2.05	2.01	1.96	1.94	1.91	1.89	1.86	1.83	1.80
15	3.07	2.70	2.49	2.36	2.27	2.21	2.16	2.12	2.09	2.06	2.02	1.97	1.92	1.90	1.87	1.85	1.82	1.79	1.76
16	3.05	2.67	2.46	2.33	2.24	2.18	2.13	2.09	2.06	2.03	1.99	1.94	1.89	1.87	1.84	1.81	1.78	1.75	1.72
17	3.03	2.64	2.44	2.31	2.22	2.15	2.10	2.06	2.03	2.00	1.96	1.91	1.86	1.84	1.81	1.78	1.75	1.72	1.69
18	3.01	2.62	2.42	2.29	2.20	2.13	2.08	2.04	2.00	1.98	1.93	1.89	1.84	1.81	1.78	1.75	1.72	1.69	1.66
19	2.99	2.61	2.40	2.27	2.18	2.11	2.06	2.02	1.98	1.96	1.91	1.86	1.81	1.79	1.76	1.73	1.70	1.67	1.63
20	2.97	2.59	2.38	2.25	2.16	2.09	2.04	2.00	1.96	1.94	1.89	1.84	1.79	1.77	1.74	1.71	1.68	1.64	1.61
21	2.96	2.57	2.36	2.23	2.14	2.08	2.02	1.98	1.95	1.92	1.87	1.83	1.78	1.75	1.72	1.69	1.66	1.62	1.59
22	2.95	2.56	2.35	2.22	2.13	2.06	2.01	1.97	1.93	1.90	1.86	1.81	1.76	1.73	1.70	1.67	1.64	1.60	1.57
23	2.94	2.55	2.34	2.21	2.11	2.05	1.99	1.95	1.92	1.89	1.84	1.80	1.74	1.72	1.69	1.66	1.62	1.59	1.55
24	2.93	2.54	2.33	2.19	2.10	2.04	1.98	1.94	1.91	1.88	1.83	1.78	1.73	1.70	1.67	1.64	1.61	1.57	1.53
25	2.92	2.53	2.32	2.18	2.09	2.02	1.97	1.93	1.89	1.87	1.82	1.77	1.72	1.69	1.66	1.63	1.59	1.56	1.52
26	2.91	2.52	2.31	2.17	2.08	2.01	1.96	1.92	1.88	1.86	1.81	1.76	1.71	1.68	1.65	1.61	1.58	1.54	1.50
27	2.90	2.51	2.30	2.17	2.07	2.00	1.95	1.91	1.87	1.85	1.80	1.75	1.70	1.67	1.64	1.60	1.57	1.53	1.49
28	2.89	2.50	2.29	2.16	2.06	2.00	1.94	1.90	1.87	1.84	1.79	1.74	1.69	1.66	1.63	1.59	1.56	1.52	1.48
29	2.89	2.50	2.28	2.15	2.06	1.99	1.93	1.89	1.86	1.83	1.78	1.73	1.68	1.65	1.62	1.58	1.55	1.51	1.47
30	2.88	2.49	2.28	2.14	2.05	1.98	1.93	1.88	1.85	1.82	1.77	1.72	1.67	1.64	1.61	1.57	1.54	1.50	1.46
40	2.84	2.44	2.23	2.09	2.00	1.93	1.87	1.83	1.79	1.76	1.71	1.66	1.61	1.57	1.54	1.51	1.47	1.42	1.38
60	2.79	2.39	2.18	2.04	1.95	1.87	1.82	1.77	1.74	1.71	1.66	1.60	1.54	1.51	1.48	1.44	1.40	1.35	1.29
120	2.75	2.35	2.13	1.99	1.90	1.82	1.77	1.72	1.68	1.65	1.60	1.55	1.48	1.45	1.41	1.37	1.32	1.26	1.19
∞	2.71	2.30	2.08	1.94	1.85	1.77	1.72	1.67	1.63	1.60	1.55	1.49	1.42	1.38	1.34	1.30	1.24	1.17	1.00

From M. Merrington and C. M. Thompson (1943). Table of Percentage Points of the Inverted Beta (F) Distribution. *Biometrika 33*, pp. 74–87. Reprinted with permission from Biometrika.

Table I — Critical Values for PPMC

Reject H_0: $\rho = 0$ if the absolute value of r is greater than the value given in the table. The values are for a two-tailed test; d.f. $= n - 2$.

d.f	$\alpha = 0.05$	$\alpha = 0.01$
1	0.999	0.999
2	0.950	0.999
3	0.878	0.959
4	0.811	0.917
5	0.754	0.875
6	0.707	0.834
7	0.666	0.798
8	0.632	0.765
9	0.602	0.735
10	0.576	0.708
11	0.553	0.684
12	0.532	0.661
13	0.514	0.641
14	0.497	0.623
15	0.482	0.606
16	0.468	0.590
17	0.456	0.575
18	0.444	0.561
19	0.433	0.549
20	0.423	0.537
25	0.381	0.487
30	0.349	0.449
35	0.325	0.418
40	0.304	0.393
45	0.288	0.372
50	0.273	0.354
60	0.250	0.325
70	0.232	0.302
80	0.217	0.283
90	0.205	0.267
100	0.195	0.254

Source: From *Biometrika Tables for Statisticians,* vol. 1 (1962), p. 138. Reprinted with permission.

Table J — Critical Values for the Sign Test

Reject the null hypothesis if the smaller number of positive or negative signs is less than or equal to the value in the table.

n	One-tailed, $\alpha = 0.005$ / Two-tailed, $\alpha = 0.01$	$\alpha = 0.01$ / $\alpha = 0.02$	$\alpha = 0.025$ / $\alpha = 0.05$	$\alpha = 0.05$ / $\alpha = 0.10$
8	0	0	0	1
9	0	0	1	1
10	0	0	1	1
11	0	1	1	2
12	1	1	2	2
13	1	1	2	3
14	1	2	3	3
15	2	2	3	3
16	2	2	3	4
17	2	3	4	4
18	3	3	4	5
19	3	4	4	5
20	3	4	5	5
21	4	4	5	6
22	4	5	5	6
23	4	5	6	7
24	5	5	6	7
25	5	6	6	7

Note: Table J is for one-tailed or two-tailed tests. The term n represents the total number of positive and negative signs. The test value is the number of less frequent signs.

Source: From *Journal of American Statistical Association,* vol. 41 (1946), pp. 557–66. W. J. Dixon and A. M. Mood.

Table K	Critical Values for the Wilcoxon Signed-Rank Test

Reject the null hypothesis if the test value is less than or equal to the value given in the table.

	One-tailed, $\alpha = 0.05$	$\alpha = 0.025$	$\alpha = 0.01$	$\alpha = 0.005$
n	Two-tailed, $\alpha = 0.10$	$\alpha = 0.05$	$\alpha = 0.02$	$\alpha = 0.01$
5	1			
6	2	1		
7	4	2	0	
8	6	4	2	0
9	8	6	3	2
10	11	8	5	3
11	14	11	7	5
12	17	14	10	7
13	21	17	13	10
14	26	21	16	13
15	30	25	20	16
16	36	30	24	19
17	41	35	28	23
18	47	40	33	28
19	54	46	38	32
20	60	52	43	37
21	68	59	49	43
22	75	66	56	49
23	83	73	62	55
24	92	81	69	61
25	101	90	77	68
26	110	98	85	76
27	120	107	93	84
28	130	117	102	92
29	141	127	111	100
30	152	137	120	109

Source: From *Some Rapid Approximate Statistical Procedures,* Copyright 1949, 1964 Lerderle Laboratories, American Cyanamid Co., Wayne, N.J. Reprinted with permission.

Table L	Critical Values for the Rank Correlation Coefficient

Reject H_0: $\rho = 0$ if the absolute value of r_S is greater than the value given in the table.

n	$\alpha = 0.10$	$\alpha = 0.05$	$\alpha = 0.02$	$\alpha = 0.01$
5	0.900	—	—	—
6	0.829	0.886	0.943	—
7	0.714	0.786	0.893	0.929
8	0.643	0.738	0.833	0.881
9	0.600	0.700	0.783	0.833
10	0.564	0.648	0.745	0.794
11	0.536	0.618	0.709	0.818
12	0.497	0.591	0.703	0.780
13	0.475	0.566	0.673	0.745
14	0.457	0.545	0.646	0.716
15	0.441	0.525	0.623	0.689
16	0.425	0.507	0.601	0.666
17	0.412	0.490	0.582	0.645
18	0.399	0.476	0.564	0.625
19	0.388	0.462	0.549	0.608
20	0.377	0.450	0.534	0.591
21	0.368	0.438	0.521	0.576
22	0.359	0.428	0.508	0.562
23	0.351	0.418	0.496	0.549
24	0.343	0.409	0.485	0.537
25	0.336	0.400	0.475	0.526
26	0.329	0.392	0.465	0.515
27	0.323	0.385	0.456	0.505
28	0.317	0.377	0.488	0.496
29	0.311	0.370	0.440	0.487
30	0.305	0.364	0.432	0.478

Source: From N. L. Johnson and F. C. Leone, *Statistical and Experimental Design,* vol. I (1964), p. 412. Reprinted with permission from the Institute of Mathematical Statistics.

Table M Critical Values for the Number of Runs

This table gives the critical values at $\alpha = 0.05$ for a two-tailed test. Reject the null hypothesis if the number of runs is less than or equal to the smaller value or greater than or equal to the larger value.

Value of n_1	Value of n_2																		
	2	3	4	5	6	7	8	9	10	11	12	13	14	15	16	17	18	19	20
2	1	1	1	1	1	1	1	1	1	1	2	2	2	2	2	2	2	2	2
	6	6	6	6	6	6	6	6	6	6	6	6	6	6	6	6	6	6	6
3	1	1	1	1	2	2	2	2	2	2	2	2	2	3	3	3	3	3	3
	6	8	8	8	8	8	8	8	8	8	8	8	8	8	8	8	8	8	8
4	1	1	1	2	2	2	3	3	3	3	3	3	3	3	4	4	4	4	4
	6	8	9	9	9	10	10	10	10	10	10	10	10	10	10	10	10	10	10
5	1	1	2	2	3	3	3	3	3	4	4	4	4	4	4	4	5	5	5
	6	8	9	10	10	11	11	12	12	12	12	12	12	12	12	12	12	12	12
6	1	2	2	3	3	3	3	4	4	4	4	5	5	5	5	5	5	6	6
	6	8	9	10	11	12	12	13	13	13	13	14	14	14	14	14	14	14	14
7	1	2	2	3	3	3	4	4	5	5	5	5	5	6	6	6	6	6	6
	6	8	10	11	12	13	13	14	14	14	14	15	15	15	16	16	16	16	16
8	1	2	3	3	3	4	4	5	5	5	6	6	6	6	6	7	7	7	7
	6	8	10	11	12	13	14	14	15	15	16	16	16	16	17	17	17	17	17
9	1	2	3	3	4	4	5	5	5	6	6	6	7	7	7	7	8	8	8
	6	8	10	12	13	14	14	15	16	16	16	17	17	18	18	18	18	18	18
10	1	2	3	3	4	5	5	5	6	6	7	7	7	7	8	8	8	8	9
	6	8	10	12	13	14	15	16	16	17	17	18	18	18	19	19	19	20	20
11	1	2	3	4	4	5	5	6	6	7	7	7	8	8	8	9	9	9	9
	6	8	10	12	13	14	15	16	17	17	18	19	19	19	20	20	20	21	21
12	2	2	3	4	4	5	6	6	7	7	7	8	8	8	9	9	9	10	10
	6	8	10	12	13	14	16	16	17	18	19	19	20	20	21	21	21	22	22
13	2	2	3	4	5	5	6	6	7	7	8	8	9	9	9	10	10	10	10
	6	8	10	12	14	15	16	17	18	19	19	20	20	21	21	22	22	23	23
14	2	2	3	4	5	5	6	7	7	8	8	9	9	9	10	10	10	11	11
	6	8	10	12	14	15	16	17	18	19	20	20	21	22	22	23	23	23	24
15	2	3	3	4	5	6	6	7	7	8	8	9	9	10	10	11	11	11	12
	6	8	10	12	14	15	16	18	18	19	20	21	22	22	23	23	24	24	25
16	2	3	4	4	5	6	6	7	8	8	9	9	10	10	11	11	11	12	12
	6	8	10	12	14	16	17	18	19	20	21	21	22	23	23	24	25	25	25
17	2	3	4	4	5	6	7	7	8	9	9	10	10	11	11	11	12	12	13
	6	8	10	12	14	16	17	18	19	20	21	22	23	23	24	25	25	26	26
18	2	3	4	5	5	6	7	8	8	9	9	10	10	11	11	12	12	13	13
	6	8	10	12	14	16	17	18	19	20	21	22	23	24	25	25	26	26	27
19	2	3	4	5	6	6	7	8	8	9	10	10	11	11	12	12	13	13	13
	6	8	10	12	14	16	17	18	20	21	22	23	23	24	25	26	26	27	27
20	2	3	4	5	6	6	7	8	9	9	10	10	11	12	12	13	13	13	14
	6	8	10	12	14	16	17	18	20	21	22	23	24	25	25	26	27	27	28

Source: Adapted from C. Eisenhardt and F. Swed, "Tables for Testing Randomness of Grouping in a Sequence of Alternatives," *The Annals of Statistics,* vol. 14 (1943), pp. 83–86. Reprinted with permission of the Institute of Mathematical Statistics and of the Benjamin/Cummings Publishing Company, in whose publication, *Elementary Statistics,* 3rd ed. (1989), by Mario F. Triola, this table appears.

Table N   Critical Values for the Tukey Test

$\alpha = 0.01$

k \ v	2	3	4	5	6	7	8	9	10	11	12	13	14	15	16	17	18	19	20
1	90.03	135.0	164.3	185.6	202.2	215.8	227.2	237.0	245.6	253.2	260.0	266.2	271.8	277.0	281.8	286.3	290.4	294.3	298.0
2	14.04	19.02	22.29	24.72	26.63	28.20	29.53	30.68	31.69	32.59	33.40	34.13	34.81	35.43	36.00	36.53	37.03	37.50	37.95
3	8.26	10.62	12.17	13.33	14.24	15.00	15.64	16.20	16.69	17.13	17.53	17.89	18.22	18.52	18.81	19.07	19.32	19.55	19.77
4	6.51	8.12	9.17	9.96	10.58	11.10	11.55	11.93	12.27	12.57	12.84	13.09	13.32	13.53	13.73	13.91	14.08	14.24	14.40
5	5.70	6.98	7.80	8.42	8.91	9.32	9.67	9.97	10.24	10.48	10.70	10.89	11.08	11.24	11.40	11.55	11.68	11.81	11.93
6	5.24	6.33	7.03	7.56	7.97	8.32	8.61	8.87	9.10	9.30	9.48	9.65	9.81	9.95	10.08	10.21	10.32	10.43	10.54
7	4.95	5.92	6.54	7.01	7.37	7.68	7.94	8.17	8.37	8.55	8.71	8.86	9.00	9.12	9.24	9.35	9.46	9.55	9.65
8	4.75	5.64	6.20	6.62	6.96	7.24	7.47	7.68	7.86	8.03	8.18	8.31	8.44	8.55	8.66	8.76	8.85	8.94	9.03
9	4.60	5.43	5.96	6.35	6.66	6.91	7.13	7.33	7.49	7.65	7.78	7.91	8.03	8.13	8.23	8.33	8.41	8.49	8.57
10	4.48	5.27	5.77	6.14	6.43	6.67	6.87	7.05	7.21	7.36	7.49	7.60	7.71	7.81	7.91	7.99	8.08	8.15	8.23
11	4.39	5.15	5.62	5.97	6.25	6.48	6.67	6.84	6.99	7.13	7.25	7.36	7.46	7.56	7.65	7.73	7.81	7.88	7.95
12	4.32	5.05	5.50	5.84	6.10	6.32	6.51	6.67	6.81	6.94	7.06	7.17	7.26	7.36	7.44	7.52	7.59	7.66	7.73
13	4.26	4.96	5.40	5.73	5.98	6.19	6.37	6.53	6.67	6.79	6.90	7.01	7.10	7.19	7.27	7.35	7.42	7.48	7.55
14	4.21	4.89	5.32	5.63	5.88	6.08	6.26	6.41	6.54	6.66	6.77	6.87	6.96	7.05	7.13	7.20	7.27	7.33	7.39
15	4.17	4.84	5.25	5.56	5.80	5.99	6.16	6.31	6.44	6.55	6.66	6.76	6.84	6.93	7.00	7.07	7.14	7.20	7.26
16	4.13	4.79	5.19	5.49	5.72	5.92	6.08	6.22	6.35	6.46	6.56	6.66	6.74	6.82	6.90	6.97	7.03	7.09	7.15
17	4.10	4.74	5.14	5.43	5.66	5.85	6.01	6.15	6.27	6.38	6.48	6.57	6.66	6.73	6.81	6.87	6.94	7.00	7.05
18	4.07	4.70	5.09	5.38	5.60	5.79	5.94	6.08	6.20	6.31	6.41	6.50	6.58	6.65	6.73	6.79	6.85	6.91	6.97
19	4.05	4.67	5.05	5.33	5.55	5.73	5.89	6.02	6.14	6.25	6.34	6.43	6.51	6.58	6.65	6.72	6.78	6.84	6.89
20	4.02	4.64	5.02	5.29	5.51	5.69	5.84	5.97	6.09	6.19	6.28	6.37	6.45	6.52	6.59	6.65	6.71	6.77	6.82
24	3.96	4.55	4.91	5.17	5.37	5.54	5.69	5.81	5.92	6.02	6.11	6.19	6.26	6.33	6.39	6.45	6.51	6.56	6.61
30	3.89	4.45	4.80	5.05	5.24	5.40	5.54	5.65	5.76	5.85	5.93	6.01	6.08	6.14	6.20	6.26	6.31	6.36	6.41
40	3.82	4.37	4.70	4.93	5.11	5.26	5.39	5.50	5.60	5.69	5.76	5.83	5.90	5.96	6.02	6.07	6.12	6.16	6.21
60	3.76	4.28	4.59	4.82	4.99	5.13	5.25	5.36	5.45	5.53	5.60	5.67	5.73	5.78	5.84	5.89	5.93	5.97	6.01
120	3.70	4.20	4.50	4.71	4.87	5.01	5.12	5.21	5.30	5.37	5.44	5.50	5.56	5.61	5.66	5.71	5.75	5.79	5.83
∞	3.64	4.12	4.40	4.60	4.76	4.88	4.99	5.08	5.16	5.23	5.29	5.35	5.40	5.45	5.49	5.54	5.57	5.61	5.65

Table N *(continued)*

$\alpha = 0.05$

k \ v	2	3	4	5	6	7	8	9	10	11	12	13	14	15	16	17	18	19	20
1	17.97	26.98	32.82	37.08	40.41	43.12	45.40	47.36	49.07	50.59	51.96	53.20	54.33	55.36	56.32	57.22	58.04	58.83	59.56
2	6.08	8.33	9.80	10.88	11.74	12.44	13.03	13.54	13.99	14.39	14.75	15.08	15.38	15.65	15.91	16.14	16.37	16.57	16.77
3	4.50	5.91	6.82	7.50	8.04	8.48	8.85	9.18	9.46	9.72	9.95	10.15	10.35	10.53	10.69	10.84	10.98	11.11	11.24
4	3.93	5.04	5.76	6.29	6.71	7.05	7.35	7.60	7.83	8.03	8.21	8.37	8.52	8.66	8.79	8.91	9.03	9.13	9.23
5	3.64	4.60	5.22	5.67	6.03	6.33	6.58	6.80	6.99	7.17	7.32	7.47	7.60	7.72	7.83	7.93	8.03	8.12	8.21
6	3.46	4.34	4.90	5.30	5.63	5.90	6.12	6.32	6.49	6.65	6.79	6.92	7.03	7.14	7.24	7.34	7.43	7.51	7.59
7	3.34	4.16	4.68	5.06	5.36	5.61	5.82	6.00	6.16	6.30	6.43	6.55	6.66	6.76	6.85	6.94	7.02	7.10	7.17
8	3.26	4.04	4.53	4.89	5.17	5.40	5.60	5.77	5.92	6.05	6.18	6.29	6.39	6.48	6.57	6.65	6.73	6.80	6.87
9	3.20	3.95	4.41	4.76	5.02	5.24	5.43	5.59	5.74	5.87	5.98	6.09	6.19	6.28	6.36	6.44	6.51	6.58	6.64
10	3.15	3.88	4.33	4.65	4.91	5.12	5.30	5.46	5.60	5.72	5.83	5.93	6.03	6.11	6.19	6.27	6.34	6.40	6.47
11	3.11	3.82	4.26	4.57	4.82	5.03	5.20	5.35	5.49	5.61	5.71	5.81	5.90	5.98	6.06	6.13	6.20	6.27	6.33
12	3.08	3.77	4.20	4.51	4.75	4.95	5.12	5.27	5.39	5.51	5.61	5.71	5.80	5.88	5.95	6.02	6.09	6.15	6.21
13	3.06	3.73	4.15	4.45	4.69	4.88	5.05	5.19	5.32	5.43	5.53	5.63	5.71	5.79	5.86	5.93	5.99	6.05	6.11
14	3.03	3.70	4.11	4.41	4.64	4.83	4.99	5.13	5.25	5.36	5.46	5.55	5.64	5.71	5.79	5.85	5.91	5.97	6.03
15	3.01	3.67	4.08	4.37	4.59	4.78	4.94	5.08	5.20	5.31	5.40	5.49	5.57	5.65	5.72	5.78	5.85	5.90	5.96
16	3.00	3.65	4.05	4.33	4.56	4.74	4.90	5.03	5.15	5.26	5.35	5.44	5.52	5.59	5.66	5.73	5.79	5.84	5.90
17	2.98	3.63	4.02	4.30	4.52	4.70	4.86	4.99	5.11	5.21	5.31	5.39	5.47	5.54	5.61	5.67	5.73	5.79	5.84
18	2.97	3.61	4.00	4.28	4.49	4.67	4.82	4.96	5.07	5.17	5.27	5.35	5.43	5.50	5.57	5.63	5.69	5.74	5.79
19	2.96	3.59	3.98	4.25	4.47	4.65	4.79	4.92	5.04	5.14	5.23	5.31	5.39	5.46	5.53	5.59	5.65	5.70	5.75
20	2.95	3.58	3.96	4.23	4.45	4.62	4.77	4.90	5.01	5.11	5.20	5.28	5.36	5.43	5.49	5.55	5.61	5.66	5.71
24	2.92	3.53	3.90	4.17	4.37	4.54	4.68	4.81	4.92	5.01	5.10	5.18	5.25	5.32	5.38	5.44	5.49	5.55	5.59
30	2.89	3.49	3.85	4.10	4.30	4.46	4.60	4.72	4.82	4.92	5.00	5.08	5.15	5.21	5.27	5.33	5.38	5.43	5.47
40	2.86	3.44	3.79	4.04	4.23	4.39	4.52	4.63	4.73	4.82	4.90	4.98	5.04	5.11	5.16	5.22	5.27	5.31	5.36
60	2.83	3.40	3.74	3.98	4.16	4.31	4.44	4.55	4.65	4.73	4.81	4.88	4.94	5.00	5.06	5.11	5.15	5.20	5.24
120	2.80	3.36	3.68	3.92	4.10	4.24	4.36	4.47	4.56	4.64	4.71	4.78	4.84	4.90	4.95	5.00	5.04	5.09	5.13
∞	2.77	3.31	3.63	3.86	4.03	4.17	4.29	4.39	4.47	4.55	4.62	4.68	4.74	4.80	4.85	4.89	4.93	4.97	5.01

Table N *(concluded)*

$\alpha = 0.10$

k v	2	3	4	5	6	7	8	9	10	11	12	13	14	15	16	17	18	19	20
1	8.93	13.44	16.36	18.49	20.15	21.51	22.64	23.62	24.48	25.24	25.92	26.54	27.10	27.62	28.10	28.54	28.96	29.35	29.71
2	4.13	5.73	6.77	7.54	8.14	8.63	9.05	9.41	9.72	10.01	10.26	10.49	10.70	10.89	11.07	11.24	11.39	11.54	11.68
3	3.33	4.47	5.20	5.74	6.16	6.51	6.81	7.06	7.29	7.49	7.67	7.83	7.98	8.12	8.25	8.37	8.48	8.58	8.68
4	3.01	3.98	4.59	5.03	5.39	5.68	5.93	6.14	6.33	6.49	6.65	6.78	6.91	7.02	7.13	7.23	7.33	7.41	7.50
5	2.85	3.72	4.26	4.66	4.98	5.24	5.46	5.65	5.82	5.97	6.10	6.22	6.34	6.44	6.54	6.63	6.71	6.79	6.86
6	2.75	3.56	4.07	4.44	4.73	4.97	5.17	5.34	5.50	5.64	5.76	5.87	5.98	6.07	6.16	6.25	6.32	6.40	6.47
7	2.68	3.45	3.93	4.28	4.55	4.78	4.97	5.14	5.28	5.41	5.53	5.64	5.74	5.83	5.91	5.99	6.06	6.13	6.19
8	2.63	3.37	3.83	4.17	4.43	4.65	4.83	4.99	5.13	5.25	5.36	5.46	5.56	5.64	5.72	5.80	5.87	5.93	6.00
9	2.59	3.32	3.76	4.08	4.34	4.54	4.72	4.87	5.01	5.13	5.23	5.33	5.42	5.51	5.58	5.66	5.72	5.79	5.85
10	2.56	3.27	3.70	4.02	4.26	4.47	4.64	4.78	4.91	5.03	5.13	5.23	5.32	5.40	5.47	5.54	5.61	5.67	5.73
11	2.54	3.23	3.66	3.96	4.20	4.40	4.57	4.71	4.84	4.95	5.05	5.15	5.23	5.31	5.38	5.45	5.51	5.57	5.63
12	2.52	3.20	3.62	3.92	4.16	4.35	4.51	4.65	4.78	4.89	4.99	5.08	5.16	5.24	5.31	5.37	5.44	5.49	5.55
13	2.50	3.18	3.59	3.88	4.12	4.30	4.46	4.60	4.72	4.83	4.93	5.02	5.10	5.18	5.25	5.31	5.37	5.43	5.48
14	2.49	3.16	3.56	3.85	4.08	4.27	4.42	4.56	4.68	4.79	4.88	4.97	5.05	5.12	5.19	5.26	5.32	5.37	5.43
15	2.48	3.14	3.54	3.83	4.05	4.23	4.39	4.52	4.64	4.75	4.84	4.93	5.01	5.08	5.15	5.21	5.27	5.32	5.38
16	2.47	3.12	3.52	3.80	4.03	4.21	4.36	4.49	4.61	4.71	4.81	4.89	4.97	5.04	5.11	5.17	5.23	5.28	5.33
17	2.46	3.11	3.50	3.78	4.00	4.18	4.33	4.46	4.58	4.68	4.77	4.86	4.93	5.01	5.07	5.13	5.19	5.24	5.30
18	2.45	3.10	3.49	3.77	3.98	4.16	4.31	4.44	4.55	4.65	4.75	4.83	4.90	4.98	5.04	5.10	5.16	5.21	5.26
19	2.45	3.09	3.47	3.75	3.97	4.14	4.29	4.42	4.53	4.63	4.72	4.80	4.88	4.95	5.01	5.07	5.13	5.18	5.23
20	2.44	3.08	3.46	3.74	3.95	4.12	4.27	4.40	4.51	4.61	4.70	4.78	4.85	4.92	4.99	5.05	5.10	5.16	5.20
24	2.42	3.05	3.42	3.69	3.90	4.07	4.21	4.34	4.44	4.54	4.63	4.71	4.78	4.85	4.91	4.97	5.02	5.07	5.12
30	2.40	3.02	3.39	3.65	3.85	4.02	4.16	4.28	4.38	4.47	4.56	4.64	4.71	4.77	4.83	4.89	4.94	4.99	5.03
40	2.38	2.99	3.35	3.60	3.80	3.96	4.10	4.21	4.32	4.41	4.49	4.56	4.63	4.69	4.75	4.81	4.86	4.90	4.95
60	2.36	2.96	3.31	3.56	3.75	3.91	4.04	4.16	4.25	4.34	4.42	4.49	4.56	4.62	4.67	4.73	4.78	4.82	4.86
120	2.34	2.93	3.28	3.52	3.71	3.86	3.99	4.10	4.19	4.28	4.35	4.42	4.48	4.54	4.60	4.65	4.69	4.74	4.78
∞	2.33	2.90	3.24	3.48	3.66	3.81	3.93	4.04	4.13	4.21	4.28	4.35	4.41	4.47	4.52	4.57	4.61	4.65	4.69

Source: "Tables of Range and Studentized Range," *Annals of Mathematical Statistics*, vol. 31, no. 4. Reprinted with permission of the Institute of Mathematical Sciences.

Appendix D

Data Bank

Data Bank Values

This list explains the values given for the categories in the Data Bank.

1. "Age" is given in years.

2. "Educational level" values are defined as follows:

   0 = no high school degree   2 = college graduate
   1 = high school graduate   3 = graduate degree

3. "Smoking status" values are defined as follows:

   0 = does not smoke
   1 = smokes less than one pack per day
   2 = smokes one or more than one pack per day

4. "Exercise" values are defined as follows:

   0 = none   2 = moderate
   1 = light   3 = heavy

5. "Weight" is given in pounds.

6. "Serum cholesterol" is given in milligram percent (mg%).

7. "Systolic pressure" is given in millimeters of mercury (mm Hg).

8. "IQ" is given in standard IQ test score values.

9. "Sodium" is given in milliequivalents per liter (mEq/1).

10. "Gender" is listed as male (M) or female (F).

11. "Marital status" values are defined as follows:

    M = married   S = single
    W = widowed   D = divorced

Data Bank

ID number	Age	Educational level	Smoking status	Exercise	Weight	Serum cholesterol	Systolic pressure	IQ	Sodium	Gender	Marital status
01	27	2	1	1	120	193	126	118	136	F	M
02	18	1	0	1	145	210	120	105	137	M	S
03	32	2	0	0	118	196	128	115	135	F	M
04	24	2	0	1	162	208	129	108	142	M	M
05	19	1	2	0	106	188	119	106	133	F	S
06	56	1	0	0	143	206	136	111	138	F	W
07	65	1	2	0	160	240	131	99	140	M	W
08	36	2	1	0	215	215	163	106	151	M	D
09	43	1	0	1	127	201	132	111	134	F	M
10	47	1	1	1	132	215	138	109	135	F	D

Data Bank   (continued)

ID number	Age	Educational level	Smoking status	Exercise	Weight	Serum cholesterol	Systolic pressure	IQ	Sodium	Gender	Marital status
11	48	3	1	2	196	199	148	115	146	M	D
12	25	2	2	3	109	210	115	114	141	F	S
13	63	0	1	0	170	242	149	101	152	F	D
14	37	2	0	3	187	193	142	109	144	M	M
15	40	0	1	1	234	208	156	98	147	M	M
16	25	1	2	1	199	253	135	103	148	M	S
17	72	0	0	0	143	288	156	103	145	F	M
18	56	1	1	0	156	164	153	99	144	F	D
19	37	2	0	2	142	214	122	110	135	M	M
20	41	1	1	1	123	220	142	108	134	F	M
21	33	2	1	1	165	194	122	112	137	M	S
22	52	1	0	1	157	205	119	106	134	M	D
23	44	2	0	1	121	223	135	116	133	F	M
24	53	1	0	0	131	199	133	121	136	F	M
25	19	1	0	3	128	206	118	122	132	M	S
26	25	1	0	0	143	200	118	103	135	M	M
27	31	2	1	1	152	204	120	119	136	M	M
28	28	2	0	0	119	203	118	116	138	F	M
29	23	1	0	0	111	240	120	105	135	F	S
30	47	2	1	0	149	199	132	123	136	F	M
31	47	2	1	0	179	235	131	113	139	M	M
32	59	1	2	0	206	260	151	99	143	M	W
33	36	2	1	0	191	201	148	118	145	M	D
34	59	0	1	1	156	235	142	100	132	F	W
35	35	1	0	0	122	232	131	106	135	F	M
36	29	2	0	2	175	195	129	121	148	M	M
37	43	3	0	3	194	211	138	129	146	M	M
38	44	1	2	0	132	240	130	109	132	F	S
39	63	2	2	1	188	255	156	121	145	M	M
40	36	2	1	1	125	220	126	117	140	F	S
41	21	1	0	1	109	206	114	102	136	F	M
42	31	2	0	2	112	201	116	123	133	F	M
43	57	1	1	1	167	213	141	103	143	M	W
44	20	1	2	3	101	194	110	111	125	F	S
45	24	2	1	3	106	188	113	114	127	F	D
46	42	1	0	1	148	206	136	107	140	M	S
47	55	1	0	0	170	257	152	106	130	F	M
48	23	0	0	1	152	204	116	95	142	M	M
49	32	2	0	0	191	210	132	115	147	M	M
50	28	1	0	1	148	222	135	100	135	M	M
51	67	0	0	0	160	250	141	116	146	F	W
52	22	1	1	1	109	220	121	103	144	F	M
53	19	1	1	1	131	231	117	112	133	M	S
54	25	2	0	2	153	212	121	119	149	M	D
55	41	3	2	2	165	236	130	131	152	M	M

Data Bank (concluded)

ID number	Age	Educational level	Smoking status	Exercise	Weight	Serum cholesterol	Systolic pressure	IQ	Sodium	Gender	Marital status
56	24	2	0	3	112	205	118	100	132	F	S
57	32	2	0	1	115	187	115	109	136	F	S
58	50	3	0	1	173	203	136	126	146	M	M
59	32	2	1	0	186	248	119	122	149	M	M
60	26	2	0	1	181	207	123	121	142	M	S
61	36	1	1	0	112	188	117	98	135	F	D
62	40	1	1	0	130	201	121	105	136	F	D
63	19	1	1	1	132	237	115	111	137	M	S
64	37	2	0	2	179	228	141	127	141	F	M
65	65	3	2	1	212	220	158	129	148	M	M
66	21	1	2	2	99	191	117	103	131	F	S
67	25	2	2	1	128	195	120	121	131	F	S
68	68	0	0	0	167	210	142	98	140	M	W
69	18	1	1	2	121	198	123	113	136	F	S
70	26	0	1	1	163	235	128	99	140	M	M
71	45	1	1	1	185	229	125	101	143	M	M
72	44	3	0	0	130	215	128	128	137	F	M
73	50	1	0	0	142	232	135	104	138	F	M
74	63	0	0	0	166	271	143	103	147	F	W
75	48	1	0	3	163	203	131	103	144	M	M
76	27	2	0	3	147	186	118	114	134	M	M
77	31	3	1	1	152	228	116	126	138	M	D
78	28	2	0	2	112	197	120	123	133	F	M
79	36	2	1	2	190	226	123	121	147	M	M
80	43	3	2	0	179	252	127	131	145	M	D
81	21	1	0	1	117	185	116	105	137	F	S
82	32	2	1	0	125	193	123	119	135	F	M
83	29	2	1	0	123	192	131	116	131	F	D
84	49	2	2	1	185	190	129	127	144	M	M
85	24	1	1	1	133	237	121	114	129	M	M
86	36	2	0	2	163	195	115	119	139	M	M
87	34	1	2	0	135	199	133	117	135	F	M
88	36	0	0	1	142	216	138	88	137	F	M
89	29	1	1	1	155	214	120	98	135	M	S
90	42	0	0	2	169	201	123	96	137	M	D
91	41	1	1	1	136	214	133	102	141	F	D
92	29	1	1	0	112	205	120	102	130	F	M
93	43	1	1	0	185	208	127	100	143	M	M
94	61	1	2	0	173	248	142	101	141	M	M
95	21	1	1	3	106	210	111	105	131	F	S
96	56	0	0	0	149	232	142	103	141	F	M
97	63	0	1	0	192	193	163	95	147	M	M
98	74	1	0	0	162	247	151	99	151	F	W
99	35	2	0	1	151	251	147	113	145	F	M
100	28	2	0	3	161	199	129	116	138	M	M

Data Set I   Record Temperatures

Record high temperatures by state in degrees Fahrenheit

112	100	128	120	134
118	106	110	115	109
112	100	118	117	116
118	121	114	114	105
109	107	112	114	115
118	117	118	125	106
110	122	108	110	121
113	120	119	111	104
111	120	113	120	117
105	110	118	112	114

Record low temperatures by state in degrees Fahrenheit

−27	−80	−40	−29	−45
−61	−32	−17	−66	−2
−17	12	−60	−36	−36
−47	−40	−37	−16	−48
−40	−35	−51	−60	−19
−40	−70	−47	−50	−47
−39	−50	−52	−34	−60
−19	−27	−54	−42	−25
−50	−58	−32	−23	−69
	−30	−48	−37	−55

Data Set II   Identity Theft Complaints

The data values show the number of complaints of identity theft for 50 selected cities in the year 2002.

2609	1202	2730	483	655
626	393	1268	279	663
817	1165	551	2654	592
128	189	424	585	78
1836	154	248	239	5888
574	75	226	28	205
176	372	84	229	15
148	117	22	211	31
77	41	200	35	30
88	20	84	465	136

Data Set III   Length of Major North American Rivers

729	610	325	392	524
1459	450	465	605	330
950	906	329	290	1000
600	1450	862	532	890
407	525	720	1243	850
649	730	352	390	420
710	340	693	306	250
470	724	332	259	2340
560	1060	774	332	3710

Data Set III   Length of Major North American Rivers   *(continued)*

2315	2540	618	1171	460
431	800	605	410	1310
500	790	531	981	460
926	375	1290	1210	1310
383	380	300	310	411
1900	434	420	545	569
425	800	865	380	445
538	1038	424	350	377
540	659	652	314	360
301	512	500	313	610
360	430	682	886	447
338	485	625	722	525
800	309	435		

Data Set IV   Heights (in Feet) of 80 Tallest Buildings in New York City

1250	861	1046	952	552
915	778	856	850	927
729	745	757	752	814
750	697	743	739	750
700	670	716	707	730
682	648	687	687	705
650	634	664	674	685
640	628	630	653	673
625	620	628	645	650
615	592	620	630	630
595	580	614	618	629
587	575	590	609	615
575	572	580	588	603
574	563	575	577	587
565	555	562	570	576
557	570	555	561	574

Heights (in Feet) of 25 Tallest Buildings in Calgary, Alberta

689	530	460	410
645	525	449	410
645	507	441	408
626	500	435	407
608	469	435	
580	468	432	
530	463	420	

Data Set V   School Suspensions

The data values show the number of suspensions and the number of students enrolled in 40 local school districts in southwestern Pennsylvania.

Suspensions	Enrollment	Suspensions	Enrollment
37	1316	63	1588
29	1337	500	6046
106	4904	5	3610
47	5301	117	4329
51	1380	13	1908
46	1670	8	1341
65	3446	71	5582
223	1010	57	1869
10	795	16	1697
60	2094	60	2269
15	926	51	2307
198	1950	48	1564
56	3005	20	4147
72	4575	80	3182
110	4329	43	2982
6	3238	15	3313
37	3064	187	6090
26	2638	182	4874
140	4949	76	8286
39	3354	37	539
42	3547		

Source: U.S. Department of Education, *Pittsburgh Tribune-Review,* March 14, 2004.

Data Set VI   Acreage of U.S. National Parks, in Thousands of Acres

41	66	233	775	169
36	338	223	46	64
183	4724	61	1449	7075
1013	3225	1181	308	77
520	77	27	217	5
539	3575	650	462	1670
2574	106	52	52	236
505	913	94	75	265
402	196	70	13	132
28	7656	2220	760	143

Source: The Universal Almanac 1995, p. 45.

Data Set VII   Acreage Owned by 35 Municipalities in Southwestern Pennsylvania

384	44	62	218	250
198	60	306	105	600
10	38	87	227	340
48	70	58	223	3700
22	78	165	150	160
130	120	100	234	1200
4200	402	180	200	200

Source: Pittsburgh Tribune-Review.

Data Set VIII   Oceans of the World

Ocean	Area (thousands of square miles)	Maximum depth (feet)
Arctic	5,400	17,881
Caribbean Sea	1,063	25,197
Mediterranean Sea	967	16,470
Norwegian Sea	597	13,189
Gulf of Mexico	596	14,370
Hudson Bay	475	850
Greenland Sea	465	15,899
North Sea	222	2,170
Black Sea	178	7,360
Baltic Sea	163	1,440
Atlantic Ocean	31,830	30,246
South China Sea	1,331	18,241
Sea of Okhotsk	610	11,063
Bering Sea	876	13,750
Sea of Japan	389	12,280
East China Sea	290	9,126
Yellow Sea	161	300
Pacific Ocean	63,800	36,200
Arabian Sea	1,492	19,029
Bay of Bengal	839	17,251
Red Sea	169	7,370
Indian Ocean	28,360	24,442

Source: The Universal Almanac 1995, p. 330.

Data Set IX   Commuter and Rapid Rail Systems in the United States

System	Stations	Miles	Vehicles operated
Long Island RR	134	638.2	947
N.Y. Metro North	108	535.9	702
New Jersey Transit	158	926.0	582
Chicago RTA	117	417.0	358
Chicago & NW Transit	62	309.4	277
Boston Amtrak/MBTA	101	529.8	291
Chicago, Burlington, Northern	27	75.0	139
NW Indiana CTD	18	134.8	39
New York City TA	469	492.9	4923
Washington Metro Area TA	70	162.1	534
Metro Boston TA	53	76.7	368
Chicago TA	137	191.0	924
Philadelphia SEPTA	76	75.8	300
San Francisco BART	34	142.0	415
Metro Atlantic RTA	29	67.0	136
New York PATH	13	28.6	282
Miami/Dade Co TA	21	42.2	82
Baltimore MTA	12	26.6	48
Philadelphia PATCO	13	31.5	102
Cleveland RTA	18	38.2	30
New York, Staten Island RT	22	28.6	36

Source: The Universal Almanac 1995, p. 287.

Data Set X Keystone Jackpot Analysis*

Ball	Times drawn	Ball	Times drawn	Ball	Times drawn
1	11	12	10	23	7
2	5	13	11	24	8
3	10	14	5	25	13
4	11	15	8	26	11
5	7	16	14	27	7
6	13	17	8	28	10
7	8	18	11	29	11
8	10	19	10	30	5
9	16	20	7	31	7
10	12	21	11	32	8
11	10	22	6	33	11

*Times each number has been selected in the regular drawings of the Pennsylvania Lottery since November 4, 1995.

Source: Copyright *Pittsburgh Post-Gazette,* October 31, 1996, all rights reserved. Reprinted with permission.

Data Set XI Pages in Statistics Books

The data values represent the number of pages found in statistics textbooks.

616	578	569	511	468
493	564	801	483	847
525	881	757	272	703
741	556	500	668	967
608	465	739	669	651
495	613	774	274	542
739	488	601	727	556
589	724	731	662	680
589	435	742	567	574
733	576	526	443	478
586	282			

Source: Allan G. Bluman, 2004.

Data Set XII Fifty Top Grossing Movies–2000

The data values represent the gross income in millions of dollars for the fifty top movies for the year 2000.

253.4	123.3	90.2	61.3	57.3
215.4	122.8	90.0	61.3	57.2
186.7	117.6	89.1	60.9	56.9
182.6	115.8	77.1	60.8	56.0
161.3	113.7	73.2	60.6	53.3
157.3	113.3	71.2	60.1	53.3
157.0	109.7	70.3	60.0	51.9
155.4	106.8	69.7	59.1	50.9
137.7	101.6	68.5	58.3	50.8
126.6	90.6	68.4	58.1	50.2

Source: Reprinted with permission from the *World Almanac and Book of Facts 2002.* Copyright © 2002 K-III Reference Corporation. All rights reserved.

Data Set XIII Hospital Data*

Number	Number of beds	Admissions	Payroll ($000)	Personnel
1	235	6,559	18,190	722
2	205	6,237	17,603	692
3	371	8,915	27,278	1,187
4	342	8,659	26,722	1,156
5	61	1,779	5,187	237
6	55	2,261	7,519	247
7	109	2,102	5,817	245
8	74	2,065	5,418	223
9	74	3,204	7,614	326
10	137	2,638	7,862	362
11	428	18,168	70,518	2,461
12	260	12,821	40,780	1,422
13	159	4,176	11,376	465
14	142	3,952	11,057	450
15	45	1,179	3,370	145
16	42	1,402	4,119	211
17	92	1,539	3,520	158
18	28	503	1,172	72
19	56	1,780	4,892	195
20	68	2,072	6,161	243
21	206	9,868	30,995	1,142
22	93	3,642	7,912	305
23	68	1,558	3,929	180
24	330	7,611	33,377	1,116
25	127	4,716	13,966	498
26	87	2,432	6,322	240
27	577	19,973	60,934	1,822
28	310	11,055	31,362	981
29	49	1,775	3,987	180
30	449	17,929	53,240	1,899
31	530	15,423	50,127	1,669
32	498	15,176	49,375	1,549
33	60	565	5,527	251
34	350	11,793	34,133	1,207
35	381	13,133	49,641	1,731
36	585	22,762	71,232	2,608
37	286	8,749	28,645	1,194
38	151	2,607	12,737	377
39	98	2,518	10,731	352
40	53	1,848	4,791	185
41	142	3,658	11,051	421
42	73	3,393	9,712	385
43	624	20,410	72,630	2,326
44	78	1,107	4,946	139
45	85	2,114	4,522	221

(continued)

Data Set XIII Hospital Data* *(continued)*

Number	Number of beds	Admissions	Payroll ($000)	Personnel
46	120	3,435	11,479	417
47	84	1,768	4,360	184
48	667	22,375	74,810	2,461
49	36	1,008	2,311	131
50	598	21,259	113,972	4,010
51	1,021	40,879	165,917	6,264
52	233	4,467	22,572	558
53	205	4,162	21,766	527
54	80	469	8,254	280
55	350	7,676	58,341	1,525
56	290	7,499	57,298	1,502
57	890	31,812	134,752	3,933
58	880	31,703	133,836	3,914
59	67	2,020	8,533	280
60	317	14,595	68,264	2,772
61	123	4,225	12,161	504
62	285	7,562	25,930	952
63	51	1,932	6,412	472
64	34	1,591	4,393	205
65	194	5,111	19,367	753
66	191	6,729	21,889	946
67	227	5,862	18,285	731
68	172	5,509	17,222	680
69	285	9,855	27,848	1,180
70	230	7,619	29,147	1,216
71	206	7,368	28,592	1,185
72	102	3,255	9,214	359
73	76	1,409	3,302	198
74	540	396	22,327	788
75	110	3,170	9,756	409
76	142	4,984	13,550	552
77	380	335	11,675	543
78	256	8,749	23,132	907
79	235	8,676	22,849	883
80	580	1,967	33,004	1,059
81	86	2,477	7,507	309
82	102	2,200	6,894	225
83	190	6,375	17,283	618
84	85	3,506	8,854	380
85	42	1,516	3,525	166
86	60	1,573	15,608	236
87	485	16,676	51,348	1,559
88	455	16,285	50,786	1,537
89	266	9,134	26,145	939
90	107	3,497	10,255	431
91	122	5,013	17,092	589

Data Set XIII Hospital Data* *(continued)*

Number	Number of beds	Admissions	Payroll ($000)	Personnel
92	36	519	1,526	80
93	34	615	1,342	74
94	37	1,123	2,712	123
95	100	2,478	6,448	265
96	65	2,252	5,955	237
97	58	1,649	4,144	203
98	55	2,049	3,515	152
99	109	1,816	4,163	194
100	64	1,719	3,696	167
101	73	1,682	5,581	240
102	52	1,644	5,291	222
103	326	10,207	29,031	1,074
104	268	10,182	28,108	1,030
105	49	1,365	4,461	215
106	52	763	2,615	125
107	106	4,629	10,549	456
108	73	2,579	6,533	240
109	163	201	5,015	260
110	32	34	2,880	124
111	385	14,553	52,572	1,724
112	95	3,267	9,928	366
113	339	12,021	54,163	1,607
114	50	1,548	3,278	156
115	55	1,274	2,822	162
116	278	6,323	15,697	722
117	298	11,736	40,610	1,606
118	136	2,099	7,136	255
119	97	1,831	6,448	222
120	369	12,378	35,879	1,312
121	288	10,807	29,972	1,263
122	262	10,394	29,408	1,237
123	94	2,143	7,593	323
124	98	3,465	9,376	371
125	136	2,768	7,412	390
126	70	824	4,741	208
127	35	883	2,505	142
128	52	1,279	3,212	158

*This information was obtained from a sample of hospitals in a selected state. The hospitals are identified by number instead of name.

Glossary

adjusted R^2 used in multiple regression when n and k are approximately equal, to provide a more realistic value of R^2

alpha the probability of a type I error, represented by the Greek letter α

alternative hypothesis a statistical hypothesis that states a difference between a parameter and a specific value or states that there is a difference between two parameters

analysis of variance (ANOVA) a statistical technique used to test a hypothesis concerning the means of three or more populations

ANOVA summary table the table used to summarize the results of an ANOVA test

Bayes' theorem a theorem that allows one to compute the revised probability of an event that occurred before another event when the events are dependent

beta the probability of a type II error, represented by the Greek letter β

between-group variance a variance estimate using the means of the groups or between the groups in an F test

biased sample a sample for which some type of systematic error has been made in the selection of subjects for the sample

bimodal a data set with two modes

binomial distribution the outcomes of a binomial experiment and the corresponding probabilities of these outcomes

binomial experiment a probability experiment in which each trial has only two outcomes, there are a fixed number of trials, the outcomes of the trials are independent, and the probability of success remains the same for each trial

boxplot a graph used to represent a data set when the data set contains a small number of values

categorical frequency distribution a frequency distribution used when the data are categorical (nominal)

central limit theorem a theorem that states that as the sample size increases, the shape of the distribution of the sample means taken from the population with mean μ and standard deviation σ will approach a normal distribution; the distribution will have a mean μ and a standard deviation $\sigma/\sqrt{n}$

Chebyshev's theorem a theorem that states that the proportion of values from a data set that fall within k standard deviations of the mean will be at least $1 - 1/k^2$, where k is a number greater than 1

chi-square distribution a probability distribution obtained from the values of $(n - 1)s^2/\sigma^2$ when random samples are selected from a normally distributed population whose variance is σ^2

class boundaries the upper and lower values of a class for a grouped frequency distribution whose values have one additional decimal place more than the data and end in the digit 5

class midpoint a value for a class in a frequency distribution obtained by adding the lower and upper class boundaries (or the lower and upper class limits) and dividing by 2

class width the difference between the upper class boundary and the lower class boundary for a class in a frequency distribution

classical probability the type of probability that uses sample spaces to determine the numerical probability that an event will happen

cluster sample a sample obtained by selecting a preexisting or natural group, called a cluster, and using the members in the cluster for the sample

coefficient of determination a measure of the variation of the dependent variable that is explained by the regression line and the independent variable; the ratio of the explained variation to the total variation

coefficient of variation the standard deviation divided by the mean; the result is expressed as a percentage

combination a selection of objects without regard to order

complement of an event the set of outcomes in the sample space that are not among the outcomes of the event itself

compound event an event that consists of two or more outcomes or simple events

conditional probability the probability that an event B occurs after an event A has already occurred

confidence interval a specific interval estimate of a parameter determined by using data obtained from a sample and the specific confidence level of the estimate

confidence level the probability that a parameter lies within the specified interval estimate of the parameter

confounding variable a variable that influences the outcome variable but cannot be separated from the other variables that influence the outcome variable

consistent estimator an estimator whose value approaches the value of the parameter estimated as the sample size increases

contingency table data arranged in table form for the chi-square independence test, with R rows and C columns

continuous variable a variable that can assume all values between any two specific values; a variable obtained by measuring

control group a group in an experimental study that is not given any special treatment

convenience sample sample of subjects used because they are convenient and available

correction for continuity a correction employed when a continuous distribution is used to approximate a discrete distribution

correlation a statistical method used to determine whether a linear relationship exists between variables

correlation coefficient a statistic or parameter that measures the strength and direction of a linear relationship between two variables

critical or **rejection region** the range of values of the test value that indicates that there is a significant difference and the null hypothesis should be rejected in a hypothesis test

critical value (C.V.) a value that separates the critical region from the noncritical region in a hypothesis test

cumulative frequency the sum of the frequencies accumulated up to the upper boundary of a class in a frequency distribution

data measurements or observations for a variable

data array a data set that has been ordered

data set a collection of data values

data value or **datum** a value in a data set

decile a location measure of a data value; it divides the distribution into 10 groups

degrees of freedom the number of values that are free to vary after a sample statistic has been computed; used when a distribution (such as the t distribution) consists of a family of curves

dependent events events for which the outcome or occurrence of the first event affects the outcome or occurrence of the second event in such a way that the probability is changed

dependent samples samples in which the subjects are paired or matched in some way; i.e., the samples are related

dependent variable a variable in correlation and regression analysis that cannot be controlled or manipulated

descriptive statistics a branch of statistics that consists of the collection, organization, summarization, and presentation of data

discrete variable a variable that assumes values that can be counted

disordinal interaction an interaction between variables in ANOVA, indicated when the graphs of the lines connecting the mean intersect

distribution-free statistics *see* nonparametric statistics

double sampling a sampling method in which a very large population is given a questionnaire to determine those who meet the qualifications for a study; the questionnaire is reviewed, a second smaller population is defined, and a sample is selected from this group

empirical probability the type of probability that uses frequency distributions based on observations to determine numerical probabilities of events

empirical rule a rule that states that when a distribution is bell-shaped (normal), approximately 68% of the data values will fall within 1 standard deviation of the mean; approximately 95% of the data values will fall within 2 standard deviations of the mean; and approximately 99.7% of the data values will fall within 3 standard deviations of the mean

equally likely events the events in the sample space that have the same probability of occurring

estimation the process of estimating the value of a parameter from information obtained from a sample

estimator a statistic used to estimate a parameter

event outcome of a probability experiment

expected frequency the frequency obtained by calculation (as if there were no preference) and used in the chi-square test

expected value the theoretical average of a variable that has a probability distribution

experimental study a study in which the researcher manipulates one of the variables and tries to determine how the manipulation influences other variables

explanatory variable a variable that is being manipulated by the researcher to see if it affects the outcome variable

exploratory data analysis the act of analyzing data to determine what information can be obtained by using stem and leaf plots, medians, interquartile ranges, and boxplots

extrapolation use of the equation for the regression line to predict y' for a value of x which is beyond the range of the data values of x

F distribution the sampling distribution of the variances when two independent samples are selected from two normally distributed populations in which the variances are equal and the variances s_1^2 and s_2^2 are compared as $s_1^2 \div s_2^2$

F test a statistical test used to compare two variances or three or more means

factors the independent variables in ANOVA tests

finite population correction factor a correction factor used to correct the standard error of the mean when the sample size is greater than 5% of the population size

five-number summary five specific values for a data set that consist of the lowest and highest values, Q_1 and Q_3, and the median

frequency the number of values in a specific class of a frequency distribution

frequency distribution an organization of raw data in table form, using classes and frequencies

frequency polygon a graph that displays the data by using lines that connect points plotted for the frequencies at the midpoints of the classes

goodness-of-fit test a chi-square test used to see whether a frequency distribution fits a specific pattern

grouped frequency distribution a distribution used when the range is large and classes of several units in width are needed

Hawthorne effect an effect on an outcome variable caused by the fact that subjects of the study know that they are participating in the study

histogram a graph that displays the data by using vertical bars of various heights to represent the frequencies of a distribution

homogeneity of proportions test a test used to determine the equality of three or more proportions

hypergeometric distribution the distribution of a variable that has two outcomes when sampling is done without replacement

hypothesis testing a decision-making process for evaluating claims about a population

independence test a chi-square test used to test the independence of two variables when data are tabulated in table form in terms of frequencies

independent events events for which the probability of the first occurring does not affect the probability of the second occurring

independent samples samples that are not related

independent variable a variable in correlation and regression analysis that can be controlled or manipulated

inferential statistics a branch of statistics that consists of generalizing from samples to populations, performing hypothesis testing, determining relationships among variables, and making predictions

influential observation an observation which when removed from the data values would markedly change the position of the regression line

interaction effect the effect of two or more variables on each other in a two-way ANOVA study

interquartile range $Q_3 - Q_1$

interval estimate a range of values used to estimate a parameter

interval level of measurement a measurement level that ranks data and in which precise differences between units of measure exist. *See also* nominal, ordinal, and ratio levels of measurement

Kruskal-Wallis test a nonparametric test used to compare three or more means

law of large numbers when a probability experiment is repeated a large number of times, the relative frequency probability of an outcome will approach its theoretical probability

least-squares line another name for the regression line

left-tailed test a test used on a hypothesis when the critical region is on the left side of the distribution

level a treatment in ANOVA for a variable

level of significance the maximum probability of committing a type I error in hypothesis testing

lower class limit the lower value of a class in a frequency distribution that has the same decimal place value as the data

lurking variable a variable that influences the relationship between x and y, but was not considered in the study

main effect the effect of the factors or independent variables when there is a nonsignificant interaction effect in a two-way ANOVA study

marginal change the magnitude of the change in the dependent variable when the independent variable changes 1 unit

maximum error of estimate the maximum likely difference between the point estimate of a parameter and the actual value of the parameter

mean the sum of the values, divided by the total number of values

mean square the variance found by dividing the sum of the squares of a variable by the corresponding degrees of freedom; used in ANOVA

measurement scales a type of classification that tells how variables are categorized, counted, or measured; the four types of scales are nominal, ordinal, interval, and ratio

median the midpoint of a data array

midrange the sum of the lowest and highest data values, divided by 2

modal class the class with the largest frequency

mode the value that occurs most often in a data set

Monte Carlo method a simulation technique using random numbers

multimodal a data set with three or more modes

multinomial distribution a probability distribution for an experiment in which each trial has more than two outcomes

multiple correlation coefficient a measure of the strength of the relationship between the independent variables and the dependent variable in a multiple regression study

multiple regression a study that seeks to determine if several independent variables are related to a dependent variable

multiple relationship a relationship in which many variables are under study

multistage sampling a sampling technique that uses a combination of sampling methods

mutually exclusive events probability events that cannot occur at the same time

negative relationship a relationship between variables such that as one variable increases, the other variable decreases, and vice versa

negatively skewed or left-skewed distribution a distribution in which the majority of the data values fall to the right of the mean

nominal level of measurement a measurement level that classifies data into mutually exclusive (nonoverlapping) exhaustive categories in which no order or ranking can be imposed on them. *See also* interval, ordinal, and ratio levels of measurement

noncritical or nonrejection region the range of values of the test value that indicates that the difference was probably due to chance and the null hypothesis should not be rejected

nonparametric statistics a branch of statistics for use when the population from which the samples are selected is not normally distributed and for use in testing hypotheses that do not involve specific population parameters

nonrejection region *see* noncritical region

normal distribution a continuous, symmetric, bell-shaped distribution of a variable

normal quantile plot graphical plot used to determine whether a variable is approximately normally distributed

null hypothesis a statistical hypothesis that states that there is no difference between a parameter and a specific value or that there is no difference between two parameters

observational study a study in which the researcher merely observes what is happening or what has happened in the past and draws conclusions based on these observations

observed frequency the actual frequency value obtained from a sample and used in the chi-square test

ogive a graph that represents the cumulative frequencies for the classes in a frequency distribution

one-tailed test a test that indicates that the null hypothesis should be rejected when the test statistic value is in the critical region on one side of the mean

one-way ANOVA a study used to test for differences among means for a single independent variable when there are three or more groups

open-ended distribution a frequency distribution that has no specific beginning value or no specific ending value

ordinal interaction an interaction between variables in ANOVA, indicated when the graphs of the lines connecting the means do not intersect

ordinal level of measurement a measurement level that classifies data into categories that can be ranked; however, precise differences between the ranks do not exist. *See also* interval, nominal, and ratio levels of measurement

outcome the result of a single trial of a probability experiment

outcome variable a variable that is studied to see if it has changed significantly due to the manipulation of the explanatory variable

outlier an extreme value in a data set; it is omitted from a boxplot

parameter a characteristic or measure obtained by using all the data values for a specific population

parametric tests statistical tests for population parameters such as means, variances, and proportions that involve assumptions about the populations from which the samples were selected

Pareto chart chart that uses vertical bars to represent frequencies for a categorical variable

Pearson product moment correlation coefficient (PPMCC) a statistic used to determine the strength of a relationship when the variables are normally distributed

Pearson's index of skewness value used to determine the degree of skewness of a variable

percentile a location measure of a data value; it divides the distribution into 100 groups

permutation an arrangement of n objects in a specific order

pie graph a circle that is divided into sections or wedges according to the percentage of frequencies in each category of the distribution

point estimate a specific numerical value estimate of a parameter

Poisson distribution a probability distribution used when n is large and p is small and when the independent variables occur over a period of time

pooled estimate of the variance a weighted average of the variance using the two sample variances and their respective degrees of freedom as the weights

population the totality of all subjects possessing certain common characteristics that are being studied

population correlation coefficient the value of the correlation coefficient computed by using all possible pairs of data values (x, y) taken from a population

positive relationship a relationship between two variables such that as one variable increases, the other variable increases or as one variable decreases, the other decreases

positively skewed or right-skewed distribution a distribution in which the majority of the data values fall to the left of the mean

power of a test the probability of rejecting the null hypothesis when it is false

prediction interval a confidence interval for a predicted value y

probability the chance of an event occurring

probability distribution the values a random variable can assume and the corresponding probabilities of the values

probability experiment a chance process that leads to well-defined results called outcomes

proportion a part of a whole, represented by a fraction, a decimal, or a percentage

P-value the actual probability of getting the sample mean value if the null hypothesis is true

qualitative variable a variable that can be placed into distinct categories, according to some characteristic or attribute

quantiles values that separate the data set into approximately equal groups

quantitative variable a variable that is numerical in nature and that can be ordered or ranked

quartile a location measure of a data value; it divides the distribution into four groups

quasi-experimental study a study that uses intact groups rather than random assignment of subjects to groups

random sample a sample obtained by using random or chance methods; a sample for which every member of the population has an equal chance of being selected

random variable a variable whose values are determined by chance

range the highest data value minus the lowest data value

range rule of thumb dividing the range by 4 given an approximation of the standard deviation

ranking the positioning of a data value in a data array according to some rating scale

ratio level of measurement a measurement level that possesses all the characteristics of interval measurement and a true zero; it also has true ratios between different units of measure. *See also* interval, nominal, and ordinal levels of measurement

raw data data collected in original form

regression a statistical method used to describe the nature of the relationship between variables, that is, a positive or negative, linear or nonlinear relationship

regression line the line of best fit of the data

rejection region *see* critical region

relative frequency graph a graph using proportions instead of raw data as frequencies

relatively efficient estimator an estimator that has the smallest variance from among all the statistics that can be used to estimate a parameter

residual the difference between the actual value of y and the predicted value y' for a specific value of x

resistant statistic a statistic that is not affected by an extremely skewed distribution

right-tailed test a test used on a hypothesis when the critical region is on the right side of the distribution

run a succession of identical letters preceded by or followed by a different letter or no letter at all, such as the beginning or end of the succession

runs test a nonparametric test used to determine whether data are random

sample a group of subjects selected from the population

sample space the set of all possible outcomes of a probability experiment

sampling distribution of sample means a distribution obtained by using the means computed from random samples taken from a population

sampling error the difference between the sample measure and the corresponding population measure due to the fact that the sample is not a perfect representation of the population

scatter plot a graph of the independent and dependent variables in regression and correlation analysis

Scheffé test a test used after ANOVA, if the null hypothesis is rejected, to locate significant differences in the means

sequence sampling a sampling technique used in quality control in which successive units are taken from production lines and tested to see whether they meet the standards set by the manufacturing company

sign test a nonparametric test used to test the value of the median for a specific sample or to test sample means in a comparison of two dependent samples

simple event an outcome that results from a single trial of a probability experiment

simple relationship a relationship in which only two variables are under study

simulation techniques techniques that use probability experiments to mimic real-life situations

Spearman rank correlation coefficient the nonparametric equivalent to the correlation coefficient, used when the data are ranked

standard deviation the square root of the variance

standard error of the estimate the standard deviation of the observed y values about the predicted y' values in regression and correlation analysis

standard error of the mean the standard deviation of the sample means for samples taken from the same population

standard normal distribution a normal distribution for which the mean is equal to 0 and the standard deviation is equal to 1

standard score the difference between a data value and the mean, divided by the standard deviation

statistic a characteristic or measure obtained by using the data values from a sample

statistical hypothesis a conjecture about a population parameter, which may or may not be true

statistical test a test that uses data obtained from a sample to make a decision about whether the null hypothesis should be rejected

statistics the science of conducting studies to collect, organize, summarize, analyze, and draw conclusions from data

stem and leaf plot a data plot that uses part of a data value as the stem and part of the data value as the leaf to form groups or classes

stratified sample a sample obtained by dividing the population into subgroups, called strata, according to various homogeneous characteristics and then selecting members from each stratum

subjective probability the type of probability that uses a probability value based on an educated guess or estimate, employing opinions and inexact information

sum of squares between groups a statistic computed in the numerator of the fraction used to find the between-group variance in ANOVA

sum of squares within groups a statistic computed in the numerator of the fraction used to find the within-group variance in ANOVA

symmetric distribution a distribution in which the data values are uniformly distributed about the mean

systematic sample a sample obtained by numbering each element in the population and then selecting every kth number from the population to be included in the sample

t distribution a family of bell-shaped curves based on degrees of freedom, similar to the standard normal distribution with the exception that the variance is greater than 1; used when one is testing small samples and when the population standard deviation is unknown

t test a statistical test for the mean of a population, used when the population is normally distributed, the population standard deviation is unknown, and the sample size is less than 30

test value the numerical value obtained from a statistical test, computed from (observed value − expected value) ÷ standard error

time series graph a graph that represents data that occur over a specific time

treatment group a group in an experimental study that has received some type of treatment

treatment groups the groups used in an ANOVA study

tree diagram a device used to list all possibilities of a sequence of events in a systematic way

Tukey test a test used to make pairwise comparisons of means in an ANOVA study when samples are the same size

two-tailed test a test that indicates that the null hypothesis should be rejected when the test value is in either of the two critical regions

two-way ANOVA a study used to test the effects of two or more independent variables and the possible interaction between them

type I error the error that occurs if one rejects the null hypothesis when it is true

type II error the error that occurs if one does not reject the null hypothesis when it is false

unbiased estimator an estimator whose value approximates the expected value of a population parameter, used for the variance or standard deviation when the sample size is less than 30; an estimator whose expected value or mean must be equal to the mean of the parameter being estimated

unbiased sample a sample chosen at random from the population that is, for the most part, representative of the population

ungrouped frequency distribution a distribution that uses individual data and has a small range of data

uniform distribution a distribution whose values are evenly distributed over its range

upper class limit the upper value of a class in a frequency distribution that has the same decimal place value as the data

variable a characteristic or attribute that can assume different values

variance the average of the squares of the distance that each value is from the mean

Venn diagram a diagram used as a pictorial representative for a probability concept or rule

weighted mean the mean found by multiplying each value by its corresponding weight and dividing by the sum of the weights

Wilcoxon rank sum test a nonparametric test used to test independent samples and compare distributions

Wilcoxon signed-rank test a nonparametric test used to test dependent samples and compare distributions

within-group variance a variance estimate using all the sample data for an F test; it is not affected by differences in the means

z **distribution** *see* standard normal distribution

z **score** *see* standard score

z **test** a statistical test for means and proportions of a population, used when the population is normally distributed and the population standard deviation is known or the sample size is 30 or more

z **value** same as z score

Glossary of Symbols

a	y intercept of a line
α	Probability of a type I error
b	Slope of a line
β	Probability of a type II error
C	Column frequency
cf	Cumulative frequency
$_nC_r$	Number of combinations of n objects taking r objects at a time
C.V.	Critical value
CVar	Coefficient of variation
D	Difference; decile
$\bar{D}$	Mean of the differences
d.f.	Degrees of freedom
d.f.N.	Degrees of freedom, numerator
d.f.D.	Degrees of freedom, denominator
E	Event; expected frequency; maximum error of estimate
$\bar{E}$	Complement of an event
e	Euler's constant ≈ 2.7183
$E(X)$	Expected value
f	Frequency
F	F test value; failure
F'	Critical value for the Scheffé test
MD	Median
MR	Midrange
MS_B	Mean square between groups
MS_W	Mean square within groups (error)
n	Sample size
N	Population size
$n(E)$	Number of ways E can occur
$n(S)$	Number of outcomes in the sample space
O	Observed frequency
P	Percentile; probability
p	Probability; population proportion
$\hat{p}$	Sample proportion
$\bar{p}$	Weighted estimate of p
$P(B|A)$	Conditional probability
$P(E)$	Probability of an event E
$P(\bar{E})$	Probability of the complement of E
$_nP_r$	Number of permutations of n objects taking r objects at a time
π	Pi ≈ 3.14
Q	Quartile
q	$1 - p$; test value for Tukey test
$\hat{q}$	$1 - \hat{p}$
$\bar{q}$	$1 - \bar{p}$
R	Range; rank sum

F_S	Scheffé test value
GM	Geometric mean
H	Kruskal-Wallis test value
H_0	Null hypothesis
H_1	Alternative hypothesis
HM	Harmonic mean
k	Number of samples
λ	Number of occurrences for the Poisson distribution
s_D	Standard deviation of the differences
s_{est}	Standard error of estimate
SS_B	Sum of squares between groups
SS_W	Sum of squares within groups
s_B^2	Between-group variance
s_W^2	Within-group variance
t	t test value
$t_{\alpha/2}$	Two-tailed t critical value
μ	Population mean
μ_D	Mean of the population differences
$\mu_{\bar{X}}$	Mean of the sample means
w	Class width; weight
r	Sample correlation coefficient
R	Multiple correlation coefficient
r^2	Coefficient of determination
ρ	Population correlation coefficient
r_S	Spearman rank correlation coefficient
S	Sample space; success
s	Sample standard deviation
s^2	Sample variance
σ	Population standard deviation
σ^2	Population variance
$\sigma_{\bar{X}}$	Standard error of the mean
Σ	Summation notation
w_s	Smaller sum of signed ranks, Wilcoxon signed-rank test
X	Data value; number of successes for a binomial distribution
$\bar{X}$	Sample mean
x	Independent variable in regression
$\bar{X}_{\text{GM}}$	Grand mean
X_m	Midpoint of a class
χ^2	Chi-square
y	Dependent variable in regression
y'	Predicted y value
z	z test value or z score
$z_{\alpha/2}$	Two-tailed critical z value
!	Factorial

Appendix F

Bibliography

Aczel, Amir D. *Complete Business Statistics,* 3rd ed. Chicago: Irwin, 1996.

Beyer, William H. *CRC Handbook of Tables for Probability and Statistics,* 2nd ed. Boca Raton, Fla.: CRC Press, 1986.

Brase, Charles, and Corrinne P. Brase. *Understanding Statistics,* 5th ed. Lexington, Mass.: D.C. Heath, 1995.

Chao, Lincoln L. *Introduction to Statistics.* Monterey, Calif.: Brooks/Cole, 1980.

Daniel, Wayne W., and James C. Terrell. *Business Statistics,* 4th ed. Boston: Houghton Mifflin, 1986.

Edwards, Allan L. *An Introduction to Linear Regression and Correlation,* 2nd ed. New York: Freeman, 1984.

Eves, Howard. *An Introduction to the History of Mathematics,* 3rd ed. New York: Holt, Rinehart and Winston, 1969.

Famighetti, Robert, ed. *The World Almanac and Book of Facts 1996.* New York: Pharos Books, 1995.

Freund, John E., and Gary Simon. *Statistics—A First Course,* 6th ed. Englewood Cliffs, N.J.: Prentice-Hall, 1995.

Gibson, Henry R. *Elementary Statistics.* Dubuque, Iowa: Wm. C. Brown Publishers, 1994.

Glass, Gene V., and Kenneth D. Hopkins. *Statistical Methods in Education and Psychology,* 2nd ed. Englewood Cliffs, N.J.: Prentice-Hall, 1984.

Guilford, J. P. *Fundamental Statistics in Psychology and Education,* 4th ed. New York: McGraw-Hill, 1965.

Haack, Dennis G. *Statistical Literacy: A Guide to Interpretation.* Boston: Duxbury Press, 1979.

Hartwig, Frederick, with Brian Dearing. *Exploratory Data Analysis.* Newbury Park, Calif.: Sage Publications, 1979.

Henry, Gary T. *Graphing Data: Techniques for Display and Analysis.* Thousand Oaks, Calif.: Sage Publications, 1995.

Isaac, Stephen, and William B. Michael. *Handbook in Research and Evaluation,* 2nd ed. San Diego: EdITS, 1990.

Johnson, Robert. *Elementary Statistics,* 6th ed. Boston: PWS–Kent, 1992.

Kachigan, Sam Kash. *Statistical Analysis.* New York: Radius Press, 1986.

Khazanie, Ramakant. *Elementary Statistics in a World of Applications,* 3rd ed. Glenview, Ill.: Scott, Foresman, 1990.

Kuzma, Jan W. *Basic Statistics for the Health Sciences.* Mountain View, Calif.: Mayfield, 1984.

Lapham, Lewis H., Michael Pollan, and Eric Ethridge. *The Harper's Index Book.* New York: Henry Holt, 1987.

Lipschultz, Seymour. *Schaum's Outline of Theory and Problems of Probability.* New York: McGraw-Hill, 1968.

Marascuilo, Leonard A., and Maryellen McSweeney. *Nonparametric and Distribution-Free Methods for the Social Sciences.* Monterey, Calif.: Brooks/Cole, 1977.

Marzillier, Leon F. *Elementary Statistics.* Dubuque, Iowa: Wm. C. Brown Publishers, 1990.

Mason, Robert D., Douglas A. Lind, and William G. Marchal. *Statistics: An Introduction.* New York: Harcourt Brace Jovanovich, 1988.

MINITAB. *MINITAB Reference Manual.* State College, Pa.: MINITAB, Inc., 1994.

Minium, Edward W. *Statistical Reasoning in Psychology and Education.* New York: Wiley, 1970.

Moore, David S. *The Basic Practice of Statistics.* New York: W. H. Freeman and Co., 1995.

Moore, Davis S., and George P. McCabe. *Introduction to the Practice of Statistics,* 3rd ed. New York: W. H. Freeman, 1999.

Newmark, Joseph. *Statistics and Probability in Modern Life.* New York: Saunders, 1988.

Pagano, Robert R. *Understanding Statistics,* 3rd ed. New York: West, 1990.

Phillips, John L., Jr. *How to Think about Statistics.* New York: Freeman, 1988.

Reinhardt, Howard E., and Don O. Loftsgaarden. *Elementary Probability and Statistical Reasoning.* Lexington, Mass.: Heath, 1977.

Roscoe, John T. *Fundamental Research Statistics for the Behavioral Sciences,* 2nd ed. New York: Holt, Rinehart and Winston, 1975.

Rossman, Allan J. *Workshop Statistics, Discovery with Data.* New York: Springer, 1996.

Runyon, Richard P., and Audrey Haber. *Fundamentals of Behavioral Statistics,* 6th ed. New York: Random House, 1988.

Shulte, Albert P., 1981 yearbook editor, and James R. Smart, general yearbook editor. *Teaching Statistics and Probability, 1981 Yearbook.* Reston, Va.: National Council of Teachers of Mathematics, 1981.

Smith, Gary. *Statistical Reasoning.* Boston: Allyn and Bacon, 1985.

Spiegel, Murray R. *Schaum's Outline of Theory and Problems of Statistics.* New York: McGraw-Hill, 1961.

Texas Instruments. *TI-83 Graphing Calculator Guidebook.* Temple, Tex.: Texas Instruments, 1996.

Triola, Mario G. *Elementary Statistics,* 7th ed. Reading, Mass.: Addison-Wesley, 1998.

Wardrop, Robert L. *Statistics: Learning in the Presence of Variation.* Dubuque, Iowa: Wm. C. Brown Publishers, 1995.

Warwick, Donald P., and Charles A. Lininger. *The Sample Survey: Theory and Practice.* New York: McGraw-Hill, 1975.

Weiss, Daniel Evan. *100% American.* New York: Poseidon Press, 1988.

Williams, Jack. *The USA Today Weather Almanac 1995.* New York: Vintage Books, 1994.

Wright, John W., ed. *The Universal Almanac 1995.* Kansas City, Mo.: Andrews & McMeel, 1994.

Photo Credits

Design Element Photos

World icon: © Vol. 34/PhotoDisc; CD icon: © Vol. OS40/PhotoDisc.

Chapter 1

Opener (spreadsheet): © Vol. 14/PhotoDisc, (bar graph): © Jack Star/PhotoLink/Getty Images, p. 2: © Vol. 102/Corbis; p. 10 © Vol. 253/Corbis.

Chapter 2

Opener: Copyright 2005 Nexus Energy Software Inc. All Rights Reserved. Used with permission, p. 34: © PhotoDisc; p. 74: © Vol. 14/PhotoDisc.

Chapter 3

Opener: © Michael Newman/Photo Edit; p. 96(top): © PhotoDisc; p. 96(bottom): © Vol. 599/Corbis.

Chapter 4

Opener: © Richard Heinzen/SuperStock; p. 172: © Vol. 122/Corbis; p. 217: © Vol. 579/Corbis; p. 223: © Betts Anderson/Unicorn Stock Photos.

Chapter 5

Opener: © Robin Sachs/Photo Edit; p. 238: © Vol. 41/Corbis; p. 242: © Phil Schermeister/Corbis; p. 256: © Vol. 10/PhotoDisc.

Chapter 6

Opener: Library of Congress; p. 286: © Vol. 106/Corbis; p. 309: © Vol. 202/Corbis.

Chapter 7

Opener: USDA; p. 348: © Vol. 124/Corbis; p. 373: © Brand X Pictures/Getty Images; p. 378: © C. Orrico/Superstock.

Chapter 8

Opener: © PhotoDisc; p. 392: © Vol. 79/Corbis; p. 425: © Jose Luis Pelaez, Inc./Corbis; p. 449: © Vol. 48/PhotoDisc.

Chapter 9

Opener (both): © Bob Shirtz/SuperStock, p. 464: © Michelle Bridwell/PhotoEdit; p. 505: © Antonio Reeve/Photo Researchers, Inc.

Chapter 10

Opener: © PhotoDisc; p. 528: © Tony Freeman/PhotoEdit; p. 540: © Vol. 79/PhotoDisc; p. 566: © Michael Kagan.

Chapter 11

Opener: © Vol. DV513/PhotoDisc; p. 584: © Vol. 130/Corbis; p. 609: © Vol. 56/PhotoDisc.

Chapter 12

Opener: © PhotoDisc; p. 620: © Jim Pickerell/Stock Connection; p. 626: © Joe Sohm/Unicorn Stock Photos; p. 636: © Martha McBride/Unicorn Stock Photos.

Chapter 13

Opener: © Rick Gayle Studio/Corbis; p. 660: © Tony Freeman/PhotoEdit.

Chapter 14

Opener: © Jeffrey Greenberg/Photo Researchers, Inc., p. 708: Courtesy of Hastos-Hall Productions, p. 713: © Vol. 31 PhotoDisc.

Selected Answers*

Chapter 1

Review Exercises

1. Descriptive statistics describe the data set. Inferential statistics use the data to draw conclusions about the population.

3. Answers will vary.

5. Samples are used to save time and money when the population is large and when the units must be destroyed to gain information.

6. *a.* Inferential   *e.* Inferential
   b. Descriptive   *f.* Inferential
   c. Descriptive   *g.* Descriptive
   d. Descriptive   *h.* Inferential

7. *a.* Ratio   *f.* Ordinal
   b. Ordinal   *g.* Ratio
   c. Ratio   *h.* Ratio
   d. Interval   *i.* Nominal
   e. Ratio   *j.* Ratio

9. *a.* Discrete   *e.* Discrete
   b. Continuous   *f.* Discrete
   c. Continuous   *g.* Continuous
   d. Continuous

11. Random, systematic, stratified, cluster

12. *a.* Cluster   *c.* Random   *e.* Stratified
    b. Systematic   *d.* Systematic

13. Answers will vary.   15. Answers will vary.

17. *a.* Experimental   *c.* Observational
    b. Observational   *d.* Experimental

19. Possible answers:
    a. Workplace of subjects, smoking habits, etc.
    b. Gender, age, etc.
    c. Diet, type of job, etc.
    d. Exercise, heredity, age, etc.

21. The only time claims can be proved is when the entire population is used.

23. Since the results are not typical, the advertisers selected only a few people for whom the weight loss product worked extremely well.

25. "74% more calories" than what? No comparison group is stated.

27. What is meant by "24 hours of acid control"?

29. Possible answer: It could be the amount of caffeine in the coffee or tea. It could have been the brewing method.

Chapter Quiz

1. True   2. False
3. False   4. False
5. False   6. True
7. False   8. *c*
9. *b*   10. *d*
11. *a*   12. *c*
13. *a*   14. Descriptive, inferential
15. Gambling, insurance   16. Population
17. Sample
18. *a.* Saves time   *c.* Use when population is infinite
    b. Saves money
19. *a.* Random   *c.* Cluster
    b. Systematic   *d.* Stratified
20. Quasi-experimental   21. Random
22. *a.* Descriptive   *d.* Inferential
    b. Inferential   *e.* Inferential
    c. Descriptive
23. *a.* Nominal   *d.* Interval
    b. Ratio   *e.* Ratio
    c. Ordinal

*Answers may vary due to rounding or use of technology.

Note: These answers to odd-numbered and selected even-numbered exercises include *all* quiz answers.

24. *a.* Continuous *d.* Continuous
    b. Discrete *e.* Discrete
    c. Continuous

25. *a.* 47.5–48.5 seconds
    b. 0.555–0.565 centimeter
    c. 9.05–9.15 quarts
    d. 13.65–13.75 pounds
    e. 6.5–7.5 feet

Chapter 2

Exercises 2–2

1. To organize data in a meaningful way, to determine the shape of the distribution, to facilitate computational procedures for statistics, to make it easier to draw charts and graphs, to make comparisons among different sets of data

3. *a.* 11.5–18.5; 15; 7
    b. 55.5–74.5; 65; 19
    c. 694.5–705.5; 700; 11
    d. 13.55–14.75; 14.15; 1.2
    e. 2.145–3.935; 3.04; 1.79

5. *a.* Class width is not uniform.
    b. Class limits overlap, and class width is not uniform.
    c. A class has been omitted.
    d. Class width is not uniform.

7.

Class	Tally	Frequency	Percent
A	////	4	10
M	7HL 7HL 7HL 7HL 7HL ///	28	70
H	7HL /	6	15
S	//	2	5
		40	100

9.

Limits	Boundaries	f	cf
19–21	18.5–21.5	2	2
22–24	21.5–24.5	13	15
25–27	24.5–27.5	11	26
28–30	27.5–30.5	3	29
31–33	30.5–33.5	1	30
		30	

The average speed is about 24.5 miles per hour.

11.

Limits	Boundaries	f	cf
745–751	744.5–751.5	4	4
752–758	751.5–758.5	5	9
759–765	758.5–765.5	7	16
766–772	765.5–772.5	11	27
773–779	772.5–779.5	2	29
780–786	779.5–786.5	1	30
		30	

13.

Limits	Boundaries	f	cf
27–33	26.5–33.5	7	7
34–40	33.5–40.5	14	21
41–47	40.5–47.5	15	36
48–54	47.5–54.5	11	47
55–61	54.5–61.5	3	50
62–68	61.5–68.5	3	53
69–75	68.5–75.5	2	55
		55	

15.

Limits	Boundaries	f	cf
31–39	30.5–39.5	4	4
40–48	39.5–48.5	5	9
49–57	48.5–57.5	5	14
58–66	57.5–66.5	12	26
67–75	66.5–75.5	13	39
76–84	75.5–84.5	5	44
85–93	84.5–93.5	3	47
		47	

17.

Limits	Boundaries	f	cf
150–1,276	149.5–1,276.5	2	2
1,277–2,403	1,276.5–2,403.5	2	4
2,404–3,530	2,403.5–3,530.5	5	9
3,531–4,657	3,530.5–4,657.5	8	17
4,658–5,784	4,657.5–5,784.5	7	24
5,785–6,911	5,784.5–6,911.5	3	27
6,912–8,038	6,911.5–8,038.5	7	34
8,039–9,165	8,038.5–9,165.5	3	37
9,166–10,292	9,165.5–10,292.5	3	40
10,293–11,419	10,292.5–11,419.5	2	42
		42	

19. The percents sum to 101. They should sum to 100% unless rounding was used.

Exercises 2–3

1. Eighty applicants do not need to enroll in the developmental programs.

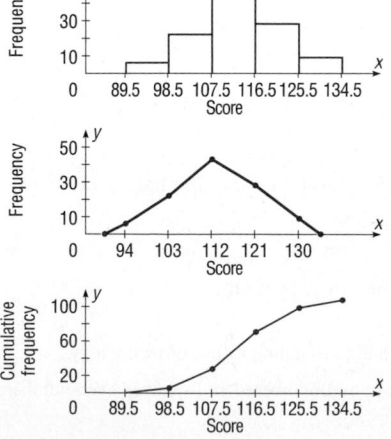

3. The distribution is slightly left skewed.

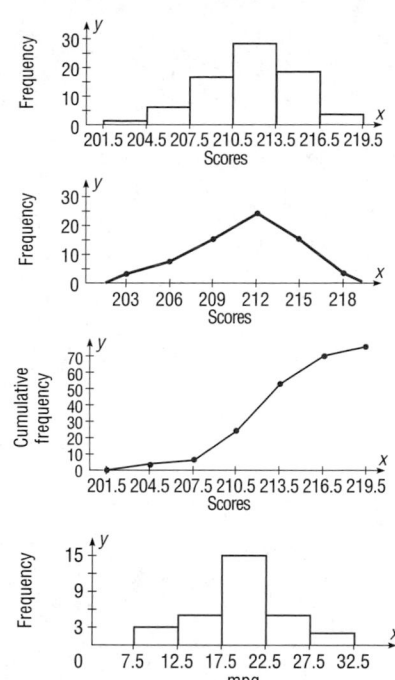

5.

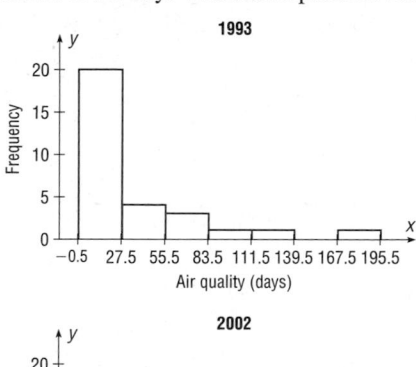

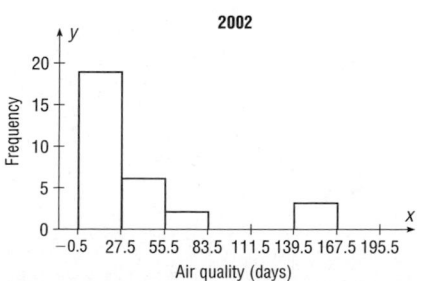

7. Both graphs are similar in that they are positively skewed. Also, it looks as if the air quality has improved somewhat in that there are slightly smaller values in 2002, which means fewer days with unacceptable levels of pollution.

9. The histogram has a peak at the class of 66.5–75.5 and is somewhat negatively skewed.

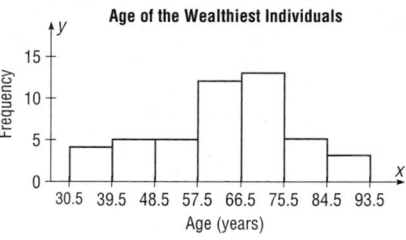

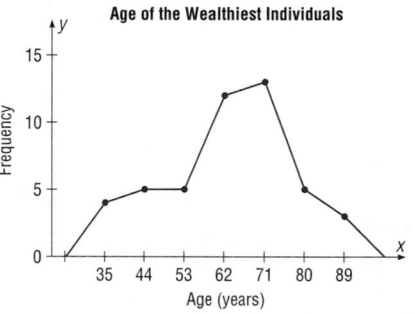

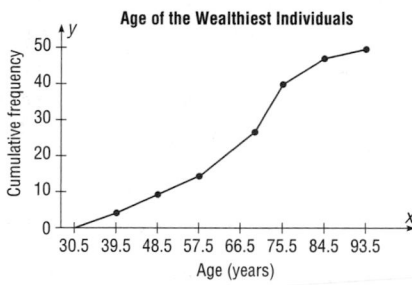

11. The peak is in the first class, and then the histogram is rather uniform after the first class. Most of the parks have less than 101.5 thousand acres as compared with any other class of values.

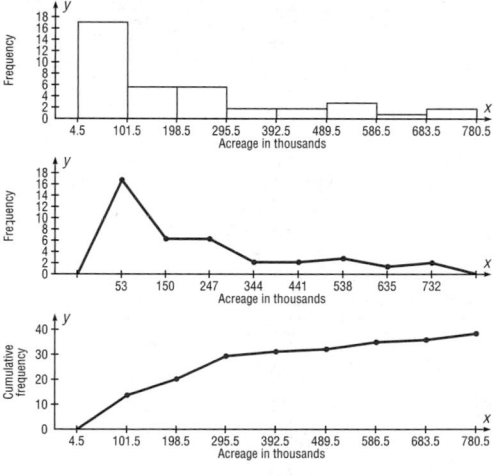

13. The proportion of applicants who need to enroll in the developmental program is about 0.26.

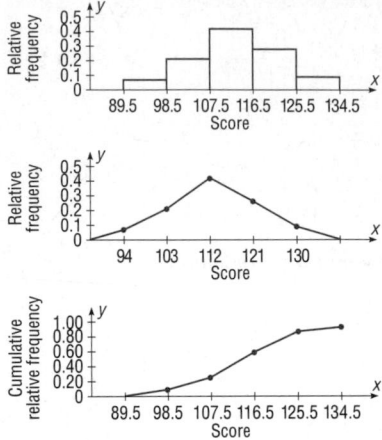

15.

Class boundaries	rf	crf
79.5–108.5	0.17	0.17
108.5–137.5	0.28	0.45
137.5–166.5	0.04	0.49
166.5–195.5	0.20	0.69
195.5–224.5	0.22	0.91
224.5–253.5	0.04	0.95
253.5–282.5	0.04	0.99*
	0.99	

*Due to rounding.

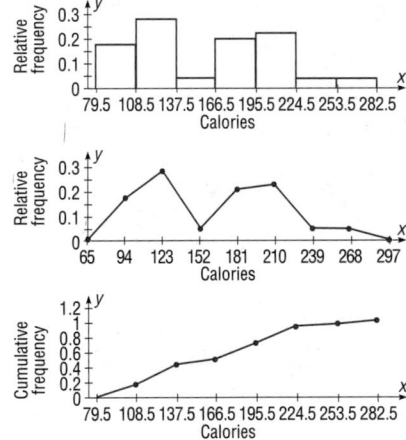

The histogram has two peaks.

17.

Class boundaries	rf	crf
−0.5–27.5	0.63	0.63
27.5–55.5	0.20	0.83
55.5–83.5	0.07	0.90
83.5–111.5	0.00	0.90
111.5–139.5	0.00	0.90
139.5–167.5	0.10	1.00
167.5–195.5	0.00	1.00
	100.00	

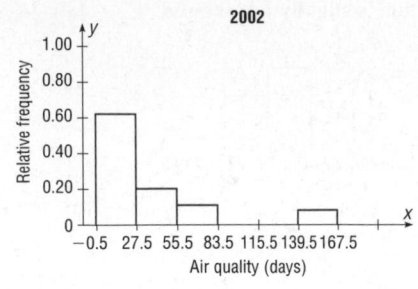

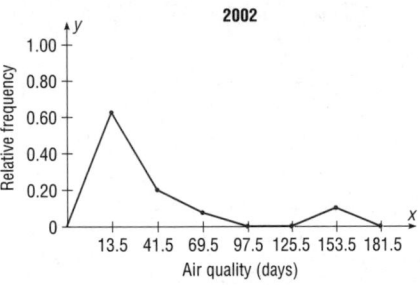

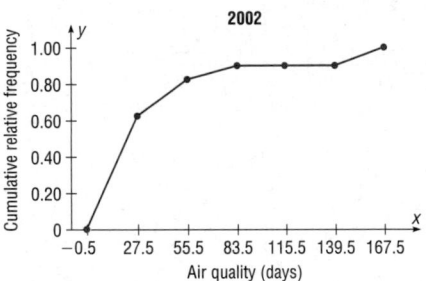

19. *a.*

Limits	Boundaries	Midpoints	f	cf
22–24	21.5–24.5	23	1	1
25–27	24.5–27.5	26	3	4
28–30	27.5–30.5	29	0	4
31–33	30.5–33.5	32	6	10
34–36	33.5–36.5	35	5	15
37–39	36.5–39.5	38	3	18
40–42	39.5–42.5	41	2	20

b.

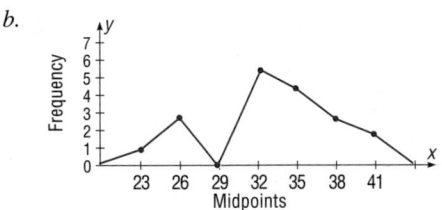

c.

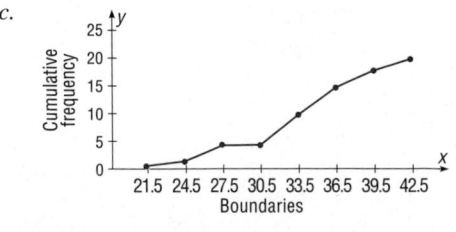

Exercises 2–4

1. The majority of money should be spent for drug rehabilitation.

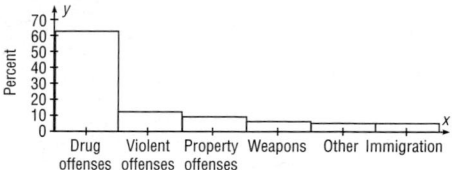

3. The best place to market products would be to the home viewers.

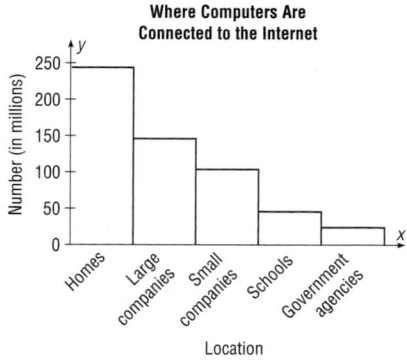

5.

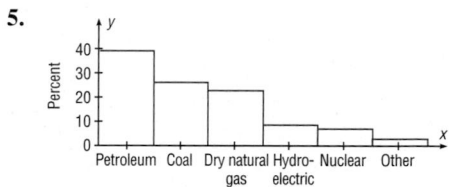

7. There is a steady increase in consumption of tobacco products.

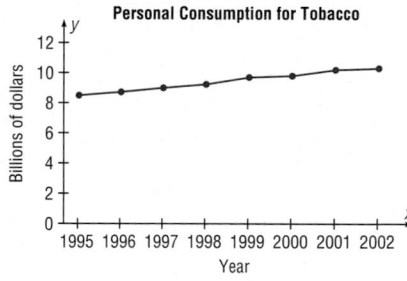

9. The graph shows a decline in the percentages of registered voters voting in presidential elections.

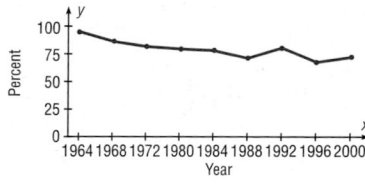

11.

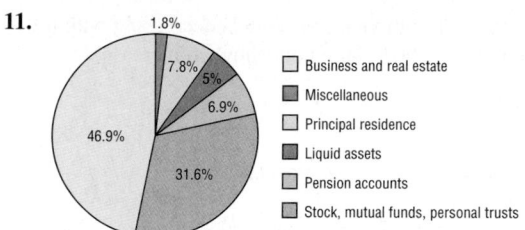

13. The pie graph better represents the data since we are looking at parts of a whole.

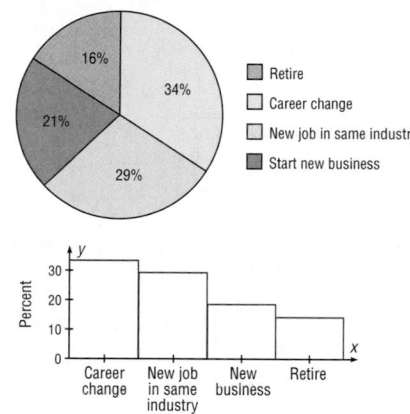

15. The distribution is somewhat symmetric, unimodal and has a peak in the 50's.

```
4 | 2 3
4 | 6 6 7 8 9 9
5 | 0 1 1 1 1 2 2 4 4 4 4 4
5 | 5 5 5 5 6 6 6 7 7 7 7 8
6 | 0 1 1 1 2 4 4
6 | 5 8 9
```

17.

Variety 1		Variety 2
2	1	3 8
3 0	2	5
9 8 8 5 2	3	6 8
3 3 1	4	1 2 5 5
9 9 8 5 3 3 2 1 0	5	0 3 5 5 6 7 9
	6	2 2

The distributions are somewhat similar in their shapes; however, the variation of the data for variety 2 is slightly larger than the variation of the data for variety 1.

19.
```
1 | 3 4 8 9
2 | 5 8 9
3 | 2 8
4 | 1
```

21. Production of both veal and lamb is decreasing with the exception of 1990, where both show an increase.

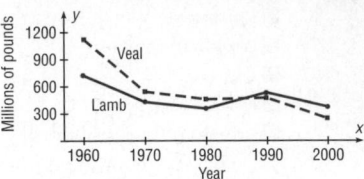

23.

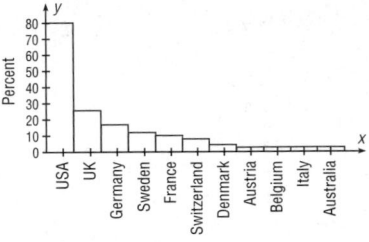

25. The values on the *y* axis start at 3.5. Also there are no data values shown for the years 2004 through 2011.

Review Exercises

1.

Class	*f*
Newspaper	10
Television	16
Radio	12
Internet	12
	50

3.

Class	*f*
Baseballs	4
Golf balls	5
Tennis balls	6
Soccer balls	5
Footballs	5
	25

5.

Class	*f*	cf
11	1	1
12	2	3
13	2	5
14	2	7
15	1	8
16	2	10
17	4	14
18	2	16
19	2	18
20	1	19
21	0	19
22	1	20
	20	

7.

Class limits	Class boundaries	*f*	cf
85–105	84.5–105.5	4	4
106–126	105.5–126.5	7	11
127–147	126.5–147.5	9	20
148–168	147.5–168.5	10	30
169–189	168.5–189.5	9	39
190–210	189.5–210.5	1	40
		40	

9.

Class limits	Class boundaries	*f*	cf
170–188	169.5–188.5	11	11
189–207	188.5–207.5	9	20
208–226	207.5–226.5	4	24
227–245	226.5–245.5	5	29
246–264	245.5–264.5	0	29
265–283	264.5–283.5	0	29
284–302	283.5–302.5	0	29
303–321	302.5–321.5	1	30
		30	

11.

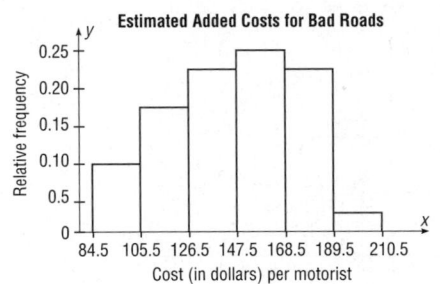

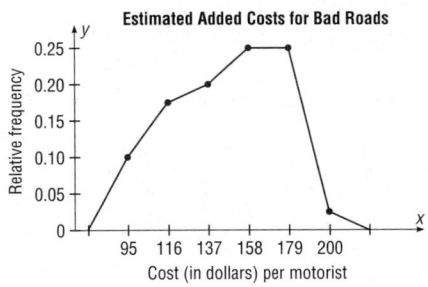

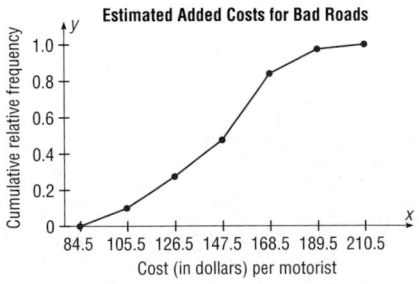

13.

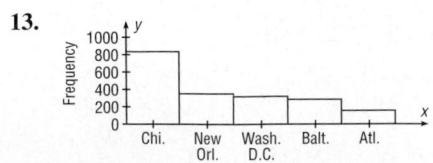

15. Over time the wage has increased.

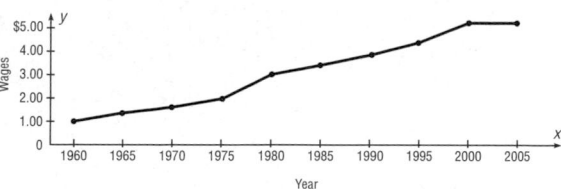

17. About the same number of people watched the first and second debates in 1992 and 1996. After that more people watched the first debate than watched the second debate.

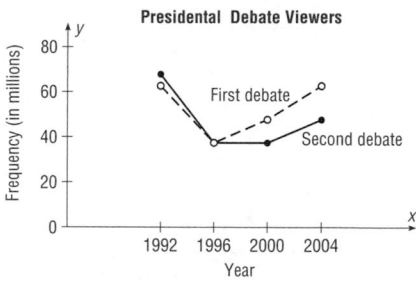

19.

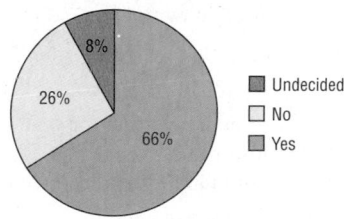

21. The peak of the distribution is in the range of 25–29.

```
1 | 2 4
1 | 6 7 8 8 9
2 | 0 2 3 4
2 | 5 5 5 6 6 9 9
3 | 2 3
3 | 5 7 8 8 9
```

Chapter Quiz

1. False
2. False
3. False
4. True
5. True
6. False
7. False
8. *c*
9. *c*
10. *b*
11. *b*
12. Categorical, ungrouped, grouped
13. 5, 20
14. Categorical
15. Time series
16. Stem and leaf plot
17. Vertical or *y*

18.

Class	*f*	cf
H	6	6
A	5	11
M	6	17
C	8	25
	25	

19.

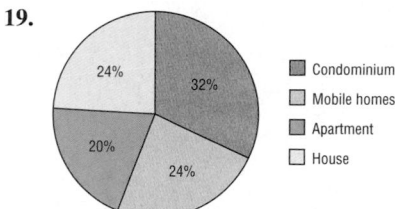

20.

Class boundaries	*f*	cf
0.5–1.5	1	1
1.5–2.5	5	6
2.5–3.5	3	9
3.5–4.5	4	13
4.5–5.5	2	15
5.5–6.5	6	21
6.5–7.5	2	23
7.5–8.5	3	26
8.5–9.5	4	30
	30	

21.

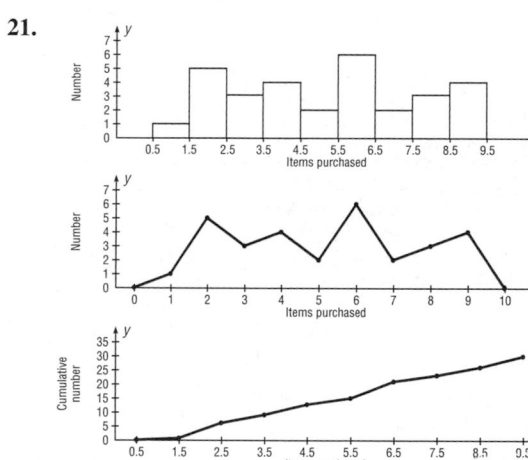

22.

Class limits	*f*	Class boundaries
27–90	13	26.5–90.5
91–154	2	90.5–154.5
155–218	0	154.5–218.5
219–282	5	218.5–282.5
283–346	0	282.5–346.5
347–410	2	346.5–410.5
411–474	0	410.5–474.5
475–538	1	474.5–538.5
539–602	2	538.5–602.5
	25	

23. The distribution is positively skewed with one more than one-half of the data values in the lowest class.

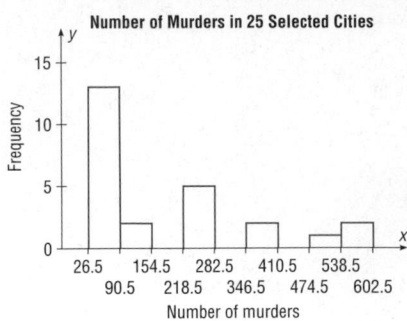

Number of Murders in 25 Selected Cities

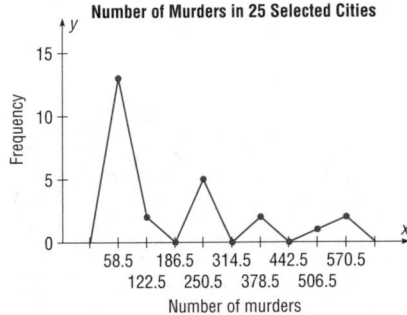

Number of Murders in 25 Selected Cities

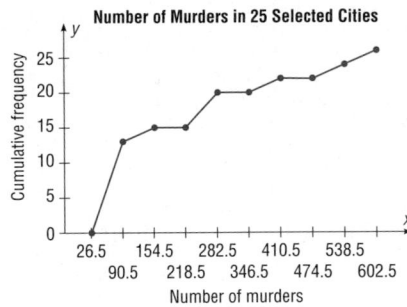

Number of Murders in 25 Selected Cities

24.

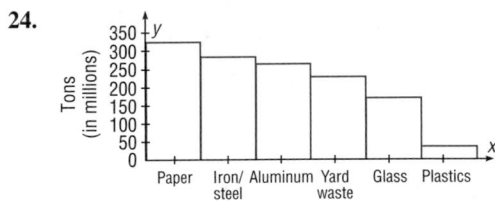

25. The fatalities decreased in 1999 and then increased the next two years.

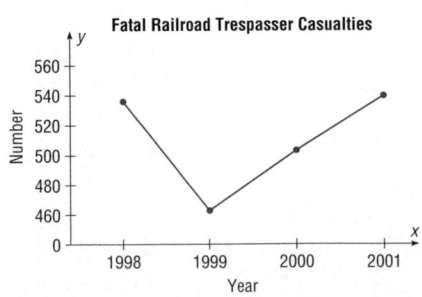

Fatal Railroad Trespasser Casualties

26.

1	5 9
2	6 8
3	1 5 8 8 9
4	1 7 8
5	3 3 4
6	2 3 7 8
7	6 9
8	6 8 9
9	8

Chapter 3

Exercises 3-2

1. *a.* 3.724   *b.* 3.73   *c.* 3.74 and 3.70   *d.* 3.715

3. *a.* 15.1   *b.* 7   *c.* 3   *d.* 31
Answers will vary.

5. 270.8; 209; no mode; 369. It seems that the average number of identity thefts is not higher than 300.

7. *a.* 6.63   *b.* 6.45   *c.* None
d. 6.7; answers will vary

9. *a.* 5678.9   *b.* 5342   *c.* 4450   *d.* 5781.5
The distribution is skewed to the right.

11. Year 1: *a.* 922.6   *b.* 527   *c.* None   *d.* 2130.5
Year 2: *a.* 911.7   *b.* 485   *c.* 1430   *d.* 2055
The mean, median, and midrange data for year 2 are somewhat less than those values for the data for year 1, indicating the number of fatalities has decreased.

13. *a.* 211.6   *b.* 211–213

15. 188.48; 34–96. Since most of the data are in the lowest class, the mean is probably not the best measure of the average. If the individual data values are available, the median may be a better measure of the average. A procedure for finding the approximate median for grouped data is found in Exercise 42 of this section.

17. *a.* 85.1   *b.* 74.5–85.5

19. *a.* 33.8   *b.* 27–33

21. *a.* 23.7   *b.* 21.5–24.5

23. 44.8; 40.5–47.5

25. 63; 66.5–75.5   **27.** 2.896

29. $545,666.67   **31.** 82.7

33. *a.* Median   *c.* Mode   *e.* Mode
b. Mean   *d.* Mode   *f.* Mean

35. Both could be true since one may be using the mean for the average salary and the other may be using the mode for the average.

37. 6

39. *a.* 36 mph   *b.* 30.77 mph   *c.* $16.67

41. 5.48

Exercises 3–3

1. The square root of the variance is the standard deviation.

3. σ^2; σ

5. When the sample size is less than 30, the formula for the variance of the sample will underestimate the population variance.

7. 48; 254.7; 15.9 (rounded to 16)
The data vary widely.

9.

	Temp. (°F)	Precip. (inches)
Range	32	4
Variance	147.6	1.89
Standard deviation	12.15	1.373

The temperatures are more variable.

11. For St. Paul: 30; 77.1; 8.8. For Chicago: 43; 237.1; 15.4. The data for Chicago is more variable since the standard deviation is much larger.

13. 21; 38.1; 6.2. According to the range rule of thumb, $s \approx 5.25$. The actual standard deviation is 6.2. The estimate is close.

15. For 1995:   For 1996:
$R = 4123$ $R = 3970$
$s^2 = 1,030,817.6$ $s^2 = 1,019,853.8$
$s = 1015.3$ $s = 1009.9$

The data for 1995 are more variable.

17. 11,263; 7436, 475.0; 2727.0

19. 133.6; 11.6

21. 27,941.46; 167.2

23. 211.2; 14.5

25. 211.2; 14.5; no, the variability of the lifetimes of the batteries is quite large.

27. 11.7; 3.4

29. For the East, CVar = 0.373. For the West, CVar = 0.494. The data for the West is more variable.

31. 23.1%; 12.9%; age is more variable

33. *a.* 96%   *b.* 93.75%

35. \$4.84–\$5.20

37. 89–101

39. 86%

41. 16%

43. All the data values fall within 2 standard deviations of the mean.

45. 56%; 75%; 84%; 88.89%; 92%

47. 4.36

49. It must be an incorrect data value, since it is beyond the range using the formula $s\sqrt{n-1}$.

Exercises 3–4

1. A z score tells how many standard deviations the data value is above or below the mean.

3. A percentile is a relative measurement of position; a percentage is an absolute measure of the part to the total.

5. $Q_1 = P_{25}$; $Q_2 = P_{50}$; $Q_3 = P_{75}$

7. $D_1 = P_{10}$; $D_2 = P_{20}$; $D_3 = P_{30}$; etc.

9. *a.* 1   *b.* −2   *c.* −2.5   *d.* −1.5   *e.* 0.67

11. *a.* 0.75   *b.* −1.25   *c.* 2.25   *d.* −2   *e.* −0.5

13. *a.* 0.75   *b.* 1.67   The score for part *b* is higher.

15. *a.* −0.93   *b.* −0.85
c. −1.4; score in part *b* is highest

17. *a.* 21st   *b.* 58th   *c.* 77th   *d.* 33rd

18. *a.* 7   *b.* 25   *c.* 64   *d.* 76   *e.* 93

19. *a.* 235   *b.* 255   *c.* 261   *d.* 275   *e.* 283

20. *a.* 376   *b.* 389   *c.* 432   *d.* 473   *e.* 498

21. *a.* 17th   *b.* 39th   *c.* 53rd   *d.* 79th   *e.* 91st

23. 82   25. 47

27. 2.1   29. 12

31. *a.* 12; 20.5; 32; 22; 20   *b.* 62; 94; 99; 80.5; 37

Exercises 3–5

1. 6, 8, 19, 32, 54; 24

3. 188, 192, 339, 437, 589; 245

5. 14.6, 15.05, 16.3, 19, 19.8; 3.95

7. 11, 3, 8, 5, 9, 4

9. 95, 55, 70, 65, 90, 25

11. The distribution is positively skewed.

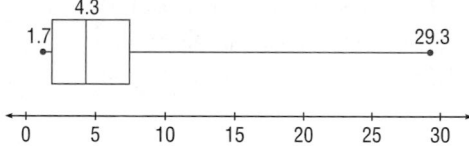

13.

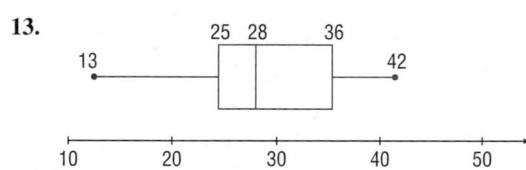

15. The distribution is positively skewed.

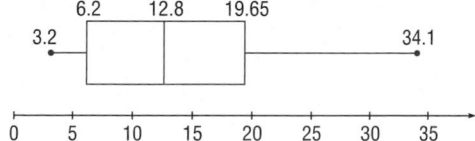

17. *a.* May   *b.* 1999

c.

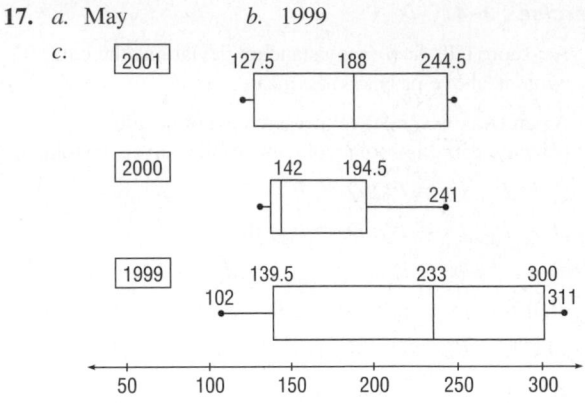

The data for the year 2000 is the least variable and has the smallest median.

Review Exercises

1. *a.* 109.9   *c.* 114   *e.* 169   *g.* 44.8
    b. 97   *d.* 144.5   *f.* 2007.1

3. *a.* 7.3   *b.* 7–9   *c.* 10.0   *d.* 3.2

5. *a.* 55.5   *b.* 57.5–72.5   *c.* 566.1   *d.* 23.8

7. 1.1   **9.** 6

11. Magazine variance: 0.214; year variance: 0.417; years are more variable

13. *a.*

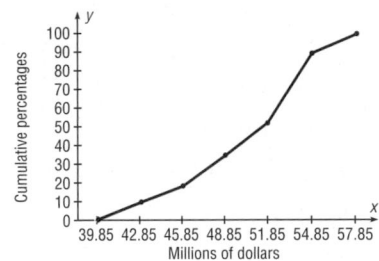

b. 49, 52, 53 (answers are approximate)

c. 15th; 33rd; 91st (answers are approximate)

15. $0.26–$0.38   **17.** 56%

19. 88.89%

21. The employees worked more hours before Christmas than after Christmas. Also, the range and variability of the distribution of hours are greater before Christmas.

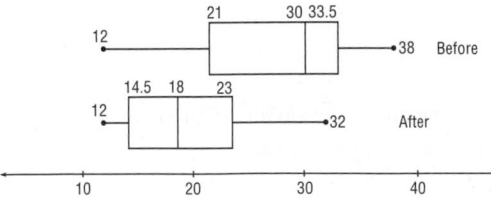

Chapter Quiz

1. True   **2.** True

3. False   **4.** False

5. False   **6.** False

7. False   **8.** False

9. False   **10.** *c*

11. *c*   **12.** *a* and *b*

13. *b*   **14.** *d*

15. *b*   **16.** Statistic

17. Parameters, statistics   **18.** Standard deviation

19. σ **20.** Midrange

21. Positively   **22.** Outlier

23. *a.* 15.3   *c.* 15, 16, and 17   *e.* 6   *g.* 1.9
    b. 15.5   *d.* 15   *f.* 3.61

24. *a.* 6.4   *b.* 6–8   *c.* 11.6   *d.* 3.4

25. *a.* 51.4   *b.* 35.5–50.5   *c.* 451.5   *d.* 21.2

26. *a.* 8.2   *b.* 7–9   *c.* 21.6   *d.* 4.6

27. 1.6   **28.** 4.5

29. 0.33; 0.162; newspapers   **30.** 0.3125; 0.229; brands

31. −0.75; −1.67; science

32. *a.* 0.5   *b.* 1.6   *c.* 15, *c* is highest

33. *a.* 56.25; 43.75; 81.25; 31.25; 93.75; 18.75; 6.25; 68.75
    b. 0.9
    c.

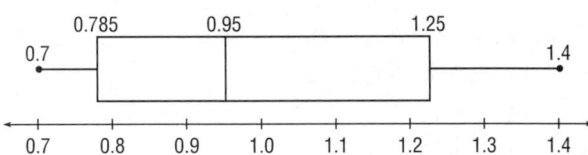

34. *a.*

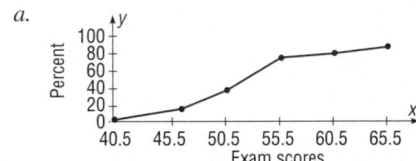

b. 47; 53; 65

c. 60th, 6th, 98th percentiles

35. The cost of prebuy gas is much less than that of the return without filling gas. The variability of the return without filling gas is larger than the variability of the prebuy gas.

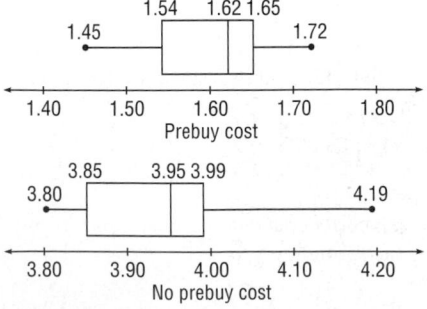

36. 16%, 97.5%

Chapter 4

Exercises 4–2

1. A probability experiment is a chance process that leads to well-defined outcomes.

3. An outcome is the result of a single trial of a probability experiment, but an event can consist of more than one outcome.

5. The range of values is 0 to 1 inclusive.

7. 0

9. 0.80 Since the probability that it won't rain is 80%, you could leave your umbrella at home and be fairly safe.

11. *a.* Empirical   *d.* Classical   *f.* Empirical
    b. Classical   *e.* Empirical   *g.* Subjective
    c. Empirical

12. *a.* $\frac{1}{6}$ *c.* $\frac{1}{3}$ *e.* 1   *g.* $\frac{1}{6}$
    b. $\frac{1}{2}$ *d.* 1   *f.* $\frac{5}{6}$

13. *a.* $\frac{5}{36}$ *b.* $\frac{1}{6}$ *c.* $\frac{2}{9}$ *d.* $\frac{1}{6}$ *e.* $\frac{1}{6}$

14. *a.* $\frac{1}{13}$ *c.* $\frac{1}{52}$ *e.* $\frac{4}{13}$ *g.* $\frac{1}{2}$ *i.* $\frac{7}{13}$
    b. $\frac{1}{4}$ *d.* $\frac{2}{13}$ *f.* $\frac{4}{13}$ *h.* $\frac{1}{26}$ *j.* $\frac{1}{26}$

15. *a.* $\frac{1}{6}$ *b.* $\frac{1}{2}$ *c.* $\frac{1}{2}$

17. *a.* 0.44   *b.* 0.04   *c.* 0.56

19. *a.* 0.26, or 26%   *b.* 0.29, or 29%   *c.* 0.9, or 90%
    d. The event in part *c* is most likely to occur since it has the highest probability of occurring.

21. *a.* $\frac{1}{8}$ *b.* $\frac{1}{4}$ *c.* $\frac{3}{4}$ *d.* $\frac{3}{4}$

23. $\frac{1}{9}$

25. *a.* $\frac{9}{19}$ *b.* $\frac{9}{38}$ *c.* $\frac{5}{38}$
    d. The event in part *a* is most likely to occur since it has the highest probability of occurring.

27. 0.54

29. *a.* Sample space

	1	2	3	4	5	6
1	1	2	3	4	5	6
2	2	4	6	8	10	12
3	3	6	9	12	15	18
4	4	8	12	16	20	24
5	5	10	15	20	25	30
6	6	12	18	24	30	36

    b. $\frac{5}{12}$ *c.* $\frac{17}{36}$

31.

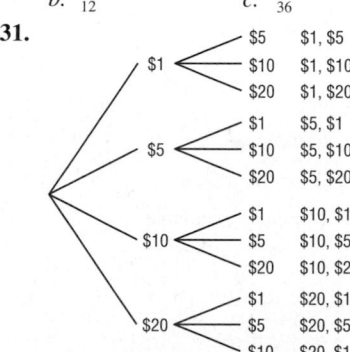

33.

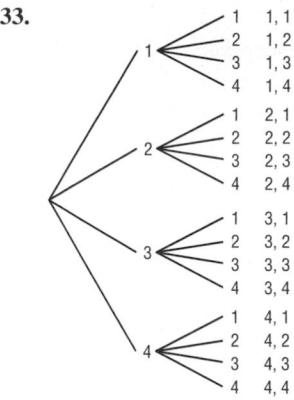

35.

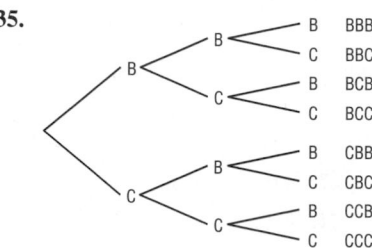

37. *a.* 0.08   *b.* 0.01   *c.* 0.35   *d.* 0.36

39. The statement is probably not based on empirical probability, and is probably not true.

41. Answers will vary.

43. *a.* 1:5, 5:1   *d.* 1:1, 1:1   *g.* 1:1, 1:1
    b. 1:1, 1:1   *e.* 1:12, 12:1
    c. 1:3, 3:1   *f.* 1:3, 3:1

Exercises 4–3

1. Two events are mutually exclusive if they cannot occur at the same time (i.e., they have no outcomes in common). Examples will vary.

3. $\frac{1}{6}$; $\frac{61}{365}$ (excluding leap year) Using the days makes the probability slightly larger than using months.

5. $\frac{11}{19}$

7. *a.* $\frac{7}{25}$ *b.* $\frac{3}{8}$ *c.* $\frac{17}{100}$
    d. The event in part *b* has the highest probability so it is most likely to occur.

9. 0.55

11. *a.* $\frac{6}{7}$ *b.* $\frac{4}{7}$ *c.* 1

13. *a.* 0.7   *b.* 0.0667   *c.* 0.7667

15. *a.* $\frac{38}{45}$ *b.* $\frac{22}{45}$ *c.* $\frac{2}{3}$

17. *a.* $\frac{14}{31}$ *b.* $\frac{23}{31}$ *c.* $\frac{19}{31}$

19. *a.* $\frac{1}{15}$ *b.* $\frac{1}{3}$ *c.* $\frac{5}{6}$ *d.* $\frac{5}{6}$ *e.* $\frac{1}{3}$

21. *a.* $\frac{5}{12}$ *b.* $\frac{1}{8}$ *c.* $\frac{2}{3}$ *d.* $\frac{23}{24}$

23. *a.* $\frac{3}{13}$ *b.* $\frac{3}{4}$ *c.* $\frac{19}{52}$ *d.* $\frac{7}{13}$ *e.* $\frac{15}{26}$

25. $\frac{7}{10}$ **27.** 0.06

29. 0.30

Exercises 4–4

1. *a.* Independent *e.* Independent
   b. Dependent   *f.* Dependent
   c. Dependent   *g.* Dependent
   d. Dependent   *h.* Independent

3. 0.706; the event is likely to occur since the probability is greater than 0.5.

5. 0.373; the event is unlikely to occur since the probability is less than 0.5.

7. *a.* 0.0954   *b.* 0.9046   *c.* 0.1601

9. 0.5139   11. 0.0298

13. 0.0002   15. $\frac{243}{1024}$

17. $\frac{5}{28}$

19. 0.210; the event is unlikely to occur since the probability is less than 0.50.

21. 0.116   23. 0.03

25. $\frac{49}{72}$ 27. 0.6

29. 89%   31. 70%

33. 82%

35. *a.* 0.4712   *b.* 0.0786

37. *a.* 0.1717   *b.* 0.8283

39. 0.8073   41. 0.9869

43. $\frac{14,498}{20,825}$ 45. 26.6%

47. $\frac{31}{32}$

49. 0.721; the event is likely to occur since the probability is about 72%.

51. $\frac{7}{8}$

53. No, since $P(A \cap B) = 0$ and does not equal $P(A) \cdot P(B)$.

55. Enrollment and meeting with DW and meeting with MH are dependent. Since meeting with MH has a low probability and meeting with LP has no effect, all students, if possible, should meet with DW.

Exercises 4–5

1. 100,000; 30,240   3. 5040

5. 40,320   7. 120

9. 1000; 72   11. 10

13. *a.* 40,320   *c.* 1   *e.* 2520   *g.* 60   *i.* 120
    b. 3,628,800   *d.* 1   *f.* 11,880   *h.* 1   *j.* 30

15. 24   17. 120

19. 840   21. 151,200

23. 5,527,200   25. 300

27. *a.* 10   *c.* 35   *e.* 15   *g.* 1   *i.* 66
    b. 56   *d.* 15   *f.* 1   *h.* 36   *j.* 4

29. 120   31. 210

33. 15,504   35. 3080

37. 495; 210; 420   39. 200

41. 2970   43. 136

45. 330   47. 125,970

49. 15

51. *a.* 48   *b.* 60   *c.* 72

Exercises 4–6

1. $\frac{11}{221}$

3. *a.* $\frac{4}{35}$ *b.* $\frac{1}{35}$ *c.* $\frac{12}{35}$ *d.* $\frac{18}{35}$

5. 0.0003; 0.0021   7. $\frac{1}{1225}$

9. *a.* $\frac{10}{143}$ *b.* $\frac{60}{143}$ *c.* $\frac{15}{1001}$ *d.* $\frac{160}{1001}$ *e.* $\frac{48}{143}$

11. *a.* 0.3216   *b.* 0.1637   *c.* 0.5146
    d. It probably got lost in the wash!

13. $\frac{5}{72}$ 15. $\frac{1}{60}$

Review Exercises

1. *a.* $\frac{1}{6}$ *b.* $\frac{1}{6}$ *c.* $\frac{2}{3}$

3. $\frac{16}{45}$ 5. $\frac{17}{30}$

7. *a.* $\frac{1}{10}$ *b.* $\frac{11}{30}$ *c.* $\frac{13}{15}$ *d.* $\frac{13}{15}$

9. 0.98   11. 28.9%

13. *a.* $\frac{2}{17}$ *b.* $\frac{11}{850}$ *c.* $\frac{1}{5525}$

15. $\frac{5}{13}$ 17. 0.4

19. 0.51   21. 57.3%

23. *a.* $\frac{19}{44}$ *b.* $\frac{1}{4}$ 25. $\frac{31}{32}$

27. 175,760,000; 78,624,000; 88,583,040

29. 350   31. 45

33. 26,000   35. 495

37. 15,504   39. 175,760,000; 0.0000114

41. $\frac{2}{7}$

Chapter Quiz

1. False   2. False

3. True   4. False

5. False   6. False

7. True   8. False

9. *b*   10. *b* and *d*

11. *d*   12. *b*

13. *c*   14. *b*

15. *d*   16. *b*

17. *b*   18. Sample space

19. 0, 1   20. 0

21. 1   22. Mutually exclusive

23. *a.* $\frac{1}{13}$ *b.* $\frac{1}{13}$ *c.* $\frac{4}{13}$

24. *a.* $\frac{1}{4}$ *b.* $\frac{4}{13}$ *c.* $\frac{1}{52}$ *d.* $\frac{1}{13}$ *e.* $\frac{1}{2}$

25. *a.* $\frac{12}{31}$ *b.* $\frac{12}{31}$ *c.* $\frac{27}{31}$ *d.* $\frac{24}{31}$

26. *a.* $\frac{11}{36}$ *b.* $\frac{5}{18}$ *c.* $\frac{11}{36}$ *d.* $\frac{1}{3}$ *e.* 0 *f.* $\frac{11}{12}$

27. 0.68   **28.** 0.002

29. *a.* $\frac{253}{9996}$ *b.* $\frac{33}{66,640}$ *c.* 0

30. 0.54   **31.** 0.53

32. 0.81   **33.** 0.056

34. *a.* $\frac{1}{2}$ *b.* $\frac{3}{7}$

35. 0.99   **36.** 0.518

37. 0.9999886   **38.** 2646

39. 40,320   **40.** 1365

41. 1,188,137,600; 710,424,000

42. 720   **43.** 33,554,432

44. 56   **45.** $\frac{1}{4}$

46. $\frac{3}{14}$ **47.** $\frac{12}{55}$

48.

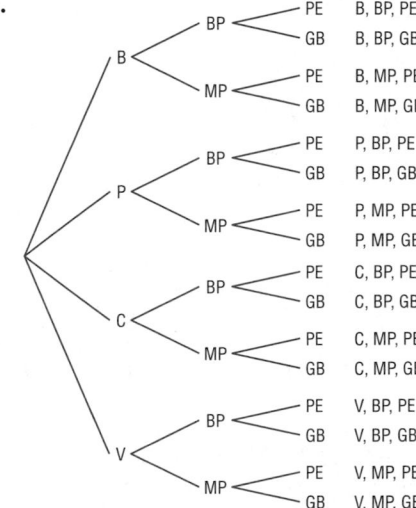

Chapter 5

Exercises 5–2

1. A random variable is a variable whose values are determined by chance. Examples will vary.

3. The number of commercials a radio station plays during each hour. The number of times a student uses his or her calculator during a mathematics exam. The number of leaves on a specific type of tree.

5. A probability distribution is a distribution that consists of the values a random variable can assume along with the corresponding probabilities of these values.

7. No; probabilities cannot be negative and the sum of the probabilities is not 1.

9. Yes

11. No, the probability values cannot be greater than 1.

13. Discrete   **15.** Continuous

17. Discrete

19.

X	0	1	2	3
$P(X)$	$\frac{6}{15}$	$\frac{5}{15}$	$\frac{3}{15}$	$\frac{1}{15}$

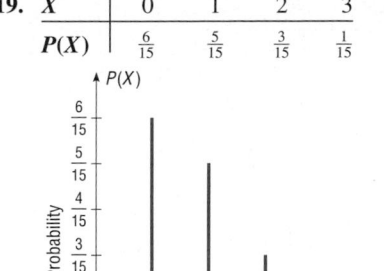
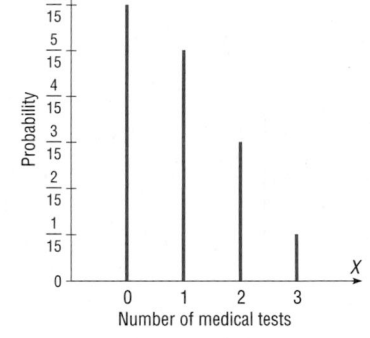

21.

X	0	1	2	3	4	5
$P(X)$	0.75	0.17	0.04	0.025	0.01	0.005

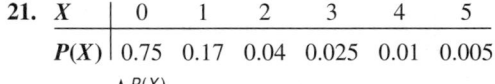
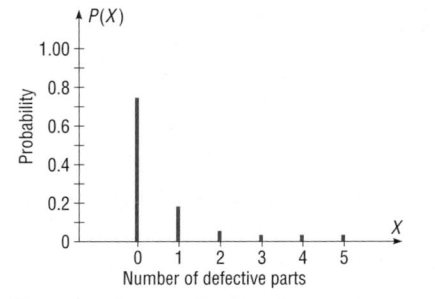

23.

X	1	2	3	4	5	6
$P(X)$	$\frac{1}{2}$	$\frac{1}{6}$	$\frac{1}{12}$	$\frac{1}{12}$	$\frac{1}{12}$	$\frac{1}{12}$

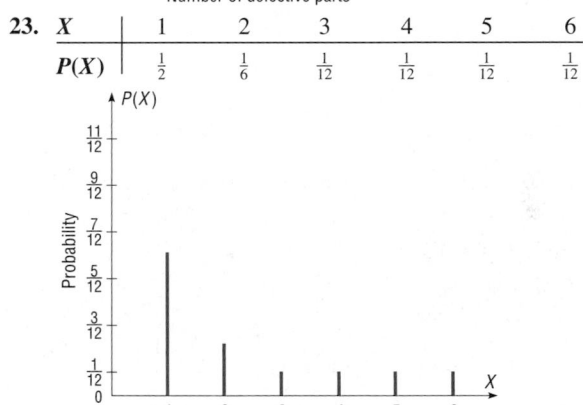
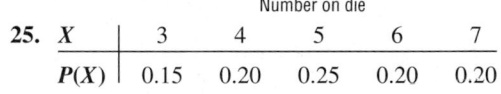

25.

X	3	4	5	6	7
$P(X)$	0.15	0.20	0.25	0.20	0.20

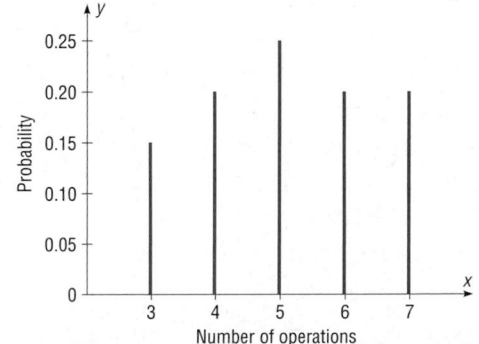

27.

X	$1	$5	$10	$20
P(X)	$\frac{2}{9}$	$\frac{1}{3}$	$\frac{1}{9}$	$\frac{1}{3}$

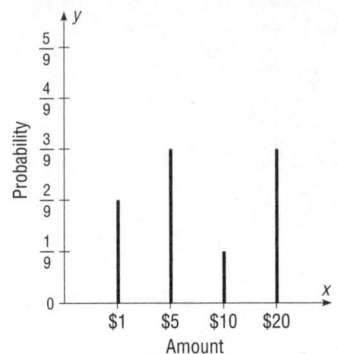

29.

X	1	2	3	4
P(X)	$\frac{1}{4}$	$\frac{1}{4}$	$\frac{3}{8}$	$\frac{1}{8}$

31.

X	1	2	3
P(X)	$\frac{1}{6}$	$\frac{1}{3}$	$\frac{1}{2}$

Yes

33.

X	3	4	7
P(X)	$\frac{3}{6}$	$\frac{4}{6}$	$\frac{7}{6}$

No; The sum of probabilities is greater than one

35.

X	1	2	4
P(X)	$\frac{1}{7}$	$\frac{2}{7}$	$\frac{4}{7}$

Yes

Exercises 5–3

1. 0.2; 0.3; 0.6; 2

3. 1.3, 0.9, 1. No; on average, each person has about one credit card.

5. 2.0; 1.6; 1.3; $200   **7.** 6.6; 1.3; 1.1

9. 13.9; 1.3; 1.1   **11.** $260

13. $0.83   **15.** −$1.00

17. −$0.50, −$0.52   **19.** $39,000; yes

21. 10.5   **23.** Answers will vary.

25. Answers will vary.

Exercises 5–4

1. *a.* Yes *c.* Yes *e.* No *g.* Yes *i.* No
    b. Yes *d.* No *f.* Yes *h.* Yes *j.* Yes

2. *a.* 0.420 *c.* 0.590 *e.* 0.000 *g.* 0.418 *i.* 0.246
    b. 0.346 *d.* 0.251 *f.* 0.250 *h.* 0.176

3. *a.* 0.0005 *c.* 0.342 *e.* 0.173
    b. 0.131 *d.* 0.007

5. 0.021; no, it's only about a 2% chance.

7. 0.267

9. 0.071

11. *a.* 0.346 *b.* 0.913 *c.* 0.663 *d.* 0.683

13. *a.* 0.121 *b.* 0.088 *c.* 0.967
    d. Event *c* is most likely to occur since it has the highest probability.

14. *a.* 75; 18.8; 4.3   *e.* 100; 90; 9.5
    b. 90; 63; 7.9   *f.* 125; 93.8; 9.7
    c. 10; 5; 2.2   *g.* 20; 12; 3.5
    d. 8; 1.6; 1.3   *h.* 6; 5; 2.2

15. 8; 7.9; 2.8   **17.** 9; 8.73; 2.95

19. 210; 165.9; 12.9   **21.** 0.199

23. 0.559   **25.** 0.018

27. 0.770; yes. The probability is high, 77%.

29.

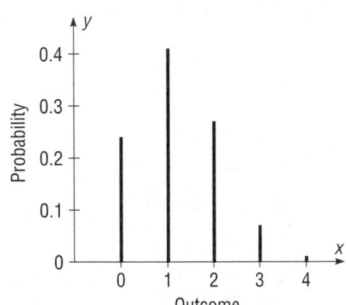

Exercises 5–5

1. *a.* 0.135 *c.* 0.0096 *e.* 0.0112
    b. 0.0324 *d.* 0.18

3. 0.06

5. $\frac{1}{108}$

7. *a.* 0.1563 *c.* 0.0504 *e.* 0.1241
    b. 0.1465 *d.* 0.071

9. *a.* 0.0183 *b.* 0.0733 *c.* 0.1465 *d.* 0.7619

11. 0.3554   **13.** 0.9502

15. 0.1563   **17.** 0.38

19. 0.13   **21.** 0.597

Review Exercises

1. Yes

3. No; the sum of the probabilities is greater than 1.

5.

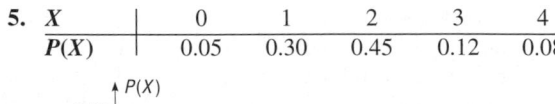

X	0	1	2	3	4
P(X)	0.05	0.30	0.45	0.12	0.08

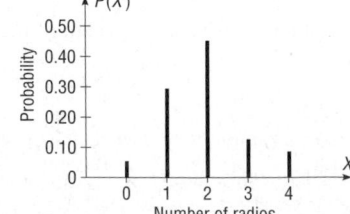

7.

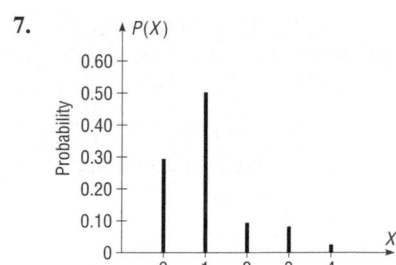

9. 15.2; 1.6; 1.3

11. 24.2; 1.5; 1.2

13. $7.23; $7.23

15. *a.* 0.122   *b.* 0.989   *c.* 0.043

17. 135; 33.8; 5.8

19. 0.886

21. 0.190

23. 0.008

25. 0.050

27. *a.* 0.5543   *b.* 0.8488   *c.* 0.4457

29. 0.27

31. *a.* $\frac{21}{44}$ *b.* $\frac{1}{22}$ *c.* $\frac{7}{22}$

Chapter Quiz

1. True   **2.** False

3. False   **4.** True

5. chance   **6.** $n \cdot p$

7. 1   **8.** *c*

9. *c*   **10.** *d*

11. No, since $\Sigma P(X) > 1$ **12.** Yes

13. Yes   **14.** Yes

15.

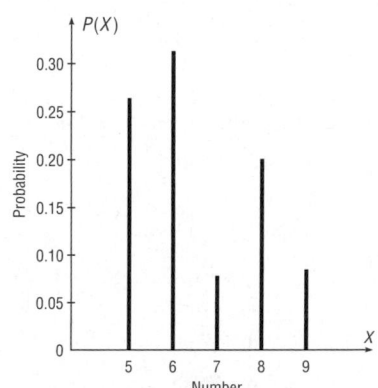

16.

X	0	1	2	3	4
P(X)	0.02	0.3	0.48	0.13	0.07

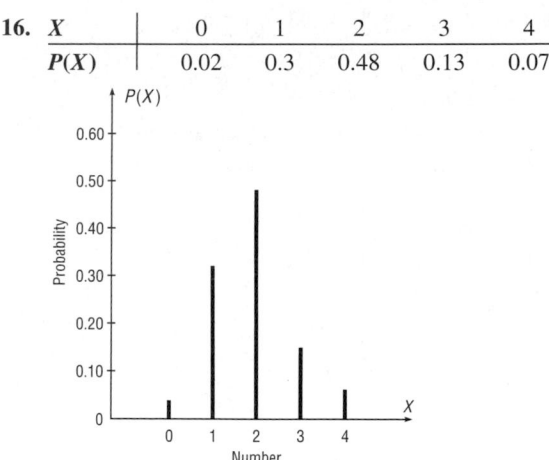

17. 2.0; 1.3; 1.1   **18.** 32.2; 1.1; 1.0

19. 5.2   **20.** $9.65

21. 0.124

22. *a.* 0.075   *b.* 0.872   *c.* 0.125

23. 240; 48; 6.9   **24.** 9; 7.9; 2.8

25. 0.008   **26.** 0.0003

27. 0.061   **28.** 0.122

29. *a.* 0.5470   *b.* 0.9863   *c.* 0.4529

30. 0.128

31. *a.* 0.160   *b.* 0.42   *c.* 0.07

Chapter 6

Exercises 6–3

1. The characteristics of the normal distribution are as follows:

   a. It is bell-shaped.

   b. It is symmetric about the mean.

   c. Its mean, median, and mode are equal.

   d. It is continuous.

   e. It never touches the *x* axis.

   f. The area under the curve is equal to 1.

   g. It is unimodal.

3. 1, or 100%   **5.** 68%; 95%; 99.7%

7. 0.2734   **9.** 0.4808

11. 0.4090   **13.** 0.0764

15. 0.1145   **17.** 0.0258

19. 0.8417   **21.** 0.9826

23. 0.5596   **25.** 0.3574

27. 0.2486   **29.** 0.4418

31. 0.0023   **33.** 0.1131

35. 0.9522   **37.** 0.0706

39. 0.9222   **41.** −1.94

43. -2.13 **45.** -1.26

47. *a.* -2.28 *b.* -0.92 *c.* -0.27

49. *a.* $z = +1.96$ and $z = -1.96$

    b. $z = +1.65$ and $z = -1.65$, approximately

    c. $z = +2.58$ and $z = -2.58$, approximately

51. 0.6827; 0.9545; 0.9973; they are very close.

53. 2.10 **55.** -1.45 and 0.11

57. $y = \dfrac{e^{-X^2/2}}{\sqrt{2\pi}}$

Exercises 6–4

1. 0.0012

3. *a.* 0.0516 *b.* 0.3503

5. *a.* 0.3023 *b.* 0.0062

7. *a.* 0.3557 *b.* 0.8389

9. *a.* 0.0012 *b.* 0.3432 *c.* 0.500

    d. About 0.12% of the microwave ovens would be replaced.

11. *a.* 0.9803 *b.* 0.2514 *c.* 0.3434

13. *a.* 0.9699 *b.* 0.8264

    c. Use the range rule of thumb. The range is about $4 \times 16 = 64$ minutes.

15. *a.* 0.3281 *b.* 0.4002 *c.* Not usually

17. $\$5518.25 < \mu < \$7465.75$; yes

19. The maximum size is 1927.76 square feet; the minimum size is 1692.24 square feet.

21. 0.006; $\$821$

23. The maximum price is $9222, and the minimum price is $7290.

25. 4.13; 7.04 **27.** $\$18,840.48$

29. 18.6 months

31. *a.* $\mu = 120, \sigma = 20$ *b.* $\mu = 15, \sigma = 2.5$

    c. $\mu = 30, \sigma = 5$

33. There are several mathematics tests that can be used.

35. 1.05 **37.** $\mu = 45, \sigma = 1.34$

39. Not normal   **41.** Not normal

Exercises 6–5

1. The distribution is called the sampling distribution of sample means.

3. The mean of the sample means is equal to the population mean.

5. The distribution will be approximately normal when the sample size is large.

7. $z = \dfrac{\bar{X} - \mu}{\sigma / \sqrt{n}}$ **9.** 0.0045

11. 0.1112; no, since the average weight of the group is within 2 standard deviations (standard errors) of the mean.

13. 0.0571

15. *a.* 0.3859 *b.* 0.1841

    c. Individual values are more variable than means.

17. 0.4176 **19.** 0.1254

21. *a.* 0.3446 *b.* 0.0023

    c. Yes, since it is within 1 standard deviation of the mean.

    d. Very unlikely

23. *a.* 0.3707 *b.* 0.0475

25. 0.9153 **27.** 0.0143

29. $\sigma_{\bar{X}} = 1.5, n = 25$

Exercises 6–6

1. When p is approximately 0.5, as n increases, the shape of the binomial distribution becomes similar to that of the normal distribution. The conditions are that $n \cdot p$ and $n \cdot q$ are both ≥ 5. The correction is necessary because the normal distribution is continuous and the binomial distribution is discrete.

2. *a.* 0.0811 *c.* 0.1052 *e.* 0.2327

    b. 0.0516 *d.* 0.1711 *f.* 0.9988

3. *a.* Yes   *c.* No   *e.* Yes

    b. No   *d.* Yes   *f.* No

5. 0.166 **7.** 0.0043

9. 0.0793 **11.** 0.4761

13. 0.9951; yes

Review Exercises

1. *a.* 0.4744 *e.* 0.2139 *i.* 0.0183

    b. 0.1443 *f.* 0.8284 *j.* 0.9535

    c. 0.0590 *g.* 0.0233

    d. 0.8329 *h.* 0.9131

3. *a.* 0.5000 *b.* 0.1515

    c. Actually about 85% of the auto mechanics would be making more than $25,000.

5. 40.13%; 12.92%

7. *a.* 0.7054 *b.* 0.8869

9. 92.2–107.8

11. 0.0023; yes, since the probability is less than 1%.

13. 0.2033 **15.** 0.0465

17. Not normal

Chapter Quiz

1. False   **2.** True

3. True   **4.** True

5. False   **6.** False

7. *a*

9. *b*

11. *c*

13. Sampling error

14. The population mean

15. Standard error of the mean

16. 5

17. 5%

18. *a.* 0.4332   *d.* 0.1029   *g.* 0.0401   *j.* 0.9131
    b. 0.3944   *e.* 0.2912   *h.* 0.8997
    c. 0.0344   *f.* 0.8284   *i.* 0.017

19. *a.* 0.4846   *d.* 0.0188   *g.* 0.0089   *j.* 0.8461
    b. 0.4693   *e.* 0.7461   *h.* 0.9582
    c. 0.9334   *f.* 0.0384   *i.* 0.9788

20. *a.* 0.7734   *b.* 0.0516   *c.* 0.3837
    d. Any rainfall above 65 inches could be considered an extremely wet year since this value is 2 standard deviations above the mean.

21. *a.* 0.0668   *b.* 0.0228   *c.* 0.4649   *d.* 0.0934

22. *a.* 0.4525   *b.* 0.3707   *c.* 0.3707   *d.* 0.019

23. *a.* 0.0013   *b.* 0.5   *c.* 0.0081   *d.* 0.5511

24. *a.* 0.0037   *b.* 0.0228   *c.* 0.5   *d.* 0.3232

25. 8.804 centimeters

26. 121.24 is the lowest acceptable score.

27. 0.015   **28.** 0.9738

29. 0.0495; no   **30.** 0.0630

31. 0.8577   **32.** 0.0495

33. Not normal   **34.** Approximately normal

Chapter 7

Exercises 7–2

1. A point estimate of a parameter specifies a particular value, such as $\mu = 87$; an interval estimate specifies a range of values for the parameter, such as $84 < \mu < 90$. The advantage of an interval estimate is that a specific confidence level (say 95%) can be selected, and one can be 95% confident that the interval contains the parameter that is being estimated.

3. The maximum error of estimate is the likely range of values to the right or left of the statistic which may contain the parameter.

5. A good estimator should be unbiased, consistent, and relatively efficient.

7. For one to be able to determine sample size, the maximum error of estimate and the degree of confidence must be specified and the population standard deviation must be known.

8. *a*

9. *a.* 2.58   *c.* 1.96   *e.* 1.88
    b. 2.33   *d.* 1.65

10. *b*

11. *a.* 82   *b.* $77 < \mu < 87$ *c.* $75 < \mu < 89$
    d. The 99% confidence interval is larger because the confidence level is larger.

12. 0.5

13. *a.* 12.6   *b.* $11.9 < \mu < 13.3$
    c. It would be highly unlikely, since this is far larger than 13.3.

15. $145,030 < \mu < 154,970$

17. $4913 < \mu < 5087$; 4000 hours does not seem reasonable since it is outside the interval.

19. $59.5 < \mu < 62.9$ **21.** 114

23. 25   **25.** 147

Exercises 7–3

1. The characteristics of the *t* distribution are: It is bell-shaped, it is symmetric about the mean, and it never touches the *x* axis. The mean, median, and mode are equal to 0 and are located at the center of the distribution. The variance is greater than 1. The *t* distribution is a family of curves based on degrees of freedom. As a sample size increases, the *t* distribution approaches the standard normal distribution.

3. The *t* distribution should be used when σ is unknown and $n < 30$.

4. *a.* 2.898   *c.* 2.624   *e.* 2.093
    b. 2.074   *d.* 1.833

5. $15 < \mu < 17$

7. $\bar{X} = 33.4$; $s = 28.7$; $21.2 < \mu < 45.6$; the point estimate is 33.4, and it is close to 32. Also, the interval does indeed contain $\mu = 32$. The data value 132 is unusually large (an outlier). The mean may not be the best estimate in this case.

9. $266 < \mu < 286$; it is highly unlikely.

11. $13.5 < \mu < 15.1$; about 30 minutes.

13. $\$17.29 < \mu < \$19.77$

15. $109 < \mu < 121$

17. $38.8 < \mu < 44.4$

19. $56,466 < \mu < 56,970$; $56,466

21. $\bar{X} = 2.175$; $s = 0.585$; $\mu > \$1.95$ means one can be 95% confident that the mean revenue is greater than $1.95; $\mu < \$2.40$ means one can be 95% confident that the mean revenue is less than $2.40.

Exercises 7–4

1. *a.* 0.5, 0.5   *c.* 0.46, 0.54   *e.* 0.45, 0.55
    b. 0.45, 0.55   *d.* 0.58, 0.42

2. *a.* $\hat{p} = 0.15$; $\hat{q} = 0.85$ *d.* $\hat{p} = 0.51$; $\hat{q} = 0.49$
    b. $\hat{p} = 0.37$; $\hat{q} = 0.63$ *e.* $\hat{p} = 0.79$; $\hat{q} = 0.21$
    c. $\hat{p} = 0.71$; $\hat{q} = 0.29$

3. $0.365 < p < 0.415$

5. $0.092 < p < 0.153$; 11% is contained in the confidence interval

7. $0.797 < p < 0.883$ 9. $0.153 < p < 0.307$

11. $0.286 < p < 0.562$. It would not be considered somewhat larger since 0.52 is in the interval.

13. $0.419 < p < 0.481$

15. *a.* 3121   *b.* 4161

17. *a.* 99   *b.* 273

19. 1068   21. 95%

Exercises 7–5

1. Chi-square

3. *a.* 3.816; 21.920   *d.* 0.412; 16.750
   b. 10.117; 30.144   *e.* 26.509; 55.758
   c. 13.844; 41.923

5. $8.64 < \sigma^2 < 28.93$
   $2.94 < \sigma < 5.38$
   Yes, it takes between 3 and 5 minutes.

7. $0.40 < \sigma^2 < 2.25$
   $0.63 < \sigma < 1.50$

9. $232.1 < \sigma^2 < 691.6$; $15.2 < \sigma < 26.3$

11. $4.1 < \sigma < 7.1$

13. $16.2 < \sigma < 19.8$

Review Exercises

1. 7.8; $7.64 < \mu < 7.97$; about 7.64 miles per hour

3. 7.5; $7.46 < \mu < 7.54$

5. $25 < \mu < 31$

7. 28

9. $0.434 < p < 0.660$; yes, it looks as if up to 66% of the people are dissatisfied.

11. 460

13. $0.218 < \sigma < 0.435$. Yes. It seems that there is a large standard deviation.

15. $5.1 < \sigma^2 < 18.3$

Chapter Quiz

1. True   2. True

3. False   4. True

5. *b*   6. *a*

7. *b*

8. Unbiased, consistent, relatively efficient

9. Maximum error of estimate

10. Point   11. 90; 95; 99

12. $23.45; $22.79 < \mu < 24.11

13. $44.80; $43.15 < \mu < 46.45

14. 4150; $3954 < \mu < 4346$

15. $45.7 < \mu < 51.5$

16. $418 < \mu < 458$ 17. $26 < \mu < 36$

18. 180   19. 25

20. $0.604 < p < 0.810$ 21. $0.295 < p < 0.425$

22. $0.342 < p < 0.547$ 23. 545

24. $7 < \sigma < 13$

25. $30.9 < \sigma^2 < 78.2$ 26. $1.8 < \sigma < 3.2$
   $5.6 < \sigma < 8.8$

Chapter 8

Exercises 8–2

Note: For Chapters 8–13, specific *P*-values are given in parentheses after the *P*-value intervals. When the specific *P*-value is extremely small, it is not given.

1. The null hypothesis states that there is no difference between a parameter and a specific value or that there is no difference between two parameters. The alternative hypothesis states that there is a specific difference between a parameter and a specific value or that there is a difference between two parameters. Examples will vary.

3. A statistical test uses the data obtained from a sample to make a decision about whether the null hypothesis should be rejected.

5. The critical region is the range of values of the test statistic that indicates that there is a significant difference and the null hypothesis should be rejected. The noncritical region is the range of values of the test statistic that indicates that the difference was probably due to chance and the null hypothesis should not be rejected.

7. α, β

9. A one-tailed test should be used when a specific direction, such as greater than or less than, is being hypothesized; when no direction is specified, a two-tailed test should be used.

11. Hypotheses can be proved true only when the entire population is used to compute the test statistic. In most cases, this is impossible.

12. *a.* ±1.96   *d.* +2.33   *g.* +1.65   *i.* −1.75
   b. −2.33   *e.* −1.65   *h.* ±2.58   *j.* +2.05
   c. +2.58   *f.* −2.05

13. *a.* $H_0: \mu = 24.6$ and $H_1: \mu \neq 24.6$
   b. $H_0: \mu = \$51,497$ and $H_1: \mu \neq \$51,497$
   c. $H_0: \mu \leq 25.4$ and $H_1: \mu > 25.4$
   d. $H_0: \mu \geq 88$ and $H_1: \mu < 88$
   e. $H_0: \mu \geq 70$ and $H_1: \mu < 70$
   f. $H_0: \mu = \$79.95$ and $H_1: \mu \neq \$79.95$
   g. $H_0: \mu = 8.2$ and $H_1: \mu \neq 8.2$

Exercises 8–3

1. H_0: $\mu = \$69.21$ (claim) and H_1: $\mu \neq \$69.21$; C.V. $= \pm1.96$; $z = -1.15$; do not reject. There is not enough evidence to reject the claim that the average cost of a hotel stay in Atlanta is $69.21.

3. H_0: $\mu \leq \$24$ billion and H_1: $\mu > \$24$ billion (claim); C.V. $= 1.65$; $z = 1.85$; reject. There is enough evidence to support the claim that the average revenue is greater than $24 billion.

5. H_0: $\mu \geq 14$ and H_1: $\mu < 14$ (claim); C.V. $= -2.33$; $z = -4.89$; reject. There is enough evidence to support the claim that the average age of the planes in the executive's airline is less than the national average.

7. H_0: $\mu = 29$ and H_1: $\mu \neq 29$ (claim); C.V. $= \pm1.96$; $z = 0.944$; do not reject; there is not enough evidence to say that the average height differs from 29 inches.

9. H_0: $\mu \leq \$19,410$ and H_1: $\mu > \$19,410$ (claim); C.V. $= 2.33$; $z = 2.81$; reject. There is enough evidence to support the claim that the cost of attendance has increased.

11. H_0: $\mu = 125$ and H_1: $\mu \neq 125$ (claim); C.V. $= \pm2.58$; $z = -2.96$; reject. There is enough evidence to support the claim that the average number of guests differs from 125.

13. H_0: $\mu = \$24.44$ and H_1: $\mu \neq \$24.44$ (claim); C.V. $= \pm2.33$; $z = -2.28$; do not reject. There is not enough evidence to support the claim that the amount spent at a local mall is not equal to the national average of $24.44.

15. a. Do not reject. d. Reject.
b. Reject. e. Reject.
c. Do not reject.

17. H_0: $\mu \geq 264$ and H_1: $\mu < 264$ (claim); $z = -2.53$; P-value $= 0.0057$; reject. There is enough evidence to support the claim that the average stopping distance is less than 264 ft.

19. H_0: $\mu \geq 546$ and H_1: $\mu < 546$ (claim); $z = -2.4$; P-value $= 0.008$. Yes, it can be concluded that the number of calories burned is less than originally thought.

21. H_0: $\mu \geq 47.1$ and H_1: $\mu < 47.1$ (claim); $z = -3.20$; P-value < 0.01; reject. There is enough evidence to support the claim that the average farm size is smaller. Yes, they should consider updating their information.

23. H_0: $\mu = 30,000$ (claim) and H_1: $\mu \neq 30,000$; $z = 1.71$; P-value $= 0.0872$; reject. There is enough evidence to reject the claim that the customers are adhering to the recommendation. Yes, the 0.10 level is appropriate.

25. H_0: $\mu \geq 10$ and H_1: $\mu < 10$ (claim); $z = -8.67$; P-value < 0.0001; since P-value < 0.05, reject. Yes, there is enough evidence to support the claim that the average number of days missed per year is less than 10.

27. H_0: $\mu = 8.65$ (claim) and H_1: $\mu \neq 8.65$; C.V. $= \pm1.96$; $z = -1.35$; do not reject. Yes; there is not enough evidence to reject the claim that the average hourly wage of the employees is $8.65.

Exercises 8–4

1. It is bell-shaped, it is symmetric about the mean, and it never touches the x axis. The mean, median, and mode are all equal to 0, and they are located at the center of the distribution. The t distribution differs from the standard normal distribution in that it is a family of curves and the variance is greater than 1; and as the degrees of freedom increase, the t distribution approaches the standard normal distribution.

3. a. $+1.833$ c. -3.365 e. ±2.145 g. ±2.771
b. ±1.740 d. $+2.306$ f. -2.819 h. ±2.583

4. Specific P-values are in parentheses.
a. $0.01 < P$-value < 0.025 (0.018)
b. $0.05 < P$-value < 0.10 (0.062)
c. $0.10 < P$-value < 0.25 (0.123)
d. $0.10 < P$-value < 0.20 (0.138)
e. P-value < 0.005 (0.003)
f. $0.10 < P$-value < 0.25 (0.158)
g. P-value $= 0.05$ (0.05)
h. P-value > 0.25 (0.261)

5. H_0: $\mu \geq 11.52$ and H_1: $\mu < 11.52$ (claim); C.V. $= -1.833$; d.f. $= 9$; $t = -9.97$; reject. There is enough evidence to support the claim that the amount of rainfall is below average.

6. H_0: $\mu \geq 2000$ and H_1: $\mu < 2000$ (claim); C.V. $= -3.747$; d.f. $= 4$; $t = -0.104$; do not reject. There is not enough evidence to support the claim that the average number of acres is less than 2000.

7. H_0: $\mu = \$40,000$ and H_1: $\mu \neq \$40,000$ (claim); C.V. $= \pm2.093$; d.f. $= 19$; $t = 3.61$; reject. Yes; there is enough evidence to support the claim that the average salary is not $40,000.

9. H_0: $\mu \geq 700$ (claim) and H_1: $\mu < 700$; C.V. $= -2.262$; d.f. $= 9$; $t = -2.71$; reject. There is enough evidence to reject the claim that the average height of the buildings is at least 700 feet.

11. H_0: $\mu \leq \$13,252$ and H_1: $\mu > \$13,252$ (claim); C.V. $= 2.539$; d.f. $= 19$; $t = 2.949$; reject. Yes; there is enough evidence to support the claim that the mean cost has increased.

13. H_0: $\mu \leq \$54.8$ million and H_1: $\mu > \$54.8$ million (claim); C.V. $= 1.761$; d.f. $= 14$; $t = 3.058$; reject. Yes. There is enough evidence to support the claim that the average cost of an action movie is greater than $54.8 million.

15. H_0: $\mu = 132$ (claim) and H_1: $\mu \neq 132$; C.V. $= \pm2.365$; d.f. $= 7$; $t = -1.7999$; do not reject. Yes. There is not

enough evidence to reject the claim that the average time is 132 minutes.

17. H_0: $\mu = 5.8$ and H_1: $\mu \neq 5.8$ (claim); d.f. = 19; $t = -3.462$; P-value < 0.01; reject. There is enough evidence to support the claim that the mean number of times has changed.

19. H_0: $\mu = \$15{,}000$ and H_1: $\mu \neq \$15{,}000$; d.f. = 11; $t = -1.10$; C.V. $= \pm 2.201$; do not reject. There is not enough evidence to conclude that the average stipend differs from $15,000.

Exercises 8–5

1. Answers will vary.

3. $np \geq 5$ and $nq \geq 5$

5. H_0: $p = 0.647$ and H_1: $p \neq 0.647$ (claim); C.V. $= \pm 2.58$; $z = -0.86$; do not reject. No. There is not enough evidence to support the claim that the proportion is different from 0.647.

7. H_0: $p = 0.40$ and H_1: $p \neq 0.40$ (claim); C.V. $= \pm 2.58$; $z = -1.07$; do not reject. No. There is not enough evidence to support the claim that the proportion is different from 0.40.

9. H_0: $p = 0.63$ (claim) and H_1: $p \neq 0.63$; C.V. $= \pm 1.96$; $z = -0.88$; do not reject. There is not enough evidence to reject the claim that the percentage is the same.

11. H_0: $p = 0.54$ (claim) and H_1: $p \neq 0.54$; C.V. $= \pm 1.96$; $z = -0.81$; do not reject. No. There is not enough evidence to reject the claim that 54% of fatal car/truck accidents are caused by driver error.

13. H_0: $p = 0.54$ (claim) and H_1: $p \neq 0.54$; $z = 0.93$; P-value $= 0.3524$; do not reject. There is not enough evidence to reject the claim that the proportion is 0.54. Yes, a healthy snack should be made available for children to eat after school.

15. H_0: $p = 0.18$ (claim) and H_1: $p \neq 0.18$; $z = -0.60$; P-value $= 0.5486$; since P-value > 0.05, do not reject. There is not enough evidence to reject the claim that 18% of all high school students smoke at least a pack of cigarettes a day.

17. H_0: $p = 0.67$ and H_1: $p \neq 0.67$ (claim); C.V. $= \pm 1.96$; $z = 3.19$; reject. Yes. There is enough evidence to support the claim that the percentage is not 67%.

19. H_0: $p \geq 0.576$ and H_1: $p < 0.576$ (claim); C.V. $= -1.65$; $z = -1.26$; do not reject. There is not enough evidence to support the claim that the proportion is less than 0.576.

21. No

23. $z = \dfrac{X - \mu}{\sigma}$

$z = \dfrac{X - np}{\sqrt{npq}}$ since $\mu = np$ and $\sigma = \sqrt{npq}$

$z = \dfrac{X/n - np/n}{\sqrt{npq/n}}$

$z = \dfrac{X/n - np/n}{\sqrt{npq/n^2}}$

$z = \dfrac{\hat{p} - p}{\sqrt{pq/n}}$ since $\hat{p} = X/n$

Exercises 8–6

1. a. H_0: $\sigma^2 \leq 225$ and H_1: $\sigma^2 > 225$; C.V. $= 27.587$; d.f. = 17

   b. H_0: $\sigma^2 \geq 225$ and H_1: $\sigma^2 < 225$; C.V. $= 14.042$; d.f. = 22

   c. H_0: $\sigma^2 = 225$ and H_1: $\sigma^2 \neq 225$; C.V. $= 5.629$; 26.119; d.f. = 14

   d. H_0: $\sigma^2 = 225$ and H_1: $\sigma^2 \neq 225$; C.V. $= 2.167$; 14.067; d.f. = 7

   e. H_0: $\sigma^2 \leq 225$ and H_1: $\sigma^2 > 225$; C.V. $= 32.000$; d.f. = 16

   f. H_0: $\sigma^2 \geq 225$ and H_1: $\sigma^2 < 225$; C.V. $= 8.907$; d.f. = 19

   g. H_0: $\sigma^2 = 225$ and H_1: $\sigma^2 \neq 225$; C.V. $= 3.074$; 28.299; d.f. = 12

   h. H_0: $\sigma^2 \geq 225$ and H_1: $\sigma^2 < 225$; C.V. $= 15.308$; d.f. = 28

2. a. $0.01 < P\text{-value} < 0.025\ (0.015)$

   b. $0.005 < P\text{-value} < 0.01\ (0.006)$

   c. $0.01 < P\text{-value} < 0.025\ (0.012)$

   d. $P\text{-value} < 0.005\ (0.003)$

   e. $0.025 < P\text{-value} < 0.05\ (0.037)$

   f. $0.05 < P\text{-value} < 0.10\ (0.088)$

   g. $0.05 < P\text{-value} < 0.10\ (0.066)$

   h. $P\text{-value} < 0.01\ (0.007)$

3. H_0: $\sigma = 60$ (claim) and H_1: $\sigma \neq 60$; C.V. $= 8.672$; 27.587; d.f. = 17; $\chi^2 = 19.707$; do not reject. There is not enough evidence to reject the claim that the standard deviation is 60.

5. H_0: $\sigma \geq 15$ and H_1: $\sigma < 15$ (claim); C.V. $= 4.575$; d.f. = 11; $\chi^2 = 9.0425$; do not reject. There is not enough evidence to support the claim that the standard deviation is less than 15.

7. H_0: $\sigma \leq 1.2$ (claim) and H_1: $\sigma > 1.2$; $\alpha = 0.01$; d.f. = 14; $\chi^2 = 31.5$; P-value $< 0.005\ (0.0047)$; since P-value < 0.01, reject. There is enough evidence to reject the claim that the standard deviation is less than or equal to 1.2 minutes.

9. H_0: $\sigma \leq 20$ and H_1: $\sigma > 20$ (claim); C.V. $= 36.191$; d.f. = 19; $\chi^2 = 58.5502$; reject. There is enough evidence to support the claim that the standard deviation is greater than 20.

11. H_0: $\sigma \geq 35$ and H_1: $\sigma < 35$ (claim); C.V. $= 3.940$; d.f. = 10; $\chi^2 = 8.359$; do not reject. There is not enough evidence to support the claim that the standard deviation is less than 35.

13. H_0: $\sigma^2 \le 25$ and H_1: $\sigma^2 > 25$ (claim); C.V. = 22.362; d.f. = 13; $\chi^2 = 23.622$; reject. There is enough evidence to support the claim that the variance is greater than 25.

Exercises 8–7

1. H_0: $\mu = 1800$ (claim) and H_1: $\mu \ne 1800$; C.V. = ± 1.96; $z = 0.47$; $1706.04 < \mu < 1953.96$; do not reject. There is not enough evidence to reject the claim that the average of the sales is $1800.

3. H_0: $\mu = 86$ (claim) and H_1: $\mu \ne 86$; C.V. = ± 2.58; $z = -1.29$; $80.00 < \mu < 88.00$; do not reject. There is not enough evidence to reject the claim that the average monthly maintenance is $86.

5. H_0: $\mu = 22$ and H_1: $\mu \ne 22$ (claim); C.V. = ± 2.58; $z = -2.32$; $19.47 < \mu < 22.13$; do not reject. There is not enough evidence to support the claim that the average has changed.

7. The power of a statistical test is the probability of rejecting the null hypothesis when it is false.

9. The power of a test can be increased by increasing α or selecting a larger sample size.

Review Exercises

1. H_0: $\mu = 98°$ (claim) and H_1: $\mu \ne 98°$; C.V. = ± 1.96; $z = -2.02$; reject. There is enough evidence to reject the claim that the average high temperature in the United States is 98°.

3. H_0: $\mu \le \$40,000$ and H_1: $\mu > \$40,000$ (claim); C.V. = 1.65; $z = 2.00$; reject. There is enough evidence to support the claim that the average salary is greater than $40,000.

5. H_0: $\mu \le 67$ and H_1: $\mu > 67$ (claim); C.V. = 1.383; d.f. = 9; $t = 7.47$; reject. There is enough evidence to support the claim that 1995 was warmer than average.

7. H_0: $\mu = 6$ and H_1: $\mu \ne 6$ (claim); $t = 1.835$; d.f. = 9; C.V. = ± 2.821; do not reject. There is not enough evidence to support the claim that the attendance has changed.

9. H_0: $p \le 0.602$ and H_1: $p > 0.602$ (claim); C.V. = 1.65; $z = 1.96$; reject. Yes. There is enough evidence to support the claim that the proportion is greater than 0.602.

11. H_0: $p = 0.65$ (claim) and H_1: $p \ne 0.65$; $z = 1.17$; P-value = 0.242; since P-value > 0.05, do not reject. There is not enough evidence to reject the claim that 65% of teenagers own their own radios.

13. H_0: $\mu \ge 10$ and H_1: $\mu < 10$ (claim); $z = -2.22$; P-value = 0.0132; reject. There is enough evidence to support the claim that the average time is less than 10 minutes.

15. H_0: $\sigma \ge 4.3$ (claim) and H_1: $\sigma < 4.3$; d.f. = 19; $\chi^2 = 6.95$; $0.005 < P$-value < 0.01 (0.006); since P-value < 0.05, reject. Yes, there is enough evidence to reject the claim that the standard deviation is greater than or equal to 4.3 miles per gallon.

17. H_0: $\sigma = 18$ (claim) and H_1: $\sigma \ne 18$; C.V. = 11.143 and 0.484; d.f. = 4; $\chi^2 = 5.44$; do not reject. There is not enough evidence to reject the claim that the standard deviation is 18 minutes.

19. H_0: $\mu = 4$ and H_1: $\mu \ne 4$ (claim); C.V. = ± 2.58; $z = 1.49$; $3.85 < \mu < 4.55$; do not reject. There is not enough evidence to support the claim that the growth has changed.

Chapter Quiz

1. True   **2.** True

3. False   **4.** True

5. False   **6.** b

7. d **8.** c

9. b **10.** Type I

11. β **12.** Statistical hypothesis

13. Right   **14.** $n - 1$

15. H_0: $\mu = 28.6$ (claim) and H_1: $\mu \ne 28.6$; $z = 2.14$; C.V. = ± 1.96; reject. There is enough evidence to reject the claim that the average age of the mothers is 28.6 years.

16. H_0: $\mu = \$6500$ (claim) and H_1: $\mu \ne \$6500$; $z = 5.27$; C.V. = ± 1.96; reject. There is enough evidence to reject the agent's claim.

17. H_0: $\mu \le 8$ and H_1: $\mu > 8$ (claim); $z = 6$; C.V. = 1.65; reject. There is enough evidence to support the claim that the average is greater than 8.

18. H_0: $\mu = 500$ (claim) and H_1: $\mu \ne 500$; d.f. = 6; $t = -0.571$; C.V. = ± 3.707; do not reject. There is not enough evidence to reject the claim that the mean is 500.

19. H_0: $\mu \ge 67$ and H_1: $\mu < 67$ (claim); $t = -3.1568$; P-value < 0.005 (0.003); since P-value < 0.05, reject. There is enough evidence to support the claim that the average height is less than 67 inches.

20. H_0: $\mu \ge 12.4$ and H_1: $\mu < 12.4$ (claim); $t = -2.324$; C.V. = -1.345; reject. There is enough evidence to support the claim that the average is less than the company claimed.

21. H_0: $\mu \le 63.5$ and H_1: $\mu > 63.5$ (claim); $t = 0.47075$; P-value > 0.25 (0.322); since P-value > 0.05, do not reject. There is not enough evidence to support the claim that the average is greater than 63.5.

22. H_0: $\mu = 26$ (claim) and H_1: $\mu \ne 26$; $t = -1.5$; C.V. = ± 2.492; do not reject. There is not enough evidence to reject the claim that the average is 26.

23. H_0: $p = 0.39$ (claim) and H_1: $p \ne 0.39$; C.V. = ± 1.96; $z = -0.62$; do not reject. There is not enough evidence to reject the claim that 39% took supplements. The study supports the results of the previous study.

24. H_0: $p \ge 0.55$ (claim) and H_1: $p < 0.55$; $z = -0.8989$; C.V. = -1.28; do not reject. There is not enough evidence to reject the survey's claim.

25. H_0: $p = 0.35$ (claim) and H_1: $p \neq 0.35$; C.V. $= \pm 2.33$; $z = 0.666$; do not reject. There is not enough evidence to reject the claim that the proportion is 35%.

26. H_0: $p = 0.75$ (claim) and H_1: $p \neq 0.75$; $z = 2.6833$; C.V. $= \pm 2.58$; reject. There is enough evidence to reject the claim.

27. P-value $= 0.0323$

28. P-value $= 0.0001$

29. H_0: $\sigma \leq 6$ and H_1: $\sigma > 6$ (claim); $\chi^2 = 54$; C.V. $= 36.415$; reject. There is enough evidence to support the claim.

30. H_0: $\sigma = 8$ (claim) and H_1: $\sigma \neq 8$; $\chi^2 = 33.2$; C.V. $= 27.991$, 79.490; do not reject. There is not enough evidence to reject the claim that $\sigma = 8$.

31. H_0: $\sigma \geq 2.3$ and H_1: $\sigma < 2.3$ (claim); $\chi^2 = 13$; C.V. $= 10.117$; do not reject. There is not enough evidence to support the claim that the standard deviation is less than 2.3.

32. H_0: $\sigma = 9$ (claim) and H_1: $\sigma \neq 9$; $\chi^2 = 13.4$; P-value > 0.20 (0.291); since P-value > 0.05, do not reject. There is not enough evidence to reject the claim that $\sigma = 9$.

33. $28.9 < \mu < 31.2$; no

34. $\$6562.81 < \mu < \$6637.19$; no

Chapter 9

Exercises 9–2

1. Testing a single mean involves comparing a sample mean to a specific value such as $\mu = 100$; testing the difference between two means involves comparing the means of two samples, such as $\mu_1 = \mu_2$.

3. The populations must be independent of each other, and they must be normally distributed; s_1 and s_2 can be used in place of σ_1 and σ_2 when σ_1 and σ_2 are unknown and both samples are each greater than or equal to 30.

5. H_0: $\mu_1 = \mu_2$ (claim) and H_1: $\mu_1 \neq \mu_2$; C.V. $= \pm 2.58$; $z = -0.856$; do not reject. There is not enough evidence to reject the claim that the average lengths of the major rivers are the same.

7. H_0: $\mu_1 \leq \mu_2$ and H_1: $\mu_1 > \mu_2$ (claim); C.V. $= +1.65$; $z = 2.56$; reject. Yes, there is enough evidence to support the claim that pulse rates of smokers are higher than pulse rates of nonsmokers.

9. H_0: $\mu_1 \leq \mu_2$ and H_1: $\mu_1 > \mu_2$ (claim); C.V. $= +2.05$; $z = 1.12$; do not reject. There is not enough evidence to support the claim that the noise levels in the corridors are higher than the noise levels in the clinics.

11. H_0: $\mu_1 \geq \mu_2$ and H_1: $\mu_1 < \mu_2$ (claim); C.V. $= -1.65$; $z = -2.01$; reject. There is enough evidence to support the claim that the stayers had a higher grade point average.

13. H_0: $\mu_1 \leq \mu_2$ and H_1: $\mu_1 > \mu_2$ (claim); C.V. $= +2.33$; $z = +1.09$; do not reject. There is not enough evidence to support the claim that colleges spend more money on male sports than they spend on female sports.

15. H_0: $\mu_1 = \mu_2$ and H_1: $\mu_1 \neq \mu_2$ (claim); $z = 1.01$; P-value $= 0.3124$; do not reject. There is not enough evidence to support the claim that there is a difference in self-esteem scores.

17. $2.8 < \mu_1 - \mu_2 < 6.0$ **19.** $-7.3 < \mu_1 - \mu_2 < -1.3$

21. H_0: $\mu_1 - \mu_2 \leq 8$ (claim) and H_1: $\mu_1 - \mu_2 > 8$; C.V. $= +1.65$; $z = -0.73$; do not reject. There is not enough evidence to reject the claim that private school students have exam scores that are at most 8 points higher than those of students in public schools.

Exercises 9–3

1. The variance in the numerator should be the larger of the two variances.

3. One degree of freedom is used for the variance associated with the numerator, and one is used for the variance associated with the denominator.

5. *a.* d.f.N. $= 15$, d.f.D. $= 22$; C.V. $= 3.36$
   b. d.f.N. $= 24$, d.f.D. $= 13$; C.V. $= 3.59$
   c. d.f.N. $= 45$, d.f.D. $= 29$; C.V. $= 2.03$
   d. d.f.N. $= 20$, d.f.D. $= 16$; C.V. $= 2.28$
   e. d.f.N. $= 10$, d.f.D. $= 10$; C.V. $= 2.98$

6. Specific P-values are in parentheses.
   a. $0.025 < P$-value < 0.05 (0.033)
   b. $0.05 < P$-value < 0.10 (0.072)
   c. P-value $= 0.05$
   d. $0.005 < P$-value < 0.01 (0.006)
   e. P-value $= 0.05$
   f. P-value > 0.10 (0.112)
   g. $0.05 < P$-value < 0.10 (0.068)
   h. $0.01 < P$-value < 0.02 (0.015)

7. H_0: $\sigma_1^2 = \sigma_2^2$ and H_1: $\sigma_1^2 \neq \sigma_2^2$ (claim); C.V. $= 2.53$; d.f.N. $= 14$; d.f.D. $= 14$; $F = 4.52$; reject. There is enough evidence to support the claim that there is a difference in the variances.

9. H_0: $\sigma_1^2 = \sigma_2^2$ and H_1: $\sigma_1^2 \neq \sigma_2^2$ (claim); C.V. $= 2.86$; d.f.N. $= 15$; d.f.D. $= 15$; $F = 7.85$; reject. There is enough evidence to support the claim that the variances are different. Since both data sets vary greatly from normality, the results are suspect.

11. H_0: $\sigma_1^2 = \sigma_2^2$ and H_1: $\sigma_1^2 \neq \sigma_2^2$ (claim); C.V. $= 4.99$; d.f.N. $= 7$; d.f.D. $= 7$; $F = 1$; do not reject. There is not enough evidence to support the claim that there is a difference in the variances.

13. H_0: $\sigma_1 = \sigma_2$ and H_1: $\sigma_1 \neq \sigma_2$ (claim); C.V. $= 2.27$; $F = 4.52$; reject. There is enough evidence to support the claim that the standard deviations of the ages are different. One reason is that there are many more people who play the slot machines than people who play roulette. This could possibly account for the larger standard deviation in the ages of the players.

15. H_0: $\sigma_1^2 = \sigma_2^2$ and H_1: $\sigma_1^2 \neq \sigma_2^2$ (claim); C.V. = 4.03; d.f.N. = 9; d.f.D. = 9; $F = 1.1026$; do not reject. There is not enough evidence to support the claim that the variances are not equal.

17. H_0: $\sigma_1^2 = \sigma_2^2$ (claim) and H_1: $\sigma_1^2 \neq \sigma_2^2$; C.V. = 3.87; d.f.N. = 6; d.f.D. = 7; $F = 3.18$; do not reject. There is not enough evidence to reject the claim that the variances of the heights are equal.

19. H_0: $\sigma_1^2 = \sigma_2^2$ (claim) and H_1: $\sigma_1^2 \neq \sigma_2^2$; $F = 5.32$; d.f.N. = 14; d.f.D. = 14; P-value < 0.01 (0.004); reject. There is enough evidence to reject the claim that the variances of the weights are equal.

Exercises 9–4

1. H_0: $\sigma_1^2 = \sigma_2^2$; C.V. = 4.03; $F = 1.93$; do not reject. H_0: $\mu_1 = \mu_2$ and H_1: $\mu_1 \neq \mu_2$ (claim); C.V. = ±2.101; d.f. = 18; $t = -4.02$; reject. There is enough evidence to support the claim that there is a significant difference in the values of the homes based on the appraisers' values. $-\$7762 < \mu_1 - \mu_2 < -\$2434$

3. H_0: $\sigma_1^2 = \sigma_2^2$; C.V. = 3.05; $F = 1$; do not reject. H_0: $\mu_1 = \mu_2$ (claim) and H_1: $\mu_1 \neq \mu_2$; C.V. = ±2.048; $t = -1.61$; do not reject. There is not enough evidence to reject the claim that the means are equal.

5. H_0: $\sigma_1^2 = \sigma_2^2$; C.V. = 14.94; $F = 1.41$; do not reject. H_0: $\mu_1 \leq \mu_2$ and H_1: $\mu_1 > \mu_2$ (claim); C.V. = 2.764; d.f. = 10; $t = 1.45$; do not reject. There is not enough evidence to support the claim that the average number of family day-care homes is greater than the average number of day-care centers.

7. H_0: $\sigma_1^2 = \sigma_2^2$; C.V. = 4.19; $F = 1.70$; do not reject. H_0: $\mu_1 = \mu_2$ and H_1: $\mu_1 \neq \mu_2$ (claim); C.V. = ±2.508; d.f. = 22; $t = -2.97$; reject. There is enough evidence to support the claim that there is a difference in the average times of the two groups. $-11.1 < \mu_1 - \mu_2 < -0.93$

9. H_0: $\sigma_1^2 = \sigma_2^2$; C.V. = 1.98; $F = 1.55$; do not reject. H_0: $\mu_1 \leq \mu_2$ and H_1: $\mu_1 > \mu_2$ (claim); C.V. = 1.282; d.f.= 48; $t = 11.427$; reject. There is enough evidence to support the claim that the average cost of a movie ticket in London is greater than the average cost of a movie ticket in New York City. One reason for the difference is the rate of exchange of the money.

11. H_0: $\sigma_1^2 = \sigma_2^2$; C.V. = 7.15; $F = 1.23$; do not reject. H_0: $\mu_1 = \mu_2$ and H_1: $\mu_1 \neq \mu_2$ (claim); C.V. = ±2.228; d.f. = 10; $t = 0.119$; do not reject. There is not enough evidence to support the claim that the color of the mice made a difference. $-5.9 < \mu_1 - \mu_2 < 6.5$

13. $\$2626.60 < \mu_1 - \mu_2 < \$11,589.00$

Exercises 9–5

1. *a.* Dependent   *d.* Dependent
b. Dependent   *e.* Independent
c. Independent

3. H_0: $\mu_D \geq 0$ and H_1: $\mu_D < 0$ (claim); C.V. = -1.397; d.f. = 8; $t = -2.8$; reject. There is enough evidence to support the claim that the seminar increased the number of hours students studied.

5. H_0: $\mu_D = 0$ and H_1: $\mu_D \neq 0$ (claim); C.V. = 2.365; d.f. = 7; $t = 1.6583$; do not reject. There is not enough evidence to support the claim that the means are different.

7. H_0: $\mu_D \leq 0$ and H_1: $\mu_D > 0$ (claim); C.V. = 2.571; d.f. = 5; $t = 2.24$; do not reject. There is not enough evidence to support the claim that the errors have been reduced.

9. H_0: $\mu_D = 0$ and H_1: $\mu_D \neq 0$ (claim); d.f. = 7; $t = 0.978$; $0.20 < P$-value < 0.50 (0.361). Do not reject since P-value > 0.01. There is not enough evidence to support the claim that there is a difference in the pulse rates. $-3.23 < \mu_D < 5.73$

11. Using the previous problem $\overline{D} = -1.5625$, whereas the mean of the before values is 95.375 and the mean of the after values is 96.9375; hence, $\overline{D} = 95.375 - 96.9375 = -1.5625$.

Exercises 9–6

1a. *a.* $\hat{p} = \frac{34}{48}, \hat{q} = \frac{14}{48}$ *d.* $\hat{p} = \frac{6}{24}, \hat{q} = \frac{18}{24}$
b. $\hat{p} = \frac{28}{75}, \hat{q} = \frac{47}{75}$ *e.* $\hat{p} = \frac{12}{144}, \hat{q} = \frac{132}{144}$
c. $\hat{p} = \frac{50}{100}, \hat{q} = \frac{50}{100}$

1b. *a.* 16   *b.* 4   *c.* 48
d. 104   *e.* 30

3. $\hat{p}_1 = 0.533$; $\hat{p}_2 = 0.3$; $\overline{p} = 0.44$; $\overline{q} = 0.56$; H_0: $p_1 = p_2$ and H_1: $p_1 \neq p_2$ (claim); C.V. = ±1.96; $z = 3.64$; reject. There is enough evidence to support the claim that there is a significant difference in the proportions.

5. $\hat{p}_1 = 0.747$; $\hat{p}_2 = 0.75$; $\overline{p} = 0.749$; $\overline{q} = 0.251$; H_0: $p_1 = p_2$ and H_1: $p_1 \neq p_2$ (claim); C.V. = ±1.96; $z = -0.07$; do not reject. There is not enough evidence to support the claim that the proportions are not equal.

7. $\hat{p}_1 = 0.83$; $\hat{p}_2 = 0.75$; $\overline{p} = 0.79$; $\overline{q} = 0.21$; H_0: $p_1 = p_2$ (claim) and H_1: $p_1 \neq p_2$; C.V. = ±1.96; $z = 1.39$; do not reject. There is not enough evidence to reject the claim that the proportions are equal. $-0.032 < p_1 - p_2 < 0.192$

9. $\hat{p}_1 = 0.55$; $\hat{p}_2 = 0.45$; $\overline{p} = 0.497$; $\overline{q} = 0.503$; H_0: $p_1 = p_2$ and H_1: $p_1 \neq p_2$ (claim); C.V. = ±2.58; $z = 1.302$; do not reject. There is not enough evidence to support the claim that the proportions are different. $-0.097 < p_1 - p_2 < 0.297$

11. $\hat{p}_1 = 0.347$; $\hat{p}_2 = 0.433$; $\overline{p} = 0.385$; $\overline{q} = 0.615$; H_0: $p_1 = p_2$ and H_1: $p_1 \neq p_2$ (claim); C.V. = ±1.96; $z = -1.03$; do not reject. There is not enough evidence to say that the proportion of dog owners has changed

$(-0.252 < p_1 - p_2 < 0.079)$. Yes, the confidence interval contains 0. This is another way to conclude that there is no difference in the proportions.

13. $\hat{p}_1 = 0.25$; $\hat{p}_2 = 0.31$; $\overline{p} = 0.286$; $\overline{q} = 0.714$; $H_0: p_1 = p_2$ and $H_1: p_1 \neq p_2$ (claim); C.V. $= \pm 2.58$; $z = -1.45$; do not reject. There is not enough evidence to support the claim that the proportions are different. $-0.165 < p_1 - p_2 < 0.045$

15. $0.077 < p_1 - p_2 < 0.323$

17. $\hat{p}_1 = 0.43$; $\hat{p}_2 = 0.58$; $\overline{p} = 0.505$; $\overline{q} = 0.495$; $H_0: p_1 = p_2$ and $H_1: p_1 \neq p_2$ (claim); C.V. $= \pm 1.96$; $z = -2.12$; reject; there is enough evidence to support the claim that the proportions are different.

19. $-0.0631 < p_1 - p_2 < 0.0667$. It does agree with the *Almanac* statistics stating a difference of -0.042 since -0.042 is contained in the interval.

Review Exercises

1. $H_0: \mu_1 \leq \mu_2$ and $H_1: \mu_1 > \mu_2$ (claim); C.V. $= 2.33$; $z = 0.59$; do not reject. There is not enough evidence to support the claim that single drivers do more pleasure driving than married drivers.

3. $H_0: \sigma_1 = \sigma_2$ and $H_1: \sigma_1 \neq \sigma_2$ (claim); C.V. $= 2.77$; $\alpha = 0.10$; d.f.N. $= 23$; d.f.D. $= 10$; $F = 10.365$; reject. There is enough evidence to support the claim that there is a difference in the standard deviations.

5. $H_0: \sigma_1^2 \leq \sigma_2^2$ and $H_1: \sigma_1^2 > \sigma_2^2$ (claim); $\alpha = 0.05$; d.f.N. $= 9$; d.f.D. $= 9$; $F = 5.06$. The P-value for the F test is $0.01 < P$-value < 0.025 (0.012); reject since P-value < 0.05. There is enough evidence to support the claim that the variance of the number of speeding tickets issued on Route 19 is greater than the variance of the number of speeding tickets issued on Route 22.

7. $H_0: \sigma_1^2 \leq \sigma_2^2$ and $H_1: \sigma_1^2 > \sigma_2^2$ (claim); C.V. $= 1.47$; $\alpha = 0.10$; d.f.N. $= 64$; d.f.D. $= 41$; $F = 2.32$; reject. There is enough evidence to support the claim that the variation in the number of days factory workers miss per year due to illness is greater than the variation in the number of days hospital workers miss per year.

9. $H_0: \sigma_1^2 = \sigma_2^2$; C.V. $= 1.98$; $F = 1.11$; do not reject. $H_0: \mu_1 \leq \mu_2$ and $H_1: \mu_1 > \mu_2$ (claim); C.V. $= 1.28$; $t = 1.31$; reject. There is enough evidence to support the claim that it is warmer in Birmingham.

11. $H_0: \sigma_1^2 = \sigma_2^2$; $F = 1.12$; do not reject since $p > 0.10$; $H_0: \mu_1 = \mu_2$ and $H_1: \mu_1 \neq \mu_2$ (claim); d.f. $= 7$; $t = -0.828$; do not reject. Since $p > 0.10$, there is not enough evidence to support the claim that the means are different. A cafeteria manager would want to know the results to make a decision on which beverage to serve.

13. $H_0: \mu_D \geq 0$ and $H_1: \mu_D < 0$ (claim); C.V. $= -1.895$; d.f. $= 7$; $t = -2.73$; reject. There is enough evidence to support the claim that the music has increased production.

15. $\hat{p}_1 = 0.15$; $\hat{p}_2 = 0.05$; $\overline{p} = 0.104$; $\overline{q} = 0.896$; $H_0: p_1 = p_2$ and $H_1: p_1 \neq p_2$ (claim); C.V. $= \pm 1.96$; $z = 2.41$; reject. There is enough evidence to support the claim that the proportion has changed. $0.023 < p_1 - p_2 < 0.177$. The confidence level does not contain 0; hence, the null hypothesis is rejected.

Chapter Quiz

1. False
2. False
3. True
4. False
5. d
6. a
7. c
8. b
9. $\mu_1 = \mu_2$
10. Pooled
11. Normal
12. Negative
13. $\dfrac{s_1^2}{s_2^2}$

14. $H_0: \mu_1 = \mu_2$ and $H_1: \mu \neq \mu_2$ (claim); $z = -3.69$; C.V. $= \pm 2.58$; reject. There is enough evidence to support the claim that there is a difference in the cholesterol levels of the two groups. $-10.2 < \mu_1 - \mu_2 < -1.8$

15. $H_0: \mu_1 \leq \mu_2$ and $H_1: \mu > \mu_2$ (claim); C.V. $= 1.28$; $z = 1.60$; reject. There is enough evidence to support the claim that the average rental fees for the apartments in the East are greater than the average rental fees for the apartments in the West.

16. $H_0: \sigma_1^2 = \sigma_2^2$ and $H_1: \sigma_1^2 \neq \sigma_2^2$ (claim); $F = 1.637$; d.f.N. $= 17$; d.f.D. $= 14$; P-value > 0.20 (0.357). Do not reject since P-value > 0.05. There is not enough evidence to support the claim that the variances are different.

17. $H_0: \sigma_1^2 = \sigma_2^2$ and $H_1: \sigma_1^2 \neq \sigma_2^2$ (claim); $F = 1.296$; C.V. $= 1.90$; do not reject. There is not enough evidence to support the claim that the variances are different.

18. $H_0: \sigma_1 = \sigma_2$ (claim) and $H_1: \sigma_1 \neq \sigma_2$; d.f.N. $= 11$; d.f.D. $= 11$; C.V. $= 3.53$; $F = 1.13$; do not reject. There is not enough evidence to reject the claim that the standard deviations of the number of hours of television viewing are the same.

19. $H_0: \sigma_1^2 = \sigma_2^2$ and $H_1: \sigma_1^2 \neq \sigma_2^2$ (claim); $F = 1.94$; C.V. $= 3.01$; do not reject. There is not enough evidence to support the claim that the variances are not equal.

20. $H_0: \sigma_1^2 \leq \sigma_2^2$ and $H_1: \sigma_1^2 > \sigma_2^2$ (claim); $F = 1.08$; C.V. $= 5.05$; do not reject. There is not enough evidence to support the claim that the variance of the number of murders committed on the East Coast is greater than the variance of the number of murders committed on the West Coast. One factor that could influence the results is the populations of the cities that were selected.

21. $H_0: \sigma_1 = \sigma_2$ and $H_1: \sigma_1 \neq \sigma_2$ (claim); $F = 1.65$; C.V. $= 2.46$; do not reject. There is not enough evidence to support the claim that the standard deviations are different.

22. H_0: $\sigma_1^2 = \sigma_2^2$; C.V. = 5.05; $F = 1.23$; do not reject.
H_0: $\mu_1 = \mu_2$ and H_1: $\mu_1 \neq \mu_2$ (claim); $t = 10.922$;
C.V. = ±2.779; reject. There is enough evidence to
support the claim that the average prices are different.
$0.298 < \mu_1 - \mu_2 < 0.502$

23. H_0: $\sigma_1^2 = \sigma_2^2$; C.V. = 9.60; $F = 5.71$; do not reject.
H_0: $\mu_1 \geq \mu_2$ and H_1: $\mu_1 < \mu_2$ (claim); C.V. = -1.860;
d.f. = 8; $t = -4.05$; reject. There is enough evidence to
support the claim that accidents have increased.

24. H_0: $\sigma_1^2 = \sigma_2^2$; C.V. = 4.02; $F = 6.155$; reject.
H_0: $\mu_1 = \mu_2$ and H_1: $\mu_1 \neq \mu_2$ (claim); $t = 9.807$;
C.V. = ±2.718; reject. There is enough evidence
to support the claim that the salaries are different.
$\$6653 < \mu_1 - \mu_2 < \$11,757$

25. H_0: $\sigma_1^2 = \sigma_2^2$; $F = 23.08$; reject since P-value < 0.05.
H_0: $\mu_1 \leq \mu_2$ and H_1: $\mu_1 > \mu_2$ (claim); d.f. = 10;
$t = 0.874$; $0.10 < P$-value < 0.25 (0.198); do not reject
since P-value > 0.05. There is not enough evidence to
support the claim that the incomes of city residents are
greater than the incomes of rural residents.

26. H_0: $\mu_1 \geq \mu_2$ and H_1: $\mu_1 < \mu_2$ (claim); $t = -4.172$;
C.V. = -2.821; reject. There is enough evidence to
support the claim that the sessions improved math skills.

27. H_0: $\mu_1 \geq \mu_2$ and H_1: $\mu_1 < \mu_2$ (claim); $t = -1.714$;
C.V. = -1.833; do not reject. There is not enough
evidence to support the claim that egg production was
increased.

28. H_0: $p_1 = p_2$ and H_1: $p_1 \neq p_2$ (claim); $z = -0.69$;
C.V. = ±1.65; do not reject. There is not enough
evidence to support the claim that the proportions are
different. $-0.105 < p_1 - p_2 < 0.045$

29. H_0: $p_1 = p_2$ and H_1: $p_1 \neq p_2$ (claim); C.V. = ±1.96;
$z = 0.544$; do not reject. There is not enough evidence
to support the claim that the proportions have changed.
$-0.026 < p_1 - p_2 < 0.0460$. Yes, the confidence interval
contains 0; hence, the null hypothesis is not rejected.

Chapter 10

Exercises 10–3

1. Two variables are related when a discernible pattern exists
between them.

3. r, ρ (rho)

5. A positive relationship means that as x increases,
y increases. A negative relationship means that as
x increases, y decreases.

7. Answers will vary.

9. Pearson product moment correlation coefficient

11. There are many other possibilities, such as chance, or
relationship to a third variable.

13. H_0: $\rho = 0$ and H_1: $\rho \neq 0$; $r = -0.832$; C.V. = ±0.811;
reject. There is a significant relationship between a
person's age and the number of hours he or she exercises.

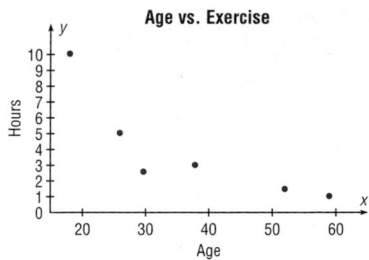

Age vs. Exercise

15. H_0: $\rho = 0$ and H_1: $\rho \neq 0$; $r = -0.883$; C.V. = ±0.811;
reject. There is a significant relationship between the
number of years a person has been out of school and his
or her contribution.

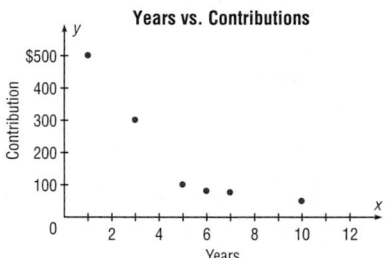

Years vs. Contributions

17. H_0: $\rho = 0$ and H_1: $\rho \neq 0$; $r = 0.104$; C.V. = ±0.754;
do not reject. There is no significant linear relationship
between the number of larceny crimes and the number of
vandalism crimes committed on college campuses in
southwestern Pennsylvania.

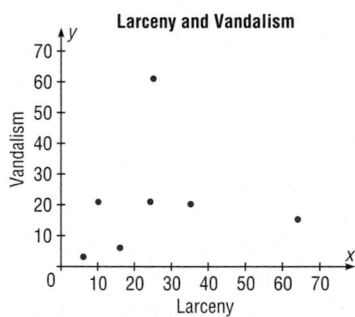

Larceny and Vandalism

19. H_0: $\rho = 0$ and H_1: $\rho \neq 0$; $r = 0.580$; C.V. = ±0.632;
do not reject. There is no significant linear relationship
between the number of tornadoes per year and the number
of deaths per year from these tornadoes.

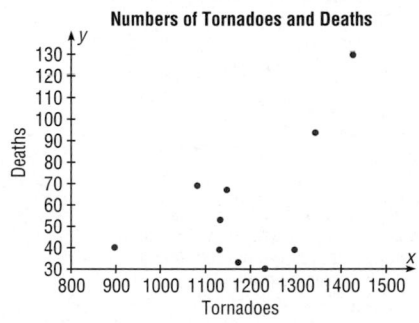

Numbers of Tornadoes and Deaths

21. H_0: $\rho = 0$ and H_1: $\rho \neq 0$; $r = 0.997$; C.V. = ±0.811;
reject. There is a significant linear relationship between

the number of people under age 5 and the number of people who are age 65 and older in the U.S. cities.

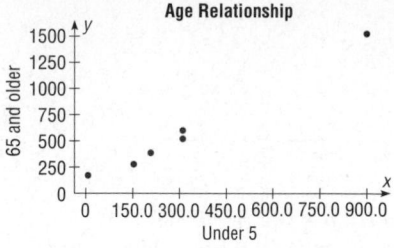

23. $H_0: \rho = 0$ and $H_1: \rho \neq 0$; $r = 0.883$; C.V. $= \pm 0.754$; reject. There is a significant linear relationship between the average daily temperature and the average monthly precipitation.

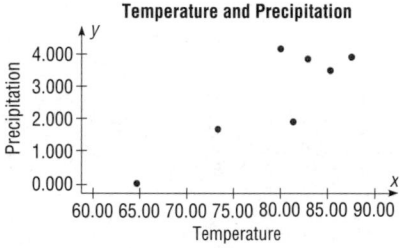

25. $H_0: \rho = 0$ and $H_1: \rho \neq 0$; $r = 0.725$; C.V. $= \pm 0.754$; do not reject. There is no significant linear relationship between the number of calories and the cholesterol content of fast-food chicken sandwiches.

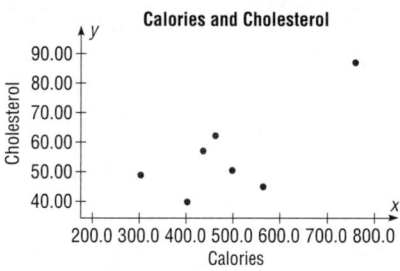

27. $H_0: \rho = 0$ and $H_1: \rho \neq 0$; $r = 0.831$; C.V. $= \pm 0.754$; reject. There is a significant linear relationship between the number of licensed beds in a hospital and the number of staffed beds.

29. $r = 1.00$: All values fall in a straight line. $r = 1.00$: The value of r between x and y is the same when x and y are interchanged.

Exercises 10–4

1. A scatter plot should be drawn, and the value of the correlation coefficient should be tested to see whether it is significant.

3. $y' = a + bx$

5. It is the line that is drawn through the points on the scatter plot such that the sum of the squares of the vertical distances from each point to the line is a minimum.

7. When r is positive, b will be positive. When r is negative, b will be negative.

9. The closer r is to $+1$ or -1, the more accurate the predicted value will be.

11. When r is not significant, the mean of the y values should be used to predict y.

13. $y' = 10.499 - 0.18x$; 4.2

15. $y' = 453.176 - 50.439x$; 251.42

17. Since r is not significant, no regression should be done.

19. Since r is not significant, no regression should be done.

21. $y' = 14.165 + 1.685x$; 351

23. $y' = -8.994 + 0.1448x$; 1.1

25. Since r is not significant, no regression should be done.

27. $y' = 22.659 + 0.582x$; 48.267

29. $H_0: \rho = 0$ and $H_1: \rho \neq 0$; $r = +0.956$; C.V. $= \pm 0.754$; reject; d.f. $= 5$; $y' = -10.944 + 1.969x$; when $x = 30$, $y' = 48.126$. There is a significant relationship between the amount of lung damage and the number of years a person has been smoking.

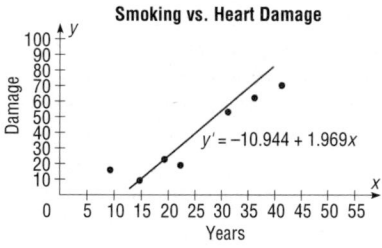

31. $H_0: \rho = 0$; $H_1: \rho \neq 0$; $r = 0.970$; C.V. $= \pm 0.707$; reject; $y' = 34.852 + 0.140x$; when $x = 500$, $y' = 104.9$. There is a significant relationship between the number of tons of coal produced and the number of employees.

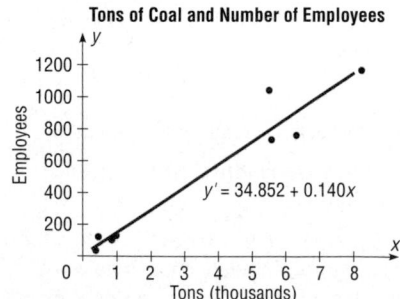

33. H_0: $\rho = 0$ and H_1: $\rho \neq 0$; $r = -0.981$; C.V. $= \pm 0.811$; reject. There is a significant relationship between the number of absences and the final grade; $y' = 96.784 - 2.668x$.

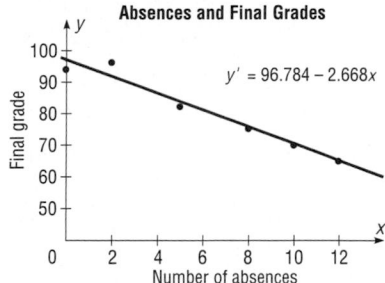

Absences and Final Grades

35. H_0: $\rho = 0$ and H_1: $\rho \neq 0$; $r = -0.265$; P-value > 0.05 (0.459); do not reject. There is no significant linear relationship between the ages of billionaires and their net worth. No regression should be done.

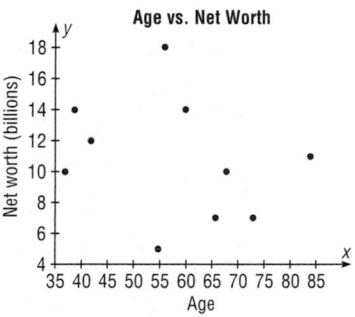

Age vs. Net Worth

37. 453.173; 21.1

Exercises 10–5

1. Explained variation is the variation due to the relationship. It is computed by $\Sigma(y' - \bar{y})^2$.

3. Total variation is the sum of the squares of the vertical distances of the points from the mean. It is computed by $\Sigma(y - \bar{y})^2$.

5. The coefficient of determination is found by squaring the value of the correlation coefficient.

7. The coefficient of nondetermination is found by subtracting r^2 from 1.

9. $r^2 = 0.49$; 49% of the variation of y is due to the variation of x; 51% of the variation of y is due to chance.

11. $r^2 = 0.1369$; 13.69% of the variation of y is due to the variation of x; 86.31% of the variation of y is due to chance.

13. $r^2 = 0.0025$; 0.25% of the variation of y is due to the variation of x; 99.75% of the variation of y is due to chance.

15. 2.092*   **17.** 94.22*

19. $1.59 < y < 12.21$*   **21.** $30.46 < y < 472.38$*

*Answers may vary due to rounding.

Exercises 10–6

1. Simple regression has one dependent variable and one independent variable. Multiple regression has one dependent variable and two or more independent variables.

3. The relationship would include all variables in one equation.

5. They will all be smaller.

7. 3.48 or 3   **9.** $39,390

11. R is the strength of the relationship between the dependent variable and all the independent variables.

13. R^2 is the coefficient of multiple determination. R^2_{adj} is adjusted for sample size and number of predictors.

15. The F test

Review Exercises

1. H_0: $\rho = 0$ and H_1: $\rho \neq 0$; $r = 0.682$; C.V. $= \pm 0.641$; reject. There is a significant linear relationship between the number of hits and the number of strikeouts. $y' = 7.22 + 0.388x$; 18.9.

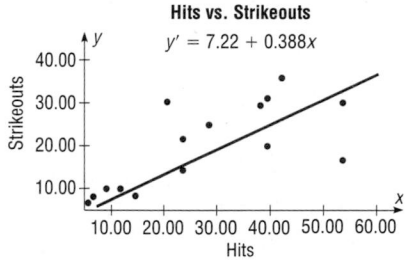

Hits vs. Strikeouts

3. H_0: $\rho = 0$ and H_1: $\rho \neq 0$; $r = 0.873$; C.V. $= \pm 0.834$; d.f. $= 6$; reject. There is a significant relationship between the mother's age and the number of children she has; $y' = -2.457 + 0.187x$; $y' = 3.9$.

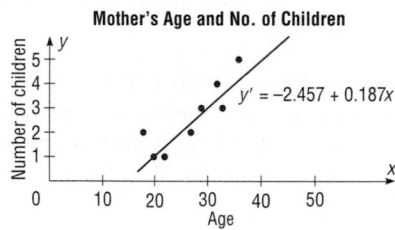

Mother's Age and No. of Children

5. H_0: $\rho = 0$ and H_1: $\rho \neq 0$; $r = -0.974$; C.V. $= \pm 0.708$; d.f. $= 10$; reject. There is a significant relationship between speed and time; $y' = 14.086 - 0.137x$; $y' = 4.222$.

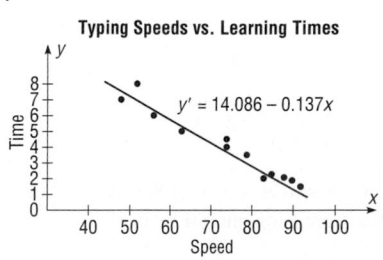

Typing Speeds vs. Learning Times

7. H_0: $\rho = 0$ and H_1: $\rho \neq 0$; $r = 0.999$; C.V. $= \pm 0.917$; reject. There is a significant relationship between the number of cows and the number of pounds of milk produced in the counties located in southwestern Pennsylvania. $y' = 0.8760 + 1.494x$.

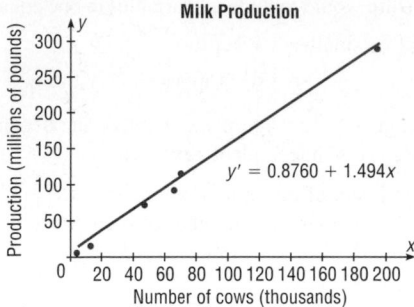

9. 0.468* (TI value 0.513)

11. $3.34 < y < 5.10$*

13. 22.01*

15. $R^2_{\text{adj}} = 0.643$*

*Answers may vary due to rounding.

Chapter Quiz

1. False   **2.** True

3. True   **4.** False

5. False   **6.** False

7. *a*   **8.** *a*

9. *d*   **10.** *c*

11. *b*   **12.** Scatter plot

13. Independent   **14.** $-1, +1$

15. *b*   **16.** Line of best fit

17. $+1, -1$

18. H_0: $\rho = 0$ and H_1: $\rho \neq 0$; $r = 0.600$; C.V. $= \pm 0.754$; do not reject. There is no significant linear relationship between the price of the same drugs in the United States and in Australia. No regression should be done.

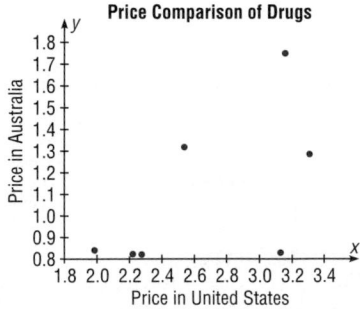

19. H_0: $\rho = 0$ and H_1: $\rho \neq 0$; $r = -0.078$; C.V. $= \pm 0.754$; do not reject. No regression should be done.

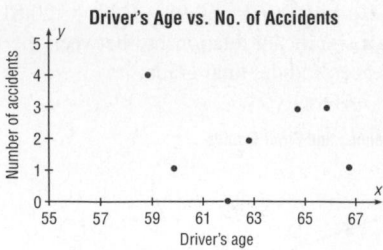

20. H_0: $\rho = 0$ and H_1: $\rho \neq 0$; $r = 0.842$; C.V. $= \pm 0.811$; reject. $y' = -1.918 + 0.551x$; 4.14 or 4.

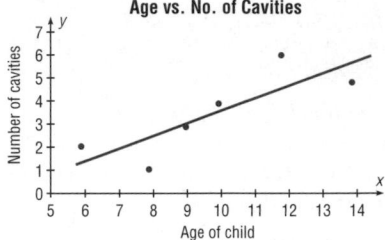

21. H_0: $\rho = 0$ and H_1: $\rho \neq 0$; $r = 0.602$; C.V. $= \pm 0.707$; do not reject. No regression should be done.

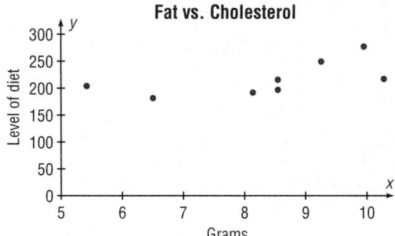

22. 1.129*

23. 29.5* For calculation purposes only. No regression should be done.

24. $0 < y < 5$*

25. 217.5 (average of y' values is used since there is no significant relationship)

26. 119.9*   **27.** $R = 0.729$*

28. $R^2_{\text{adj}} = 0.439$*

*These answers may vary due to the method of calculation or rounding.

Chapter 11

Exercises 11–2

1. The variance test compares a sample variance with a hypothesized population variance; the goodness-of-fit test compares a distribution obtained from a sample with a hypothesized distribution.

3. The expected values are computed on the basis of what the null hypothesis states about the distribution.

5. H_0: The ages of automobiles are equally distributed over the three categories (claim). H_1: The ages of automobiles are not equally distributed over the three categories.

C.V. = 5.991; d.f. = 2; χ^2 = 0.8; do not reject. There is not enough evidence to reject the claim that the average age of automobiles is equally distributed over the three categories. Tire manufacturers need to make enough tires to fit automobiles of all ages.

7. H_0: 28.1% purchased a small car, 47.8% purchased a mid-sized car, 7% purchased a large car, and 17.1% purchased a luxury car. H_1: The distribution is not the same as stated in the null hypothesis (claim). C.V. = 7.815; d.f. = 3; χ^2 = 1.9869; do not reject. There is not enough evidence to support the claim that the distribution is not the same.

9. H_0: 35% feel that genetically modified food is safe to eat, 52% feel that genetically modified food is not safe to eat, and 13% have no opinion. H_1: The distribution is not the same as stated in the null hypothesis (claim). C.V. = 9.210; d.f. = 2; χ^2 = 1.4286; do not reject. There is not enough evidence to support the claim that the proportions are different from those reported in the survey.

11. H_0: The types of loans are distributed as follows: 21% for home mortgages, 39% for automobile purchases, 20% for credit card, 12% for real estate, and 8% for miscellaneous (claim). H_1: The distribution is different from that stated in the null hypothesis. C.V. = 9.488; d.f. = 4; χ^2 = 4.786; do not reject. There is not enough evidence to reject the claim that the distribution is the same as reported in the newspaper.

13. H_0: The methods of payments of adult shoppers for purchases are distributed as follows: 53% pay cash, 30% use checks, 16% use credit cards, and 1% have no preference (claim). H_1: The distribution is not the same as stated in the null hypothesis. C.V. = 11.345; d.f. = 3; χ^2 = 36.8897; reject. There is enough evidence to reject the claim that the distribution at the large store is the same as in the survey.

15. H_0: The populations of prisoners in state prisons for serious offenses are distributed as follows: 29.5%, violent offenses; 29%, property offenses; 30.2%, drug offenses; 10.6%, weapons offenses; and 0.7%, other offenses. H_1: The distribution is not the same as stated in the null hypothesis. C.V. = 9.488; d.f. = 4; χ^2 = 15.31; reject. There is enough evidence to reject the claim stated in the null hypothesis.

17. H_0: The distribution of the ways people pay for their prescriptions is as follows: 60% used personal funds, 25% used insurance, and 15% used Medicare (claim). H_1: The distribution is not the same as stated in the null hypothesis. The d.f. = 2; α = 0.05; χ^2 = 0.667; do not reject since P-value > 0.05. There is not enough evidence to reject the claim that the distribution is the same as stated in the null hypothesis. An implication of the results is that the majority of people are using their own money to pay for medications. Maybe the medication should be less expensive to help out these people.

19. Answers will vary.

Exercises 11–3

1. The independence test and the goodness-of-fit test both use the same formula for computing the test value. However, the independence test uses a contingency table, whereas the goodness-of-fit test does not.

3. H_0: The variables are independent (or not related). H_1: The variables are dependent (or related).

5. The expected values are computed as (row total $\times$ column total) $\div$ grand total.

7. H_0: $p_1 = p_2 = p_3 = p_4 = \cdots = p_n$. H_1: At least one proportion is different from the others.

9. H_0: The type of pet owned is independent of the income of the owners. H_1: The type of pet is dependent upon the income of the owners (claim). C.V. = 21.026; d.f. = 12; χ^2 = 35.1774; reject. There is enough evidence to support the claim that the type of pet is dependent upon the income of the owner.

11. H_0: The composition of the legislature (House of Representatives) is independent of the state. H_1: The composition of the legislature is dependent upon the state (claim). C.V. = 7.815; d.f. = 3; χ^2 = 48.7521; reject. There is enough evidence to support the claim that the composition of the legislature is dependent upon the state.

13. H_0: The number of ads people think that they've seen or heard in the media is independent of the gender of the individual. H_1: The number of ads people think that they've seen or heard in the media is dependent on the gender of the individual (claim). C.V. = 13.277; d.f. = 4; χ^2 = 9.562; do not reject. There is not enough evidence to support the claim that the number of ads people think they've seen or heard is related to the gender of the individual.

15. H_0: The type of practice of an attorney is independent of the gender of the attorney (claim). H_1: The type of practice of an attorney is dependent upon the gender of the attorney. C.V. = 5.991; d.f. = 2; χ^2 = 7.963; reject. There is enough evidence to reject the claim that the type of practice is independent of the gender of the attorney.

17. H_0: The type of video rented is independent of the person's age. H_1: The type of video rented is dependent on the person's age (claim). C.V. = 13.362; d.f. = 8; χ^2 = 46.733; reject. Yes, there is enough evidence to support the claim that the type of movie selected is related to the age of the customer.

19. H_0: The type of snack purchased is independent of the gender of the consumer (claim). H_1: The type of snack purchased is dependent upon the gender of the consumer. C.V. = 4.605; d.f. = 2; χ^2 = 6.342; reject. There is enough evidence to reject the claim that the type of snack is independent of the gender of the consumer.

21. H_0: The type of book purchased by an individual is independent of the gender of the individual (claim). H_1: The type of book purchased by an individual is

dependent on the gender of the individual. The d.f. = 2; $\alpha = 0.05$; $\chi^2 = 19.43$; P-value < 0.005; reject since P-value < 0.05. There is enough evidence to reject the claim that the type of book purchased by an individual is independent of the gender of the individual.

23. H_0: $p_1 = p_2 = p_3 = p_4$ (claim). H_1: At least one proportion is different. C.V. = 7.815; d.f. = 3; $\chi^2 = 5.317$; do not reject. There is not enough evidence to reject the claim that the proportions are equal.

25. H_0: $p_1 = p_2 = p_3 = p_4$ (claim). H_1: At least one of the proportions is different from the others. C.V. = 7.815; d.f. = 3; $\chi^2 = 1.172$; do not reject. There is not enough evidence to reject the claim that the proportions are equal. Since the survey was done in Pennsylvania, it is doubtful that it can be generalized to the population of the United States.

27. H_0: $p_1 = p_2 = p_3 = p_4$ (claim). H_1: At least one proportion is different. C.V. = 6.251; d.f. = 3; $\chi^2 = 12.755$; reject. There is enough evidence to reject the claim that the proportions are equal.

29. H_0: $p_1 = p_2 = p_3 = p_4$ (claim). H_1: At least one proportion is different. The d.f. = 3; $\chi^2 = 1.734$; $\alpha = 0.05$; P-value > 0.10 (0.629); do not reject since P-value > 0.05. There is not enough evidence to reject the claim that the proportions are equal.

31. H_0: $p_1 = p_2 = p_3$ (claim). H_1: At least one proportion is different. C.V. = 4.605; d.f. = 2; $\chi^2 = 2.401$; do not reject. There is not enough evidence to reject the claim that the proportions are equal.

33. $\chi^2 = 1.075$

Review Exercises

1. H_0: People show no preference for the day of the week that they do their shopping (claim). H_1: People show a preference for the day of the week that they do their shopping. C.V. = 12.592; d.f. = 6; $\chi^2 = 237.15$; reject. There is enough evidence to reject the claim that shoppers have no preference for the day of the week that they do their shopping. Retail merchants should probably plan for more shoppers on Fridays and Saturdays than they will have on the other days of the week.

3. H_0: Opinion is independent of gender. H_1: Opinion is dependent on gender (claim). C.V. = 4.605; d.f. = 2; $\chi^2 = 6.163$; reject. There is enough evidence to support the claim that opinion is dependent on gender.

5. H_0: The type of investment is independent of the age of the investor. H_1: The type of investment is dependent on the age of the investor (claim). C.V. = 9.488; d.f. = 4; $\chi^2 = 28.0$; reject. There is enough evidence to support the claim that the type of investment is dependent on the age of the investor.

7. H_0: $p_1 = p_2 = p_3$ (claim). H_1: At least one proportion is different. $\chi^2 = 4.912$; d.f. = 2; $\alpha = 0.01$; $0.05 < P$-value < 0.10 (0.086); do not reject since P-value > 0.01.

There is not enough evidence to reject the claim that the proportions are equal.

9. H_0: $p_1 = p_2 = p_3 = p_4$ (claim). H_1: At least one proportion is different. C.V. = 6.251; d.f. = 3; $\chi^2 = 6.70$; reject. There is enough evidence to reject the claim that the proportions are equal.

Chapter Quiz

1. False   2. True
3. False   4. c
5. b 6. d
7. 6   8. Independent
9. Right   10. At least five

11. H_0: The reasons why people lost their jobs are equally distributed (claim). H_1: The reasons why people lost their jobs are not equally distributed. C.V. = 5.991; d.f. = $\chi^2 = 2.334$; do not reject. There is not enough evidence to reject the claim that the reasons why people lost their jobs are equally distributed. The results could have been different 10 years ago since different factors of the economy existed then.

12. H_0: Takeout food is consumed according to the following distribution: 53% at home, 19% in the car, 14% at work, and 14% at other places (claim). H_1: The distribution is different from that stated in the null hypothesis. C.V. = 11.345; d.f. = 3; $\chi^2 = 5.271$; do not reject. There is not enough evidence to reject the claim that the distribution is as stated. Fast-food restaurants may want to make their advertisements appeal to those who like to take their food home to eat.

13. H_0: College students show the same preference for shopping channels as those surveyed. H_1: College students show a different preference for shopping channels (claim). C.V. = 7.815; d.f. = 3; $\alpha = 0.05$; $\chi^2 = 21.789$; reject. There is enough evidence to support the claim that college students show a different preference for shopping channels.

14. H_0: The number of commuters is distributed as follows: 75.7%, alone; 12.2%, carpooling; 4.7%, public transportation; 2.9%, walking; 1.2%, other; and 3.3%, working at home. H_1: The proportion of workers using each type of transportation differs from the stated proportions. C.V. = 11.071; d.f. = 5; $\chi^2 = 41.269$; reject. There is enough evidence to support the claim that the distribution is different from the one stated in the null hypothesis.

15. H_0: Ice cream flavor is independent of the gender of the purchaser (claim). H_1: Ice cream flavor is dependent upon the gender of the purchaser. C.V. = 7.815; d.f. = 3; $\chi^2 = 7.198$; do not reject. There is not enough evidence to reject the claim that ice cream flavor is independent of the gender of the purchaser.

16. H_0: The type of pizza ordered is independent of the age of the individual who purchases it. H_1: The type of pizza

ordered is dependent on the age of the individual who purchases it (claim). $\chi^2 = 107.3$; d.f. = 9; $\alpha = 0.10$; P-value < 0.005; reject since P-value < 0.10. There is enough evidence to support the claim that the pizza purchased is related to the age of the purchaser.

17. H_0: The color of the pennant purchased is independent of the gender of the purchaser (claim). H_1: The color of the pennant purchased is dependent on the gender of the purchaser. $\chi^2 = 5.632$; C.V. = 4.605; reject. There is enough evidence to reject the claim that the color of the pennant purchased is independent of the gender of the purchaser.

18. H_0: The opinion of the children on the use of the tax credit is independent of the gender of the children. H_1: The opinion of the children on the use of the tax credit is dependent upon the gender of the children (claim). C.V. = 4.605; d.f. = 2; $\chi^2 = 1.534$; do not reject. There is not enough evidence to support the claim that the opinion of the children on the use of the tax is dependent on their gender.

19. H_0: $p_1 = p_2 = p_3$ (claim). H_1: At least one proportion is different from the others. C.V. = 4.605; d.f. = 2; $\chi^2 = 6.711$; reject. There is enough evidence to reject the claim that the proportions are equal. It seems that more women are undecided about their jobs. Perhaps they want better income or greater chances of advancement.

Chapter 12

Exercises 12–3

1. The analysis of variance using the F test can be employed to compare three or more means.

3. The populations from which the samples were obtained must be normally distributed. The samples must be independent of each other. The variances of the populations must be equal.

5. $F = \dfrac{S_B^2}{S_W^2}$

7. The Scheffé test and the Tukey test

9. H_0: $\mu_1 = \mu_2 = \mu_3$. H_1: At least one mean is different from the others (claim). C.V. = 3.47; $\alpha = 0.05$; d.f.N. = 2; d.f.D. = 21; $F = 1.9912$; do not reject. There is not enough evidence to support the claim that at least one mean is different from the others.

11. H_0: $\mu_1 = \mu_2 = \mu_3$. H_1: At least one mean is different from the others (claim). C.V. = 3.98; $\alpha = 0.05$; d.f.N. = 2; d.f.D. = 11; $F = 2.7313$; do not reject. There is not enough evidence to support the claim that at least one mean is different from the others.

13. H_0: $\mu_1 = \mu_2 = \mu_3$. H_1: At least one mean is different from the others (claim). C.V. = 3.68; $\alpha = 0.05$; d.f.N. = 2; d.f.D. = 15; $F = 8.14$; reject. There is enough evidence to support the claim that at least one mean is different from the others. Tukey test: C.V. = 3.29; $\overline{X}_1 = 7.0$;

$\overline{X}_2 = 8.12$; $\overline{X}_3 = 5.23$; $\overline{X}_1$ versus $\overline{X}_2$, $q = -2.35$; $\overline{X}_1$ versus $\overline{X}_3$, $q = 3.47$; $\overline{X}_2$ versus $\overline{X}_3$, $q = -6.35$. There is a significant difference between $\overline{X}_1$ and $\overline{X}_3$, and $\overline{X}_2$ and $\overline{X}_3$. One reason for the difference might be that the students are enrolled in cyber schools with different fees.

15. H_0: $\mu_1 = \mu_2 = \mu_3$ (claim). H_1: At least one mean is different from the others. C.V. = 4.10; $\alpha = 0.05$; d.f.N. = 2; d.f.D. = 10; $F = 3.9487$; do not reject. There is not enough evidence to reject the claim that the means are equal.

17. H_0: $\mu_1 = \mu_2 = \mu_3$. H_1: At least one mean is different from the others (claim). $F = 10.118$; P-value = 0.00102; reject. Scheffé test: C.V. = 5.22; $\overline{X}_1$ versus $\overline{X}_2$: $F = 2.91$; $\overline{X}_1$ versus $\overline{X}_3$: $F = 19.3$; $\overline{X}_2$ versus $\overline{X}_3$: $F = 8.40$. There is a significant difference between $\overline{X}_1$ and $\overline{X}_3$, and $\overline{X}_2$ and $\overline{X}_3$.

19. H_0: $\mu_1 = \mu_2 = \mu_3 = \mu_4$. H_1: At least one mean is different from the others (claim). C.V. = 5.29; $\alpha = 0.01$; d.f.N. = 3; d.f.D. = 16; $F = 0.636$; do not reject. There is not enough evidence to support that at least one mean is different from the others. Students may have had discipline problems. Parents may not like the regular school district, etc.

Exercises 12–4

1. The two-way ANOVA allows the researcher to test the effects of two independent variables and a possible interaction effect. The one-way ANOVA can test the effects of only one independent variable.

3. The mean square values are computed by dividing the sum of squares by the corresponding degrees of freedom.

5. *a.* For factor A, d.f.$_A$ = 2

   b. For factor B, d.f.$_B$ = 1

   c. d.f.$_{A\times B}$ = 2

   d. d.f.$_{\text{within}}$ = 24

7. The two types of interactions that can occur are ordinal and disordinal.

9. *a.* The lines will be parallel or approximately parallel. They may also coincide.

   b. The lines will not intersect and they will not be parallel.

   c. The lines will intersect.

11. H_0: There is no interaction effect between the time of day and the type of diet on a person's sodium level. H_1: There is an interaction effect between the time of day and the type of diet on a person's sodium level.

    H_0: There is no difference between the means for the sodium level for the times of day. H_1: There is a difference between the means for the sodium level for the times of day.

    H_0: There is no difference between the means for the sodium level for the type of diet. H_1: There is a difference between the means for the sodium level for the type of diet.

ANOVA Summary Table

Source	SS	d.f.	MS	F
Time	1800.0	1	1800.000	25.806
Diet	242.0	1	242.000	3.470
Interaction	264.5	1	264.500	3.792
Within	279.0	4	69.750	
Total	2585.5	7		

The critical value at $\alpha = 0.05$ with d.f.N. = 1 and d.f.D. = 4 is 7.71 for F_A, F_B, and $F_{A \times B}$. Since the only F test value that exceeds 7.71 is the one for the time, 25.806, it can be concluded that there is a difference in the means for the sodium level taken at two different times.

13. H_0: There is no interaction effect between the gender of the individual and the duration of the training on the test scores. H_1: There is an interaction effect between the gender of the individual and the duration of the training on the test scores.

H_0: There is no difference between the means of the test scores for the males and females. H_1: There is a difference between the means of the test scores for the males and females.

H_0: There is no difference between the means of the test scores for the two different durations. H_1: There is a difference between the means of the test scores for the two different durations.

ANOVA Summary Table

Source	SS	d.f.	MS	F
Gender	57.042	1	57.042	0.835
Duration	7.042	1	7.042	0.103
Interaction	3978.375	1	3978.375	58.270
Within	1365.500	20	68.275	
Total	5407.959	23		

The critical value at $\alpha = 0.10$ with d.f.N. = 1 and d.f.D. = 20 is 2.97. Since the F test value for the interaction is greater than the critical value, it can be concluded that gender affects test scores differently for the duration levels.

15. H_0: There is no interaction effect between the ages of the salespeople and the products they sell on the monthly sales. H_1: There is an interaction effect between the ages of the salespeople and the products they sell on the monthly sales.

H_0: There is no difference in the means of the monthly sales of the two age groups. H_1: There is a difference in the means of the monthly sales of the two age groups.

H_0: There is no difference among the means of the sales for the different products. H_1: There is a difference among the means of the sales for the different products.

ANOVA Summary Table

Source	SS	d.f.	MS	F
Age	168.033	1	168.033	1.567
Product	1,762.067	2	881.034	8.215
Interaction	7,955.267	2	3,977.634	37.087
Error	2,574.000	24	107.250	
Total	12,459.367	29		

At $\alpha = 0.05$, the critical values are: for age, d.f.N. = 1, d.f.D. = 24, C.V. = 4.26; for product and interaction, d.f.N. = 2 and d.f.D. = 24; C.V. = 3.40. The null hypotheses for the interaction effect and for the type of product sold are rejected since the F test values exceed the critical value 3.40. The cell means are as follows:

Age \ Product	Pools	Spas	Saunas
Over 30	38.8	28.6	55.4
30 and under	21.2	68.6	18.8

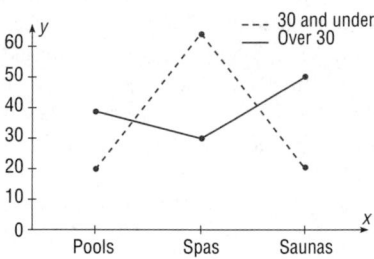

Since the lines cross, there is a disordinal interaction; hence, there is an interaction effect between the ages of salespeople and the type of products sold.

Review Exercises

1. H_0: $\mu_1 = \mu_2 = \mu_3$ (claim). H_1: At least one mean is different from the others. C.V. = 5.39; d.f.N. = 2; d.f.D. = 33; $\alpha = 0.01$; $F = 6.94$; reject. Tukey test: C.V. = 4.45; $\overline{X}_1$ versus $\overline{X}_2$: $q = 0.342$; $\overline{X}_1$ versus $\overline{X}_3$: $q = 4.72$; $\overline{X}_2$ versus $\overline{X}_3$: $q = 4.38$. There is a significant difference between $\overline{X}_1$ and $\overline{X}_3$.

3. H_0: $\mu_1 = \mu_2 = \mu_3$. H_1: At least one mean is different from the others (claim). C.V. = 3.55; $\alpha = 0.05$; d.f.N. = 2; d.f.D. = 18; $F = 0.0408$; do not reject. There is not enough evidence to support the claim that at least one mean is different from the others.

5. H_0: $\mu_1 = \mu_2 = \mu_3$. H_1: At least one mean is different from the others (claim). C.V. = 2.61; $\alpha = 0.10$; d.f.N. = 2; d.f.D. = 19; $F = 0.4876$; do not reject. There is not enough evidence to support the claim that at least one mean is different from the others.

7. H_0: $\mu_1 = \mu_2 = \mu_3 = \mu_4$. H_1: At least one mean is different from the others (claim). C.V. = 3.59; $\alpha = 0.05$;

d.f.N. = 3; d.f.D. = 11; F = 0.182; do not reject. There is not enough evidence to support the claim that at least one mean is different from the others.

9. H_0: There is no interaction effect between the type of exercise program and the type of diet on a person's glucose level. H_1: There is an interaction effect between type of exercise program and the type of diet on a person's glucose level.

H_0: There is no difference in the means for the glucose levels of the people in the two exercise programs. H_1: There is a difference in the means for the glucose levels of the people in the two exercise programs.

H_0: There is no difference in the means for the glucose levels of the people in the two diet programs. H_1: There is a difference in the means for the glucose levels of the people in the two diet programs.

ANOVA Summary Table

Source	SS	d.f.	MS	F
Exercise	816.750	1	816.750	60.50
Diet	102.083	1	102.083	7.56
Interaction	444.083	1	444.083	32.90
Within	108.000	8	13.500	
Total	1470.916	11		

At α = 0.05, d.f.N. = 1, d.f.D. = 8, and the critical value is 5.32 for each F_A, F_B, and $F_{A \times B}$. Hence, all three null hypotheses are rejected. The cell means should be calculated.

Exercise \ Diet	A	B
I	64.000	57.667
II	68.333	86.333

Since the means for exercise program I are both smaller than those for exercise program II and the vertical differences are not the same, the interaction is ordinal. Hence one can say that there is a difference for exercise and diet, and that an interaction effect is present.

Chapter Quiz

1. False
2. False
3. False
4. True
5. d
6. a
7. a
8. c
9. ANOVA
10. Tukey
11. Two
12. H_0: $\mu_1 = \mu_2 = \mu_3 = \mu_4$. H_1: At least one mean is different from the others (claim). C.V. = 3.49; α = 0.05; d.f.N. = 3; d.f.D. = 12; F = 3.23; do not reject. There is not enough evidence to support the claim that there is a difference in the means.

13. H_0: $\mu_1 = \mu_2 = \mu_3$. H_1: At least one mean is different from the others (claim). C.V. = 6.93; α = 0.01; d.f.N. = 2; d.f.D. = 12; F = 3.49. There is not enough evidence to support the claim that at least one mean is different from the others. Writers would want to target their material to the age group of the viewers.

14. H_0: $\mu_1 = \mu_2 = \mu_3$. H_1: At least one mean is different from the others (claim). C.V. = 3.68; α = 0.05; d.f.N. = 2; d.f.D. = 15; F = 10.494; reject. Tukey test: C.V. = 3.67; $\bar{X}_1$ = 47.67; $\bar{X}_2$ = 63; $\bar{X}_3$ = 43.83; $\bar{X}_1$ versus $\bar{X}_2$, q = -4.90; $\bar{X}_1$ versus $\bar{X}_3$, q = 1.23; $\bar{X}_2$ versus $\bar{X}_3$, q = 6.12. There is a significant difference between $\bar{X}_1$ and $\bar{X}_2$ and between $\bar{X}_2$ and $\bar{X}_3$.

15. H_0: $\mu_1 = \mu_2 = \mu_3$ (claim). H_1: At least one mean is different from the others. C.V. = 2.70; α = 0.10; d.f.N. = 2; d.f.D. = 15; F = 0.0509; do not reject. There is not enough evidence to reject the claim that the means are equal.

16. H_0: $\mu_1 = \mu_2 = \mu_3 = \mu_4$. H_1: At least one mean is different from the others (claim). C.V. = 3.07; α = 0.05; d.f.N. = 3; d.f.D. = 21; F = 0.4564; do not reject. There is not enough evidence to support the claim that at least one mean is different from the others.

17. *a.* Two-way ANOVA

b. Diet and exercise program

c. 2

d. H_0: There is no interaction effect between the type of exercise program and the type of diet on a person's weight loss. H_1: There is an interaction effect between the type of exercise program and the type of diet on a person's weight loss.

H_0: There is no difference in the means of the weight losses of people in the exercise programs. H_1: There is a difference in the means of the weight losses of people in the exercise programs.

H_0: There is no difference in the means of the weight losses of people in the diet programs. H_1: There is a difference in the means of the weight losses of people in the diet programs.

e. Diet: F = 21.0, significant; exercise program: F = 0.429, not significant; interaction: F = 0.429, not significant

f. Reject the null hypothesis for the diets.

Chapter 13

Exercises 13–2

1. *Nonparametric* means hypotheses other than those using population parameters can be tested; *distribution-free* means no assumptions about the population distributions have to be satisfied.

3. Nonparametric methods have the following advantages:

a. They can be used to test population parameters when the variable is not normally distributed.

b. They can be used when data are nominal or ordinal.

c. They can be used to test hypotheses other than those involving population parameters.

d. The computations are easier in some cases than the computations of the parametric counterparts.

e. They are easier to understand.

The disadvantages are as follows:

a. They are less sensitive than their parametric counterparts.

b. They tend to use less information than their parametric counterparts.

c. They are less efficient than their parametric counterparts.

5.

Data	22	32	34	43	43	65	66	71
Rank	1	2	3	4.5	4.5	6	7	8

7.

Data	4	6	6	7	8	9	9	10	13	15	16	18
Rank	1	2.5	2.5	4	5	6.5	6.5	8	9	10	11	12

9.

Data	188	188	197	256	321	530	763
Rank	1.5	1.5	3	4	5	6	7

Exercises 13–3

1. The sign test uses only positive or negative signs.

3. The smaller number of positive or negative signs.

5. H_0: median = 2.8 (claim) and H_1: median $\neq$ 2.8; test value = 8; C.V. = 4; do not reject. There is not enough evidence to reject the claim that the median is 2.8.

7. H_0: median = 25 (claim) and H_1: median $\neq$ 25; test value = 7; C.V. = 4; do not reject. There is not enough evidence to reject the claim that the median is 25. School boards could use the median to plan for the costs of cyber school enrollments.

9. H_0: median = $10.86 (claim) and H_1: median $\neq$ $10.86; C.V. = +1.96; $z = -0.77$; do not reject. There is not enough evidence to reject the claim that the median is $10.86. Home buyers could estimate the yearly cost of their gas bills.

11. H_0: $p \leq 0.5$ and H_1: $p > 0.5$ (claim); $z = -0.99$; P-value = 0.1611; do not reject. There is not enough evidence to support the claim that more than 50% of the students favor single-room dormitories.

13. H_0: median = 50 (claim) and H_1: median $\neq$ 50; $z = -2.3$; P-value = 0.0214; reject. There is enough evidence to reject the claim that 50% of the students are against extending the school year.

15. H_0: The medication has no effect on weight loss and H_1: The medication affects weight loss (claim); C.V. = 0; test value = 1; do not reject. There is not enough evidence to support the claim that the medication affects weight loss.

17. H_0: Reasoning ability will not be affected by the course and H_1: Reasoning ability increased after the course (claim); C.V. = 1; test value = 1; reject. There is enough

evidence to support the claim that reasoning ability has increased after the course.

19. H_0: The number of viewers is the same as last year (claim); H_1: The number of viewers is not the same as last year; C.V. = 0; test value = 2; do not reject. There is not enough evidence to reject the claim that the number of viewers is the same as last year.

21. $6 \leq$ median ≤ 22

23. $4.7 \leq$ median ≤ 9.3

25. $17 \leq$ median ≤ 33

Exercises 13–4

1. n_1 and n_2 are each greater than or equal to 10.

3. The standard normal distribution

5. H_0: There is no difference in the number of jetplanes grounded for repairs and the number of jetplanes in storage (claim). H_1: There is a difference in the number of planes in each group. C.V. = ± 1.96; $z = -2.65$; reject. There is enough evidence to reject the claim that there is no difference in the number of planes grounded and the number of planes in storage. The airline should plan for more storage facilities.

7. H_0: There is no difference between the stopping distances of the two types of automobiles (claim); H_1: There is a difference between the stopping distances of the two types of automobiles. C.V. = ± 1.65; $z = -2.72$; reject. There is not enough evidence to reject the claim that there is no difference in the stopping distances of the automobiles. In this case, mid size cars have a smaller stopping distance.

9. H_0: There is no difference in the number of hunting accidents in the two geographic areas. H_1: There is a difference in the number of hunting accidents (claim). C.V. = ± 1.96; $z = 2.57$; reject. There is enough evidence to support the claim that there is a difference in the number of accidents in the two areas. The number of accidents may be related to the number of hunters in the areas.

11. H_0: There is no difference in the times needed to assemble the product. H_1: There is a difference in the times needed to assemble the product (claim). C.V. = ± 1.96; $z = +3.56$; reject. There is enough evidence to support the claim that there is a difference in the productivity of the two groups.

Exercises 13–5

1. The t test for dependent samples

3. Sum of minus ranks is -6; sum of plus ranks is $+15$. The test value is 6.

5. C.V. = 20; reject

7. C.V. = 130; do not reject

9. H_0: The human dose is less than or equal to the animal dose. H_1: The human dose is more than the animal dose (claim). C.V. = 6; $w_s = 2$; reject. There is enough evidence to support the claim that the human dose costs

more than the equivalent animal dose. One reason is that some people might not be inclined to pay a lot of money for their pets' medication.

11. H_0: The number of people who watched the Presidential debates is less than or equal to the number of people who watched the Vice Presidential debates. H_1: The number of people who watched the Presidential debates is greater than the number of people who watched the Vice Presidential debates (claim). C.V. = 0; $w_s = 0$; reject. There is enough evidence to support the claim that the number of people who watched the Presidential debates is greater than the number of people who watched the Vice Presidential debates. Yes, a Vice President is only a heartbeat away from being President, so his opinions are valuable.

13. H_0: The prices of prescription drugs in the United States are greater than or equal to the prices in Canada. H_1: The drugs sold in Canada are cheaper. C.V. = 11; $w_s = 3$; reject. There is enough evidence to support the claim that the drugs are less expensive in Canada.

Exercises 13–6

1. H_0: There is no difference in the number of calories. H_1: There is a difference in the number of calories (claim). C.V. = 7.815; $H = 2.842$; do not reject. There is not enough evidence to support the claim that there is a difference in the number of calories.

3. H_0: There is no difference in the prices of the three types of lawnmowers. H_1: There is a difference in the prices of the three types of lawnmowers (claim). C.V. = 4.605; $H = 1.07$; do not reject. There is not enough evidence to support the claim that the prices are different. No, price is not a factor. Results are suspect since one sample is less than 5.

5. H_0: There is no difference in the number of carbohydrates in one serving of each of the three types of food. H_1: There is a difference in the number of carbohydrates in each of the three types of food (claim). C.V. = 9.210; $H = 11.58$; reject. There is enough evidence to say that there is a difference in the number of carbohydrates in the three foods. You should recommend the ice cream.

7. H_0: There is no difference in the number of deaths due to severe weather. H_1: There is a difference in the number of deaths due to severe weather (claim). C.V. = 6.251; $H = 5.537$; do not reject. There is not enough evidence to support the claim that there is a difference in the number of deaths due to severe weather.

9. H_0: There is no difference in the number of crimes in the five precincts. H_1: There is a difference in the number of crimes in the five precincts (claim). C.V. = 13.277; $H = 20.753$; reject. There is enough evidence to support the claim that there is a difference in the number of crimes in the five precincts.

11. H_0: There is no difference in the final exam scores of the three groups. H_1: There is a difference in the final exam

scores of the three groups (claim). $H = 1.710$; P-value > 0.10 (0.425); do not reject. There is not enough evidence to support the claim that there is a difference in the final exam scores of the three groups.

Exercises 13–7

1. 0.716

3. 0.648

5. $r_s = 0.612$; H_0: $\rho = 0$ and H_1: $\rho \neq 0$; C.V. = ±0.564; reject. There is a significant relationship between the number of temperatures and the record high temperatures.

7. $r_s = 0.943$; H_0: $\rho = 0$ and H_1: $\rho \neq 0$; C.V. = ±0.886; reject. There is a significant relationship between the sentence and the time period.

9. $r_s = 0.5$; H_0: $\rho = 0$ and H_1: $\rho \neq 0$; C.V. = ±0.738; do not reject. There is not enough evidence to say that a significant correlation exists.

11. $r_s = 0.624$; H_0: $\rho = 0$ and H_1: $\rho \neq 0$; C.V. = 0.700; do not reject. There is no significant relationship between gasoline prices paid to the car rental agency and regular gasoline prices. One would wonder how the car rental agencies determine their prices.

13. $r_s = -0.10$; H_0: $\rho = 0$ and H_1: $\rho \neq 0$; C.V. = ±0.900; do not reject. There is no significant relationship between the number of cyber school students and the cost per pupil. In this case, the cost per pupil is different in each district.

15. H_0: The number of cavities in a person occurs at random. H_1: The null hypothesis is not true. There are 21 runs; the expected number of runs is between 10 and 22. Therefore, do not reject the null hypothesis; the number of cavities in a person occurs at random.

17. H_0: The Lotto numbers occur at random. H_1: The null hypothesis is not true. There are 14 runs, and this value is between 7 and 18. Hence, do not reject the null hypothesis; the Lotto numbers occur at random.

19. H_0: The number of defective cassette cases manufactured by the machine occurs at random. H_1: The null hypothesis is not true. There are 10 runs; and since this value is between 9 and 20, do not reject the null hypothesis. The number of defective cassette cases manufactured by a machine occurs at random.

21. H_0: The number of absences of employees occurs at random over a 30-day period. H_1: The null hypothesis is not true. There are only 6 runs, and this value does not fall within the 9-to-21 range. Hence, the null hypothesis is rejected; the absences do not occur at random.

23. H_0: The IQs of students who complete the tests are random. H_1: The null hypothesis is not true. Do not reject, since there are 7 runs, and this value is within the 6-to-16 range; the IQs occur at random.

25. ±0.479

27. ±0.215

Review Exercises

1. H_0: median = 5 (claim) and H_1: median ≠ 5; test value = 6; C.V. = 2. Do not reject; there is not enough evidence to reject the claim that the median is 5.

3. H_0: The special diet has no effect on weight and H_1: The diet increases weight (claim); C.V. = 1; test value = 3. Do not reject; there is not enough evidence to support the claim that there was an increase in weight.

5. H_0: There is no difference in the amount of money the groups spent for the textbook. H_1: There is a difference in the amount of money the groups spent for the textbook (claim). C.V. = ±1.65; $z = -3.59$. Reject; there is enough evidence to support the claim that there is a difference in the amount of money the groups spent on the textbooks.

7. H_0: The number of sick days workers used was not reduced. H_1: The number of sick days workers used was reduced (claim). C.V. = 4; $w_s = 8.5$. Do not reject; there is not enough evidence to support the claim that the number of sick days was reduced.

9. H_0: The diet has no effect on learning. H_1: The diet affects learning (claim). C.V. = 5.991; $H = 8.5$. Reject; there is enough evidence to support the claim that the diets do affect learning.

11. $r_s = 0.891$; H_0: $\rho = 0$ and H_1: $\rho \neq 0$; C.V. = ±0.648; reject. There is a significant relationship in the average number of people who are watching the television shows for both years.

13. H_0: The grades of students who finish the exam occur at random. H_1: The null hypothesis is not true. Since there are 8 runs and this value does not fall in the 9-to-21 interval, the null hypothesis is rejected. The grades do not occur at random.

Chapter Quiz

1. False
2. False
3. True
4. True
5. *a*
6. *c*
7. *d*
8. *b*
9. Nonparametric
10. Nominal, ordinal
11. Sign
12. Sensitive

13. H_0: median = 300 (claim) and H_1: median ≠ 300. There are 7 plus signs. Do not reject since 7 is greater than the critical value of 5. There is not enough evidence to reject the claim that the median is 300.

14. H_0: median = 1200 (claim) and H_1: median ≠ 1200. There are 10 minus signs. Do not reject since the 10 is greater than the critical value 6. There is not enough evidence to reject the claim that the median is 1200.

15. H_0: There will be no change in the weight of the turkeys after the special diet. H_1: The turkeys will weigh more after the special diet (claim). There is 1 plus sign; hence, the null hypothesis is rejected. There is enough evidence to support the claim that the turkeys gained weight on the special diet.

16. H_0: The distributions are the same. H_1: The distributions are different (claim). $z = -0.05$; C.V. = ±1.96; do not reject the null hypothesis. There is not enough evidence to reject the claim that the distributions are the same.

17. H_0: The distributions are the same. H_1: The distributions are different (claim). $z = -0.14434$; C.V. = ±1.65; do not reject the null hypothesis. There is not enough evidence to support the claim that the distributions are different.

18. H_0: There is no difference in the GPA of the students before and after the workshop. H_1: There is a difference in the GPA of the students before and after the workshop (claim). Test statistic = 0; C.V. = 2; reject the null hypothesis. There is enough evidence to support the claim that there is a difference in the GPAs of the students.

19. H_0: There is no difference in the breaking strengths of the tapes. H_1: There is a difference in the breaking strengths of the tapes (claim). $H = 29.25$; $\chi^2 = 5.991$; reject the null hypothesis. There is enough evidence to support the claim that there is a difference in the breaking strengths of the tapes.

20. H_0: There is no difference in the reaction times of the monkeys. H_1: There is a difference in the reaction times of the monkeys (claim). $H = 6.9$; $0.025 < P\text{-value} < 0.05$ (0.032); reject the null hypothesis. There is enough evidence to support the claim that there is a difference in the reaction times of the monkeys.

21. $r_s = 0.683$; H_0: $\rho = 0$ and H_1: $\rho \neq 0$; C.V. = ±0.600; reject. There is enough evidence to say that there is a significant relationship between the drug prices.

22. $r_s = 0.943$; H_0: $\rho = 0$ and H_1: $\rho \neq 0$; C.V. = ±0.829; reject. There is a significant relationship between the amount of money spent on Head Start and the number of students enrolled in the program.

23. H_0: The births of babies occur at random according to gender. H_1: The null hypothesis is not true. There are 10 runs, and since this is between 8 and 19, the null hypothesis is not rejected. There is not enough evidence to reject the null hypothesis that the gender occurs at random.

24. H_0: There is no difference in the rpm of the motors before and after the reconditioning. H_1: There is a difference in the rpm of the motors before and after the reconditioning (claim). Test statistic = 0; C.V. = 6; do not reject the null hypothesis. There is not enough evidence to support the claim that there is a difference in the rpm of the motors before and after reconditioning.

25. H_0: The numbers occur at random. H_1: The null hypothesis is not true. There are 16 runs, and since this is between 9 and 21, the null hypothesis is not rejected. There is not enough evidence to reject the null hypothesis that the numbers occur at random.

Chapter 14

Exercises 14–2

1. Random, systematic, stratified, cluster

3. A sample must be randomly selected.

5. Talking to people on the street, calling people on the phone, and asking one's friends are three incorrect ways of obtaining a sample.

7. Random sampling has the advantage that each unit of the population has an equal chance of being selected. One disadvantage is that the units of the population must be numbered; if the population is large, this could be somewhat time-consuming.

9. An advantage of stratified sampling is that it ensures representation for the groups used in stratification; however, it is virtually impossible to stratify the population so that all groups are represented.

11. Answers will vary. 13. Answers will vary.

15. Answers will vary. 17. Answers will vary.

19. Answers will vary.

Exercises 14–3

1. Flaw—asking a biased question. Do you think XYZ Department Store should carry brand name merchandise?

3. Flaw—asking a biased question. Should banks charge a fee to balance their customers' checkbooks?

5. Flaw—confusing words. How many hours did you study for this exam?

7. Flaw—confusing words. If a plane were to crash on the border of New York and New Jersey, where should the victims be buried?

9. Answers will vary.

Exercises 14–5

1. Simulation involves setting up probability experiments that mimic the behavior of real-life events.

3. John Von Neumann and Stanislaw Ulam

5. The steps are as follows:
   a. List all possible outcomes.
   b. Determine the probability of each outcome.
   c. Set up a correspondence between the outcomes and the random numbers.
   d. Conduct the experiment by using random numbers.
   e. Repeat the experiment and tally the outcomes.
   f. Compute any statistics and state the conclusions.

7. When the repetitions increase, there is a higher probability that the simulation will yield more precise answers.

9. Use two-digit random numbers: 01 through 70 is considered a success and 71 through 99 and 00 a failure. One-digit random numbers can also be used: 1 through 7 as a success and 8, 9, and 0 a miss.

11. Use one-digit random numbers: 1 through 8 would be a hit and 9 and 0 a miss.

13. Let an odd number represent heads and an even number represent tails. Then each person selects a digit at random.

15. Answers will vary. 17. Answers will vary.

19. Answers will vary. 21. Answers will vary.

23. Answers will vary.

Review Exercises

1. Answers will vary. 3. Answers will vary.

5. Answers will vary. 7. Answers will vary.

9. Use one-digit random numbers 1 through 4 for a strikeout and 5 through 9 and 0 represent anything other than a strikeout.

11. In this case, a one-digit random number is selected. Numbers 1 through 6 represent the numbers on the face. Ignore 7, 8, 9, and 0 and select another number.

13. Let the digits 1–3 represent rock, let 4–6 represent paper, let 7–9 represent scissors, and omit 0.

15. Answers will vary. 17. Answers will vary.

19. Flaw—asking a biased question. Have you ever driven through a red light?

21. Flaw—asking a double-barreled question. Do you think all automobiles should have heavy-duty bumpers?

Chapter Quiz

1. True 2. True
3. False 4. True
5. *a* 6. *c*
7. *c* 8. Larger
9. Biased 10. Cluster

11.–14. Answers will vary.

15. Use two-digit random numbers: 01 through 45 means the player wins. Any other two-digit random number means the player loses.

16. Use two-digit random numbers: 01 through 05 means a cancellation. Any other two-digit random number means the person shows up.

17. The random numbers 01 through 13 represent the 13 cards in hearts. The random numbers 14 through 26 represent the 13 cards in diamonds. The random numbers 27 through 39 represent the 13 spades, and 40 through 52 represent the 13 clubs. Any number over 52 is ignored.

18. Use two-digit random numbers to represent the spots on the face of the dice. Ignore any two-digit random numbers with 7, 8, 9, or 0. For cards, use two-digit random numbers between 01 and 13.

19. Use two-digit random numbers. The first digit represents the first player, and the second digit represents the second player. If both numbers are odd or even, player 1 wins. If a digit is odd and the other digit is even, player 2 wins.

20.–24. Answers will vary.

Appendix A

A–1. 362,880

A–3. 120

A–5. 1

A–7. 1320

A–9. 20

A–11. 126

A–13. 70

A–15. 1

A–17. 560

A–19. 2520

A–21. 121; 2181; 14,641; 716.9

A–23. 32; 258; 1024; 53.2

A–25. 328; 22,678; 107,584; 1161.2

A–27. 693; 50,511; 480,249; 2486.1

A–29. 318; 20,150; 101,124; 3296

A–31.

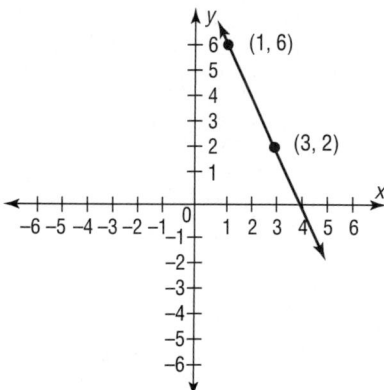

A–33.

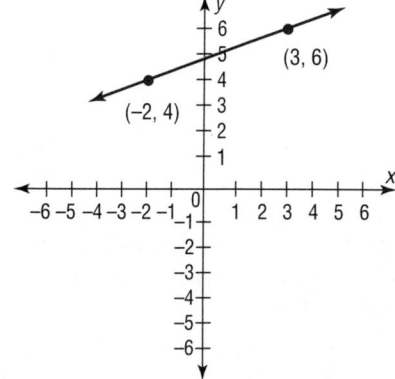

A–35.

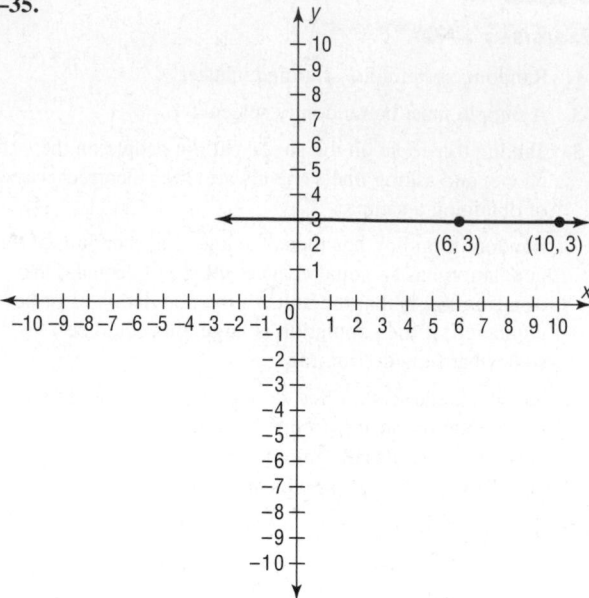

A–37.

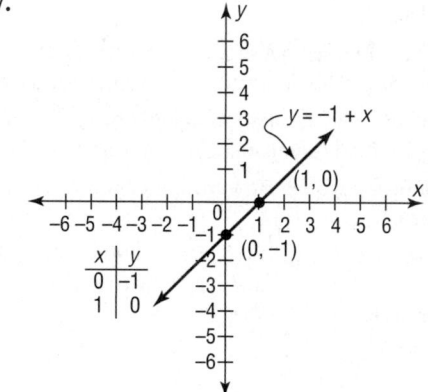

A–39.

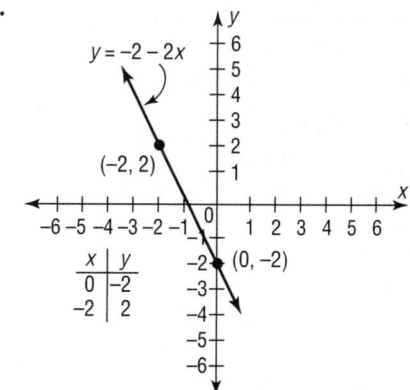

Appendix B–2

B–1. 0.65

B–3. 0.653

B–5. 0.379

B–7. $\frac{1}{4}$

B–9. 0.64

B–11. 0.857

Index

95%	98%	99%
0.025	0.01	0.005
0.05	0.02	0.01
12.706	31.821	63.657
4.303	6.965	9.925
3.182	4.541	5.841
2.776	3.747	4.604
2.571	3.365	4.032
2.447	3.143	3.707
2.365	2.998	3.499
2.306	2.896	3.355
2.262	2.821	3.250
2.228	2.764	3.169
2.201	2.718	3.106
2.179	2.681	3.055
2.160	2.650	3.012
2.145	2.624	2.977
2.131	2.602	2.947
2.120	2.583	2.921
2.110	2.567	2.898
2.101	2.552	2.878
2.093	2.539	2.861
2.086	2.528	2.845
2.080	2.518	2.831
2.074	2.508	2.819
2.069	2.500	2.807
2.064	2.492	2.797
2.060	2.485	2.787
2.056	2.479	2.779
2.052	2.473	2.771
2.048	2.467	2.763
1.960	2.326^c	2.576^d

One tail   Two tails

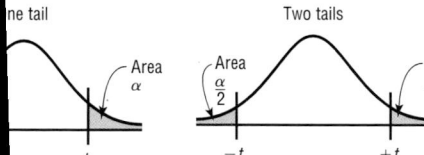

Table G The Chi-Square Distribution

Degrees of freedom	α									
	0.995	0.99	0.975	0.95	0.90	0.10	0.05	0.025	0.01	0.005
1	—	—	0.001	0.004	0.016	2.706	3.841	5.024	6.635	7.879
2	0.010	0.020	0.051	0.103	0.211	4.605	5.991	7.378	9.210	10.597
3	0.072	0.115	0.216	0.352	0.584	6.251	7.815	9.348	11.345	12.838
4	0.207	0.297	0.484	0.711	1.064	7.779	9.488	11.143	13.277	14.860
5	0.412	0.554	0.831	1.145	1.610	9.236	11.071	12.833	15.086	16.750
6	0.676	0.872	1.237	1.635	2.204	10.645	12.592	14.449	16.812	18.548
7	0.989	1.239	1.690	2.167	2.833	12.017	14.067	16.013	18.475	20.278
8	1.344	1.646	2.180	2.733	3.490	13.362	15.507	17.535	20.090	21.955
9	1.735	2.088	2.700	3.325	4.168	14.684	16.919	19.023	21.666	23.589
10	2.156	2.558	3.247	3.940	4.865	15.987	18.307	20.483	23.209	25.188
11	2.603	3.053	3.816	4.575	5.578	17.275	19.675	21.920	24.725	26.757
12	3.074	3.571	4.404	5.226	6.304	18.549	21.026	23.337	26.217	28.299
13	3.565	4.107	5.009	5.892	7.042	19.812	22.362	24.736	27.688	29.819
14	4.075	4.660	5.629	6.571	7.790	21.064	23.685	26.119	29.141	31.319
15	4.601	5.229	6.262	7.261	8.547	22.307	24.996	27.488	30.578	32.801
16	5.142	5.812	6.908	7.962	9.312	23.542	26.296	28.845	32.000	34.267
17	5.697	6.408	7.564	8.672	10.085	24.769	27.587	30.191	33.409	35.718
18	6.265	7.015	8.231	9.390	10.865	25.989	28.869	31.526	34.805	37.156
19	6.844	7.633	8.907	10.117	11.651	27.204	30.144	32.852	36.191	38.582
20	7.434	8.260	9.591	10.851	12.443	28.412	31.410	34.170	37.566	39.997
21	8.034	8.897	10.283	11.591	13.240	29.615	32.671	35.479	38.932	41.401
22	8.643	9.542	10.982	12.338	14.042	30.813	33.924	36.781	40.289	42.796
23	9.262	10.196	11.689	13.091	14.848	32.007	35.172	38.076	41.638	44.181
24	9.886	10.856	12.401	13.848	15.659	33.196	36.415	39.364	42.980	45.559
25	10.520	11.524	13.120	14.611	16.473	34.382	37.652	40.646	44.314	46.928
26	11.160	12.198	13.844	15.379	17.292	35.563	38.885	41.923	45.642	48.290
27	11.808	12.879	14.573	16.151	18.114	36.741	40.113	43.194	46.963	49.645
28	12.461	13.565	15.308	16.928	18.939	37.916	41.337	44.461	48.278	50.993
29	13.121	14.257	16.047	17.708	19.768	39.087	42.557	45.722	49.588	52.336
30	13.787	14.954	16.791	18.493	20.599	40.256	43.773	46.979	50.892	53.672
40	20.707	22.164	24.433	26.509	29.051	51.805	55.758	59.342	63.691	66.766
50	27.991	29.707	32.357	34.764	37.689	63.167	67.505	71.420	76.154	79.490
60	35.534	37.485	40.482	43.188	46.459	74.397	79.082	83.298	88.379	91.952
70	43.275	45.442	48.758	51.739	55.329	85.527	90.531	95.023	100.425	104.215
80	51.172	53.540	57.153	60.391	64.278	96.578	101.879	106.629	112.329	116.321
90	59.196	61.754	65.647	69.126	73.291	107.565	113.145	118.136	124.116	128.299
100	67.328	70.065	74.222	77.929	82.358	118.498	124.342	129.561	135.807	140.169

Source: Donald B. Owen. *Handbook of Statistics Tables,* © 1962, by Addison-Wesley Publishing Co., Inc. Reading, Massachusetts. Table A–5. Reprinted with permission of Addison-Wesley Longman Publishing Company, Inc.

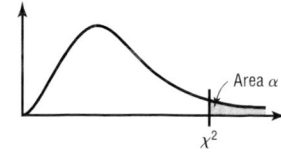

Table I Critical Values for PPMC

Reject H_0: $\rho = 0$ if the absolute value of r is greater than the value given in the table. The values are for a two-tailed test; d.f. $= n - 2$.

d.f	$\alpha = 0.05$	$\alpha = 0.01$
1	0.999	0.999
2	0.950	0.999
3	0.878	0.959
4	0.811	0.917
5	0.754	0.875
6	0.707	0.834
7	0.666	0.798
8	0.632	0.765
9	0.602	0.735
10	0.576	0.708
11	0.553	0.684
12	0.532	0.661
13	0.514	0.641
14	0.497	0.623
15	0.482	0.606
16	0.468	0.590
17	0.456	0.575
18	0.444	0.561
19	0.433	0.549
20	0.423	0.537
25	0.381	0.487
30	0.349	0.449
35	0.325	0.418
40	0.304	0.393
45	0.288	0.372
50	0.273	0.354
60	0.250	0.325
70	0.232	0.302
80	0.217	0.283
90	0.205	0.267
100	0.195	0.254

Source: From *Biometrika Tables for Statisticians,* Vol. 1 (1962), p. 138. Reprinted with permission.

Important Formulas

Chapter 3 Data Description

Mean for individual data: $\bar{X} = \dfrac{\sum X}{n}$

Mean for grouped data: $\bar{X} = \dfrac{\sum f \cdot X_m}{n}$

Standard deviation for a sample:

$$s = \sqrt{\frac{\sum X^2 - [(\sum X)^2/n]}{n-1}}$$

Standard deviation for grouped data:

$$s = \sqrt{\frac{\sum f \cdot X_m^2 - [(\sum f \cdot X_m)^2/n]}{n-1}}$$

Range rule of thumb: $s \approx \dfrac{\text{range}}{4}$

Chapter 4 Probability and Counting Rules

Addition rule 1 (mutually exclusive events):

$$P(A \text{ or } B) = P(A) + P(B)$$

Addition rule 2 (events not mutually exclusive):

$$P(A \text{ or } B) = P(A) + P(B) - P(A \text{ and } B)$$

Multiplication rule 1 (independent events):

$$P(A \text{ and } B) = P(A) \cdot P(B)$$

Multiplication rule 2 (dependent events):

$$P(A \text{ and } B) = P(A) \cdot P(B \mid A)$$

Conditional probability: $P(B \mid A) = \dfrac{P(A \text{ and } B)}{P(A)}$

Complementary events: $P(\bar{E}) = 1 - P(E)$

Fundamental counting rule: Total number of outcomes of a sequence when each event has a different number of possibilities: $k_1 \cdot k_2 \cdot k_3 \cdots k_n$

Permutation rule: Number of permutations of n objects taking r at a time is $_nP_r = \dfrac{n!}{(n-r)!}$

Combination rule: Number of combinations of r objects selected from n objects is $_nC_r = \dfrac{n!}{(n-r)!r!}$

Chapter 5 Discrete Probability Distributions

Mean for a probability distribution: $\mu = \sum[X \cdot P(X)]$

Variance and standard deviation for a probability distribution:

$$\sigma^2 = \sum[X^2 \cdot P(X)] - \mu^2$$
$$\sigma = \sqrt{\sum[X^2 \cdot P(X)] - \mu^2}$$

Expectation: $E(X) = \sum[X \cdot P(X)]$

Binomial probability: $P(X) = \dfrac{n!}{(n-X)!X!} \cdot p^X \cdot q^{n-X}$

Mean for binomial distribution: $\mu = n \cdot p$

Variance and standard deviation for the binomial distribution:

$$\sigma^2 = n \cdot p \cdot q \qquad \sigma = \sqrt{n \cdot p \cdot q}$$

Multinomial probability:

$$P(X) = \frac{n!}{X_1!X_2!X_3! \cdots X_k!} \cdot p_1^{X_1} \cdot p_2^{X_2} \cdot p_3^{X_3} \cdots p_k^{X_k}$$

Poisson probability: $P(X; \lambda) = \dfrac{e^{-\lambda}\lambda^X}{X!}$ where

$$X = 0, 1, 2, \ldots$$

Hypergeometric probability: $P(X) = \dfrac{_aC_X \cdot _bC_{n-X}}{_{a+b}C_n}$

Chapter 6 The Normal Distribution

Standard score $z = \dfrac{X - \mu}{\sigma}$ or $\dfrac{X - \bar{X}}{s}$

Mean of sample means: $\mu_{\bar{X}} = \mu$

Standard error of the mean: $\sigma_{\bar{X}} = \dfrac{\sigma}{\sqrt{n}}$

Central limit theorem formula: $z = \dfrac{\bar{X} - \mu}{\sigma/\sqrt{n}}$

Chapter 7 Confidence Intervals and Sample Size

z confidence interval for means:

$$\bar{X} - z_{\alpha/2}\left(\frac{\sigma}{\sqrt{n}}\right) < \mu < \bar{X} + z_{\alpha/2}\left(\frac{\sigma}{\sqrt{n}}\right)$$

t confidence interval for means:

$$\bar{X} - t_{\alpha/2}\left(\frac{s}{\sqrt{n}}\right) < \mu < \bar{X} + t_{\alpha/2}\left(\frac{s}{\sqrt{n}}\right)$$

Sample size for means: $n = \left(\dfrac{z_{\alpha/2} \cdot \sigma}{E}\right)^2$ where E is the maximum error of estimate

Confidence interval for a proportion:

$$\hat{p} - (z_{\alpha/2})\sqrt{\frac{\hat{p}\hat{q}}{n}} < p < \hat{p} + (z_{\alpha/2})\sqrt{\frac{\hat{p}\hat{q}}{n}}$$

Sample size for a proportion: $n = \hat{p}\hat{q}\left(\dfrac{z_{\alpha/2}}{E}\right)^2$

where $\hat{p} = \dfrac{X}{n}$ and $\hat{q} = 1 - \hat{p}$

Confidence interval for variance:

$$\frac{(n-1)s^2}{\chi^2_{\text{right}}} < \sigma^2 < \frac{(n-1)s^2}{\chi^2_{\text{left}}}$$

Confidence interval for standard deviation:

$$\sqrt{\frac{(n-1)s^2}{\chi^2_{\text{right}}}} < \sigma < \sqrt{\frac{(n-1)s^2}{\chi^2_{\text{left}}}}$$

Chapter 8 Hypothesis Testing

z test: $z = \dfrac{\bar{X} - \mu}{\sigma/\sqrt{n}}$ for any value n. If $n < 30$,

population must be normally distributed.

$$z = \frac{\bar{X} - \mu}{s/\sqrt{n}} \qquad \text{for } \sigma \text{ unknown and } n \geq 30$$

t test: $t = \dfrac{\bar{X} - \mu}{s/\sqrt{n}}$ for $n < 30$ (d.f. $= n - 1$)

z test for proportions: $z = \dfrac{\hat{p} - p}{\sqrt{pq/n}}$

Chi-square test for a single variance: $\chi^2 = \dfrac{(n-1)s^2}{\sigma^2}$

(d.f. $= n - 1$)

Chapter 9 Testing the Difference between Two Means, Two Variances, and Two Proportions

z test for comparing two means (independent samples):

$$z = \frac{(\bar{X}_1 - \bar{X}_2) - (\mu_1 - \mu_2)}{\sqrt{\dfrac{\sigma_1^2}{n_1} + \dfrac{\sigma_2^2}{n_2}}}$$

Formula for the confidence interval for difference of two means (large samples):

$$(\bar{X}_1 - \bar{X}_2) - z_{\alpha/2}\sqrt{\frac{\sigma_1^2}{n_1} + \frac{\sigma_2^2}{n_2}} < \mu_1 - \mu_2$$

$$< (\bar{X}_1 - \bar{X}_2) + z_{\alpha/2}\sqrt{\frac{\sigma_1^2}{n_1} + \frac{\sigma_2^2}{n_2}}$$

Note: s_1^2 and s_2^2 can be used when $n_1 \geq 30$ and $n_2 \geq 30$.

F test for comparing two variances: $F = \dfrac{s_1^2}{s_2^2}$

where s_1^2 is the larger variance and
d.f.N. $= n_1 - 1$, d.f.D. $= n_2 - 1$

t test for comparing two means (independent samples, variances not equal):

$$t = \frac{(\bar{X}_1 - \bar{X}_2) - (\mu_1 - \mu_2)}{\sqrt{\dfrac{s_1^2}{n_1} + \dfrac{s_2^2}{n_2}}}$$

(d.f. = the smaller of $n_1 - 1$ or $n_2 - 1$)

Formula for the confidence interval for difference of two means (small independent samples, variance unequal):

$$(\bar{X}_1 - \bar{X}_2) - t_{\alpha/2}\sqrt{\frac{s_1^2}{n_1} + \frac{s_2^2}{n_2}} < \mu_1 - \mu_2$$

$$< (\bar{X}_1 - \bar{X}_2) + t_{\alpha/2}\sqrt{\frac{s_1^2}{n_1} + \frac{s_2^2}{n_2}}$$

(d.f. = smaller of $n_1 - 1$ and $n_2 - 1$)

t test for comparing two means (independent samples, variances equal):

$$t = \frac{(\bar{X}_1 - \bar{X}_2) - (\mu_1 - \mu_2)}{\sqrt{\dfrac{(n_1 - 1)s_1^2 + (n_2 - 1)s_2^2}{(n_1 + n_2 - 2)}}\sqrt{\dfrac{1}{n_1} + \dfrac{1}{n_2}}}$$

(d.f. $= n_1 + n_2 - 2$)

Formula for the confidence interval for difference of two means (small independent samples, variances equal):

$$(\bar{X}_1 - \bar{X}_2) - t_{\alpha/2}\sqrt{\frac{(n_1 - 1)s_1^2 + (n_2 - 1)s_2^2}{n_1 + n_2 - 2}} \cdot \sqrt{\frac{1}{n_1} + \frac{1}{n_2}}$$

$$< \mu_1 - \mu_2 <$$

$$(\bar{X}_1 - \bar{X}_2) + t_{\alpha/2}\sqrt{\frac{(n_1 - 1)s_1^2 + (n_2 - 1)s_2^2}{(n_1 + n_2 - 2)}} \cdot \sqrt{\frac{1}{n_1} + \frac{1}{n_2}}$$

and d.f. $= n_1 + n_2 - 2$.

t test for comparing two means for dependent samples:

$$t = \frac{\bar{D} - \mu_D}{s_D/\sqrt{n}} \qquad \text{where} \qquad \bar{D} = \frac{\sum D}{n} \qquad \text{and}$$

$$s_D = \sqrt{\frac{\sum D^2 - [(\sum D)^2/n]}{n-1}} \qquad \text{(d.f. } = n - 1)$$

Formula for confidence interval for the mean of the difference for dependent samples:

$$\bar{D} - t_{\alpha/2}\frac{S_D}{\sqrt{n}} < \mu_D < \bar{D} + t_{\alpha/2}\frac{S_D}{\sqrt{n}}$$

(d.f. $= n - 1$)

z test for comparing two proportions:

$$z = \frac{(\hat{p}_1 - \hat{p}_2) - (p_1 - p_2)}{\sqrt{\bar{p}\bar{q}\left(\dfrac{1}{n_1} + \dfrac{1}{n_2}\right)}}$$

where $\bar{p} = \dfrac{X_1 + X_2}{n_1 + n_2}$ $\quad \hat{p}_1 = \dfrac{X_1}{n_1}$

$\bar{q} = 1 - \bar{p}$ $\quad \hat{p}_2 = \dfrac{X_2}{n_2}$

Formula for the confidence interval for the difference proportions:

$$(\hat{p}_1 - \hat{p}_2) - z_{\alpha/2}\sqrt{\frac{\hat{p}_1\hat{q}_1}{n_1} + \frac{\hat{p}_2\hat{q}_2}{n_2}} < p_1 - p_2$$

$$< (\hat{p}_1 - \hat{p}_2) + z_{\alpha/2}\sqrt{\ }$$

Chapter 10 Correlation and Regression

Correlation coefficient:

$$r = \frac{n(\sum xy) - (\sum x)(\sum y)}{\sqrt{[n(\sum x^2) - (\sum x)^2][n(\sum y^2) - (\sum y)^2]}}$$

t test for correlation coefficient: $t = r\sqrt{\dfrac{n-2}{1-r^2}}$

(d.f. $= n - 2$)

The regression line equation: $y' = a + bx$

where $a = \dfrac{(\sum y)(\sum x^2) - (\sum x)(\sum xy)}{n(\sum x^2) - (\sum x)^2}$

$b = \dfrac{n(\sum xy) - (\sum x)(\sum y)}{n(\sum x^2) - (\sum x)^2}$

Coefficient of determination: $r^2 = \dfrac{\text{explained variation}}{\text{total variation}}$

Standard error of estimate:

$$s_{\text{est}} = \sqrt{\frac{\sum y^2 - a\sum y - b\sum xy}{n-2}}$$

Prediction interval for y:

$$y' - t_{\alpha/2}s_{\text{est}}\sqrt{1 + \frac{1}{n} + \frac{n(x - \bar{X})^2}{n\sum x^2 - (\sum x)^2}}$$

$$< y < y' + t_{\alpha/2}s_{\text{est}}\sqrt{1 + \frac{1}{n} + \frac{n(x - \bar{X})^2}{n\sum x^2 - (\sum x)^2}}$$

(d.f. $= n - 2$)

Formula for the multiple correlation coefficient:

$$R = \sqrt{\frac{r_{yx_1}^2 + r_{yx_2}^2 - 2r_{yx_1} \cdot r_{yx_2} \cdot r_{x_1x_2}}{1 - r_{x_1x_2}^2}}$$